CW01198469

Vee's For Victory!

The Story of the Allison V-1710 Aircraft Engine 1929-1948

Vee's For Victory!

The Story of the Allison V-1710 Aircraft Engine 1929-1948

Daniel D. Whitney

Schiffer Military History
Atglen, PA

ACKNOWLEDGMENTS

A book like this is a project that requires a special set of circumstances to bring it to reality. Assembling the information is certainly necessary, but even more important is the support of family and friends that provide the environment and encouragement needed to carry it through. Collecting material for this book has been a 20 year undertaking and has involved searches in many archives, both public and private. It has been a treasure hunt, and quite a lot of fun. Meeting the people who have become a part of this project has been the most enjoyable part. Cooperation and friendship have been provided from every corner, and it is greatly appreciated. I hope that each of you will see your contributions and that you will be pleased with the result.

There are a few particular individuals that played important roles in bringing this together. Much of the most important technical material that formed the basis of the presentation came about from a request for information on the Allison V-1710 placed in a national aircraft history magazine. A letter from Bob Martin, who turned out to live only ten miles away from my home in Sacramento, offered some help and suggested that I contact Bob Berryessa, also living in Sacramento. Bob had original Allison design reports on the V-1710, given to him by John Hubbard, who had joined Allison in 1940 as a test engineer and was another Sacramento resident. John's collection of manuals and design documents was invaluable in assuring technical correctness in the descriptions of the engine models and how they evolved. During the course of the project all three of these fine men passed away. May this book be a tribute to them.

Serendipity led to a contact with a current Allison engineer, and he said "you need to talk to Joan Zigmunt." Joan had recently retired from Allison, but had been there during the war and had personally worked with the many personalities who had brought the V-1710 though its development and introduction to mass production. She arranged and hosted a delightful day of interviews with other Allison people who had been in engineering, production, operation, and field service during the war. Their stories gave perspective and detail on the people who made the engine a reality.

In a project like this it really helps to have advice from others who have been down the path before. Author Birch Matthews has been a valuable source of important information on all of the Bell airplanes. His experience, encouragement, and hospitality are greatly appreciated. In addition the following individuals have all contributed materials, guidance, and/or critical review and I greatly appreciate their efforts: Rick Leyes, Graham White, Peter Bowers, Warren Bodie, Hal Andrews, Jim Doll, Dick Askren, Charles Daugherty, John Kline, Don Wright, Frank Losonsky, Dave Ostrowski, and the late Gerhard Neuman.

While this project was not sponsored in any way by the Allison Engine Company, it could not have been accomplished in a comprehensive way without access to their historical archives. They also provided access to many of the photos that illustrate the story of the company and its engines.

To the people who built and flew the Allison engine in the cause of freedom.

Book Design by Robert Biondi.

Library of Congress Catalog Number: 97-81401.

Printed in the United States of America.
ISBN: 0-7643-0561-1

We are interested in hearing from authors with book ideas on related topics.

Published by Schiffer Publishing Ltd.
4880 Lower Valley Road
Atglen, PA 19310
Phone: (610) 593-1777
FAX: (610) 593-2002
E-mail: schifferbk@aol.com.
Please write for a free catalog.
This book may be purchased from the publisher.
Please include $3.95 postage.
Try your bookstore first.

Contents

Introduction

The 1920s and 1930s were a time of radical change in the field of aviation. WWI had just ended and not only were there thousands of surplus airplanes and engines around, there were thousands of recently trained, and available aviators to fly them. During the War the public saw the wonder of these new warriors of the air, and many thought of them as the chivalrous knights of a new age. Everyone seemed to be interested in aviation, and many were getting into it any way that they could.

Creativity had been stimulated, and the record is filled with inventors and developers, each having a better way to triumph in the air age. The reality of the era was that a science was developing in aerodynamics, structures, instruments, and engines. But the skilled engineers and technicians needed to be supported by adequate financing, and they had to do their part before the intrepid pilot could take to the air. We all know the story of Lindbergh's epic solo flight across the Atlantic. What many have not appreciated was that it was the preparation and work that he did before the flight that made it possible. He organized his financing, selected a technically capable aircraft builder, and decided upon an engine that was probably the most reliable of its day, an engine whose crankshaft was riding on Allison steel backed bearings.[1] Every detail was figured and planned. All he had to do was be expert, and lucky, in carrying it off, but it was not a spur-of-the-moment scheme to get rich!

The Allison Engineering Company was a bit player in the founding days of aviation, contributing specialty equipment and power transmission machinery. Their reputation was equal to their skills as engineers and machinists. They became known as a firm willing to take on the jobs that others shied away from. Building on this foundation, Norm Gilman, Allison's Chief Engineer, conceived and designed a much improved, high temperature liquid-cooled V-12 engine, the V-1710. The early development of the engine was difficult for a variety of reasons, but Allison was ultimately successful in being the first to pass the stringent Army 150 hour Type Test at 1000 bhp. Using this new engine an entire generation of advanced Pursuit aircraft was developed, beginning with the Curtiss P-37, then the immortal Lockheed P-38, Bell P-39, Curtiss P-40, and North American P-51. These were the fighter aircraft that fought the war for the U.S. during the early years of WWII. Allison kept these airplanes in the battle by continually supplying better and more powerful models of the V-1710. At the time Pearl Harbor the standard V-1710 Military rating was 1150 bhp, but demands from the combat Theaters soon resulted in Allison introducing War Emergency Ratings of nearly 1500 bhp from these same engines. With such power the combat performance of the airplanes greatly improved.

Allison quickly developed a series of improved base-line models offering Military ratings of first 1325 bhp, then 1425, and finally 1750 bhp. All the while the reliability and durability of the engine was improving. The number of models proliferated, mainly because of the demands of the military and the needs of the aircraft manufacturers, but this was fairly easily accomplished because of the building-block design of the engine that allowed different configurations to be constructed by simply fitting the appropriate accessories and reduction gear configurations to fit a particular need.

The V-1710 is often criticized for being a "low-altitude" engine, but this was actually an Army requirement as their intent was to use the engines capability for turbosupercharging when high altitude capability was needed. When the need arose for a two-stage mechanically supercharged V-1710, Allison responded with their Auxiliary Stage Supercharger and the engine was able to match the vaulted two-stage Rolls-Royce Merlin in both horsepower and altitude. Here, the story of the Allison V-1710 is comprehensively told and the similarities and differences in the Merlin and Allison are detailed as never before.

While this is the story of the V-1710 it also details the many odd, unusual and modified airplanes that flew, or were intended to be powered by the V-1710. Some of the most interesting airplanes and engine configurations ever conceived. One of these is the Allison Double-Vee V-3420, the first U.S. engine able to deliver over 2300 bhp. While never making it to the front lines, it was developed as a backstop for the Boeing B-29 when that program was in serious trouble due to the early Wright R-3350's.

This book provides information gathered from numerous archives and files, and present it in such a way that the reader can form opinions about the circumstances that surrounded the Allison V-1710 engine. I intend this book to be interesting and informative, as well as a reference to the circumstances and events that had a lot to do with the tactics and utilization of fighter aircraft during the prosecution of the war by the U.S. Army Air Forces.

The story of the Allison V-1710 is also the story of people. You will meet some interesting personalities who individually contributed to the engine, the airplanes it powered, and certainly the use and support of the

[1] Allison Division, GMC 1950.

airplanes in combat. This includes the thousands of men and women who built and supported the engines around the world. In some small way I hope that this book can be seen as a tribute to their dedication and commitment in the fight for the freedom of our country.

Of necessity this book provides technical information and details on the Allison V-1710 family of engines, but that information is in no way to be considered suitable for use in the maintenance or operation of engines discussed in the text. The reader is cautioned to not attempt to use any of the material in such a manner. This book is written for the enjoyment of the reader and those interested in the historical development of technology, and an important aircraft engine.

In the final analysis the Allison V-1710 is often compared to the similar, and contemporary, Rolls-Royce V-12 Merlin. The stories of the development of these two fine machines have many parallels, and in fact cross paths many times in the 1930s and 1940s. Allison and Rolls always were friendly competitors, joining in jet engine development and production programs in the 1960s and 1970s. Finally, on Monday November 24, 1994 it was announced at the Allison Plant that Rolls-Royce Inc. had agreed to purchase the Allison Engine Company. The joint future of these two fine engine manufacturers builds on a long and distinguished heritage that was begun with their independently developed V-12's.

While the text is written to be readable, and intended to be free of jargon, it has been necessary at times to use certain technical terms. These are defined in the Glossary, but because of frequent use, *ihp* and *bhp* are explained here. Indicated Horse Power, (ihp) is the power actually produced within the engines cylinders, the maximum available from the fuel. Unfortunately, this is not the power available to the pilot or the propeller. This results because both the friction within the engine and the power to drive the supercharger and accessories must be extracted, along with any power consumed in "pumping" the exhaust gases out through the exhaust nozzles or turbosupercharger. What remains after these adjustments is the Brake Horse Power, (bhp) the power actually available to the propeller, or on a dynamometer, the "brake."

1

The Story of Allison and Its Piston Engines

The Allison family arrived on the Indiana frontier in 1840 and immediately began to establish themselves in a variety of businesses. Over the years these included river boating, wholesale groceries and textile manufacturing. In the late 1800s they went on to develop the very successful *Allison Coupon Company*, an operation that stayed in the family until 1963. The coupons were a new concept and Allison Coupon books were used in many ventures, including being a part of the pay for workmen building the Panama Canal between 1904 and 1914. In 1922 General Motors Acceptance Corporation began using Allison Payment books for their customers who were taking out automobile loans. All of these enterprises resulted in a very well connected and influential family.

In 1904 James A. Allison joined with Carl G. Fisher and Percy C. Avery to form the *Concentrated Acetylene Company*. As the story goes, Allison and Fisher each borrowed one hundred dollars to start this precursor to the *Prest-O-Lite Company.*[1] Fisher, who had been interested in bicycle racing and was once the owner of a bicycle shop, joined the trend of the times and opened one of the first automobile garages in the area.[2] By 1904 he was building a network of automobile sales franchises, and was excited by Avery, who had an idea for marketing compressed acetylene for use in powering automobile headlights, this being in the days before automobiles had electric lights. This was timely, as automobiles were then using the troublesome and messy calcium carbide acetylene generator to power their headlights. It was the right product at the right time, and there was soon a tremendous demand for their compressed acetylene powered headlights.

In 1905 several automobile enthusiasts in the Indianapolis area became excited about automobile racing and decided to stage a 24-hour race, with nighttime illumination to be provided by large acetylene lamps. The event was a rousing success, with a new 24-hour distance record being established at 1,094 miles, overshadowing the old mark of 789 miles. It was during that November event that the idea of having a paved 2-1/2 mile banked track evolved. It took four years to develop, but in 1909 Allison, Fisher, Frank Wheeler and Arthur Newby founded the *Motor Speedway*, and on the following Decoration Day (May 30, 1911) the first *Indianapolis Speedway 500-Mile International Sweepstakes Race* was held at what was to become known as the "Brickyard." The event continued and still excites and inspires the automobile fraternity with similar fervor.

Meanwhile, the *Prest-O-Lite* acetylene business continued to prosper at its downtown Indianapolis location. To support their growing involve-

Jim Allison was a businessman adventurer willing to invest his money in high quality race cars. This quest led to financing a machine shop able to meet his exacting standards. Taking on the tough jobs led to developing the skills needed to build engines. He directly influenced the philosophy of the company that was able to go on to become a major manufacturer of horsepower to this day. (Allison)

Car #3 is Jim Allison's, early in the 1919 Indianapolis 500 race. This is the car that won that year's event. (Allison)

ment in automobile racing, Fisher and Allison setup a small machine shop in a building next door to the gas plant. One day in 1915 Prest-O-Lite had a considerable factory explosion and their immediate neighbors created a necessity for the company to move from the area! This was what caused them to relocate to a site in Speedway City, where they built a new facility with 300,000 sq. ft on land adjacent to the Speedway, all at a cost of $500,000.

When the European phase of what became WWI began in 1914 it had a constraining effect on the number of foreign race cars and teams participating in the annual 500-Mile race. Fisher and Allison became concerned about having a sufficient number of cars to conduct a viable race.[3] As a consequence, on September 14, 1915 Allison, Fisher, Theodore E. Myers, plus Wheeler and Newby, founded the *Indianapolis Speedway Team Company*, the first of three proposed racing teams.[4] Two Peugeot race cars were purchased. It is on this date that the corporate life of what later became the Allison Engine Company began.[5]

The team owners felt that since they were also the track owners and promoters, fostering cars and the race events was poor business practice and the arrangement was not publicized. Fisher and Allison, acting as the Indianapolis Motor Speedway Team Company, also arranged for the lease of four special Maxwell race cars from the Maxwell Motor Car Company. These cars were found to be far from durable. Typically only two were able to compete in any one race, the other two were in the shop being reworked for the next racing event. To compensate the team obtained three "Premier Special" race cars, a model having engines almost identical to those in the Peugeots.

Possibly to compensate of the lack of durability in his race cars, the 1916 Indianapolis race was run as a 300 mile event, rather than the expected 500 miles, with two 25-mile and a 100-mile races added to the race card. The Fisher/Allison team, captained by Johnnie Aitken, fielded four cars for the events, the two Peugeots and two of the Premier Specials.[6]

The first team manager, chief engineer, and ranking driver, was Allison's driver and friend, Johnnie Aitken. Prest-O-Lite also fielded a team, captained by their lead driver Eddie Rickenbacker, who was to later become the leading U.S. ace of WWI.

After a couple of years of trying to win a race Allison asked Johnnie just what it would take to win a race on his own race track, and with one of his own cars? Johnnie replied, "that if he had a machine shop where he could turn out precision parts, instead of using stock car parts, he could win a race." Up to this point both teams had been working out of a garage in downtown Indianapolis. Allison was further frustrated by the three mile trip from the garage to the track where testing was done, and suggested to his partners that he take over the team and build a dedicated speed shop out near the track. His partners agreed and Jim Allison became the sole owner of the Indianapolis Speedway Team Company. Fisher kept the Prest-O-Lite team. He then purchased a "fleet" of race cars, three Peugeots and two Bremiers. These cars needed a lot of work to adapt and run on American tracks. They also needed a source for parts, since spares were otherwise in far off Europe.[7] Allison hired his race driver, Johnnie Aitken, to be the head of his race team.[8] Shortly thereafter, in 1916, Allison constructed a small building to house his machine shop on the site of what later became Allison Plant No.1, also in Speedway.[9]

Developing the Company and its Products

On January 1, 1917 the company moved into its new shop in Speedway. Mr. Allison built this precision machine shop under the name of the *Allison Speedway Team Company*. His new shop was equipped with the best tools and machines available, and he hired the finest craftsmen to do the work in the 20 man operation. All of this was solely intended to support his race team. At the same time Norman H. Gilman, a practical visionary, came over from National Motors and joined Allison as Chief Engineer of the Allison Speedway Team Company.[10]

Mr. Allison had taken a liking to Aitken and due to the danger, forbid him from further racing. Instead he put him in charge, granting him the title of General Manager of the Allison Speedway Team Company.[11] Mr. Allison made himself President of the company. While the company focused on Mr. Allison's cars, there was considerable work available to support the many other race teams. Many turned to Allison for special parts and services, and a solid business was established.[12]

Gilman's task was to satisfy the perfectionist demands of James Allison, who was once heard telling a workman, "I want you to remember that this is not just another machine shop. Whatever leaves this shop over my name must be the finest work possible."[13] Later in 1917, Allison sold the controlling interest in Prest-O-Lite, and spent more of his time with the

Allison facilities during WWI when the 250 epicyclic Liberty 12-B engines for the Navy were being assembled. (Allison)

Speedway operation. The company's dogged determination to produce high-quality, world-class engines evolved from James A. Allison's seminal dream of excellence, and it is his signature that adorns the company logo to this day.[14] At the height of Mr. Allison's success he had built his fortune to well over $17,000,000.[15]

On February 23, 1917 the company reorganized, this time as the *Allison Experimental Company*, largely to reflect its high quality machine work which was being recognized in fields beyond the preparation and repair of race cars.[16,17]

When the U.S. entered WWI on April 17, 1917, Jim Allison directed his staff to stop work on cars, but to keep the operation together, as he felt it would be inappropriate to continue the races until the war was over. He then directed Gilman to, "Go find out how we can get war orders rolling. Take any jobs you like, especially the ones other fellows can't do, anything that will help us get started. Don't figure costs or wait to quote prices. We'll take care of that later."[18] Not much later, the Indianapolis 500 was canceled for the duration of the war.[19]

With war work pouring in, employment soon exceeded 300. Belying its future significance, was a project assigned by the Nordyke & Marmon Motor Company of Indianapolis, builder of the then popular Marmon automobile. They had received government contracts to build Hall-Scott aircraft engines, a water cooled 60 degree V-12. Very shortly thereafter the Liberty V-12 aircraft engine was designed by Packard's Jesse G. Vincent and E.J. Hall, an engine drawing heavily upon the Hall-Scott features and similarly sized. Contracts were then placed with a number of automobile manufacturers for Liberty's, with Nordyke & Marmon, among them. They in turn gave Allison the job of building the great majority of parts for their first Liberty engine. Allison also played a part in tool design and tooling fabrication for the project.[20] The practice of the day was for a manufacture to use a precise "master model" against which production parts were compared to assure that the parts would be interchangeable in the field. Allison constructed two master model Liberty's from Production Board drawings,

For some reason the epicyclic Liberty 12-B engines built by Allison during WWI never show up on period production lists. As many as 200-250 of these engines were built during the war. The 1.667:1 epicyclic reduction gears show well on these engines. (Allison)

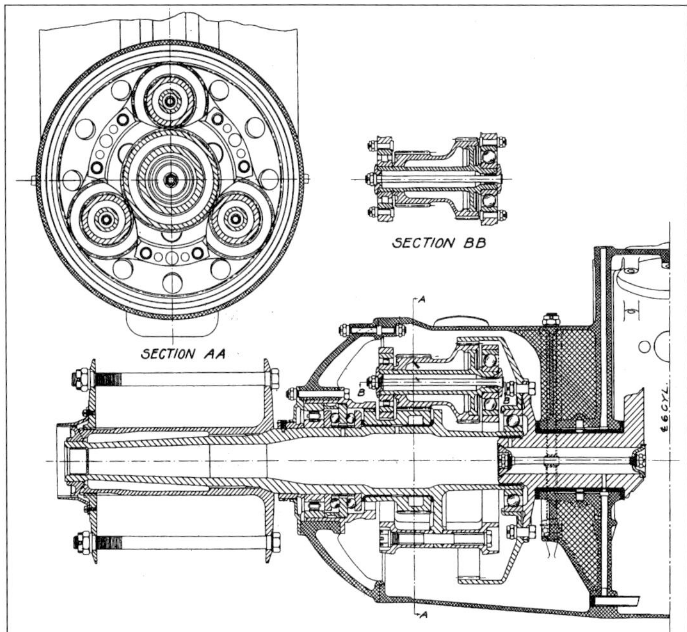

This is the epicyclic reduction gear that Col. Vincent designed for the Army Liberty. Allison fabricated four of these units for testing during WW I and then went on to construct engines incorporating the unit as the Liberty 12-B for the Navy. (Air Service)

along with a set of the necessary tools, jigs, dies, fixtures, and gages.[21] By the end of July 1918 19,500 high compression Liberty 12's were on order for the Army while 2,500 low compression models had been ordered for the Navy.[22] Nordyke & Marmon is believed to have built 1,000 of the engines while total production of the Liberty 12, by all manufactures prior to Armistice was 13,574, and 17,935 by December 31, 1918. There was a fairly lengthy "wind-down" in production resulting in a total of 20,278, 20,428 or 20,478 engines being built, depending upon the source.[23]

Late in 1917 Allison became aware that the Navy was intending to purchase some airplanes for trans-Atlantic flight that would require a Liberty engine equipped with reduction gearing.[24] Gilman and his associates worked all night, New Year's Eve 1917, to prepare a proposal. Early on New Year's day 1918, he left for Washington, and when he returned, had with him a contract for 250 epicyclic geared Liberty 12-B engines.[25,26] This represented the start of a new career for the Allison Experimental Company.[27] When the war ended the company was employing 350 draftsmen, engineers and skilled machinists.

By January 1918 the Army was working on providing gearing for the Liberty 12 as well. Their goal was to provide more power by having the engine run at higher speed, and allowing for a more efficient propeller by slowing its rotating speed. One of the Liberty's designers, Lt. Col. J.G. Vincent studied the options and chose a simplified epicyclic reduction gear design based on the one being used by Rolls-Royce on their 360 bhp V-12 Eagle. He believed that his epicyclic design would weigh 50-75 pounds less than a similarly rated spur gear design.[28] The Army's Airplane Engineering Department at McCook Field, then under the direction of Col. Vincent, designed both a 1.667:1 ratio epicyclic and a 1.50:1 ratio spur type reduction gear for the Army Liberty 12 engines. The engines were identified as the Liberty 12-B and Liberty 12-C models respectively.[29] The Army carefully analyzed this new epicyclic reduction gear in comparison to their own design which was being installed on six experimental Liberty's, three of which were being built by Allison. Noting that the Rolls-Royce

1.667:1 gear was designed for a higher speed than the Liberty, but for less power, the analysis concluded that the Liberty design was more than adequate for the intended service.[30] The Liberty gear was also built for 1.667:1 reduction ratio.[31]

The Allison Experimental Company built three epicyclic gear units in July 1918 and then received a second order for one more in October, evidently to replace one which was defective on an order for three similar units from the Willys-Morrow Co. All were for installation on six special Liberty's built by the Packard Motor Car Company, testing to be done by the Army at McCook Field.[32] While these were beautifully built and operated quite well, the Army felt that neither type, epicyclic or spur gear, significantly improved the associated engine's performance. In fact, considerable trouble was initially experienced with the epicyclic unit due to overheating of the planet gear bearings, which resulted in delays. The Army abandoned the development effort, though the Navy did receive at least two of the epicyclic configured engines in anticipation of their entire order for 250 Liberty 12-B's.[33]

The Navy was particularly interested in the epicyclic geared engine, and in August 1918 had requested 2,000 of them. The Army was concerned that production commitments for direct drive Liberty engines would suffer and therefore did not accommodate the request. The Navy could not understand why the Air Service had trouble in recognizing the benefits of the geared engine. Using the first two engines equipped with Allison epicyclic reduction gears a Navy F-5 flying boat, which did not have properly matched propellers, was able to become airborne in 23 seconds. On average, the direct drive powered model took 35 seconds.[34]

Wars End

In the 1919 Indianapolis 500-Mile Sweepstakes Race, the first following the war, the Allison racing team was successful in winning with their reworked Peugeot; driven by Howdy Wilcox, with Leo Banks as the riding machanician. Another of Allison's cars finished third.[35] This success, and the obvious high standards of the firm, went a long way to establish the Indianapolis 500 as not just a race, but as a proving ground where automobile technology was both developed and demonstrated.

With his racing goals fulfilled Mr. Allison sold all of his race cars to a movie star and race enthusiast, believed to be Ruth Long.[36] He then went on to concentrate on other activities and gradually shaped an organization of 200 men, largely skilled mechanics, which continued to earn a wide reputation for quality and skilled workmanship.[37]

On January 4, 1921, with the change in corporate direction following the withdrawal from racing, Norman Gilman was instrumental in changing the name of the firm to *Allison Engineering Company*.[38] As such the firm was a respected, going concern, making a contribution to the booming economy of the 1920s. It reflected the tradition of American private enterprise – growth by achievement.[39] Its ownership was a one-man show.

Allison Engineering Plant No.1 in Speedway during the early 1930s when it was the only Allison plant. (Allison)

Carl Fisher's 80 foot yacht *Sea Horse* could make 30 mph, powered by two 400 bhp rated Allison Marine-12 engines. Jim Allison had his engines running at 500 bhp in *Altonia II*. (Allison)

One hundred shares of $100 par value was the sole issue of stock during its lifetime, and those were all owned by James A. Allison with one share assigned to each of the four other directors.[40]

After the War James Allison moved to Florida and became involved in real estate development, but both he and Gilman shared a feeling of obligation to the men who had come together during the war and created the dynamic firm. It was their goal to keep the company together and thriving by seeking out new fields to conquer.[41] One of these fields involved the interest Allison shared with his friends for yachting. Unable to find a suitable engine for their yachts, Allison asked Gilman to build a marine engine worthy of the Allison name.

Borrowing heavily from Liberty practices the firm designed a sturdy marine V-12, of which an initial quantity of twelve were to be built. Bore and stroke were both increased over the Liberty by a half inch, as was the angle of the Vee, from the Liberty's 45 degrees to the smoother, and more appropriate, 60 degrees preferred in a V-12. The overhead valve gear was totally enclosed and the engine made extensive use of cast-iron and manganese-bronze. It developed 425 bhp at 1500 rpm, all for a price of $25,000.[42] Four engines went to Allison, two to Arthur Newby, and two were built for Carl Fisher. The other four were offered for sale to the public. It is believed that only the first eight were actually built. With twin Allison-12's, Fisher's *Sea Horse* could run at 26 knots. Allison used two engines in his *Altonia II*, giving it an ability to achieve 30 knots. The engines were known for their reliability and smoothness of operation. The benefit to the Allison Engineering Company was the experience of developing an indigenous design-to-purpose, water-cooled V-12 engine.

Post War Military Work

The Liberty 12 was the U.S. standard aircraft engine. It was an excellent engine by contemporary measures, but suffered from a comparatively short operating life; usually as a result of failure of the connecting rod and/or main bearings. For example, the 1924 round the world flight by five single-engined *Douglas World Cruisers* covered 26,345 miles in 371 hours 11 minutes flying time. Little known is that a total of 17 engines were consumed in the accomplishment. Thirty-five engines were specially prepared for the flight, with all but four distributed around the world for use as spares. Engine changes occurred on average about every 66 hours, for a total of about 4,794 miles per engine. Average airspeed for the trip was only 72.5 mph. Normally scheduled Time Between Overhaul (TBO) on these early Liberty's was 100 hours.[43] The engines had cost the Government an average of $4,000 each, and with well over 10,000 new engines

A view of an Allison angle head drive as installed on the Macon and Akron. Note the coolant radiators installed just inboard of the propeller arc and the exhaust water condensers on the ships surface above the access port. These recovered ballast water to compensate for the consumption of fuel. The V-1710-4(B2R) engines were intended for these airships. The angle gearbox could be rotated 90 degrees to move the ship vertically. (Allison)

available in warehouses, it was clear that the Army was not likely to get funds for new engines until this investment was reduced.[44]

Allison was never daunted by difficult technical problems. An example being the construction of the Liberty 12-B engines for the Navy during the war. These were unique for the time in that the Navy wanted them to be equipped with epicyclic type reduction gears. Engineering handbooks of the day clearly demonstrated, benefits aside, that the resulting higher speed of the engine would destroy the gearing due to vibration, shock and heat, while the consequential weight needed to resolve these concerns would make the engine too heavy. Allison accepted and met the challenge by "writing" a new handbook. They relied not on mass and brute strength, but upon balanced design and precision workmanship. Following the war Allison continued with a variety of geared propeller drives, among them a two-speed Liberty transmission; another with two engines to one propeller; also four engines to one propeller, with synchronizing clutches for each engine; reduction gears for early Wright engines and Army-built W-18 engine; regular production (about 200 assemblies) for the geared Packard A-1500 and A-2500 liquid-cooled V-12 engines, and many others.[45] The USS *Shenandoah* airship (1923-1924) required reduction and reverse gearing with clutches, brakes, extension shafts and vibration dampers, all of which Allison designed and furnished.[46] The most elaborate of all were the transmissions for the *Akron* and *Macon* airships (1929-1933). These had swiveling drive heads by which any or all of the eight propeller shafts could be directed horizontally or vertically, at will; so that propeller thrust was available to lift or depress the ship, as well as to drive it forward or backward. These transmissions also pioneered in the successful use of bevel gearing for transmitting large amounts of power, and very long extension shafting. These transmissions were also notable in the fact that Allison scored a clean victory in engineering, design, workmanship, performance and delivery over the German Zeppelin gear works, who had a parallel development contract.[47]

During the 1920s Allison also manufactured many interesting and important aviation related items, such as Roots-type superchargers, transmissions, reduction gears, and their mainstay in the 1924 to 1927 period, upgrading, overhaul and modernizing some 2,000-3,000 Liberty 12-A's for the government.[48] The Roots superchargers for airplane engines were for customers from all over the world, many being used to establish world altitude records that stood for many years.[49] In the process they demonstrated the practicality of high-speed Roots Blowers for use in two-cycle Diesel engines, but that came a decade later.[50]

Another early 1920s engine project for Allison was the construction of the single experimental X-4520 engine for the Army Air Corps. The project began in 1924 and continued for nearly a decade. The Army specified the design to have four banks of six air-cooled cylinders arranged in an "X" configuration. At 1200 bhp this was one of the first aircraft engines in the world able to deliver over 1000 bhp. In fact, its power was so great that when run on the experimental block it blew out the end of the building! It was impressive enough for display even ten years later at the *Chicago Century of Progress Exposition*. However, it was never installed in an airplane.[51]

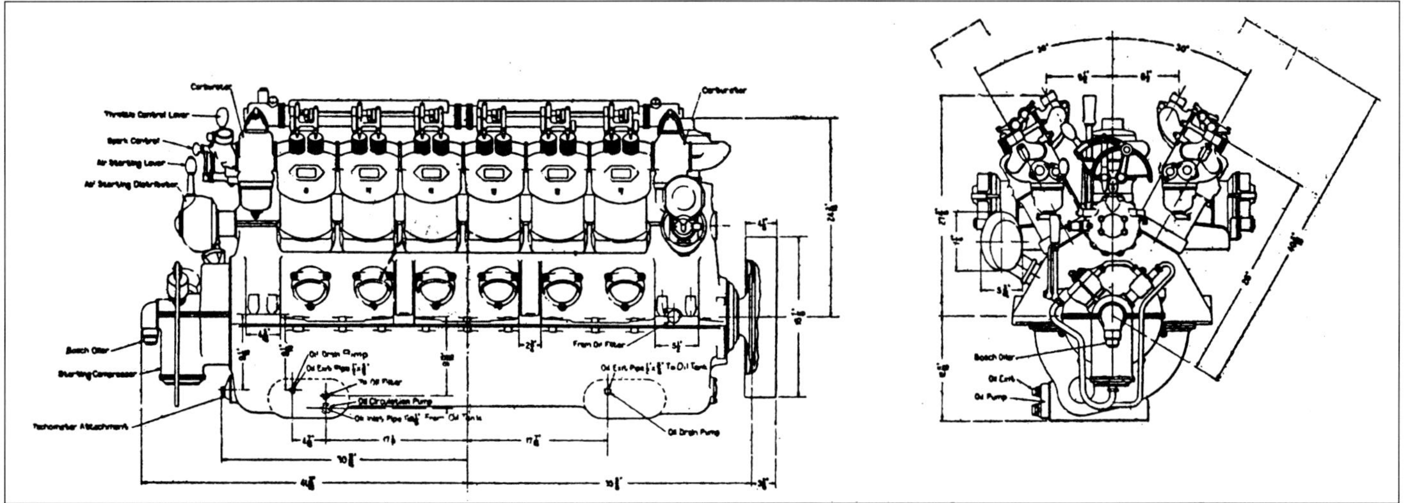

The German Maybach-Zeppelin 480 bhp reversible LV2 was a 2055 cubic inch liquid-cooled V-12 weighing 2,450 pounds was used in large airships including the "Los Angeles" and "Graf-Zeppelin". For the later "Macon" and "Akron" the engines were rated at 600 bhp. Though German engineered and expected to operate 2,000 hours between overhauls, Navy experience was that they were unreliable and expensive to operate. Connecting rods were unusual in that they were fitted with roller bearings, a major source of trouble.

The Allison Engineering Company also did a fair amount of work for the U.S. Navy during the 1920s that later influenced V-1710 engineering features and concepts. Work on the unique swiveling angle drives and extension shafts for airship propellers provided experience and a basis for the extension shafts Allison designed for use with the later V-1710-D and V-1710-E series engines. In addition, at Navy request, work was begun in 1927 to provide a 2-cycle in-line, water-cooled, six-cylinder diesel engine to replace the 2,500 pound, 600 bhp, German Maybach engines then powering the Navy's rigid airships. The Allison diesel was intended to develop 900 bhp and weigh less than 3,000 pounds. As a coincidence, this project attracted the attention of Charles F. Kettering of General Motors, who was working on a Diesel engine involving similar principles. He became a visitor to the shop, beginning an association that was to last for decades.[52] After building and testing the first engine the project was stopped because of the difficulty in recovering ballast water from the exhaust of a two-stroke diesel was recognized as an undesirable feature.[53] When Allison proposed the concept of a 1000 bhp, 4-cycle gasoline V-12 engine with a weight of only 1,000 pounds, to the Navy in 1930 it is not surprising that they were very amenable to sponsoring its development. They were looking toward obtaining a modern lightweight U.S. built engine specifically for airship use.[54]

As a further measure of their manufacturing prowess, Allison was building the prototype high-speed superchargers being developed by the General Electric Company for the Army Air Corps.[55] From this experience came the analytic and manufacturing expertise that allowed Allison to subsequently design and manufacture its own superchargers for the many thousands of V-1710's.

Changes in the Wind

By the end of the 1920s it was apparent to Gilman that Allison needed a new product line for the 1930s. Airplanes were getting larger and faster, and the current practice of simply hanging more Liberty's on them was no longer viable or competitive. The Air Corps was finally responded to the poor situation they had created for themselves when on July 1, 1929 they ordered that no more new aircraft designs would be powered by the venerable Liberty.[56] Two years earlier Lindbergh had crossed the Atlantic powered by a Wright J-5 air-cooled radial, equipped with Allison steel backed lead-bronze bearings. Pratt & Whitney was aggressively challenging Wright's dominance in the air-cooled radial engine field with impressive performance from their new R-1340 *Wasp*. Although the U.S. had dropped out of the competition following the 1925 race, the biannual Schneider Cup Sea Plane Races were capturing the imagination of aviation interests around the world, demonstrating the prodigious capability of highly tuned, water-cooled, in-line engines installed in highly streamlined airframes. Times were right for a major change in aviation technology.

A Change in Ownership

In 1927 Eddie Rickenbacker, pre-WWI driver and captain of the Prest-O-Lite race team and subsequently the U.S. leading WWI Ace, was looking for a business opportunity. He knew that his old friend Jim Allison was partially retired and living in Florida. Reflecting on his frequent trips to Indianapolis, Rickenbacker relates as to how he had been impressed by the Allison Engineering Company, which he described as having "a little gem of a plant." So he went to Jim Allison and asked to buy the company. Allison responded, "Well, Eddie, I'll tell you, it's true I spend most of my time in Florida these days, but when I do come home for the summer I like to have a desk I can put my feet on. But I have a better idea, one tailor-made for you. Why don't you take over the Speedway?" Allison related that, "Of the four men who had built the Speedway, Carl G. Fisher, Frank H. Wheeler, Arthur C. Newby; Frank Wheeler is dead, Art Newby is no longer interested and Carl Fisher is spending all his time developing Miami Beach. I'd like to see it in younger, capable hands."[57]

Allison gave Rickenbacker a price of $700,000 and a thirty-day option. Rickenbacker felt the price was fair, but he neither had the money himself or was unable to raise the money in Indianapolis within the allotted time. Allison then granted him a thirty-day extension, with the looming alternative being to relinquish the Speedway to real estate developers anxious to subdivide the property. Working with the Guardian Trust Company in Detroit, Rickenbacker was finally able to raise the $700,000 in a 6-1/2 percent bond sale. Accomplishing the feat only one hour before the option to purchase was to expire. On November 1, 1927 Eddie Rickenbacker took possession, holding 51 percent of the common stock as a bonus, while the bank held 49 percent for its effort in marketing the bond issue.[58]

Then an unexpected turn of events occurred when on August 4, 1928, James Allison died after a short bout with bronchial pneumonia. He was only fifty-five. Control of the business was left in the hands of executors of the estate. These Indianapolis interests set out to look for a new owner that would continue the business in Indianapolis.[59] The executors advised Rickenbacker that the Allison Engineering Company would soon be sold to the highest bidder.[60] Just about everybody in the aircraft industry, or who wanted to get into the aircraft industry, was after the company, though the requirement to continue the business in Indianapolis for at least ten years considerably reduced the bid prices.[61]

Wright Aeronautical, the leading engine maker of the time, was somewhat interested. After looking it over they decided there was not a sufficiently good profit record to justify operating the business, so they would only take the people and patents. Consolidated Aircraft also took a look, and came to a similar conclusion.[62] Through his associations, Rickenbacker was kept informed of the discounted prices being offered by the few bidders and kept raising his bid accordingly. He finally bid $5,000 more than the previous highest bidder, and the Allison Engineering Company was his, for a price of $90,000. Though he did not have the money, Rickenbacker managed to get a loan from an Indianapolis bank. In addition, he optioned the land around the company's Speedway facility for future expansion.[63]

According to Rickenbacker's autobiography, after acquiring the company he realized that he had neither the time to run it properly, nor the money or credit that would be required for proper expansion over the next few years. Still, during his months of ownership he claims to have been the one to arrange with the Navy to build the prototype GV-1710-A, though more specific records give May 7, 1929 as the date when Norm Gilman and his staff began to sketch and design the VG-1710, as it was known within Allison at the time.[64] It is entirely possible that Rickenbacker did have conversations with Navy personnel that encouraged Gilman that such an engine might find a buyer, but the timing suggests they could not have been conclusive.

Rickenbacker further relates that, at that time, the Fisher Brothers Investment Trust was buying into the aviation industry and they agreed to take the Allison Engineering Company "off his hands."[65] Effective January 1, 1929, the Allison Engineering Company was purchased by the Fisher Brothers, who also happened to be sitting on the General Motors Board; Eddie Rickenbacker was named president of the Fisher operation.

An important action attributed to Larry P. Fisher during his short tenure as owner, was his direction to Norm Gilman to build a small six-cylinder, in-line, liquid-cooled engine for a "family plane." He was to build this engine around the high temperature liquid-cooled cylinder Gilman had designed in 1928.

At the time General Motors was also investigating investments in the aviation business, and their findings were somewhat different than those of Wright and Consolidated. General Motors appreciated Allison's good shop and reputation, and they also gave value to the Allison bearing business and were encouraged by the two-cycle Diesel engine developments, which were ahead of a similar development project of their own. But the bottom line to General Motors was that Allison had demonstrated that it could develop and produce horsepower in many fields, and above all else, General Motors was interested in horsepower producers.[66]

At the General Motors Executive Committee meeting of March 1, 1929 "there was a lengthy discussion regarding the desirability of GMC making some effort to get into the airplane business, and finally it was unanimously:

> "RESOLVED that the...GMC invest up to $5,000,000 in the airplane business, particularly in connections with the manufacture of instruments, engines and other accessories for airplanes, and also investigate fuselage construction if a favorable opportunity develops..."

At the March 15, 1929 meeting of the General Motors Finance Committee a lengthy discussion was held regarding aviation interests that might be appropriate to GMC purposes, this list included the Allison Engineering Company. As a result, at the May 3, 1929 Executive Committee meeting the following was proposed and approved:

> "General Motors Corporation Appropriation $800,000 (Project A-1209) Purchase of Allison Engineering Company."

Of the appropriation, $600,000 was to cover the price paid for Allison by Fisher and Company plus 6 percent interest during the time they held the investment, the actual amount paid being $592,168.[67] The remaining $200,000 was to be used for required additions and improvements needed during the calendar year. The purchase was effective April 1, 1929 and was greeted with great enthusiasm in Indianapolis, while providing considerable relief to the Allison employees and others interested in continuity and employment in Speedway.[68] General Motors promised to continue operation of the business for at least ten years.[69]

When purchased, the firm employed fewer than 200 employees and had its manufacturing space in about 50,000 sq. ft.. The principal business at that time was modernization of Liberty engines for the Government and the sale of Allison bearings.[70] While not a big operation by GM standards it was important for their planned expansion into the aviation field.[71]

Reflections

It is curious that Rickenbacker could purchase the Allison Engineering Company, "Lock-Stock-and-Barrel" for $90,000 and four months later sell it to the Fisher Brothers for $500,000! A going engineering and manufacturing enterprise of Allison's size should have brought, on the open market, a figure nearer to $500,000 than $90,000! As confirmation, the Allison Engineering Company Balance Sheet of January 31, 1929 shows a net worth of $492,866.88. One might conclude that events surrounding Mr. Allison's scandalous divorce from his long-time wife, and the subsequent marriage to his secretary Miss Lucille Musset, which occurred only five days before his death, were factors.

Allison's estate was estimated at between three and six million dollars, which legally was to go to his new bride, for there was no will. Two days following his death his first wife filed a $2 million suit claiming "allenation of affections" against the new Mrs. Allison. Ultimately, his 80 year old mother, Mrs. Myra J. Allison received most of the $3 million estate.[72]

Evidently the executors (long-time friends of both Rickenbacker and Jim Allison) did not care to see the new wife profiting from the tragedy, and by the expedient of requiring the firm to stay in Indianapolis, caused the firm to be heavily discounted in the eyes of the other bidders. Rickenbacker was in the right place at the right time, and had the inside track.

Early Allison Engine Development Projects

The V-1710, while the most famous and numerously produced Allison piston engine, it was not Allison's first or only piston engine. In fact, they had a history of crafting innovative engines and unique power transmission components for diverse aircraft. Here are the milestone projects that had direct bearing on the ultimate development of the V-1710 and its features.

The Allison Twelve Marine Engine

In April 1919, with the end of WWI in sight and with the coming need to return to commercial and private endeavors, Jim Allison directed Norm Gilman to build a 400 bhp marine V-12 for use in his yacht. It was hoped that such an engine could become a profitable line of business. He planned to shakedown two of these in his own yacht and authorized initial construction of eight engines.[73]

Not surprisingly, the new marine engine incorporated and improved upon many of the attributes of the Liberty 12. The resulting engine had the same 400 bhp as the Liberty, but it differed in most other particulars. The first engine was assembled in December 1919 and connected to a marine drive gear assembly purchased from others.

It is apparent that Allison had big hopes for marketing the engine. A very handsome brochure was produced, complete with art work and photographs of the engine and its major components.[74] As an aside, Jim Allison was unhappy with the available auxiliary power plants being used for yacht service. They were all too noisy and vibrated excessively for his liking. To this end he had his company produce the *Allison Lite*, a 3 to 4 kW electric generator running at 1200 rpm. It was a four cycle, four cylinder L-head engine having a 2-1/2 inch bore and 3-1/2 inch stroke. The engine weighed 530 pounds, while the associated electrical generator controls added an-

Performance testing one of the Allison 12 Marine engines. (Allison)

Immediately following WWI Jim Allison had his company design and build the 425 bhp Marine 12 and 3-4kW Allison Lite electricity generator, both for yachts. The firm was still working on the order for epicyclic Liberty 12-B engines as well. All three are shown in this portrait. (Allison)

other 104 pounds. The general design was quite sophisticated for such an engine.[75]

Significantly, the *Allison Twelve Marine Engine* (sometimes referred to as the *Miami Engine*), was a 60 degree V-12, with dual intake and exhaust valves in the head, operating as a 4-cycle on gasoline.[76] Each cylinder was 5-1/2 inches in diameter and had a stroke of 7-1/2 inches. Displacement was 2,138 cubic inches, and used a compression ratio of 4.9:1. The Liberty cylinder was one-half inch less in both dimensions, which gave 1,650 cubic inches of displacement while utilizing only a single intake and exhaust valve in each cylinder. Main and connecting rod bearings were 7/16 inch thick bronze lined with a high grade of babbitt.

The *Miami* engine rating was 300 bhp at 1000 rpm, 400 bhp at 1400 rpm and 425 bhp at 1500 rpm. As a marine engine, light weight was not a requirement, and the engines' dry weight was a rugged 4,400 pounds, approximately eleven pounds per horsepower! Quite a difference from the later V-1710 at under one pound per horsepower.

A major improvement over the Liberty was the valve gear. Each bank had a single hollow ground camshaft with internal pressure oiling. The cams operated the dual inlet and exhaust valves, which had dual valve springs, through rocker arms with rollers operating on the cams.[77] This setup was carried over onto the V-1710, as was the total enclosure of the valve gear under close fitting valve covers.

The ignition system was unique in that dual plugs in each cylinder were fired by a magneto, and in addition, there was a third sparkplug in each cylinder operated by a battery ignition system. Variable spark advance was also provided.

The crankcase of the engine was of the dry sump type and included a built-in clutch and marine reversing gear purchased from an outside source.[78]

It is believed that only the original order for eight engines was completed, all in the 1919 and 1921 period. They were offered for sale at a price of $25,000 each.[79] With two each going into the yachts owned by Allison, Fisher (installed in his 80-foot yacht *Sea Horse*) and Newby. Mr. Allison had one of his engines (engine #5) specially prepared and tested for racing. Running on gasoline mixed with one third Benzol, at 1700 rpm, it developed 500 bhp. This was in late 1921.

These engines were well known in East Coast Yachting circles for their reliability and smoothness of operation. Considering the low compression ratio, they were also quite economical at about 0.53 pounds of fuel per horsepower per hour. The design and manufacture of these en-

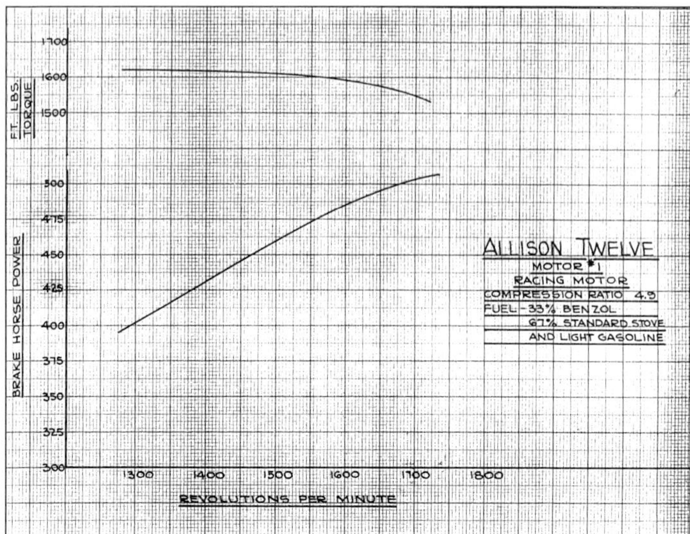

Jim Allison's interest in yachts extended to racing as well. With standard 4.6:1 compression ratio the engine delivered 425 bhp at 1500 rpm on gasoline. This curve shows that Mr. Allison's 'Miami' high compression 4.9:1 racing motor produced over 500 bhp at 1700 rpm, using 1/3 Benzol mixed with the gasoline. (Allison)

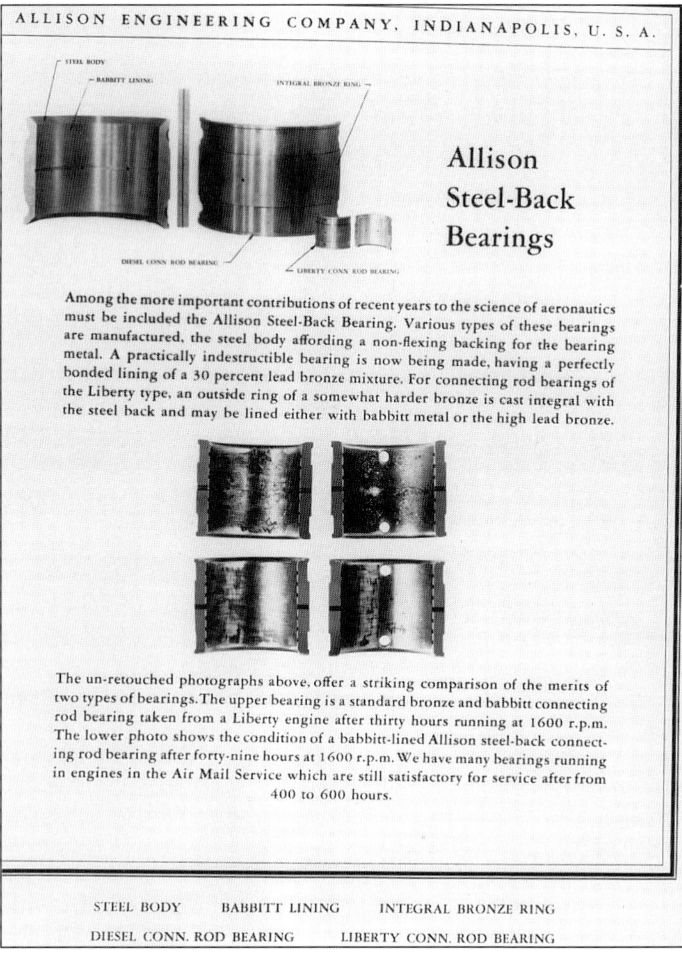

ALLISON ENGINEERING COMPANY, INDIANAPOLIS, U. S. A.

Allison Steel-Back Bearings

Among the more important contributions of recent years to the science of aeronautics must be included the Allison Steel-Back Bearing. Various types of these bearings are manufactured, the steel body affording a non-flexing backing for the bearing metal. A practically indestructible bearing is now being made, having a perfectly bonded lining of a 30 percent lead bronze mixture. For connecting rod bearings of the Liberty type, an outside ring of a somewhat harder bronze is cast integral with the steel back and may be lined either with babbitt metal or the high lead bronze.

The un-retouched photographs above, offer a striking comparison of the merits of two types of bearings. The upper bearing is a standard bronze and babbitt connecting rod bearing taken from a Liberty engine after thirty hours running at 1600 r.p.m. The lower photo shows the condition of a babbitt-lined Allison steel-back connecting rod bearing after forty-nine hours at 1600 r.p.m. We have many bearings running in engines in the Air Mail Service which are still satisfactory for service after from 400 to 600 hours.

STEEL BODY — BABBITT LINING — INTEGRAL BRONZE RING

DIESEL CONN. ROD BEARING — LIBERTY CONN. ROD BEARING

This early advertisement gives testimony to the benefits of Allison Steel Backed bearings. These were the mainstay of the company for many years. (Allison)

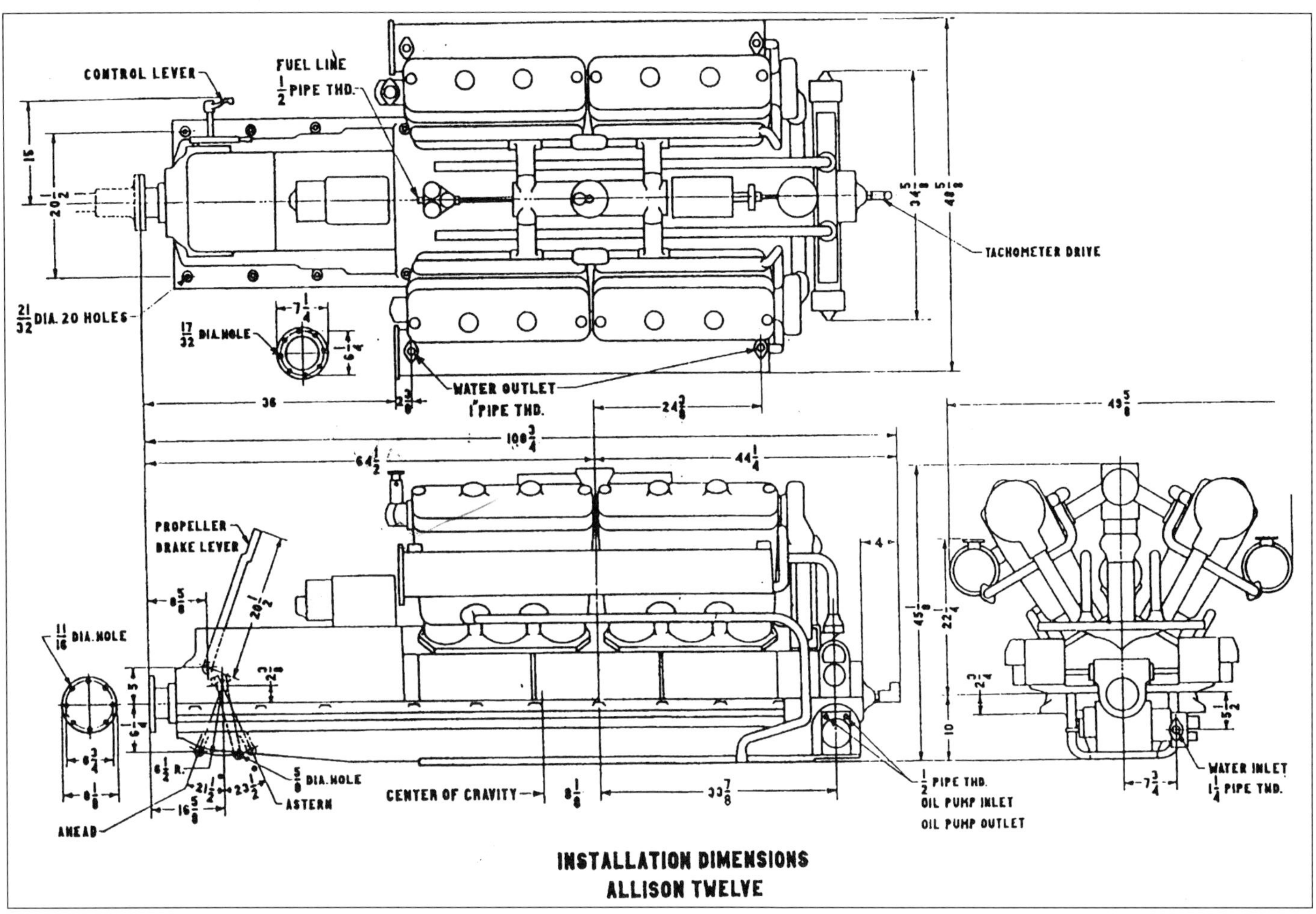

The Allison Marine 12 improved on many features seen in the Liberty 12, important ones being the use of 60 degrees between the cylinder banks, jacketed cylinder blocks and enclosed valve gear. Displacement was 2,138 cubic inches. (Allison)

gines contributed to Allison's reputation and confidence in undertaking other complex engineering assignments.

The engines gave excellent performance and operation, but the gears and clutches required further development. Allison found that the market for the engine was very limited and the production program was dropped in late 1920.[80] Allison did not again become involved with marine applications for nearly 20 years. Marine adaptations of the V-1710 are described in Appendix 11.

The Allison Steel Backed Sleeve Bearing

In the early 1920s Allison was contracted by the military to rebuild and test a Liberty 12 for hi-power by using 20 percent Benzol mixed with aviation fuel. The goal was to pull 570 bhp from the 425 bhp rated engine. Norman Gilman and his crew rebuilt the engine and ran it for 31 hours, at which point the connecting rods failed, severely damaging the engine. This incident had a far reaching effect on Gilman and the Allison firm. First, he became convinced that the water-cooled engine had potential as a high powered airplane engine. To him it appeared to overcome the limitations he saw in the air-cooled engines of the day. He determined to build a 1000 bhp water-cooled engine, but first he would have to develop a bearing that could withstand the necessarily greater loads and heat. Gilman's solution would not only be used on the remanufactured Liberty, but on all high-powered reciprocating engines in general. This developed into an entire product line for Allison, known as the "steel backed bronze sleeve bearing."[81] It is this type of bearing that are used on practically all reciprocating engines to this day.

The problem as Gilman saw it was that the Liberty's connecting rod bearings were rigidly clamped into the forked rod and worked laterally on the outside of a bronze shell that distorted under heavy loads and temperature changes. Without dimensional stability, the bearing stresses were unevenly applied and the bearing would be quickly pounded apart. Gilman calculated that a steel shell, with twice the strength and three times the stiffness of bronze, would resist the distortion and allow the bearing to work. Building the steel bearing shell was simple; the required advance in manufacturing technology was in how to get the necessarily soft lead-bronze bearing surface to adhere to the shell. The solution he and his team developed was to heat the shell to red-hot, pour the bronze around it, and to quickly quench to obtain the desired bonding and grain structure. The surplus material was then simply machined away.[82]

It wasn't all quite that simple. Rolls-Royce in England heard of the improvements and obtained a set of such bearings for one of their engines. When they tested it they observed significantly improved longevity. When they tried to manufacture the bearings themselves they found that the bronze would separate in operation and ruin the engine.[83] Rolls-Royce came to Allison and obtained a license to use the Allison developed process, which included the essential step of quenching the bronze soon after pouring.

With the new bearings it was not unusual for a Liberty 12 to successfully operate ten times as long between overhauls as the original models. By 1927 manufacture of the Gilman bearing was a mainstay of the Allison operation and was a major factor in providing the cash needed during the Depression years to continue with the development of the V-1710. The process was kept secret for years, though a patent application was filed on June 16, 1923, resulting in patent number 1,581,083 being issued April 13, 1926. From this point onward, Allison provided the bearings for almost every major aircraft engine manufactured in the U.S. well into the 1940s.[84] The story of the development of these bearings is described in detail in Appendix 10.[85]

Allison X-4520 Air Cooled Engine

In 1924 Allison was asked by McCook Field to refine their design and to build a large air-cooled inline engine for a planned large single-engined biplane.[86] The engine was contracted with Fiscal Year 1925 funds and built in short order. This was a 1200 bhp, 1800 rpm, 24 cylinder air-cooled engine using four banks of six cylinders arranged in an "X" configuration with 90 degrees between the banks.[87] Cylinder bore was 5-3/4 inches with a 7-1/4 inch stroke, and used a compression ratio of 4.9:1. It incorporated a 2.0:1 reduction gear as well as rotary induction using 5.0:1 step-up gears. A single crankshaft was used with blade and fork connecting rods operating on each of the six throws, one pair behind the other. The two lower banks of cylinders were staggered to the rear and theirs were the rear mounted pairs of connecting rods. Connecting rod bearings were the same type of steel backed bronze babbitted bearings as Allison had developed for the Liberty, while the main bearings were Hoffman roller bearings. A single camshaft served each bank of cylinders, and driven from it was a distributor providing dual ignition to each of the cylinders on that bank. Dry weight was approximately 2,800 pounds, length 108 inches, width 60 inches, and height 53 inches. The engine incorporated a number of advanced features, many of which would be seen on the later V-1710. Exhaust valves were cooled by an internal charge of a mixture of sodium nitrate and potassium nitrate salts.

Table 1-1
McCook Field Hi-Power Engine Projects

	X-4520	V-4100	Two ea 2A-1500
RPM	2000	2000	2500
BHP	1500	1333	1230
BMEP, psi	130	130	130
Propeller Diameter, ft-in	21-6	21-0	12-9
Engine Weight, pounds	2970	2400	1900
Cooling System, pounds	0	770	710
Propeller, pounds	400	350	270
Frontal Area, sq.ft.	16.2	20.7	16.5

The McCook Field Power Plant Section showed considerable interest in the engine: not only funding the prototype, but in 1926 they were considering it for further development. They produced a study that compared the X-4520 to a hypothetical 1200 bhp V-12 of 4100 cu.in. displacement, and built along the lines of the Packard 2A-1500, as well as a tandem arrangement of two 600 bhp Packard 2A-1500's. Tabulating the characteristics of these engines gives an interesting view of the state-of-the-art for engines at the time. Assuming further development and "rotary induction" (i.e. moderate supercharging) systems good to 10,000 feet, the ratings would have been as shown in the adjacent table. When delivered by Allison in 1927, the X-4520 was rated for 1200 bhp at 1800 rpm.[88]

McCook Field engineers designed the 1200 bhp air-cooled 24 cylinder, 4-bank X-4520 for large single engined bi-planes. They contracted with Allison for its detailed design and fabrication in 1925 as Army AC 25-521, Allison s/n 1. Delivered in 1927 it sat idle for several years awaiting a suitable dynamometer for testing. Cooling problems with the undeveloped cylinder soon ended the effort though the engine did make its intended power. (Allison)

McCook Field was also encouraging construction of an even larger engine, an X-5340 that would be rated at 1500 bhp with the same 130 psi BMEP. It was to use 24 cylinders having 6-1/4 inch bore and a 7-1/4 inch stroke. They did not expect much of an increase in weight, though obtaining an adequate fixed pitch propeller would have been quite a feat![89]

The engine, Army AC25-521, Allison s/n 1, was not tested until several years after delivery as a consequence of neither the Air Service nor Allison having facilities adequate for a proper test.[90] The Air Service finally did prepare a Technical Report on its testing, but not until January of 1932.[91] The engine had developed 1323 bhp at 1900 rpm, but unfortunately the testing was terminated early because of sticking of the number 4 piston in Bank D. They concluded that in its present form, the engine was unsatisfactory as a result of inadequate cylinder barrel and piston cooling.

The ubiquitous Ford Tri-motor monoplane had in the meantime made large single engined bi-planes obsolete.[92] The Air Corps soon lost interest to the point that when the X-4520 stuck a piston early in its test program, they did not even tear it down to confirm the cause.[93]

A handsome piece of machinery, showing the craftsmanship of the Allison designers and machinists, the X-4520 was by far the largest engine of its day. It was significant enough that it was displayed at Chicago's Century of Progress Exposition in 1934.[94] The engine then went into the Air Corps Museum at McCook Field, where it remained until disposed in the 1970s. Fortunately it was saved from the scrap heap and is now in the collection of the New England Air Museum, though somewhat worse for wear due to outside storage by previous owners.

In August 1942 the Ordinance Department of the Army developed a critical need for an unsupercharged 600 bhp in-line air-cooled engine to operate on 80 octane fuel and power a tank. They requested that the Materiel Command take the engine from its Museum and make the X-4520 available to them, along with all technical data on it and the V-1410 air-cooled Liberty. They intended to operate and test the engine as a V-12, while also taking one of the cylinders for use on a single cylinder test engine and another to be cut-up for inspection.[95] The Materiel Command responded that the X-4520 cylinder was not a "developed" unit and that they doubted that the engine could be reworked to place it in an operable condition. In an effort to support the Ordinance Department they did ship the engine to the Chrysler Corporation, the Departments contractor, for inspection, along with guidance on how to obtain the requested reports and drawings. In the same correspondence they recommended that the Department consider using either the Wright R-1820 or R-3350 air-cooled cylinders, which were believed to be much more appropriate for the intended purpose and readily available as both were then in mass production.[96] Viewing the engine today shows that it was not cut up when inspected by Chrysler in 1942.

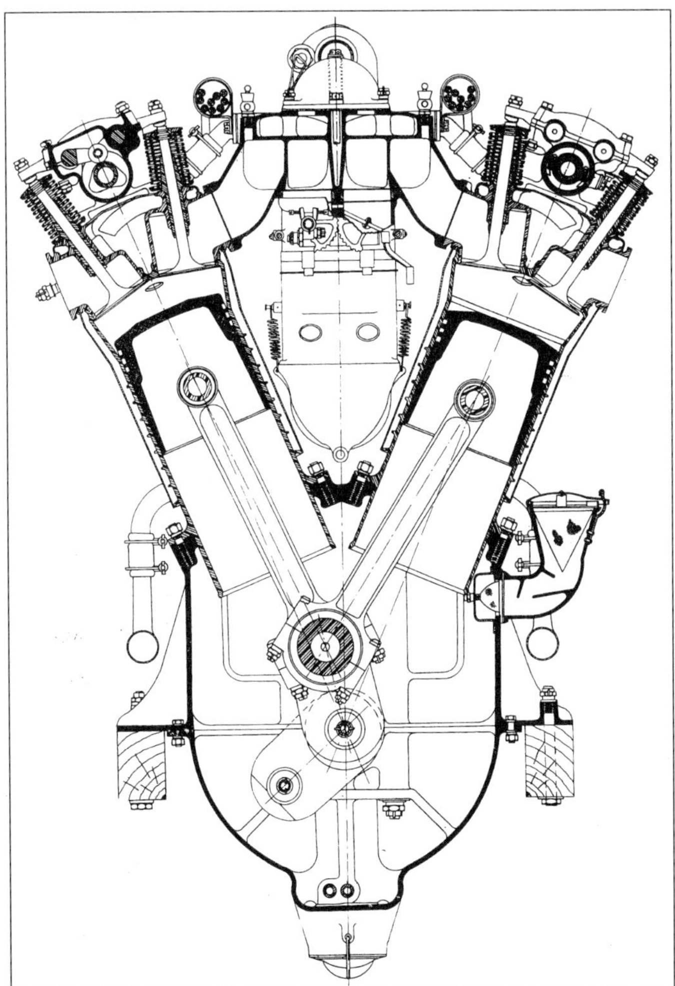

This cross-section of the 45 degree Liberty 12-A gives a good perspective of the engine. While Allison's experience on the Liberty aided in the ultimate development of the V-1710 there are only conceptual similarities. Allison developed their steel-backed bronze bearings to solve problems with distortion of the original bronze bearings used in the blade and fork connecting rods in this engine. (Air Service)

Overhaul, Rework and Modernizing Liberty 12's

During WWI Allison had built 250 Liberty 12-B epicyclic geared engines, as well as done considerable other work in support of the Liberty manufacturing program.[97] In the mid-1920s the Air Corps began to look for ways to improve the Liberty, and to use it in the variety of new aircraft designs being proposed. Allison became a major participant in these programs because of its previous experience and reputation as a quality engineering and machine facility. The original designs for the work performed by Allison on these Liberty engines were largely provided by the Engineering Department at McCook Field, though Allison contributed greatly to the success of the resulting engines as a result of their innovative manufacturing and engineering adaptations of the designs. For example, we shall see that the concept of the sleeve bearing was proposed by the Air Corps, but it was Allison who then made a practical bearing. In fact, one much improved over what the Air Corps had proposed. Likewise, the Air Corps designed the V-1410 air-cooled Liberty, but in an upright configuration. It was Allison who successfully developed the engine in the preferred inverted position. These projects were assigned to Allison as a consequence of their success in competitive bidding for the work. This work became the mainstay of the firm during the decade of the 1920s.

Allison V-1410 and GV-1410 Air-Cooled Liberty Engines

Under Army contracts, Allison upgraded standard Liberty 12-A engines while also developing two new versions of the Liberty. This was done by modifying some of the thousands of engines available from government warehouses.[98] The two new models were inverted, one being a standard sized 1650 cu.in. water-cooled model. The other air-cooled, whose capacity was reduced to 1410 cu.in., necessitated by the need to reduce the bore to allow fitting enough cooling fin area around each cylinder.

In late 1924 the Air Service assembled the first air-cooled V-1410 Liberty 12 derivative, which was an *upright,* direct drive engine. It was extensively tested resulting in Allison receiving a production order for *in-*

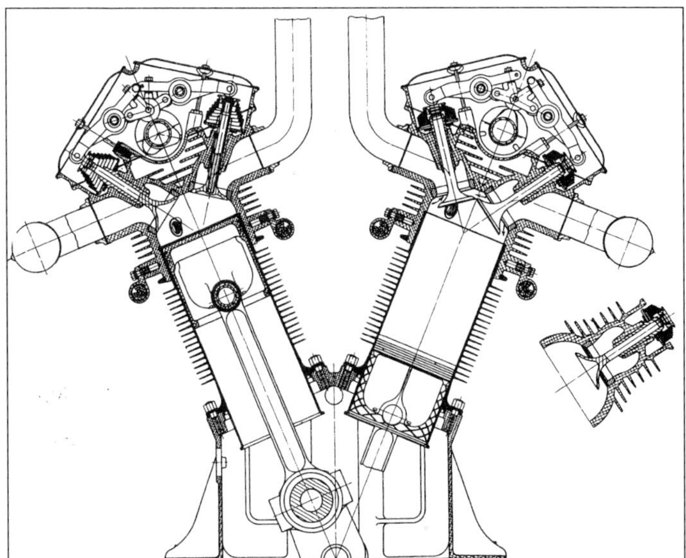

This drawing is of the Sam Heron of McCook Field designed V-1410 air-cooled derivative of the Liberty 12-A. The Army built this first engine as an upright. They then had Allison deliver small quantities of inverted versions of the engine. These differed primarily by utilizing features from the 1918 designed inverted cylinders of the X-24 Liberty. Note the unusual mixture flow from the outside, with exhaust stacks between the vees, and the use of valve springs made from coiled strips of steel, not the more typical coils of wire used by Allison. (Air Service Circular #551)

Shown here is the first Curtiss P-1 Hawk, AC25-410, when re-engined with an inverted direct drive Allison V-1410 in 1926. With this engine it was raced at Kansas City. The airplane had originally been delivered to McCook Field on July 25, 1925, powered by a Curtiss V-1150(D-12) and operated as McCook P-400. (Air Corps)

verted versions of the engine in 1925.[99] Allison produced the engine in both direct and geared models. All of these Allison built air-cooled Liberty's were inverted to secure the perceived advantages of the inverted position: better visibility, lower center of gravity, a cleaner and quieter cockpit, reduction of fire hazard and ease of servicing and maintenance. Wright Field laboratory test experience with the engine found it to be the only air-cooled engine they had tested that was able to be satisfactorily cooled by the blast from a flight propeller. All radial types they had tested needed special cooling "clubs" to obtain adequate cooling flow when on the torque stand.[100]

The first V-1410 Allison built (of a development batch of three, including two with epicyclic gears) was built as a direct drive unit and was test flown in an XCO-6 airplane in October 1925. Jimmy Doolittle, a renowned personality and aviation leader during the 1920s and 1930s, related his first experience with Allison. While he was flight testing the first inverted V-1410, immediately after the first takeoff the engine cut out cold and dropped him into a baseball diamond just east of old McCook Field. No damage was done, but the airplane had to be disassembled and hauled back to the field. Such occurrences were not unusual in the early days.[101]

Early in 1925 the Air Service became interested in having a number of air-cooled inverted and geared Liberty's for use in powering transports and multi-engined bombers. Allison requested the Navy to turn over 100 of the surplus Liberty 12-B epicyclic geared engines that had never been used, these being from the batch of 200-250 Allison had built during the war. This transfer was approved in March 1926, and the evidence suggests that these engines were then delivered to Allison for conversion to the inverted air-cooled configuration as VG-1410's.[102]

In 1926 Allison received an order for twelve inverted air-cooled Liberty engines, eight to be direct drive and four with epicyclic reduction gears.[103]

Except for the crankcase and crankshafts, and of course the reduction gear, the direct and geared engines were identical. For the geared engine Allison had designed a 5:3 (1.667:1) epicyclic propeller reduction gear on the VG-1410, which was quite advanced in that it incorporated a "spring" coupling which was effective in preventing breakage and excessive wear on the gear teeth by damping torsional vibrations. The direct drive engine used the standard heavy-type Liberty 12-A crankcase, crankshaft and connecting rods, but utilized the improved Allison steel backed bronze plain bearings. The geared engine required a special crankcase and crankshaft. Propeller and crankshaft rotation were both in the same direction. Practically all other parts were new and unique to the V-1410, particularly the air-cooled cylinders, improved valve gear and the rotary induction system. It was necessary to reduce the cylinder bore from 5 inches to 4-5/8 inches to gain space for air flow and room for the cooling fins to extend between the cylinders. This allowed a lighter piston that in turn permitted rating the engine at a higher speed. In this way the engines could equal or exceed the usual 400 bhp rating of a Liberty 12-A.

The intake system of the V-1410 was entirely different from the conventional Liberty, incorporating the "rotary induction" system. This feature gave a slight positive boost to manifold pressure and power. This was the terminology of the day and described a centrifugal fan intended to pulverize the fuel drops coming from the carburetor and thereby deliver a uniform mixture to the manifolds and cylinders. This also provided a slight supercharge that more than overcame the resistance of the manifolds and valves at high speed. The valves themselves had higher lift and extended timing. With these changes and the higher speed, the V-1410 produced more power than the standard Liberty 12-A, 410 bhp at 1800 rpm (31.75 inHgA) and 430 bhp at 1900 rpm (32.25 inHgA), with a gross weight reduction for the installation. Guaranteed fuel consumption remained at 0.53 pounds per horsepower per hour.[104]

Cylinder barrels were made of steel, with integrally machined cooling fins. Aluminum alloy heads were screwed on with a shrink fit and provided with a steel band shrunk over the joint. Significantly, Allison used salt cooled, cobalt-chrome exhaust valves. The ends of the valve stems and springs operated in an oil bath that aided in cooling the valves. While the valves were the same size as those on the Liberty, the increased lift, timing and improved port design provided for greater breathing capacity. The camshaft housing and other features of the lubrication system were similar to those on the contemporary inverted Allison V-1650 Liberty 12.

Cooling of the cylinders was accomplished by ducting air from the propeller slip stream into the Vee of the engine where a metal sheet formed a wedge shape directing the air across each cylinder. Allison claimed that this arrangement resulted in a frontal area that was materially less than that of a radial engine of the same power.[105]

Allison built and offered both the V-1410 and GV-1410 models of the air-cooled Liberty. The direct drive engines used in the XLB-3 and XCOA-1 airplanes were designated as V-1410-1 by the Air Corps, though no other specific models have been validated. Several of the 5:3 epicyclic geared engines were built, and flown, but it appears that all of the other applications only referred to them as "V-1410's."

The Liberty X-24 was an expedient way for McCook Field to obtain a 750 bhp engine during the middle of WW I. The War ended before the engine was ready for flight, but following the war the development of the inverted portion of the crankcase and cylinders became the basis for the Inverted Liberty 12. (Air Service)

V-1410 Particulars:

Bore 4-5/8 inches, Stroke 7 inches, Compression Ratio 5.3:1, 12 cylinders with 45° between the banks, 1410 cubic inch displacement. A single intake and exhaust valve were provided in each cylinder along with battery ignited dual sparkplugs. Single Stromberg N.A.S-8E Carburetor and centrifugal rotary induction. High pressure lubrication used along with a dry sump arrangement. Dry weight was 1,125 to 1,160 pounds for the VG-1410, and its width, 17 inches, height, 46-3/4 inches, overall length, 85-1/8 inches. Dry weight for the direct drive V-1410 was 1,000 to 1,025 pounds, its other dimensions were the same, except the overall length being only 78-1/2 inches.

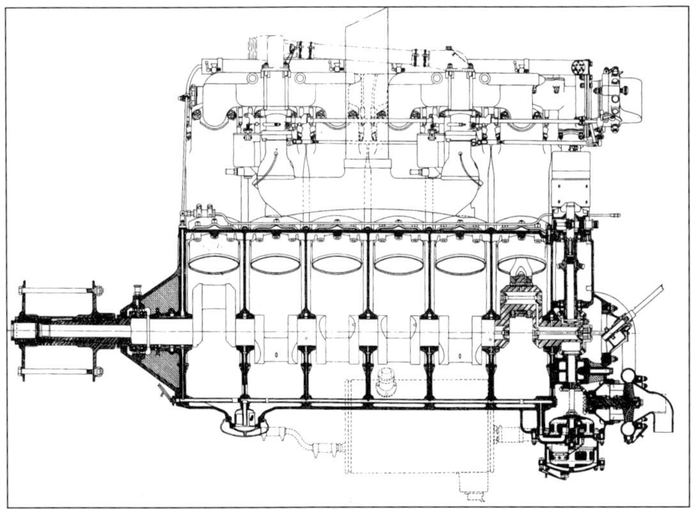

The Liberty-12 engine was conceived during WWI and became one of the great manufacturing and engineering triumphs of the period. For the day it was not only the most powerful aircraft engine, but established new standards for reliability. It evolved from and extended concepts already built into pre-war Hall-Scott engines as well as benefiting from the genius of its designers. At 400 bhp from 1,650 cubic inches it was a big engine for the day. Allison built the original "Pattern" engines for Nordyke & Marmon a Liberty production contractor. (Signal Corps)

Liberty 12's as the Allison V-1650

By 1925 the Army still had over 10,000 new Liberty 12's in its inventory.[106] In 1927 the Army Air Corps decided that the surplus Liberty engines needed to be modernized, being seriously dangerous for the peacetime mission of the day. The Army asked Allison to take on the job of rebuilding some of them to incorporate Allison's new steel backed connecting rod bearings to improve durability.[107] Allison has reported that the initial order was for 1,800 engines, but this grew to between two and three thousand Liberty's finally modernized by Allison.[108] Chief Design Engineer on this project was O.E. Hunt, who went on to become Vice President of Engineering for General Motors.[109] All of the engines that were modernized were the 5.4:1 Hi-Compression models, not the 4.9:1 "Navy" Liberty engines. Consequently, the rating on these engines were for 420 bhp, rather than the 400 bhp available from the low compression engines. A different account of the total number of engines reconditioned by Allison is given by the Chief of Air Corps, a total of 3,660 over a four year period.

In December 1918, during the middle of WWI, an "inverted" Liberty had been attempted as a component in an experimental four bank X-24 Liberty. Although the X-24 Liberty project did not continue, the interest in an inverted direct-drive Liberty did. In February 1919 a successful test of such an engine showed it's feasibility, providing that suitable oil scavenging and water flow could be arranged. In 1924 McCook Field successfully modified a water-cooled Liberty for inverted operation.[110]

In October 1925 the Air Service issued a Circular Proposal for the reconditioning and remodeling of ten standard water-cooled Liberty 12-A engines for inverted operation. Allison was the successful bidder for the work and the result was the Allison Liberty V-1650-13.[111] This was an *inverted* direct drive Liberty engine, made from a standard Liberty 12-A, with modifications to the mounting, lubrication system, carburetors and other minor parts to allow inverted installation and operation. In reconstructing the engines for inverted service a number of modern improvements were made, especially in the bearings, gears, and induction manifolds. The Army paid Allison $1,472 for each of the conversions. Twenty engines were provided in fiscal 1925 for the Loening Amphibian (COA-1), ten more were produced for the Army in 1926 as the OA-1A, with an

Table 1-2 Reconditioning Liberty 12 Engines					
Fiscal Year	Remodeled	Unit Cost	Reconditioned	Unit Cost	Comments
1928	-	-	235	$636.00	Steel Products Engineering Co, Springfield, OH
1928	167	$736.99	558	$712.41	Allison Engineering Co.
1929	105	$532.43	1,240	$594.39	Also $85,000 for 1,300 sets of Steel Backed Bearings
1930	1,500	$450.40	-	-	Allison Engineering Co.
1931	90	$509.85	-	-	Allison Engineering Co.

additional five for the Coast Guard. Total conversions by Allison are believed to be these thirty-five.[112]

The Army felt that the principle advantage of the inverted position was to provide increased visibility to the pilot in a tractor installation, as the cylinders would then project below the line of vision. In addition, inverted engines were easier to maintain from the ground, provided a lower center of gravity, and made for a cleaner and quieter cockpit, since the exhausts exited underneath the aircraft. This was believed to reduce the hazard of fire because only short air-cooled exhaust stacks were required.

Allison's improved steel backed, plain lead-bronze bearings were used on both the crankshaft journals and connecting rods. With these bearings the Liberty could operate 600-800 hours without attention according to Allison. The original plain bronze bearings usually achieved less than 75 hours.[113]

The V-1650 developed 420 bhp at 1700 rpm and 450 bhp at 1800 rpm. These ratings were about 20 bhp higher than the original Liberty 12 could produce, primarily due to the much improved fuel distribution resulting from the new Allison intake manifold and the higher rpm allowed by the new bearings. Fuel consumption was a respectable 0.53 pound per horsepower per hour. All V-1650 Liberty engines used 5.42:1 Compression Ratio pistons.

Other improvements involved replacing the auxiliary drive gears with a heavier stub-tooth design, thereby resolving what had been an occasional problem in the original Liberty. Enclosing the valve gear with an oil tight camshaft housing and internal oil scavenging represented a major improvement. Another improvement was realized in cooling the cylinder heads and combustion chambers as the cooling water in the inverted engine entered at the hottest point, the heads, and was then directed to flow upward around the cylinder wall jackets.

When long distance flight records were being set in 1927 the Allison V-1650 was there. The five Loening OA-1A Amphibians that made the Pan-American goodwill flight were all powered by the inverted Allison V-1650-13. According to Allison, "The performance of the engines on this trip was beyond criticism."[114]

During WWI a Liberty 12-C with 3:2 ratio plain spur reduction gears was tested and flown in an experimental airplane. This photo is believed to be a 420 bhp Allison V-1650-5 [V-1650-E] with 2:1 plain spur reduction gears built during the reconditioning program in the 1920s. (Allison)

In the Fall of 1926 a trading company, representing the USSR, contacted Allison and attempted to purchase either 100 new V-1650's or V-1410 air-cooled engines. Air Corps authorities established a price to Allison for old unmodified Liberty's at $2,000 each, noting that Allison would

Table 1-3 Allison V-1650 Liberty-12 Models						
Air Service Model	Manuf Model	Upright/Inverted	Reduction Gear Ratio	Takeoff Rating, bhp/rpm/alt	Continuous Rating, bhp/rpm/alt	Usage/Comments
V-1650- 1	V-1650-A	Upright	Direct	420/1700/SL	420/1700/SL	C-1, C1-C, XO-1, XO-2, O-2, O-2A, XO-4, O-5, O-6, XO-11, O-11
V-1650- 3	V-1650-D	Upright	Direct	420/1700/SL	420/1700/SL	O-1A, XLB-5, LB-5, LB-5A
V-1650- 5	V-1650-E	Upright	2.00:1 Spur	420/1700/SL	420/1700/SL	C-1A w/T-S/C
V-1650- 7	V-1650-F	Upright	5:3 Epicyclic	420/1700/SL	420/1700/SL	
V-1650- 9	V-1650-HD	Upright	Direct	420/1700/SL	420/1700/SL	Heavy Crankshaft
V-1650-11	V-1650-N	Upright	Direct	420/1700/SL	420/1700/SL	Reconditioned
V-1650-13	IV-1650	Inverted	Direct	420/1700/SL	420/1700/SL	COA-1, OA-1A/B/C

have to take "run of the depot" crankcases, and could not have any of the more desirable *Ford* built engines. It took nearly a year for the Government to finally decide, and deny, the sale to Russia. The reason being that Russia was not then recognized by the U.S. Government.

About two years later, May 1929, the Russians again sought Liberty-12's, this time the epicyclic units originally built for the Navy as Liberty 12-B's during the war many of which still remained. In fact, 113 of them had been returned to the Air Corps and 100 of these were still in their crates. These engines were to be declared surplus in the Fall of 1929, with disposal through normal channels, which would allow Allison to bid on them.[115] Once the engines were surplus, and assuming the government would issue an export license, they would have been a potentially lucrative business for Allison. It is not clear that the project ever progressed.

V-1650 Liberty Particulars:

Bore 5.0 inches, Stroke 7.0 inches, Compression Ratio 5.42:1, 12 cylinders, 1650 cubic inch displacement, direct drive. A single intake and exhaust valve were provided in each cylinder along with battery ignited dual sparkplugs. Two Stromberg N.A.Y-5 Carburetors were used and high pressure lubrication was followed by a dry sump arrangement. No supercharger. Dry weight was 885 pounds. Width, 17 inches, Height, 46-1/2 inches, overall length, 74-1/4 inches.

Many of the featured improvements Allison incorporated into this engine were further refined and subsequently utilized in the V-1710.

Allison Diesel Airship Engines

In 1927 Allison began work on a large two-cycle diesel engine having six inline cylinders and a Roots Blower providing positive induction and exhaust scavenging. The cylinder configuration was somewhat unique in that the intake ports were at the bottom of the cylinder, with four exhaust valves in the head. The engine first ran in November 1929, though it was not to achieve its rated power until 1934, and then only in single cylinder form. The intended rating was for 900 bhp and specified to weigh less than 3,000 pounds. The Navy's goal was to replace the German Maybach 600 bhp engines then powering their large rigid airships, USS *Los Angeles*, USS *Akron*, and USS *Macon*, with an improved engine.[116]

Allison operated the engine, but it could not be successfully used in the airships because of the difficulty of reclaiming water, needed as ballast, from the exhaust of the two-stroke cycle. The engine was abandoned because it had become apparent that a great deal more work and money would be required to develop it for satisfactory operation. Costs had already far exceeded the contract price.[117] With the acquisition of Allison by General Motors in 1929 the Diesel engine and every phase of its development was moved to Detroit to become part of the GM Diesel development program. This ultimately lead to the formation of the GM Detroit Diesel Division, and much later, both it and the Allison Division were rejoined into a single Division of GM.[118]

In 1927, Allison independently was attempting a specific power output in a Diesel engine which still had not been attained in any practical Diesel engine by 1950.

Allison/GM Research Lab "U" Engine

The "U" engine project was active in the 1932-1935 period and was unrelated to any other Allison or GM project. At the time two-cycle engines had caught the interest of several manufacturers, as well as Wright Field. General Motors Research was promoting a two-cycle engine, originally intended for automobiles. Allison undertook, with GM cooperation, expansion of their basic design into two practical engines: one to be a large flat engine for the Army, and the other a small radial for the $700 Flivver airplane. A two cylinder test model of the Army engine was built, demonstrating the two-cycle principle.[119]

Although this activity occurred while Allison was heavily involved in the early development of the V-1710, this unique 2-cycle, gasoline, spark ignition, liquid-cooled 2-cylinder engine demonstrated that it was not ready for practical application.

The Allison engine was known as the U-250, referring to its "U" type twin cylinder horizontally opposed cylinder layout. The engine was purchased under contract W535-ac-6192 and assigned Air Corps serial number AC34-1 (which is of interest because the XV-1710-3, also purchased that year, was AC34-4).[120] Displacement was 250 cu.in. and it was intended to construct a 10 or 12 cylinder version once development testing of the two cylinder version was completed. At a rated speed of 2500 rpm the supercharged engine produced 200 bhp. The "U" cylinders were arranged such that one barrel contained the intake ports and the other the exhaust ports, with a common combustion chamber. The result was a uni-directional flow of gasses which assured complete scavenging of the cylinder on each cycle or crankshaft revolution. Articulated connecting rods were used to provide a degree of differential timing between the intake and exhaust pistons and as assist to scavenging.

The engine suffered from excessive induction temperatures, a common problem when a Roots type blower is used as was the case with the 2-cylinder engine. Much of the research and development work on the engine and its components went on at the General Motors Research Laboratory, where the project was moved soon after GM's acquisition of Allison. The intent of GM was to ultimately use the engine in a light weight rear-engine drive passenger car. The version which received the bulk of the development work was for aircraft and would have included a centrifugal compressor, had a full scale engine been constructed. With the demands of redesigning the later XV-1710-3 and completing its Type Test, interest in the 2-cycle engine declined. The entire project was dropped in 1935, and the excellent team of engineers including the Project Engineer Ron Hazen, were sent to Allison where they immediately went to work on the all out push to get the V-1710-C through its Type Test.[121]

Extension Shafts and Gear Boxes

Allison's experience with propeller shafting and gearing dates from the early 1920s when they built the Army Air Corps a drive gear designed to couple four Liberty engines to one very large propeller. The device passed a successful test on the stand of about 80 hours at full power, but the process consumed 15 unimproved Liberty engines! Perhaps this explains why the complex transmission was never actually used.

Soon after acquiring Allison in 1929 General Motors commissioned a committee to look into developing product lines appropriate to Allison and its capabilities. One area to be developed was light weight reduction gearing and extension shafting. The most elaborate application of these concepts occurred in the period between 1929 to 1933, when they designed the long extension shafts and gearboxes for the airships USS *Akron* and USS *Macon*.[122] In these Allison demonstrated its skill in the art of design by producing a system that successfully operated with the 12 cylinder German engines. These engines proved a challenge as they had eight measurable harmonics, coupled with the shaft system that had three vibration modes of its own, giving a total of 24 possible resonant speeds! Allison had a command of the essential elements for smooth operation and engineered flight-weight transmission systems, utilizing stiff and well-balanced shafts with properly located bearings and dampers. All of this required the utmost in manufacturing excellence and is a measure of the standard of their engineering and manufacturing capability.[123]

For the airships *Macon* and *Akron* Allison built these Goodyear-Allison reduction gearboxes able to translate 90 degrees, enabling the airships to use their propellers for both vertical lift or longitudinal thrust. (Allison)

There was also an element of competition in the Airship transmission project, for the U.S. Government had given a development contract to the German Zeppelin Gear Works for parallel development. It was the Allison system which flew.[124] Significant in Allison's mastery of vibration were the contributions made by John L. Goldthwaite, hired by Gilman during the 1920s to the position of Assistant Chief Engineer. Goldthwaite became highly expert in the solution of vibration problems and it was he who in later years was mainly responsible for the development of the ingenious means of vibration damping used within the V-1710.[125]

Air versus Liquid-Cooled Engines, The Controversy

In 1928 the trend for aircraft engines in the U.S. was toward air-cooling, though its success was not so entirely unqualified as to warrant abandonment of the water-cooled type. This was due to the fact that there was no definite assurance that high-speed, supercharged engine could be successfully cooled with air, nor was the development of the air-cooled type at the stage where it could supplant the water-cooled engine in high speed aircraft. These reasons, together with excellent prospects of improving performance of water-cooled power plants through higher crankshaft speeds, supercharging and higher coolant temperatures warranted continuance of experimental water-cooled developments.[126]

Even so, the trend toward air-cooling had been advanced considerably by the 1927 announcement by the U.S. Navy Bureau of Aeronautics that in the future they would only use air-cooled engines on their one and two place aircraft.[127] The apparent reasons for this position being that they felt that such engines were better adapted to carrier stowage.[128] It was about this time that the Army and Navy (both claim original credit) were experimenting with high-temperature liquid-cooling, using ethylene glycol in place of water.[129] With this chemical coolant it was found that only a comparatively small radiator, one-forth the usual size, was required. This in turn permitted a more streamlined installation, and a considerable reduction in drag. Concurrently, aircraft speeds were increasing to the point where the drag of air-cooled engines was a serious limitation.[130]

Other factors were also making themselves felt. It was believed by many that air-cooling of radial engines could only be successful in single-row machines, there being insufficient cool air for directing across the rear bank of cylinders. The Army had been doing considerable testing of in-line air-cooled engines, such as the Liberty-12 derived Allison V-1410 and the new, but similar, Wright V-1460. These offered air-cooling and little adverse effect on the frontal area of the installation. Unfortunately they were not able to significantly increase the available power. For these reasons many in the military services believed that 800 bhp was about the practical limit for service engines, and these were likely to be water-cooled.

S. D. Heron, Mechanical Engineer for the Materiel Division of the Air Corps and a leading force in the development of the subsequent successful air-cooled cylinder, stated in 1929 that the experience of the Air Corps was that high-temperature liquid-cooled engines will stand much more abuse than any air-cooled engine yet observed. This was attributed to the uniformity of cooling and the large reserve of heat-storage capacity provided by the latent heat of evaporation of the coolant. Furthermore, at that point the Air Corps had not found any air-cooled engine which would equal the low fuel-consumption provided by high-temperature liquid-cooled engines.[131]

According to the Chief of the Air Corps, the year 1930 witnessed a remarkable expansion of the air-cooled engine field, principally the radial

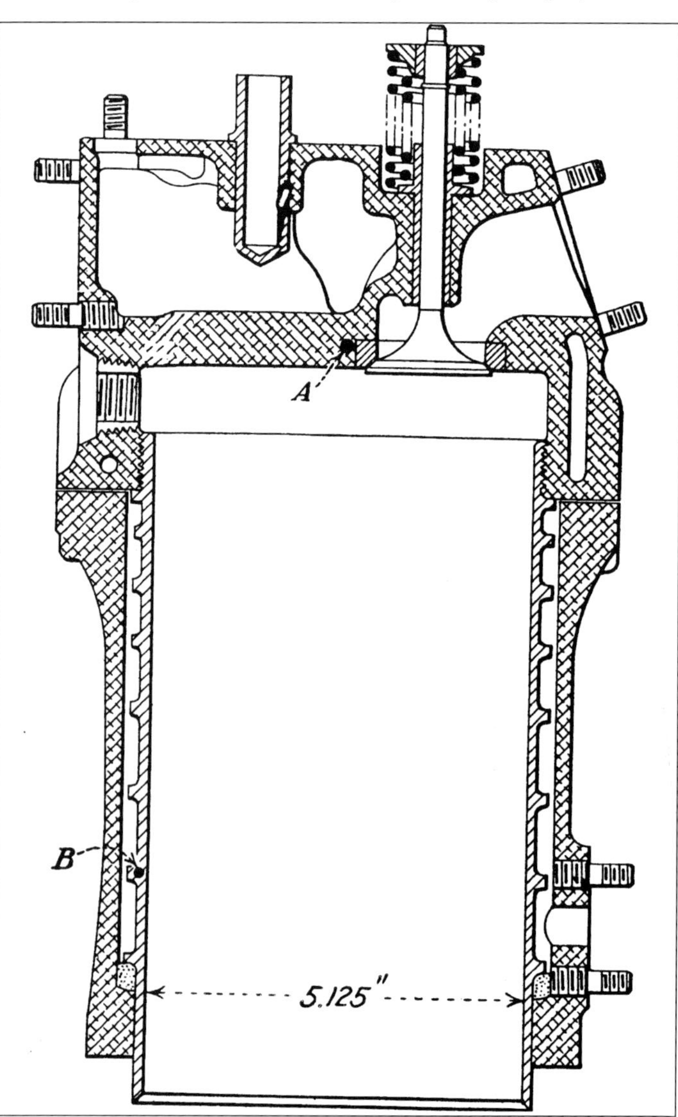

Curtiss used a number of different methods for attaching the cylinder liners to the head on various models of their D-12 and V-1570 engines. This was the improved arrangement used with the V-1570 tested for high temperature liquid-cooling, and it leaked. (Air Corps)

type. Problems arising in connection with liquid-cooled vee type engine temporarily placed the type on the defensive, although its inherent advantages over the radial type were still extant. High temperature cooling with Prestone received a temporary setback when applied to present-day engines whose parts, designed for water-cooling, proved incapable of withstanding operation at the higher temperatures and pressures possible with this type of cooling. Aside from increased output, reduced frontal area and installed weight of liquid-cooled installations held forth great promise. This was demonstrated in a standard Pursuit airplane equipped with a high compression D-12 Prestone-cooled engine. It's high speed at sea-level and service ceiling were increased 11 mph and 3,000 feet, respectively, and time to climb to 20,000 feet improved by 14 minutes.[132]

One of the drawbacks to liquid-cooling of existing engines was the tendency for ethylene glycol to leak through even the smallest passages. It wasn't so much of a problem in the plumbing attached to the engine. Rather, the problem was internal.[133] Specifically, engines like the Curtiss D-12 used a steel cylinder liner screwed into an aluminum jacket for their structure, served to contain the coolant. Hot coolant would frequently work its way through such joints, no matter how carefully assembled. It is for this reason that Gilman came up with carefully designed joints between the jackets, heads and cylinder liners. From the outset, the V-1710 was designed to remain leak-tight under all conditions. He further designed the cylinder blocks so that they could be leak-tested at high temperature prior to assembly on an engine. This assured that even if the parts were not properly assembled, the problem could be found at the component level and would not result in a defective engine.

To say that there was confusion and uncertainty about the best type of aircraft engine is an understatement. Packard had terminated its series of water-cooled V-12's and Pratt & Whitney was developing a fine series of air-cooled radial engines to battle the established Wright air-cooled engines. Curtiss had set the standard for water-cooled engines with their D-12 and Conqueror series, but the company was in financial trouble and were purchased by Wright in 1929. The end result was that Wright terminated developments of the V-1570 Conqueror as well as their own T-3 series of liquid-cooled engines. Allison had developed the air-cooled X-4520 and inverted V-1410 Liberty engines. They also designed a six-cylinder air-cooled in-line engine intended for the private plane market, which everyone forecast as being "just around the corner."

It appeared that the air-cooled engine was the preferred engine in all but the most high speed aircraft, such as the Schneider Cup racers.

Complicating the discussion was the timely development by the NACA of their "NACA Cowl." This simple device took control of the cooling airflow going past the cylinders in a radial engine and was able, in a properly designed installation, to cause the cooling air to exit the engine compartment in a manner which contributed enough thrust to overcome a major portion of the drag caused by the larger cross-section of the radial engine.

Many established players were abandoning the water-cooled engine, just as its cooling limitations were being resolved. This coupled with the considerable speed advantages being posted regularly on the world stage by streamlined aircraft built around high performance V-12 inline engines, Norm Gilman saw an opportunity to fill a gap. The liquid-cooled V-1710 became the result.

All out performance records were being set by aircraft with liquid-cooled engines, exceeding speeds of 400 mph. The standard was set by the Messerschmitt Bf 109R, and its spécially prepared, liquid-cooled, Daimler-Benz 601R V-12 "Sprint" engine when it set the world speed record at 469.549 mph on April 26, 1939. The air-cooled school countered with claims of greater vulnerability for the liquid-cooled engine, which could be put out of action by a single machine gun bullet to any part of the cooling system. The argument waged without resolve, either way, until the turbojet engines came along and settled the question by obsoleting both types of reciprocating engines! Still, during the war there were many instances of fighters with liquid-cooled engines making their way home with the cooling system shot-up completely.[134]

The Allison engine, and the Air Corps emphasis on the liquid-cooled Pursuit, became a hot topic early in 1941 when a Capital Hill newsletter *PM* apparently attempted to start a fight over the question. General Arnold believed that the magazine was simply "attempting to build up circulation by getting a fight started on this Allison engine matter."[135] The battle continued to the point that the Assistant Secretary of War for Air, Mr. R.A. Lovett challenged the situation at the Air Council in early May 1941. In response General Arnold presented arguments to Mr. Lovett which helped soothe the controversy, and provided us some insight into the decisions and circumstances facing the Army at the time:[136]

- Both the Army's and the Navy's Bureau of Aeronautics peacetime engine development programs were based upon the need for both liquid and air-cooled engines.
- Development of air-cooled engines for commercial aviation, plus the efforts of NACA and the military services had demonstrated the superiority of air-cooled engines for multi-engine aircraft. In practice, the record of the liquid-cooled engine had shown it to be the superior power plant for pursuit aircraft.
- Without the benefit of commercial engine sales, development of liquid-cooled engines were constrained by the limited military appropriations typical prior to 1939.
- Beginning in 1930 the Army began a determined effort to interest several manufacturers in liquid-cooled engines. Some $600,000 was expended with General Motors for the Allison, a very small amount in light of what was paid to the air-cooled manufacturers P&W and Wright Aeronautical. Still the Allison engine appeared equal or superior to any of the foreign liquid-cooled engines.
- Prior to the first quantity order for the V-1710 in June 1939, a total of only 81 engines had been procured for test purposes. The engine had passed a development test at 1000 bhp as early as December 1934, and 150 hours of Type Testing in December 1936. On this basis the engine was accepted as a basic type for Air Corps use in procurement proposals from the aircraft industry.
- In the Congressional Act of 1926 the Air Corps procurement of airplanes were required to be based upon guarantees of the manufacturers, rather than estimates made by Army engineers.
- In the 1938 Pursuit Competition and procurement, the P-40 won by nature of demonstrated superior performance over air-cooled engined airplanes offered by the Curtiss and Seversky Companies. In awarding the contracts the Chief of the Air Corps approved the purchase of the P-40 only after management of the General Motors Corporation warranted acceptance of guarantees to deliver the specified engines, and that in any future National Defense emergency their efforts would largely be directed at engine production.
- In the later 1939 pursuit competition, wherein manufacturers were allowed to submit designs only, that the liquid-cooled engine in general was specified as offering better performance and military utility over the air-cooled type. Even so, General Arnold had approved purchase of air-cooled single engine and two-engine models for experimental development (P-44 and XP-50) in addition to the primary purchases (P-39C and P-38).

- Beginning in the fall of 1939 the French found from combat experience that the air-cooled Curtiss pursuit version of the basic P-40 was out-performed by the liquid-cooled German Messerschmitt. When the P-40 was made available for export, France and Britain both selected it, even though improved P-36's, the Seversky P-44 and the Navy's Grumman's and Voughts were available and offered.
- With the additional procurement of the "18,000 Airplane Program" in the spring of 1940, the Air Corps collaborated with the National Advisory Committee, chaired by Dr. George J. Mead, the man largely responsible for the basic design of the Pratt & Whitney engines. He informed the Materiel Division that he had reviewed the case history of the Allison engine and suggested a series of tests which he believed would ensure correction of the defects that the engine had experienced. This action resulted in the P-40 Accelerated Service Test and in the assignment of Mr. O.E. Hunt, Chief Engineer for General Motors, taking personal charge of the engine work at Allison.
- In collaboration with the Advisory Commission of the Council of National Defense, and the British Purchasing Commission, it was determined that the requirements for 1000 bhp liquid-cooled engine could not be satisfied by the Allison Company alone. The British represented through the President that they desired to build the Rolls-Royce Merlin engine in the U.S. On June 17, 1941 the Chief of the Air Corps advised Dr. Mead that the Air Corps would collaborate in production of the Merlin provided the Army would receive engines for 3,000 airplanes over those then included in existing programs. Furthermore, the development and production of the 1700 bhp liquid-cooled Continental would be expedited and the experimental Rolls-Royce Griffin engines would be developed in this country as a backup, should the Continental not meet expectations. Subsequently the British recommended that the Saber 2000 bhp liquid-cooled engine be produced in lieu of the Griffon and that its production be at the expense of productive capacity for air-cooled engines.
- Even so, the Army had on order 866 Republic P-47B aircraft with air-cooled engines.
- The General summarized the situation by stating, "As the situation now stands, it appears that the American production ingenuity must be called upon to assist three leading manufacturers who have *failed in their guarantees to the U.S. Government* to deliver airplanes and engines selected in accordance with the Act of 1926. These manufacturers are the General Motors Corp., the Republic Aviation Corporation and the Lockheed Aircraft Company. There are indications that there will be other manufacturers (added) including the builders of air-cooled engines."

What the General was saying was that there were an abundance of problems as the aircraft and engine industries prepared for the coming war. Some of these problems could be traced to previous circumstances and methods by both the industry and the Government, but he also identified that through leadership and ingenuity a successful outcome was possible. In fact most of the points in his statement were successfully resolved, excepting the Continental engine and the intent to build either the Griffon or Saber in the U.S. Significantly, the statement also lends insight to the question of how and why the Merlin was brought on-shore. The General was able to get another 3,000 airplanes, as well as a access to a second supply of an already developed liquid-cooled V-12.

During the early days of the war, while Allison was in a period of production build-up and receiving its first combat experience with the fledgling V-1710, a heated controversy among "experts" and public alike was waging in the public press over relative merits of air-cooled versus liquid-cooled engines. So heated became the exchange over this new liquid-cooled engine that Army Air Force leaders became alarmed over the effects on morale of pilots and parents. Combat reports were carefully screened for news which could be released to offset the battering the liquid-cooled engine was taking from the armchair strategists. General Motors, with its pride and its investment in Allison, had a highly partisan interest in the controversy.[137] One of the reasons for the problem was the degree of exposure of green pilots to Allison powered aircraft. Not only were the front line fighters powered by the V-1710, but the only "advanced" trainers available to fighter units working up for departure to the combat theaters were often Allison powered early models of the P-39 and P-40.

Table 1-4
Liquid vs. Air-Cooled Installed Powerplant Weight

Component	Liquid-Cooled, pounds per bhp	Air-Cooled, pounds per bhp
Bare Engine	1.00	1.15
Engine with Cooling Systems	1.30	1.28
Total Power Plant, less Propellers	1.52	1.50

The direct resolution of the question of air vs. liquid-cooling has never occurred. In fact it remains to this day. But the acute problem during the early days of the war was resolved when neophyte pilots and ground-crews finally learned how to maintain and operate such advanced aircraft as the high altitude turbo-supercharged P-38 *Lightning*.

With the great diversity of aircraft types and the corresponding range of engines developed for them during the early 1940s, it should not be surprising that the differences, on a macroscopic scale, tended toward similarity. Table 1-4 from 1944, gives a summary of factors making up the weight of typical engine installations.[138]

Allison's Inline Engine for Flying *"Flivvers"*

With the excitement emanating from Lindbergh's 1927 trans-Atlantic flight it seemed that everyone wanted to get into aviation, or at least, be able to fly. Contemporary writers and authors dreamed of the "Flivver" airplane which would do for aviation what the Model T Ford "Flivver" had done for the automobile. The Commerce Department had even adopted a goal of a $700 family airplane to replace the family car! Allison responded to this enthusiasm and laid out a small air-cooled six-cylinder, in-line engine for the anticipated light plane market.[139] Stimulated by this market, and appreciating the limitations of the Liberty motors Allison was modernizing, and noting the interest in high temperature chemical cooling, Gilman saw an opportunity. He devised an entirely new cylinder design able to utilize chemical cooling and meet the future need for much more powerful engines. While focusing on the benefits to be had in a large 1000 bhp engine, he realized that a small engine for private planes and trainers was going to be needed for what seemed to be a waiting market. The air-cooled Flivver engine was dropped and the small liquid-cooled engine became the Allison 400-IG. It followed the VG-1710 concept and was a way to gain experience with the new cylinder since the big engine found no customers. Allison designed the new Flivver engines to be 150 bhp inverted and geared six-cylinder liquid-cooled in-line of 400 cubic inch displacement. As a result, two engines, the VG-1710 and 400-IG, were both being developed at the same time.[140]

Another of GM's early actions was to transfer the 400-IG project, and several of its designers, to the Cadillac plant at Detroit where GM is believed to have planned to establish an aircraft engine production line. Allison built and tested one or two of these engines, but the entire project died with the onset of the Depression and the Allison people returned to Indianapolis in the spring of 1930.[141] The engine development did prove some of the features later built into the V-1710, especially the cylinder block construction and valve gear.[142]

Basis for a Modern Liquid-Cooled In-line Engine

In October 1917 the War Department set up an engineering laboratory at Dayton, Ohio on property owned by the Dayton-Wright Corporation.[143] The Army named it McCook Field (after Gen. Anson McCook, a Civil War officer). The unit was established as the *Airplane Engineering Department, Signal Corps' Equipment Division*. Its object was to develop adequate aviation equipment for wartime purposes by coordinating and focusing the services aeronautical experiments, testing and research. Construction was started November 5, 1917, and throughout the First World War it did much valuable work. Later the organization's name was changed to, *"Engineering Division, Air Service, U.S. Army."* By January 1, 1919, during the height of U.S. involvement in WW-I, McCook Field had a 254-acre landing area and 69 buildings, including hangars, shops, laboratories, offices, hospital and a wind tunnel. Personnel consisted of 56 officers, 322 enlisted men and 1,096 civilians.

In 1926, on the recommendation of the Morrow Board, the Air Service became the Air Corps. Engineering Division functions were broadened to include supply, procurement and maintenance of all Army aircraft. Again the name was changed, this time to *Materiel Division*. As a consequence of McCook Field's world-acclaimed work it became obvious that an even larger experimental aeronautical research laboratory was needed. To keep the Army's air center in Dayton, home of the Wright brothers, the community raised $425,000 to buy 4,500 acres of land lying about 5-1/2 miles east of the city. This was given to the Government as a permanent home for the Materiel Division. The new facility was Officially opened on Columbus Day, 1927 as *Wright Field, Materiel Division*, USAAC. Developments coming from Materiel Division are legion, including the Wright *Whirlwind* engine, first of the successful air-cooled radials, development of superchargers, and in cooperation with the petroleum industry, improved fuels which had raised the octane rating of aviation fuel from a value of 50 in the 1920s to 100 by 1940.[144]

In the mid-1920s the Army Air Corps Materiel Division began investigating "chemically cooled" versions of what were called "water" cooled engines as a way to reduce aircraft cooling drag by utilizing the higher temperatures. If accomplished, it would make it possible to reduce radiator size by some 60 percent, and still have the same capability to cool the engine. Since water-cooled 1920s aircraft usually had their radiators simply hung out in the slipstream, sizable reductions in drag were anticipated. By 1928 the Air Corps was putting its first aircraft with "liquid", or "Prestone" (the commercial product name for Ethylene Glycol), cooled engines into service.[145] In 1930 the National Advisory Committee for Aeronautics, (NACA, predecessor to today's NASA), stated in its annual report to Congress that,

> "With the engineering materials and fuels now available, the maximum power of radials (air-cooled radial engines then common in aircraft) seems to be limited to about 600 hp. There is a real demand, both in the military services and in the Commercial field, for higher powered engines. This demand will no doubt be met in the coming year by development of liquid-cooled engines of 1,000 to 1,500 hp.[146]

Role of General Motors in Creating the V-1710

On May 14, 1929 the Executive Committee of General Motors met and appointed a committee of C. E. Wilson (VP of GMC), O.E. Hunt (VP of Engineering at GMC), C.F. Kettering and Mr. Norm Gilman "to consider the airplane engine program and formulate a plan for the Allison Engineering Company to pursue; this committee [was] to decide what type or types of engines for aviation we should build."[147]

One of the types suggested at the meeting was an engine for the small airplane to sell for around $700 that was being aggressively promoted by the U.S. Department of Commerce.[148] The engine program committee investigated the options and apparently found no active interest in the *Flivver* engine but there was interest in what became the U-250 two-cycle gasoline engine program.

In the field of higher horsepower aircraft engines they found the new 450 bhp Pratt & Whitney, and similar rated Wright air-cooled engines along with the water-cooled Curtiss "*Conqueror*", which Wright Field wanted to have converted to ethylene-glycol cooling. About this time, Wright brought out a new 750 bhp air-cooled engine, and at the same time, merged with Curtiss. The C-W interest in the *Conqueror* quickly faded, and its development was terminated.[149]

With the above knowledge obviously in hand, Norm Gilman and his team, in May 1929, began the design of a modern V-12 engine utilizing ethylene-glycol for cooling. They incorporated the latest innovations to make it capable of 1000 bhp.[150] Almost as a mark of this action, on July 1, 1929, the Air Corps finally prohibited the use of Liberty engines in any of their new aircraft.[151] Even so, the Air Corps still was supporting approximately 950 Liberties.[152]

The result of the Allison design was the VG-1710. Its existence was a direct outgrowth of the survey initiated by General Motors for new product development by their Allison acquisition. Gilman had led the survey team in focusing on the perceived need for a 1000 bhp, high temperature, liquid-cooled Vee type 12 cylinder engine.[153] With GM's concurrence he presented his design to both the Army Air Corps and Navy, but could stir up no interest or support from either. In the meantime Allison began work on the 400-IG and further refinement of the VG-1710 design.[154]

The decision to design the new engine as a liquid-cooled, V-12 in-line, was not a casual or pre-conceived commitment. Harold Caminez, a man with extensive experience as an engine designer, was brought in specifically to do the detailed design and analysis of the new engine. He subsequently published an ASME paper titled *Large Engine Development* in which he succinctly describes the rational for sizing and laying out the features of what became the V-1710.[155] The output of 1000 bhp was selected because aircraft designers were beginning to use multiple 700-750 bhp engines to obtain sufficient performance; he argues that a single 1000 bhp engine would offer the same performance. He also relates that previously rated "large" engines were required to demonstrate only the most marginal levels of reliability. During WWI, engines were required to deliver rated power for only 30 minutes, and that by the mid-1930s engines in commercial service were being rated on results of a 60-hour run. The new engine was going to be designed to have considerably improved reliability and be the first to satisfy the new Air Corps requirement to complete a rigorous 150 hour Type Test.

So that the engine would be competitive and attractive to users, it was decided to design the engine not only for 1000 bhp, but with a weight not to exceed 1.3 pounds per horsepower. Furthermore, it was desired to take advantage of the inherent high efficiency offered by liquid cooling, and provide an engine with specific fuel consumption during half power cruising conditions of 0.4 pounds per horsepower per hour.

The physical dimensions of the engine were selected in recognition of the cooling problems with the cylinder and piston if the diameter of the piston exceeds about 6 inches. Since Allison was interested in an efficient engine, which required good volumetric efficiency, the stroke/bore ratio was determined to have to lie in the range of about 83-125 percent, resulting in a stroke of less than 7-1/2 inches for a 6 inch cylinder. These dimensions result in a single cylinder of 230 cubic inches, with a peak rating of about 100 bhp. Caminez believed that such a cylinder was too large for a viable design, particularly if it was air-cooled, but felt that if liquid-cooled a reduction in dimensions would still allow developing 100 bhp, though a slight amount of supercharging would be needed.

Ten of these cylinders would deliver the desired 1000 bhp, but a ten cylinder arrangement is not desirable for a number of good mechanical reasons. Caminez believed that a more suitable number would be either twelve or fourteen cylinders. With these numbers, he then considered four different cylinder arrangements, the so-called "Hex", having two banks of either six or seven cylinders, radially arranged. Alternatively, a "Flat" 12-cylinder arrangement with 180 degrees between two banks of six cylinders, or the more usual arrangement where the two cylinder banks are set 60 degrees to one another in a "Vee." Caminez lamented, the "actual choice of cylinder arrangement from the four available is a difficult one, especially when the engineer's judgment is not biased by his organization's previous policy and practice. Each of the arrangements has definite advantages peculiar to itself."[156]

Caminez felt that the 12-cylinder Vee arrangement was well known and an imposing amount of research and investigation had been done on it, concluding that the development of such an engine, "would undoubtedly prove less costly than either of the radial arrangements."[157]

In the matter of cooling an engine Caminez relates that there are four coolants; vapor, air, water or a high temperature coolant such as ethylene-glycol. Water-cooling was eliminated because of altitude and aircraft drag considerations.[158] Air-cooling is usually believed to provide (1) low overall weight of the installation, and (2) less likelihood of trouble due to plumbing failures. He agreed with the first reason, but noted that the advantage was not so great as usually believed when the ducting, baffles and larger oil cooler radiator, were included. The difference was usually the result of only the weight of the quantity of coolant, about 0.2 pounds per bhp in a well designed installation.

Caminez also argues that the danger of coolant leaks is only about one-third as real as it appears. A liquid-cooled engine has three plumbing systems: fuel, lubricant and coolant. An air-cooled engine eliminates only the coolant, and remains vulnerable to failure should either the fuel or lubricant systems be damaged. Vibration is a continuous risk and cause of liquid system failures. In this regard a properly designed V-12 is the smoothest of the various reciprocating engine arrangements.

Caminez identifies the advantages of the liquid-cooled engine as being: (1) controlled cylinder head temperatures, permitting high compression ratios, (2) simple cowling, (3) extremely low drag, (4) flexible radiation surface or cooling drag matched to flight conditions, and (5) easy simulation of flight conditions during development testing on the torque stand.

Once it had been settled that the engine would have 12 cylinders, Caminez reduced the cylinder dimensions to the 5.5 inch bore and 6.0 inch stroke which defined the V-1710. Such an engine, with a compression ratio of 6.0:1 and running at 2500-2600 rpm with some boosting to improve volumetric efficiency, would be able to deliver the requisite 1000 bhp, and exhibit commendable fuel economy. Given the high crankshaft speed, he proposed that the engine would be equipped with 3:2 or 2:1 spur type reduction gears because of their simplicity. He also considered the trend toward high altitude operation, and as an aid to achievement of a vibrationless engine with the desired low fuel consumption, he argued "it is probable that some type of manifold [fuel] injection device should be used instead of a carburetor."[159] This engine, as described in Caminez's 1935 paper, is exactly the engine then coming off of the Allison drawing boards and being constructed as the fuel injected XV-1710-5(C2).

In the spring of 1930 Messrs. C.E. Wilson and O.E. Hunt, then Vice Presidents of General Motors, came to Indianapolis to review Allison's future programs, if any, in the aircraft business. They again reviewed Norm Gilman's strong convictions in support of the V-1710 and concluded that some further layout, design, calculations and presentations should be made and Allison's proposal re-reviewed by the Military.[160] The Army apparently did not have much interest or funds, but they suggested that the Navy might have a need and ability to support development. Navy interest was found to exist, but it was for four 750 bhp airship engines.[161] Still, in 1930 they did contract with Allison for one VG-1710, as it was already designed. Moreover, its construction and testing was expected to expedite subsequent development of the desired airship engine.

As the Navy GV-1710-A, this engine was built and tested in 1931 at 650 bhp. When delivered to the Navy on March 12, 1932 as the GV-1710-A1, it was rated to operate at 750 bhp.[162] The U.S. Army Air Corps closely monitored the engine's development, to the point that they had representatives present during all of the Navy testing.

Based on the success of the Navy engine, the Air Corps began negotiations with Allison in January 1931. These resulted in a March 1932 contract for a single XV-1710-1. This engine was based on the Navy engine and designated as Allison model V-1710-A2.[163] It was to be similar in all major aspects to the Navy GV-1710-A1, except that the propeller shaft and supporting nose case were to be approximately 12 inches longer. This was intended to accommodate streamlined aircraft nacelle installations. Furthermore, it was to be designed to allow operation at 1000 bhp and 2800 rpm, though the XV-1710-1 itself was to be rated at 800 bhp and 2400 rpm with Allison guaranteeing 750 bhp at 2400 rpm. The engine, after initial testing by Allison, was delivered for experimental tests to the Air Corps at Wright Field on July 6, 1933.

Meanwhile, Allison was hard at work on the V-1710-B Navy airship engine. This was a naturally aspirated, liquid-cooled V-12, directly developed from the GV-1710-A1, but equipped with devices to allow it to fully reverse direction and return to full power within eight seconds. Interest-

McCook Field 50-hour endurance test of the XV-1710-1(A2). The test began on August 30, 1932 and was completed September 9, 1932, using a Hartzell four bladed wooden test propeller. (Air Corps)

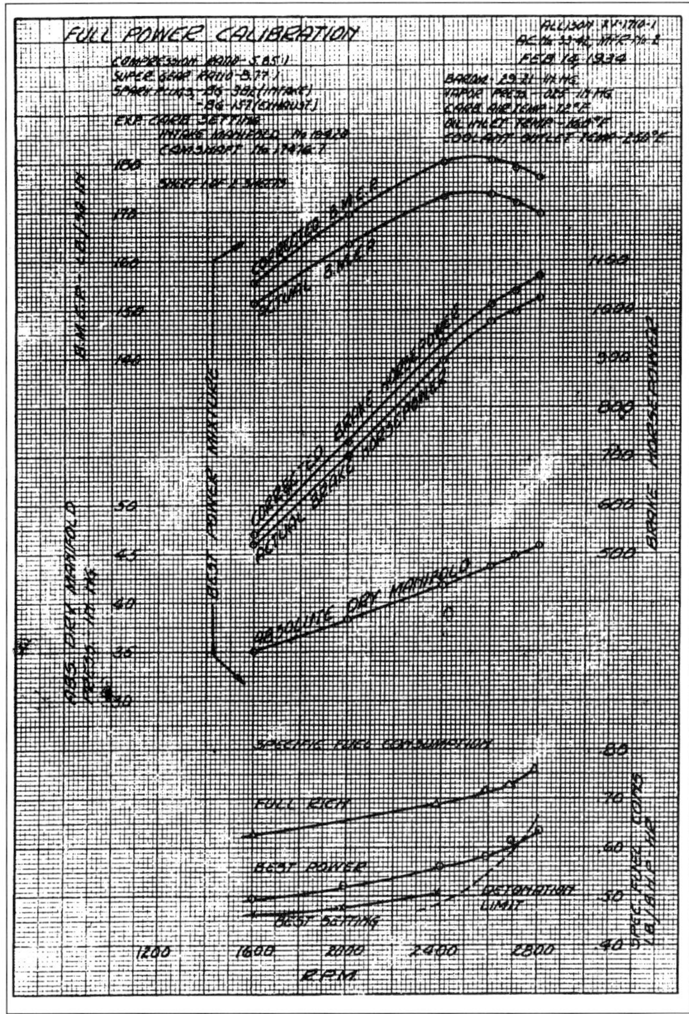

This performance curve was prepared from XV-1710-1(A2) data taken after the change to 250° F coolant temperature. The engine was rated or 750 bhp at 2400 rpm, but demonstrated 1070 bhp at 2800 rpm. (Air Corps)

ingly, the Navy had suggested that Allison should support the crankshaft and connecting rods on roller bearings like those used in the German Maybach engines then powering the dirigibles Shenandoah and Los Angeles.[164] Skeptically, Allison engineers looked at the situation. They found that more than half of the new Maybach engines in the Navy hangar at Akron were already out of use because of bearing trouble. They stayed with their "plain" type main bearing.[165]

The following statement is taken from the Air Corps Technical Report detailing the results of testing the XV-1710-1.[166] It offers insight into what was considered state of the art, along with a statement of military strategy of the day.

> At the time the first Allison XV-1710-1 engine was procured by the Air Corps, certain of its features made it of special interest as a possible military type. Among these was the fact that, being liquid-cooled, it could be turbo-supercharged to maintain its rated power at high altitudes. For such operation cooling can be obtained by radiators enlarged within practical limits, whereas increased cooling area on air-cooled engines can be provided with great difficulty, if at all. Another, was the fact that it was the first large engine type to be designed from its inception for operation with a high coolant temperature, and could therefore be expected to function satisfactorily at temperatures which would enable airplane performance to be improved by the use of small radiators.
>
> Other features of merit were its small frontal area, roughly six square feet, and its low specific weight, which, although not especially attractive at a rating of 750 or 800 horsepower, would become so if, as was expected, the engine could be developed to a rating of 1000 horsepower with small increase in weight beyond its initial 1,122 pounds.
>
> In general, design of the engine was judged to be outstanding. Accessibility of the various accessories was unusually good. The construction permitted rapid disassembly and reassembly, and workmanship in manufacture was excellent.

Development and testing of the V-1710-C, follow-on model to the XV-1710-1 for the Army, was a long and tedious affair for Allison. There were many problems, and several of them were interrelated. Since the Army would only pay for an engine upon completion of the engine acceptance test, it would sometimes be years before payment would occur. Fortunately, General Motors continued to support developments through this troublesome Depression Era.

After a fairly straight forward development and test program on the XV-1710-1(A2) the XV-1710-3(C1) was purchased with the intent of having it perform the specified 150-hour Army Type Test at 1000 bhp. This engine ran hundreds of hours in development and attempts at the Type Tests. While often close, it never satisfactorily completed the rigorous Type Test. General Motors reasons for continuing support of the protracted project was probably the order for ten YV-1710-3's by the Army in 1934, though their manufacture was pending completion of the Type Test. These were intended to power a new generation of experimental bombers. Compounding the effort were the distractions surrounding the program including: the Army insisting that the YV-1710-3's be designed and built with fuel injection, the Navy program for the V-1710-B reversible engine, the design and production of the complex drive systems for the USS *Akron* and USS *Macon* dirigibles, and designing the new X-3420 engine for Army long-range bombers.

Carl Weinbrect, Supervisor of Tests, Jud Butteim, Chief Testor and Ron Hazen, Chief Engineer with the final Type Test engine YV-1710-7 (C4/C6) at Indianapolis upon the occasion of completing the Type Test penalty runs, April 1937.

By late 1935 General Motors realized that the V-1710 program was in big trouble for want of a successful Type Test. Air Corps programs such as the Martin XB-16 were being canceled for want of V-1710's. The engine continued to regularly break its major parts and there seemed little hope for major production orders during the Depression. Still, General Motors had a lot of its own money in the project, much more than the U.S. Government, and its corporate leaders could see the war clouds rising around the world. GM corporate management decided to refocus its resources. They terminated several programs such as the U-250 engine. This brought additional talent, including Ron Hazen who had been leading the U-250 program, into the V-1710 project.

One of the nearly immediate benefits, was Hazen's recognition that the real trouble with the engine. Poor fuel distribution was badly overloading some of the cylinders and starving others. This resulted in rough operation, excessive vibration, local overheating, and detonation. He was then instrumental in throwing out all further work on fuel injection and other distracting activities, and concentrating on one model. He then designed an entirely new intake manifold in addition to substantial increases in strength and weight of the crankshaft and major castings. The result was the successful Type Test for the V-1710, completed by the YV-1710-7(C4/C6) in April 1937.[167]

People Behind the V-1710

Mr. Norman H. Gilman was long the President and Chief Engineer of the Allison Engineering Company, and continued in this capacity after the Company became a wholly owned subsidiary of the General Motors Corporation. His management and creative engineering abilities were responsible for the successful operation of the Company even before Jim Allison's retirement. In fact, he was the one who prepared Howdy Wilcox's Peugeot which won the 1919 Indy. Furthermore, it was he who was primarily responsible for the invention and development of the steel backed lead bronze bearing that remained a profitable product for the Company for many years.[168] He was also the key man behind the V-1710, but it took a lot of support and the talents of many people to get the engine into service and to keep it on the front lines through WWII. A key personality and leader in this effort was *Ronald M. Hazen*, who took over the role of Chief Engineer in 1935, allowing Gilman to retire.

Ron Hazen, a native of Wahpeton, North Dakota, had been enrolled at the University of North Dakota when WWI interrupted his studies in 1917. Drafted into the Army, he first served as a master mechanic at the 7th Aviation Instruction Center in France, then was promoted to Sergeant in charge of engines for the 96th Bombardment Squadron. He was subsequently commissioned a Lieutenant and trained as a pilot in the U.S. Army Air Service. After the war he entered the University of Michigan, from where he graduated in 1922 with a BS in Mechanical Engineering. He then took a position with the General Motors Research Corporation in Dayton, Ohio. They allowed him to take the graduate engineering course at the University of Minnesota, which he completed in 1923. He then began a stint as an instructor and Assistant Professor of Mechanical Engineering at the University, where he completed postgraduate work while spending his summers working on special engine projects at McCook Field. In 1927 he returned to full time aircraft engine work as special projects engineer for the Wright Aeronautical Corporation in New Jersey, where he stayed for two years. He then became Assistant Chief Engineer of Fairchild Engine Corporation where, during his nearly six years with the firm, he supervised the design and development of the "Ranger" six and twelve cylinder air-cooled inline aircraft engines. During this period he was promoted to Chief Engineer.[169]

It was this background and experience that Hazen offered to Norm Gilman in early 1933 when he applied for a position at Allison. Gilman hired him, but the General Motors Research Laboratories in Detroit asked that he work with them and apply his knowledge of air-cooled engines to their projects, particularly the two-cycle "U" engine. This engine was intended to be liquid-cooled in its final form. By March 1933 Hazen was busy with the design of a six-cylinder "U" type radial engine at the GM Research Laboratory, although he was still working as an Allison employee.[170] In early 1936, with the compounding problems in getting the V-1710 through its Type Test, and with Gilman planning to retire from the position of Allison's Chief Engineer, Gilman arranged for Hazen to return to Allison and named him Chief Engineer. In the process Mr. Harold Caminez, who had been brought into the V-1710 project in 1930 and had risen to the position of Project Engineer on the V-1710, left the employ of the Company. It had been he who had done the original design analysis and provided the analytical basis for the V-1710.[171] Gilman notified the Air Corps that Ron M. Hazen had been named Allison's Chief Engineer on March 6, 1936.[172]

Ron Hazen immediately threw himself into the V-1710 development and resolution of its problems. He determined that some parts of the engine were too light. In the interest of meeting the Air Corps demands for light weight, many parts of the engine were simply not up to the demands being placed on them, particularly when the engines were run in test stands simulating in-flight conditions. Engines which ran fine on rigid dynamometers would quickly fail when driving a propeller! Hazen believed a few pounds of metal in the right places would solve these problems. The other problem was due to the poor mixture distribution and delivery between the six cylinders on each bank. The "Log Type" manifolds were delivering too much fuel to cylinders 1 and 6, and starving cylinders 3 and 5. The lean cylinders tended to detonate and/or run hot. This damaged valves, pistons and often times, the crankshaft due to stresses from the uneven firing pulses.[173]

He also realized that the small Allison team was being pulled in too many directions, trying to get the XV-1710-3 through its Type Test, while designing and testing the XV-1710-5 fuel injection system and engine, as well as work on the 1600 bhp X-3420. At the same time they were being distracted by airplane manufactures seeking the perfect engine for their unique projects. The V-1710 program was six years old and it was now time to complete the Type Test and get on with putting the engine in the air and into production. Ron Hazen knew he had a good team; what was needed was to focus their efforts and move forward.

Wright Field, the customer, was demanding that fuel injection be used on all future V-1710's, and the X-3420 as well.[174] Hazen believed that was impractical given that there was no suitable fuel metering pump then available. He decided that the fuel injection project had to be terminated. Anticipating a major battle, Hazen lined up his support, including GM Engineering Vice President Ormand E. Hunt. At the pivotal meeting in March 1936, Hazen told the Air Corps that, if allowed to proceed with his plans, they would have a qualified V-1710 ready for production in a matter of months! It has been reported that the Air Corps people turned skeptically to Hunt and asked, "Can we believe Mr. Hazen is peaking with any authority?" To which Hunt responded, "Mr. Hazen is our Chief Engineer. What he says, General Motors will back up."[175]

The Air Corps had received the carbureted XV-1710-7(C2) in December 1935, intending to use it to power the Bell Aircraft modified A-11A flying test-bed. It was being fitted into the aircraft when its sibling the XV-1710-3(C1) Type Test was declared a failure. At the same time, and with the support and concurrence of Norm Gilman, Hazen and his team dove into the further redesign and construction of the V-1710-C3, to be known as the XV-1710-7 even though it combined the features needed to enable it to be converted to a fuel injected XV-1710-5.[176] On June 13, 1936, after a span of only thirteen weeks, Allison delivered the new engine to Wright Field. Not only had major parts of the engine been strengthened, but Ron Hazen had designed an entirely new intake manifold. This became a classic component of the V-1710. Known as the "Rams Horn" manifolds, they were able to solve the maldistribution problem and at the same time reduce internal pressure losses. In the next major version of the engine, the V-1710-11(C7), power was increased to 1150 bhp while the supercharger gear ratio was significantly reduced, largely due to improvements in the induction system efficiency.

Ron Hazen was very proud of his contribution to the V-1710 and always considered this manifold the "heart" of the engine.[177] The manifold went through a number of changes during the production life of the engine. These specifically included a change from magnesium to aluminum and a improvement in efficiency by changing the shape of the internal passages to further streamline the flow into the cylinders and eliminate the need for backfire screens. With his technical and managerial skills clearly established he went on to a long and productive career that took Allison well into the jet age. He was formally recognized in 1940 with his appointment to the Aircraft Powerplant Committee of the National Advisory Committee for Aeronautics.[178]

With Ron Hazen firmly in place as Chief Engineer, Norm Gilman finally retired on December 31, 1936. Mr. O.T. (Pop) Kreusser of General Motors was then named General Manager of the Division and elected President of the Allison Company.[179] General Motors Corporation served as an important source of key people necessary to successfully plan and carry out the expansion of V-1710 production preceding America's entry into WWII. *Charles J. McDowall* was another key force in the ultimate success of the V-1710. He served as Chief Design Engineer throughout the production life of the engine.[180]

Mr. W.H. McCoy of the General Motors Research Division, was assigned the responsibility for tooling and planning the four engines per day plant expansion program of late 1938. He then became the Production Manager of Plant #3 when it was completed in 1940 and quickly ramped up to producing more than ten engines per day! During 1940 other experienced production executives were transferred from other GM Divisions and they formed the core of the Allison management structure of the day.

Mr. W.G. Guthrie, formerly Works Manager of the Opel Division of GM became the Allison Works Manager in June 1940. *Mr. B. Conway* took over as Plant #3 Superintendent in March. He was previously General Master Mechanic at GM's Pontiac Division. *Mr. C.M. Jessup*, formerly Production Manager of GM's Delco-Remy Division, took over the Allison Material Control Department in May. Allison's Chief Inspector was formerly Inspection Superintendent at A.C. Sparkplug. In addition, a large number of other experienced design and production engineers were transferred in from other GM Divisions.

With a second round of expansion and real production requirements beginning in 1940, *Mr. F.C. Kroeger*, General Manager of Delco-Remy Division and one of the most successful manufacturing managers within the General Motors Divisions, was appointed General Manager of the Allison Division.[181]

Until the expansion program began in 1939 Allison's focus had been on experimental engineering, engine development and bearing manufacture. It had developed strong skills in the engineering and manufacture of prototype and unique engines and power transmission equipment, but there had not been a need to develop a deep management structure. Fortunately, General Motors was ready to commit the necessary depth, as well as the support of its manufacturing facilities.

This is not to say that everything at Allison was rosy. The rapid expansion resulted in protracted Model Testing of the new production models as well as quality control problems. On 1 July 1940 *Major James "Jimmy" Doolittle*, Army Air Corps was assigned as the Assistant District Supervisor of the Central Air Corps Procurement District, with duty as the Army Air Corps representative at the Allison Indianapolis Plant. He completed his assignment in November 1940 and submitted a report stating that the problems with Allison were largely managerial. He noted that many personnel changes had taken place and how he felt that certain top managers were more interested in quantity than quality. He credited *O.E. Hunt* with straightening out the mess, and acknowledged that progress was being made toward turning out reliable engines based on Hunt's "knowledge, skill, patience, honesty, and never-failing cooperation."[182]

In Doolittle's opinion, the large expansion at Allison had resulted in an organization unaccustomed to making aeronautical equipment, and that as such, they could not produce a satisfactory product until they acquired the "*aviation viewpoint*" and [the culture to] work to aeronautical instead of commercial standards. He stated that, "The bigger and more successful an organization is (i.e. General Motors), the harder it is for them to realize this."[183]

From the historical perspective, Doolittle's comments were certainly reflective of the circumstances as he found and viewed them. Important to the discussion is the unprecedented scope of the expansion which was underway, and significantly, the commitment of General Motors to the success of the V-1710. That Doolittle's findings and concerns were soon put to rest. Evidence showed that large quantities of quality engines were soon being produced. This is clear from the operational record and accomplishments of the tens of thousands of Allison V-1710 powered aircraft during the coming war.

But who gets the credit for "inventing" the V-1710? Norman H. Gilman, in 1940, was voted by his associates and honored by the American Manufactures Association, as a "Modern Pioneer on the Frontiers of American Industry." He was bestowed with the credit for having conceived and developed the design of the Allison engine. But clearly the whole undertaking was vastly greater than a one man job. As Gilman himself pointed out at the award ceremony, "I cannot tell how I got any particular idea. You talk to one man and that starts you off on another track. Bye and bye your thoughts start crystallizing and finally you get a definite idea. But I do know that if it had not been for Jim Allison I would never have built that engine."[184]

That is undoubtedly true. And so it is that into the development of the V-1710 went the spirit of Jim Allison and his associates and the men with whom Norman Gilman came in contact. Many of them were the outstanding men of their time. These men were laying the groundwork for the world's greatest industry – the American automobile. This group included:

Arthur Newby of National Motor Car and Vehicle Corporation,
Carl Fisher, an optimist whose eyes sparkled like diamonds when he got a big idea,
Frank Wheeler of Wheeler Schebler Carburetor Company,
Daredevil drivers, Barry Oldfield, Eddie Rickenbacker, Johnny Aiken, Joe Dawson

These and many others set the tempo of the period when the Allison engine idea was born. They undoubtedly made their imprint on Norm Gilman.

But when the Allison engine idea had grown into a definite production project, it was only the tremendous power of a far flung organization that allowed it's success. No small organization could have financed the project, could have furnished the research facilities required, or would have had available the kind of engineering and manufacturing talent needed. The men of General Motors can be proud of the job they have done. And to Norman H. Gilman goes the happiness, satisfaction and honor of having pioneered the creation of the world's finest and most powerful liquid-cooled airplane engine.[185]

The Mechanics, The People Doing the Work

Thousands of men and women working together actually made the V-1710 happen. Their stories are just as unique and important to the success of the engine as were those of the principals. A personal view of Allison during its early growth period is offered by a participant and worker, Don Wright.

Don came to Allison a young man right out of school and went into Plant 2, where he was to learn the "ropes" of engine building. His education was of the old method, where one effectively learned as an apprentice for a considerable period before taking on positions of responsibility. He speaks fondly of the people in the shop. "We had some of the most outstanding mechanics, and when I went there it was [just] a machine shop with an assembly and test area. Within those particular groups you had master mechanics all over the place, guys that [had] put race cars together, and lots of them had lifetime jobs from Jim Allison. That's something that I think I ought to mention, lots of them had lifetime jobs from Jim Allison. When he sold the outfit in '29 he made an arrangement with General Motors that the guys who worked there at that time would have a job for life: Burt Lang, Marty Becker, Shorty Stockdale, Erni Ceise, Slim Anderson. Of course I got to know all of them because I was the kid around there. I was waiting to go off to GMI [General Motors Institute] the next year around.[186] When I got in there they said, well, we can't sponsor anymore because we're only a small Division and we can only send so many. By the time I got around to going I'd been through the assembly and test operations and was working in final assembly when I was about 19, and then when I was 20 they sent me to California, and I thought I was in heaven. I was living in the Hollywood Roosevelt Hotel, in an apartment, and I had a Ford convertible with black and red leather upholstery. But those guys, like George Lishlion, when it came to building an engine they were expert. Every place that you would go there was a guy that was an expert. Russ Wright, he was in charge of experimental operations when I went to work there they wouldn't let me work in there. The first day that Virgil Grimes was foreman was the day I went to work. Carl Weinbrect was the superintendent. So this fellow Russ Wright, a rather eccentric individual...he and I got along fine...other people he didn't get along with.[187,188]

Development of the Allison Division of GMC

Table 1-5 recaps the early evolution of the corporate history of Allison. Following acquisition by General Motors, the Allison Engineering Company remained fairly stable through 1936, when roughly half its personnel were applied in the manufacture of aircraft engine bearings and the other half to the development of the V-1710 and other military projects. One major change occurred on January 1, 1934 when the Allison Engineering Company became merely a sales organization for convenience in handling existing Government contracts. The manufacturing organization became the Allison Division of General Motors Corporation. The Allison Engineering Company was gradually phased out as new contracts took effect. The Allison Division became a major supplier of aircraft engines, and remained a full Division of General Motors Corporation until it was sold in 1994, 60 years later.

The successful completion of the Type Test of the YV-1710-7(C4) in April 1937, and the promising performances of the V-1710 powered prototype Curtiss XP-37 and Bell XFM-1 during that year, resulted in the Air Corps purchasing service test quantities of these airplanes.[189] Allison then received orders for 22 V-1710's in 1937 and another 43 in 1938 to support the YP-37/YFM-1 and other Air Corps experimental aircraft projects, the XP-38 and XP-39. For more information see Chapter 17 on Production and Manufacturing for details on delivery of these engines.

Given the circumstances of the Depression, getting the V-1710 through Type Testing would probably not have been possible without the corporate resources of General Motors and its steady support of the Allison Division. Through 1937 GMC had supported development with about $900,000 of its own money, while the Navy and Army together had paid about $1 million between 1930 and 1937.[190] For perspective, the Air Corps appropria-

Table 1-5
Corporate Evolution of Allison

Corporate Name	Date	Products
Indianapolis Speedway Team Co.	1913	Race Car Designs and Models
Allison Speedway Team Co.	1917	Superchargers, Tank Tracks, Whippet Tanks, Hi-Speed Tractors, Tool-Jigs for Liberty Engines, Liberty Master Models
Allison Experimental Co.	1918	Same
Allison Engineering Co.	1919	Same, Race Car modifications
Allison Engineering Co.	1927	Roots Blowers, A/C Reduction Gears, Hi-Speed Superchargers, Steel backed bearings, Rebuilding Liberty Engines
Allison Engineering Co., purchased by GMC.	1929	A/C Reduction Gears, Hi-Speed Superchargers, Allison Bearings, V-1710
Allison Division, GMC	1934	A/C Reduction Gears, Hi-Speed Superchargers, Allison Bearings, V-1710
Allison Engineering Division, GMC	1937	V-1710, Aircraft Engine Bearings
Allison Division, G.M.	1939	V-1710, Aircraft Engine Bearings
Allison Division, G.M.	1945	V-1710, V-3420, J33 Jet Engines, Shock Absorbers, Aircraft Engine Bearings
Allison Engine Company	1993	Military and Commercial jet and turbine engines
Allison Engine Company of the Rolls-Royce Aerospace Group	1994	On November 21, 1994 Allison was purchased by Rolls-Royce. Products will continue to be helicopter, turboprop and military jet engines along with a line of Industrial turbine engines.

Table 1-6
Early Contracts, Orders and Production

Date	Contract	No. Ordered	Comments/Deliveries
6-26-30	Navy C-17952	1	GV-1710-A
1-5-33	W535-ac-05592	1	XV-1710-1(A2)
1-24-33	Navy C-29907	3	V-1710-4(B1R/2R) Reversible, Airships
1934	W535-ac-06551	1	XV-1710-3(C1), Type Test
6-15-34	W535-ac-06795	10	YV-1710-3, Del as various Models
3-1937	W535-ac-09678	1	XV-3420-1(A1), AC38-119
1937	W535-ac-10291	1	XV-1710-15(C9), AC38-120
1938	W535-ac-10830	1	XV-1710-17(E2)
1938	W535-ac-11279	1	XV-1710-17(E2)
1-5-38	W535-ac-10660	59	V-1710-21(C10) and V-1710-23(D2)
2-18-39	Navy C-65197	2	XV-1710-6(E1) for Bell XFL-1
6-27-39	W535-ac-12553	969	4 Models, includes 132 for P-38's
11-8-39	W535-ac-13420	81	New E-4 Model for P-39C's
12-8-39	F-207(French)	115	V-1710-C15, 15 were spares
2-1-40	F-223(French)	669	700 ordered as C-15, 31 to China
5-25-40	A-196(British)	3,500	879 C-15 delivered, plus 69 to China 810 as V-1710-E4 1,416 as V-1710-F3R 163 as V-1710-F5R 163 as V-1710-F5L
12-12-40	W535-ac-16323	3,691	Delivered to AAC as various models
2-6-41	China	50	V-1710-C15A, spare engines
	China	31	C-15's from F-223, French order
	China	69	C-15's from A-196, British order
8-16-41	ac-67 (Lend-Lease)	2,160	AAF for Lend-Lease
10-29-41	W535-ac-21623	6,953	Delivered to AAF as various models
11-29-41	ac-628 (Lend-Lease)	932	AAF for Lend-Lease
12-9-41	W535-ac-22831	810	Delivered to AAF as various models

Production and Delivery Summary
Prototype and Development, 16
1937 Deliveries, 7
1938 Deliveries, 12
1939 Deliveries, 48
1940/1155 total (w/705 Foreign)
1941/6437 total (w/3645 Foreign)
1942 Deliveries, 15,319
1943 Deliveries, 21,381
1944 Deliveries, 20,302
1945 Deliveries, 5,501
1946 Deliveries, 17
1947 Deliveries, 603
1948 Deliveries, 145

Note: These deliveries total 70,943, a slightly larger number than from other sources.
Significantly, In the 1938-1941 period prior to Pearl Harbor, Allison built 7,652 engines and was ahead of its contract schedules.

tions for 1932-33 were only $25,673,236, while the total sales of military planes and engines dropped from $17,167,794 in 1932 to $14,494,798 in 1933.[191] These trends and levels of expenditure need to be remembered when the pace of V-1710 development is assessed. This also explains why engines were built and re-built so many times. Moreover, the military would only pay for engines that were "accepted"; that is, ones which had completed their contract acceptance tests or requirements. This meant that most of the government funds were not actually delivered to Allison until long after the original expense.

By the last half of the 1930s, Allison became somewhat desperate for sales of its engine. Plant 2 had been constructed in 1936 to provide facilities for manufacturing as many as 100 engines per year, yet by 1939 production was quickly coming to an end on the total of only 79 engines which had been ordered since 1934. Allison was working closely with aircraft manufacturers seeking applications for the engine. This was a large part of their reason to seek approval for application of a non-turbosupercharged engine. Allison also sought foreign sales. To this end, the 1937 vintage V-1710-C6, had been proposed to a Swiss delegation in November 1938. It was a commercial offering that never received a military designation. On February 2, 1939 Allison requested approval from the Materiel Division at Wright Field to release specifications and installation drawings on the stated "obsolete" engine to the Swiss. The request was passed to the Chief of the Air Corps. In response, the Air Corps noted that while the offered engine was obsolete, "any engines contemplated for export sale would be to current production specifications." Furthermore, they stated that, "the Allison V-1710 type engines are specified for certain of the types of airplanes in the Emergency Procurement Program..." with the result that the Chief of the Air Corps objected to furnishing the material inasmuch as the capacity of the Allison Engineering Company will be completely absorbed by proposed procurement by the U.S. Army for a period of at least two years.[192] This was before the results of the delayed 1938 Pursuit Competition were determined and still did not result in any tangible orders for the V-1710.

It was not until the January of 1939 when the V-1710 powered Curtiss XP-40 airplane won the rescheduled 1938 Pursuit Competition that Allison was clearly in the picture as a production source for military aircraft engines. Even at that time the quantities for the next fiscal year (FY-40) were planned on a basis of two or three hundred engines *per year.* The rapid development of the war in Europe made rather sudden changes in the level of proposed requirements. In June 1939 the Air Corps issued a contract for 837 engines to Allison.[193] Unfortunately from a production standpoint, this order included four models of the V-1710, each differing from the other. A fifth model was added when another contract for 81 engines was received that December. These were to power the Bell P-45 (P-39C).

Early in 1939, following the Pursuit Competition, Secretary of War Louis Johnson called W.S. Knudsen, President of GM, and requested they build a plant in Indianapolis suitable for production quantities of the V-1710. The military was still faced with a critical shortage of funds and he was only able to promise a contract to follow for 836 engines.[194] Even so, General Motors elected to honor the verbal request.

Immediately following the 1939 Armistice Day 500 mile race at the Indianapolis Speedway, General Motors, Allison, and military officials drove the short distance from the race track to an old pear orchard on 10th Street in Speedway.[195] There they participated in ground breaking ceremonies for a new factory and office building. This was before the actual receipt of the first production contract from the Air Corps. Total floor space was to be 310,000 sq.ft. and Allison immediately started ordering the necessary machine equipment and tools for mass producing the V-1710. The production plans at that time were more than ample to take care of all contract requirements placed, or anticipated, in 1939 or 1940. Production in the new facility, known as Plant #3, started in February 1940 and engines were first completed at this plant in June. Back in December 1939 an order for 115 V-1710-C15 engines had been received from the French, though events required that these engines be delivered to the British after the fall of France in the Spring of 1940. While some accounts criticize Allison for having been slow to deliver these early engines, the facts are that more engines were delivered during the year 1940 than were on order January 1, 1940.[196]

From another perspective, in the Summer of 1940 the U.S. AAC placed orders for 3,000 Rolls-Royce Merlin's, to be built by Packard in Detroit. That October the Air Corps increased its orders to Allison, but still only for a total of 1,050 engines. Even so, one reason given by the Air Corps for the Merlin order was to free Allison production capacity and insure sufficient V-1710's were available for the Lockheed P-38 and Bell P-39.

Additional contracts were received on February 1, 1940 for 700 C-15 engines for the French[197], 3,500 in five models for the British[198] on May 25, 1940, and 3,691 for five different models from the U.S. Army Air Corps on December 11, 1940. Allison anticipated these orders and increased its factory floor space three-fold during the year. In addition the production facilities of other General Motors Divisions, notably Cadillac and Delco Remy, were made available and committed to production of major V-1710 components. This expansion in conjunction with the development and proof testing of new models, along with obtaining and training personnel was an achievement which would not have been possible without the resources, manufacturing experience and personnel reservoir of General Motors Corporation.

Where the rate of engine production had been about two engines a month at the end of 1938, by the end of 1939 it was eight per month, a 400 % increase. By the end of 1940 Allison and its component suppliers were producing 300 finished engines per month! During 1941 this was again dramatically increased and reached 1,000 engines per month by the time of Pearl Harbor and America's entry into WWII. The 1941 expansion was complicated by the requirement to complete production of the C-15 engine and get four new models into production during the year. Factors effecting 1940 production were an Air Corps contract changing the number of C-15 engines from 776 to 300 and a substitution of F-3R engines, a model which had not yet completed development. At the same time the major changes in the P-39 program involved revision of the turbosupercharged E-2 engines from sea-level to altitude rated E-5's, a type not then completely developed.[199] Consequently, total deliveries to the Air Corps in 1940 were only 342. Another 833 C-15's were delivered to the British as well.[200]

In recognition of the changing conditions in Europe following the invasion of Poland in September 1939, the U.S. was working to find ways to aide their allies. At the very time France was being over-run by Germany in the spring of 1940, the Materiel Command was proposing to divert one of their V-1710-33's and carry it through for final French Air Commission acceptance for delivery. This was intended to expedite their first engine for Curtiss' use in connection with setting up the prototype on the French contract.[201] With the fall of France and the British takeover of their contracts, this was no longer required.

Role of Wright Field and the Materiel Division in WWII

Following the First World War there was the usual radical reduction in military expenditures and downsizing of the standing Army and Navy. During the 1920s, budgets for military development had been small. With the beginning of the Depression in October 1929, there was even less money to be spent on anything more than simply maintaining a minimal military structure. Consequently, the Air Corps had consolidated its activities to achieve the maximum of efficiency and productivity. Wright Field, located at Dayton, Ohio, was the home of the Materiel Division, and was clearly the center of the Air Corps technical activities. It had evolved into this role from the mission of the earlier, and similar, McCook Field of WWI fame. Wright Field, though fairly impoverished, was where technical progress was being accomplished and sponsored, even though often the sponsorship was only supported with enthusiasm. Most of the true technical accomplishments within the aviation industry during the period were done using corporate and investor funds, though often with significant kibitzing from the on-looking officers and engineers of the Materiel Division.

Tactics for future air war were being argued and developed by the strategic military thinkers of the day. One vigorous advocate of a superior interceptor airplane was the Air Corps Tactical School instructor Capt. Claire L. Chennault. His outspoken opposition to the multiseat fighter concept, exemplified by the then forthcoming Bell XFM-1 *Airacuda*, eventually led the Captain to retire on disability in 1937. Alternatively, Capt. Ross G. Hoyt, long with the 1st Pursuit Group at Selfridge Field, advocated adoption of an "Interceptor Pursuit" and a long-range multiplace fighter.

It is surprising to realize the comparatively low rank of many of the major contributors to the development of the aircraft and engines which proved so appropriate in the conflict to come. A key example being Lt. Ben Kelsey, who ran the Pursuit Projects Office at Wright Field for the last half of the 1930s. He not only prepared the specifications for the competitions held to select the next Air Corps pursuits, but actually performed the first flights of several of the new aircraft, including the Bell XFM-1 and Lockheed XP-38 as neither firm had pilots qualified in such high performance aircraft.[202]

Meager as it was, the aviation industry still provided new aircraft and engines to Wright Field for evaluation, and hopefully, acceptance by the Materiel Division. When a manufacturer received an order it could be assured that it would be for a "bare-bones" design, and at a constrained price. The first three V-1710 engine orders Allison received were for one engine each, and payment was not forthcoming until the contracted performance and longevity was demonstrated! The Army demanded value for its funds, and the record shows that they prevailed.

Wright Field procured all of the aircraft, and aircraft accessories, including engines, used by the Army. They rigorously applied the Congressionally mandated procurement regulations, a process that required competitive bidding and performance contracts. Aircraft related contracts were administered from Wright Field. Each was assigned a unique and sequential number, though except for Lend-Lease contracts, all had a W535-ac prefix. A partial listing of contracts interesting to the history of the V-1710 is provided in Appendix 8.

Acceptance testing was done by the Propulsion Branch of the Materiel Division at Wright Field, utilizing their personnel, facilities and equipment. Consequently, development required very close cooperation and involvement of all parties. Sometimes Wright Field, which had a very qualified core of leading engine designers and engineers, would direct manufacturers to incorporate features or changes not supported by the manufacturer. There are several instances of this in the V-1710 record. Interestingly, Allison usually won out on these technical issues in the long run, but still the side-track and diversion of resources was often taken.

A Chronology of the Early V-1710 Engines

The Army had been interested for some time in an inline engine designed to accommodate "chemical" cooling at 300°F. They intending it primarily as a high altitude bomber engine due to the types inherent capacity for turbosupercharging. But without funds, and with Curtiss-Wright forgoing further development of the *Conqueror*, the Air Corps was reduced to an observer of industry events. Concurrent with the testing of the Navy GV-1710-A Allison began work on the engine the Navy really wanted, the reversible airship engine. The V-1710-B (Navy V-1710-4) was conceived, designed and developed during the 1932-1935 period.[203] This effort was then in parallel with development of the Army "long nose" V-1710. The V-1710-B was a considerably different engine from the original GV-1710-A in that it featured the capability to reverse direction of rotation, from full power to full power, in eight seconds.

Since Airships operated close to sea level the engine was not equipped with an internal supercharger. Instead it had two carburetors located within the vee formed by the cylinder blocks. This engine was tested and rated at 650 bhp. Three engines were built, the first test unit was Allison Engineering Company (AEC) #4. The first of the two pre-production YV-1710-4(B2R) engines was being prepared for shipment at the time the U.S.S. *Macon* was lost off the California coast. This was the last of the Navy's airships intended for the engine, and its loss on the night of February 12/13, 1935 soon ended the program.

From an Allison concept and designation perspective, the "B" series followed the "A", with the "C" for the Army following both of them. As for actual production, the "A1" for the Navy was followed by the 'long nosed' XV-1710-1 "A2" for the Air Corps then the three "B's." The "B" model involved a significant amount of engineering to incorporate the reversing feature, which required shifting cams and revised drives and gearing for the ignition. The Navy "A1" was for "proof-of-concept", as the Navy had no direct application in mind, their focus always being the airship engine.

The Model Test engine was the V-1710-B1R. It successfully demonstrated 650 bhp, naturally aspirated, the ability to shift from full power in one direction to the other in under eight seconds, and contributed to the improvement of the cylinder blocks being used in the C-Model engines. (Allison)

When the Army specified the XV-1710-1 in late 1932, they also defined the XV-1710-3, which was to be a model to power bomber aircraft, specifically the Boeing XB-15 and Martin XB-16. The XV-1710-3 was intended to deliver 1000 bhp and be similar to the XV-1710-1 except that it would have 3:2 reduction gearing instead of 2:1, along with 8.0:1 supercharger gearing in place of the 8.77:1 gears in the XV-1710-1. It was also to have crankcase mounting pads like those on the Navy GV-1710-A, and be suitable for heavy aircraft installations. In fact, when produced the XV-1710-3 (Allison Model V-1710-C1) became an improved XV-1710-1, retaining the same 2:1 reduction gearing, but with 8.0:1 supercharger gears and designed for rating at 1000 bhp. Following delivery in 1934, the single XV-1710-3 started through a protracted test and development period that failed to cumulate in a satisfactory 150-hour Military Type Test. To initiate the flight test program, the first of 10 "Pre-production" YV-1710-3 engines was constructed as the XV-1710-7(C2), and assigned to the A-11A being built by Bell Aircraft. The actual production of the remaining engines was to be delayed until the completion of a satisfactory Type Test.

In the early 1930s the Army was highly interested in having an engine with fuel injection and Allison had designed the VG-1710 to accommodate the feature. In 1934 the Army directed that the first two engines of the YV-1710-3 pre-production batch were to be designed and built as fuel injected XV-1710-5's. In fact, even though two sets of parts were fabricated, to allow constructing then in either carbureted or fuel injected forms, events conspired such that both were built as carbureted engines. Because of problems impeding the completion of the XV-1710-3 Type Test, the second pre-production engine, AC34-6, which had been intended to be assembled as a fuel injected -5, was redesigned with improved cylinder blocks and the new Hazen intake manifolds and delivered as the XV-1710-7(C3) Type Test engine. Even in this form one more round of upgrades was required before the 150-hour Type Test at 1000 bhp was successfully completed. This was accomplished with the YV-1710-7(C6) AC34-8 in April 1937, the first engine in the U.S. to accomplish the feat. Other engines had previously run at over 1000 bhp, but none had successfully completed the severe 150-hour Type Test as defined by the Army Air Corps.[204]

Allison envisioned a commercial version of the type tested engine. With Army support, they submitted the results to the Department of Commerce for Type Approval as the 1000 bhp V-1710-C4. Contemporary press releases and photographs show the actual Type Test engine and describe it as the Allison V-1710-C6. This is technically correct in that the final configuration of the engine on the test stand was described by Allison Specification 104-A, the spec describing the C-6. The commercial C-6 was to have differed from the tested V-1710-C4/C6 in that the supercharger drive ratio was to be reduced to 6.75:1 rather than the tested 8.00:1. This was appropriate given the anticipated lower operating altitudes to be encountered in commercial operations of the day. No commercial C-4 or C-6 engines were ever sold. The military designation remained C-4 or C-6 for the V-1710-7.

The V-1710 Takes to the Air

It had been intended that the first pre-production YV-1710-3 would be used for initial flights of the V-1710. When the XV-1710-7(C2), AC34-5, was delivered it was configured the same as the final build of the XV-

From the firewall forward the Consolidated A-11 AC33-208 was redesigned by Bell as their Model 2 to mount the XV-1710-7(C2), designated by the Air Corps as the A-11A. The C-2 engine (AC34-5) was only used for initial fitup and systems tests, being replaced by YV-1710-7(C4), AEC #9. First flight occurred on December 14, 1936 at Wright Field. AEC s/n 9 flew in this airplane for a total of 304 hours before it was overhauled. The large 3-bladed propeller and short landing gear required 3-point landings and takeoffs. (Allison)

1710-3(C1) Type Test engine. When that engine was declared as having failed the Type Test, Wright Field determined that AC34-5 should not be flown. Bell then completed the A-11A and installed AC34-5 as the XV-1710-7(C2) for systems checkout and ground operation, while Allison redesigned the engine in preparation for another attempt at the Type Test.

Wright Field had gone through an extensive effort to find a suitable aircraft in their inventory that could properly flight test the new V-1710. They finally selected the Consolidated A-11 for conversion as the flying test bed.

The first Allison V-1710 to fly was a contemporary of the YV-1710-C6 that finally completed the Type Test. Specifically it was Air Corps model YV-1710-7 serial number AC34-7, Allison model V-1710-C4 serial number AEC#9. It was installed in the Consolidated A-11, number 33-208, which had been modified by the fledgling Bell Aircraft Co. as it's Model 2, the A-11A test bed for the V-1710 engine. It first flew on December 14, 1936. It then entered an "accelerated" test and demonstration period during which it accumulated 304 hours of flight in 23 months. The engine was in pretty good shape after 304 hours of flight, overhaul cost to the Air Corps for Allison to repair the cylinder blocks was $174.66, including $66.46 for labor. After overhaul following the intended 300 hour Time Between Overhaul (TBO) interval, #9 was reinstalled in the A-11A and flew another 7 hours before it and the A-11A, were retired to mechanic training duties at Chanute Field, Illinois in September 1939.

The first "built-to-purpose" V-1710 powered aircraft was the Curtiss XP-37. This aircraft was intended to investigate the combination of the V-1710 and the turbosupercharger developed for it, a package on which the Air Corps was basing much of its technological future. This arrangement goes back to 1933 when the Air Corps commissioned the General Electric Company to develop one of its turbosuperchargers to work with the V-1710, intending it to deliver 1000 bhp at 20,000 feet.[205]

The XP-37, basically a new fuselage with a P-36 wing and empennage made its first official flight at Wright Field on June 16, 1937 powered by a 1000 bhp YV-1710-7(C4). In 1938 this engine was rebuilt and upgraded to become a V-1710-11(C7). The -11 could produce 1150 bhp for takeoff, and was internally similar to the V-1710-13 being used on the contemporary Bell XFM-1 pusher. The significant improvements included the reduction of the gear ratio driving the engine stage supercharger, along with strengthening several components, increasing the cylinder compres-

Captain S. R. Harris of the Wright Field Test Section stands in front of the Curtiss XP-37 when powered by the YV-1710-7(C4), July 23, 1937. Scoop on side of fuselage is air intake for radiator and intercooler, slot over wing is hot air discharge. Slot in leading edge of wing is for induction air to turbosupercharger. Oil cooler inlets directly behind Capt. Harris. (Allison)

When first flown in September 1937 the Bell XFM-1 was a giant technological step forward. It's unconventional configuration featured V-1710-9(D1) pusher engines and GE Form F-10 turbosuperchargers. (Allison)

sion ratio and incorporating a new torsional vibration dampers on the crankshaft to allow rating at the higher specified rpm. The reduced supercharger gear ratio improved the efficiency of the engine and allowed the associated turbosupercharger to be better utilized.

At this time the Air Corps major program for the V-1710 was as a "pusher" for the Bell XFM-1. As such, the engine reduction gear assembly was redesigned in the form of a five foot propeller speed "extension shaft" to the propeller, hence defining the Allison Model "D." The engine was the V-1710-D1, Army designation V-1710-9, and had the same features and 1000 bhp takeoff rating as the YV-1710-7(C4). The Bell XFM-1 really required more power, in fact initial studies were done assuming 1250 bhp V-1710 fuel injected engines. Allison was able to accommodate 1150 bhp at takeoff by configuring the engine internally as per the V-1710-11(C7). This resulted in the V-1710-13, still identified by Allison as the V-1710-D1. The Army Materiel Command performed a 150-hour Type Test on the -13 engine, the test results being assumed as applicable to the V-1710-11 as well since the V-1710-13 Type Test engine had originally been built to perform the V-1710-11 Type Test.

In June of 1937 the Lockheed P-38 won the Wright Field X-608 twin engine interceptor design competition, the specification of which included the requirement that the winner use the Allison V-1710-C7 engine. Since the P-38 was a twin engine configuration it was necessary to develop an engine with the ability to "feather" its propeller. This was the V-1710-C8, still known within the Air Corps as the V-1710-11. Another unique feature of the XP-38 was the requirement for "opposite" handed or rotating propellers, a feature that resulted in canceling the effects of propeller torque and thereby improving aircraft handling. Allison responded with their left-hand turning V-1710-C9, built from C-8 components, and subsequently designated by the Air Corps as the XV-1710-15.

The low thrust line of the V-1710-11/15(C8/C9) made for an extremely sleek installation in the prototype Lockheed XP-38. Photo taken at March Field, California at time of first flight. Adjustable inlet scoops for oil coolers are visible just below props. (Allison)

Table 1-7
Scheduled Deliveries on Service Test Contracts

	M-38	A	M	J	J	A	S	O	N	D	J-39	F	M	A	M	J	J	A	S	O	N	D	J-40	Total
V-1710-11/XP-37/8																								2
V-1710-19/XP-40							1																	1
V-1710-21/YP-37		1	1	4	4	4	4	1																17
V-1710-23/YFM-1				X16						2	3	5	5	7	8	6	3							39
XP-37/XP-38 A/C																								1/1
YFM-1 Airplanes	X20													1			3	3	3	1			2	13
YP-37 Airplanes								1					12											13
XP-40 Airplane								1																1

Notes:
Contract W535-ac-10660 for 20 engines for the YP-37's was issued as of January 5, 1938, then amended on June 16 to include the 39 engines for the YFM-1's.
"X20" denotes date of contract. Following values are number of aircraft or engines to be delivered within month.

With the advent of the Bendix-Stromberg pressure "carburetor" it was possible to effectively eliminate the problems of carburetor icing, and at the same time provide improved operation free of the "cutout" phenomena typical of float type carburetors in negative g. This device used a somewhat conventional carburetor arrangement to measure airflow and to locate the throttle plates. What was unique was the subsequent control of a pressurized stream of fuel which was sprayed directly into the eye of the integral engine-stage supercharger. Such devices were soon nearly universal on all high powered aircraft engines. Adaptation by Allison resulted in the V-1710-C10, and similar V-1710-D2 engine. When purchased by the Army to power the "service test" YP-37's and YFM-1's, these models were designated as V-1710-21 and V-1710-23 respectively. The V-1710-21's were also the first engines to utilize cylinder blocks which were interchangeable with the new V-3420 and all subsequent V-1710 models.

By the time the V-1710-C10 model was going into development at 1150 bhp, some 3000 hours of Type Testing or equivalent development running on "C" type engines had been completed at the 1000 bhp level.[206]

With thirteen each of the YP-37 and twin-engined YFM-1 airplanes, plus the needed spares, Allison received it largest engine order yet, for 59 engines, dated January 5, 1938. Twelve of these were delivered during 1938, and another 40 during 1939. Production was finally moving after nearly ten years of effort.[207]

This is a wooden mockup of the 1600 bhp Allison X-3420. Note the use of Hazen era Fuel Injection plenum located between each pair of cylinder banks, and the pairing of the banks on the single crankshaft. (Allison)

V-1710's for Production Aircraft

The Army's interest in a "Very Long Range" bomber, and the consequential need for an engine with over 1600 bhp, caused them to request Allison to develop such a powerplant, but using V-1710 components and technology to expedite its availability. This request officially started in 1935, but preliminary work was done in 1934. In 1936 efforts were suspended when the V-1710 Type Test became overriding. This doubled V-1710 engine was the X-3420, not the later XV-3420. With the XV-1710-7(C3) appearing to be successful with the Type Test, attention returned to the 1600 bhp engine. Given Allison's limited production and engineering capacity in late 1936, Allison argued for minimum changes and maximum commonality between the engines and was successful in getting Wright Field to agree that the new engine was to be built with two crankshafts and four cylinder blocks to be interchangeable with those on the V-1710. New crankcase and accessories sections were developed which defined the configuration of the V-3420.

With the frequent need to redesign the V-1710 to meet a multiplicity of aircraft installation requirements and configurations, an effort was made in the design of the V-3420 to insure maximum component and subassembly interchangeability. For example, to simplify the task of reversing engine rotation the crankshaft was designed with identical flanges on each end. Reversing the direction of rotation only required swapping the crankshaft end-for-end, and putting a "idler" gear into the accessories drive to maintain proper direction of rotation for the supercharger and camshafts. This effort immediately paid off in a reverse technology transfer, that is, back to the V-1710 engines configured to power the revolutionary Bell XP-39 and the P-38's ordered after the XP-38.

The numerous operating troubles encountered with the turbosupercharged installations in the XP-37 and XFM-1 convinced the Allison Company that they were in a poor position to continue their engine development as long as it was tied to the turbo. They deemed this as being extremely hazardous to safe engine operation in its current state of development. As it did not appear to them that the situation would be improving rapidly, they concluded that it was essential to provide an altitude rated V-1710.[208]

The Bell XP-39 required the engine to be mounted behind the pilot, with an extension shaft driving a remote reduction gear mounted in the nose of the aircraft. This feature was deemed necessary to accommodate a

The design of the XV-3420-1 was begun in 1936 at 2000 bhp, but when it began running in April 1938 it was rated for 2300 bhp. Development of the engine took longer than expected as a consequence of the demands for putting the V-1710 into mass production and the decision by the Military to defer development as they no longer had a use for it. When the Wright R-3350 introductory problems threatened the Boeing B-29 program in 1941 the V-3420 program was hurriedly restarted. Note the early PN34550 form of the Hazen intake manifolds. (Allison)

37 mm cannon mounted to fire through the hub of the propeller. When produced, the configuration was identified as the Allison V-1710-E series. The Navy was interested in the concept as well and sponsored the Bell XFL-1 *Airabonita*, powered by an Allison V-1710-E1, military XV-1710-6. This was an "altitude" rated engine. The Army wanted the XP-39 to be turbosupercharged, so Allison built the sea-level rated V-1710-E2 (Army designation XV-1710-17) in this form.

In cooperation with the leading aircraft manufactures, Allison became interested in selling an "altitude" rated version of the V-1710 as a way to advance sales in view of the problems with turbo installations. The difference between the engines was that higher internal supercharger gear ratios would be used, giving critical altitudes in the 10,000 to 14,000 foot range. To this end Allison proposed to the Army their models V-1710-C11 and C-12, neither of which was ever purchased or issued a military model number.

Allison was repeatedly advised by the Air Corps that they had no interest in even an experimental airplane with an altitude rated engine. However, Allison was finally successful in convincing the Curtiss Company of the necessity of protecting both of their interests for the upcoming 1938 Pursuit Competition by offering an airplane with an altitude rated engine. Time was short, for the competitors were to deliver flight articles that November.[209] Concurrently the Air Corps was interested in the inherent differences between an air-cooled versus liquid-cooled installation in otherwise identical airframes. When Curtiss offered a liquid-cooled adaptation of its air-cooled and radially engined P-36 the Air Corps was quick to authorize the configuration, but not for the reasons being promoted by Allison and Curtiss.

Even though documents have been found which provide the foundation for this course of events, the participants still cannot agree on the reasons for the changes, delays, and finally the approval to use an altitude rated engine.

Ben Kelsey, who as the Fighter Projects Officer at the time, and who was in a position to know these arguments, relates in his book that he discussed the origins of the XP-40 with its designer, Donovan Berlin, in the 1970s. As Chief Engineer at Curtiss, Berlin had been responsible for both the air-cooled P-36, and its liquid-cooled development, the P-40. Kelsey relates as how he was certain that the Air Corps had requested Curtiss to modify a P-36 by installing an Allison engine to provide full-scale evidence of the benefits or disadvantages of liquid versus air-cooling. Berlin, on the other hand, was convinced that Curtiss had initiated the program to get Allison to increase the altitude performance of its engine and thus to provide a plane that would win the next procurement competition. These apparently conflicting positions were reconciled easily when Berlin said, "I've always wondered why our proposal was approved so quickly." Probably, Allison was also prepared to pursue any modification that promised the possibility of quantity procurement to offset its large development investment. As Kelsey notes, regardless of how the results were achieved, the quality of the aviation products that were developed represented competitive performance plus adaptability, versatility, and utility that were unique for the time.[210]

For the 1938 fighter competition Curtiss took the "bait" and selected the Allison V-1710-C13 altitude rated engine for their re-engined and Allison powered P-36A Hawk. Allison received an order for the engine in June and delivered the engine to Curtiss in September. The competition was then delayed to early 1939 for lack of competitors, which was fortuitous as it gave time to considerably refine the XP-40. The result was that the XP-40 won the competition. Its V-1710-C13, which had a three barreled carburetor and used the 8.77:1 supercharger gears as first used on the XV-1710-1. The engine was assigned the Air Corps designation XV-1710-19. The Army then ordered 134 P-40's, and 234 V-1710-19's to power and support them using FY-39 funds. Allison subsequently made further improvements in the supercharger inlet elbow, carburetor and engine, and the actual deliveries of the production engines were as V-1710-C15's, Air Corps model V-1710-33. This was the first production contract for the Allison V-1710 and occurred ten years after Allison began the project.

Improved V-1710's

Throughout the development period for the V-1710-C/D there had been concern about the propeller reduction gear. Originally intended as a weight saving feature, and subsequently to enhance streamlining, the design was

This is the initial October 1938 form of the XP-40. By the time of the 1938 Pursuit Competition the radiator had been relocated to the chin and the XV-1710-19(C13) further refined. (Allison)

In its turbosupercharged form the Bell XP-39 used the new XV-1710-17(E2). Unfortunately the installation suffered from an excessive amount of drag and the airplane was unable to demonstrate its guaranteed 400 mph speed. (Allison)

unique in utilizing an "internal" spur gear arrangement. The outer rim of the requisite "ring" gear was supported by a "plain" steel backed lead-bronze bearing of the type that was Allison's stock in trade. While all of the calculations showed this to be a satisfactory arrangement, it was prone to miss-alignment and was not reliable at over 1150 bhp. Consequently, Allison took advantage of the redesigned and interchangeable components coming from the V-3420 and V-1710-E programs and created the V-1710-F series. This engine had a more conventional "external" spur type reduction gear nose case, but otherwise utilized the other components of the new "E" series engines. This degree of interchangeability and commonality meant that the two models could safely and economically be built on the same assembly line. The V-1710-F1, Army model V-1710-25, was used for development only, the first production models being those for the pre-production and early P-38's, the V-1710-27/29's. These were Allison model V-1710-F2R/F2L, and marked the first time the "L", designating "left-turning" engines, was used.[211]

Problems with the external and internal drag of the turbosupercharger installation on the XP-39 had led the NACA to recommend its removal in September 1939. This required Allison to rapidly develop an altitude rated model of the E-2 engine as a replacement. The result was an interim designation of the original engines when modified as the V-1710-E2A, ["A" for "Altitude"], and given the Air Corps designation V-1710-31. The actual altitude engines were in fact rebuilds of the two XV-1710-17's, using new supercharger drive step-up gears and the carburetor from the similarly configured and altitude rated V-1710-19/33(C13/C15) for the Curtiss P-40's. The original intent had been for the 12 turbosupercharged YP-39's to use the improved V-1710-17(E3), and the one YP-39A to demonstrate the altitude rated V-1710-31(E2A). With the decision to remove the turbosupercharger from the XP-39, creating the altitude rated XP-39B, the entire batch of thirteen service test P-39's, were redesignated as YP-39's and were to use the improved altitude rated V-1710-E4 engine, which was to be the production version of the engine for the YP-39A. When finally produced the V-1710-E4 had difficulty in passing the 150-hour Model Test at the desired 1150 bhp, the crankcases kept failing. Consequently, the need to power the thirteen altitude rated YP-39's resulted in another interim engine, the V-1710-E5, of which sixteen were built on contract W535-ac-12553 as V-1710-37's. The two original XV-1710-17's were continuously revised until they ended up configured as the prototype E-5's. The V-1710-E5 engines were operated with the same 1090 bhp ratings as the production V-1710-33(C15). With an improved crankcase the V-1710-35(E4) was subsequently qualified at 1150 bhp and became the definitive early P-39 production engine. Thousands were built.

The V-1710-33(C15) was the last of the "C" models and the V-1710-23(D2) was the end of production for the pushers [excepting four modified as altitude rated V-1710-41(D2A)'s], though Allison was working on left hand turning -D models for the stillborn production Bell FM-1. There was also a V-1710-D3 tractor type extension shaft engine, proposed to power the Bell Model 3 prior to the redesign to provide a cannon firing through the propeller. A modified version of this engine was the first to be equipped with a remote reduction gear driven by a crankshaft speed extension shaft. This V-1710-D3 Mod A was the engine used in both the Bell Model 3 and Model 4 submittals that resulted in the XP-39.

The D-3 Mod A was a unique engine that came about during the preparation of proposals for the 1937 X-609 single-engine Interceptor competition, that had required proposers to use the latest Allison, the 1150 bhp V-1710-C7. Both of the competing Bell proposals, the Model 3 and Model 4, as well as the Curtiss Design 80A, elected to incorporate an extension shaft engine revised to accommodate mounting a cannon or machine guns in the nose. After selection of the Bell Model 4 as the winning design it was decided to redesign this extension shaft engine to utilize improved components coming from the V-3420 project. The result was the "E" model V-1710. The "E" model went through a long list of variants as modifications and improvements necessary to support the Bell P-39, P-63, as well as other follow-on programs. Most significant was the adaptation of an Auxiliary Stage Supercharger, a mechanically driven second stage supercharger, giving the P-63 a high altitude capability. This feature was later adapted to several "F" model engines as well, along with the ultimate Allison V-1710 series, the "G." Late E/F and all of the "G" models used a 12 counterweight crankshaft and could be rated for maximum power at 3200 rpm. The cylinder compression ratio was reduced in most of the "G" engines to 6.0:1 to accommodate higher manifold pressures. With water injection they could exceed 2200 bhp and 110 inHgA MAP.

Table 1-8
Scheduled Production on First Production Contract

	A-39	M	J	J	A	S	O	N	D	J-40	F	M	A	M	J	J	A	S	O	N	D
V-1710-17/YP-39			X27								1	2	2	4	4	3					
V-1710-19/P-40			X27				1			2	15	15	15	30	35	50	75	75	75	75	75
V-1710-27/YP-38			X27						1	1	2	3	3	3	3	3	3	1			
V-1710-29/YP-38			X27						1	1	2	3	3	3	3	3	3				
YP-39A Airplane	X27										1										
YP-39 Airplanes	X27														2		3	3	4		
P-40 Airplanes	X25											3				15	20	30	40	45	45
YP-38 Airplanes	X27										1	1	1	2	2	3	3				
P-38 Airplanes						X16									2	6	9	10	14	14	11

	J-41	F	M	A	M	J	J	A	S	O	N	D	Total	Comments
V-1710-17/YP-39													16	Delivered as V-1710-37
V-1710-19/P-40	75	75	75	13									776	Delivered as V-1710-33
V-1710-27/YP-38													23	
V-1710-29/YP-38													22	
YP-39A Airplane													1	Delivered as YP-39
YP-39 Airplanes													12	
P-40 Airplanes	45	45	45	45	45	45	45	11					524	
YP-38 Airplanes													13	
P-38 Airplanes													66	

Notes:
1. A total of 837 engines were originally ordered on this first production contract, W535-ac-12553. A change order later added 132 engines for 66 P-38's at no additional cost. This was due to the large British orders causing a reduction in the unit cost of the engine. The last 542 of the V-1710-19 engines were to be procured with FY-40 funds.
2. "X25" denotes date of contract. Following values are number of aircraft or engines to be delivered within month.
3. A total of 776 V-1710-19 engines were ordered, but switched to V-1710-33(C15) in November 1939. Only 300 C-15's were delivered on this contract, the balance becoming V-1710-F3R's for the P-40D/E.
4. This information obtained from Status of Deliveries of Engines and Aircraft on 1939 Production Program, RG18, File 452.1 to 452.17, Box 130 "Bulky" at National Archives.

Production Problems

The first production contract for V-1710's was signed in June 1939, and required delivery of the first two V-1710-19 engines in January 1940. This was to be followed by 15 each for several months, and then stepping up to 75 per month by August 1940. In parallel, limited production of the V-1710-F engines for the YP-38 and "E" engines for the YP-39 was to occur. Allison determined they would produce the large order for V-1710-C engines in the new Plant 3, and would build the initial quantities of E/F engines in Plant 2, what was effectively the "prototype" shop. Construction of Plant 3 was for all intents on schedule, as was the provisioning with tools, fixtures and jigs. Even with this aggressive effort and the support of the larger General Motors Corporation, initial deliveries were found to not be up to the guaranteed standards for either quantity or quality.

With Bell, Curtiss and Lockheed producing airframes to the original schedule the situation quickly developed into a crisis. In March 1940 Allison provided the following forecast of deliveries against U.S. and Foreign contract requirements:

Army reports were skeptical of these numbers, suspecting only a total of 20 engines would be delivered within the first two months, not 37. The chief cause for delay was in the production of suitable quality non-ferrous castings.

Table 1-9
Initial Forecast of 1940 V-1710 Production Deliveries
for US/Foreign

1940	V-1710-33 for P-40	V-1710-27/29 for P-38	V-1710-37/35 for P-39	Total for US
March	11/33	2/14	0/4	13
April	25/15	8/8	4/5	37
May	35/30	8/26	4/6	47
June	50/35	18/36	6/1	74
July	80/50	34/36	2/1	116
August	120/75	40/34	6/6	166
September	180/75	64/23	6/6	250
October	292/75	None	8/8	300
November	342/75	None	8/8	350
December	390/75	None	10/10	400

Note: Allison forecast of deliveries as of 3-11-40. NARA RG18, File 452.8, Box 808.

A major concern of General Motors in 1939 was the limited production commitment by the Air Corps. In order to provide the contracted number of engines in the shortest possible time, Allison was building facilities capable of producing over 300 engines per month, yet just over 1,000 engines were on order. At this point Allison was successful in getting approval for foreign sale of later engine models. The Air Corps was ever the tough negotiator, and in exchange for the approval was able to get Allison/GMC to forgive the accumulated one million dollars in development expenses paid by Allison and yet to be reimbursed by the Army.[212] Allison also agreed to an accelerated schedule for furnishing the government one improved model free, and to redesign and accelerate development of the V-3420. In turn Allison was intending to be able to profit from foreign sales of the E/F model V-1710.[213]

By July 1940 the situation had further deteriorated due to protracted Model Testing of the V-1710-33 that delayed the deliveries to the airframe manufacturer. The situation was bad enough that General Brett, Chief of the Materiel Division, sent a rather scathing letter to Allison detailing several problems with the "C" engine specifically, and noting that the new "F" engine was showing similar traits. He noted that as of July 15, 1940 Allison was to have delivered 138 V-1710-33 engines to Curtiss, but that they had delivered only 77, and that while 102 V-1710-27/29 engines were to have been produced for the Lockheed P-38, only two had been delivered. With a considerable portion of important combat airplanes scheduled for production in Fiscal Years 1941 and 1942 using Allison engines, he believed that dramatic actions needed to be taken to insure that the Munitions Program not be jeopardized.[214]

General Brett went on to state that it was an Army and Navy requirements that when a part failed on Type Test it must be redesigned before a second official test is undertaken. Army policy was to have the manufacturer initiate and use his own judgment in effecting the correction of such inherent features of design. He concluded that in this case the Air Corps felt justified on insisting that Allison take immediate actions to correct the deficiencies that testing and analysis had indicated needed to be changed to provide satisfactory performance of the engine in service. This included the crankcase, main bearings, connecting rods, thrust bearing lock and distributor bearings.

In June 1940 France fell to the Nazi invasion and on July 1, 1940 Jimmy Doolittle was recalled to the Air Corps as a Major. His first assignment was as Assistant Supervisor, Central Air Corps Procurement Center, on special assignment from General Arnold to function as his trouble shooter at the Allison Indianapolis plant.[215] While he certainly did not single-handedly turn the situation around, Doolittle worked with senior Allison and GM management, guiding them in the needs of the Army and aviation requirements in general.

By November 8 a total of only 277 "C" and 4 "E" engines had been delivered, yet by the end of the year a total of 1,175 engines were delivered.[216] Allison had finally gotten its production line in order and most of the bottlenecks cleared. Significant in this respect was the shortage of test stands for both production and experimental use.[217] Having at least six different models in development at the same time, and usually able to only run one of each due to the shortage of stands, constrained both development and production.

After getting back on schedule at the end of 1940 things once again deteriorated in early 1941. As an example, during December 1940 out of a batch of 36 engines, 14 were returned for penalty runs prior to final acceptance due to damaged bearings and scored parts caused by foreign matter in the test stand lubrication system, none of the rejections were due to lack of performance.[218] By April 1941 Allison was in arrears on the British order by some 1,000 engines. The situation was so serious that the Under Secretary of War, Robert P. Patterson, wrote to Allison demanding that positive action be taken to meet the delivery obligations. The problem was attributed to a repetition of the delays that had previously been encountered, primarily as a result of the rapid expansion coupled with the lack of high quality workmanship required for aircraft engines.[219] The trouble was brought about by the simultaneous production of the V-1710-C15 engine and introduction into mass production of the E-4, F-2R/L and F-3R engines. These later models were quite different from the C-15 and required major revisions to the production flow and processes, as well as many new and/or upgraded parts from new suppliers. In June 1941 the Assistant Secretary of War for Air, Robert A. Lovett, asked the Air Corps if production of the "C" model could be accelerated so as to complete its orders and permit concentration on the E/F types.[220] This was accomplished, with the last V-1710-C15 being assembled in July 1941.[221] By the time of Pearl Harbor, and during the AVG battles in China, the C-15 model engines for their P-40's had been out of production for over six months!

Table 1-10
Allocation of V-1710 Production, Fall 1941

	Engines/mo
North American P-51	60
Bell P-39	180
Curtiss P-40	250
Lockheed P-38	320

The effort to expand production to a rate of 1,000 engines per month by December 1941, was impacted as Allison was unable to obtain sufficient machine tools to fill the available floor space. As of September 15, 1941 they had some 200 machine tools undelivered, even though they held a priority rating of A-1-C. As such, Allison was listed behind 525 other prime contractors on the master priority list! The needed tools limited production to about 800 engines per month and as a result the Air Corps was rationing engines to the airframe manufacturers.[222] A combination of revised priorities and management attention was successful in achieving the 1,000 engines per month rate in December 1941. Given the considerable problems incurred, that standard was achieved in 23 months from the beginning of production from Plant 3, a facility intended to have a maximum capacity of 300 engines per month!

Improving Performance of the V-1710

With the aircraft manufacturers and Wright Field always pressing for more performance of their aircraft, Allison was placed in the position of needing to continuously improve, upgrade and push the engine to deliver more power, and to do it at a higher altitude. Over the years there have been many offering criticism or allegations that Allison was reluctant to advance the engine. This criticism is somewhat surprising since it was on Allison's own initiative that they conceived and developed a 1000 bhp engine at a time when the military was intending to stay with multiple smaller engines on its aircraft.

Early engines were all designed to deliver 1000 bhp, but the first customer to build an aircraft around the engine, Bell with their XFM-1, needed 1250 bhp from each engine. Thus from the outset it was recognized that there was a need to aggressively pursue increasing the power available from the basic engine. Allison believed that the configuration and structure of the basic V-1710-C limited the model to no more than 1150 bhp in turbosupercharged installations and 1090 bhp when altitude rated. In both cases the indicated horsepower being developed was the same, the differences were in the amount of power required to drive the internal supercharger.

The Army became quite interested in the output potential of the V-1710 soon after they ordered the 1090 bhp V-1710-19 powered Curtiss P-40 into production. The contract for the P-40's had been awarded to Curtiss on April 25, 1939, but the first aircraft was not scheduled to be delivered for nearly a year, March 1940.

In August 1939 Major General H.H. Arnold, Chief of the Air Corps, initiated a program to determine what was being done to increase the rating of the V-1710, with the intention being to obtain an engine rating in the 1500 to 1800 bhp range, "...as (is) now being obtained from foreign engines of a similar type and of practically the same weight."[223]

Significantly, this intent was reflected in the CP 39-770 single engine pursuit competition where the aircraft manufacturers using V-1710's included performance estimates assuming 1500 bhp from the altitude rated models, and 1600 bhp from the sea-level rated models.

The Materiel Division promptly responded by establishing a project to convert a spare V-1710-21 into a V-1710-19 with all of Allison's latest features, and then systematically investigate its capability to demonstrate 5-minute ratings above 1500 bhp.[224] This effort was the subject of a letter to the Materiel Division from General Arnold dated August 29, 1939 in which he authorized the tests and directed that his office "will be kept advised as to the progress being made on this program."[225]

- First Priority-Increase the power output of the type of Allison engine that is to be installed in the P-40 airplanes. The changes indicated necessary to accomplish this will be made on engines now under contract as early in the contract as possible.
- Second Priority-Increase the power output of all Allison engines, both sea level and altitude rated types, with a view of making available for future procurement an improved V-1710 with a rating of not less than 1500 horsepower.

By September 21, 1939 the modified V-1710-19 had been delivered to Wright Field and was being tested. Separately, but at the same time, a development program had been laid out for obtaining 1090 bhp at 14,500 feet in the altitude rated XP-39B. It was also determined that a similar program for turbo installations would have a second priority and be accomplished only when time and facilities permitted. Significantly, it was at this point that the Air Corps began discussions with Allison regarding investigations into the development of two-stage gear driven superchargers for special application in high performance interceptors.[226]

On September 1, 1939 Germany invaded Poland, the action that initiated what was to become known as World War II. While U.S. aircraft and manufacturing capacity was being expanded, it was far from being ready for the conflict.

General Arnold was not one to tarry. On September 5, 1939 he instigated a letter from the War Department to William S. Knudsen, President of General Motors Corporation, in which he states,

> "...concerning our relatively inferior position in regard to engines of high performance, as compared to England and Germany...I was alarmed to learn that not only was the rate of development of the Allison engine lagging behind that of other aircraft engines in this country, but also was being far outdistanced by foreign engines, such as the Rolls-Royce *Merlin*."

Arnold noted that construction of the new Allison plant at Indianapolis was progressing satisfactorily, but stated that "little effort appears to be directed by your company and your engineers toward improvement and advancement over the present model." He goes on to state that "The present engine will not be satisfactory for further procurement unless greater emphasis is placed...toward improving this engine until it at least equals performance attained and demonstrated by other countries.[227] At this point in time the Army Air Corps had placed only a single production order with Allison, and that for 837 engines.

In early October Allison's General Manager, Mr. Kreusser, provided a rebuttal to the Air Corps protest that Allison was lagging in engine development as compared to progress being reported by European engine manufacturers. The lengthy list of points made are interesting and substantive in view of subsequent actions and developments by the various parties.[228]

- Horsepower ratings are not properly qualified as to how they were obtained.
- The U.S. Air Corps Type Test requirements are more stringent than those of the British or any other nation.
- Horsepower expressed in terms of "per cubic inch" means very little. The curve included in the letter to Mr. Knudsen was considered as misleading because it was based on what are considered "flash" performance rather than Type Test.
- Allison argued that they had been actively expanding their manufacturing facilities and personnel consistent with sound logic, (a reference to the scale of orders received from the Air Corps-Author).
- Allison contended that frontal area was an important consideration not included in the Air Corps assessment. The following table was provided.

Table 1-11
Frontal Area of Contemporary V-12 Engines (circa 1940)

	R-R Merlin, Used in Spitfires	V-1710-C Used in P-40's	V-1710-F Used in P-38's
Frontal Area, square feet	5.85	5.49	5.19
Weight, pounds	1,395	1,325	1,260
Military Rating, bhp	1160	1090	1150
Critical Altitude, feet	16,730	13,700	12,000
bhp/sq.ft. at Military rating	198.2	198.4	221.5
Pounds/bhp, at Military rating	1.202	1.216	1.095
Takeoff bhp	1320	1090	1150
Pounds/bhp, at Takeoff rating	1.058	1.216	1.095

At the same time Allison made note of the considerable and extensive development program that was then underway. Its major features were:

- Durability tests of the V-1710-19 at 1090 bhp at 13,200 feet in support of releasing the engine for production.
- Conversion of four V-1710-23 engines to sea-level V-1710-41's for the Bell YFM-1B aircraft.
- Endurance and performance development of the V-1710-F2R/L engines to meet Lockheed P-38 production schedules.
- Completion of V-1710-E development for the Bell P-39, including resolution of the torsional vibration problems with the ten foot long extension shaft.
- Company sponsored development testing in anticipation of an Air Corps experimental program involving four V-1710-F3R high horsepower at high critical altitude engines.
- Development of bevel drive gears for use with the V-3420.

- Development of the V-3420 in a wide range of gear ratios and configurations. Allison was awaiting a decision to commit to a single experimental aircraft to lend direction to the program.
- Several Navy projects were aggressively underway.
- Studies of possible configurations and engines able to deliver 4000 bhp and greater.
- Single cylinder development program supporting higher output, improved combustion chambers, pistons, valves, rings, fuels, etc.
- Development of manufacturing processes and materials needed to support advanced and improved engines.

These various issues and concerns were resolved by General Arnold to the point that at the end of January 1940 he authored a letter for Louis Johnson, Assistant Secretary of War, to send to Wm. S. Knudsen, President of General Motors. It identified the necessity of immediately modifying the V-1710-E engine to be capable of delivering at least 1350 bhp, and to be in production no later than January 1941. The reason behind this action was to simply insure that superior engines would be available to U.S. Forces. At the time the War Department was considering releasing the present 1150 bhp powered models and aircraft for foreign sales and obtaining Congressional approval of such sales depended upon not releasing the most advanced U.S. equipment.[229]

That these efforts were successful is reflected in the appearance of the 1325 bhp V-1710-E6 engine in 1941, replacing the 1150 bhp rated V-1710-E4 used in the P-39C/D/Q and defining the improved performance P-39K/L. Similar improvements were made on the "F" engines at the same time, where the 1150 bhp V-1710-F2R/L and V-1710-F3R, respectively powering the P-38D/E, P-40D/E and P-51A, were replaced by 1325 bhp rated V-1710-F5R/L and V-1710-F4R in the P-38F and P-40K.

Concurrent with getting these new models into mass production Allison was at work on the two-stage V-1710-E9 and V-1710-F7 engines as well. The story of the two-stage evolution is related elsewhere in this text, but suffice to say that its development time line, and power ratings, paralleled those of the contemporary two-stage Rolls-Royce Merlin.

Beginning on December 7, 1941 the U.S. was directly involved in WWII and Allison was soon able to achieve a considerable increase in rated power of its engines by establishing War Emergency Ratings. It took some effort to persuade conservative interests within the Materiel Division to allow rating their engines in such a way, but Allison argued that it better reflected the way the engines were being used in the combat theaters, with or without the Materiel Division approved rating. The British and Germans had been rating their engines in this way for a long time. By making it official, a better comparison of the Allison and foreign engines was be achieved. The WER rating was assigned only after successfully completing a 7-1/2 hour demonstration at the claimed rating. These ratings were also retroactively applied to earlier models after qualification testing.

Table 1-12
V-1710 Military and War Emergency Power Ratings

	Military, bhp	WER, bhp
V-1710-35(E4)	1150	1490
V-1710-63(E6)	1325	1580
V-1710-85(E19)	1125	1410
V-1710-39(F3R)	1150	1490
V-1710-73(F4R)	1325	1580
V-1710-49/53(F5R/L)	1325	1325
V-1710-51/55(F10R/L)	1325	1450
V-1710-111/113(F30R/L)	1425	1600

By 1943 the "easy" increases in power had been achieved. The engines were fundamentally strong at their current ratings, including WER. Still there was considerable demand for further power increases, and Allison exploited every pathway to provide greater power. This included switching coolant from nearly pure Ethylene-Glycol to pressurized water, with 30 percent Ethylene-Glycol added as antifreeze. This coolant cooled hot spots within the heads better and supported additional power by deferring the tendency of fuels to detonate. Detonation was usually the phenomena that established a given rating, not structural considerations.

Fuels were another problem, and limitation. In March 1943 Allison complained that the Grade 130 fuel they were provided by the Government was evidently inferior to that being used in the combat zones, but still they had no choice and thereby the established ratings suffered to some degree. This was later somewhat resolved when AN-F-28 replaced AN-F-27 fuel, but in 1943 it was a limitation. Another problem for Allison's development program was expressed by Ron Hazen to General Echols, Commander of Materiel Command, as;

> "There is apparently a larger and larger number of people in the Army Air Forces, War Production Board and other Government Agencies who seem to feel that each item which they propose is the Number One job for the Allison Division. The problem is getting serious enough so that I can see the approach of a complete breakdown of our forward engineering program on higher outputs if we are not relieved of a large number of these items by the issuance of a broad policy of priority on what the engine manufacturer should do first."[230]

Some of the "priority one" items being requested at the time included:

- 80-20 glycol-water cooling
- Ceramic coating of the complete engine internal coolant system
- Centrifugally cast cylinder barrels
- Chrome plating of worn cylinder barrels
- Material substitutions in a number of locations
- Variations in fuels

The Army quickly responded with the needed definition of priorities and made increasing War Emergency Ratings as the program to be given highest priority, though cautioning that to the degree the offered list of priority tasks effected WER ratings, they were to be included.[231]

In June 1943 the Materiel Command credited Allison with maximum ratings for the V-1710, as demonstrated by the V-1710-97(G1R), as 1725 bhp at 3400 rpm, with the two-stage E-11 being able to carry 1150 bhp at 3000 rpm up to 22,400 feet. From these ratings they notified Allison that they were interested in the maximum possible extension of both horsepower and altitude ratings that, "would definitely put the Allison V-1710 type engine 'way out in front' relative to advanced designs of liquid-cooled engines of comparable displacement." They wanted to exploit the potentialities of combining all of the most advanced features of engine design in a high-output high-altitude model. Features under consideration included either integral two-stage or external Auxiliary Supercharger type of two-stage supercharging, relocated carburetor, intercooler, aftercooler, or both, water injection, increased speed ratings and improvements in mechanical design for increased output.[232]

In the spirit of the Lend-Lease program, the Rolls-Royce representative to Packard for the Merlin, Mr. James Ellor, made it known to General Echols through the U.S. Air Attaché in London, that his services were avail-

With the expectation of the coming end of the war both the Army Air Forces and U.S. interests wanted an all-U.S. Mustang, for once the war was over Rolls-Royce was to be the only source of Merlin's. The North American XP-51J lightweight Mustang was to use the V-1710-119(F32R) and be able to equal or exceed the performance of any Merlin powered P-51. (NAA via SDAM)

able to the Air Corps. He stated that he was willing to work with anyone they desired in connection with the development or installation of airplane engines in the U.S.[233] General Echols then proposed to Allison that Mr. Ellor be consulted to insure the Rolls-Royce experience in growing the Merlin was considered in the development of the V-1710.

At this juncture, a very poor situation developed. The Materiel Command Power Plant Laboratory alleged that Allison was reluctant to incorporate "needed changes" into the engine to "exploit the potentialities of the V-1710 engine to the utmost."[234] Per Wright Field, discussions and conferences with Allison representatives "have been a series of explanations of the engine's shortcomings, tending to blame the major part of their difficulties on the airplane installation and on the quality of the fuel." Army Air Forces investigations found that the P-63 installation did have certain defects, namely high coolant temperatures, but the fuel was the same, or better, than that being used by other manufacturers. The AAF attempted to break the developing impasse by bringing in consultants to study and advise on engine development. They were Professor C. Fayette Taylor of the Massachusetts Institute of Technology and Mr. Ellor of Rolls-Royce. Mr. Ellor's services had been offered to Allison several weeks earlier and no follow-up had yet occurred. This was not setting well with the Air Force.[235] What they wanted was an engine with a higher horsepower rating at increased altitudes. For such an engine intercooling or aftercooling was considered to be essential.[236]

The two-stage V-1710-119(F32R) was equipped with an aftercooler to remove half the heat of compression. In this way the engine was to be able to deliver maximum power to quite high altitudes. Development was paced by the new Bendix Stromberg Speed Density fuel metering device, a necessity as there was insufficient space for an engine mounted carburetor. (Allison via Jack Wetzler)

Mr. Hunt, on behalf of Allison, responded to the General on September 8, 1943 noting that they had now met with Mr. Ellor and were in complete accord with the program proposals the Army had put forward. He acknowledged that Allison was appreciative of the Air Force for having arranged to install a two-stage V-1710 in a North American P-51F type airplane, and that they were working on aftercooling such an engine.[237]

The result of this request was the V-1710-119(F32R) aftercooled engine equipped with a speed density "carburetor" and an Auxiliary Stage Supercharger fitted into the lightweight XP-51J Mustang.

Summary

Allison's V-1710 has received a lot of criticism at various times, and has been summarized by some as only a "so-so" engine. Careful investigation shows that much of the criticism was simply due to the conditions in which the engines were operated and maintained, or even, downright misstatements. One such incident occurred when charges were made in May 1943 that "defective Allison engines have cost the lives of 5,000 pilot trainees." As a consequence General Arnold was caused to respond in a memorandum to the President. In it he stated that the charges were, "so contrary to fact that I do not anticipate any serious discussion of the subject."[238]

The P-38's, P-39's and P-40's were the only modern fighter aircraft the U.S. Army had in the early going during WWII. Dirt airstrips in far-off, hot, high and humid corners of the world were hard on engines, airplanes, equipment and people. Fuel was often of poor quality so detonation and engine damage often was the result. Spare parts and necessary specialized tools were often few and far between. Initially compounding these conditions were the deficiencies in aircraft design, aircrew and maintenance personnel training; with the result that systems and components were often pushed to their limits, and beyond. An example being the wing leading edge intercoolers on the early P-38's. They simply were not adequate to cool the turbosupercharged induction air for the new higher powered Allison engines being installed in the latest combat models. Still, Allison powered aircraft carried the bulk of the fighter war for the first three years, that is, until the Merlin Mustang and P-47 were available and in place. At the end of the war the late model P-38's were the match, and often the master, of any challenger. The U.S.'s highest scoring ace, Lt. Richard Bong, exclusively flew P-38's.

A letter from the Joint Production Board to the Commanding General, AAF, dated May 28, 1945 noted that in light of the latest revised production schedules for P-63 and P-38 aircraft, deliveries of V-1710-111/113, -117, -109 and -109A engines were in excess of requirements. It stated that action had been taken to discontinue manufacture of these types of engines as of that date.[239]

So ended the series production of the Allison V-1710, except for special models unique to the several development programs still underway. The only engine to see further series production was the V-1710-143/145(G6R/L) for the post-war North American Aviation built P-82E/F.

After producing nearly 70,000 V-1710's during the war, the massive production effort by the Allison Division was stopped. Still, the V-1710 was soldiering on at the front lines, just as it had been at the beginning of the war four years before. A lot of people had done a lot of work, and put a lot of heart into a lot of V-1710's. It was a commitment and contribution like so many others at the time, and it was a part of what it took to bring the war to a successful conclusion.

A concluding view from the fighting front relative to the Allison V-1710 was offered by Col. Robert L. Scott in his epic tome *God is My Co-Pilot*, written upon his return from a combat tour in China with General Chennault of *Flying Tiger* fame:

"Some day, when the war is over and our sturdy American engines driving great American ships have won victory with air power, I hope and pray – with all fighter pilots who have faced our enemies in aerial combat, from the hot sands of Libya to the cold tundra of the Aleutians, from the jungle heat of Guadalcanal to those torrential rains of the Burmese Monsoons – that some understanding group of citizens will go to Kitty Hawk, North Carolina. There, beside the statue that commemorates the first flight of the Wright Brothers, I hope that they will build a monument to the Curtiss P-40 with its *Allison* Engine.[240]"

The V-1710 was a well designed and strong engine. Post-war hydroplane racers have modified them to run at over 4,000 rpm and with manifold pressures well over 100 inHgA, conditions which compute to over 4000 bhp! At Reno we find that the Unlimited Air Racers who push their Packard Built Rolls-Royce Merlin's the hardest are using Allison V-1710 connecting rods. All-in-all, the Allison people made quite an engine, and a considerable contribution to the war effort. The V-1710 existed because of Allison foresight and the support of the General Motors Corporation during the lean Depression years. When the war came, it was there, and it remained on the front lines to the end of the piston powered fighter era.

NOTES

[1] Mr. Worden had been the bookkeeper for Prest-O-Lite, and then went to work directly for Mr. Allison on February 1, 1918. Over several years he became quite close to Mr. Allison. He relates that Mr. Allison never finished high school; however, after becoming successful in business he began studying very hard and acquired a brilliant education. Worden 1962.

[2] Allison Division, GMC 1962 Draft, 2.

[3] Bosler, T.C. 1952. A letter from T.C. Bosler of Allison to O. T. Kreusser relating discussions with "Pop" Myers of the Indianapolis Speedway Office.

[4] Worden 1962.

[5] The Indianapolis Speedway Team Company incorporated as of September 14, 1915, incorporators were listed as James A. Allison, Carl G. Fisher, Theodore E. Myers, Arthur C. Newby and Frank H. Wheeler. The corporation was dissolved November 12, 1919. Marion County Clerk's Office, 1915.

[6] Bosler, T.C. 1952.

[7] Allison Motors, 1920, 23.

[8] Worden 1962.

[9] Bosler, T.C. 1952.

[10] Allison Division, GMC 1950, 11.

[11] During the fall of 1918 Johnnie Aitken became one of the many who died in the terrible epidemic of influenza that swept the nation. Mr. Allison felt the loss deeply. Worden 1962.

[12] Allison Division, GMC 1962 Draft, 3.

[13] Sonnenburg, Paul, and William Schoneberger. 1990, 22.

[14] Raflo, Diane. 1994, 40.

[15] Worden 1962.

[16] The Allison Experimental Company was incorporated February 23, 1917 with only three names on the papers, Quincy A. Myers, Edward E. Gates and Everett F. McCoy, and note that, "For the first year of the existence of the corporation there are to be only three Stockholder-Directors. The three called a special meeting on October 4, 1920 in the Allison Experimental Co. office to change the articles of Incorporation and By-Laws, and to increase the number of Directors. James A. Allison, owner of 98 shares presided, the other stockholders present were Norman H. Gilman, one share, and L.M. Langston, one share.

[17] Hazen, R.M. 1957, 3.

[18] Sonnenburg, Paul, and William Schoneberger. 1990, 22.

[19] Fleming, Roger 1950, 1.

[20] Hazen, R.M. 1957, 3.

[21] Allison Division, GMC 1950, 11.

[22] *Forecast of Production of Planes and Engines for Week Ending July 26, 1918*. Library of Congress.

[23] Number of Liberty-12 engines:

Packard	6,500
Ford	3,950
General Motors (Cadillac & Buick)	2,528
Lincoln	6,500
Nordyke & Marmon	1,000
Total	20,478

Sources: Dickey, Philip S. III. 1968, 67, and Boyne, Walter J. 1984.

Note that Boyne gives GM production as 2,328 for a total of 20,278. According to *U.S. Army Aircraft Production Facts*, May 31, 1919, p.30, during the 18 months of the war production was 15,572. Available at Stanford University, W87.2:Ai7[2]

Note: This tabulation does not include the 250 Liberty-12B epicyclic geared engines built by Allison for the US Navy. It is possible that these engines were conversions of production by others, but since an entirely new crankcase and crankshaft was required, it is more likely that the Allison production has simply been overlooked in the tabulations done since WWI. Author.

[24] *Memorandum of Questions Discussed in Conference Between Mr. Nash, Mr. Landon, Colonel Vincent, et al*, August 16, 1918. H.H. Arnold manuscript, Roll 197, Library of Congress.

[25] Dickey, Philip S. III, 1968, 31.

[26] It is believed by your author that these engines were conversions of Liberty 12-A's built by others. The changes were fairly substantial, requiring a new crankcase and crankshaft upon which Liberty cylinders and the new reduction gear were mounted. It may also be that the variation in number of Liberty's built may involve how these Allison models were counted, i.e. as conversions or "total".

[27] Allison Division, GMC early 1940s, 6.

Table 1-13
Early V-1710 Engine Models

AEC Model	AEC S/N	Mil S/N'	Mil Model	Delivery Date	W535-ac-Contracts	Usage/Comments
A-1Bld1	1		GV-1710-A	8-13-31	C-17952	GV-1710-A1, Navy Contracted 6-26-30, final delivery 3-23-32.
A-1Bld2	1		XV-1710-2	3-12-32		GV-1710-A1 Build #2, Re-delivery, test complete Sept 1932.
A-1Bld3	1		XV-1710-2	1934		Air Corps used as Fuel Injection test-bed.
A-2	2	33-42	XV-1710-1	6-28-33	5592	First Army "Long Nose", used for feasibility testing.
B-1R	3	BU___	XV-1710-4	10-30-34	C-29907	29907 A-1 Prop Shaft. Reversible for Airships. Contract of 1-24-33.
B-2R	4	BU___	YV-1710-4	3-08-35	C-29907	With Airship extension drive shaft coupling
B-2R	5	BU___	YV-1710-4	5-27-35	C-29907	Had Airship extension drive shaft coupling.
C-1	6	34-4	XV-1710-3	6-20-34	6551	First Type Test engine, through 4-13-36, 328 hrs Total Time.
						On June 18, 1934 the following ten engines contracted as YV-1710-3's.
C-2	7	34-5	XV-1710-7	12-24-35	6795	Intended as 1st Flight engine for A-11A. Used for A-11A pre-flight trials.
						On 12-15-36 it was shipped to Curtiss for use in Propeller Testing as a C-6.
C-2	7	34-5	XV-1710-5	NA	6795	Was to be rebuilt for fuel injection in Fall 1936, but FI effort was terminated.
C-3	8	34-6	XV-1710-7	6-13-36	6795	Hazen redesign, 2nd Type Test, ran 245 hrs, 6-15-36 to 1-5-37.
						Remaining eight engines released for construction Sept 22, 1936
C-4	9	34-7	YV-1710-7	9-25-36	6795	1st Flt engine, A-11A AC33-208, 12-14-36, accumulated 304 hrs.
C-4	10	34-8	YV-1710-7	1-19-37	6795	Built from YV-1710-9(D1), completed 150-hr TT, 4-23-37.
D-1	11	34-9	YV-1710-9	2-27-37	6795	First Flight XFM-1, LH position. Later modified to YV-1710-13.
D-1/C-7	12	34-10	YV-1710-9	3-27-37	6795	Built for -9 TT, completed -11 TT.
C-4	13	34-13	YV-1710-7	2-08-37	6795	First Flight engine for XP-37, later upgraded to -11(C7).
D-1	14	34-14	XV-1710-13	6-28-37	6795	Built for -11(C7) TT, modified & ran -13 TT through 11-19-37.
D-1	15	34-11	YV-1710-9	4-09-37	6795	Began -9 TT, Modified to YV-1710-13, flew in XFM-1 from 1-38.
D-1	16	34-12	YV-1710-9	5-25-37	6795	First Flight XFM-1, RH position. Later modified to YV-1710-13.
						Pre-Production Engines
C-9	17	38-120	XV-1710-15	4-04-38	10291	LH Turning "C8" for XP-38. Later used for training at Chanute.
						20 V-1710-11's ordered for YP-37's, Dec 1937, most delivered as -21(C10), Contract date 1-5-38. Contract amended June 16, 1938 to add 39 V-1710-13's, that were delivered as -23(D2)'s.
C-8	18	38-581	V-1710-11	4-04-38	10660	RH Turning engine for XP-38. Later used in Hi-Power -19 test.
C-8	19	38-582	V-1710-11		10660	Damaged in XP-37 when it crashed Summer 1938.
C-10	20-1	None	V-1710-21			1st engine to attempt the V-1710-21 TT, 4-20-38 to 5-17-38.
C-10	20-2	None	V-1710-21			2nd engine to attempt the V-1710-21 TT, 10-26-38 to 11-28-38.
C-13	23	38-585?	XV-1710-19	8-15-38	10660	First flight engine for XP-40, built from V-1710-21.
E-2	30	38-644	XV-1710-17	11-11-38	10830	Initial engine for XP-39, modified to prototype E-5.
C-10	31	38-592	V-1710-21		10660	3rd engine to attempt the V-1710-21 TT, 1-26-39 to 3-2-39.
E-2	33	38-931	V-1710-17		11279	First Flight engine for XP-39, modified to prototype E-5.
C-13	37	38-597	V-1710-19		10660	-21 configured as -19 for Hi-Power tests, 9-22-39 to 1-11-40.
D-2	61	38-620	V-1710-23		10660	Engine for Bell YFM-1 pusher.
C-15	84-1	None	V-1710-19/33	8-11-39		Made 3 attempts at V-1710-33 Model Test, 8-11-39 to 9-10-39.
C-15	84-2	None	V-1710-33			Second engine to attempt V-1710-33 MT, 10-10-39 to 11-11-39.
C-15	84-3	None	V-1710-33			3rd engine to attempt V-1710-33 MT, 1-26-40 to 1-29-40.
E-1	88	BU 4310	XV-1710-6	1-04-40	C-65197	First flight engine for XFL-1 Airabonita. Contracted 3-3-39.
E-1	89	BU 4311	XV-1710-6	3-28-40	C-65197	Performed Altitude Calibration and used as spare for XFL-1.
E-5	137	40-571M	V-1710-37			Ran 124.5 hrs of 150-hr Type Test, Feb 1941.
C-15	150	39-948?	V-1710-33			4th engine to attempt V-1710-33 MT, started 5-11-40.
F-3R	301	40-4395	V-1710-39	10-07-40		First flight engine for NA-73X Mustang.
F-3R	305		V-1710-39			Type Test
C-15	360?	39-1125	V-1710-33			5th engine, completed V-1710-33 MT, started 8-15-40.

References: History, by J.H. Hunt of Allison, with comments by J.L. Goldthwaite, 1942, and Confidential Allison History, dated January 15, 1941.
Abbreviations: TT = Type Test, MT = Model Test

[28] Dickey, Philip S. III, 1968, 31.
[29] Engineering Division, 1920, Vol. II, No. 143, 11-13.
[30] Engineering Division, 1918, Vol.1, No. 2, 53-85.
[31] Engineering Division, 1918, Vol.1, No. 2, 53-85.
[32] Engineering Division, 1918, Vol.1, No. 3, 108-116. First Army Procurement Order to Allison for three units was CS-54, a second order for three more was received on CS-203.
[33] Air Service Information Circular No.143, 1920, 34.
[34] Dickey, Philip S. III, 1968, 31.
[35] Allison Motors, 1921, 23.
[36] Worden 1962.
[37] Allison Division, GMC 1950.
[38] The Allison Experimental Company filed a petition in Marion Circuit Court on January 4, 1921 to change the name to Allison Engineering Company.
[39] Worden 1962.
[40] Allison Division, GMC 1950, 1.
[41] Allison Division, GMC 1962 Draft, 4.
[42] In 1995 dollars or value, a similar 400 bhp, but provided by a gas turbine (Allison 250) yacht engine, would cost about $250,000. Adjusting for inflation, in terms of 1920 dollars, 400 bhp yacht engines still cost about $25,000!
[43] Bowers, Peter M. 1974, 8-50.
[44] Norton, William 1981, 55.
[45] Fleming, Roger 1950, 2.
[46] Allison Division, GMC 1950, 13.
[47] Goldthwaite, John L. 1950, 2.
[48] Hazen, R.M. 1957, 3.
[49] Allison built Roots-type blowers, or superchargers, for many different customers and applications, beginning in the early 1920's and continuing at least through the mid-1930's. In the early 1920's there was considerable debate about whether the Roots or Centrifugal types were the better for aircraft, in fact the Air Service performed a fly-off between the two types using a DH-4MS in a landmark development effort. One installation was on the Packard A-2500 water-cooled V-12, which used two Allison offered at least these two Roots blowers:

No. 2 size- Impellers 9-1/2 inches in diameter by 8-1/4 inches long. Displacement, 660 cubic inches per revolution, weight 88 pounds including gearing.

No. 3 size- Impellers 9-1/2 inches in diameter by 4 inches long. Displacement, 320 cubic inches per revolution, weight 60 pounds including gearing.

See Aeronautical Engine Laboratory, 1935, 22.

[50] Goldthwaite, John L. 1950, 2.
[51] *History of Allison Division in the Aircraft Engine Industry*, p.3.
[52] Allison Division, GMC 1950, 13.
[53] Schlaifer, R. and S.D. Heron, 1950, 274.
[54] Page', Victor W. 1929, 1850-1858.
[55] Rickenbacker, Edward V. 1967, 150.
[56] Dickey, Philip S. III, 1968, 76.
[57] Rickenbacker, Edward V. 1967, 151.
[58] Rickenbacker, Edward V. 1967, 150-160.
[59] Allison Division, GMC 1962 Draft, 2.
[60] Rickenbacker, Edward V. 1967, 160.
[61] Allison Division, GMC 1950.
[62] Allison Division, GMC 1950, 14.
[63] Rickenbacker, Edward V. 1967, 160.
[64] Rickenbacker, Edward V. 1967, 160.
[65] Rickenbacker, Edward V. 1967, 160.
[66] Allison Division, GMC 1950, 14.
[67] Allison Engineering Company Balance Sheet of January 31, 1929, which would be at the end of the first month of Fisher Brothers ownership, shows a Net Worth of $492,866.88.
[68] Allison Division, GMC 1950, 14.
[69] Charles E. Wilson was then the president of General Motors, and had become well acquainted with the skill of the Allison organization while he was general manager of the Delco Remy Division of GM, located in Anderson Indiana, and had recognized that the organization represented an asset which GM could use while continuing the Indianapolis operation. On his recommendation the firm was purchased for $592,168. Fleming, Roger 1950, 3.
[70] Fleming, Roger 1950, 4.
[71] *General Motors and the Aviation Industry*, for The General Motors Story, DRAFT of March 6, 1957, p.18.
[72] *Campus Heritage*, History of the Indianapolis campus of Marian College.
[73] *Aircraft Engine Information*, data provided by R.M. Hazen, Technical Assistant to General Manager, for use in Allison History prepared in 1957. Letter of July 24, 1957, page 4.
[74] Allison Motors, 1920.
[75] Allison Motors, 1920.
[76] Jim Allison used these engines in his 72-foot yacht *Altonia II,* which he often based in Miami.
[77] This may well have been one of the first applications of rollers being used as cam followers in high speed engines.
[78] Allison Engineering Company, 1920 Specification.
[79] Allison Division, GMC 1950, 12.
[80] Hazen, R.M. 1957, 4.
[81] Nordenholt, G.F. 1941, 218-223.
[82] Sonnenburg, Paul, and William Schoneberger. 1990, 30.
[83] Rolls-Royce licensing agreement. See Sonnenburg, Paul, and William Schoneberger. 1990, 31.
[84] Allison was granted a number of patents in the course of establishing themselves, and in developing the V-1710. The most important of these were often identified on a dataplate affixed to the engine crankcase. A V-1710-F10R School Engine belonging to the Air Force Museum, lists the following Patents: 1948479, 2001866, 2011855, 2024334. Patent 1,948,479 covers the Relief valve arrangement in the Lubrication system, first used on XV-1710-1. 2,024,334 covers certain features of the valve mechanism, first used on the XV-1710-1.
[85] Cruzans, Mr. 1947.
[86] Fleming, Roger 1950, 3.
[87] Allison X-4520, Air Corps Contract 836, Serial Number 25-521, Bore 5-3/4 inches, Stroke 7-1/4 inch, Displacement 4520 cu.in. From the data plate on the engine, in storage at the New England Air Museum, Bradley International Airport, Connecticut, April 1995. (Calculated displacement is 4518.3 in.cu).
[88] Allison Division, GMC 1950, 13.
[89] Wright Field, Chief of Power Plant Section letter to Allison Chief Engineer, *X-4520 Engine*, May 15, 1926.
[90] When Allison began the V-1710 program they again started with manufacturers serial number 1, which was the GV-1710-A, the first Navy engine.
[91] Air Corps Technical Report Number 3585, 1932.
[92] *Allison General History, 1962*, internal Allison report, p. 5.
[93] Air Corps Technical Report Number 3585, 1932.
[94] Sonnenburg, Paul, and William Schoneberger. 1990, 43.
[95] *In-Line Air-Cooled Engines for Tanks*, Ordnance Department letter to Materiel Command at Wright Field, August 29, 1942. NARA RG18, File 452.8, Box 807.
[96] *In-Line Air-Cooled Engines for Tanks*, Wright Field letter to Chief of Ordnance Department, September 26, 1942. NARA RG18, File 452.8, Box 807.
[97] Allison reports differ on the number built, references exist supporting both 200 and 250.
[98] In 1924 the Air Service had 11,801 Liberty engines stored at Little Rock Air Intermediate Depot. Norton, William 1981, 80.
[99] Air Service Information Circular No.551, 1926.
[100] Frank, Gerhardt W. 1929, 445-459.
[101] Doolittle, Major J.H. 1941.
[102] Dickey, Philip S. III. 1968, 31.
[103] Technical Bulletin No.46, 1926, 30-32.
[104] Gilman, N.H. 1927, 1468-1470.
[105] *Allison V-1410 and VG-1410 Engine Specification*, Commercial, no date, probably 1927.
[106] Kelsey, Benjamin S. B.G. USAF (Ret.) 1982, 29.
[107] Allison Division, GMC 1962 Draft, 5.
[108] *History of Allison Division in the Aircraft Engine Industry*, p.3.
[109] Atkinson, Robert P. 1989.
[110] Dickey, Philip S. III. 1968, 28-29.
[111] Technical Bulletin No.46, 1926, 31.
[112] Dickey, Philip S. III. 1968, 30.
[113] Christy, Joe, 1971, 56.
[114] *Allison V-1650 Engine Specification*, no date, probably 1927.
[115] NARA RG 342, RD3480, Box 6896.
[116] Sonnenburg, Paul, and William Schoneberger. 1990, 44.
[117] Fleming, Roger 1950, 2.
[118] Allison Division, GMC 1950, 16.
[119] Fleming, Roger 1950, 4.
[120] The XV-1710-3 purchased on contract W535-ac-6551 was AC s/n 34-4, while the ten YV-1710-3's purchased on contract W535-ac-6795 were assigned s/n's 34-5 through 34-14.
[121] Allison Division, GMC 1950, 16 & 19.
[122] Letter from John C. Felli to Mr. O. A. Lundin, Allison Comptroller regarding the material going into *The General Motors Story-General Motors and the Aviation Industry*, prepared in 1957. Letter of March 14, 1957.
[123] Goldthwaite, J.L. 1932.
[124] Allison Division, GMC 1950.
[125] Nordenholt, G.F. 1941, 218-223.
[126] *Annual Report of the Chief of the Air Corps, 1928*, 59.
[127] Schlaifer, R. and S.D. Heron, 1950, 260.
[128] Goldthwaite, John L. 1950, 4.
[129] The first known attempt to cool an aircraft engine cylinder by means of a high-boiling-point cooling-liquid was made by Dr. A.H. Gibson at the Royal Air-

craft Establishment in 1916. He used aniline and achieved jacket temperatures as high as 314 °F. The first work at McCook Field was done in March 1923 at the suggestion of S. D. Heron, who also recommended Ethylene-Glycol as the coolant. The first flight test of an engine modified for high-temperature chemical cooling was a Liberty-12 powering an Air Corps TP-1 airplane on February 25, 1924. The Navy also investigated high-temperature cooling in a Wright E-4 V-8 and then in a 600 bhp Wright T-3 V-12. Frank, Gerhardt W., 1929, 445-459.

[130]Fleming, Roger 1950, 3.

[131]Frank, Gerhardt W., 1929, 343.

[132]*Annual Report of the Chief of the Air Corps, 1930*, 54-55.

[133]Ordinary 1928 vintage Air Corps specification radiator hose was not suitable as the coolant temperature was above the vulcanizing temperature of the rubber lining, causing it to swell and loosen. Early flight installations used 100 psig steam hose, which was not entirely satisfactory because of its stiffness causing it to be difficult to clamp securely. The development of high-temperature liquid-cooling is an example of how one advance required another supporting development, all of which then allowed (in this case) the successful development and use of the V-1710. Frank, Gerhardt W., 1929, 339.

[134]*Allison General History*, draft prepared in 1962, p.15.

[135]Arnold, H.H.,Major General, Chief of the Air Corps letter to Mr. Louis A. Johnson of Clarksberg, West Virginia, January 31, 1941. NARA RG18, File 452.8, Box 807.

[136]*Allison Engine Situation*, 1941.

[137]Allison Division, GMC 1950, 25.

[138]Bourdon, M.W. 1944.

[139]Goldthwaite, John L. 1950, 3.

[140]Allison Division, GMC 1950, 13.

[141]Article in Allison News.

[142]Fleming, Roger 1950, 3-4.

[143]As an interesting sidebar to the creation of McCook Field, the site had originally belonged to the predecessor of the Dayton-Wright Corporation who had used it in the early days of aviation as a flying field. During WWI it was the largest tract of land used as a flying field. In 1929 General Motors purchased a 40 percent interest in Fokker Aircraft Corporation of America for $7,782,000, thereby acquiring the capital stock of the Dayton-Wright Corporation subsidiary of Fokker. The assets included patents and cash worth $6,500,000, and McCook Field, valued at $1,282,000. This was following the relocation of military activities and interest to the new Wright Field site some 5-1/2 miles out of the city, which had occurred in 1927. *The General Motors Story*, GMC letter from J.C. Felli to Roger C. Fleming, Director of Public Relations, Allison Division, July 3, 1956.

[144]Echols, Maj. Gen O. P. 1941, 83.

[145]*15th Annual Report of the NACA*, 1929, 45 & 86.

[146]*16th Annual Report of the NACA*, 1930, 64.

[147]Allison Division, GMC 1950.

[148]Allison Division, GMC 1950, 16.

[149]Allison Division, GMC 1950, 17.

[150]According to John L. Goldthwaite, "I think Gilman's most important single contribution was his insistence on the 1000 bhp idea. In this he stood alone at the time, against all advice from the Army, Navy and others. It was urged that the limit of usefulness for single-engine planes was around 750 bhp (the variable pitch propeller would soon change this picture); and that larger planes would simply use more engines, not larger ones (air power dreams were modest in those days). Goldthwaite, John L. 1950, 3.

[151]Dickey, Philip S. III. 1968, 76, which refers to a McCook Field memo, Chief, Air Corps, from Commander, January 24, 1929.

[152]*Annual Report of the Chief of the Air Corps, 1932*, 77.

[153]General Motors Corporation. 1957, 21.

[154]The 400-IG engine was rated for 150 bhp at 2800 rpm. It featured two valves per cylinder having a lift of 7/16 inch, each valve was somewhat larger than those in the GV-1710, which was rated for 2400 rpm and had a lift of .5 inch. A major difference was that the 400-IG used separate cams for each valve where the GV-1710 operated two valves (exhaust or intake) with a double rocker arm. Caminez, H., Report No. 28, 1930.

[155]Caminez, Harold and F.N.M. Brown. 1935.

[156]Caminez, Harold and F.N.M. Brown. 1935.

[157]Caminez, Harold and F.N.M. Brown. 1935.

[158] This is interesting, and reflects the 1934/35 version of the dilemma, for early in the War the V-1710 was switched from pure ethylene-glycol to pressurized water, with ethylene-glycol only as an antifreeze. Cooling effectiveness was actually improved, along with a reduction in "cooling drag".

[159]Caminez, Harold and F.N.M. Brown. 1935.

[160]Hazen, R.M. 1957, 2.

[161]Allison Division, GMC 1950, 17.

[162]Sonnenburg, Paul, and William Schoneberger. 1990, 48.

[163]Procured on Wright Field contract W535-ac-5592.

[164]The German Maybach engines were officially considered to be very fine engines, actually they were not so good. *Allison Engineer J.L. Goldthwaite*

[165]*Allison General History*, draft prepared in 1962, p.9.

[166]Air Corps Technical Report No. 4038, 1937, 48.

[167]Goldthwaite, John L. 1950, 5.

[168]Allison Division, GMC, 1941, 4.

[169]Resume' in Allison history files, Ronald M. Hazen.

[170]Hazen, R.M. 1933.

[171]Harold Caminez had been in charge of engine design for the Engineering Division of the U.S. Army Air Corps at McCook Field, Dayton, Ohio during and after WW-I. While there he made a study of types of power plants most suited to the needs of aviation, and as a result, developed the design for an engine with a unique "cam" drive. He patented the design and the Engineering Division adapted an engine to test the cam drive principle. These tests proved the high mechanical efficiency of the cam mechanism and showed the possibility of obtaining high power in a compact engine by its use. Sherman M. Fairchild recognized the value of this invention and the Fairchild Caminez Engine Corporation was formed in January 1925 to construct cam type aircraft engines. Caminez was named vice-president and Chief Engineer of the Fairchild Caminez Engine Corporation. The Company completed its first engine that year, and submitted it to 50 hours of testing at McCook Field, followed by installation in a Fairchild aircraft. The first flight of the engine was in April 1926. From this experience the Company developed its Model 447-C engine, which was offered commercially in 1927. This 447 cubic inch four-cycle, air-cooled, four cylinder radial engine weighed 350 pounds and developed 145 bhp at 1100 rpm, the highest power output per cubic inch demonstrated as of that time in a non-supercharged, air-cooled engine. Caminez obviously had considerable talent and experience as an engine designer and engineer, as well as experience as a corporate manager and developer. In joining Allison in 1930, to engineer the designs of Norm Gilman, it was likely that he saw considerable opportunity for advancement. When Gilman did not name him as his replacement as Allison's Chief Engineer in 1936, he evidently saw it as a serious vote of non-confidence. His resignation from Allison would not be surprising under the circumstances. See *Fairchild Caminez Aircraft Engine Model 447-C* sales brochure, 1927 for more details on the Caminez Cam Engine.

[172]Allison letter, Gilman to Col. O. P. Echols, Wright Field, 3-6-36.

[173]Hazen, R.M. 1941.

[174]Development work done by Wright Field in 1930 had found that mechanical metering of fuel, using the Marvel fuel pump, promised a marked reduction in specific fuel consumption and commensurate increase in horsepower over a carburetor equipped engine at the same manifold pressure. *Annual Report of the Chief of the Air Corps, 1930*, 55.

[175]Sonnenburg, Paul, and William Schoneberger. 1990, 51.

[176]Allison Division, GMC, 1942 news release.

[177]Wright, Donald F. 1994.

[178]Atkinson, Robert P. 1989.

[179]Kreusser had returned to GM after establishing the Chicago Museum of Science and Industry. He had previously been with GM for 13 years in various executive engineering roles, including being the director of the GM Proving Grounds. Fleming, Roger, 1957, 4.

[180]Atkinson, Robert P. 1989.

[181]Allison Division, GMC, 1941, 4.

[182]Doolittle, Major J.H. 1941.

[183]Doolittle, J.H. and Carroll V. Glines, 1992, 200-201.

[184]Nordenholt, G.F. 1941, 218-223.

[185]Nordenholt, G.F. 1941, 223.

[186]Each Division within General Motors was annually allocated a number of positions in this prestigious post-graduate GM training center. Being selected was not only an honor, but did a lot to insure future success within the GM organization. Due to its small size prior to the War, Allison received very few slots.

[187]Dick Askren, an engineer in Allison at the time, relates about Wright that, "He was strict and severe, and he was good." Askren, R.W. "Dick" 1994.

[188]Wright, Donald F. 1994.

[189]On April 23, 1937, when the YV-1710-7(C4/C6) completed its Type Test at 1000 bhp, and the XV-1710-11(C8) was ready for Type Testing, the Wright R-1820 was rated at 575 bhp, and the Pratt & Whitney R-1830 was at 800 bhp. It was not long before both engines, the major competition for the V-1710 at the time, were developed for 1000 bhp ratings. Allison Division, GMC 1950, 29.

[190]Christy, Joe, 1971, 61.

[191]Kelsey, Benjamin S. B.G. USAF (Ret.) 1982, 39-40.

[192]Letters between Allison General Manager and Air Corps, located in NARA RG 342, RD 3480, Box 6896.

[193]Allison memo of January 15, 1941 identifies 857 engines as having been the aggregate of the early Army purchases. Referenced in Allison memo to Mr. E.B. Newill, March 7, 1957.

[194]Sloan, Alfred P., Jr. 1964, 371. Sloan says that they built the plant with orders for only 836 engines.

[195] Hazen, R.M. 1945 Report.

[196] Hazen, R.M. 1945 Report.

[197] Some references give this French order as 669 engines, which is the number actually delivered to the British from the French contract F-223, the other 31 were diverted to the Chinese for the AVG.

[198] Sixty nine British C-15's were diverted to the Chinese for the AVG. The breakout of engine models was: 948 C-15, 810 E-4, 1,416 F-3R, 163 F-5R, 163 F-5L.

[199] Allison Division, GMC, 1941, 2-3.

[200] Doolittle, Major J.H. 1941.

[201] Materiel Division letter proposing diversion of a single Air Corps contract W535-ac-12553 production V-1710-33 to French Air Commission as first engine on French contract, May 15, 1940. NARA RG 342, RD 3480, Box 6896.

[202] Bodie, Warren M., 1991, 20.

[203] Allison's Manufactures' model numbers, e.g. V-1710-C4, described the configuration, as defined by their model specifications and reflected the development status, or unique features incorporated in each of their engines. These were of little interest to the military, who issued model "dash" numbers to denote major changes in engine features or ratings and often to define uniqueness required by a particular aircraft installation, and then only for the engines they actually purchased. Consistent throughout the military was a convention that the Army would assign odd "dash" numbers to its engines, and the Navy even dash numbers. In both cases the dash number followed the cubic inch displacement, rounded to the nearest 5 cubic inches, and a prefix denoting the cylinder configuration. Allison also used its model numbers for any commercial or non-Army models, i.e. the V-1710-C15's (not V-1710-33's as in U.S. Army nomenclature) that powered the British *Tomahawks* prior to Lend-Lease.

Throughout this text every attempt has been made to present nomenclature in the vernacular used by the parties during the life of the V-1710. When using the Allison model numbers, in both writing and discussion, persons familiar with the engines would use the form C-4, C-15, E-2 or G-6 rather than the more formal "V-1710-C4", etc. The hyphenated form was used in all but the most formal of discussions.

[204] A description of the various engine serial numbering schemes used by the military and Allison are given in Appendix 1.

[205] Superchargers. NARA RG 342.

[206] Allison Division, GMC, 1941, 10.

[207] Allison general history, draft prepared in 1962, p.15.

[208] Allison Division, GMC, 1941, 8.

[209] Allison Division, GMC, 1941, 8.

[210] Kelsey, Benjamin S. B.G. USAF (Ret.) 1982, 69-70.

[211] Appendix 9 details the significant differences between the C-15 and F-2R engines. Except for the reduction gear the E/F engines were identical.

[212] In July 1937 the Air Corps was complaining internally about the cost of the 60 pre-production V-1710 engines, those for the YP-37 and YFM-1, (engine costs proposed as $23,922 each). The President of the Allison Engineering Company had been to Wright Field to justify the price and stated that the Company had un-absorbed development costs of $900,000 and additional production tooling and shop equipment of $250,000 to procure in a rising market for labor and material to contend with before the company could go into quantity production. Allison stated that in prorating unabsorbed development costs it was assumed that 300 engines would be ordered over a two year period. The result was that the Materiel Division decided to audit the Allison books before agreeing to the engine price. *Estimate of Funds Required for the Procurement of Airplanes in the Fiscal Year 1938*, Materiel Division letter to the Chief of the Air Corps, July 15, 1937. NARA, RG18, Entry 187D, Box 48.

The Army was clearly in arrears in supporting Allison in developing the V-1710. As of December 31, 1936 total costs by the Army, including their own direct labor, materiel, travel and overhead amounted to $506,361.19, of which only $419,423.70 had been paid to Allison. Engineering Section Memo Report, E-57-680-14, 1937.

[213] Priority Teletype DHQ-T-527 from General Echols, Chief, Materiel Division, April 17, 1940. NARA RG18, File 452.8, Box 808.

[214] Brett, General George H. July 25, 1940.

[215] Thomas, Lowell and Edward Jablonski, 1976, 151-153.

[216] *Status of Allison Engines*, internal memo to Chief of Air Corps, November 29, 1940. NARA RG18, File 452.8, Box807.

[217] Doolittle, Major J.H. 1941.

[218] Teletype INSP-T-688 from Army Inspectors to General Arnold, January 9, 1941. NARA RG18, File 452.8, Box 807.

[219] Letter from Robert P. Patterson, Under Secretary of War, to Allison Engineering Company, April 11, 1941. NARA RG18, File 452.8, Box807.

[220] Memorandum for General Echols from the Assistant Secretary of War for Air, Robert A. Lovett, June 18, 1941. NARA RG18, File452.8, Box807.

[221] Telegram from F.C. Kroeger of Allison to AAC Production Engineering Section, September 10, 1941. NARA RG18, File452.8, Box807.

[222] *Allison Engines versus Airframes*, memorandum to Robert A. Lovett, Assistant Secretary of War for Air, from Brig. General O. P. Echols, September 12, 1941. NARA RG18, File452.8, Box807.

[223] Correspondence from Chief of Air Corps to Chief, Supply Division, August 3, 1939, "The CAC recently asked that a letter be written to the Materiel Division to determine whether or not the Allison Company had been contacted in an effort to bring the present Allison type up to a 1600 h.p. rating...He thought it imperative that early steps be taken...". NARA RG18, File 452.8, Box 808.

[224] Engineering Section Memo Report Serial No. E-57-4272-1, 1939.

[225] *Allison Engine*, Letter to Chief, Materiel Division, Air Corps from Major General H.H. Arnold, Chief of the Air Corps. NARA RG18, File 452.8, Box 808.

[226] First Endorsement, by Chief of the Materiel Division, in response to General Arnold's letter of August 29, 1939, dated September 21, 1939. NARA RG18, File 452.8, Box 808.

[227] Letter from Louis Johnson, Assistant Secretary of War, to Wm. S. Knudsen, September 5, 1939. NARA RG18, File 452.8, Box 808.

[228] *Allison Engineering Company's Development Progress*, Materiel Division letter to General Arnold dated December 29, 1939. NARA RG18, File 452.8, Box 808.

[229] Quoting from the reference, "...so that consideration can be given for the release for foreign sale of the types of V-1710 engines now installed in the more advanced pursuit airplanes under construction for the aviation expansion program. Release of these types for foreign sale should immediately make it necessary for all our advanced pursuit airplanes to be retrofitted with the improved model V-1710." Johnson to Knudsen letter, January 31, 1940. NARA RG18, File 452.8, Box 808.

[230] Allison letter X2C20AE, to General O. P. Echols, Commanding General, Materiel Command, Army Air Forces from R.M. Hazen, Allison Chief Engineer, March 20, 1943. NARA RG18, File 452.8, Box 808.

[231] Letter from Materiel Division to R.M. Hazen, March 31, 1943. NARA RG18, File 452.8, Box 808.

[232] *Development of Increased B.H.P. and Altitude for the Allison V-1710 Engine*, letter from Chief of Procurement Division to Allison Division, June 14, 1943. NARA RG18, File 452.8, Box 808.

[233] Letter from Lt.Col. Thomas Hitchcock, Asst. Military Air Attaché, American Embassy, London, England to Major General O. P. Echols, Materiel Command, June 28, 1943. NARA RG18, File 452.13-M, Box 756.

[234] Letter to O.E. Hunt, Vice President GMC, Allison Division from Major General C. E. Branshaw, Commander of Materiel Command, August 13, 1943. NARA RG18, File 452.13-M, Box 756.

[235] *Development of Horse Power for the Allison V-1710 Engine*, memorandum for Major General O. P. Echols from Major General C. E. Branshaw, August 13, 1943. NARA RG18, File 452.13-M, Box 756.

[236] Letter to O.E. Hunt, Vice President GMC, Allison Division from Major General C. E. Branshaw, Commander of Materiel Command, August 13, 1943. NARA RG18, File 452.13-M, Box 756.

[237] Letter from O.E. Hunt, Executive Vice President, GMC to Brigadier General C. E. Branshaw, September 8, 1943. NARA RG18, File 452.13-M, Box 756.

[238] Arnold stated that pilot trainees did not fly in Allison equipped planes in the schools. During the period 7-1-42 to 1-31-43, Allison powered planes aggregated 283,783 flying hours, incurring 0.051 fatal accidents per 1,000 flying hours, a total of 14 or 15 pilots, a rate in line with the general Air Corps. He noted the improvements that had been made. In May 1940, production rate was 12 engines per month, and since that date over 26,000 had been produced and the present rate was 1,700 per month. At the same time the average service life of the engine had increased from 55 hours between overhauls, to a then current average of 200 hours. Horsepower had increased in the same period from 1000 bhp to 1452 bhp. He closed noting that "Reports from theater commanders indicate very satisfactory performance of this engine and full pilot confidence in it." H.H. Arnold, Commanding General, AAF, *Memorandum For The President: Performance of Allison Engines*, May 8, 1940. H.H. Arnold manuscript, Roll 165, Library of Congress.

[239] *Termination of V-1710 Type Aircraft Engines*, File 452.8- V-1710, ProdSec/ProcDiv to CG, AAF, Washington DC, dated May 28, 1945.

[240] Scott, Robert L., Colonel, Air Corps, U.S. Army, 1944, 260.

2

First Flight for the V-1710

In early 1935 the Air Corps began in earnest to plan for the initial flight testing of the Allison V-1710, intending that the first flight engine would be the first of the ten YV-1710-3's purchased in 1934. All that was needed was a release for manufacture, though that release was pending the requirement to successfully complete the 150-hour Type Test at the specified 1000 bhp. In anticipation of the XV-1710-3(C1) successfully completing the Type Test, XV-1710-7(C2) [AC34-5] was delivered on December 24, 1935, and shipped to Bell Aircraft for installation in the A-11 they were modifying as the A-11A for initial flight testing of the V-1710.

Selecting an Aircraft

The Materiel Division at Wright Field conducted a comprehensive review of aircraft that might be suitable to serve as the V-1710 flying test bed, envisioning that it would require an aircraft with special capabilities, and one that could be simply modified to accommodate the big Allison. The finalists in this process were the Bellanca C-27A, Kreider-Reisner (Fairchild) XC-31 and the Douglas O-43A.[1]

Each of these airplanes was found to be less than ideal when considered as a testbed for the V-1710. The Bellanca C-27A would have been some 375 pounds over its design weight of 9,600 pounds, and not insignificantly, somewhat nose-heavy. This could have been accommodated by repositioning of internal aircraft equipment and cargo, so it was believed that the C-27A could be a satisfactory testbed.

The 1934 vintage Bellanca C-27A was powered by an air-cooled Wright R-1820 and one can only imagine how it would have looked with the longer V-1710 had it been modified as the testbed. (Hal Andrews)

The single Kreider-Reisner/Fairchild XC-31 was the first U.S. aircraft built specifically for cargo.[2] It was powered by a Wright *Cyclone* R-1820-25 of 750 bhp, and believed to still be the largest single engined cargo plane ever built.[3] It would have been 520 pounds over its design gross weight of 12,400 pounds with the YV-1710-3 installed. As with the C-27A, a scheme for bringing it into balance and within gross weight was devised, but not attempted on the C-27 either. However, the type had only completed its first flight on September 22, 1934, and had yet to complete its own test program.[4] It was therefore deemed to be unavailable for use as the test bed.

The Douglas O-43A, powered by a Curtiss 650 bhp V-1570-59 Conqueror liquid-cooled V-12, at first looked like a good candidate for conversion. Investigation found that the aircraft would be 156 pounds over its design maximum weight even after all unnecessary equipment was removed. With a maximum allowable gross weight of only 5,300 pounds there was no margin. Furthermore, the heavy Allison moved the center of gravity too far forward, hence Wright Field concluded that the O-43A was unsatisfactory as the testbed.

At this point the fledgling Bell Aircraft Company enters the picture. Consolidated Aircraft had just relocated its operations to Southern California, vacating facilities it had been occupying in Buffalo, New York. Larry Bell was able to find enough financing to start his own aircraft firm, using the vacant Consolidated facilities and some of the ex-Consolidated staff who wished to remain in the area. Bell had been working with Consolidated's Bob Woods, the capable young designer of the YP-25 and P-30, on starting this venture for a couple of years. Together they hoped to be able to introduce advanced aircraft concepts and emerge as a successful new aircraft company. This was a tall order in the middle of the Great Depression, even for visionary upstarts.

Bell began operations in the summer of 1935 manufacturing airframe components, mostly for the relocated Consolidated, and was in definite need of additional work. Larry Bell was in close contact with the Wright Field people, and was having some success in selling them on the concept of the multi-place fighter or bomber/destroyer. His original 1934 plans for the aircraft showed a twin-engined pusher, using the new Allison V-1710. He was clearly aware of the status of the engine, and of the Air Corps need for a suitable flying testbed. With his ties to Consolidated, and the avail-

The XC-31 being prepared for its first flight, piloted by W.N. DeWald, Hagerstown Airport, Maryland, September 22, 1934. (Photo by Lyle S. Mitchell, provided by Kent A. Mitchell)

ability of key ex-Consolidated employees, Bell was able to convince the Air Corps that the Allison testbed should be converted from the Consolidated A-11 then being flight tested at Wright Field. Only four of the A-11's were built, and aircraft developed from the Curtiss Conqueror powered Consolidated P-25, but without that aircraft's turbosupercharger.

The conversion turned out to be a pretty good match, and a suitable aircraft for the purpose of putting the Allison through its paces. Wright Field issued Bell Aircraft an order in August 1935 for the conversion, resulting in contract W535-ac-7949 for $24,995.[5] This amount is interesting in that contracts over $25,000 were deemed "major" contracts and required an extensive process of review and approval. Expediting the $24,995 to Bell had the effect of keeping the new firm, and its' multi-place fighter concept, alive a while longer. Bell designated the conversion project as their "Model 2." The "Model 1" was already associated for the first all-Bell design, the multi-place fighter.

The resulting aircraft was known as the Consolidated A-11A, though "Consolidated" had no direct involvement with the conversion. Some references to this airplane identify it as the XA-11A, but that designation is never seen in original source documents describing the aircraft, its modification, or operation.[6] Historians should also note that there was an earlier "A-11A" proposed by Consolidated, dating from August 1934.[7] It would have been an A-11 powered by either the Wright R-1510 or P&W R-1535, no Air Corps procurement action followed. The airframe that actually became the A-11A was Consolidated A-11 Air Corps AC33-208, the first of the four A-11's[8], and an aircraft that had been assigned to Wright Field since delivery to the Air Corps in August 1934. It had flown a total of 118

Powered by the liquid-cooled 600 bhp Curtiss Conqueror V-12, the Douglas O-43A was a strong candidate for conversion to the larger Allison V-1710. (San Diego Aerospace Museum)

Wright Field shipped Bell only the fuselage of the Consolidated A-11 AC33-208 for reworking as the A-11A testbed for the V-1710. Bell installed the intended first flight engine, the 1000 bhp XV-1710-7(C2) [AEC s/n 7] and mounted the fuselage on a crude undercarriage for ground running and system checkout in April 1936. This engine did not fly. (Bell Textron via Peter Bowers)

The Bell modified A-11A being prepared for shipping to Wright Field. The XV-1710-7(C2) is installed. (Bell via Bowers)

hours up to July 1935 when it was flown to Bell at Buffalo for modification.[9] The aircraft was modified and fitted with XV-1710-7(C2), AEC# 7, AC34-5, prior to being shipped to Wright Field by surface transportation on May 14, 1936. This engine was intended as the first flight engine until April 13, 1936, when the similarly configured XV-1710-3(C1) Type Test was declared as failed.[10]

At this time the only other V-1710-C engine was the XV-1710-3(C1), (AEC# 6, AC34-4), and it had just been torn down following its failure at the attempted 150-hour Type Test. The XV-1710-7(C2) engine was used only for fit-up and ground running to test the aircraft's cooling, oil, fuel and control systems. The A-11A never flew with this engine. The engine was subsequently rebuilt by Allison as a C-3 and sent to Curtiss Electric for use in testing propellers intended for use with the V-1710.

The failure of the Type Test engine initiated a 13 week crash program to redesign the engine to solve the mixture distribution problem and generally strengthen it throughout. In June, the considerably revised AC34-6, AEC# 8, was delivered as the XV-1710-7(C3) to restart the Type Testing.

The first flight engine was the further revised YV-1710-7(C4), [AC34-7, AEC# 9] that was not officially released for production until September 22, 1936, although the decision to release the final eight engines (from the batch of 10 YV-1710-3's) had been reached at a conference on September 3. In anticipation Allison had already built the engine and was able to ship it on September 24, 1936, even though the XV-1710-7(C3) attempting the 150-hour Type Test, was not to accumulate 150 hours until October 27.[11] The C-4 engine was installed in the A-11A after both were shipped to Wright Field in the fall of 1936.[12]

On this basis, and appreciating that after three years of testing the V-1710 had yet to fly, the decision was made to proceed. While engine #8 had run some 245 hours, the Air Corps deemed it too as having failed the Type Test, due to the many component failures that occurred during the protracted testing. Because of the difficulties in completing the Type Test, Wright Field restricted engine power output in the A-11A during its first 30 hours of flight.

The aircraft first flew on Allison power on December 14, 1936.[13] Wright Field test pilot Air Corps Captain Frank Irvin made the first flight. First Lt. Benjamin S. Kelsey, as Officer in Charge of the Fighter Development Office was not only instrumental in the conception and development of advanced pursuit aircraft, but he personally flew the V-1710 powered A-11A for over 100 hours during the flight test program.[14] During the first 100 hours on the new engine, Kelsey flew the A-11A, which was not turbosupercharged, to an altitude of 26,400 feet.[15] He subsequently performed the first flights on the Allison V-1710 powered Bell XFM-1 and Lockheed XP-38, and in addition logged time in the YV-1710-7(C4) powered Curtiss XP-37.

Allison supervised the installation of the engine and then observed and checked flight test data for an additional 9 to 12 months.[16] Allison's representative was T.S. McCrae. In late February 1937 the Wright Field Engineering Section asked the War Department if Mr. McCrae could go

A-11A in Buffalo, NY fitted with the XV-1710-7(C2), AEC#7, AC34-5. Note the pre-Hazen intake manifolds are just visible above the top of the engine, identifying it as the single XV-1710-7(C2), intended as the first flight engine until April 1936. (Bell photo taken April 24, 1936, via B. Matthews)

aloft in the two-seat A-11A on a 30 minute flight "...as Engine Manufacturer has no flight test facilities and desires their representative to check engine operation and smoothness in flight, while completing manufacturers record on engine performance under all conditions of operation." We do not know if the flight was authorized, but if so, he would have been the first Allison man to fly behind the V-1710.[17]

As installed in the A-11A the YV-1710-7(C4) was equipped with a two-position controllable pitch propeller. The engine was rated at 1000 bhp at 2600 rpm at sea-level, though as installed it would deliver 1060 bhp at 2600 rpm at full throttle. For the first thirty hours of flight, to insure proper break-in and reliability, the power from the Allison was limited by a throttle-stop on the quadrant. During the first 20 hours it was 700 bhp at 2300 rpm, then for the next five hours the limit was raised to 800 bhp at 2420 rpm, and then five hours set at 900 bhp at 2520 rpm. The limiting ratings for the YV-1710-7(C4) powered A-11A are given in Table 2-1.[18,19]

Table 2-1
Restricted Ratings for YV-1710-7(C4) in A-11A

	Before Hours	Horsepower	RPM	Max MAP, inHgA	Fuel Burn, GPH
Takeoff/Rated	<20	700	2300	32.0	NA
Takeoff/Rated	20-25	800	2420	35.0	NA
Takeoff/Rated	25-30	900	2520	38.0	NA
Takeoff/Rated	>30	1000	2600	41.0	NA
Cruising	<20	700	2020	26.5	NA
Cruising	20-25	800	2120	28.0	@ 45.5
Cruising	25-30	900	2200	29.5	@ 50.8
Cruising	>30	1000	2280	31.5	NA

Early in the test program the Air Corps performed its standard battery of flight tests done with new aircraft, though the reports have not been located. According to Lt. Mark Bradley (later, General) who was one of the pilots flying the A-11A in its "fly, refuel, fly" program, the airplane was "not at all pleasant to fly because it had a large-diameter propeller, forcing the pilot to land and takeoff from the 3-point position every time or pay the price."[20]

The goal of the flight test program was to put the Allison through the intended full overhaul cycle life of 300 hours. The program seems to have gone fairly well, with 20-30 hours flown per month during the first four months of 1937, all in the winter. For the next twelve months only 49 hours were flown, and in three of the months no time was accumulated. Sketchy records suggest that the aircraft may have been involved in one or two minor incidents, or accidents during the 300 hour test program. The engine may have been a problem in one of them, for it was necessary to replace the accessory drive extension shaft, at a cost of $58.50 in June 1937, and no flight time was recorded that July.[21] The problem did not require its return to the factory. In May 1938, the program was intensified, with 63 hours flown in three months, followed that July by a complete teardown and inspection of the engine after completing 200 hours. It was found to be in excellent condition. After reassembly an order must have been issued to complete the program, for the final 83 hours were flown during October, bringing the total time on the engine to 304 hours.[22] For safety reasons the aircraft had been restricted to only local flying once 275 hours had been accumulated.[23]

One problem that appeared during the flight tests was a slight coolant leak between the coolant jacket and the cylinder head at the attaching studs. Leakage amounted to about one quart of coolant in three days. The leak was due to a design defect and had already been resolved on new production engines.[24]

Wright Field then disassembled the engine for inspection. The cylinder blocks were returned to Allison where they were repaired at a cost of $174.66, including $66.46 in labor. The work involved installing new cyl-

This is YV-1710-7(C4), AEC #9, the first V-1710 to fly. It was fitted with gun synchronizer type distributors. (Allison)

inder coolant jackets, gaskets, sealing washers and most of the 1/4 inch diameter studs and nuts. The major components, including the cylinder barrels and valves did not require replacement. The reduction gear assembly was also returned to Allison, primarily for inspection, and was refurbished as well.[25]

The airplane flew again in July 1939, though likely only as a "station hack" at Wright Field, for it only accumulated seven hours in four months. Its last flight occurred on September 22, 1939 when it was delivered to Chanute Field, Illinois for duty as a Class 26 maintenance trainer.

On March 28, 1944, the occasion of Allison's delivery to the Army Air Force of its 50,000th engine, the Air Force presented Allison with "ole-number 9" the first flight engine, resplendent in a new coat of paint.

NOTES

[1] *Installation of Allison V-1710-3 Engine in an Airplane*, Wright Field letter AD-51-309, February 20, 1935. NARA RG 342, V-1710 1932/5 file.

[2] The Kreider-Reisner Aircraft Corporation was a subsidiary of the Fairchild Corporation. The XC-31 was known by the manufacture as the Fairchild Model 95.

[3] Mitchell, Kent A. 1992, 70-75.

[4] Mitchell, Kent A. 1992, 70-75.

[5] Matthews, Birch. 1996, 26. Contract date was September 22, 1935.

[6] The summary document "*Model Designations of Aircraft Engines, USAF*" published January 1, 1949, and earlier version of that document, do identify the airplane as the XA-11A, but not the documents produced for the airplane itself.

[7] *Three View A-11A Bi-Place Attack Airplane, Wright 1510 & P&W 1535 Engine*, Consolidated Drawing 27Z008, August 17, 1934. *SDAM*

[8] Wagner, Ray, 1960, 60.

[9] According to the aircraft history card the airplane *was transferred to Consolidated Corporation on October 4, 1935 for modification to A-11A*. Evidently some Air Corps people did not know that Bell had become an independent contractor during the summer of 1935. The A-11, AC33-208, was equipped with a special high compression (7.25:1) Curtiss V-1570-59 of 675 bhp (Manuf No. 20241, AC33-54).

[10] Engineering Section Memorandum Report E-57-541-27, 1938.

[11] Hunt, J.H. 1942, A-11A.

[12] The non-original data plate affixed to the engine for historical purposes in 1944 states, *"Sold to the Army Air Corps October 16, 1936. Placed in the Consolidated A-11A airplane at Wright Field and was first flown during the winter of 1936. It remained in the airplane for 300 hours of flying."* Information obtained by Rick Leyes, Curator for Aero Propulsion, NASM.

[13] Appendix 6 lists the first flights of many Allison powered aircraft.

[14] Allison News, 2nd March Issue, Vol.III, No. 17.

[15] Allison PR Release by Mr. O. T. Kreusser, General Manager of the Allison Engineering Division of GM, 1937.

[16] Hunt, J.H. 1942, A-11A.

[17] Telegram, War Department to Materiel Division, 2-26-37.

[18] *Operating Instructions for the Allison XV-1710-7 Engine in the A-11A Airplane*, Wright Field Engineering Section Memo Report No. R-57-311-5, November 9, 1936. Note that neither Allison nor the Air Corps were at all consistent in the nomenclature for this engine. Documents exist identifying it as both an XV-1710-7 and YV-1710-7.

[19] *Supplementary Operating Instructions for the Allison XV-1710-7 Engine in the A-11A Airplane*, Wright Field Engineering Section Memo Report No. R-57-311-8, January 8, 1937.

[20] Bodie, Warren M., 1991, 15.

[21] *Replacement Parts for V-1710-3 Engine AC34-7*, Allison letter to Chief, Materiel Division, Wright Field, June 25, 1937.

[22] Aircraft History Card for Consolidated A-11A AC33-208 with Allison YV-1710-7 AC34-7.

[23] Engineering Section Memo Report No. R-57-311-9, 1938.

[24] Engineering Section Memo Report No. R-57-311-9, 1938.

[25] *Repair of Cylinder Blocks for V-1710 Engine-A.C. 34-7*, Allison letter to Chief, Materiel Division, December 12, 1938.

3

A New Generation of Fighter Aircraft

Following World War I the public was enamored with the glamour of airplanes and flying in general. At the same time, military aviation was in the doldrums. Overwhelming quantities of aircraft, engines and accessories from that conflict, along with the cutbacks in budgets and little perceived need for a standing Army, saddled the military with leftover equipment from the war. Alternatively, during the 1920s aircraft designers created a generous supply of ideas for improved aircraft, but those purchased by the military were usually required to be powered by the obsolete but plentiful 400 bhp Liberty-12 engines of WWI fame. Aircraft companies came and went as the Army often purchased an aircraft design, and then had the airplane produced after a separate competition.

Entrepreneurs and designers did have some success in chipping away at the domain of the Liberty. In 1922 Curtiss introduced their 435 bhp D-12 water-cooled V-12 engine with 1145 cubic inch displacement. It was significant for introducing true mono-block construction. It revolutionized the strength and reliability of the in-line water-cooled V-12 engine. The D-12 powered the high speed aircraft of the period, and paved the way for limited quantities of new engines for the military. A total of 1,192 D-12's were built during its 1922-1932 production run.[1] Through the 1920s others attempted to replace the Liberty, including new radial engines by both Wright and the fledgling Pratt & Whitney firms, along with an enlarged D-12, the 600 bhp Curtiss Conqueror V-1570. Attempting to compete with these firms was Packard and their series of large water-cooled V-12's and the contemporary Wright T-2 and T-3 V-12's for the Navy. None of these engines ever received extensive production orders. Concurrent with these efforts, Allison developed a good business in upgrading Air Corps Liberty's, while also developing derivative models such as the inverted water and air-cooled Liberty engines. They also built special equipment for others, such as the reduction gear assemblies for Packard V-12's. By 1930 the Air Corps and Congress were both finished with the Liberty engine and biplanes. They were ready for the next step-up in power: to go from 600-700 bhp to 1000 bhp or more.[2] Such an engine in a monoplane pursuit would for the first time allow standard military aircraft to exceed 300 mph, an important tactical advantage and a goal of Air Corps planners and strategists of the day.

Featuring mono-block cylinder banks, the 435 bhp Curtiss D-12 was really the first "modern" V-12. It had the best power to weight ratio of the early engines. In the 1920s, its strength and durability set records. It was naturally aspirated and direct drive. (Author)

A companion effort that had been started as a crash program during W.W.I, was the Air Corps continuation of funding to develop the turbosupercharger. Depending upon perspective, the turbosupercharger promised to provide an altitude advantage in fighter combat, or to many of the Air Corps' bomber advocates, would allow bombers to fly above and faster than enemy pursuits. United States military strategy at the time was one of isolation, providing only for defensive armaments. The resulting doctrine allowed fighters only to intercept high flying enemy bombers after they would reach U.S. shores. With this strategy in mind the Army was charged with defense over the Continental United States, the Navy being responsible for stopping any invasion from the sea. Army planners turned their requirements into the need for "1000 bhp" pursuits to be equipped with the turbosupercharger as a means to reach potential penetrations of high flying bombers.[3] Fortunately, from the work begun for the War Department, Bureau of Aircraft Production, in 1917 General Electric had succeeded in bringing the turbosupercharger to a state of development such that it could be incorporated into this new generation of aircraft.[4] Development of the turbosupercharger had been slow and fraught with technical challenges in design and materials, but GE development success gave the U.S. a commanding lead in the demanding technology.

Lockheed's proposal in the X-604 competition for a Multi-Place-Fighter was designated as the XFM-2, though not built. It had been developed as Wright Field classified project M-10-35. The pair of tractor configured V-1710's would have been contemporaries to the V-1710 pushers used on the Bell XFM-1. Note the location of the turbosuperchargers, on top of the nacelles, just as in the later XP-38. (Bell via Matthews)

Serendipity: Air Corps Planning and the Industrial Entrepreneurs

In 1935 a young MIT graduate and Air Corps Officer, Lt. Benjamin S. Kelsey, was assigned as Officer-in-Charge of the Fighter Projects Office of the Materiel Division at Wright Field. Lt. Kelsey was an outstanding and confident pilot, as well as an engineer with a passion for exceptional aircraft. Although endowed with a lengthy job title, he was in effect a "one-man" operation in support of Pursuits, surrounded by the "Bomber Mentality" that permeated the Air Corps at the time. Army aviation had set its priorities as: bombers for defense and long range observation, observation aircraft for battlefield surveillance, and attack aircraft to clear the battlefield. With the U.S. Navy charged with protecting the sea lanes and the shores of the continent, popular opinion was that overseas aggressors would not be able to penetrate to our shores. Pursuit aircraft were deemed useful only for encouraging public and political support for funding the Air Corps.[5]

Lt. Kelsey developed the opinion that the two-seat fighter, a concept that had been advancing during the preceding ten years, was archaic and in need of overhaul. The unsolicited Bell Aircraft Company proposal of a turbosupercharged high-altitude fast bomber-destroyer, armed with heavy cannon, was in line with his vision for the future of fighter aircraft. His office soon issued a circular calling for such a heavily armed fighter able to destroy high flying bombers operating at speeds greater than those of any bomber then flying.

The resulting design competition received proposals from Bell and Lockheed, which the Air Corps designated as the XFM-1 and XFM-2 respectively (eXperimental-Fighter-Multiplace).[6] Bell won the competition by 0.4 points out of a possible 100, attributed to the heavier forward cannon fire of their proposal.[7] Lt. Kelsey not only served as the project officer, but when the aircraft first flew in 1937 he was the primary test pilot. At the time he was current on practically every plane in the Air Corps inventory, and Bell had no one qualified to fly the advanced prototype.

It appears that even before the first flight Kelsey realized that the *Airacuda*, as the Bell XFM-1 was named, would fail to achieve its goals. He developed the belief that fighters must be compact single-seaters capable of uncompromised performance. His objective was a high performance aircraft able to carry twice the then standard 500 pounds of armament. Performance was to be achieved by using a 1500 bhp engine, although there was no such engine even on the drawing boards at the time.[8]

Lt. Kelsey participated in Air Corps studies done in 1935 and 1936 that determined that after 1940, fighters should be capable of: maximum speeds over 400 mph, operating altitudes of 20,000 feet and better, and

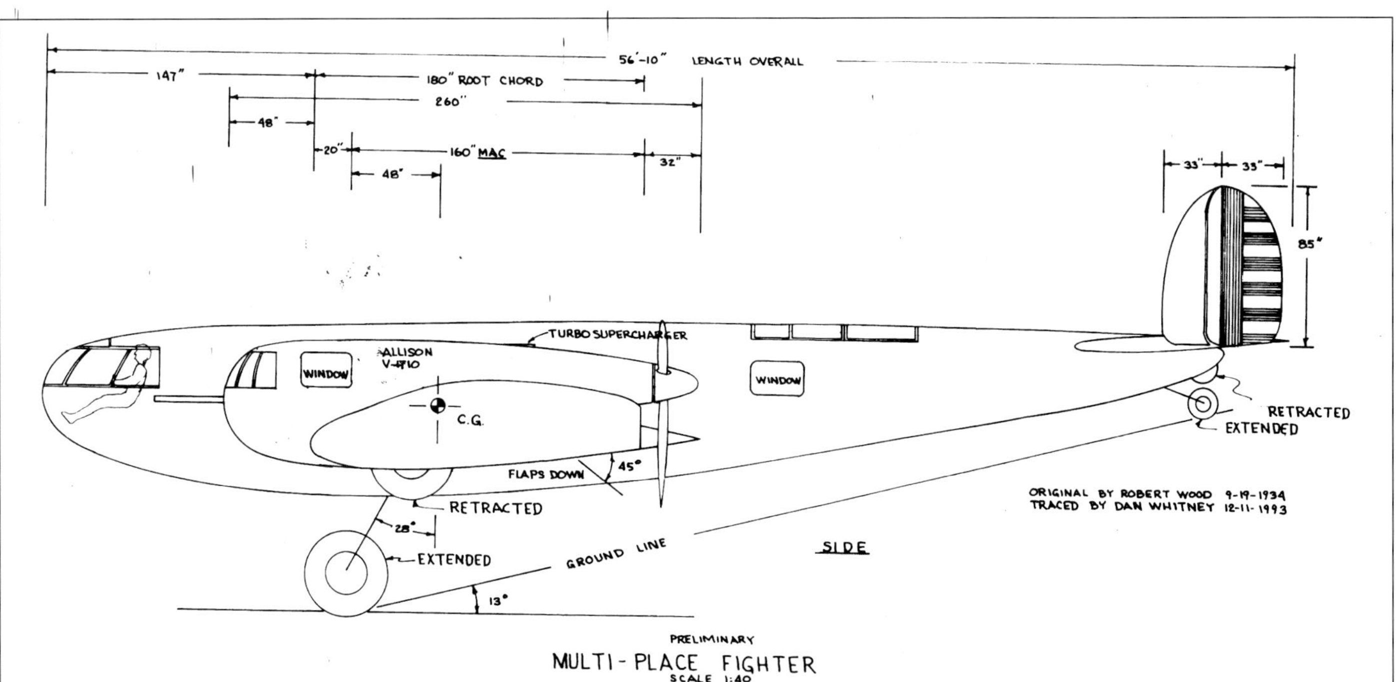

This drawing shows the original configuration of the Bell Multi-Place-Fighter as drawn by designer Robert Wood in 1934, more than a year before Bell was formed and the Air Corps issued its specification X-604 defining the Multi-Place-Fighter. At the time of this drawing the Air Corps was just beginning to test the XV-1710-3, and already Wood was assuming he would have a turbosupercharged V-1710 configured as a pusher for his airplane. Note the use of twin tail fins, a feature not found on the actual Bell XFM-1. (Author)

able to carry armament loads roughly four times the then standard. Bombers, by the same criteria, should have speeds roughly twice current capability, also operate at 20,000 feet or more, and carry two to four times the load twice the distance. Such airplanes would need four instead of two engines, with each able to produce 2000 bhp.[9] But this was 1936, and the country was in the middle of the Great Depression. Funds for research and development of Kelsey's desired pursuit aircraft simply were not available.

Kelsey then combined talents with Lt. Gordon Saville, a like thinking member of the Army Procurement Board, and they conspired to issue a procurement circular for an "Interceptor" type aircraft.[10] As an Interceptor it bypassed the old concept of "Pursuits." While not defining an exact tactical mission, the circular described an aircraft having the performance wanted by Lt. Kelsey, along with the then unorthodox feature of a retractable tricycle landing gear.[11] The engine specified was the most powerful engine expected to be ready for production in 1938, the 1150 bhp Allison V-1710-C7. High altitude performance was to be gained by exploiting the Allison's designed-in suitability for turbosupercharging. In this respect the Circular specified use of the Form F-10 turbo being specifically developed by the General Electric Company for the V-1710. It was this procurement circular that motivated GE to continue with the development of the turbo. Otherwise they may well have terminated the effort, a casualty of the Great Depression. As a matter of perspective, including those for both test and production purposes fewer than 100 turbosuperchargers had been built in 20 years of effort.[12]

The finalized Air Corps requirements for the new *Interceptor* design competition were issued in February 1937 as Circular Proposal Specification X-608 (twin-engined interceptor) and X-609 (single-engined interceptor). X-608 replaced earlier preliminary design studies done under Project M-12-36, while X-609 consolidated like work done under Project M-13-36.[13] Contract awards were made in June and October 1937 respectively. Lockheed won the X-608 competition over the proposed Vultee XP1015 with the twin engined Model 22.[14] The winning X-608 airplane was designated by the Air Corps as the XP-38. Bell won the single-engined X-609 competition with their cannon carrying Model 4, designated by the Air Corps as the XP-39. It was selected over the proposed Bell Model 3, the Curtiss Designs 80 and 80A, and several preliminary versions of what later became the Seversky XP-41.[15]

When the Air Corps began supporting V-1710 development they purchased two engines, the XV-1710-1 and XV-1710-3. The first for development in December 1932, and the second for Type Testing, ordered in January 1934. On June 18, 1934 they issued Allison contract W535-ac-10660 for ten "service test" YV-1710-3's, intending them to power experimental and service test models of the first of this new generation of 1000 bhp aircraft.[16] Initially this purchase was to provide engines for the revolutionary four engined Boeing XB-15 and Martin XB-16 bombers. The Boeing was switched to air-cooled radials when the V-1710-3 was late in completing the Type Test. The Martin evolved into a large twin-boom aircraft that was ultimately configured to use six 1000 bhp V-1710-3's, four as tractors and two as pushers. The Air Corps purchased the engineering design, but canceled the XB-16 project before the aircraft was constructed.[17] Ultimately, all ten of these engines were built and delivered to later standards. In the event, they did accomplish the original purpose of powering the first of this new generation of aircraft, though in the form of 300 mph fighters, the XP-37 and XFM-1.

By the late 1930s the Air Corps was using Allison liquid-cooled engines in most of their experimental planes. This included the XFM-1, XP-37, XP-38, XP-39 and XP-40. As a measure of the importance placed by the Air Corps on the new V-1710 powered pursuits, the following comments were directed to Allison by Wright Field on December 7, 1938, just three years before Pearl Harbor:[18]

> 1. It is imperative that first priority in your development work be given to the V-1710 engines for the below listed airplane types.
> 2. It is intended to continue the use in these airplanes of engine models essentially similar to those used in the experimental and/or service test articles, as follows:
>
> (a) For the Curtiss P-37 a right-hand tractor model with long reduction gear housing rated for use with an exhaust turbine supercharger.
>
> (b) For the Lockheed P-38, a model or models identical to that for the Curtiss P-37 but of such construction that each airplane may be equipped with two engines of opposite propeller rotation.
>
> (c) For the Bell P-39, a right-hand tractor model with crankshaft speed extension shaft and outboard reduction gear, rated for use with an exhaust turbine supercharger.
>
> (d) For the Curtiss P-40, a right-hand tractor model with long reduction gear housing, rated at the highest practicable critical altitude which should, if possible, exceed the military rating of 1000 hp or more at 15,000 feet or higher. Which Mr. Hazen of Allison stated would be available within two months.
>
> (e) For the Bell FM-1, a left-hand pusher model with propeller extension shaft, rated for use with an exhaust turbine supercharger.

This long list of priorities encompassed all of the Air Corps high performance aircraft projects at the time. The diversity of the engine models complicated the development effort for the fledgling Allison engine program. The ultimate success of each model is a tribute to the soundness of Gilman's original design and Hazen's development team, the path to success was long and arduous. With the 1939 procurement of 134 P-40's service use was begun.[19]

Bell X/YFM-1 *"Airacuda"*

When Consolidated Aircraft was still in upstate New York, and as early as 1934, Robert Woods was at work (probably in his off-hours) conceiving the multiplace-fighter/bomber-destroyer. It was a project that was to come

An elaborate fuselage/wing/nacelle test stand for the XFM-1 was setup at Wright Field. It was used to Type Test the drive. Both the XV-1710-9 and XV-1710-13 Type Test attempts used it. (Air Corps)

to life only when he later became the Chief Engineer at the to be formed Bell Aircraft Co. His concept envisioned a cannon firing aerial platform, with turbosupercharged engines, able to seek out high flying bomber formations and bring them down with superior firepower. With this idea in mind Larry Bell was successful in convincing the Air Corps to issue a Request For Proposals, their Specification X-604, for such an aircraft in 1936. This followed the decision of Consolidated Aircraft to relocate to southern California, a decision that was not entirely satisfactory to many Consolidated employees. Amiably, Larry Bell was able to organize a new firm under his name, and acquire Consolidated's New York facilities as well as the services of many employees of the former Consolidated Aircraft operation.

Woods' concept of a bomber destroyer put the cannons in front of the aircraft where they were operated by gunners occupying the location usually occupied by the engines in typical tractor arrangements. Woods solved the resulting propulsion dilemma by mounting the engines as "pushers", located directly behind the gunners and cannons. Apparently, little thought was given as to how the gunners might safely exit the aircraft in an emergency, for the propellers were directly in their egress path! Allison, with their extensive experience in design and manufacturing of gearing and shafting, met the requirements of the installation by simply turning the V-1710 around and driving the propeller via an extension bolted to the reduction gear output shaft. This required revising the coolant plumbing to adapt to the new position and development of a five-foot extension shaft to connect the engine reduction gear to the propeller and remote thrust bearing that was rigidly attached to the airframe. Allison's first official record of this program is an engine layout dated January 23, 1936, and used by Bell in their original XFM-1 specification of March 7, 1936.[20] It envisioned use of the 1250 bhp fuel injected version of the YV-1710-9(D1) pusher engine, along with the General Electric Form F-10 turbosupercharger.[21]

The aircraft was loaded with advanced features, one of which was detailed in the original XFM-1 specification as Appendix II, *New and Novel Features of the XFM-1 Airplane*:

> Extension Drive Shafts are provided for the pusher propellers. By the development of this arrangement considerable variation may be provided in engine location on the airplane. On this airplane exceptional mounting and service facilities are provided by mounting the engine directly on the main wing beams. Some propeller-engine load troubles should also be avoided by the full floating type propeller bearing which is supported in the nacelle independently from the engine. A tubular torque shaft which *carries only torque* (emphasis added) and which is equipped with self aligning connections, floats between the engine and propeller bearing assembly. This mechanism will be developed by Allison together with the V-1710-9 engine and represents an especially important innovation. One of the features of this installation is that the engine may be removed without disturbing the propeller installation, an especially important feature when complicated propeller pitch controls are involved.[22]

Wright Field used this XFM-1 nacelle mockup to Type Test first the XV-1710-9(D1), and then the YV-1710-13(D1). Note the complete lack of components associated with a turbo. The V-1710 had not been tested with the turbo prior to the first flight of the XFM-1. (Allison)

The importance of the Multi-Place Fighter to the Air Corps in 1936 can be appreciated in that five of the eight available service test YV-1710's were assigned to the Bell XFM-1 program. Likewise the program was important to Allison, for at the time they employed fewer than 25 engineers and a little over 100 persons in the experimental shop.[23]

The V-1710 first specified for the XFM-1 was to be based on the 1000 bhp fuel injected XV-1710-5(C2), but to be rated at 1250 bhp by increasing the supercharger step-up gear ratio, using a higher cylinder compression ratio and running on 100 octane fuel. This was revised on December 15, 1936 so that the actual V-1710-9(D1) was a contemporary of the carbureted YV-1710-7(C4), the 1000 bhp engine that the Air Corps was then aggressively pushing through the 150-hour Type Test program at Wright Field. In fact, when the XV-1710-7(C3) failed its attempt at the Type Test it was YV-1710-9(D1) AC34-8, the engine that had been prepared to do the pusher engine Type Test, that was hurriedly reconfigured as a YV-1710-7(C4). This was the engine that officially completed the 150-hour Type Test at 1000 bhp on April 23, 1937. As a result, YV-1710-9's AC34-9 and AC34-12 were released for the initial flights of the XFM-1, qualified by association with the Type Test engine. A second engine (AC34-11) was then prepared as the V-1710-9(D1) Type Test engine. Wright Field oper-

This is XV-1710-9(D1) AEC #10, AC34-8, as built- to perform the Type Test for the "D" engine at Wright Field prior to the first flight of the XFM-1. It was this engine that was re configured as the final Type Test YV-1710 7(C4). Upon success of that test, and the similar ity of the "D", it was determined to not comp ete a Model Test prior to first flight of th- e XFM-1 with the D-1 engines. The propell r speed extension shaft and remote thrust bear ing are included. Note the coolant outlet - elbow is at the accessories end of the engine a location unique to the pusher engines for the X/YFM-1. (Allison)

Table 3-1
Evolution of Bell XFM-1 Design

A/C Model	XFM-1	XFM-1	XFM-1	XFM-1	XFM-1
Bell Model	1	1	1	1	1
Number Built	(1)	(1)	(1)	1	(1)
Contract Number	ac-8347	ac-8347	ac-8347	ac-8347	ac-8347
High Speed at Critical Alt, mph Mil Power	316.9/20,000	316.9/20,000	316.9/20,000	287.0/20,000	301.3/20,000
Time to Climb, Guaranteed, min.	10.0/15,000	7.5/15,000	7.5/15,000	7.82/15,000	6.43/15,000
Wing Area, sq.ft.	688	688	688	688	688
Empty Weight, pounds	11,123.0	11,971.5	11,971.5	13,130.0	13,130.0
Gross Weight, pounds	16,340	17,333.5	17,333.5	18,492.0	18,492.0
Design Load Factors	+8.5, -4.25	+8.5, -4.25	+8.5, -4.25	+8.5, -4.25	+8.5, -4.25
Engine Model	V-1710-9	V-1710-9	V-1710-D1	V-1710-9(D1)	V-1710-13(D1)
Allison Engine Specification	103	103	106	106	109-C1**
Turbosupercharger	F-10 Turbo	F-10 Turbo	F-10 Turbo	F-10 Turbo	F-10 Turbo
Performance at Takeoff, bhp/rpm	1250/2800	1250/2800	1000/2800	1000/2600	1150/2800
Critical Altitude, Military, ft.	20,000	20,000	20,000	20,000	20,000
Fuel Octane	100	100	92		
Aircraft Specification	1Y012	1Y012-A	1Y012-B	1Y012-B	1Y012-B
Date of Revision to Aircraft Specification	3-7-1936	7-28-1936	12-15-1936	9-20-1937	9-10-1937
Government Specification	X-604/M-9-35	X-604/M-9-35	X-604/M-9-35	X-604/M-9-35	X-604/M-9-35
Date of Government Specification	12-30-1935	12-30-1935	12-30-1935	12-30-1935	12-30-1935
Army Min. Hi-Speed at Critical Alt, mph	330/20,000	330/20,000	330/20,000	330/20,000	330/20,000
Army Min. Time to Critical Alt, in	12/15,000	12/15,000	12/15,000	12/15,000	12/15,000
NACA Wing Section at Root	2218	2218	2318	2318	2318
NACA Wing Section at Tip	2209	2209	2309	2309	2309
Forward Cannons	2 ea, 20 mm	2 ea, 37 mm	2 ea, 37 mm	2 ea, 37 mm	2 ea, 37 mm
Comments	Fuel Injected	Fuel Injected	Carbureted D1	Carbureted D1	Carbureted D1
* Actual by Test					
** Engines AC34-9, 34-11 & 34-12 only	To be right or left rotation, AAC Option	To be right and left rotation@ AAC Option	Both To be right hand rotation	At time of first flight	From about 4th flight

XFM-1 used 110v, 800 Hz AC electrical system, but in 1Y012-A spec was changed to be 800 Hz or 360 Hz at Air Corps option. Specification 1Y012-B resolved to use the 800 Hz system, but later YFM-1 used 24v DC only Electrical System.

ated it for only 18 hours in their XFM-1 test nacelle before the spring type crankshaft vibration damper failed, requiring a complete redesign of the engine's torsional vibration system. At that point the engine was replaced by the uprated and appropriately modified XV-1710-13(D1), AC34-14, and another complete Type Test begun.[24] This engine was tested from June 28 to November 19, 1937, during which it completed 150-hours of Type Testing, only to suffer a broken crankshaft at approximately 171 hours total time during subsequent propeller testing. Still the test qualified the engine for flight in the XFM-1.[25]

Bell Aircraft and the Air Corps were continually asking for more power for the XFM-1. The original specification had called for 1250 bhp from fuel injected V-1710-9's, yet here is the first aircraft, about to fly, but with only 1000 bhp available from each of its carbureted YV-1710-9's. With the redesign of the basic engine underway for the V-3420 project, and with the desire to increase engine efficiency by raising the compression ratio to 6.65:1 from the then current 6.00:1, newly redesigned cylinder blocks were available and used as the basis for a new model able to give 1150 bhp at takeoff, the V-1710-11(C7). The V-1710-13(D1) [built to Allison Specification 106] was a companion engine, configured as a "pusher" and specified for later installation in the XFM-1. The new engine models had improved induction systems, a reduced supercharger gear ratio, a higher compression ratio, improved vibration dampers and numerous other detail improvements.

XFM-1 Flies

On September 1, 1937, as Lt. Ben Kelsey lifted the YV-1710-9(D1) powered XFM-1 off the Buffalo Municipal Airport runway on its first flight, the left YV-1710-9(D1) engine (AC34-9) backfired violently, damaging the intercooler ducting and engine accessories housing.[26] Fortunately, the engine continued to operate through the abbreviated 20 minute flight. Subsequent investigation showed that the turbo regulator had stuck, causing the turbo to run at high speed and raising the carburetor inlet pressure to 36-37 inHgA, as compared to the intended 30 inHgA at full throttle.[27] This caused the turbo to deliver unusually hot air to the carburetor. Another significant cause of the backfire was the fact that the new selsen type engine tachometers were miss-calibrated and the engines were later determined to have been running at more than 3500 rpm. It was estimated that the engine was producing over 1350 bhp, under conditions ideal for a backfire! It was fortunate that the skill of the pilot and the basic ruggedness of the Allison could keep the plane in the air and allow for a safe landing. Following repairs, and installation of the spare YV-1710-9 engine, the XFM-1 then made its second flight on September 24. Again Lt. Kelsey had an

Outboard side of the starboard nacelle on the XFM-1 gives a good view of how the GE Form F-10 turbo was installed. Notice the air intake on the right side of the photo. This turbo used combustion air flowing across the bearings to cool them. The horizontal silver shroud is the exhaust manifold delivering hot exhaust to the turbine. The waste gate was at the nozzle on the left end of the manifold. (Bell via B. Matthews)

adventure, when the right hand F-10 turbo failed, slinging turbine blades and causing him to feather the propeller just prior to landing. Then at touchdown the right main landing gear did not properly lock down, with the result that it collapsed at touchdown, damaging the right wing and propeller.[28] During the resulting stand-down for repairs, the engines were replaced with the newly upgraded YV-1710-13's, and the turbos changed to the improved Form F-13.[28A] Things then progressed quickly, for the repaired aircraft made another ten flights before being flown to Wright Field on October 21, 1937.[29]

During the period while the YFM-1 airplanes were being designed and built the XFM-1 engine installations were modified to develop those to be used in the new airplanes. The most significant change was relocation of the turbo from the side of the nacelle, above the leading edge of the wing, to a position below the nacelle and behind the rear wing spar. The considerably improved and revised Form F-13/Type B-1 turbo was used. In this configuration the XFM-1 remained in service until August 1940 until the follow-on YFM-1's were on strength. It accumulated a total of 103 hours of flight, not high by modern standards, but for the Multi-Place Fighter program it was the high-time aircraft.

Engine Installation and Cooling

The initial nacelle arrangement on the XFM-1 was considerably different than on the later YFM-1, a configuration to which it was modified during its operational life. Most noticeable is the turbosupercharger, which was installed on the outboard side of each nacelle, ahead of the engine. This and the earlier installation of the Form F-10 turbo in the XP-37 soon demonstrated the inadequacies of that unit. Its physical arrangement of components made it prone to bearing failures, while the ducting of combustion air over the bearings made installation complicated and inefficient. Photos of the installation show the combustion air intake being just ahead of the turbine wheel, with the intercooler intake scoop just above the glazed gunners compartment. Flow through the air-to-air intercooler was controlled by a flap on the outlet just behind the inlet and on top of the nacelle. The radiator was in the wing, with air being drawn in through the leading edge and exiting on the top surface, just ahead of the rear spar. The oil cooler was mounted in a separate housing completely under the wing and aft of the rear spar.[30]

When the airplane was on the ground the pusher mounted engines required that the coolant distribution header to the cylinders have reversed metering orifices to provide the proper flow to each. Likewise the coolant header tank had to be located high and to the "rear" of the engine to function properly. A continuing problem during ground operation was the total lack of airflow though the oil and coolant radiators, the result of their being ahead of the slipstream of the propellers. As a consequence, it was usually necessary to tow the airplane to the runway before engine start, and then wait only for temperatures to reach minimums before beginning the takeoff.

As with any good development program, the prototype article goes through a number of changes, and that was certainly true of the XFM-1. Soon after the first flights the 1150 bhp engines were installed. After extensive testing in this configuration the airplane was modified to become a prototype for the YFM-1 engine installation. These changes completely revised the nacelles, with the new GE Type B-6 turbos being mounted under the wing fed by a single exhaust manifold connected on the outboard side of the nacelle. The B-6 had two inlets to its nozzle box. Only one was used on each nacelle and the other was blanked off. This allowed a single turbo to be used on either side of the airplane.

The YFM-1 Program

In May 1938, nine months following the first flight of the XFM-1, the Air Corps contracted with Bell for 13 Service-Test YFM-1's. The first ten airframes were to be conventional "tail-draggers", with the final three featuring tricycle landing gear and designated as YFM-1A. This feature was likely the direct result of Lt. Kelsey's influence.[31] Thirty-nine engines were purchased on an amendment to the FY-1938 contract with Allison, W535-ac-10660, that had been issued to obtain 20 V-1710-11 engines for the thirteen Curtiss YP-37's. This change order provided two engines for each of the 13 YFM-1 aircraft. When the order was issued it was intended that V-1710-13's would be provided, but this was soon changed, and all of the YFM-1/1A's were delivered with further improved V-1710-23(D2)'s. The major difference between the two models were the -23's use of revised valve timing from the C-8 and use of the new two-barrel Bendix-Stromberg pressure discharge carburetor instead of the earlier four barrel float type. Delivery of these engines for the most part followed those for the YP-37's. All carried FY-38 serial numbers.

On the YFM-1's, the turbosupercharger was changed to the new GE Type B-6, providing a dual inlet nozzle box specifically configured for use in the YFM-1. This greatly simplified ducting of the hot exhaust gases to the turbo, that was now mounted below the engine much in the fashion to be used with the coming Bell XP-39.

As a group, the YFM-1 aircraft did not fare particularly well. The airplane was a comparatively complex machine of an unconventional design and incorporated many technical advances in propulsion, armament, fire control and in its electric systems. Many of these advanced features were specified and required by Wright Field, not the contractor. For example, the XFM-1 was the first aircraft to use a high frequency alternating current power system. In this instance it operated at 800 Hz, and was troublesome to the point that it was not continued in the YFM-1's. These advances caused problems for both the crews operating and maintaining the aircraft.

At least six of the YFM-1's were wrecked in one way or another, including the #7 aircraft that was lost during spin testing before delivery. In just over two years, the thirteen airplanes accumulated a total of 539 hours of flight time among them, an average of only about 41 hours each!

Table 3-2
Evolution of Bell YFM-1 Design

A/C Model	YFM-1	YFM-1A	YFM-1B	YFM-1C
Bell Model	7	8	7A	17
Number Built	10-(2)	3	2	0
Contract Number	ac-11122	ac-11122	ac-11122	ac-11122
High Speed at Critical Altitude, mph Mil Power	305/20,000	305/20,000	268/12,600	404/13,200
Time to Climb, Guaranteed, minutes	10.0/15,000	10.0/15,000	10.6/15,000	5.99/15,000
Wing Area, sq.ft.	688	688	688	706
Empty Weight, pounds	11,955.5	12,314.0	13,118.2	NA
Gross Weight, pounds	17,281.5	17,640.0	18,444.2	NA
Design Load Factors	+8.5, -4.25	+8.5, -4.25	+8.5, -4.25	+4, -2
Engine Model	V-1710-13(D2)*	V-1710-13(D2)*	V-1710-41(D2A)	XH-2470
Allison Engine Specification	109-F	109-F	114-C Revised	Lycoming H-24
Turbosupercharger	Type B-6	Type B-6	None	None
Performance at Takeoff, bhp/rpm	1150/2950	1150/2950	1090/3000	1800
Critical Altitude, Military, ft.	20,000	20,000	13,200	15,000
Fuel Octane				
Aircraft Specification	1Y012-D	1Y012-Dsup1	1Y012-Dsup2	1Y012-Dsup3
Date of Revision to Aircraft Specification	4-15-1938	10-5-1938	10-19-1939	6-6-1940
Government Specification	X-604/M-9-35	X-604/M-9-35	X-604/M-9-35	X-604/M-9-35
Date of Government Specification	12-30-1935	12-30-1935	12-30-1935	12-30-1935
Army Min High Speed at Critical Alt, mph/ft	330/20,000	330/20,000	330/20,000	330/20,000
Army Min. Time to Climb to Critical Alt, min/ft,	12/15,000	12/15,000	12/15,000	12/15,000
NACA Wing Section at Root	23018	23018	23018	23020-Mod
NACA Wing Section at Tip	23009	23009	23009	23009
Nose Cannon	2 ea, 37 mm	2 ea, 37 mm	2 ea, 37 mm	2 ea, 37 mm
Comments			PT-13E Carb	2 engines
* Delivered as V-1710-23(D2) with PD-12	Spec wanted	Tricycle Gear	Altitude Rated	Tricycle Gear
Pressure type carburetors, Spec 114-C.	1250 bhp		Engines	

The three tricycle gear YFM-1A airplanes were equipped with the same 1150 bhp turbosupercharged V-1710-23(D2) engines as the YFM-1 models. (Bell, Aug. 1940)

The three tricycle airplanes had more than their share of problems. All were lost in crashes, and the one fatality of the program occurred when YFM-1A #12, AC38-497, suffered an in-flight fire caused by a broken oil line. This was induced by airframe vibrations and occurred on a flight from Chanute Field, Illinois to Keesler Field in Mississippi. Both the crew chief and the pilot were forced to parachute from the burning plane. The pilot was killed when his parachute did not open. Total time on the aircraft was only 15:25 hours. The accident report sums up the cause as:

> "Inherent defects in (aircraft) design caused constant maintenance difficulties and the flying of this type had been very limited."[32]

The YFM-1 Airacuda was a difficult airplane to handle. The Pilot's Manual included the following warning:[33]

> Due to close proximity of propeller to tail surfaces, a sudden reduction of power of one engine either through an engine failure or excessive movement of one throttle will result in a much more violent and immediate control reaction than on multi-engine tractor type airplanes. Failure of one engine may result in a spin unless the other engine is retarded, or trim tab control adjusted immediately.

In April 1942 the remaining flyable YFM-1's were grounded. A few were assigned to the mechanic training programs at Keesler and Chanute. None were to fly again.

Altitude Rated YFM-1B

As originally contracted, all of the YFM-1's were to have turbosuperchargers, giving the airplanes a respectable altitude capability. In an effort to reduce problems associated with the operation, maintenance and control of the turbosuperchargers, two aircraft, YFM-1's #4 and #5 (AC38-489 and AC38-490) were modified on the production line to the altitude rated YFM-1B configuration. Per contract change dated October 19, 1939, the turbos, associated ducting, and controls were all removed and their engines modified to the V-1710-41(D2A) "Altitude" Rated configuration. These were V-1710-23's with higher supercharger gear ratios (8.77:1 versus 6.23:1) allowing them to develop 1090 bhp takeoff power to 13,200 feet.[34] With this change the engines were using the same components and authorized the same ratings and performance as the Altitude Rated V-1710-33(C15) then being flown in the Curtiss XP-40. It appears that Allison did not assign a new Specification number for the modified engines. Only the four engines for the two aircraft were converted to this configuration, no spares. Allison was paid an additional $1,690 for each of the conversions.[35]

The altitude rated models enjoyed somewhat reduced weight, and as a result the YFM-1B performed similar to the earlier models up to its somewhat reduced critical altitude. YFM-1B AC38-490 was the last flyable Airacuda. It went to Class 26 disposal status at Langley Field, Virginia in April 1942. Total airframe time was 45.7 hours.

Follow-on FM-1's

As a part of the Aviation Expansion Program a need was defined for procurement of 43 Two-Engine MultiPlace Fighter airplanes, and in January 1939 the Air Corps produced Type Specification C-618 describing it. This resulted in Circular Proposal No. 39-780 that was released to 82 aircraft manufacturers on March 11, 1939. When bids were received that July, only the Bell Aircraft Corporation responded, and that was with an aircraft essentially similar to the YFM-1 that they were then building.

Bell offered two versions of their Model 7, a turbo equipped model and an altitude rated V-1710-D4 powered model, both described in Table 3-3.

The evaluation Board determined that the turbosupercharged model was the winner, but that both models were acceptable for quantity production if "necessary" changes were incorporated. The Board did take exception to the claimed high speed of the "B" model, noting that based upon flight tests of the similar XFM-1 they were of the opinion that a high speed closer to 300 mph would be achieved. The specification required a 10 hour endurance at a reduced cruise power of 0.40 lbm/bhp/hr. The Board believed that the aircraft could only achieve 235 mph at this condition.

The Board went on to note that "testimony before the Board indicates that this design represents the peak of development of this model...and that substantial improvements in performance by the installation of existing engines of greater horsepower will call for several major changes in the structure." Procurement of the 13 YFM-1's then on order was deemed "sufficient to work out a number of important technical and tactical problems confronting the Air Corps. These were mainly: the development of a satisfactory flying platform for cannon; suitable cannon mounts; cannon fire control systems; questions concerning the efficiency of pusher propellers and extension drive shafts; turbo supercharger installations on airplanes of this type and size."

Higher headquarters assessed the relatively low figures of merit as a deficiency in performance and noted that, "the basic design precludes incorporating improved type of high power air-cooled engines, therefore the airplane is totally unsuitable for procurement."[36] The comparison was made to the performance of contemporary light bombardment airplanes such as the Douglas A-20, which far outdistanced the FM-1. They went on to note that the procurement of the 13 YFM-1 airplanes would be sufficient to

Table 3-3
Specification C-618 Requirements and Guarantees

Performance	Air Corps Minimum	Air Corps Desired	Bell 7Y003-B, Guaranteed	Bell 7Y003-C, Guaranteed
High Speed at 20,000 feet	300	375	325	315
Operating Speed at 15,000 feet	255	320	256	256
Time to Climb to 15,000 ft, min	12.0	8.0	8.0	8.0
Endurance at Operating Speed, hrs	10.0	10.0	10.0	10.0
Engine Model	NA	NA	V-1710-23	V-1710-D4
Supercharging	NA	NA	Turbo	Blower
Engine Specification	NA	NA	127-A	131
Figure of Merit	NA	NA	620.62	571.42

determine practicability of fire control apparatus and the efficiency of novel engineering features incorporated in the design.

General Arnold agreed with the evaluation and decided to hold the announcement of the results of the evaluation until August 28, 1939, the date that the results for the Single Engine Pursuit proposals under CP 39-770 were due to be announced. He intended to then shift the CP 39-780 funds reserved for the 43 airplanes to the purchase of additional Single Engine Pursuits. In his penciled comments he emphasizes, "Give Bell as many as he can produce."[37] The winner of the CP 39-770 competition was the Seversky Model AP4L, that resulted in the award for 80 P-44 airplanes. Bell placed second with their Model 4F, which became the P-39C,[38] and was awarded an order for 37 airplanes, though this was augmented by the 43 canceled airplanes from the CP 39-780 competition, making a total of 80 for Bell as well.[39] In this regard the War Department satisfied their goal for number of aircraft, and saved some money in the process.

As late as mid-1941, Brig. Gen. George C. Kenney, then Assistant Chief of the Materiel Division at Wright Field, said that the YFM-1 *Airacuda*,

> "...is by no means through as far as the Army is concerned. There has been a lull in development, perhaps, but that has been for a period in which we had to iron out the bugs. After all, the ship is of radical design and many new changes developed from the original model before we began to get the kind of performance out of it that we wanted."[40]

Modernizing the Bell FM-1 would have required more horsepower and possibly the use of "handed" engines to improve in-flight handling. The airplane was always underpowered, having been intended to have 1250 bhp available from each engine. Although considerable effort was expended in looking at alternative propulsion schemes, 1150 bhp was the highest power rating achieved from any engine during the life of the program.

In March 1938 the Air Corps authorized Allison to release information on the new V-3420 to Bell Aircraft. Hazen's letter to Robert Woods, Bell's Chief Engineer, describes the XV-3420-1(A1) as a 2300 bhp sea-level rated tractor engine, but notes that it could be provided as a pusher

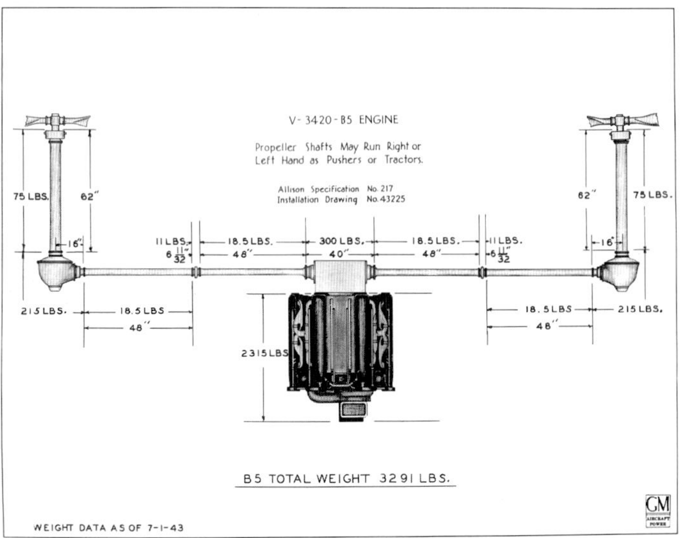

Even after the demise of the Bell FM-1 the Air Corps had Allison continue development of the twin-pusher version of the V-3420 Double Vee for similar installations. This was the V-3420-7(B5), first specified in August 1940, and updated in May 1942 to deliver 2600 bhp. (Allison)

and/or equipped with extension shafts operating at either crankshaft or propeller speed.[41] During the balance of the year, a lively discussion transpired between Wright Field and Allison regarding the feasibility and type of extension shaft system that should be developed for such an installation.

Several other engines were considered for the *Airacuda*. One version got as far as getting a designation assigned: YFM-1C (preliminary specification June 6, 1940). The YFM-1C was intended to expand on the mission that the *Airacuda* had originally been assigned. The primary tactical mission was now to form close-in responsive support of the ground forces by the attack and destruction by bombs and gunfire, ground or naval personnel and light materiel targets. It was to also be capable of the *Airacuda's* earlier mission, i.e. carrying out both precision and area bombing missions and sustained attack of hostile aircraft in flight.[42]

The YFM-1C would have been powered by two 1800-2300 bhp Lycoming XH-2470 engines, mounted in the fuselage and driving the pusher propellers through extension shafts and right angle gear boxes. An alternate proposal was the sleeve valved, liquid-cooled, two-stage and aftercooled, H-2600 being developed by P&W as their X-1800A-2G. That engine was rated for 2000 bhp at takeoff, with 1850 bhp available up to 20,000 feet.[43] That proposal ended when P&W quit working on the X-1800 in favor of developing the XR-4360 in November 1940.[44] Another installation Bell considered in 1940 was a version powered by a pusher arrangement using the 1940 version of Allison's V-3420-7(B5).[45] The proposed 2300 bhp engine would have been located in the fuselage driving extension shafts through angle gear boxes to outboard 2.50:1 reduction gearboxes configured for pusher operation.[46] These in turn were to drive the propellers through propeller speed extension shafts, the same as those used with the original V-1710-D engines. This would have made an interesting aircraft, but still only capable of delivering 1150 bhp to each propeller.

On November 22, 1939 both the P-39 and FM-1 were granted a release for export as Bell Models P-400 and Model P-300 (sometimes identified as the Bell FB-300) respectively.[47] There is no record of any foreign governments expressing substantive interest in the FM-1 aircraft, but the occasion of its release to the export market is noteworthy.

Summary

It appears that events overtook the Multiplace Fighter and caused its demise. Lt. Ben Kelseys close association and involvement probably had a lot to do with it as well. Events in Europe were showing that it was already obsolete as a concept. As early as September 1939 General Arnold sent a letter to the Assistant Secretary of War comparing the performance of the P-38 and FM-1 airplanes based upon the manufacturers guaranteed performance from Circular Proposals 39-775 and 39-780 respectively. He showed that the P-38 was nearly 100 mph faster at 20,000 feet, required a crew of one versus five for the FM-1, and had the same armament except for having only one instead of two 37 mm cannons.[48] The YFM-1's were being delivered to the Air Corps from February through October 1940, the period spanning the Battle of Britain. European combat experience with other multiplace-fighters or bomber-destroyers, such as the British Boulton-Paul Defiant and German Messerschmitt Bf 110 that were then in combat, made it clear to observers on both sides that such aircraft required single-place fighter escorts just to get into combat. Even then, not all of them made it back. Coupled with the teething problems of a complex new design, and the concurrent demands being placed upon Bell for the P-39, there just was not enough interest or resources available to improve the aircraft. Furthermore, by 1940 entirely new designs were being brought forward that were incorporating the lessons being learned in European combat. These later designs were the aircraft that would carry the U.S. into and through most of the coming world wide airwar.

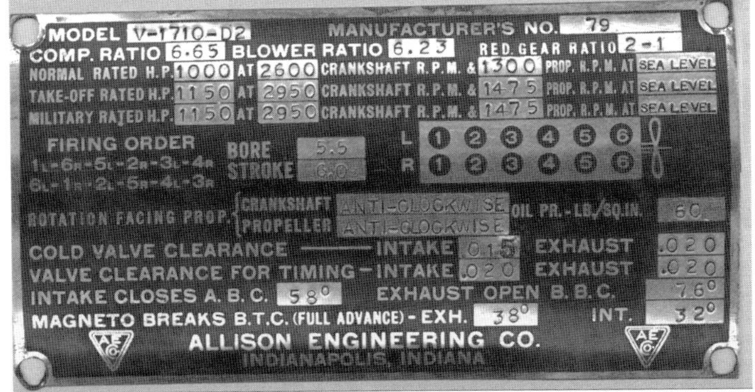

This data plate from V-1710-D2, AEC #79 was removed from YFM-1 AC38-478 while it sat in the scrap yard at Keesler Field, Mississippi, 1944. Note how the intake valve clearance had been restamped for the new 0.015 inch setting. This simple change resolved the early V-1710 backfire problem. (Author)

The X/YFM-1 program should be viewed as constructive and worthwhile. The interest and priority given it during the Depression sustained the young Bell Aircraft Company while at the same time focused Air Corps strategic and materiel thinking. Such action resulted from a recognition of changing needs and requirements, concluding with cancellation of the X/YFM-1 program. New methods of design, new systems and equipment as well as improved maintenance and test programs were developed by the Air Corps because of the program. These were in place to guide the rapid development and introduction of the coming generation of even more complex aircraft, and the methods for operating and supporting them. It also provided the bulk of the orders for the V-1710 prior to the full series production of the V-1710-33(C15) that began in 1940.

Today there are few if any artifacts or relics from the X/YFM-1 program. The adjacent photo is of the data plate from Allison V-1710-23(D2), AEC #79. It was removed from the port engine of YFM-1 AC38-478 while it sat on the scrap yard/disposal pile at Keesler Field, Mississippi in 1944. The late Bob Berreyesa, then a cadet in training there, retrieved the dataplate. It is unlikely that there are many other components from the X/YFM-1 program remaining.

Curtiss X/YP-37

In the mid-1930s the European success with liquid-cooled in-line engines in high speed aircraft caused American aircraft manufacturers, as well as the Army and Navy, to rethink the performance needed in modern military fighter aircraft. Curtiss, long the preeminent supplier of advanced aircraft designs to both the Army and Navy, was then asked by the Army to provide an aircraft to demonstrate the performance of their new Allison V-1710 and GE turbosupercharger combination.

Curtiss was constrained by the Depression, and an evolving corporate propensity to minimize costs and risks by creating each new model from the model then in production. As such, it is no surprise that Curtiss' Chief Engineer, Don Berlin, accepted encouragement from the Air Corps in 1936 to adapt their radial powered P-36 airframe to the new 1000 bhp Allison in-line engine. Furthermore, the Air Corps specified the craft was to incorporate the new General Electric Form F-8 turbosupercharger that had been designed specifically for the V-1710. This would be the first installation powered by the paired V-1710 and GE turbosupercharger. As such, it was intended to demonstrate the capabilities of a turbosupercharged aircraft, the primary reason for the Air Corps original interest in the V-1710. The Air Corps wanted to get these two new power plant components into a modern airframe and thereby investigate the potential of the turbo. Curtiss was awarded a contract on February 16, 1937 for the single XP-37.

Because of the experimental nature of the engine installation in the XP-37 the aircraft was procured without performance guarantees. Consequently, the modified airframe that became the XP-37 may at first appear as a hodgepodge of equipment and ideas, but it was in fact a highly significant milestone in American aviation. It was to be the final stepping stone to a generation of similarly configured and powered fighters and bombers. As such it was the first to use the actual equipment that was to make U.S. aircraft a dominant force in the coming decade.

The XP-37 airframe was that of the original prototype Curtiss Design 75, constructor's number 11923. This was the same airframe that had first been powered by the short-lived 900 bhp Wright SCR-1670-G5 when entered in the 1935 Pursuit competition. For the April 1936 Pursuit Competition it was re-engined with a 675 bhp Wright R-1820F Cyclone as the Curtiss Design 75B, losing to the Seversky 1-XP, which became the P-35. Airframe 11923 was then rebuilt as the Design 75D, its final radial engined form. As a matter of historical convenience the radial engined airplane has

XP-37 undergoing maintenance while in its original YV-1710-7 powered form. (SDAM)

The XP-37 (Curtiss Design 75I) as originally delivered to Wright Field powered by 1000 bhp YV-1710-7(C4) AEC #13, AC34-13. Inlet to turbo/carb is rectangular scoop just inboard of landing gear, oil coolers fed from scoops behind prop spinner while large scoop on port side supplied the intercooler and radiators, with the hot air exiting the fuselage side above landing gear. Exhaust collector is visible, though routed internally to turbo located in black painted area on fuselage centerline. Oval cutout behind spinner supplied cooling air to the shroud around the exhaust pipe connecting to the GE Form F-10 turbo. (Air Corps Photo)

retroactively been referred to as the "XP-36", but it was always privately owned and never delivered to the military until as the Curtiss Design 75I, Air Corps XP-37 (AC37-375), and still identified as C/N 11923.[49,50]

The length of the new Allison engine required a considerable redesign of the P-36 type fuselage. Whether for balance, or because the radiators and intercooler were located in the previous cockpit area, the pilots' compartment was relocated far to the rear, much like the *Gee Bee* racer. Measured from the wing attach points, the tail of the XP-37 remained in the same location as on the YP-36.[51] From the pilot's perspective this aft location proved to be a major disadvantage in a potential combat aircraft, as the wing and fuselage effectively blanked out much view ahead. Still the aircraft held considerable promise in that it was expected to be capable of over 340 mph at its critical altitude of 20,000 feet. This was to make it the first U.S. pursuit to be able to exceed 300 mph. This was a big im-

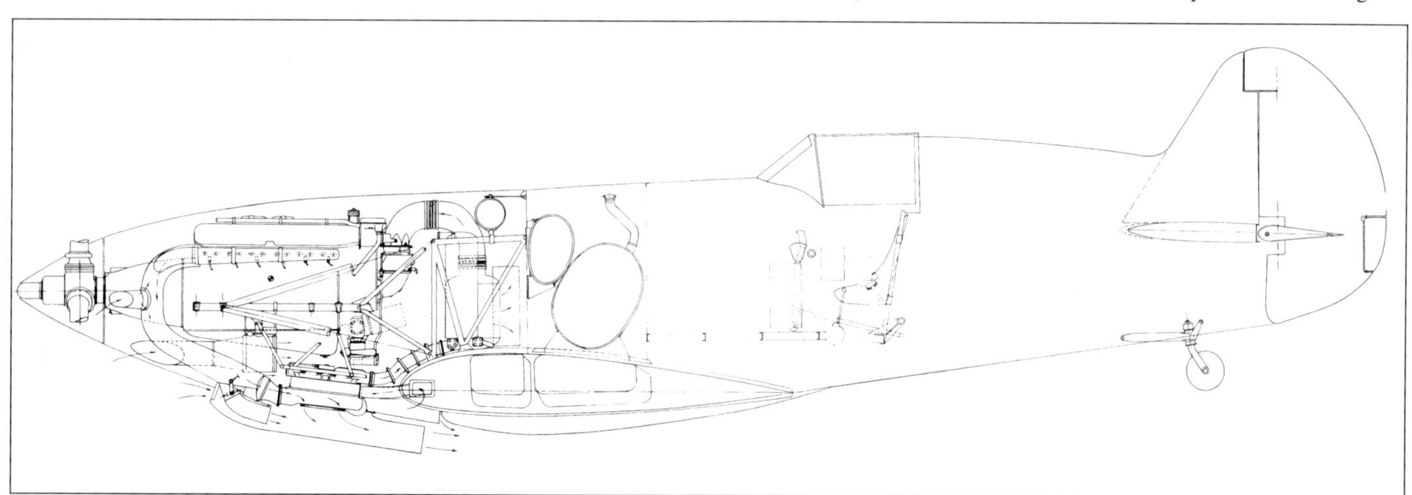

This is the XP-37, Curtiss Design 75I, as it appeared when delivered to the Air Corps at Wright Field in 1937. The engine was the Type Tested 1000 bhp YV-1710-7(C4) with induction air provided by a GE Form F-10 turbocharger rated for 20,000 feet. No photos have been found showing the aircraft equipped with the shroud over the turbo as shown above. (Author)

provement over the 293 mph top speed of the Curtiss Y1P-36, which was limited to a considerably lower ceiling. Although cramped, the revamped nose of the airplane included streamlined baffles to improve ducting into the radiator and intercooler. It also was fitted with the improved and newly developed GE Form F-10 turbosupercharger, as well as the usual accessories and necessary armament.

Engine Installation and Cooling the X/YP-37

In the XP-37 the intercooler was placed on the aircraft centerline at the rear of the engine compartment. There were two radiator sections, one on each side of the intercooler. Scoops on the sides of the fuselage drew air into the compartment, with a slot in the fuselage skin directly ahead of the firewall allowing the heated air to exit. If additional cooling was required a small door or shutter could be opened by the pilot on each side of the fuselage. In the later, but similar, YP-37 this arrangement was somewhat improved with internal sheetmetal ducting to direct and control the airflow through the coolers. In that installation the intercooler was located on the left side of the fuselage and an enlarged single radiator on the right. The intercooler was fed by a 50 square inch scoop on the left side of the fuselage while a 120 square inch scoop on the right side directed air into the radiator. The radiator was provided with a streamlined duct directing the heated air to the exit door on the right side of the fuselage. The intercooler discharge, along with that from the oil coolers, continued to exit through the slot around the fuselage.

Because the radiators and intercooler were directly in front of the cockpit, the pilot was literally in a "hot box." It was reported that cockpit heat could be almost unbearable, although the pilots flying the YP-37's at Ladd Field, Alaska reported they never noticed heat to be a problem. Another problem found during the early flight testing was that the pilot had to remember to close the radiator outlet shutters prior to landing. Otherwise, the warm outflow caused tail buffeting on the XP-37 when in the landing configuration.[52] This may be the reason the fuselage behind the pilot was lengthened 1'-6" in the YP-37's, a change intended to move the elevators out of the path of the turbulence. If so, it was not successful, as the Official Operating Instructions for the YP-37 cautioned the pilot to close the radiator shutter when in the landing configuration.[53]

In all of the P-37's the turbosupercharger was mounted on the aircraft centerline, directly below the engine. While this was compact and provided for directing a blast of cooling air across the hot turbine wheel, it was not ideal from an aerodynamic perspective.[54] Consequently, early aircraft like the XP-37 and its contemporaries that used similar turbo installations were not always able to live up to their proponents performance expectations. Significantly, a factor common to most of the "high-performance" non-turbo designs as well.[55]

The XP-37 when being prepared for return to Curtiss for repairs following its being damaged at Wright Field soon after delivery. (SDAM)

When the XP-37 Design 75I was damaged at Wright Field in an emergency landing with a turbo fire it was expedient to simply cut the wings off and rush it back to Curtiss for repair in April 1937. Early Hazen designed intake manifolds are clearly visible, confirming YV-1710-7(C4) is installed. (USAF Museum)

Operating History

While comprehensive documentation on the early history of the XP-37 is still missing, it is known that the aircraft was damaged in an emergency landing due to a turbo fire upon delivery to Wright Field on April 20, 1937.[56,57] Consequently, the only Allison powered entry in the 1937 Pursuit Competition was unable to compete.[58] As a result of the damage the aircraft was shipped back to Curtiss for repairs, being redelivered to Wright Field, June 16, 1937.[59] To expedite the return to Curtiss the wings were sawn off just outboard of the landing gear, an action probably reflecting the degree of damage and the need to replace them anyway.[60] The initial engine installed in the aircraft was Allison YV-1710-7(C4), AC34-13, a sea-level engine rated for 1000 bhp at 2600 rpm to 20,000 feet. This was the engine installed in the aircraft when it was finally and officially accepted from the contractor for its official first flight at Wright Field, July 23, 1937.[61,62]

Wright Field soon found that the initial turbo installation was prone to overheating and fire. They made somewhat crude changes to obtain sufficient safety to allow continued flight testing. Some six or seven forced landings for various reasons necessitated immediate solution of the functional problems before the performance capability of the aircraft could be investigated. These investigations focused on severe buffeting and vibration of the airframe and required considerable time to resolve. During this period it was determined that radical changes in the induction system were also necessary. These were made in a makeshift manner to check the principles involved. When the first altitude runs were finally begun the F-10 turbosupercharger failed, abruptly ending the performance evaluation. At this point the aircraft was to be returned to Curtiss for incorporation of needed and desirable modifications.[63]

The above occurred during the initial five months of flight testing and it was then decided to rebuild and improve the XP-37 to be a competitor in the upcoming 1938 Pursuit Competition. In anticipation of this change the aircraft remained at Wright Field and was down for modifications needed to prepare it for baseline performance testing from November 1937 through February 1938. This testing finally occurred in early March 1938 when another 1-1/2 hours were added to the previous 28 hours that had been flown at Wright Field. The early March 1938 flights were to complete the required contract acceptance tests and were to be flown to 20,000 feet to demonstrate the promised top speed of 340 mph, but the F-10 turbo again failed prior to completing the tests.[64] Data had been obtained using 720 bhp, which the Air Corps then used to estimate that at the rated 1000 bhp from the engine the airplane would be able to deliver as much as 317 mph at 10,000 feet.[65] The aircraft was then prepared for return to Curtiss for the scheduled rework.

Mid-afternoon on March 10, 1938, Wright Field test pilot Capt. Samuel R. Harris, Jr. was taxiing the Design 75I aircraft across a field to the runway in preparation for the flight to Curtiss in Buffalo, NY, when the left landing gear strut folded up, dropping the wing onto the ground.[66] The official accident investigation concluded that the rough ground caused the tail wheel to oscillate up and down, which in turn cycled hydraulic pressures throughout the system enough so that the left main landing gear unlocked. Although considerable damage was done to the left wing and aileron, both assemblies were repairable.[67] At the time of the incident both the aircraft and engine had logged only 29-1/2 hours of flight in Army service.

During the reconstruction the 75I was converted to Design 80 form. As such it functioned as the prototype for the proposed Model 81, the design that became the YP-37. To provide the improved performance expected of the Design 80, the original V-1710-7(C4) engine was upgraded to the new V-1710-11(C7) standard.[68] It had first been intended to configure the rebuilt aircraft with the military version of the C-4 Type Test engine, the 1000 bhp V-1710-C6 [Allison Specification 104-A], while retaining the Form F-10 turbo. Alternatively, Allison was able to offer the 1150 bhp V-1710-11(C7/C8) in time for the rebuild, and that model engine was used instead.[69] When returned to flight in the summer of 1938, AC37-375 was quite a different aircraft than when in its original Design 75I form.

The second example of the V-1710-11(C8), AC38-582, was assigned to power the XP-37 in its appearance as the Curtiss Design 80. Curtiss performed preliminary tests on the revised airplane that indicated satisfactory freedom from vibration and buffet and reasonable power plant functioning. On September 17, 1938, during ground operations at Curtiss, another landing gear failure occurred; this time resulting in the prop striking the ground and breaking the prop and gearcase on the new engine. The damaged reduction gearcase was sent to Allison were it was repaired and

The XP-37, believed to be during the period when powered by the 1150 bhp V-1710-C7/C8. (Air Corps)

returned to Wright Field for reinstallation on AC38-582, which then became the "spare" engine.[70] The original engine, AC34-13, having previously been upgraded to V-1710-11(C7) standard when overhauled following the first 29-1/2 hours in the XP-37, had already been installed to expedite the program. During this standdown it was necessary to also rework the turbo to prevent failures which had been experienced in other installations of the Form F-13 turbo, recently redesignated as the Type B-1 turbo. It was December 1938 before all of the conversion work, repairs, and ground running on the Design 80 were complete. Curtiss then began a flight test program on the aircraft that accumulated 12 hours of flight time by the end of the year.[71]

The new 1150 bhp V-1710-11(C7/C8), along with the new and considerably revised GE Form F-13 turbosupercharger, required extensive changes in the routing of exhaust manifolds and other plumbing within the engine compartment. In addition vibration isolating engine mounts had been incorporated along with modifications intended to remove buffeting and directional instability. For all practical purposes, the C-7 and C-8 engines were identical except for the C-8's being built on the production contract for the YP-37 engines though it also included provisions for propeller feathering, along with other minor internal improvements and revised valve timing.

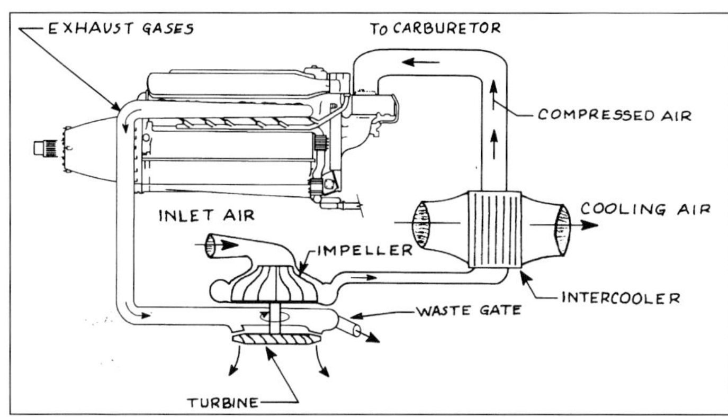

Schematic of Turbosupercharger installation for the X/YP-37. (Author)

The Engine

Some references, including the Official Aircraft Record Card for the XP-37, identify the first engine as a GV-1710-7 (for "G"eared "V"-12). That nomenclature was a carry-over by Curtiss of the nomenclature they used on their previous generation of large liquid-cooled V-12's, the Curtiss GV-1570 Conqueror. The "G" was their way of distinguishing between the geared and direct drive Conquerors. On the V-1710, such a designation was officially used only on the first Navy engine, the GV-1710-A.

In summary, the XP-37 actually existed in two different versions, distinguished by different engine/turbo installations. First in its original Curtiss Design 75I configuration, when powered by the YV-1710-7(C4), AC34-13, and a GE Form F-10 turbosupercharger.[72] In March 1938, this was all changed when it was rebuilt as the Curtiss Design 80, with an 1150 bhp V-1710-11(C8), AC38-582, installed. At the same time a considerably revised turbo, the new GE Form F-13, was incorporated. This unit provided for streamlining the internal ducting while offering an increase in the critical altitude to 25,000 feet. The original engine (AC34-13) was then rebuilt by Allison as a C-7, and held as a spare. In this form both engines used the new 6.65:1 compression ratio cylinder blocks and improved induction system. Also incorporated was a lower supercharger gear ratio and an improved crankshaft vibration damper. These changes allowed an increase in rated engine speed and power to 1150 bhp at 2950 rpm.

The Curtiss XP-37 in its 1938 Design 80 configuration in flight near Wright Field. Powered by the 1150 bhp V-1710-11(C8) and GE Form F-13 turbo. In this form it competed for the delayed 1938 Pursuit Competition as the prototype for the similarly powered YP-37. (Air Corps)

The Turbo

The original General Electric Form F-10 turbosupercharger, the evolved form of the Form F-8 unit the Air Corps had GE develop specifically for the 1000 bhp V-1710, was quite different from the later, and subsequently well known, GE Type B family of turbosuperchargers. Both the turbine and supercharger wheels were of the same dimensions as in the later Type B units, but in the Form F-10 the impeller took its air from a collar located between the turbine and the compressor. This somewhat awkward arrangement was used to provide a steady flow of cooling air over the bearing housing, with that air then being drawn into the compressor and supplied to the engine. The configuration required an outboard bearing to support the end of the comparatively long compressor shaft. This bearing was a major source of trouble in that it tended to fail due to its position and associated problems in providing adequate lubrication. While the F-10 turbo had been specified for a number of early aircraft it was soon found to be a failure and was quickly removed from both the XP-37 and XFM-1, the only aircraft to actually fly with it. GE developed the Form F-13 turbo in 1937 to resolve these problems, and to improve the overall efficiency and ease of installation of the unit. The Form F-13 differed by having the impeller reversed so that the air intake was from the shaft end rather than from the center of the assembly. This in turn cantilevered the impeller from the central bearing housing and allowed a shorter and stiffer shaft, while at the same time eliminating the troublesome outboard bearing. This configuration has been adopted for nearly all of the millions of turbosuperchargers since built for both aircraft and automobiles. One consequence of this arrangement was to require use of a cooled oil lubrication system, as there was otherwise no cooling for the bearings.[73] On or about January 1, 1939, this model was redesignated by the Air Corps as the GE Type B-1 turbosupercharger. As such, it defined the configuration, dimensions and rotating components for the hundreds of thousands of General Electric Type B turbos to follow.

The combustion air was ducted to the supercharger from slots in the leading edge of both wings on the Design 75I. This was revised on the Design 80 to a pair of intakes located just below the propeller near the fuselage centerline. These provided air to the oil coolers as well as the new center inlet Form F-13 turbo. The follow-on Model 81 YP-37's were originally specified to use the GE Type B-1 turbo, but by the time they were being delivered the GE Type B-2 turbo was available and used instead.[74] The air intake for both the oil coolers and turbo were consolidated into a

The GE Type B-2 turbosupercharger as installed in the YP-37. Duct for combustion air into the compressor is just below the engine crankcase. (Army Air Corps)

The XP-37 when rebuilt as the Design 80 with the new 1150 bhp V-1710-11(C8) and Form F-13 turbo for the 1938 Pursuit Competition. Note revised intercooler scoop, relocated carburetor air and oil cooler inlet scoops. (Air Corps)

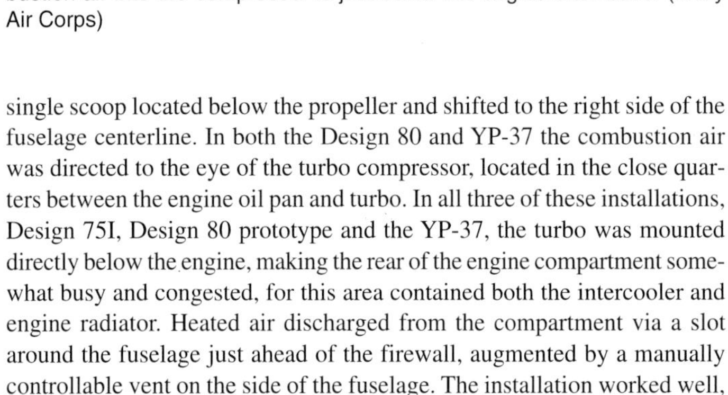

single scoop located below the propeller and shifted to the right side of the fuselage centerline. In both the Design 80 and YP-37 the combustion air was directed to the eye of the turbo compressor, located in the close quarters between the engine oil pan and turbo. In all three of these installations, Design 75I, Design 80 prototype and the YP-37, the turbo was mounted directly below the engine, making the rear of the engine compartment somewhat busy and congested, for this area contained both the intercooler and engine radiator. Heated air discharged from the compartment via a slot around the fuselage just ahead of the firewall, augmented by a manually controllable vent on the side of the fuselage. The installation worked well, but the state of the art of turbosuperchargers, and particularly their controls, caused ongoing problems, including failed bearings and frequently, turbine overspeeding, manifold pressure excursions and broken turbine buckets.

Performance of the XP-37, in both forms, was somewhat disappointing, apparently due to the high drag of the turbosupercharger, intercooler, and radiator installations. Although the engine was badly abused during testing to improve supercharger controls, and from overspeeding by the somewhat experimental Curtiss-Electric controllable propeller, the engine still gave satisfactory service.[75]

General Mark E. Bradley, Commanding General of the USAF Logistics Command, was then a Lieutenant and a test pilot at Wright Field. He picked up the Design 80 XP-37 from Curtiss on January 24, 1939 and returned it to Wright Field in a 2:05 hour delivery flight.[76,77] He remem-

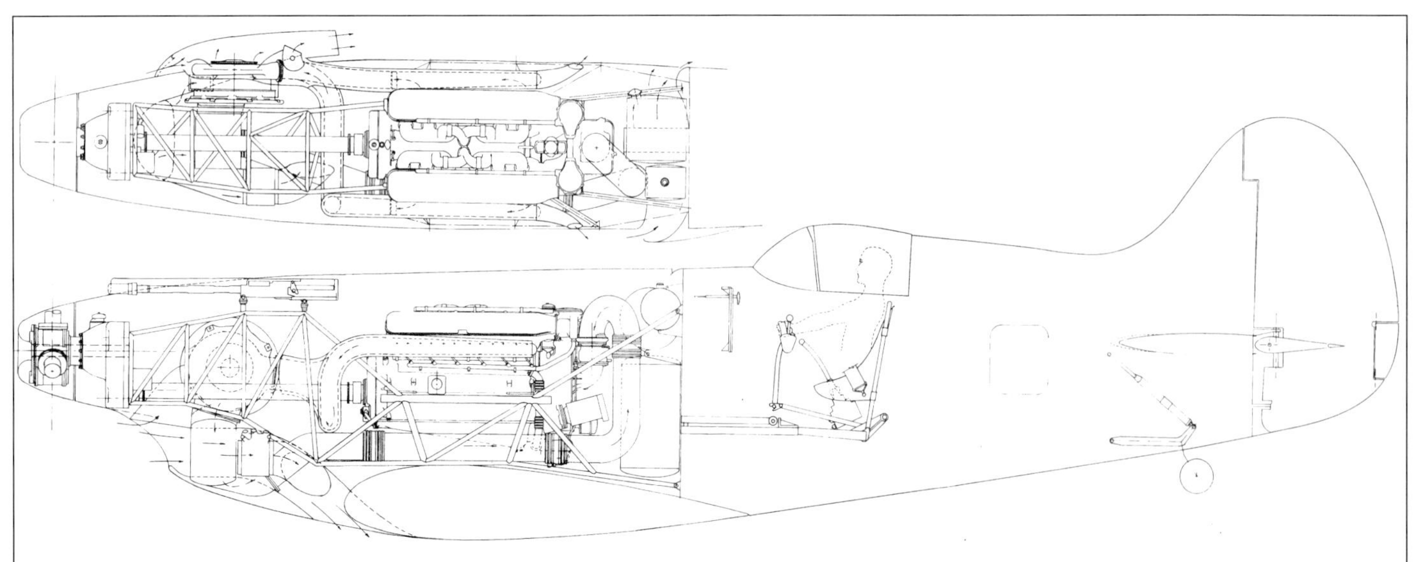

The proposed Curtiss Design 80A would have used the Allison V-1710-D3 Mod B, an adaptation of the Army specified V-1710-C7 engine. The D-3 Mod A/B were the first V-1710 to have a crankshaft speed extension shaft to a remote reduction gear, which was done to allow installation of a 23mm Madsen cannon (not shown for clarity) arranged to fire through the propeller hub. Notice in the plan view above how the turbo was installed alongside the extension shaft. This aircraft would have been the same length as the XP-37; it was not built. (Author)

bered the flight not only for the extremely poor weather,[78] "...but principally because I was flying one of the most, if not the most for the period, advanced fighter planes in the world. A piece of engineering that was a shadow of things to come."[79] The aircraft was delivered just in time for the delayed 1938 Pursuit competition.

The Curtiss Design 80 lost in the competition to the Curtiss Design 75P (XP-40), powered by the first altitude rated Allison, the 1090 bhp XV-1710-19(C13). This did not effect the thirteen service test YP-37's that had been ordered back in December 1937, for they were already under construction at Curtiss.[80] The YP-37 service test aircraft were delivered in the period of April 29, 1939 to December 5, 1939.

The XP-37 soldiered on as the prototype for the YP-37's, and remained in service at Wright Field until November 15, 1941. At that point it was flown to Chanute Field, Illinois and used as a Class 26 maintenance trainer. Its flying days were over after having accumulated a total of 162 hours of flight.

Engine AC38-582 was later overhauled and equipped with the new PD-12 pressure carburetor, thus configuring it as a V-1710-21(C10). No records show whether this engine was flown in the XP-37 in this form, though the engine was in this form when later used for exhaust jet ejector performance testing by Wright Field.[81]

Curtiss Design 80A

As an aside, Curtiss seriously considered rebuilding the XP-37 as its Design 80A rather than as the Design 80. This aircraft would have had a radically altered engine compartment in that it would have mounted the Allison V-1710-D3 Modification "B" extension shaft engine with remote reduction gear. This was an engine derived from the 1938 Pursuit Competition specified V-1710-C7, but differing by having a 56-inch extension to a remote reduction gear and fitted to accommodate the Danish 23mm Madsen cannon. With this arrangement a cannon could fire through the hollow propeller hub. The design was considered seriously enough that Allison configured and successfully operated their EX-2 "Workhorse" engine with the 56 inch extension shaft and remote reduction gearbox. The turbosupercharger would also have been relocated, from below the engine to alongside the extension shaft and cannon. The result was an aircraft with a very strange profile, though the overall length of the aircraft, and the distance from the pilot to the propeller was not changed from the original Design 75I.

The XP-37 airframe certainly had a varied life. It was variously a Design 75, 75B, 75D, 75I and 80.[82] It operated with the Wright R-1670 and R-1820, as well as the turbosupercharged Allison C-4, C-7, C-8 and possibly C-10 engine models, while using both the GE Form F-10 and Form F-13 turbosuperchargers in its Design 75I and Design 80 configurations respectively.

Curtiss YP-37

The thirteen YP-37's were ordered in December 1937. They were intended to be powered by the 1150 bhp V-1710-11(C8) engine, though only the first two engines of the twenty ordered were delivered as such, one going to the XP-38 and the other becoming the first engine for the XP-37 Design 80. The engines delivered for the YP-37's were built as V-1710-21(C10)'s. The main difference between the two engine models being the use of a Bendix-Stromberg pressure carburetor in the C-10 instead of the original four barrel float type carburetor in the C-7/C-8, though internally many detail improvements coming from ongoing Type and Model testing were also incorporated. Besides a different engine, the YP-37's were longer overall by 18 inches, the result of lengthening the fuselage aft of the cockpit.[83] They also used the production standard GE Type B-2 turbosupercharger and a single coolant radiator.

According to Curtiss designer Don Berlin, the turbosupercharger on the YP-37 was still not working. There were turbo regulator and control problems as well as turbine bucket life. Curtiss did not have time to develop it along with the airplane and the new P-40, winner of the recent 1938/39 Pursuit Competition.[84]

In hindsight this points out a major problem with aircraft manufacturing and design concepts as overseen by the Air Corps Materiel Division at the time. There was little, if any, of what today would be called "systems integration" being done. Engines, turbosuperchargers, propellers, aircraft accessories, armament, and communications equipment were "GFE," that is, Government Furnished Equipment. The aircraft designers had to bring it all together, but no *one* was responsible to see that the systems and equipment, along with their controls, operated harmoniously. For example, the Allison V-1710-C was specified and designed to work with a turbosupercharger, but no Allison engine had ever been run connected to a turbo before those installed in the Curtiss XP-37 and Bell XFM-1 airframes! Furthermore, it was not clear who between Allison, General Electric, Curtiss and Wright Field had the responsibility to integrate these systems into a

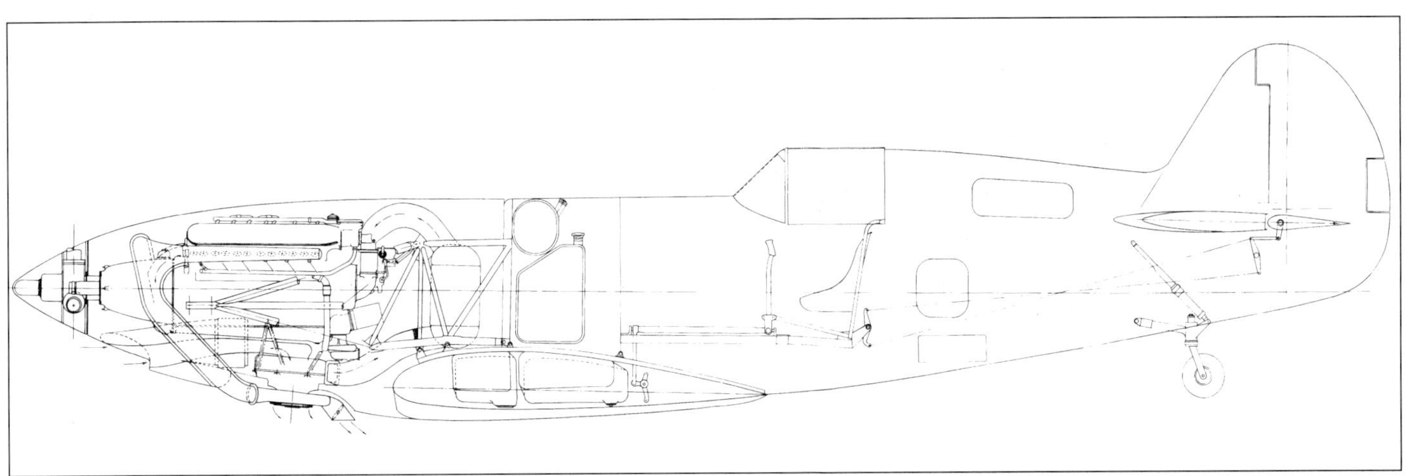

The 13 YP-37's, Curtiss Model 81, built with the 1150 bhp V-1710-21(C10) that featured new Bendix injection carburetor fed by a new and improved GE Type B-2 turbosupercharger rated for a critical altitude of 25,000 feet. (Author)

cohesive and effective package.[85] GE finally solved the turbosupercharger control problem, but not until nearly five years later, when W.W.II was nearly half over. In another example of poor systems management, Curtiss Electric Propellers nearly grounded the entire P-40 fleet just before Pearl Harbor because of inherent overspeeding due to inadequate propeller governors. As the records show, these early installations had problems that were never satisfactorily resolved, and in fact, were primary reasons that the YP-37 and YFM-1 never received further orders or development.

This was the situation in the late 1930s, and reflects the undeveloped state of art for both new technological equipment and supporting management processes. Programs such as the X/YFM-1, X/YP-37 and X-608/9 procurements that were responsible for many of the improvements in equipment, testing, design, manufacturing, operations and management that were revealed during the coming war. Still, the availability and success of the later aircraft suggests that the lessons were learned. A good example was the close relationship between Allison, Bell and Curtiss that resulted in altitude rated V-1710's being available and competitive in the later Pursuit Competitions.

Operating History of the YP-37's

The first YP-37, AC38-472, spent the first 17 months of its operational career at Curtiss, having been retained for test and development work. The number of flight hours accumulated during this period is unknown. It was officially delivered to Wright Field in June 1941, and flew one to three hours a month until transferred to Chanute Field in Illinois, where it flew occasionally until March 1942. By then it was the last flying YP-37.

The balance of the YP-37's were sent to Langley Field where the squadrons of the 36th Pursuit Group operated them from November 1939 until the spring of 1941. In August 1940, AC38-477 and 38-481 were flown to the Sacramento Air Depot where they were winterized for operation and testing in Alaska. That October they arrived at Ladd Field, Fairbanks, Alaska, where 38-481 operated until November 1941, 38-477 having been lost while landing in a white-out the previous November.[86] These, and two B-17B bombers, were the first aircraft systematically tested for prolonged operation in the adverse arctic climate. Many lessons were learned and were immediately incorporated into military specifications, greatly improving the reliability and performance of the coming wartime aircraft. For example, they learned from the bush pilots to drain the oil on all engines while it was still warm. Wright Field then developed an oil dilution system in which small amounts of fuel were mixed with the oil just before shutting down the engine. This reduced the viscosity of the oil and allowed a cold engine to be cranked even after it had cold soaked. At first oleo struts on the landing gear would fail because of rubber O-rings shrinking and cracking in the cold. Likewise, rubber hoses broke like kindling wood. Wright Field soon provided neoprene O-rings and hoses that saved the day. Another change was in the coolant, as a mixture of 50:50 Ethylene Glycol and water was used in deference to the cold weather.[87] Nearly pure Ethylene Glycol, the specified coolant, freezes at about 10°F while a 50:50 mixture protects to about -40°F. The liquid-cooled Allison engines in the YP-37 airplanes presented no special problems in keeping the operating temperatures up in the recommended range, though Air Corps mechanics did use felt to blank off various amounts of the Prestone and oil radiators. The Prestone temperature could be controlled very little by adjustment of the usual shutters.[88]

Table 3-5
YP-37 Performance

YP-37 with V-1710-21 and Type B-2 Turbo	Restricted, Sept. 1940	Original, Unrestricted
Takeoff, bhp/rpm/map, inHgA	1000/2770/33.4	1150/2950/37.7
Normal	880/2600/31.0	1000/2600/34.2
Speed at Crit. Alt., 34.3 inHgA	346 mph @ 24,000'	NA/25,000
Cruise at 2280 rpm, 27.4 inHgA	308 mph @ 24,000'	NA

Following a year and half with the 36th Pursuit Group, the Langley YP-37's were delivered to Chanute Field, except for 38-474 which was assigned to the NACA for test and evaluation at their Langley, Virginia facility, and 38-475 that went to the armament school at Lowry Field in Colorado.[89] The aircraft at Chanute saw a little flying time soon after their arrival, but with the increased demand for modern aircraft upon which student mechanics could hone their skills, all were soon assigned as Class 26 Maintenance Trainers. Prior to reclassification several of the YP's were flown to schools at Keesler Field and Sheppard Field where they also finished out their days as classroom trainers.

When originally delivered, the V-1710-21 in the YP-37 was rated for Takeoff and Military power at the intended 1150 bhp, running at 2950 rpm with 37.7 inHgA and using 92 Octane fuel. Normal power was 1000 bhp at 2600 rpm at up to the 25,000 foot critical altitude of the turbosupercharger. In September 1940 this was considerably revised when the Materiel Division issued new operating instructions that reduced Takeoff power to 1000 bhp at 2770 rpm with 33.4 inHgA. Normal rating was reduced to 880 bhp at 2600 rpm up to the 25,000 foot critical altitude.[90] The reason for this rerating was the same as that being applied to all V-1710's then in service and pending Modernization to the final C-15 Model Test configuration. Although the V-1710-21(C10) had been Model Tested and approved for rating at 1150 bhp, subsequent efforts to complete the Model Test on the V-1710-33(C15) had shown it was necessary to strengthen the crankcase. Due to the similarity of the various engine models, the restricted rating that was issued for the early P-40's was also applied to the V-1710-21. It appears that there was no effort to Modernize the C-10 engines, given their few numbers and no intent to extend the operational life of the YP-37.

Table 3-4
Summary of Early Allison Powered Curtiss Pursuits

Air Corps	Curtiss	Engine	GE Turbo	1st Flight	Comments
XP-37	Design 75I	YV-1710-7(C4)	Form F-10	3-1-1937?	From Design 75, P-36
XP-37	Design 80	V-1710-11(C8)	Form F-13	Aug. 1938	Rebuilt in 1938
XP-37	Design 80A	V-1710-D3/Mod B	Form F-13	None	Proposal Only
XP-40	Design 75P	XV-1710-19(C13)	None	10-1-1938	From 10th P-36
YP-37	Model 81	V-1710-21(C10)	Type B-2	1-24-1939	13 Built
P-40	Model 81A	V-1710-33(C15)	None	4-4-1940	Production

Likewise, it was a conscious decision by the Air Corps to not Modernize the similar D-2 engines for the Bell YFM-1 airplanes.[91]

None of the P-37's were involved in fatal accidents and only 38-477 had to be surveyed due to crash damage, and that was primarily due to its remote location in Alaska. The average flying time for the 14 aircraft was 168 hours, with 38-481 in Alaska being the high time aircraft at 279.3 hours. Engine performance was fairly good, with most of the aircraft keeping their original engine until removed for overhaul at 150 hours.

The X/YP-37's did not go into series production even though their performance was actually better than the competition. A major reason being the aft pilot position, needed to balance the supercharger and forward location of the radiators and intercooler. This arrangement was considered unsuitable for a combat aircraft. They did serve to confirm the contention that small fighter aircraft needed to have higher power and higher altitude ratings built into their engines, a lesson incorporated into the coming generation of successful combat fighters that came into service after 1940.[92]

As early as October 1937, well before the YP-37 was contracted, Curtiss was at work on a follow-on to the YP-37. This was to be the Y1P-37A, which appears from the limited information available to have been a refined aircraft incorporating a new wing with inward retracting 30-inch diameter wheels. The engine and turbosupercharger would have been those in the YP-37.[93] The project did not proceed.

Lockheed X/YP-38 *"Twin Engine Interceptor Pursuit*

In his quest for high performance, Lt. Kelsey desired a single engined fighter powered by a 1500 bhp engine. As no such engine was available in the mid-1930s, the 1937 design competition defined by Circulars X-609 and X-608 sought to compare the best possible performance from a single engined aircraft, using the most powerful engine available, against similarly powered twin engine aircraft. By doubling the horsepower it was hoped to approximate the performance of the sought 1500 bhp single engine fighter.

When the Materiel Division issued Circular X-608 for a twin engine interceptor in February 1937 it specified that the proposals must use the turbosupercharged 1150 bhp Allison V-1710-C7. This specification was issued prior to the Air Corps completing their 150 hour test on the 1000 bhp V-1710-C4/C6 Type Test engine, and prior to the availability of the improved 1150 bhp V-1710-11(C10).

Lockheed's X-608 Model 22 proposal. Details were quite different than in the later XP-38. 2. Flush landing gear doors, 3. Radio antenna, 5. Turbo waste gate, 6. GE Form F-10 turbo, 7. Engine air intake. (Lockheed)

Circular X-608 specified a minimum speed of 360 mph and formalized work that had been underway since February 1936 on classified project M-12-36. Three aircraft firms prepared proposals for the V-1710-C7 powered Twin Engine Interceptor Pursuit; Lockheed with their Model 22, Vultee's XP1015 and a design from Curtiss. Lockheed believed that their design was capable of 417 mph at 20,000 feet, and though many of the officers at Wright Field were skeptical, the design was the clear winner. Lockheed was notified that they had won on June 23, 1937, and soon signed contract W535-ac-9974 for the single XP-38.[94]

In an effort to get even more mileage out of the Model 22 effort, Lockheed offered their Model 24 to the Navy. A sketch of the general arrangement of that aircraft shows it to be a hook equipped Model 22, complete with V-1710-C model engines as intended for the Model 22.[95] Turbosuperchargers are not shown on the outboard side of the boom/nacelle, but then this is only a sketch.

The Model 22 proposed by Lockheed for the X-608 competition in 1937 was quite different from the Model 22 aircraft that flew as the XP-38

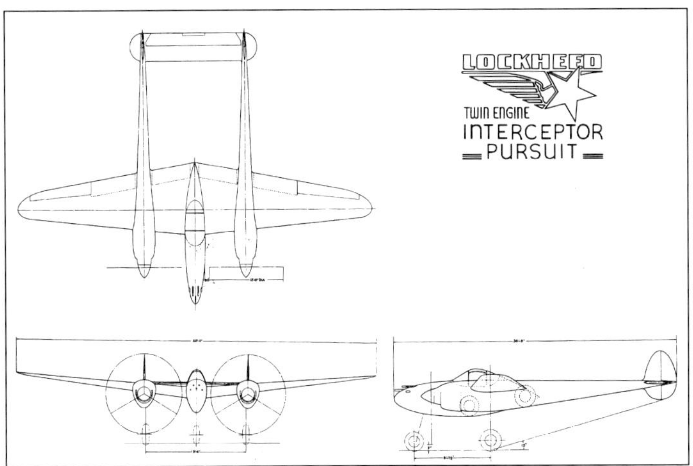

Lockheed Model 22 as submitted to Wright Field for X-608 Competition in April 1937. Note the chin mounted radiators and clean booms. (Lockheed)

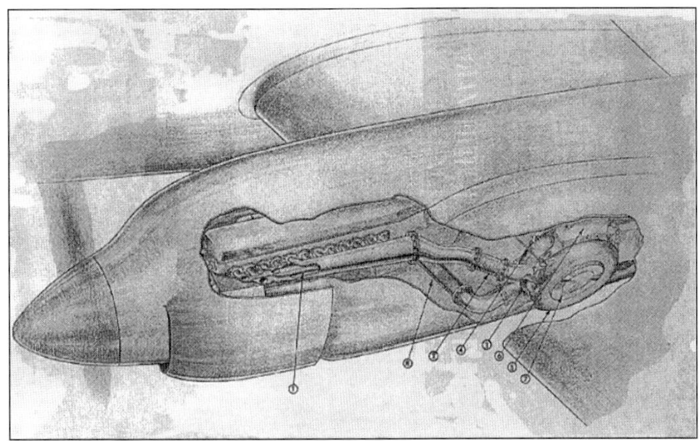

The original Lockheed Model 22 submitted for the X-608 competition incorporated a considerably different turbosupercharger installation than on the later XP-38. Exhaust gases were to be routed to the turbo, located below the wing on the outboard side of the nacelle. 1. Solid baffle in exhaust manifold, 2. Flexible exhaust joint, 3. Waste gate, 4. Control rod to turbo regulator, 5. Form F-10 turbo, 6. Supercharger air intake, 7. Exhaust exits through turbine wheel, 8. Exhaust cross-over from inboard cylinders. (Lockheed)

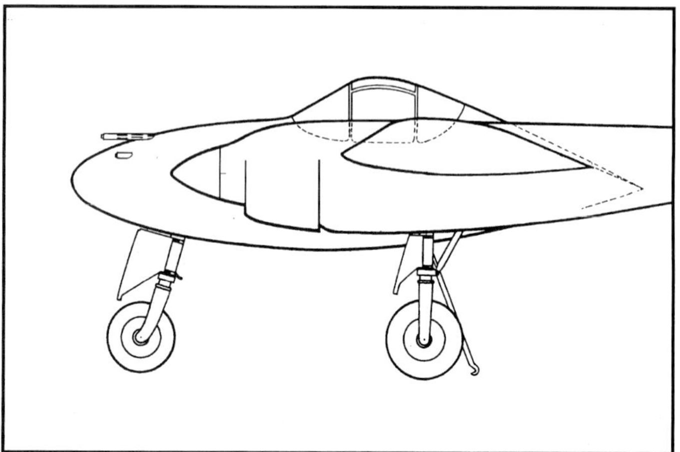

Lockheed offered their Model 24 to the Navy. The aircraft is an obvious adaptation of the liquid-cooled Model 22, but equipped with a hook. (Matthews)

The Model 22 proposal located kidney shaped radiators in a section of NACA type cowling. The engine driven coolant pump delivered Prestone to the integral cylinder block coolant headers. Hot coolant exited the front of each block and was directed into the adjacent radiator segment. Each of the 9 inch deep radiator cores was connected to the expansion tank over the engine reduction gear. A special oil cooler was to be mounted below the engine in the space between the radiator segments. (Lockheed)

in January 1939. While the overall configuration of the proposed airframe contained the now familiar characteristics of the P-38, there were considerable differences in the engine, turbosupercharger, intercooler installations and radiator arrangements.[96]

The original Model 22 was designed to use the required Allison V-1710-C7 engine and General Electric Form F-10 turbosupercharger, the same units likewise specified for the contemporary X-609 Bell Model 4, (XP-39) and soon to be flying on the Curtiss XP-37 and Bell XFM-1. This turbo had the compressor "reversed" on the drive shaft, as compared to the later and better known Type B units. As a result, the air intake to the turbo was drawn into the region between the turbine and compressor wheels. This is clearly evident in the accompanying drawing from the Lockheed Model 22 submittal. Most interesting is the entirely different location of the turbo, on the outboard side of the engine nacelles, under the wing and adjacent to the position of the retracted main landing gear wheel. This

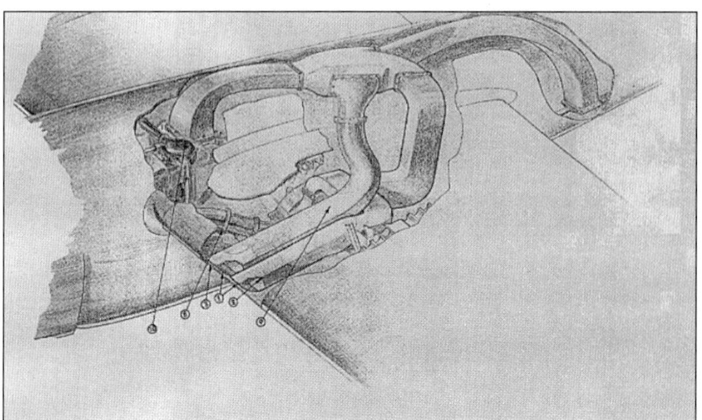

Lockheed originally intended the Model 22 to have a separate air-to-air intercooler mounted in the nacelle just behind the wing spar. Compared to this arrangement the elegance of the later integral leading edge intercoolers on the actual aircraft is apparent. 1. Wire screens covered the leading edge intakes, 2. Supercharger air inlet, 3. Cooling air to Intercooler, 8. Flexible connection of hot exhaust to turbo, 9. Ducts passed through Main Wing Spar, 10. Compressed air into carburetor. (Lockheed)

turbo had the combustion air intake arranged as a scoop just behind the turbine wheel, and extending out into the air stream. As shown in the drawing, this resulted in a comparatively clean installation with fairly straight forward routing of the exhaust manifolds and compressor discharge ducting to the intercooler and engine carburetor.

It is noteworthy to note that this installation incorporated a fairly compact air-to-air intercooler, with the cooling air drawn in through a slot in the wing leading edge and subsequently exiting into the "dirty" airstream far back on the bottom of the boom. While this alone would have been a good installation from an aerodynamic viewpoint, the actual XP-38 showed an extra measure of design elegance by integrating the intercooler directly into the structure of the leading edge.

Study of the proposed installation suggests reasons for this design to have been abandoned in favor of the arrangement used on the real P-38's, with the turbo mounted turbine wheel up on the top surface of the nacelle just behind the rear wing spar. Having the turbo in the side of the nacelle probably cut dramatically into the space in the boom for the wheel well, though the turbo was located just ahead of the retracted main wheel. At the same time, the intercooler was a large device located in the approximate position ultimately occupied by the turbo. By using the wing leading edge as an intercooler there was a considerable reduction in the quantity of air brought onboard for intercooler cooling, while at the same time freeing the space for installation of the turbo and elimination of the weight of the intercooler.

Another factor that may have had a significant role in selecting the final arrangement was the fact that both the air supply duct for the intercooler and the combustion air into the turbo required adjacent cutouts in the main wing spar. The necessary strengthening of this vital component probably caused adverse weight increase in the airframe and contributed to the desire to redesign when detailing the XP-38.

The radiator installation was likewise unique, particularly to American practice. Two 9 inch deep radiator core segments were positioned on each side of the engine, just ahead of the cylinder banks. A special oil cooler was to be placed on the engine nacelle centerline, directly under the reduction gear. The air for the radiators and oil cooler was to be taken in through a horseshoe shaped annular duct extending 2/3 the way up the sides of the nacelle, and exiting through a streamlined slot just ahead of the wing. In all, the arrangement appeared much like a segment of the recently developed NACA cowl that was to do so much to allow air-cooled radial engines to become competitive on the next generation of aircraft.

Apparent in the engine installation in the proposed Model 22 is the sophistication of the internal aerodynamics, this at a time prior to systematic NACA investigations into the matter. Considering that this design was a contemporary of the Curtiss XP-37 with a V-1710 and the same GE Form F-10 turbo, the differences are night and day. For example, the airframe designer was responsible for the engine exhaust system. The engine manufacturers responsibility traditionally ended at the exhaust flange fitting. Lockheed proposed a unique arrangement in which the front port of each pair of exhaust ports was angled up into a duct that incorporated a solid horizontal dividing baffle, below which the aft port of each cylinder pair exhausted. The intent was to minimize turbulence within the collectors, thereby providing the maximum exhaust energy to the downstream turbo. In truth this was probably more trouble that it was worth and apparently was never used.

The final design for the new aircraft was completed on September 23, 1937 and specified for the first time "handed" engines. Allison engine models V-1710-C7/C9 specified.[97] Interestingly, Allison had only prepared their Model Specification 111 for the C-9 "left-hand" turning engine on September 1, 1937, the date of the first flight of the Bell XFM-1. Things were happening rapidly as Lockheed moved to construct the XP-38.

The X-608 competition winning XP-38 resulted in the Air Corps deciding in December 1938 to purchase three YP-38's, two to be powered by the new V-1710-27/29(F2R/L) and the other by the P&W R-1830, with the contract specifying an option for seven more aircraft. These seven were to be ordered only after evaluation of the first YP-38 versus a similar number of YP-39's.[98] This order was soon revised to a total of 13 Allison powered YP-38's, and the radial engined version dropped. When the XP-38 (AC37-457) made its first flight on January 27, 1939 it was powered by Allison V-1710-11/15(C8/C9) engines, the C-8 being a C-7 equipped for propeller feathering and incorporating other detail improvements, including revised valve timing. The C-9 was built using C-8 components modified for left-hand rotation. These turbosupercharged sea-level engines were rated at 1150 bhp at 2950 rpm to 25,000 feet. Turbosupercharger air inlets, located under each wing, led to the GE Type B-2 turbosuperchargers, installed on top of the engine nacelle booms, just aft of the rear wing spar. The leading edges of the wing were ingeniously constructed to act as intercoolers. They not only cooled the induction air, but also providing a measure of leading edge de-icing. Exhaust gas collection manifolds routed the hot gasses from the engine to the turbo and/or waste gate, both located on top of the boom and behind the wing spar. The small scoop on top of the XP-38 nacelle was to provide cooling air for the red-hot exhaust manifold in the close cowled engine nacelles. Scoops leading to the coolant radiator extended from the sides of the booms midway to the tail. It is interesting to note their relatively small size, as compared to similar installations on later model P-38's, and realize that these radiator inlets do not have boundary layer splitters. This reflects their being sized for the 1150 bhp Allison's and the unique location of the radiator core within the boom, an arrangement used only in the XP-38. The heated air exited through a chute on top of the boom. Later models of the P-38 required larger radiators able to support higher powered V-1710 models rated for 1425 and 1600 bhp. The tightly cowled engines show the advantage the original Air Corps concept for the highly streamlined cowling available with the V-1710-C engines and their low centerline and extended propeller shaft.

By the fourth flight of the XP-38, Lockheed aerodynamicists had determined that elevator buffeting at high speed was more than likely the result of high-activity "prop wash" and that a change from inboard (at the top) propeller rotation to outboard rotation would solve the problem.[99] The solution was to interchange the right and left engines, which *was not done* on the prototype before its crash in New York on February 11, 1939. The aircraft only had about five hours in the air by the time Lt. Kelsey flew the XP-38 on its record breaking cross country delivery flight. Changes were already being made. For example the exhaust manifold cooling slot had been provided with a closed cover having only a small scoop for manifold cooling air. The engine oil cooler scoop is shown on the bottom of each nacelle.

XP-38 at March Field on or about the time of its first flight. Note the radiator exit chute is open, top of boom (in front of far rudder), also the open oil cooler inlet scoop just behind and below spinner, the vent slot in nacelle under wing, just ahead of turbo/carb air scoop under wing trailing edge. The scoop on top of the nacelle, above the leading edge of wing, supplies cooling air to the turbo manifold shroud. (Air Corps)

The two first flight engines were procured in different ways, and were the first two engines Allison built after the sixteen development engines. The V-1710-11(C8) was taken from the batch of 20 procured for the Curtiss YP-37's, and was the first engine of the batch, AC38-581 (AEC#18). The left-hand turning V-1710-15(C9) engine was AC38-120 (AEC#17), procured on a separate contract, W535-ac-10291 and actually built before any of the C-8's.[100] Both engines were severely damaged in the crash at Mitchel Field, but each went on to interesting further service.

The left-hand engine was probably the only such turning engine in the Air Corps at the time. As such there were not a lot of uses available for it, and in such instances it was Air Corps practice to send unique aircraft hardware to Chanute Field, Illinois, where they could be used in the training of aircraft mechanics. The engine was repaired, equipped with a new propeller and setup on a test stand for training mechanics on how to operate the new V-1710 engine.[101]

The right-hand V-1710-11 was not repaired, though it was held as a source of spare parts at Wright Field. The crankcase, crankshaft, and connecting rods were subsequently pressed into service in the special V-1710-19(C13) type engine [AC38-597, AEC #37] built for the High-Powered V-1710-C test series performed in late 1939. When the original test engine was found to have a cracked crankshaft midway through the project, replacement components were taken from 38-581 and used to complete the tests. It was with these parts that the peak output of 1575 bhp for the V-1710-C series was produced.

Following the loss of the XP-38, Lt. Kelsey sat down and prepared a report based upon the partial data from the few hours of testing that had been completed. He estimated that when using 1150 bhp from both engines the XP-38 would have been able to cruise at 394 mph at 20,000 feet. Armed with this information, he joined with General H. Arnold and toured the halls of Congress and the War Department in Washington DC. They were successful in salvaging the P-38 program. The result was an April 27, 1939 order for 13 V-1710-F2R/L powered YP-38's. The order for $2,180,725 to Lockheed was a strong endorsement for the advanced features and performance demonstrated by the XP-38 in its short operational life.[102]

Engine Installation and Cooling

When constructed, the XP-38 (Model 22) had considerably different systems than those proposed in the original submittal. The air-to-air package type intercoolers had been eliminated in favor of the integral leading edge devices and the turbos relocated to the top of the booms. The engine radiators were now in the booms as well, with the oil coolers streamlined into the nacelles, just below the engine. While the airplane was carefully designed and wind tunnel tested to insure the best installations of each of these components, there was little growth margin included in the design. For example, the radiator and oil cooling systems were designed to accommodate only 1000 bhp from each engine, the power rating needed for continuous flight at 20,000 feet with a speed of 312 mph.[103]

To achieve the promised 400 mph at 20,000 feet the XP-38 was to use the full 1150 bhp available from each engine. Lockheed engineers hoped that there was enough "latent" heat capacity with the systems, along with a fair amount of heat lost throughout the plumbing, to allow five minutes at the 1150 bhp rating.[104] This suggests that the leading edge intercoolers were likewise constrained for continuous duty to no more than 1000 bhp.

The excellence of the XP-38 design can be quantified by comparison with similar components of drag of the XP-39, as shown in Table 3-6.

When the V-1710-C8 engine was running at 1000 bhp the oil cooling load was 75 bhp, with 110 pounds of oil circulating per minute. To cool this oil Lockheed used two standard Air Corps 8 inch diameter oil coolers mounted below the engine, with cooling air provided by a variable position scoop and similarly moveable exit chute. Lockheed did not like the high drag resulting from this installation amounting to 84 percent of the drag of the boom mounted Prestone radiators. They subsequently redesigned the cooler intake into the streamlined nacelle, as can be seen in photos of the YP-38's and on through the Model 322 aircraft.

Table 3-6
Comparison of Drag Between XP-38 and XP-39

	XP-38 Speed Loss, mph	XP-38 C_D	XP-39 C_D
Prestone Radiator	10.7	0.0025	0.0024
Oil Radiator	8.9	0.0021	0.0040
Engine Cowl Gills	3.4	0.0008	-
Supercharger and Exhausts	0.8	0.0002	0.0047
Carburetor Air Scoop	3.0	0.0007	0.0019
Intercooler	-	-	0.0007
Excessive Cooling Drag	-	-	0.0015
Total Engine Accessories	-	0.0063	0.0152
Total Aircraft @ 400 mph	-	0.0252	0.0329

The Prestone radiators were installed in a unique fashion on the XP-38 as compared to all later P-38's. Although located at the same station on the booms, there was only a single radiator core that was buried within the boom and fed cooling air by scoops extending out into the slipstream. Hot air was then exhausted through a chute, able to be positioned to control the flow, located on the top of the boom. Structural limitations kept Lockheed from increasing the chute exit area, the obvious way to provide additional cooling. According to Allison the C-8 engine required 410 hp of cooling capacity when operating at its rated 1150 bhp, and 360 hp when the engine was at 1000 bhp. The XP-38 design could only accommodate 315 hp in each radiator, though it was hoped that conservatism in the analysis, normal heat losses in the plumbing, and the benefits of the heated discharge air would makeup the additional needed capacity. The limitations on cooling were most constraining during maximum power climbs. Subsequent models of the P-38 used a pair of "kidney" shaped radiators placed in short ducts appended to the sides of the booms, thereby simplifying the installation and eliminating the structural constraints on radiator exit area.

For high speed flight at 20,000 feet Lockheed determined that the radiator chutes had to be open about 30 degrees, which produced a drag coefficient of 0.00346 and consumed 234 of the 2000 bhp being provided by the engines. While this installation was incorporated on the Model 22, Lockheed was not happy with it; particularly when they found that recent NACA work had shown other radiator installations, utilizing streamline shaped ducts, were more than four times as efficient.[105] These and other improvements coming from the shortened XP-38 flight test program showed that a redesigned radiator installation would add 8 mph to the plane, and a non-retracting oil cooler scoop would add another 3 mph while changes to the turbo installation would add 4 mph.[106] All of these features showed up on the YP-38 and subsequent models of the P-38.

YP-38

The YP-38's were quite different than the prototype XP-38. First they were designed to be amenable to mass production, which resulted in considerably revised detailing within the aircraft structure. Probably the most discernible change, and certainly the one readily apparent to observers, was the revised cowling lines on the engine nacelles, necessitated by the incorporation of the new high thrust-line V-1710-F model engines.

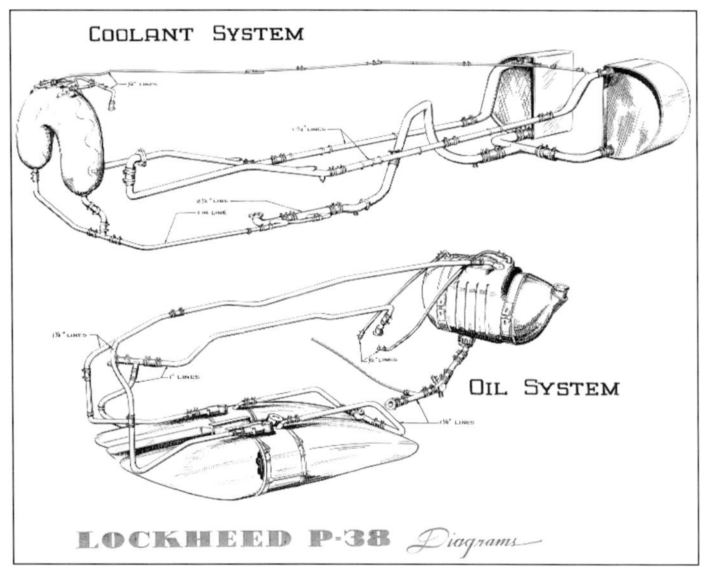

These were the basic coolant and oil cooling systems used in the various models from the Lockheed YP-38 on. Sizes of the coolers and internal flow rates were increased to meet the needs of the higher powered V-1710's. Lockheed also made considerable improvements in reducing cooling drag and to improve heat removal over the years. (Allison)

As is detailed elsewhere in this text, by 1938 Allison was of the opinion that the V-1710-C engine was at the limit of its development potential as a consequence of the design limitations of the reduction gear. With the design opportunity provided by the V-3420 project they began to redesign and standardize the components in the V-1710 and V-3420 engines. The result was the V-1710-E extension shaft engine and a companion powersection designed to drive an integral external spur type reduction gear, the V-1710-F. Work on this engine dates from 1938, and so it was only natural that it would be incorporated from the outset in the improved YP-38's, actually begun prior to the first flight of the XP-38.

With the P-38 requiring "handed" engines, Allison ingeniously designed their new engine models to all use the same crankshaft. By the simple expediency of end-for-end reversing of the crankshaft during assembly, and providing appropriate connections of the ignition wires to the sparkplugs along with the proper positioning of several gears in the accessories section, the engine could be operated either right-, or left-handed. The YP-38's were the first to fly with the new V-1710-F, using the F-2R/L models as the Air Corps V-1710-27/29 respectively. These provided the same 1150 bhp as the earlier "C" model engines in the XP-38, but offered considerably improved growth capability and ease of manufacture.

The engines in the YP-38's seemed to have operated quite well for a new engine in a new airframe, their only symptomatic problem being the tendency to backfire. This problem was common to all early V-1710's during the initial introduction into squadron service. The problem was resolved after a series of changes and modifications that are detailed in Chapter 4. Suffice to say that while new and stronger induction manifolds were provided, along with improved backfire screens, the substantive change was to simply increase the tappet clearance adjustment on the intake valves from 0.010 to 0.015 inches. Retrofitting sodium cooled intake valves into the early V-1710's also helped, but that occurred somewhat later.

The same engine installation as on the YP-38's was used on all of the follow-on P-38 models up through the P-38E, a total of 388 aircraft. When the P-38F model of the aircraft was introduced, it used the V-1710-49/53(F5R/L) with its higher internal supercharger gear ratio, allowing increasing power to 1325 bhp for Takeoff and Military. The balance of the installation remained the same. These later combat ready P-38's are detailed in Chapter 5.

Summary

The P-38 was not just another airplane. It embodied advances in aerodynamics, propulsion and systems that were years ahead of both foreign and domestic competition. It became the measure of other airplanes, and it is significant that it was still in production at the end of the war. It was not the fastest airplane of WWII, but it could compete with any U.S. or foreign aircraft at any point during the war. That it was available, and breaking ground in so many areas, was significant in the prosecution of the war.

The XP-38 and YP-38's were blazing a path in every aspect of aviation when they were first introduced.

Table 3-7
Specification X-608 and X-609 Requirements and Guarantees

Performance	1936 Minimum	1939 Minimum	1939 Desired	XP-38, Guaranteed	XP-39, Guaranteed
High Speed at 20,000 feet	360	400	450	417	400
High Speed at Sea Level	270	310	350	393	330
Time to Climb to 20,000 ft, min.	6.0	6.0	4.0	4.5	5.0
Endurance at Operating Speed, hrs.	1.00	1.00	NA	1.75	1.00
Range at Cruising Speed, miles	NA	700	NA	1,386	NA

Bell X/YP-39 *"Airacobra"*

Bell submitted their Models 3 and 4 to the Air Corps in response to Type Specification X-609 that was issued on March 19, 1937, a few days before the YV-1710-C4/C6 completed the Type Test. In this specification Lt. Kelsey had defined the requirements for single engined interceptors. Submittals were to have the capability for "the tactical mission of interception and attack of hostile aircraft at high altitude." Minimum airspeed at altitude was to be 360 mph and the aircraft was to be able to reach 20,000 feet within six minutes. Submittals were also required to use the only engine known to be capable of such performance at the time, the Allison V-1710-C7. The X-609 specification used the Allison nomenclature since Air Corps practice was to assign engine model numbers only after an engine was actually contracted. In this case the C-7 ultimately became the V-1710-11, though the innovative Bell designs required a much different configuration and was assigned another designation by both Allison and the Air Corps.

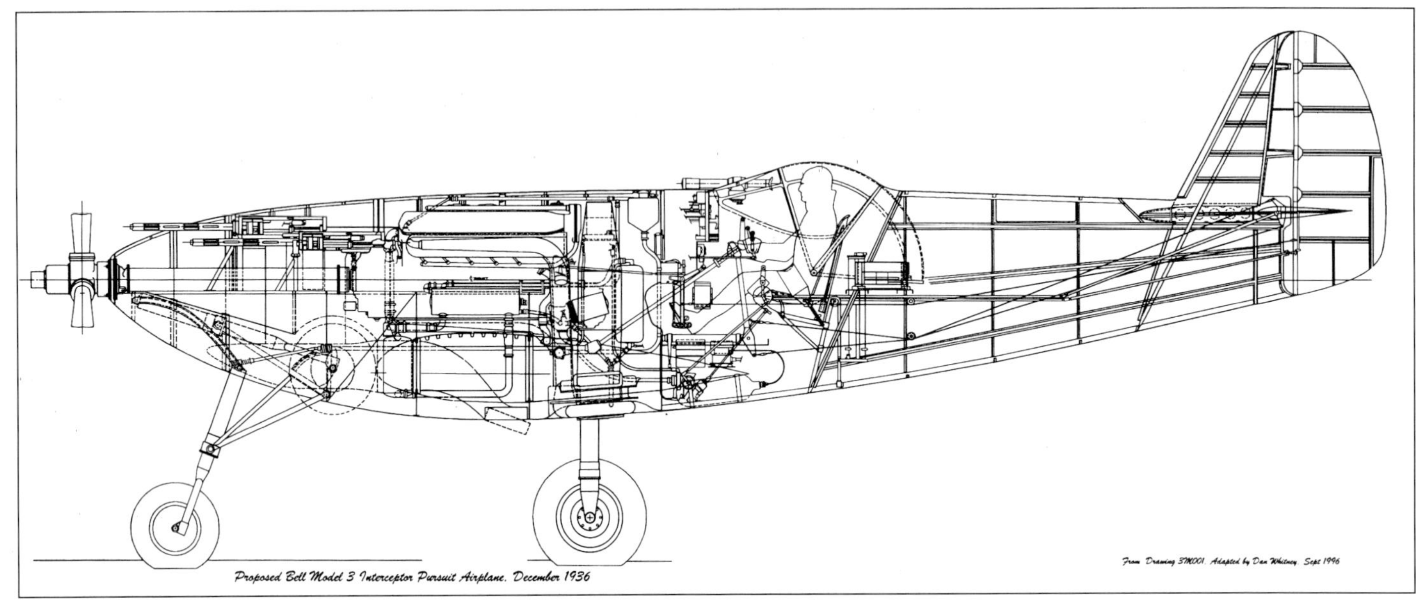

The initial December 1936 configuration of the Bell Model 3 used the 1250 bhp rated V-1710-D3 propeller speed tractor engine. A Form F-10 turbo was mounted behind the rear wing spar. This was all revised in May 1937 so that the proposed Model 3 and Model 4 were submitted for the X-609 Interceptor Pursuit competition with both models incorporating the Modified D-3 featuring a remote reduction gear able to mount a cannon. (Author)

Bell Model 3

The Bell interceptor project began early in 1936 and was initially identified by the Classified project number M-13-36. This was quite an advanced aircraft. Not only was the engine installation like none other before it, there were many new and unique features, such as incorporating the radio mast into the leading edge of the vertical stabilizer and having the capability to use a rocket to boost acceleration in the interceptor role. As with most aircraft development programs the concept evolved considerably overtime. The accompanying Model 3 Inboard Profile dates from December 1936 and shows an aircraft with a turbosupercharged tractor V-1710-D3, but with a propeller speed extension shaft and expected to deliver 1250 bhp.[107] Such a configuration would obviously not be able to accommodate a cannon firing through the propeller hub, and in fact only two .50 cal machine guns are mounted, synchronized to fire through the propeller. Note that the engine is mounted directly over the wing and that the exhaust manifolds are routed internally to the Form F-10 turbosupercharger, which is mounted on the fuselage centerline and directly behind the rear wing spar. An interesting observation is the attractive streamlining of the fuselage that results.

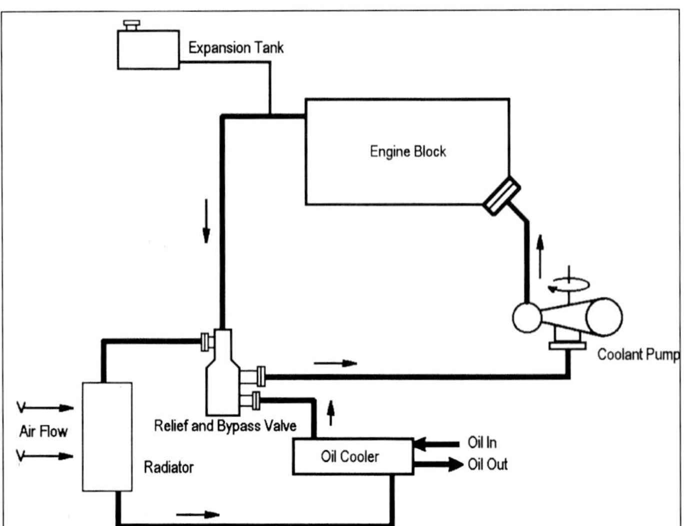

The Bell Model 3 integrated the oil cooler into the engine liquid-cooling system. An advanced concept, it did not carry over onto the XP-39 though a similar system was used by North American in the XP-51J. (Author)

This engine rating of 1250 bhp was also intended for the Bell XFM-1 at the time, but was never a rating used in flight. Achieving this rating would have required a degree of ground level boosting from the GE Form F-10 turbosupercharger. In the case of the Model 3 the result was a reduction in the critical altitude to 15,000 feet, where the airplane was expected to achieve 351 mph and a rate of climb of over 4,400 feet per minute in its interceptor role.[108]

The Model 3 configuration actually submitted in March 1937 has quite a different profile, one that shows that the higher thrust line of the new external spur gear type of reduction gear was simply grafted onto the earlier design for the Model 3 airframe. The reason of course being that it featured the installation of a 25 mm cannon, arranged to fire through a three inch diameter hole in the propeller hub. As a consequence of this change Allison designed a new engine, based upon the D-3, but featuring a crankshaft speed extension shaft to drive a 5:3 reduction gear mounted in the nose. This was the V-1710-D3 "Mod A." The pilot was seated behind the engine, in a position much like that in the Curtiss XP-37 Design 80A. Another consequence was that the turbo was moved forward to a position directly below the engine.

Bell Model 4

The reasons for completely revising the Model 3 as a new model probably resulted from the desire to better integrate the newly added cannon, which had severely degraded the pilots view over the nose. By putting the pilot in front of the engine, visibility was considerably improved. The installation of the Form F-10 turbo was simplified, now fitting completely behind the wing spar. The Model 4 was a newly designed airframe, but mounted the same equipment as in the final configuration of the Model 3, including the 1150 bhp V-1710-D3 "Mod A" engine, but it did require an additional five feet of extension shafting to reach the reduction gear. With the engine at

As submitted for the X-609 competition the Bell Model 3 fitted the D-3 "Mod A" extension shaft engine enabling the installation of a cannon mounted to fire through the propeller hub. (Bell Aircraft)

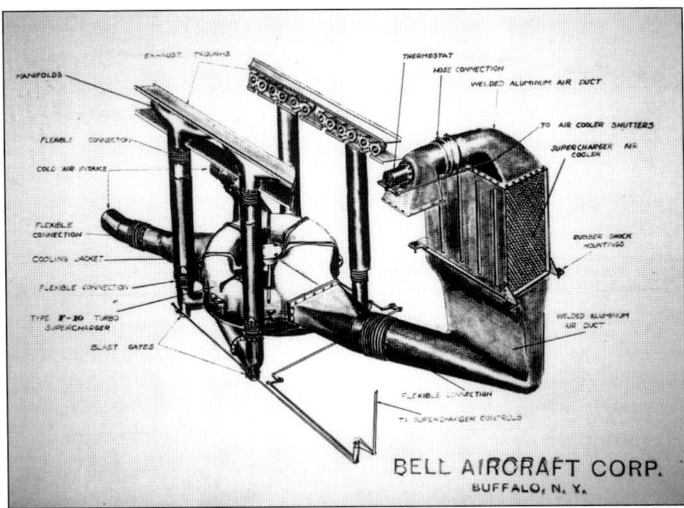

In their final forms both the Bell Model 3 and Model 4 used the same arrangement for routing the hot exhaust gasses to the Form F-10 turbosupercharger. The arrangement was simplicity itself, but it turned out to be an aerodynamic disaster and resulted in the turbo being deleted from the P-39. (NARA)

near the center of gravity it was expected to be a highly maneuverable aircraft.

This cannon armament was a deciding factor in favor of the Bell submittals, just as it had been a year earlier in the Multi-Place Fighter competition they had won with the cannon armed XFM-1 Airacuda.

Using this engine and turbo, combined with the light weight of the winning Bell Model 4, a high rate of climb was assured.[109] The Bell Model 4 was selected on promised performance of over 400 mph. It triumphed over their Model 3 (which had "won" on a cost adjusted basis), the Curtiss YP-37 Designs 80 and 80A, as well as several Seversky AP-4 models. The Air Corps designated the winner as the XP-39.[110]

Modifying the V-1710-C7 for the Bell Model 3 and Model 4

The Air Corps specified that X-609 competitors were to use the 1150 bhp Allison V-1710-C7. Its streamlined integral nose case, would not have worked in either the proposed Bell Model 3 or 4 with their arrangements for remote installation of the reduction gear needed to accommodate the nose mounted cannon and machine guns. Clearly the engine would have to be redesigned. The reduction gears would have to be separated and mounted well ahead of the engine power section. In this way space and provisions for mounting the cannon to fire through the propeller hub were accomplished. With their extensive experience in experimental engineering and power shafting, no firm was better qualified than Allison to undertake the innovative endeavor. Allison proposed to provide a modified version of their original "extension shaft" V-1710-D2 engine, that they had offered in a tractor configuration for the initial Model 3 design studies as the V-1710-D3.[111]

In its first incarnation (December 1936) the Model 3 was to be powered by a propeller speed extension shaft V-1710-D3 rated for 1250 bhp at 2900 rpm, and boosted with a Form F-10 turbo for a critical altitude of 15,000 feet.[112] In March 1937 the engine performance was revised and no longer required turbo boosting at sea level. When the Model 3 and Model

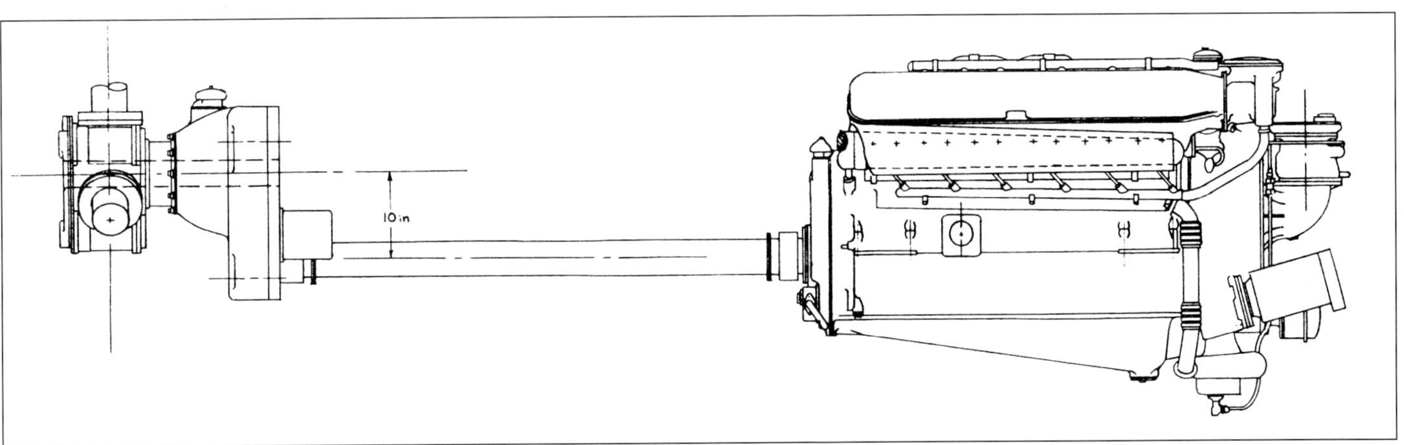

This is the V-1710-D3 engine modified with the first remote reduction gear and a crankshaft speed extension shaft. The hollow propeller shaft would accommodate a cannon, the reason for adopting the feature. Mod "A" and Mod "B" of this arrangement were offered respectively for use in the competing Bell and Curtiss proposals in the Army's 1937 X-609 competitions. They differed only in length of the extension shafts. (Author)

4 designs were submitted for the X-609 competition, the remote reduction gear versions of the D-3 were specified and given a sea-level rating of 1150 bhp at 2950 rpm, with the turbo boost being used to provide a critical altitude of 20,000 feet. Allison issued these revised specifications for these three versions of the D-3 in May 1937, all of which used the C-7 power section.[113] The result was a "straight" D-3, as a propeller speed tractor, and two models having different length crankshaft speed extension shafts to remote reduction gears able to accommodate a cannon within the propeller shaft. Once developed, the resulting extension shaft and remote reduction gears were amazingly trouble-free during the following years of production and service.[114]

Events Surrounding the First "E" Engine for the XP-39

Even before the X-609 design competition Allison was working on a continuing Air Corps request to "pair" the V-1710 and thereby double the available horsepower. The Army was now interested in ways to achieve 2000 bhp or more, up from their 1935 requirement for a "doubled" X-3420 engine of 1600 bhp. This power was going to be required for their long desired Very Long Range (VLR) bomber project.[115] With the sage guidance of Allison's then new Chief Engineer Ron Hazen, and likely influence of the General Motors Corporation, Allison engineers were ever mindful of the advantages and necessity of maximizing commonality in their products. The need for an extension shaft engine, coupled with the requirement to design the double vee (X-3420) bomber engine, along with the recent experience in incrementally upgrading the V-1710-C series to get it through the 1000 bhp Type Test, as well as the demand for handed engines, gave Allison the motivation to meet all of these objectives at once. They designed both of the new engines for maximum commonality: Cylinder banks, pistons, connecting rods, crankshafts and many accessories would be interchangeable on these and all subsequent Allison reciprocating engines.

Design and development of the V-1710-E engines, with their extension drives to remote reduction gears, progressed rapidly.[116] The details of how, and which, engine came first, and how they were related, is both interesting and important to the development of the V-1710. The story also resolves misconceptions perpetuated in earlier accounts.

In September 1937 Allison completed its design study for the new XV-3420 "Double Vee" engine, a much improved arrangement over the previous X-3420 effort. It design used four V-1710 cylinder blocks mounted on a single crankcase. It also included the innovative feature of using two identical crankshafts that were mirror images around the center thrust bearing. In this way, by simply installing the crankshaft end-for-end, the engine could be timed to run in the opposite direction. This was a desirable feature as the V-3420 was intended to be offered with left/right handed propellers, or with counter-rotating propellers. This feature also allowed a V-1710 engine, using this crankshaft, to be operated either as a right-hand, or left-hand turning engine as well. The feature offered a number of advantages in both manufacture and operation, and it may be more than coincidental that it was designed at the same time Allison was offering a left-hand turning V-1710 for Lockheed to use in their X-608 proposal. Allison was anticipating a considerable demand for "handed" engines.

While Allison was busy preparing proposals for the aircraft manufacturers entering the Air Corps X-608 and X-609 design competitions, they were also meeting with the Navy on the possibility of a Bell built fighter for carrier service. Although the Air Corps had specified the 1150 bhp turbosupercharged V-1710-C7 was to be used in their competition, the Navy was just as adamant about wanting 1150 bhp, but without the turbosupercharger. This was the first formal requirement to Allison for any type of altitude rated V-1710 and resulted in definition of the Allison V-1710-E1, Specification #112 dated September 17, 1937. With all of these different configuration demands it is not surprising that Allison used the opportunity for the design of the XV-3420 to design an engine having unprecedented flexibility for adopting it and its components.

At this point an interesting contradiction occurs regarding who was paying for the design of the V-3420. Ron Hazen later reported that the V-3420 engine was designed using corporate funds, though we know that contract W535-ac-9678 was authorized in May 1937 to cover the XV-3420 study, and ultimately the first engine.[117] It is likely that both are correct. Corporate funds were used to expedite the effort, and then, somewhat after the effort was well underway, the government funds finally became available. Certainly the cost was greater than the bargain price paid by the government, hence Hazen is correct to the extent that the company paid the majority of the expenses.

As is told elsewhere in this book, Allison was eager, even desperate at the time, to get a production order for the V-1710. They were now seven years into the V-1710 program and had delivered a total of 15 engines, the proceeds of which did not cover even half of the expenses to date. Feeling that a share of the problem was the Air Corps focus on requiring turbosupercharging, a component that was still not adequately developed, Allison became somewhat aggressive in seeking applications for altitude rated engines. To this end they issued a second specification for an altitude rated engine, Specification #113 that defined the V-1710-C11 on September 21, 1937, less than a week following that for the E-1 Navy engine.

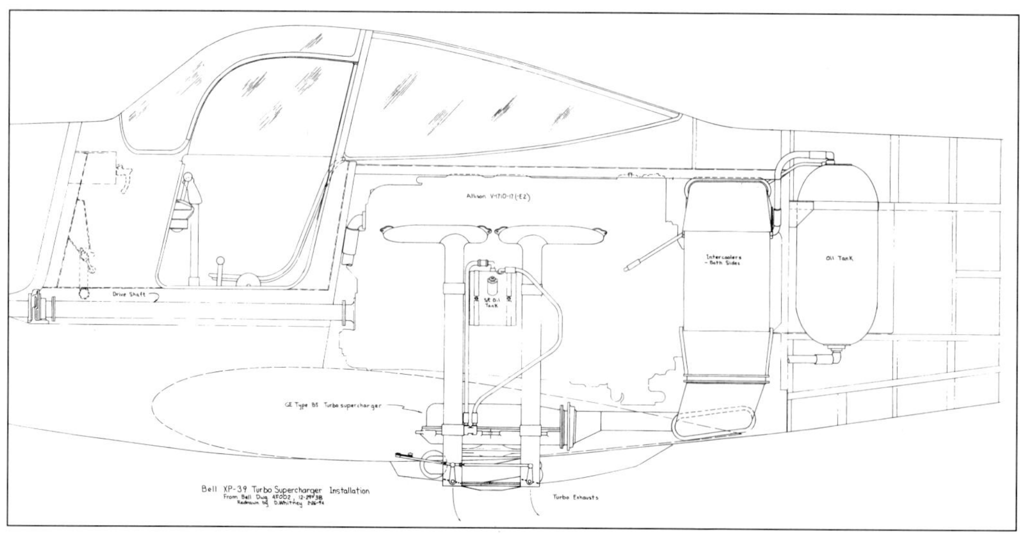

When the Model 4 won the X-609 contract the turbo was the GE Form F-10. When the XP-39 was built the improved GE Type B-5 was used in a similar installation. (Author)

On October 10, 1937 Bell was awarded a contract for their winning Model 4, now designated as the XP-39. During the summer, while the proposals were being evaluated, Allison configured their EX-2 "workhorse" engine with the 56 inch long crankshaft speed extension shaft and remote reduction gearbox, proceeding to investigate the vibration characteristics of the new arrangement. If so, this would have been the V-1710-D3 "Mod B" configuration offered to Curtiss for the XP-37 Design 80A. It was not until November 24, 1937 that Allison issued Specification 115 describing the sea-level rated V-1710-E2 engine for the Bell XP-39, a design that certainly benefited for the interim work done on the house D-3, Mod B.[118]

The above explains why the XP-39 ended up with its "E" series engine designated as the E-2, even though its engine was the first actual "E" model to be built. That first engine (AC38-644, AEC#30, purchased on contract W535-ac-10830) for the XP-39, was delivered November 11, 1938.[119] The Army also separately purchased a spare -17 (AC38-931, AEC#33, on Contract W535-ac-11279). The Navy did not contract for their XV-1710-6(E1) engines until March 3, 1939, purchasing two of them to power the Bell Model 5, Navy XFL-1 derivative of the XP-39. Those engines were delivered on January 4 and March 28, 1940.

To expedite development, Allison built what was in truth the first "E" model engine, their EX-3 "workhorse" engine. It was configured as an E-2 and used extensively for development work on the type.[120] This early work was in direct support of both the E-2 and E-1 models.

Subsequent to the October contract with Bell for the XP-39, some 27 pages of revisions were accumulated before the end of the year. These were incorporated into the aircraft specification as of January 12, 1938.[121] Chief among the changes was the specification of the V-1710-17(E2) engine rather than the V-1710-C7 derived V-1710-D3 Mod "A", stated in the original proposal. In addition, the powerplant was to be equipped with the GE Type B-1 turbosupercharger. The Type B-1 being the redesignated Form F-13 turbo that had replaced the GE Form F-10 specified in the X-609 Circular.[122] Both units used the same dimensions for the impeller and turbine wheels, but the physical arrangement of the Type B was improved by cantilevering the impeller and taking the intake air from the front of the unit rather than from between the turbine and compressor. In addition, all of the bearings were now located in a central housing, which required a separate bearing oil cooling capability be added. Furthermore, the Air Corps had recently changed the designation system relative to turbosuperchargers, using the term "Type" instead of "Form", and an alphabetical prefix to identify the horsepower class of the unit. By December 1938 the specified turbo had again been changed, this time to the GE Type B-5, a built-to-purpose unit having four inlets to the turbine nozzlebox, a unique configuration specifically for the XP-39 installation. This greatly simplified the collection and routing of the exhaust gases, as the manifolds were separated into groups of three adjacent cylinders, collected and piped to the supercharger mounted directly below the engine. Each pipe was provided with a waste gate for turbo control, the four being mechanically linked together. Unfortunately, the four waste gates and their discharge nozzles were exposed to the slipstream in a manner that contributed measurably to the drag of the installation.

Engine Development in Support of XP-39

Vibration problems occurred in the XP-39 during its ground running in early 1939, and it took several months to sort those out. Allison initially attributed the problem to the lack of adequate vibration damping in the V-1710-E which was allowing torsional oscillations to be passed through the original friction disk type vibration damper into the accessories section causing the camshafts, distributors and magnetos not to run smoothly. (The solution was to design a combination hydraulic vibration damper and accessories drive coupling to replace the plate type friction damper that had been located between the rear of the crankshaft, driving the accessories.) This worked exceptionally well on the higher order vibrations, and was one of the reasons that the Allison V-1710 has always been revered as a very smooth running engine. Given the nature of torsional vibrations inherent in extension shaft systems, this hydraulic damper was critical to the success of the installation and proved itself in a wide range of other innovative installations in subsequent years. Still, in early May 1939 Allison concluded that the new damper had not resolved the fundamental vibration problem, which they then determined was due to 1/2-order vibrations coming from the light-weight [2-1/2"x5/32" wall] extension shafts. Allison committed to resolve the problem by designing and manufacturing extension shafts that were 30 percent stiffer, and that would fit in the narrow space available in the XP-39. New extension shafts were manufactured and installed in late July 1939 in Allison's EX-3 workhorse engine [configured as a V-1710-17(E2)] and mounted in their rigid test stand to perform a contract required 35-hour test. The test was completed on July 31, 1939 with the new 2-5/8 inch OD extension shafts having 0.200" thick walls and driving a 1.8:1 reduction gear. With this configuration all torsional vibrations were less than +/-1 degree within the operating range. Allison ran this engine for a total of approximately 345 hours in the course of preparing for the 35-hour test of the E-2. These changes solved the torsional vibration problem once and for all.[123] These new extension shafts were not available to install in the XP-39 until it was rebuilt in its XP-39B configuration.

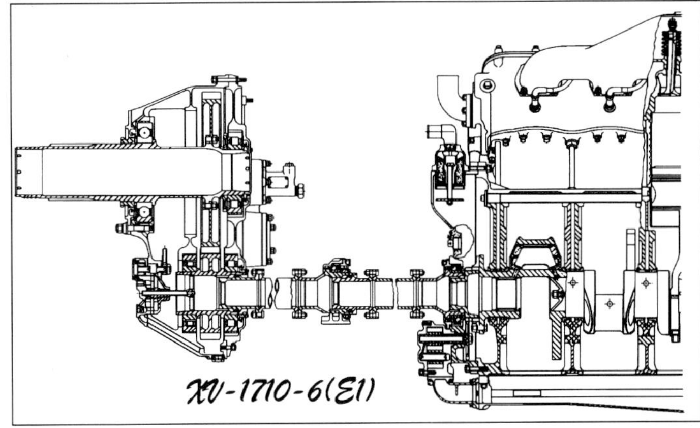

This section of the XV-1710-6(E1) typifies the early extension shaft "E" series engines with the extension shaft to the 1.80:1 reduction gear. The hollow #60 propeller shaft would accommodate the barrel of a 37 mm cannon. Gun synchronizers and other aircraft accessories were mounted on the rear of the reduction gear box. The extension shaft sections were identical to those in the P-39 and coupled through an intermediate bearing assembly. For the E-1 the center coupling was shorter than in the P-39 as the Navy insisted upon an extended front coupling on the engine. In this way the engine could be removed without removing the extension shaft itself. (Author)

With the decision to remove the turbosupercharger from the XP-39, it became necessary for Allison to develop and Model Test a suitable Altitude rated engine. This was done by first converting the two sea-level rated XV-1710-17(E2)'s, with their 6.44:1 supercharger gears, to V-1710-31(E2A)'s. This was accomplished by installing 8.80:1 supercharger step-up gears and the larger PT-13B carburetor as was being used on the altitude rated and contemporary V-1710-19(C13) for the XP-40. No actual V-1710-31's were produced, and in fact, other changes were made that resulted in these two engines being further modified until finally they were configured and served as prototypes for the V-1710-E5. While this was an

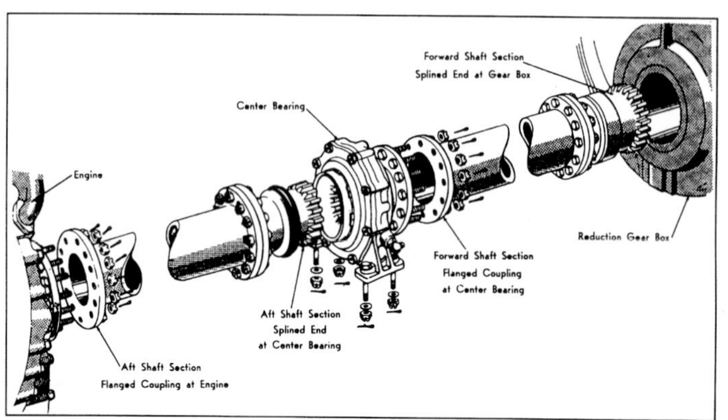

Detail design and arrangement of the extension shaft was fairly conventional, though it incorporated many features to allow for airframe deflection and torsional vibration. The center bearing was commercially available but mounted in a special housing. The system proved to be reliable and able to handle considerably greater loads than the original 1000 bhp. (Matthews)

expedient to get the XP-39B into the air, the engine necessarily suffered a reduction in power from the previous 1150 bhp due to the additional power demanded by the higher speed supercharger it was now driving. The first modified engine was scheduled for delivery to Bell in mid-October 1939, rated for 1090 bhp at 13,200 feet, while the EX-3 workhorse engine was continuing torsional vibration and endurance testing in Indianapolis. This work was to be completed by November 1, at which time 9.60:1 gears were to be tested in EX-3. It was expected that these gears would provide 1090 bhp at up to 15,000 feet.[124]

Allison responded with the desired 1150 bhp in the follow-on V-1710-35(E4) by increasing the engine rated speed from 2950 to 3000 rpm. Unfortunately this model engine was unable to complete the necessary 150-hour Model Test in time for the YP-39's and in the interim, the V-1710-37(E5) was produced specifically for these preproduction aircraft. The E-5 was basically the E-2 crankcase and cylinder blocks derated to a safe 1090 bhp at 3000 rpm for takeoff and military operations up to 13,200 feet. Sixteen of these engines were built to support the thirteen YP-39's. Achieving the 1150 bhp rating required Allison to again strengthen the crankcase in order to successfully pass the 150-hour Model Test necessary before committing the engine to series production. At overhaul, many if not all, of the E-5's were upgraded with the new components, thereby qualifying them for the increased ratings of the E-4. These Modernized engines had an "M" appended to their Air Corps serial numbers.

First Flight of the XP-39

After initial assembly at the Bell factory in Buffalo, the XP-39 was disassembled and shipped by rail to Wright Field, intending to be a participant in the delayed 1938 Pursuit Competition. Delivery was by December 27, 1938, and in time for the completion now scheduled for January 1939.[125] Assembly and final fitting of detail components were complicated by the need to resolve problems with adjustments to the engine and leakage of the integral fuel tanks. These and subsequent problems kept the aircraft from ever getting into the competition.

Taxi testing finally began on March 1, 1939, and then the engine had to be removed for upgrading by Allison.[126] It is believed that it was at this point that the original friction plate type vibration damper in the accessories drive was replaced with the first version of the hydraulic vibration damper. Finally on March 25 both the engine and the turbosupercharger were installed and test running of the installation began. During an engine run on April 5 a slight roughness was noted in the low speed range of 570 to 1400 rpm, and only 2650 rpm could be reached, though rated speed was 3000 rpm. This was found to be due to 1/2-order torsional vibrations being passed into the accessories section and magneto drive when the engine was installed in the aircraft and driving the long flexible extension shaft.

The XP-39 was quite an attractive little airplane. The intercooler housing shows clearly. It was found to be a source of considerable drag, as were the turbo wheel and exhaust waste gates that show in silhouette in photo. Note the "cuffed" prop and high drag tall canopy. (Air Corps)

Overall, it had taken several months for Allison to both literally invent and develop the new type of hydraulic vibration damper sufficiently to complete the contract required engine qualification tests prior to authorizing the first flight. Using the first model of the vibration damper, work went ahead in preparation for the first flight. Active development of the hydraulic damper for the XP-39 continued until July 31, 1939 when the 35 hour qualification test was successfully completed.

After identifying the 1/2-order vibration from the long extension shaft, Allison at first argued that the Army should proceed with flight tests of the airplane and simply minimize operation in the high speed range where the vibration became objectionable. Wright Field objected, noting that should a single cylinder short-out during high power operation that 1/12 of the power of the engine would be contained in the vibration, a condition they maintained as unsafe. This was confirmed by Allison on their workhorse engine when they demonstrated a violent shaking when the event was simulated. As a result Allison designed new extension shafts that were 30 percent stiffer by increasing them from 2-1/2 inch diameter and 5/32 inch (0.156 inch) wall to 2-5/8 inch diameter and 0.200 inch wall. This, in combination with the further improved hydraulic damper, raised the system natural frequency and better dampened torsional vibrations so that there were no serious resonant speeds for torsional vibration encountered from 600 to 3100 rpm.[127] The new extension shafts were not available prior to the airplane being sent to the NACA for wind tunnel testing, further suggesting that the airplane was never allowed to attempt a full power performance demonstration of its capability before being rebuilt as the XP-39B.

Finally, on April 6, 1939, the XP-39 flew for twenty minutes with Bell test pilot James Taylor, USNR, at the controls. The engine incorporated the hydraulic vibration damper, but not the stiffened extension shaft, so care was taken to minimize operation of the engine in the high vibration ranges. John Kline, the Allison man assigned to the XP-39 relates that, "We had the G.. Damn'st Party that night at the Biltmore that you ever saw!"[128]

XP-39 Performance

It is clear from recently located NACA test reports on the XP-39 that it was not meeting the contracted performance guarantees.[129] While it has been reported that the aircraft was able to climb from takeoff to 20,000 feet in five minutes, and that the maximum speed at that altitude was 390 mph, with the airplane weighing 5,550 pounds, the data does not show it. Birch Matthews, aviation historian and author who has researched Bell Aircraft and the P-39 extensively, reports that he has never found a source document confirming this maximum speed. Furthermore, given that General Arnold was hurriedly arranging to have the NACA put the airplane in its wind tunnel for drag reduction tests only a month after the first flight, suggests that all was not well. During the initial flight testing several problems were identified and quick fixes incorporated for the engine and oil cooling systems by adding/revising scoops for the inlets, as well as enlarging the turbosupercharger inlet.

Another factor in the poor performance equation was that the airplane was also overweight. The original contracted weight was 5,550 pounds, which had increased to 5,855 pounds even before the airplane was finished. When the Air Corps officially weighed it in at Wright Field it grossed 6,104 pounds. Being ten percent overweight had to hurt performance.[130]

As a consequence of these experiences General Arnold directed on April 21, 1939 that:

> "... pursuit airplanes being procured in the Emergency Program must reach the maximum performance attainable in order to approach foreign developments in pursuit, and have a sufficient speed differential over foreign bombardment airplanes. All elements which reduce performance without a commensurate (improvement) in war effectiveness must be eliminated from the structure. With the above in mind, the following controlling factors will govern in the modification of present models such as the P-39 and any other models or design submitted in response to Specification C-616[131]:

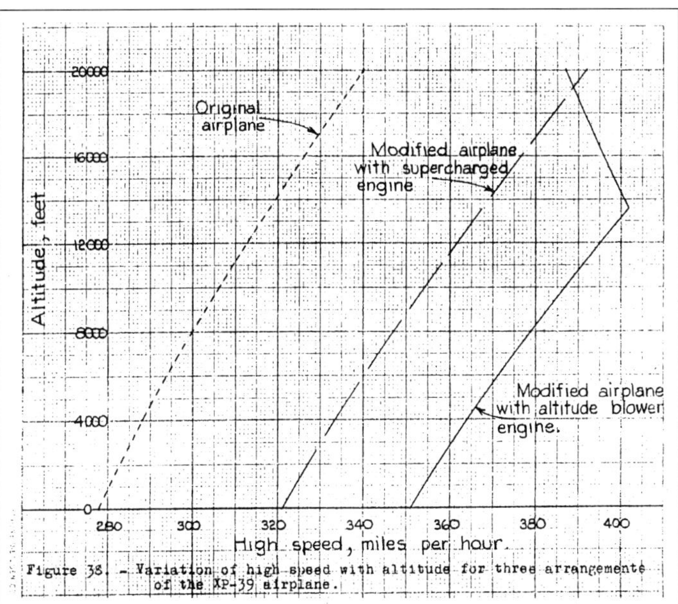

Published in September 1939, the NACA test results on the XP-39 gave convincing evidence that the only hope for reaching General Arnold's goal, and Bell's guarantee, of 400 mph was with the new 1150 bhp Altitude Rated Engine. (NACA)

The XP-39 was placed in the NACA Full Scale wind tunnel not long after its first flights. Following testing as received each drag producing proturbance was cleaned up. The result was the recommendation to delete the drag producing systems for the turbo and the turbo itself. When this photo was taken a considerably reduced intercooler inlet was being tried. (NARA)

a. The fuselage will be streamlined to accommodate a pilot not over five feet eight inches tall and weighing not more than 160 pounds.

b. The critical altitude of the airplane will be approximately 15,000 feet. This is the approximate critical altitude of the bulk of foreign airplanes, and as a general guide, our pursuit will have its optimum performance at the critical altitude of the bombers of our potential enemies.

c. No consideration will be given to the baggage compartments if this adversely affects performance.

d. The design fuel load will be for one hour at full throttle at the critical altitude.

e. Eliminate any requirement for individually lighted instruments.

f. Flaps will be manually operated.

Furthermore, every effort must be made to obtain a high speed at 15,000 feet in excess of 400 miles per hour. To this end the (X)P-39 will be placed in the N.A.C.A. tunnel at Langley Field, Virginia, immediately upon completion of its acceptance tests, which tests will be expedited."[132]

In June 1939 the XP-39 was transferred to the NACA full scale wind tunnel at Langley Field, where NACA engineers did considerable work to first quantify the sources of drag, both internal and external, and then modify the aircraft and recommended changes so that it could achieve its guaranteed performance.

Key findings were that the Prestone radiator, which was installed in a duct in the left wing, had inlets and outlets which were obstructed by wing spar web members and the outlet was further restricted by a bump formed for the wheel well and landing gear actuating linkage. The 13 inch diameter oil cooler was installed in a duct on the right side of the airplane, and the Army had fitted a sharp edged inlet to enable the ship to be flown safely. The intercooler was originally in a duct on the left side of the fuselage, but the shape of the inlet to the duct was abrupt and sharp edged, factors relating to the high drag count. Tuft observations made in the wind tunnel showed that the oil cooler, intercooler, carburetor air intake and radiator duct produced disturbed flow in the wing roots. The stall occurred in the left wing root behind the radiator duct several degrees of angle of attack before occurring on the right wing root.

Flight testing had identified considerable problems in achieving adequate cooling of oil, Prestone and the intercooler. NACA found that though the V-1710-E2 required 10,250 scfm to properly cool the radiator, the original installation provided only 7,880 scfm when climbing at 160 mph, and 16,900 scfm at 350 mph. Too little during climb, far too much at high speed. The intercooler cooling flow was only 1,600 scfm at high speed, while 5,000 to 7,000 was needed if the engine was to achieve full power without detonation. As received, the NACA determined that the maximum speed of the aircraft was estimated to be 340 mph at 20,000 feet, with the Allison putting out 1150 bhp.

Table 3-8
XP-39 Drag Distribution

Component	Drag Count
Turbosupercharger	0.0033
Waste Gates/exhausts	0.0014
Air Intake	0.0019
Intercoolers	0.0007
Oil Cooler	0.0040
Prestone Radiator	0.0024
Landing Gear	0.0019
Excessive Cooling Drag	0.0015
Gun Ports	0.0006
Canopy	0.0004
Air Leaks	0.0004
Subtotal	0.0185
Total Aircraft	0.0329

Following the several modifications tested by NACA they were able to considerably reduce the drag to the point where the turbosupercharged airplane was estimated to be capable of 392 mph at 20,000 feet, again using 1150 bhp from the engine.

The NACA went on to consider using the new Allison V-1710-E4 altitude rated engine, again with 1150 bhp but at a lower critical altitude. This would allow elimination of the turbosupercharger and intercooler installation and make possible a material reduction in the airplane drag. They did not take into account the effects of reducing the weight that removing the turbo would produce, and still the high speed was now expected to be 402 mph, at 13,200 feet. Furthermore, this configuration at 20,000 feet was expected to give about the same performance as the NACA modified turbosupercharger installation would have.[133] The major NACA recommendations were:[134]

- Improve streamlining of wheel well doors.
- Lower the canopy to improve streamlining.
- Remove the turbosupercharger due to high installation drag.
- Relocate carburetor scoop from left side of fuselage to just behind the canopy.
- Install the altitude rated engine.

In light of the directions from General Arnold, coupled with the critical and immediate need for fighters able to at least match foreign aircraft performance at about 15,000 feet, it should be no surprise why the decision was made to eliminate the turbo from the P-39 and get on with production.

The above findings and recommendations were unfortunate for the turbosupercharged XP-39, but probably saved the production program. The original concept of Specification X-609 was negated by a series of decisions by the Bell design team and the new Air Corps Project Officer assigned to relieve Lt. Kelsey of the concurrent responsibility for both the P-38 and P-39, fighter projects. It has been reported that Capt. Ben Kelsey, as Chief of the Pursuit Branch, was vociferous in his arguments *against* removing the turbo from the XP-39.[135] He then seemed to have had a change in heart, for in May 1940 he wrote a letter regarding the future use of Exhaust Turbosuperchargers on Pursuit Aircraft.[136] In it he noted that while some work had been done, and more was proposed, to study the fundamentals of their proper location within an aircraft and by that minimize drag and weight of the installation. Furthermore, he argued that future pursuit aircraft were going to weigh approximately 10,000 pounds and be powered by 2000 bhp engines; thus, their performance even above their critical altitude would still be substantial, even without the turbo. He also states that the turbine supercharger is "exceptionally vulnerable and the normal installation distributes this vulnerability over a large part of the airplane volume." He recommends that "more returns might be realized by a more thorough investigation of other means of supercharging."[137]

To the credit of the subsequent altitude rated models of the P-39, they were quite successful at the low and medium altitudes where their altitude rated Allison's could press the attack to the advantage of their heavy aerial cannon. Conceived as they were as interceptors, and given their small size demanded by the original time-to-climb criteria, the limited fuel and am-

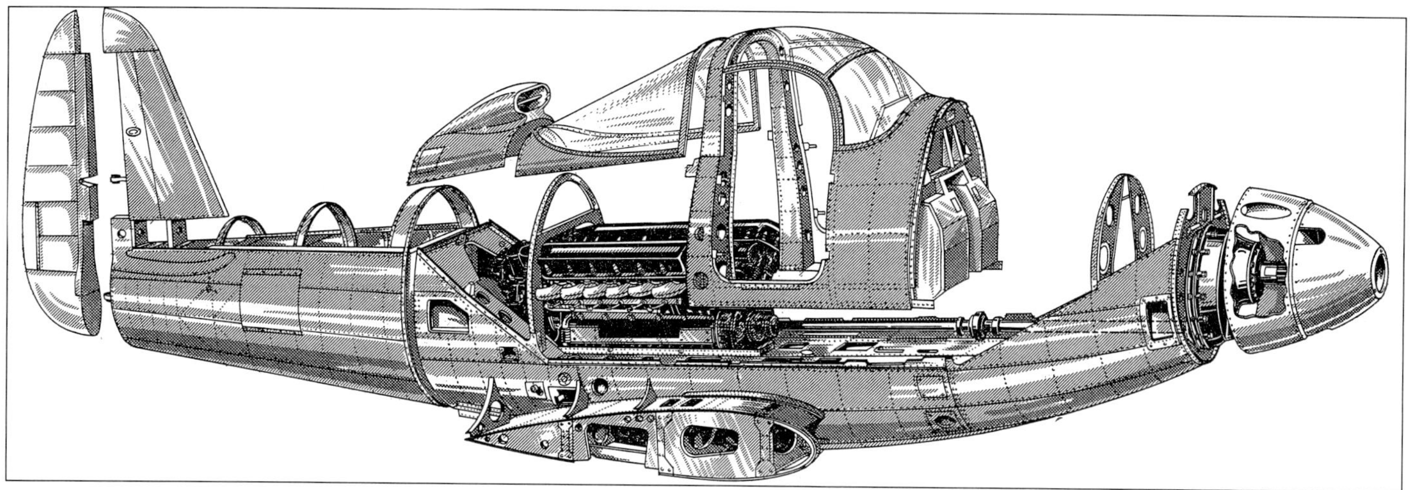

The P-39 fuselage was a lightweight though robust structure able to support the engine and extension shaft and minimize vibrations. Even though designed to limit drive line deflections to less than +/- 1 degree, the propeller centerline could move through a range of 3.64 inches. The pilot was well protected in the event of a crash by the robust structure. (AVIATION, Feb. 42)

munition loads restricted their use to Army ground support and local area air supremacy. Even so, they served well in this capacity in the Southwest Pacific Area and Russian Front until larger and more powerful aircraft could replace them in the later years of W.W.II.

P-39 Engine Installation

The use of the extension shaft arrangement created some unique challenges to Bell in the supporting structure they designed for the P-39. Specifically, they intended to avoid the vibration problems that had plagued their X/YFM-1 by building a robust structure. Power shafting is always subject to vibration, particularly if there is any miss-alignment, imbalance or excessive flexibility in the system. Consequently Bell designed a very stiff structure to support the engine and to minimize deflection under severe loads on the reduction gear, some ten feet ahead of the engine. The design could keep deflections to within +/-1 degree of the crankshaft centerline. Even so, this allowed to a total travel of 3.64 inches! Furthermore, the long shaft would "twist" 5-3/4 degrees when the engine was operating at its full 1150 bhp. The actual extension shaft used two identical sections 48-9/16 inches long with a 7-7/32 inch long center support/coupling mounted to the airframe structure. In their developed form each shaft was 2-5/8 inches in diameter and had a 0.200 inch thick wall. Overall length of the extension shaft was 104-11/32 inches.

The pilot sat astride the shaft housing and often had well over 1000 bhp passing between his knees. Despite fears expressed by many engineers during its introduction, the extension shaft of the *Airacobra* and later *Kingcobra* is not known to have suffered a single failure from torsional vibration, the usual Achilles' heel of such installations.[138]

Three sets of splined couplings, and a commercially available double row self-aligning ball bearing on the center bearing mount, took care of any misalignment. A splined "floating" coupling was employed to connect the extension shaft to the crankshaft. The other two splined couplings were located at the center bearing and at the reduction gear pinion. Self aligning bearings were mounted in the front cover plate of the engine and at the mid-span coupling.

Design and reality are not always the same thing. The case in point is how the actual airframe had to be adjusted and the engine and extension aligned to have the system run true in operation. John Kline, the Allison engineer assigned to Bell for the XP-39, relates:

> "On the initial XP-39 we made the alignment of the engine and extension shafting with the airplane on its load points while sitting on jacks, in the hangar. Everything checked out perfect. So they put it down on its gear and it was completely out of line! So the engineers had to figure out how much deflection to take into account when you set it up in the factory so that it would be just right when it was down on its wheels. Essentially what they did was to 'miss-align' it."[139]

The Service Test YP-39

On December 14, 1938 General Arnold sent a Secret communication to the Chief of the Materiel Division at Wright Field authorizing immediate action to purchase three each P-38 and P-39 airplanes. Each purchase was to include an option for an additional seven airplanes, but only one of the two types. The selection was to be based upon the results of tests of the three experimental articles.[140]

Engine compartment of Bell Model 12, YP-39, November 1940. The engine is the V-1710-37(E5). (Bell via Matthews)

The Air Corps defined its requirements for the Service Test YP-39's in Specification C-616, which because of the rapidly changing world conditions and the Congressionally sponsored "Emergency Program" for accelerating aircraft purchases, resulted in an increase in the "Service Test" quantity to be thirteen airplanes. The Air Corps issued contract W535-ac-12635 for $1,073,445 on April 27, 1939, for 12 YP-39's and one YP-39A. These were to be powered by the turbosupercharged V-1710-E3 and V-1710-31(E2A) altitude rated engine (without a turbosupercharger) respectively. The YP-39A was intended to be the first of the service test aircraft to be delivered.[141] With the subsequent decision to revamp the turbosupercharged XP-39, the 12 YP-39's, along with the single YP-39A, were all reconfigured to use the production version of the altitude rated "E" engine, the V-1710-E4.

Allison then had considerable trouble in getting the E-4 engine through its Model Test at 1150 bhp. The problems were primarily due to the need to reinforce specific portions of the crankcase to prevent cracking. As a result of the need for engines for the YP-39's, an interim model was defined, the V-1710-37(E5). It had all of the features of the E-4, but was rated at 1090 bhp, the same as the contemporary V-1710-33(C15).

Engine Installation and Cooling

Both the XP-39 and YP-39 were intended to have minimum drag streamlined inlet and exits for the various cooling systems for the engine. In the original XP-39/XP-39B the coolant radiator was fed cooling air from a single large leading edge scoop in the left wing, and there was a separate scoop for the oil cooler in the right wing. This was revised in the YP-39's and subsequent production P-39's where there were two separate leading edge inlets in each wing, adjacent to the fuselage. The inboard scoops fed a single coolant radiator mounted on the aircraft centerline and exiting through the bottom of the fuselage. Each outboard scoop fed a separate engine oil cooler, and also exited beneath the fuselage. Shutters on the outlet of each of the three ducts controlled the flow of air through the ducts, and thereby the temperature of the fluids being cooled.

The turbosupercharger intercooler was unique to the XP-39, and had been found to be a major source of drag during the NACA wind tunnel tests. Those tests determined that the Bell installation created a relatively high pressure drop across the intercooler radiator. This was a consequence of its small size, a factor intended to reduce the weight of the installation. As a result of this approach the NACA found that the required airflow varied from 5,000 to 7,000 cubic feet per minute, and that the cooler would be unable to meet the cooling load. Most intercoolers of the day were intended to remove half of the heat of compression added by the turbo, but in the XP-39 the intercooler was only able to remove 25 percent of the heat added during high speed flight, and about 12 percent during climb.[142] The result of this reduced performance would have been high induction temperatures and the danger of detonation damage to the engine during high power operation. This issue alone, and the difficulty overcoming it within the small airframe, may have been sufficient reason to warrant deletion of the turbosupercharger from the P-39.

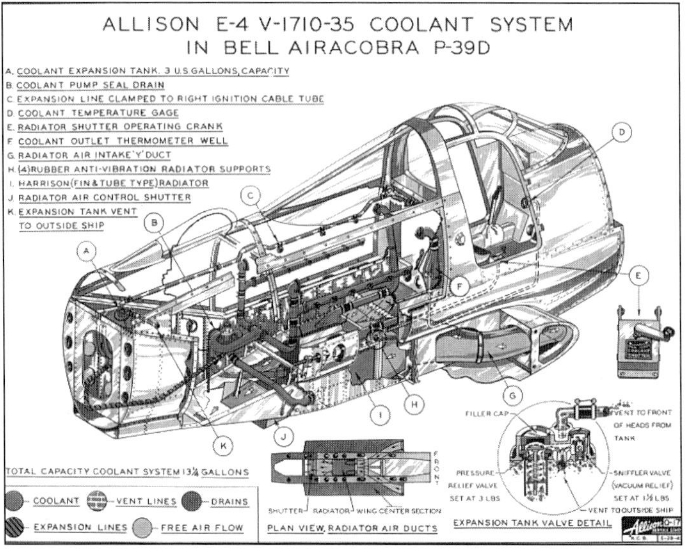

The cooling system in the P-39 was quite compact since the engine was directly above the radiators. Leading edge ducts delivered air to the radiators, with a shutter on the bottom of the aircraft controlling airflow, and thereby, coolant temperature. (Allison Service School)

Summary

Equipped with an altitude rated engine the P-39 was not the aircraft originally intended in the Interceptor circular, X-609. It was the maneuverable aircraft originally intended, but its small size was a problem when the tactical circumstances involving long range and heavy combat loads were imposed. The steadily increased ratings available with later V-1710's were hard pressed to compensate for the increased weight of armor and armament that were included in the later models of the aircraft. Still, for missions within its capability, the aircraft did quite well. Those missions were not those intended in X-609, but rather low and medium altitude support of ground forces, most notably on the Eastern Front in Russia.

There were subsequent efforts made to turbosupercharge the P-39, though none of the proposals were an enhancement of the basic aerodynamics of the aircraft. In mid-1941 Bell built a flying mockup of two configurations that consisted of a saddle package installed on top of the fuselage aft of the cockpit. Neither offered any promise and in fact, cost in terms of high speed. As the Air Corps and Bell were busy working on the forthcoming XP-39E and XP-63, powered by the new two-stage Allison engine, a turbosupercharged P-39 became irrelevant.[143]

The P-63 Kingcobra was to resolve the shortcomings of the low altitude rated engine and build on the strengths of the P-39. It was larger, giving more range and load carrying capability. It used a modern low drag laminar airfoil, and incorporated a two-stage supercharged Allison engine that provided the desired high altitude capability. The story of its development is beyond the scope of this chapter, but suffice to say that it met its operational objectives. Thousands were employed by the Russians on the Eastern Front in W.W.II.

A few P-39's remained flyable after the war and demonstrated new uses for the V-1710. This included those used in the stripped down P-39 Cobra racers, where Cobra II succeeded in winning the 1946 Thompson against Mustangs, Lightning's and Corsair's, at a new closed course record speed of 373.9 mph.

Bell XFL-1 *"Airabonita"*

Official Navy interest in a liquid-cooled inline engined aircraft first surfaced in a Bureau of Aeronautics, memorandum in November 1937. BuAer was becoming concerned that the high speed limit had just about been reached for radial engined fighters.[144] Larry Bell was ever the salesman and his actual role in these events is only speculative, but he appears to have convinced the U.S. Navy to allow a liquid-cooled engine into their September 1938 specification SD-240 fighter competition. It was probably not too difficult to get the Navy's interest, given the performance being demonstrated overseas by the then current generation of inline engined fighters, and with the new Curtiss XP-37 and Bell XFM-1 already in the air.[145]

XFL-1 Fuselage during construction with the first XV-1710-6, BuNo. 4310, AEC #88 in-place for initial fitup. Note the collector type exhaust manifolds. (Bell photo via B. Matthews)

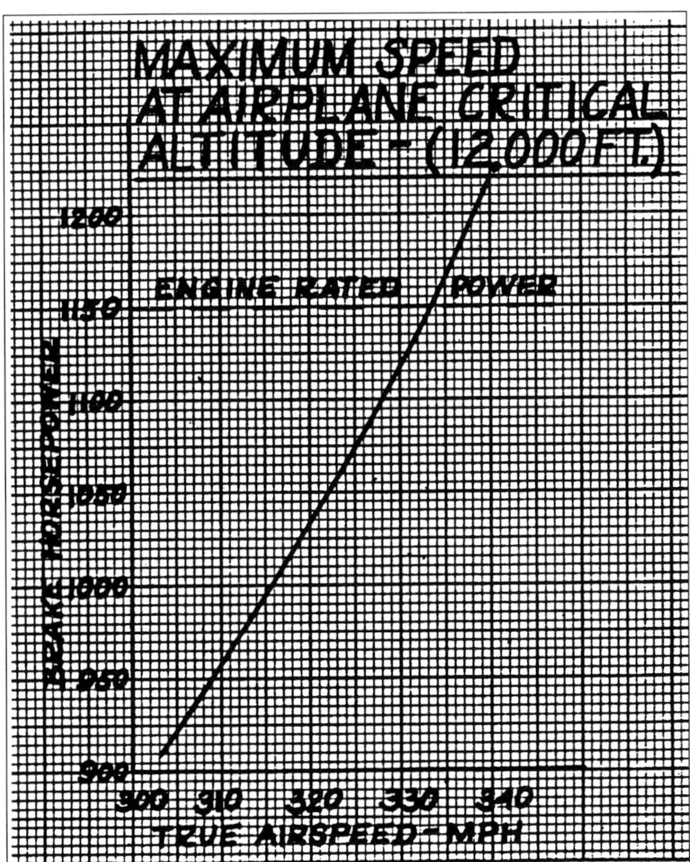

Compare the speed at Critical Altitude expected from the XFL-1 to that from the XP-39B when both were equipped with 1150 bhp altitude rated engines. (Navy via Matthews)

When specification SD 240 was issued it required a maximum power-off stall speed of 70 mph, 1000 mile range, three gun armament and offered "consideration...to a liquid-cooled engine installation provided a material increase in performance over the air-cooled engine can be shown."[146] Bell then became one of five companies submitting proposals in response for a new type VF (heavier-than-air, Fighter) single-seat fighter airplane. After a thorough evaluation of the ten designs that were submitted, the Bureau of Aeronautics carefully defined rating system placed the Vought "B" design (which was subsequently developed into the F4U Corsair of W.W.II fame) as the winning design. An Allison V-1710 powered development of the Brewster *Buffalo* came in second, and a Grumman design was third. The Allison powered Bell Model 5 finished sixth overall.

BuAer documents state, "While the disadvantages attending the introduction of a liquid-cooled engine into the Naval service are realized, it is our opinion that, if high speed in single-engine VF can be obtained only by use of liquid-cooled engines, this step must...be thoroughly explored."[147]

Following initial analysis of performance, design features, deck handling, price, and other factors, the Assistant Chief of BuAer on May 10, 1938 reported in a letter, the decision to award Vought a contract for their "B" proposal and award the twin-engined VF design to Grumman, to become the XF5F-1. The letter also recommended that the Navy drop consideration of the Bell single-engined design due to the low performance of the Bell with the present Allison engine. This conference decision apparently did not prevail in the final award process, as an earlier May 1938 directive from the Plans Division to the Bureau Chief set the tone for the ultimate determination of the 1938 VF experimental program. The Plans Division fully agreed with the first two choices of Vought and Grumman, but said, in addition, that "favorable consideration should be given to an award for the Bell single-engined design using the new Allison 15,000 foot engine."[148, 149] The Bureau Chief then set in motion the creation of three new Navy carrier fighters in mid-1938: The Vought XF4U-1, Grumman

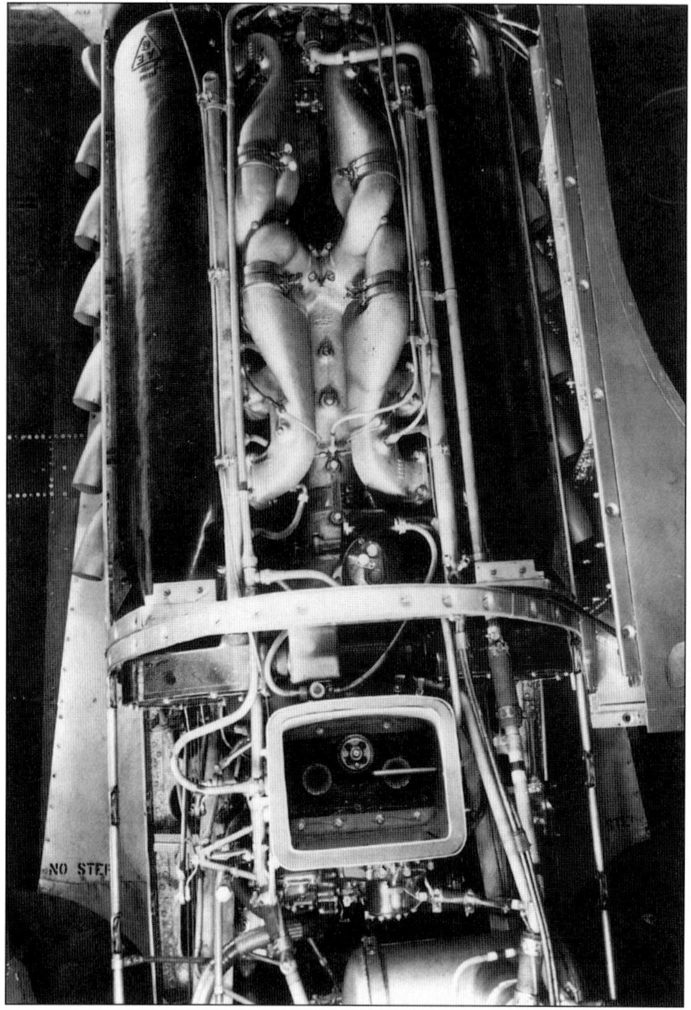

Unique view of XV-1710-6, AEC #89, installed in XFL-1, BuNo 1588. Note use of individual ejector exhaust stubs that first appeared when this engine was installed. (Navy via H. Andrews)

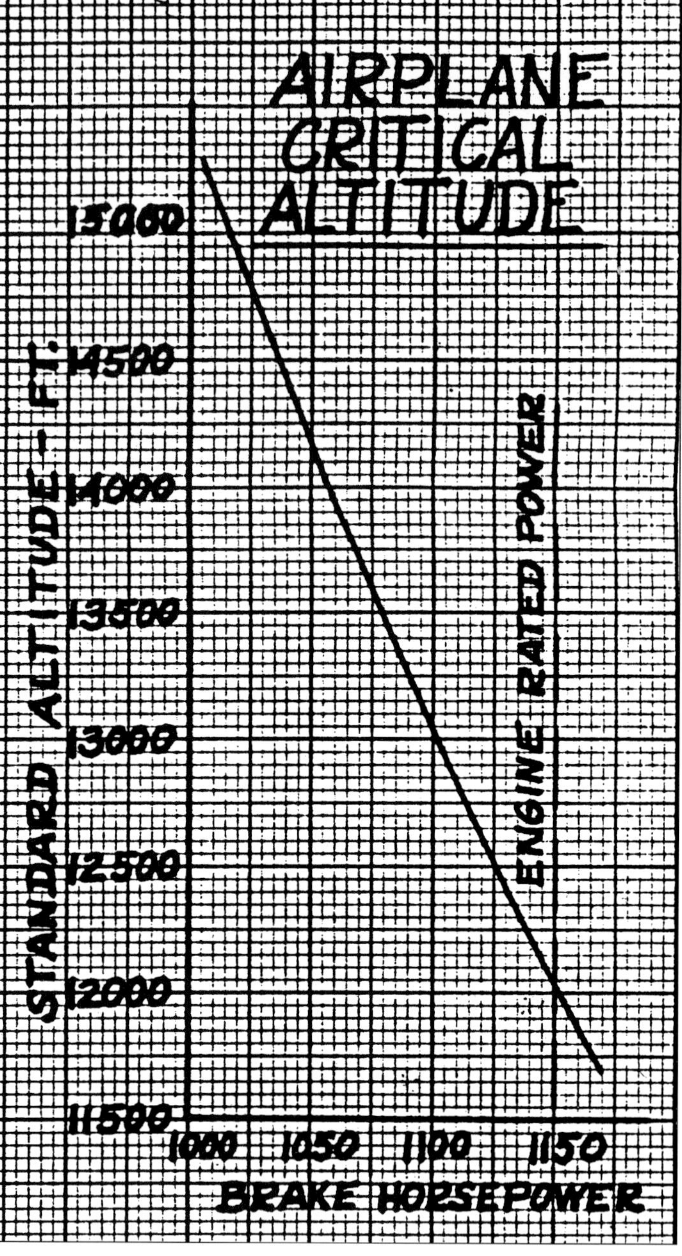

At above the 12,000 foot Critical Altitude for the XFL-1 considerably less power was available from the XV-1710-6(E1)and performance was impacted. (Navy via Matthews)

XF5F-1, and finally, the undeniable longshot, the very sleek Bell Model 5, which they designated as the XFL-1 (BuNo 1588).[150]

The Navy finally agreed to contract for a single aircraft for testing on their contract Number 63629, dated November 8, 1938. The contract was contingent upon equipping the aircraft with conventional undercarriage, and having the wing area increased over that of the P-39 to assure a landing speed not to exceed 70 mph.[151] The XFL-1 wing measured 220 square feet when it began flight testing.[152]

On December 19/20, 1938 a U.S. Navy team evaluated Bell's full scale mock-up of the XFL-1, noting many minor improvements to be incorporated. The XFL-1 design and fabrication progressed slowly, and the aircraft weight increased measurably. Furthermore, as the project developed it became evident that the contract was funded well below the realistic needs for development by a company working to develop three new and advanced fighters at the same time.[153]

In December 1939 it was determined that the aircraft was considerably overweight, even without including an extra 25 pounds from the engine and remote gearbox. At the same time there was not space in the airframe to accommodate a more powerful two-stage Allison engine to compensate.[154] This burden essentially dealt the XFL-1 a fatal blow concerning any production plans for the aircraft.

The specified Allison XV-1710-6(E1) engine was a major problem for the design team. The first engine, AEC #88, BuNo. 4310, was not delivered to Bell Aircraft until January 4, 1940, more than four months after the contract called for delivery of the finished prototype airplane.[155]

The Navy had originally wanted to purchase three engines for the XFL-1; one for installation, one as a spare, and one for calibration and Type Test. This latter engine was needed as the Air Corps had no similar model scheduled for calibration and Type Testing. When Navy Contract No. 65197 was let to Allison on March 3, 1939, it was only for two XV-

The XFL-1 was a Pretty Bird, and appears in photos to be quite different than the baseline XP-39B, but the centerline distance from prop to engine were the same on both airplanes. Much of the difference in appearance is due to different cooling arrangements. Because of conventional wheel layout the XFL-1 had to adapt radiators in ducts below the wing. (Allison)

1710-6(E1) altitude rated engines for the XFL-1 program. Guaranteed static performance was 1150 bhp at 2950 rpm up to 9,000 feet. When tested 1150 bhp was obtained with 3000 rpm at up to 12,000 feet, including the effect of high speed ram air.[156]

It had been decided that due to cost (Remember, 1939 was still during the Depression. In fact these events occurred during a further downturn known as the "Roosevelt Depression") and other considerations, to not perform a Type Test on this engine until the airplane design was evaluated and a decision made regarding its use in production quantities.[157]

Due to the bonus/penalty clause in the contract, the Navy felt that it was very desirable to obtain an altitude calibration of the engine in order to be on a more sound basis in evaluating the airplane performance at the critical altitude. At the time the contract was being finalized it was felt that the desired calibration could be accomplished in a period of about two weeks by using the "spare" engine. This engine, AEC #89, BuNo. 4311, was delivered to the Navy on March 28, 1940 and run through a complete calibration by the Aeronautical Engine Laboratory, located at the Naval Aircraft Factory, Philadelphia, PA.[158] Testing occurred between May 9, 1940 and July 5, 1940, far exceeding the original two week estimate. During the calibration and testing the engine accumulated about 109 hours total time, and had been completely disassembled for inspection at the end of the test. Inspection showed the engine to be in generally excellent condition, so it was overhauled and returned to "spare" status.[159]

Many changes to the basic E-1 engine had been necessary to meet the original Navy needs and requirements. Compared to contemporary altitude rated Air Corps E-4/E-5 models being used in the P-39; the Navy used the same 8.80:1 supercharger gear ratio, but a larger PD-12G1 carburetor in place of the PD-12B4, different valve timing and a Cartridge type starter. Another change was in the overall length of the extension shaft itself. The XFL-1 used two of the same 48-9/16 inch long sections as the P-39, but the center support and coupling was revised to be only 6-11/32 inches long, making the overall length of the extension shaft 103-15/32 inches.[160] This was done to make engine changes possible without first removing the extension shaft, and in turn it required an extended coupling face to be mounted on the engine crankshaft. The total distance between the engine's vertical centerline and propeller were the same on both the P-39 and XFL-1.

An interesting observation by the Navy test personnel was a comparison of the V-1710 characteristics in contrast to the radial air-cooled engines they usually dealt with. They found that the "peaking" tendencies of air flow and power at the higher speeds (as prevalent with 2-valve type radial engines) were not present, due principally to minimum restriction of intake air and exhaust gas flow afforded by the 4-valve cylinder heads, large intake manifolds, and exceptionally good efficiency of the supercharger at the higher speeds. Furthermore, adequate cooling of the cylinders permitted running at near "best power" fuel/air ratios (requiring no excess fuel for cylinder cooling, as is usually necessary with air-cooled engines) at or near rated and take-off conditions as well as at the lower power conditions.[161]

First Flight

On May 13, 1940, during a high speed taxi test at Buffalo a sudden gust of wind lifted the XFL-1 from the runway on what became an unintended first flight. A flight considerably complicated by the inadvertent inflation of the "flotation" bags intended for safety if forced down at sea. On August 15, 1940, after accumulating a total of 32-1/4 hours of flight, the aircraft was grounded to investigate an oil leak, which had been coming from the front bearing cover for some time.[162] As a result, Allison installed new bearings, probably needed as a result of poor oil cooling. The engine had accumulated a total of 58 hours and 28 minutes of operation up to this point. Flight testing resumed on October 3, and following a flight on October 4, the 57th flight and one that had gone to 35,000 feet, it was discovered that the Cuno filter contained a copious quantity of bronze flakes from the engine bearings. The new bearings had been ground to a fine bronze dust, necessitating an engine change.[163] The engine had accumulated a total of

This photo of the XFL-1 shows the aircraft at the time when XV-1710-6, AEC#89 was being installed. (Matthews)

83 hours, 55 minutes, of which 33 hours and 25 minutes were flight time.[164] The recently overhauled AEC#89 engine was then installed.[165] During the standdown for installation of the new engine the radiator shutters were reworked. Flight testing resumed at Buffalo Airport on January 1, 1941.

Engine Installation and Cooling

Initial ground testing of the aircraft in early May 1940 revealed engine cooling difficulties that were to plague the aircraft throughout its four year career. Much of the focus was on engine oil cooling and attempted fixes including increasing the size of the coolers, though to no avail.

The Navy requirement to build the *Airabonita* with conventional landing gear necessitated retracting the wheels into the area of the wing where the cooling ducts were located on the P-39. The cooling radiators were then mounted in tunnels below the wing, an arrangement being used in contemporary models of the Supermarine Spitfire and Messerschmitt Bf 109. The cooling problem was found to be disturbed airflow in front of the radiator ducts. Conditions were so bad that reverse flow was found near the outboard side of each tunnel, a condition that was accentuated in the starboard unit due to rotation of the slipstream from the propeller. Guide vanes were tried as a way to improve the flow, but for ground running it was usually necessary to spray water into the radiators. Several months were spent altering the inlet and outlet geometry of the cooling tunnels in an effort to seek a permanent solution.[166]

The original arrangement had the below wing tunnels smoothly faired into the lower surface of the wing, just below the flap hinge line, and the heated air exiting through large ports on the upper wing surface. Each tunnel had a splitter to divide the inboard section for the oil cooler from the larger outboard radiator section. Separate controllable outlets were provided for each. A series of modifications were effected that by October 1940 had resulted in deepened tunnels discharging straight through, all on the lower wing surface.

In July 1940 the coolant tunnels were revised to allow a portion of the oil cooler air to discharge through slots in the upper surface of the wing, with the balance passing straight through the coolers, the same as the coolant radiators. This arrangement was changed in August when the Army standard 9 inch diameter, 12 inch long oil coolers were used in reshaped and deeper tunnels. Satisfactory oil cooling was finally demonstrated in October of the year using this configuration.

The original concept for engine exhaust manifolds had all twelve ports on each bank connecting into a single manifold and discharge port. This was not how it was built. Rather, the airplane was first delivered with collector manifolds for each group of three cylinders. By October 1940 these had been replaced by the more usual P-39 type of ejector exhausts, one for each cylinder.

Another area receiving considerable evolution during the flight testing was the carburetor air intake for the engine. When first flown a very simple, and low profile, scoop was mounted behind the cockpit glazing and over the carburetor. This was found to be quite inadequate and was soon replaced by the production P-39 unit, which was fairly sophisticated in that its shape provided a necessary boundary layer splitter function and considerably improved recovery of ram pressure into the carburetor.

Summary

There were a lot of problems with the *Airabonita* project, the result being that the XFL-1 never achieved the promise Bell or the Navy had for it. The success of the Bell P-39 project, and lack of adequate capital for development and expansion of Bell's engineering and manufacturing capacity had to effect the program. The Navy's insistence on a conventional landing

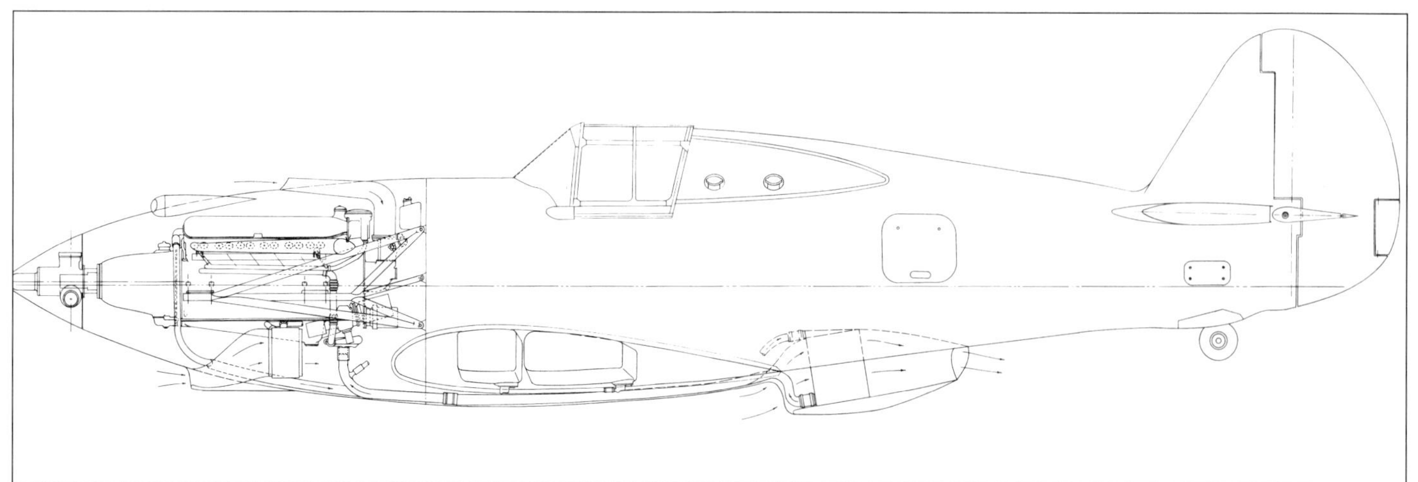

XP-40, as first flown with rear located radiator. It was powered by the 23rd V-1710 built, the XV-1710-19(C13), and had a Military altitude rating for 1150 bhp up to 10,000 feet and 1000 bhp up to 13,500 feet, using 2950 rpm. The altitude rating resulted from using 8.77:1 supercharger gears and the larger three barrel PT-13B carburetor from the 2300 bhp XV-3420-1. (Author)

This is the XP-40 as configured at the time of its first flight. Note the aft radiator and oil cooler inlet below the engine. The short carburetor air intake is just visible on top of the engine and the single nozzle from the collector exhaust manifold is apparent. (Air Corps)

gear arrangement, coupled with Robert Woods success in convincing the BuAer that as detail design progressed a known aft center of gravity condition would be resolved, resulted in an airplane that was unmanageable unless heavily ballasted. This situation was further aggravated by the requirement of the Navy to use the 23mm Madsen cannon rather than Air Corps 37mm cannon. The resulting weight shift had a serious detrimental effect on performance.[167]

The XFL-1 represented a transition in Naval thinking. Not just in consideration of a liquid-cooled engine, but importantly in getting the Navy to consider the use of tricycle landing gear on a carrier aircraft. Bell's lobbying in this regard resulted in a BuAer memo to Bell, dated October 12, 1939 stating "The contractor is requested to submit to the Bureau a study of the practicability of incorporating tricycle landing gear in airplanes of this basic design." Bell's initial response was a hook equipped XP-39B to expedite testing. Investigations found the airframe structure of the XP-39B was not adequately stressed for carrier use. This resulted in the *Airabonita* having a considerably different airframe structure from that of the XP-39B.[168] This revised structure was needed to accommodate the conventional landing gear arrangement that contributed significantly to the endemic cooling problems of the XFL-1. There simply was insufficient internal volume in the airframe to accommodate the radiators and oil coolers, with the result that they were housed under the wings, similar as to those on some European aircraft. As provided, it was found that the details of the design were restricting sufficient flow to the coolers. Resolving these problems required most of the protracted development effort spent on the project.[169]

The XP-40 inflight during the early fall of 1938, while in its first configuration. The short carburetor scoop is clearly visible, as are the collector type exhaust manifolds, as well as the ventral location of the coolant radiator. (Air Corps)

But Bell was not yet through with the Navy. On February 17, 1941 Bell sent the BuAer a specification for a new Navy fighter, the Bell Model 21. This proposal not only included tricycle landing gear (to resolve the persistent center of gravity problem of the XFL-1) and folding wings, but also the new two-stage supercharged Allison V-1710-E11 engine.[170] This would describe a transitional model along the lines of a navalized XP-39E. The XP-39E was Bell's Model 23, the first to mount the two-stage Allison. The XP-63 was their Model 24, the first Bell aircraft with a laminar flow wing, and also the two-stage engine.[171] The proposed Navy project did not proceed.

Curtiss XP-40 and P-40/B/C/G *"Tomahawk"*

Don Berlin, Chief Designer at Curtiss for the P-36 and P-37 aircraft, was frustrated by the continuing problems with the turbosupercharged Curtiss XP-37 [Models 75I and 80], and the lack of potential for the P-36. Given the urgency of the upcoming 1938 Pursuit Competition, he obtained from Allison an estimate of cost and performance of an "altitude" rated V-1710 for use in a P-36 derivative.[172]

In February 1938 Allison wrote to the Chief of the Materiel Division at Wright Field noting that for some time they had been receiving requests from aircraft designers about the possibility of an Altitude Rated V-1710. As such they had proposed, as early as September 1937, providing 1150 bhp from their altitude rated V-1710-C11 for fighter/pursuit airplanes. It would have had 8.77:1 supercharger gears and be rated for 9,000 feet. In January 1938 the V-1710-C12 was proposed for attack planes. It was to have 7.33:1 supercharger gears and a critical altitude rating of 4,000 feet. In both cases the engines would have been converted from the contracted V-1710-C8 engines. The cost of converting the first engine to C-11 configuration was estimated at $7,205.[173] At the same time Allison proposed investigating the altitude improvement that could be achieved by utilizing the large three barrel carburetor then being used on the XV-3420-1 double vee engine. They believed that with this carburetor and an improved supercharger inlet elbow, the rated altitude could be raised to between 10,500 to 11,000 feet.[174]

Using this information, and the promise of 1050 bhp, on March 3, 1938 Curtiss submitted to the Materiel Division a proposal for a P-36 airframe mated to this engine. It stated in part, "Wind tunnel tests indicate that (a P-36) with the Allison V-1710 (altitude rated) engine, a high speed of 350 mph at 15,000 feet is possible. This is based on a modified V-1710 having a gear-driven supercharger giving 1000 bhp at 2,600 rpm at 10,500 feet. It is estimated that this engine will develop 1050 bhp at 2950 rpm at 15,000 feet with carburetor ram air..." Wright Field immediately approved Allison to proceed with the design change to the engine, and for Curtiss to proceed with the installation of the engine in the number ten P-36A airframe. Thus the P-40 series was born in the form of the Curtiss Design 75P, Air Corps designation XP-40 (AC38-010).

The reasons for sponsoring the XP-40 varied considerably, depending upon the interests of the party. The Army Air Corps states that they sponsored the XP-40 for the sole purpose of comparing the differences in liquid-cooled inline powered aircraft versus an air-cooled radial engine in an otherwise identical airframe.[175] Curtiss was looking for a quick fix for the P-36 as a way to win the 1938 Pursuit competition. Allison was desperate for a production order as a way to justify their considerable investment, and to support continued development of the V-1710.

The selection by the Army of the Curtiss P-36A for the conversion was not accidental, and only minimal changes were made to adapt the inline Allison engine. The P-36 fuselage was noticeably narrower just behind the engine, so that the narrower inline fit nicely within the width of the cockpit, sized to the width of the pilots shoulders. Keeping with the minimum change dictate, the only new structure was forward of the firewall, and fitting of the radiator below the airframe so that the structure of the P-36 fuselage was not effected. Even the coolant lines to/from the radiator were externally mounted on the centerline of the lower fuselage.

The official order for the XP-40 was issued in July 1938. The plane was first flown on October 10, 1938, powered by the new Allison XV-1710-19(C13) that evolved until rated at 1090 bhp at 2950 rpm up to 10,000 feet at the time of the coming competition.[176] The pilot for this first flight was Mr. E. Elliot, Curtiss Airplane Division Assistant Chief Test Pilot.[177] Other accounts state that Lt. W. M. Morgan was the first pilot.[178] Most likely both made flights on the first day, Lt. Morgan being the first military pilot.

All was not a bed of roses for the new pursuit. Testing through October 1938 produced the comparison to the P-36 given in Table 3-9.

For all intents and purposes the XP-40 in its initial configuration was performing about as well as a P-36A with an assumed 1150 bhp engine. In an effort to improve the initial performance, Curtiss relocated the radiator to a position beneath the engine, and it was in this form that the XP-40 won the January 1939 Competition. Even so, a contract was not implemented as the XP-40 did not meet the performance guarantees.[178A] The Air Corps noted that the liquid-cooled installation in the XP-40 was not particularly clean aerodynamically and they authorized a program to considerably improve it. To this end the Air Corps requested NACA to test the aircraft in their full-scale wind tunnel at Langley Field. Those tests were summarized in a May 1939 report that stated that the emphasis was directed toward reducing the drag of the radiator installation without decreasing the quantity of cooling air passing though it. Compared to the airframe with a faired over radiator, the XP-40 radiator increased the drag of the airframe by 18.9 percent. Only about 16 percent of the drag was attributed to passing the air through the radiator core, the balance being consumed by the frontal area of the scoop.[179] As a result the radiator was relocated to a position just below the engine crankcase. In this position the radiator scoop was wholly within the original fairing line of the smooth nose cowling. This configuration reduced drag to a figure of 10.2 percent over that with a faired fuselage. In this arrangement the total amount of radiator or cooling drag was less than 5 percent of the total airplane drag.[180] This course of events should lay to rest the story told by some authors that the move as a result of a request "by Curtiss-Wright Sales Representatives" and their perception that it was more aesthetically appealing.[181]

Table 3-9
Liquid vs. Air-Cooled P-36

	High Speed, mph	Horsepower	Altitude, ft
XP-40	326	1150	8,800
P-36A	313	1050	8,500
P-36A (Calculated)	325 (Calc'd)	1150 (Assumed)	8,800

Airframe modifications incorporated in the XP-40 prototype as a result of the Langley testing, and that defined the form of the P-40/B/C were:

- Change the carburetor airscoop to eliminate the large scoop on top of the cowling between the machine gun fairings. The original had been made wide and low so as to not obstruct vision, with the result that it was inefficient and created turbulent flow over the cockpit enclosure.
- The radiator was attached to the outside of the P-36A fuselage with the least possible alteration to the structure, and to maintain airplane balance and low drag. The type of radiator was then changed and the assembly partially buried within the fuselage in an effort to further reduce drag. Still later the unit was relocated to just below the engine.
- The initial collector type exhaust manifolds created excessive backpressure at full power at altitude. These were replaced with Rolls-Royce jet type exhaust stacks similar to those being used on the Hurricane and Spitfire. The hope was that they would produce a similar 7 mph increase in speed as claimed by those aircraft.
- The workmanship on the initial cowlings was relatively poor.

Even at this early date the use of the V-1710-F, with its higher critical altitude, lighter weight, higher thrust line and shorter nose was being considered for any follow-on P-40's.

During the period when the airplane was being worked up in preparation for the Pursuit Competition, the engine was also improved. Just prior to the January competition, Ron Hazen and his staff at Allison decided upon a modification to the impeller and supercharger then installed in the XP-40. This they believed would improve the efficiency and capacity of

This is how the XP-40 was configured following the December 1938 mods done to prepare for the delayed 1938 Pursuit Competition. The supercharger intake was improved on the XV-1710-19. The ventral radiator is gone, relocated to just below the engine. Inlets for the oil cooler are in the wing roots. The carburetor scoop was eliminated, replaced by oversized gun ports, which were ducted to the rear of the engine. At this point the half wheel covers were still in place. (Air Corps)

the supercharger. They worked all night to perfect the modification and took it over to Wright Field, arriving there on the morning of the Pursuit Competition with just enough time to install it in the competition engine. With this improvement the XP-40 was found to be some 40 mph faster than the previous Pursuit competition winner. This in an era when the Army Air Corps was used to seeing 10-15 mph improvements in the then annual Pursuit Competition.[182]

With most of these changes incorporated, along with a much improved supercharger and induction system the XP-40 handily won the delayed 1938 Pursuit Competition (held in January 1939) against the XP-38 (it had crashed on its record setting delivery flight and was therefore unable to compete!), the yet to fly XP-39, the Seversky AP-4, the Curtiss XP-37 (Design 80) and other proposed YP-37 derivatives.[183]

1938 Pursuit Competition

In 1938 Pursuit aviation was in a transitional period. The X-608/X-609 competition had been held and the considerably advanced, but experimental, XP-38 and XP-39 Interceptors were expected to be delivered in the Fall of the year. Still there was a critical need to get new aircraft into the expanding squadrons. In the Fall of 1937 the Air Corps announced that the FY-1938 Pursuit competition would be held November 15, 1938 for *flyable* aircraft complying with Type Specification 98-610. The intent was to procure 147 single-engine, single-place Pursuits having a high speed at 15,000 feet of 310 to 370 mph with fuel for two hours of flight, and carrying two .50 cal guns.[184]

Cutaway of an early C-15 used by the Air Corps for instruction of P-40 ground crew and pilots. Use of such training aids made the thousands of parts understandable. (Air Corps)

As it turned out, none of the likely competitors, including the rebuilt Curtiss XP-37 (Design 80 prototype), XP-38, XP-39 or XP-40, were ready or available for the scheduled competition. Consequently the competition was rescheduled for January 1939. Unfortunately both the XP-38 and XP-39 were still unavailable, so the competition centered upon the Curtiss and Seversky entries. Of these the XP-40 was the clear winner, and received the major FY-39 Pursuit production order, revised down to 134 aircraft. Interestingly, prior to the competition the Air Corps had intended to order 13 service test XP-38 *or* XP-39 aircraft, but due to the promise of the two designs decided to order 13 of each. This second batch of 13 was "taken" from the procurement authorization for 147 FY-39 Pursuits, setting the P-40 purchase at 134. When the contract was actually put in place it was for multi-year procurement of a total of 524 P-40's. It was to provide the engines for these aircraft that gave Allison its first series production contract.

The P-40 and Its Engine

As a result of the competition the Air Corps awarded contract W535-ac-12414 to Curtiss on April 26, 1939 for 524 P-40's, contract amount $12,872,898. Allison received an initial order for 393 engines, though this was increased to 837 V-1710's of various models by the time contract W535-ac-12553 was in-place, all at a cost of $15,000,000.[185] This contract was later revised to deliver 969 engines at the same total cost. Each contract was the largest of its type let by the War Department since W.W.I.

Curtiss elected to designate these early P-40's as their Model 81A, which is interesting since the aircraft was considerably different from the Model 81 YP-37. This may have been due to the wing and empennage components, since those on both aircraft had Model 81 part numbers.[186]

The first block of the production engine order to Allison was for 234 engines to support the block of 134 P-40 aircraft ordered with FY-39 funds. These engines were ordered as V-1710-19's, and were to be identical to the engine in the prototype XP-40. The first of these engines, AC39-897 (AEC #84), was sent to Wright Field to perform the obligatory military 150-hour Model Test for a "normal" rating of 960 bhp at 2600 rpm and a takeoff rating of 1090 bhp at 3000 rpm, a test that had been successfully accomplished on the earlier pre-production engine models. The remaining 390 aircraft of this order were to be built using FY-40 funds, and were constructed as 66 P-40's, 22 P-40D and 301 P-40E's, though these later models were not delivered until mid/late 1941. All of the aircraft up to the P-40D were powered by the V-1710-33(C15), an internally strengthened C-13. As an interim measure to keep the production lines going while waiting for the V-1710-F3R powered P-40D/E models, a further 131 P-40B and 193 P-40C V-1710-33 powered models were purchased using FY-41 funds. All of these aircraft were delivered during the first half of 1941. Adding up the numbers results in a total of 524 V-1710-33 powered P-40's for the U.S. Air Corps, the same number as originally ordered, but the models, funding sources and serial numbers became mixed in the process.[187] Included in these numbers were those rebuilt as P-40G models. These were V-1710-33 powered airframes that had the export Model H81-A2 wing equipped with two .30 cal guns in each panel. A total of 46 airframes were so fitted and designated as P-40G's in August/September 1941.[187A] Two were subsequently returned to the P-40 configuration in 1942. Only the last of the first 200 P-40's, AC39-221 was built as a P-40G, though an

Development of the XP-40 continued while production lines were being setup. As shown here AC38-010 has been reworked as the prototype of the P-40 Tomahawk. The classic lines of the nose are fixed; the single carburetor scoop has reappeared and been extended, the radiator and oil coolers are now streamlined into the nose, temperature control gills are fitted on the outlet and individual ejector exhaust stubs are being used for the first time on a V-1710. The engine remained the XV-1710-19(C13) until it began flight testing the V-1710-33(C15) with its improved supercharger inlet. Photo taken at Wright Field in fall/winter 1939/1940, WWII had already begun in Europe. The U.S. had only this one P-40. (Air Corps)

additional 44 were created by refitting the new wing onto original P-40's.[188] According to Allison, 779 V-1710-C15's were built for these early "long nose" Army P-40's, though a later check of the records suggests that 776 were "ordered", an order which was changed after 300 were delivered, with the F-3R substituted for the balance.[189]

By far the largest number of *Tomahawks*[189A] (a total of 1,180) were built for the French and British. All were powered by the Allison V-1710-C15, an engine having identical performance and interchangeable with the Air Corps V-1710-33(C15), although the export version did often have different accessories.

Engine Installation and Cooling

When first flown at the Curtiss factory the XP-40 radiator was housed in a ventral duct located just aft of the trailing edge of the wing. The installation contributed significantly to the drag of the aircraft and was first partially buried, and then in December 1938, soon relocated to a chin position beneath the engine. The oval air intake served the radiator, with the oil coolers now being supplied with air taken from the wing roots, and discharging beneath the firewall. The carburetor air scoop was covered over, with air now being taken from gun ports, each oversized to provide 16.5 sq.in. of flow area. Ram air from these inlets was ducted down to the carburetor inlet.[189B] The airplane was in this form when the Pursuit Competition was won.

Following the Pursuit Competition the aircraft was again modified, this time to prototype the configuration planned for the production P-40/B/C. Here the radiator inlet was enlarged and extended forward about half way to the spinner. The duct was then segmented, one for each of the two radiator cores and with the center section feeding the oil cooler. At the same time the carburetor air inlet was extended forward to a position where the full effect of the slipstream could be captured. These revisions were quite satisfactory and remained unchanged for the entire production run of V-1710-C15 powered models. One important benefit of this final configu-

Table 3-10
P-40 Tomahawk Models

Model	Airplanes Built	Engine Model	Customer	Comments
XP-40	1	XV-1710-19(C13)	US Army Air Corps	Converted P-36A
P-40	200	V-1710-33(C15)	US Army Air Corps	FY-39/40, 21 to Russia*
P-40B	131	V-1710-33(C15)	US Army Air Corps	Same as H81-A2
P-40C	193	V-1710-33(C15)	US Army Air Corps	Improved P-40B
P-40G	(46)	V-1710-33(C15)	US Army Air Corps	From earlier models
Total U.S. P-40's	**525**			
Tomahawk H81-A1	100	V-1710-C15	France	Delivered to Britain
Tomahawk H81-A1/2/3	1,080	V-1710-C15	Britain	Ordered by Britain
Tomahawk	(200)	V-1710-C15	Russia	Diverted from US/UK
Tomahawk H81-A2	(100)	V-1710-C15	China	Diverted from Britain
Total P-40 Tomahawks	**1,705**			

ration was that the entire engine installation and cooling systems could be assembled as a "power egg" for easy assembly and installation on the aircraft.

In the first configuration of the XP-40 a streamlined shroud covered the cowled collector type exhaust manifold, discharging thorough a single port on each side of the fuselage. When the radiator was moved forward the exhaust manifold was deleted and jet type ejector exhaust stacks installed for the first time on a U.S. aircraft, one for each cylinder. At high speed these could improve the thrust available to the aircraft by as much as 10 percent, and provide an additional 7 mph in top speed.

Summary

For a variety of reasons, explained in Chapter 9, the V-1710-33(C15) for the P-40 suffered considerable delay in completing the Army required Model Test. As a consequence Allison produced some 228 engines at their own risk, meaning that they agreed to rework them to the final Model Test standard at their own cost. In the interim the Army restricted these early engines from the intended power ratings pending completion of the testing and Modernization program.

A total of 1,705 V-1710-33(C15) powered Curtiss P-40 type aircraft were built, including the XP-40 that was re-engined with a C-15 in the fall of 1939. This breaks down to 525 for the U.S. Air Corps and 1,180 for the British, who accepted all of those originally purchased by the French, and also includes the 100 diverted to China for use by the American Volunteer Group. Engines delivered on foreign contracts have dataplates identifying them as Allison Model V-1710-C15, the U.S. Army model number is blank. Given the wisdom of the day, which required nominally fifty percent spare engines, the total number of V-1710-C15's built was 2,550, including those built for the non-turbosupercharged Lockheed P-322. Specific problems and developments surrounding the introduction of the engine into service in the P-40 are discussed in Chapter 4.

Curtiss entered their model CP 39-13 in the CP 39-770 pursuit competition for a lightweight fighter. As the XP-46, it and the Republic XP-47 were to compete with the Douglas XP-48 in a fly-off. The XP-48 was a true lightweight, powered by a Ranger SGV-770, compared to the 1150 bhp V-1710-39 to be used by the XP-46 and XP-47. (Curtiss, via SDAM)

Curtiss XP-46 and XP-46A

Often considered as a footnote to aviation history and a non-effective redesign of the P-40, the Curtiss XP-46 was actually a unique aircraft that played an important role in advancing the technology of U.S. fighter aircraft available in W.W.II. The P-40 had won the 1939 major production contracts not only because it was marginally the best pursuit flying at the time, but importantly, it could be had in quantity at a critical juncture in world history. According to General Ben Kelsey, USAF Ret, who in 1939 as Lt. Kelsey was heading up the Pursuit Project Office at Wright Field, the aircraft he and his boss General Echols really wanted was the V-1710-39(F3R) powered XP-46,(Curtiss Model 86).[190,191]

The XP-46A was a clean and compact addition to the Curtiss Hawk family. It and the XP-46 were powered by the new 1150 bhp V-1710-39(F3R) engine. Unfortunately the 8-gun XP-46 was unable to best the performance of its similarly powered Curtiss P-40D stablemate. (Allison)

Following the German invasion of Poland on September 1, 1939, the British and French sent purchasing commissions to the U.S. with orders to purchase practically everything that could be thrown into the war. While the P-40 was then being readied for mass production, it was in reality a streamlined adaptation of the four-year old P-36 airframe, re-engined with the Allison V-1710-C15. Meanwhile, Curtiss had been hard at work on an entirely new "Hawk" for almost two years, and had done a considerable amount of wind-tunnel work on it. On July 24, 1939, Curtiss submitted their CP 39-13 and CP 39-13a designs to the Air Corps in response to the CP 39-770 Pursuit competition. These differed only in selection of the engine, those being the V-1710-F3R and V-1710-C15 respectively. The CP 39-13a "won" over the -13 with a score of 752.50 versus 752.40 points, and was recommended by the Review Board for procurement on an experimental contract.[192] The design placed third, with first going to the Republic P-44 with an order for 80 aircraft powered by the P&W R-2180 and second place to an improved Bell P-39.[193] The improved P-39 was first known as the Bell P-45, but by the time the contract was in place it specified an order for 80 altitude rated V-1710-35(E4) powered P-39C's. Two examples of the third place Curtiss design were ordered as the XP-46/A on September 29, 1939. Soon thereafter two similarly configured Republic AP-10's were ordered as the XP-47/A. These aircraft were built to Kelsey's requirements for a lightweight fighter powered by an altitude rated and improved 1150 bhp Allison V-1710-F3R engine. In this fashion Kelsey was attempting to achieve the performance expected of a conventionally powered 1600 bhp aircraft by proportionally limiting the weight to the available 1150 bhp.[194] Kelsey and Echols approached General Hap Arnold with a request to slow early P-40 production to allow introduction of the P-46. The demand for P-40's for the Allies, and as mounts for the training and commissioning of new U.S. Army combat groups was just too great. "...Every P-40 you could get out of that plant was already obligated and running late."[195]

The P-46 program was still not finished. Curtiss made a formal Foreign Release Agreement with the Army on April 18, 1940 that would have ended P-40 deliveries to the U.S. at 200 planes by October 1940, and released the P-46 for export.[196] The purpose of this agreement being to allow Curtiss to concentrate on delivering contracted export H81-A *Tomahawks* through the end of the year and then to provide P-46's to Britain in 1941. In exchange, the Army was to receive the balance of its order for 524 P-40's as 324 combat ready P-46's, to be delivered in the period from March 1941 though February 1942. At the same time, Curtiss would design and develop the P-53, a P-40 fuselage fitted with a laminar flow wing and believed capable of 450 mph.[197] For a while the P-53 was to have been powered by the new Allison V-1710-F11R, a unique engine intended to use a new and highly efficient Birmann designed supercharger driven through a two-speed gear drive.[198]

As initially rolled out the XP-46A appears true to the concept shown in the CP 39-13 mockup. While the ventral radiator duct appears to be similar to that later used by North American in their NA-73X (P-51), it clearly did not incorporate the same degree of aerodynamic sophistication. (Curtiss via SDAM)

Table 3-11
Comparison of XP-46 and XP-51

	XP-46	**XP-51**
Span	34 ft, 3-1/4 in	37 ft. 0-3/8 in.
Length	30 ft, 2 in	32 ft, 2-5/8 in.
Height	10 ft, 1 in	NA
Wing Area, sq.ft.	208	235.75
Empty Wt, pounds	5,625	5,990
Gross Weight, pounds	7,665	8,633
Max Speed, mph	355 mph at 12,200 ft.	382 mph at 13,700 ft.
Max. Speed, mph	NA	328 mph at 1,000 ft.

General Arnold evidently became the "spoil sport" in this arrangement with his concern for increasing production of the P-40. On May 24, 1940 Curtiss proposed discarding P-46 production and instead use the same 1150 bhp V-1710-39(F3R) engine in a reworked P-40, the P-40D. This aircraft would include heavier guns, armor, and self-sealing fuel tanks. The Air Corps accepted the improved P-40D design as of June 17, 1940 and plans for the P-46 were then finally dropped. Given all that was going on, there is the appearance that there was little coordination occurring between various branches of the Army, such as Kelsey's Fighter Projects Office, and the War Department which was overseeing Anglo-French Purchasing Commission activities. As a measure of the importance of the P-40 at the time, mention should be made that General Arnold had a study initiated on April 22, 1940 on building P-40's from either plastic or wood.[199] He was very concerned about the strategic situation and obviously had a better grasp of the desperate circumstances facing the Allies and U.S. than did the Anglo-French Purchasing Commission, or the rank and file in the Army and/or government.

When approached by the British Purchasing Commission General Echols suggested that they find a manufacturer who wasn't already bogged down in high priority work, and he promised that Curtiss-Wright and the Air Corps would make available all of the material they had on the XP-46 to help them build a new fighter. This was Kelsey's and Echol's way of getting a P-46 type fighter anyway.[200]

The British then approached North American Aviation to either license build P-40's, or design a new fighter for them using the Curtiss wind tunnel data. According to Kelsey, North American purchased the XP-46 data from Curtiss in May 1940 for $56,000, and then signed a contract with the British on May 29, 1940 to deliver the NA-73X prototype in four months.[201] As a condition of the agreement, the Air Corps had required that two of the initial North American production aircraft would be given to the Air Corps for evaluation as XP-51's. According to Kelsey the XP-51 wing area, landing gear arrangement, aft radiator location (configured to produce net thrust), and aircraft empty weight, were as in the XP-46.[202] The record shows clearly that such was not the case.

Engine Installation and Cooling

Contract W535-ac-12553 became the first production source for the V-1710-F3R when 476 C-15's were diverted to the later standard, but these were preceded by five specially ordered F-3R engines obtained on a sepa-

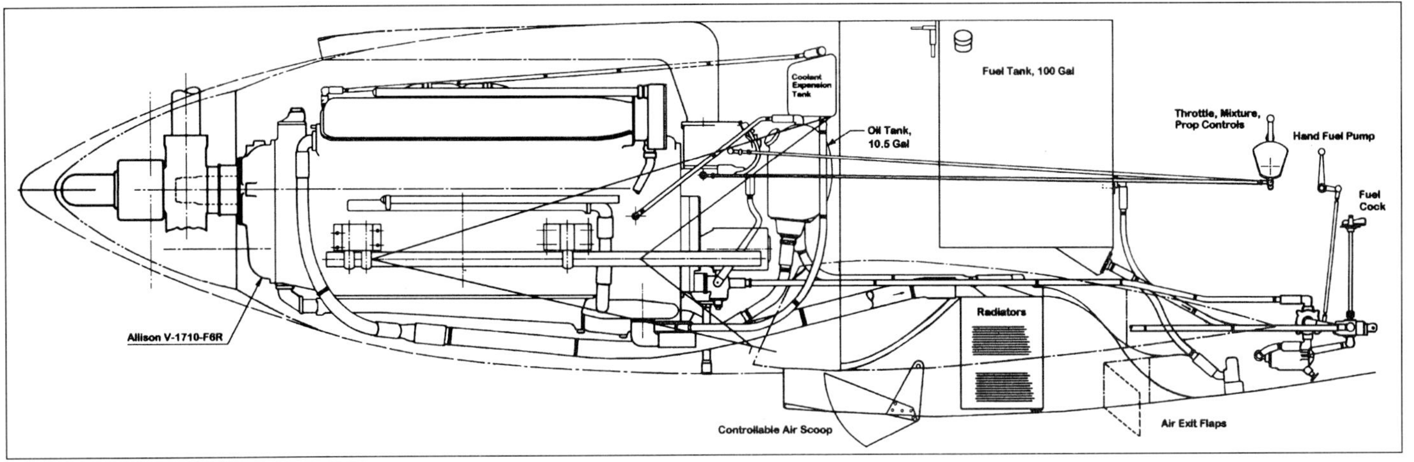

Curtiss Design 39-13 won one of the 1939 lightweight fighter contracts and became the XP-46. As a result of the considerable wind tunnel work done on the airplane, Curtiss submitted this design as their CP 40-1 proposal for the 1940 R-40C competition. A high altitude V-1710-F6R, with 9:6:1 supercharger gears was to be used. As submitted a 3-blade propeller was to be fitted, but as shown on the drawing, Curtiss offered a smooth cowl line having the propeller moved forward 5 inches. The final Allison design of the F-6R describes an engine fitted with a counter-rotating propeller. Such an installation would likely have matched the extended cowl line. Note how the radiator installation is configured, positioned, and uses a controlled position air scoop on the inlet. The radiator installation of the NA-73X/P-51 is not much different. (Author)

rate contract specifically for the experimental XP-46 and XP-47. These were ordered on contract W535-ac-13841, dated January 27, 1940. By the time the airframes were ready to receive them, circumstances had changed. The XP-47/A had been canceled in favor of the radically different and much larger XP-47B. The P-47 was no longer to be Allison powered. As a result the first engine (originally intended to perform the Model Test) had already been installed as the first flight engine in the North American NA-73X, and then suffered crash damage. The two engines for the XP-46's were evidently available when the delayed airframes needed them, and of the last two engines, one became the replacement for the repaired NA-73X and the other was installed in the prototype P-40D.

The radiator installation on the XP-46 returned to the ventral position as first used on the XP-40, though drawings suggest that the XP-46 adapted a better layout and improved internal ducting to smooth and control the cooling air flow. Still it remained for the later Curtiss CP 40-1 proposal to suggest internal ducting likely to offer a degree of thrust recovery. Even so, neither Curtiss offering included the key feature of a boundary layer splitter so critical to the success of the NA-73X and P-51 radiator installation.

Engine mounting in the XP-46 used a tubular truss similar to those Curtiss had used with the V-1710's in the earlier X/YP-37 and XP-40. As usual these supported the oil cooler as well.

Summary

After all of the corporate and government intrigue surrounding the airplane, and without a likely production commitment, Curtiss was allowed to delay completion of the XP-46A and XP-46 "lightweight" fighters. The airplane featured an overall clean design built around the new V-1710-39(F3R) engine. It was about the smallest airplane that could handle the 1150 bhp and carry a heavy brace of Pursuit weapons. The airframe incorporated a number of advanced features: the landing gear was wide-set and inward retracting, while the radiator was buried in the rear of the fuselage with a diffuser duct inlet that enabled it to recover some thrust from the radiator heat. In addition, automatic leading edge slats were provided on the outer portions of the wing to improve low speed, high-g turning ability and performance.

The first of the two P-46's to fly was the XP-46A. It weighed 5,471 pounds, without armor and self-sealing fuel tanks, when delivered on February 21, 1941. It was able to make 410 mph with its 1150 bhp Allison V-1710-39. The second aircraft, the XP-46, had other military improvements in addition to carrying two .50-caliber machine guns in the nose, and eight .30-caliber machine guns in the wing when delivered September 22, 1941. Top speed was down to 355 mph, a speed that could be exceeded by the Curtiss P-40D.[203] Since the contract had specified a top speed of 410 mph, the Army penalized Curtiss $14,995 for not meeting guarantees.

Curtiss later updated the design as the CP 40-1 and submitted it in the Army's 1940 R-40C Pursuit competition. This time intending to utilize the altitude rated Allison V-1710-F6R engine [1125 bhp to 15,000 feet] with counter-rotating propellers, along with considerably improved radiator ducting.[204] The design was not selected for development by the Air Corps.

Mockup for Republic AP-10 proposal built for the CP 39-770 specification, and became the XP-47 when contracted. Note how the V-1710-F3R defined the shape of the very light aircraft that mounted a total of eight machine guns. (Photo provided by Warren Bodie)

Profile of the XP-47 with its liquid-cooled V-1710-39(F3R) appears much like that of the later and much larger XP-47H with its turbosupercharged liquid-cooled Chrysler IV-2220. (Warren Bodie)

The Chrysler IV-2220 liquid-cooled V-16 was a good match for the huge turbocharged XP-47H. The similarity with the profile of the 1939 P-47A with the V-1710 is quite apparent. (SDAM)

Republic XP-47 and XP-47A

Another result of the 1939 single engine pursuit competition (CP 39-770) was the Republic AP-10. The design was submitted in August 1939 by the new Republic Aviation Corporation, successor to Severski, and their Chief Designer, Alex Kartveli. As specified, this was a "light weight" fighter that was to utilize the Allison V-1710-39(F3R). Though configured very much like the competing Curtiss CP 39-13 (XP-46), its initial form was considerably lighter, weighing only 4,600 pounds.

The Air Corps Review Board determined that the design was unsuitable for mass production, but recommended that further investigation was merited as an experimental project. When Materiel Division changes were incorporated the weight increased to 4,900 pounds. A contract (W535-ac-13436) was proposed for construction of two of the revised AP-10's as the XP-47 and XP-47A. In November 1939 the War Department disagreed with the specifics of the proposed design and did not approve the contract. By again reworking the design to further increase the wing area and armament, a new contract (W535-ac-13817) was approved and put in place authorizing the XP-47 and XP-47A on January 17, 1940. The weight had now increased to 6,900 pounds.

The lightweight Republic XP-47/XP-47A project ended before the first airframe was completed, a casualty of the Emmons Board report, produced in the summer of 1940. That Board criticized the lack of diversity of Air Corps aircraft and engines.[205] The Air Corps responded with an effort to diversify the types of engines being used in its Pursuit aircraft. Republic was then given new specifications for increasing the engine power in its R-2800 powered P-44, and at the same time told to terminate work on the lightweight XP-47. The result was to evolve into an entirely new airframe built around the Pratt & Whitney R-2800-11 and designated XP-47B, an aircraft that was to become the immortal *Thunderbolt* in W.W.II. Termination of the Allison powered models was officially completed by contract change on August 15, 1940.[206]

With the termination of the XP-47 and XP-47A in favor of the radial engined P-47B, the two F-3R engines procured specifically for the CP 39-770 competition airplanes were put to good use. One powered the NA-73X when it returned to flight after its untimely crash and the other the P-40D prototype.

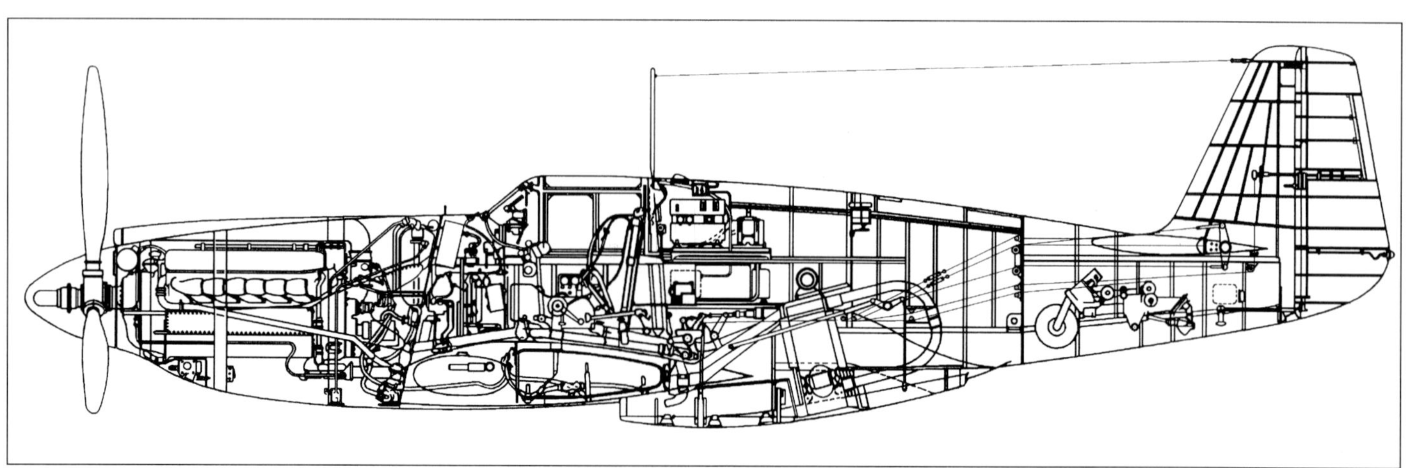

Powered by the 1490 bhp WER rated V-1710-39(F3R), the P-51 (Model 91) was provided as Defense Aid to Britain. This was the same engine that powered the NA-73X (NX19998). (San Diego Aerospace Museum collection, adapted by author)

North American NA-73X/XP-51 *"Mustang"*

The story of the P-51, known variously as the *Mustang*, *Apache*, *Invader*, is considered well known. In April 1940 the British Purchasing Commission asked the North American Aviation Company to license build Curtiss P-40's. North American responded that they could build a superior aircraft, and do it in the same 120 day period it would take them to tool-up and produce their first P-40.[207] On May 4, 1940 the British Air Purchasing Commission officials approved the preliminary design and began to define the contractual agreement. Since North American had never produced a fighter aircraft, a condition of the agreement was that they were required to obtain all current data from the Curtiss design team on their latest developments. A unit price of $50,000 was established and the British placed an order for 320 NA-73 aircraft on May 29, 1940. The clock had started.[208]

Design of the Airplane

Lack of time, money, and non-spectacular performance, were the reasons that the Curtiss P-46 was not produced in quantity for the U.S. Air Corps. Lt. Kelsey had private conversations with key individuals at NACA Langley about the need to salvage a P-46 type aircraft, but does not relate any knowledge of the interactions between Langley and North American about the incorporation of the laminar flow airfoil into the P-51.

While many histories report that when the British contract for the North American NA-73X was signed the design did *not* include the laminar flow airfoil, a review of wind tunnel work done at Cal Tech for North American suggests otherwise. On May 21 and 22, 1940, Dr. Clark Millikan, Cal Tech's world renowned aerodynamist, directed testing of two North American Aviation supplied 1/4 scale wings for the NA-73. These wings were identical except that the first used a NAA High Speed section they had derived from the new NACA 45-100 family of laminar flow profiles. The second used a conventional NACA 23017 profile at the root, tapering to 23008 at the tip.[209] Kelsey later applauded North American's genius and guts to be the first to adopt the new airfoil for production as well as designing an extremely clean airframe that was easy to manufacture and maintain.[210]

All P-51 models were pleasing to the eye, but the Allison powered models like this XP-51 were really sleek, largely as a result of locating both the oil cooler and radiator in the thrust producing duct in the bottom of the fuselage. (Allison)

Providing an Engine

This engine was one of five in a batch purchased on contract W535-ac-13841 in January 1940, specifically for the Curtiss and Republic lightweight fighter projects that came out of the CP 39-770 competition. North American was able to get the first engine from the batch, the one originally intended to perform the F-3R Model Test. In fact the engine was the first of the V-1710-39 models to be built and it is believed to be only the third flyable V-1710-F, following the two F-2's in the first YP-38.[211] Since Allison was aggressively producing engines from two factories at this point, it is important to look at what was actually going on.

This was a very busy period for Allison, as they were not only having problems rapidly increasing production of the C-15 in their new Plant 3, but also providing and supporting four new engine models, all of which had their first flights within a six week period! The first YP-39 flew on September 13, 1940, powered by the new V-1710-37(E5). The first YP-38 flew on September 17, 1940, powered by the new V-1710-27/29(F2R/L). The NA-73X first flew on October 26, 1940, the first flight by a V-1710-39(F3R)! Not only were the F-2R/L, F-3R and E-5 engines all in the design, Model Testing and flight qualification phases at the same time, but the V-1710-33(C15) still had not completed its Model Test, while the Air Corps had only just finished the demanding accelerated flight test program of their first C-15 powered P-40's!

As it was, North American received the first V-1710-39(F3R), Allison s/n 301, while AEC #302 was then used for the F-3R Model Test. Although the first contracted F-2R engine was used for that series Model Test, it bore Allison s/n 305, which would give it a production date *after* that of the F-3R in the NA-73X![212] Although the V-1710-27(F2R) engine completed 150 hours of testing on October 31, 1940, the V-1710-39(F3R) Model Test was much delayed, and was not able to begin until February 1941, nearly five months after the first flight of the NA-73X.[213] That test had been terminated after 112 hours and was successfully rerun in March. Any problems Allison had in providing the engine for the NA-73X were due to its being the first of the type, and not because of corporate inattention to the needs of its customer, North American.

Prior to releasing AEC #301 for flight, Allison ran a 50 hour Full Power Development Test on their EX-13 "Workhorse" engine, configured as a V-1710-F3R. As of August 25, 1940 the test engine had accumulated only 30 hours, delivering 1108 bhp at 3000 rpm from 44.6 inHgA at 0.680 LBM/bhp/hr. Meanwhile, the NA-73X was nearing completion.

While the "F" series was a contemporary of the "E" series, with both sharing cylinder blocks and crankshafts from the double vee V-3420 engine, the reduction gear assembly for the "E and F" engines differed from that in the V-1710-C's in that they were of the more conventional "external spur gear" variety. As executed this gave a higher thrust line and in the "F", and close coupled the propeller to the front of the engine. Mechanically the early F-3R engine, except for the propeller reduction gear, was a contemporary of the V-1710-37(E5), with which it shared the same 8.80:1 supercharger gear ratio, carburetor and crankcase.

Engine Installation and Cooling

The engine installation in the Mustang was particularly clean, and advanced. The man responsible for this was Art Chester, a well known race pilot of the 1930s who had joined North American in 1939 as a result of the hiatus in air racing during the war years. Although not formally trained as an engineer, he became the project engineer for the P-51 power plant installation as a result of his extensive experience in building state of the art race planes. Particularly effective was the cantilever structure built-up from flat aluminum stock and a few small castings that Chester designed to support the engine.[214] This provided improved rigidity for mounting the engine while allowing for the attachment of all cowling, ducting, and the two .50 caliber machine guns. The design was driven by the need to improve the producability of the airplane. The entire engine compartment, with all of its accessories and cowling could be built-up for installation as an assembly on the fuselage at the proper place on the manufacturing line.[215]

On the matter of the thrust recovery by the radiator, Edgar Schmued designer of the NA-73X, is quoted in *Mustang Designer*,[216] "that Atwood's claim about radiator design was incorrect, and that nothing was owed to the XP-46 design."[217] This may be entirely true, as a number of major changes were incorporated into the radiator duct in the various models of the P-51. But Ray Wagner author of *Mustang Designer* notes that Britain's Sir Henry Self had been authorized by London to buy P-40's, and relates how Self told Atwood to get P-40 data from Curtiss so he could tell London that North American was utilizing P-40 experience. Evidently Atwood made a quick train trip to Buffalo and Curtiss agreed to sell him a box of data. In a May 1, 1940 letter Atwood said that the Curtiss people "are furnishing data covering a series of wind tunnel, cooling and performance tests of a *similar airplane*, which data will assist us..."[218] It is likely that the data in question was from the XP-46 effort and that its major contribution was to have a goal of using the heated air from the radiator to provide positive thrust over the normal "cooling drag", a goal that the P-51 was able to achieve. That the XP-46 radiator installation was not fully developed is apparent when the details in the follow-on Curtiss CP 40-1 design are studied. Those appear more like what was already being built into the NA-73X.

Epilogue

As of November 20, 1940, 3:20 hours of flight testing had been accumulated on the NA-73X. Then during the fifth flight, the engine lost power and the contract test pilot, Paul Balfour, attempted a gear-down dead-stick landing into a plowed field. The usual resulted: the aircraft over on its back, and broken.

A comprehensive analysis of the engine was begun by Allison on November 28, 1940 with the result that the engine was found to be in perfect order, excepting crash damage that included complete separation of the propeller and reduction gear.[219] The crash was attributed to the pilots unfamiliarity with the operation of the aircraft's fuel system and resulted

The damage done to AEC #301 when the NA-73X crashed due to fuel starvation, was concentrated in the nose and reduction gear case. As bad as it was, the need for F-3R's was so great that the engine was repaired and powered the first Air Corps XP-51 during its first two years of testing. (Allison)

from simply running his selected fuel tank dry after only 15 minutes of flight![220]

As a sidelight to this episode, one that shows the critical need for V-1710's at the time, AEC#301 was repaired, modernized as AC40-4395M, and returned to service. In fact, it was installed in the forth production Mustang, the first XP-51, when it was delivered to the Air Corps. The engine was still installed when that aircraft was turned over to the NACA at Langley Field on December 27, 1941, over a year later. That aircraft remained at Langley for nearly another year, and sometime during that period the engine was changed.[221]

The V-1710 was a rugged engine and its modular construction allowed major repairs to be effected by simply replacing the damaged components. In contrast, the Rolls-Royce Merlin had the reduction gear case cast integral with the crankcase. Gear-up landings or nose-overs often broke or cracked the casting between the reduction gear and the crankcase necessitating a heavy external brace to be installed to effect a repair.

After the war the NACA returned AC41-38 to the Army Air Force with 290 hours total time. They placed it in storage at Orchard Airport, Park Ridge, Illinois, now the site of Chicago's O'Hare Field, along with many other aircraft retained for a future museum. On January 3, 1949 much of the collection was turned over to the Smithsonian Institution, including Mustang AC41-38. It was subsequently dismantled and stored at the NASM facility at Silver Hill, Maryland.

In 1975 the XP-51 was traded to the Experimental Aircraft Association Air Museum for a historic and rare Northrop Alpha. The XP-51 was then the subject of an extensive restoration to flying status by Darrell Skurich for the EAA, being shown for the first time following restoration at Oshkosh-76. In addition to a complete restoration, the F-3R engine was overhauled and upgraded to a later dash number by installing a heavier crankshaft and late style rods and pistons. Skurich then flew the aircraft, and for a time was the only person current in both the XP-51 and more common P-51D Mustang. He compared the two as a race car to a Rolls-Royce. The XP being "...fast, noisy, uncomfortable, and hot; the D has room, you can see out of it-just like being in the lap of luxury after flying the XP. It's a lot faster than the D. At the same power settings it is about 45 mph faster. I flew along side...a D...and had ample opportunity to compare the two. The XP changes speed instantly...It's uncomfortable...*but it sure is fun!*"[222]

Wallace-Martin NF-2

With the National Emergency developing as a consequence of the widening war in Europe many firms were eager to get involved in the production of war materiel. The example had been set in all previous U.S. wars, being in the right place at the right time could lead to lucrative contracts and corporate success. Apparently the Wallace-Martin Aircraft Corporation of Long Island City, New York was such a contender. They had desires to build their NF-2 shipboard fighter for demonstration to the Navy, and intended to power it with a 1150 bhp Allison V-1710-F3R engine.

Late in July 1940 Wallace-Martin issued a purchase order to Allison for a single F-3R engine. Allison returned the order noting that it was not possible for them to make a direct sale of the "special engine" requested. Only the Air Corps could authorize such a transaction.[223] This response began a series of actions revolving around getting an early F-3R for the project by marshaling congressional support. In this way Wright-Martin brought pressure on the Navy, causing them to state that they believed the proposed development was worthy of consideration. In August 1940 Captain M.A. Mitscher, USN requested that the Chief of the Air Corps authorize the release of one of these engines for sale by Allison to Wallace-Martin.[224]

The Air Corps responded that the V-1710-F3R was a new type and the first article would not be ready for Model Tests until the latter part of September at the earliest, and that sufficient engines of the type would not be available to meet Air Corps requirements until late 1942. As such they could not make an engine available before the end of 1940, though they did offer to release a C-15 for immediate sale.[225] Significant to the overall story of the V-1710 are the circumstances surrounding the F-3R at this point, for only five of these engines were being readied for the initial installations, and the first of these was being rushed for installation in the North American Aviation NA-73X prototype. That engine was actually the first F-3R, for the "prototype" was a Allison "workhorse" engine and had accumulated only 30 hours of operation by August 25. The engine for the NA-73X was delivered on October 7, 1940. Wallace-Martin wanted an engine at a time when they simply did not exist.

As a compromise the Air Corps allowed Allison to provide Wallace-Martin with drawings and a mockup of the F-3R. Things then began to get even more interesting when the Chief of Air Corps Intelligence sent a letter to the Materiel Division assessing the interests of the Wallace-Martin as, "they are attempting to sell a "pig in the poke" to the Navy. They have no outstanding development to offer, but want to get in on the "gravy train" in building airplanes for the Government and they have made such financial guarantees that the Navy can no longer refuse to give them a chance to try their hand. They have even contacted the President's Office."[226]

In this situation Wallace-Martin began an "all fronts" assault on the reluctance of the Air Corps to let them have an engine. In a letter to General Arnold they asked that Allison be authorized to loan them an engine on the basis that, "the design and construction of this experimental pursuit plane is so far in advance of present designs that the release of this engine is justified under the National Defense Program."[227] The next day the Wallace-Martin General Manager wrote to Sol Bloom, their Congressman,[228] claiming that they had designed the airplane around the V-1710-F3R at the suggestion of the Navy, and now with the Air Corps opposed to providing them an engine, their assets had been frozen. They claimed to have spent some $225,000 of their own funds since June 1 to develop a new experimental shop and staff it with 45 mechanics, 20 engineers and draftsmen for the purpose of designing an interceptor fighter-type airplane. Their request to the Congressman was for his support to get necessary experimental contracts, not only for the engine but for the project and accessories as well![229]

As usual with Congressional Inquiries, the Air Corps carefully reviewed the situation and responded, noting that they "considered the development of the Wallace-Martin airplane to be a purely Navy project" and since they "still had a shortage of motors, cannot release a motor at this time."[230]

The Congressman followed up with the Air Corps in December to see if conditions may have improved in the interim.[231] The Air Corps response was that the shortages still existed, and that they could not release a motor "unless urgently requested by the Navy."[232] If that was all it would take the Navy was obliging: on the same day they sent an official request to the Air Corps.[233]

In the summer of 1940 the Wallace-Martin Aircraft Company aggressively pursued the development of their NF-2 shipboard fighter for demonstration to the Navy. The engine they wanted was the new V-1710-F3R. Curtiss and North American were also waiting for this engine. Everybody got involved, the Air Corps, Navy and Congress, but it didn't fly. (NARA)

Things had changed. The Air Corps responded to the Navy that "diversion of an engine would be on the basis of an outright sale by Allison to Wallace-Martin", but that an engine would not be available before at least March 1, 1941 as "production of the "F" type engine had been suspended due to difficulties encountered in model tests and was not contemplated to be resumed before February 1941."[234] In their authorization to Allison for the sale the Materiel Division was emphatic in noting that allowing the diversion of an engine "does not alter the total number of engines contracted for by the Air Corps. An additional engine will have to be constructed as a replacement."[235]

At this point the story appears to end. At least we know that while all of this was going on the Allison powered Bell XFL-1 ship-board fighter was nearing completion. There were proponents within the Navy to allow fighters with liquid-cooled engines aboard its carriers, but then none were operated as such by the Navy. In truth the XFL-1 was most likely an effort by the Navy to "hedge its bets", while the Wallace-Martin seems to have been powered by pure politics. The same methods are still practiced today-not just the political approach, but by bureaucratic organizations as well.

Table 3-12 **Performance of Early V-1710 Powered Aircraft**								
Aircraft	First Flight	Engine	TO bhp	Max. T.O. Weight, #	Max mph	Critical Alt, ft	Climb Time, min.	Service Ceiling, ft
X-608/ X-609	SPEC	V-1710-C7	1150	NA	360	20,000	6.0@20k	NA
XFM-1	9-1-1937	V-1710- 9(D1)	1000	18,492	287	20,000	7.82@15k	NA
XFM-1	10-1937	V-1710-13(D1)	1150	18,492	301	20,000	6.43@15k	30,500
YFM-1	9-28-1939	V-1710-23(D2)	1150	19,000	270	12,600	10.3@15k	30,500
YFM-1A	-1939	V-1710-23(D2)	1150	19,000	270	12,600	10.3@15k	30,500
YFM-1B	-1940	V-1710-41(D2A)	1090	19,000	268	12,600	10.6@15k	30,500
XP-37(75I)	4-20-1937	YV-1710-7(C4)	1000	6,643	<340	20,000	7.1@20k	35,000
XP-37(80)	Sept. 1938	V-1710-11(C7)	1150	NA	370[1]	25,000	NA@25k	NA
YP-37	1-24-1939	V-1710-21(C10)	1150	6,889	331	25,000	8.5@20k	34,000
XP-38	1-27-1939	V-1710-11/15(C8/C9)	1150	15,416	413	20,000	6.5@20k	38,000
YP-38	9-17-1940	V-1710-27/29(F2R/L)	1150	14,348	405	20,000	6.0@20k	38,000
P-38E	1941	V-1710-27/29(F2R/L)	1150	15,482	395	25,000	8.0@20k	39,000
XP-39	4-6-1939	XV-1710-17(E2)	1150	6,204	390[2]	20,000	5.0@20k	32,000
XP-39B	11-25-1939	V-1710-37(E5)	1090	6,450	375	15,000	7.5@20k	36,000
YP-39	9-13-1940	V-1710-37(E5)	1090	7,235	368	13,600	7.3@20k	33,300
P-39C	1940	V-1710-35(E4)	1150	7,300	379	13,000	3.9@12k	33,200
P-39D	1941	V-1710-35(E4)	1150	8,200	368	12,000	5.7@15k	32,100
XFL-1	6-13-1940	XV-1710-6(E1)	1150	7,212	338	11,000	3.8@10k	30,900
XP-40	10-14-1938	XV-1710-19(C13)	1090	6,870	327	12,000	3.2@12k	31,000
P-40	4-4-1940	V-1710-33(C15)	1040	7,215	357	15,000	5.3@15k	32,750
P-40D	5-22-1941	V-1710-39(F3R)	1150	8,809	350	15,000	6.4@15k	30,600
XP-46	9-29-1941	V-1710-39(F3R)	1150	7,665	355	12,200	5.0@12.3k	29,500
XP-46A	2-2-1941	V-1710-39(F3R)	1150	7,322	410	12,200	NA	29,500
NA-73X	10-26-1940	V-1710-39(F3R)	1150	8,400	382	13,000	10.9@20k	30,800
XP-51	5-1-1941	V-1710-39(F3R)	1150	8,400	382	13,000	10.9@20k	30,800
A-36A	Sept. 1942	V-1710-87(F21R)	1325	10,000	360	20,000	___@__k	30,000
P-51A	1942	V-1710-81(F20R)	1200	9,000	390	20,000	9.1@20k	31,350

1. 370 mph figure has not been confirmed in a primary reference.
2. 390 mph was guarantee, 340 mph was best determined in full scale wind tunnel.

NOTES

[1] Byttebier, Hugo T. 1972, 74.

[2] Dickey, Philip S. III. 1968, 76, which refers to a McCook Field memo, Chief, Air Corps, from Commander, January 24, 1929.

[3] Bodie, Warren M., 1991, 15-20.

[4] Airplane Engineering Division, January 1919, 5.

[5] Bodie, Warren M., 1991, 12.

[6] Matthews, Birch J. 1996, 31. The Lockheed design was based upon their Model 10 Electra transport, but using two turbosupercharged tractor type V-1710's.

[7] Bodie, Warren M., 1991, 14.

[8] Bodie, Warren M., 1991, 14.

[9] Kelsey, Benjamin S. B.G. USAF (Ret.) 1982, 87.

[10] Bodie, Warren M., 1991, 14.

[11] Kelsey certainly wanted the performance improvements inherent by incorporating tricycle landing gear, but performance alone was insufficient to overcome the internal Air Corps opposition to incorporating the feature. It was felt by many that even retractable landing gear should not be incorporated on the grounds of maintenance, complexity and weight. His reasoning for tricycle gear was improved stability on the ground, particularly during landing and taking off with heavy and fast airplanes, making them much easier and safer to operate. Newly trained pilots could match performances that required high skills and considerable performance in time of war. Kelsey, Benjamin S. B.G. USAF (Ret.) 1982, 82.

[12] Bodie, Warren M., 1991, 17.

[13] See Appendix 8 for Wright Field Contract Chronology.

[14] Carter, Dusty, Spring 1988, 18-22. Engines were to be buried in the wing with air supplied by leading edge inlets.

[15] Bodie, Warren M., 1991, 16-20.

[16] Army order for ten 800 bhp at 2400 rpm YV-1710-3's, to be based on the detailed engine design done for 1000 bhp at 2800 rpm, on Contract W535-ac-6795 dated June 18, 1934. Price was $200,000. By Engineering Order, dated July 27, 1935, the engines ratings were changed to 1000 bhp at 2600 rpm.

[17] Wright Field Report No. TSEST-A2, 1946.

[18] *Priority of V-1710 Engine Development*, Letter to Allison Engineering Company from Chief, Materiel Division, CC-CEN-MIM-57, December 7, 1938.

[19] Kelsey, Benjamin S. B.G. USAF (Ret.) 1982, 31.

[20] Hunt, J.H. 1942, XFM-1. Allison internal report, *Allison XFM-1 History*, with comments by J.L. Goldthwaithe.

[21] Bell Aircraft Preliminary Specification, Report No. 1Y012, 1936, 2. The Bell specification indicates that V-1710-9 engines had a "military rating of 1250 bhp at 2800 rpm and a normal rating of 1000 bhp at 2600 rpm at sea level and 20,000 feet altitude, geared 2 to 1, equipped with a type F-10 turbosupercharger." These would have been the early fuel injected versions of the V-1710-9, not the 1000 bhp carbureted version actually delivered.

[22] Bell Aircraft Preliminary Specification, Report No. 1Y012, 1936, 67.

[23] Schlaifer, R. and S.D. Heron, 1950, 278.

[24] Engineering Section Memo Report Serial No. E-57-362-1, 1938.

[25] Engineering Section Memo Report Serial No. E-57-1406-3, 1939.

[26] Bell test operations with the XFM-1 and subsequent models were conducted from the Buffalo Municipal Airport, where the company leased hangar space. The Airacuda were towed tail first, using a truck, to the airport where wing installation and final assembly occurred prior to flight testing. Per correspondence with Birch Matthews, September 1995.

[27] *Damage to and Repair of Model YV-1710-9 Engine A.C. No. 34-9, as Result of Backfire in XFM-1 Airplane*, Hazen of Allison letter to Chief, Materiel Division, Wright Field, September 8, 1937.

[28] Matthews, Birch. 1996, 49.
[28A] At this time the recommendation was made to use the Form F-13 (Type B-1) turbo, which necessitated a complete rearrangement of the induction system to mount it below the wing. These changes were made some time later, at which point the SFM-1 nacelles were prototypes for the YFM-1 installation. Modifications to upgrade to the YFM-1 engine configuration cost $38,104. *Skytiger, History's Forgotten Fighter*, Boyne, Walt, *AIRPOWER*, Vol.1, No.1, Sept. 1971, 20-62.

[29] Matthews, Birch. 1996, 49

[30] Matthews, Birch. 1996, 34-52.

[31] Douglas Aircraft had modified a single OA-4C Dolphin amphibian with a prototype tri-cycle landing gear. Col. O.P. Echols sent Lt. Kelsey and Jake Harmon (Bomber Projects Officer) out to get experience with it. Both were quite impressed with the improvement in handling, with the result being a desire to incorporate the feature into future aircraft. Kelsey included the feature in the coming X-608 and X-609 Interceptor circulars by offering contractors several extra merit points if fully retracting tricycle landing gear was included in their design. Bodie, Warren M., 1991, 14-15. In addition, the opportunity of the YFM-1 contract allowed for another test of the concept.

[32] Technical Report of Aircraft Accident Classification Committee, January 24, 1942, *Chanute Field investigation of YFM-1A A.C. No.38-497*, Pilot 1st Lt. James O. Reed.

[33] *Handbook of Instruction of Operation and Flight Instructions for the Models YFM-1 and YFM-1B Multi-Place Fighter Airplanes*, Section IV, p.21.

[34] *Supplement 2 to Bell Specification 1Y012-D*, dated October 19, 1939 states the following:

> 2 Engines, Allison V-1710-23 Converted to ratings equivalent to those of the V-1710-33 engine, per Allison Specification 114-C revised to show ratings equivalent to Allison Spec 126-B Revised, June 1, 1939.

[35] Change No.7 to Contract W535-ac-10660, December 7, 1939. NARA RG18, File 452.8, Box 808.

[36] *Airplanes, Multiplace Fighter, Circular Proposal 39-780*, letter to Chief of the Air Corps, August 8, 1939. NARA RG18, Bulky, Box 711.

[37] Note from Major General H.H. Arnold, Chief of the Air Corps. NARA RG18, Bulky, Box 711.

[38] These aircraft were initially identified as P-45's, but this was soon changed to P-39C to minimize confusion. Only the first 20 were delivered as P-39C, the balance being the combat ready P-39D model.
[39] *Circular Proposal 39-770, Single Engine Interceptor Pursuit Airplanes*, Routing and Record Sheet to General Arnold, September 2, 1939. NARA RG18-Bulky, File 452.1 Airplanes, CP 39-770, Box 710.

[40] Vandenberg, Maj. Hoyt, 1941, 54-62.

[41] Hazen letter to Mr. Robert J. Woods, Chief Engineer at Bell Aircraft, March 10, 1938. V-3420 file at NARA.

[42] *Appendix IIIA to Bell Specification 1Y012-D, Model YFM-1C*, June 6, 1940.

[43] Information provided by author Birch Matthews in telecon with author, 5-18-1994.

[44] X-1800 is 24 cylinder liquid-cooled sleeve valve, vertical "H" configuration with integral two stage supercharger and separate aftercoolers to each half of engine. Rated for 2000 bhp. Design started in December 1939 and discontinued November 1940. Description data on H-2600 P&W X-1800 obtained from data tags on engine X-95 at P&W East Hartford, April 7, 1995.

[45] *From Weight Data Including Dimensions for V-1710 and V-3420 Engines*, Allison Engineering Publication, July 21, 1943.

[46] Allison Model Specification No. 217, 1940. Model V-3420-7(B5). This specification was updated as 217-A on May 19, 1942, which was after the end of the Bell FM-1 program. This suggests that the latter model may have been considered for the McDonnell P-67.

[47] *Report on Airplanes, Engines and Propellers on Foreign Order in the Aircraft Industry*, as of December 31, 1939. NARA RG18, Entry 293B, Box 250.

[48] *Requirements for Pursuit Airplanes, Aviation Expansion Program*, memo for The Assistant Secretary of War from Chief of the Air Corps, September 13, 1939. NARA RG18, Bulky, File 452.1-Aircraft Misc., Box 717.

[49] According to Aviation Historian Hal Andrews of Arlington, Virginia. With respect to the XP-37 designation, the proper Curtiss nomenclature is 75I, not the H-75I as is often shown. Neither the name Hawk, nor the "H" prefix, as is often applied to the Curtiss Model 75's, came about until after the export, fixed gear Model 75 was named the Hawk 75 in 1938. It was then followed by the retractable gear Hawk 75A for the French, which was clearly known as the H-75A as contrasted to the Design 75A for one of the prototype configurations. Correspondence with the author, 2-16-1995 and 5-29-97.

[50] Bowers, Peter M. 1979, 348-365.

[51] Curtiss drawing SK-1090, 1937.

[52] Beauchamp, Gerry, To be published in 1995, 46.

[53] Wright Field file D52.1/1422 Curtiss. 1940.

[54] It was not until the late 1930's that the NACA began systematic investigations into interference drag, due to mixing of airflows around an airframe, that the importance of carefully designing inlets and exhausts, along with venting of cooling air, was understood.

[55] An observation by aviation historian Hal Andrews, May 1997.

[56] Since the Air Corps normally included contractors flying time on the aircraft record card only after official delivery, it is believed that the first flight was actually done by Curtiss at Buffalo, and that the aircraft was damaged on the delivery flight to Wright Field on April 20, 1937.

[57] Allison Division, GMC, 1941, 7.

[58] Allison Division, GMC, 1941, 7.

[59] Beauchamp, Gerry, 1977, 12-25. Note that Shamburger, Page and Joe Christy, 1972, say the XP-37 was delivered by surface freight to Wright Field on April 1, 1937, "though it was damaged in the process and redelivered after repairs on June 16, 1937." No tangible documents have been found to verify this later story, and it is not believed likely, based on the photos of the damaged aircraft.

[60] Wright Field photographs #56944/5, taken May 4, 1937.

[61] Wright Field Aircraft Status Report, 1937.

[62] Delivery date for the XP-37, AC37-375 was July 23, 1937, with Allison YV-1710-7 AAC34-13 installed. Information contained in *Technical Report of Aircraft Accident Classification Committee*, dated March 11, 1938, pilot Capt. Samual R. Harris, Jr. Kirtland AFB, Personal Operations Records.

[63] *Performance of XFM-1, XP-37 and XP-40 Airplanes*, Letter from the Chief of Materiel Division to Chief of the Air Corps, November 4, 1938. NARA RG18, Entry 167C, Box 27.

[64] Shamburger, Page and Joe Christy, 1972, 112.

[65] *Performance of XFM-1, XP-37 and XP-40 Airplanes*, Letter from the Chief of Materiel Division to Chief of the Air Corps, November 4, 1938. NARA RG18, Entry 167C, Box 27.

[66] Captain Harris was a prolific flyer with considerable experience since earning his wings in 1927. At time of the March 10, 1938 incident his total time was 3645:20 hours, and he had logged 5:25 hours, out of the aircraft's total of 29:30 hours, in the XP-37. He had also logged considerable recent time in the P-36, A-17 and PB-2A. Source, March 10, 1938 Mishap Report.

[67] Wright Field document B71/6062, *XP-37 AC37-375 March 10, 1938 Incident Report*, notes taken by Jefferies from report at Air Force Museum, September 1956.

[68] The "Design 80" is usually missing from all available lists of Curtiss design studies and/or production aircraft. Detail drawings of the original XP-37 Design 75I show liberal use of components direct from the 75 series. The drawings for rebuilding the XP-37 in its role of becoming the prototype for the YP-37's rely on unique components, which were assembled as Design 80, except for the horizontal stabilizer and elevator which were from the 75. The YP-37 was Curtiss Design 81, with the majority of it constructed from Model 81 components, excepting again the Design 75 horizontal stabilizer and elevator. Significantly, the aberration for the rebuilt Design 80A, with the extension shaft engine, and the improved XP-37 were identified as the Design 80 (dwg 81906). This design dates from May 25, 1937 (which was prior to the official delivery of the 75I) and uses the V-1710-C7 engine. Both the Design 80 and 80A were to have inward retracting landing gear, a feature not incorporated into the actual aircraft. The new production series, which began with the YP-37, was actually the first to use the Model 81 designation, which was followed by the production P-40 and P-40B/C (Models 81-A1 through 81-A3 respectively).

[69] *Curtiss Design 80 Engine Installation Drawing*, see Curtiss microfilm roll "N", frame 119, at NASM.

[70] Allison letter to Chief, Engineering Section, Materiel Division, Air Corps, Wright Field, *Reduction Gear Assembly Repair*, October 6, 1938 and response letter from J.P. Richter, Major, Air Corps, dated October 20, 1938.

[71] XP-37 AAC37-375 Aircraft History card.

[72] More explicit documents show the first flight engine as the YV-1710-7(C4), AAC s/n 34-13, a contemporary of the 1000 bhp Type Test engine.

[73] *Preliminary Data for General Electric Exhaust-Driven Turbine Supercharger, Form F-13*, attached to GE Letter from S. Moss to Mr. T. Jacocks, Navy Department Bureau of Aeronautics, October 15, 1937. NASM File B3006100.

[74] The Type B-1 and Type B-2 units differed only in that the B-1 had provisions for cabin pressurization, a feature omitted on the B-2, as well as other minor detail.

[75] Allison Division, GMC, 1941, 7.

[76] Bodie, Warren M. 1994, 72-73.

[77] Some references give the date of this flight as January 20, 1939.

[78] Beauchamp, Gerry, To be published in 1995, manuscript p.49.

[79] Beauchamp, Gerry, 1977, 18.

[80] This was done largely to provide Allison and GE with production experience with the engine and turbosupercharger pending the larger orders likely to be required due to the worsening world conditions.

[81] Materiel Division Memorandum Report, Serial No. EXP-M-57-535-3, 1941.

[82] Originally built as the company Design 75 demonstrator, the many subsequent developments caused Curtiss to assign suffixes to the various projects. According to Bowers, Peter M. 1979:

Design 75: Original designation of c/n 11923, powered by the twin row 900 bhp Wright SCR-1670-G5. This installation was found unsatisfactory and a P&W R-1535 was temporarily installed.

Curtiss 75B: Design 75 c/n 11923 repowered by a 675 bhp Wright R-1820F Cyclone for the April 1936 Pursuit Competition. This was changed to a 850 bhp Wright SGR-1820-G5 for the final competition.

Curtiss 75D: This has been identified by some as a retroactive designation of the c/n 11923 airframe when in its original Design 75 form, and sometimes (unofficially) referred to as the "XP-36". Curtiss historian Hal Andrews believes this is incorrect as there are later pictures of a 75D powered by an R-1820.

Curtiss 75I: Airframe c/n 11923 reworked and designated XP-37 AC37-375, when powered by the turbosupercharged 1000 bhp, Allison YV-1710-7(C4), AAC34-13.

Curtiss 80: The XP-37 airframe c/n 11923 when again reworked, this time as the YP-37 prototype, and powered by the 1150 bhp V-1710-11(C8) AAC38-582 and V-1710-11(C7) AAC34-13, and new GE Form F-13 turbosupercharger.

[83] This table was compiled from original Curtiss drawings available on Microfilm at NASM, [roll-frame]: XP-37, [R-138]; X/YP-37 Design 80, [N-124]; YP-37 Design 80A, [N-121]; YP-37, [N-41]. The foreword cockpit windshield framing was considerably changed for the YP-37, leading to the minor deviations shown in the table. Other authors have reported that the fuselage on the YP-37 was lengthened ahead of the cockpit. Clearly this is not the case. Except for the "Actual Length" column, these values should not be taken to be exact given that they were scaled from the drawings. Incidentally, comparing the overall length given in reputable published sources gives no consistency for the YP-37 length either! Author.

	Actual Length from Drawings, Ft.	Overall Lengt, Nose/Tail Ft.	CL Prop 10 Top Windshield Frame, Ft.	Top Windshield Frame to Rudder Hinge, Ft.	CL Prop 10 Rudder Hinge, Ft.
XP-37 Design 751	30.67	31.3	16.9	10.5	27.4
XP-37 Design 80	30.67	30.8	16.2	10.5	26.7
XP-37 Design 80A	30.67	30.4	16.8	10.2	27.0
XP-37 Model 81	33.21	32.8	17.0	11.9	29.0

[84] Shamburger, Page and Joe Christy, 1972, 112.

[85] John D. Kline (Allison Installation Engineer on XP-37, XP-38, XP-39, XP-40 and NA-73X), during an interview in November 1994 concurred that there was no one person or agency actually responsible to bring an aircraft and all of its systems together in a cohesive way. He says this resulted in a lot of "finger pointing". Unfortunately, this is one aspect of history that repeats itself, for it continues today.

[86] On November 19, 1940 Colonel Gaffney was flying YP-37 from Ladd Field to Elmendorf Field at Anchorage. The weather conditions enroute were subject to very rapid and complete change, and such occurred with the weather closing the pass before and behind the Colonel's line of flight and forcing him to seek an emergency landing field near the foot of Mt. McKinley, at the Golden Zone Mine, Colorado, Alaska. The landing was accomplished but the high speed and fast landing pursuit ship nosed over in the deep snow and was damaged beyond repair. It had operated in Alaska for only 24 hours. The pilot was saved from serious injury. *History of the Air Corps Station, Ladd Field, Fairbanks, Alaska*. H.H. Arnold manuscript, Roll 183, Library of Congress.

[87] Correspondence from Col. Ancil D. Baker, USAF(Ret), who accumulated 26:10 hours flying the YP-37's at Ladd Field in the period October 29, 1940 to July 25, 1941.

[88] *History of the Air Corps Station, Ladd Field, Fairbanks, Alaska,* p.45. H.H. Arnold manuscript, Roll 183, Library of Congress.

[89] This aircraft was the last YP-37 to be surveyed, February 1, 1943. The NACA flew #38-474 over 21 hours each of the first two months they had it, but according to the Record Card, after February 1941 the aircraft was not flown.

[90] Wright Field file D52.1/1422 Curtiss. 1940. The NACA flew #38-474 over 21 hours each of the first two months they had it, but according to the Record Card, after February 1941 the aircraft was not flown.

Wright Field file D52.1/1422 Curtiss. 1940.

[91] Teletype TEX-T-409 from Wright Field Experimental Engineering Section to General Echols, *Pertaining to Status of Allison Engine Model Tests*, January 23, 1941. NARA RG18, File 452.8, Box 807.

[92] Kelsey, Benjamin S. B.G. USAF (Ret.) 1982, 120.

[93] The designation Y1P-37A suggests a considerable amount of government interest, though it may have been simply wishful thinking by Curtiss, for the Y"1"P- designator suggests that supplemental government funds were to be provided for the project, along with an apparently official designation of the new model. See Curtiss sketch SK 1066 and associated drawings, Curtiss Microfilm at NASM Roll L, Frames 3, 4 and 96, dated 10-20-1937.

[94] Bodie, Warren M., 1991, 16-20.

[95] Correspondence from Birch Matthews, July 6, 1997.

[96] *Lockheed Model 22 Twin Engine Interceptor Pursuit, Manufacturing Plan*, submitted to Air Corps for X-608 competition in 1937.

[97] Bodie, Warren M., 1991, 20; reference to Lockheed Report 1160, September 23, 1937.

[98] *Purchase of Pursuit Airplanes (Interceptor Type)*, Immediate Action SECRET correspondence from H.H. Arnold, Chief of the Air Corps to Chief, Materiel Division, Wright Field, December 14, 1938.

[99] Bodie, Warren M., 1991, 34.

[100] *XP-38 Mishap Report*, Technical Report of Aircraft Accident Classification Committee, February 12, 1939, Mitchel Field, Long Island, NY.

[101] Wright Field Memo Report EXP-M-57-503-126, 1940.

[102] Bodie, Warren M., 1991, 42.

[103] White, R.J. 1938, 4.

[104] White, R.J. 1938, 2.

[105] White, R.J. 1938, 37-38.

[106] Bodie, Warren M., 1991, 39.

[107] Bell Drawing 3M001, March 1936.

[108] Bell Drawing 3M001, December 1936.

[109] In their original Model 4 specification, Bell listed "*Model V-1710 with extension drive shaft, independent reduction gear box and type F-10 turbo supercharger*," Bell Aircraft Corporation, Report No. 4Y003, 1937, 3.

[110] Matthews, Birch. 1996, 73.

[111] Allison issued Specification #110, March 20, 1937, that included the "Mod A" configuration of the V-1710-C7 power section. This model mated a crankshaft speed extension shaft to a remote reduction gear able to mount a cannon firing through the propeller hub. This engine was featured in the Bell Model 3 and Model 4 aircraft proposals that were submitted to the Air Corps on June 3, 1937.

[112] Bell Drawing 3M001, December 1936.

[113] Allison Model Specification No. 110, 1937.

[114] Appendix 5 provides a complete list of Allison Engine Specifications.
[115] Kelsey, Benjamin S. B.G. USAF (Ret.) 1982, 88.

[116] "E" was the next major design to follow the "D" pusher engines, and only coincidentally could it be considered to imply "extension" shaft.
[117] Hazen, R.M. 1945 Report, 2.

[118] Following the designation of the left-hand turning XV-1710-15 for the Lockheed XP-38 following the X-608 competition, it is likely that the Air Corps designation XV-1710-17 was assigned to the engine powering the winner of the X-609 competition, meaning that it may have applied to the V-1710-D3 'Mod A' for a time.
[119] Allison Division, GMC, 1941, 9.

[120] *V-1710-E and F Dive Test to Determine the Advisability of Decreasing Oil Flow by Decreasing Main Bearing Clearance*, Allison Experimental Department Report No. 311, by J.C. Schmid, 1-6-1940, p.2. File H-10.

[121] Bell Aircraft Corporation, Report No. 4Y003, 1937.

[122] Correspondence from Robert J. Woods to Chief, Engineering Section, Materiel Division, Wright Field, *Revision to Specification 4Y003, Contract AC-10341, 12 January 1938*, p.2. Mention is made of the engine specification being Allison Engineering Company Specification No. 115-B, Revision B, dated 10 December 1937.

[123] Emmick, Wm. G. 1939.

[124] Teletype message S-290, dated September 19, 1939, from Wright Field Experimental Engineering Section. NARA RG18, File 452.8, Box 808.

[125] Date per September 30, 1995 letter to author from Bell historian and author Birch Matthews.

[126] Bell Aircraft Corporation log of XP-39 activities at Wright Field, January-April 1939, per Birch Matthews.

[127] Emmick, Wm. G. 1939.

[128] Kline, John 1994.

[129] Silverstein, Abe and F.R. Nickle, 1939.

[130] Comments provided by Birch Matthews in correspondence with author, September 30, 1995.

[131] Air Corps Type Specification C-616 resulted in the purchase of the 13 YP-39's.
[132] Letter of April 21, 1939 from H.H. Arnold, Chief of the Air Corps, to Chief, Materiel Division, Wright Field. NARA RG18, Entry 293B, Box 249, file 452.1-Pursuits.

[133] Silverstein, Abe and F.R. Nickle, 1939.

[134] NACA Wartime Report WR L-489, Table I.

[135] Comments provided by Birch Matthews in correspondence with author, September 30, 1995.

[136] Inter Office Memo, *Exhaust Turbine Superchargers on Pursuit Aircraft*, Capt. B. S. Kelsey, Chief Pursuit Branch, May 11, 1940. NARA Record Group 342, Supercharger-Engines, 1940.

[137] While these activities resulted in removing the turbo from the P-39, the Air Corps was far from through with the turbosupercharger. In October 1941 Wright Field summarized its findings in a memo, *Aerodynamic Problems Introduced by the Use of Turbo Superchargers*, which reported on work which had been underway since April of 1940. It relates that it was well known at the time the early installations were developed, that they were experimental and that space limitations and mechanical difficulties took precedence over aerodynamic considerations. When service type airplanes were equipped with turbos it was desirable to reduce drag, but this was difficult for a number of factors, including:

a. Difficulty in getting reliable inflight test data.
b. Drag of intercoolers.
c. Cooling drag of the exposed rotor casing.
d. Drag caused by the discharge of hot exhaust gasses.

Some improvement occurred by installing as much of the equipment as possible within the airplane or nacelle, and it was intended to make a significant improvement by submerging the turbo entirely within the aircraft. To make this possible Wright Field designed a "radiant" cooling cap which cooled the supercharger rotor without mixing the cooling air and the hot exhaust gas. This eliminated the dangerous and undesirable "afterburning" which otherwise occurred at the rotor. In this manner completely shrouded installations were feasible that would then allow recovery of some of the exhaust gas thrust by directing the gasses through a properly shaped nozzle. Source: Materiel Division Memo Report, *Aerodynamic Problems Introduced by the Use of Turbo Superchargers*, October 20, 1941. NARA RG 342, RD3466, Box 6841.

[138] McClarren, Robert, *Douglas XB-42*, an article in the files of SDAM.

[139] Kline, John 1994.

[140] *Purchase of Pursuit Airplanes (Interceptor Type)*, Immediate Action SECRET correspondence from H.H. Arnold, Chief of the Air Corps to Chief, Materiel Division, Wright Field, December 14, 1938.

[141] *Status of Deliveries of Aircraft on 1940 Procurement Program*, October 1, 1939. RG18, File 452.1 to 452.17, Box 130 'Bulky'.

[142] Matthews, Birch. 1996, 95.

[143] Comments provided by Birch Matthews in correspondence with author, September 30, 1995.

[144] Plans Division memorandum Aer-PL-3-EMN from Commander A.C. Davis to Chief of the Bureau of Aeronautics, 16 November 1937, p.1.

[145] Materiel Division Memo Report, *Aerodynamic Problems Introduced by the Use of Turbo Superchargers*, October 20, 1941. NARA RG 342, RD3466, Box 6841.

[146] Specification allowing liquid-cooled engines, *Class VF Airplanes-Request for Designs and Informal Proposals*, dated 1 February 1938.

[147] Martin, Bob, 1993, 206.

[148] This would be typical of the single-stage engines when driving the supercharger through 9.60:1 step-up gears.

[149] *High Altitude with Geared Supercharger*, in a letter from Allison Engineering to the Materiel Division, dated 6-6-1938, Allison noted that the practical limit on maximum altitude rating for a single-stage supercharger was about 15,000 feet. Two-stage arrangements using turbosuperchargers for the first stage were recommended if higher altitudes were desired for the V-1710-19 or V-3420 engines.

[150] Martin, Bob, 1993. 206.

[151] The wing area of the Proposed Bell Model 4 was originally 200 square feet, but by the time it flew as the XP-39 it was up to 213 square feet.

[152] *Detail Specification for Proposed Navy Fighter - Bell Aircraft Model 5*, Bell Report No. 5Y005.

[153] Matthews, Birch J. 1996, 124-147.

[154] At this date Allison was just starting the design of the Auxiliary Stage Supercharger, which resulted in an engine about 20 inches longer. This in turn required a lengthened airframe, which the Air Force accomplished in the P-63, though it was years later.

[155] V-1710-E/F production was suspended in the fall of 1939 to allow quality upgrades in manufacture of components found deficient by Type/Model testing of the first E/F's. Although a few E/F's were build in mid-1940, production was again delayed in the fall of 1940 to generally strengthen critical parts.

[156] Sanwald, G.L. 1940.

[157] NARA RG72, Contract 65197 file, BAU Memo, Feb 20, 1939.

[158] NARA RG72, Contract 65197 file, Allison Letter of April 8, 1940.

[159] Sanwald, G.L. 1940.

[160] Sanwald, G.L. 1940.

[161] Sanwald, G.L. 1940.

[162] *Bell XFL-1 Pilot's Report*, by R.M. Stanley, Bell Report 5Y045, August 15, 1940.

[163] Martin, Bob, 1993, 206.

[164] *Bell XFL-1 Pilot's Report*, by R.M. Stanley, Bell Report 5Y045, October 4, 1940.

[165] NARA RG72, Contract 65197 file, Inspector of Naval Aircraft Restricted letter of Oct 8, 1940.

[166] Matthews, Birch. 1996, 141.

[167] Comments provided by Birch Matthews in correspondence with author, September 30, 1995.

[168] Comments provided by Birch Matthews in correspondence with author, September 30, 1995.

[169] Matthews, Birch. 1996, 137.

[170] The actual timing of developments suggests that the proposal would have referenced the V-1710-47(E9), which during its development period evolved to become the V-1710-93(E11).

[171] Noted Naval aviation historian Hal Andrews notes that "the Navy was not a proponent of the laminar flow wing", so we can assume that the Model 21 was more like the Model 23 than the Model 24. Correspondence with author, 3-30-1996.

[172] Shamburger, Page and Joe Christy, 1972, 113.

[173] *Altitude Ratings for Allison V-1710 Engines*, Allison General Manager O.T. Kreusser letter to Chief, Materiel Division, February 11, 1938. The cost of a basic V-1710-C8 was $26,560. The $7,205 for the proposed V-1710-C11 altitude rated engine covered 250 man-hours to design the new supercharger drive and inlet, $550 for patterns for the housing, diffuser and intake elbow, $600 for a PD-12 carburetor, 500 man-hours of shop labor and 50 hours of testing. Quite a bargain by today's standards!

[174] *Altitude Rating for Allison V-1710 Engines*, CONFIDENTIAL letter from Allison General Manager, O.T. Kreusser to Chief, Materiel Division at Wright Field, February 19, 1938. NOTE: There is an error in this letter in that the gear ratio of the C-11 and C-12 are reversed relative to the values and applications given in Allison Specifications 113-B and 116 respectively, Author.

[175] *Engineering Comparison of Hawker Hurricane, Supermarine Spitfire and Curtiss XP-40 Airplane*, letter from Chief of Materiel Division to Chief of the Air Corps, November 21, 1938. NARA RG18, Entry 167C, Box 27.

[176] Engine ratings can be very confusing, with many excellent sources giving different values for seemingly identical engines and installations. In the case of the XV-1710-19 for the XP-40, following its initial flights and the 1938/39 Pursuit Competition it was re-rated to 1090 bhp, and the rated speed increased from 2950 to 3000 rpm. The net effect was to retain a rated critical altitude of 13,500 feet. Allison recommended the re-rating because they had qualified the engine based upon testing of earlier 1150 bhp rated engines which used 6.23:1 supercharger gears. With the 8.77:1 gears requiring additional power for the supercharger, Allison recommended that the engine be limited to the same indicated power as previously qualified. The V-1710-33 production version of the engine was initially rated for 1090 bhp for both Takeoff and Military at up to 13,200 feet. The ratings were taken from identical performance curves, the only difference being the selection of a higher altitude for rating. From, *Release for Export Sale of P-40 Type Airplane*, Materiel Division Memo of November 8, 1939.

[177] Tydon, Walter, 1997, 140.

[178] *Allison News*, 2nd March Issue, 1944, Vol.III, No.17.

[178A] Hart, Eric H., XP-40 Marsupial Coolant System, comments on Walter Tydon article in *AAHS Journal*, Summer, 1997, comments in *Journal*, Winter 1997, page 320.

[179] Tydon, Walter, 1997, 140.

[180] Tydon, Walter, 1997, 141.

[181] Shamburger, Page and Joe Christy, 1972, 114.

[182] Hazen, R.M. 1945 Report, 4.

[183] Knott, James E. 1968.

[184] Letter to Chief of Materiel Division at Wright Field on Pursuit Airplanes, March 26, 1938. NARA RG18, Entry 293B, Box 249.

[185] Researching creditable records gives a wide range to the number of engines promised, or intended to be ordered from the new Allison plant. Contract W535-ac-12553 is variously noted for either 924 or 969 engines in either four or five models, while some accounts say 857 engines were ordered. Further confusion is due to the way blocks of engines were ordered under this one contract using funds from different Fiscal Years. The important fact remains that when General Motors authorized Allison to build Plant #3 there was no promise of more than 1000 engines needed to meet future government requirements. General Motors took the

chance in the spring of 1939, and built a plant designed to produce 1000 engines per year. The actual quantity was an initial order for 134 in FY-39 for the initial block of P-40's, which was increased to cover the FY-40 requirements giving a total of 837. This was further increased when the 66 P-38's from the CP 39-775 competition were ordered. While the total was now 969, there was no change in the total contract dollar amount. The consequence of foreign orders for the V-1710 had reduced the unit cost to the point that 132 engines for the P-38's were provided at no additional cost. Government contract records, NARA.

[186] Shamburger, Page and Joe Christy, 1972, 116. There is some confusion regarding Curtiss Model numbers. Shamburger & Christy say that the first Production P-40 was the H81, and that there was no H81-A, but that the French model was the H81-A1, and the British received the H81-A2/A3. The H81-B was the P-40B/C.

[187] Shamburger, Page and Joe Christy, 1972, 248.

[187A] Hart, Eric H., Curtiss P-40 Historian, personal correspondence with Author, January 23, 1998.

[188] Shamburger, Page and Joe Christy, 1972, 248.

[189] *Allison "Firsts"*, notes compiled in 1942 by J.L. Goldthwaite of Allison.

[189A] Hart, Eric H., The names *Tomahawk* and *Kittyhawk* were assigned by the British to the Hawk 81A, the names were not adopted by the Army. In about October 1941 the USAAF did assign the name *Warhawk*, which applied to all U.S. P-40's. Correspondence with Author, January 23, 1998.

[189B] Beauchamp, Gerry, *The Phantom Prototype*, Wings, Vol. 4, No.6, December, 1942, 16-27.

[190] In his book The Dragon's Teeth?, Kelsey gives special notice to role played by Oliver Echols during the 1930's. From 1934 to 1940, he was Chief Engineer and then Assistant Chief of the Materiel Division at Wright Field. During this time, he was the key figure in the initiation of all the weapons that became available for the dramatic expansion of production. A review of relevant correspondence of the period shows his close involvement and leadership in advancing technology and bringing practical aircraft into the inventory. Kelsey, Benjamin S. B.G. USAF (Ret.) 1982, 142.

[191] Ethell, Jeff, 1st Quarter 1981, 65.

[192] *Evaluation-Circular Proposal 39-770 Review Board.* NARA RG18, Bulky, Box 711 CP 39-770.

[193] Wagner, Ray 1991.

[194] Ethell, Jeff, 1st Quarter 1981, 69 and 80.

[195] Ethell, Jeff, 1st Quarter 1981, 66.

[196] *Army/Curtiss Foreign Release Agreement*, April 18, 1940. NARA, RG 18, 452.1C, Box 266.

[197] Wagner, Ray 1991.

[198] Allison Photolab Log, photos #03435/40, July 1941.

[199] General Arnold correspondence of 4-22-40 provided by Ray Wagner, Archivist, San Diego Aerospace Museum, 1992.

[200] Ethell, Jeff, 1st Quarter 1981, 69.

[201] Ethell, Jeff, 1st Quarter 1981, 69.

[202] Ethell, Jeff, 1st Quarter 1981, 69.

[203] Shamburger, Page and Joe Christy, 1972, 127-128.

[204] Curtiss Drawing P-3064 dated 4-8-1940, see NASM Curtiss Microfilm roll/frame L-49 and the Power Plant Installation Drawing P-3071 dated 4-10-1940, microfilm frame L-55.

[205] Bodie, Warren M. 1994, 113. The Emmons Board (named for its chairman, M/Gen. Delos C. Emmons, Chief of the GHQ Air Force) was delving into every facet of the experimental fighter program. On 19 June 1940, the Board issued a critical report, which was endorsed by M/Gen. Henry H. Arnold, Chief of the Air Corps. One key target of the report was that too much of America's pursuit airplane program hinged on "one type liquid-cooled engine." The real purpose of the Board was to come up with recommendations about which fighters should be produced in quantity for combat operations. The Board strongly recommended expedited development of fighters powered by air-cooled radial engines and, if possible, having other sources to develop and produce liquid-cooled engines.

[206] Bodie, Warren M. 1994, 113-116.

[207] Not everyone realizes that General Motors owned 30 percent of NAA stock up until 1948. During that time they were well represented on the NAA Board and directly contributed GM manufacturing and product positioning guidance.
[208] Gruenhagen, Robert W. 1980, 35 & 39.

[209] The following table provides some of the more critical parameters from the tests reported by Millikan, Clark B. 1940:

	45-100 Mod	23017/008
Airfoil Efficiency Factor	0.96	0.91
Minimum Profile Drag Coefficient	0.0068	0.0086
Optimum Coefficient for Lift	0.15	0.20
Maximum Lift Coefficient	1.12	1.37

The new laminar flow airfoil had a thickness ratio of 15.1% at the 39% cord line at the wing root, and 11.4% at the 50% chord line at the tip. According to Gruenhagen, Robert W. 1980, 40, this resulted in a wing with the maximum thickness at 40% from the leading edge, and a slight negative pressure gradient over the leading 50-60 percent of the wing surface. The new airfoil demonstrated laminar flow over the forward 30 percent of the wing, a factor which was significant in the outstanding performance of the P-51, though it was found that production wings in the field seldom could demonstrate laminar flow. Of interest to historians should be the fact that the NAA drawing number for the NACA 45 series wing was 73-01001, while that for the alternative 23 series wing was X73-01003, suggesting that the NAA focus was on the laminar profile from the beginning. The commitment to the original wing layout is demonstrated by comparing the original design, as submitted for the wind tunnel tests, to that on the mature P-51D:

	73-01001	P-61D
Area, square feet	233.42	233.75
Span, Feet	37.03	37.03
Aspect Ratio	5.875	5.815
Chord at Center of Airplane, inches	104	103.99
Chord at Wing Tip, inches	50	50
Taper Ratio	2.16:1	2.16:1
Dihedral, degrees	5	5
Sweepback at Leadin Edge, deg-min-sec	3-36-0	3-35-32

The additional wing surface area on the actual airplane was the result of having to sweep the leading edge foreword adjacent to the fuselage to provide space for the main landing wheels, a feature not included on the wind tunnel models.

Conversely, Curtiss retained a conventional section on the XP-46, as well as their CP-40-1 derivative of the P-46 which was being designed in April 1940 for submittal in response to the Army's Circular Proposal for the 1940 pursuit competition. Curtiss proposed to use a 4 gun wing of only 150 square feet having a NACA 2418 root section tapering to 2409 at the tip. Empty weight was to be 4,902 pounds, maximum gross weight being 6,963 pounds, including the V-1710-F engine. See NASM Curtiss Microfilm Roll L, Frame 48. This wing section was little different than the NACA 2215 root, 2209 tip used for the 236 square foot wing of the P-36/P-37/P-40, the Curtiss standard for some five years. See NASM Curtiss Microfilm Roll N, Frame 40.

[210] Ethell, Jeff, 1st Quarter 1981, 71.

[211] The reasoning being that the first V-1710-F2 purchased for the YP-38's was AC39-1131, Allison #305, and it was used for the F-2 Model Test.

[212] Allison Report No. A2-7, 1941. The 150-hour test was accomplished in the period October 21 to October 31, 1940, using engine AEC#305, AC39-1131.

[213] Priority Telegram PROD-T-662 to Chief, Materiel Division, March 13, 1941. NARA RG18, File 452.8, Box 807.

[214] Wagner, Ray. 1990, 52 and 59.

[215] Gruenhagen, Robert W. 1980, 42.

[216] Wagner, Ray. 1990, 230.

[217] "Atwood" is Lee Atwood, Vice President of North American.

[218] Wagner, Ray. 1990, 53-54.

[219] Allison Service Engine Report No. A2-6, 1940.

[220] Wagner, Ray. 1990, 66.

[221] *Airplane and Engine Records* attached to letter from NASA Langley Research Center, Richard T. Layman Historical Program Manager, November 28, 1994.

[222] *The Restoration of the XP-51*, by Jack Cox, *Sport Aviation*, December 1976, p.58/64.

[223] Allison letter to Wallace-Martin Aircraft Corporation, July 29, 1940. NARA RG18, File 452.8, Box 808.

[224] *Request for Engine for Wallace-Martin Aircraft Corporation*, letter from Chief of the Bureau of Aeronautics to the Chief of the Air Corps, August 1, 1940. Signed by Captain M.A. Mitscher, USN. NARA RG18, File 452.8, Box 808.

[225] *Request for Engine for Wallace-Martin Aircraft Corporation*, letter to Chief of the Bureau of Aeronautics from Chief of the Materiel Division, August 29, 1940. NARA RG18, File 452.8, Box 808.

[226] *Request for Engine for Wallace-Martin Aircraft Corporation*, letter to Chief, Materiel Division from Chief, Air Corps Intelligence Section, October 2, 1940. NARA RG18, File 452.8, Box 808.

[227] Wallace-Martin Aircraft Corporation letter to General H.H. Arnold, Chief of Air Corps from Alastair Bradley Martin, President, October 7, 1940. NARA RG18, File 452.8, Box 808.

[228] The Honorable Sol Bloom, Chairman, Foreign Affairs Committee, House of Representatives.

[229] Wallace-Martin Aircraft Corporation letter to The Honorable Sol Bloom, Chairman, Foreign Affairs Committee, House of Representatives, October 8, 1940. NARA RG18, File 452.8, Box 808.

[230] Draft response by General Arnold to Congressman Bloom, October 21, 1940. NARA RG18, File 452.8, Box 808.

[231] Letter from Congressman Sol Bloom to Asst, Chief of the Air Corps, December 10, 1940. NARA RG18, File 452.8, Box 807.

[232] Air Corps response to Congressman Bloom, December 13, 1940. NARA RG18, File 452.8, Box 807.

[233] *Engine for Wallace-Martin Aircraft Corporation*, letter to Chief, Air Corps from Chief, Bureau of Aeronautics, December 13, 1940. NARA RG18, File 452.8, Box 807.

[234] *Engine for Wallace-Martin Aircraft Corporation*, letter to Chief, Bureau of Aeronautics from Air Corps, December 19, 1940. NARA RG18, File 452.8, Box 807.

[235] Letter from Materiel Division to Allison Engineering Company, January 10, 1941. NARA RG18, File 452.8, Box 807.

4

Into Service with the Curtiss P-40 "Tomahawk"

Allison started on the V-1710 in 1930 and by mid-1937 had delivered all 16 of the engines which had been ordered to that point. Late in 1937 they received an order for 20 V-1710-11's for the YP-37's, and then early in 1938 that order was amended to add 39 V-1710-13's for the YFM-1 program, all of these engines being rated for 1150 bhp. Allison was in the position of having one of the most advanced engines in the world at the time, but they were not considered as a major engine manufacturer. At the same time the considerably worsening world political situation was making the U.S. desperate for fighters and aircraft able to hold their own should a conflict develop.

Although not everything the Air Corps wanted in terms of performance and firepower, the selection of the Curtiss P-40 with its altitude rated V-1710 in the spring of 1939 was still a considerable advance in the standard of U.S. warplanes.[1] Development of the pre-production engines had been a long and protracted effort, but it served to refine the engine. It was the first to pass the strenuous 150-hour Military Type Test at 1000 bhp, and went on to be rated at 1150 bhp for the service test aircraft. Following this success Allison received its first real production commitment for 234 engines for the 134 P-40's purchased with FY-39 funds.

By 1938/39, things were beginning to look good for Allison. All of the Air Corps high performance Pursuit programs were committed to Allison V-1710 power, including the Curtiss P-37, Lockheed P-38, Bell P-39, Curtiss P-40, Curtiss P-46/P-46A, and Republic P-47/P-47A. There were other fighter projects being developed as competitors and intended as successors, but none of them were ever to enjoy major procurement or a significant role as a Army fighter. In fact, the only follow-on Army models to become major production programs during the war were the P-39 derived (and V-1710 powered) Bell P-63, the re-designed and radial engined Republic P-47B, and of course, the North American P-51B/C/D with the two-stage Packard-built Rolls-Royce Merlin, developed from the 1940 design of the V-1710 powered British Mustang (NA-73).

The Curtiss P-40 was Allison's first major production program. As such, it was the program to endure the pains associated with introducing the engine into squadron service. This was no small endeavor considering variability's and new demands coming from being away from the factory and laboratory. In this respect the Allison V-1710 was no different than any new engine thrust into service. Even today, all new engines have "teething" problems to varying degrees. This was emphasized during the war with the introduction of the various models of the P&W R-2800, the Curtiss-Wright R-3350 and P&W R-4360. Each went through crisis periods. The introductory problems with the V-1710 focused on backfiring and integration with the airframe and propeller. All were satisfactorily resolved before the U.S. entered the war.

As it turned out the Allison powered P-40 was in the thick of the war from the outset, with even a few P-40's able to rise in an uncoordinated effort to defend Pearl Harbor on December 7, 1941. Though several victories were claimed, they were insufficient to turn the attack. Four hours later a P-40B shot down the first aircraft in the defense of the Philippines. Then on December 20, the A.V.G. began their rise to fame in their V-1710

The early V-1710-33 powered P-40 *Tomahawk's* were quickly distributed around the country and to the major U.S. outposts in the Panama Canal, Hawaii and the Philippines. Squadrons were soon issuing Unsatisfactory Reports detailing backfires and a few broken crankshafts, The Accelerated Service Test was to resolve questions about the effectiveness of Allison fixes. It also established the operational requirements for service use of the aircraft. (R. Besecrer Collection)

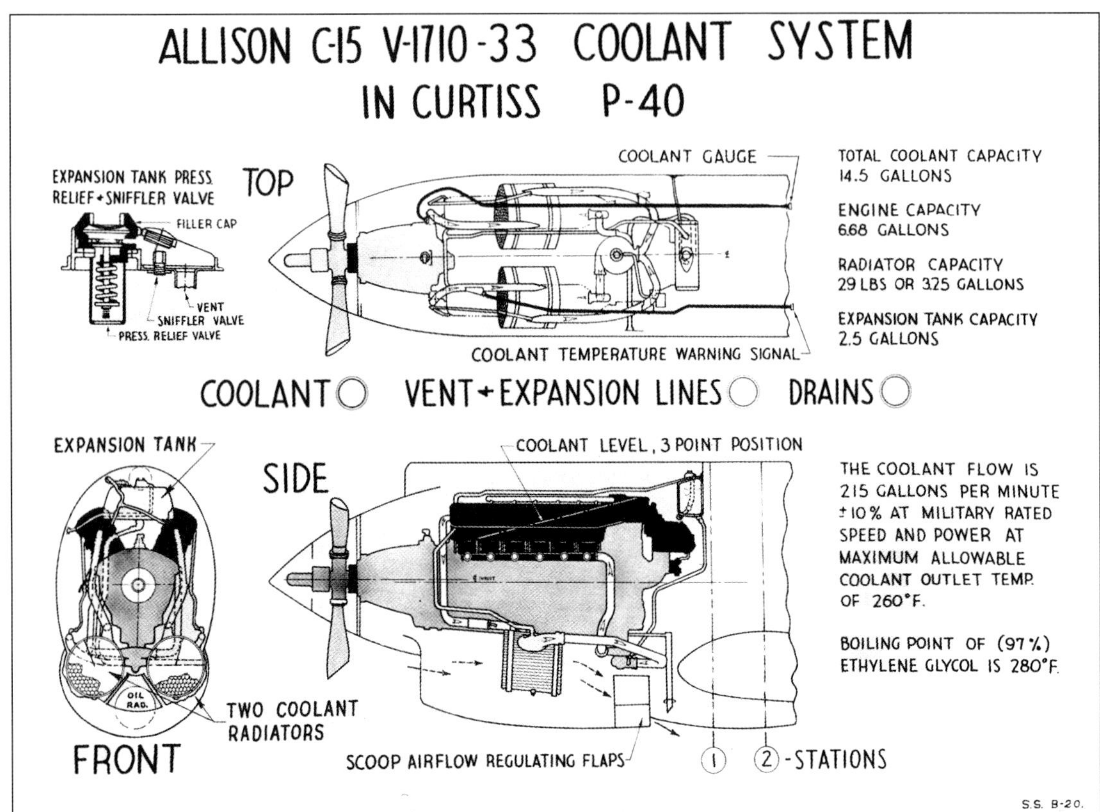

The important cooling system for the P-40 was compactly installed within the engine compartment. (Allison)

powered *Tomahawks* when they repelled the Japanese attempting to bomb Kunming, China.[2] The following discussions relate the technical aspects, procurement and operational impacts of these and related problems with introduction and early prosecution of the war.

Accelerated Service Test

In the late Spring of 1940 Curtiss, Allison and the Air Corps were both in a state of high consternation. The Nazis were inflicting war in western Europe with the Blitzkrieg into Northern Europe and France. Conditions were bleak for the Allies and the one and only U.S. Army Pursuit remotely capable of meeting the onslot, the Curtiss P-40, was just then getting into production. To make matters worse, the Allison engines for these planes were behind schedule. Not only was the Army without its P-40's, the large British and French orders for the aircraft were behind schedule as well.

There were abundant reasons for the engine situation. Allison had broken ground only a year earlier for the new factory in which to produce the engines. Furthermore, at the same time they had to recruit and train workers to operate the machinery and build the needed quantities of the V-1710-33. The engine itself was still embroiled in a torturous series of Model Tests, in fact five engines were to be used before the engine was finally able to complete the 150-hour test. There were many reasons for these problems, one being the insistence of the Materiel Division that the engine be tested at 1090 bhp, when Allison recommended 1040 bhp, along with a number of problems with the available test stands and test procedures. These conditions resulted in Allison delivering some 228 engines prior to completing the Model Test. They agreed to redeliver them after upgrading to the final configuration, all at their own expense. Conditions were to the point by mid-June 1940 that the situation demanded the attention of the highest levels of the companies, agencies, and governments involved. At this point Dr. George J. Mead, Chief of the Aeronautics Section of the National Advisory Commission to the Council of National Defense, was named to oversee the resolution of the situation.

On June 17, 1940 a conference was held at Wright Field, attended by Major General Arnold, Brigadier General Brett, Colonel Echols for the Air Corps, and the senior management of the Allison Company, Messrs. Hunt, Evans, Kreusser, and Hazen. They reported to Dr. Mead that the engine was being released for unlimited production and delivery to the Curtiss Company, but with the flight rating limited to 950 bhp. In addition, six P-40's were being ordered to Patterson Field, Ohio for an accelerated service test of one hundred fifty hours. During the service test, flight operations were to be scheduled as closely as possible to mirror the test stand operating schedule. That schedule itself was established to mirror a "normal" flight with a warm-up period at idle, followed by a period at takeoff power, then climb power followed by a cruise period, then idle prior to repetition until the required number of hours were achieved. Concurrent with the service test Allison was to run the Model Test at a rating of 1040 bhp. The purpose of all of this being to develop a "Yard Stick" so that ground test and flight test would be more closely correlated. It was hoped that a consequence of this approach would result in expedited rating tests and a better determination of the optimum rated power for the engine, while establishing a baseline for the attainment of higher ratings for use in future engines.[3]

The Accelerated Service Test of the six P-40's was completed in less than a month, July 17, 1940. Seven airplanes were actually involved in the test, and five of them successfully completed the required 150 hours.[4] One crankshaft failed and one nose section was found cracked. On the basis of the accelerated service test, and concurrent Model Test, the P-40's entered service with a 950 bhp rating, though this was to change within a year when the Allison recommended 1040 bhp rating was approved.

John Kline, Allison Tech Rep at Patterson Field for the Accelerated Service Test, remembers that this was the first such test the Army had ever run with such intensity, or on more than one airplane. He said,

> "basically, it was to shake down the engine, to find out all that was wrong with it. I used to tell Bill Watson this, because he was (Curtiss) project Engineer on the P-40, "You know Bill, strange thing, we're running the accelerated service test on the engine, but we found more things wrong with the airplane than we did with the engine!" We'd find little "nut and bolt" items on the engine, that was it, nothing major. That worked out pretty well."[5]

The failed crankshaft was found to be a problem unique to the single failed unit. The result of the accelerated service test was a considerable improvement in confidence in the V-1710-33 and the P-40. Production continued and was quickly accelerating to the point that by mid-September Allison was ahead of the original delivery schedule.

The accelerated service test procedure was subsequently used on many of the Army's new aircraft. The P-39 was the next one to undergo the process, also at Patterson Field. These tests also answered the moot question as to how many of the Model Test failures were the fault of the engine and how many the fault of the test stand. According to Jimmy Doolittle in his report to General Arnold, they were both guilty, but the engine was being unduly penalized. On the whole, Doolittle concluded that the engine comported itself very well during these tests.[6]

Accelerated Service Testing was found to be a rapid and effective way to shakedown a new aircraft and at the same time determine any unique servicing or operating problems likely to occur in subsequent service in a combat theater. By doing the test early, necessary changes could be integrated into production. This improved the standardization and performance of the aircraft and also focused the needs for supporting the maintenance and supply of the aircraft.

Going to Sea in a P-40

Early on the Air Corps could see the need to be able to fly its Pursuits off the decks of Navy carriers. The first demonstration of the capability came on October 14, 1940 when the Eighth Pursuit Group took off from the USS *Wasp* while at sea. Twenty-four P-40's had been flown to Chambers Field in Virginia on October 12, and then were taxied to the pier alongside the carrier. Hoisting aboard went quickly, a mere 3-1/2 minutes being required to load each aircraft. All were spotted on the flight deck, though the dimensions of the P-40 were such that the carrier could have moved them below decks via the elevators if desired.

Pilots had previously visited the carrier to be instructed in carrier take-off procedures. Whitewash lines were then painted on a runway to represent the carrier flight deck, and each pilot made several practice take-offs to familiarize himself with the procedure and to determine the best method of obtaining maximum performance.

Pilots went aboard the USS *Wasp* on the night of October 13 and the Wasp embarked early the next morning and steamed to a position approximately 80 miles east of Langley Field. With the carrier steaming into the wind and making about 30 knots across the deck, the P-40's were spotted 390 feet from the bow. Their engines were run-up against the brakes until the tail tended to lift, which occurred at about 30 inHgA manifold pressure. No difficulties were encountered. All of the airplanes flew off the deck, usually being airborne in less than the available 390 feet. They then assembled in formation and returned direct to Langley.[7]

Twenty-four P-40 *Tomahawk's* and eight O-47A's were deck loaded on the USS *Wasp* in October 1940 and successfully launched at sea in a twenty-three minute period. A minimum of 400 feet of deck was available. With a 30 knot wind across the deck the typical takeoff run was 300 feet, using full throttle and 15 degrees of flaps. The space required for takeoff was about the same as for the Navy SB2U-2. (SDAM)

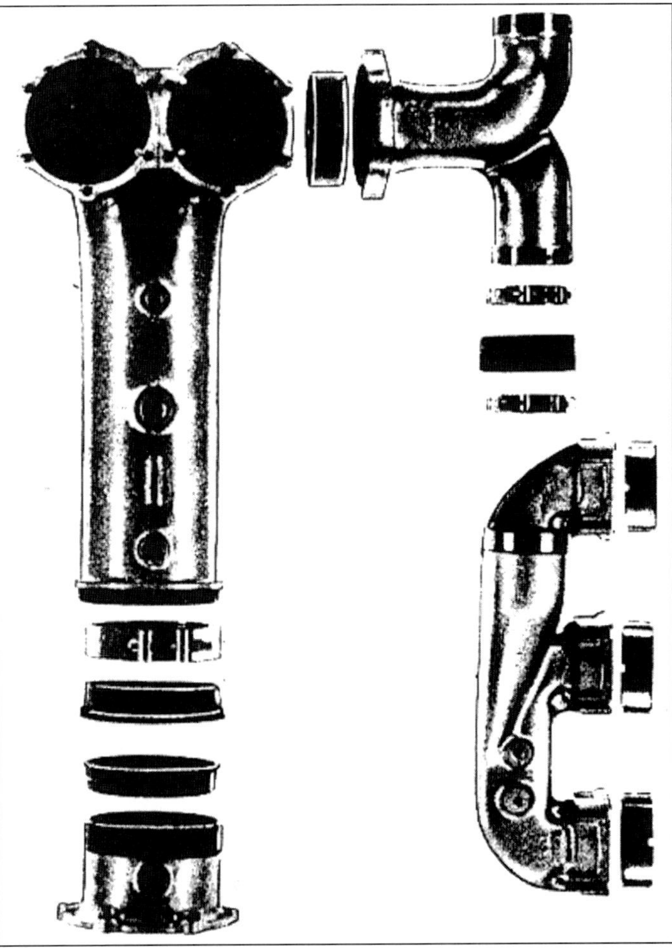

This parts breakdown of the induction manifold clearly shows how the flow distribution backfire screen between the gas pipe and manifold tee was fitted, along with the individual port filters set into the wide flange PN40772/3 manifolds. (Air Corps)

This capability was exercised a number of times during the war when an aircraft carrier was pressed into service as a way to transport a complete fighting unit to a position where they could join with invasion forces.

Backfiring

From its inception the Allison V-1710 was recognized as a high performance engine. It was the first production engine to be rated at near 1 horsepower per pound, as well as developing power approaching 1 horsepower per cubic inch. This was at a time when the best radial engines were hard pressed to achieve 0.5 bhp per cubic inch. The V-1710 was also the first engine designed to operate with high coolant temperatures, as well as having a comparatively high BMEP, or specific power. The XV-1710-1, the first Air Corps V-1710, was designed with sodium cooled *intake* valves as a way to lessen the likelihood that a hot intake valve would ignite the charge filled induction system and cause backfiring. While sodium filled *exhaust* valves had been used in high powered engines for several years, the V-1710 was the first to apply the feature to both the intake and exhaust valves. Unfortunately, that first engine suffered from "sticking" of the intake valves, which then caused backfiring. Consequently, to expedite development, conventional "hollow-stem" intake valves were used throughout the V-1710 development period and on into the first production engine, the V-1710-33 for the Curtiss P-40's.

Backfiring can be initiated from a number of causes, including: overrevving, malfunctioning carburetor, ice in the intake, excessive carburetor air temperature, malfunctioning carburetor heat, an overspeeding turbosupercharger, fuel tank runout, oil in the induction manifold, valve sticking, burned or broken intake valves, or often, improper ignition wiring or timing. In fact, the tendency toward backfiring is aggravated, or caused by, anything that creates a lean fuel/air mixture.

Backfiring is a term difficult to comprehensively define, for it can vary from simple burning of the mixture in the induction system all of the way to violent explosions, sometimes with catastrophic results to the engine. During the development program Allison remained concerned about backfiring and consequently incorporated "Backfire Screens" in the induction system of the early engines. These were redesigned in the XV-1710-7(C3) and subsequent engines to be of a circular cross-section steel frame with alternate coiled sheets of flat and corrugated bronze. They were approximately one inch wide and formed into a coarse mesh set crosswise to the induction flow. They were intended to act much like the screen on a Bunsen burner, or in a miners lamp; that is, to hold the flame on the downstream face and prevent its' propagation into the mixture filled supercharger and carburetor. There were two screens, one located on each branch of the manifold tee where the intake trunk splits for the left and right banks of the induction manifolds. In this location they were not only intended to prevent backfire flames from reaching the supercharger, but were necessary to insure proper mixture distribution and flow straightening within the manifold passages.[8] During the development program the only backfires that occurred were as a result of specific valve and/or ignition failures. It was not believed to be a problem requiring further investigation or resolution. In general it was believed that the screens were satisfactory and that the engine was not particularity prone to backfiring.

On the first flight of the Bell XFM-1, just as the aircraft broke ground, its left-hand engine suffered a violent backfire, and considerable damage was done to the engine and induction system. The flight was successfully concluded some 20 minutes later, with the engine continuing to operate throughout the flight. This was only the third airplane to be V-1710 powered, following the Consolidated A-11A test bed and Curtiss XP-37, both within the previous nine months. The likely cause of the backfire was determined to be an over-speeding engine, due to miss-calibrated tachometers, and a stuck turbo regulator that overboosted the engine. This created high induction temperatures along with a lean mixture; all conditions conducive to backfiring. Consequently, neither Allison nor the Air Corps Materiel Division had reason to view the incident as a precursor to subsequent backfire events. With this experience, confirmed by the generally satisfactory operation of the engines in the XP-37 and XFM-1, the Air Corps moved ahead with procurement of the P-40 in January 1939.

By April 1939 the Air Corps orders for V-1710-C engines for the P-40 program had increased to 393, with orders for P-40's then totaling 200. The next order was for 131 P-40B's that were ordered in FY-41, the order for engines was increased by 131, to a total of 524 V-1710-33's, with spares to be drawn from the earlier production. These engines were part of the total of 969 engines delivered under the first production contract W535-ac-12553, which also provided the first V-1710-27, -29, -37 and -39's.

It was a year later before deliveries of the P-40 to the Air Corps actually began on June 1, 1940, the summer of the *Battle of Britain*. The first batch of 200 (these were "straight" P-40's, not a later subtype) were completed by October 15, 1940, overlapping with deliveries of British P-40's (Model H-81A "*Tomahawks*" in British nomenclature) which continued in production through various Marks to August 21, 1941. The overall status of equipment for the Air Corps at this time was dismal. No combat airplane in the service was equipped with leakproof fuel tanks, and only three P-39's were considered to offer adequate fire power. Specifically, as of January 31, 1941 GHQ Air Force could send into combat against a modern Air Force only those Groups shown in Table 4-1.[9]

Another 131 P-40B and 193 P-40C's, both V-1710-33 powered, were delivered to the Air Corps in the period of January 3, 1941 to May 20, 1941.[10] By the summer of 1941 the Air Corps had received its entire compliment of 524 Curtiss P-40's.[11] Many had been dispatched to the Philippine Islands, Hawaii and the Canal Zone, where Allison was supporting 43, 142 and 12 engines respectively. These were the first line of defense. With this significant effort for the Air Corps expansion program, the P-40 became its most numerous combat type. It was the core of Army Pursuit aviation when the attack on Pearl Harbor sparked U.S. entry into what is now known as WW2.[12]

Table 4-1 U.S. Combat Strength, January 1941

Number of Groups	Number of Aircraft	Type of Aircraft
1-1/2 Groups Heavy Bombers	57	B-17
_ Group Medium Bombers	33	B-23
2 Groups Pursuits	173, 3	P-40, P-39
1 Squadron Light Bombers	13	A-20A

By the end of 1940, the P-40's in squadron service were reporting a considerable numbers of backfires, some with dire consequences. Numerous "Unsatisfactory Reports" (UR's) began to describe the problems and incidents, which occurred under a wide variety of operating conditions. There was no clear pattern or obvious single cause. Particularly troublesome was the tendency of the lightweight magnesium intake manifolds to rupture and/or catch fire following the backfire. The high intensity of such a fire led to numerous lost aircraft, injuries, and some fatalities. Clearly it was a problem requiring immediate resolution, as shown by the following excerpts from Air Corps Squadron Reports:

> Hamilton Field radiogram: "While flying squadron engineering mission on January 14, 1941 in P-40 type airplane, Air Corps No. 39-254, pilot dived from 22,000 feet and pulled out at 15,000 feet, air speed 400, carburetor automatic rich, with throttle approximately 25

Accessibility on early P-40 was good and the Accelerated Service Test showed that for a new engine and airplane the reliability was satisfactory by the standards of the day. Safety remained an issue until the Backfire problem was resolved. (Allison)

> inHgA. Pilot noticed cowling assembly, left and right hand, rip open (along) with cover leading edge gun faring, left hand, breaking loose, hitting and cracking left hand glass windshield. Engine began to run slightly rough and pilot returned to field. On inspection the top right and left hand cowlings were found damaged. Intake air duct on cowl shattered. Screen carburetor air scoop was bulged outward, indicating violent backfire. Also webs on supercharger inlet cover assembly were cracked, (as well as) a 1/2 inch crack in the accessory drive housing. A similar event occurred in P-40 airplane Air Corps No. 39-242 on January 28, 1941. While flying at 3,000 feet the pilot reported that the engine began to run rough and began to backfire and (he) returned to field. Condition of cowl assembly and blower section very similar to those listed above.[13]
>
> Selfridge Field letter of April 24, 1941, Subject: *Wreck of P-40 Airplane No. 40-343, engine A.C. No. 40-85.* A preliminary superficial examination revealed that the fire in this engine was confined to the rear section of the engine. The intake manifolds and intake pipes leading to the manifolds were completely destroyed by fire. Large pieces of ash containing metal that were apparently the result of burning of the destroyed parts were identified as part of the intake manifold system. In addition the crankshaft was broken at No. 5 crankpin. Both the No. 6 connecting rods had failed and the rod bearing showed evidence of having been subjected to excessive heat. It was Allison's contention, upon inspection of the engine, that the primary cause of the failure was seizure of No. 6 main rod bearing. This is not concurred in by the Power Plant Laboratory in as much as there was no evidence whatever of scoring or galling of the rod bearing or crankpin journal.[14]

Both the Materiel Division at Wright Field and Allison in Indianapolis focused on the problem and how to resolve it, though they did not always agree on either the cause or the corrective action. When John Kline, who had been intimately involved in the early introduction of the V-1710 into service, was asked during an interview with your author he reacted, "Its been a nice day up to now!"[15] Obviously the experience was traumatic for everyone involved.

The initial focus was on the backfire screens themselves.[16] They created a real dilemma, for their very presence adversely affected engine performance. The tee screens were believed necessary for proper mixture distribution between the banks. Still, they significantly reduced the critical altitude of the engine, and if possible, everyone would have agreed to their removal as a solution.

The first approach was to develop a screen with a finer mesh, thus intending to improve its effectiveness in limiting the passage of the backfire into the whole induction system. This was done even though it would further restrict high altitude performance. Allison did not prefer this approach, for it led to another very unique and serious problem.

During Type Testing and development in the mid-1930s it was found that the original screens would often foul after about 50 to 85 hours of operation. The fouling was caused by accumulations of the dye used in "ethyl" fuel additive to identify the resulting high octane fuel as containing lead. The dye had been required by the U.S. Surgeon General as a matter of health safety.

Accumulations could be fairly heavy. In fact the XV-1710-7 on Type Test suffered at times with as much as 8 inches mercury pressure drop across the backfire screens. Trying to develop takeoff power with this much loss would cause excessively high temperatures in the induction system, alone enough of a factor to lead to detonation and possibly backfiring. The proposed new backfire screens made of finer mesh caused a further reduction in critical altitude of 300 to 500 feet, even when clean. It was found that they tended to foul even faster.

In December 1940 Allison introduced a new induction manifold that featured an extension of the manifold divider wall down to the surface of the backfire screen. This was intended to prevent a backfire from spreading from one end of the bank to the other. Known as the "divided tee" manifold, it was put into service on the second production contract engines, i.e. all engines on Air Corps contract W535-ac-16323, and was retrofit onto V-1710's already in service (none of the 300 on the first production contract W535-ac-12553 had been built with the divided tee). The manifold was to be used with a "fine" mesh backfire screen.

One finding during the investigations was that the entire induction system was quite dirty and contaminated with oil.[17] It was determined that the oil was coming from the supercharger shaft oil seal and efforts were immediately begun by Allison to come up with an improved oil seal. This was done and a very effective "labyrinth" type seal developed. Upon introduction into service, on the first V-1710-27/29's powering the YP-38's, it was found that the problem of fouled backfire screens became even worse! It was then determined that engine oil in the induction system had been helping to "minimize" the accumulation of dye on the backfire screens; without it they fouled in 8 to 10 hours!

In the Spring of 1941, Allison again redesigned the oil seal in an effort to control the flow of oil vapors within the engine. Concurrently an extensive test and development program to improve engine venting on the P-40 was performed by both Curtiss and the Wright Field Power Plant Sections. Pressure surveys were taken inside and around the engine compartment during flight. The result was that the engine crankcase breather ports were relocated, one to exit the cowling just ahead of the right side exhaust stacks, the other just behind the right side stacks.[18] These changes effectively removed combustible vapors from the engine compartment. The supercharger oil seal vent was then routed into the engine compartment and pressurized with air from the spark plug cooling blast tube. These changes improved the venting of the engine, but really had little effect on resolving either the fouling of backfire screens or the backfire problem.

One proposed fix, given serious consideration, was to add an amount of engine oil to the fuel equal to the amount of ethyl fluid, about 4 cc's per gallon. This was rejected because of the difficulty in assuring proper addition and mixing as well as the known adverse impact on the fuels' octane rating. The V-1710-33(C15) was rated on 100 octane fuel, and as an "altitude" rated engine, needed every bit of octane benefit it could derive from the fuel to minimize the likelihood of destructive engine detonation at high altitudes. The interim resolution was to again revise the supercharger oil seal, this time intending to allow a nominal amount of oil to leak into the induction system, and thereby minimize the fouling of the manifold tee backfire screens.

Another tact attempted was to get the dye removed from all ethyl doped fuels. This was a monumentally difficult undertaking, but the effort was begun.[19] Using a new dye without the tendency to foul was investigated, and such a dye was developed. In the event, this approach was not successful as the long term effects of the new dye on fuel lines, fuel bladders and fuel system components was not known. At this point, with the military buildup for what became World War II well underway, there was neither time nor support for the development effort that would be required. Allison was stuck with dye in high octane fuel.

In-flight problems continued. The pilot would note a drop in power and see gray/white smoke coming from one bank or the other, an indication of a backfire. While not usually of the explosive variety, this internal burning would cause serious damage to the backfire screens, often times burning them away, as well as leading to magnesium fires if not stopped in time. More violent backfires could split the manifolds, as well as result in fires. Damage to the screens created other problems, specifically an increase in the pressure drop across them and they could become the source of metal fragments that would then flow into the intake valves and cylinders. Heated to incondensence, these particles could cause detonation and further backfires. At the minimum they could, and often did, lead to valves sticking open, as well as eroding the valve and cylinder wall surfaces. All of this, coupled with a susceptibility of some intake valves to lead attack and corrosion, led Allison on a program to develop an improved intake valve, as well as further improvements in the intake manifolds.

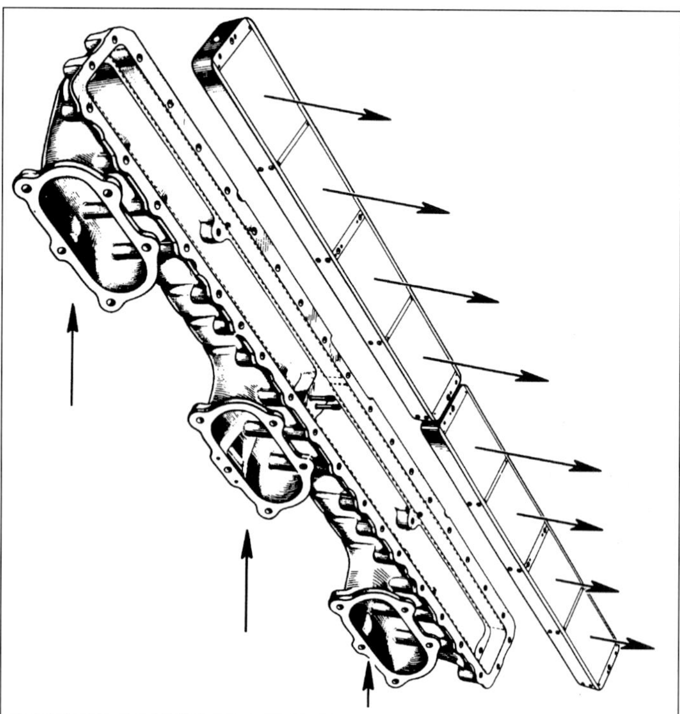

View looking up into the manifold spanning a single cylinder bank, and showing the backfire screens. Rolls-Royce had to revise the manifold to include a web in the middle of the manifold to limit any backfire to a single group of three cylinders. Wright Field wanted Allison to use these same plenum type backfire screens in the V-1710. The result would have been a loss of ram helping to pack mixture into each cylinder. (Courtesy of RRHT)

The Materiel Division at Wright Field felt that the major problem was that the screens were too far away from the intake ports.[20] They believed that the existing arrangement, with all six cylinders of one bank being "protected" by a single screen located at the tee in the distribution trunk, was increasing the chances of a backfire. They took two avenues for resolving the problem. First they designed a manifold with individual intake port screens, which required a total of 12 screens per bank, as a screen had to be placed in front of each of the dual intake ports serving the six cylinders on each bank. Modified manifolds, and enough screens to equip test aircraft, were flight tested in February 1941 to determine their effectiveness at controlling the backfires. The second tact was to design another intake manifold that could accommodate the backfire screen then being used in the Packard built Rolls-Royce Merlin. Two Merlin screens spanned the intake valves for a bank of six cylinders and had the effect of creating "plenums" behind both the backfire screen and the six pairs of intake valves.[21] Allison of course was reluctant to adopt this later arrangement. Their reason (other than the obvious one of being a competitors design) was that the V-1710 induction manifold was developed to provide a large amount of "ramming", accomplished by having individual intake trunks tuned to recover the maximum amount of momentum available in the pulsating flow into the cylinders. These Ron Hazen developed ramming manifolds were particularity effective in this regard.

In April 1941 a conference on the backfire situation was held between Allison and Wright Field representatives. Note was made of the number of significant issues with which they were dealing, and comment was included commending the cooperative attitude and results being obtained. One action item of significance from this meeting was recognition that in-service engines were showing a tendency to decrease intake valve clearance over time. Allison felt that this could be a major contributor to the backfire problem since insufficient clearance would cause intake valves to be "cracked open" when they should be tightly closed, thus leaking hot combustion gasses back into the induction manifold and initiating backfires. Field testing with the intake valve clearance set at 0.015 inches, instead of the then current handbook value of 0.010 inches, along with 300 hours of Allison factory testing, had shown improvement in reducing the number of backfires, as well as the tendency for the clearance to be reduced in service. The recommendation from the conference was for immediate implementation of the revised tappet settings by Tech Order Change to all V-1710-33's.[22]

On May 5, 1941 the Materiel Division notified Allison that they were canceling the divided tee and fine mesh manifold retrofit program in favor of saving their money and expediting the availability of the new manifold with individual intake port screens.[23] Allison believed that this action was taking a significant chance relative to the cause of backfires. Furthermore, the Materiel Division only intended to obtain enough of these manifolds to install on 168 engines in P-40's scheduled for overseas duty; their intent being to move on to the Rolls-Royce type backfire screens for all of the other existing, and subsequent production engines. At this point the Unsatisfactory Reports on the backfire problem were continuing to pour in. The situation was very serious.

The reason for the Air Corps to change its decision about adopting the divided tee manifold and fine mesh screen was a fire in an V-1710-27 engine in a YP-38 that had just received the new divided tee manifolds with medium mesh backfire screens. On May 10, 1941 this decision was again revised to once again require installation of divided tee manifolds on all engines of the first production batch, (those procured on W535-ac-12553) as well as changing the screens in all engines to be the "medium" mesh screens then being used in the V-1710-27/29's powering the early P-38's.[24]

The British Air Commission responded to a questionnaire from the Chief of the Air Corps on May 8, 1941 about their experience in operating the Allison V-1710-C15 powered *Tomahawk* in the deserts of North Africa. The British reported that they had "no trouble due to backfires or rusting airscrew thrust bearings, nor oil pressure trouble at high altitudes." Regarding the engines in general, "requires less maintenance than British types. Most unserviceability due to lack of spares. Position not improving."[25] The British V-1710-C15's were built with the original magnesium manifolds with manifold tee type backfire screens. They had reset their intake valve clearance at 0.015 inches and were using hollow stem intake valves with the engines operating at a rated speed of 3000 rpm. Significantly, they also had the "pendulum type" oil tank pickups and had overcome the initial overspeeding troubles with the Curtiss Electric propellers. This experience gave confidence to Allison that the backfire problems could be resolved without resorting to the redundant backfire screens and associated restrictions to flow in the manifolds being directed by Wright Field.

On May 15, 1941 Wright Field summarized all of the different recommendations and opinions, along with the various directives to date applicable to Army Air Corps P-40's:[26]

- Install new aluminum intake manifolds with individual port screens in combination with the Allison tee screens.
- Install pendulum type oil tank to insure oil supply in negative "g".
- Install changes to insure more reliable propeller operation.
- Set intake valve clearance at 0.015 inches.
- Install ceramic spark plugs in place of 2-piece plugs.[27]
- Relocate crankcase breathers to right side of cowling, fore and aft of Connect the supercharger oil seal vent to the spark plug cooling air blast tube.

On June 9, 1941 Air Corps issued Technical Order 02-5AC-9 with Specific Operating Instructions for V-1710-33's that were to be equipped with individual port type backfire screen intake manifolds. These were manifolds with P/N's 40772 and 40773 installed in Air Corps P-40, P-40B and P-40C airplanes.[28]

By September 19, 1941 the situation in the field had not materially changed, and the Chief of Experimental Engineering at Wright Field wrote to Allison demanding, "...that all V-1710's be provided with backfire screens near the intake valves as that should completely and effectively eliminate backfires, unless the Materiel Division is provided with an alternative that they can test that indicates that there is no danger from fires or damage to the engine and its induction system when backfire screens are not employed."[29]

An interesting perspective on the situation! The direction for resolving the problems is contained in comments made to the Chief of Air Staff by O.P. Echols, Chief of the Materiel Division on September 10, 1941, when he notes that, "Aluminum alloy manifolds will be furnished on future V-1710-C15, F-3R and E-4 engines, as soon as present supply of magnesium manifolds is depleted and aluminum type are available." He goes on to say:[30]

> The aluminum alloy intake manifolds for use with the port backfire screens are not furnished on any British engines. The port type manifolds and screens were incorporated in Air Corps V-1710 series engines to reduce or prevent fires which had been experienced in flight. Airplane fires in flight on all P-40 series airplanes have been eliminated as a result of the above change plus other modifications, as evidenced by complete absence of unsatisfactory reports. The British engines do not have these port backfire screens because their use was set up as a requirement for Air Corps engines at the insistence of this Division, and as this organization has no jurisdiction over British engines, the requirement of the Air Corps were not made applicable to British procurement.

At this point in time Allison had delivered to the British, 1,650 V-1710-C15, 396 V-1710-F3R and 236 V-1710-F2R/L engines, so there was a considerable base from which to establish in flight operating conditions.

As of October 1941 Allison was still contending that port backfire screens were unnecessary. They acknowledged that the new manifolds, those with port backfire screens, had significantly reduced the occurrence of backfires and explosions, but maintained that the change to stronger manifolds, made of aluminum rather than magnesium, was the essential factor. To this end a test was begun in late October using two P-40's, both having the new manifolds, but in one aircraft the port screens were to be removed. In the meantime, Air Corps engines were being fitted with manifolds having both tee and the port type backfire screens.

According to Don Wright, then in Allison's Service Department, the screens were viewed "almost [as] an option, most people didn't want backfire screens. You lost some power. What would happen, the manifold there where it joined with the supercharger would get [a slight] offset, so we put a little bumper [on] it to hold it in line, finally we got it aligned pretty well and it would do OK, unless you got a backfire from a bad plug or something, it really raised hell with the intake. So you figured the backfire screens were going to take care of that. Really, what took care of it was cleaning up the engine, it didn't need those things, but as I remember it, most of the time [that] those backfire screens were in the inlet ports [they] finally just disintegrated if you left them in there long enough.[31]

The American Volunteer Group, the AVG or *Flying Tigers*, began operating their *Tomahawks* in Burma during the late summer and fall of 1941.[32] They had precious few spares, other than the 150 engines purchased to support the 100 airframes. Just prior to Pearl Harbor, December 7, 1941, the AVG had Pan Am air-freight them 400 backfire screens. This suggests that they were being consumed at a high rate. This number would have provided two spare sets for each aircraft originally on strength. War for the Flying Tigers started following Pearl Harbor, and by March 1942 the Allison's in the *Tomahawks* were getting tired, worn and badly in need of overhaul. Many engines were no longer able to deliver full power and detonation was a frequent complaint, though a review of AVG files does not show problems of epidemic proportions. Their Allison Tech Rep, Mr. Arne Butteburg, determined that many of the performance problems were the result of poor maintenance technique, and most specifically, that carburetors were often "cleaned" while installed by simply pouring raw gasoline through them. Coupled with operations in dusty conditions, the result was plugged sensing ports in the carburetors, along with heavy deposits of dirt in the backfire screens. In one case the backfire screens were so dirty a "Shark" was only able to develop 25 inHgA at "takeoff", when they should have been producing 41 inHgA. In mid-March 1942 an aggressive and thorough cleaning and maintenance program was instituted that restored power and operability.[33]

In July 1942, after an extensive development and test program, Allison was successful in getting Materiel Command approval to *remove* both the tee and port backfire screens from the altitude rated engines in the P-40 and Bell P-39.[34] Key to this decision was another new Allison intake manifold tee, PN42790, this one designed to provide constant velocity of the charge through its entirety. Made of aluminum, and with its passages having a smooth oval cross-section, it was known as the "streamlined" manifold. It achieved the proper mixture distribution without the troublesome tee backfire screen. This eliminated the backfire screen pressure drop and the requirement be cleaned every 8 to 10 flight hours. P-40 and P-39 aircraft with the new streamlined manifolds gained 2,400 feet in critical altitude, where they also enjoyed an increase in speed of 7 to 10 mph. To qualify the new manifold and provide a basis for removing the screens, a P-39 was flown and approximately 80 backfire events induced using the old manifolds and screens. While there were no fires, the screens were destroyed to the point that the engine had to be replaced. The new engine, with streamlined manifolds and no screens was then flown, and some 120 backfires purposely induced. This testing slightly bulged the elbow in the intake manifold system, but no fires or serious damage resulted. Another 2,000 backfires were accomplished under a variety of conditions on engines in test stands with similar results.

The likelihood of backfiring had been reduced by the previous reintroduction of sodium cooled intake valves, and adoption of the increased intake valve clearance. A further reduction in likelihood of backfiring was expected by not having the screens and the known problems that occurred when they deteriorated. With the removal of the backfire screens it was also possible to go back to the oil tight supercharger oil seal that would significantly reduce the overall engine oil consumption while improving spark plug life.

Expedited special testing of the new manifolds in the P-38 then cleared them for universal application in September 1942, when the Materiel Command approved removing the backfire screens from engines equipped with the streamlined manifolds in turbosupercharger installations, i.e. the P-38's.[35]

In early 1943 another spat of backfiring occurred in some P-39F, P-40D and P-40K airplanes. This time the trouble was traced to malfunctioning Automatic Mixture Controls, and/or conditions in that the flow through the carburetor was so low that it would cause lean mixtures. These conditions were likely during gliding flight and low power operation. Improved training in piloting technique was instrumental in solving the backfire problem this time.[36]

Backfiring again became a problem following the War, with the installation of the high-powered, two-stage V-1710-143/145's in the North American P-82E/F Twin Mustangs. These engines had a Takeoff rating of 1600 bhp and under War Emergency Ratings could achieve 2250 bhp when water injected, but some claim that those engines did not enjoy a well supported development program. The allegation being that the emphasis was then on building jet engines. Problems with backfiring delayed entry into squadron service and ended the V-1710 program on a somewhat sour note in the eyes of some observers. As a replay of previous actions, North American designed a new intake manifold that incorporated port type backfire screens, and Allison adamantly opposed and prevented their adoption.[37] A lot of pilot engine operating technique and induction system maintenance was necessary to have adequate and reliable performance in this high performance engine. Also, since the aircraft were not then expected to be flying combat missions, the authorization to use the 2250 bhp WER rating was withdrawn.

Intake Valves

The reason for the valve clearance to "close-up" during service was believed by Allison to be caused by "over-revving" of the engines. Several reasons for over-revving were believed to be present, including poor piloting technique, but mainly the cause was poor performance of the Curtiss Electric propeller governor in the early P-40. To this end Ron Hazen, Allison's Chief Engineer, proposed on May 9, 1941 that the Air Corps change the maximum rated rpm for the V-1710 from 3000 to 2800 rpm.[38] The belief was that over-revving caused the valves to "pound" their seats, effectively increasing the length of the valve (by shortening the "height" of the aluminum cylinder head), with the result being observed as a reduction in valve clearance and an increased tendency to backfire due to slightly open intake valves, even when the valve was "closed." Technical Order 02-5A-13 was issued to implement the change from 0.010 inches to 0.015 inches, in May 1941 as a way to provide a margin of safety.

In June 1941 Allison determined that a new sodium cooled intake valve, PN41164, though having a thicker "stem", did not adversely effect flow through the cylinder intake ports and that they would incorporate it into production "E" and "F" engines. It was also feasible to retrofit into the "C" engines, such as the V-1710-33. On September 11, 1941 Wright Field directed Allison to deliver 50,400 sodium cooled intake valves for installation in the 2,100 engines expected to have been delivered to the Air Corps as of October 1, 1941.[39] The Air Corps insisted that these be provided at no cost under the "latent defect" clause of the Allison engine contract. These valves were not retrofit into the engines purchased by the British, though subsequent deliveries of Lend-Lease engines were so equipped.

All engines then in Air Corps service, as well as the subsequent production engines, received sodium cooled intake valves. While this significantly improved the safety of the V-1710-C engine, indications were occasionally still found in some engines of "pounding" of the aluminum-bronze valve seat inserts for the intake valves. This was resolved on the V-1710-E/F engines by providing steel valve seat inserts.[40]

Changing Engine Ratings

Allison maintained that the poor performance of the Curtiss Electric propeller and governor resulted in frequent engine overspeeding. The result was often backfiring as well as failure of the heavily loaded number 6 main crankshaft bearing (forward end of the crankshaft). With the concurrence of the Wright Field Propeller Laboratory, the Materiel Division authorized the rerating of the V-1710 at 2800 rpm for takeoff as of June 12, 1941.[41] This direction was issued by Wright Field in service operating instructions for V-1710-33, -27, -29, -35 and -37 engines as early as January 1941 noting that,

> "operation of the propeller governor of the Curtiss propeller used on this engine series is not satisfactory under some conditions. Repeated instances have been reported from...where the propeller governor is set of 3000 rpm during high power, when sudden opening of the throttle may cause the engine to over-rev as high as 3600 rpm before the propeller governor acts to increase pitch and reduce rpm."[42]

Until such time as very fast operating propellers became available on all fighter aircraft, Allison requested that operating instructions should limit maximum engine speed to 3120 rpm. The adoption of these new take-off ratings coincided with the installation of the new intake manifolds with both port type and Tee backfire screens. The result of all of this was that at 2800 rpm, the same MAP was able to deliver the same power as before, but with a considerable margin before overspeeding. Military ratings continued to be specified at 3000 rpm, necessary to maintain the desired critical altitude.[43] The reduced speed gave the same power as before because of the reduced power needed to drive the supercharger, along with a small reduction in engine friction at the lower speed.

To further resolve the problem, in July 1941 the Air Corps issued a Technical Order to install a new type of propeller governor on all P-40 airplanes. In addition, the commutator brushes and holders in the Curtiss Electric propeller were to be replaced with a new "impregnated" type by that September, as well as fitting a low-pitch stop to the props to minimize the possibility of overspeeding.[44] As a further precaution, the Wright Field Inspection Division issued requirements that engine operating instructions were to be issued including the following precautionary instructions:[45]

> (1) V-1710-33 engines now in service must continue to be operated under present restrictions until such time as they are brought up to model test standards.
> (2) Instructions must be issued to operating personnel specifying that a maximum engine speed of 3120 rpm will not be exceeded during take off or in flight.
> (3) For normal peace time operation from fields in good condition a propeller governor setting of 2770 rpm will give excellent take off characteristics, and eliminate danger of over-revving.

Separately, on April 8, 1941, the Power Plant Laboratory had recommended that the remaining unmodernized V-1710-33's continue to be restricted as shown in Table 4-2.[46]

Table 4-2
Unmodernized V-1710-33 Ratings

Takeoff	950 bhp at 2770 rpm at Sea Level
Military	950 bhp at 2770 rpm at 8,000 feet
Normal	838 bhp at 2600 rpm at 8,000 feet

This is a very early V-1710-33(C15), AEC #155, AC 39-853M, being readied for flight in the 1990s. It was delivered 6-28-1940, not long after AEC#150, the final Model Test engine. The "M" in the s/n shows that it was later "Modernized" to bring it up to the Model Test standard. Detailed inspection shows that modern builders have significantly improved the engine by using many later model components. Including intake manifolds, coolant piping, ignition system without gun synchronizers and liberal use of improved internal components should make for a reliable engine. (Author)

The definitive basis for rating the engines at 2770 rpm instead of the recommended 2800 rpm has not been determined, though two reasons are possible. The first is the governmental way of providing "margin" on the 2800 rpm recommendation from Allison, i.e. the 2770 value is 1 % under the "limit" and should therefore preserve any guarantees offered by the manufacturer. The other reason is that the "C" engine entered the 4-1/2 order torsional vibration region at 2800 rpm. This of course was mitigated by the pendulum damper on the crankshaft, and should not have been a problem even at 3000 rpm, though with the concerns about the crankshafts in unmodernized engines, this was likely the constraint. The setting of limits is often the consequence of factors entirely separate from engineering requirements.

Allison continued to work to improve the main and connecting rod bearings with the result that by mid-1941 production bearings allowed safe operation into the range of 3700-3800 rpm during dives with V-1710-E/F engines. Since the engines ran very smoothly in the 3000-3500 rpm range it was not unusual for the pilot to be unaware that the engine was actually overspeeding. As usual, implementing such changes took time. For example, the V-1710-85(E19) takeoff rating in the P-39 was changed from 1200/2800 to 1200/3000/51.5 inHgA, but not until November, 1942.[47]

Consequently when one encounters handbook values on the ratings of various Allison engine models it is easy to get quite different ratings for the same model and aircraft installation, depending upon whether or not backfire screens were installed, the rated rpm, and whether or not takeoff or WER ratings were given. The values depend upon *when* the reference was written. Since altitude rated engines were required to be "throttled" for sea-level takeoff conditions, full power was still available at the reduced rpm. It was just that the manifold pressure would have to be slightly higher. At altitude, the effect was to reduce the maximum height at which the engine could still produce rated power. This was not too much of a hindrance, for few if any, of these rerated engine installations actually got into combat. In any case, Allison guaranteed that the same power would be available for takeoff at the lower rpm by operating at the slightly higher manifold pressure. Furthermore, during the summer of 1942 a program of re-rating all models to provide for War Emergency Rating (WER) power levels was under way by Allison, providing much increased performance in combat conditions.[48]

The restrictions were lifted on modernized engines. Modernization was required on the first 228 V-1710-33 engines, those whose manufacture preceded the completion of the successful 150-hour Model Test.[49] During modernization they were fitted with the stronger crankcase then being used in production.

Other differences in the W535-ac-12553 engines included their having plain copper-lead main bearings, able to operate at up to 3500+/-100 rpm during dives, providing that an uninterrupted oil supply was maintained at the pump inlet. The follow-on contract W535-ac-16323 engines had an additional lead coating on their main bearings and would stand 3600+/-100 rpm. None of these early engines had automatic boost regulators.

Reduction Gear Limitations

From the beginning the power limiting component in the V-1710-C engine was the propeller reduction gear assembly. In the effort to provide the maximum amount of streamlining, (and it was believed to get the lightest reduction gear), Allison had utilized an "internal" spur gear design driven by an "over-hung" crankshaft pinion. The "driven gear" was set in a necessarily large diameter ring gear with "internal" teeth. This in turn was supported by a large "plain" bearing, fourteen inches in diameter. With the engine qualified to run for short periods at up to 3700 rpm, the sliding

SECTION III
SPECIFIC OPERATING INSTRUCTIONS

ENGINE: V-1710-33
P-40, P-40B, P-40C

RESTRICTED

CONDITION	FUEL PRESSURE LB/IN.^{2}	OIL PRESSURE LB/IN.^{2}	OIL TEMP. °C	COOLANT TEMP. °C
DESIRED	12-16	60-70	60-80	105-115
MAXIMUM	16	85	95	125
MINIMUM	12	55		85
IDLING	9	15		

MAX. PERMISSIBLE ENGINE OVER SPEED: 3120 R.P.M.

MAX. ALLOWABLE OIL CONSUMPTION AT:
NORMAL RATED POWER 12.4 QTS./HR.
MAXIMUM CRUISING 9.3 QTS./HR.
MINIMUM SPECIFIC FUEL FLOW 5-7 QTS./HR.

FUEL GRADE 100 OCTANE

⊕ OPERATING CONDITION	HORSE POWER	R.P.M.	MAN. PRESS. (IN. HG)	PRESSURE ALTITUDE (IN FEET)	BLOWER CONTROL POSITION	USE LOW BLOWER BELOW	MIXTURE CONTROL POSITION	MIN. F/A RATIO	FUEL FLOW GAL/HR	MAX. CYL. HD. TEMP. °C	REMARKS
TAKE-OFF	1040	2800	40.6	Sea Level	-	-	Auto-Rich	-	104	-	5 Minute Operation Only
MILITARY RATED POWER	1040	3000	37.2	14,300	-	-	Auto-Rich	-	115	-	5 Minute Operation Only
⊙ NORMAL RATED POWER (100%)	930	2600	33.7	12,800	-	-	Auto-Rich	-	90	-	-
MAX. CRUISING (75%)	697	2280	27.9	12,800	-	-	Auto-Rich	-	59	-	-
DESIRED CRUISE (67%)	623	2280	25.4	12,800	-	-	Auto-Rich Auto-Lean	-	53 48	-	-
DESIRED CRUISE (60%)	558	2190	23.9	12,800	-	-	Auto-Rich Auto-Lean	-	47 42	-	-
CRUISE FOR MIN. SPECIFIC FUEL FLOW	400 450 495 520 450	1950 1950 1950 1950 1950	24.2 24.2 24.2 24.2 FT.	Sea Level 5,000 10,000 15,000 20,000	- - - - -	- - - - -	Auto-Lean Auto-Lean Auto-Lean Auto-Lean Auto-Lean	- - - - -	30 34 37 39 34	- - - - -	- - - - -

⊕ REFER TO T.O. NO. 00-10 FOR DEFINITION OF EACH OPERATING CONDITION ⊙ MAXIMUM PERMISSIBLE CONTINUOUS HORSE POWER REVISED 12-1-41

T.O. NO. 02-5AC-1

This page from the pilots operating manual clearly shows the 1040 bhp rating as approved for use by the Air Corps.

velocity in this bearing was an extremely high 13,561 feet per minute, or 154 mph. To keep everything working smoothly Allison directed almost 1/3 of the engine oil supply, at a pressure of about 75 psig, into the bearing. Still it was enough of a concern that early development engines were equipped with a bearing temperature indicator. If the temperature increased abnormally it was time to significantly reduce power and soon land. All of this meant that there was a limit to the amount of power that could be reliably transmitted. The value was set at 1150 bhp, though the V-1710-19(C13) Hi-BHP test successfully operated at up to 1575 bhp at 3000 rpm.

Allison was quite adamant about not overpowering the V-1710-C15 engine, stating that "on account [of] extreme over-load on [the] over-hung pinion at the higher horsepower's involved" in operating at takeoff manifold pressures up to the critical altitude of the engine.[50] The limitation was not the bearing, but rather the bending stresses developed in the front cheek of the crankshaft as a result of the high loads transmitted to the crankshaft extension carrying the pinion gear.

In early 1935 Wright Field asked Allison to investigate the feasibility of designing a reduction gear with conventional bearings for the engine. Such a unit was designed, and included provisions for a bearing to support the outboard end of the crankshaft and thereby eliminate the troublesome bending loads on the front crankcheek. Unfortunately, considerable redesign of the engine would have been required, including not only the new reduction gear assembly, but significantly, new upper and lower crankcases, new crankshaft, accessory drive system and considerable revision to other detail parts.[51] While the design was proposed for fitting to the X/YV-1710-3 and XV-1710-5 engines, it was determined to not be practical during the Type Test period. With the opportunity to completely redesign the engine following the Type Test in 1937, these features were incorporated in the "E" and "F" series engines, which used straight cut "external" spur gears. As developed, the E/F design followed closely the 1935 concept, and resulted in lighter and more compact reduction gears than the original "C" gear set. This created a higher "thrust line" that reduced frontal area. Allison got their start in the engine business by building high powered gearing and drives, and these reduction gears benefited from all of that experience.

In September 1940 the Air Corps issued a Technical Order specific to the early production V-1710-C/D engines. It required inspection and reworking of the reduction gears on overhauled engines that had had any of their gears that mated with the large internally cut reduction gear replaced. Allison was having trouble maintaining tooth profile on the internally cut teeth. If galling or burring was found, the fix required hand stoning of the damaged tooth faces following by further running the engine, disassembly and re-inspection.[52] Evidently a similar procedure had been followed during original manufacture an assembly of each engine.

Within months of introduction into service on the P-40, reports of rust and corrosion in the V-1710-33(C15) reduction gear began coming in. The problems were in the Propeller Thrust Bearing and also between the two propeller shafts. As early as January 1941 Allison acknowledged that excessive pitting and rusting of the Propeller Thrust Bearing was occurring.

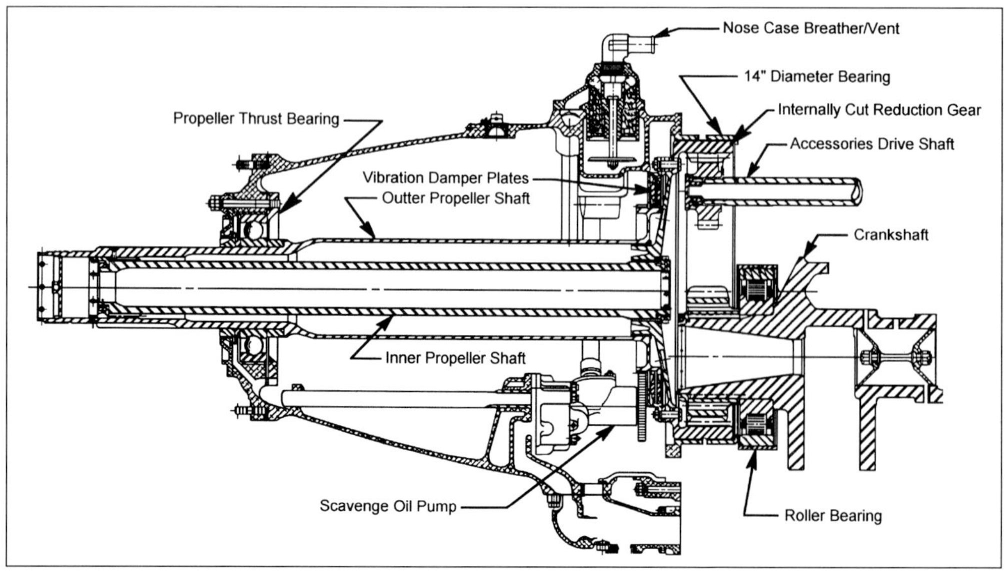

The V-1710-C15 reduction gear rode on a 14 inch diameter plain bearing and was driven by an overhung pinion gear on the end of the crankshaft. The accessories drive was taken from the same internally cut ring gear. The dual propeller shafts were adopted to incorporate the vibration damper. As a result the outer shaft carried the weight and/or thrust of the propeller while the inner shaft supplied the torque. The propeller thrust bearing was not initially provided with an oil supply adequate to insure it was coated when the engine was shutdown. About 5 cfm of air was normally drawn into the engine through the thrust bearing causing it to rust if the air was humid and the bearing not coated with oil. Allison solved the problem by fitting a spray nozzle to direct a small stream of oil onto the backside of the bearing. (Author)

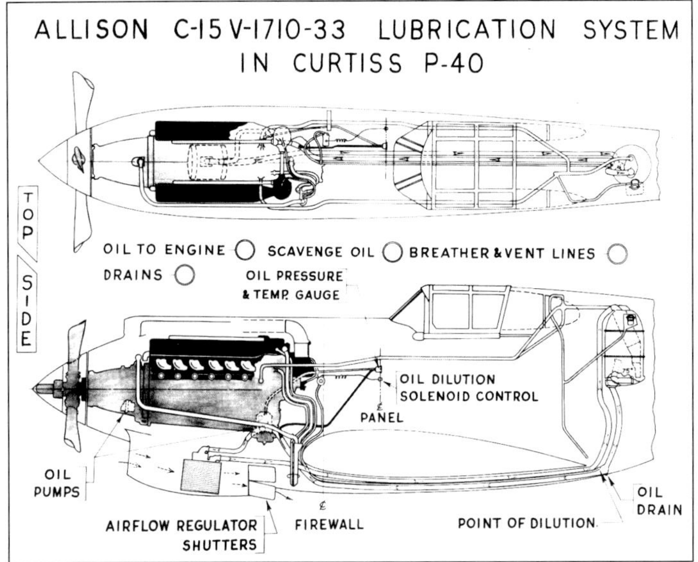

The Tomahawk lubrication system was fairly straight forward, and included features such as dilution to aid in cold weather starting. Hot oil was collected within the engine by the scavenge pumps and directed through the oil cooler to the oil tank. Note the drawing shows the oil tank pickup needed for inverted flight. The oil pressure pump drew from the tank and delivered 60 psi oil to the bearings. (Allison via Hubbard)

The problem appeared to be due to the location of the gear case vent, and the resulting high crankcase vacuum.[53] The consequence was that it was impossible for any oil vapor to reach this bearing, coupled with humid air being drawn in as the engine cooled following normal operation. It was found that lubricating oil was not adequately wetting all of the surfaces of the propeller shaft and rear face of the thrust bearing. In early March 1941 Allison conducted a test program of different nozzles that would direct a stream of oil onto the surfaces prone to rusting. They devised a field mountable external tube and nozzle assembly that obtained its oil supply from the pressure line in the governor pad, and for production engines, a similar nozzle was mounted inside the reduction gear assembly accomplishing the same thing.[54]

No effective fix had been implemented by the time the engines for the AVG were built. Consequently, when the AVG began their operations in the humid jungles of Burma they immediately began to observe rusting. During October 1941 the Allison representative to the AVG devised his own fix and tested it on the first AVG assembled airplane, P-8124 with V-1710-C15 AEC #946. This engine had already developed a rusty thrust bearing. He installed a stainless steel dam in the bottom of the gear case. The purpose was to retain a pool of oil sufficient to insure everything was oil covered, though it required hand rotation of the prop after cooldown to completely wet the bearing. This worked quite well for the AVG, but required the ground crew to hand rotate the prop at the end of each day, as well as prior to startup every day.[55] Allison also incorporated a production procedure to cover the two concentric propeller shafts with soft grease. This succeeded in solving that portion of the rust problem.

Hopper Oil Tank and #6 Main Bearing Failures

The Army Air Corps' original P-40's, and some of the early P-40B's, were built with "hopper" type oil tanks. These were adequate for positive "g" operations, but negative "g" would deprive the engine oil pump of its source of oil, as all of the oil would be at the "top" of the hopper. The highly loaded bearings in the V-1710 demanded a constant supply of high pressure oil. Allison maintained that when the oil supply was interrupted the heavily loaded number 6 main bearing (the one at the front of the engine) would quickly "wipe", allowing metal-to-metal contact and soon result in a failed bearing and broken crankshaft.[56,57] The British *Tomahawks* had a different oil tank, one with a pendulum type pickup that was free to "flop" from the bottom to the top of the tank during periods of negative "g". These tanks were to be retrofitted onto the Air Corps airplanes, but the work was slow to be completed. Fortunately the British in North Africa and the AVG in China-Burma all used British contract *Tomahawks* and did not have the problem.

Lockheed P-38 models prior to the P-38H were originally equipped with hopper type oil tanks and would also suffer damaging loss of oil pressure during certain maneuvers. Many of these early P-38's were being used within the U.S. in training roles, and as late as January 1943 there were no plans to retrofit the pendulum equipped tanks. The result that the loss rate of engines and aircraft was unnecessarily high.[58]

Summary of Early Operational Problems in P-40 Aircraft

In January 1941, nearly a year before the beginning of the war, Allison summarized the problems with operation of the V-1710-33 in the Curtiss P-40. These troubles were felt to be more or less serious in nature in that they often resulted in engine failure, whether due to the engine or the installation. The following are presented in the general order of their first appearance.[59]

Exhaust Ignition Lead Sleeve Failures
Allison provided new sleeves made of phanolic for all engines in service, although the Air Corps was not always prompt in their installation.

Oil Pressure Failures Resulting in Burned Out Bearings
Three or four engines were affected. The trouble was attributed to the viscosity valve on the oil cooler in combination with the plumbing arrangement of the oil hopper tank. At the time of the report only two or three of the Air Corps planes had been equipped with the "export" tank, which would resolve the problem. Experience by the 79th Pursuit Squadron at Hamilton Field had found that operating the engine for more than 30 seconds with oil pressure below 50 psig would result in bearing failure.[60]

Roller Bearing Press on the Crankshaft
Only the first few P-40's were effected as Allison had early-on increased the press of the bearing on the crankshaft.

Front Crankshaft Roller Bearing Lock Ring
A couple of instances of the lock ring loosening resulted in an immediate change in the design to minimize likelihood of improper assembly.

Over-revving Due to Unsatisfactory Propeller Governing
Two cases had occurred where engines were washed-out due to governor malfunction allowing the engine to exceed 3500 rpm. There were many instances where the engine speed increased into the 3400-3500 rpm range, that did not result in reports of damage to the engine at the time.

Oil Dilution Valve Leakage
A large number of reports had been received complaining of bad leakage during operation, resulting in very low oil pressure. The problem

was that the valves supplied by the airframe manufacture were not designed for use with a pressure carburetor.

Thrust Bearing Rusting and Failure
During the first cold weather operation of the P-40's a number of failures, or near failures, occurred due to the high crankcase vacuum, a consequence of the location of the breather vent port on the aircraft. The pressure was preventing any oil vapor from reaching the bearing. Allison responded by immediately providing a spray nozzle to insure positive flow of oil to the bearing. Wright Field was continuing to work on a revised vent location. This problem remained for units in the field such as the AVG.

Prestone Leaks
Several UR's were received reporting coolant leakage from faulty castings. Foundry and inspection methods were improved as a result. Some hose and hose clamp trouble was experienced on early installations, but improved hose and hose clamps, which required a change in the Air Corps Specifications, resolved the problem.

Faulty Assembly
During the accelerated service testing of P-40's at Fairfield Army Air Depot, (located at Patterson Field, Dayton, OH), one case of a rag, and another of a rubber washer blocking oil supply to a bearing were found. Allison changed its assembly and inspection methods to resolve the problem. One step was to do final assembly in a separate room where rags were issued and controlled. This resolved the problem.

Crankshafts
Two crankshafts had failed, both occurring in the #11 cheek in front of the #6 bearing at the very front of the engine. The cause was due to bending resulting from use of a test stand during the final factory run-in that had undesirable vibration characteristics with a critical 1-1/2 order vibration occurring at 2000 rpm. Use of the questionable test stand was discontinued and additional attention was paid to insure a smooth running test propeller was used. The problem did not occur during flight operations.

Icing in Carburetor Spray Nozzle
Thought to be impossible with a pressure carburetor, the problem was found to be due to icing in the vent space of the spray nozzle. Improved venting resolved the problem.

Generator Drive Gear Failures

World War II between the U.S. Allies and the Axis powers officially began for the U.S. on December 7, 1941, but at the same time an international confrontation was occurring between the Allies over failures of the generator drives on the V-1710-C15 in the Curtiss *Tomahawks*. In a Priority Teletype from London General Arnold was notified "Generator Drive Failures have happened and situation, with imminent possibility of eventual grounding of all *Tomahawks*, is assuming serious proportions. Generator coupling has not enough damping characteristics. We are still experiencing difficulties in spite of (modification) efforts."[61] Averell Harriman, U.S. Ambassador, wrote to General Arnold on December 8, 1941 that "since British received first *Tomahawk* continuous trouble has been experienced with C-15 Allison engine generator drive as a result of torsional vibration.

The majority of the Tomahawks were in service with the British in North Africa, where the first confirmed enemy aircraft, an Italian bomber, was destroyed by a Tomahawk on June 8, 1941. The aircraft was still going through its introductory service trials with the U.S. Army at the time. While there were systems and specific mechanical problems during this period, for the most part the aircraft and its V-1710-C15 engine performed at least as well as the other V-12's in the Theater, the Rolls-Royce Merlin in the Spitfire V's and Daimler-Benz DB601 in the German Messerschmitt. (Photo taken September 9, 1941. Allison)

This V-1710-C15 is in the Capetti Collection of the Politecnico di Torino Museum in Turin, Italy. It is a very early British C-15 with Allison s/n 479, brought to Italy for test and evaluation following capture during the North African campaign. Evidently a controllable Curtiss-Electric propeller was not available so a fixed pitch wooden propeller was fitted, along with non-standard exhaust stacks. (Politecnico di Torino Museum)

All C-15 generator drives now have new (coupling) rubbers and elongated slots...still the failures are occurring. New drive gears are required to replace those damaged prior to replacing the couplings." Continuing, *Tomahawks* are distributed approximately as follows: 500 in the Middle East, 300 in England, 200 in Russia and 100 in China.[62]

The situation was particularly acute for the Russians, who in the midst of a severe winter reported that all of their P-40's were grounded for want of generator drives. A measure of the critical situation in Russia, which was fighting a Nazi invasion, was a report from Harriman that, "in some instances Russians have removed generators and are attempting to fly their *Tomahawks* on batteries, but if batteries are run down, flaps and landing gear must be operated by hand. Guns cannot be fired and propeller pitch cannot be controlled."

The problem had been known for some time, and Allison had believed that it had been resolved by revising the couplings and providing new rubber dampers.[63] When the problem resurfaced Allison designed heavier gears and provided same to the Allies for field installation. Unfortunately many of these were not promptly delivered, for example 168 sets were being held at the British Embassy in Washington, while they awaited delivery instructions. To resolve the problem Allison began immediate production of an additional 500 sets of gears and couplings to reissue to the field.[64]

With continuing failures, and the observation that the new coupling was not much of an improvement in sub-zero temperatures, an investigation was begun to completely resolve the problem. It was found that the originally specified C-15 Model Test for the British had been run with a 500 watt generator. The British separately specified that the minimum size generator for use on the *Tomahawk* was to be rated for greater than 900 watts, but the unit they had procured was rated at 1,500 watts. The result was that the drive was dramatically undersized for the greater torsional loads being imposed upon it. The final resolution was for the British to obtain a generator better suited to the installation and needs of the *Tomahawk*. In mid-January 1942 they began receiving Leese Neville M1 generators which had been found to work without problems. These were rapidly dispatched to the British acquired *Tomahawks* around the world, including those in Russia, England, the Middle East and China. The generators were individually packaged for shipment and included a generator, new control panel and a set of replacement gears. Seventy-five sets were shipped to China for installation in the remaining AVG *Tomahawks*.[65]

Service in China with the AVG, *The Flying Tigers*

On June 24, 1941 production was completed on 150 V-1710 engines for General Chennault's First American Volunteer Group (AVG) of China *(The Flying Tigers)* ...and so usually begins *and ends*, the story of the Allison engines that powered the "Sharks" in their epic combats of 1941 and 1942 over China and Burma. In fact there was a lot more. It involved the men and women of Allison, and the tireless technicians, mechanics, and pilots who maintained and operated their Allison's under the most primitive of conditions.

When Allison received the request from China in January 1941 for engines to power their batch of *Tomahawks* for the planned AVG, no production capacity was available. The British had previously reserved all available production.[66,67] Still Allison negotiated an agreement for the engines with the Universal Trading Corporation of New York City, the Chinese representative in the U.S. The transaction itself required approval of the U.S. Government, who in March 1941 was busy notifying Allison that they would have to go through proper channels to obtain approval for any such sale.[68] The supply of engines was critical and the need great during this period.

Table 4-3 **Ratings for V-1710-33(C15) Engines**
1040 bhp at 3000 rpm for Takeoff, with 41.0 inHgA MAP
1090 bhp at 3000 rpm to 13,200 ft, with 38.9 inHgA MAP
960 bhp at 2600 rpm to 12,000 ft, with 35.0 inHgA MAP

The British had, in effect, "traded" the 100 *Tomahawks*, without engines, in exchange for a larger number of *Kittyhawks* supplied by the U.S. as Defense Aid. The arrangement was that the Chinese needed to obtain their own supply of engines and spares. Someone at Allison remembered that they had a warehouse of "off-dimension" parts, those that did not meet either U.S. Army or British contract specifications, but that were otherwise sound. They proposed that by hand fitting, matching and repairing these parts, suitable engines for the spare order could be provided. The salvage and repair techniques used involved fitting steel inserts that were plated to fit oversize tapped holes, connecting rod bearings altered to fit slightly undersize crankshafts, and dozens of other similar fixes were made. When tested, these engines developed more horsepower and used less fuel than the standard engines that the U.S. military would accept. The Chinese purchasers were delighted, and the needed 100 engines were soon hand fitted by the Allison experts. The balance of the spares order came from later production. One retired engineer remembered that "in the field, those engines, made out of matched parts that had been rejected on Army standards, had a better field record than the standard engines."[69]

Why did the Chinese engines perform so well? Dick Askren, who was at the time an engineer with Allison, states simply, "Because they were done by some of the best mechanics we had and they were fussier than hell about the way they put it together, every engine!"[70]

The engines for the China contract were built to Allison Specification 145A, rather than Spec 126D as for the C-15, and identified by Allison as their Model V-1710-C15A. There was no "military" designation as the engines were the result of a "commercial" sale to a foreign nation, the only way Congress allowed the transfer of military materials prior to Lend-

Lease. These engines were actually assembled on a separate "Chinese" assembly line in Indianapolis to insure that none of the previously "rejected" parts would accidentally be installed in U.S. or British engines.[71] Ratings for the V-1710-C15A for the AVG's Curtiss H81-A2 *Tomahawks* were the same as for the British V-1710-C15 and Army V-1710-33's.[72] Like the British engines, these used the early magnesium type induction manifolds, and did not result in any particular problems for the AVG.[73]

The AVG was a rough and ready outfit not obsessed with operating by the book. Still as experienced military aviators, many with extensive experience in fighters, they were wise in the ways to get the most from their airplanes, and engines. To this end there are conflicting reports as to the ratings the AVG pilots used in actual combat. While most report staying within the handbook limits, others note that the engine had additional capability and that in the heat of combat most were not watching the gages.

Frank Losonsky, a crew chief for the 3rd Squadron, AVG, relates that, "the pilots naturally wanted lots of manifold pressure, but too much destroyed the engines, which were normally set to operate at 58 inHgA, the "take-off" position. The pilots wanted 62 inHgA, I recall we could make one small adjustment to improve engine performance. We could lengthen the "regulator-to-carburetor" rod by a certain number of turns, supposedly to "correct small manifold pressures." This "unauthorized" adjustment increased the aircraft speed and performance, just a little."[74] As well it should, for as shown on the accompanying Allison performance chart for the C-15, 58 inHgA would allow 1600 bhp at up to about 1,700 feet when running at 3000 rpm. If pilots were getting 62 inHgA they were doing it by running the engine up to about 3200 rpm, where about 1700 bhp would be available.[75] It is very likely that these power levels were used at times, and seldom if ever officially acknowledged. In the heat of combat, and with the pilot's life in the balance, using the engine for all it was worth was acceptable. Even so, it appears that the engines took the abuse and continued to operate successfully.

Allison and the AVG

When provisioning the First American Volunteer Group the Chinese felt that it would be advisable to send trained technical personal to China to assist in the training of their officers and enlisted personnel in maintenance and overhaul of their new airplanes.[76] In general this component of the AVG was to act as technical advisors to both the AVG and Chinese tactical units. In this regard the Chinese had the foresight to come to Allison early in the deal and formulate plans for this technical support. Along with the engines, China's U.S. based purchasing agent, Universal Trading Corporation of New York, placed an order for a few hundred thousand dollars worth of well selected Allison spare parts, maintenance and overhaul tools. The magnitude of the effort was such that O.T. Kreusser, Director of Allison

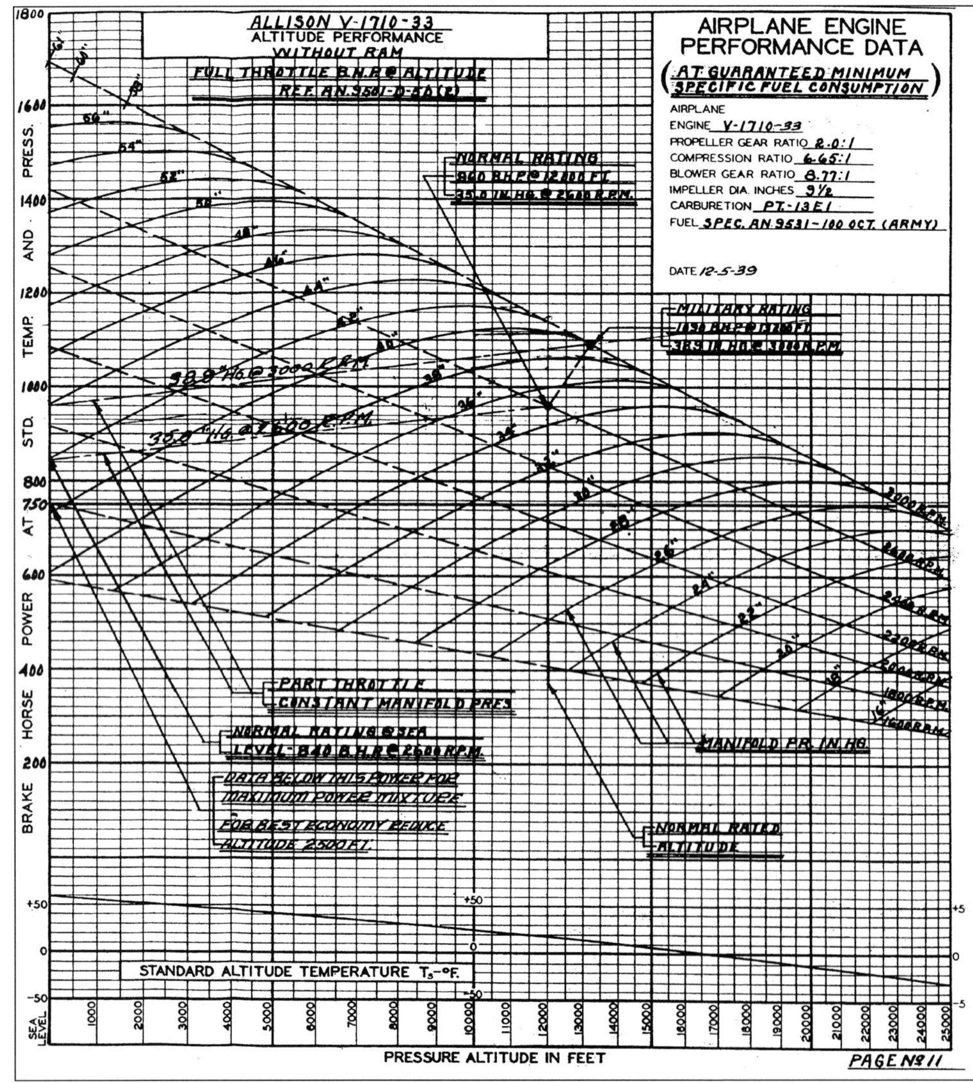

Performance capabilities of the altitude rated V-1710-33(C15) were validated by the Hi-Power Test Program in the fall of 1939. Allison offered Bell a C-15 rated for 1500 bhp for their 1939 Model 11 proposal, though the rating was never approved. At the reported maximum of 58 inHgA the engine would have been developing 500 bhp more than authorized by the factory! (Allison)

Service and Training at the time, took Arne Butteberg and Tye M. Lett, Jr., both with considerable field service experience, to Washington D.C. to confer with the Chinese Aeronautical Mission to the U.S.[77] This was at the same time as Chennault was touring the country seeking volunteers for the flying and maintenance arms of the First American Volunteer Group.

In a meeting with Chennault, it was arranged that a group of the crew chiefs, which he recruited, would take a month-long intensive training course at the Allison factory prior to departing for China. The selected men received the training and actually arrived in China in July 1941, where they became the nucleus of the AVG ground crew. Later, eleven Chinese, most of them American born, who had aeronautical experience either in college or practical operations, took six weeks of training at Allison. It was these men, who were an industrious and competent bunch, that Allison relied upon to support the far off AVG. During their training they were at the factory and had contact with the assembly and acceptance of the spare engines for the AVG. Only nine of these men were actually able to get through to China and into the fight, where they became the backbone of the AVG third and forth echelon maintenance and overhaul crews at Kunming and Loiwing. One of these men, Teh Chang Koo, was the son of Dr. Wellington Koo, Chinese Ambassador to the Court of St. James, and a key player in support of the AVG and Allison. He was a graduate of the Cornell Engineering School, and started out at Allison as a "grease ball." He learned about the engine the hard way, ending up as an inspector for accepting the Chinese spare engines. He then went to Burma and supported the Allison activities for the AVG.[78]

Butteberg and Lett departed Indianapolis for China on May 12, 1941. They left San Francisco on the first commercial flight of Pan Am's Boeing 314 *California Clipper*, headed for Manila. From there they traveled by Dutch steamer to Hong Kong where they obtained transport on Chinese National Aeronautical Corporation aircraft to Lashio, Burma via Chungking and Kunming. They then drove to Mandalay and caught a train to Rangoon, Burma. The trip took 27 days and covered 13,000 miles. At its end they were at the Inter Continent facilities where the AVG *Tomahawks* were being uncrated and tested. The flying was being done from the adjacent Mingaladon airport, the main Rangoon landing field. It was not until the end of July that the first contingent of AVG personnel arrived and setup operations at the British airfield at Toungoo.

Training at Toungoo, Burma for the First AVG was anything but routine. The P-40 was the most advanced pursuit being produced in the U.S. and operating it in primitive conditions was a new experience for everyone. On September 26, 1941 Bob Little's plane ran off into the boondocks and "nosed-up". While looking like a washout, it typically took about 6 hours to replace the prop and oil cooler and another 6 hours to repair the landing gear. The AVG mechanics got a lot of practice at this. (Chennault Collection, Hoover Institution Archives)

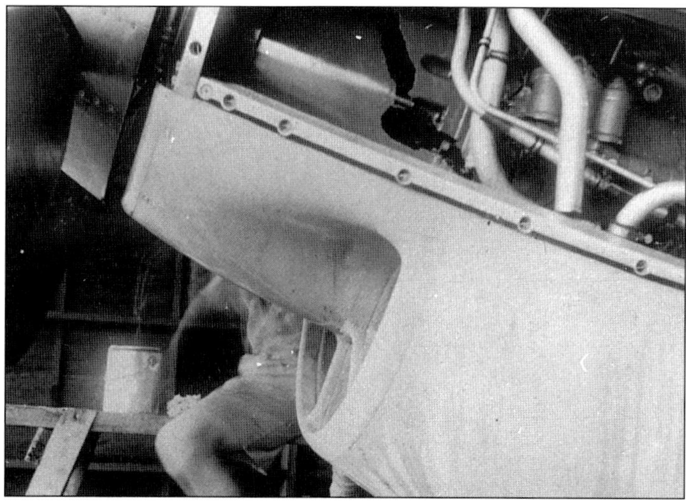

Not every nose over during training at Toungoo was so simple. Here an AVG Tomahawk suffered a broken nose case. (Pentecost Collection, SDAM)

Chennault clearly focused the AVG to be a tactical fighting unit. He directed that only simple ground servicing operations should occur on the operational fields. If operations were expected to exceed fifteen minutes in duration, (that is if a plane could not be made flight ready in 15 minutes), then it was to be withdrawn to a dispersal point for servicing by other maintenance personnel. With the Chinese air raid warning system able to give the AVG 20 minute warnings of coming Japanese attacks, the AVG was usually able to have its available aircraft serviced and airborne prior to their arrival. Planes that were not ready to fly after 15 minutes were hidden in the jungle.

Following the first bombing attack on the AVG, in Rangoon on December 23, 1941, it became apparent to Crew Chief Henry Fox that it was going to be essential to mobilize the AVG ground crew. He instigated a meeting with Squadron Leader John Newkirk and Allison's Tye Lett, with Newkirk authorizing action. Lett found himself in a position to exert some influence in resolving the situation, for General Motors had been appointed by the American Government as the agents for handling Lend-Lease shipments of military vehicles to Burma. Many vehicles were sitting on the Rangoon docks, and Lett was able to appeal for help from Lieut. Col. Twitty, Commanding Officer of the American Military Mission in Rangoon. With Twitty's help the AVG was able to obtain the vehicles needed by the AVG in Burma.[79]

Many factors contributed to increase the difficulty of maintaining the AVG P-40's. First insufficient spare parts had been delivered to the AVG, meaning that they had to cannibalize damaged planes for practically everything. Serendipitously, there were a lot of them available. Operating conditions and airfields were primitive and rough and these factors contributed to numerous wrecks, particularly at Toungoo during the AVG workup period. The P-40 would punish an unwary pilot who stepped on the brakes too hard, and end up on its nose. This usually required replacing the oil cooler, cooling flaps and often the prop. It would take about a half day to complete the job, assuming the necessary parts were available.[80]

Spare parts and tools for the aircraft and engines were a constant headache. During the first three months in Burma there was only one small shipment of parts received. Fortunately it included one set of major engine repair tools that had only been thrown in at the last moment as a result of

Here is the same Tomahawk prepared for engine removal. Details of the installation are clearly shown. (Pentecost Collection, SDAM)

This photo was taken during AVG training at Toungoo, well before the Flying Tigers entered combat or deployed to forward airfields. Already heavy build-ups of dirt are apparent on the radiators and coolers. (Pentecost Collection, SDAM)

Lett and Butteberg's pleading with the Allison Spare Parts Department in Indianapolis. Consequently the time honored practice of "cannibalization" was the only way to keep the operation going, a result of the high standards and interchangeability of components. In this regard Arne Butteberg earned a reputation as a mechanical genius, it was said, "just give him some bailing wire and a rusty file and he could run a major engine overhaul!"[81]

The V-1710-C15A's in the AVG *Tomahawks* acquitted themselves very well. In fact, no pilot in the AVG is believed to have lost his life due to engine failure. The credit not only goes to the engines, but in the view of Tye Lett, to the loyal ground crew who were a painstaking bunch of lads.[82]

Lett, Butteberg and Koo were the key men responsible for this accomplishment in the field. It was they who managed the maintenance and repairs of the engines powering the *Tomahawks*. Furthermore, as the only manufacturer's representatives with the AVG, it also fell to them to train the Chinese to maintain not only the Allison engine but the Curtiss-Wright electric propellers and governors, Bendix-Stromberg pressure carburetors, Pesco electric fuel pumps and the remaining accessories, gadgets and systems on the P-40's. They established an instruction course for an initial batch of 22 Chinese mechanics who had previously worked on an earlier generation of aircraft powered by the V-12 Hispano-Suiza liquid-cooled engine. This instruction involved translation of the technical materials from English to Chinese by Koo, a very difficult accomplishment. After this first class they went on to instruct officers of the Chinese Air Force attached to aircraft factories, base depots and flying fields. They in turn instructed their subordinates. Though the approach and materials were comparatively crude, this effort succeeded in building a pyramid of trained Allison and Curtiss-Wright P-40 service personnel in China. Koo provided the translations of the various technical materials necessary to train the many Chinese technicians.[83]

Service of the AVG V-1710-C15A Engines

Considering their remote location and the primitive operating conditions experienced by the First American Volunteer Group, Chennault's *Flying Tigers* in China seem to have had few endemic engine problems.[84] Their engines were mainly randomly produced, (see above), beginning with Allison serial number 829 and spreading through 1628. Thirty one of the engines originally installed in the airframes came from French contract F-223 (taken over by the British), while the remaining 69 were from British contract A-196.[85] An additional 50 engines were purchased as spares and delivered on a separate order from China. All were designated as model C-15A. Most, if not all, of the actual spare engines were sequential around AEC production number 3000. It is believed that the V-1710-C15A designation was applied to all 150 engines, as Allison provided the AVG only with copies of their Specification 145, not the otherwise similar 126-D. It is believed that between 50 and 100 engines were built from the hand fitted parts, with the balance coming from the standard V-1710-C15 production line. From the spare engine serial numbers we know that they were produced *after* the V-1710-33(C15)'s for the U.S. Air Corps P-40's, and near or following the end of the production run for the British. As such, they all certainly benefited from the previous operating experience and improvements resulting from the service experience of the first 2000 or so engines. Still the AVG had maintenance problems, though often times simply due to the lack of sufficient spare parts, special tools and the length of the supply line to the remote combat theater.

The AVG was initially able to obtain a limited quantity of spare parts by delivery from the States via Pan Am Clipper, but this ended when the

A new V-1710-C15A being prepared for installation in an AVG Tomahawk at Toungoo. Note the mostly Chinese crew. (Pentecost Collection, SDAM)

This V-1710-C15A engine (Allison 899, Air Ministry Number, A 200012) was installed by Curtiss in the Model 81-A2 Tomahawk that became P-8127 in the American Volunteer Group, the Flying Tigers. By December 2, 1941 both it and P-8127 were assigned to the 1st Squadron as AVG #14, flown by John Farrell. Latter P-8127 was assigned to the 2nd Squadron as #47 and flown by Flight Leader John Petach. By March 1942 he had earned 1-1/3 victories. The engine was found in a Sacramento area scrap yard and purchased by Jim Appleby for $50 in 1960. It was then acquired by Ben Giebeler from whom it was obtained for the C-W Historical Association to use in a Tomahawk restoration. (Giebeler)

war started. On December 4, 1941 2,300 pounds of critically needed tires, ailerons and quantities of engine parts left Hawaii bound for delivery to Chennault. Another 1,000 pounds was to leave San Francisco by Clipper on December 7.[86] The first Clipper had made it as far as Wake Island when the Japanese attacked Pearl Harbor. As it was considered a sitting duck in the Wake Island lagoon, the cargo was unloaded and the Clipper returned to Hawaii with priority passengers. As far as the AVG was concerned, the critical parts were lost.

Sparkplugs were always a problem for the AVG, primarily due to their chronic short supply. Another item that early-on caused a lot of downtime, was the electrical generator. The AVG *Tomahawk*'s differed from standard U.S. Army P-40's in that the larger generators were installed. As a consequence, the gear train was adversely affected if the engines did not run smoothly. During the rainy season in Burma the engines often ran rough because of the 100 percent humidity affected the entire ignition system, plugs and harnesses.[87] The result was that initial operations saw frequent failures of the generator drive gears, shafts and the rubber drive coupling. An improved coupling was sent out from Allison, but generators remained a problem. Finally, in late March 1942, 75 new generators, complete with new control panels, braces and associated parts were provided, for installation in all P-40's in China.[88] No proof has been found that they were ever installed, though crew chiefs did report having to measure the drive splines for wear.[89]

Some of the operations for the Flying Tigers were quite demanding on the engines. On January 18, 1942 Squadron Leader R.J. Sandell of the 1st Pursuit Squadron took off from Kunming with three P-40's, one of which had been modified as a photo ship, for reconnaissance of Hanoi in French Indo-China. After a 35 minute flight toward Hanoi they landed and refueled, then began the long flight, cruising at an altitude of 24,000 feet. Upon their return from the mission Sandell reported that his airplane, P-40 #11 with engine AEC#1432, had burned 127 gallons of gas in 2:45 hours, cruising at 21 inHgA and 2300 rpm and making a ground speed of 270 mph.[90] Tragically, Sandell was lost in this airplane on February 7, 1942. The airplane and engine had accumulated only 54 hours total time.

Ground operations exposed the engines to clouds of dust everywhere the AVG operated. It clogged the carburetors and backfire screens to the point that it was dangerous to increase manifold pressure for fear the engine would quit cold.[91] Cleaning the entire induction system would restore the engines' power and smooth operation, but after a single day's operations it would often be just as dirty again. Backfire screens could become plugged in as little as an hour of operation![92] Things were bad enough that the AVG's Fritz Wolf reported that when eight P-40's attempted to takeoff for an air raid on Rangoon, three were forced to abort because of fouled carburetors and/or induction systems. In February 1942 Chennault sent a telegram to Bob Neale down in Rangoon, directing Henry Fox, the 1st Squadron's Line Chief, "to remove carburetors and clean butterfly valves, seats and impact tubes (as) this prevents rapid fouling of spark plugs."[93] The problem came to a head in early March 1942 when Mr. Arne Butteberg, one of the Allison technical representatives to the AVG, investigated reports that two "Sharks" that had just come in from Burma. It was reported that they could only develop 25 inHgA manifold pressure at takeoff. He found all of the important sensing ports and screens in the entire induction and carburetion system plugged solid with dirt, as well as dirt being packed into the coolant and oil radiators. This in turn would cause excessively high engine temperatures.[94] Buttebergs' recommendations to Chennault resulted in the immediate issuance of orders requiring frequent disassembly and cleaning of induction system components by AVG ground crew. Furthermore, it prohibited the previous practice of simply pouring raw gasoline down the carburetor. That practice may have "cleaned" the carburetor duct, but caused the loosened dirt to be deposited onto the backfire screens and other critical ports and passages within the carburetor. The revised maintenance procedures resulted in restoring the critically needed performance of the AVG *Tomahawks*. An interesting observation about Col. Chennault, and how he built the *First AVG* into the success that it was, is apparent in his letter back to Butteburg acknowledging the recommenda-

This is one of the V-1710-C15A engines hand built from non-spec parts for the AVG. Allison #899, Air Ministry Number, A 200012 was one of the engines sold to the Chinese. Note the external oil piping on the reduction gear used to direct a spray of oil onto the thrust bearing, and a portion of the Curtiss-Electric propeller control still installed. This engine appears to have been simply removed from the aircraft and never overhauled. (Giebeler)

tions and the action he was taking. He states, "As you know, you are entitled to all the privileges of a member of the Group."[95] Chennault knew people, and how to motivate and mold them into a fighting force.

Between September 1, 1941 and April 19, 1942 the AVG lost 39 of the original 99 P-40B's [with V-1710-C15A], along with two of the 17 replacement P-40E's [powered by the V-1710-39(F3R)]. As of April 19, an additional 47 aircraft were awaiting major overhaul or repair, repairs such as needing a new wing or two. Only 28 aircraft were "in-commission", and of those, a further 20 percent were out-of-service for minor maintenance and inspection on a daily basis. Tires, propellers, engines and engine accessories were the critical components, as most replacement parts came from damaged or out-of-service aircraft.[96] Interestingly, no spare tires or propellers were originally purchased, or delivered along with the airplanes.[97]

The Allison was held in high regard by the AVG, as attested in a cable sent by Col. C.L. Chennault, Commanding the AVG, to Allison workers on April 8, 1942:

> "Our pilots flying Curtiss P-40 pursuit airplanes equipped with Allison liquid cooled engines have been extremely successful in flight operations against the invading Japanese Air Force...Performance of these liquid cooled engines has been absolutely amazing under the most grueling wartime fighting conditions...We have destroyed over two hundred Japanese planes in the air and on the ground without, to our knowledge, losing a pilot or airplane due to engine failure. You keep producing'em and we will keep'em flying-Lets go America!"[98]

A key component of the engine maintenance program was to establish an overhaul capability. The early V-1710-C15 engines were scheduled to undergo an overhaul after 200 hours of flight operation. It is believed that none of the AVG engines were changed for excessive operating time. Rather, they were operated until some type of major maintenance was required. For the last several months of operation the AVG only had about 25 aircraft combat ready. Even so, the actual number of flying hours was surprisingly small. Review of maintenance records shows many cases of airplanes with only 50 hours total time as of February 1942, and that by the end of the AVG period, July 4, 1942, few of the airplanes had accumulated 200 hours.

This view of Flying Tiger V-1710-C15A engine, Allison 899, Air Ministry Number, A 200012, shows the gun synchronizing distributors have been cannibalized, along with the generator and starter. (Giebeler)

Throughout the AVG period (the AVG was disbanded and became the 23rd Fighter Group on July 4, 1942) efforts were underway to establish an overhaul capability by CAMCO, the longtime American run aircraft repair and assembly company that had been supporting the China Air Force since the mid-1930s. Some progress was made in this regard, but sufficient tools and a lack of necessary spare parts hampered the effort. Consequently, it was mid-March 1942 before the first engine overhaul was completed, an overhaul done without benefit of the two sets of overhaul tools originally ordered by the AVG. As late as April 18, 1942 Chennault was still sending cables to Washington seeking them, and supplies such as valve grinding compound.[99] It was in February 1942 when it was realized that a fair number of new engines, at least 14 located at the CAMCO Lungtao repair facility, were available and that they could each be installed in an aircraft in about 2-1/2 days. Walter Pentecost, who oversaw the Allison overhaul depot and was the AVG advance man who had supervised the erection of the boxed P-40's in Burma, recommended this to Chennault as an expedient way of improving the "Sharks" performance over China.[100] After 90 days in combat the AVG's P-40's were "war weary." It is believed that not all of the spare engines were used prior to overhauled engines becoming available.[101]

First Hand Experience

Gerhard Neumann, line chief for the AVG's 1st Squadron, stationed early on in Kunming, China, relates that he never heard of one of their engines backfiring. This would be consistent with the British experience with their engines installed in sister ships to the AVG's *Tomahawks*, but operating in North Africa. Both agreed that they were pleased with their Allison V-1710-C15's, that they "... were fine engines operating under difficult conditions."[102] Neumann goes on to establish that their major maintenance problems, other than a lack of spare parts, were with the propeller governors, spark plug fouling, and failed distributor rotors. In fact, one time he even fashioned a distributor rotor from a piece of Water Buffalo horn as a replacement for the recently introduced Bakelite part. It lasted a year, as long as the airplane itself did![103]

Insight to the tendency of the V-1710-C to backfire is explained by Erik Shilling, AVG Pilot. When the AVG pilots first gathered in Burma, he notes that fewer then 20 had ever flown a fighter, and most of them had not flown the P-40. He explains that the whole matter was a question of technique. That the AVG experienced only minor problems with backfiring can be attributed to their previous extensive flying experience and willingness to learn from each other while the AVG was forming.

> "We did have a backfire problem in the beginning, but I always associated it with too rapid a throttle movement, and I think most of the fellows learned this quite early. As a matter of fact, I felt the Allison was able to take a great deal of manifold pressure—50/60 inches—without a problem, but no over revving at all. As I recall, the max over-rev was 3120 rpm and takeoff was 3000 rpm. Rapid throttle movement was a no-no, since it would accelerate like mad. If not careful, the Allison would over-rev in seconds.
>
> Tactical approaches were helpful in reducing power to prevent backfiring. In other words, in making the decent power could be pulled back to minimum cruise power. Come in low over the field, 50/100 feet, pull up sharply, pulling lots of G's, make a steep turn also pulling G's (high induced drag to slow down), when the speed dropped to max gear down speed, drop gear, and then start pulling power.

The Allison didn't backfire if power was pulled off gradually as speed diminished, insuring the engine drove the prop. When the power was less, so that the speed drove the prop, the engine would backfire.

I personally didn't have backfiring because I was used to flying behind the Allison (in the YP-37 and YFM-1), and I soon learned how to prevent the engine from backfiring. Power could not be handled like a radial."

A 3rd Squadron AVG crew chief, and youngest member of the group, Frank Losonsky, went to work for Allison in the 1950s, and relates his experiences that had been gained under the most adverse conditions. He was personally responsible for up to four aircraft at a time, and notes that spark plugs, magnetos and oil coolers were their major problems. He relates that they had removed the manifold tee backfire screens because they were disintegrating and fouling sparkplugs and valves. Conflicting with Neumann's account, he notes that backfiring had become a considerable problem when they began operating from Kunming. A consequence of the high field altitude effecting the fuel mixture during engine starts and low power operation.[104] Evidently the "fix" for rusting reduction gear thrust bearings was not universally effective on all of the AVG planes because he remembers pulling the ball bearings and cleaning the rust off of them with crocus cloth, then putting them back into the engine![105]

AVG P-40 Maintenance

The AVG was notorious for lack of military protocol in its operations, which can be understood given their mercenary environment and remote location. Still it should not be surprising that they did adhere to the essentials needed to field an effective fighting force, including in-air discipline and rigorous maintenance and preparation of the Sharks.

Each day, before beginning flight operations every aircraft was inspected by its assigned crew chief. Special attention was given to spark plug wires and condition of fuel lines within the engine compartment. Each five hours the carburetor fuel filter was cleaned and engine idle speed adjusted to about 625 rpm. After 25 hours the plugs were pulled and inspected for fouling. At 50 hours the distributor rotor was pulled, cleaned and any pits ground out. The plugs were also inspected and regapped. At the same time the Prestone coolant was drained and replaced.[106] At 100 hours the spark plugs were replaced, if any new plugs were available, and the slip ring contacts for the Curtiss-Electric propeller were checked and cleaned.[107] Similar inspections and service were done to the airframe as a whole, with specific attention to the various seals, bearings and fittings used in the landing gear and flying controls. Even allowing for the primitive conditions, the skilled and dedicated ground crews were able to keep a viable fighting force in the air.

In preparation for a scheduled flight, or alert duty, the crew chief would also warm-up his ships before the pilots arrived. After completing the daily engine inspection and determining the condition of the flight controls and tires, the fuel sumps were drained to remove any water or collected crud and the fuel tanks were checked for quantity. Prior to the first start of the day the ignition switch would be confirmed "off" and with the throttle opened, someone in the ground crew would pull the propeller through for at least two revolutions, the purpose being to circulate the oil and get the gas flowing. Checking that the carburetor heat was "off" and the radiator flaps "open" to insure the engine would not overheat on the ground. The throttle would then be "cracked" open, mixture set to "idle cut off", propeller switch "on" and set to low-pitch (only minimum load on the engine during the start), and a fuel tank selected. The actual start was accomplished by operating the "wobble pump" to obtain an indication of fuel pressure to the carburetor, taking care to not "over-prime" for fear of a backfire, and then toggling the "starter" switch. This energized an electric motor that spun a flywheel which would automatically engage the engine when it had reached sufficient speed. After 2 or 3 revolutions the engine would usually fire and the starter switch would be released. At the same time the mixture control would be set to "automatic rich" and the engine would quickly accelerate to idle speed. Fuel and oil pressure would be carefully monitored, and if minimum oil pressure for idling (15 psig) had not been achieved with 30 seconds the mixture control had to be positioned at "idle cut-off" and the ignition cut to prevent damage to the bearings.

By monitoring the color and texture of the flames coming from the exhaust stacks during the start it was possible to determine over priming

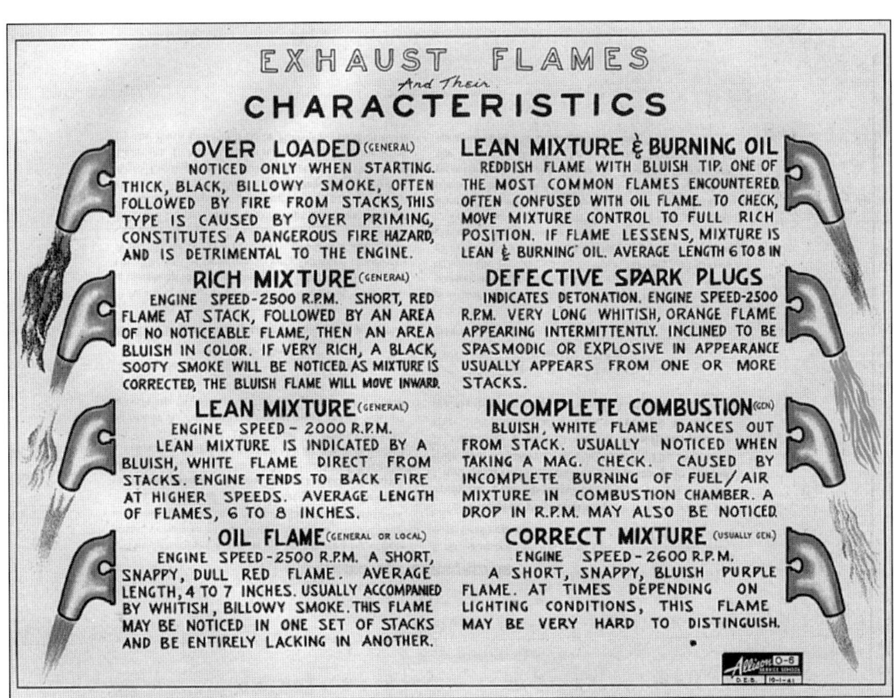

A skilled crew chief or pilot could tell a lot about the condition of an engine by simply observing the flames exiting the individual exhaust stacks. (Allison Service School)

and general condition of each of the cylinders. This was a major safety consideration provided by an experienced crew chief or pilot.

After a successful start it was necessary to cycle the propeller governor and observe the correct decrease in engine speed as the blades took a bigger bite of air, and then do a runup at 2300 rpm, but to not more than 30 inHgA (to prevent lifting the tail) and check that each set of spark plugs and the magneto was working properly. A standard engine start took about five minutes, but the entire preflight procedure required about 30 minutes.[108]

The propeller governor controls were mounted directly aft of the reduction gear box and stuck up into the vee of the engine. It was often necessary to replace the unit on hot engines just returning from combat, and to do so within the 20-25 minutes before the Japanese retaliation fighters or bombers arrived. A most trying job![109]

Spare engines were a problem at the China end of the AVG supply line as well. According to Neumann, who was there through at least October 1942, he was unaware of any "new" spare engines at Kunming, and furthermore, the Kunming Squadron never had more than two spare engines. When they needed another they used engines salvaged from crashes in Chinese territory. Under such conditions he performed three engine changes for the 1st Squadron during the early months of the war. He believes that no AVG engine exceeded 150 hours before the plane was either shot-down, bombed on the ground, or became a "hangar queen", i.e. a source of parts. Crew chief Frank Losonsky relates that most engine changes were the result of finding metal contamination in the oil.[110]

From a review of the engine serial numbers listed on the AVG reports, it is entirely possible that the entire lot of 50 spare engines were never used by the AVG. One hundred came installed in the P-40 *Tomahawk* airframes, and a sequential production block of 22 more have been identified that were in the field in Burma. An inventory of spare engines done on November 29, 1941, barely a week before the start of the war, shows that both Kunming and Loiwing had 15 new engines on hand, but that 12 new engines were still in Rangoon.[111] It is possible that the other eight engines had been used during training, though other factors make this unlikely.[112] There is a good chance that some of these engines were still in Rangoon when the city was evacuated and burned prior to being overrun by the Japanese in March 1942. This suggests that at most, the AVG had only 138, or fewer, engines to work with.

Getting the engines from Rangoon into China would have been a logistics problem. Each of the engine overseas shipping boxes were 38x57x111 inches (about 4x5x10 feet) and weighed 2,240 pounds (a long ton) complete with an engine.[113] Each would have had to have been loaded on motor trucks and hauled up the torturous Burma Road, during a time when available transport was in critically short supply. In fact, Neumann led one of the first supply convoys for the Chinese from Kunming to the Burma border back in 1940. The trip covered the 580 miles and required over eight days each way. They delivered tin and wolfram for eventual shipment to the U.S. and returned with aviation fuel, machine gun ammunition and 500 pound bombs.[114]

Fuel for use by the AVG in their extensive theater of operations was a considerable problem. In the summer of 1941 the U.S. Finance Department signed a contract with Texaco for 4,685,000 gallons of 100 octane gas to be delivered at Rangoon, Burma. Texaco was to begin monthly deliveries that September of 20,000 drums, equivalent to 1,060,000 gallons.[115] Rangoon was a long way from Kunming and central China, and the cost of delivery, in terms of losses of fuel, was high. As of January 2, 1942 there was a total of 1,063,847 gallons of 100 octane in China, scattered over some 20 different locations.[116] This was only a one month supply for the AVG *Tomahawks*, assuming two hours of flight per day.[117] Even so, the real problem was fuel quality, for not only did conditions contribute to dirt and contamination, but a lot of the fuel had been stored and stockpiled in barrels for several years. In addition, as a consequence of handling a lot was lost when filtering, pumping or pouring from awkward 50 gallon barrels.

Considerable fuel related trouble was occurring with the P-40's based in Burma, those using fuel supplied by the British. Then in November 1941, stocks of fuel were found in Burma that were believed to contain 4.8 cc of TEL per gallon, compared to the maximum of 3.6 cc's specified by Allison for the V-1710's.[118] At the time of the first air raid on Rangoon the AVG was attempting to assess the viability of using RAF fuel arriving from Sumatra. This Netherlands East Indies gas had the same octane rating as that proscribed for the V-1710's, but it used a different base.[119] Everything was a compromise. Things were no better in China where Chennault was trying to find a laboratory able to test the fuel and assist in resolving performance problems, but none was available.[120] Dirt in the fuel was a continuous concern and often required that it be passed through a chamois as it was being poured into the P-40's fuel tanks. For whatever reason, the fuel in China has been reported as being a "dirty" color.[121] The age and natural degradation of the available fuel stocks probably meant that little, if any, of the fuel being used could actually deliver the required 100 octane performance. No specific reports have been found of fuel dye clogging the backfire screens in the AVG engines, a problem of significant proportions to the P-40's being operated in the U.S. at the time. Even so, Frank Losonsky reports that the backfire screens were removed from the AVG engines.[122] This was probably due to the lack of sufficient spares, the excessive workload to change and clean them if used, and importantly, finding that they actually caused backfires when they deteriorated.

The effect of lower grade fuel would be a problem during takeoff and combat, times when maximum performance of the engine was critical to the safety and success of the pilot. In this regard a measure of credit probably should go to the pilots, most of whom had considerable military flying experience prior to joining the AVG. With experience, a pilot could operate his engine on the edge of detonation and thereby achieve peak performance. With degraded fuel and dirty engines this might be at power levels far less than what the Allison had produced on the factory test bed. Wise use and care of your engine was essential to completing a combat mission over the jungles of Burma or the mountain peaks and remoteness of China.

Service of the P-40E with the AVG

From the beginning of the AVG Chennault was concerned about expanding his forces and obtaining replacements of both personnel and aircraft. The Chinese had obtained a compliment of Douglas DB-7 light bombers that were on their way to China when the war started. These were intended for the Second American Volunteer Group, but were unceremoniously unloaded in Australia and taken over by Allied forces. The Chinese were able to acquire three Curtiss CW-22 light-weight Pursuits, but they were all lost by the AVG on delivery flights into China early in 1942. Madam Chiang Kai-shek was instrumental in obtaining 30 Vultee P-48 fighters that arrived in Rangoon during December 1941. These were to be used by the Chinese Air Force, but because of the large number of P-40's that were not to be repaired for several months, Chennault requested that the Chinese allocate them to the AVG. This transfer did not occur, possibly because the airplanes had to be trucked to Magwe from Rangoon and then assembled.[123] It is likely that many of these aircraft simply fell into the custody of the Japanese when they overran the Burma peninsula.

At this point efforts from the Chinese delegation in Washington D.C. were successful in getting the Army to direct a number of P-40E airplanes to Chennault. These airplanes were delivered to the Gold Coast of Africa

Captain Eddie Rickenbacker views a C-15 returned to Allison by the British after safely completing a mission over the North African desert and receiving 14 bullet holes. The similar AVG engines were taken from the same British and French production contracts. (Allison)

in crates were they were assembled and prepared for AVG pilots to fly them on to China, an epic journey in itself. AVG pilot G.B. McMillan provided Chennault with a narrative diary of his experiences on one of the delivery trips. The effort and timing of the excursion from China is representative of the conditions under which the AVG was laboring.

McMillan left Kunming on February 16, 1942 via the CNAC airline along with five other AVG pilots, and were flown to Calcutta, India. The next day they prepared to leave by B.O.A.C. flying boat. They had no priority on B.O.A.C. with the result that the trip to Cairo, Egypt took until February 23 and involved many stops along the way. They left Cairo on the 25th by Pan American Airways Africa, and again after numerous stops arrived in Accra, Gold Coast on the 27th. The group checked out their new P-40E's, finding several things wrong with them which required additional maintenance before they could begin their return flight.

Departure for China was on March 2. On the first leg of the trip one plane had a bad tire and another bad spark plugs, causing two of the pilots to have to wait for repairs. Of the group of four planes making the next day's flight to French Equatorial Africa, three were having plug trouble and there were no spare plugs available in Accra nor enroute. With no chance of repairs until they reached Cairo, the group pressed on, even though one of the airplanes had now developed propeller problems. On March 5 they landed at an emergency desert airfield due to a bad dust storm, which became so bad that they backtracked to an earlier field. In the course of the flight the entire electrical system went out on R.T. Smith's airplane and it had to be left behind. The three remaining airplanes reached an airfield on the Suez canal about 50 miles from Cairo on March 7. This field had U.S. mechanics who went to work on the airplanes, all of which were in bad need of repairs and inspections after 3,900 miles and no maintenance. On March 10 Flight Leader Smith caught up with them, and on the 13th they were joined by three other P-40's, two piloted by Pan Am contract pilots.

On March 15 the first group of four planes departed for the east. With the exception of one airplane developing a bad fuel leak that required repair, the group covered another 4,000 miles to arrive in Karachi, India. The airplanes were again in bad shape from lack of maintenance; in particular: one generator, one starter, two hydraulic systems, spark plugs and the gas tank leak. At the same time the guns were removed, cleaned, bore-sighted and loaded. They left on March 21 and arrived in Kunming, China late in the day on the 22nd, with the "generator not functioning on Wing Man Green's airplane and the usual amount of plug trouble." Total flight time for the return trip was 53:45 hours.[124]

The trip had required nearly five weeks, and denied Chennault the combat services of one pilot for every airplane delivered this way. With the

AVG losing two to six airplanes and pilots each month, the effort to replace his losses was in effect doubling them.

The V-1710-39(F3R) powered P-40E's were very important to the AVG as they were equipped with six .50 cal machine guns and could be fitted with bombs and dropable fuel tanks. Very quickly they were pressed into service as the preferred model to take into combat, although most formations routinely mixed the P-40B and "E" models. For a while the AVG assigned all of the "E" model airplanes to the First Squadron.

When the AVG was inducted into the U.S. Army, July 4, 1942, not a lot changed. The new 23 Fighter Group continued to be short of staff and airplanes for a considerable period. The remaining P-40's fought on as well. When the last of the AVG *Tomahawks* flew combat is not known for sure, but loss reports from the China Air Task Force note that as of September 1942 there were ten P-40B's grounded for want of maintenance and repairs.[125] While they continued in operation, on October 22, 1942 the Army placed its P-40/P-40B airplanes in the "Restricted" class, meaning that they were no longer considered to be combat airplanes.[126]

Pilots' Comments on the Tomahawk

What did the AVG pilots think about the engines in the P-40's? While no one pilot can speak of the varied experiences of over 100 men in combat, your author has put the question directly to several surviving AVG pilots. "Buster" Keeton, 2nd Squadron, relates that he was originally afraid they were going to have trouble with the Allison, based upon his previous experience only on air-cooled radials. He then goes on, "but after I got to flying them I really loved them. A very rugged engine. Once the mechanics learned how to take care of them they were great!"[127]

Ed Rector, 2nd Squadron AVG, remarked about the dependability of the engine and noted that he had not heard of any of the AVG Allison's having any destructive backfire problems. It was nothing like the problems being experienced by the P-40's in U.S. Army Air Corps service. He did say that you had to be careful operating the engine while on the ground, for too much ground running would foul the plugs and could be a problem on takeoff.[128]

Rector goes on to commend the ground crews, and Allison Tech Rep Arne Butteberg in particular. "Butteberg knew that engine A to Z. He worked just like one of the crew chiefs, out there at all hours. Dedicated. I remember him with great fondness."

Dick Rossi, 1st Pursuit Squadron, AVG and a 6-1/4 victory Ace, comments that the AVG pilots did not normally overboost or abuse their engines. He believes that he never overboosted his engine, even in the heat of combat. He does remember an incident where Greg Boyington, 1st Pursuit Squadron, overboosted his engine on a go-around during training at Toungoo.[129,130] Jim Cross was the next guy to fly the airplane and had the engine fail on takeoff, ending up in a rice paddy. One problem he does remember was what the mechanics called "fuseing" of the ignition harness, which could cause a serious loss of power, particularly at takeoff. He remembers this specifically because he was trying to takeoff from Magwe in R.T. Smith's lucky No. 77 when this happened to him.[131] He ended up wiping the gear off the plane, ending its flying days. He says R.T. still hasn't forgiven him![132]

Operation of the Curtiss *Tomahawks* by the AVG on the frontier of the war was difficult given the primitive conditions and lack of supporting infrastructure. Still they acquitted themselves marvelously. They were there, and they triumphed in spite of the conditions and "hand-me-down" equipment. As an integral part of the whole, the Allison V-1710 did all that was asked of it. It had plenty of power, and ruggedly soldiered on to help write a sterling chapter in the annals of the war.[133]

NOTES

[1] Hazen, R.M. 1941.

[2] *Allison History*, 1962, 19.

[3] *Decisions Relating to the Allison V-1710-33 Engines*, Telegram to Dr. George J. Mead from General Arnold, June 17, 1940. NARA RG18, File 452.8, Box 808.

[4] *Allison Engine Situation*, 1941.

[5] Interviews with Retired Early Allison Personalities, made in Indianapolis on November 28, 1994 by Dan Whitney.

[6] *Allison Engines*, report compiled at the request of Chief of the Air Corps, Major General H.H. Arnold by Major J.H. Doolittle, Asst. Supervisor, Central Procurement District, Detroit, MI, January 19, 1941. NARA RG18, File 452.8, Box 807.

[7] *Embarkation of Army Aircraft in Aircraft Carriers*, letter to Commanding Officer, Langley Field, VA from Eighth Pursuit Group (F), October 14, 1940. NARA RG18, Entry 293B, Box 249.

[8] *Commercial Operation and Maintenance Handbook for Allison V-1710-C15 Engines*, March 7, 1940, p.18.

[9] Memorandum for Mr. Lovett, February 10, 1941. See H.H. Arnold manuscript, Roll 197, Library of Congress.

[10] The P-40A was the version purchased by the French. With the fall of France in the spring of 1940, the British took over the order and all were delivered as the *Tomahawk Mk I*, Curtiss H81-A1.

[11] As an example of the rapidly changing technology and production standard, the P-40/B were assigned "Restricted" status on October 22, 1942. As RP-40/RP-40B they were not to be assigned or used in combat theaters. *Model Designations of Army Aircraft, 11th Edition*, January 1945.

[12] The Air Corps expansion program was to provide 25 Combat Groups by 4-1-1941, and 54 Groups by 4-1-42. As of January 1941 plans and reality had provided the following, where Pursuit airplanes included the obsolescent P-26, P-35, and P-36. Only the 198 P-40's, without self-sealing tanks and modern armament were considered Modern:
From Memorandum for the Chief of Staff, *Status of Pursuit Airplanes in the Air Corps*, January 7, 1941. H.H. Arnold manuscript, Roll 197, Library of Congress.

[13] *Failure of V-1710-33 Engines, Air Corps No. 39-997 and 39-999*, Wright Field, Service Division Letter to AAC Representative at Allison Engineering Company, February 11, 1941.

[14] *Investigation of Fire and Backfire in V-1710-33 Engines Installed in P-40 Airplanes*, Wright Field Routing and Record Sheet, Experimental Engineering Section, May 2, 1941.

[15] Kline, John 1994.

[16] Backfire screens were constructed from corrugated strips of copper about one inch wide. When properly assembled they acted first as a "flow straightener", then if a flame was to "backfire" to them, it would be held on the downstream face of the screen, just as occurs in a "miners lamp" or "bunsen burner". The pulsing flow, relatively high temperatures, and vibrations inherent in being located near the center of the engine contributed to often damage the thin metal strips. In any case they acted as a filter and restricted the flow, though Allison depended upon them to smooth

and balance flow between the banks of the early Rams Horn type manifolds. They adversely effected the performance and critical altitude of the engine as the supercharger had to develop more pressure, and heat, in order to overcome the restriction.

[17] Leakage of oil into the induction remained a problem for some time. In October 1941 Curtiss Wright complained that Leakage of Oil into Supercharger Housings of V-1710 Engines was causing fouling of spark plugs in new P-40D/E aircraft. Their tests showed that by eliminating the line providing high pressure air to he supercharger seal vent improved the condition and recommended Allison re-configure the engines accordingly on all service engines. The proposed change included routing the spark plug blast tube to the seal vent line, thereby supplying ram air to the vent. Contract file W535-ac-12414, cross reference of 10-4-41 regarding C-W letter of 9-23-1941, NARA RG342, RD3466, Box 6841.

[18] *Cause of Fires in P-40 Airplane*, T.S. McCrae, Assistant Chief Engineer/Allison to Assistant Chief Materiel Division, Wright Field, April 14, 1941.

[19] *Deposits on Tee Backfire Screens on all V-1710 Engines*, Hazen of Allison letter to Commanding General, Materiel Center, Wright Field, May 25, 1942.

[20] *Backfire Protection in Allison V-1710 Series Engines*, letter to Allison from F.O. Carroll, Lt. Col, Chief, Experimental Engineering Section, September 19, 1941, NARA, RG 342, Allison-Backfiring.

[21] Not until after the first 24 RM14SM engines had been built was the plenum on the valve side of the flame trap divided by partitions between the cylinders. This was very near the end of Merlin production, 1944. Harvey-Bailey, Alec and Dave Piggott, 1993, 49 & 51, 84-86.

[22] *Conferences at Wright Field on April 8 and April 29, 1941 on the Allison V-1710-33 Engines and Curtiss P-40 Airplanes*, Materiel Division Inter-Office Memo, May 2, 1941.

[23] Transcript of telephone conversation between Mr. Hazen of Allison and Captain R.J. O'Keefe, Maintenance Branch, F.S.S., May 5, 1941.

[24] *Difficulties Encountered in Allison Engines, Particularly Fire and Backfiring*, Air Corps letter to Allison, May 10, 1941 reporting on recent Wright Field Conferences with Allison.

[25] British Air Commission, Roderic Hill Air Marshal, response letter to questionnaire, Office of the Chief of the Air Corps, August 11, 1941. NASM, Wright Field file D52.1/630 Curtiss.

[26] *Modification of P-40 Airplanes to Reduce Fire Hazards*, Wright Field Inter-Office Memo, Chief, Experimental Engineering Section to Chief, Field Service Section, May 15, 1941.

[27] This change is confirmed by Dick Askren, Allison Tech Rep on Wright Field Accelerated P-40 test program, mid/late 1941. Also, correspondence from John Kline, Allison Tech Rep, 1-14-95.dw

[28] Appendix 7 provides a partial listing of V-1710 related Technical Orders issued by Wright Field.

[29] *Backfire Protection in Allison V-1710 Series Engines*, letter to Allison from F.O. Carroll, Lt. Col, Chief, Experimental Engineering Section, September 19, 1941, NARA, RG 342, Allison-Backfiring.

[30] *P-40 Difficulties*, endorsement on communication to Chief of Air Staff by O.P. Echols, Chief of the Materiel Division on September 10, 1941. NARA RG18, File 452.8, Box 807.

[31] Wright, Donald F. 1994.

[32] These *Tomahawks* came from a block built for the British, but they used Allison V-1710-C15A engines which was a model specifically built for the Chinese. Specified performance was the same as that for the British V-1710-C15.

[33] *Study of Poor Performance of P-40*, letter, A. Butteberg Allison Engineering Representative to General C.L. Chennault, Commanding, First A.V.G., March 11, 1942. Chennault Papers, Box 4.

[34] *Backfire Screens, Allison Engines*, Air Corps report on conference held at Materiel Center documenting decision to remove both the port and tee backfire screens and install the constant velocity manifolds on all V-1710's, June 24, 1942.

[35] *Results of Test of Constant Velocity Manifold on P-38 Installations*, Teletype from Air Corps Production Engineering Section to Engineering Division, August 31, 1942.

[36] *Carburetor Intake Scoops for Eliminating Backfires in Glide and the Development of Double Accelerating Pumps*, Allison letter to Materiel Command, Wright Field, April 5, 1944.

[37] Wagner, Ray. 1990, 159.

[38] *Difficulties Encountered in Allison Engines, Particularly Fire and Backfiring*, Air Corps letter to Allison, May 10, 1941 reporting on recent Wright Field Conferences with Allison.

[39] *Replacement of Intake Valves, V-1710 Series Engines*, Air Corps letter to Allison Engineering Company, September 11, 1941.

[40] *Difficulties Encountered in Allison Engines, Particularly Fire and Backfiring*, Allison Engineering Co. letter from Assistant Chief Engineer T.S. McCrae to Wright Field, June 17, 1941.

[41] *Reduced Take-Off Rating RPM on all Allison V-1710 Engines*, Materiel Division memo, Power Plant Laboratory to Propeller Lab, 6-21-1941.

[42] Wright Field Memo Report Insp-M-41-5-E, 1941.

[43] *XV-1710-6 Engine, Maximum Allowable Overspeed For*, Hazen of Allison letter to Navy Department, August 5, 1941.

[44] *Difficulties Encountered in Allison Engines, Particularly Fire and Backfiring*, letter from Chief, Field Service Section to Allison Division/GMC, July 21, 1941. NARA RG18, File 452.8, Box 807.

[45] Wright Field Memo Report Insp-M-41-5-E, 1941.

[46] *Operating Instructions for the Allison V-1710-33 Engines Installed in Curtiss P-40 and P-40B Airplanes at Manufacturer's Ratings*, Wright Field Routing and Record Sheet, Chief, Field Service Section, April 8, 1941.

[47] *Allison Model Specification No. 165-B*, for V-1710-85 Aircraft Engine, November 2, 1942.

[48] Allison had no real objection to operating their engines at higher levels, providing that the proper maintenance and care was given the engine. The impetuous for the WER Rating program came from field reports of how the pilots were actually operating their engines in combat. They were pulling WER a long time before it was authorized! The engines were quite capable of higher power, it was only the conservative Army rating policies that held them down. Conversely, the British practice was to rate an engine as high as reliability would allow, and as improvements were incorporated, to continue to uprate the engine. Planned engine removal rates were typically higher for British than for AAC/AAF engines.

[49] *Operating Instructions for the Allison V-1710-33 Engines Installed in Curtiss P-40 and P-40B Airplanes at Manufacturer's Ratings*, Wright Field Routing and Record Sheet, Chief Field Service Section, April 2, 1941.

[50] Cable Regarding Re-Rating Allison V-1710-C15 Engine, from Hazen to Dixon at Wright Field, August 28, 1941. Wright Field file D52.41/81 Allison.

[51] *Proposed V-1710 Engine Reduction Gear Bearing Redesign*, H. Caminez of Allison letter to Power Plant Branch, Wright Field, February 11, 1935.

[52] *Inspection and Reworking of Propeller Shaft Reduction Gearing—V-1710-11,-21,-23,-33 and -41*, TO 02-5A-8, September 1940.

[53] Allison Division, GMC, 1941.

[54] *Investigation of a Method of Getting Oil to the Forward End of Propeller Shaft and Thrust Bearing of the V-1710 C-10 and C-15 Allison Engines*, Allison Test Department Report No. P3-2, J. Hubbard, March 12, 1941.

[55] Information provided to author by Frank Losonsky, Crew Chief, 3rd Squadron, AVG, September 1995.

[56] *Cause of Fires in P-40 Airplanes*, recap of conference held at Wright Field April 8, 1941, 19 different open items discussed and resolved. Allison Engineering Company letter from Asst. Chief Engineer, T.S. McCrae to Wright Field, April 14, 1941.

[57] Retired Allison Tech Rep Don Wright related the following incident concerning Ben Kelsey while on a visit to the Lockheed facilities in California during the war:

Before he left he wanted to show them how he could fly a P-38, so he takes off, goes over the field at Glendale upside down, and down to Mines Field and flies it upside down [too]. So I get down to Mines Field the next day and this airplane he's been flying is from Mines Field, and they said, "Boy you should have seen this demonstration we had! I say, "What happened?" They told me, and I said, "How long was he upside down?" "Oh, way from out there to over here, close to the ocean." I said, "You better pull the Cuno strainers on those engines because I think he probably knocked all of the bearings out of it!" The Engineering Officer almost fell on his face, but they checked, and sure enough, "Here's all of the bearings!" You can't run them upside down [very long], you run out of oil, [for without oil pressure the bearings are lost immediately]!

[58] *Recent Engine Failures*, record of telegram from Commanding General 4th Air Force to Materiel Command, January 2, 1943. NARA RG18, File 452.13, Box 1260.

[59] Allison Division, GMC, 1941, 22.

[60] *Endorsements Regarding Forced Landing Report Originated by the 79th Pursuit Squadron*, by Chief, Field Service Section, September 5, 1941. NARA RG18, File 452.8, Box 807.

[61] Priority Teletype E-784 to Wright Field Power Plant Laboratory, December 18, 1941. NARA RG18, File 452.8, Box 807.

[62] Averell Harriman, in London, letter of December 8, 1941 for Hopkins and Stettinius, attachment to letter to Major General H.H. Arnold from E.R. Stettinius, Jr., Administrator of the Office of Lend-Lease Administration, Washington D.C., December 27, 1941. NARA RG18, File 452.8, Box 807.

[63] As early as July 1941 the Air Corps P-40's at Hamilton Field, California, had been having generator problems, but all of their engines were equipped with Eclipse units. The Eclipse was a 12 volt, 25 ampere unit rated for only 300 watts, which was not sufficient to carry all the operations in the P-40 if performed at one time, i.e., flaps, wheels, lights, and radio. Memo from General Brett to Colonel Eaker, August 5, 1941. When the 33rd Squadron was dispatched to Reykjavik, Iceland for Project Indigo that summer, they were determined to not have a generator problem for their generators were manufactured by Leese-Neville and were giving satisfactory service. Memo to General Stratemeyer from Lt. Col. Bennett E. Meyers, August 7, 1941. Both memos from H.H. Arnold manuscript, Roll 129, Library of Congress.

[64] *Replacement Parts for Allison Engines*, memo to Mr. E.R. Stettinius, Jr., from John L. Pratt, December 19, 1941. NARA RG18, File 452.8, Box 807.

[65] *Allison Generator Drives*, letter from British Air Commission to Ministry of Aircraft Production, January 9, 1942. NARA RG18, File 452.8, Box 807.

[66] Great Britain had purchased 140 *Tomahawk Mk I*, the ex-French *Tomahawks*, and 110 *Tomahawk Mk II/IIA* aircraft which had been delivered during the Fall of 1940. When discussions began in Washington about providing P-40's to China, Britain had 630 Mk IIB's on order, (RAF serials AH991/999, AK100/570 and AM370/519, see Shamburger, Page and Joe Christy, 1972, 252) and unknown to China, Curtiss-Wright had sufficient surplus components on hand or available for 300 more that could be built in the Spring of 1941. Curtiss was interested in producing these aircraft to keep production flowing while the P-40D *Kittyhawk* was prepared for production. In December 1940 it was determined that if Britain would let China have 100 from those on order, they would be replaced two for one. Curtiss identified the unique model for the Chinese as their H81-A3, though all references by the AVG suggests that the airframes were identified as H81-A2's. Deliveries were to be 50 in January, 25 in February, 25 in March, 1941, the last of the 100 aircraft was delivered March 6, 1941. All were randomly taken from the AM370/519 block at a unit cost of $36,745, w/o GFE. As a consequence, Britain received all of the 300 additional airframes. Ford, Daniel, 1991, 46-51.

[67] U.S. Army practice was to purchase the airframe from the manufacture, and then provide the bulk of the operational equipment, including the engine, propeller, armament, radios and much of the ancillary equipment itself as GFE, <u>G</u>overnment <u>F</u>urnished <u>E</u>quipment. As a result China had to fend for itself for these items, they were not automatically available just because they had purchased the "airplanes".

[68] *Foreign Sale of Allison Engines*, letter from Air Corps Factory Representative to Asst. Chief, Materiel Division, March 8, 1941, and response dated March 24, 1941. NARA RG18, File 452.8, Box 807.

[69] *Allison-Power of Excellence, 1915-1990*, P. Sonnenberg and William A Schoneberger, Coastline Publishers, Malibu CA, 1990, p.76.

[70] Interview with R. W. "Dick" Askren, Engineer involved with engineering the early models of the V-1710, later with the Service Department, November 28, 1994, Indianapolis, IN.

[71] *Allison General History, 1962*, p.18.

[72] *Specifications and Operation Data-Allison Model V-1710-C15 Engines Used in Curtiss Model 81A Aeroplanes*, Allison Engine Field Service Memorandum Far Eastern No. 1, Rangoon, Burma, letter Lett, Tye M. Jr. (Far Eastern Representative, Allison Division, GMC) to Col. C.T. Chien, September 25, 1941.

[73] Information provided to author by Frank Losonsky, Crew Chief, 3rd Squadron, AVG, September 1995.

[74] Losonsky, Frank S. and Terry M. Losonsky, 1996, 31.

[75] In the later P-40K, with the V-1710-73(F4R) engine, a factory WER 5-minute rating was established at 60 inHgA, giving 1580 bhp at 3000 rpm. The AVG had been out of business for over a year before these models were available.
[76] Lett, Tye M. Jr. 1942.

[77] Arne Butteberg, along with Tye M. Lett, Jr., were assigned as the Allison Service Representatives to the AVG. Butteberg came to Allison in 1939 with a long history and much experience in aviation. He had been trained as a Naval pilot in the Norwegian Air Force in 1919 and went on to technical training in Norwegian and German schools, as well as numerous jobs within various aviation factories. Beginning in May 1941, he spent 21 months with the AVG in China. Under the most primitive of conditions, and with a minimum of tools and spare parts, he succeeded in establishing the maintenance facilities and training technicians (both Chinese and American) needed by the AVG and follow-on Army and Chinese aviation forces. He established a record of innovative methods and techniques for providing critically needed parts and components. <u>Tail Spins</u>, Vol 1, No. 1, 1943, published by Service Department, Allison Division, GMC.

[78] Lett, Tye M. Jr. 1942, 3.

[79] Lett, Tye M. Jr. 1942, 14.

[80] Losonsky, Frank S. and Terry M. Losonsky, 1996, 44.

[81] Lett, Tye M. Jr. 1942, 17.

[82] Lett, Tye M. Jr. 1942, 19.

[83] Lett, Tye M. Jr. 1942, 20.

[84] Three American Volunteer Groups were authorized to be formed. The 1st AVG was to be a fighter outfit (Curtiss P-40's), the 2nd AVG was to be a bomber unit and 33 Douglas DB-7B's (A-20 to the AAC) were procured for it. The 3rd AVG was to be another fighter unit (a shipment of Vultee Vanguard P-66's was on its way when Chennault traded them to the Chinese for their P-43's). The 2nd AVG and its aircraft were midway across the Pacific when the war started. All were diverted to Australia when the Japanese attacked Pearl Harbor on December 7, 1941. On December 8 action was taken in Washington D.C. to terminate the "volunteer" program and return the personnel to the US Military forces. The project was therefore terminated because of US defense needs. Ford, Daniel, 1991, 93-94.

[85] In <u>*Flying Tiger-A Crew Chief's Story*</u>, Frank Losonsky relates how some of the Sharks had the French style throttle, full throttle rearward, compared to U.S. practice of foreword for full throttle. This would suggest that the British had only taken over the French order, as is. The AVG quickly modified these airplanes to have standard throttles after several nearly disastrous incidents.
[86] Ford, Daniel, 1991, 93-94.

[87] Lett, Tye M. Jr. 1942, 18.

[88] Letter to BG C. L. Chennault in Kunming from China Defense Supplies, Inc., Calcutta, March 6, 1942. Hoover Institute, Chennault papers, Box 2-9.

[89] Information provided to author by Frank Losonsky, Crew Chief, 3rd Squadron, AVG, September 1995.

[90] Report of Reconnaissance of Hanoi, French Indo-China, Intelligence Report of Squadron Leader R.J. Sandell, 1st Pursuit Squadron, January 18, 1942. Chennault Papers, Box 6, Folder 2.

[91] 1st Squadron pilot Fritz Wolf report to Col. Chennault on conditions at Rangoon. Ford, Daniel, 1991, 233.

[92] Information provided to author by Frank Losonsky, Crew Chief, 3rd Squadron, AVG, September 1995.

[93] Chennault to Neale Telegram, No. 1277-KR-90 (0840), February 13, 1942. Chennault Papers, Box 5, Folder 2.

[94] Letter to General C. L. Chennault from A. Butteberg, Allison Engineering Representative, Subject: *Study of Poor Performance of P-40*, March 11, 1942. Chennault Papers, Box 4.

[95] Letter from C. L. Chennault, Commanding, A.V.G. to Mr. A. Butteburg, March 12, 1942.

[96] Lett, Tye M. Jr. December 1942, 104/106+.

[97] Handwritten notes by General C.L. Chennault, Commanding, First A.V.G. Chennault Papers, Box 2, Folder 9.

[98] Cable to Otto T. Kreusser, Allison Division, General Motors Corporation, from General C. L. Chennault, Commanding, First A.V.G., April 20, 1942. Chennault Papers, Box 4.

[99] Letter from C. L. Chennault requesting a cable be sent to Gen T.H. Shen, China Defense Supplies, Washington, April 18, 1942. Chennault papers, Box 2-9.

[100] Letter from Walter Pentergast to Col. Chennault, February 26, 1942. Chennault papers, Box 1-12.

[101] Information provided to author by Frank Losonsky, Crew Chief, 3rd Squadron, AVG, September 1995.

[102] Correspondence from Gerhard Neumann to author, July 1994.

[103] *Herman The German*, autobiography of Gerhard Neumann, AVG crew chief, 1st Squadron, p.110.

[104] Telecon between author and Frank Losonsky, 9-24-1995.

[105] *Outstanding Memories of Allison Retires-Frank Losonsky*, Allison employee February 4, 1952 to March 1, 1981.

[106] According to Frank Losonsky the AVG had no shortage of Prestone, though they never threw any away.

[107] Losonsky, Frank S. and Terry M. Losonsky, 1996, 35.

[108] Losonsky, Frank S. and Terry M. Losonsky, 1996, 29-30 and 50-51.

[109] Letter from Gerhard Neumann to author, July 25, 1994.

[110] Information provided to author by Frank Losonsky, Crew Chief, 3rd Squadron, AVG, September 1995.

[111] *Observations of Facilities Available and Recommendation for Maintenance and Overhaul Work on Allison Engines*, November 29, 1941, Tye M. Lett, Jr, Allison Engineering Company Representative. Document in Chennault Papers, Box 4.

[112] An engine inventory done at the CAMCO factory in Lungtao, done on January 19, 1942 noted they had 15 new engines, AEC serial numbers: 2998, 2999, 3001, 3003, 3005, 3006, 3007, 3016, 3017, 3018, 3019, 3020, 3021, 3022, 3023. They also had seven engines awaiting overhaul. Ten more, requiring repair and most stripped of useable accessories, were received from Toungoo on January 26, 1942. On February 3rd 30 wings were received from Toungoo, most were badly damaged.

[113] Tech Order No. 02-5AC-2, *Handbook of Service Instructions, Model V-1710-33 Engine and Associated Models*, January 20, 1941, p.24.

[114] *Herman The German, Enemy Alien U.S. Army Master Sergeant #10500000*, by Gerhard Neumann, 1984, p.69.

[115] Description of U.S. Finance Department contract with Texaco to supply the noted quantities of fuel. Telegram from General Shen Teh-tsi with information for Col. Chennault, August 12, 1941. Chennault Papers, Box 6, Folder 3.

[116] Vehicles, Oil, and Cartridge Depots Under the 5th Route Commanding Headquarters, January 2, 1942. Chennault Papers, Box 4.

[117] On average a "Shark" could be expected to consume about 100 gallons per hour of flight, and one hundred aircraft would consume one million gallons in one hundred hours of flight. The available fuel also had to fuel support and transient aircraft, including flight of air transports to carrying some of it into China. Of course the AVG never had 100 aircraft on strength, so on balance it seems fair to assume that only enough fuel was available, on average, to support no more that two hours of flight for each *Tomahawk*, each day!

[118] This amount of lead should have given a PN/Grade 155 rating, quite satisfactory as compared to the PN/Grade 150 for the standard, or it may have been used to dope substandard fuel stocks to achieve a fuel Grade 100 rating. The concern was effect on plugs and valves within the engine. In this respect the AVG wanted to get enough information so that Allison could authorize its use. Chennault letter of December 2, 1941 from Toungoo to Mr. R.C. Chen in Rangoon. Chennault Papers, Box 5, Folder 2.

[119] Lett, Tye M. Jr. 1942, 24.

[120] Letter from S.T. Young, Head of the Dept of Chemistry, National Southwest Associated University, Kunming, China to Colonel C. L. Chennault, Commanding FAVG, Kunming, dated March 25, 1942. Chennault Papers, Box 5, Folder 2.

[121] Information provided to author by Frank Losonsky, Crew Chief, 3rd Squadron, AVG, September 1995.

[122] Losonsky, Frank S. and Terry M. Losonsky, 1996, 75.

[123] Letter from General Chennault to General Chow, December 28, 1941. Chennault Papers, Box 7-5.

[124] *Pilot G.B. McMillan's Narrative Diary of Trip to and from the Gold Coast, Africa*. Chennault Papers, Box 7-8.

[125] *Recapitulation of Losses, China Air Task Force and Enemy*, letter from Office of the Operations Officer, Kunming, China, February 6, 1943. H.H. Arnold manuscript, Roll 183, Library of Congress.

[126] *Model Designations of Army Aircraft, 11th Edition*, January 1945.

[127] Discussion with "Buster" Keeton, pilot, AVG 2nd Squadron, occurred at *Friends of the American Fighter Aces, Northern California Chapter, Spring Meeting,* held at McClellan AFB, CA, March 11, 1995.

[128] Discussion with Colonel Edward F. "Ed" Rector, pilot, AVG 2nd Squadron, occurred at *Friends of the American Fighter Aces, Northern California Chapter, Spring Meeting*, held at McClellan AFB, CA, March 11, 1995.

[129] The following is from Greg Boyington's book Baa Baa Black Sheep, Bantam edition, February 1977, pages 30-31:

One day after I had been given a cockpit checkout by a qualified pilot, I had my first ride. The P-40 didn't feel too strange, considering that I had never flown one before, and had been inactive for three months.

Everything went okay until I came in to touch down for a landing. Having been accustomed to three-point landings in my Marine Corps flying, I tried to set this P-40 down the same way, even though I had been instructed to land this plane on its main gear only. I bounced to high heaven as a result of my stubbornness, and I started to swerve off the runway. So I slammed the throttle on, making a go-around. In my nervousness I had put on so many inches of mercury so quickly that the glass covering the manifold pressure gauge cracked into a thousand pieces. After I had landed in the proper manner on the second try, I was informed in no uncertain words: "You can't slam the throttle around like you did in those God-damned Navy air-cooled engines."

Jim Cross took the same P-40 up after lunch and the engine blew up. When this happened, Jim was lucky to make a wheels-up landing in a nearby rice paddy. Even though Jim wasn't hurt, I felt very bad about it, as they were forced to use this P-40 for spare parts.

[130] The V-1710 in the Curtiss *Tomahawk* drove a Curtiss Electric propeller. Normal procedure would have the pilot bring the engine rpm up to the "takeoff" rpm setting while on approach, so that in the event of a "go-round" simply opening the throttle was all that was required to make power. The Allison could then make additional power very rapidly as the supercharger was already developing maximum pressure ratio, with only the airflow constrained by the carburetor throttle on the inlet. Curtiss had serious problems with their early propellers because of the relatively slow speed at which the electric motors repositioned the blades, thereby loading the propeller and preventing overspeed. Initial field experience with overspeeding of US Army P-40 engines was so bad, that Wright Field had the engines re-rated for takeoff using only 2800 rpm instead of 3000. It is hard to say how fast Boyington's engine actually went, but to break the gage would suggest that manifold pressure was probably well over 60 inHgA, and the speed for this in the V-1710-C15 would probably have been over 3600 rpm. The momentary power level was probably in the range of 1800 bhp, though likely limited by detonation! There should be little surprise that the engine failed on its next flight!

[131] Ford, Daniel, 1991, 255.

[132] Discussion with J.R. "Dick" Rossi, pilot, AVG 1st Squadron, occurred at *Friends of the American Fighter Aces*, Northern California Chapter, Spring Meeting, held at McClellan AFB, March 11, 1995.

[133] Information provided to author by Frank Losonsky, Crew Chief, 3rd Squadron, AVG, September 1995.

5

The V-1710 in Combat Aircraft

For the USA the shooting phase of World War II officially began at Pearl Harbor on December 7, 1941. The country had been preparing for the eventuality for some time, certainly since September 1, 1939, the occasion of the Nazi invasion of Poland. Where previously there had not been whole hearted support for the U.S. to participate in the war, after the attack on Pearl Harbor there was an enormous outpouring of commitment to right a grievous wrong.

It is hard, looking back more than over 50 years, to understand the situation facing the common people, their leaders, and the peacetime reared military. While advertised by President F. D. Roosevelt as the "Arsenal of Democracy", the truth is that at Pearl Harbor the U.S. lost most of its offensive Naval punch and that its Army and the Air Corps were equipped with weapons that were decidedly second rate. The only reason that we had even a fighting chance of defending ourselves was that for nearly two years American factories had been producing defense materiel for sale to the British, French and their allies. This was indeed fortunate for the course of history, for much of the infrastructure necessary to support a large fighting force was already in place and at work. This was particularly true in the field of aviation.

During the heart of the 1940 Battle of Britain, with hundreds of Spitfires and Hurricanes battling hordes of Messerschmitt Bf 109 and Bf 110's, Heinkel He 111's and Junkers Ju 87 Stukas, the U.S. Army Air Corps took delivery of its most advanced pursuit aircraft, the thirteen Curtiss YP-37's. These were significant improvements over the Sikorsky P-35 and Curtiss P-36's which then constituted the core of the front line fighter force. The first major pursuit production program to number in the hundreds was that of the Curtiss P-40. Delivery of the first batch of 199 non-combatant ready P-40's occurred from June through October 1940, followed by 131 combat ready P-40B's, delivered January through April 1941.[1]

With the world situation deteriorating in 1939, the U.S. Army Air Corps had flown the XP-38, XP-39 and XP-40. The XP-40 had won the Pursuit Competition that January and during the year sizable orders were placed, as well as pre-production batches of the XP-38 and XP-39 being ordered. Things were beginning to move, for each of these designs was a significant improvement on previous U.S. "combat" aircraft. Even so, as the events in Europe were to soon prove, none of the early models of these aircraft was up to the standard necessary to survive or be successful in modern combat. Also in 1939, the Air Corps issued Circular Proposals to the aircraft industry for production quantities of single and twin-engined interceptors, as well as a Multi-Place fighter.

As combat reports began accumulating and the lessons of this round of aerial combat digested, it became clear that the aircraft the U.S. was developing were woefully inadequate and would have to be significantly improved. When Congress finally gave the European purchasing commissions access to U.S. manufacturers the available "advanced" designs were revised and improved to qualify them as combat capable aircraft. The result was serial modification and incremental improvement in each of the available models, along with the initiation of a few new designs such as the *Mustang* from North American Aviation and P-47B *Thunderbolt* from Republic.

During 1940 many government and military leaders believed that it was no longer "if", but rather "when", the U.S. would be drawn into the widening conflict. Consequently, the acquisition of aircraft, and efforts to improve their combat capability, took on entirely new significance well before Pearl Harbor. By early 1941 the first of these "combat" qualified aircraft began to enter U.S. Army Air Corps service. Initially their numbers were not large, but they were quickly evolving into machines capable of engaging contemporary Axis aircraft. Concurrently the Axis were improving their own combat capability based upon two years or more of experience and development.

Creating a modern fighting force on such comparatively short notice was not an easy undertaking. Much of the credit goes to the men and women of the nation who saw a common threat, then pulled together as a production and combat team the likes of which the world had never seen. Enormous accomplishments occurred seemingly overnight, a tribute to the sacrifice and commitments at home and by the forces overseas.

As an indication of the scope and complexity of the task, this chapter presents a description of the various models of combat aircraft employed by the U.S. that were powered by the Allison V-1710 aircraft engine. The focus is on the features of the engines that were unique to a particular aircraft model, its purpose and mission as effected by the changing demands from the combat theaters, and strategy and tactics for the war. It is also important to recognize how specific models, and their capabilities, were developed to counter a particular threat such as the Mitsubishi A6M *ZerSen "0"* or Focke Wulf Fw 190A.

Lockheed P-38 *"Lightning"*

The winner of the 1937 Air Corps Type Specification X-608 design competition was the Lockheed Model 22, designated Air Corps Model XP-38. Following its loss on the delivery flight the whole program hinged solely upon the promise of the design. Fortunately, Lockheed was awarded a contract for a service test batch of thirteen YP-38's. These aircraft differed in detail from the XP-38, with the most significant change being use of the new V-1710-F series engines. In this respect the YP-38's were the first to mount an engine that would provide the P-38 with steadily increasing power, rising from 1150 bhp to 1725 bhp in later models. While the design of the aircraft showed the genius of Lockheed and its Chief Designer Clarence L. "Kelly" Johnson, it was an airframe of uncommon adaptability and capability. Significant to its many accomplishments was the incorporation of turbosuperchargers, that enabled the Allison' to maintain their takeoff and War Emergency ratings to extremely high altitudes.[2] The combination of the turbo and the engine-stage supercharger made it a "two-stage" installation, with intercooling. In this section we look at the evolution of the unique features of the engine and turbo installations that paced the development of advanced capabilities of the airplane.

The first YP-38 standing for a publicity photo at Lockheed Burbank Airport. (Lockheed via SDAM)

Early P-38's

As Lockheed's Model 122 the YP-38's were powered by 1150 bhp V-1710-27/29(F2R/L) engines having intercooled induction air supplied by General Electric Type B-2 turbosuperchargers. Induction air was ingeniously cooled by an intercooler formed by the leading edge section of the outer wing panels. So as to eliminate the effects of propeller torque, along with tail buffeting encountered on the XP-38, the engines were installed such that the "right" turning engine was installed on the right wing, and the "left" turning engine on the left. This arrangement was continued for all subsequent P-38's.

Table 5-1
Early P-38 Development Milestones

Date	Number	Comment
6-23-37	1	XP-38 ordered
4-27-39	13	YP-38 ordered
1-27-39	-	XP-38 makes First Flight
8-9-39	66	Ordered P-38 (Model 222-62-01)
6-5-40	667	Model 322-61-04 ordered, British Contract A-242
8-30-40	607	AAF orders for P-38E/F
9-17-40	-	First Flight of a YP-38, #39-689
6-20-41	-	First Production P-38, #40-744 delivered
7-1-42	-	First P-38D delivered
8-1941	-	First British 322-61-04 delivered, AE978
10-1941	-	Second 322-61-04 delivered, AE979
11-16-41	-	First P-38E delivered

As a component of the Emergency Procurement Program of 1939 the Air Corps developed Type Specification C-615 describing the desired characteristics for procurement of a twin engine service aircraft. Bids were invited from 87 aircraft manufacturing firms under the associated Circular Proposal 39-775, issued on March 11, 1939. Only four competitors offered airplanes when the bids were opened on July 8, 1939. From these the Allison powered Lockheed Model 222 was the clear winner, and received an order for 66 aircraft to be delivered within 15 months. These were actually delivered as P-38 and P-38D models, there being 29 and 36 produced respectively, with the odd aircraft being the airframe diverted for modification as the pressurized XP-38A.

There were also two competing aircraft from the CP 39-775 competition selected for experimental development. These were the second place airplane, Lockheed Model 222-A (to be powered by turbosupercharged P&W R-1830-C3G engines)[3] and the Wright R-1820-G666-1/2 powered Grumman Model 41 (later designated XP-50). Only the Grumman made it to flight status. The other competitors were the Burnelli Model XBP-1, also to be powered by the Allison V-1710-F2R/L, and Bellanca Model 33-220 to be powered by Allison V-1710-33's.

The 29 "straight" P-38 aircraft were built as the Lockheed Model 222-62-02.[4] These were effectively YP-38's with offensive armament, though they were not up to combat standards. Most served in test and training rolls.

Engines for these 66 aircraft were obtained at no cost to the government by taking advantage of the unit cost reductions occurring at Allison due to the large orders being received from the French and British. As a result the U.S. engine contract (ac-12553) totals were increased from 837 to 969 in September 1940, but the total contract cost was unchanged.

A single Model 622 was built as the XP-38A, and was unique in having a pressurized cockpit. It was powered by the V-1710-27/29's as were the Model 222-62-02's from which it was derived. The next models, P-38B and P-38C were proposals only and were never built.

Lockheed Model 222/322 P-38D/E/F/G *"Lightning"*

P-38D Built as the Lockheed Model 222-62-08D, the P-38D was the first truly combat capable *Lightning*. These were the remaining 36 aircraft on the CP 39-775 order for 66 P-38's, and incorporated a number of improvements deemed essential as a result of lessons learned in European combat. These aircraft entered production in June 1941, and were powered by the same V-1710-27/29(F2R/L) and GE Type B-2 turbos as had been used in the 13 YP-38's and 29 P-38's.

P-38E In September 1941 the P-38E, also powered by the V-1710-27/29, entered production and 210 were built.[5] This was the first major "production" version of the aircraft, though the model was still not considered to be really combat ready, the Type B-2 turbosupercharger was retained. These aircraft led varied service lives doing test and training duties, primarily within the Continental U.S.

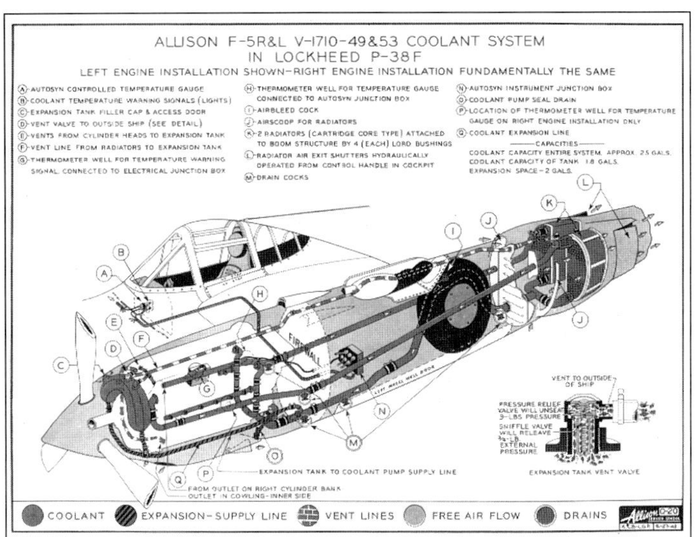

Here are the details of the coolant system on the early P-38. The streamlined nacelles and locating the radiators aft of the wing resulted in lots of plumbing. (TO 30-5A-1)

Table 5-2 **Lockheed P-38F Recommended Ratings**				
	Engine bhp	MAP, inHgA	RPM	Altitude, Ft
Ratings as of March 13, 1942				
Takeoff	1150	41.5	3000	SL
Military	1150	41.5	3000	SL to 25,000
Normal Cruise	1000		2600	SL to 25,000
Ratings as of October 23, 1942				
Takeoff	1325	47.5	3000	SL to 4,000
Military	1240		2800	4,000 to 25,000
Normal	1000		2600	SL to 27,000

MAP was to be reduced 1-1/2 inHgA per 1000 feet above critical altitude, while maximum carburetor air inlet temperature was to be less than 135 oF. Fuel was to be AN-VV-F-781, Amendment 5. There was no War Emergency Rating officially established at this time.

P-38F Entering production in March 1942, the P-38F became the first truly combat capable P-38 due to many changes in equipment. Significant among these being its improved engine, the V-1710-49/53(F5R/L). It took over the production lines and remained the primary model until introduction of the P-38G in August 1942. The turbos remained the Type B-2. Concurrent with production of the P-38F were a further 99 similar aircraft that were factory built as photo-reconnaissance craft, designated by the AAF as the F-4-1-LO. These were the first mass produced reconnaissance aircraft in the U.S.

The key change incorporated into the V-1710-F5 was an improved engine-stage supercharger. It had rotating steel inlet guide vanes mounted on the impeller, and a higher step-up gear ratio driving the supercharger. These, and other supporting improvements that strengthened the internals of the engine allowed it to be rated for 1325 bhp from takeoff up to the critical altitude of the turbosupercharged aircraft. At the higher altitudes this rating was actually more than could be accommodated by the integral wing leading edge intercoolers.[6]

This is V-1710-49(F5R), AEC#8286, AC41-33720, recovered from under the Greenland icecap, with P-38F "Glacier Girl". The long exposure to water had corroded the pistons within the nitrided steel cylinder liners to the point that it took 1000 psi of hydraulic pressure to push them out. The engine is being rebuilt to fly by Al Boushea for the Lost Squadron project. About 40 percent of the parts have been replaced to make it air worthy. (Author)

On March 13, 1942 Operating Instructions were issued for the P-38F and F-4 aircraft. This was practically coincident with their introduction and reduced allowable takeoff and military ratings from 1325 bhp to 1150 bhp because of difficulties being experienced with the integral wing leading edge intercoolers on the airplanes. In order to obtain maximum performance, Lockheed then ran extensive flight tests to determine optimum engine operation in the P-38F. The Wright Field Materiel Center concurred with their findings and recommended the ratings given in Table 5-2.[7]

In response to the restricted ratings, by the end of May 1942 Lockheed had performed extensive flight tests to determine optimum engine operating conditions in the P-38F. The critical parameter was to limit power if the temperature of air coming from the intercoolers exceeded 135 °F. Even so, it took Wright Field until the end of October to agree to restoring the 1325 bhp rating, and even longer to get the official directives into the field. In this environment the 8th Air Force Fighter Command took matters into its own hands and with the testing and guidance of Col. Kelsey was using even more aggressive ratings.

Table 5-3 **P-38F Combat Ratings by 8th Fighter Command**				
	Engine bhp	MAP, inHgA	RPM	Altitude, Ft
Military, after 8-15-1942	1300	47.0	3000	SL to 20,000
Military, after 8-23-1942	1250	45.0	3000	15 to 25,000
WER, Proposal by 8th FC	1450	52.0	3000	Below 11,000

After two months operating with these ratings no engine failures had been reported.

Lockheed Model 322 Lightning's

The Model 322 for Export

In April 1940 the Anglo-French Purchasing Committee contracted for 667 Model 322 aircraft, the French to receive the Model 322-F, and the British the Model 322-B. This was a significant order, for the total number of P-38's that had been ordered by the Army was only 80 at this point. The engine was selected with care to be the V-1710-C15, the simple reason being to standardize on the altitude rated engine being used in the Curtiss H-81A *Tomahawk* that both countries had already purchased.[8] With the collapse of France in June 1940, long before the first 322-F was built, the British took over the entire order. They then named their Model 322-B the *Lightning Mk I*, and amended the contract to provide the final 524 aircraft on the contract with V-1710-F5R/L's as their *Lightning Mk II*.[9]

Only three of the British 322-B models had been delivered by the time of Pearl Harbor, though production was underway on the entire order. With U.S. entry into the war circumstances were considerably changed, the most dramatic being the takeover of *Lightning* production by the U.S. Army. The Army then identified the V-1710-C15 powered airplanes as the P-322-I (still the Lockheed Model 322-61-04). With the Army takeover of the order, they wanted the aircraft to have handed engines, so Lockheed defined the Model 322-62-18, Army Model P-322-II. This aircraft was powered by the altitude rated V-1710-27/29(F2R/L), but otherwise similar to the P-322-I and likewise without turbosuperchargers. They were converted from P-332-I's at a Lockheed modification center, where the engine and cowling changes were made. As the V-1710-F2's retained their original 6.44:1 supercharger gears (as intended for turbosupercharged installations), the P-322-II had a considerably reduced critical altitude. These aircraft served primarily as advanced trainers for the Air Corps where they were effective in the low level mission.

AE979 was the second P-322-B on the British order. The lower thrust line of the V-1710-C15's shows clearly, as do the YP-38 style radiator ducts. Exhaust collectors were simply routed from each cylinder bank to exit above the nacelle just ahead of the front wing spar. These "Castrated Lightnings" were attractive airplanes. (Lockheed via SDAM)

Two *Lightning Mk I's*, AF105 and AF106 (V-1710-C15 powered Lockheed Model 322-61-04, Constructors Numbers 3028 and 3029) were delivered in January 1942 and shipped to England that March. The first P-322-B, Lockheed constructor's number 3001, British serial AE978, was also the first aircraft to be converted as a Model 322-62-18. As such, it was also shipped to England in that March, where the three aircraft were tested, primarily by the Royal Aircraft Establishment at Boscombe Down. These were the only P-322's to reach England.

It has sometimes been reported that when the RAE tested these aircraft the result was cancellation of the remaining production order, but that is not likely the case. The actual reasons for cancellation appear to have been several.

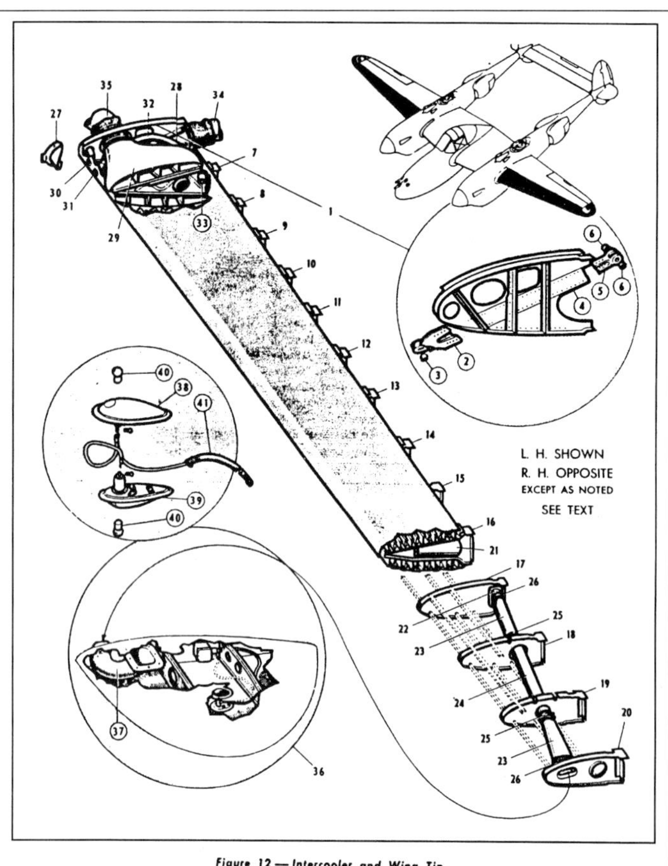

For the XP-38 through the P-38H Lightning the wing leading edges functioned as pressurized induction air intercoolers. These were sized to remove turbo compression heat equivalent to 1000 bhp, the Normal engine rating, not Military. This was a limitation for the early Lightning's. The last three feet was a section used to "turn the air from the lower to upper channels. Hot turbo air entered at 34 and discharged to the carburetor at 35. Internal cooling air exited at 37, the wingtip. The inboard section of the intercooler was nearly two feet deep, tapering to less than one foot at the wingtip. (Lockheed via Bob Cardin)

This view of P-38F AF41-2314 shows that the radiator housings were the same as those on the P-322. (Allison)

First the war was now nearly two years past the Battle of Britain and circumstances were considerable different, including the condition of the Exchequer and the need to redirect their limited funds. Second, it was recognized that without handed engines the aircraft was difficult to handle and did not have the performance of the P-38. The reasons for not originally providing a left handed C-15 are not clear, for the P-38 certainly needed the torque canceling benefits of the feature. It is likely that with the 1939/1940 push to produce sufficient numbers of C-15's, and at the same time develop the follow-on V-1710-E/F series, there were simply insufficient resources to commit to the effort. A left-handed C-15 would have been little different than the changes that had been incorporated into the C-9 for the XP-38, but introducing it into production would have required considerable duplication of spare parts to support such an engine in the field. Given that the specification of C-15's in the first place was for commonality with the C-15 powered Hawk 81A, commonality was probably the primary reason handed engines were not provided. The need for handed engines was the likely reason for the U.S. Army to have immediately made the conversion to altitude rated V-1710-F2R/L engines.

The deletion of the turbos may also have been a factor, for by 1942 combat over Europe was being entered at greater altitudes. Back in 1940 competing Spitfire, Hurricane and German aircraft like the Messerschmitt Bf 109, were all equipped with similarly capable single-stage engines. Combat was therefore likely to be limited to moderate altitudes and the savings in weight, cost and complexity of the turbo were probably the overriding considerations for its deletion by the British. When delivered in 1942 the "*Castrated Lightnings*" were definitely obsolete as combat capable aircraft.

In June 1941 the U.S. Joint Aircraft Committee decided that only 143 Model 322-61-04 P-322's would be built. Production of the "Castrated P-38" on the British order was ended by their decision to switch all P-38 types to the Allison "F" engines with turbosuperchargers, thereby providing for early elimination of the "C" engine manufacture and accelerating production of both the Allison "F" engine and the P-38 airplane.[10]

Another factor was the timing. The P-322 was being built parallel to the P-38E and F-4 reconnaissance models of the P-38, the types in production following Pearl Harbor. The U.S. Military was seriously concerned about invasion and/or attack of the West Coast and was eager for any capable defensive aircraft. Meanwhile the P-322-I production was being placed into open storage pending RAE testing of the first articles and a decision by the British on taking delivery.

For a variety of reasons 23 of the aircraft were not converted to the F-2 powered P-322-II configuration (Lockheed Model 322-62-18) upon delivery to the U.S. AAC.[11] The 120 conversions to Model P-322-II were all delivered to the Air Corps by July 1942. The Air Corps purchased the remaining 524 aircraft on the British contract and had them finished in unique production blocks as P-38F-13/15 *Lightning I* and P-38G-13/15 *Lightning II*.

P-38G The distinguishing feature of the P-38G was introduction and use of the V-1710-51/55(F10R/L). Its significant difference from the earlier V-1710-F5 engines was the adaptation of a larger carburetor, increased power ratings, and use of a higher supercharger gear ratio. The major structural and reciprocating components remained as on the earlier series engine. After the first block of P-38G-1 aircraft the turbo was upgraded from the Type B-2 to the GE Type B-13.[12] The benefit of the new turbo was that it had a higher critical altitude from being rated for a higher maximum speed.

During the Summer of 1942, when the need to establish War Emergency Ratings for aircraft to be used in combat theaters was recognized. The earlier Model 222 *Lightning* did not fare well. After considering the limitations of the wing leading edge intercooler Allison recommended that War Emergency Ratings *not* be established for any models prior to the P-38H.[13]

This P-38G was named "THUMPER" when used for testing skis, evidently there were problems. Of interest are the openings in the wing leading edge that fed cooling air into the leading edge intercooler. The large scoop aft of the prop spinner fed cooling air into the exhaust manifold shroud, the small scoop just below supplied the blast tube for cooling the exhaust spark plugs. Carburetor intake are came from the scoop just below the wing. The form of the bent prop tips shows the engine was making power when the airplane crash landed at Wright Field, January 13, 1944. (AAF via SDAM)

P-38H The P-38H was a transitional model of the *Lightning*. It retained the appearance of the Lockheed Model 322 (P-38G), including its integral wing leading edge intercoolers and sleek nacelles, but was powered by the improved V-1710-F17 model engines intended for the Model 422-81-14 (P-38J) and rated for 1425 bhp. Specifically the P-38H was Lockheed's Model 422-81-20, clearly conceived and designed after the P-38J (LAC Model 422-81-14) and reflecting its production as a transition model.[14] The visible difference between the 322/422 Models is apparent on the P-38J airframes that used core type intercoolers located in the chin of the nacelle, directly below the engine. Starting with the later P-38J-5-LO, two 55-US gallon fuel tanks took the place of the leading edge intercool-

With the introduction of the high powered F-17 and F-30 engines it was necessary to improve the cooling capacity of the P-38. This P-38L shows that larger radiator cores were fitted in new housings, and a boundary layer splitter was also provided. This improved the effectiveness of the entire installation. (Author)

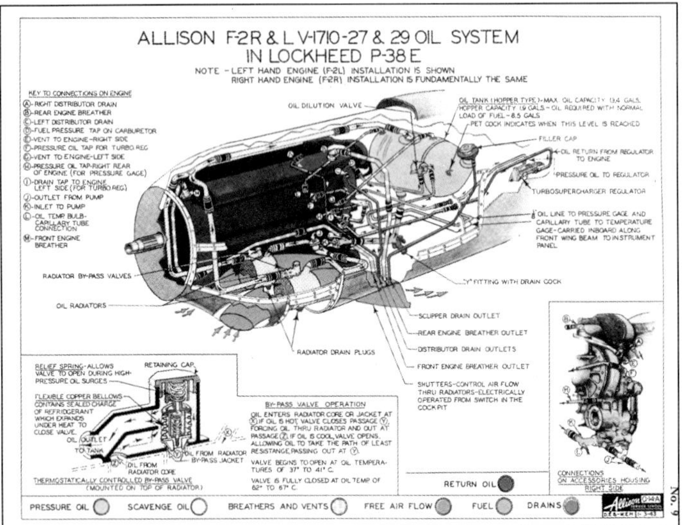

Lube oil cooling for all models of the P-38 was via coolers mounted directly below the engines. These were compact and effective installations. (TO 30-5A-1)

ers.[15] Another significant change introduced on the P-38H, and one difficult to discern, was use of enlarged coolant radiators and scoops on the sides of the booms. These incorporated boundary layer bleeds as another new feature to further improve cooling. Additional radiator capacity was required because of the higher power available with the V-1710-F17 engines.[16]

All Lockheed P-38 models prior to the P-38H were built with "hopper" type oil tanks and would suffer damaging loss of oil pressure during negative "g" flight maneuvers. Many of these early P-38's were being used within the U.S. in training roles, and as late as January 1943 there were no plans to retrofit them with the pendulum equipped tanks. The result was that the loss rate of engines and aircraft was unnecessarily high.[17]

While Lockheed was ready to change the production line to the P-38J, it was delayed due to a lack of suitable core type intercoolers.[18] Lockheed built three preproduction P-38J's before the single preproduction P-38H, followed by the single P-38K and then the two blocks of production P-38H's, 226 as the P-38H-1 with Type B-13 turbos, and 375 as the P-38H-5, which used the higher rated Type B-33 turbosupercharger.[19] All were built on the same contract that produced the production P-38J's. The reason for the military designation suffixes to be sequenced different than the Lockheed model numbers has to do with the military practice of not assigning designations until the actual procurement contracts were in place.

For the V-1710-89/91(F17R/L) engine, Allison had established a War Emergency Rating of 1600 bhp, available when running at 3000 rpm with 59 inHgA manifold pressure. This rating required the full capacity of the turbosupercharger to deliver 31-1/2 inHgA to the carburetor deck, and in addition, for the intercooler to cool the induction air to below 80 °F when operating at the 25,000 foot critical altitude of the installation. In the airplanes with leading edge intercoolers 40 to 45 inHgA[20] backpressure would have been required for Type B-13 turbos to deliver the required flow and pressure at the critical altitude. Allison feared that field operations with this amount of backpressure limited intercooling capacity were likely to cause detonation. Consequently, WER was reduced to 1425 bhp, which was still risky given the limited capacity of the turbo and leading edge intercoolers.[21]

Interestingly, at the same power settings, configurations and altitudes, the P-38H was marginally faster than the P-38J/L.[22] This can be attributed to the better streamlining of the nacelles.

P-38H's were produced from May through December 1943, and large numbers of them reached England that fall and winter. They were the workhorse long range fighter of the 8th Air Force well into 1944, when first augmented, and then replaced, by the Merlin powered North American P-51B/C Mustang and the new P-38J.

Lockheed Model 422 *Lightning's*, the P-38J/L/M

The initial production P-38 was the Lockheed Model 222, that in its various forms (P-38D/E/F/G) resulted in 1,660 *Lightning's*. A further 667 similar Model 322's were built on British orders, though most were actually delivered to the USAAF as P-38F and P-38G models. The ultimate *Lightning* was the Lockheed Model 422, 7,694 of which were produced for the AAF variously as their models P-38H/J/L/M.

P-38J The P-38H (LAC 422-81-20) and P-38J (LAC 422-81-14) models were both powered by the V-1710-89/91(F17R/L), but the P-38J was the first to adapt the nacelle chin location for the installation of the high capacity core type intercoolers for the Type B-33 turbosuperchargers. These nacelles were the visual feature distinguishing the Model 422 *Lightning* from earlier models. The P-38J was the major production model from August 1943 through June 1944, when the much more sophisticated P-38L entered production.

P-38 Long Range Performance

Model	RPM	MAP, inHgA	Total gph	IAS, mph	TAS, mph	Alt,ft	A/C Weight, pounds
With two 165 Gallon Tanks, Maximum Continuous:							
P-38G							LATER
P-38H	2600	44	245	232	360	30,000	18,100 to 13,500
P-38J/L	2600	44	245	227	354	30,000	19,400 to 14,500
With tank supports only, Maximum Continuous:							
P-38G							LATER
P-38H	2600	44	245	238	396	30,000	16,100 to 13,500
P-38J/L	2600	44	245	252	388	30,000	17,400 to 13,500
With two 165 Gallon Tanks, Maximum Range:							
P-38G							LATER
P-38H	1850	30	85	178	269	25,000	18,100 to 13,500
P-38J/L	1850	31	89	170	260	25,000	19,400 to 14,500

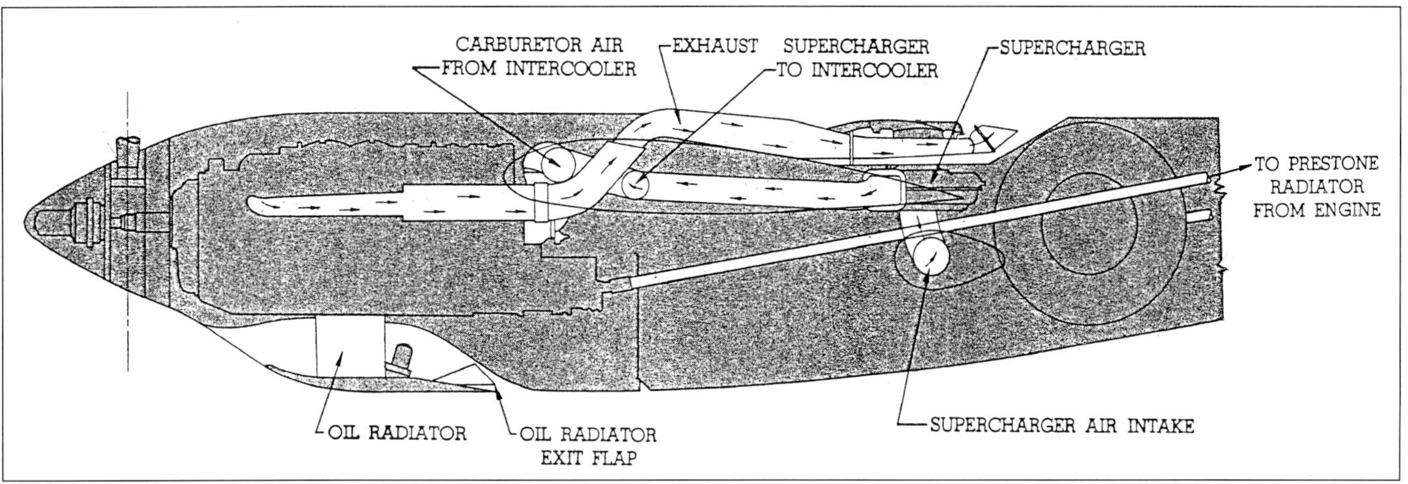

Internal ducting of the P-38F typified the nacelles of P-38's with leading edge intercoolers, those prior to the Model 422. (Lockheed)

The distinguishing feature of the P-38J/L airplanes were the deep nacelles mounting both oil coolers and intercooler cores directly below the engines. (Lockheed via SDAM)

With "handed" engines the propeller torque on the aircraft was canceled. This gave an advantage in maneuverability to the *Lightning* pilot. On all but the XP-38, the left turning propeller/engine was installed on the left wing nacelle and the right turning unit on the right.[23] Compared to the early F-2L and F-5L engines, these F-17 engines had wider and stronger gears in the accessories drive that improved their reliability, particularly that of the left turning engine. Except for the positioning of these drive gears the R/L engines were identical, and could, if necessary, be changed from Right to Left turning in the field.

Allison Time Bomb

When the V-1710-89/91 powered P-38H entered combat it was limited to a WER rating of 1425 bhp because of insufficient intercooler capacity. With core type intercoolers, the P-38J was able to use the full WER rating of 1600 bhp. This lead to an immediate rash of problems for the 8th Fighter Command, for they suddenly found numerous cases where Allison's were reported to be simply "blowing-up." This caused some to refer to the engine as the "Allison Time Bomb." Subsequent investigations have not found a single cause, but rather suggest that a number of factors conspired to cause a crisis of sorts. As best can be determined, they were:

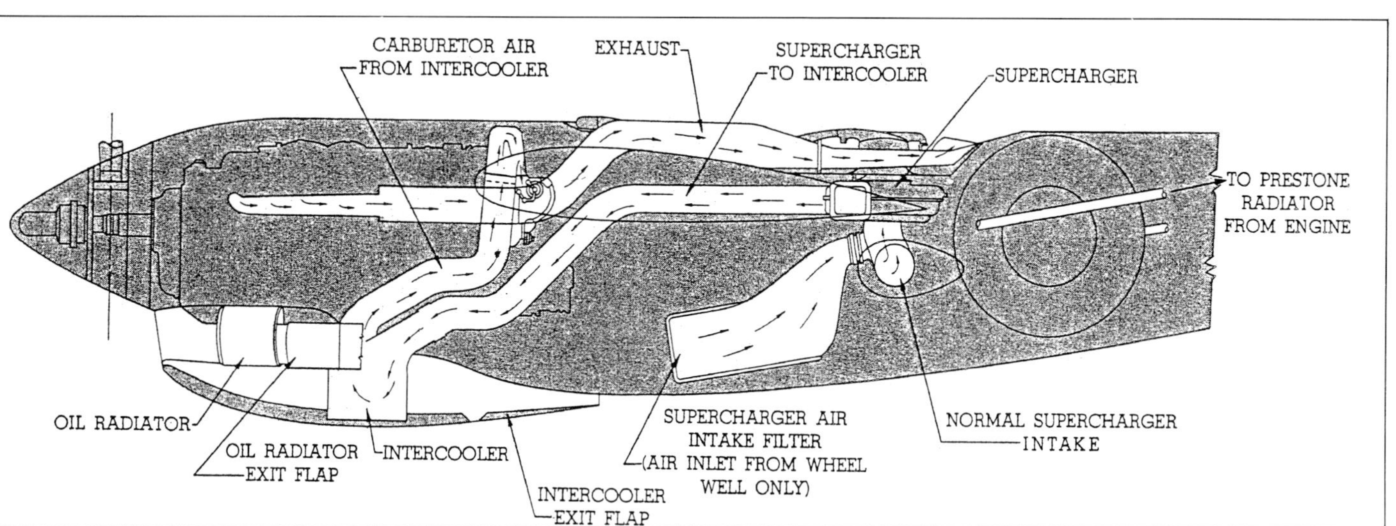

This drawing clearly shows the major changes made within the nacelles when Lockheed changed to core type intercoolers in the P-38J. (Lockheed)

- Coincident with the introduction of the P-38J's was the arrival in England of many new and relatively inexperienced pilots.[24]
- Engines were routinely being operated at the new 1600 bhp WER rating that was obviously closer to the detonation limits than with the earlier restriction to 1425 bhp.
- At 1600 bhp the engine maintenance requirements were exacting. Exhaust plugs were to have been changed after every flight that WER was pulled. This was not always done, leading to lead fouling and increased likelihood of detonation during subsequent operations at WER.
- Pilots were arriving from the U.S. where they had been trained to use high rpm and low MAP when cruising on combat missions.[25] This was very hard on engines and *not* consistent with either Lockheed or Allison technical instructions.
- It appears that the fuel being provided in England for the P-38's during the winter of 1942/43 was not entirely adequate, as the TEL would condense in the manifolds, particularly during cruise, and lead to destructive detonation.[26,27]
- The improved intercoolers were providing considerably lower manifold temperatures, which allowed TEL condensation during cruise, as well as increasing the likelihood of plug fouling.

Another perspective on the situation is related by Tony LeVier, renowned Lockheed Test Pilot who was personally flew many of the milestone P-38 tests. In May 1944 he was on a special assignment to the 8th Fighter Command in England and involved with the introduction of the P-38J. He relates how General Doolittle had assigned four P-38's to them for a special program to improve the combat capability of the P-38, including investigation of the exploding engine problem. Along with Lockheed's Ward Bernan they decided upon their own to "supercharge" the crankcase of a V-1710 to well above the normal 20 inches of water pressure-in fact to 20 feet of water-and run a Torture Test to see the effect. Twenty "inches of water" is a little less than 1 psi of positive pressure greater than the local atmospheric pressure. Twenty "feet of water" is 7.2 psi, and if the P-38 was at 17,500 feet or less, the pressure inside the crankcase would be at least equivalent to that at sea-level.[28] The supercharging pressure came from the "blow-by" of hot combustion gasses past the piston rings that is normal in any engine. By restricting the size of the outlet vent the pressure could be raised to the higher level.

LeVeir reported that he made two flights in the modified P-38, in each the modified engine was run continuously at Military power, 3200 rpm and 60 inHgA, at an altitude of 35,000 feet until the fuel was almost gone.[29] He reported that the engine "purred along smooth as silk" and after each of these punishing flights the engine oil sumps and magnetic plugs were inspected for signs of internal engine damage, he reported they were "clean as a whistle." A compression check by the Allison Rep found the twelve cylinders to have the highest values he had ever seen. These results were reported to Lockheed in California, though no reports of consequential changes followed.

While it cannot be denied that LeVier did what he reported, nor that he observed what he did, we do need to consider the likely effects of the modification and whether or not the higher crankcase pressure had any effect on the durability of the engine.

The higher crankcase pressure could be expected to upset the internal distribution of pressure throughout the engine, the likely effect being on the oil seals into the supercharger, propeller shaft and accessories. These effects should be comparatively minor, accept for a possible increase in oil consumption through leakage. It needs to be remembered that the higher crankcase pressures were considered in the design of the engine, for they often occurred as a consequence of piston ring wear, sticking or failure. In such cases the crankcase breather provided the vent path for the resulting high pressure, high temperature blow-by gasses.

Some might believe that the higher pressure in the crankcase would provide a "cushion" of air to smooth the deceleration of the pistons at the bottom of each stroke, and thereby reduce any tendency to "pound" the bearings to failure. Close tolerances between the bearings, connecting rods and crankshaft, while maintaining a steady supply of lubricating oil was sufficient to handle the deceleration forces. Those forces are considerable-over 1,000 "g's" when running at Military rpm. The air "cushion" by comparison would not have any effect. Remember that as the twelve pistons reciprocate, the "average" volume of the crankcase never changes, hence there is no "compression cushion" being provided.

Another point is to note that even with 20 "feet of water" pressure in the crankcase, at 35,000 feet the absolute pressure in the crankcase would be equivalent to the atmospheric pressure at about 9,000 feet, 21 inHgA. This is considerably less than the reported manifold pressure of 60 inHgA that was being run and further suggests that any cushioning effect was negligible. The major effect of the restricted vent path was to simply reduce the quantity of blow by gas. Depending upon the condition of the piston rings this may or may not have improved their functioning.

It is too bad that LeVeir did not run the remaining test: returning the vent path and crankcase pressure to normal and then repeating the demonstration. Given the preliminaries, one should expect the engine to have continued to run smoothly and show no ill effects of the sustained high altitude high power running.

The ultimate fix took awhile to define and implement. According to John Kline, then of the Allison Service Department and assigned to Lockheed Engineering Flight Test, Lockheed finally [solved the problem by] injecting a little more heat [done by reducing the amount of intercooling] into the carburetor inlet on the flights in the real cold conditions successfully correcting the problem.[30] Allison was working on a fix as well. Their assessment of the problem was that the low temperatures at high altitudes were causing fuel condensation within the intake pipe between the supercharger and distribution tee. One consequence of this effect was uneven concentrations of TEL in the mixture, a sure cause of lean mixture and detonation. Allison developed an intake pipe fitted with a boost venturi, similar to those in a carburetor, that would reatomize any fuel collecting within the pipe. This device was retrofit onto most V-1710's by an official Tech Order modification.

Resolving these conditions took several months, and in the meantime Major General Jimmy Doolittle, Commanding General 8th Air Force during 1944, felt that he couldn't wait and made his decision to rely upon the Merlin powered P-51 Mustang as the primary 8th Air Force long range escort fighter. While the P-38's and Republic P-47's were never fully replaced, they had reached their zenith within the 8th Air Force.[31] Even so, Doolittle often flew a P-38 while commanding the 8th Air Force, both for commuting between stations and sometimes over Europe. In fact, he chose a P-38 as a platform from which to observe the action over Normandy on D-Day.[32]

Early in 1944 a crisis developed in the Mediterranean Theater where these engines were being operated in photo-recon F-5B aircraft on long range high altitude missions. Engines were failing due to exhaustion of the oil supply, coupled with oil fouling of the sparkplugs. The problem was serious enough that all of the aircraft of one wing were at one time grounded.[33] Engine teardowns in the Theater showed evidence of leakage from the blower section into the crankcase around the labyrinth shaft seal. The seal was still effective in retaining oil, but excessive crankcase pressure at altitudes above 25,000 feet when using high boost, was believed to

Table 5-5
Lockheed P-38 Lightning Engines

Aircraft Model	No. Built	Engine Model	Reduction Gear Ratio	S/C Gear Ratio	Turbo Model	Takeoff, bhp	Military, bhp/alt/MAP	War Emergency, bhp/alt/MAP
XP-38	1	V-1710- 11/15(C8/C9)	2.00:1	6.23:1	B- 2	1150	1000/25,000/33.4	1150/25,000/37.5
YP-38	13	V-1710- 27/29(F2R/L)	2.00:1	6.44:1	B- 2	1150	1150/25,000/39.4	None
XP-38A	1	V-1710- 27/29(F2R/L)	2.00:1	6.44:1	B- 2	1150	1150/25,000/39.4	None
P-38	29	V-1710- 27/29(F2R/L)	2.00:1	6.44:1	B- 2	1150	1150/25,000/39.4	None
P-38D	36	V-1710- 27/29(F2R/L)	2.00:1	6.44:1	B- 2	1150	1150/25,000/39.4	None
P-38E	210	V-1710- 27/29(F2R/L)	2.00:1	6.44:1	B- 2	1150	1150/25,000/39.4	None
F-4	99	V-1710- 27/29(F2R/L)	2.00:1	6.44:1	B- 2	1150	1150/25,000/39.4	None
P-38F	526	V-1710- 49/53(F5R/L)	2.00:1	7.48:1	B- 2	1325	1325/25,000/47.0	None
F-4A	20	V-1710- 49/53(F5R/L)	2.00:1	7.48:1	B- 2	1325	1325/25,000/47.0	None
P-38G	1,082	V-1710- 51/55(F10R/L)	2.00:1	7.48:1	B-13	1325	1325/25,000/47.0	None
F-5A	181	V-1710- 51/55(F10R/L)	2.00:1	7.48:1	B-13	1325	1325/25,000/47.0	None
P-38K	1	V-1710- 75/77(F15R/L)	2.36:1	8.10:1	B-14	1425	1425/27,000/54.0	1600/27,000/60.0
P-38H-1	226	V-1710- 89/91(F17R/L)	2.00:1	8.10:1	B-13	1425	1425/24,900/54.0	1425/24,900/54.0
P-38H-5	375	V-1710- 89/91(F17R/L)	2.00:1	8.10:1	B-33	1425	1425/24,900/54.0	1425/24,900/54.0
P-38J	2,970	V-1710- 89/91(F17R/L)	2.00:1	8.10:1	B-33	1425	1425/24,900/54.0	1600/10,000/60.0
F-5B	200	V-1710- 89/91(F17R/L)	2.00:1	8.10:1	B-33	1425	1425/24,900/54.0	1600/10,000/60.0
P-38L	3,811	V-1710-111/113(F30R/L)	2.00:1	8.10:1	B-33	1425	1425/24,900/54.0	1600/10,000/60.0
P-322-I	23	V-1710- 33(C15)	2.00:1	8.77:1	None	1040	1040/14,300/41.0	1090/13,200/38.9
P-322-II	120	V-1710- 27/29(F2R/L)	2.00:1	6.44:1	None	1150	1150/NA/39.4	None
Total	9,924							

In addition, 113 P-38L-5-VN were built by Consolidated-Vultee at Nashville. Note, the XP-49 is not included in total.
* All ratings given with the engines at 3000 rpm.

be causing oil to vent overboard. Long missions were emptying the reservoir and resulting in failed engines. The situation was acute enough that the Theater Command was recommending substituting V-1710-51 and -55 engines in place of the -89/-91.[34] The resolution was somewhat surprising. Based upon flight tests of P-38J airplanes at Lockheed it was found that the way in which the ends of the engine crankcase breather vent hoses were scarfed (cutoff and shaped) and their position in the airplane slipstream resulted in oil being lost during high altitude operation. Corrective action was issued by T.O. 01-75-56. The bulk of the oil loss through the breathers was attributed to worn piston rings causing high backpressure within the crankcase and upon the scavenge pump. This resulted in a high oil level within the engine, which with an improperly scarfed breather hose and operation at high altitudes, was sufficient to causc oil to be discharged overboard.[35] A difficult and complex problem, but as is often the case, one that was easily resolved.

P-38L Production was from June 1944 through August 1945, effectively to the end of the war. Not only did they incorporate improvements such as the critical dive flaps, hydraulically boosted ailerons and additional fuel capacity, they also utilized the further improved V-1710-111/113(F30R/L) engines, while continuing to use the Type B-33 turbosupercharger.

The F-30 engine was still rated as the F-17's, but incorporated many internal improvements, most notably the 12-counterweight crankshaft. As a consequence it could be operated up to 3200 rpm. Using Grade 150 fuel it could deliver 1725 bhp under WER conditions.

P-38M A number of P-38's were modified for other missions and given other designations. Significant among them were the 75 P-38M-6 models that were built as two-seat night fighters by conversion of late model P-38L airframes. Engines and turbo installations were unchanged.

F-5 Photo-reconnaissance A fairly large number of Model 222 and 422 airframes were modified to function as photo-reconnaissance aircraft. Cameras replacing the majority of the guns. These aircraft were known as F-5's. Model 222 airframes were designated as the F-5A, with subsequent suffixes identifying the photo ship as being derived from the later series of P-38's. Again, the modified aircraft retained their engines and turbo installations as they came from the parent P-38. All but the F-5A's were pro-

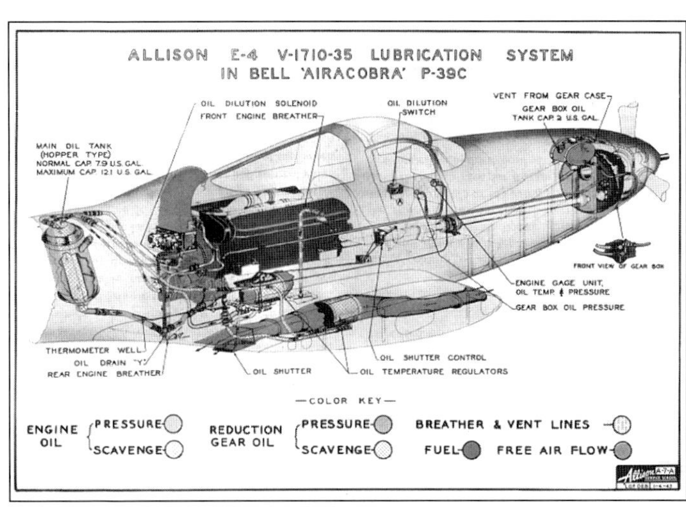

The central location of the engine in the P-39 allowed for compact cooling systems. Here the lube oil system is shown, including the NACA refined internal ducting to improve cooling effectiveness. (TO 30-5A-1)

duced at modification centers using stock airplanes, not on the Lockheed production line itself.

Bell P-39C/D/F/M/N/Q and P-400 *"Airacobra"*

In its original XP-39 form the Bell *Airacobra* had been intended to be a turbosupercharged interceptor. When the powers to be, with the aid of NACA, decided to remove the turbo the *Airacobra* became a single-stage aircraft, it was relegated to combat at the lower altitudes. There simply was insufficient volume in the aircraft to accommodate other engine/supercharger combinations, had they even been available at the time, which they were not. As a result, the early *Airacobra* models used the specially designed single-stage V-1710-E with an induction system derived from the altitude rated V-1710-33(C15) powering the early Curtiss P-40's.

All of the early *Airacobra* combat aircraft used the 1,400 pound V-1710-35(E4) engine, rated for 1150 bhp for takeoff and up to 12,000 feet, and with a War Emergency rating of 1490 bhp after 1942. These engine power sections were contemporaries of the V-1710-39(F3R), though equipped with the extension shaft and remote reduction gear unique to the P-39 and later P-63 installations.

The P-39 entered combat to mixed reviews. Along with the Curtiss P-40 it was the only available pursuit at the beginning of the war, and it was a situation that it must be used. A June 1942 report to General Arnold on P-39 combat experience summarized engagements against the Japanese Zero as 46 P-39's shot down against 47 Zero's. Noted in the report is that most P-39's shot down were hit in the engine and coolant system from the rear, and that a new piece of armor plate weighing 35 pounds was being installed immediately behind the engine. This change was expected to significantly reduce the loss rate. A weight reduction program was also to occur, with the expected improvement in performance to make the airplane a better match against the Zero.[36]

Some of the model designations used on different P-39 models were unique only in that they involved a change in the type of propeller used. The change usually being a necessity as a result of propeller production shortages. As a way to resolve the problem another source of propellers was sought. To this end, the Aeroproducts Division of General Motors was formed. It was a huge endeavor and a great deal was riding on its success. That it was ultimately successful is well known, but as with most engineering developments "Gremlins" had a way to gum up the works. The story below tells of what must have been a terrible moment for those involved:

"I could tell you a good story about the P-39. When we had progressed a little, into '41 or '42, the Air Force knew that we had to have more sources for propellers. So, GM setup a plant at Vandalia, Ohio. Well they had hired a couple of guys from Hamilton Standard, [and] made their first prototype propeller. It was better than gold! Captain [Signa] Gilkey, who was in charge of Accelerated Service Test at Wright Field, put that [propeller] over in the corner [of the hangar] and put a full time 24 hour guard on it. When the time comes, Capt. Gilkey said he was the one who was going to fly it. He got to be a General you know. Comes the day to fly it, everybody is standing around, you know, a big show, and he goes out, gets into the airplane, fires it up, [and] pretty soon clickity-click-click, the blades hit the concrete! The nose wheel [had] folded! There went that precious Aeroproducts propeller, and he came out of there, and I'll give the guy credit, he didn't blame anybody. He said, "that damned horn was blowing in my ear and I couldn't hear it!" [The horn for when the gear was to be down and locked.] He ran a good show!"

P-39C Bell's Model 13 was the first production *Airacobra*. It was procured as the P-45 when selected as the 2nd place finisher behind the Republic P-44 in the CP 39-770 Pursuit Competition of 1939. The improvements warranting a production contract over the YP-39 were the incorporation of the developed Allison V-1710-35(E4). Due to the similarities to the earlier YP-39's it was subsequently determined to identify the airplane as the P-39C submodel as a way to reduce confusion. Initially 80 of the aircraft were ordered, but due to inadequate protective equipment for combat only 20 were delivered as P-39C's, the balance of the contract being produced as the combat qualified P-39D. Production of the P-39C occurred while Allison was in the major expansion of its facilities and capacity to build engines. Many of the twenty airframes were completed as of December 1940, except for their V-1710-35 engines. The delay in engine delivery was due to Allison having to reprocess engines in which their inspectors had found aluminum chips left from manufacturing. These problems were soon overcome, but not before receiving considerable adverse attention.[37]

P-400 The French Purchasing Commission ordered 165 Bell Model 14 *Airacobras*. Following the collapse of France in June 1940 the order was taken over by the British, though by then the quantity on order had been increased to 170. In all the British ordered a total of 675 airplanes on

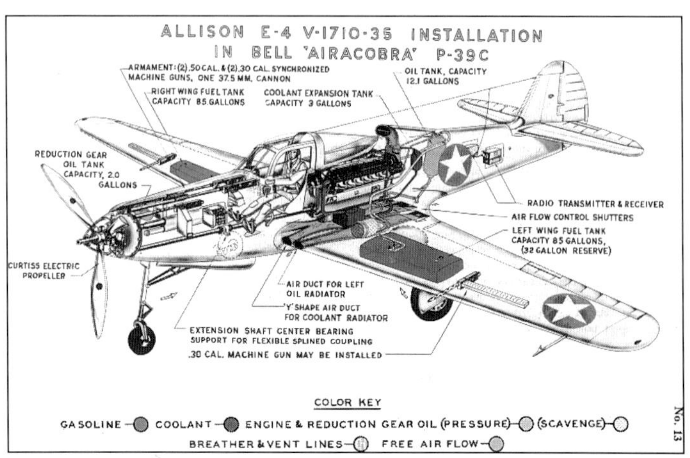

The P-39 was a compact airplane with the center location of the engine making for very close coupled plumbing. (TO 30-5A-1)

The Air Force Museum has setup an entire P-39 drive system, including the cannon. The V-1710-85(E19) is driving a 2.23:1 reduction gear as in the P-39Q-1. (Author)

An early V-1710-35(E4) powered P-39C airplane over the winter countryside. (Bell)

This is the second XP-39E, the first actually fitted with the two-stage V-1710-E9. Note the enlarged dorsal carburetor air intake as compared to the contemporary P-39. (Materiel Division)

three contracts, naming them the *Caribou*.[38] Although the P-400's incorporated features intended to qualify them for combat, the British canceled the order after receiving only a few, declaring the aircraft to be unsuitable.[39] A total of 212 were then diverted to the Soviet Union as part of the British contribution to the Lend-Lease program.[40] Of these, 54 were lost at sea during delivery. The U.S. Army commandeered 179 from the order. These were also shipped around the world, many ending up in the early battles in North Africa and the South Pacific.[41] All were powered by the V-1710-E4.

P-39D When the P-39C was revised to make the aircraft combat ready the aircraft was identified by Bell as their Model 15, designated as the P-39D by the Air Corps. These aircraft retained the V-1710-35(E4) while incorporating protective armor and self-sealing fuel tanks needed for combat.

XP-39E The XP-39E (Bell Model 23) program was separate from the P-63 project, though both were begun at about the same time and were to use the new Allison two-stage engine. The XP-39E was derived from a P-39D fuselage modified to accommodate the V-1710-47(E9) two-stage engine, intended to once again give the aircraft a high altitude capability. Reports that the aircraft was fitted with a laminar flow wing are erroneous. That was a distinction left to the P-63. The XP-39E did have a larger wing than other P-39's, and used a three percent thicker airfoil section at the wing root.[42] As the two-stage engine was still in development when the first aircraft was otherwise ready for flight, a single-stage V-1710-35(E4) was used for the flight test program on airplane AF41-19501. The first flight of the aircraft was on February 21, 1942. Once the E-9 engine was cleared for flight it was fitted in the second aircraft (AF41-19502), and began its flight test program on April 4, 1942.

Early in 1942, after the first airplane had been flown by the contractor for a short period of time it was decided that the design did not warrant procurement of a production quantity and that the remaining two airplanes would be used to develop the Allison two-stage engine for use in the P-63 series airplanes.[43] This decision caused some reshuffling within the deployed commands. For example, ten P-39E's had been designated for shipment to the 33rd Pursuit Squadron at Reykjavik, Iceland on a convoy leav-

Although the XP-39E was a derivative of the P-39 its appearance was quite different. Internally extensive changes were made, not only to accommodate the new two-stage V-1710-47(E9) engine, but to the cooling systems as well. Note the dual radiator intakes now in the leading edges of the wings. (Materiel Division)

Table 5-6
Early Experimental Models of the Bell P-39

A/C Model	XP-39	XP-39B	YP-39	YP-39A	XP-39E
Bell Model	4	4	12	4-D	23
Number Built	(1)	1	13	(1)	3
Contract Number	ac-10341	ac-10341			ac-18373
High Speed at Critical Altitude, mph Mil Power	400	394	375*		386/22,400
High Speed at Low Altitude, mph	330/SL	321.5/SL	326		NA
Time to Climb to Critical Alt, min	5.0	5.35	4.5*		9.3/20,000
Wing Area, sq.ft.	200	213.6	213	213	235.6
Empty Weight, pounds	4,295.6	4,289.7	4,955		7,631.1
Gross Weight, pounds	5,855	5,845	7,024		8,917.9
Design Load Factors	+8.0, -6.0				NA
Engine Model	V-1710-D3	V-1710-17(E2)	V-1710-37(E5)	V-1710-E4	V-1710-47(E9)
Allison Engine Specification	110-ModA	115-B	121-E	130	137-A/G
Turbosupercharger	Type F-10	Type B-1	None	None	Aux S/C*
Performance at Takeoff	1150/2950	1150	1090/3000	1150/3000	1325/3000
Critical Altitude, Military	20,000	20,000	13,200	12,000	1150/22,400
Aircraft Specification	4Y003	4Y003	4Y003-C	4Y003-D	23-947-001
Date of Original Aircraft Specification	6-3-1937	1-12-1938	3-1-1939		3-1-1941
Government Specification	X-609	X-609	C-616		
Date of Government Specification	3-19-1937	3-19-1937	1-25-1939		
Army Min High Speed at Critical Alt, mph	360	360	360	360	
Army Min High Speed at Low Altitude, mph	290/SL	290/SL	300	300	
Army Min Time to Climb to Critical Alt, min	6.0	6.0	6.0	6.0	
Army Minimum Load Factors	+8.0, -4.0	+8.0, -4.0	+8.0, -4.0	+8.0, -4.0	+8.0, -4.0
NACA Wing Section at Root	0015		0015	0015	0018
NACA Wing Section at Tip	0009		23009	23009	23009
Nose Cannon	25 mm	37 mm			
* Actual by Test	Rebuilt as XP-39B				* Intercooled

ing New York in January 1942. The squadron was standing a lone watch over the North Atlantic and had seriously depleted their compliment of Curtiss P-40C airplanes.[44]

Concurrent improvements in what became the single-stage P-39N discounted the advantages intended with the P-39E.[45] The first aircraft had been lost during spin tests after only a month and 14:55 hours of flight testing. As a result of the poor spin characteristics of the aircraft it was decided to lengthen the fuselage of the somewhat similar P-63.

The second and third XP-39E aircraft were both bailed to Allison in the summer of 1943 for development flight testing of features to be introduced on the P-63. This testing included automatic engine controls, improved pistons and rings, use of water injection, functioning of the engine hydraulic coupling, use of an engine aftercooler, and studies of carburetion, including a relocation of the carburetor from the inlet of the auxiliary stage to the inlet of the engine stage supercharger.[46] During the testing and development of the V-1710-47(E9)'s they were continually modified, functioning as prototypes for the V-1710-E11. As such their original 7.48:1 engine stage supercharger gears were replaced by the 8.10:1 gears used in the E-11. For this reason the performance of the XP-39E is usually tabulated assuming the ratings of the standard the V-1710-93(E11).

P-39F The P-39F (Bell Model 15B) was the P-39D revised to use the new General Motors Aeroproducts Hydromatic Propeller mounted on its V-1710-E4 instead of the earlier Curtiss Electric.[47] This model came about because of the critical shortage of Curtiss Electric propellers that were also being used on the P-40 and P-38's.

P-39J In an attempt to improve the altitude performance of the P-39D an early effort was made to equip the E-4 engine with an improved supercharger impeller having 47 degree rotating inlet guide vanes and 9.60:1 drive gears, as well as an automatic boost control. Twenty-five aircraft were built as P-39J's with the new V-1710-59(E12). It was soon found that service life of the modified engines, rated for 1100 bhp to 13,800 feet, was too short. Consequently the project was dropped, and along with it, an order for a similar number of Curtiss P-40J aircraft powered by the similarly configured V-1710-61(F14R).[48]

P-39K The next *Airacobra* was the V-1710-63(E6) powered P-39K; also based on the P-39D, but "improved" to the extent that it was some 800 pounds heavier. It was powered by a strengthened engine and the new supercharger impeller from the -59 giving it higher ratings, a contemporary of the V-1710-73(F4R). The E-6 was rated for 1325 bhp for takeoff and up to its Military rated critical altitude of 11,800 feet. It used a 2.00:1 reduction gear ratio instead of the 1.80:1 fitted on the V-1710-E4. In addition it could deliver 1580 bhp when operated at its War Emergency rating. The critical altitude of the engine remained as for the V-1710-E4 since both used the same supercharger gears, maximum rpm, and carburetor.

P-39L Similar to the P-39K, a batch of V-1710-63 powered aircraft were built as P-39L's, the only substantive difference being that they returned to the Curtiss Electric propeller. Use of the Curtiss Electric propeller required a plug in the oil supply line from the reduction gear through the propeller shaft.

Table 5-7
Bell Single Seat Pursuit Proposals and Early Production P-39's

A/C Model	Proposal	XFL-1	Proposal	P-39C	P-39D
Bell Model	3	5	11	12	15
Number Built	0	1	0	20	60
Contract Number	None/X609		CP 39-770	ac-13383	ac-
High Speed at Critical Altitude, mph Mil Power		346.0/ 9,000'	411/15,000'	398	395
High Speed at Low Altitude, mph			358/ 5,000'	353	355
Time to Climb to Critical Alt, min		4.088/15,000'	3.9/15,000'	4.0	5.5
Wing Area, sq.ft.		220	166.5	213	213
Empty Weight, pounds		4,350	4,352	NA	NA
Gross Weight, pounds		5,960	5,581	7,075	7,406
Design Load Factors			NA	+7.5, -3.75	+8.0, -4.0
Engine Model	V-1710-D3	XV-1710-6(E1)	V-1710-33(C15)	V-1710-35(E4)	V-1710-35(E4)
Allison Engine Specification	110-Mod A	112	126-B	130-D	130-D
Turbosupercharger	Type F-10	None	None	None	None
Performance at Takeoff	1150/2950	1150/2950		1150/3000	1150/3000
Critical Altitude, Military	20,000	9,000	15,000	12,000	12,000
Aircraft Specification		5Y005	11Y002	4Y003-F	4Y003-G
Date of Original Aircraft Specification			Bell/6-21-39	7-1-1940	7-21-1940
Government Specification			C-619/CP39-770	C-619/CP39-770	CP39-770
Date of Government Specification			6-24-1939	6-24-1939	6-24-1939
Army Min High Speed at Critical Alt, mph			365	365	365
Army Min High Speed at Low Altitude, mph			308	308	308
Army Min Time to Climb to Critical Alt, min			6.5	6.5	6.5
Army Minimum Load Factors			+8.0, -4.0	+8.0, -4.0	+8.0, -4.0
NACA Wing Section at Root		0015	27-218		
NACA Wing Section at Tip		23009*	27-209		
Nose Cannon		37 mm	None		
* Actual by Test		* or modified	441 mph @	Circular	
		0009	15,000'/1500 hp	39-770	

Table 5-8
Bell P-39 Aircraft and Engines

Aircraft Model	No. Built	Engine Model	Red Gear Ratio	S/C Gear Ratio	Turbo Model	Takeoff, bhp	Military, bhp/alt/MAP	War Emergency, bhp/alt/MAP
XP-39	1	V-1710-17(E2)	1.80:1	6.44:1	B- 5	1150	1150/25,000/41.6	None
XP-39B	(1)	V-1710-37(E5)	1.80:1	8.80:1	None	1150	1150/12,000/42.0	None
YP-39	13	V-1710-37(E5)	1.80:1	8.80:1	None	1150	1150/12,000/42.0	None
P-39C	20	V-1710-35(E4)	1.80:1	8.80:1	None	1150	1150/12,000/42.0	1490/4,300/56.0
P-400	675	V-1710-E4	1.80:1	8.80:1	None	1150	1150/12,000/42.0	1490/4,300/56.0
P-39D	923	V-1710-35(E4)	1.80:1	8.80:1	None	1150	1150/12,000/42.0	1490/4,300/56.0
XP-39E	3	V-1710-47(E9)	2.23:1	7.48:1	None	1150	1150/ 21,000/49.7	None
P-39F	229	V-1710-35(E4)	1.80:1	8.80:1	None	1150	1150/12,000/42.0	1490/4,300/56.0
P-39J	25	V-1710-59(E12)	1.80:1	9.60:1	None	1100	1100/13,800/44.2	None
P-39K	210	V-1710-63(E6)	2.00:1	8.80:1	None	1325	1150/11,800/	1580/2,500/60.0
P-39L	250	V-1710-63(E6)	2.00:1	8.80:1	None	1325	1150/11,800/	1580/2,500/60.0
P-39M	>1	V-1710-63(E6)	2.00:1	8.80:1	None	1325	1150/11,800/	1580/2,500/60.0
P-39M-1	<240	V-1710-83(E18)	2.00:1	9.60:1	None	1200	1125/15,500/44.5	1410/9,500/57.0
P-39N	2,095	V-1710-85(E19)	2.23:1	9.60:1	None	1200	1125/15,500/44.5	1410/9,500/57.0
P-39Q	4,905	V-1710-85(E19)	2.23:1	9.60:1	None	1200	1125/15,500/44.5	1410/9,500/57.0
Total	9,589							

Source: Index of Serial Numbers Assigned to AAF and DA Aircraft, June 1946.

P-39M Based on reports from the combat theaters, Bell determined that an improved P-39 was needed. Their goal was that the *Airacobra* would fulfill its purpose much better if it could climb faster and if the top of the engine and the accessory compartment were protected from gunfire. Many *Airacobras* that had been shot down in combat had been lost simply because of one bullet penetrating the engine manifold, carburetor, or camshaft covers. The objective was to provide maximum armor protection for these vital areas. In order to achieve a weight reduction a great number of more-or-less useful, but not necessary, items were removed from the aircraft. The gear box armor was removed inasmuch as it was no longer required from a balance standpoint and Bell service representatives knew of no actual instance when this armor plate had ever been struck by a bullet in battle. The first of these aircraft, a P-39M-1, became known at Bell as "Old Ironsides."[49]

The P-39M reverted to the Aeroproducts propeller and some were powered by the V-1710-63(E6), but in most cases they used the V-1710-83(E18) engine. The E-6 was rated as the E-4 for 1360 bhp at Takeoff as it used the same 8.80:1 supercharger gears.

The P-39M-1 was built with the V-1710-83(E18) engine. It had been strengthened to accommodate 9.60:1 supercharger gears and resolve problems that occurred with the -59(E12). Consequently, the Takeoff rating was reduced to 1200 bhp, but the new gears gave an improved Military altitude rating of 1150 bhp at 15,500 feet, and a War Emergency rating of 1410 bhp at 9,500 feet.

P-39N This was the first large scale production model of the P-39, with 2,095 built.[50] Again a different engine was used, this time the V-1710-85(E19), which was really the V-1710-83 power section driving through a 2.23:1 reduction gear in place of the earlier 2.00:1 unit. The performance ratings of the engine were the same as those for the V-1710-83(E18).

P-39Q This was the major production model with 4,905 built.[51] Like the P-39N it was powered by the V-1710-85(E19).

The Allison Powered Curtiss Hawks

Curtiss P-40D/E *"Kittyhawk*
Even before the V-1710-C15 was selected for production as the first series produced V-1710, Allison was hard at work to bring the V-1710-E/F series of engines through their development and Model Testing programs. The V-1710-F incorporated all of the improvements coming from the V-1710-C engines and the V-3420 project as well as the lessons being learned from the early P-39 development and operation of the V-1710-E. In fact, the first "E" engines were built along side the beginning production of the C-10 engines for the Curtiss YP-37's. The first "F's" for the Lockheed YP-38's, were built along side of the last of the D-2's for the Bell YFM-1's. Since the "F" was to be the engine for the production P-38, and was configured as a compact tractor engine, it was also a candidate for an improved P-40. The propeller shaft on the "F" had been relocated to approximate the geometric center of the engine. The engine selected for the new P-40 was the V-1710-F3R, the same engine as selected for the Curtiss XP-46 (Curtiss Model 86) and Republic XP-47, and that would be used on the North American NA-73X. The Air Corps had ordered five of these engines in 1939 using FY-40 funds, one for a Model Test, and it is believed, one each for the XP-46/XP-46A and the XP-47/XP-47A.[52]

P-40D The P-40D airplane was an evolution of the Curtiss *Tomahawk*, though most parts were updated and received Curtiss "Model 87" numbers. The wing was redesigned to carry four .50 caliber machine guns and to mount two podded 20 mm cannon. Both of the rifle caliber fuselage guns, and the need for gun synchronizers on the engines were deleted, though the V-1710-39 specification provided them anyway. The bulk of

This is the second P-40D (AC40-359), when assigned to Allison Flight Test at Weir Cook Airport, Indianapolis, October 1940. Even with additional weight and features needed to enter combat the new V-1710-39 gave the airplane superior performance to that of the Curtiss XP-46. All in all a very satisfactory airplane that was heavily used in the early stages of the war. (Allison)

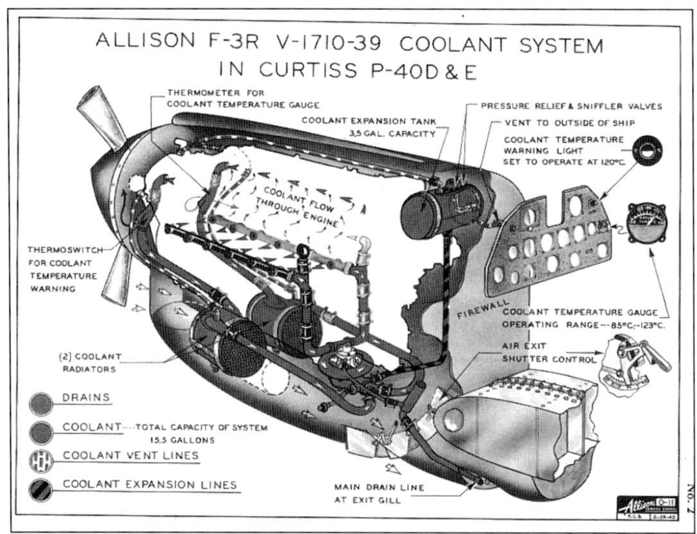

When the V-1710-F was installed in the P-40 the chin radiators were retained, but all was completely revised to support the higher powered engine. (TO 30-5A-1)

P-40 Kittyhawk radiator and oil cooler inlets dominate the chin area, with the carburetor air intake directly above the propeller spinner. (Author)

the fuselage remained as in the Model 81 *Tomahawk*, though from the firewall forward everything was revised so as to better accommodate the short nosed V-1710-F. Actual work on the P-40D began in the spring of 1940 and followed that on the XP-46. The engine installations in the two airplanes had a lot in common, except for the obvious difference in location of the radiator and oil coolers to a ventral position on the XP-46.

With the 1150 bhp V-1710-39(F3R) engine additional radiator area was necessary. This was accommodated by installing a pair of enlarged radiator cores and an oil cooler, each in its own duct. All were located just below the engine. Since the "F" engine was some six inches shorter than the "C", the result was the definitive boxy nose that became a hallmark of subsequent production P-40's. As a result of the higher thrust line this arrangement also eliminated the space previously used for the nose guns,

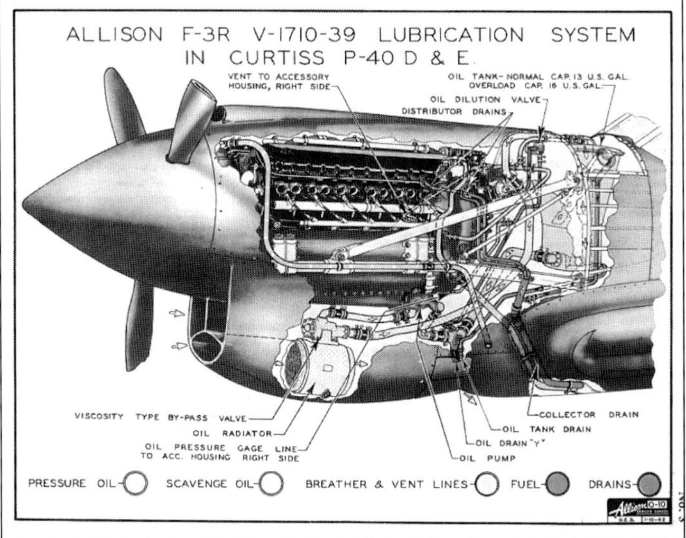

Cooling the lube oil in the P-40 was rather straight forward. Whenever an engine was changed due to internal failure it was necessary to also replace the oil cooler. Small bits of metal would otherwise be pumped into the new engine and ruin its bearings. (TO 30-5A-1)

which were deleted. Twenty four of these airplanes were ordered by the Air Corps, though one was rebuilt and delivered as the P-40E and another became the XP-40F. Hence only 22 of these were delivered as the P-40D to the Army Air Corps, though the British received an additional 560.[53] The British identified these Curtiss Model H87-A2 aircraft as the *Kittyhawk Mk I*.

The service experience of these early aircraft shows the critical need for aircraft in general, offering perspective on the circumstances of the day. The first P-40D (AC40-358) was delivered to the Air Corps in May 1941 and then tested by both Curtiss in New York and the Air Corps at Wright Field.[54] This airplane was soon reworked to become the first P-40E, which was its form when delivered to the Army. The second P-40D (AC40-359) was delivered to Wright Field and was extensively tested by them during the summer of 1940. In October it was assigned to Allison and operated by them for engine testing at Indianapolis before being returned at the end of the year. The second airplane suffered an engine failure at Patterson Field in August 1941, but then was repaired and served in Umnak, Alaska into 1944. The third P-40D (AC40-360) was commandeered and fitted with a Rolls-Royce Merlin XX as the XP-40F. The airplane delivered as the third P-40D accumulated 296 hours of flight in training duty at Tuskegee, Alabama before suffering engine failure in September 1942. The balance of these early F-3R powered airplanes served with only the usual problems of aircraft in training rolls at the time.

The V-1710-F3R was rated for 1150 bhp at Takeoff and Military, up to a critical altitude of 11,700 feet. This was a marginal improvement over the earlier C-15, but the new package benefited from the experience gained on the *Tomahawk* and resulted in an overall improvement in reliability and load carrying ability. Even though the weight of the aircraft had increased, the weight of the new higher powered engine was reduced by 15 pounds.

P-40E The P-40E was very similar to the P-40D, and used the same V-1710-39 engine and systems, though the aircraft was fitted with six .50 caliber machine guns in the wing instead of the four in the "D." The provisions for the cannon were deleted.[55] A total of 2,320 P-40E's were built as Curtiss Model H87-A3/A4's.[56] The first P-40E was the first P-40D (40-358) fitted with the new six gun wing. The airplane was not equipped with air cleaners, given satisfactory operation of the engines even in the dusty

Middle East. This was attributed to the location of the carburetor air intake directed to the down-draught carburetor.[57]

P-40F While the P-40F *Kittyhawk* was powered by a Packard built V-1650-1 Merlin it is important to the story of the P-40 and mentioned here. To expedite development of the model, P-40D airframe 40-360 was converted to Rolls-Royce Merlin XX power as the XP-40F and was flying in October 1941, some three months before the first Packard powered P-40F, AC41-13600. The P-40D/E/F *Kittyhawks* were all developed at the same time, with the P-40D only a few months ahead of the E. Maximum speed of the V-1710-39 powered P-40E was 354 mph at 15,000 feet. The P-40F with the single-stage Merlin weighed 240 pounds more than the P-40E, but could make 364 mph at its critical altitude of 20,000 feet. Operational experience showed a critical need for retrofitting these airplanes with air cleaners. The high consumption rate of V-1650-1 engines may have been attributed to lack of this feature. This engine used an up-draught carburetor fed from the chin intake.[58]

Kittyhawk Mk I A total of 560 Curtiss H87-A2 aircraft, similar to the AAC P-40D were delivered to Britain. Although these were Lend-Lease aircraft they did not receive U.S. AAC serial numbers and were delivered directly to Britain. Delivered August though December 1941.

Kittyhawk Mk IA A total of 1,500 Lend-Lease P-40E aircraft produced as Curtiss models H87-A3 and H87-A4. Delivered September 1941 through June 1942.

Kittyhawk Mk III These were P-40K and P-40M "Warhawk" models provided to Britain under Lend-Lease and all 364 are counted under the AAF totals. Deliveries were from November 1942 through January 1943.

Kittyhawk Mk IV These "Warhawk" models were also Lend-Lease aircraft from various blocks of P-40N, with 458 delivered. Deliveries were from March 1943 through January 1944.

Curtiss P-40J/K/M/N/Q/R *"Warhawk"*

P-40J The stillborn P-40J was intended to be a much improved medium altitude fighter by the simple technique of installing an engine having a higher critical altitude. This was to be achieved by using 9.60:1 supercharger gears in a V-1710-F3R with numerous detail improvements, designated V-1710-61(F14R). The engine began Model Testing during the week following Pearl Harbor and both Allison and Curtiss were working to quickly put the improved P-40J into production early in 1942. Unfortunately the introduction of the higher ratio gears caused many problems with the reliability of the supercharger drive.

This project was a companion to the similar production of 25 Bell P-39J aircraft that had like configured V-1710-59(E12) engines. These P-39J's were built, but suffered poor reliability as a consequence of the high loads being imposed on the supercharger drive gears. As a result all were converted back to their original standard and the program to utilize 9.60:1 gears went back to the test benches. Considerable development work then occurred, finally resulting in a series of engines that successfully used the 9.60:1 gears in the P-40M/N series. Those engine models were V-1710-81/99/115, Allison models F-20R/F-26R and F-31R. The P-40J did not reach flight status.

P-40K The P-40K was similar to the P-40E, followed it in production and was similarly intermingled in production with the single-stage Merlin

ENGINE: V-1710-61
P-40J

CONDITION	FUEL PRESSURE LB/IN.²	OIL PRESSURE LB/IN.²	OIL TEMP. °C	COOLANT TEMP. °C
DESIRED	12-16	60-70	60-80	105-115
MAXIMUM	16	85	95	125
MINIMUM	12	55		85
IDLING	9	15		

MAX. PERMISSIBLE ENGINE OVER SPEED: 3120 R.P.M.

MAX. ALLOWABLE OIL CONSUMPTION AT:
NORMAL RATED POWER 13.6 QTS./HR.
MAXIMUM CRUISING 10 QTS./HR.
MINIMUM SPECIFIC FUEL FLOW 5-7 QTS./HR.

FUEL GRADE 100 OCTANE

⊕ OPERATING CONDITION	HORSE POWER	R.P.M.	MAN. PRESS. (IN. HG)	PRESSURE ALTITUDE (IN FEET)	BLOWER CONTROL POSITION	USE LOW BLOWER BELOW	MIXTURE CONTROL POSITION	MIN. F/A RATIO	FUEL FLOW GAL/HR	MAX. CYL. HD. TEMP. °C	REMARKS
TAKE-OFF	1100	2800	46.5	Sea Level			Auto Rich		119		5 Minute Operation Only
MILITARY RATED POWER	1100	3000	44.2	13800			Auto Rich		136		5 Minute Operation Only
⊙ NORMAL RATED POWER (100%)	950	2600	37.5	13700			Auto Rich		103		
MAX. CRUISING (75%)	712	2280	29.8	13700			Auto Rich		66		
DESIRED CRUISE (67%)	636	2280	27.2	13700			Auto Rich Auto Lean		59 48		
DESIRED CRUISE (60%)	570	2190	25.2	13700			Auto Rich Auto Lean		50 43		
CRUISE FOR MIN. SPECIFIC FUEL FLOW	375 415 455 485 490	1950 1950 1950 1950 1950	23.0 23.0 23.0 23.0 FT.	Sea Level 5000 10000 15000 20000			Auto Lean Auto Lean Auto Lean Auto Lean Auto Lean		28 31 34 36 37		

⊕ REFER TO T.O. NO. 00-10 FOR DEFINITION EACH OPERATING CONDITION ⊙ MAXIMUM PERMISSIBLE CONTINUOUS HORSE POWER REVISED 12-1-41

T.O. NO. 02-5AB-1B

For a time the P-40J was a high priority project and intended to become a major production model. This page from the pilot's manual gives the parameters for operating the V-1710-61(F14R). (Air Corps)

powered P-40F *Warhawks*. The aircraft was the heaviest P-40 model and suffered from poor handling on takeoff and in high power dives due to the higher engine power developed by its V-1710-73(F4R) engine, a contemporary of the similarly rated V-1710-63(E6). While it has been reported that large fillets were added to the vertical fins of the P-40K-1 and -5 models to counter the P-40K handling problem, those fillets first appeared about two-thirds through the production run of the V-1710-39 powered P-40E, including some P-40E-1's.[59] The lengthening of the fuselage by another 20 inches was developed for the Merlin powered P-40F and incorporated on the P-40K-10 and -15 models to resolve the handling problem.[60] This was a way to improve the effectiveness of the vertical tail without having to increase its dimensions. This change resulted in placing the rudder-hinge line behind the elevator-hinge line, an arrangement and fuselage that was retained in the subsequent P-40M/N models.[61]

The V-1710-73(F4R) was provided with an Automatic Manifold Pressure Regulator as well as incorporating a number of internal improvements allowing an increase in the take-off rating to 1325 bhp instead of the 1150

Externally the V-1710-61(F14R) looks much like the F-3R from which it was derived. Internally it featured an improved supercharger with 9.6:1 step-up gears to improve the altitude capability, but the extra load on the gears caused several Model Test failures. Use of the 9.6:1 ratio had to await a new accessories section featuring wider gears in the F-20R. (Allison)

RESTRICTED

SECTION III
SPECIFIC OPERATING INSTRUCTIONS

ENGINE: V-1710-33
P-40, P-40B, P-40C

CONDITION	FUEL PRESSURE LB/IN.^{2}	OIL PRESSURE LB/IN.^{2}	OIL TEMP. °C	COOLANT TEMP. °C
DESIRED	12-16	60-70	60-80	105-115
MAXIMUM	16	85	95	125
MINIMUM	12	55		85
IDLING	9	15		

MAX. PERMISSIBLE ENGINE OVER SPEED: 3120 R.P.M.

MAX. ALLOWABLE OIL CONSUMPTION AT:
NORMAL RATED POWER 12.4 QTS./HR.
MAXIMUM CRUISING 9.3 QTS./HR.
MINIMUM SPECIFIC FUEL FLOW 5-7 QTS./HR.

FUEL GRADE 100 OCTANE

⊕ OPERATING CONDITION	HORSE POWER	R.P.M.	MAN. PRESS. (IN. HG)	PRESSURE ALTITUDE (IN FEET)	BLOWER CONTROL POSITION	USE LOW BLOWER BELOW	MIXTURE CONTROL POSITION	MIN. F/A RATIO	FUEL FLOW GAL/HR	MAX. CYL. HD. TEMP. °C	REMARKS
TAKE-OFF	1040	2800	40.6	Sea Level	-	-	Auto-Rich	-	104	-	5 Minute Operation Only
MILITARY RATED POWER	1040	3000	37.2	14,300	-	-	Auto-Rich	-	115	-	5 Minute Operation Only
⊙ NORMAL RATED POWER (100%)	930	2600	33.7	12,800	-	-	Auto-Rich	-	90	-	-
MAX. CRUISING (75%)	697	2280	27.9	12,800	-	-	Auto-Rich	-	59	-	-
DESIRED CRUISE (67%)	623	2280	25.4	12,800	-	-	Auto-Rich Auto-Lean	-	53 48	-	-
DESIRED CRUISE (60%)	558	2190	23.9	12,800	-	-	Auto-Rich Auto-Lean	-	47 42	-	-
CRUISE FOR MIN. SPECIFIC FUEL FLOW	400 450 495 520 450	1950 1950 1950 1950 1950	24.2 24.2 24.2 24.2 FT.	Sea Level 5,000 10,000 15,000 20,000	- - - - -	- - - - -	Auto-Lean Auto-Lean Auto-Lean Auto-Lean Auto-Lean	- - - - -	30 34 37 39 34	- - - - -	- - - - -

⊕ REFER TO T.O. NO. 00-10 FOR DEFINITION OF EACH OPERATING CONDITION ⊙ MAXIMUM PERMISSIBLE CONTINUOUS HORSE POWER REVISED 12-1-41

T.O. NO. 02-5AC-1

The operating instructions contained in the Pilots manual provide a lot of information about the engine and how to safely operate it. These were revised fairly frequently, reflecting changes such as removal of backfire screens and authorization of War Emergency Rated power. The V-1710-73(F4R) offered 1550 bhp for five minutes of WER in the P-40K.

bhp available with the V-1710-39(F3R). This was accomplished by general strengthening of the engine components, as the same supercharger gear ratio of 8.80:1 was used. With the engine weight increased by 35 pounds to 1,345 pounds, the engine was delivering nearly one horsepower per pound of weight at its Military rating. The War Emergency Rating was increased to 1580 bhp at 60 inHgA. In the early days of the war, before the Army officially established WER ratings, V-1710's in the combat theaters were often being overboosted. Allison field representatives reported cases where V-1710-73's were routinely operated at up to 66 inHgA in combat, and one representative in Australia noted cases were 70 inHgA were being used on P-40K and P-39D/K/L aircraft.[62] Such indications show that the engines were being greatly abused. At 3000 rpm and at sea level the engine was only capable of about 62 inHgA, at which point it would be producing 1760 bhp. Achieving 66 inches would have required running the engine to at least 3200 rpm, and 70 inches would have meant at least 3400 rpm. While the engine was capable of safely exceeding 4000 rpm at this point in its development, the valve springs and/or fuel Grade were probably acting as the power limiters. Later models necessarily used stiffer springs and higher grade fuel when 60-65 inHgA was to be exceeded.

One important change that supported these increased ratings was the adoption of a pressurized mixture of 30 percent Ethylene Glycol and 70 percent water as the coolant. This replaced the previously used 97 percent Ethylene Glycol solution, and significantly improved the removal of heat from the cylinder heads by taking advantage of the better cooling capability of pressurized water. The Glycol was needed only for freeze protection.

P-40M The P-40M's used the lengthened P-40K-10 airframe and featured built-in carburetor air filters as well as being fitted with the V-1710-81(F20R) engine.[63] This engine was similar to the V-1710-73, but with 9.60:1 supercharger gears in place of the earlier 8.80:1, giving a 2,500 foot higher altitude for the Military rating. Because of the additional power required to drive the supercharger, the rated power available to the propeller was reduced to 1200 bhp at Takeoff and 1410 bhp for War Emergency, though at nearly four times the altitude: 9,500 feet versus 2,500 feet.

P-40N Curtiss built the P-40N as a "lightweight" P-40, eliminating two of the wing guns and making extensive use of aluminum for components such as the oil and coolant radiators. The engine for the early blocks, P-40N-1, -5, -10 and -15 was the V-1710-81. As a result this model, at 378 mph, was the fastest of the *Warhawks*.[64] The P-40N-20 introduced the V-1710-99(F26R) with its Auto Engine Regulator, a device to further simplify the pilots task of managing the engine and manifold pressure. Otherwise the engine was similar to and rated as the V-1710-81.

Due to the obvious obsolescence of the P-40 design and reduced pilot training requirements, an order for 1,000 P-40N-40's was terminated in November 1944 after only 220 of this block were delivered.[65] They used the V-1710-115(F31R) engine, but had the same ratings as the V-1710-99. The new model engine reflected internal differences in that it used the improved 12 counterweight crankshaft, a strengthened accessories section and wider accessories gears. All of these changes had the effect of improving engine durability. This added some 30 pounds of engine weight, 26 pounds being due to the new crankshaft.

XP-40Q Curtiss did get a two-stage P-40 when they installed the V-1710-101(F27R) in three heavily modified earlier P-40's. These airplanes went through numerous modifications during the course of their test programs, with many of the changes being with respect to particulars of the engine itself. See Chapter 6, 'Odds and Mods' for details of these airplanes.

P-40R A total of about 600 P-40F's and P-40L's, originally powered by the Packard built single-stage Rolls-Royce Merlin V-1650-1, were re-engined with V-1710-81's due to a severe shortage of the original V-1650-1 power plant and associated spare parts. This situation developed because of two conditions: Army Production Engineering had specified only 12 percent spare parts for the V-1650-1, while events showed that they needed about 32 percent. Secondly, the production of the single-stage Merlin V-1650-1 was terminated following initial completion of the original contract for 3,000 engines so that Packard could devote all of its resources to building two-stage Merlins. With the heavy demand for P-40's in the combat theaters in early 1943 the only solution was to switch 1,500 Merlin P-

Table 5-9
Curtiss P-40 Kittyhawk and Warhawk Engines

Aircraft Model	No. Built	Engine Model	Red Gear Ratio	S/C Gear Ratio	Aux S/C Ratio	Takeoff, bhp	Military, bhp/alt/MAP	War Emergency, bhp/alt/MAP
XP-40	1	V-1710- 19(C13)	2.00:1	8.77:1	None	1060	1150/10,000/41.2	None
P-40	199	V-1710- 33(C15)	2.00:1	8.77:1	None	1040	1090/13,200/38.9	None
P-40B	131	V-1710- 33(C15)	2.00:1	8.77:1	None	1040	1090/13,200/38.9	None
P-40C	193	V-1710- 33(C15)	2.00:1	8.77:1	None	1040	1090/13,200/38.9	None
Tomahawk	1,180	V-1710- C15	2.00:1	8.77:1	None	1040	1090/13,200/38.9	None
P-40D**	582	V-1710- 39(F3R)	2.00:1	8.80:1	None	1150	1150/11,700/44.6	1490/4,300/56.0
P-40E	2,320	V-1710- 39(F3R)	2.00:1	8.80:1	None	1150	1150/11,700/44.6	1490/4,300/56.0
P-40G	1	V-1710- 33(C15)	2.00:1	8.77:1	None	1040	1090/13,200/38.9	None
P-40K	1,300	V-1710- 73(F4R)	2.00:1	8.80:1	None	1325	1150/12,000/42.0	1580/ 2,500/60.0
P-40M	600	V-1710- 81(F20R)	2.00:1	9.60:1	None	1200	1125/15,500/44.5	1410/ 9,500/57.0
P-40N-1/15	1,977	V-1710- 81(F20R)	2.00:1	9.60:1	None	1200	1125/15,500/44.5	1410/ 9,500/57.0
P-40N-20/35	3,022	V-1710- 99(F26R)	2.00:1	9.60:1	None	1200	1125/15,500/44.5	1360/ SL /57.0
P-40N-40	220	V-1710-115(F31R)	2.00:1	9.60:1	None	1200	1125/15,500/44.5	1360/ SL /57.0
XP-40Q-1	(1)***	V-1710-101(F27R)	2.36:1	8.10:1	6.85:1	1325	1100/28,500/50.0	1220/25,000/56.0
XP-40Q-2	(1)	V-1710-101(F27R)	2.36:1	8.10:1	6.85:1	1325	1250/12,000/50.0	1500/ 6,000/60.0
XP-40Q-3	(1)	V-1710-121(F28R)	2.36:1	8.10:1	7.23:1	1425	1100/28,000/____	1700/26,000/____
Total	11,726							

* All ratings given with the engine at 3000 rpm.
** Includes 560 Delivered to Britain as Kittyhawk Mk I.
*** XP-40Q were all conversions of earlier airframes.

40's planned for production to V-1710's. Even so, the British still gave up 600 of their single-stage Merlin's to become spare parts for the Merlin powered P-40's.[66] All of this is made even more fascinating when it is recognized that one of the original motivations for building the single-stage Merlin P-40's was to free-up Allison production so that additional V-1710's could be allocated to Lockheed P-38's and Bell P-63's![67] The P-40F's (a total of 1,312 were built) became the P-40R-1 while the P-40L's (a total of 650 were built) became the P-40R-2.[68,69] The conversion amounted to installation of all components and structure of the P-40K "power egg" from the firewall forward.

Modernizing the P-40

Don Berlin, designer of the P-40, was never satisfied with the limited altitude performance of the aircraft. In June 1941, he expressed a desire to develop a turbosupercharger installation on a P-40D, with a view toward incorporation of turbos on future models of the P-40. While the P-40D was being updated on the production line to become the P-40E, he was at work on a turbosupercharged version to be known as the XP-40H. The goal always being to recover the loss in altitude performance caused by use of the medium altitude ratings of the single-stage V-1710's. The Air Corps approved of this activity and sent one of their GE Type B-2 turbos to Curtiss for the effort.[70] By October 1941 Curtiss had decided to not proceed with a turbosupercharged P-40, but rather to incorporate the feature into their new P-60A, then being designed as the Curtiss Model 95.

In an attempt to update the P-40 aerodynamics Curtiss had fitted a P-40 fuselage with a laminar flow wing, resulting in their Model 90 (XP-60).[71] This project combined the earlier efforts to incorporate a laminar flow wing and other improvements on what was to have been the Continental XIV-1430 powered Curtiss XP-53.[72] When originally conceived as Wright Field project MX-69, the XP-60 was to be powered by the first Allison two-stage engine, the V-1710-45(F7R). When development of that engine was put on hold during 1941, while Allison increased its production capacity and qualified its engines for combat, Curtiss went ahead with the Merlin engine as a part of the effort to free Allison production for the P-38, P-39 and P-51. As a result the XP-60 was powered by a single-stage Packard built Merlin V-1650-1, and the aircraft did prove to be faster than the P-40. This resulted in a large production order for 1,950 P-60A's, though these were canceled on December 20, 1941, just after the start of the war. Curtiss was awarded a consolation contract for production of Republic P-47D's as the P-47G-1-CU. Apparently the Army wanted Curtiss to concentrate on production of the critically needed P-40's, and not expend extensive effort on the development and manufacture of the entirely new P-60A.[73] The effort to produce the P-47 by Curtiss was a disappointment, for Curtiss had only produced 354 Thunderbolts by the time the contract was canceled in March 1944.[74]

The Curtiss P-60 had potential. In 1943 the P-60E was flown against the Merlin powered North American P-51B and found to be clearly faster at 20,000 feet. The report recommended reconsideration of the decision not to build the P-60, which elicited a response from higher headquarters that "(the recommendation) should not in any way be interpreted as suggesting reconsideration of the production of the P-60. That airplane has been definitely canceled."[75]

North American P-51 and P-51A *Mustang, Mustang Mk I/II* and A-36A

North American Aviation built their NA-73X prototype fighter specifically for the British Purchasing Commission. It made its first flight on October 26, 1940 powered by the first Allison V-1710-39(F3R). This was the first altitude rated "F" model and employed a considerably improved supercharger as compared to previous V-1710's, including the V-1710-F2's powering the YP-38's, the only other "F" models flying at the time. Though this was a British project, the prototypes' engine was provided by the Army

North American Mustang IA, Model NA-91, was provided by Lend-Lease as the Army P-51. Powered by the V-1710-39(F3R) it demonstrated outstanding performance at low and medium altitudes. The outstanding contours and finish of the airplane were keys to its performance. Even though above the engine critical altitude, the similar Model NA-99 P-51A could reach 409 mph, the only combat aircraft in the world at the time capable of 400 mph at 20,000 feet. (NAA via SDAM)

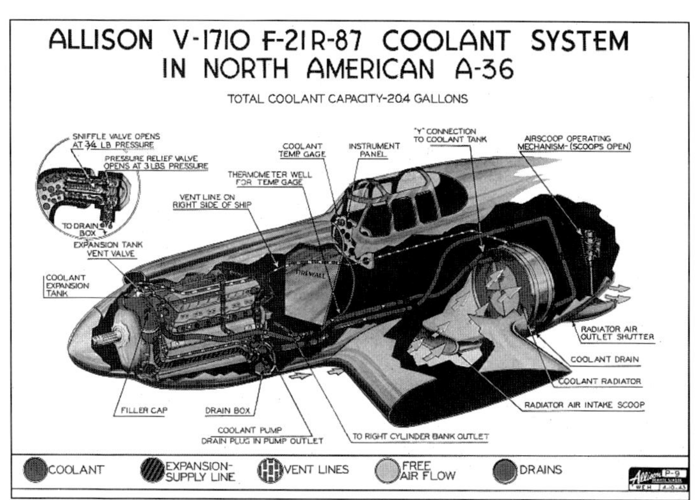

The ground support mission of the A-36 introduced the V-1710-87(F21R) low level engine, and included a new radiator air scoop that was quite different than on the P-51. (TO 30-5A-1)

Air Corps, it being the first of a batch of five V-1710-39's intended for the two new experimental XP-46/A and two XP-47/A aircraft.

Based upon the success promised by the initial testing of the prototype, Britain then ordered 320 NA-73 aircraft as their *Mustang Mk I.* Two of these were required to be delivered to the U.S. Army Air Corps for evaluation. The Army in turn designated them as XP-51's. Interestingly, the first of these, AC41-038, was powered by the repaired V-1710-39 that had powered the prototype NA-73X when it crashed due to fuel starvation, (V-1710-F3R AEC#301). The British then purchased a further 300 similar *Mustang I*'s, these being NAA model NA-83's. With the advent of Lend-Lease the U.S. Army began its purchase of P-51's, acquiring 150 aircraft for Britain as *Mustang IA*'s, North American Model NA-91. With the attack on Pearl Harbor and U.S. entry into WWII, 53 of these aircraft were

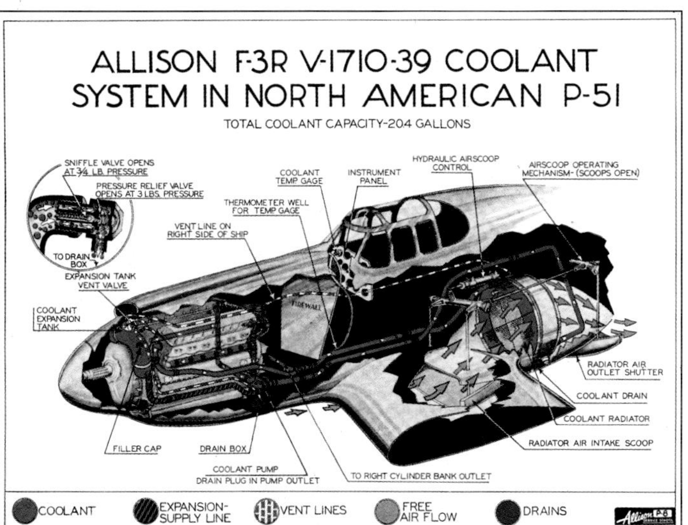

The early F-3R powered P-51 Mustangs used a variable area scoop to feed the radiator duct. Otherwise the coolant system was typical for an Allison. (TO 30-5A-1)

retained for photo-recon purposes and another two were allocated for conversion to two-stage Merlin power by North American.[76,77] All of these models were powered by the V-1710-39, with some 4,694 being built in total. The engine not only powered the *Mustang Mk I/IA* and P-51, but was also used by numerous models of the Curtiss P-40 *Kittyhawk* and both Curtiss XP-46's.

The first purchase of the North American craft for use by the U.S. was as the A-36A *Invader.* A batch of 500 of these dive bomber adaptations were contracted in April 1942 as the North American Model NA-97. These were to be popularly known as the "*Invader*", and were the first of the series to utilize a new engine optimized for low level missions, the V-1710-87(F21R). A total of 835 of these engines were built. With its intended low level bombing and army support mission it was not desirable to have a "high altitude" engine, so the supercharger gears were changed to 7.48:1. This reduced the critical altitude, but gave a boost to engine output at low levels. The result was a takeoff and Military rating of 1325 bhp being available at up to 2,500 feet. War Emergency rating of the engine was 1500 bhp, using 52 inHgA at up to its critical altitude of 5,400 feet. All of this resulted in a sleek aircraft that was considered a "hot rod" at the lower altitudes.

The A-36A, (Model NA-97) while appearing much like the P-51, was considerably different than the earlier P-51 and *Mustang I*, particularly in the way the air scoop was shaped and fitted. It was the first Mustang to be tested in the NACA full scale wind tunnel, and as a consequence incorporated a number of detail improvements throughout. In addition to the different engine, it had dive brakes and four .50 cal wing guns. The first flight of the A-36 was in October 1942.[78]

In June 1942 the Army Air Forces ordered 310 P-51A aircraft, the North American Model NA-99.[79] These were powered by an improved F-4R, the V-1710-81(F20R) which was tailored for medium altitude fighter duties by being the first Allison to successfully employ 9.60:1 supercharger step-up gears. With this engine the P-51A had 1200 bhp available for takeoff and a Military rating of 1125 bhp up to 15,500 feet. Under WER conditions the engine was good for 1410 bhp at up to 9,500 feet using 57.0 inHgA. This engine saw wide use, not only in Mustangs, but in the Curtiss

At one time or another the Allison Flight Test department operated one or more of almost every aircraft powered by the V-1710. Their P-51A was AF43-6012, the tenth one built. The Air Forces had done extensive performance tests on it at Eglin Field, Florida prior to bailing it to Allison. (Allison)

P-40 Warhawk series as well, with nearly 3000 V-1710-81's being built. The major difference between the -81 and -39 being the strengthened 9.60:1 internal supercharger gears and improved supercharger inlet guide vanes that gave a critical altitude of 15,500 feet compared to the earlier 11,700 feet.

Prior to P-51A AF43-6012 being assigned to Allison, it and AF43-6013 were performance tested at Eglin Field, Florida. The following graphs show the engine power and airplane performance from sea level up to 30,000 feet, with bomb racks installed.

Following the Eglin testing, P-51A AF43-6012 was issued to Allison for use in its various flight testing and engine development programs. It continued in this role through the war. In April 1945 Allison was authorized to flight test the new V-1710-97(G1R) engine in the airplane.[80] With the new engine, rated for 1725 bhp at 3400 rpm, the airplane should have been a real performer.

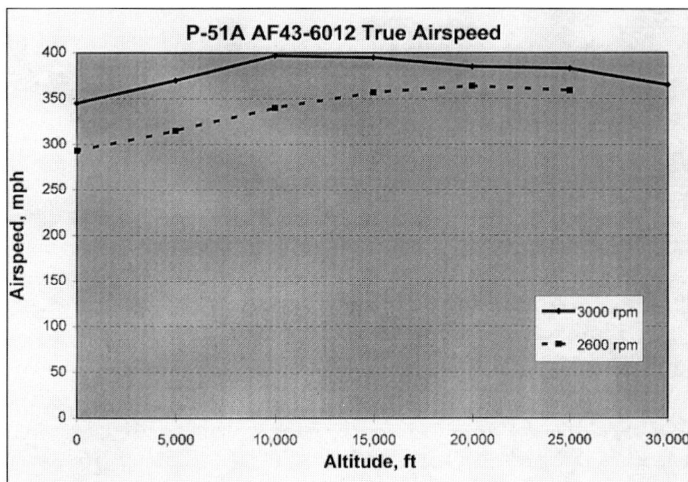

Using 3000 rpm, P-51A AF43-6012 maintained a speed of nearly 400 mph up to 25,000 feet when tested at Eglin Field, Florida, March 31, 1943. At 2600 rpm maximum continuous cruise speeds peaked at 20,000 feet, well above the 15,000 foot critical altitude for the engine. (Author)

Improved Mustangs

In February 1941, three months before the first flight of an Air Corps XP-51, Lee Atwood, VP of North American Aviation, wrote to General Motors and the Air Corps recommending that Allison develop a *two-speed* supercharged V-1710 instead of continuing to work on two-stage versions. This

Table 5-10 Engines in North American P-51 Mustangs
V-1710 Allison Powered Mustangs*

Aircraft Model	No. Built	Engine Model	Red Gear Ratio	S/C Gear Ratio	Aux S/C Ratio	Takeoff, bhp	Military, bhp/alt/MAP	War Emergency, bhp/alt/MAP
NA-73X	1	V-1710- 39(F3R)	2.00:1	8.80:1	None	1150	1150/11,700/44.6	1490/4,300/56.0
Mustang Mk I/73	320	V-1710- 39(F3R)	2.00:1	8.80:1	None	1150	1150/11,700/44.6	1490/4,300/56.0
XP-51	(2)	V-1710- 39(F3R)	2.00:1	8.80:1	None	1150	1150/11,700/44.6	1490/4,300/56.0
Mustang Mk I/83	300	V-1710- 39(F3R)	2.00:1	8.80:1	None	1150	1150/11,700/44.6	1490/4,300/56.0
P-51/Mustang IA/91	150	V-1710- 39(F3R)	2.00.1	8.80:1	Nonc	1150	1150/11,700/44.6	1490/4,300/56.0
A-36A/NA Model 97	500	V-1710- 87(F21R)	2.00:1	7.48:1	None	1325	1325/ 2,500/____	1500/ 5,400/52.0
P-51A/NA Model 99	310	V-1710- 81(F20R)	2.00:1	9.60:1	None	1200	1125/15,500/44.5	1410/ 9,500/57.0
XP-51J/Model 105	2	V-1710-119(F32R)	2.36:1	8.10:1	7.64:1	1500/58"	1200/30,000/52.0	1900/ SL /78.0
Total	1,583							

* All ratings given with the engine at 3000 rpm, accept for F-32R at 3200 rpm. The Auxiliary Stage, or first-stage impeller, was 12-3/16 inches in diameter, the engine-stage was 9.5 inches.

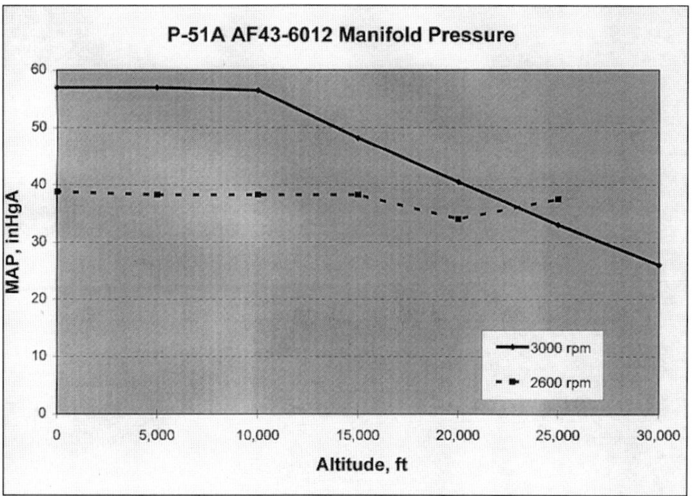

At 3000 rpm, P-51A AF43-6012 maintained WER rating of 1360 bhp at 57 inHgA on the V-1710-81(F20R) up to a critical altitude of 10,000 feet. The Military rating of 1150 bhp at 3000 rpm was available from 44.2 inHgA at up to 14,600 feet. At 2600 rpm maximum continuous cruise, 1000 bhp was available from 38.3 inHgA. (Author)

request was the result of a NAA study considering the two-stage Continental V-1430 in the NA-73. As a high boost two-stage engine it required an intercooler, a device the NAA engineers were unable to accommodate within the airframe without considerable increase in weight and drag.[81] If the two-stage engine was to proceed he believed it would be "a long time" before it could be successfully installed in production airplanes. NAA engineers anticipated that a two-speed V-1710 would raise the critical altitude by 5,000 feet and add 20 mph to the already high speed.

Still the mark of the day was being established by the Allison powered Mustang. British testing and experience demonstrated that the V-1710-F3R powered aircraft was "an excellent low and medium altitude fighter and certainly the best American fighter that has so far reached this Country." In comparisons with the Spitfire Vb, the Mustang was faster than the single-stage, two-speed Merlin powered Vb at all altitudes up to 25,000 feet. When tested against a captured Focke-Wulf 190 the Allison Mustang did the best in all respects, except rate of climb. Even so, the North American representative in England reported to them that as of September 1942 the Mustang had the lowest priority that could be granted to a production airplane in the U.S.[82]

By June 1942 the results were in from the Rolls-Royce analysis of a two-stage, two-speed Merlin 61 installed in the Mustang and the focus of the entire program shifted to redirecting Packard production for this engine and North American's enthusiastic introduction of the P-51B powered by that engine. Contracts were placed for 400 production P-51B-1NA on August 26, 1942, though the XP-51B was not to fly for the first time until November 30, 1942, only a month after the first flight in England of the similarly powered Rolls-Royce conversion.[83] These dates and actions bring into question the role of Rolls-Royce as being the genius of the Merlin powered Mustang.

But the P-51 wasn't finished with the Allison just yet. Following the August 17, 1943 debacle of the 8th Air Force over Regensberg for want of adequate escort fighters, General Arnold directed the Materiel Command at Wright Field to begin a design study of the P-51B around the two-stage V-1710. He stated that,

> "there is a probability that at a future date a requirement will develop for a greatly increased number of P-51 type airplanes. It is apparent that the production of the Merlin engine is definitely limited and that no appreciable increase is to be expected...it is requested, therefore, that studies of the P-51B and P-51F airplanes modified to incorporate the two-stage V-1710-93 engine."[84]

On September 8, 1943 O.E. Hunt, GMC Executive Vice President, wrote to the Materiel Command noting the Company's appreciation for having arranged to install the two-stage V-1710 in a P-51F airplane, and stating that Allison would fully cooperate with the Materiel Command and North American in the project. Allison stated that they felt the V-1710 could equal or better the horsepower rating of any comparable liquid-cooled engine under the same operating conditions, and furthermore, that its structure would stand up to such ratings in a superior fashion.[85] The goals for the program were to exceed the performance of the V-1650-3 Merlin and

P-51A AF43-6012 was bailed to Allison for flight testing engine developments. This photo was taken in August 1943 at Weir Cook Airport. One assignment was the first flight test of the new 1725 bhp V-1710-97(G1R), in April 1945. (Allison)

Table 5-11
Engines in North American P-51 Mustangs
V-1650 Merlin Powered Mustang Ratings

Aircraft Model,	No. Built	Engine Model/ Weight, pounds	Red Gear Ratio	S/C Gear Ratio	Takeoff, bhp	Military, bhp/alt/MAP	War Emergency, bhp/alt/MAP
P-51B/C	3,738	V-1650- 3/1690#	2.088:1	8.094:1-Hi	NA	NA	1330/23,000/67"
P-51B/C	-	V-1650- 3	2.088:1	6.391:1-Lo	1380/61"-Lo	1490/13,750/61"	1600/11,800/67"
P-51D/K	9,802	V-1650- 7/1690#	2.088:1	7.349:1-Hi	NA	NA	1505/19,300/67"
P-51D/K	-	V-1650- 7	2.088:1	5.802:1-Lo	1490/61"-Lo	1590/ 8,500/61"	1720/ 6,200/67"
P-51H	555	V-1650- 9/1745#	2.088:1	8.094:1-Hi	1830/80"Wet	NA	1505/19,300/67"
P-51H	-	V-1650- 9	2.088:1	6.391:1-Lo	NA	1490/13,750/61"	1720/ 6,200/67"
XP-51F	3	V-1650- 3/1690#	2.088:1	8.094:1-Hi	1380/61"-Lo	NA	1330/23,000/67"
XP-51G	2	RM.14SM/1690#	2.381:1	7.349:1-Hi	NA	w/Grade 150	1850/22,800/81"
XP-51G	-	RM.14SM	2.381:1	5.802:1-Lo	NA	w/Grade 150	2200/ 2,000/81"
XP-51G	-	RM.14SM	2.381:1	5.802:1-Lo	1675/66"-Lo	NA	1720/ 6,200/67"
XP-51G	-	RM.14SM	2.381:1	5.802:1-Lo	NA	w/Grade 130	1850/ 6,000/71"
XP-51G	-	RM.14SM	2.381:1	7.349:1-Hi	NA	w/Grade 130	1600/18,000/71"
P-51L	2	V-1650-11/1690#	2.088:1	8.094:1-Hi	NA	NA	1860/19,400/90"Wet
P-51L	-	V-1650-11	2.088:1	8.094:1-Hi	NA	NA	1780/22,500/80"Wet
P-51L	-	V-1650-11	2.088:1	8.094:1-Hi	NA	NA	1405/25,000/70"Dry
P-51L	-	V-1650-11	2.088:1	6.391:1-Lo	1380/61"-Lo	NA	2270/ 4,000/90"Wet
Total	14,102						

Rolls-Royce RM.14SM used the new British Skinner-Union Speed Density Carburetor, based on the same principles as the Bendix Speed Density unit on the V-1710-143/145.
All of these two-stage Merlin's used a first stage impeller 12.0 inches in diameter, the second-stage was 10.1 inches. All ratings at 3000 rpm and on Grade 100/130 fuel unless otherwise noted.

at least equal the performance expected from the coming Rolls-Royce RM14SM engine.[86] These objectives were then the focus of the light-weight NAA XP-51J airplane and its Allison V-1710-119(F32R) engine, both of which are discussed in the next chapter.

Although the Merlin Mustang, P-51B/C/D, claimed fame as the ultimate piston engined fighter of WWII, the early Allison powered *Mustang* was its match at low and medium altitudes. The sleek profile of the early Mustang was possible because of its close cowled Allison V-1710, a very handsome and smooth running aircraft that can be appreciated in the few P-51A's and A-36A's still flying 50 years later.

Bell P-63 *"Kingcobra"*

General Arnold had been promised, and expected, great things from the XP-39. It was to have been a 400 mph, fast climbing, interceptor. In this it failed, but the General was still fixed on improving its performance with a new model. In June 1942 he noted that "many times I have been informed that the Allison Engine Company (two-stage) supercharger is getting along all right and will be in production very shortly, but as yet I have seen no supercharger." He then produced a set of questions regarding the new supercharger and the airplane to use it. The response was that the two-stage engine would enter production in December 1942 and that the Bell P-63 would be powered by it. This airplane was to enter production in April 1943 and soon thereafter supplant the P-39.[87]

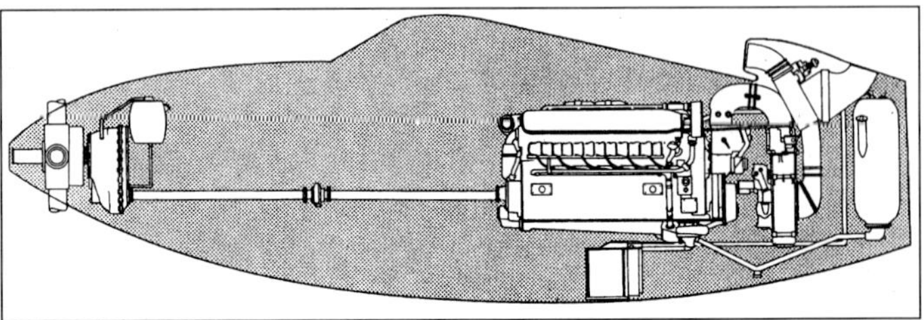

With the necessity to lengthen the fuselage of the P-63 for stability the Auxiliary Stage was easily accommodated. This drawing gives perspective on the compact installation of the two-stage V-1710-E11. (Matthews)

The Bell P-63 was one of two aircraft intended by the War Department to meet three principal needs: "low" and "high" altitude capability for quick climb, maneuverability, and fire power. The XP-63 was to meet the low altitude requirement (22,000 feet) while the Republic XP-47E was to be the high altitude airplane, for which it was to incorporate a pressurized cabin to enable sustained operations around 40,000 feet.[88]

Bell Aircraft's *Kingcobra* claims the distinction of being the only American fighter that had its maiden flight after the war started that was actually built in comparatively large numbers.[89] The concept was to build a fighter benefiting from the significant advances occurring in aviation, and in the process make up for the shortcomings of the P-39 as an interceptor once its turbosupercharger had been removed. The mission was to intercept Fw 190 and Zero fighters, scourges of the Allies far-flung forces.[90] Though the aircraft was obviously derived from the P-39, it was an enlarged aircraft built to utilize the new high altitude Allison V-1710 with its two-stage mechanically driven supercharger. As a result of the poor spin characteristics of the somewhat similar two-stage Allison powered XP-39E aircraft, it was decided to further lengthen the fuselage of the P-63.[91] The design also adopted a laminar flow airfoil, improved internal aerodynamics and ducting, and improved overall detail.

Allison had begun work on a two-stage engine as early as 1938, and was then spurred on by General Arnold to match or exceed single-stage, two-speed, Merlin ratings being offered in the Fall of 1939.[92] The Allison concept for a two-stage engine was to continue with their "building block" approach to the engine. This approach was intended to minimize changes in manufacturing and require little or no change to the power section of the engine.

The new supercharger stage would be built as a separate unit and act as a first-stage supercharger providing sea-level air pressure to the engine stage supercharger. As such, it would be driven by a power takeoff coupling fitted to the starter gear. Initially the unit was directly driven via gearing from the starter, but this had the disadvantage of requiring considerable throttling to maintain manifold pressure to acceptable limits when below the critical altitude. Since throttling is inherently wasteful of power, Allison devised a hydraulic coupling that operated much like the torque converter in a modern automobile automatic transmission. This had the advantage of giving a continuously variable speed capability to the first-stage, meaning that it would operate the optimum speed for the conditions of flight. Another advantage was to isolate the new supercharger from the torsional vibrations inherent in a drive coming directly from the crankshaft. The only other WWII era engines to have this feature were the large liquid-cooled Daimler-Benz V-12's powering German aircraft such as the Messerschmitt Bf 109.

Because of the comparative high efficiency of the Allison supercharger, combined with the advantages of the hydraulic coupling and the space constraints in the P-63, Allison did not provide an intercooler for its two-stage "E" series production engines, although considerable engineering and testing was spent on such an installation.[93] In order to take advantage of the high manifold boost capability from the two-stage setup, Allison elected to develop and adopt water injection as a means of limiting detonation that would otherwise occur during War Emergency operation.

Even so, with the design of the P-63 increased internal volume was available, so once again consideration was given to turbosupercharging its V-1710. In this regard Allison identified the V-1710-E20 sea-level rated single-stage engine, which would have been unique in that it would have been supplied by the Wright Model 800 TSBA-1 turbosupercharger. The Wright turbo was somewhat similar to the more usual GE turbo's in configuration, but had a considerably different design for the turbine wheel in that the blades were air-cooled. As proposed it would have made a very clean installation.

The P-63 never was included in the U.S. order of battle during WWII, though three squadrons were formed on the P-63 with the intention of using them for the invasion of Europe. There were two reasons for the U.S. not using the P-63 in the European campaigns. First, Air Force tests of the P-63A-1 and A-5 models at Eglin Field, Florida concluded that the airplane had a decided lack of combat radius. In fact, it had the shortest combat radius of any contemporary first line Army fighter of the day (P-38, P-47 and P-51). The short legs were a consequence of having the least internal fuel capacity per installed horsepower (takeoff) of *any* Army or Navy fighter that saw action during the war.[94] The second reason was more practical: there simply wasn't time to equip, train, and deploy *Kingcobra* squadrons to England in time for the June 1944 invasion.[95]

However all was not lost. The Soviets were faced with an entirely different set of circumstances. Furthermore, they did not require a great combat radius. What they wanted was a powerful and maneuverable fighter with heavy firepower. With its' 37 mm cannon the P-63 excelled as a "tank buster" for the Soviets on the Eastern Front, but it also had the capability to combat German aircraft at high altitudes. A total of 3,303 P-63's were built, with 2,421 being dispatched under Lend-Lease to the USSR. A total of 2,400 were actually delivered, as the other 21 were lost in transit. While it is often reported that another 300 P-63's were provided to the Free French Air Force, research by Allain Pellitier of France has determined that something less than 200 airplanes were actually received.[96] The French used them mainly in Southeast Asia following WWII.

XP-39E Contrary to many accounts, the P-63 program was separate from the XP-39E (Bell Model 23) project, though both were begun about the same time and used the same new V-1710-47(E9) two-stage engine. Reports that it was fitted with a laminar flow wing are erroneous, that was a distinction left to the P-63.[97]

XP-63 The XP-63 was originally intended to be powered by the upright two-stage Continental V-1430 engine, but in February 1942 this was deleted and the two-stage Allison V-1710-47(E9) fitted instead.[98,99]

Two of these Bell Model 24's were built, AF41-19511/19512. They were powered by the V-1710-47(E9) two-stage engine, though during the test program numerous modifications were incorporated. This is seen in Allison Specification No. 137 that went from revision A through G during the course of development, all using the 7.48:1 engine stage gears. By the time the P-63A went into production, the differences between the E-9 flight test engines and the production V-1710-93(E11) were slight, as in their final form they even incorporated the same engine-stage 8.10:1 supercharger gears.[100]

The original concept for the aircraft was as a more drastic improvement of the standard P-39 than offered by the XP-39E. The XP-63 embodied a new low-drag airfoil, larger wheels, increased fuel capacity and a number of other improvements. The airplane had been stripped of weight so that it was envisioned as a stop-gap, fast climbing interceptor that could be in production at an early date. The need for such an interceptor was to counter pressure being felt by the Materiel Division for a competitor to the Japanese Zero and German Focke Wulf Fw 190.[101] Development took longer than intended as difficulty was encountered in getting the desired degree of reliability from the early models of the Auxiliary Stage Supercharger, as well as introduction of associated improvements in the V-1710 power section to accommodate the higher indicated power required with the Aux Stage.[102]

XP-63A This was the third prototype (AF42-78015) of the *Kingcobra*, purchased out of sequence mainly to replace the first XP-63, it had been lost in January 1943 after only one month of testing. The XP-63A made its first flight on April 26, 1943, powered by the V-1710-93.[103] With this airplane, Bell finally had achieved what the P-39 was never able to do: 400 mph at high altitudes.[104]

P-63A Deliveries of 200 of these V-1710-93(E11) powered aircraft began in October 1943. As the P-63A-8, they were the first production aircraft in the U.S. to be equipped with ADI injection.[105] With ADI and WER ratings these aircraft could climb to 20,000 feet in less than five minutes, a considerable improvement over the 5.5 minutes needed by the

This two-stage V-1710-93(E11) powered P-63A-9 was destined for service on the Russian Front. Military Rating of 1150 bhp was available up to 22,400 feet, with WER of 1390 to 18,000 feet where the airplane made 422 mph. (Bell via SDAM)

Table 5-12
Early P-63 Performance Specifications

A/C Model	XP-63	XP-63A	P-63A-8	P-63A-10	TP-63A-10
Bell Model	24	24A	33A-8	33A-10	38
Number Built	2	1	200	730	NA
Contract Number	ac-18966	ac-18966	ac-29318	ac-29318	ac-
Speed at Critical Alt, mph Mil Power	407/22,400	407/22,400	417/24,000	417/24,000	NA
Military Climb to Critical Alt, min	7.0/22,400	5.47/20,000	6.8/22,400	7.2/22,400	NA
Wing Area, sq.ft.	248	248	248	248	248
Empty Weight, pounds	6,053.8	6,184.8	6,531.9	6,693.7	NA
Gross Weight, pounds	7,524.7	7,655.7	8,213.2	8,264.7	NA
Design Load Factors	+8.0, -4.0	+8.0, -4.0	+8.0, -4.0	+8.0, -4.0	+8.0, -4.0
Engine Model	V-1710-47	V-1710-93	V-1710-93	V-1710-93	V-1710-93
Allison Engine Specification	137-G	142-C	142-C	142-C	142-C
Two-Stage Supercharger	Auxiliary	Auxiliary	Auxiliary	Auxiliary	Auxiliary
Performance at Takeoff, bhp/rpm	1325/3000	1325/3000	1325/3000	1325/3000	1325/3000
Mil Rating at Critical Alt, bhp/rpm	1150/3000	1150/3000	1150/3000	1150/3000	1150/3000
Military Rating, Critical Altitude	22,400	22,400	22,400	22,400	22,400
War Emergency Rating, bhp/rpm	None	1150/3000	1390/3000	1390/3000	None
War Emergency mph/altitude, ft	None	416/25,400	422/18,000	422/18,000	None
War Emerg Time to Climb/Altitude, ft	None	6.5/22,400	4.8/20,000	4.9/20,000	None
Aircraft Specification	24-947-001	24-947-001-1	24-947-001-1	33-947-001-10	38-947-001
Date of Original Aircraft Specification	4-1-1941	10-15-1942	4-1-1944	4-15-1944	12-15-1943
Army Minimum Load Factors	+8.0, -4.0	+8.0, -4.0	+8.0, -4.0	+8.0, -4.0	+8.0, -4.0
NACA Wing Section at Root	66,2X-116	66,2X-116	66,2X-116	66,2X-116	66,2X-116
NACA Wing Section at Tip	66,2X-216	66,2X-216	66,2X-216	66,2X-216	66,2X-216
Nose Cannon	37mm	37mm	37mm	37mm	37mm
Comments					

contemporary P-39's to reach only 12,000 feet. This was accomplished even though the P-63A was nearly 1,000 pounds heavier. Early models of the V-1710-93 were often troubled with failures of the bearings supporting the drive to the hydraulic coupling. This occurred both on the Model Test and some in-service engines, once resolved, the Auxiliary Stage equipped engines were quite reliable.[106] There were also a number of early P-63A's lost to a mysterious gradual loss of power. Finally an intact aircraft survived an incident and the investigation determined that an improperly designed carburetor float was being used in the large Stromberg carburetor. Fixes were devised and all *Kingcobras* were grounded for a time throughout Flying Training Command while the carburetors were retrofitted. This solved the problem once and for all.[107]

The higher power ratings of the -93 engine required a new type of piston ring by used. These new "keystone" piston rings were fitted from the 13th aircraft on. Engine and airframe production was ahead of a successful Model Test of the engine, so the first 125 or so airplanes were assigned within the continental limits of the U.S. and limited to 45 inHgA, pending completion of the V-1710-93 Model Test.[108]

XP-63B The P-63 was originally conceived as an improved P-39 having a two-stage engine with high altitude capability, and was initially intended to use the new Continental V-1430. The XP-63B was conceived early in 1943 by General Echols, Commander of the Materiel Command, as a way to utilize the new two-stage Rolls-Royce Merlin engines to be built by Packard. He had personally directed that the Merlin be tried out in the P-63.[109] The result was the Bell Model 34, XP-63B. As of May 1943 the documents describe an aircraft to be built around a modified Packard V-1650-3. This was revised that October to utilize a V-1650-5, a model with a new front crankcase that was able to mate with the Allison extension shaft system as utilized by the P-39 and P-63. Two aircraft were ordered, but neither were completed. While the reasons for cancellation are only speculative, it appears that there was little, if any, performance advantage over the contemporary Allison powered P-63A. With the two-stage Merlin, the XP-63B would have been some 400 pounds heavier than the P-63A. Although the Merlin critical altitude was about 1,000 feet higher than the E-11, the P-63A demonstrated a considerably better time to climb to 20,000 feet: 4.9 minutes versus 6.0. Given that the intended mission for

The air and mixture flow paths within the V-1710-93(E11) and -117(E21) are clearly shown in this diagram. Note how the carburetor is mounted on the intake to the Auxiliary Stage Supercharger. (Allison)

This is a diagram of the induction airflow for the V-1710-109(E22). It achieved a higher critical altitude by deleting the flow restriction caused by the carburetor on the inlet to the Aux Stage. The carburetor used was that used on the turbosupercharged engines, and likewise mounted on the inlet to the engine stage supercharger. The gain of about 3,000 feet in critical altitude would have been put to good use in the coming P-63E. (Allison)

the P-63 at the time was to intercept Zero's and Fw 190's, this difference was considerable.

P-63C Bell shifted production to the P-63C in January 1945 and proceeded to deliver a total of 1,227 prior to contract termination following the end of the European war in May. Allison delivered a total of 2,237 two-stage V-1710-117(E21)'s to the Air Force to power and support these aircraft. These engines differed from the E-11 by having the 12-counterweight crankshaft and a slightly higher Aux Stage step-up gear ratio that raised the Military rated critical altitude to 25,000 feet, an increase of 3,500 feet.

The E-11 and E-21 models were both arranged so that the carburetor was installed directly above the inlet elbow to the Auxiliary Stage Supercharger, though the fuel was delivered into the eye of the Engine-stage Supercharger impeller as on all other V-1710's. There was a considerable disadvantage inherent in this location of the carburetor. Even though it was a large three barrel unit, it caused a one to two inch pressure loss at the inlet to the impeller.[110] This alone would reduce the critical altitude of the aircraft by over 2,000 feet.

In their V-1710-109(E22) for the P-63D/E, Allison revised the arrangement so that maximum ram air from the intake scoop would be delivered directly to the Auxiliary Stage Supercharger impeller, where it would then boost the pressure to one to two inches of mercury *above* the pressure at sea-level. This air was then delivered to a two barrel carburetor that was installed in the same position as on turbosupercharged V-1710's, just above the inlet elbow to the engine-stage supercharger.[111] At Military ratings the result was an increase of 3,000 feet in the critical altitude of the airplane.

Performance of the airplane far exceeded the limited roll assigned the *Kingcobra* by the Army, as demonstrated in the following testing. Bell Chief Test Pilot Jack Woolams and Army Lieut. Jesse Beitman flew simulated dogfights in P-63C-5's against Captain M.G. Hebrard, an outstanding French pilot who was flying a Republic P-47 *Thunderbolt*. In separate "combats" over the Niagara Falls Airport on March 4, 1945, the Bell pilots outperformed and outmaneuvered the P-47 with extreme ease. Captain Hebrard, who was at Bell for the specific purpose of testing the latest P-47

Table 5-13
P-63B/C Performance Specifications

A/C Model	XP-63B	XP-63B	P-63C-1	P-63C-5	P-63C-5
Bell Model	34	34	33C-1	33C-5	33C-5
Number Built (Contracted)	0	2, Canceled	215	585	426(675)
Contract Number, W535-	Proposal	W33-038-ac-621	ac-29318	ac-29318	ac-40041
Speed at Critical Alt, mph Mil Power	420/24,200	420/25,800	413/26,700	418/28,590	418/28,590
Military Climb to Critical Alt, min	7.5/24,200	8.5/25,800	8.8/25,000	8.6/25,000	8.6/25,000
Wing Area, sq. ft.	248	248	248	248	248
Empty Weight, pounds	6,809	7,105.9	6,798.5	6,855.7	6,855.7
Gross Weight, pounds	8,264.4	8,676.3	8,448.4	8,640.2	8,640.2
Design Load Factors	+8.0, -4.0	+8.0, -4.0	+8.0, -4.0	+8.0, -4.0	+8.0, -4.0
Engine Model	V-1650-3	V-1650-5	V-1710-117	V-1710-117	V-1710-117
Allison Engine Specification	AC1035	AC1050	178-A	178-A	178-A
Two-Stage Supercharger	2-Stage	2-Stage	Auxiliary	Auxiliary	Auxiliary
Performance at Takeoff, bhp/rpm	1375/3000	1390/3000	1325/3000	1325/3000	1325/3000
Mil Power at Critical Alt, bhp/rpm	1300/3000	1238/3000	1100/3000	1100/3000	1100/3000
Military Rating, Critical Altitude	24,200	29,500	25,000	25,000	25,000
War Emergency Rating, bhp/rpm	NA	1360/3000	1440/3000	1565/3000	1565/3000
War Emergency mph/altitude, ft	NA	426/23,400	425/18,500	436/18,500	436/18,500
War Emerg Time to Climb/Altitude, ft	NA	6.0/20,000	5.0/20,000	5.0/20,000	5.0/20,000
Aircraft Specification	34-947-001	34-947-001	33-947-005	33-947-007	33-947-007
Date of Original Aircraft Specification	5-15-1943	8-20-1943	1-15-1944	11-15-1944	11-15-1944
Army Minimum Load Factors	+8.0, -4.0	+8.0, -4.0	+8.0, -4.0	+8.0, -4.0	+8.0, -4.0
NACA Wing Section at Root	66,2X-116	66,2X-116	66,2X-116	66,2X-116	66,2X-116
NACA Wing Section at Tip	66,2X-216	66,2X-216	66,2X-216	66,2X-216	66,2X-216
Nose Cannon	37mm	37mm	37mm	37mm	37mm
Comments					249 Canceled

The single Bell XP-63D, AF43-11718 used the V-1710-E22 engine with the between stage carburetor. Allison Test Flight, Weir Cook Airport, Indianapolis. (Allison)

Table 5-14
P-63D/E/F Performance Specifications

A/C Model	P-63D-1	P-63E-1	P-63E-1	P-63F-1
Bell Model	37	41E-1	41E-1	43
Number Built (Contracted)	1	13(1574)	(425)	2
Contract Number	ac-	ac-29318	ac-40041	ac-29318
Speed at Critical Alt, mph Mil Power	437/30,000	439/31,800	439/31,800	414/29,800
Military Climb to Critical Alt, min	11.2/28,000	8.7/28,000	8.7/28,000	7.7/27,500
Wing Area, sq.ft.	255	255	255	255
Empty Weight, pounds	7,076.0	7,088.1	7,088.1	7,111.9
Gross Weight, pounds	9,053.7	8,897.3	8,897.3	8,911.8
Design Load Factors	+8.0, -4.0	+8.0, -4.0	+8.0, -4.0	+8.0, -4.0
Engine Model	V-1710-109	V-1710-109	V-1710-109	V-1710-133
Allison Engine Specification	179-C	179-E	179-E	E-30
Two-Stage Supercharger	Auxiliary	Auxiliary	Auxiliary	Auxiliary
Performance at Takeoff, bhp/rpm	1425/3000	1425/3000	1425/3000	1500/3000
Mil Power at Critical Alt, bhp/rpm	1100/3000	1100/3000	1100/3000	1150/3000
Military Rating, Critical Altitude	28,000	28,000	28,000	27,500
War Emergency Rating, bhp/rpm	1300/3200	1340/3200	1340/3200	2200/3200
War Emergency mph/altitude, ft	452/27,800	452/27,500	452/27,500	449/13,900
War Emerg Time to Climb/Altitude, ft	4.4/20,000	4.5/20,000	4.5/20,000	3.0/20,000
Aircraft Specification	37-947-001	41-947-001	41-947-001	43-947-001
Date of Original Aircraft Specification	12-15-1943	1-15-1945	1-15-1945	10-15-1945
Army Minimum Load Factors	+8.0, -4.0	+8.0, -4.0	+8.0, -4.0	+8.0, -4.0
NACA Wing Section at Root	66,2X-116	66,2X-116	66,2X-116	66,2X-116
NACA Wing Section at Tip	66,2X-216	66,2X-216	66,2X-216	66,2X-216
Nose Cannon	37mm	37mm	37mm	37mm
Comments				

Table 5-15
Gunfire Protection for the RP-63C-2

Description	Weight, pounds	
External Armor Plate:		
Gear Box Cowl	41.7	
Fuselage forward of station 153	124.1	
Nosewheel Doors	24.5	
Cabin Superstructure (Upper)	20.0	
Cabin Doors	47.2	
Cabin Superstructure (Lower)	41.9	
Fuselage, station 153 to 231-1/2	15.0	
Main Wheel Fairing and Flipper Doors	27.7	
Aft Fuselage, aft of station 231-1/2	0.2	
Wing	681.0	
Center Wing Section	139.0	
Shutters and Other on Bottom	46.9	
Leading Edge of Fin and Stabilizer	7.0	
Total External Armor Plate		1,216.2
Internal Armor Plate:		
Gear Box	56.1	
Fume tight bulkheads	33.1	
Aft of Pilot's Head	11.4	
Hinge Protectors	1.4	
Total Internal Armor Plate		102.0
Armor Glass:		
Windshield	28.4	
Doors	50.5	
Cabin	90.6	
Total Armor Glass		169.5
Total Gunfire Protection		1,487.7
Empty Weight of Aircraft, pounds		9,071.0
Gross Weight of Aircraft, pounds		10,200.0

against the P-63 in simulated combat, had a record of more than a dozen kills in actual combat. Notwithstanding his experience and ability, he was unable to keep Woolams or Beitman in their individual missions against him from piloting their *Kingcobras* so that they easily flew on the tail of the P-47 amply long enough for a victory. At no time was the P-47 a challenge to the swifter climbing, more maneuverable P-63's. Captain Hebrard said when he landed that he had gone up with the intention of trying all of the tricks he had learned in actual war.[112]

P-63D Only one Bell Model 37, P-63D (AF43-11718) was built. It is believed to have been built upon a P-63A fuselage modified to demonstrate the bubble canopy planned for the P-63E-5. It also incorporated the larger wing designed for the P-63E/F airplanes.[113] The two-stage engine was an early version of the V-1710-109(E22) [Allison Specification 179-C], with the between stages carburetor, as used in the later P-63E [Allison Specification 179-E]. This was the fastest of the P-63's, being capable of 452 mph at its WER critical altitude of 27,800 feet.[114]

P-63E This Bell Model 41 airframe was reported to be similar to the P-63D and used the same V-1710-E22 engine, the only obvious difference being the retention of the original "automobile" door style cockpit. Although 2,943 were ordered for the Russians, all but 13 were canceled with the end of the war in Europe. The P-63E (AF43-11720) first flew in the spring of 1945. Maximum speed was 439 mph at its considerable critical altitude of 31,800 feet.[115] The P-63E-5 model was to have been the major production version and would also have incorporated the NACA improved tail demonstrated on the P-63F airplane as well as the bubble canopy from the P-63D.[116] In April of 1945 Allison delivered one of the next generation V-1710-E30 engines for installation in the P-63E. It was estimated that this engine would increase the speed of the airplane by 31 miles per hour at 20,000 feet and by 19 mph at 30,000 feet. The additional power was expected to increase the rate of climb of the Kingcobra by 700 feet per minute at sea level, 800 fpm at 20,000 feet and 500 fpm at 30,000 feet.[117]

P-63F The first of two P-63F's was AF43-11719, making it difficult to determine if it was a P-63D or P-63E derivative. Suffice to say that it was similar since its serial number was between the D and E models. The second prototype was 43-11722, an airframe taken from the batch of P-63E's. The P-63F used the next generation two-stage engine, the V-1710-133(E30).[118] This engine provided 1500 bhp for takeoff, and with water injection could deliver 2200 bhp at 13,900 feet using 3200 rpm with a top speed of 449 mph. The airplane really performed in climbing, capable of reaching 20,000 feet in only 3.0 minutes. The aircraft was distinguished by its taller, and somewhat pointed, vertical tail, a feature intended for production on the P-63E-5. Like the P-63E its production contract was canceled following V.E. Day.[119]

"Pinball" RP-63's These were developed and used in the frangible bullet gunnery training program. A total of 100 P-63A's were converted as models RP-63A-11 and RP-63A-12, the "R" noting "Restricted from Combat." With the success of the early program another 200 "Pinball's" were built from P-63C airframes, and designated RP-63C-2. An order for production of 450 RP-63G's began deliveries in September 1945, but was terminated upon the cessation of hostilities during that month after only 32 were completed.[120] The RP-63's were not outstanding performers, as performance was restricted by the extra weight of the heavy aluminum plates used to protect the aircraft and pilot. Gross weight of the RP-63A was 10,063 pounds, reaching 10,573 for the RP-63G, typically some 1,600 to 1,800 pounds heavier than their fighter configured brethren. With performance not an issue, the "Pinball" airplanes were all rated for 300 mph using Military power.

The detail on the armor plating built into the P-63 is given in the following table. For comparison, the empty weight of a standard P-63C was 6,799 pounds and its allowable maximum gross weight a respectable 8,448 pounds, which was more than 600 pounds less than the *empty* weight of the Pinball.

RP-63G The final production model of the *Kingcobra*, the RP-63G was restricted from combat as it was a "Pinball" target plane for gunner training. Consequently it could dispense with the Auxiliary Stage Supercharger entirely and utilize the V-1710-135(E31)[121] in its comparatively low level mission.[122] The power section of the engine was the slightly heavier version typical of the late model engines as it incorporated the 12 counterweight crankshaft that allowed it to operate at higher rpm without overloading the bearings. Some 258 of these engines were reportedly built, far more than needed for the few aircraft that were actually available to use them. All of these engines were converted by Allison from other models. Consequently, many of these engines soldiered on years later as the powerplant of choice on the Unlimited Hydroplane Circuit, and as the basis for engines to power Warbird restorations such as the Planes of Fame P-51A *Mustang*, though in that case it was configured with a V-1710-F reduction gear nose case.[123]

Ivan Hickman, who in WWII piloted the P-63, writes in his book Operation Pinball, that the P-63 was undeniably a fine fighter, attested to by those who flew and compared it with the P-51, P-47 or the P-38. Up to

Table 5-16 RP-63 "Pinball" Performance Specifications					
A/C Model	RP-63A-11	RP-63A-12	RP-63C-2	RP-63G-1	RP-63G-1
Bell Model	33A-11	33A-12	33C-2	41G	41G
Number Built	5	95	200	2	30
Contract Number	ac-29318	ac-29318	ac-29318	ac-29318	ac-11724
Speed at Critical Alt, mph Mil Power	300	300	300	300/20,000	300/20,000
Military Climb to Critical Alt, min	NA	NA	NA	NA	NA
Wing Area, sq. ft.	248	248	248	255	255
Empty Weight, pounds	8,945.8	8,985.8	9,071.0	9,422.2	9,422.2
Gross Weight, pounds	10,063.0	10,103.0	10,200.0	10,572.9	10,572.9
Design Load Factors	NA	NA	NA	NA	NA
Engine Model	V-1710-93	V-1710-93	V-1710-117	V-1710-135	V-1710-135
Allison Engine Specification	142-C	142-C	178-A	183-A	183-A
Two-Stage Supercharger	Auxiliary	Auxiliary	Auxiliary	Engine Only	Engine Only
Performance at Takeoff, bhp/rpm	1325/3000	1325/3000	1325/3000	1200/3000	1200/3000
Mil Rating at Critical Altitude, bhp/rpm	1100/3000	1100/3000	1100/3000	1125/3000	1125/3000
Military Rating, Critical Altitude	22,400	22,400	25,000	15,000	15,000
War Emergency Rating, bhp/rpm	None	None	None	None	None
War Emergency mph/altitude, ft	None	None	None	None	None
War Emerg Time to Climb/Altitude, ft	None	None	None	None	None
Aircraft Specification	33-947-001-11	33-947-001-12	33-947-006	41-947-005	41-947-005
Date of Original Aircraft Specification	7-27-1944	10-15-1944	11-15-1944	5-1-1945	5-1-1945
Army Minimum Load Factors			+8.0, -4.0	+8.0, -4.0	+8.0, -4.0
NACA Wing Section at Root	66,2X-116	66,2X-116	66,2X-116	66,2X-116	66,2X-116
NACA Wing Section at Tip	66,2X-216	66,2X-216	66,2X-216	66,2X-216	66,2X-216
Nose Cannon	Ballast	Ballast	Ballast	Ballast	Ballast

15,000 feet it would outperform the P-51 in at least two categories, climb and maneuverability. In regard to the latter, the Kingcobra was particularly outstanding. P-63's actually turned inside AT-6 training planes, a considerably slower and supposedly tighter turning airplane. It is interesting to contemplate how it would have fared against the famed Japanese Zero, a contest that was never to be.

Hickman goes on to identify the major shortcoming of the P-63 as its 132 U.S. gallon internal fuel capacity. When developed as an interceptor this was an advantage. By the time the *Kingcobra* was available in numbers sufficient for use in combat the strategic situation had shifted away from the defensive interceptor to the need for extremely long range escort fighters. Consequently the USAAF never employed the P-63 in a combat role. Rather, some 1,952 Kingcobras were extensively and successfully

Table 5-17 Bell P-63 Aircraft and Engines								
Aircraft Model	Number Built	Engine Model	Reduction Gear	Engine S/C Ratio	Aux S/C Ratio	Takeoff, bhp	Military Rating, bhp/alt/MAP	War Emergency, bhp/alt/MAP
XP-63	2	V-1710- 47(E9)	2.227:1	7.48:1	6.85:1	1325	1150/21,400/49.7	None
XP-63A	1	V-1710- 93(E11)	2.227:1	8.10:1	6.85:1	1325	1150/22,400/50.0	1825/SeaLev/75.0 Wet
P-63A	1,725	V-1710- 93(E11)	2.227:1	8.10:1	6.85:1	1325	1150/22,400/50.0	1825/SeaLev/75.0 Wet
RP-63A	100	V-1710- 93(E11)	2.227:1	8.10:1	6.85:1	1325	1150/22,400/50.0	ADI Disconnected
P-63C	1,227	V-1710-117(E21)	2.227:1	8.10:1	7.23:1	1325	1100/25,000/52.0	1800/24,000/76.0 Wet
RP-63C	200	V-1710-117(E21)	2.227:1	8.10:1	7.23:1	1325	1100/25,000/52.0	ADI Disconnected
P-63D	1	V-1710-109(E22)	2.227:1	8.10:1	7.23:1	1425*	1100/28,000/50.7	1750/SeaLev/75.0 Wet
P-63E	13	V-1710-109(E22)	2.227:1	8.10:1	7.23:1	1425*	1100/28,000/50.7	1750/SeaLev/75.0 Wet
P-63F	2	V-1710-133(E30)	2.227:1	8.10:1	7.64:1	1500*	1150/27,500/58.0	2200/SL/80/85 Wet
RP-63G	32	V-1710-135(E31)	2.227:1	9.60:1	None	1200	1125/15,000/43.2	ADI Disconnected
XP-63H	(1)	V-1710-127(E27)	2.48:1	8.10:1	7.23:1	2825	2430/17,000/85.0	2980/11,000/100.0 Wet
Total	3,303							

* All ratings given with the engine at 3000 rpm except for V-1710-109 , -127 and -133 WER ratings at 3200 rpm .
Note: The approximately 200 P-63's for the French were delivered from above production.

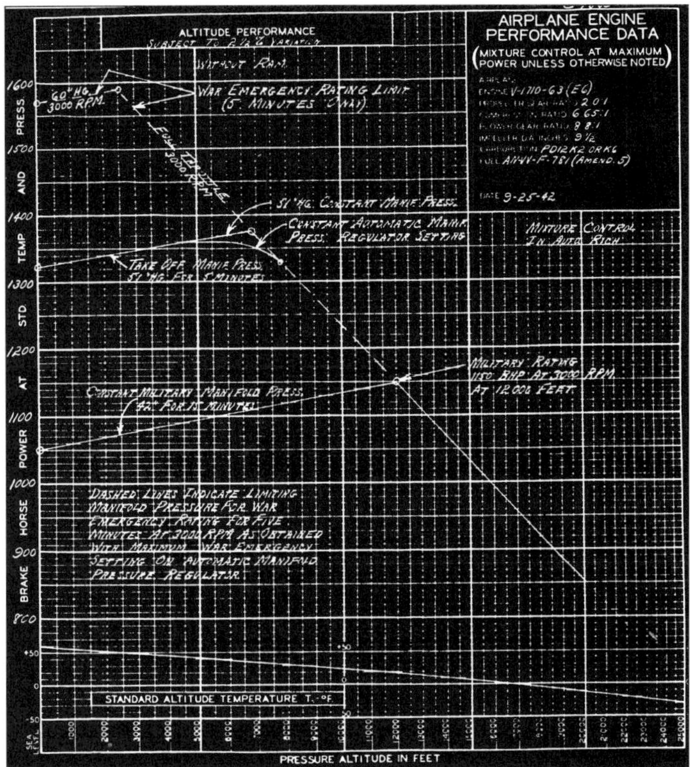

Wright Field required standard engine performance curves to assume no forward airspeed, i.e. no ram air, the result being conservative estimates of power during flight. These full throttle, maximum rpm WER curves show the altitude at which constant manifold pressure can no longer be maintained. These ratings were established by Allison in September 1942. The rise in power until the critical altitude is reached is due to the reduction in air temperature with altitude, which results in a greater mass of air being drawn into the engine.

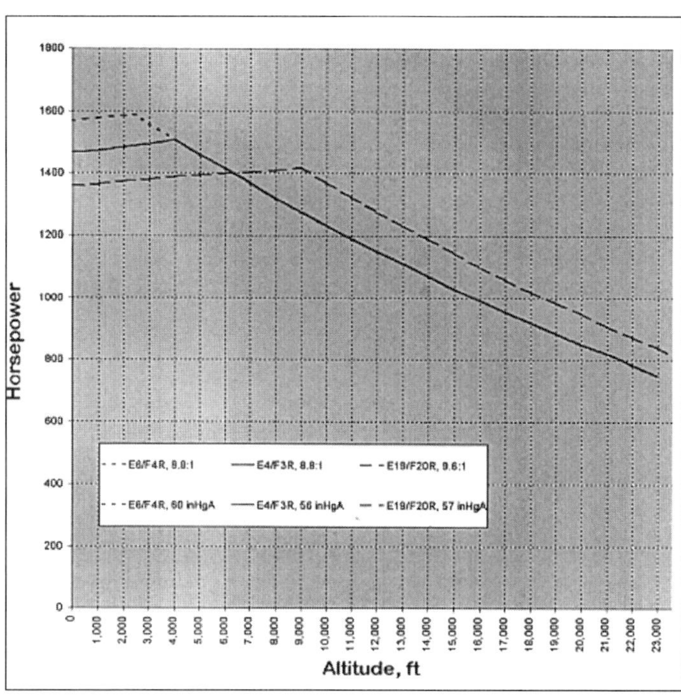

Establishing the performance of an engine requires fixing the rpm and the supercharger drive ratio, which sets the power available at any altitude. The maximum power is determined either by the strength of the engine or by the ability of the fuel to resist detonation. This is clear when the E4/F3R are compared to the physically stronger E6/F4R. Otherwise these engines were identical and all rated on the same AN-F-781 100/125 Octane fuel.

used by the Russians in a ground support role for their army's as they reclaimed eastern Europe from the Nazi.

War Emergency Power Ratings, WER

It is unfortunate that most published information presented on Allison V-1710 powered aircraft only refer to the "takeoff" rating of the engine in that aircraft. The implication is that it represents the maximum amount of power available to the pilot. Such was definitely not the case since WER ratings were being developed by Allison within the first three months of the war.

Sometimes these WER ratings could not be fully utilized because of the inability of the host aircraft to provide the necessary support for the engine to function at the higher power. An example being the Lockheed P-38F/G/H with their inadequately sized leading edge intercoolers, and low capacity turbosupercharger.[124] The power and the reliability were there, but it took a long time for the Army Air Forces to provide the necessary grade of fuel, disseminate information and train its pilots and support personnel to properly use the power and engine.

In September 1942 the U.S. Army prepared its second report on *Allison Engines Under Wartime Conditions*. The report detailed experience to date with engines then powering British aircraft, namely the *Mustang, Tomahawk, Kittyhawk* and *Airacobra*.[125] While the report is nominally complimentary to the V-1710, it criticizes the conservative maximum ratings being imposed by the USAAF. Significantly, it shows the differing philosophies in wartime Britain regarding the ratings and use of a fighter engine, as compared to those in the U.S.:

> "The F-3R engine would well stand rating up to considerably higher boosts for combat purposes, and this step is strongly recommended."
>
> "In Britain if an engine becomes reliable, it is immediately rated up for combat purposes. The engine manufacturers feel that in view of the risks the pilots have to take, manufacturers should be prepared to risk their engines as well. Engine failure nowadays is only a minor cause of casualties. Even in peace time, competition between engine builders and the military purpose for which engines are intended, also ensures the same rapid progress in ratings in Britain."

At the time, the AAF rated its engines based upon the maximum power they could sustain during the specified 150-hour Model Test regime, as defined by the performance of the fuel they were using. Typically, this was a test to specifically define the "Takeoff" and "Military" performance ratings. There was no "sprint" or peak ratings defined, allowed or considered by Wright Field. With the introduction to combat, pilots were pushing their engines to the limits. Since few of these early installations utilized manifold pressure regulators it was easy for British pilots to pull more than "rated" power. The consequence of unconstrained operation was a serious possibility of engine damage or reduced reliability. The experience of British pilots was that many had pulled 50-56 inHgA (45 inHgA was the handbook maximum for Military Power) from their F-3R's for 5 to 15 minutes while over enemy territory without trouble.[126] As U.S. pilots entered the combat theaters, they were asking for authorization to use this same capability from their Allison's.

The U.S. Government did not allow "combat" power ratings until December 1942, until the war was a year along. Allison helped the Air Forces authorities to write the War Emergency or Combat Rating procedures after assisting in determining what constituted a suitable test. The result was the 7-1/2 hour test used to establish the five minute emergency rating; the test required a minimum of 5 hours at the intended power level. Table 5-18 shows the increases in rated power achieved by V-1710's on standard Army specification fuel of the day, and in some cases with water-alcohol injection.[127]

Table 5-18
Evolution of Maximum Ratings
(Current Production Engines)

Date	Rated Takeoff, bhp	War Emergency, bhp
March 1941	1150	None
October 1942	1325	1500
January 1943	1425	1600
November 1943	1725	1725

Allison and Wright Field became quite embroiled over the quality of the fuel upon which they were rating the engines. During most of the war the fuel was specification AN-F-28, rated as Grade 100/130. As the war progressed demands for this fuel exceeded the supply of premium feedstocks resulting in a need to use substandard stocks and then supplement them with additional TEL and/or other blending agents. The result was particularly troublesome for the V-1710 in cold weather, for the specified volatility of the fuel was decreased. As a consequence this led to appreciable fuel/mixture maldistribution within the long intake trunk and manifolds. In addition to the distribution concern the heavier "ends" of the fuel blend are naturally the ones containing the most effective anti-knock components. Without complete vaporization of these constituents the engine could easily experience detonation at high power. Allison's response was to introduce the "Madam Queen" air intake pipe that incorporated a venturi to reatomize any liquid fuel collecting in the manifold.[128]

Allison was willing to rate their engines at the maximum detonation free manifold pressure that the available fuel would allow. During the fall of 1943, development of increased WER ratings were somewhat setback by the above mentioned fuel formula change. The later formula caused piston ring sticking within the engine.[129] This was the direct cause of the development of what became known as the "keystone" piston ring. These rings were mandatory on any of the later highly rated engines, particularly the two-stage models.

In March 1944, improved AN-F-33 Grade 115/145 fuel was available for special uses and Allison ran test engines on it that achieved WER ratings of 2000 bhp. During the summer of 1944 some experimental tests were run using the exotic fuel Triptane, PN Grade 200/300. Engines normally rated for operation at 1500 bhp were able to qualify at 2400-2700 bhp on this fuel! Allison ran tests of Triptane with 4.6 cc of TEL/gallon and were able to support detonation free operation without ADI up to 430 psi IMEP at a fuel/air ratio of 0.10. At 3000 rpm in a V-1710, this delivered 2785 ihp, and at 3400 rpm, a total of 3157 ihp.

Performance limits

Any engine is limited in its performance by several interrelated, and often conflicting, requirements. Reliability is the result of the strength of the engines' components as adapted to the specifics of the design and balanced against the stresses imposed upon the thousands of parts. For a given engine it is the stresses related to speed and power that define the amount of time it can operate before failure of the weakest component. Destructive stresses are produced by detonation, so when the absolute maximum amount of power is required it is detonation that defines the limits for the engine. Detonation cannot be tolerated in any engine for more than the briefest of moments.

Onset of detonation is actually determined by the characteristics of the fuel being used, and the density (as defined by the MAP and temperature) of the fuel/air mixture going into the cylinders. Consequently, for a given fuel and mixture temperature, there is a MAP at which detonation begins based upon the pressure being generated in the cylinders. Adjustments are made to provide a margin of safety, thereby establishing the War Emergency Power Rating.

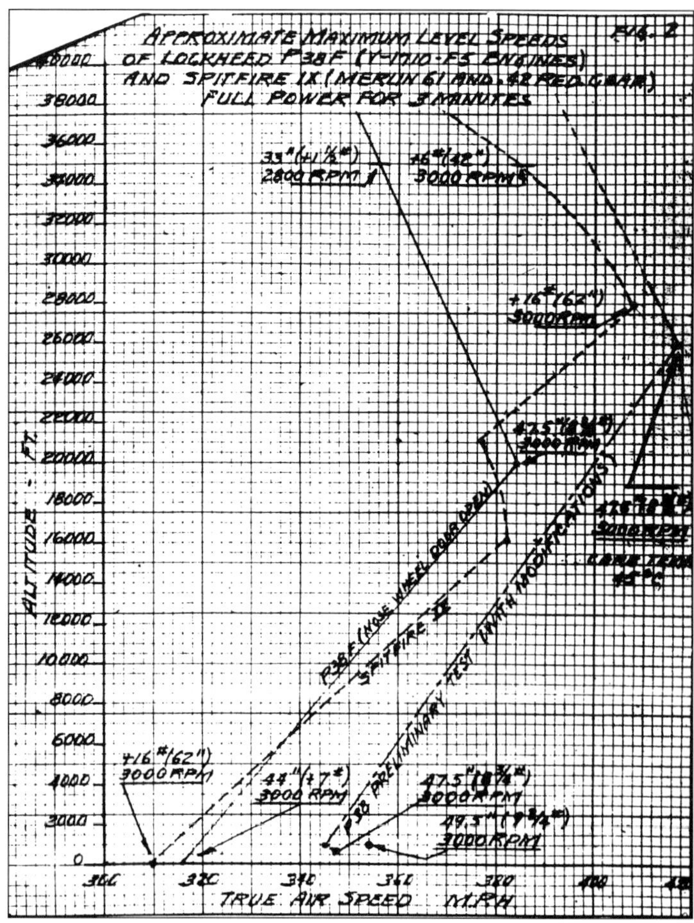

The 8th Fighter Command Technical Section under Col. Kelsey took on the task of improving the performance of the P-38F, intending to increase its speed compared to the Spitfire. Tests were run using the 8th Fighter Command ratings authorized in August 1942, 47.5 inHgA for 1325 bhp for their V-1710-F5 engines. Allison had done supporting work to authorize this rating, though Materiel Command at the time limited these engines to 1150 bhp. The figure shows the effect of the increased power ratings compared to the then new Spitfire Mk IX, the first airplane to be equipped with the new two-stage Merlin 61. (NARA)

Table 5-19
Allison Recommended War Emergency Ratings

	WER, bhp*	MAP, inHgA	Altitude, ft	S/C Ratio	Comments
Altitude Rated:					
V-1710-35, -39	1490	56.0	4,300	8.80:1	<5 min.
V-1710-63, -73	1580	60.0	2,500	8.80:1	<5 min., Stronger Const.
V-1710-81, -83, -85, -99	1410	57.0	9,000	9.60:1	<5 min.
Sea-Level Rated:					
V-1710-51, -55, -89, -91	1600	60.0	Turbo	8.10:1	<5 min.
V-1710-111, -113	1700	70.0	Turbo	8.10:1	P-38L, Grade 150
V-1710-111, -113	2000	75.0	Turbo	8.10:1	P-38L, Grade 150

*From Allison Performance Curves. Generally, listed handbook values are 20-50 bhp less for the same MAP. All values given at 3000 rpm with Grade 100/130 Fuel.

Allison V-1710 engines models were fundamentally all the same, except for their supercharger drive gear ratios. The higher the ratio, the more power required to drive the supercharger at full throttle. Since all models could operate at similar ihp, the differences in WER horsepower ratings shown in Table 5-19 are primarily due to the power consumed by the engine driven supercharger stages. With the introduction of the two-stage engine, higher MAP was required to produce the previous bhp's since more power was being taken from the engine to drive both superchargers. This required that either an intercooler be provided, or as was done for the Bell P-63, water/alcohol anti-detonant (ADI) fluid be injected into the induction mixture to provide a cooling effect and thereby delay detonation to a higher IHP plateau.

The problem facing the engineers at home and pilots in the combat theaters was to determine the power ratings they could safely use without damaging their engines. They did not have the luxury of laboratory facilities and knowledge of the inherent limitations within the engine. All they could do was "push the envelope", carefully. Support by the Army brass in the states to increase ratings was slow during the first year or more of the war. To their credit, Allison had already provided aggressive WER rating recommendations to the Army by the fall of 1942.

In the late summer of 1942 the Eighth Air Force was just getting its feet wet in England, and the issue of performance of U.S. fighter aircraft was of primary importance to everyone involved. The senior on-scene Allison man at the time was R.L. Jahnke, Zone Manager, British Isles, and he reported on a number of official and adhoc programs underway to increase the power available from the V-1710's, primarily those in the P-38F (powered by the V-1710-F5R/L, the primary 8th Air Force fighter of the day).

One of these efforts was under the direction of Major Hough, USAAF Chief of 8th AF Fighter Command, Technical Section, a respected engineer and P-38 pilot. In an attempt to exceed the performance of the new Spitfire Mk IX, the first fighter to be equipped with the two-stage Merlin 61, he had modified the air intakes on the P-38 and improved the streamlining of the guns while using increased power from the engines. His efforts were successful in besting the Spitfire Mk IX.

The Eighth Fighter Command then took things into their own hands, and on August 15, 1942 issued new operating limits for all P-38's in England.[130] These limits were decided in a conference with Col. B.S. Kelsey, then assigned to the 8th AF Air Technical Section at Bovingdon, England.[131] While it was recognized that engine life may have been reduced at the new limits, it was established that this was not of primary importance under the circumstances. Furthermore, if these ratings appeared satisfactory, they were to be systematically raised again if possible.

At the time Allison was also working to establish appropriate WER ratings for the various V-1710 models, an activity directly spurred by the above events in the field. In response to a conference held in Indianapolis on September 5, 1942, Ron Hazen, Allison's Chief Engineer, recommended that the War Emergency Ratings shown above be established.

These ratings were recommended by Allison based upon their previous endurance testing experience with these particular engine models, coupled with the reports and experiences coming in from combat theaters. The lower ratings on the -35 and -39 as compared to the -63 and -73 en-

Table 5-20
8th Fighter Command Established P-38 Combat Ratings

	Takeoff	MAP	Altitude,	Comments
P-38F Handbook	1150	41.5	25,000	From 3-13-1942 to 8-15-42
8th FC, after 8-15-1942	1300	47.0	10,000	No time limit specified
8th FC, after 8-15-1942	1250	45.0	20,000	No time limit specified
Proposed by Col. Kelsey*	1325	47.5	20,000	<5 min w/CAT <50 degC
8th FC, after 8-23-1942*	1250	45.0	25,000	above 15,000'
Future Proposal by 8th FC	1450	52.0	11,000	

* After two months operating with these ratings no engine failures had been reported.

Table 5-21 7-1/2 Hour Approval Test for War Emergency Ratings						
	WER, bhp	MAP, inHgA	rpm	Altitude, ft	S/C Ratio	Comments
V-1710-89/91	1600	60.0	3000	10,000	8.10:1	<5 min., Dry
V-1710-93	1825	75.0	3000	Sea-level	8.10:1	840#/hr ADI,2-stage S/C
V-1710-133	_NA_	100.0	3200	Sea-level	8.10:1*	w/ADI, tested Fall 1946

* Two-stage engine tested on AN-F-33 fuel and ADI. Others based on AN-F-28, Amend #2, PN Grade 100/130 Fuel

gines are based upon the numerous durability related improvements, and safety factors that had been incorporated into production subsequent to when most of the -35 and -39 engines had been built.

It was important that specific indication of use of these ratings be known to the crew chiefs of airplanes that had used WER ratings. Devices such as gates or seals were to be used to provide positive indication that WER ratings had been used. Additionally, the AAF was to relax its assumptions on engine lifetime on affected engines. Furthermore, Allison recommended retaining the boost controls on the affected engines, and that they be reset to maintain the WER values on aircraft requiring combat performance.

It evidently took some time before these recommendations reached the combat forces. While the 8th Fighter Command was working up their own combat ratings, that to their credit were being methodically developed, they did not approach the comparatively aggressive ratings being recommended by Allison. As of the end of November 1942, Allison Service representatives in England advised that no official information was available there for the use of WER on Allison engines, while others reported that in the Middle East automatic manifold pressure regulators were being removed from some V-1710-73 engines (Curtiss P-40K) because there were no operating instructions on their proper operation and adjustment. Service Representatives documented cases where up to 66 inHgA were being used in combat on the V-1710-73. Reports from Australia noted cases were 70 inHgA was being used on the V-1710-73 (P-40K) and -63 engines (Bell P-39D/K/L).[132]

Altitude rated V-1710's were usually provided with short "jet" type exhaust stacks or stubs. They not only directed the exhaust in the desired direction, but provided significant power and thrust at high speeds. At speed and altitude, a Mustang would be obtaining as much as 10 to 14 percent of its net propulsive thrust from these exhaust jets. Allison found that with the introduction of WER ratings their previous recommendations on the size of the ejector jet nozzle was inadequate to pass the additional exhaust flow without creating excessive back pressure and potentially causing detonation. They had previously recommended that stack areas of 3.2 to 3.6 square inches per cylinder were adequate for up to 1325 bhp at takeoff and 1150 bhp up to 12,000 feet. At the new ratings they recommended 4.2 to 4.4 square inches per cylinder. As an expedient, Allison recommended that field forces could simply saw off the ends of the stacks to achieve the desired area.[133] It was the responsibility of each of the airplane manufactures to design and test the shape and function of the stacks on their particular airplanes.[134] Allison only supplied the flange to which the aircraft manufacturer's stack was welded, and even it had to be procured separately from Allison.[135]

In the spring of 1943, as a result of the need to qualify engines for a specific WER rating, Wright Field formalized a "7-1/2 hour Approval Test" to demonstrate satisfactory performance at the claimed level.[136] This development was instituted during the summer/fall of 1943, with the V-1710-89/91(F17R/L) and V-1710-93(E11) being the first engines to officially attempt and achieve the rating.

WER ratings were definitely hard on engines, particularly those in flight testing and development. Consequently, Allison provided the following guidance for such engines as a good safety measure:[137]

- Engines should be removed and replaced at the end of ten hours total time at War Emergency Rating.
- New exhaust spark plugs should be installed prior to any expected WER operation to avoid running plugs which may have previously become partially lead fouled. Such plugs may form lead globules which lead to immediate pre-ignition which causes piston and ring failure.
- Intake plugs should be replaced at the end of one hour of WER operation, or after ten hours of total operation during which at any time, the engine operated at WER ratings.

Summary

With the introduction of WER ratings, the Allison was able to provide performance equivalent to any engine then in service, and exceed that of many. Still the engine was limited in the altitude at which it could produce this power. This was a constraint only in the single-stage installations, and it is necessary to make comparisons to like equipped competitors. When high altitude two-stage equipped aircraft entered the mid and lower altitudes the single-stage V-1710 powered aircraft were their masters when using WER ratings.

Table 5-22
Aircraft Performance at War Emergency Ratings

Airplane	Engine	WER Rating, bhp/Alt/inHgA	Max, mph @ SL	Max at Altitude, mph/Alt/MAP	Max SL Climb, Ft/min	Gross Weight, pounds	Time to Critical Alt, minutes	Service Ceiling, ft
P-38F	V-1710-49/53	1325/25,000/47	NA	395/25,000/47"	NA	15,900	8.8 to 20k'	39,000
P-38G	V-1710-51/55	1325/25,000/47	386	400/25,000/	NA	15,800	8.5 to 20k'	39,000
P-38J-25	V-1710-89/91	1612/10,000/60.8	345	422/25,000/___	4000	17,500	6.19	40,000
P-39D	V-1710- 35	1490/ 4,300/56	NA	368/12,000/	3750	7,500	5.7 to 15k'	32,100
P-39K	V-1710- 63	1580/ 2,500/60	NA	368/13,800/	NA	7,600	5.7 to 15k'	32,000
P-39N	V-1710- 85	1410/ 9,500/57	NA	399/15,000/	NA	7,600	3.8 to 15k'	38,500
P-40B	V-1710- 33	1090/13,200/38.9	NA	352/15,000/___	2860	7,325	5.1 to 15k'	32,000
P-40E	V-1710- 39	1490/ 4,300/56	NA	366/15,000/___	2050	8,280	7.6 to 15k'	29,000
P-40F	V-1650- 1	1435/11,000/63	NA	364/20,000/___	3250	8,500	7.6 to 15k'	34,400
P-40K	V-1710- 73	1580/ 2,500/60	NA	370/20,000/___	2000	8,400	7.5 to 15k'	28,000
XP-40Q-2	V-1710-121	1700/26,000/__	NA	422/______/___	NA	NA	NA	NA
XP-47J	R-2800- 57	2800/__/72Wet	NA	507/34,300/___	4900	NA	6.75	NA
A-36A	V-1710- 87	1500/ 3,500/52	366	368/14,000/35"	NA	8,370	NA	27,000
P-51A	V-1710- 39	1330/10,000/57	374	409/20,000/52"	NA	8,600	10.1 to 25k'	34,000
P-51B/C	V-1650- 3	1330/23,500/67	NA	431/30,000/44"	NA	9,800	9.1 to 25k'	42,000
P-51D/K	V-1650- 7	1505/19,300/67	NA	438/25,000/43"	NA	10,100	9.6 to 25k'	41,900
P-51H	V-1650- 9	1630/23,500/80Wet	NA	487/25,000/77"	NA	9,500	9.1 to 25k'	41,600
XP-51J	V-1710-119	1900/SeaLev/78Dry	424	491/27,400/	NA	7,550	3.6 to 20k'	43,700
XP-58	V-3420-11/13	3000/28,000/51.5	NA	436/25,000	NA	39,192	12.1 to 25k'	38,400
XP-63A	V-1710- 47	NA	NA	421/24,100	NA	7,525	5.8 to 20k'	45,500
P-63A	V-1710- 93	1825/SeaLev/75Wet	NA	410/25,000/	NA	8,800	7.3 to 25k'	43,000
P-63C	V-1710-117	1800/24,000/76Wet	NA	410/25,000/	NA	8,800	8.6 to 25k'	38,600
P-63F	V-1710-133	2200/SeaLev/80Wet	NA	NA	NA	NA	NA	NA
XP-82	V-1650-23/25	2270/ 4,000/90Wet	NA	468/22,800	4900	19,100	6.4 to 25k'	40,000
P-82E	V-1710-143/5	2250/SeaLev/100Wet	NA	465/21,000	4020	20,684		40,000
Fw 190A-3	BMW 801D-2	1700/SeaLev/	312	418/21,000	2830	8,770	12 to 26k'	34,775
Bf 109F-4	DB 601E-1	1300/18,045/	321	358/ 9,840	3860	6,393	5.2 to 16k'	36,090
Bf 109G-2	DB 605A-1	1355/18,700/	317	406/28,540	4590	6,834	7.6 to 33k'	39,370
Spitfire VB	Merlin 45	1350/12,000/62.5	332	357/19,500	4750	6,750	7.0 to 20k'	35,500
Spitfire IX	Merlin 66	1580/16,000/67	312	408/25,000	4100	7,500	5.7 to 20k'	43,000
Spitfire XIV	Griffon 65	1935/15,500/67	357	439/24,500	4580	8,375	7.0 to 20k'	43,000

Notes:
1. Allison V-1710 is a liquid-cooled, supercharged, upright V-12 of 1710 cubic inch displacement.
2. BMW 801 was an air-cooled, supercharged two row radial with 14 cylinders and 2551 cubic inch displacement.
3. DB 601 was a liquid-cooled, supercharged, inverted V-12 of 2071 cubic inch displacement.
4. DB 605 was a liquid cooled, supercharged, inverted V-12 of 2182 cubic inch displacement.
5. Rolls-Royce Merlin is a liquid cooled, supercharged, upright V-12 of 1650 cubic inch displacement.
6. Rolls-Royce Griffon is a liquid cooled, supercharged, upright V-12 of 2239 cubic inch displacement.
7. It is often not known what power settings were used to establish maximum performance, but it can be assumed that maximum available was used.Often this would be less that the given WER values, particularly when above the rated WER altitude.
8. Then it can be assumed that engine was at full throttle and making maximum available power. Time to Climb to Critical Altitude done using best climb power, not WER. V-1710 climbing at 2600 rpm, V-1650 climbing at 2700 rpm.

[1] Shamburger, Page and Joe Christy, 1972, 248.

[2] Lockheed's P-38 as designed by Kelly Johnson was a classic case of serendipity. Johnson was a designer uncharistically founded in the analytical and practical aspects of aerodynamics, structures and manufacturing, all wrapped in the classic American work ethic. After graduation from the University of Michigan in 1932 he was unable to find a job in aviation. Taking the advice of Detroit Aircraft's (predecessor to Lockheed Aircraft) then Chief Engineer Richard von Hake, Kelly returned to the graduate program at the University of Michigan and obtained a Master of Science degree. Significantly, his major was in supercharging of engines and boundary layer control! Both of these subjects were obvious in his early work at Lockheed, evidenced on the Electra and the experimental XC-35, the first pressurized fuselage aircraft. It used exhaust driven turbosuperchargers to pressurize the cabin. In his design of the XP-38 he clearly had no qualms about including these two new technologies, along with his skill with advanced aerodynamics. See Johnson, C.L."Kelly" with Maggie Smith, 1985, 20.

[3] There is some confusion within the CP 39-775 documents regarding the designation of this airplane. Lockheed identified it as their model 222-A, but the Air Corps always refers to it as the Model 222-B.

[4] Aviation Historian Birch Matthews has investigated early Lockheed Model Numbers and offers that the basic aircraft model number was followed by a number identifying the engine model used. This was later upgraded with a third number representing the interior arrangement. The XP-38 (22-64-02) was the basic Model 22 equipped with the Allison C-9(64). Lockheed designated the Allison engines as; 60(F5R/L), 61(C15), 62(F2R/L), 63(XV-3420-1), 64(C9), 65(F3R), 68(F10R/L), 69(F4), 81(F17R/L), 85(F15R/L), 86(V-3420-A16R/L), 87(F30R/L).

[5] Bodie, Warren M., 1991, 246-251.

[6] The critical altitude of a supercharger is limited primarily by the speed at which it is turning. The GE Type B turbos used in the P-38 were speed limited by the strength of the materials used in the high temperature turbine wheel. In 1942 GE introduced a new material that allowed rating at higher speeds, hence raising the critical altitude for the installation, and the P-38.

[7] *Operating Instructions for Allison V-1710-49 and -53 Engines Installed in the Lockheed P-38F Airplane*, Letter from Col. F.O. Carroll, Chief Experimental Engineering Section, Wright Field to Allison, October 23, 1942. NARA RG 342, RD 3774.

[8] Engines for aircraft purchased directly by foreign governments were considered as "commercial" sales and as such did not receive or use the US Air Corps engine model number, even though the engines might be identical.

[9] Francillon, Rene J. 1982, 162.

[10] *Discontinuance of Allison "C" Engine Production to Improve Production of "E" and "F" Models*, memorandum for Mr. Robert A. Lovett, Assistant Secretary of War for Air from the Chief of the Air Corps, June 21, 1941. NARA RG18, file 452.8, Box 807.

[11] Unfortunately there is not unanimity in the airframe and model count, even between well researched sources. In his authoritative The Lockheed P-38 Lightning, Warren Bodie found 22 P-322-I's and 121 Model P-322-II's, as well as a total of all models at 9,925 P-38's built by Lockheed.

[12] Francillon, Rene J. 1982, 166.

[13] *War Emergency Ratings for Allison V-1710-49, -51, -53, -55 and -87 Engines*, Confidential Allison letter to Commanding General AAF, Materiel Center, Wright Field, November 12, 1942.

[14] Bodie, Warren M., 1991, 194.

[15] Francillon, Rene J. 1982, 167.

[16] Not everyone realizes that for liquid-cooled engines, a good rule of thumb is that the engine cooling system will have to remove heat equivalent to that going into useful work, about 20 percent of the energy supplied by the fuel. Another 10 percent is removed by the oil cooler, and the balance, 50 percent, goes out as heat in the exhaust gasses. This is the reason why turbosupercharging is so beneficial, for it relieves the engine of doing a major portion of the work of compression while using energy otherwise going to waste.

[17] *Recent Engine Failures*, record of telegram from Commanding General 4th Air Force to Materiel Command, January 2, 1943. NARA RG18, File 452.13, Box 1260.

[18] Production of the 225 P-38H-1's ended in August 1943 and the 375 P-38H-5's begun. At the same time limited production of P-38J-5's began, with P-38J-10 production beginning in October 1943. Both P-38H-10's and P-38J-10's were being produced through December. Bodie, Warren M., 1991, 194.

[19] Francillon, Rene J. 1982, 167.

[20] Sea Level static pressure is 29.92 inHgA. General Electric designed most of their turbosuperchargers to deliver 31-1/2 inHgA at the critical altitude, which was determined by the maximum speed capability of the turbine wheel. The approximately 1-1/2 inHg over sea level pressure was to compensate for the pressure losses through the ducting, intercooler and carburetor. So that 1600 bhp could be delivered by the engine it was necessary to increase the speed of the turbo so that the entire 31-1/2 inHgA was available **after** losses through the system had been overcome. GE had made sufficient improvements in the turbo so that the higher speed could be provided, and Lockheed could now take advantage of the full capability of the V-1710-F17 WER rating.

[21] *War Emergency Ratings for V-1710-89, -91 and -93 Engines*, Confidential Allison letter, Chief Engineer R. Hazen to Commanding General AAF, Materiel Center, Wright Field, November 14, 1942.

[22] *Pilots Flight Operating Instructions for P-38H Series, P-38J Series, P-38L-1, L-5 and F-5B Airplanes.*

[23] As viewed from the pilots seat, facing foreword, the propellers turned "outboard".

[24] Bodie, Warren M., 1991, 199.

[25] Tony Le Vier, Lockheed Engineering Test Pilot, filed the following in a report on his four-month mission to England, January-June 1944, during which he helped introduce the new P-38J.

[26] During the winter of 1943 Allison was hard at work on the problem, and claimed that recent changes in the specification for AN-F-28 Grade 100/130 fuel had allowed use of heavier components to obtain the desired Grade, but in the process a considerable reduction in fuel volatility resulted. Consequently during cold weather operations fuel would condense in the manifold and considerably magnify the fuel distribution problems between cylinders. This new fuel also caused ring sticking. The fuel change resulted in Allison developing the "Madam Queen" air intake pipe with its venturi to revaporize any condensed fuel, as well as the "Keystone" piston ring. See R.M. Hazen, Allison Chief Engineer, letter #BC7C19AE, to H.M. Poyer, Director of Engineering, Bell Aircraft Corporation, March 19, 1943. NARA RG18, File 452.13, Box 1260.

[27] According to Kelsey, who was at the time commanding the 8th Air Force Operational Engineering Section, the V-1710 seemed to behave poorly with the aromatic fuels then available in England. Kelsey, Benjamin S. B.G. USAF (Ret.) 1982, 134.

[28] From the Lockheed published LADC *Star* of May 19, 1944.

[29] The V-1710-F17 engines in this airplane were rated for Takeoff and Military as 1425 bhp at 3000 rpm and 54 inHgA. LeVier is evidently referring to the War Emergency Rating of 1600 bhp. This was available from 3000 rpm (3200 rpm was not authorized for the F-17) and 60 inHgA.

[30] Kline, John 1994.

[31] Bodie, Warren M., 1991, 199.

[32] Doolittle, J.H. with Carroll V. Glines, 1992, 373.

[33] Teletype TSO 18 from Caserta, Italy, February 2, 1944. NARA RG18, File 452.131-A, Box 756.

[34] Classified Message Center NRSC 0379 from Allied AFSC MTO to Patterson Field, March 13, 1944. NARA RG18, File 452.131-A, Box 756.

[35] *Supercharger Seal Leakage in V-1710 Engines*, communication from Chief, Engineering Division to Commanding General, AAF, April 10, 1944. NARA RG18, File 452.131-A, Box 756.

[36] General H.H. Arnold, *Summary of P-39 Combat Experience*, Routing and Record Sheet, June 30, 1942. H.H. Arnold manuscript, Roll 165, Library of Congress.

[37] Matthews, Birch. 1996, 161.

[38] Matthews, Birch. 1996. Contract quantities were 170, 205 and 300 for the 675 total. These quantities were not included in the other P-39 production totals.

[39] Matthews, Birch. 1996.

[40] Green, William and Gordon Swanborough, 1977, 6.

[41] Mitchell, Rick, 1992, 17-18.

[42] Comments provided by Birch Matthews in correspondence with author, September 30, 1995.

[43] Technical Report No. 5022, 1943.

[44] Memo to Chief of the Army Air Forces, *Supply of 33rd Pursuit Squadron at Raykjavik, Iceland.* H.H. Arnold manuscript, Roll 129, Library of Congress.

[45] Experimental Engineering Section, CONFIDENTIAL Memorandum, 1942, 2.

[46] Technical Report No. 5022, 1943, 18.

[47] Angelucci, Enzo with Peter Bowers, 1985, 47.

[48] Wright Field Files D52.41/114 and D52.41/125, *Model Testing Allison V-1710-61.* Files at NASM, Silver Hill.

[49] *Description of Heavily Armored P-39*, Bell Report 26-922-001, 3-19-1943.

[50] Mitchell, Rick, 1992, 30-36.

[51] Angelucci, Enzo with Peter Bowers, 1985, 41-47.

[52] Five V-1710-25(F1) engines were upgraded at delivery to a similar number of V-1710-39(F3R)'s, all on contract W535-ac-13841.

[53] Shamburger, Page and Joe Christy, 1972, 248.

[54] *List of Curtiss-Wright Reports Pertaining to P-40 Series Weight and Performance*, Curtiss-Wright Document available from SDAM.

[55] Technical Order 01-25CF-1, 1943.

[56] Shamburger, Page and Joe Christy, 1972, 248.

[57] Memo to General Harmon, *Air Cleaners*, from Lt. Col. F.I. Ordway, Jr., April 22, 1942. H.H. Arnold manuscript collection, Roll 129, Library of Congress.

[58] Memo to General Harmon, *Air Cleaners*, from Lt. Col. F.I. Ordway, Jr., April 22, 1942. H.H. Arnold manuscript collection, Roll 129, Library of Congress.

[59] Correspondence with aviation historian Hal Andrews, 3-10-1996.

[60] Correspondence with aviation historian Hal Andrews, 3-24-1996.

[61] Shamburger, Page and Joe Christy, 1972, 212-213.

[62] *War Emergency Rating Operating Instructions for Allison V-1710-35, -39, -63, -73, -81 and -85 Engines*, Allison letter, Chief Engineer R. Hazen to Commanding General AAF, Materiel Center, Wright Field, December 14, 1942.

[63] Shamburger, Page and Joe Christy, 1972, 217.

[64] Shamburger, Page and Joe Christy, 1972, 217.

[65] Historian Hal Andrews had photos of last P-40N (AAF44-47968) on the production line, not 44-47964 as reported in Shamburger, Page and Joe Christy, 1972, 250.

[66] *Case History of the V-1650 Engine*, 1945. Transcript of Conversation between Colonel Sessums and Colonel Johnson, March 11, 1943.

[67] Letter 6-161, dated February 27, 1943, from Chief of Staff to B.E. Meyers, Brig. General, U.S.A., Maxwell AFB Archives microfilm roll #A2073, frame 1106.

[68] Shamburger, Page and Joe Christy, 1972, 251.

[69] Bowers, Peter M. 1979, 492.

[70] Teletype PROD-T-665 from Materiel Division, Wright Field, dated 6-12-41.

[71] Teletypes PROD-T-719 and PROD-T-732 from Materiel Division, Production Engineering Branch, Wright Field, dated 10-17-41.

[72] Correspondence with aviation historian Hal Andrews, 3-24-1996.

[73] Bowers, Peter M. 1979, 437.

[74] Bodie, Warren M. 1994, 140.

[75] *P-51B Aircraft*, memo from Requirements Division to Materiel Division, July 6, 1943. NARA RG18, File 452.1, Box 698.

[76] North American converted the first aircraft, designated as P-51-1-NA, the balance were depot modified and designated as P-51-2-NA. The aircraft was known as "*Apache*" until July of 1942, after which it was the "*Mustang*". These aircraft were subsequently redesignated as F-6A.

[77] Gruenhagen, Robert W. 1980, 195.

[78] Gruenhagen, Robert W. 1980, 61.

[79] Gruenhagen, Robert W. 1980, 196.

[80] Telegram from Wright Field Power Plant Laboratory to Allison, April 12, 1945. NARA RG342, RD3775, Box7411.

[81] Letter from J.L. Atwood, Vice President of North American Aviation, to O.E. Hunt, General Motors Corporation, February 236, 1941. NARA RG18, File452.8, Box807.

[82] *History of the Mustang P-51 Aircraft*, Memo Report from Thomas Hitchcock, Assistant Military Air Attaché, Army Air Corps, London, October 8, 1942. NARA RG18, File 452.13, Box 756.

[83] Birch, David, 1987, 64.

[84] *P-51B Airplane, Design Study Around the Allison V-1710-93 Engine*, letter from Assistant Chief of Air Staff to Commanding General, Materiel Command, September 1, 1943. NARA RG18, File 452.1, Box 698.

[85] Letter from O.E. Hunt, Executive Vice President, GMC, to Brigadier General C.E. Branshaw, September 8, 1943. NARA RG18, File 452.13-M, Box 756.

[86] Letter from Chief, Power Plant Laboratory, Wright Field, to Allison Division, September 25, 1943. NARA RG18, File 452.13, Box 1260.

[87] General H.H. Arnold, *Summary of P-39 Combat Experience*, Routing and Record Sheet, June 30, 1942. H.H. Arnold manuscript, Roll 165, Library of Congress.

[88] *Procurement Activities of the Air Corps, Memorandum for the Chief of Staff*, War Department, 1942. See H.H. Arnold manuscript, AAF Plans Folder, Roll 199, Library of Congress.

[89] Angelucci, Enzo with Peter Bowers, 1985, 52.

[90] Experimental Engineering Section, CONFIDENTIAL Memorandum, 1942, 2.

[91] Technical Report No. 5022, 1943, 12.

[92] Letter from Major General H.H. Arnold, Chief of the Air Corps, to Mr. William Knudsen, President of General Motors regarding the *Relative Horsepower Output of the Present Allison engine and the Rolls-Royce Merlin*, September 11, 1939. See *V-1650 Case History*, Maxwell AFB, Microfilm Roll #A2073.

[93] In the late fall of 1943 Allison was aggressively working on an intercooler for the two-stage V-1710-93 for the P-63. Several intercoolers were fabricated and tested, both with and without an engine. Considerable difficulty was encountered getting the Materiel Division to process the fabrication orders and contracts to procure the specially built units from the Harrison Radiator Division of General Motors. Release of production orders had still not occurred as late as May 22, 1944, though testing had been showing the expected good results. See *Progress Reports on Two-Stage Allison V-1710 Engine Project*, NARA RG 342, RD 3971, Box 8146.

[94] Comments provided by Birch Matthews in correspondence with author, September 30, 1995.

[95] Matthews, Birch. 1996.

[96] Comments provided by Birch Matthews in correspondence with author, September 30, 1995.

[97] Comments provided by Birch Matthews in correspondence with author, September 30, 1995.

[98] Technical Report No. 5022, 1943.

[99] Bell Report 24-947-001-1 dated 10-15-1942, Supplement 1, Summary of Change Orders, No. 6465 #2, 2-24-42.

[100] Angelucci, Enzo with Peter Bowers, 1985, 52.

[101] Experimental Engineering Section, CONFIDENTIAL Memorandum, 1942, 2.

[102] Technical Report No. 5022, 1943.

[103] Angelucci, Enzo with Peter Bowers, 1985, 52.

[104] Matthews, Birch. 1996, 187.

[105] Comments provided by Birch Matthews in correspondence with author, September 30, 1995.

[106] Engineering Division Memo Report, Serial No. ENG-57-503-1217, 1944, 10.

[107] Hickman, Ivan, 1990, 75.

[108] *Engines for P-63A-1 Airplanes*, memo to Commanding General, AAF, Operations from Chief, Aircraft Distribution Office, October 21, 1943. NARA RG18, File452.13, Box 1260.

[109] *Case History of the V-1650 Engine*, 1945.

[110] Pressure loss as indicated by a decrease in the height of a column of mercury the pressure would support, by say 2 inches. Pressure indications in "inHg" can be converted to pressure in terms of "pounds per square inch" by multiplying by the ratio of standard sea-level pressure: 14.69 psi/29.92 inHgA. A 2 inHg pressure drop is approximately a 1 psi pressure decrease.

[111] Technical Order AN 02-5AD-2, 1945.

[112] Bell Aircraft Corp., Niagara Frontier Division, First Issue of Confidential Management Letter, March 7, 1945.

[113] Comments provided by Birch Matthews in correspondence with author, October 14, 1995.

[114] Waag, Robert J. 1973, 16-23.

[115] Waag, Robert J. 1973, 16-23.

[116] Comments provided by Birch Matthews in correspondence with author, October 14, 1995.

[117] Bell Aircraft Corp., Niagara Frontier Division, Confidential Management Letter of April 14, 1945.

[118] Matthews, Birch. 1996.

[119] Waag, Robert J. 1973, 16-23.

[120] Hickman, Ivan, 1990, 62-64.

[121] This engine is often incorrectly identified as the V-1710-135(G4).

[122] Comments provided by Birch Matthews in correspondence with author, October 14, 1995.

[123] Mormillo, Frank B. 1982, 64-67.

[124] In deference to the outstanding Lockheed P-38 design team, the aircraft was only intended to have 1150 bhp engines, and they designed an elegant intercooler into the structure of the wing. When larger engines became available the choices were to either enlarge the wing or restrict the power rating. They finally elected to redesign the nacelles to incorporate core type intercoolers, and further turned adversity to success by turning the wing leading edges into fuel tanks.

[125] Allison Engines Under Wartime Conditions, 1942.

[126] Allison Engines Under Wartime Conditions, 1942, 2.

[127] Jahnke, R.L. 1942.

[128] Hazen of Allison letter to Mr. H.M. Poyer, Director of Engineering, Bell Aircraft Corporation, March 19, 1943. NARA RG18, File 452.13, Box 1260.

[129] *Development of Increased Horsepower and Altitude for the Allison V-1710 Engines (Particularly for N.A.A. P-51 Airplane)*, Allison letter C1L2AE from Hazen to Commanding General, AAF Materiel Command, November 2, 1943. NARA RG18, File 452.13, Box 1260.

[130] Jahnke, R.L. 1942.

[131] Col. Ben Kelsey and Major Cass Hough were both highly qualified pilots and engineers who played major roles in the development and successful combat deployment of the P-38, both were among the first to fly P-38's directly to England in Operation BOLERO. Once there B/Gen. Frank Hunter, commander of the new 8th Fighter Command asked them to setup an Air Technical Section at Bovingdon, an assignment that began in January 1943, and was important to the innovations and tactics that were developed for the P-38. Bodie, Warren M. 1994, 134.

[132] *War Emergency Rating Operating Instructions for Allison V-1710-35, -39, -63, -73, -81 and -85 Engines*, Allison letter, Chief Engineer R. Hazen to Commanding General AAF, Materiel Center, Wright Field, December 14, 1942.

[133] Hazen, R.M. 1943.

[134] Hazen, R.M. 1943.

[135] Allison Model Specification No. 230, 1943, paragraph E-32a.

[136] Materiel Command Memorandum Report No. ENG-57-503-848, 1943.

[137] *Proposed Time Limit for War Emergency Rating Operation on Engines Operated in United States*, Allison letter, Chief Engineer R. Hazen to Commanding General AAF, Materiel Center, Wright Field, February 19, 1944. NARA RG 342, RD 3774.

6

Odds and Mods Powered by the Allison V-1710

The Allison V-1710 was developed to be mass produceable for a wide variety of different installations and arrangements. As such it was both available and adapted for use in a wide range of aircraft, some quite unique. A good share of the impetus for these developments came from Wright Field, where the Army was instigating a number of interesting advanced concepts in 1934/35. Significant to this effort was initiation of the advanced bomber Project 'D' on February 5, 1935, followed by issuance of Type Specification X-203 to a limited number of aircraft manufacturers. The breakout of the projects is given in Table 6-1.[1]

In this chapter we investigate installations that resulted in development of flight hardware, but never achieved series production status.

Table 6-1
Early Experimental Bomber Projects

Restricted Project	Description
M-2-35	Development of 1600 bhp Allison X-3420 for the Project 'D' Airplanes
M-3-35	Project 'A' airplane, the V-1710-3 powered Boeing XBLR-1 (XB-15)
M-4-35	Project 'D' airplane that became Sikorsky XBLR-3, not built
M-5-35	Project 'D' airplane that became Douglas XBLR-2 (XB-19)

Bombers

The early 1930s were a time when the streamlining of vehicles was a passion. Cars, trains and planes were the obvious benefactors, but even buildings and appliances were afforded the treatment. One concept applied to aircraft was that of burying the engines within the wing and/or fuselage, with extensions shafts to drive propellers streamlined to the wing for minimum drag. The compact liquid-cooled engine was believed to be particularly adaptable to such installations.

Concurrently the "bomber boys" were dominating the development of military airpower, and describing very long range bombers with supercharged engines able to fly higher and faster than any opponent. These theories were to be brought to reality in the mid-1930s in the form of the Boeing XB-15 and Martin XB-16 bombers.

At the same time the Army bought their first V-1710, the XV-1710-1(A2) in 1932, they also specified the XV-1710-3 for bombers. Like the XV-1710-1, this engine was to be designed for 1000 bhp, but initially rated at 750 bhp while retaining the 3:2 reduction gear of the GV-1710-A1, though otherwise improved. The engine was never contracted nor developed, instead the focus shifted to the 1000 bhp "C" pursuit and "D" extension shaft models. The XV-1710-3 would not have had the streamlined nosecase, as it, as well as latter engines, were intended to be suitable for buried installations. Both the XB-15 and XB-16 were initially intended to be powered by this engine. In the event, the delays in completing the V-1710 Type Test at

For years the Boeing XB-15 was the largest airplane in the world. It was to have been powered by Type Test delayed1000 bhpYV-1710-3 engines. Note the similarity to the Martin Model S-145-1. The air-cooled radial replacement engines were never able to deliver as much power as the large airplane needed. (Hal Andrews)

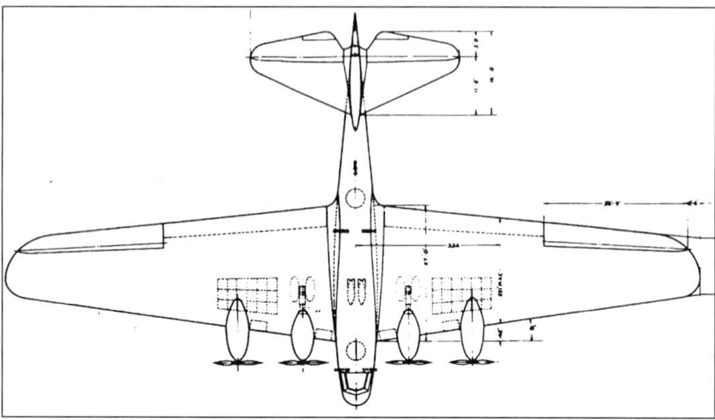
The Air Corps provided Martin with a description and drawing of this B-15 bomber and asked Martin to submit a proposal. This was done as their Model S-145-1 proposal, dated June 12, 1934. (Glen L. Martin Museum via Dave Ostrowski)

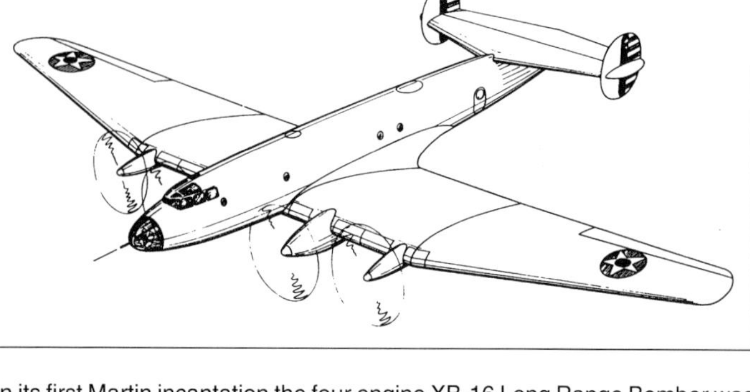
In its first Martin incantation the four engine XB-16 Long Range Bomber was a very attractive airplane. Notice the streamlined nacelles, a result of having the turbocharged V-1710 engines buried within the wing. (Glen L. Martin Museum via Dave Ostrowski)

1000 bhp resulted in the XB-15 being switched to air-cooled radial engines and the XB-16 being canceled outright. Even so, the military planners continued to consider ways to achieve the advantages of the liquid-cooled engine in a bomber. The following Bomber projects were built around the Allison V-1710.

Boeing XB-15

The V-1710 was originally intended as a bomber engine. When the initial build of the XV-1710-3(C1) passed its fifty hour experimental test at 1000 bhp in the fall of 1934, it had already been selected as the power plant for the Air Corps Project "A" bomber, the Boeing XB-15, largely because it was the only 1000 bhp engine available.[2] Because of the continuing problems in satisfactorily completing the 150-hour Type Test at 1000 bhp with that engine, the Materiel Division began studies in November of 1935 to substitute P&W R-1830-B engines, although they resulted in a reduction in the performance of the airplane.[3] The replacement R-1830-11's were rated for 1000 bhp at takeoff, and 850 bhp at 5,000 feet, all when using 100 octane fuel.[4]

Martin XB-16

This was a Secret and highly important Air Corps project in the early 1930s that like the Boeing XB-15, was designed to Project Specification X-200. While many of the details remain sketchy, recently obtained drawings from 1934 by the Glen L. Martin Company provide some interesting background. It appears that Wright Field held a competition for production of the B-15. Martin drawing S-145-1 notes that "This was not a Martin design but was traced directly from an Air Corps drawing and submitted." The drawing shows a 63,351 pound airplane with a 2,700 sq.ft. wing strikingly similar to the airplane that later flew as the Boeing XB-15. At the date of the drawing the four engines were rated for 1000 bhp up to 7,000 feet. The reduction gear ratio was to be "approximately 0.6:1." The Allison XV-1710-3 was to be rated for 1000 bhp, and without a turbosupercharger would have

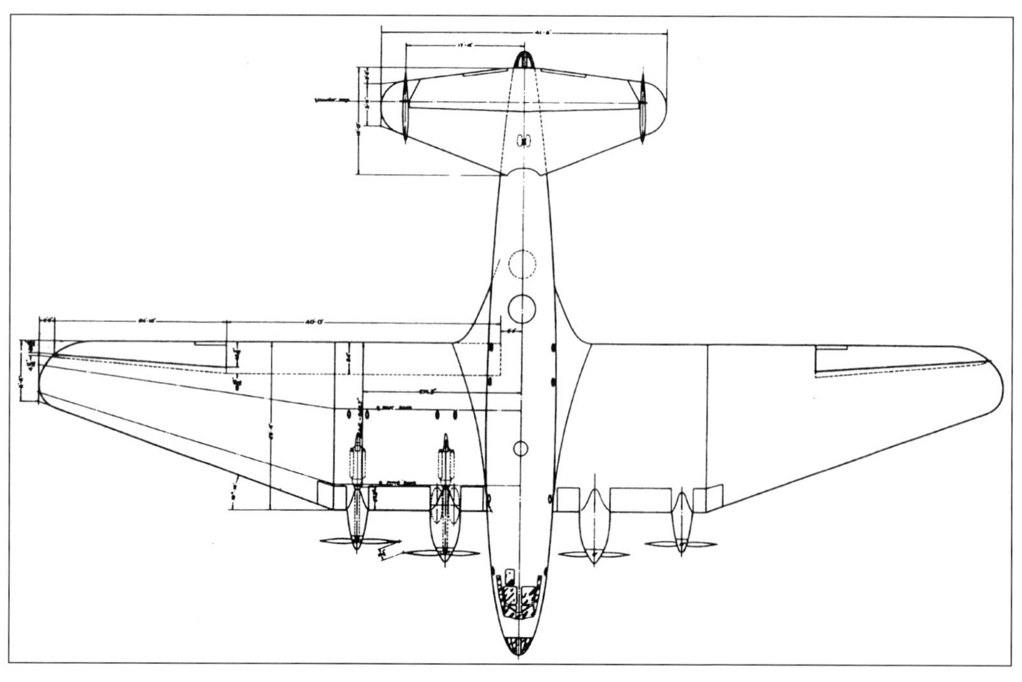
Martin buried the V-1710-3's in the 19.3% thick wing and Allison was to provide propeller speed extension shafts approximately 8 feet long to reach the propellers on the Model 145-A XB-16. (Glen L. Martin Museum via Dave Ostrowski)

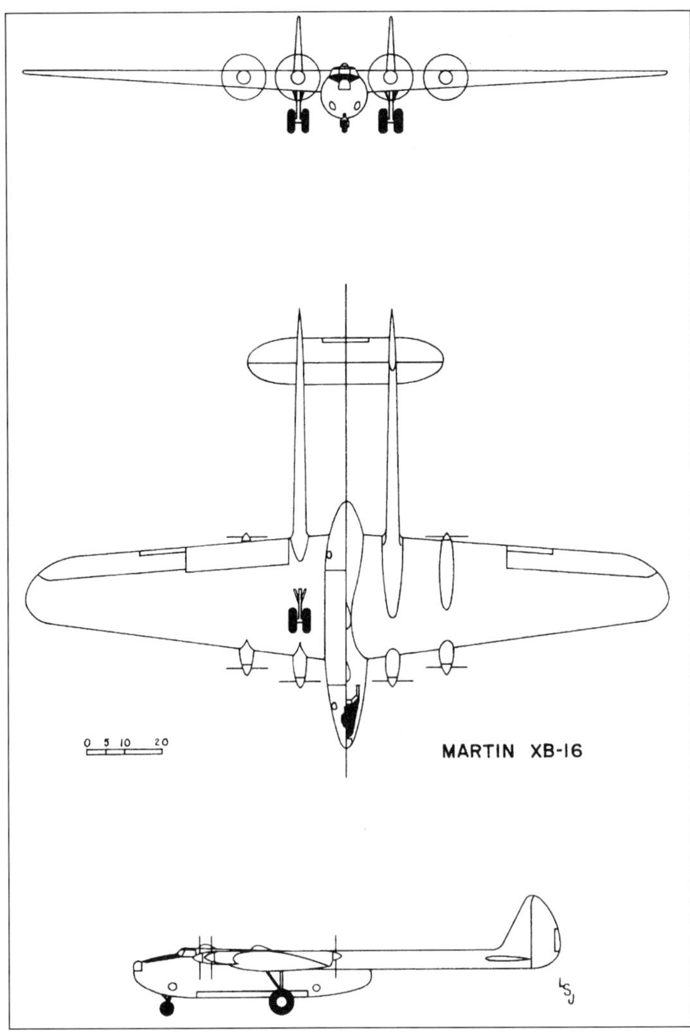

The Martin XB-16 design continued to grow and finally was to require six YV-1710-3's to get its bulk into the air. (Jones)

had a critical altitude of about 7,000 feet. Its gear ratio would have been equivalent to 0.67:1.

Feedback from the Army must have been encouraging for Martin went ahead with a design of their own. That airplane was identified as the Martin Model 145-A Long Range Bomber, and had already received the designation XB-16 when first shown on drawing S-145-1A, dated November 27, 1934.

The airplane was to be a high-speed, long range, heavy bomber utilizing four tractor type 1000 bhp V-1710-3 engines buried entirely in the wings and driving 12 foot 3 inch diameter propellers via extension shafts. Type F turbosuperchargers gave a critical altitude of 20,000 feet, where the aircraft was to have a top speed of 237 mph and an endurance of 42 hours when cruising at 120 mph with its crew of ten. This would have been a large, conventionally configured aircraft, having a wing span of 140 feet offering 2,600 square feet of lifting area for a gross weight of 65,000 pounds.[5]

As is often the case during the development phase, the aircraft grew and became a quite different configuration and airplane. The final design shows an aircraft with a wing span of 173 feet, distributed to provide 4,256 square feet of area to lift the new gross weight of 104,880 pounds. Additional power was also required, which was obtained by burying two more

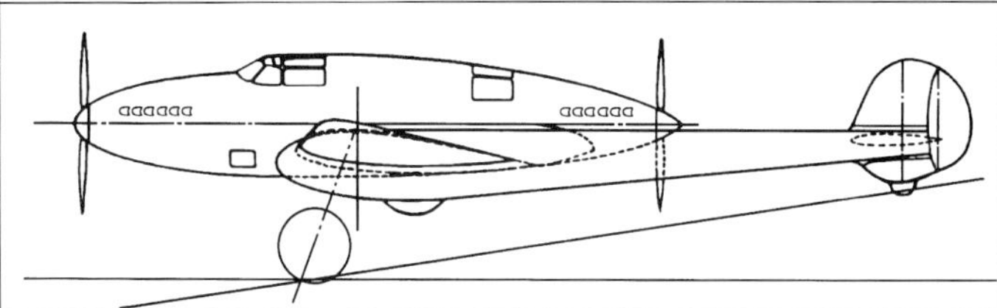

Figure 6-1 In 1936 Martin proposed their twin boom Model 151K be powered by tractor/pusher V-1710-C engines rated for 1000 bhp at 2800 rpm. (Glen L. Martin Museum via Dave Ostrowski)

V-1710-3 engines in the large wing, arranged as pushers and positioned directly behind the outboard tractors. The configuration was also dramatically changed, adopting tricycle landing gear and long slender booms to support the tail. The resulting configuration was somewhat like the much later Fairchild C-119 "Boxcar" in appearance.

Although the engineering design of the XB-16 was purchased by the Air Corps, the decision was made not to proceed with construction of the actual aircraft. Many factors such as the availability of engines and the effects of the Depression probably had a lot to do with the decision to cancel the project. The Boeing B-17 was the actual winner over the XB-16, not the XB-15.[6]

Martin Twin Bomber

The 1930s were a prolific period for proposals of new aircraft. All of the manufacturers kept their designers busy trying to best the competition and get the attention of the Army or Navy. One of Martin's preliminary 1936 designs was their Model 151K-1. As shown in the Figure 6-1 sketch, the light twin-engined bomber was to use early V-1710-C engines in a tractor/pusher configuration. The available information suggests it was an unsupercharged installation mounting 1000 bhp engines rated at 2800 rpm. This would be equivalent to the rating of the V-1710-C4/C6. The airplane was to have an empty weight of 15,000 pounds, with an overload gross weight of 30,000 pounds. It is entirely possible that this airplane was intended as a competitor in the XFM-1/2 competition.

Vega XB-38

As noted above, the major reason for initial Air Corps interest in the V-1710 was to power high altitude bombers. In the early 1930s it was believed that only high-temperature liquid-cooled engines would allow turbosupercharging for high power at altitude, exactly the objective of the

Vega built the XB-38 from an early B-17E. The V-1710-89's made for a most attractive airplane photographed here under Southern California camouflage netting. (Bodie)

The XB-38 engine installation on this test run appears simple due to there being no turbo or exhaust ducting. Note the unique structural support for the engine, and the lack of any attempt to streamline the nacelle firewall to take full advantage of the V-12 form. (Bodie)

The Vega XB-38 in its finished form was an attractive aircraft from every angle. The high powered and streamlined V-1710-89's made it a very streamlined aircraft. (SDAM)

bomber strategists in the Air Corps. Streamlined installations were simply a benefit. The four-engined Boeing XB-15 and Martin XB-16 bombers were to be powered by the most powerful engine available, the 1000 bhp Allison V-1710-3. Delays in completing the V-1710 Type Test meant canceling the XB-16 and switching the Boeing bomber to air-cooled radials. While the liquid-cooled engine was missing from the ranks of this new generation of heavy bombers the interest in using, the V-1710 as a bomber engine remained. It finally became a reality in the XB-38.

The XB-38 resulted from the Wright Field MX-234 Classified project and was conceived as a possible answer to a number of questions facing the Army Air Corps prior to U.S. entry into WWII in late 1941. Specifically, they were very interested in increasing the performance of the Boeing B-17, particularly with respect to takeoff, climb, service ceiling and top speed. In addition, there was considerable interest in comparing liquid-cooled versus the standard air-cooled engine installations in the same type airframe, it being argued that liquid-cooled engines provided better cooling at high altitudes. Another issue of the day was the desire to provide an alternate source of engines for the critical B-17 program, a bomber then being ordered in unprecedented numbers. As a follow-on, in April/May 1942 Allison prepared sketches and a preliminary design for equipping the Consolidated B-24 with V-1710's, but that project did not proceed.[7]

When fitted for flight and all of the cooling and exhaust systems were installed the engine compartments were busy. The exhaust ducts and shrouds appear fairly straight forward and well planned, yet one of these failed and the result was a fatal fire and loss of the aircraft. (Bodie)

As a part of the B-D-V (Boeing-Douglas-Vega) manufacturing triumphant for the Boeing B-17, the Vega Aircraft Corporation was asked by the AAF in early 1942 to build, test and deliver one Vega Model V-134-1 (XB-38) within ten months of contract approval. The aircraft was to be a modification of the Boeing B-17F with four 1425 bhp Allison V-1710-89 engines substituted for the original 1200 bhp air-cooled Wright R-1820-97's. All other major features of the aircraft, except the engines, superchargers, radiators and propellers, were to remain as in the standard B-17.

It was decided to convert the Boeing B-17E assigned to Vega as a "pattern" aircraft to assist in the startup of their production of the B-17.[8] The modification was to install the -89's along with a new model General Electric turbosupercharger, the Type B-17, as well as new three-bladed Hamilton Standard 12'-1" diameter propellers. This installation was to provide 1425 bhp up to the critical altitude of 25,000 feet.[9]

The engine installation was somewhat unique in that it departed from the usual tubular steel V-1710 engine mount that required that some of the lateral engine loads be carried through the engine crankcase. For the XB-38, a rigid light-weight, box-beam structure was built-up from aluminum sheet. It not only mounted the engine, but freed the engine crankcase from carrying the side loads.[10] The structure was designed to tie into the existing mounts for the radial engines, located on the nacelle firewall. No attempt was made to further streamline or reduce the cross-section area from that of the original radial installation.

The oil coolers were mounted directly below the engine support structure so that the likelihood of "coring", and loss of cooling effectiveness would be minimized. Coring is caused by oil congealing in the center of the oil cooler due to the low temperatures at high altitudes and it renders

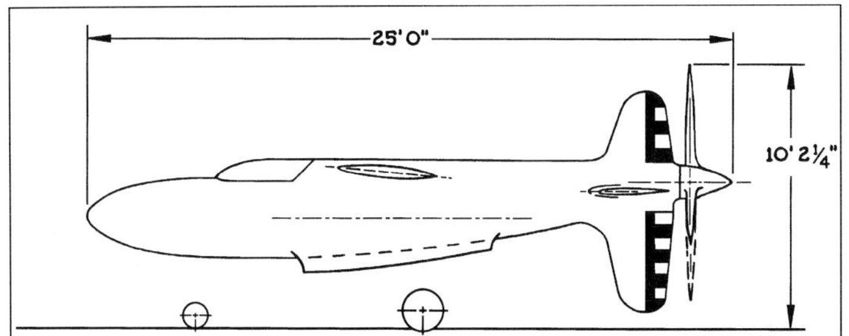

New NACA buried engine and advanced aerodynamic features were incorporated by Douglas as in their proposed light-weight DS-312 entry in the CP 39-770 pursuit competition of 1939. It featured an inverted turbosupercharged Ranger SGV-770 and a single 9'0" pusher propeller. (NARA)

the oil cooler practically useless. This in turn can lead to engine failure due to over-heating of the remaining oil. The rectangular coolant radiators for the engines were located in the leading edges of the wing, between the engine nacelles. Vega elected to retain the exhaust, carburetor and intercooler air ducts as on the B-17, but they increased the sizes to accommodate the higher airflows needed to supply the high powered V-1710's.

Although work was progressing rapidly on the XB-38, the project was scaled back in December of 1942 while Vega focused on higher priority work. As a result, the conversion was not completed until May 1943, with its first flight occurring on the 19th of the month.

The first five flights went well for an experimental aircraft, though on the last of these an exhaust manifold leak was discovered. It took nearly three weeks to revise the exhaust manifold's flexible joints, and return the aircraft to flight status. Then on its ninth test flight on June 15, 1943, during a speed run at 25,000 feet and with a total time of about 12 hours as the XB-38, a fire broke out in the inboard rear section of the No. 3 engine compartment. The flight crew was unable to extinguish the blaze and had to abandon the aircraft.[11]

This was not the end of the B-38 program, for in July 1943 Washington was considering having Vega switch all of their production to an "escort" version of the B-38. This would have been similar to the YB-40 armed escort versions of the B-17 then being field tested by the 8th Air Force in England. In early August General Arnold held a review of the entire B-38 project. They considered the recent unsatisfactory experience with the escort YB-40's, the improved supply of R-1820's, and the need for as many B-17G's as possible. The decision was reached to terminate the B-38 project. It is speculative, but had the review been held a week later things might have been different, for on August 17, 1943 the first Schweinfurt/Regensburg mission was flown. Without escorts, the 8th Air Force lost 60 B-17's on that single mission.[12] That mission, and another 60 lost on the next mission to the same targets that October, which became known as "Black Thursday", led to a crash program to provide suitable long-range escort fighters.[13] The B-38 was not to become a part of that program.

Douglas XB-42/XB-42A *"Mixmaster"*

In 1939 Douglas proposed their model DS-312, a small and lightweight high performance fighter, as an entry in the Air Corps CP 39-770 single-engine pursuit competition.[14] The aircraft was based on new aerodynamic principles sponsored by the NACA and was potentially capable of speeds in the 500 mph bracket, though the offered model was expected to achieve only 412 mph at 20,000 feet. In consideration of the features the new and novel design offered, the Air Corps evaluators recommended that it be developed for future experimental procurement in 1942. It was to be powered by a turbosupercharged Ranger SGV-770 inverted air-cooled V-12 of approximately 550 bhp. Yet it was to be competitive with conventional 1150 bhp fighters, primarily because of its small size, a gross weight of only 3,900 pounds and a single four-bladed Hamilton-Standard pusher propeller. What distinguished the aircraft was its configuration and profile, establishing the form and features of the later XB-42.

The result of CP 39-770 was negotiation of a contract for the development of several advanced designs for future experimental procurement. This included the single Douglas DS-312 along with one each of the Curtiss CP 39-13a (XP-46) and Seversky AP-10 (XP-47).

During the contract negotiation and definition process the DS-312 became two different airplanes. The light-weight fighter became the XP-48, but only after having been reconfigured into a conventional tractor arrangement while still using the originally proposed turbosupercharged SGV-770 Ranger engine and high aspect ratio wing.[15]

The original configuration, with the engine buried in the fuselage, grew in size and adopted a larger engine. The revised design was known as the DS-312A and used the same turbosupercharged Allison V-1710 liquid-cooled engine and extension shafts as in the XP-39.[16] The engine was to be located amidship and connected to a remote reduction gear mounting a pair of 9 foot diameter three-bladed contra-rotating pusher-propellers at the extreme rear end of the fuselage.[17] This configuration was not only

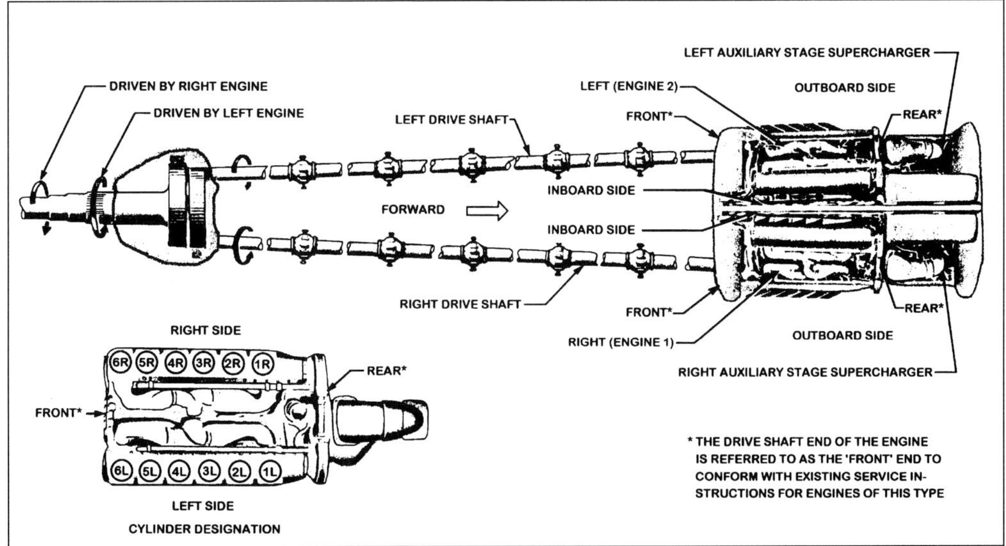

This figure, adapted from the XB-42A Flight Manual, gives a good idea of the power system layout, and attempts to resolve the confusion in nomenclature inherent in pusher configurations. (Author)

aerodynamically clean, but allowed installation of heavy cannon and concentration of armament in the nose. Neither the DS-312 or DS-312A left the drawing board, but the original configuration did become the baseline for a number of follow-on design concepts by Douglas.[18] The one design that did reach flight status was the scaled-up XA-42 *Mixmaster*, which flew as the XB-42.

The Specification for the Douglas DS-312A identifies the engine to be a V-1710-E5, with a GE Type B Turbosupercharger having a critical altitude of 25,000 feet. It was to be an extension shaft engine driving a counter-rotating pusher propeller. While the specifics are clear, the likelihood is that this reference to the E-5 was for convenience only. Allison did not develop a specification for this installation as was the usual case for serious proposals. Rather, the E-5 was probably the model in development/production at the time the aircraft was formulated and the fact that it was Altitude Rated was not considered important.

Douglas Aircraft began internal studies in early 1943 for a radically new aircraft design intended to meet the Army Air Force's recurring need during the war for a smaller, more efficient, more economical, speedy, longer-range tactical bombardment aircraft. Initially intended as an attack aircraft and designated by the Army as the XA-42, it was redesignated XB-42 on November 25, 1943 in a programmatic change by the AAF toward light/medium bombers. This change was due to General H.H. Arnold, General of the Air Force, who believed light bombers would be safer as a consequence of their tailored bomb loads rather than reliance on heavy machine guns typical of "Attack" aircraft. The light bomber would thereby avoid using costly strategic bombers in strictly tactical situations.[19] With the advent of the Heavy Bomber B-29 class of aircraft, the strategic role for the Boeing B-17 was being supplanted and Allison was informed that one of the objectives of the XB-42 project was to provide a replacement for the B-17.[20] The original contract, W535-ac-40188, calling for two purely experimental aircraft and one static test airframe was approved on June 25, 1943.[21]

In the record time of less than ten months from the May 1943 date of their unsolicited and informal proposal, Douglas designed and built the first aircraft (AF43-50224, with the "bug eye" cockpit canopies). Powered by a pair of V-1710-103's, Douglas pilot Bob Brush made a 22 minute first flight on May 6, 1944.[22] By September 30, 1944 Douglas had conducted 42 flights, and accumulated 34:35 hours, without major difficulties being encountered. The second aircraft, AF43-50225, first flew on August 1, 1944 powered by V-1710-129's.[23] Not long after this first flight it was fitted with a conventional "green house" cockpit canopy.

The XB-42 was a fairly small aircraft with an empty weight of 20,888 pounds and a maximum takeoff weight of 35,000 pounds. With its compact design, internal engines and cooling radiators, it promised very good performance and a top speed of 386 mph, along with a range of over 1,800 miles.

The power plant installation was by far the most radical single item in the XB-42. Ed Burton, Chief Engineer at Douglas Santa Monica, was originator of the XB-42 buried engine concept and foremost proponent of extension shaft aircraft.[24] In the view of Carlos Wood (Chief of Douglas' Preliminary Design Section at the time) it presented surprising few difficult problems, primarily because of the deliberate attempt to make the installation merely an unconventional arrangement of conventional components. In this regard the design utilized two late model V-1710-E extension shaft engines driving a remote counter-rotating reduction gearbox via six sections of extension shafting borrowed from the Bell P-39.[25] The engines were installed, side-by-side, just behind the cockpit, with the vertical centerline of each inclined about 20 degrees and with the crankshafts toed in a few degrees to follow a straight line from the extension shafts to the reduction gearbox mounted behind the tail.

Given the experimental nature of the program, and the considerable technical developments that become available for the engines in the XB-42's, there were a number of different engines proposed for and/or used by the two airplanes. Tabulated they are:

By some accounts the engines for the XB-42A were V-1710-133(E30)'s, but the actual aircraft history cards show that such was not the case. The first aircraft, AF43-50224, was powered by two different sets V-1710-103's when in its XB-42 form, and then a single pair of V-1710-137(E23C)'s when modified to become the XB-42A.[26] The discrepancy is likely due to the specifics of the V-1710-137(E23C), for these engines were originally built by Allison as their E-30 engine. As such, it is not surprising that the engines in the XB-42A would be described as V-1710-133(E30)'s. The major components of the engines were the same, though the accessories, appurtenances and controls unique to the XB-42A were enough to identify them as a separate model. The E-30 engine was used in other types of aircraft as well, another reason for a XB-42A specific designation.

According to Allison production records, ten engines were supplied for the XB-42 program. Five of each model, V-1710-103 and V-1710-133, were originally delivered.[27] Clearly the numerous changes occurring during development and testing created the evolution of models described above. It was typical for development programs such as this to purchase one engine for model testing, and then the flight engine and a spare. Since this was a twin-engine aircraft this would mean five engines for each aircraft, since each used a different engine model.

As originally proposed, the XB-42 aircraft were to use two V-1710-93(E11) two-stage engines, rated for takeoff at 1325 bhp from 3000 rpm and capable of 1500 bhp at War Emergency ratings. High altitude cruise ratings were 1200 bhp, available up to 22,500 feet as a result of the E-11's Auxiliary Stage Supercharger.[28] The counter-rotating reduction gear was taken from the V-3420-B, being used in the Fisher XP-75, and modified so that flight on a single propeller was possible. In addition, the reduction gear ratio was uniquely tailored to this installation at 2.772:1. The initial

Table 6-2
Engine Usage in XB-42/XB-42A Aircraft

Engine Model	XB-42/A #43-50224	XB-42 #43-50225	Comments
V-1710- 93(E11)	Did Not Use	Did Not Use	Proposal Only
V-1710-103(E23)	A-042208 & A-042212	Did Not Use	A-042208 rebuilt as -129's
V-1710-125(E23A)	Not Flown	Not Flown	Test and Spares
V-1710-129(E23B)	Did Not Use	A-063331/2	Flight Engines
V-1710-137(E23C)	A-071175/6	Did Not Use	In XB-42A Configuration Only
V-1710-133(E30)	Not Flown	Not Flown	Not configured for XB-42
V-1710-E24	Not Flown	Not Flown	w/2.46:1 RG for Martin AC

A low pass by the XB-42A following a high speed test flight at Muroc Army Base, pilot is likely Capt. Glenn Edwards. The black soot above the wing is indicative of a high power run using water injection, the black streak below the wing is oil spew from the left turning engine, a problem never completely solved when running at 3200 rpm and WER power of 1900 bhp from 75 inHgA.(SDAM)

design configuration intended to mount a pair of three bladed Hamilton-Standard pusher propellers of 13'-2" and 13'-0" diameter.[29]

When the first XB-42 was flown, it was powered by a pair of Allison V-1710-103(E23) engines rated for takeoff at 1325 bhp from 3000 rpm. It was also able to deliver 1820 bhp under War Emergency conditions. The propellers were of the counter-rotating design and operated through the 2.772:1 reduction gears. They were built by Curtiss Electric as their Model 3-C-2 and were provided with individual governing and separate feathering controls. According to Douglas designer Wood, no trouble was encountered from the engine and drive system and the bearings and shafts were in excellent condition after more than 600 hours on the test stand and over 150 hours of flight.

Early flight testing revealed that when open, the wake from bomb bay doors caused considerable propeller vibration. The vibration problem was minimized by using dynafocal mounts on the aft located reduction gearbox and revising the bomb bay doors for snap action, thereby minimizing the time they were open. Less significant propeller vibrations occurred when the radiator coolant doors were open or the flaps were down for landing. These vibrations were determined to not constitute a danger to the propellers.

Typical of pusher type propeller driven aircraft, engine cooling during ground operations was a problem due to the lack of any slipstream to induce airflow through the radiators. The problem was solved by installing electric fans downstream from the radiators and oil coolers. These were run anytime there was weight on the nose gear and the engine oil pressure was greater that 15 psi. Basically, when running an engine while on the ground. Another problem was with the original narrow inlet to the radiator cooling duct. It created a very objectionable buffeting of the elevator during any approach to stall, or during tight turns. This problem was subsequently reduced in the No. 1 airplane by the installation of a wider inlet and the use of internal ducting built with larger radius curves.[30]

On the whole, the engine installation was extremely clean, incorporating for the first time flush-type air scoops with boundary layer bleeds for the engine induction system, as well as having well designed internal ducting to the radiators mounted in their between-the-wing-spars location. All of this had the desired effect of minimizing installation and cooling drag.

As typical in experimental aircraft projects, the first and second XB-42's were different in many details, such as the above mentioned cooling ducts. Another difference was the use of lighter propeller blades on the second aircraft, resulting in much greater vibration when power was reduced on the forward propeller.[31]

Douglas had completed 150 flights on the first aircraft prior to turning it over to the Air Force, who agreed to curtail its flight test program and assign all of the performance guarantee requirements to the second airplane, which had the desired final configuration. Unfortunately the second XB-42 was unable to meet the contractor's guarantees of high speed and range. The first plane was then to run additional tests to get comparative data for use following its intended conversion to XB-42A configuration. For this purpose the aircraft was conditionally accepted by the Air Force, pending installation of satisfactory dynafocal shock mounts for the reduction gearbox.

The required tests were then accomplished in four flights performed at the Douglas Clover Field, Santa Monica, California facility by Captain Edwards in March 1946. About 20 minutes into the last of these flights, and after three minutes with both V-1710-103 units running at 3000 rpm and 52 inHgA, the left unit began running exceedingly rough and backfired violently at 24,000 feet. Captain Edwards throttled back both engines, since he did not know at the time which unit was giving trouble, and began a decent. Shortly thereafter he noticed that the left hand oil quantity gauge was approaching zero, as well as the left hand unit oil pressure gauge. He then feathered the engine and made a normal single engine landing. The left engine had suffered connecting rod failures on the front six cylinders. That in turn led to the fracture of the upper and lower crankcase and oil pan in the plane of the number three crankthrow, as well as fractures at the No. 1 crankthrow. The damage was caused by the continuing problem of oil spewing out of the left hand unit breather when running at full power for extended periods. At the conclusion of its operation as an XB-42, 43-50224 had made 154 contractor flights for 129:55 duration, and the Air Force had made 14 flights adding 14:30 hours. Total time on the airframe being 144:25 hours, all powered by the V-1710-103.[32]

The problems with "oil spew" on the left-hand engine unit tended to occur whenever running for more than five minutes at takeoff and/or WER conditions above 75 inHgA, conditions under which each power unit could produce 1900 bhp at 3200 rpm. Allison finally resolved the problem by redesigning the front cover and crankcase breather on the high powered left turning "E" model engines.[33]

After the fairly extensive flight testing by Douglas, the two XB-42's were officially inspected by the Air Force at the contractor's Santa Monica plant during the first week of September 1945. While nothing major was found, there remained a number of small items to resolve before official acceptance. These were accomplished on the second aircraft first, since it was the one configured as desired by the Air Force. Its final acceptance occurred December 8, 1945, the date it was flown from Long Beach, California to Washington, D.C., setting a new transcontinental speed record in the process. The co-pilot on the flight was the same Lt. Glen Edwards mentioned above, and it is for him that Edwards AFB is named.[34] The National Aeronautical Association supervised record distance was 2,295 miles, and the time, which was reduced due to the effects of a 60 mph tail wind, was 5 hours, 17 minutes, and 34 seconds for an average speed of 433.6 mph. Douglas adjusted the figures and calculated that the true air speed for the flight was about 375 mph, during which it consumed 1,153 gallons of gasoline, resulting in an average of about 2 miles per gallon. Very efficient for the day, but still not up to expectations.

Table 6-3
Operating Time for XB-42 AF43-50225 Engines

	Flight Time, hrs	Total Time, hrs
Left Engine, V-1710-129L, s/n A-063332	81:10	136:20
Right Engine, V-1710-129R, s/n A-063331	83:40	142:00
Gear Box	62:10	93:35

The record setting XB-42 AF43-50225, with its V-1710-129's, was temporarily stationed at Bolling Field, Washington, D.C., prior to being flown to Schenectady, New York where General Electric was to install radar equipment prior to scheduled tactical testing during January 1946 at Eglin Field, Florida. On December 16, 1945, an engineering test flight was scheduled following a complete spark plug change, and to investigate operation of the right engine in the Auto-Lean condition as well as check the hydraulic system, which had been having some difficulty. While all of the objectives of the flight were satisfactorily accomplished, the aircraft was unfortunately lost while returning to Bolling Field.

The incident that lead to the crash was failure of the main landing gear to extend due to a faulty diaphragm in a hydraulic accumulator. While trouble shooting this problem the left engine began running rough, as though overheated or seizing. The pilot then noted high coolant temperatures on both engines, the result of his shutting the coolant flaps upon entering the landing configuration. At this point the left engine stopped and he feathered its' propeller. Now at 1800 feet, and five miles from the runway, with the landing gear only partially extended, and with the one remaining engine at about 50 inHgA and 3000 rpm, the aircraft would not maintain altitude. The right engine then stopped producing power, and could not be restarted, even though the pilot switched to other fuel tanks. The pilot then ordered the two observers on the flight to bail-out, and was in the process of exiting the craft himself when he remembered to jettison the propellers. When this was done by the installed explosives, the aircraft immediately pitched nose down and the pilot was only able to get free about 400 feet above the ground. Fortunately, no one was injured in the incident.

A subsequent investigation could find no problems with the engines, other than that associated with crash damage.[35,36] This led to the likely cause being fuel starvation due to running the selected tank dry and not allowing sufficient time for the selected reserve tank to come on-line.[37] At the time of the accident the airplane had 118:20 hours total time, with the major components having operating time as shown in Table 6-3.

The XB-42 Engines

It appears that Allison and the Army Air Force departed somewhat in their engine designation schemes when it came to the engines for the XB-42. Usual practice was for Allison to indicate in their designation the direction of rotation as either R/L, and for the military to assign a separate dash number for each. In the XB-42 program they seem to have mixed the practices along lines intended for the V-3420, which the configuration mirrored with its extension shafts into the counter-rotating gearbox. Both "power sections" received only a single designation by both the military and Allison; furthermore, Allison deviated from their practice by using A-B-C suffixes to identify progressive developments of the basic series, some of which also received new military designations.

The designations of the paired engines in AF43-50224 were somewhat unique. With its initial V-1710-103 power sections, each carried the same engine serial number, A-042208. The right hand power section (as viewed from the pilots seat facing forward) was further identified as "Unit #1", the left hand being "Unit #2." The right hand engine, that drove the rear 13'-0" diameter propeller on the #50 spline shaft, rotated counter-clockwise as viewed from the rear of the aircraft, while its power section turned clockwise. The left hand power section drove the forward propeller that was mounted on a #70 spline shaft and turned its 13'-0" propeller clockwise, as viewed from the rear, its power section turned counter-clockwise. Engine(s) A-042208 was installed for the first 47:10 hours of flight, but at that point they were replaced by V-1710-103 s/n A-042212, that remained for the next 95:50 hours of flight. The A-042208 power sections were then overhauled, and converted in the process to V-1710-129 configuration, meaning that they had increased Auxiliary Stage drive ratios of 7.64:1 in place of the earlier 6.85:1 gears.[38]

Evidently the engine designations scheme for the XB-42 engines had become too confusing, so with the manufacture of new V-1710-129(E23B)'s for the second XB-42 (AF43-50225), and the later V-1710-137(E23C) powered XB-42A, each power section carried a unique serial number, independent of the "Unit 1/2" scheme. On the XB-42A, the right side power section, driving the aft propeller on the #50 spline, was s/n A-071175. The left side power section, driving the front propeller on the #70 spline, was A-071176. As of June 26, 1946, when the aircraft completed modifications to become the XB-42A, V-1710-137 engines were installed. Total time on the aircraft had reached 143:00 hours.

The XB-42A

In September 1944 the Army Air Force requested Douglas to develop a proposal to install auxiliary jet engines on the XB-42, and in addition, that the static test airframe be made airworthy as an all-jet powered aircraft, subsequently to be known as the XB-43. In response, Douglas submitted its proposal on February 23, 1945 calling for the installation of two 1,600 pound thrust Westinghouse Model 19XB-2A jet engines, one under each wing.[39] The Air Force finally approved a revised specification on March 8, 1946, though final contract approval was not received until March 31, 1947, much of the delay being due to the unavailability of the jet engines from the Navy.[40]

With the loss of the No. 2 XB-42, selection of AF43-50224 for conversion to the XB-42A configuration was fairly straight forward. During the considerable down-time while awaiting the Westinghouse jet engines, improved Allison V-1710-137(E23C) engines were installed in June 1946, along with improved dynafocal mounts for the reduction gear box. The reworked aircraft was flown for the first time as the XB-42A on May 27, 1947 and added another 23 flights and 18:20 hours to the 144:25 hours it had ac-

With "two-turnin and two-burnin" the XB-42A prepares for a test flight. Delays in delivery of satisfactory Westinghouse 19XB-2A engines delayed the development and interest of the project, favoring instead the XB-43 all jet version of the aircraft. (SDAM)

Table 6-4
Engines for the XB-42, XB-42A

Engine Model	Engine S/C Ratio	Aux S/C Ratio	Takeoff, bhp	Military Rating, bhp/alt/MAP	War Emergency, bhp/rpm/alt/MAP	Comments
V-1710- 93(E11)	8.10:1	6.85:1	1325/3000/54.0	1180/21,500/52.0	1825/3000/SL/75.0 Wet	Proposal Only[1]
V-1710-103(E23)	8.10:1	6.85:1/6.80:1	1325/3000/54.0	1150/22,400/___	1820/3200/SL/___ Wet	AC#224, 4-44/6-46[2]
V-1710-125(E23A)	8.10:1	6.85:1/6.80:1	1675/3200/75.0	1100/28,000/___	1900/3200/__/75.0 Wet	[3]
V-1710-129(E23B)	8.10:1	7.23:1/7.18:1	1675/3200/75.0	1100/28,000/51.0	1900/3200/__/75.0 Wet	AC#225 1st Flt[5]
V-1710-137(E23C)	8.10:1	7.64:1/7.59:1	1790/3200/100.0	1150/27,500/57.0	____/____/__/100.0 Wet	6.00:1 CR[7]
V-1710-E24	8.10:1	7.23:1/7.18:1	1500/3200/____	1325/3200/20,000	Not Established	Not Used[4]
V-1710-133(E30)	8.10:1	7.64:1	1500/3200/____	1150/27,500/____	1800/3200/SL/75.0 Wet	6.00:1 CR[6]

NOTES:
1. Original Design Proposal for XB-42.
2. These engines used during first six months of flight testing the first XB-42, AF43-50224.
3. Similar to -103 but with 12 counter-weight crankshaft and provisions for H2O/Alcohol injection and spark retard devices. Left Hand engine had a problem with "oil spew" from breather after 5 minutes at 75 inHgA.
4. Proposed engines with Aftercoolers and 2.458:1 reduction gears configured as in XB-42, believed to have been intended for a Martin Aircraft project, 1943. None built.
5. Improved E-23/E-23A engines with many components from V-1710-133 and spark retard distributors. Initial engines for the second XB-42, AF43-50225, 7.18:1 Aux S/C drive on Left engine.
6. Reduced compression ratio to allow higher manifold pressure and power without detonation. Increased Aux S/C drive ratio to increase critical altitude. Both S/C impellers used "parabolic" shaped guide vanes. Tested for War Emergency rating of 100-110 inHgA, wet, approximately 2300 bhp. Also, "Left Hand" engine drove Aux S/C through 7.59:1 power takeoff gear train. No record of ever being installed in either XB-42, though many references say that they were. The four E-23C engines were originally built as E-30's.
7. E-30 engines configured for use in XB-42A, 7.59:1 Aux S/C drive on Left engine.
8. All engines with 6.65:1 Compression Ratio, except as commented.

quired as a XB-42.[41,42] In this form the aircraft received very little flight time as a result of the considerable amount of maintenance being required, and the delays in receiving spares and support for the jet engines.[43] The XB-42A was finally accepted by the Air Force upon project close-out, August 8, 1948. On October 5, 1948 Major R.L. Cardenas delivered the aircraft to the National Air Museum Storage Activity, located at Orchard Place Airport, Park Ridge, Illinois.[44] In April 1959 the fuselage of the XB-42A was relocated, and now rests in the collection of the NASM, Smithsonian Institution, Suitland Annex, Silver Hill, Maryland.[45]

Transports

Douglas DC-8 "*Skybus*", The Transport of Tomorrow

In October of 1945 Douglas Aircraft announced their post-war successor to the DC-3 transport, the DC-8 *Skybus*.[46] The new low-winged commercial aircraft was to have a 39,500 pound gross weight, carry 48 passengers in a five abreast configuration, offer a pressurized cabin, and to be powered by two 1630 bhp Allison V-1710's. This was an effort to capitalize on the development of the XB-42, but in fact it described a considerably different aircraft. The DC-8 offered nearly twice the weight and installed the engines in the below floor cargo compartment instead of directly behind the cockpit. The DC-8 was to be 50 percent faster and carry twice as many passengers for a direct cost only one-half that of the standard Douglas DC-3 twin-engine transport.[47]

The engines were to be quite different than the two-stage units used in the XB-42. Rather than be fitted with Auxiliary Stage Superchargers, they would have been the first V-1710 production model to have had a two-speed drive to the engine single-stage supercharger.[48] It is believed that the engine intended for the DC-8 was to be the Allison V-1710-E29, similar to the V-1710-E22 but incorporating the features developed for the V-1710-131(G3R).[49] Curtiss Electric propellers were to be used, though the diameter was to be increased to 15 feet as compared to the 13 foot units on the XB-42.[50] These propellers each had their own pitch-changing capability and reverse thrust was available for in-flight aerodynamic braking or feathering. Since the engines were mounted below the floor, the extension drive shafts had to be angled up to the tail mounted reduction gear box. This feature required introduction of a bevel gearbox with a one-to-one ratio and its own lubrication system.

Douglas claimed this arrangement greatly improved single engine performance with no off-set thrust in case of engine failure, as well as a drag coefficient 25 percent less than in a conventional aircraft. Performance for the day would have been a great improvement over the competition at 270 mph at 10,000 feet and less than 4,000 feet required for either takeoff or landing, all with a useful load of 15,585 pounds.[51]

Introduced at the same time as the conventionally configured Convair 240 and Martin 2-0-2, and faced with development costs and a sales price that was substantially higher, the aircraft did not receive the necessary airline orders. The airlines had expressed concerns about added maintenance costs stemming from the use of long extension shafts and the general non-conventional arrangement. Consequently the project lost out to the competition and the *DC-8* designation was recalled to await the dawn of the jet age a decade later.[52] One other attempt to salvage the concept was in the

Cloudster II. It was basically an executive configuration of the XB-42 airframe. In the post-WWII market swamped with low priced conventional aircraft it was unable to obtain any sales and the project was canceled after completion of one prototype.[53]

Douglas XC-114 and YC-116

In an effort to streamline and increase the performance of the Douglas C-54 transport, military version of the four engined commercial DC-4, it was proposed to build the C-54E model with liquid-cooled engines. As contemplated by Douglas in December 1944 the new aircraft would be able to use alternatively the Rolls-Royce V-1650 Merlin, Allison V-1710, or the Wright R-2180 liquid-cooled 42 cylinder engines.[54] Historically, Douglas went on to later build 125 C-54E's, but all were powered by a model of the P&W R-2000 air-cooled radial that powered earlier production C-54's.[55]

Wright Field went ahead with plans to use liquid-cooled engines on the aircraft, but elected to assign new model numbers. The first of these was to be the Douglas XC-114, AF45-874, that was to be powered by four V-1710-131(G3R) engines. The airframe was similar to that of the delivered C-54E's, but with the fuselage lengthened 81 inches, in fact it was the same fuselage to be used on the soon to appear Douglas DC-6's that had been prototyped on the Douglas XC-112A (AF45-873), a P&W R-2800 engine powered aircraft.[56]

The G-3 engine model was somewhat unique for Allison in that it was equipped with a two-speed drive to its single-stage, engine driven supercharger. It is clear that these features would be desirable in an aircraft with a mission that required heavy weight takeoffs and cruising at only medium altitudes. As executed, the supercharger impeller was the enlarged 10.25 inch diameter unit used on most of the "G" series engines. These engines provided 1600 bhp at takeoff, from 3200 rpm, quite an improvement over the 1350 bhp available from the C-54's usual R-2000 air-cooled engines.[57]

Douglas was also to build the YC-116, AF45-875, an aircraft that was similar to the XC-114 and likewise powered by the Allison V-1710-131, but equipped with a revised wing structure to allow thermal de-icing instead of the more usual pneumatic boots on the wing leading edges.

These engine installations were an improvement over those developed by Canadair for their C-54 derivative, the Rolls-Royce Merlin 620 powered "North Star." Instead of the "annular radiators" used on the Merlin installations, Douglas opted for a streamlined version of the chin radiator arrangement similar to that used previously by the Curtiss P-40. Unfortunately, maximum advantage could not be taken of the slim V-12 engine, for Douglas retained the circular firewall used to interface with the earlier radial engined wing. Still the performance of the aircraft was improved considerably in all areas as compared to the R-2000 powered models.

Early in 1946 these aircraft were discontinued. The four nacelles built for the XC-114 were retained for future testing on the XC-112A.[58]

In late 1946 Douglas was actively offering this arrangement with V-1710-G2R/L engines to the airlines. Discussions were serious enough for Allison to assign their Specification 251 to a proposal offered to American Airlines.[59]

Installation of liquid-cooled Allison V-1710-G2R/L two-speed engines added to the trim appearance of the Douglas DC-4 passenger airplane, according to this artist's conception. Engineers expected substantially improved airplane performance and greater passenger comfort from this type of engine installation. (Allison)

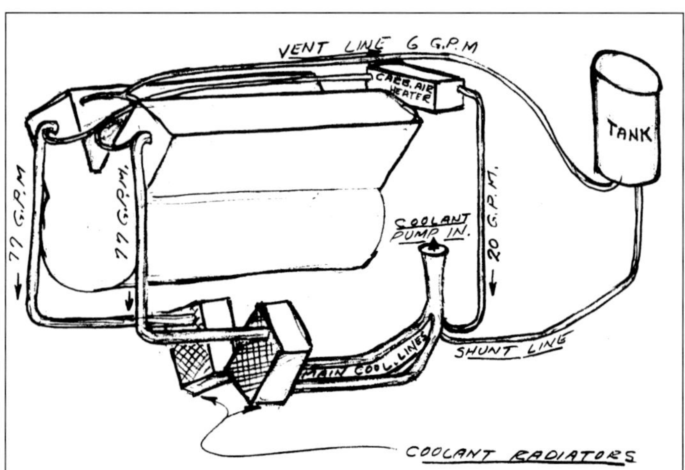

This January 1945 sketch was made by Allison engineer John Hubbard and shows the proposed cooling system layout for the Allison powered C-54 derivative. Note that 20 gpm of hot coolant was directed through a carburetor air heater. During low-power cruise conditions the single-stage supercharger would not otherwise provide enough heat to prevent formation of carburetor ice or evaporate the fuel. (Hubbard)

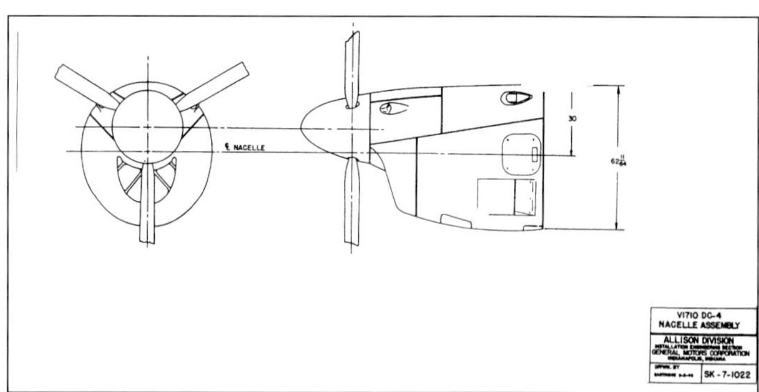

The Allison designed DC-4/C-114 nacelle for the G-2 was very compact, incorporating the radiator and oil coolers. The overall cross section still fit directly to the firewall intended for the R-2000 radial engine, so much of the benefit of a streamlined in-line engine was lost. (Allison)

Table 6-5
Allison Engines For Douglas 4-Engine Transports

Engine Model	Engine S/C Ratio	Aux S/C Ratio	Compression Ratio	Takeoff, bhp	Military Rating, bhp/rpm/alt/MAP	Weight, Pounds
V-1710-G2R/L	NA	None	NA	NA	NA	1,470
V-1710-131(G3R)	7.48:1-Lo	None	6.50:1	1600/3200/61.7	1360/3000/ 4,750/____	1,475
V-1710-131(G3R)	9.60:1-Hi	None	6.50:1	Not Allowed	1220/3000/15,000/____	-
R-2000-7(2SD-G)	7.15:1-Lo	None	6.50:1	1350/2700/____	1350/2700/ 2,000/____	1,570
R-2000-7(2SD-G)	8.47:1-Hi	None	6.50:1	Not Allowed	1000/2550/14,000/____	-

NOTES:
1. All engine weights are "dry", and do not include radiators or oil coolers.
2. Above engines all rated on 100/130 Octane fuel.
3. R-2000-7 ratings and specifics from P&W Aircraft Twin Wasp 2SD-G Engine Specification No. 9004, Single-Stage Two-Speed, dated 7-1-1944.

Fighters

Brewster Model P-22

In the late 1930s nearly everyone building airplanes had a project intended to incorporate the new Allison. Included in this group was the Brewster Aeronautical Corporation. Their proposed Model P-22 was a midwing monoplane pursuit for the Army dating from February 1939. The airplane was based on their F2A-1 *Buffalo* but featured a two-stage V-1710. It would have had a gross weight of 6,240 pounds with an empty weight of 4,620 pounds. With a wing area of 209 square feet, it would have been similar in size to a P-39. The wing section was to be NACA 2300 series, 9 percent thick at the tip and 18 percent at the root. These are not particularly aggressive figures, and they resulted in a maximum speed of 307 mph at sea level, 347 mph at the 12,000 foot critical altitude. Very little information is available on the engine selection, but it appears to have been an unidentified model with an auxiliary stage supercharger. It was to be rated for 1200 bhp from 3000 rpm at up to 12,000 feet. At 20,000 feet, where the airplane was expected to reach 363 mph, the Military rating was 1150 bhp.[60]

In March 1940 Brewster offered a very similar airplane, based upon the F2A-2 as their proposal P-29, Model 140, to the French Purchasing Commission. No contracts were issued and it is not clear which model V-1710 would have been used, though it is likely that the French would have required them to use the single stage V-1710-C15.

Bell Model 11

In response to the 1939 Army Type Specification C-619 and the resulting Circular CP 39-770, Bell submitted five different proposals. Two were based on the turbosupercharged XP-39 (Model 4E), and two were similarly based upon the single-stage V-1710-E4 powered YP-39 derivative (Model 4F). The fifth proposal was their Model 11, an airplane considerably different than the baseline P-39.

Drawings for this airplane predate the Air Corps issuance its Circular for the competition, again suggesting that Larry Bell was working closely with Wright Field. This airplane would have been much more conventional, configured as a tail-dragger and powered by a tractor mounted V-1710-33(C15). It's appearance was much like the competing Allison V-1710-F3R powered CP 39-770 designs from Curtiss and Seversky that became the XP-46/XP-46A and XP-47/XP-47A respectively. Critical altitude was expected to be 15,000 feet, where it was to clock 411 mph, as against the YP-39 derivatives making 400-407 mph at the same altitude. Empty weight was to be 4,352 pounds and gross weight was 5,581 pounds. The Model 4F placed second in the overall competition, receiving an order for 80 P-39C (initially the Army P-45) aircraft. Of the 19 designs offered in the competition the Model 11 finished number 11, receiving a figure of merit score of 701.65, compared to the top score of 797.25 for the Republic P-44, and 781.05 for the P-39C.[61]

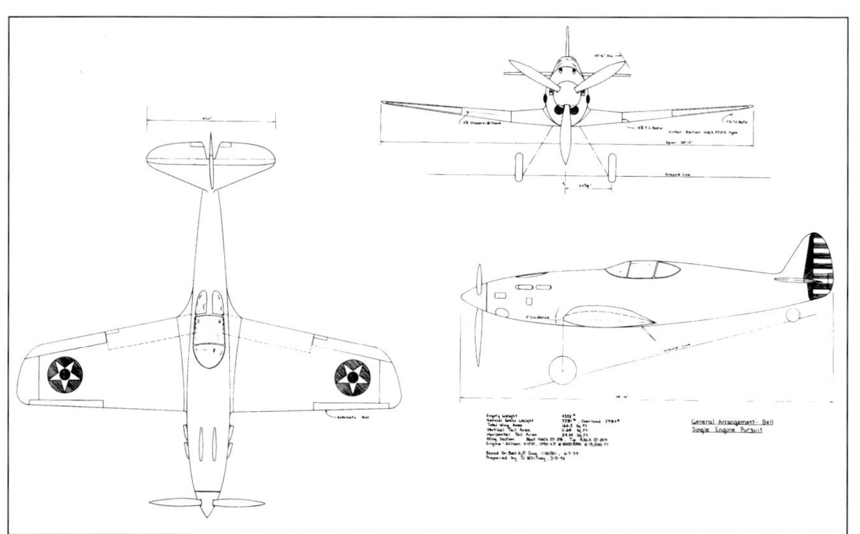

In the 1939 Pursuit Competition Bell offered a comparatively conventional airplane as an alternative to the extension shaft equipped P-39. The selected engine was the V-1710-33(C15) with a special rating offering 1500 bhp at 15,000 feet, where the Model 11 was to achieve a top speed of 441 mph. (Author)

All of the Bell proposals in the CP 39-770 competition included performance estimates based upon considerably more engine power from their Allison's than was authorized prior to the introduction of WER ratings. In the case of the V-1710-33 powered Model 11, it was to have 1500 bhp available at 15,000 feet. This power was to result in a maximum speed of 441 mph. The turbosupercharged XP-39 derivatives were to have 1600 bhp available and result in a top speed of 447 mph at 20,000 feet.

Bell XP-52 (Model 13-A/B)

Bell Aircraft Corporation submitted their Model 13-B under the 1940 Circular Proposal R-40C solicitation for advanced and unconventional aircraft. Four different Model 13's were offered. Two had upright Continental V-1430 engines: the Model 13A (to be powered by the single-stage V-1710-E8), and the Model 13B, which would have been the first instance of Bell attempting to regain the altitude advantage by using a two-stage V-1710-E9. The airplane was similar in size to a P-39, but with the wing area reduced from 213 square feet to 143 square feet.[62] The Air Corps liked the design but rather than funding it, negotiated for the XP-52, based upon the Model 13B, in exchange for release of the P-39 for foreign sale, and a payment to Bell of $50.00.

The aircraft was to be powered by a pusher version of the Allison V-1710-E9 two-stage engine with a military rating of 1225 bhp up to its critical altitude of 18,000 feet, where it was expected to achieve 443 mph. Empty weight was to be 5,430 pounds and gross weight was 7,104 pounds. Armament would have been 1-37mm cannon, 2-.50 cal synchronized guns, and 4-.30 cal guns mounted in the wings.[63] Later reports on the design ignore the Allison engine connection and report that it was to be powered by either the Lycoming XH-2470 or Continental I-1430-5 engine.

The XP-52 was abandoned in favor of the larger and later Bell Model 20 (XP-59), that was also configured as a twin boom pusher. It to was abandoned when Bell was asked to use the XP-59 designation as a cover for the first U.S. jet airplane, the Bell XP-59A.

Curtiss XP-40H

The Curtiss XP-40H was to have been an Allison powered and turbosupercharged version of the P-40E.[64] At a conference at Wright Field on June 10, 1941 Mr. Don Berlin of Curtiss received authority to begin such a project and the Materiel Division immediately shipped one GE Type B-2 turbosupercharger to Curtiss at Buffalo, New York.[65] By that October the decision had been made to not turbosupercharge the P-40, but instead incorporate the feature into the coming Curtiss P-60. The turbo was to be installed behind the pilot in the XP-60.[66]

The XP-40K incorporated extensive sheet metal and plumbing changes in an attempt to dramatically streamline the P-40. (SDAM)

This frontal view gives an idea of how extensive the relocation of oil and coolant radiators was on the XP-40K. Performance figures for the aircraft have not been located. (SDAM)

Curtiss XP-40K

In an effort to streamline and modernize the obsolete P-40, P-40K AF42-10219 was extensively modified by relocating the radiator cores and oil cooler from below the engine. These were re-shaped and then installed in an airfoil shaped "glove" fitted between the landing gear struts. It is not apparent that there was much if any reduction in form drag or streamlining of the aircraft. In any case, the benefits, if they existed at all, were insufficient to warrant interrupting the production lines.[67] The engine remained the standard V-1710.

Curtiss XP-40Q

This program began with an early two-stage version of the V-1710 fitted into the first XP-40Q. That engine is believed to have been the V-1710-45(F7R), but has not been confirmed. In May 1943 Allison was able to make available an improved engine, the V-1710-101(F27R), which was a two-stage engine having the carburetor mounted between the stages. Curtiss believed that with this engine, "a substantial performance increase in speed and altitude will result." This improved engine was desired as other airframe manufacturers (notably Bell) had been having difficulties with the early two-stage engines similar to the one then installed.[68] In addition to the two-stage engine, the XP-40Q had been modernized by giving it a laminar flow wing section and a bubble canopy.

The basic engine was an "F" configuration of the V-1710-E11 as developed for the P-63 airplanes, though the various models went through considerable development and incorporation of advanced components. For example, by November 1943 an engine with the carburetor located between the supercharger stages was being flown with the 6.85:1 Auxiliary Stage gears and obtaining 54 inHgA at altitudes up to 30,000 feet. When the 7.23:1 gears, identifying the V-1710-121(F28R), were installed the performance was improved to the point that at high-altitude it equaled the performance of the Rolls-Royce RM14SM high performance two-stage Merlin.[69]

In January 1944 one of the XP-40Q's was flown to Eglin Field, Florida for Army tactical and performance tests against other late model Army fighters. For these critical tests the Auxiliary Stage Supercharger drive ratio was changed to 7.23:1 (making it in effect a V-1710-121(F28R)) with Allison authorizing the modified V-1710-101 to operate with a 3200 rpm WER rating of 63 inHgA at sea-level.[70] A new control schedule for the hydraulic coupling biased the low-level performance of the Aux Stage while giving a greater critical altitude.

The V-1710-101(F27R) two-stage engine was based on the V-1710-89(F17R), but equipped with the 2.36:1 reduction gear and rated for 3200 rpm. The carburetor was located between the supercharger stages. Military rating of 1150 bhp at up to 22,400 feet, with WER of 1500 bhp at up to 6,000 feet from 60 inHgA without ADI. (Allison)

As the Curtiss Model 87X and the culmination of the XP-40Q program, P-40N-25 AC43-24571 was the third and final XP-40Q. It featured a bubble canopy and a cut-down aft fuselage deck as had been previously tested on the XP-40N. The extended nose was to accommodate the engine with its Auxiliary Stage Supercharger. The modifications made for a long and fairly sleek aircraft. The XP-40Q-2 version, with the clipped wing tips and utilizing water injection, demonstrated a maximum speed of 422 mph at 20,500 feet. In a time-to-climb trial it demonstrated a climb to 20,000 feet in 4.8 minutes. This made it the fastest and highest flying of all the P-40's.[71,72]

Specifics of the three XP-40Q's are given below.

XP-40Q-1 P-40K-10 AC42-9987. At various times this aircraft had either a single-stage V-1710-81(F20R), or one of several different two-stage V-1710-101(F27R)'s during development. It did not have the bubble canopy and the semi-flush, low-drag coolant radiators, which faired into the leading edges of the wing and extended across the bottom of the fuselage. Carburetor air was taken from the top of

XP-40Q-2 in flight shows some of the improvements, streamlined engine cowl, clipped wing tips, four guns instead of six and 360 degree bubble canopy. Radiator capacity was increased by leading edge inlets and radiators in place of the removed guns. With the two-stage V-1710-101(F27R) engine and light weight it was quite a performer, 422 mph. (SDAM)

The XP-40Q program lasted almost three years, involved three different airplanes, and several engine models. The longer two-stage installation was balanced by a lengthened fuselage. This was the highest flying and fastest of all of the P-40's. Each of the airplanes went through considerable evolution of configuration, engines and cooling systems, all intended to improve performance. This is believed to be the #3 airplane when at Allison. (Allison)

After the war XP-40Q-2 was surplused and raced in the 1947 Thompson as Race 82. (SDAM)

the engine nacelle as in most previous P-40's. The wing retained the rounded wingtips, but mounted only four .50 caliber guns. In June 1944 the aircraft was assigned to "Post Operations" at Wright Field, suggesting it was no longer needed for test and development.[73] In July 1944, Captain Abel ground-looped the aircraft while landing at Wright Field, and it was recommended to be surveyed.[74] In August, the Materiel Command requested that the airplane be given priority for repair, as it was needed for engine experimentation by Allison at Indianapolis.[75]

XP-40Q-2 P-40K-1 AC42-45722, the earlier XP-40N cutdown rear fuselage and bubble canopy were incorporated. The coolant radiators remained below the nose, but in a tighter and more streamlined cowling. The oil coolers were relocated to the positions in the wings vacated by removing two guns. Air was provided by leading edge inlets. The wing tips were squared and had been clipped to 35 feet 3 inches. With the two-stage V-1710-101 engine rated for WER power, and a weight of 8,363 pounds, the airplane could climb at 3,800 ft/min at 13,000 feet, reaching 21,000 feet in six minutes.[76] It is believed to have flown with both the F-27R and F-28R engines. The airplane had been loaned to Allison for flight investigation of sparkplug life, when on March 15, 1946 Allison requested a six-month extension of the loan. The AAF responded that the aircraft had been declared surplus and therefore they could no longer loan it. Allison then offered $1,250 for the airplane and spare parts then on-hand. It is not clear whether or not Allison then was able to procure the aircraft, though it was on the civil registry in time for the coming Thompson Trophy race. In any case, registered as NX300B, the aircraft was an unofficial starter in the 1947 Thompson Trophy Race. It was in 4th place when it caught fire and was abandoned in-flight.

XP-40Q-3 P-40N-25 AC43-24571 was fitted out like the XP-40Q-2 but incorporated increased capacity brakes, a flat bullet-proof windshield, a four bladed propeller and was fitted with the Allison V-1710-121(F28R). This engine with its auxiliary stage supercharger developed 1425 bhp at 3000 rpm for takeoff, used the 12-counterweight crankshaft that allowed War Emergency Power of 1700 bhp at 26,000 feet to be developed at 3200 rpm. This aircraft was the only XP-40Q delivered to the AAF as such, though not until early 1945. The others retained their original P-40K delivery status.[77]

In the evolution of the XP-40Q a number of airframes and engines were used. Again demonstrating the flexibility of the V-1710. When a nose-over of one of the airplanes broke the reduction gear off of the only available V-1710-101, Allison shipped Curtiss a new 2.36:1 reduction gear and the Materiel Division provided a spare V-1710-89(F17R). The Allison Tech Rep at Curtiss then installed the accessories section, carburetor and Auxiliary Stage Supercharger from the damaged V-1710-101 on the new engine, along with the new reduction gear, and the aircraft was returned to service. This same engine was subsequently converted to a V-1710-121 by changing the Aux Stage drive gears to 7.23:1 and rescheduling the control for the hydraulic drive coupling.

Lockheed P-38K

Impetus for this aircraft was likely the result of studies done by the Wright Field Propeller Laboratory in October 1942 to determine the effects of the proposed WER ratings on propeller performance. They considered a P-38H powered by the new Allison V-1710-75(F15) using either a 3-Blade, 12'-6" or a 4-Blade 12'-0" unit. With the latter propeller the increased engine ratings were better utilized, and the advantage increased with altitude.[78]

Only one of these aircraft was built, and it did not fly for nearly a year after being recommended. The stated purpose for the project to investigate a new wide chord high altitude Hamilton-Standard Hydraulic 3-Blade propeller with "high activity" blades. These were to be driven by higher powered V-1710-75/77(F15R/L) engines rather than the V-1710-89/91(F17R/L) engines then in the P-38H/J's.[79] The aircraft, AF42-13558, was delivered in September 1943 and demonstrated a considerable degree of im-

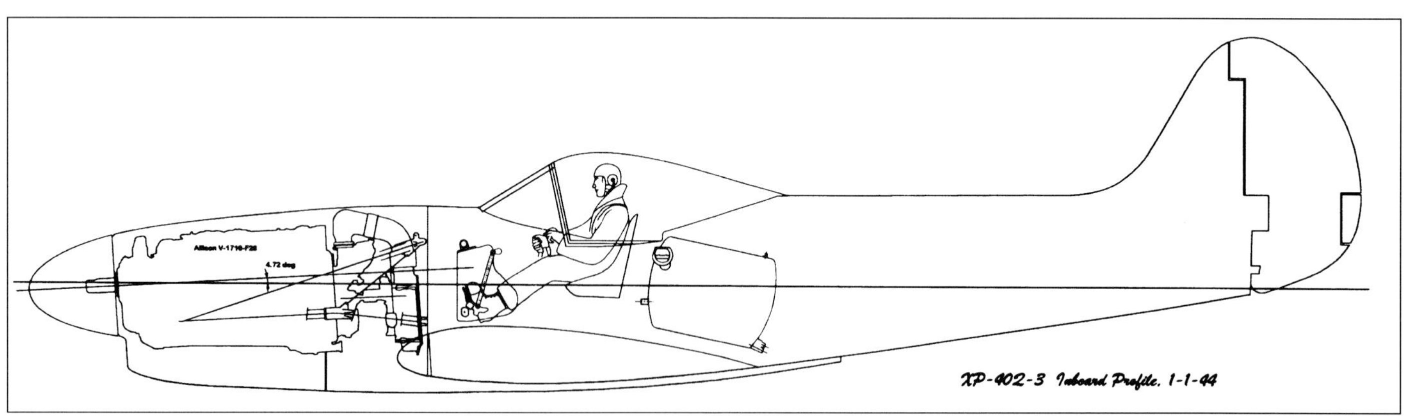

This inboard profile of the XP-40Q-3 gives a good idea of the numerous changes incorporated into this much revised P-40. The two-stage V-1710-121(F28R) provided 1700 bhp up to 26,000 feet in this aircraft. (Author)

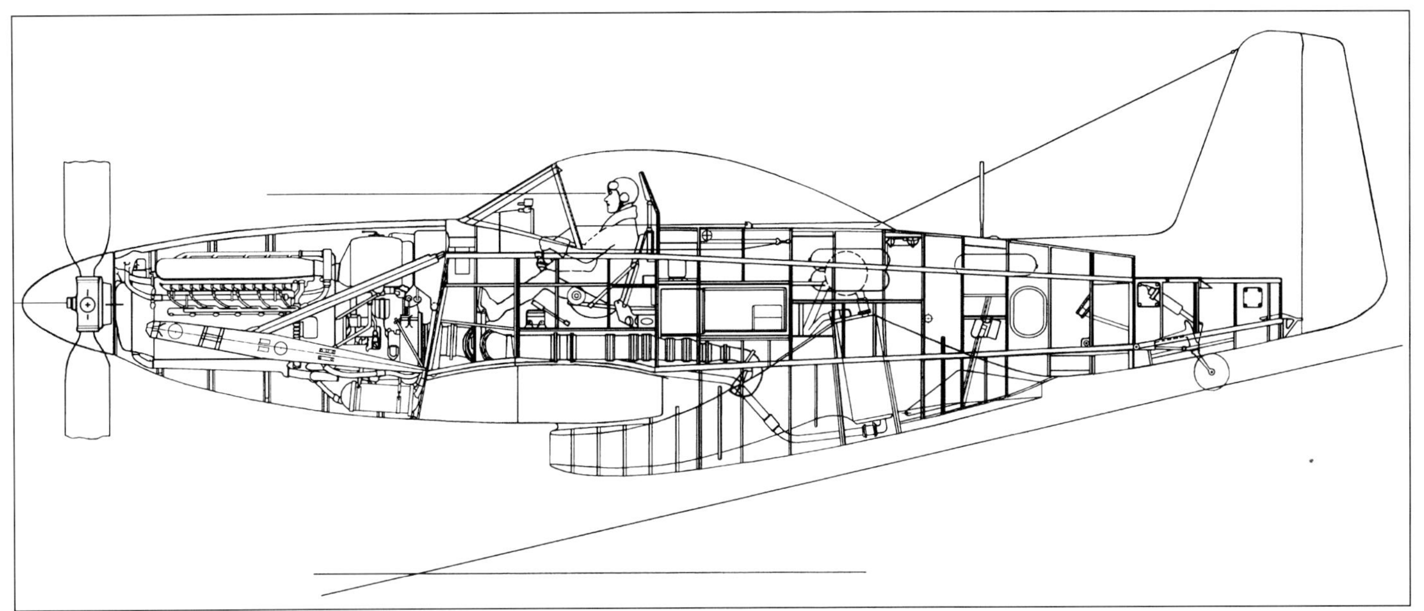

The XP-51J was the Allison powered version of the light-weight Mustang. Power was from a new two-stage V-1710-119(F32R) complete with Aftercooler and Speed Density 'carburetor'. The Aftercooler is plainly visible, located just aft of the cylinder heads of the engine. Note carburetor air taken from the cooling duct. (Inboard Profile, NAA Dwg 105B-00003, October 12, 1944, adapted by Author)

proved performance over the P-38J's then going into serial production, though the aircraft used the chin located core type intercoolers as on the P-38J.

This was accomplished by using larger propellers, which required a higher reduction gear ratio of 2.36:1, rather than the standard 2.00:1 used with the usual Curtiss Electric props. This kept the blade tips from overspeeding. Another unique feature was the incorporation of water injection on the engines, making this the only P-38 so equipped. While it has not been confirmed, there is reason to believe that this may have been necessary as a result of the considerably higher critical altitude of 29,600 feet and service ceiling of 46,000 feet available from the new and higher capacity GE Type B-14 turbosupercharger. At these high altitudes such a unit would require additional induction air cooling which was beyond the capacity of the standard core type intercoolers.

While it appears that the P-38K could, and probably should, have gone into series production, the reasons given for not doing so relate to the disruption in production necessary to accommodate the larger propeller, and its need for the prop spinner diameter to be increased by one inch. The entire nacelle and cowl would also have had to be designed and built anew to interface with oil cooler and intercooler air intakes while accommodating the larger 2.36:1 reduction gear.[80] This dilemma reflects the critical need to increase intercooler capacity, which was resolved in the P-38J. Lockheed was just introducing it into production and any revision would have further disrupted and delayed production of the critically needed P-38's.

North American XP-51J

In January 1943 North American Aviation proposed to Wright Field a completely revised Mustang, built to British design standards rather than those specified by the U.S. Army. The resulting "light weight" Mustangs became the XP-51F and XP-51G models, with the Merlin V-1650-3 powered XP-51F flying for the first time on February 14, 1944. Five of the aircraft were procured, with the last two being identified as the XP-51G, and powered by "reverse lend-lease" supplied Rolls-Royce RM14SM Merlin engines.

According to the USAF Report on the program, "Due to the fact that no new fighter type aircraft were being built incorporating the Allison V-1710 engine, a study was made on the advisability of installing the Allison engine in a P-51F type aircraft."[81] Alternatively, following the 8th Air Force disaster at Regensburg, where the bombers were well beyond the range of fighter escort, General Arnold had directed the study because of the likely greatly increased demand for P-51's and the limitations on availability of two-stage Packard Built Merlin V-1650's.[82] As a result of this study, the procurement of two Allison powered XP-51J aircraft was authorized in

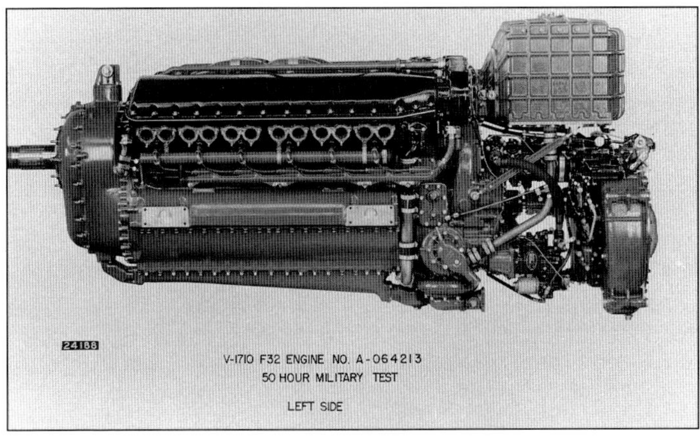

The V-1710-119(F32R) was intended to be the U.S. answer to the two-stage Rolls-Royce Merlin. The variable speed Auxiliary Stage Supercharger fed through the new Bendix SD-400 Speed Density 'carburetor' to the standard 9-1/2 inch diameter engine stage supercharger. It discharged through an aftercooler sized to remove half of the heat of compression. Rated at 3200 rpm, 1200 bhp was available to 30,000 feet from 52 inHgA. WER ratings were 1900 bhp at sea-level from 78 inHgA-wet, and 1720 bhp at 20,700 feet. The advanced RM14SM Merlin in the XP-51G delivered 1690 bhp at 18,500 feet. (Allison)

Portrait of XP-51J at Mines Field, LAX, April 1945. This V-1710-119 powered lightweight Mustang would have been the best performing Mustang in the stable. But single engine Mustang production was coming to an end, as was the war, and the "J" did not become a production model. (NAA via SDAM)

November 1943. The Model Specification for Allison V-1710-119 powered airplanes was dated January 1, 1944, but it was not until June 30, 1944 that the contract to procure two additional "light weight" XP-51J *Mustangs* was signed.

The engine selected was a V-1710 equipped with an Auxiliary Stage Supercharger and an Aftercooler sized to remove 50 percent of the supercharger compression heat. This engine was intended to provide performance at least as good as the two-stage Merlin then going into production P-51B/C Mustangs. The engine also introduced a new type of "carburetor", the Bendix "Speed Density" unit that measured the mass flow of air into the engine, then computed and controlled the quantity of fuel being injected. This change was needed to provide the space in which to install the large aftercooler.

As of November 1943 Allison expected to have the first engine running on the test bed in March 1944, with a flight engine available approximately May 1, 1944. In contrast to the situation at Packard and Continental, who were unable to produce the number of Merlins required, Allison had considerable excess manufacturing capacity, as their production orders were using only a little over half of their capacity. Between August and December 1944 Allison reported they could have produced over 10,000 additional engines. With the unending requirements for P-51's it is no wonder that an Allison two-stage powerplant would have given a big boost to meeting demand.[83]

On June 30, 1944 the AAF officially ordered two XP-51J (AF44-76027/8) "light weight" Mustang aircraft as proposed by North American in their Model Specification, dated January 1, 1944. The primary purpose of the aircraft was to produce a lightweight fighter with exceptional high speed, maneuverability, and performance characteristics, with priority being given to rate of climb and maneuverability.[84] In the process they were to test the new Allison V-1710-119(F32R) two-stage aftercooled engine. Allison put the first F-32R on test in July 1944. First flight of the aircraft was not until April 23, 1945, piloted by North American's Joe Barton.

The aircraft was different from the Merlin powered XP-51F only from the firewall forward. The F-32R developed 1500 bhp for takeoff and 1720 bhp at War Emergency conditions up to 20,700 feet. At sea-level, 2100 bhp was available. At 27,400 feet, the aircraft was to be capable of 491 mph, be able to climb to 20,000 feet in five minutes (6,600 ft/min using 150 Grade fuel) and have a 43,700 foot service ceiling. It has not been made clear why, but it has been reported that the XP-51J was not popular with the North American flight test people. Only seven flights were made on the first aircraft and two on the second before the aircraft were turned over to the AAF on February 15, 1946.[85] The airplanes were then turned over to Allison for engine development purposes.[86] At least one was used to flight test development models of the V-1710-143(G6R).[87]

Accommodating the elongated two-stage Allison in an existing airframe was usually a problem for the aircraft designer given the necessity of maintaining longitudinal balance. This was not the case in this installation as the propeller and overall dimensions stayed in the same positions as on the XP-51F. The extra length was provided by moving the firewall back about 9-10 inches. Induction air was obtained from within the duct to the coolant radiator, and then ducted under the cockpit floor to the intake for the Aux Stage Supercharger. Without the chin carburetor air intake required for the updraft Merlin powered P-51H, the aircraft was extremely sleek.

The No. 1 XP-51J made its initial flight April 23, 1945, and preliminary flight tests indicated that the V-1710-119 was not developed suffi-

All Mustangs were sleek, but the lightweight V-1710-119(F32R) powered XP-51J exemplified the type. The clean cowling is due to the carburetor intake being located in the cooling duct. (NAA via SDAM)

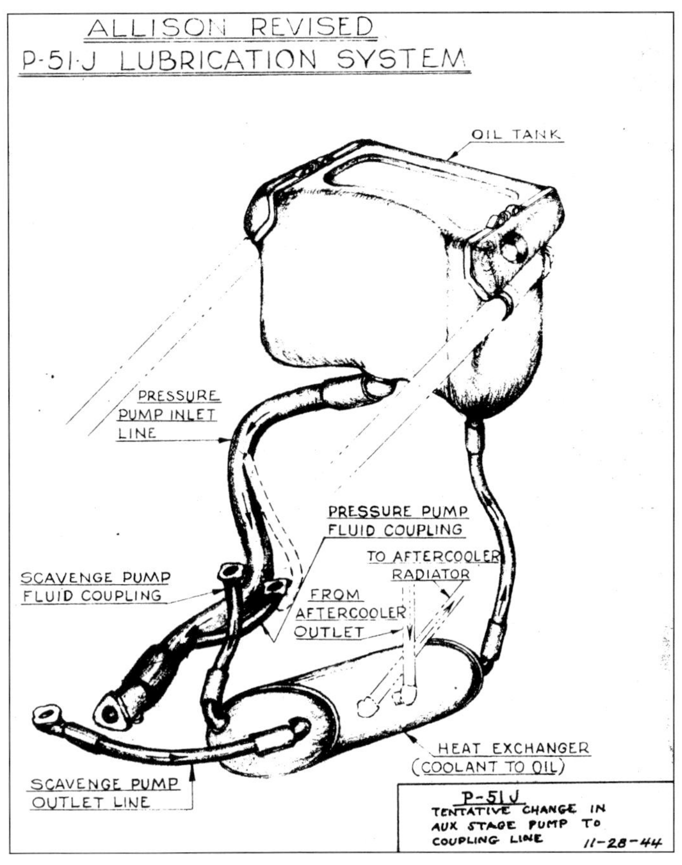

The lube oil system for the F-32R in the XP-51J was fairly sophisticated in that it used an oil to water heat exchanger cooled by the aftercooler circuit rather than a radiator. (Allison via Hubbard)

ciently for an airplane installation. The problem was that the V-1710-119 incorporated a fuel metering system in place of a carburetor, and it was this feature that was not sufficiently developed. Manifold pressure on this particular engine during the manufacturer's flights was limited to only 54 inHgA, which would produce 1150 bhp at 2700 rpm. The airplane, AF44-76027, was soon loaned to Allison on a Bailment Contract for further engine development. The second airplane, AF44-76028, was given to Allison for use in obtaining spare parts for the flying airplane.[88]

The V-1710-119 engine was also to be used in the XP-82A, but again it was not to be. The problem with this engine was not it's two-stage supercharger or aftercooler, but rather the use of the new Bendix Speed Density Carburetor. The unit was just not sufficiently developed for the engine. It was not until the V-1710-143/145's, developed for the P-82E, that the device came into its own. Later on, one of the XP-51J's was used for testing the V-1710-143 prior to its being installed in the P-82's. In this way they again demonstrated the easy re-configuration of seemingly quite different Allison's. The engine tested was actually a V-1710-133(E30) converted "...as nearly as practicable..." to represent the coming V-1710-143.[89]

Curtiss XP-55 *"Ascender"*

Not content with the progress then being made in the advancement of Pursuit aviation, on November 27, 1939 the Air Corps issued Specification XC-622, that resulted in the R-40C Request for Data issued February 20, 1940.[90] The result was a procurement program for experimental aircraft that featured the pusher configured airplanes, the Consolidated Vultee XP-54, Curtiss XP-55 and Northrop XP-56.[91] This program set very stringent performance requirements and called for the use of the forthcoming Pratt & Whitney liquid-cooled X-1800-A3G engine, a two-stage H-2600 sleeve valve engine rated at 2000 bhp.[92] The specification placed a premium on low overall drag for the aircraft as well as improved pilot visibility and offensive armament placement. Overall, the specification was seeking air-

Table 6-6
Lightweight Mustangs

	XP-51F	XP-51G	P-51H	XP-51J
North American Model	NA-105	NA-105A	NA-126	NA-105B
Number Built	3	2	555	2
Contract Number	ac-37857	ac-37857	ac-1752	ac-37857
Speed at Critical Alt, mph Mil Power	___/21,500	/30,000	/30,000	470/36,000
Military Climb to Critical Alt, min	/20,000	___28,000	___28,000	5.0/20,000
Wing Area, sq.ft.	235	235	235	235.7
Empty Weight, pounds	5,634	5,750	7,148	6,030
Gross Weight, pounds	7,340	NA	9,250	7,540
Design Load Factors				+7.33/-3.67
Engine Model	V-1650-3	RM14SM	V-1650-9	V-1710-119(F32R)
Engine Specification				184-C
Two-Stage Supercharger	Integral	Integral	Integral	Auxiliary
Performance at Takeoff, bhp/rpm	1380/3000	1675/3000/66"	1380/3000/61"	1500/3200/58"
Mil Power at Critical Alt, bhp/rpm	1330/3000/23,300	NA	NA	1200/3200/30,000
Military Rating, Critical Altitude	1210/25,700	NA	NA	NA
War Emergency Rating, bhp/rpm/alt/MAP	1790/3000	1690/3000/18,500/70"	1630/3000/23,500/80" Wet	1720/3200/20,700 Wet
War Emergency mph/altitude, ft	491/21,500	NA	487/	491/27,400
War Emerg Time to Climb/Altitude, ft	___/20,000	NA	NA	3.6/20,000
NACA Wing Section at Root	66,2-18155	66,2-18155	66,2-18155	66,2-18155
NACA Wing Section at Tip	66,2-1812	66,2-1812	66,2-1812	66,2-1812
First Flight Date	2-14-1944	8-9-1944	NA	4-23-1945
Comments				

This is the second XP-55 and shows the streamlined aft location of the single V-1710-95(F23R) pusher engine. (AAF via SDAM)

craft with radical design innovations that might provide significant performance advances.

Curtiss-Wright of St. Louis, MO was awarded a contract to develop its proposed Model CW-24 *Ascender*, a "tail-first" canard design of what was actually a flying wing, driven by a "pusher" propeller. Because of the radical design and the need to do both wind tunnel and scale model testing, a contract for three full scale XP-55 aircraft was not issued until July 10, 1942. Since the P&W X-1800 was long dead by that time, the design was revised and proposed to be powered by the Allison V-1710-F16 on the basis of reliability and availability. The F-16 would have been unique as the only V-1710 to have a counter-rotating gear box. The pusher propeller would have been jetisonable to permit safe bail-outs by the pilot.

Because of the many pressing demands being placed on Allison early in the war, the Air Corps elected to not proceed with the counter-rotating V-1710 and instead to fit a pusher configured V-1710-95(F23R). This single-stage engine had a 1275 bhp takeoff rating and was able to deliver 1125 bhp at a critical altitude of 15,500 feet.[93]

Allison provided the special V-1710-95(F23R) pusher engine for the XP-55. This photo shows to advantage the oil cooler mounted above and radiator cores below the engine, as well as the cowl flaps controlling the air exiting the compartment. (Bob Martin)

Detail of the exhaust manifolds on the first XP-55 shows a scoop to cool the exhaust port area and how the front stack was a single ejector rather than the dual ports on the rest of the cylinders. (SDAM)

Maximum level-flight speed of the *Ascender* was 390 mph at 19,300 feet, with a climb to 20,000 feet in 7.1 minutes. Service ceiling was 34,600 feet and empty weight was 5,325 pounds. The Air Corps expected good things from the aircraft in the interceptor roll.[94] A total of three XP-55's were built, with the first one flying on July 13, 1943.[95]

Curtiss-Wright was not able to push the development of the XP-55 because of demands being placed on it for other aircraft developments. This included delays due to the higher priority of their existing production programs and the need to develop the all-wood transport. As early as November 1942 the Air Corps was considering the design for conversion to jet power, it being a "natural" for conversion because of the lack of a tail.[96] The advances made during the early years of the war by conventionally configured aircraft, along with the coming of turbojet power, resulted in no follow-on contacts or interest in procuring the P-55 as a combat type.

Curtiss XP-60A

The Curtiss P-60 series was intended as a modernization of the dated P-40 airframe and configuration. It began as the Wright Field Classified Project No. MX-69. While it continued the Curtiss practice of evolution of its designs, the original concept was to build the aircraft to utilize the new Allison V-1710-45(F7R) two-stage engine with aftercooler. Contracts for the development of this engine were first established in December 1940, and then on December 2, 1941 and again on December 8, 1941 a second and third F-7R were purchased for the MX-69 airplane.[97]

The basic P-60 was to utilize a NACA laminar flow airfoil, have flush mounted inward retracting landing gear. A wide variety of engines were considered for installation. This included, in addition to the V-1710-45, a turbosupercharged Allison, and either a Continental or Chrysler engine.[98] Almost every available and/or developing liquid-cooled engine was considered at one time or another, but of the five P-60's built, only the Allison powered XP-60A (Curtiss Model 95A) with the single stage V-1710-75(F15R) is considered here in detail, as the F-7R powered model was never built. The demise of this model was probably a consequence of the subsequent Army decision to terminate development airplanes that were interfering with the critical need for production aircraft and engines at the onset of the war.

After the demise of the earlier Curtiss XP-53 project, the XP-60 became the Curtiss program intended to get a laminar flow wing into the air.

The second XP-53 airframe was powered by a single-stage Packard V-1650-1 Merlin and designated the XP-60. It featured a laminar flow airfoil and inward retracting landing gear. Note the intake to the updraft carburetor located in the cooling duct. (SDAM)

The XP-53 airplane was to have been based upon the P-40, but with a new NACA laminar flow wing and inward retracting landing gear, while mounting eight .50 caliber machine guns, and powered by the 1600 bhp Continental XIV-1430-3. Two XP-53's had been ordered in October 1940, though only two months into the project the Air Corps requested that the second prototype (AC41-19508) be revised to be powered by a single-stage Packard Built Rolls-Royce Merlin V-1650-1, hence becoming the XP-60. To expedite construction of both aircraft, Curtiss incorporated the rear fuselage and tail from the P-40D. The XP-60, Curtiss Model 90, made its first flight on September 18, 1941. Flight testing revealed changes were needed in the empennage, the modified aircraft then being identified as the Curtiss Model 90A.[99]

On October 31, 1941 Curtiss received a production contract for 1,950 P-60A's. They were to be different than the prototype by utilizing the 1425 bhp takeoff rated Allison V-1710-75(F15R) and a GE Type B-14 turbosupercharger.[100] This was the same engine and turbosupercharger combination to be used on the one-off Lockheed P-38K. Apparently the two-stage F-7R development was lagging to the extent that it was not considered for the production order.

Following the December 7, 1941 attack on Pearl Harbor, the Army determined to concentrate production and development on proven types. Consequently the eight gun P-60A's were canceled as of December 20, 1941. Work continued on several of the YP-60A's that were intended as prototypes suitable for later production. To utilize production capacity freed by the cancellation, Curtiss was contracted to become a second source of the Republic P-47D, designated as P-47G-1-CU.

The XP-60A, AC42-79423, first flew on November 11, 1942, but without its GE Type B-14 turbo. The initially planned turbo installation was considered to be a fire hazard. When the turbo was later installed the maximum level-flight speed of the XP-60A was found to be 420 mph at 29,000 feet, with a climb to 25,000 feet in 12.4 minutes. Service ceiling was 34,600 feet. Empty weight 7,806 pounds. With this performance the Air Corps expected good things from the aircraft in the interceptor roll.[101]

There was also to be an Allison V-1710-75 powered XP-60B (AC42-79425) aircraft. It differed from the XP-60A in that it was to use a Wright turbosupercharger in place of the more usual GE Type B unit. The aircraft was not completed with either this engine or the Wright turbo.

According to the Army, "The Curtiss XP-60A, B and C are what might be termed a consolation prize given to Curtiss when the decision was made for Curtiss to go into production on P-47 airplanes. Because of the relatively low performance of these three airplanes, the first of which was to have an Allison engine with General Electric turbo, the second an Allison engine with Wright turbo, and the third to be held for the later installation of the Chrysler liquid-cooled engine, the Pursuit Projects Unit did not feel that these airplanes had much to offer and recommended cancellation of

This is the XP-60A, the first of an order to Curtiss for 1,950 with V-1710-75(F15R) engines and incorporating a GE Type B-14 turbosupercharger. A lot was expected of this P-40 derivative with a laminar flow wing. The attack on Pearl Harbor terminated development in preference to increased production of existing types. (SDAM)

This is the prototype V-1710-127(E27) photographed September 20, 1945. Note the induction air intake between the Auxiliary Stage Supercharger and the General Electric CT-1 power turbine. It drove through the Aux Stage drive into the Accessories Section of the engine. The carburetor was the between stage PD-12K15, starter and generator were mounted on the remote propeller reduction gearbox. The modified P-38 type exhaust manifolds contained up to 100 inHgA of exhaust gasses exceeding 1725 °F. War Emergency Rating was 2980 bhp at 3200 rpm from 100 inHgA up to 11,000 feet on Grade 115/145 fuel. Continuous Cruise BSFC was an outstanding 0.405 #/bhp-hr at 1200 bhp and 26,000 feet. (Allison)

the contract. However, in view of the fact that the engineering was so far along and the cancellation of the contract would result only in saving of shop assembly time, this recommendation was never forwarded. Recently, when the pressure was exerted for a fast climbing interceptor as a competitor to the Jap Zero and the Focke Wulf 190, it was decided to utilize the XP-60C airplane for a quick installation of the two-stage R-2800 engine..."[102] The Curtiss P-60 had potential. In 1943 the P-60E was flown against the Merlin powered North American P-51B and found to be clearly faster at 20,000 feet.

Bell XP-63H, with Turbo-Compound V-1710-127(E27)

By 1944 Allison had its mass production problems under control and reliable engines were being delivered in large numbers. Allison engineers were then able to turn their attention to the increasing demands for improving the performance of Allison powered aircraft. Military tactics required improved takeoff performance, to carry ever greater quantities of fuel for the long range escort missions, and at the same time provide more power at ever higher altitudes, necessary to provide the tactical advantage in combat.

It was well known that recovery of some of the nearly 50 percent of fuel energy wasted in the exhaust gases could go a long way in meeting these objectives. So in 1944 Allison launched a major effort to improve the V-1710 and satisfy the demands from the operating theaters. The path selected was to develop a "turbo-compound" engine, a concept where the hot exhaust gasses power a turbine geared to the engine crankshaft where it can contribute to the power delivered to the propeller. This was an ambitious development for at the time no one had ever flown such a machine or ever demonstrated a practical engine/turbine setup.

The project went through several development and test phases. In these, component parts were subjected to extensive performance testing under conditions similar to those the final V-1710-127(E27) turbo-compound engine would see. The actual engine utilized components from the latest Allison standard: the V-1710-G that was developed and rated for operation at 3200 rpm, with high manifold pressures and able to operate at sustained high power. Its major component differences from earlier models involved using the 12 counterweight crankshaft and pistons having reduced compression ratio of 6.00:1 instead of the previous standard of 6.65:1.

The concept followed Allison's basic design approach of interfacing stand-alone components with minimum redesign of each, and thereby im-

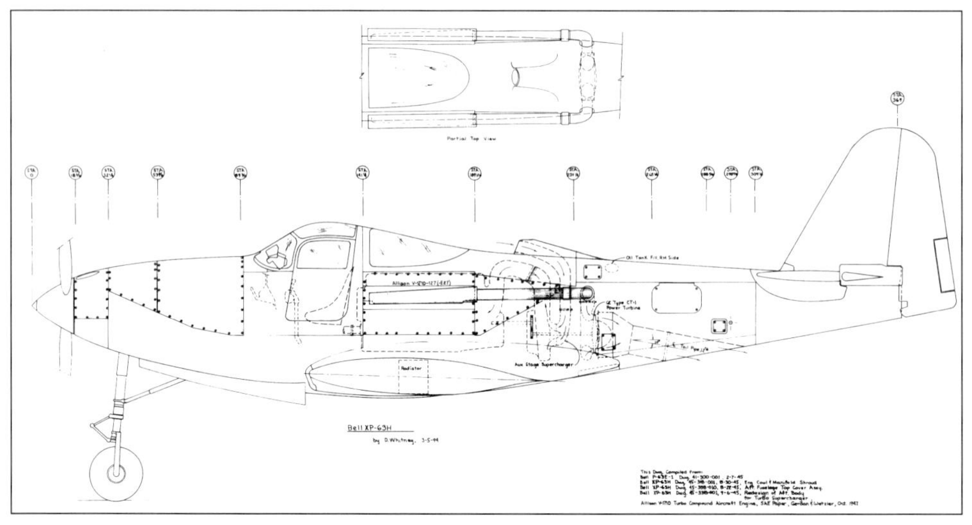

The turbocompound XP-63H would have been the first such aircraft in the world. With the V-1710-127(E27) able to crank out a maximum of 3090 bhp it should have been quite a performer. (Author)

posing minimum impact upon the production line. The engine was configured as a "E" model, for driving a remote reduction gear box, and provided with the usual integral engine supercharging stage. It was also mechanically coupled to an Auxiliary Stage Supercharger, giving the configuration a two-stage breathing capability and a high critical altitude. The Auxiliary Supercharger was modified to allow direct connection to the power recovery exhaust gas turbine. This device was derived from the GE Type CH-5 turbosupercharger, as used in 2800 bhp late model P-47's and R-4360 powered bombers.

The turbine was provided by GE as their CT-1 power turbine. Three of these were obtained.[103] It differed from those in the Type CH-5 turbosuperchargers in that the nozzles and buckets were arranged to drive the wheel in the opposite direction. It had a similar nozzle flow area of 16.5 square inches and operated with an efficiency of about 62 percent. No waste gate was provided, as it was intended to operate the turbine at all times and the sizing of the nozzle area defined the maximum turbine power and engine backpressure. GE established the maximum allowed exhaust gas temperature into the turbine at 1725 degrees Fahrenheit, this being significantly higher than the nominal 1600 degrees seen in normal turbosupercharger operations. It was clearly at the upper limit for the design and available turbine materials. This temperature limitation was the "Achilles Heel" of the project.

The project did not really get underway until Contract W-33-038-ac-7191 was approved in January 1945. At that point, the engine was described a having a 2.48:1 reduction gear and was to be suitable for use in the P-63 airplane with a total weight of 1,983 pounds. Three of the engines were to be built from existing V-1710-109(E22) cores, though only one and a working test mockup were actually built. The test mockup was completed in June 1945 and the first full engine that September. Military ratings were on Grade 100/130 fuel are shown in Table 6-7.

Table 6-7
V-1710-E27 Initial Ratings

Horsepower	rpm	Altitude, ft
1450	3000	Sea Level
1600	3000	25,000
1550	3200	29,000

Allison proposed adding an aftercooler to remove 50 percent of the compression heat, which would improve the performance as shown in Table 6-8.[104]

It was found by testing that with the engine at 3200 rpm, and at 11,000 feet, with water injection and 100 inHgA manifold pressure, the turbine caused a back pressure of 70 inHgA and delivered 890 bhp. At this manifold pressure, and with the turbine coupled to the engine, 2980 bhp was being delivered to the propeller! The turbine was directly connected to the engine through 5.953:1 gears contained in the Auxiliary Supercharger gearbox. The connecting shaft was concentric with the Aux Supercharger impeller, but the impeller continued to be driven through its hydraulic coupling. In this way the Aux Supercharger only consumed the amount of power necessary to provide the demanded pressure at the inlet to the engine-stage supercharger. This minimized the temperature of the air delivered to the engine and the likelihood of detonation.

Table 6-8
Aftercooled E-27 Performance

Horsepower	rpm	Altitude, ft	Fuel Grade
Military:			
1450	3200	Sea Level	100/130
1620	3200	30,000	100/130
War Emergency:			
1900	3200	Sea Level	100/130
2030	3200	24,000	100/130
2370	3200	Sea Level	150
2430	3200	18,000	150

The XP-63H project got far enough along that the turbo-compound V-1710-127(E27) did get installed. (Bell via Matthews)

Ducting of the hot exhaust gases from the engine to the power turbine was accomplished by using late model P-38 manifolds and fittings. These were suitable up to the maximum back pressures, which could reach 100 inHg, nearly 50 pounds per square inch at 1725 °F.

The one-off Bell XP-63H began as a P-63E-1, with modifications to accommodate the turbo-compound Allison engine and function as a flying test bed.[105] The fuselage was not lengthened since the P-63 series was already "stretched" to accommodate the two-stage Allisons, i.e. those with the Auxiliary Stage Supercharger. At first it appears that the installation was somewhat crude, in that the exhaust manifolds were simply run external to the fuselage. In fact, this was an attempt to accomplish a degree of exhaust gas cooling prior to passing it to the turbine, with its critical temperature limitations. The P-38 type collector manifolds were bolted directly to the V-1710 cylinder blocks, and were covered with a simple sheet metal shroud open on both ends. This extended into the airstream and continued to just past the rear of the cylinder blocks. From this point, the circular pipes were external to the fuselage back to the point where they entered through a circular cut-out, connecting to the turbine nozzle box situated on the centerline of the aircraft.

The turbine exhaust was then collected by a shroud and ducted to a 12 inch diameter ventral exit also on the fuselage centerline. This required the elimination of the ventral stabilizing fin normally seen on late model P-63's, hence the profile of the XP-63H resembled the earlier P-63A and C models. At maximum power and altitude this exhaust duct configuration should have been able to contribute an additional 100-200 pounds of net propulsive thrust, though no exact figures have turned up. The only other significant modification was the relocation of the oil tank, details of which have not been found, though it appears to have remained near the usual P-63 location.

Table 6-9 Turbocompound V-1710-127(E27) Ratings		
Rating	BHP/RPM/Altitude, ft/MAP	Comments
War Emergency	3090/3200/28,000/100 inHgA	Wet
Military	2320/3200/28,000/ 85 inHgA	Dry
Normal	1740/3000/33,000/ 52.5inHgA	Dry
Maximum		
Cruise	1340/2700/26,000/	Dry, BSFC 0.365 #/hp-hr

All of this suggests that the XP-63H may have had a fairly "aft" center of gravity. The V-1710-127 was expected to weigh 1,950 pounds, dry. The dry weight of the V-1710-109 for the P-63E weighed in at 1,660 pounds. The CT-1 alone probably weighed about 150 pounds, and was certainly located quite far aft. Possibly, the ventral fin removal assisted in restoring balance, but a fair amount of ballast was going to have to be fitted into the nose compartment. To this end the starter and generator were mounted on the propeller reduction gearbox.

Even so, the V-1710-127(E27) was quite a performer! Allison's proposed guaranteed performance, on AN-F-33 Grade 115/145 fuel and incorporating a turbine able to operate at 1,950 °F is given in Table 6-9.

The "Achilles Heel"

Lean Fuel/Air mixtures are known to cause high exhaust gas temperatures. Allison found that even the normal "rich" mixture ratio of 0.08:1 pounds of fuel per pound of air, and water injection sufficient to depress detonation at 100 inHgA, the exhaust gas temperature would still be 1825°F.

The result was that they believed that a air-cooled power turbine able to operate with 1950 °F gas would have to be developed to realize the promise of the turbo-compound.[106] While Allison proposed to provide such a device, later in 1946 they changed their mind, believing the effort could be better directed to establishing Allison's role in the field of jet propulsion. Consequently the program was terminated and the V-1710-127 and its XP-63H testbed never flew.

In 1942 Curtiss-Wright Aeronautical Corporation had begun investigations and design studies that finally led to Model Testing the TC 18, a turbo-compound, 18 cylinder R-3350 in 1950. Wright had begun development testing in 1948, and developed a configuration achieving the desired high efficiency. This engine went on to become the mainstay of the final generation of piston powered airliners and bomber aircraft such as the Douglas DC-7, Lockheed Constellation, and Lockheed P-2 Neptune. To accommodate the high exhaust gas temperatures Wright developed air-cooled blading for their power turbine.[107] It was a good approach, but it still lagged several years behind the Allison program that developed the first turbo-compound aircraft engine, the V-1710-127.

Turbosupercharged Bell P-63

In 1943 Bell proposed their Model 33, a P-63 airframe mounting a Wright 800TSB-A1 turbosupercharger in place of the usual Allison Auxiliary Stage. The turbo would have been controlled by a GE turbo-regulator and was to incorporate a Packard type (Merlin 66) aftercooler as an intercooler between the turbo and engine stage superchargers. Allison offered the V-1710-E20 for use in this unique installation. By totally enclosing the turbo within the fuselage minimum drag should have resulted. The Wright turbo featured air-cooled turbine blades which would have been beneficial in the close coupled arrangement. The project did not proceed and no Military designation was assigned. It was sea-level rated for 1425 bhp to 27,000 feet from 53.5 inHgA.

North American P-82 "*Twin Mustang*"

Although built by North American Aviation, and appearing as if two light weight P-51 fuselages sharing a common wing and empennage, the P-82 was actually an entirely new and re-engineered aircraft. The twin-fuselage concept was not unique to the P-82, for it had contemporaries such as Germany's Heinkel He 111Z and Messerschmitt Bf 109Z "Zwilling" (meaning "Siamese Twin").[108] It is hard to determine who first conceived the arrangement, but it is clear that both the Army Air Force and North American Aviation believed the concept possessed features to resolve the specific problems being faced by Allied escort fighters over Western Europe when it was proposed in 1943/1944.

The design was tailored by North American to provide the essential features required for long-range fighter escort: high performance, the reliability of two engines, increased fuel capacity, and less pilot fatigue because of the two pilots. Such capability was critical at this point in the war, as learned by the Army Air Force in its disastrous missions to Schweinfurt, Germany on August 17 and October 14, 1943. Without escorts, over 120 heavy bombers and 1,200 aircrew had been lost on the two missions.

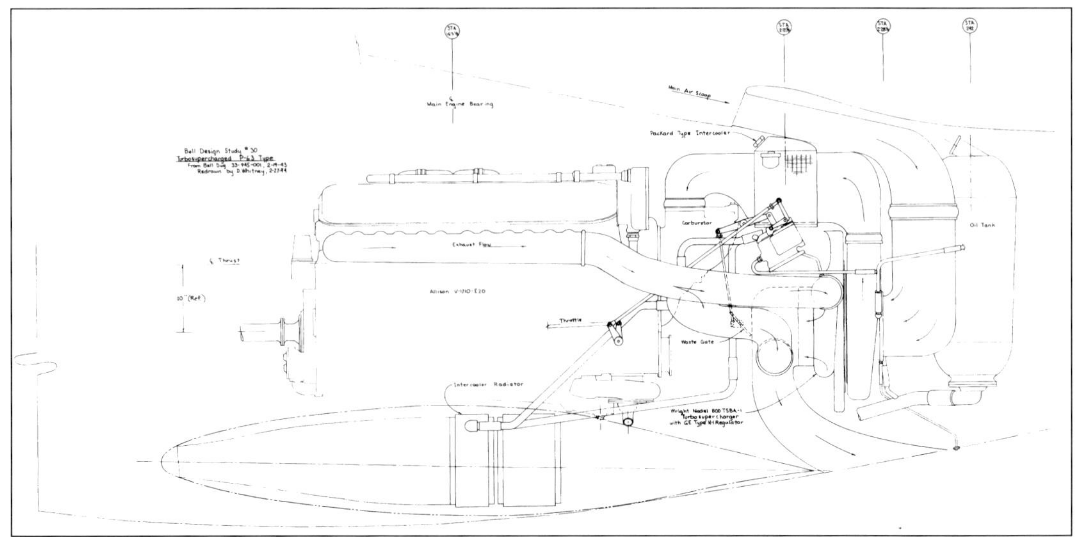

In their Model 33 version of the P-63 Bell proposed to pair the Allison V-1710-E20 with a Wright turbosupercharger having air cooled turbine blades to compensate for the high temperature exhaust gasses flowing past the highly stressed turbine blades. The configuration was to incorporate a liquid-cooled aftercooler from the two-stage Merlin 66 to control induction mixture temperatures. (Author)

The P-82E/F airplanes were sleek and offered outstanding performance. Opposite rotating V-1710-143/145(G6R/L) engines with two-stage supercharged engines gave high critical altitude, speed and long range. This is the second of 100 P-82E's. (Allison)

North American Aviation submitted its unsolicited and unique design to the Army Air Force on January 7, 1944.[109] The Air Force responded on January 15, 1944 with its Authority For Procurement, and in February issued a letter contract for four experimental P-82's and one static test airframe. The first two aircraft were to be powered by the latest versions of the Packard Merlin, V-1650-23/25, and the next two by Allison V-1710-119(F32)'s. These aircraft were designated XP-82 and XP-82A respectively.[110] Later in the month, an order was placed for 500 Merlin powered P-82B production models. Allison lost out on this early competition as they had proposed to use the two-stage, 1500 bhp for takeoff, V-1710-139/141(F33R/L) in the P-82B's.[111] The need for the aircraft was critical, so it rapidly received the Air Forces' support, as well as their concurrence with North American's recommendation to stay with the Merlin, given that the V-1710-119 was not yet ready for Model Testing.

After the end of the war in September 1945, the P-82A was canceled when Wright Field realized that the P-82Z would be in series production by the time the V-1710-119 for the P-82A would be available. The "Z" in the P-82Z was the designation given to aircraft authorized as a device for maintaining aircraft manufactures during the transition to peace-time production levels. In this case the "Z" airplanes were the twenty Merlin powered P-82B's. The first flight of a P-82B was on October 19, 1945. The first delivery was in March 1946.[112] Production of the Merlin was terminated at the end of the war and any further P-82 production would have to use an American engine.[113] The V-1710-119 was Allison's answer to the two-stage Merlin V-1650 series, as it incorporated the Auxiliary Stage Supercharger along with an Aftercooler mounted in a location similar to that on the two-stage Merlins.

Even for a firm with the skill, resources, and priorities of North American, developing a completely new aircraft in the middle of the war was a troubling and vexing undertaking. As a result only the 20 Merlin powered "Z" airplanes were produced on the original production order.

From the beginning of the program the government had always wanted to have an all-American engine power the P-82's.[114] They not only wanted it to be a stronger engine than the Merlin, but also to eliminate the $6,000 royalty required to be paid for each license built Rolls-Royce Merlin.[115] With the end of the war, Packard production of the V-1650 Merlin was immediately terminated and any question about which engine would be used in any further P-82 production.[116]

It was determined on December 7, 1945 that 250 Allison powered P-82's would be built. This was subsequently divided between 100 P-82E "Escort Fighters", to be assigned to what became the Strategic Air Command, and 150 P-82F "All Weather Interceptors" or "Night Fighters" for the fledgling Air Defense Command, though the last 59 were delivered as the P-82G.[117] All were to use V-1710-143/145(F36R/L) engines, though by May 1946 Allison had revised to specification and defined the model V-1710-G6R/L instead. The engines differed primarily as a result of changing the balance of the gear ratios between the engine and the Auxiliary Stage Superchargers.[118] The V-1710-143/145 Military designation was retained for the new model.

In anticipation of ordering engines for the P-82E's, on October 8, 1945 the AAF directed Allison to designate their Models V-1710-F36R/L as Air Force Models V-1710-143/145 respectively.[119] Later that month, Allison agreed to retain and store (at no cost to the government) parts then in the V-1710 Termination Inventory that could be subsequently used in the production of the F-36 engines.[120] In March 1946, and preceding the actual engine purchase order, the Air Force asked Allison to establish the earliest date they could deliver 750 V-1710-143/145 engines for the P-82 program. This schedule was to be used as the basis for determining the date upon which North American Aviation would commence delivery of the P-82E's.[121]

As the NA-144, the P-82E first flew on February 17, 1947, powered by Allison V-1710-143/145(G6R/G6L) engines. The marriage was not a particularly happy one.

Engine problems caused the delay in delivery of finished airframes for a year, and added substantially to the cost of the program. Edward Schmued, *Mustang Designer*, describes the central problem:

"The engine we had intended for this design was a Merlin. The United States Air Force was tired of paying a $6,000 royalty to England for each Merlin engine built in this country by Packard, on a royalty basis. So they decided, then, to substitute an Allison V-1710 for the Merlin."

"Now the Merlin engine had a very high rating, 2,270 hp with 90-inHg manifold pressure, and the Air Force told the Allison people they had to duplicate this performance. It was obvious that the way their engine was built was not suitable for these high manifold pressures. The British built a backfire screen into their engine, which made it run properly. But the Allison people refused to do that."

"To help the situation along, we actually modified an Allison engine with a backfire screen that worked fine. But then, the Secretary of Defense was powerful enough to override the Air Force and told Allison not to do anything. Which, of course, left the F-82 with a rating far below that of a good trainer. The manifold pressure was reduced to 60 inches to keep this thing from backfiring into the blower case and damaging the engines. This was a very sad situation, because it really ruined the project. The Secretary of Defense (James Forrestal) favored General Motors and, I think, he had a good idea of how to protect the Allison people."

"The Allison people were already on the way to build jet engines and did not really like to go back and build reciprocating engines...We had demonstrated the Allison engine with backfire screens, but it didn't make any difference, because Allison never agreed to install them. We just couldn't get the support from the Department of Defense."

As a counter-point, George Gehrkens, the P-82 project engineer, stated that he feels that the above characterization of the political situation by Schmued is very overdrawn.[122] According to M.J. Hardy "the Twin Mustang was well liked by its pilots and carried on its single-engined predecessor's tradition of outstanding performance."[123]

When Don Wright, one of Allison's Field Service Representatives at the time was asked about the P-82 experience, his comment was,

"It was an electricians nightmare, and it wasn't the fact that the engines wouldn't lift the airplane off or fly it, rather you could never keep it in commission. All of the systems in there were all of the latest electronic advancements from the labs, this was done in a period when the labs at Wright Field put all of these requirements into this airplane. It was just a mess all of the time, we had water injection for takeoff, we had WER, and we had a Speed Density pump instead of a carburetor. Really though, the problem with the airplane was to get it to stay in commission. I don't know the percent of time the engine caused it to be out of commission, but it seamed to always be due to electronics."[124]

For several years after the war, Allison continued to offer improved V-1710 models for both military and commercial applications. Had the move to the jet engine not developed, it is clear that these later V-1710's would have been in the leading high performance aircraft of the day.

As an indication of the concern about costs and the post-war reduction of military budgets, Allison was asked by the government to consider utilizing some of the V-1710's then in storage. These were complete engines and the Air Force wanted them used as a source of parts for the upcoming production of the needed V-1710-143/145 engines.[125]

XP-82A Only one of the two ordered aircraft was constructed, AF44-83888, being accepted in October 1945, some three months after the first XP-82. Its power plants were the V-1710-119(F32), a two-stage engine equipped with the Bendix Speed-Density carburetor and an aftercooler.[126] As such, this engine was the match of the Merlin's installed in the XP-82 and P-82B's, though in the event the engine proved troublesome as a result of the state of development of the speed-density unit.[127]

P-82E Without the rush of war, considerable time was taken to incorporate the latest technology and features into these production aircraft. While the contract for the aircraft was officially signed on October 10, 1946, some 90 change orders had accumulated by January 1950. First flight by a P-82E was on February 17, 1946, with a total of 100 aircraft ultimately being delivered. The aircraft was powered by the Allison V-1710-143/145(G6R/L), two-stage engines with water injection for takeoff and War Emergency power. In consideration of the need to get the aircraft into service, the first 200 engines were restricted from operating at War Emergency settings at the time of delivery, though this was to be rectified by Allison at a later date. For this reason, most handbooks and technical material relating to the P-82E/F do not provide War Emergency rating information. The Model Test engine was a G-6L and it successfully passed the

This is AF46-405, the first of the 150 P-82F radar equipped all-weather/ night-fighter version of the twin-Mustang. Powered by the same two-stage V-1710-143/145 as the P-82E. The night fighter was fitted with flame damping exhaust stacks to prevent the pilots from losing their night vision. (NAA via SDAM)

Table 6-10
Engines for the P/F-82's

Engine Model	Engine S/C Ratio	Aux S/C Ratio	Engine S/C Impeller Dia, in[5,6]	Weight, pounds	Takeoff, bhp/rpm/MAP	Military Rating, bhp/alt/MAP	War Emergency, bhp/rpm/alt/MAP[1]
V-1650- 23/25	6.391:1	None	10.10/12.0	1742	1380/3000/61.0	1500/3000/15,700/61.0	2270/3000/ 4,000/90
V-1650- 23/25	8.095:1	None	10.10/12.0	1742	Not Allowed	955/3000/30,000/46.0	Not Allowed
V-1710-119(F32R)	8.10:1	7.64:1	9.50	1750	1500/3200/62.5	1200/3200/30,000/52.0	2100/3200/ 4,000/__
V-1710-139(F33R)	8.10:1	7.64:1	9.50	1560	1500/3000/____	1150/3000/27,500/____	____/3200/______/__
V-1710-141(F33L)	8.10:1	7.59:1	9.50	1560	1500/3000/____	1150/3000/27,500/____	____/3200/______/__
V-1710-143(F36R)	7.48:1	8.80:1	10.25	1560+	1850/3200/Wet	1250/3000/32,600/____	2200/3200/12,700/__
V-1710-145(F36L)	7.48:1	8.80:1*	10.25	1560+	1850/3200/Wet	1250/3000/32,600/____	2200/3200/12,700/__
V-1710-143(G6R)	7.48:1	8.08:1	10.25	1595	1600/3200/74.0	1250/3200/32,700/____	2250/3200/_____/100
V-1710-145(G6L)	7.48:1	8.03:1	10.25	1595	1600/3200/74.0	1250/3200/32,700/____	2250/3200/_____/100
V-1710-147/9(G9R/L)	7.48:1	8.08/8.03	10.25	1625	1600/3200/74.0	1250/3200/32,700/68.0	2250/3200/_____/100
V-1710-G10R/L	8.10:1	NA:1	10.25	NA	2000/3200/____	2250/3200/28,500/Dry	2550/3200/20,000/Wet

NOTES on P-82 engines:
1. Ratings noted as "Wet" are with Water/Methanol Injection to control detonation.
2. Above engines all rated on Grade 100/130 fuel except "G" models on Grade 115/145.
3. All engine weights are "dry", and do not include radiators or oil coolers.
4. All engines in table use 6.00:1 compression ratios.
5. First stage Merlin supercharger impeller is 12.0 inches in diameter.
6. Allison Auxiliary Stage, or first stage, supercharger is 12-3/16 inches in diameter.
7. All of the V-1710's intended for the P-82 used 2.36:1 propeller reduction gears.
8. * Approximate ratio.

150-hour test in October 1946, but troubles resurfaced on the first four aircraft whose acceptance dates were between September and December 1946. The Air Force redesignated them as P-82A's and restricted their use to testing. Allison was asked to have two-stage carbureted V-1710-139/141(F33R/L) engines available for these airplanes, pending resolution of the V-1710-143/145 problems. It is not known if they were used.[128]

P-82F A contract change limited the number of P-82E's to 100, with the remaining 150 aircraft being configured for air defense duties as "All Weather Interceptors" and equipped with APS-4 or APG-28 radar. Ninety one were delivered as P-82F's. Like the P-82E's, these aircraft were powered by the V-1710-143/145.

P-82G Nine of the P-82F's were equipped with the SCR-720C radar that had been prototyped on the P-82C. Otherwise, they were P-82F airframes, of which North American delivered another fifty.

P-82H Five of the P-82G's and nine of the P-82F's that were specially modified for service in the cold climate of Alaska.

F-82's Upon becoming an equal Service, the new U.S. Air Force changed the designations of all of its "Pursuits", existing and forthcoming, to "F" for Fighter. Consequently, all references to the aircraft after June 1948 identify it as an "F-82." Suffixes from the earlier models were retained unchanged.

F/P-82 Program Summary

The conflicting priorities and interests of all of the involved parties resulted in giving the P-82 program a bad name. Alternatively, when the aircraft was working, it was truly a high performance fighter and one that proved itself during the early days of the Korean War where it was the aircraft to shoot down the first three Yak's during the conflict.

Production of the V-1710-143/145 was trouble plagued as a result of the drastic reduction in force at Allison that occurred at the end of the war. Furthermore, it was obvious that the era of the high performance piston engine was drawing to a close, as demonstrated by the governments' assignment to Allison of the production contract for the General Electric J33 jet engine. The conversion from recips to jets was all consuming, and it took awhile to restart production of the V-1710 and to resolve the issues relating to the use of earlier produced parts.

Development of the V-1710-143/145 was prolonged by the commitment to use the new Bendix Speed Density carburetor. Allison had been working with this unit since July 1944 on the V-1710-119(F32R) program, and its installation in the XP-51J had been essentially a failure because of the state of development of the Speed Density carburetor. Things were little better for the G-6's and their Bendix SD-400-D1 speed density carburetor, though with it, the engine successfully completed the government 150-hour Model Test.

When the engines entered Squadron service they were plagued by spark plug fouling, Auxiliary Stage Supercharger failures, oil spewing through breathers and backfiring at high and low power. In addition, they were often running rough and tended to surge.[129] Most of these problems are typical of improper fuel control, specifically running lean. The plug fouling was due to excessive oil leaking past the piston rings, a result of the changes made to the piston clearances necessary to allow running at the intended very high power ratings, as well as the generally poor piston ring designs and material available at the time.

As the F-82F, the Twin Mustang entered Air Defense service in March 1948 and were based on both coasts of the continental U.S., as well as a squadron in Alaska and another in Japan. When the Korean War suddenly erupted in 1950, only the F-82G had the necessary range to reach the battlefield, where it claimed the first aerial victory of the conflict. Soon replaced by jets, the aircraft continued to provide defense for the country. The last active duty mission was flown by F-82G AC46-377 from Ladd Field Alaska on November 12, 1953. As an example of aircraft usage, during the five year career of the aircraft it had accumulated 715 hours of flight.[130]

The Official Air Force history of the P-82 program states, "The major problem in the F-82 program was the failure of the government furnished engine."[131] Certainly it had problems, and working with its highly strung carburetion system must have been a frustration for the pilots and ground crews supporting the aircraft around the world. The Air Force and North American were requiring extremely high performance from a production engine. Unlimited Air Racers in the 1990s are not able to develop power equivalent to the War Emergency ratings on the P-82's for more than three or four 15 minute runs before their highly tuned and carefully prepared Merlin's self-destruct. To qualify for its War Emergency ratings, Allison had to successfully perform a 7-1/2 hour endurance test. The P-82's were literally running on the "racers edge"!

What Could Have Been

Allison continued to work on advanced power plants for the P-82. While the port fuel injected V-1710-147/149(G9R/L) reached the point of a military designation, the most interesting would have been a derivative of the V-1710-127(E27) turbo-compound engine designated as the V-1710-G10. In this respect Allison did design studies of a compact integral two-stage turbine feed-back engine for the P-82. The G-10 would also have used port fuel injection, but would have also been aftercooled and would have fit into the P-82E without altering the distance between the engine firewall and propeller disk. Allison offered that the G-10 model should be the successor to the G-6, rather than the already established G-9. It was expected that the Model Test of the G-10 would be completed within 24 months of go-ahead by the Air Force.[132]

Table 6-11
V-1710-G10 Turbo-Compound Performance

Horsepower	rpm	Altitude, ft	Comments
Takeoff: 2000	3200	Sea Level	Grade 115/145, Dry
Military: 2250	3200	28,500	Critical Altitude, Dry
War Emergency:			
2400	3200	Sea Level	Wet
2550	3200	20,000	Critical Altitude, Wet
Normal Cruise:			
1400	2700	Sea Level	
1700	2700	30,000	

The primary purpose to the turbo-compounded G-10 was not necessarily additional power, rather to provide considerably improved fuel economy. Discussions with North American Aviation had determined that the 3000 bhp available would have required extensive propeller changes and upset airplane stability, so the performance shown in Table 6-11 was offered.

At the same time Allison offered another advanced two-stage engine, their V-1710-H, which would have eliminated the engine-stage supercharger. It would have been replaced by an Allison designed two-stage aftercooled centrifugal compressor driven by a two-stage exhaust turbine. Like the G-10, this engine was intended to fit into the space and engine mounts of the P-82E. While the performance ratings would not have changed from the G-10, the package should have been simpler and considerably more efficient.

NOTES

[1] *Power Plants for Project "A", Restricted Project No. M-3-35, and Project "D", Restricted Projects Nos. M-4-35 and M-5-35*, Air Corps Materiel Division letter to Chief of the Air Corps, November 9, 1935. NARA RG18, Entry 167D, Box 46.

[2] This description of events and designations is contrary to reports by many authors, for example some have reported that the V-3420 was to be used in the Boeing XB-15. See endnotes for sources.

[3] Wright Field Report No. TSEST-A2, 1946.

[4] Bowers, Peter M. 1966, 199-201.

[5] Jones, Lloyd S. 1962, 39-44.

[6] *Casualty Rate on Experimental Airplanes*, June 10, 1949. NARA RG342, RD 3957 Box8096.

[7] Allison Photolab log, April/May 1942.

[8] The aircraft was the ninth production B-17E, AC41-2401.

[9] Allen, Grant. 1969, 14-17.

[10] Allen, Grant. 1969, 15.

[11] Allen, Grant. 1969, 16.

[12] Allen, Grant. 1969, 14-17.

[13] Caidin, Martin. 1960. Black Thursday,

[14] *Evaluation-Circular Proposal 39-770, Board Report*, August 19, 1939. NARA RG18, Bulky Box 711, CP 39-770.

[15] Francillon, Rene J. 1979, 712.

[16] The specification for the DS-312A identifies the engine as the V-1710-E5, but with a GE Type B turbosupercharger and driving a counter-rotating pusher gearbox via an extension shaft. Allison specifications do not detail such a configuration, but the linkage to the sea-level engine is appropriate as the E-5, an altitude rated engine as defined by Allison Spec 121-E, was the evolved from of the sea-level rated E-3 described by the earlier Spec 121-A/B. The E-5 was simply the current model at the time while the E-3 was the intended production version of the XV-1710-17(E2) in the XP-39.

[17] Ludwig, Paul A. 1996, 34.

[18] Communication with Paul A. Ludwig, author of *Hap Arnold's Ghost Fighters*. The DS-312A used the NACA 24-212 airfoil at the root, with dihedral of 0 degrees, 43 minutes, with a span of 40 feet. It was sweptback 3 degrees and had 5 degrees of incidence. Fuel capacity was 160 gallons, gross weight 6,800 pounds, length 30 feet, 1/2 inch, the tail height was 108 inches. Maximum speed at the 25,000 foot critical altitude was 500 mph, though it was to be capable of 534 mph at 30,000 feet.

[19] *XB-42 and XB-42A Mixmaster, Douglas Airplane Company, Inc.*, Appendix II, p.509-515. Document in Library of Congress, abridged copy obtained from Ben Giebler, July 1991.

[20] Atkinson, Robert P. 1989.

[21] *Case History of the B-42 Airplane Project*, 1948, 140-155.

[22] Wood, Carlos. 1947, 22-29.

[23] Francillon, Rene J. 1979, 376-7.

[24] Letter from Donald P. Frankel, Allison Field Engineer, LA Office, to Wm C. Campbell, Director of Communications, Allison Gas Turbine Division, GMC, April 2, 1990.

[25] Other accounts of the installation state that there were five sections of P-39 extension shafting used, careful analysis of recently available photos, sketches and drawings in *Pilots Handbook of Flight Operating Instructions for Model XB-42A Airplane* confirm that six sections were used.

[26] Technical Order Compliance Record for V-1710-137 AAF s/n A-071175/6, and Propeller Historical Record for XB-42A #43-50224, provided by NASM Silver Hill, 1995.

[27] *Allison Engine Production*, from Allison Service Department Notebooks. Allison Engine Company, June 1995.

[28] As detailed elsewhere in this volume, Allison achieved a "two-stage" gear driven supercharger capability in its engines by connecting a separate "Auxiliary Stage Supercharger" to the accessories section, driven from the crankshaft by a gear connected to the starter drive.

[29] NACA Wartime Report A-78, 1944.

[30] Boyne, Walt. 1973, 12.

[31] Boyne, Walt. 1973, 12.

[32] *XB-42 Converted to the XB-42A*, summary information in file at San Diego Aerospace Museum.

[33] See information contained in the section detailing description of the V-1710-E23B engine.

[34] Letter from Donald P. Frankel, Allison Field Engineer, LA Office, to Wm C. Campbell, Director of Communications, Allison Gas Turbine Division, GMC, April 2, 1990.

[35] *Report of Inspection of Parts of Douglas XB-42 Airplane, AAF #43-50225,* Col. Jean A. Jack, Bolling Field, Washington, D.C. January 23, 1946. NARA, RG 342, RD 3766, Box 3766.

[36] *Report of Major Accident: XB-42 #43-50225*, December 16, 1945, Pilot, Lt. Col. Fred J. Ascani. Includes information on engine models and serial numbers. Copy from HQ AFSA/JARF, Kirtland AFB NM 87117-5670.

[37] Ascani, Maj. General Fred J. and John M. Fitzpatrick, 1993, 30-31. General Ascani, who was the pilot on the flight, reported in his article that, "In retrospect, I later decided that I may not have waited long enough in any one position for the fuel to reach the engine."

[38] This gear change has yet to be confirmed, but altitude ratings suggest it. Author.

[39] The actual jet engines used were: Westinghouse X19XB2B, AAF #WE-000039, accepted 12-30-1947 and installed on the XB-42A 8-30-1948 with a total of 13.2 hours of operating time. Westinghouse X19XB3, AAF #WE-000056, accepted 8-29-1946 and installed on the XB-42A 1-27-1948 with a total of 26.7 hours of operating time. The Westinghouse X19 engine was subsequently assigned the designation J30, which was the first assignment in the new "J" for "Jet" classification.

[40] *XB-42 Converted to the XB-42A*, summary information in file at San Diego Aerospace Museum.

[41] *XB-42 Converted to the XB-42A*, summary information in file at San Diego Aerospace Museum.

[42] Technical Order Compliance Record for V-1710-137 AAF s/n A-071175/6, and Propeller Historical Record for XB-42A #43-50224, provided by NASM Silver Hill, 1995.

[43] *Case History of the B-42 Airplane Project*, 1948, 140-155.

[44] *XB-42 Converted to the XB-42A*, summary information in file at San Diego Aerospace Museum.

[45] *XB-42 and XB-42A Mixmaster*, Appendix II, p.509-515 of document provided by Ben Giebler, 7-29-1991.

[46] *The Transport of Tomorrow-Today*, Douglas AIRVIEW, Volume XII, Number 10, October 1945, p.6-7.

[47] Betts, Ed 1997, 110-116.

[48] *Douglas Reveals Features of the DC-8,* Aero Digest, November 1, 1945, p.60-61.

[49] This "two-speed" engine appears to be an extension shaft configured companion to the engine Allison was preparing for the Douglas XC-114 and XC-116 developments of the usually radial air-cooled engined C-54 or DC-4 transport. As such it may have been designated by Allison as the V-1710-E29, and would not have received a military "dash" number designation. In this respect it would have been an "Americanized" version of the Canidair "North Star" which was a Rolls-Royce Merlin powered C-54 derivative that was produced in fairly large numbers. The engine intended for the XC-114 and XC-116 was the V-1710-131(G3R), using new 6.50:1 compression ratio pistons and incorporating a 10.25 inch diameter engine stage supercharger impeller (in place of the here-to-fore standard 9.50 inch unit) and provided with a revised accessories section offering 7.48:1 and 9.60:1 step-up gear ratios. Engine weight was a respectable 1,475 pounds and ratings were 1600/3200/Takeoff/61.7 inHgA and 1150/2700/15,500 for Hi-gear cruising with a respectable fuel consumption of 0.43 pounds per horsepower per hour.

[50] *New U.S. Transports; The Unconventional DC-8 and Martin Model 202, FLIGHT*, October 11, 1945, p.402.

[51] *New U.S. Transports; The Unconventional DC-8 and Martin Model 202, FLIGHT*, October 11, 1945, p.402.

[52] Francillon, Rene J. 1979, 714-715.

[53] *Yesteryear.. Douglas XB-42 "Mixmaster"*, by R.E. Williams-Publications, p.35.

[54] Allison Division Inter-Office Memo, *Conferences at Douglas Aircraft Corporation on C-54E Power Plant*, January 6, 1945. Copy via J. Hubbard.

[55] Francillon, Rene J. 1979, 329.

[56] Francillon, Rene J. 1979, 410-411.

[57] Francillon, Rene J. 1979, 330-345.

[58] One official Army publication states that the engines for the XC-114 were to be V-1710-133's. That engine was the high powered two-stage E-30 model, an engine configured entirely differently than the G-3. The confusion probably comes from the fact that the E-30 was the model incorporating the latest Allison developments, and as such, it was the reference model for the later engines, including the G series. Author. Also see *Army Aircraft Model Designations*, Chief of Engineering Division, Report TSEST-A7, June 1, 1946.

[59] Allison Division drawing, *3-View V-1710 Powered DC-4*, SK-7-1022, dated 9-3-1946.

[60] *Brewster Aeronautical Corporation Model P-22 Airplane Characteristics*, February 16, 1939. Provided by Dr. George R. Inger, Ames, Iowa, January 1997.

[61] *Circular Proposal 39-770, Airplanes, Interceptor Pursuit (Single Engine) Board Recommendations*, September 2, 1939. NARA RG18, File 452.1-Airplanes, CP 39-770, Box 710 (BULKY).

[62] Matthews, Birch. 1996, 372.

[63] Confidential *Descriptions of Experimental Aircraft*, July 1, 1940. NARA RG18, Entry 293B, Box 244.

[64] *Evolution of the P-40, Aircraft and Jets*, Vol.1, No.1, June 1946, p.17.

[65] Teletype E-106 and PROD-T-665 of 6-10/12-1941 to Chief, Materiel Division, Wright Field. NARA, RG 342, RD 3466, Box 6841.

[66] Teletype messages of 10-17/18-1941 to Chief, Production Engineering Branch, Wright Field. NARA, RG 342, RD 3466, Box 6841.

[67] See photo in: *Evolution of the P-40, Aircraft and Jets*, Vol.1, No.1, June 1946, p.17/19, also San Diego Air Museum Photo file of XP-40K, as well as Allison photo.

[68] *XP-40Q-2 Airplanes, Two-Stage Engine Installation with Carburetor between Stages*, Curtiss-Wright letter to Commanding General AAF, Materiel Command, Wright Field, May 1, 1943.

[69] *Development of Increased Horsepower and Altitude for the Allison V-1710 Engines, Particularly for N.A.A. P-51 Airplane*, letter from R.M. Hazen of Allison to Commanding General, AAF Materiel Command, C1L2AE, November 2, 1943. RG18, file 452.13, Box 1260.

[70] Letter from Chief, Technical Staff, Engineering Division @ W.F. to Curtiss-Wright, December 13, 1943. NARA RG 342, RD 3834.

[71] Bowers, Peter M. 1979, 492.

[72] Green, William and Gordon Swanborough, 1977, 59-62.

[73] *Assignment of XP-40Q Airplane No. 42-9987 to Post Operations*, Letter to Commanding Officer, Wright Field from Chief, Technical Staff Engineering Division, June 26, 1944.

[74] *XP-40Q, 42-9987 Airplane*, Letter from Chief Inspector to Captain D.T. Smith, July 31, 1944.

[75] *Request for Priority Repair of XP-40Q, 42-9987*, teletype from Materiel Command to Commanding General AAF, Washington D.C., August 14, 1944.

[76] *XP-40Q-2 Performance*, Curtiss-Wright letter to Commanding General AAF, Materiel Command, Wright Field, January 10, 1944.

[77] Shamburger, Page and Joe Christy, 1972, 251.

[78] *War Emergency Ratings on Allison Engines*, letter from Propeller Laboratory to Chief, Experimental Engineering Section, Wright Field, October 16, 1942. NARA RG 342, RD 3774.

[79] Also known as "Paddle" blades for they were unusually wide for the time. They were particularly effective at high altitudes, but the primary advantage was that they allowed for a smaller diameter and could therefore be turned at a higher rpm before encountering high mach effects. This is the reason that the V-1710-F15 had the increased reduction gear ratio.

[80] Bodie, Warren M., 1991, 170-171.

[81] Technical Report No. 5653, 1947.

[82] *P-51B Airplane, Design Study Around the Allison V-1710-93 Engine*, letter from Assistant Chief of Air Staff to Commanding General, Materiel Command, September 1, 1943. NARA RG18, File 452.1, Box 698.

[83] Allison letter to Commanding General, AAF Materiel Command, Wright Field, November 5, 1943. NARA RG18, File 452.13, Box 1260.

[84] North American Aviation Report No. NA-8033, 1944.

[85] Wagner, Ray. 1990, 145-146.

[86] Gruenhagen, Robert W. 1980, 123.

[87] *Flight Test of V-1710-143 Engine in P-51J Airplane*, letter from Wright Field Power Plant Laboratory to Allison, May 31, 1946. The letter authorized converting V-1710-133 engine to -145 for test. NARA RG342, RD3775, Box7411.

[88] Technical Report No. 5653, 1947.

[89] AAF Letter to Allison, *Flight Test of V-1710-143 Engine in P-51J Airplane*, dated 5-31-46, re: Contract W33-038-AC-2123.

[90] Matthews, Birch. 1996, 104.

[91] Technical Report No. 5306, 1945.

[92] Data plate on P&W X-1800 serial number X-95. Design started December 1939, discontinued November 1940. Vertical "H" configuration with 2600 cubic inch displacement, liquid cooled 24-cylinder, sleeve valves, two-stage gear driven supercharger with integral aftercooler segments for each block of 12 cylinders. The X-1800 project was terminated in favor of developing the XR-4360 28-cylinder air-cooled radial engine. Information obtained at P&W East Hartford, CT, on 4-8-95.

[93] Green, William and Gordon Swanborough, 1977, 69-71.

[94] Bowers, Peter M. 1979, 466-469.

[95] XP-55, a description produced by Curtiss-Wright Corporation, St. Louis, Missouri Division.

[96] Memorandum for General Vananan: *A Review of the Experimental Airplane Program*, November 6, 1942, by E.O. Carroll, Col. Air Corps, Chief, Experimental Engineering Section. NARA RG 342, RD3957, Box 8096.

[97] Materiel Division contracts W535-ac-16146 dated December 3, 1940, W535-ac-22957 dated December 2, 1941 and W535-ac-22967 dated December 8, 1941. Engine deliveries were to be September 1940 for the first contract, and in April 1942 for the follow-on contracts.

[98] *A Review of the Experimental Airplane Program*, Memorandum for General Vananan, November 6, 1942. NARA RG342, BD3957, Box 8096.

[99] Angelucci, Enzo with Peter Bowers, 1985, 173-175.

[100] Green, William and Gordon Swanborough, 1977. 71-72.

[101] Bowers, Peter M. 1979, 441-442.

[102] Experimental Engineering Section, CONFIDENTIAL Memorandum, 1942, 2.

[103] Project MX-467, 1945.

[104] Project MX-467, 1945.

[105] Wright Field Report No. TSEST-A2, 1946.

[106] *Proposal for Allison Model V-1710-G10 Engines*, Allison letter to Engineering Division, Wright Field, June 6, 1946. NARA RG342, RD 3775, Box 7411.

[107] Bentele, Max, 1991, 27.

[108] The prototype Bf 109Z was completed in 1943, and was very similar to the P-82 except that the right-hand cockpit was fared over, there was no copilot. Production aircraft would have used two Junkers Jumo 213E V-12's rated for 1750 bhp at take-off and 1320 bhp at 29,530 feet. Normal loaded weight was to be 16,050 pounds, maximum speed being 462 mph at 26,250 feet. The prototype, built from Bf 109F's with DB 601E-1 V-12 engines and weighing 13,000 pounds loaded, was damaged in a bombing raid and the program was abandoned in 1944. Green, William. 1971, 118.

[109] Technical Report No.5673, 1948.

[110] Technical Report No.5673, 1948. Frame 0582.

[111] Allison Model Specification No. 198, 1945.

[112] The 10th and 11th Merlin powered P-82B's were modified, with AAF44-65169 becoming the P-82C (Night Fighter prototype with SCR-720 radar), and AAF44-65170 becoming the P-82D (Night Fighter prototype with APS-4 radar). Angelucci, Enzo with Peter Bowers, 1985, 338-341.

[113] USAF Historical Advisory Committee, 1975, 14, note 8.

[114] In deference to the "experimental or developmental" nature of the early P-82 program it is interesting to compare the prices the government was paying for similar series produced engines. According to T.O. 00-25-30, dated August 1, 1945, the Air Force paid $17,558 for the two-stage Packard built Merlin V-1650 going into the North American P-51D. At the same time, Allison was being paid $13,160 for the two-stage V-1710 going into the Bell P-63C.

[115] Wagner, Ray. 1990, 159.

[116] Wagner, Ray. 1990, 158-159.

[117] Angelucci, Enzo with Peter Bowers, 1985, 338-341.

[118] AAF Letter to Allison, *Allison V-1710-143 and -145 Engines*, 5-21-1946. Filed in Contract W33-038-AC-13849, see NARA RG 342, RD 3775, Box 7411.

[119] Telegram TSEPL-5-10-39, to Allison from Chief, Power Plant Laboratory, Propulsion & Accessories Subdivision, Engineering Division. See NARA RG 342, RD 3775, Box 7411.

[120] Letter from AAF Site Inspector at Allison to Headquarters, TSBPE-3B, dated 10-

23-1945. Record in NARA RG 342, RD 3775, Box 7411.

[121] AAF Letter to Allison dated 3-27-46, in Contract file W33-038-AC-13849. Copy in NARA RG 342, RD 3775, Box 7411.

[122] Wagner, Ray. 1990, 232.

[123] Hardy, M.J. 1979, 121.

[124] Interview on November 28, 1994 in Indianapolis with Donald F. Wright, retired from Allison Service Department.

[125] AAF Letter to Allison re: Allison V-1710-143/145 Engines, Contract W33-038-AC-13849, dated 5-21-1946. NARA RG 342, RD 3775, Box 7411.

[126] While no records show a unique military engine model equivalent to a left-hand V-1710-119, it is likely that the XP-82A was provided with V-1710-F32R/L engines, both designated V-1710-119. This type of designation deviated from previous practice, but seems to have been prevalent on the later development programs such as the XB-42.

[127] In an October 3, 1945 letter from Allison's Chief Engineer to the Director of the Air Technical Service Command, the troubles encountered with the SD-400-B1 and SD-400-D1 Bendix Speed Density Carburetors during development are reviewed. He concludes that, with incorporation of the changes made to date, and those contemplated, the Allison Division believes that the speed density carburetor is a practical method of metering fuel. This position then set the stage for the units use in the V-1710-143/145. The Speed Density Carburetor was an entirely different device from previous carburetors. Its functioning was actually more like an analog computer measuring mass air flow from engine rpm, manifold temperature and pressure. It was quite a complex device to develop. Referenced letter is in NARA RG 342, RD 3775, Box 7411.

[128] *Future Ratings For Allison Model V-1710 2-Stage Engines*, R.M. Hazen of Allison letter to Director, Air Technical Service Command, Wright Field, August 22, 1945. NARA RG342, RD 3775, Box 7411.

[129] USAF Historical Advisory Committee, 1975, 13-21.

[130] *Journal of Military Aviation*, January/February 1987, 14-20.

[131] *Case History of the F-82E, F and G*, 1951, 7, and Technical Report No.5673, 1948.

[132] *Proposal for Allison Model V-1710-G10 Engines*, Allison letter to Engineering Division, Wright Field, June 6, 1946. NARA RG342, RD 3775, Box 7411.

7

The GV-1710-A for the Army and Navy

The usual story of the early V-1710 relates that Allison first proposed Gilman's engine to the Army, who in turn expressed little interest, and no confidence in Allison's ability to successfully produce it. The story then goes on to note Navy interest in using the engine to power its rigid airships, and as they say, the rest is history. The actual events were somewhat different.

The Allison V-12 project dates from 1928 when Norm Gilman made preliminary designs of a high temperature chemically cooled cylinder. Development was delayed by the death of founder James Allison in August 1928, and the purchase of the firm by first Eddie Rickenbacker, and then sold again on January 1, 1929 by the Fisher Brothers Investment Corporation. During his tenure Lawrence P. Fisher has been credited with directing Gilman to proceed with a small six cylinder in-line engine incorporating his new cylinder. This was the 400-IG engine, intended for a small "family plane."

The Fishers resold the firm to General Motors on April 1, 1929. With an eye to the future of the aircraft engine business, the new owners directed a comprehensive study of the aircraft engine market and Allison's options within it. To this end a committee was appointed on May 14, 1929 consisting of Messrs. C.W Wilson, O.E. Hunt, and C.F. Kettering from GM and N.H. Gilman of Allison. They were to consider the airplane engine program and formulate a plan for the Allison Engineering Company to pursue, and specifically, this Committee was to decide what type or types of engines for aviation GM should build.[1] The study Committee recommended that Allison be directed to proceed with both a large V-12 and a small six-cylinder inline. Both engines were to use Gilman's new high temperature liquid-cooled cylinder. As a result, Gilman and his Allison Engineering Company team began sketching out the 12 cylinder glycol cooled engine capable of 1000 bhp as of May 7, 1929.[2] The initial engine would be limited to about 750 bhp as that was all the likely military customers could use.[3] Gilman intended the engine to be capable of accepting turbosupercharging and fuel injection. From this point on, corporate support never wavered, even during the economic Depression that began that October and continued for a decade. This was the beginning of the 1710 cubic inch Allison V-1710.

The Engine

In spite of Wright Field and Curtiss poor experience with Prestone cooling, Allison's Norm Gilman felt certain that he could develop a successful design of a 1000 bhp liquid-cooled engine. He showed initial drawings of the new Allison VG-1710 to Army engineers with the suggestion that the engine be used in a bomber, but Wright Field was not interested. Wright Field suggested to Allison that the Navy might be interested, for they were believed to be looking for such an engine for their proposed Aluminum Aircraft Company XP2H-1 flying boat.[4] He then approached the Navy, who were interested in the VG-1710 as a replacement for the German Maybach engines then powering their large airships. Although airships would not require the high altitude, high speed turbosupercharged engine Gilman had designed, the VG-1710 provided an expedient way to advance development. The resulting Navy support did give a reason to proceed.[5]

In January 1930, the Wright Field Experimental Engineering Section issued their specification X-28214 "1000 h.p. Engine" to the General Motors Corporation, as well as other potential manufacturers. In the cover letter the Air Corps stated they were interested in having a 1000 bhp aircraft engine developed following their tentative specification. They invited comments and suggestions, noting that, "while some aspects may seem drastic, all data were the result of actual accomplishments in existing equipment either in the U.S. or Europe." The Army asked that Allison express their attitude toward undertaking such a development and give as definite an outline of procedure as they would be willing to subscribe to in case GMC "deems it advisable to undertake this development." It was not proposed that the Army would underwrite or subscribe to the development "until an actual experimental engine has been built and is either being tested or has been tested."[6]

General Motors VP of Engineering, O.E. Hunt, responded on February 11, 1930 noting that, "we are developing a Prestone-cooled aviation engine…." He followed ten days later with a more detailed response to the Air Corps proposed specification, stating that they had been "out of touch for several years with the Army and Navy and their requirements for engines for military aircraft.[7] Hence they did not feel qualified to offer constructive suggestions on the specifications." He went on,

> "This lack of contact and continuing development experience has naturally placed us in a position where we feel that we cannot, at the moment, give you any assurance that we could in the immediate future contribute a development that would be of interest to you in connection with the specifications. We can say, however, that we are interested in large size, fluid cooled, aircraft engines and we hope that

we can depend upon your cooperation in our development activities in connection with them, with the idea that through such cooperation we might develop our engines along lines that would be interesting to you in the future."[8]

Concurrent with the above communications Allison was discussing their VG-1710 with the Navy, with the result that Bureau of Aeronautics Contract No. 17952 for one GV-1710 engine was signed on June 28, 1930, the engine to be rated for 650 bhp at sea-level. With work underway on the Navy GV-1710 Allison next contacted Wright Field in January 1931 to setup a meeting of Mr. Gilman and Mr. Caminez with the Materiel Division "regarding a new engine." The meeting was held later in the month and considered in detail design features that would be required by the Army. In view of the effects of the Depression, both upon General Motors and the buying public, the small six cylinder 400-IG engine project was dropped during 1931 and Allison resources focused on the V-1710.

Then came an interesting twist. In November 1931 the Air Corps sent a letter to Allison and other engine manufacturers titled "Policy-Engines Submitted for Examination and Test by the U.S. Army Air Corps." It stated in part:

"Manufacturers who desire to design and construct, at their expense and without obligation to the Government, engines to be tested by the manufacturer and witnessed by an Air Corps representative to determine their suitability as Air Corps types, will be governed by this policy."

The policy was quite detailed, and onerous to the manufacturer. The General Motors response was issued by F.O. Clements, Technical Director, Research Laboratories, GMC. He quickly acknowledged receipt of the Policy and responded,

"This particular Laboratory wishes to be perfectly foot-loose in the carrying on of an engine project. This means of course that our engine developments will be carried out at our own expense. When they reach a point where we feel that they might be of some interest to military or naval authorities, we will be glad to submit them in order to obtain your reactions thereto. It seems to me that your restrictions are too severe and will tend to prevent folks from submitting new designs of engines."

Evidently as a bridge over the policy issues, in December 1931 the Chief of the Bureau of Aeronautics requested that Wright Field have a representative witness their 50-hour test of the GV-1710-A.

In a report to the General Motors Executive Committee, dated December 8, 1931, the following circumstances were noted:[9]

"The outlook for the Allison Engineering Company seems to be particularly promising and with the development of new airplane engines, and particularly the large engine for the Navy, it would appear that the Allison Engineering Company could operate upon a very satisfactory basis.[10] It was pointed out, however, that an expenditure of approximately $413,000 will be required at Allison for the construction of a building for the manufacture of engines if we are successful in obtaining certain orders from the U.S. Navy.[11] In this connection it will be necessary for us to assure the Navy Department that we are willing to support the Allison Engineering Company in constructing the necessary plant should the Navy give Allison these orders. The committee was unanimously in favor of our giving this support."

GM's response to the needs of Allison, even during the worst days of the Depression, reflects the confidence that continued the support of the V-1710 for another eight years before a major production order was to be received.

In May 1932 Wright Field wrote to the Navy stating that for several years they had been doing considerable research on Prestone cooling of aircraft engines, and that they had made the results of this work generally available to the industry. Hence they were interested in the VG-1710 engine and were "anticipating possible procurement of one of the engines for testing." They then requested that the Navy make available cross sectional drawings and general installation information on their engine.

This led to conferences during the fall of 1932 attended by Allison's Gilman and Caminez at which a detailed specification was prepared for the Allison V-1710-1, an engine to be rated at 750 bhp at 2400 rpm, but designed with the understanding that the Air Corps would operate the engine at a rating of 1000 bhp and 2800 rpm.[12] Allison agreed to this specification on October 20, 1932, with Norm Gilman and Harold Caminez returning to Wright Field for a conference on November 3, 1932 to finalize details of the new engine, which included the requirements for turbosupercharging and fuel injection.[13] [14]

The most significant differences between the Navy and Army engines was changing the reduction gear from 3:2 to 4:2, which required a 14 inch diameter ring gear in place of the 10 inch gear in the Navy engines. The propeller shaft was also extended some 12 inches to better streamline installation of the engine. The Air Corps designated their first engine as the XV-1710-1, through to Allison it was to become known as the V-1710-A2. The Air Corps responded with an Authority for Purchase on November 12, 1932, consummating Contract W535-ac-5592 on January 5, 1933. The engine, XV-1710-1(A2), Air Corps No. 33-42, Mfr. No. 2, was completed by Allison on June 28, 1933 and delivered to the Air Corps on July 6, 1933 for a price of $38,237.32, including spare parts, special tools and data.[15,16]

The Materiel Division tested the engine for nearly a year and in the process demonstrated a maximum power of 1070 bhp at 2800 rpm from 45.7 inHgA. The conclusive test of the XV-1710-1 was performance of a 50-hour development test at 800 bhp. This was satisfactorily concluded with the Materiel Division recommending that a model of the engine be subjected to a 50-hour development test at a rating of 1000 bhp.[17] This was later done using the XV-1710-3(C1) engine.

When the changes wanted by the Army were incorporated it was clear that the engine was quite different than the original Navy model and Allison designated the subsequent Army models as their V-1710-C series.[18] Allison continued to work on both the new Army V-1710-C and the unique reversible Navy airship V-1710-B engines well into 1935, though the Navy program came to an end with the loss of the airship USS *Macon* in February 1935. The V-1710-C continued as an active program through 1941.

Summary

Allison had begun the design of the V-1710 prior to the expression of military interest in such an engine. Some time later the Army was aggressively looking for someone to provide a high temperature Prestone cooled 1000 bhp engine. While there was likely a fair amount of informal interchange between the Materiel Division engineers and Allison during the period that the Navy GV-1710-A was being built and tested, it is not reflected in the formal record. When discussions on the specification of an Army V-1710 did get underway in the fall of 1932, they were quite detailed in contrast to those for the Navy VG-1710. This was even though the Air Corps specified that their engine was to be "essentially the same as the Allison GV-1710 engine which completed the Navy 50-hour development test."[19]

While Allison did not subscribe to the stated Air Corps engine development Policy, they did comply in that they funded development independent of the Air Corps, coordinated with military representatives and inspectors, and provided detailed information on the design and progress of the engine. All of this set the stage for the rapid design, contracting, construction, and testing of the Air Corps XV-1710-1(A2) once they obtained the funds. Being the second engine and contracting in the midst of the Depression, the Air Corps paid far less than the $75,000 the Navy paid for their GV-1710.[20]

Design of the GV-1710-A:

Allison approached the GV-1710-A project with its usual methodical and analytical engineering design and development approach. Every component was designed and analyzed to insure its strength and function *prior* to there being any fabrication of parts. This is documented in a very comprehensive engineering analysis done for the 750 bhp engine by Allison's Harold Caminez, dated September 24, 1930. The methods he developed and applied for this analysis were continued and applied to each of the major engine models right on through the V-1710-G series. This analytical approach took a lot of uncertainty out of the development process.[21,22]

It is instructive to consider the actual 1930 description of this first engine,[23] a portion of which is repeated here:

> Specific features of the VG-1710 design were that it was to be a twelve cylinder, 60 degree vee, liquid cooled engine, geared 3:2 and with approximately 1710 cubic inches piston displacement. It was to be equipped with an Eclipse Series VII hand-electric inertia starter with integral booster coil; one double Scintilla magneto; Scintilla ignition switch; suitable carburetor; modified C-3 fuel pump (as GFE); priming pump and connections; size S.A.E. #40 propeller shaft; provisions for driving type G-1 generator; provisions for driving two E-4 gun synchronizers; two separate tachometer drives; two-way propeller hub for size #2 blade roots; exhaust pipes and gaskets; provisions for installing Marvel type fuel injectors and pumps; geared centrifugal supercharger; tool kit; instruction book. This engine shall develop not less than 750 propeller B.H.P. corrected, at 2400 crankshaft R.P.M. with 60 degrees F. carburetor air temperature, 29.92 inHgA barometer pressure and standard dry air, on Navy Aviation gasoline with six c.c. of ethyl fluid per gallon added; and with cooling liquid outlet temperatures not less than 300 degrees F. The engine shall weigh not more than 1000 pounds dry, without starter, ignition switch, fuel pump, priming pump and leads, propeller hub, exhaust pipes, tool kit and instruction book, but with all other equipment included.

The designer tabulated the performance of the engine in Table 7-1.

Of significance is the requirement, from the very beginning, that the engine design would be required to accommodate fuel injection. Another significant requirement is the light weight of under 1,000 pounds required for the engine, a factor in the future development of the C/D engines as every increase in ratings made it necessary to strengthen the next weakest point or component in the engine.

Table 7-1
GV-1710-A Design Performance

RPM	BMEP, psi	BHP
3000	180	1165
2700	175	1020
2400	145 w/34inHgA	750 (Rated Conditions)
2250	124	600

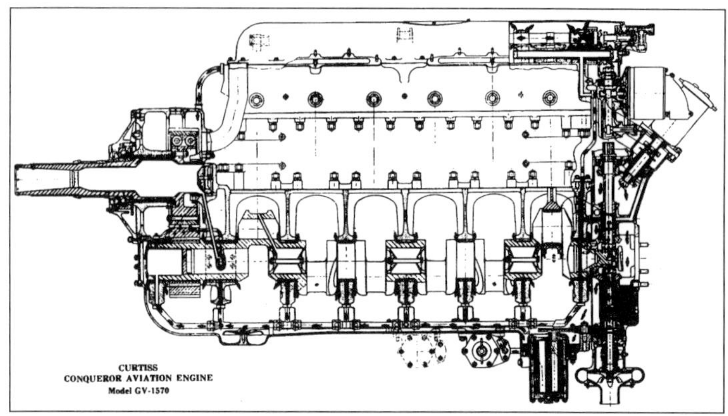

The geared 650 bhp Curtiss GV-1570 was the largest and best V-12 available at the end of the 1920s. The series went through a number of different crankshaft evolution's, always getting larger diameter bearings and a stiffer crankshaft as ways to combat bearing failures. (AAC)

Allison took great care in the design and analyses to insure that the crankshaft was conservatively designed, and then compared its loadings and stresses to those of other contemporary crankshafts. By providing generous bearing surface, loads were about 75 percent of those in the 630 bhp Curtiss Conqueror V-1570 and 87 percent of those in the 420 bhp Pratt & Whitney R-1340 Wasp. The crankshaft was supported by six plain Allison steel-backed bearings and a large roller bearing at the front. This bearing was to carry the load of the propeller reduction gear reaction as well as that from the front face of the number six throw, the one in "front."

The then recent history of the Curtiss Conqueror engine was clearly in mind by the designer, particularly its crankshaft and bearing problems. This experience was considered in the design of the VG-1710 with its 20 percent increase in output, with the result that:

The superchargers in both "A" models were quite similar in form. This is the 9-1/2 inch unit from the XV-1710-1, damaged during the first attempt at the 50-hour test when the retaining tab failed. It gives a good look at the simple design approach. More complex blade profiles and improved diffusers were necessary to get above the 50 percent efficiency of these early designs. (AAC)

The VG-1710 engine bearings sizes are better proportioned than in the V-1570 because in the former the maximum and mean load in all the bearings are more nearly equal. The center main bearing of the Curtiss engine is the troublesome bearing. In the VG-1710 this bearing has a lower specific load. The maximum and mean load of the VG-1710 intermediate main bearings exceeds those of the corresponding Curtiss V-1570 bearings but are less than on the Curtiss V-1570 center main bearing. With proper design and lubrication, the bearing sizes of the VG-1710 engine are adequate at the normal rating. Stresses within the crankshaft material were significantly below those that were being used in the old Liberty engine, consequently Allison concluded that an adequate factor of safety and that no difficulties should be experienced as long as the critical vibration speeds are outside the running range of the engine.

The primary dimensions of the bearings in the V-1710 were unchanged through its production life, even though its ratings increased from this original design for 750 bhp at 2400 rpm up to a pinnacle of over 2200 bhp at 3200 rpm. Much of the increase was realized by counterbalancing the crankshaft as a way to reduce the centrifugal forces acting on the bearings and crankcase. The analytical methods for designing the crankshaft were those established by the Air Corps at McCook Field.[24]

The VG-1710 was equipped with an Allison designed "Rotary Induction System"; that is, a supercharger, but one intended to only provide for uniform mixture distribution and delivery. It was not used to "supercharge" for additional power, which could have been done if greater power was desired. The VG-1710 rotary induction system was intended to deliver 1710 cubic inches of standard air for every two revolutions of the crankshaft, thus a volumetric efficiency of 100 percent. The impeller in the system was designed to provide only enough boost to compensate for pressure losses to and through the carburetor, intake manifold, and intake valves, as well as to compensate for density lost due to heat absorbed from the engine and manifolds. The result was a requirement to "boost" manifold pressure to 34.9 inHgA, or 5 inHg over sea-level pressure.[25]

The half shrouded twelve bladed radial vaned impeller was designed using relations taken from Marks Handbook for Mechanical Engineers, for impellers without inlet guide vanes.[26] Originally it was assumed that the unit would be 80 percent efficient, which would have allowed a 6.0:1 drive gear ratio to be used with an 8.0 inch diameter impeller. The efficiency was revised by September 1932 to an assumption of 50 percent, resulting in 8.0:1 step-up gears. Early running of the unit probably showed that the original efficiency assumptions were not realistic. In fact, it was not until Whittle designed the compressor for his first jet engine in the late 1930s that an 80 percent unit of this type was achieved, and it had inlet guide vanes and a vaned diffuser. The actual impeller for the GV-1710-A was 8.25 inches in diameter, probably to give some margin in performance. As designed, the supercharger drive required 37 bhp. At its rated speed the aluminum impeller would be rotating 19,200 rpm, and concern about its internal stresses required it to be heat treated so that it had a factor of safety greater than ten.

A centrifugal compressor must have a diffuser. Its function it is to convert the high velocity of the air streaming off of the impeller into pressure by slowing it down. By so doing the kinetic energy of the air is recovered as pressure, the whole purpose of the device. The GV-1710-A used a "vane-less" diffuser that in reality is a specially shaped narrow nozzle fixed and surrounding the periphery of the impeller. These can work quite well, but usually only over a narrow range of conditions. Consequently, more modern superchargers will have stationary vanes surrounding the impeller to form the necessary diffuser channels.

This is the original Allison designed "Rotary Induction System" supercharger, with the inlet cover removed. The straight blades are only supported to half their length. The smooth annulus around the perimeter of the impeller is the rear half of the vane-less diffuser. The step-up drive gears are also shown. (Allison)

The valve gear arrangement of the V-1710 was similar to that developed for the Allison 400-IG engine. The 400-IG was the 150 bhp Inverted and Geared liquid-cooled six cylinder engine designed around the original Gilman cylinder. The difference was primarily that the 400-IG had only two valves per cylinder while the V-1710 used four valves to better accommodate the larger cylinder. Both engines used a single cam with dual helical springs for each valve. In the V-1710 a single cam operated each pair of valves through a double rocker arm. Original cam profiles, springs and general arrangement were otherwise the same on both engines.[27] Allison had quite a lot of experience with this valve operating system, both in single cylinder and V-12 service. Significantly, this design included a roller riding on the cam and was quite rugged.

While testing the later Type Test XV-1710-3(C1) engine, it was found that the original cam profile was causing outer valve spring failures. The problem was the rapid changes in acceleration that were causing the spring to "flutter." A new profile was designed that smoothed accelerations and resolved the problem. During the long production life of the V-1710 changes in the valve gears were limited to strengthening the springs, to allow for higher manifold pressures and higher operating speeds, and the use of improved materials.

The V-1710 had very few problems with its valve gear, but the roots of a serious backfire condition that occurred during the introduction of the V-1710-33 powered Curtiss P-40, can be traced to the design of the VG-1710. The designer calculated that the tappet clearance of the intake valves would increase by 0.009 inches when the engine reached its 300 °F operating temperature. The initial cold clearance setting was therefore established

The rebuilt XV-1710-1 engine was connected to this dynamometer for calibration runs at Wright Field beginning August 25, 1932. Once completed, the engine was installed on a rigid test stand and fitted with a 70 inch diameter Hartzell four bladed wooden test propeller. The actual endurance test began on August 30, 1932 and was completed September 9, 1932. (AAC)

on the GV-1710-A as 0.015 inches for the intake and 0.020 for the exhaust valves. By the time the XV-1710-1(A2) was built the intake valve clearance had been established as 0.010 inches, probably rationalized because of the change to 250 °F coolant temperature. In service on the V-1710-33(C15), it was found that this setting was occasionally allowing hot combustion gasses to leak back into the intake manifold, particularly as the valves and seats wore. The result was often backfires. The solution was to return the setting to 0.015 inches, as initially used on the GV-1710-A. This later setting was then retrofit and established as common to all production V-1710's.

Allison's expertise with plain bearings was evident in the reduction gear on the A/B/C/D series of engines. The A and B used a 10 inch diameter ring, running on a plain bearing, to drive the 3:2 reduction gear's and absorb the crankshaft torque through an internally cut gear. In the A/B series the supercharger, camshafts, distributors and accessories were all powered by a long quill shaft that was driven from teeth cut in the exterior perimeter of the reduction gear ring gear. On the A-2 and later C/D series it was necessary to enlarge the ring gear to 14 inch diameter to provide the required 2:1 reduction ratio. Consequently the quill shaft was driven from the inside of the ring gear by the same teeth driven by the crankshaft pinion. While this arrangement was satisfactory for the original 750 bhp rating, and allowed for the upgrade to 1150 bhp in the later C/D engines, it was found that the overhung crankshaft driving pinion was a limiting feature when challenged by the demands for higher power a decade later.

The "A" model engine crankshaft was not provided with counterweights, hence the engine was limited to 2400 rpm for normal operation. This was not too much of a hindrance given that aircraft of the day could not really utilize the available 750 bhp due to their fixed pitch propellers. Even so, and allowing for the numerous changes subsequently made to improve and strengthen the crankshaft, the engine demonstrated the fundamentals of a good design. The basic design of the pistons and connecting rods was validated, though they, and all other components of the engine, were to be continually upgraded and improved during the production and development life of the subsequent higher rated engine models.

Developing the GV-1710-A

On August 20, 1931, the V-1710 had its first run. This began a series of development and demonstration tests in preparation for attempting the Navy contract required endurance test.[28] Based on early performance, the contract was revised in November 1931 to specify that the engine be built to deliver its designed power of 750 bhp at 2400 rpm, rather than the originally specified 650 bhp.

During the fall and winter of 1931-32 Allison redesigned and upgraded components for the 750 bhp rating, while continuing to operate the engine to gain experience and component performance data. The uprated engine, with its new 8.0:1 supercharger gears, began the official Navy Bureau of Aeronautics specified 50-hour endurance test on May 3, 1932. Witnessed by both a Navy and Army representative, the test was to be accomplished in ten 5-hour segments. These tests showed both promise and problems. After only a few hours of testing it was decided to redesign major components before continuing the testing, even though the engine had demonstrated 778 bhp running at 2445 rpm with a manifold pressure of 33.4 inHgA, and using 307 °F coolant temperature on 87 octane fuel.[29]

The first build of the 650 bhp engine satisfactorily completed the Contractor's Tests per the Navy contract on March 21, 1932. It was then submitted to an endurance test in accordance with the Navy specifications.

The initial attempt at the 50-hour endurance test began May 3, 1932. After 4:25 hours it was necessary to shutdown due to burned exhaust valves on three cylinders. Otherwise, the engine was found to be in good condition. A top overhaul was performed and the engine prepared to begin the endurance test anew.

A second attempt at the 50-hour endurance test was begun on May 12, 1932. After 1:02 hours the crankshaft broke at the number 6 front crank cheek. Another attempt was not to be made until an improved cylinder head and crankshaft were available. These first two attempts at the endurance test revealed defects in the initial design of the cylinder block and the

The "A" model engines did not use a counterweighted crankshaft since the 2400 rpm rated speed did not require counter balancing to maintain acceptable loads on the large plain bearings. The large roller bearing and pinion drive gear for the propeller is at the right. (AAC)

Table 7-2 GV-1710-A Engines Shipped			
Model	Number	Aircraft	Comments
GV-1710- A(A1)	1	Model Test	Prototype for Navy, later known as V-1710-2
XV-1710- 1(A2)	1	Model Test	Prototype Long-nose "C" type for Air Corps
Total	2		

front end of the crankshaft. Additional coolant passages were to be provided in the cylinder head casting to provide for adequate exhaust valve seat cooling, and the strength of the front crankcheek was to be more than doubled.

The rebuilt engine was connected to a dynamometer for calibration runs beginning August 25, 1932. Once completed, the engine was installed on a rigid test stand and fitted with a 70 inch diameter Hartzell four bladed wooden test propeller. The actual endurance test began on August 30, 1932 and was completed September 9, 1932.

Even though the third attempt at the Endurance Test went smoothly, a number of parts were found to be unserviceable by the subsequent teardown inspection. These included:

- Crankcase upper half; due to five fractures.
- Crankcase lower half; one fracture.
- Rocker Arm Assemblies; two tappet screws damaged.
- Master Rod Bearings; several indicated poor bonding of metal.
- Prop Shaft, Reduction Gear; 3 teeth showing signs of failure.
- Oil Pumps; drive shafts slightly pitted.
- Water pump; slight pitting on drive gear.
- Valves; several on left bank were leaking slightly or burned.

On August 30, 1932 the strengthened engine, with new valve timing to improve valve cooling, was returned to the test stand and calibrated in preparation for the endurance test. By comparison with the first attempt at the 50-hour endurance test, the second test was a satisfying forward march. Testing was completed on September 9, 1932 with only minor interruptions and routine servicing.[30] It is interesting to note that in the interval between the May and September tests, the designation of the engine had been changed from VG-1710 to GV-1710-A and was subsequently referred to as the Allison V-1710-A1.

Although only one GV-1710-A was built, and its development life a comparatively short two years or so, it soldiered on for a number of years. NACA Langley used it for awhile, though the purpose has not been established. It is clear that it never flew. Later, the Army paid to have it modified to test the intended port type fuel injection system being developed for their planned XV-1710-5 engine.

The Air Corps inspector reported that, "The [GV-1710-A] engine operated throughout the entire test with exceptional smoothness and the performance of the engine was satisfactory at all times. It is believed that the changes proposed by the Allison Engineering Company will eliminate failures such as were encountered during the development test."[31] With this recommendation the Air Corps proceeded to obtain an engine of its own.

Table 7-3 V-1710-A Engine Models

A-1 Build 1: The engine was purchased on Navy contract 17952, dated June 26, 1930, and designed for 750 bhp at 2400 rpm, but guaranteed for 650 bhp at 2400 rpm as of September 24, 1930. Calculated dry weight, 940.0 pounds. The crankshaft pinion gear was 5.5 pitch diameter and used 4-5 pitch 20 degree tooth face angles. Both exhaust and intake valve seats were straight cut at 45 degrees. There were no counterweights on the crankshaft and the 8-1/4 inch diameter supercharger impeller had blades that were only partially shrouded. Valve timing was a "symmetrical" 15/55/55/15, giving 250 degree duration.

Allison made two attempts at the Navy required 50-hour development test. The first attempt began on May 3, 1932 and accumulated only 4:25 hours before it was necessary to shutdown because of a burned valve on the No. 4 cylinder. The engine was given a top overhaul and another attempt begun on May 12, 1932. After only 1:02 hours of running, all at

The GV-1710-A1 Build 1 for the Navy shows nearly every feature of V-1710's built 10-15 years later, though in detail practically every feature was changed, strengthened or improved. Note the "bowtie" exhaust ports, unique to this build of the engine. (Allison)

near the rated 750 bhp, the crankshaft fractured through the front crankcheek. With this inauspicious beginning the V-1710 began a series of redesigns that were to continue for nearly a decade. An improved cylinder block was designed, intended to improve cooling of the exhaust valves, as well as an improved and strengthened crankshaft. It took about ten weeks to complete the rework of the engine.[32] The engine was first accepted March 12, 1932.

Particulars, GV-1710-A, Build #1:

Model	GV-1710-A	Valve Clearance:		Carburetor	NA-U8J
Manuf s/n	1	Intake, in	0.015	Fuel Grade	87 Octane
Air Corps s/n	na	Exhaust, in	0.020	Coolant	Ethylene Glycol
Compression Ratio	5.88:1	Valve Timing, IO	15 BTC	Weight, pounds	1030
Supercharger Ratio	8.00:1	Inlet Closes	55 ABC	Length, in	78-19/32
Reduction Gear Ratio	3:2	Exhaust Opens	55 BBC	Width, in	30-1/4
Propeller Shaft	No. 40	Exhaust Closes	15 ATC	Height, in	41-11/32
Date Delivered	3-12-1932	Ignition, Intake	34 BTC	S/C Impeller, in	8-1/4
Contract, Navy	17706	Exhaust	36 BTC		

Performance: 778bhp/2400rpm/SL/___" @ 0.530#/bhp-hr,
Comments: Crankshaft without counterweights. Tested 3-12-1932 to 5-12-1932.

A-1 Build 2: This designation reflects the second configuration of the first engine, when it completed the Navy Contract tests. Retroactively this engine was known as the V-1710-2. Estimated performance as of April 22, 1931 was 800 bhp at 2400 rpm and 1000 bhp at 3000 rpm, though the guaranteed output was 750 bhp at 2400 rpm. Calculated dry weight was up to 970.4 pounds, with the extra 30 pounds over the Build 1 version being placed in the connecting rods and piston assemblies. By the time the engine successfully completed the Navy 50 hour test its weight was up to 1035 pounds. The crankshaft pinion gear was changed to 5.71 pitch diameter and 3.5-5 pitch 20 degree tooth face angles. Intake valve seat inserts were venturi shaped with 30 degree seats. Exhaust seats had a straight passage with 45 degree faces. Valve timing was revised.

In its Build 2 form the GV-1710-A was a considerably different engine. Note the new "eye glasses" design of the exhaust ports, a feature of all subsequent V-1710's and a reflection of the considerable changes incorporated into the blocks and heads to improve cooling around the exhaust valves. The crankcase breathers on the sides were also unique. Both Builds of this engine show how the coolant piping was arranged to provide a supply to the cylinder jackets and to the heads. This split was replaced on the "C" Model engines with an enlarged supply to the jackets only, though with the higher powered E/F engines the above arrangement was used. (Allison)

Particulars, GV-1710-A, Build #2:

Model	GV-1710-A	Valve Clearance:	Carburetor	NA-U8F	
Manuf s/n	1	Intake, in		Fuel Grade	87 Octane
Air Corps s/n	na	Exhaust, in		Coolant	Ethylene Glycol
Compression Ratio	5.85:1	Valve Timing, IO	10 BTC	Weight, pounds	1035
Supercharger Ratio	8.00:1	Inlet Closes	52 ABC	Length, in	NA
Reduction Gear Ratio	3:2	Exhaust Opens	48 BBC	Width, in	NA
Propeller Shaft	No. 40	Exhaust Closes	11 ATC	Height, in	NA
Date Re-Delivered		Ignition, Intake	34 BTC	S/C Impeller, in	8-1/4
Contract,	17706	Exhaust	41 BTC		

Performance: 795bhp/2480rpm/SL/___" @ 0.500#/bhp-hr,
Comments: Crankshaft without counterweights. 50-hour Test run 8-30-1932 to 9-9-1932

A-1 Build 3: The Materiel Division at Wright Field had equipped a R-1340 nine-cylinder air-cooled engine with a mechanical fuel injection and had good results with it.[33] They desired to make a corresponding application to a twelve-cylinder vee type engine. For this purpose they initiated work in February 1931 to develop suitable manifolding and injection equipment for use on the Curtiss-Wright V-1570 engine. Tests began on this engine in August 1931, but were terminated shortly thereafter due to the necessity of correcting carburetion troubles being encountered with the V-1570 engines then in service. In FY-34 Allison agreed to design and test an induction system for the GV-1710-A (V-1710-2) and use it as a test bed for developing the port fuel injection installation intended for the Air Corps V-1710-5(C2) model engine.

Allison developed a new manifold assembly, PN31000, for the engine, but it was soon found to develop excessive turbulence in the air-stream passing the fuel discharge port. Even so it did demonstrate very consistent fuel distribution, all with 1-1/2 percent. The engine developed 783 bhp at 2400 rpm with BSFC of 0.510 #/bhp-hr. Allison then designed their PN31103 manifold to smooth airflow past each discharge nozzle. Fuel distribution remained uniform, within 1-1/2 percent, with the engine producing 741 bhp at 2400 rpm, with BSFC of 0.503 #/bhp-hr. Significantly, this first effort with fuel injection equaled the power developed when the engine was fitted with the NA-F7C carburetor, but with the BSFC improved by six percent at the maximum power mixture. Even so, Wright Field was not satisfied with either of these manifolds and suggested that new designs be developed that would be better able to take advantage of the ramming effect of properly coordinated valve timing and fuel injection.[34] A photo of the engine in this configuration is shown in Chapter 7.

A-2: The contract for this engine was issued by the Air Corps in March 1933, though the specification dated from November 3, 1932. The Air Corps intended the engine to be essentially the same as the Navy GV-1710-A, but with the propeller shaft extended 12 inches into a streamlined nose case, while the lowest point on the engine was to be raised to be at least 2 inches nearer the crankshaft. They also specified it was to be fitted with a counter-balanced crankshaft, 2:1 reduction gears and use a revised supercharger impeller of 9-1/2 inch diameter driven through 8.77:1 step-up gears. In this form the engine was to be rated for 750 bhp at 2400 rpm, but was to be designed with the understanding that the Air Corps would operate the engine at 1000 bhp and 2800 rpm.[35]

This engine was acceptance tested in June 1933 at its intended rating of 750 bhp at 2400 rpm and delivered to the Air Corps for their 50-hour development test. In view of the very large increase in horsepower asked of this engine, it was expected that a good many failures and a long development period would occur. Unfortunately, as determined later, this did not occur and the test was completed without too much trouble. The test consisted of 9-1/2 hours at 1000 bhp and 40-1/2 hours at 90 percent power and was run on a dynamometer with apparently very favorable mounting conditions.[36]

AEC s/n 2: The XV-1710-1(A2), Air Corps serial number AC33-42, was the second V-1710 built. It differed from the Navy engine by having a counter-weighted crankshaft, forged instead of cast pistons, but used the same valve timing and other details as in the Build 2 configuration of the earlier Navy GV-1710-A1. Of necessity, the plain bearing used to support the reduction gear was increased to 14 inches in diameter to accommodate the Air Corps desired 2:1 reduction gear ratio, the Navy engine used a 10 inch diameter bearing supporting gears having a 3:2 ratio. This engine was built with intake manifolds PN18579, which were replaced in October 1933 with PN19420 manifolds that provided increased flow area to reduce pressure losses. Delivered per contract W535-ac-5592 to the Air Corps on June 28, 1933, the Wright Field test program began on July 6, 1933 and was concluded June 1, 1934. Manifolds actually installed when the engine was retired are PN19408, on the right side, front and rear, and PN19409 on the left side, front and rear.

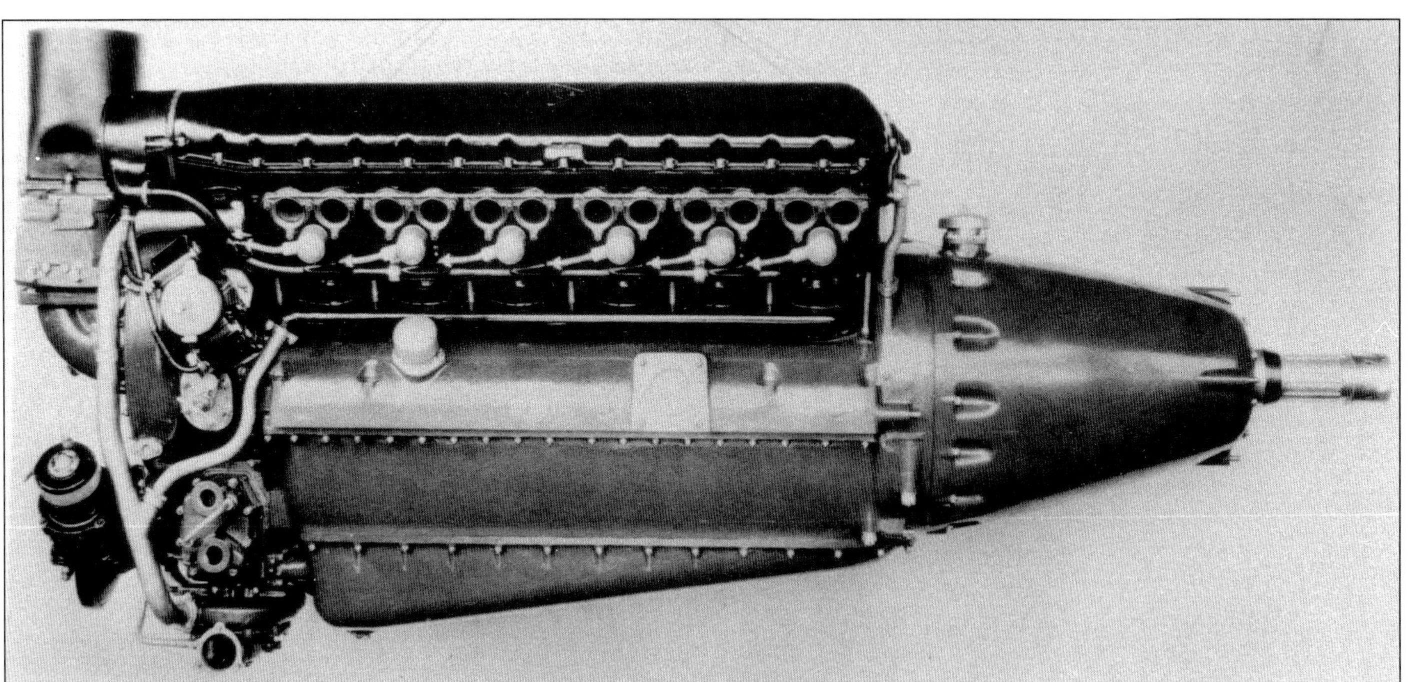

The XV-1710-1(A2) with its 2:1 reduction gear shows a considerably more streamlined and modern appearance than the A-1. Note that the crankcase is completely revised from that used for the A-1 Build 2. With the exception of the crankcase breather this configuration was to continue on through all of the service test engines. The SAE#40 propeller looks small for an engine designed to produce 1000 bhp. Allison

Particulars as of XV-1710-1(A2) at End-of-Test:					
Model	XV-1710-1	Valve Clearance:	Carburetor	NA-F7B	
Manuf s/n	2	Intake, in	0.010	Fuel Grade	92 Octane
Air Corps s/n	33-42	Exhaust, in	0.020	Coolant	Ethylene Glycol
Compression Ratio	5.75:1	Valve Timing, IO	15 BTC	Weight, pounds	1122
Supercharger Ratio	8.77:1	Inlet Closes	55 ABC	Length, in	93-25/32
Reduction Gear Ratio	2.00:1	Exhaust Opens	55 BBC	Width, in	30-25/32
Propeller Shaft	No. 40	Exhaust Closes	15 ATC	Height, in	37-1/16
Date Delivered	7-6-1933	Ignition, Intake	32 BTC	S/C Impeller, in	9-1/2
Contract, W535	ac-5592	Exhaust	38 BTC		

Performance: 1070bhp/2800rpm/SL/45.9" @ 0.632#/bhp-hr, 1011/2600/SL/43.6" @ 0.581#/bhp-hr, 932/2400/SL/41.8" @ 0.559#/bhp-hr, 534/1700/SL/31.3" @ 0.451#/bhp-hr,
Comments: Crankshaft with six counterweights. Tested 7-6-33 to 6-1-34. Exhaust valve stems were sodium filled.

The test program for this engine was intended to determine the performance characteristics and see if the engine could successfully operate for 50 hours at 1000 bhp and 2650 rpm. It took three attempts, several modifications and some redesign of components to obtain the desired performance and reliability, but at a reduced level of 800 bhp at 2400 rpm. Allison had only guaranteed 750 bhp for this engine. Wright Field operated it at up to 1070 bhp.

Principal parts effected by the upgrades during testing were the supercharger impeller and drive, the 14 inch diameter reduction gear plain bearing, enlargement of the intake manifolds[37], and improvements to the inlet and exhaust valves. During the testing the engine was initially operated at the intended 300 °F coolant outlet temperature, but it was found to be running on the edge of detonation, and that the amount of heat being carried away in the oil would require an oil cooler as large as the coolant radiator. Testing by the Air Corps in January/February 1934 defined the conditions and resulted in a complete series of tests at 250 °F. Performance was improved by reducing the coolant outlet temperature to 250 °F, and consequently this, and all subsequent V-1710's, ran at this lower coolant temperature.[38]

At about the time as the first development tests were in progress, the V-1710 type engine was being considered for use in a proposed long-range aircraft. In such an installation, it was estimated that cruise power of 50 percent of rated and the lowest possible specific fuel consumption would be desired. Tests were run and the engine found to deliver 535 bhp at 1700 rpm for 0.450 pounds of fuel per horsepower per hour. Quite commendable for the day.[39]

With these preparatory runs completed the first attempt at the 50-hour test was begun, only to end after 1-1/2 hours when the supercharger impeller was damaged, by failure of the impeller retaining nut tab washer. Following this damage, it was considered desirable to reduce the severity of the test and reserve the 1000 bhp development test for the XV-1710-3 model, for which procurement had been initiated in January 1934.

The second attempt at the 50-hour test, this time scheduled for 800 bhp, began March 29, 1934. It ended after 18-1/2 hours due to internal gas leakage into the coolant from a cracked cylinder liner. When rebuilt with stronger cylinder liners it also included new exhaust valves having a thinner wall and a more flexible head to improve seating. Only the valve stem was sodium cooled. This valve had a "tulip" or cupped head that caused the engine compression ratio to be reduced to 5.75:1 from 5.85:1. The original valve heads were larger and mushroom shaped to accommodate internal sodium cooling. The original sodium cooled inlet valves were replaced by conventional hollow stem valves to eliminate the previous instances of sticking.

The third attempt at the 50-hour test began on May 7, 1934, and was completed without a failure of any nature on May 11. Subsequent teardown found no unsatisfactory conditions in the propeller reduction gearing, supercharger drive, valve mechanism, accessory drives, or other parts of the engine. Not appreciated at the time was the precursor to later troubles with the cylinder liners incurred because of the tendency for them to go out-of-round. Four connecting rod bearings and a set of new main bearings, along with one new exhaust valve, were used in reassembling the engine for further testing. Although the test qualified the engine for operation at 800 bhp and 2400 rpm, the engine was also demonstrated to a maximum of 1070 bhp at 2800 rpm using a manifold pressure of 45.9 inHgA while consuming 0.632 pounds of fuel per horsepower-hour.[40]

Summary

The Air Corps concluded that, "The performance and reliability demonstrated in the test proved the V-1710 type engine to have potential military value", and recommended, "that a model of the V-1710 type engine be subjected to a 50-hour development test at a rating of 1,000 horsepower."[41]

At completion of testing, the Air Corps reflected on the lowered coolant temperature, noting that it had been done solely for the purpose of reducing the possibilities of detonation within the operating range of the engine. There had been no serious mechanical difficulties with leakage or thermal distortion at the higher temperature. They felt that if other means could be used to reduce the possibility of detonation, for instance, an improvement in supercharger efficiency, there was no reason why advantage should not again be taken of the opportunity to reduce radiator size by increasing the coolant temperature.[42] As a consequence, the General Electric Company was brought in to assess the design of the Allison supercharger.

In its final form, the AEC#2 supercharger impeller was a "half shrouded" (the back plate behind the blades extended only for half of their

Table 7-4
XV-1710-1 Supercharger Performance

S/C Horsepower	Pressure Ratio	Engine speed, rpm	Efficiency, %
110	1.65	2800	54
87	1.56	2600	55
66	1.48	2400	58

length), 9-1/2 inches in diameter, having 12 radial blades 0.657 inches wide/high at the perimeter. The blades were chamfered at the inlet, but not curved. The necessary initial fluid swirl being established by four curved vanes cast in the inlet elbow. Six curved diffuser vanes were supported by the inlet cover, with a scroll used to duct the discharged mixture into the trunk leading to the intake manifolds. This was a major difference from the supercharger in the GV-1710-A with its "vane-less" diffuser.[43]

The Wright Field report on the 50-hour Development Test states, "It is apparent from the data obtained that a very creditable supercharger design has been achieved, despite the manufacturer's lack of extensive experience with centrifugal compressors. It is considered that with refinement, especially of the diffuser and scroll to eliminate all possible sources of turbulence, [that] higher values [of temperature rise ratio] might be more nearly approached."[44]

Following the formal test program the Air Corps considered ways to improve the V-1710. In June 1934, they held a conference attended by Allison, representatives of the Power Plant Branch of the Materiel Division, and Dr. Sanford Moss, GE's prominent supercharger expert. It was proposed that General Electric redesign the supercharger to guarantee certain pressure and temperature rises, i.e. the efficiency, of the unit. The objective was not to increase the rated power, but rather to insure that the supercharger was of optimum design.[45] GE did propose a redesigned supercharger that was offered to provide 1000 bhp at 2600 rpm using a 7.33:1 step-up gear ratio.[46] The proposal was not adopted and Allison who continued developing superchargers of their own design.

After completing its use in the development program, AC33-42 continued to be useful. It was sent to the NACA who used it from the summer of 1936 until March 1938 in its propeller test program and for other special tests, most notably running in a wind tunnel to determine radiator cooling requirements for the V-1710.[47] It was delivered in 1938 to Randolph Field, Texas for Classroom Instructional Purposes. The engine is now in the collection of the Smithsonian's National Air and Space Museum, Gerber Facility.

NOTES

1 Minutes from GMC Operations Committee, May 14, 1929.

2 Sonnenburg, Paul, and William Schoneberger. 1990, 48.

3 Sonnenburg, Paul, and William Schoneberger. 1990, 48.

4 Schlaifer, R. and S.D. Heron, 1950, 275.

5 Nordenholt, G.F. 1941, 218-223.

6 Engineering Section Memo Report, E-57-680-14, 1937, 2.

7 Their most recent related work had been developing and manufacturing improved liquid-cooled models of the old Liberty 12-A for the Army as the Allison V-1650 Liberty. They also developed inverted air and liquid-cooled versions based upon the 12-A and 12-B. This experience, when coupled with their previous projects to design and manufacturer the marine V-12's, the air-cooled X-4520, aircraft reduction gears and various power transmission systems, provided Allison with current and extensive knowledge of the state-of-the art of engine technology.

8 Engineering Section Memo Report, E-57-680-14, 1937, 2.

9 GMC Executive Committee Report of December 8, 1931 Committee meeting.

10 This would be the expected order for twenty V-1710-B reversible airship engines.

11 It is likely that this was the cost for what became known as "Plant 2" in Speedway.

12 Engineering Section Memo Report, E-57-680-14, 1937, 7.

13 Engineering Section Memorandum Report, E-57-6, *1932*. e

14

Engineering Section Memorandum Report, E-57-6, *1932*.

E*arly Contracts for Allison Aircraft Engines,* a compilation by the Allison Division, GMC.

15

*Engineering Section Memo Report, E-57-680-1*4, 1937, 8.

Air Corps Technical Report No. 4038, 1937, 55.

16

As engine power had increased during the 1920s it was no longer feasible to simply provide larger propellers, for they developed excessive tip speeds and needed very tall landing gear for adequate ground clearance. Reduction gears were the solve tip speeds and needed very tall landing gear for adequate ground clearance. Reduction gears were the solution. The "GV-" prefix designation had come into use during the 1920s when engines were first built with reduction gears, the "G" indicating reduction "gearing", rather than the then more conventional direct drive. Examples can be found of "GR-" radial engines as well. The classic US military designation of Type-Displacement-Model, with Navy engines using "even" model numbers was not applied to the V-1710 until the V-1710-B models were ordered. As the "second" Navy model, they became the V-1710-4 and retroactively the GV-1710-A was known as the V-1710-2.

17

Allison designated the new V-12 as their model VG-1710. When the Navy purchased the first engine they designated it as the GV-1710, evidently continuing with the scheme that Curtiss had introduced on the V-1570 series. In the interval between the May and September 1932 tests references to the engine changed from VG-1710 to GV-1710-A. When the Air Cor the 1920s when engines were first built with reduction gears, the "G" indicating reduction "gearing", rather than the then more conventional direct drive. Examples can be found of "GR-" radial engines as well. The classic US military designation of Type-Displacement-Model, with Navy engines using "even" model numbers was not applied to the V-1710 until the V-1710-B models were ordered. As the "second" Navy model, they became the V-1710-4 and retroactively the GV-1710-A was known as the V-1710-2.

18

Allisone power had increased during the 1920s it was no longer feasible to simply provide larger propellers, for they developed excessive tip speeds and needed very tall landing gear for adequate ground clearance. Reduction gears were the solution. The "GV-" prefix designation had come into use during the 1920s when engines were first built with reduction gears, the "G" indicating reduction "gearing", rather than the then more conventional direct drive. Examples can be found of "GR-" radial engines as well. The classic US military designation of Type-Displacement-Model, with Navy engines using "even" model numbers was not applied to the V-1710 until the V-1710-B models were ordered. As the "second" Navy model, they became the V-1710-4 and retroactively the GV-1710-A was known as the V-1710-2.

19 Allison designated the new V-12 as their model VG-1710. When the Navy purchased the first engine they designated it as the GV-1710, evidently continuing with the scheme that Curtiss had introduced on the V-1570 series. In the interval between the May and September 1932 tests references to the engine changed from VG-1710 to GV-1710-A. When the Air Corps purchased their first engine Allison referred to it as the V-1710-1. Only when later models were begun were these first two engines referred to by Allison as their V-1710-A1 and V-1710-A2. Neither carried such on its data plate.

20 According to the 1931 Aircraft Yearbook, the Navy contract paid Allison $75,000 for this first engine during the year. Interestingly in today's terms, this was a pretty good deal. In 1990s dollars, the cost would be about $600,000, not too bad for a first of a kind development. During the 1930s the military demonstrated an exceptional ability to get value for its few dollars.

21 Caminez, H., Report No. 26, 1930.

22 Caminez, H., Report No. 28, 1930. Note that at this point in time Allison was referring to the engine as the VG-1710, not GV-1710.

23 Caminez, H., Report No. 26, 1930, 3.

24 McCook Field Report Serial No. 2038, Air Service Information Circular No. 421.

[25] Caminez, H., Report No. 43, 1932, 13.

[26] *Marks Handbook* for Mechanical Engineers, page 1638.

[27] Caminez, H., Report No. 28, 1930.

[28] *The Company Behind the Allison Engine*, internal Allison write-up, 1943 vintage, p.2.

[29] *Fifty Hour Endurance Test of Allison VG-1710 Engine*, Report of Army Air Corps Inspector to Chief, Power Plant Branch, May 21, 1932.

[30] *Report on Development Test of Allison GV-1710-A Engine*, Report of Army Air Corps Inspector to Chief, Power Plant Branch, September 30, 1932. NARA RG342, file: V-1710 1932/1935.

[31] *Report on Development Test of Allison GV-1710-A Engine*, Report of Army Air Corps Inspector to Chief, Power Plant Branch, September 30, 1932. NARA RG342, file: V-1710 1932/1935..

[32] *Fifty Hour Endurance Test of Allison VG-1710 Engine*, by Charles W. Bond, Air Corps Inspector, to Chief Power Plant Branch, May 21, 1932.

[33] Wright Field file D00W/4040.

[34] Engineering Section Memo Report, Serial No. E-57-240-4, 1934.

[35] Engineering Section Memo Report, Serial No. E-57-6, 1932, 5.

[36] Allison Division, GMC, 1941, 5.

[37] Rick Leyes, Curator for Aero Propulsion at the Smithsonian Institution inspected the XV-1710-1(A2) in January 1997, noting that it has two data plates. The Allison Data Plate says Model: V-1710-1, Manufacturers No.: 2, while the U.S. Army Air Corps Data Plate says, Type: XV-1710-1, Serial No.: A.C. 33-42, Order No.: W535 AC-5592, Date Accepted: 6-28-33. He also noted that the intake manifold casting numbers on the right side, front and rear, are PN19408 and left side, front and rear, PN19409. Communication with Author.

[38] Air Corps Technical Report No. 4038, 1937, 31-32.

[39] Air Corps Technical Report No. 4038, 1937, 55.

[40] Air Corps Technical Report No. 4038, 1937, 14.

[41] Air Corps Technical Report No. 4038, 1937, 6.

[42] Air Corps Technical Report No. 4038, 1937, 54.

[43] Air Corps Technical Report No. 4038, 1937, 88.

[44] Air Corps Technical Report No. 4038, 1937, 57.

[45] Engineering Section Memo Report, Serial No. E-57-250-9, 1934.

[46] Engineering Section Memo Report Serial No. E-57-541-1, 1934.

[47] Harold Caminez of Allison letter to Hall I. Hibbard of Lockheed Aircraft Corp., January 20, 1936.

8

The V-1710-B: Power for Navy Airships

When the Navy originally contracted with Allison for the GV-1710-A it was to be understood to be an expedient way to begin the development of the engine they really wanted, a modern lightweight and powerful engine for their rigid airships. In their original form, the airships USS *Macon* and USS *Akron* were each powered by 560 bhp German Maybach VL-2 reversible, twelve cylinder vee engines. Each ship was powered by eight of the heavy and difficult to maintain engines.[1] The plan was to procure twenty of the proposed Allison airship engines just as soon as the engine could be designed and pass the 150-hour Navy endurance test.[2,3]

Initial testing of the GV-1710-A had been successfully concluded in the Fall of 1932, preparing the way for the airship engine. On January 24, 1933 the Navy awarded Contract No. 29907 to Allison for three experimental models of the airship engine, and requiring a 150-hour endurance test of the first engine, designated as the V-1710-4. Total amount of the contract was $85,500.[4] These engines were to be considerably different from the previous GV-1710-A. Most significant, they were designed to reverse from full-power, to full-power in the opposite direction, within eight seconds.[5] Power from the engines was to be transmitted to fixed pitch propellers mounted on outriggers equipped with swiveling heads that allowed the propeller thrust to be directed either horizontally or vertically. Drive shafts sixteen feet long transmitted engine power to the swiveling head. These were the same transmission and gear arrangements that were designed and manufactured by the Allison Engineering Company in (1929-1933). They were already in service in service in the USS *Akron* and USS *Macon*.

This is the unsupercharged XV-1710-4(B1R) that performed the 150-hour Navy endurance test at 650 bhp, 2400 rpm, May 23 to June 25, 1934. Downdraft carburetors show well, along with manual controls, including levers for reversing direction of rotation from full power in only eight seconds. Note that the B-1R was fitted with a #40 propeller shaft. (Allison)

Since airships operate at near sea-level, it was prudent to eliminate the supercharger and configure the engines for normally aspirated carburetors. A considerable amount of redesign of the V-1710 was necessary to incorporate the reversing feature. Much of this effort was expended in detailed engineering and testing of the linkages necessary to shift the valve and ignition timing for reverse operation. In recognition of having the reversing feature Allison amended their designation with an "R" suffix. These engines becoming the V-1710-B1R and B2R.

Consider the problem of reversing the direction of rotation of a 60 degree V-12 having two banks, each of six cylinders:

> One bank can be considered to be mechanically indifferent as to the direction of rotation, for its pistons simply rise and fall in response to the motion of the crankshaft, independent of direction of rotation, although the firing order is altered.
>
> Things are not so simple for the other bank. When "reversed" it must operate its valves and fire its cylinders in the opposite order. Details of just how Allison accomplished all of this have not come to light. Further complicating the operation however, was the necessity of maintaining a constant direction of rotation for the coolant and oil pumps along with the engine tachometer drives. Suffice to say that Allison did accomplish the feat.

As an earlier component of the project, Allison brought its considerable expertise with shafting, gears, and power drives into play and designed and built the beautiful remote angle head propeller drives that allowed the propeller to swing through a 90 degree arc.[6] Coupled with the V-1710-B reversing feature, an airship would be able to maneuver quite well,

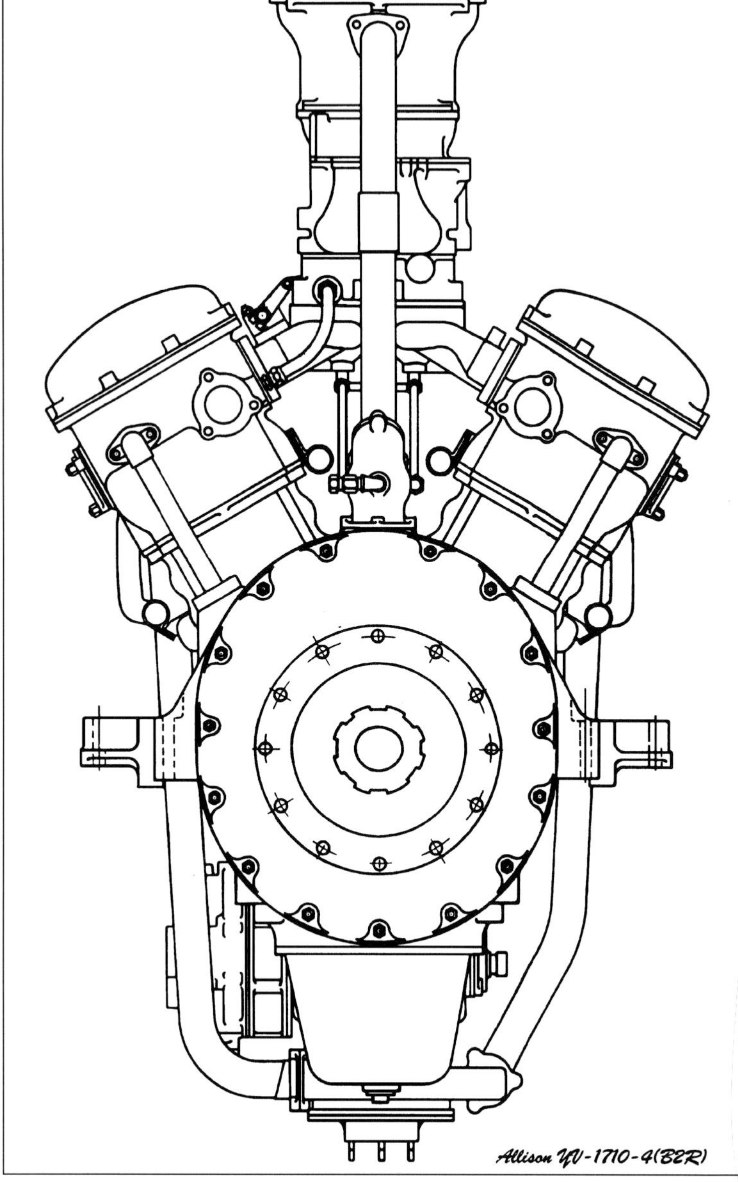

Front View of V-1710-4(B2R) Reversible Airship Engine. (Adapted by author from Allison drawing #19960)

having the ability to power itself fore or aft, up or down. All in all, this was quite an accomplishment for the small staff of engineers and machinists who were the core of the Allison operation.

With the reversible engine project well underway, tragedy struck on April 4, 1933, when the USS *Akron* (ZRS-4) crashed off of the coast of New Jersey. Since the Navy still had the USS *Macon* (ZRS-5) in operation the Allison engine project was continued. The first engine, AEC s/n 3, was constructed that September.

After over 200 hours of preliminary testing and calibration on the engine, Allison began the official 150-hour endurance test of XV-1710-4(B1), AEC s/n 3, on May 23, 1934. The test was terminated on June 25, 1934 after 120 hours as a result of Prestone leakage. Disassembly found the engine to be generally in excellent condition, though four of the cylinder barrels had cracked. These failures were attributed to concentration of bending stresses occurring in the barrels caused by the relief machined into the piston to prevent scuffing of the forged aluminum pistons at the piston pin bosses. The fix was to use pistons identical to those in the Navy GV-1710-A, which had exhibited no difficulty in 300 hours of testing. Cylinder barrel heat treatment also changed from a water to oil quench.[7]

The engine was generally renewed and prepared for another attempt at the 150-hour endurance test. New parts utilized: Cylinder Block Assemblies, crankcase assemblies, camshaft bearings, connecting rod bearings, main bearings, accessory housing casting, control tower casting, magnetos, all accessory ball bearings, propeller shaft thrust bearing, crankshaft roller bearing, magneto shifter assembly, ignition manifolds and wires, coolant pump, pistons, rings, backfire screens and miscellaneous small parts. The engine had accumulated 325 hours total time when renewed.[8]

In September 1934 the renewed engine completed the 150-hour endurance test on the dynamometer at 650 bhp and 2400 rpm, thereby clearing the two remaining engines for construction.

These engines were somewhat different than the V-1710-B1R in that they used the latest cylinder blocks from the Army V-1710-C program. The most noticeable external difference being the way that the coolant outlet connections from the blocks are arranged. On the B-1R they exited 90 degrees to the long axis of the blocks, and connecting to a common crossover pipe. The B-2R blocks have the coolant exit on the block centerline through a flush faced pad. This was a result of the improved internal coolant flow arrangement that Allison had instituted to improve cooling within the upper head.

The V-1710-4(B2R) engines were in final test and assembly in preparation for delivery to the Navy, when on February 12, 1935, the last of the two largest Navy airships, the USS *Macon*, was lost off of the California coast. With this loss the Navy ended its rigid airship development program, and with it, the need for the Allison V-1710-B.

The extensive running and testing of the V-1710-4 did contribute to the concurrent V-1710-C program for the Army, particularly the improvements in design and manufacture of the cylinder liners and bearings.

Table 8-1
Reversible Airship Engines Delivered

Model	Number	Aircraft	Comments
XV-1710- 4(B1R)	1	Model Test	No supercharger, 650 bhp, reversible
YV-1710- 4(B2R)	2	Dirigible	Pre-production, coupling in place of propeller shaft.
Total	3		

Table 8-2: V-1710-B Engine Models

B-1R - AEC s/n 3: The three V-1710's purchased for the airship program were procured on Navy Contract 29907, dated 1-24-33. As the Navy XV-1710-4, this engine performed the 150-hour Type Test for the Reversible, naturally aspirated, model. It was delivered October 30, 1934, and used two NA-Y6E Stromberg carburetors. The follow-on YV-1710-4(B2R) model was intended for flight. Testing was completed at 650 bhp in September 1934, after accumulating 325 hours total time. This engine was provided with a SAE #40 propeller shaft. The "R" in the Allison designation refers to "Reversible", not "Right" hand rotation as in later models.

Particulars as of End-of-Test:
7.00:1 CR, naturally aspirated, no supercharger, 3:2 Propeller Reduction Gear as in GV-1710-A, valve timing is not known, two NA-Y6E Carbs, SAE #40 propeller shaft, weight 1162 pounds and equipped with backfire screens. Un-weighted crankshaft. Exhaust valve stems were sodium filled. Actual test performance 690/2400/SL/-2.26"Hg Vacuum @ 0.530#/hp-hr, rated on 83 octane fuel with 2cc TEL added. Official delivery date, October 30, 1934.

B-2R - AEC s/n 4 and 5: These were the preproduction YV-1710-4 engines that were nearly ready for delivery to the Navy when the Rigid Airship program was terminated. Official delivery dates were March 8, 1935 and May 27, 1935 respectively. Essentially the same as the Type Test engine, but with all final improvements incorporated. These engines differed from the B-1R by having a coupling flange in place of the SAE #40 propeller shaft. They also used the latest evolution cylinder blocks with axial coolant outlets instead of the earlier lateral connections. Weight 1,160 pounds.

When the B-1R was tested it was necessary to include a flywheel on the drive shaft to simulate the inertia of a propeller. Note coolant "'y" piping. This was abandoned on the "C/D" engines, with all supply going into the cylinder jackets through the lower manifold. On the higher powered E/F/G the "y" was reintroduced to increase coolant supply to the cylinder heads. (Allison)

NOTES

[1] Althoff, William F., 1990, 87.

[2] Schlaifer, R. and S.D. Heron, 1950, 276.

[3] Nordenholt, G.F. 1941, 218-223.

[4] Engineering Section Memo Report, Serial No. E-57-680-14, 1937, 8.

[5] Hazen, R.M. 1941, 490.

[6] Althoff, William F., 1990, 87.

[7] Montieth, O.V. 1934, 6.

[8] Montieth, O.V. 1934. This test completed only 125 hours of the endurance test.

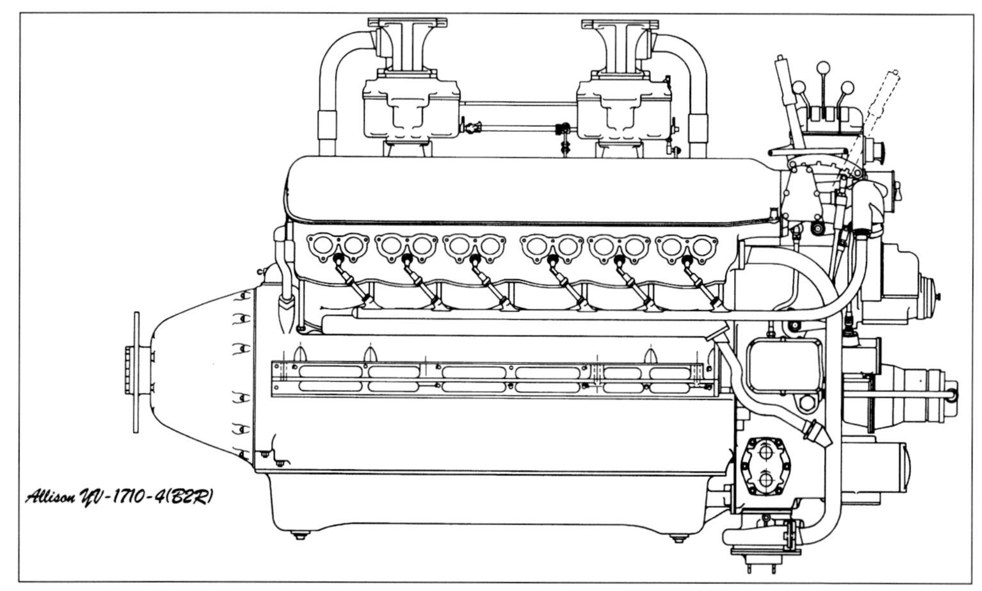

The Reversible Airship engines used a flange to couple to the extension shaft system. An engine operator would be stationed at the engine to manipulate the throttle, mixture, spark and reversing controls on orders from the airship commander. (Drawing by Author)

9

The V-1710-C "Long Nose" for Army pursuits

Background to the V-1710-C

With the introduction of the V-1710-C aircraft designers were not only provided with a 1,000 horsepower engine for the first time, but one with a fully streamlined shape. It was a truly advanced engine for the day. Consequently, the engine and its derivatives were to power practically all of the modern pursuits proposed and/or developed for the U.S. Army during the late 1930s. Lessons learned in the development and operation of these engines provided the basis for the improved and still higher powered "E", "F" and "G" series V-1710's. In total, the V-1710 powered over 60 percent of the Army's WW2 fighters, as well as the final generation of liquid-cooled pursuit aircraft, the Bell P-63C, and the North American P-51J and P-82E/F models.[1]

Following the successful demonstration by the XV-1710-1, Allison and the Air Corps undertook the development of the similar V-1710-C, intending it for use in a new generation of bombers and pursuits. It took nearly five years to develop the V-1710-C to the point that it could reliably complete the newly instituted 150-hour military Type Test. The reasons for the protracted development are many and varied, though without question the small size of the Allison team, the changing government demands and the budgetary limitations brought about by the Great Depression certainly played significant roles.

The Army had originally been reluctant to support the Allison proposal to build an aircraft engine at all, largely as a result of their critical assessment of Allison's' limited resources and minimal relevant experience. While the status as a Division of General Motors certainly worked to Allison's advantage in overcoming this Wright Field reluctance, it still remained a daunting task to invent, develop and later mass produce a successful large and modern aircraft engine.

World and economic events conspired to encourage the Army to support the Allison engine effort. The rearming of Europe, the commitment to liquid-cooled engines by the European powers, and the lack of development potential and corporate support of the only other U.S. liquid-cooled aircraft engine, the Curtiss-Wright V-1570 Conqueror, and their short-lived SGIV-1800, were the major factors.[2] In 1931-32 Wright Aeronautical had begun development of a follow-on to the Conqueror, their SGIV-1800. This V-12 engine was sponsored by the Navy, utilized individual cylinders specifically designed for high-temperature cooling, and intended to incorporate a supercharger rated for 12,000 feet.[3] It was run in late 1932 or early 1933, and in June 1934 underwent a 50-hour development test at 800 bhp. At this point the Navy lacked funds to continue and ceased support of the V-1800. The Army refused to provide any appreciable support, and the company did not wish to do further development at its own expense.[4] Earlier production of a family of water-cooled V-12's by the Packard Motor Car Company had ended when the Navy adopted a policy of using only air-cooled engines and Packard decided to focus their production and engineering capacity on manufacture of automobiles instead of further de-

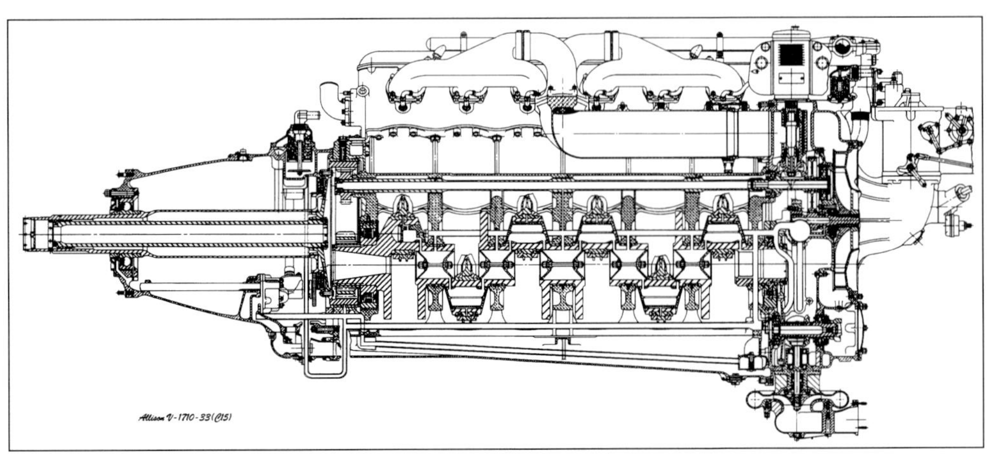

This crossection of the V-1710-33(C15) shows the detail of how all of its components and systems were integrated into a powerful and effective powerplant for an entire generation of advanced pursuit aircraft. (Author)

velopment of aircraft engines.[5] Another significant contributing factor appears to have been a disagreement with the Army over design and manufacturing standards and policies. As a result the Army completely withdrew all development assistance from the company in 1928. Without subsidies, Packard dropped all further work on its liquid-cooled aircraft engines.[6]

In this environment the early 1930s effort by Wright Field to encourage engine manufacturers to build high temperature liquid-cooled engines utilizing their new HYPER two valve cylinder found interest only with Lycoming and Continental.[7] Allison was already committed to their own four valve cylinder and associated concepts and was not interested in participation in HYPER development. The Chrysler IV-2220 inverted V-16 liquid-cooled engine used some of these features, but it was not begun until after 1939.[8]

Developments Up To 1936

It is hard to believe that so few engines were built and used in support of the Air Corps sponsored Allison V-1710 development program. Only four engines were used during the five years it took to accomplish the 150-hour Type Test, each existed one at a time. The June 1934 purchase of ten "YV-1710-3" engines was intended to meet the need for test engines and provide for use in new models of aircraft designed to specifically make use of the engine. The first of these aircraft were to be the four engine Boeing XB-15 and the Martin XB-16.

For comparison, Rolls-Royce was developing their similarly rated Merlin at the same time. Their first flight occurred on April 12, 1935 and they had four flying test beds prior to Merlin C No. 11 powering the first flight of the prototype Hurricane on November 6, 1935. All of this was in spite of the engine having failed to pass the necessary civil 50-hour provisional certification of airworthiness test. By the following February, the prototype Hurricane was flying with Merlin C No.19. A production order for 200 Merlin I engines was issued to Rolls-Royce in March 1936, a full year before the V-1710-C even completed its 150-hour Type Test.[9] Money, commitment and an experienced engine development team greatly accelerated the development of the Merlin.

There were four development engines needed to complete the Army Type Test for the V-1710. These were the XV-1710-1(A2), the XV-1710-3(C1), the XV-1710-7(C3), and the YV-1710-7(C4/C6).

The XV-1710-1(A2), a "proof of concept" engine, followed by the XV-1710-3(C1) that was purchased to perform the 1000 bhp Type Test.

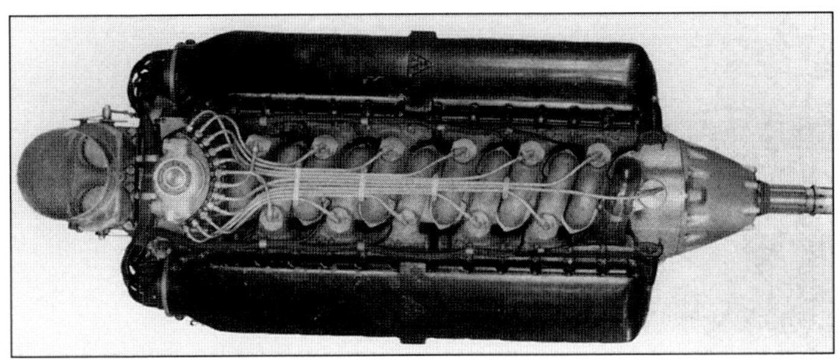

Navy GV-1710-A/Build #3, modified by the Army for fuel injection development tests. Injector pump was mounted on the vertical magneto drive shaft. Legible on fuel injector in original photo is "Marvel Carburetor Co. Flint, Michigan" and "Chandler Fuel Charger". Tubing runs, in clockwise order read: "5L-2L-3L-4L-6R-1R-1L-6L-5R-2R-3R-4R". A major problem for the fuel injection program was the lack of a mixture control able to automatically compensate for altitude. Note glass trunk on 6R (front) cylinder for viewing mixture during testing. (Allison)

After a protracted test period that involved many rebuilds and improvements, it was replaced by a completely redesigned engine in 1936, the XV-1710-7(C3). Though improved, and tested for about 250 hours, it was also ruled unsuccessful in that it was not able to go the 150 hours without major rework or repair. The YV-1710-7(C4), AEC #10, AC34-8, incorporated all of the improvements developed to date, although it was upgraded to C-6 form during the test. It demonstrated a comparatively smooth run and completed the Type Test between January 28 through April 23, 1937.

Even while the Type Testing was continuing, Allison was pressed to develop other arrangements and features into the engine. Particularly distracting was the Air Corps demand for the design and development of the XV-1710-5(C2) fuel injected engine in 1934. Concurrently, Allison began work on developing the extension shaft version of the V-1710-C. This new "D" series was to operate as a "pusher" for the Bell multi-place-fighter/bomber-destroyer. In March 1936, nearly a year before the V-1710 Type Test was successfully completed, the Air Corps combined these efforts and ordered a fuel injected V-1710-9(D1) pusher configured development of the V-1710-C engine to power the unique Bell XFM-1 aircraft.

The mid-1930s could be viewed as either a desperate or particularly creative period for aviation. Manufacturers and designers were each trying to out-do the other with innovative and unique designs. With the promise

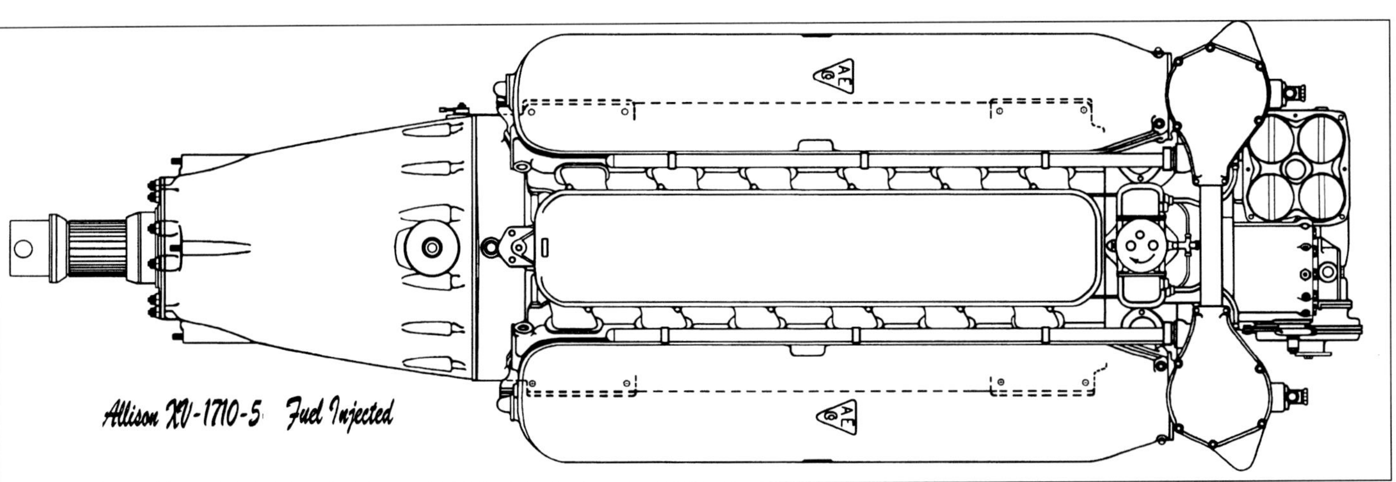

Final Configuration of the Proposed XV-1710-5(C3) Fuel Injected Engine. The induction system is as run on the Allison EX-2 Workhorse engine in May 1936. Note the air plenum between the cylinders. (Drawing by Author)

of the most advanced power plant, Allison's small staff was involved with nearly all of them.

Although the Army had ordered ten "Service-Test YV-1710-3's" in 1934, only the one for Type Testing, along with the intended first flight carbureted XV-1710-7(C2), were constructed prior to September 22, 1936, the date the Army officially released the last eight engines for manufacture. The contract had stipulated that the Type Test was to be completed prior to their manufacture. Furthermore, the Army did not pay for an engine until it was actually delivered. Allison was in a deep financial hole as a result, and General Motors clearly kept the program funded and alive. It was the first of these later engines, Allison "Number 9", that became the first V-1710 to fly.

The Fuel Injected V-1710

In July 1934 Mr. H. Caminez, V-1710 Project Engineer, delivered layouts to Wright Field for their review showing design details of the proposed fuel injector installation on the YV-1710-3 engines.[10] This was in response to one of the original XV-1710-1 contract requirement that the engine be designed to accommodate Marvel fuel injection.[11] In February 1935 the Air Corps further diverted the limited Allison engineering and Type Test engine development activities by requiring that the first two engines on the ten engine W535-ac-6795 contract be built with fuel injection into the cylinder ports, in place of the previous carburetion system. This necessitated a complete redesign of the accessory housing, accessory drives, and intake manifolding, as well as the necessity of providing cylinder heads with individual ports connecting to the manifolds in place of the previous plenum spanning groups of three cylinders. Interestingly, the original approach to intake manifold design was similar to that used by the later Rolls-Royce Merlin. In April 1935 the Air Corps further directed that all ten of the service test engines be changed over to this fuel injection system.[12] These engines were designated as the YV-1710-5, and were to be rated for 1000 bhp for both "Normal" and "Takeoff", the same as the carbureted YV-1710-3's then on order. The new intake manifold (PN33425)[13] provided individual trunks discharging into each of the twelve intake ports from a central plenum, thus accommodating the positive displacement "Chandler Fuel Charger" fuel injection pump manufactured by the Marvel Carburetor Company of Flint, Michigan. This device was designated by the Air Corps as their Type S-1 Fuel Injector.[14]

The first engine built on the contract, AEC#7 (AC34-5), had been delivered to the Air Corps in the carbureted form as the XV-1710-7(C2) on December 24, 1935, where it was to be used for flight testing in the A-11A engine test bed. It was intended that this engine would be converted to fuel injection as soon as the feature was through Type Testing.

Allison did build a fuel injected "workhorse" engine on its own, as their EX-2 development engine, and had it running in May 1936 in a "pusher" extension shaft configuration. In this form it represented the originally specified 1250 bhp V-1710-9(D1), but was probably operated with 6:1 compression ratio at no more than 1000 bhp. In addition, the original Navy GV-1710-A, when in its "Build 3" configuration, had been operated with "spaghetti" type port injection manifolds and used for an early series of fuel injection tests.

When Ron Hazen arrived at Allison he consolidated the design efforts on the fuel injection and carbureted engines and oversaw the redesign of an engine that could be built in either fuel injected or carbureted forms. Engines YV-1710-3 AC34-5 and 34-6, AEC#'s 7 and 8, the first of the batch of ten contracted Service Test engines were then designed, and had parts fabricated that would allow them to be assembled in either the YV-1710-5(C3) [fuel injected], or YV-1710-7(C3) [carbureted] forms. The Army insisted that AC34-5 be rebuilt in the new fuel injected configuration as soon as the redesign was complete and parts available. The first engine to result from the 13 week redesign effort was AEC #8, and it was built as the carbureted XV-1710-7(C3), and assigned the role of completing the Type Test.

With the emphasis by Allison and Wright Field on completing the Type Test and getting the service test engines into the aircraft being readied for them, the fuel injection program was relegated to an Allison in-house development effort using their EX-2. This resulted in recognition of the lack of a fuel metering pump with a viable mixture control. This was the primary reason given by Ron Hazen for termination of active development of the fuel injected model during the summer of 1936. The injector needed to be able to not only deliver uniform quantities of fuel, but also incorporate an effective automatic mixture control. Even with the March 1936 decision to focus efforts on the carbureted engine Type Test, the fuel injection engine development program continued. The feature was continued in the design of the improved and convertible C-3 models, at least through October 1936.

Intake Manifold Evolution

The intake manifold was the heart of the V-1710, and as such received a lot of attention during the life of the engine, particularly during the early development years. Designed from the outset as a supercharged engine, considerable thought was given to the layout of the passages for the mixture as it passed through the supercharger and into the cylinders. With the introduction of features for adoption of fuel injection in 1934, it became necessary to provide separate trunks leading directly to each cylinder to accommodate a fuel spray nozzle just ahead of each inlet port. Such an arrangement is nearly universal on production passenger automobile engines in the 1990s.

Allison had been revising the manifolds on practically every model or rebuild of the early development engines. One reason was to reduce the

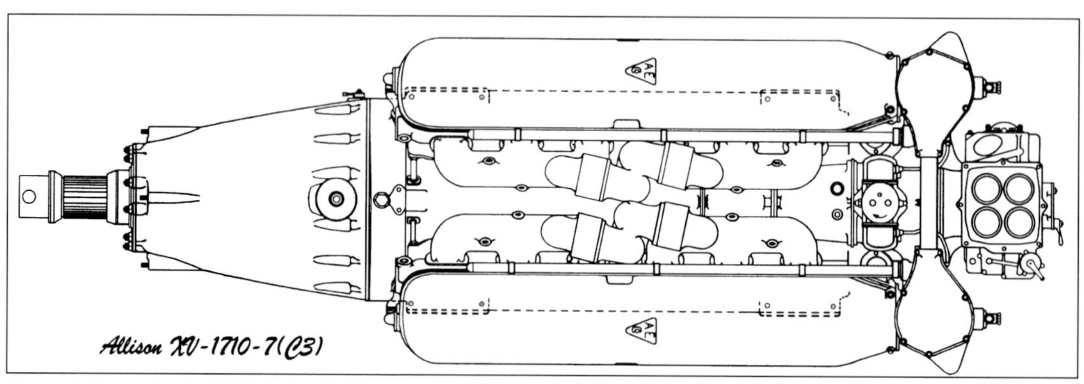

Ron Hazen introduced the tuned ramming intake manifolds, PN34054, during the redesign that resulted in the XV-1710-7(C3). When delivered in June 1936, this was the first use the new manifolds. All subsequent V-1710 and V-3420 engines used developments of this manifold. (Author)

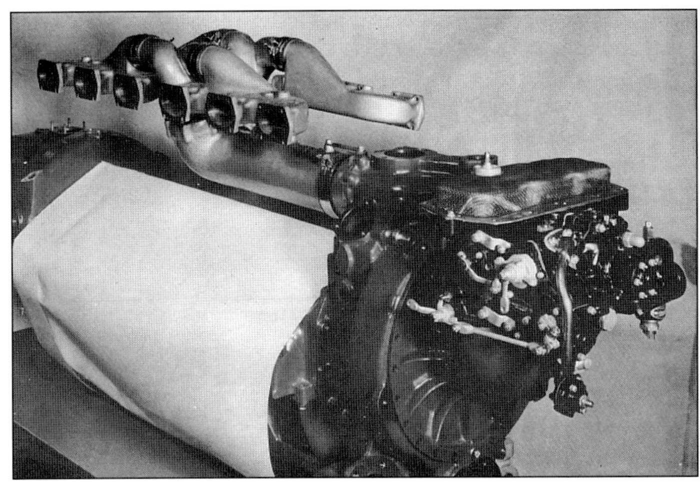

This is the induction system on the C-15 engine. As shown it uses Tee backfire screens and PN 36085/6 magnesium manifolds. (Allison via Hubbard)

flow restrictions, and thereby reduce the supercharger step-up gear ratio, this was accomplished by enlarging the passages ways and plenums spanning each group of three cylinders.

When Ron Hazen joined the project as Chief Engineer in early 1936, he undertook a critical assessment of the engine and initiated a general redesign which focused on strengthening the engine to the point that it could reliably complete the 150-hour Type Test. One likely motivation for the extent of the effort was the finding from in-house testing that there was considerable maldistribution of the mixture between the banks, and among the cylinders on each bank. It was determined that this was the source of much of the uneven torsional vibrations that were being observed, as well as the frequent burning of valves and pistons. Hazen's response was to design the "ramming" manifolds, that were effective in equalizing the distance traveled by the mixture in reaching each cylinder, while at the same time smoothing the pulses caused by charging each cylinder. Previous manifolds caused considerable imbalance due to the 60 degree delay in delivering mixture between the banks. By carrying each of the four induction trunks back to the center of the engine, not only were the lengths equalized, but each was able to take credit for the ramming inertia of the charge pulsing through each trunk. It was the PN34054 manifold that first offered this feature.

To prevent inadvertent explosions and backfires with the large volume of combustible mixture in the supercharger and manifolds, backfire screens were fitted to minimize the spread of a fire or explosion. There were many different sizes and arrangements of backfire screens used at various times, though the primary ones are given in the table below. As is told in the chapter on introduction to service of the P-40, the change from magnesium to aluminum, and carrying the web in the dividing "tee" all of the way down to the backfire screen aided in reducing the fires and destructive effects of backfires in these early engines. It was determined by Allison that resetting the intake valve clearance was the most significant change responsible for resolving the backfire problem. Even so, for a while the Air Corps directed Allison to install both the "tee" backfire screens and in addition, separate screens at each inlet port. Thousands of these manifolds were built, and rather than scrap them once Allison had resolved the backfire problem, they were fitted with a sleeve in place of the screen at the port. In this way, the flow path into the cylinders was smooth.

In 1942 Allison revised the manifolds again to allow removal of all of the backfire screens. Originally Hazen had incorporated the "Tee" screen, not only as a protection against backfires, but to aid in equalizing flow between the two banks. When removed flow distribution would again become a concern, so a "streamlined" manifold was developed. It achieved equal flow to each bank by paying careful attention to the flow cross-section, which was now elliptical in shape in comparison to the original circular section. This manifold was quite successful and provided several thousand additional feet in the rated critical altitude of the aircraft in which it flew.[15]

Engine Model Designations

An interesting aspect of developing the V-1710 was the increasing complexity of the effort, and the necessity of upgrading the management of the program in parallel with improving the engine. Not long after Ron Hazen joined the V-1710 program as Chief Engineer he notified Wright Field that Allison had created its own Model Designation system to distinguish their efforts from those represented by the Air Corps model designations.[16] This was not long after Allison had instituted its *Allison Specification* scheme, a

Table 9-1
Intake Manifold Evolution

PN_____	GV-1710-A(A1) Build 1 & 2, Central trunk w/4 branches to Plenums spanning 3 cylinders
PN18579	XV-1710-1 prior to October 1933, Plenum spanning 3 cylinders
PN19409	XV-1710-1 and Build 1 of XV-1710-3, Plenum spanning 3 cylinders
PN19420	XV-1710-1 after to October 1933, enlarged to reduce losses, Plenum spanning 3 cylinders
PN33425	Fuel Injection Manifold with individual ports, for XV-1710-5, 7-15-35, pre-Hazen
PN33550	Manifold on Carbureted XV-1710-7(C2) AEC#7, 12-24-35, pre-Hazen
PN34054	Hazen "Ramming" Manifolds w/Tee BF Screens, used on Service-Test engines, first installed on XV-1710-7(C3), AC34-6, AEC#8 also on XV-3420-1. Used until C-10 engines.
PN34162	Tee w/BF Screens used w/Hazen manifolds, dividing vane terminated about 9/16 inch above BF Screen, used PN36930 BF Screen
PN36085/6LL	Improved "Ramming" type Magnesium Manifold quadrants without Individual Port Backfire Screens, first used on C-10 (Contract W535-ac-10660) , the 969 W535-ac-12553 engines, and most of the British C-15's.
PN36085/6BB	Aluminum Manifold quadrants without Individual Port Backfire Screens, retrofit onto all U.S. engines.
PN40041	"Divided Tee" with dividing web extended down to surface of screen, Replaced PN34162.
PN40772/3	Aluminum Manifolds with Individual Port Backfire Screens-Most reworked to remove port screens.
PN43775	Streamlined Manifold Tees with no Backfire Screens, used on post-1942 engines. Connects to either 36085/6 or 40772/3 manifold quadrants.

system for defining the specifics of an engine that has continued in use by Allison for over 60 years.[17] These efforts provided structure for managing internal accounting, engineering and development as well as proposal of engines to customers, and for managing subsequent improvements in each model. In 1935 Allison started with Specification 101 describing the then current XV-1710-7(C2). The X-3420 "double vee" program started at the same time with Specification 201. All previous projects and models were simply ignored, with the result that their designations remained project specific.

A degree of confusion surrounds the designations and manufacture of the early V-1710's. Much of the confusion occurs because of the transitory nature of events at the time. For example, the flip/flapping on the fuel injected and carbureted engines were described only by changes to contract documents. Ron Hazen soon instigated the decision to completely redesign the engine, expanding the in-house designation of C-2 for the current convertible carbureted/fuel injected engine model and assigning C-3 to be the newly improved convertible engine. The previously built "C" model became the C-1 retrospectively. With the decision to defer fuel injection development a revised accessories section was designed that eliminated the fuel injection pump and saved weight, thus defining the C-4 model. Subsequent models reflected the evolution of requirements and features demanded by an ever broadening list of applications for the engines.

Ron Hazen Led Redesign and Type Testing

When Hazen arrived to take over as Chief Engineer at Allison in January 1936 he found a somewhat confusing array of projects and directions. The XV-1710-7 engine, AC34-5, mirroring the latest improvements in the XV-1710-3, had just been delivered to the Air Corps where it was intended to power the Bell A-11A flight test aircraft. He undertook a personal study of the engine and what was needed to get it to satisfactorily complete the Type Test. This resulted in directing the efforts into finishing up the design of the new fuel injected engine so that it could be built in either fuel injected or carbureted versions. This was to prove beneficial when the Air Corps declared, in April 1936, that the XV-1710-3 Type Test engine had failed the Type Test.

In March 1936 the V-1710 engine situation was thoroughly reviewed by Wright Field and Allison. Ron Hazen was adamant that the fuel injected engine program should be terminated or deferred, as no satisfactory fuel injector was available. He offered that all resources should be focused on a comprehensive redesign of the original carbureted version of the engine so as to resolve the many problems still being experienced with the Type Test engine. The parties agreed to take three months for the redesign. Still, the Army prevailed by requiring the design to be convertible to fuel injection. The result was the V-1710-5/7(C3) model of the engine having a convertible accessories section and cylinder blocks able to accommodate either the new Ramming Manifolds or Port fuel injection.[18] This occurred just as the XV-1710-7(C2) [AC34-5] was being fit into the A-11A.[19]

Ron Hazen actively lead the Allison team in the redesign and construction of the engine in the 13 weeks between March 7 and June 13, 1936. Actually, a considerable amount of work had been underway on the V-1710-5/7 models for at least six months, but the refocusing of resources and effort brought the considerably improved engine to the test bed in record time.[20] In addition to use of PN34054 manifolds in place of PN33550, they altered the coolant passages to be more effective with glycol, redesigned the combustion chamber, provided new forged pistons, and generally strengthened and/or improved numerous minor parts and components. These new manifolds were equipped with backfire screens, though they were not in the same location as those in the later C-4 with its Hazen designed ramming manifolds.[21]

In the six months following its June delivery, the new XV-1710-7(C3) engine accumulated 245 hours of Type Testing, sufficient to cause the Air Corps to officially release the eight remaining engines on the contract for construction as of September 22, 1936.

During the redesign, Hazen accommodated the Fuel Injection requirement by providing a pad to mount a Type S-1 fuel injection pump and distributor, along with a new air throttle, and necessary controls.[22] In addition, two sets of intake manifolds were designed for the engine: the new "ramming" manifolds for the C-3 carbureted version and a central plenum chamber with leads to each cylinder for the XV-1710-5. As the accompanying drawing of the XV-1710-5(C3) shows, the arrangement looks suspiciously like the manifold arrangements used successfully on high performance turbosupercharged automobile racing engines running in the Indy 500 in the 1980s and '90s'.

Allison drawings dated November 1935, show the first two engines on the contract for the ten "YV-1710-3's", AC34-5 and AC34-6, being assembled in either XV-1710-5 or XV-1710-7 configurations.[23] Parts had been fabricated for both engines to be built-up in either form. The drawings for the XV-1710-7 were revised as of February 21, 1936 and show the first application of the Hazen designed "ramming" manifold, the type that became standard on all subsequent V-1710's. With the press of continuing Type Testing using the XV-1710-7(C3) (AC34-6) to replace the XV-1710-3 (AC34-4), it was built in the improved carbureted form. Engine AC34-5 had already been delivered as the carbureted XV-1710-7(C2) when the redesign occurred and was in the A-11A during the redesign period. This is the engine that would have been the first to be built with fuel injection as the XV-1710-5 had that program continued. When Allison won a stay in the development of the fuel injected models, both the XV-1710-3(C1), AC34-4, and the XV-1710-7(C2) were rebuilt with the improved dynamically balanced crankshaft and C-4 accessories sections, making them C-6 models. Engine AC34-5 was then shipped to Curtiss for use in propeller testing.[24]

Table 9-2
Early "C" Model Engines

AEC Model	AAC Model	Serial Numbers	Comments
C-1	XV-1710-3	AC34-4, AEC #6	1st Type Test Engine, redelivered as C-6
C-2	XV-1710-7	AC34-5, AEC #7	Carbureted, Delivered 12-24-35, later redelivered as C-6
C-3	XV-1710-5	None	C3 was convertible to Fuel Injection, applied to AEC #7 & AEC #8
C-3	XV-1710-7	AC34-6, AEC #8	2nd Type Test Engine, Carbureted
C-4	YV-1710-7	AC34-7, AEC #9	1st Flight Engine, A-11A
C-4	YV-1710-7	AC34-8, AEC #10	3rd and final Type Test Engine
C-4	YV-1710-7	AC34-13, AEC #13	1st Flight Engine for XP-37

During the course of Type Testing the XV-1710-7(C3) was replaced by the YV-1710-7(C4). It was this engine that ultimately was the successful Type Test engine. During its testing it was necessary to replace the crankshaft with the dynamically balanced unit developed for the 1150 bhp engine. As a result the engine was designated the C-6 by Allison.

Features of the C Series

The major external changes insisted on by the Air Corps in their specification of the XV-1710-1 was that the propeller shaft be extended approximately 12 inches ahead of its location on the Navy GV-1710-A(A1), that the lower crankcase be streamlined to allow a smooth transition from the extended nosecase, and that the propeller reduction gear ratio be 2:1, versus the earlier 3:2.[25] This was done to allow maximum streamlining of the engine when installed in an aircraft.

The 1000 bhp requirement necessitated running at a higher rpm than the Navy engine, that in turn required a counter-balanced crankshaft to minimize the dynamic loads on the main bearings and crankcase. As on the GV-1710-A, Allison continued to use their unique "internal" spur type reduction gear with its large plain bearing. They had selected this to aid in streamlining, and they believed at the time, to reduce the weight of the reduction gear assembly. With the increase in engine speed the reduction gear ratio needed to be increased so as to not overspeed the propeller. The change required a larger ring gear, with a 14 inch OD plain bearing to support it, along with a change in the way the accessories drive "quill" shaft was driven.

In the GV-1710-A, which used a 10 inch OD ring gear, the quill shaft was driven by a gear running on the outside of the ring. In the V-1710-C the quill shaft was driven by a gear engaging the internal reduction gear itself. Such changes, though seemingly simple, can be extremely complex and require a significant amount of engineering, along with development and testing. The result of all of this was that the reduction gear plain bearing operated at very high bearing velocities and required a comparatively large oil flow to cool and lubricate its highly loaded parts. By the time the V-1710-C13 was ready to fly in the XP-40, Allison had resolved the lubrication of this bearing to the point that they no longer recommended having a thermocouple to monitor its condition, in fact, they stated that "the large reduction gear bearing used on our design is now as reliable as any other bearing in the engine."[26]

Incidentally, many aviation historians and writers persist in incorrectly identifying this reduction gear as an "epicyclic" or "planetary" gear design. It was not. This perception is probably due to the comparatively close centerlines of the propeller and crankshaft in the "long nose." They were only 3-3/16 inches apart due to the use of the "internal" spur gear arrangement, but it is clearly an error and needs to be revised in numerous accounts.

Dealing with the Air Corps, as represented by the Materiel Division at Wright Field during this period, was tough. The V-1710 was developed in the heart of the Great Depression. Money was scarce and military budgets were small. Consequently the Army demanded value, and placed every burden it could onto its contractors, even if they were losing money. A case in point is the issuance to Allison of a letter in December 1936 requiring that Allison must guarantee the satisfactory performance of the sparkplugs in the V-1710's for 500 hours of operation, or one year, whichever was completed first. This was at a time when sparkplugs were still the "two part" type and were a constant source of problems for everyone in aviation.[27]

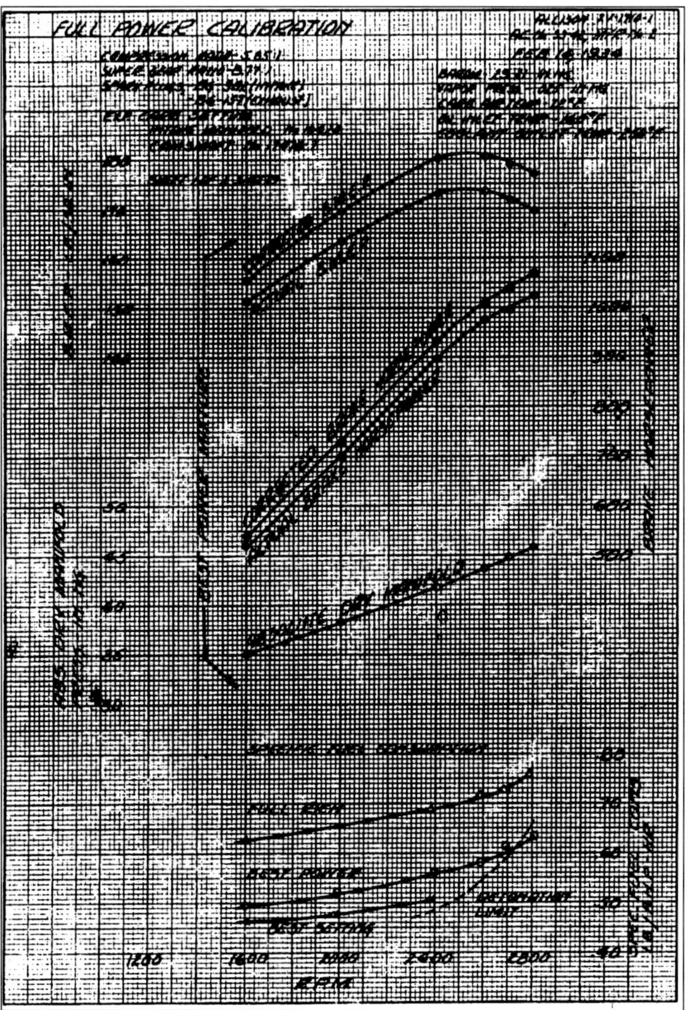

Although Allison only guaranteed the XV-1710-1(A2) to deliver 750 bhp, they knew the Army intended to qualify it at 1000 bhp and 2650 rpm for 50 hours. During the testing it was successfully run to over 1000 bhp. (Air Corps)

Summary of Development Running Time

The following is a compilation of official testing on the V-1710 done by the Air Corps at Wright Field through the end of 1936.

In addition, the YV-1710-7(C4) continued for the balance of its 150-hour Type Test, and then was returned to Allison where it ran another 116-1/2 hours prior to beginning a series of severe Dive tests that required about another ten hours, much of it at high overspeed. This total is about 1,060 hours of official test time to accomplish the Type Test.

Table 9-3
Summary of Development Running

Model	AC Serial Number	Manuf Serial No.	Test Time, Hr:Min
XV-1710-1(A2)	33-42	2	190:00
XV-1710-3(C1)	34-4	6	328:00
XV-1710-7(C2)	34-5	7	0:00
XV-1710-7(C3)	34-6	8	258:55
YV-1710-7(C4)	34-7	9	5:05
Total			782:00

Follow-on Models

By the time the Type Test was completed on the YV-1710-7(C4) one was already flying in the test-bed Consolidated/Bell A-11A and another was about ready to go for the first flight of the Curtiss XP-37. In addition, two of the service test engines, as carbureted YV-1710-9's, had been delivered for installation in the Bell XFM-1. Soon after the first flights of these later aircraft, both airframe manufacturers received orders for service test batches of thirteen aircraft, with the Air Corps contracting separately with Allison for engines with which to power them. Twenty V-1710-11's were ordered for the YP-37's and somewhat later, 39 V-1710-13's for the YFM-1's were added. The last of these engines, a V-1710-D2, was shipped in October 1940. Production capacity, for even such a small order, was extremely limited. All of the contract W535-ac-10660 engines were built in the original Allison Plant #2.

"Spare" V-1710-11's from the batch for the YP-37's were used to power other prototype aircraft, specifically the XP-37, the Lockheed XP-38, and the Curtiss XP-40. The latter was unique in that it required Allison to develop its first "Altitude Rated" engine, or conversely, Allison pursued the aircraft manufacturers to try its first altitude rated engine. This helped to gain the first production order.

An altitude rated engine is one intended to operate with excess supercharger capacity at sea-level, where it is required to be throttled to prevent overboosting. At altitude this capacity is available to extend the height at which the engine can continue to deliver its full rated power. This requires a larger carburetor and a higher internal supercharger gear ratio in order that a useful critical altitude could be achieved. With this engine the XP-40, itself a conversion of the tenth P-36A production airframe, was the winner in the 1938/9 Pursuit competition and 134 were ordered. To power them, Allison received an initial order for 234 V-1710-19's.

Only one V-1710-19(C13) was actually built on the production order, and it failed its Model Test. Improvements were quickly made to further enlarge the supercharger intake elbow and carburetor along with other internal improvements, the result was a change in designation to V-1710-33(C15). In this form the engine was further refined and finally completed the 150-hour Type Test after consuming five engines. By the time this had been done, 228 V-1710-33 engines had already been delivered.[28] At no cost to the Air Corps, Allison reworked and modernized these earlier engines to the final Model Test standard. These early engines received a "M" stamped behind their Air Corps serial number to identify them as "Modernized."

V-1710-33(C15) AEC #155, AC 39-953M is one of the first batch of production C-15's and was one of those requiring Modernization. Note remarking of the Air Corps s/n on the dataplate following Modernization. (Author)

Much of the problem with the V-1710-C15 Model Test was the Air Corps insistence that the engine be rated at 1090 bhp, while Allison recommended 1040 bhp. The final takeoff rating was indeed established at 1040 bhp. These ratings were from identical altitude power calibration curves for the C-13 and C-15, but the C-15 engine was rated at a point on the curve corresponding to a higher altitude as compared to the C-13. See Appendix 2 for details on how engines were rated.

This rating controversy was not just a matter of desiring different numbers, though everyone certainly wanted the most power possible to be available to the pilot and aircraft. Since the C-13/C-15 engines were in reality conversions of the V-1710-21(C10) YP-37 engine, it is important to understand why the later engines could not be rated at the same 1150 bhp available with the C-10. The answer is in the amount of power required to drive the internal gear driven supercharger. As a "sea-level" engine the C-10 was rated for operation with a turbosupercharger, and thus only needed comparatively low 6.23:1 ratio internal supercharger gears. At the rated 2950 rpm of the C-10, when it was producing 1150 bhp, the supercharger was consuming about 85 bhp. Alternatively, the C-15, with its' 8.77:1 step-up gears and running at 3000 rpm to develop the desired 1090 bhp, the supercharger was requiring about 175 horsepower. Both engines were actually limited to the same maximum indicated internal gas pressure allowed by available 100 octane fuel. As rated, both engines were operating at a nominal 1400 IHP, hence Allison's concern and the basis for their recommendation of 1040 bhp. The extra 50 bhp was simply not available.

Table 9-4
Modernized C-15 Ratings

Allison Rating	bhp/ rpm/Altitude, ft
Takeoff	1040/3000/Sea Level
Normal, Sea Level	840/2600/Sea Level
Normal, at Altitude	930/2600/14,600
Military	1040/3000/12,800

As a consequence of the long period and multiple tries at completing the V-1710-33 Model Test, the Air Corps restricted the engine to 950 bhp at 2770 rpm, the latter limit being due to difficulties with the crankshafts on some of the early P-40's.[29] On January 25, 1941 Wright Field cleared the V-1710-33 for production and service with ratings based on Allison's recommendations coming from the several 150-hour Model Tests of the V-1710-33, -27, -29, -35, -37 and given in Table 9-4.[30]

Until modernized with an improved crankshaft and other parts, V-1710-33's in squadron service continued to be restricted to:

The XV-1710-19 engine in the XP-40, (AC38-585**?**, AEC #23) was later replaced by a new V-1710-33 and the -19 subsequently had its details upgraded to latest production standards, but still configured as a V-1710-19.[31] The remaining engines on W535-ac-10660 were delivered as V-1710-21(C10)'s and -23(D2)'s, rather than the V-1710-11(C8) and V-1710-13(D1) models originally ordered. The difference being that the new Bendix-

Table 9-5
Early Un-Modernized C-15 Ratings

Restricted Rating	bhp/ rpm/Altitude, ft
Takeoff	950/2,770/Sea Level
Normal, at Altitude	838/2,600/8,000
Military	950/2,770/8,000

Stromberg pressure carburetor was used in place of the previous Bendix float carburetor. Another -21 (AC38-597 AEC #37) was subsequently converted to a modified V-1710-19 configuration and used for the "High Power" engine test project, that ran from November 1939 to January 1940. In this effort it finally developed 1575 bhp, certainly the maximum ever intentionally drawn from a V-1710-C/D. Interestingly, during the test program a flaw was found in the crankcase and the internals of this engine, and they were replaced by those from V-1710-11 (AC38-581), the engine that had been damaged in the delivery flight crash of the XP-38! It was with these components that testing reached the 1575 bhp level.

According to Allison Chief Engineer Ron Hazen, the V-1710-C15 engines were produced at a peak rate of 300 per month. The bulk of production going to the French and British orders, with the U.S. Army receiving only 30 to 40 percent of production.[32] Though initial AAC orders were for 776 V-1710-19 engines, this was soon changed to an even 300 C-15's, with the balance to be produced as the improved 1150 bhp V-1710-39(F3R). Most of the "C" model production was for the French and British, with the Army promoting early completion of those contracts to expedite production of the E/F models. Production of the C-15 model was completed in July 1941.[33]

A total of 2,550 V-1710-33(C15)'s were built, and used to power early Curtiss P-40's and *Tomahawks*, along with the batch of Lockheed P-322's. These aircraft were distributed all over the world in the early days of WW2. In 1941 they were the primary Allied fighter in the Desert War in North Africa, and were the front line defensive fighters with the U.S. Army in the Philippines, Hawaii and the Canal Zone. In a letter from Air Vice Marshal at the Royal Air Force, Cairo dated June 25, 1941 he states that:[34]

- "The first Allison engine in 250 Squadron has now reached 120 hours. It is running perfectly, and we have felt justified in raising the life to 180 hours.
- The general performance of the Allison has been satisfactory. No air cleaners are fitted, and no extra cooling for these conditions. Serviceability of the Squadron is 15 (out of sixteen) this morning.
- In comparison with our own trials and troubles in the development of our engines, this is a most creditable achievement."

But still, the P-40's are probably best known for being the mounts of the American Volunteer Group, the *Flying Tigers*, in China in 1941 and 1942.

Altitude Rated "C" Models

Initial Air Corps interest in the V-1710 was largely because of its ability to be turbosupercharged to high power at high altitude. The slow development of turbosuperchargers and their controls was having an adverse effect on Allison and the companies building the aircraft they were to power. It appears to have been a common desire among Allison and most of these builders to offer "Altitude Rated" aircraft powered by a single-stage V-1710. Allison offered their C-11 for use in a pursuit aircraft in September 1937, and the C-12 for a mid-altitude pursuit or attack airplane in January 1938.

Success was finally achieved when the AAC desired to see a comparison between air-cooled and liquid-cooled versions of the same airframe. The result was the 1938 Curtiss XP-40 powered by the new altitude rated C-13. In September 1937 the Navy had become the first to officially ask for an Altitude engine, the E-1 for the Bell XFL-1. Still things did not move quickly, for it not until April 1938 that the C-13 was defined for the XP-40 project. It was over a year later, June 1939, that the large V-1710-19(C13) production order for the initial P-40's was received. The C-14 was a proposed sea-level engine that was not contracted, while the C-15 became the end of the model line, and the major production version, as it was the developed C-13 and fitted into all of the P-40 *Tomahawks*.

Aircraft Powered by the V-1710-C

The V-1710-C was the primary program for Allison from 1932 into 1941. Only a handful of engines were produced prior to the large order for the Curtiss P-40, and even then the majority of the engines were procured on foreign orders. Still a number of important aircraft were powered by the engine.

Table 9-6
V-1710-C Powered Aircraft

Aircraft	Engine Model	Comments
Bell A-11A	YV-1710-7(C4)	Engine Test Bed, 1936
Curtiss X/YP-37	YV-1710-7/-21(C10)	Service Testing 1937/40
AAC X-608/X-609	V-1710-C7	Prototype Competitors
Lockheed XP-38	XV-1710-11/-15(C8/C9)	Prototype, 1938
Curtiss XP-40	XV-1710-19(C13)	Prototype, 1938
Curtiss P-40/B/C	V-1710-33(C15)	Production Pursuits, 1940
Lockheed P-322	V-1710-33(C15)	Production Pursuits, 1942

Elsewhere in this book each of these aircraft and its engine is described in detail. In addition there were many other aircraft proposed to use the engine, the most powerful high altitude engine of the day. These included the Boeing XB-15, Bell Model 11, Bellanca 17-110, Brewster P-22, Curtiss CP 39-13a and at least two foreign aircraft, the French Arsenal VG-32 prototype all-wood aircraft (expected 390 mph but it never flew) as well as a Dewoitine D-520 version to have been manufactured by Ford in Detroit.[35]

Summary

In the period between 1931 and through 1937, Allison had built and delivered sixteen V-1710 engines in ten different models. Horsepower rating was increased from 750 to 1150 bhp, while weight increased from 1,035 to 1,325 pounds. The net result was an improvement in power rating from 0.725 to 0.868 bhp/pound, though sprint values reached 1575 bhp (1.19 bhp/pound). Power was growing faster than weight, while at the same time reliability was considerably improved. In addition to the basic model, major variations during the period included the reversible dirigible engine, fuel injection models and both propeller speed and crankshaft speed extension shaft engines. It is fairly obvious that neither the Allison Company nor the military services used particularly good judgment in loading a very small engineering organization, having only meager development and test facilities, with such a wide variety of projects. However, the pressure for highly experimental variations of the V-1710 engine and rapid increase in power output continued as witnessed by the initiation of a contract on the V-3420 and the rebuilding of six of the ten engines on Contract W535-ac-6795 in two models for operation at 1150 bhp. The net result from the financial standpoint of the Allison Company was a total deferred development charge, over and above the contract prices on delivered engines, of approximately $800,000 as of January 1, 1938. But the U.S. had the engine it needed to power a new generation of advanced military aircraft.[36]

Table 9-7
V-1710-"C" Series Features

Model	Comp Ratio	S/C Ratio	Carb	Weight, lbs	Fuel Octane	Takeoff bhp/rpm/MAP	BSFC, #/bhp-hr	Comments
XV-1710-1(A2)	5.75	8.77	NA-F7B	1122	87	1010/2600/43.6	0.632	Sodium Exhaust Valves, SAE#40 Prop
XV-1710-3(C1)	5.75	8.77	NA-F7C	1160	92	1000/2650/42.4	0.609	Build#1: New Valve Timing, Sodium filled Intake and Exhaust Valves.
XV-1710-3(C1)	5.90	8.00	NA-F7C	1160	92	1000/2650/41.9	0.680	Build #2: SAE#50 w/Dampers, New Valve Timing, hollow stem intake valves.
XV-1710-3(C1)	6.05	8.00	NA-F7C	1160	92	1070/2650/43.2	0.590	Build #3: New CR.
XV-1710-3(C1)	6.05	8.00	NA-F7C	1214	92	1000/2650/43.0	0.660	Build #4: New friction vib damper
XV-1710-7(C2)	6.00	8.00	NA-F7C	1160	92	1050/2650/43.2	0.670	Pre-Hazen redesign.
XV-1710-5(C2)	7.00	8.00	S1 Inj.	1290	100	1250/2800/____	0.650	Fuel Injected, New valve timing.
XV-1710-7(C3)	6.00	8.00	NA-F7C	1245	92	1066/2600/40.1	0.525	First with Ramming Intake Manifold.
YV-1710-7(C4/C6)	6.00	8.00	NA-F7C	1265	92	1046/2600/40.7	0.535	150-hr Type Test, Pendulum Vib Dampr
V-1710-(C6)	6.00	6.75	NA-F7C	1280	92	1000/2600/42.4	0.600	AC34-4 & 34-5 rebuilt in 1937.
XV-1710-11(C7)	6.65	6.23	NA-F7C	1300	91	1150/2950/____	NA	1st With Pendulum Vib Damper.
V-1710-11(C8)	6.65	6.23	NA-F7F3	1310	91	1150/2950/37.5	0.580	Revised Valve Timing, w/Turbo-S/C.
XV-1710-15(C9)	6.65	6.23	NA-F7F3	1310	91	1150/2950/37.5	0.580	Left Hand Turning (C8), 1 built.
XV-1710-19(C13)	6.65	8.77	PT-13B1	1320	100	1060/2950/42.5	0.610	1st Altitude Rated Engine.
V-1710-21(C10)	6.65	6.23	PD-12B4	1310	92	1150/2950/37.5	0.600	20 ordered, w/PD Carb & Turbo-S/C.
V-1710-33(C15/A)	6.65	8.77	PT-13E1	1325	100	1040/3000/42.0	0.647	Total 2550 built.

Table 9-8
Long Nose Engines Shipped

Model	Number	Aircraft	Comments
XV-1710- 1(A2)	1	Development	AC33-42, AEC#2, delivered 6-28-33
XV-1710- 3(C1)	1	1st Type Test	AC34-4, AEC#6, delivered 6-20-34
XV-1710- 7(C2)	1	A-11A Ground Run only	AC34-5, AEC#7, Was intended to do 1st Flights, del 12-24-35
XV-1710- 5	0	None	AC34-5, AEC#7 was to be rebuilt to this configuration
XV-1710- 7(C3)	1	2nd Type Test	AC34-6, AEC#8, Hazen designed intake manifolds, del6-13-36
YV-1710- 7(C4)	3	3rd TT, A-11A, XP-37	Passed 150-hr Type Test @ 1000 bhp, 1st Flt Engines
V-1710-C6	(2)	1937 Rebuilds	AC34-4 when redelivered w/6.75:1 gears. AC34-5 when redelivered for Propeller testing at Curtiss.
XV-1710-11(C7)	1	Type Test	First 1150 bhp engine
V-1710-11(C8)	2	XP-37, XP-38	AC38-581/582, AEC#18/19
XV-1710-15(C9)	1	XP-38	Left-Hand Turning, AC38-120, AEC#17
V-1710-21(C10)	19	YP-37, Type Test	Two C-8's from this order were replaced with TT engines
V-1710-19(C13)	2	XP-40	First to be Altitude rated, 2nd engine delivered to France
V-1710-33(C15)	2,400	P-40, P-322	1,770 engines delivered on British contracts.
V-1710-C15A	150	P-40	100 from British/French contracts plus Spares for AVG/China.
Total	2,582		

Table 9-9
V-1710-"C" Series Ratings
BHP/RPM/Altitude,ft/MAP,inHgA

Model	Takeoff	Normal @SL	Normal @ Alt	Military	Comments
XV-1710-1(A2)	1010/2600/SL/43.5	750/2000	NA	NA	1st Long Nose
XV-1710-3(C1)	1000/2650/SL/43.0	750/2400	NA	NA	1st Type Test
XV-1710-7(C2)	1000/2600/SL/NA	700/2300	NA	1000/2600/SL	Intended for 1st Flt.
XV-1710-5(C2)	1250/2800/SL/NA	1000/2600	NA	1250/2800/turbo	Fuel Inj, Not Built
XV-1710-7(C3)	1000/2600/SL/41.0	700/2300	NA	1123/2800/SL	1st Hazen, Type Test
YV-1710-7(C4)	1000/2600/SL/41.0	700/2300	NA	1123/2800/SL	1st Flight & Type Test
V-1710-11(C8)	1150/2950/SL/37.5	880/2600	880/2600/25,000	1000/2770/25,000/33.4	w/Turbo
XV-1710-15(C9)	1150/2950/SL/37.5	880/2600	880/2600/25,000	1000/2770/25,000/33.4	LH turning C8 w/Turbo
XV-1710-19(C13)	1060/2950/SL/42.5	910/2600	1000/2600/ 1,000	1090/2950/10,000/NA	1st Altitude Rated
XV-1710-19(C13)	1060/2950/SL/42.5	910/2600	1000/2600/ 1,000	1150/2950/10,000/41.2	for XP-40
V-1710-21(C10)	1150/2950/SL/37.5	880/2600	880/2600/25,000	1000/2770/25,000/33.4	w/Turbo, PY-37
V-1710-33(C15)	950/2770/SL/NA	838/2600	930/2600/14,600	950/2770/ 8,000/NA	Un-Modernized
V-1710-33(C15)	1040/3000/SL/41.?	840/2600	930/2600/14,600	1040/3000/12,800/NA	Modernized
V-1710-33(C15)	1040/3000/SL/42.0	840/2600	960/2600/12,000	1090/3000/13,200/38.9	w/o BF Screens

Table 9-10 V-1710-C Engine Models

C-1: The V-1710-C1 designation was retroactively applied to the first "C" engine, XV-1710-3 [AEC#6, Army AC34-4]. It was contracted on March 29, 1934, with the intention that it would perform the Military Type Test for the V-1710. Confidence was such that on June 18, 1934 an order for an additional ten service test YV-1710-3's was placed, though their production was subject to satisfactory completion of the Type Test.[37] All were to be similar to the earlier XV-1710-1(A2), but designed for 1000 bhp. The amended final contract, W535-ac-6795, cost for these ten engines was $281,353.

There was also an earlier designation of a V-1710-3 engine. During the conference that defined the XV-1710-1 engine in November 1932 a V-1710-3 was also described in detail. Like the XV-1710-1 it was to be rated at 750 bhp at 2400 rpm, but differed by having a 3:2 reduction gear, and the supercharger impeller enlarged to 8-1/2 inches in diameter and driven through new 8.1:1 step-up gears. Otherwise it was similar to the XV-1710-1, having 5.8:1 compression ratio, No. 40 propeller shaft as well as gun synchronizers and sodium cooled intake valve stems.[38] This engine was never contracted nor developed, instead the focus shifted to obtaining an engine qualified for attempting the 1000 bhp Type Test.

The increase to 1000 bhp was to be accomplished by increasing the rated speed to 2600 rpm, with overspeeds allowed to 2800 rpm. This was to be accomplished by providing crankshaft counterbalancing sufficient to insure that center main bearing loads did not exceed 2000 psi at 3640 rpm.

The evolution and development of this engine model was pivotal to the future definition of the V-1710 and is therefore provided in detail below.

AEC s/n 6: The XV-1710-3(C1), Air Corps No. 34-4, was designed to be fully capable of completing a Type Test at 1000 bhp. It, like the XV-1710-1 used the PN19409 "Plenum" type intake manifolds that spanned and fed three adjacent cylinders.[39] The major changes from the XV-1710-1(A2) were:

This photo of the XV-1710-3(C1) shows it in its Build 1 or 2 configuration. Note how the early crankcase breather and mounting pads differed from the later arrangement. (Allison)

- use of more compact and neater appearing Prestone lines and ignition manifolding
- redesign of the rear case to make mounting of accessories more compact and accessible
- revised valve timing to increase overlap and improve high speed operation
- Adoption of SAE #50 Propeller Shaft
- Provision of oil system for controllable pitch propeller
- NA-F7C carburetor
- Various detail improvements throughout engine

In its "Build 1" configuration it successfully completed a preliminary 50 hour "Suitability", or development test, at 1000 bhp preparatory to attempting the first Type Test. Suitability testing occurred from 8-22-34 to 12-1-34 and included 6 hours at 1000 bhp, 2650 rpm with 0.609 BSFC and 44 hours at 900 bhp, 2560 rpm with 0.551 BSFC.

XV-1710-3(C1) "Build 1" Particulars, as run during 50-Hour Development Test:							
Model	XV-1710-3		Valve Clearance:			Carburetor	NA-F7C
Manuf s/n	6		Intake, in	0.010		Fuel Grade	92 Octane
Air Corps s/n	34-4		Exhaust, in	0.020		Coolant	Ethylene Glycol
Compression Ratio	5.7:1?		Valve Timing, IO	22 BTC		Weight, pounds	1160
Supercharger Ratio	8.77:1		Inlet Closes	56 ABC		Length, in	93.76?
Reduction Gear Ratio	2.00:1		Exhaust Opens	64 BBC		Width, in	30.58?
Propeller Shaft	No. 50		Exhaust Closes	22 ATC		Height, in	37.06?
Date Delivered	6-20-1934		Ignition, Intake	32 BTC		S/C Impeller, in	9-1/2 dia x3/4
Contract, W535	ac-6551		Exhaust	38 BTC		Mech Eff,%	86.1
Performance: 1000 bhp/2645 rpm/SL/42.4"HgA @ 0.615 #/bhp-hr, 900/2560/SL/39.3" @ 0.541 #/bhp-hr. Sodium filled inlet valve stems were replaced by hollow stem units during test. Single 3-1/4" diameter propeller shaft and without any vibration dampers. Comments: Tested 8-22-1934 to 12-1-34.							

Following the successful 50-hour suitability test the engine was rebuilt with the following major changes in preparation for the 150-Hour Type Test:

- Redesigned Cylinder Blocks, with increased radii of fillets in cylinder head castings and redesigned coolant passages.
- Redesigned pistons with higher compression ratio and more oil control rings.
- New oil seals for supercharger and accessories.
- Larger capacity coolant pump.
- Redesigned distributors and ignition wiring.
- Reduced oil flow to connecting rod bearings.
- Reduced supercharger gear ratio as a result of widening the blades from 3/4 in to 1 in at tip.

XV-1710-3(C1) "Build 2" Particulars when Submitted for Type Test:							
Model	XV-1710-3		Valve Clearance:			Carburetor	NA-F7C
Manuf s/n	6		Intake, in	0.010		Fuel Grade	92 Octane
Air Corps s/n	34-4		Exhaust, in	0.020		Coolant	Ethylene Glycol
Compression Ratio	5.90:1		Valve Timing, IO	22 BTC		Weight, pounds	1160
Supercharger Ratio	8.00:1		Inlet Closes	56 ABC		Length, in	
Reduction Gear Ratio	2.00:1		Exhaust Opens	64 BBC		Width, in	
Propeller Shaft	No. 50		Exhaust Closes	22 ATC		Height, in	
Date Re-Delivered	4-19-1935		Ignition, Intake	32 BTC		S/C Impeller, in	9-1/2x1
Contract, W535	ac-6551		Exhaust	38 BTC			
Performance: Actual 1000 bhp/2650 rpm/SL/41.35" @ 0.530#/bhp-hr, Corrected 1040/2560/SL/_" @ 0.530 BSFC. Same 3-1/4"OD propeller shaft, replaced with 4-13/16 "OD shaft in June 1935. New supercharger impeller with 1 inch wide tips in place of 3/4 inch allowing reduced step-up gear ratio. Comments: Tested 4-19-1935 to 7-29-1935. Accumulated 30-3/4 hours of Type Testing in this configuration.							

On May 14, 1935 the revised engine satisfactorily completed calibration runs in preparation for the Type Test, delivering 1000 bhp at 2650 rpm from 0.530 lbs/bhp-hr. It was then installed on a propeller torque stand that had to be considerably strengthened before the testing could begin. Torsiograms on the crankshaft were then taken that showed resonant single-node 1-1/2 order vibrations occurring at about 2400 rpm, and 6-1/2 order two-node vibrations of about +/- 1/2 degree occurring at a frequency of about 220 cycles per second when operating at 2000 rpm. Both vibrations were unacceptable for sustained operation within the 2000-2400 rpm range. By selecting a suitable controllable pitch propeller, and operating in the range of the intended rated speed of 2650 rpm, vibration levels were low enough to be considered safe for the first 50-hours of the Type Test. Allison immediately began the design of a larger propeller shaft, intended to change the fundamental frequency to a higher engine speed.[40] The first attempt at the Type Test was begun on June 20, 1935, but was soon terminated by an overheated reduction gear plain bearing. A new bearing having increased clearances was installed along with the new propeller shaft, now 4-13/16 inch OD and designed for resonance at about 3000 rpm. Torsionometer testing showed the fundamental 1-1/2 order vibration was still a factor, beginning at 2900 rpm, and that the two-node vibration at 2100 rpm was now in excess of +/- 1 degree. This was considered dangerously high for the crankshaft. Type Testing resumed on July 5, but after only one hour of running, a crack was discovered in the front end of the cast aluminum coolant jacket on the left bank. The engine was once again returned to Indianapolis. So began a long and tortuous series of starts, stops, repairs, improvements, redesigns lasting nearly a year on this engine.

On September 28, 1935 Allison again redelivered the Air Corps XV-1710-3 engine to Wright Field. It had been almost completely rebuilt (described as Build 3) with new parts and new features to resolve problems identified in previous testing. These included:

XV-1710-3(C1) "Build 3" Particulars:							
Model	XV-1710-3		Valve Clearance:			Carburetor	NA-F7C
Manuf s/n	6		Intake, in	0.010		Fuel Grade	92 Octane
Air Corps s/n	34-4		Exhaust, in	0.020		Coolant	Ethylene Glycol
Compression Ratio	6.05:1		Valve Timing, IO	22 BTC		Weight, pounds	1214
Supercharger Ratio	8.00:1		Inlet Closes	56 ABC		Length, in	
Reduction Gear Ratio	2.00:1		Exhaust Opens	64 BBC		Width, in	
Propeller Shaft	No. 50		Exhaust Closes	22 ATC		Height, in	Date Re-Delivered
9-28-1935	Ignition, Intake		32 BTC	S/C		Impeller, in	9-1/2 dia x 1
Contract, W535	ac-6551		Exhaust	38 BTC		Mech Efficiency,%	86.1

Performance: Corrected, 1070bhp/2650rpm/SL/43.2" @ 0.590#/bhp-hr, Actual 1000/2650/SL/42.2 @ 0.590 BSFC.
Incorporated dual propeller shaft, without friction damper, and spring type crankshaft harmonic vibration damper, as well as new cylinder heads having individual inlet ports and intake manifolds, and a redesigned S/C diffuser.
Comments: Tested 9-28-1935 to 4-13-1936.

- New Double Propeller Shaft
- Spring driven flywheel type Torsional Vibration Damper incorporated into back end of revised crankshaft to cancel the high frequency, two-node vibration at about 2100 rpm.[41]
- Redesigned and Heat treated upper crankcase.
- New solution heat treated cylinder blocks having individual inlet ports for adaptation of fuel injection.
- New Intake manifolds with individual ports, PN33550.
- Slightly increased CR, 6.05:1, due to new heads.
- New pistons with thicker skirts.
- Weight increased by 54 pounds to 1,214 pounds.

The early engines used an intake manifold (PN18579 and/or PN19420) with each of the four branches delivering to a common plenum supplying mixture to a group of three adjacent cylinders. This was the same approach used by the later Rolls-Royce Merlin intake manifolds and actually gave the XV-1710-3 Build 1 configuration, as used for the 50-hour test, superior performance to the Build 3 arrangement with its individual intake ports to each cylinder (PN33550).[42] The individual inlet ports were provided as an expedient for the later adaptation of port type fuel injection. These manifolds can be seen on the photos of the XV-1710-7(C2), AC34-5.

As a result of the extensive changes and new parts incorporated, the previous 30-3/4 hours of running was discredited and the engine recalibrated prior to restarting the Type Test. During calibration the engine developed 1050/2650/SL/43.2 @ 0.670 #/bhp-hr, then the 150-hour Type Test was begun anew on October 8, 1935.

After only 14-1/2 hours had been accumulated, the new spring type torsional vibration damper failed, damaging parts of the crankcase and accessories housing. At this point the entire approach to the problem of vibration damping was reconsidered, and it was decided to incorporate the solution that Allison had applied to the drives on the airship "*Shanandoah*." The result was to fit disc type friction clutches to both the crankshaft and propeller shafts. Subsequent testing showed effective reduction of vibration amplitudes throughout the entire speed range of the engine. This included both one- and two-node forms of vibration with equal effectiveness.

Testing was then continued in this Build 4 configuration, but with numerous interruptions for repairs and replacement of components. By mid-March 1936 the test seemed to be headed to a satisfactory conclusion, when on March 21,1936 with 148-1/2 hours completed, the left-hand cylinder head cracked between the No. 3 and 4 cylinders. With a used cylinder head installed the engine completed 150 hours on April 7, followed by an additional 14-1/2 hours of penalty time due to the earlier replacement of the crankshaft and vibration damper. During the penalty run, at 156-1/4 hours total time, the right-hand cylinder head was found cracked, similar to the earlier problem on the left bank. On April 13, 1936 the test was terminated after a total of 164-1/2 hours of Type Testing.

Concurrent with completion of the test, Hazen notified Wright Field of testing they had been doing on the intake system. They had developed conclusive evidence of unequal air and power distribution between cylinders accounting for the predominance of difficulties occurring on the left bank. Coupled with the revealed weaknesses in cylinder heads and crankcases, Allison's Ron Hazen proposed a complete and thorough re-design of the engine to resolve these fundamental problems. It was from this testing that he developed the "Rams Horn", or "ramming" type intake manifolds that resolved the imbalance and resulted in a much smoother running, better performing and more reliable engine. The first drawings showing the new manifolds are dated February 21, 1936, prior to the conference with Wright Field.[43] Hazen knew very well what he would be able to deliver when he went to negotiate the fate of the fuel injected V-1710 with Wright Field that March.

Although the XV-1710-3 (Build 3/4) completed 150 hours of Type Testing, plus a 14-1/2 hour penalty run on the crankshaft and vibration damper, it was deemed to have failed the Type Test because of numerous major parts failing, which included the following:[44]

6 Cylinder head coolant jackets cracked
4 Crankcases cracked
3 Crankshafts broken or cracked
3 Propeller shaft gears cracked
1 Crankshaft pinion pitted
1 Reduction gear bearing seized
1 Accessory housing cracked
13 Pistons burned or cracked
46 Exhaust valves burned
1 Reduction gear case cracked
2 Oil pans cracked
5 Distributor rotors burned

Overall, the attempted Type Testing of the XV-1710-3 occurred from 4-19-35 through 4-13-36 and accumulated 164.5 hours of Type Testing and 328 hours total run time. A major contributing cause of the failures

XV-1710-3(C1) "Build 4" when Rebuilt with Disc Type Vibration Damper					
Model	XV-1710-3	Valve Clearance:		Carburetor	NA-F7C
Manuf s/n	6	Intake, in	0.010	Fuel Grade	92 Octane
Air Corps s/n	34-4	Exhaust, in	0.020	Coolant	Ethylene Glycol
Compression Ratio	6.05:1	Valve Timing, IO	22 BTC	Weight, pounds	1214
Supercharger Ratio	8.00:1	Inlet Closes	56 ABC	Length, in	
Reduction Gear Ratio	2.00:1	Exhaust Opens	64 BBC	Width, in	
Propeller Shaft	No. 50	Exhaust Closes	22 ATC	Height, in	
Date Re-Delivered	11-26-35	Ignition, Intake	32 BTC	S/C Impeller, in	9-1/2 dia x 1
Contract, W535	ac-6551	Exhaust	38 BTC	Mech Efficiency,%	86.1

Performance: Average 1000 bhp/2650 rpm/SL/43.0" @ 0.660 #/bhp-hr
Incorporated new crankshaft disc type friction harmonic vibration damper at 14-1/2 hours.
Comments: Tested 11-29-1935 to 4-13-1936. Total Type Test time accumulated, 164-1/2 hours.

was due to lack of a suitable test stand and the Army's inexperience in testing 1000 bhp class engines on torque stands. That is, on test stands where they drove a metal flight propeller instead of a mechanical dynamometer.[45] Serious vibrations were induced by the test stand, resulting in damage to the engine structure. Everything associated with a 1000 bhp class engine had to be developed, including supporting equipment and procedures.

Engine AC34-4 was returned to Allison who then rebuilt it with the new induction system from the C-3, upgrades from the C-4, as well as with 6.75:1 supercharger gears in effect making it a V-1710-C6.

C-2: With the XV-1710-3(C1) undergoing Type Testing, the need was growing to get the engine into an airplane and begin flying. To this end Allison prepared the first of their engine model descriptions for the use of the airplane manufactures, Allison Specification 101 that described the XV-1710-7. Like the C-1 designation, the C-2 was designation was applied retroactively when Ron Hazen announced the Allison model designation system in the summer of 1936.[46]

AEC s/n 7: This was the first engine constructed of the ten in the service test batch. It was built as the XV-1710-7, a 1000 bhp engine incorporating all of the developments and improvements then incorporated into the XV-1710-3. In fact, according to Allison's Goldthwaite, it was identically configured to the final Build 4 version of the XV-1710-3. The engine had been intended to power the A-11A and was accepted December 24, 1935. Allison had been doing design and component development work for the V-1710-7 since mid-1935. One of its major features, the PN33550 intake manifolds, is shown on the photograph of the engine. This manifold was equipped with backfire screens, though they were not in the same location as those in the later C-4 with its Hazen designed ramming manifolds.[47] This was the configuration of the manifold that Ron Hazen found gave such poor mixture distribution to the various cylinders. In April 1936 this engine was with Bell Aircraft at Buffalo where it was fit into the A-11A test-bed aircraft intended to power its first flights. In the event, the engine was "grounded" by the failure of the XV-1710-3 Type Test and used only for initial runs and systems checks of the V-1710 "first flight" aircraft.

When replaced by the actual first flight engine, AC34-5 was returned to Allison who rebuilt it with the new induction system from the C-3, upgrades from the C-4, along with 6.75:1 supercharger gears before shipping it as a V-1710-C6 to Curtiss Electric on December 15, 1936 where they used it for propeller testing.[48]

This is the XV-1710-7(C2), AC34-5, the engine originally intended to power the A-11A. Note the PN33550 Induction Manifold that caused poor mixture distribution in the final build of the XV-1710-3(C1) Type Test engine. (Allison)

YV-1710-5, Built to Allison Specification 102, this was the Army's long sought fuel injected V-1710. Allison evidently never assigned a separate model designation, probably because they had terminated work on the project prior to the institution of their designation system. Work had began in earnest on project during the summer of 1934. The development was somewhat more expensive than conventional carbureted engines that Allison was offering to the government for a price of about $28,000 each. The first fuel injected engine was offered for $34,000, and the Air Corps was to provide the fuel injection system.[49] By March 1935 Allison was awaiting a contract change order to construct two XV-1710-5 fuel injected engines that were expected to be ready for delivery about four months following receipt of the order.[50] Events had developed to the point that in January 1936 Allison responded to Lockheed Aircraft with a description of a fuel injected engine based upon the XV-1710-7(C2), but with 7.0:1 compression ratio and running on 100 octane fuel and would deliver 1250 bhp at 2800 rpm. While proposing the engine, Allison was adamant that they would

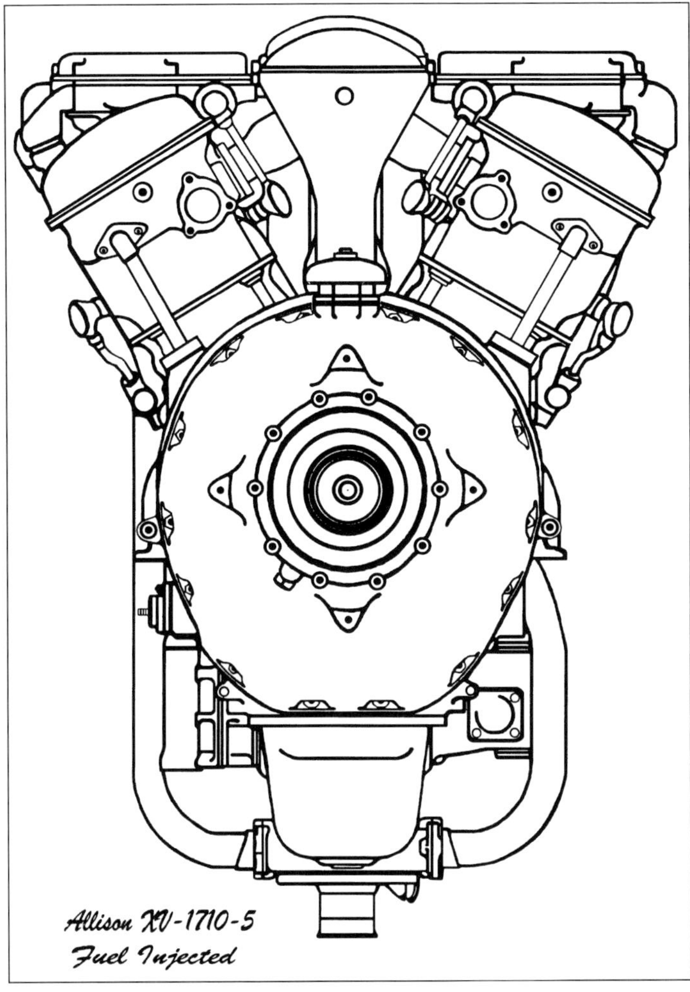

Front View of Proposed XV-1710-5(C3) Fuel Injected Engine. The manifold plenum between the cylinder banks shows well in this view. (Author)

not develop it unless they were recompensated by the customer.[51] Lockheed was well underway with the design of an aircraft intended to utilize this engine as they were seeking details regarding the thermal loads for the related coolant and oil coolers.

While all of the details are not available, it appears that one of Ron Hazen's first activities upon joining Allison in Indianapolis was to integrate the design and improvement efforts on the carbureted and fuel injected engines. The result was an entirely new accessories section able to be used on both models. With the near crisis circumstances occurring with the XV-1710-3 being Type Tested at Wright Field in March 1936, the Army agreed with Allison's proposal to focus on redesign and completing the Type Test on the carbureted engine. Subsequent to the March conference the engine was completely redesigned, but retained the features in the accessories section needed to provide fuel injection. Since AC34-5 had already been delivered as the XV-1710-7(C2), and was installed in the A-11A when the need for a new engine to carry on the carbureted Type Test occurred, AC34-6 then was selected to perform that role. In the meantime, Allison was active in progressing with the fuel injected engine. On May 4, 1936, right in the middle of the 13 week "redesign" period, they began run-in of their EX-2 workhorse engine *with the fuel injection system*.[52] Drawings of the XV-1710-5 and XV-1710-7(C3) prepared in February 1936 clearly show that as of that date, Allison was proposing to convert the two XV-1710-5 designed engines, AC34-5 and AC34-6 into the carbureted model.[53] Both engines were subsequently rebuilt and redelivered as carbureted V-1710-C6 models.[54]

An outstanding feature of the Allison Fuel Injected engine is the configuration of the induction system, for it shows that under Hazen's guidance, the early "spaghetti" type of induction trunks to each port, utilized in the Build 3 configuration of the GV-1710-A, had been abandoned for a central plenum and short riser from each port. When the Fuel Injected effort was officially abandoned in 1937 it was because of lack of an adequate fuel injection and metering pump.

Both engines, AEC #7 and AEC #8, were initially identified for construction and/or assembly as fuel injected in late 1935. In the event neither were, and no other fuel injected engines were built for nearly ten years.

C-3: The result of the 13-week redesign effort in 1936 was the much improved 1000 bhp engine intended to satisfactorily complete the Type Test. It included features so that it could be assembled in either fuel injected or carbureted forms. When Hazen instituted the new Allison model designation system in the summer of 1936 this convertible engine became the Allison V-1710-C3. To the Air Corps they were the V-1710-5 and V-1710-7, fuel injected and carbureted respectively, designations that had been on the books since at least November 1935. With the need to replace the Type Test XV-1710-3 with an improved model incorporating all of the features coming from the redesign effort, it was decided to build the next engine from the pre-production batch as a carbureted C-3 for this purpose.

AEC s/n 8: This engine, AC34-6, became the XV-1710-7(C3) Type Test engine. It was the first to incorporate the Hazen manifolds and new universal accessories section. The parts that had been prepared for its buildup in the fuel injection configuration were not used, though its new cylinder heads were designed to be able to function with either fuel injection or carbureted type of intake manifolds.

- Changes included in the C-3 engine:
- Hazen's new even distribution PN34054 intake manifolds.
- Cylinder blocks and crankcase again redesigned and strengthened.
- Revised block coolant flow path, for improved cooling.
- New cams with valve timing, increased overlap and earlier Exhaust Opening for greater power at higher speed.

The entire engine was redesigned for a final assault on the Type Test by Ron Hazen and his team. This new XV-1710-7(C3) engine was the first to have the new Hazen designed intake manifolds. (Allison)

XV-1710-7(C3) Particulars when Delivered for Type Test:					
Model	XV-1710-7	Valve Clearance:		Carburetor	NA-F7C
Manuf s/n	8	Intake, in	0.010	Fuel Grade	92 Octane
Air Corps s/n	34-6	Exhaust, in	0.020	Coolant	Ethylene Glycol
Compression Ratio	6.00:1	Valve Timing, IO	44 BTC	Weight, pounds	1,245
Supercharger Ratio	8.00:1	Inlet Closes	58 ABC	Length, in	
Reduction Gear Ratio	2.00:1	Exhaust Opens	76 BBC	Width, in	
Propeller Shaft	No. 50	Exhaust Closes	26 ATC	Height, in	
Date Delivered	6-15-1936	Ignition, Intake	32 BTC	S/C Impeller, in	9-1/2
Contract, W535	ac-6795	Exhaust	38 BTC	Mech Efficency,%	86.5

Performance: 1064 bhp/2600 rpm/SL/40.35" @ 0.526 #/bhp-hr, and 1123/2800/ at full throttle
Incorporated redesigned cylinder blocks with strengthened top deck and improved cooling, new PN34054 even distribution intake manifolds, new valve timing with earlier exhaust opening and increased overlap, accessory housing designed for easy conversion to XV-1710-5, exhaust rockers redesigned to insure both valves open at same time. Backfire screens in manifold tee.
Comments: Type Tested 7-10-1936 to 10-27-1936 when 150 hours completed. Penalty run begun 11-18-1936, at 210-3/4 hours LH Cylinder head cracked between No 3 and No 4 cylinders, replaced with obsolete design. At 245 hours LH Cylinder head again cracked, as had the RH head. Since new parts were not available the test was terminated, January 5, 1937.

- Improved accessory assembly, suitable for fuel injected XV-1710-5 model.
- Higher capacity oil and coolant pumps.
- Strengthened internal reduction gear.
- Improved seal for Distributor drive.
- Eliminated Fuel Injection pump drive and mounting.
- Improved Supercharger Scroll.
- Backfire Screens included to improve mixture distribution and provide backfire protection.

Type Testing occurred from 6-15-36 to 1-5-37 using a 13'-0" diameter 3-blade Curtiss-Electric propeller.

The XV-1710-7(C3) was not successful in completing the 150-hour Type Test because of the following major failures:

- Reduction Gear Housing cracked at 60-3/4 hours,
- Two right-hand cylinder heads cracked, both between No.3 and No.4 cylinders, at 139-1/2 and 245 hours, left cylinder head was found similarly cracked at 245 hours during the penalty run on October 27, 1936.
- Three coolant pump bodies cracked, at 25-1/2, 143-1/4 and 149-1/2 hours.

After testing, AC34-6 was later sent to Randolph Field for instructional training purposes in September 1938.[55]

C-4: A total of three of these engines were built. While they incorporated improvements identified as necessary or desirable by the XV-1710-7(C3) during its attempt at the Type Test, the distinguishing change was an again revised accessories section that eliminated the provisions of the C-3 for conversion to fuel injection.[56] The first engine to incorporate these features was AC34-7, AEC#9, the first flight engine when it was delivered September 25, 1936. The second was the YV-1710-7(C4) that had been converted from the XV-1710-9(D1) Type Test engine, AC34-8, AEC#10, used for the successful attempt on the elusive Type Test. All were operated using Curtiss-Electric propellers.

The changes incorporated into the engines, as compared to the XV-1710-7(C3), were:[57]

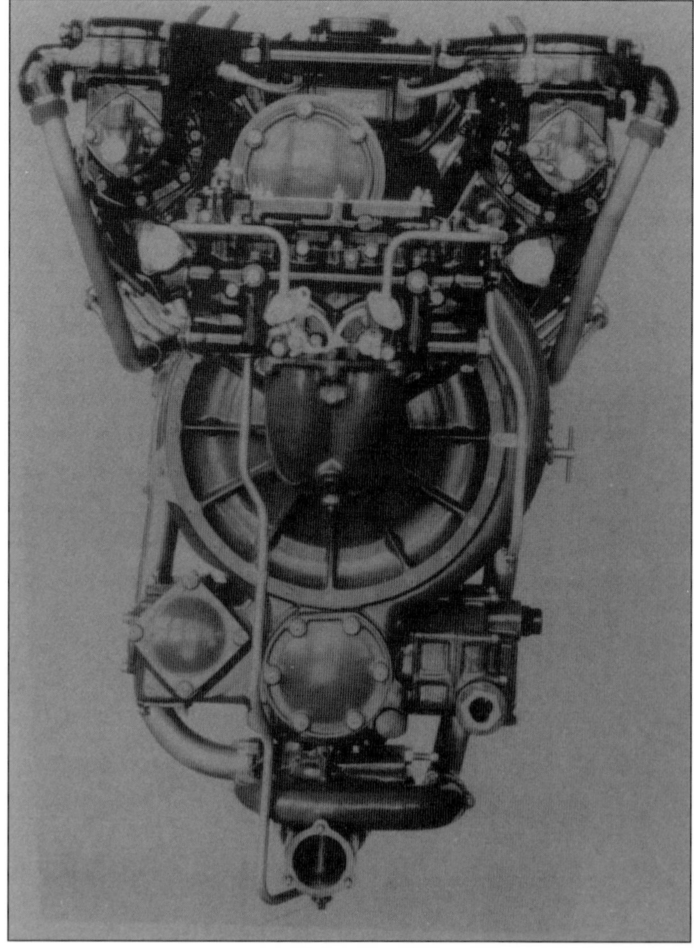

This view of the C-3 accessories section clearly shows the mounting provided for the Chandler Fuel Injection pump. (Allison)

- Cylinder heads further strengthened between #3 and #4 cylinders.
- Crankshaft pinion nut locked by rivet instead of cotter-pin.
- Accessory drive shaft bearings solid bronze instead bronze-lined steel.
- Looser thread fit on crankcase core plugs.
- Crankshaft lubrication system redesigned to improve overspeed reliability.
- Ignition distributor drives redesigned to improve oil sealing, eliminate synchronizer drives and reduce engine frontal area.
- Accessories housing redesigned to reduce weight by elimination of fuel-injector drive.
- Generator, vacuum pump and fuel pump drive ratios increased to provide sufficient speed for cruising at low engine rpm.
- Generator and fuel pump pads revised to later standard.
- Spark plug cooling tubes added.

During the final Type Test one further major change occurred. It was a result of failure of the disc type vibration damper to adequately reduce torsional vibrations within the crankshaft. At 116-1/2 hours into the test the crankshaft was found to be nearly cracked through. As a result Allison provided the crankshaft and new pendulum type harmonic vibration damper they were preparing for the 1150 bhp V-1710-11 series.[58] This worked extremely well and was an integral feature of every V-1710 subsequently built. With the fitting of this crankshaft Allison referred to the engine as the C-6, which explains the various photos of the final Type Test engine being identified as a C-6. The Air Corps never purchased any C-6's as such.

This view of the accessories end of AEC #10 was taken when it was first built as a D-1, note the coolant outlet elbows from the blocks. The C-4, C-6, and D-1 engines all used this accessories section, without features for fuel injection. (Allison)

AEC s/n 9: This was the actual first flight engine, YV-1710-7(C4), AC34-7, AEC#9, and the first built to the C-4 standard. It was cleared for flight following the first 150 hours of the XV-1710-7(C3) Type Test on AC34-6, even though that engine was still proceeding with its penalty runs. It first flew on December 14, 1936, installed in the Bell Aircraft modified Consolidated A-11A that had been shipped by rail to Wright Field, where the flight qualified YV-1710-7(C4), AEC #9, took the place of the XV-1710-7(C2), AEC#7, in the Fall of 1936. During its first 20 flight hours it was limited to 700 bhp, on its way to accumulating at least 300 flight hours, the intended time between overhauls. After 200 hours of flight the engine was torn down and given a through inspection that it passed with "flying colors." Reassembled, it flew on to accumulate 304 hours before again being torn down for inspection. This time it was overhauled though it did not require any major components to be replaced, and was soon returned to flight status. After another seven hours of flight, the difficult to handle A-11A was flown to Chanute Field in Illinois and became a mechanic trainer. The engine is now in the collection of the Smithsonian's National Air and Space Museum, Gerber Facility.

AEC s/n 10: This engine had originally been built to perform the XV-1710-9(D1) "Pusher" Type Test, so when the XV-1710-7(C3) had to be replaced it was readily available. The conversion was done by installing the reduction gear and nose case from the original XV-1710-7(C3) Type Test engine, AC34-6, rearranging the coolant flow orifices and installing the tractor type ignition harnesses. The engine was not a "true" C-4 as it retained the pusher type ignition distributors without provisions for gun interrupter's.

This was the third Type Test engine, and the one that successfully completed the 150-hour Type Test at 1000 bhp, although it actually was in the form of Allison Specification 104-A when testing was completed. As such it would be correctly identified as a V-1710-C6. Wright Field did the 150 hours of Type Testing in the comparatively brief period from 1-28-37 through 3-23-37. Allison then performed a 116.5 hour penalty run to accumulate 150 hours on the replacement crankshaft and new pendulum type harmonic vibration damper. This was completed April 23, 1937. The series of dive tests were done a few days later.

Type Testing the YV-1710-7(C4/C6)

As the final Type Test engine it was calibrated prior at the start of the test and found to produce 1046 bhp at 2600 rpm with 40.68 inHgA manifold pressure, with a "best power" specific fuel consumption of 0.535 pounds/bhp-hr and a mechanical efficiency of 84.8 percent. At the 30 hour point five cylinder block hold-down studs were found to be broken due to "resonant transverse" vibration, and were replaced by the new "muted" design that Allison had already prepared for installation in subsequent models.

At 116-1/2 hours difficulty was experienced with oil scavenging and investigation revealed the crankshaft (PN34600) to be almost completely cracked through at the #4 crankpin journal. This necessitated replacing the crankshaft and the connecting rods and bearings on the #4 journal. The replacement crankshaft was of the new design (PN36000) that incorporated the six-weight dynamic balancer (PN34747 Hub, PN34748 Weights) tuned to cancel the 4-1/2 order two-node torsional vibrations. In addition,

YV-1710-7(C4) Particulars when Delivered for Type Test:					
Model	YV-1710-7	Valve Clearance:		Carburetor	NA-F7C
Manuf s/n	10	Intake, in	0.010	Fuel Grade	92 Octane
Air Corps s/n	34-8	Exhaust, in	0.020	Coolant	Ethylene Glycol
Compression Ratio	6.00:1	Valve Timing, IO	44 BTC	Weight, pounds	1,269
Supercharger Ratio	8.00:1	Inlet Closes	58 ABC	Length, in	94-15/32
Reduction Gear Ratio	2.00:1	Exhaust Opens	76 BBC	Width, in	28-15/16
Propeller Shaft	No. 50	Exhaust Closes	26 ATC	Height, in	42-3/32
Date Delivered	1-19-1937	Ignition, Intake	32 BTC	S/C Impeller, in	9-1/2
Contract, W535	ac-6795	Exhaust	38 BTC	Mech Efficiency,%	84.8

Performance: 1046 bhp/2600 rpm/SL/40.68" @ 0.535 #/bhp-hr, and 1123/2800/ at full throttle
Incorporated strengthened cylinder heads between Nos. 3 and 4 cylinders and redesigned accessories housing from D-1 engine having reduced weight by eliminating provisions for fuel injection. Delivered with friction type vibration dampers. Backfire screens in manifold tee. Assembly drawing #34406. Following the crankshaft failure at 116-1/2 hours, due to manufacturing defect, the latest crankshaft with 2-node dynamic vibration damper and the separate 3-clutch plate friction damper in the propeller shaft was installed.
Comments: Type Tested for 150 hours at Wright Field 1-28-1937 to 3-23-1937, then Allison ran penalty run for 116-1/2 hours, completed 4-23-1937. Dive tests were completed May 1, 1937.

the propeller shaft friction damper was changed from the eight plate to four plate design (having 7 and 3 friction plates respectively). The crankshaft failure was attributed to fatigue stresses from an insufficiently damped two-node torsional vibration, compounded by an extraordinary condition of surface hardening in the #4 journal. This hardening was traced to a process failure during heat treatment at the Park Drop Forge Company plant in Cleveland, Ohio. After replacement of the crankshaft assembly, the test was completed without incident.

Wright Field completed 150 hours of Type Testing and then the engine was shipped to Allison, who had been contracted to complete a 116-1/2 hour penalty run on the replacement crankshaft, as well as perform the contract required series of 60 dive (short overspeed) tests. Wright Field was unable to perform these tests themselves because of other test program requirements. Allison began the penalty run on April 1, 1937 and had to disassemble the engine twice. Both times the top piston rings were stuck, and then at a total time of 232 hours, the 5-L piston pin failed. This resulted in wrecking the #5 connecting rods, pistons and barrels, as well as breaking the #5 counter-weight from the crankshaft. A new counterweight was welded on and new cylinder liners, pistons and connecting rods installed at the #5 position. Otherwise, the engine was serviceable and was back on the block and completed the penalty run on April 23, 1937, after 266-1/2 hours total time.

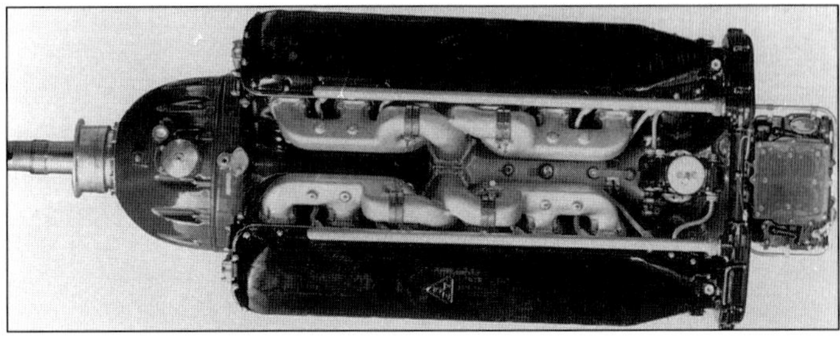

Top view of the final Type Test engine, AEC #10, when initially built as a D-1. The reduction gear from the XV-1710-7(C3) was fitted, along with revised coolant manifolds to prepare the engine for testing as the YV-1710-7(C4). When the crankshaft and dynamic dampers were replaced at 116 hours into the test the changes resulted in redesignation as Allison Model C-6. (Allison)

Following the penalty run the engine was again disassembled and inspected with only normal wear indications being found. It was then prepared for the demanding "dive" tests that involved a total of sixty "dives" of 20 seconds each, with ten at each five percent rpm plateau up to 130 percent of normal speed, all with normal rated manifold pressure. At the conclusion of the dive tests, the engine was again disassembled and no defects were revealed in any of the major parts.

With the Air Corps' cooperation, Allison submitted results from the YV-1710-7(C4/C6) Type Test to the Department of Commerce for Commercial Type Certification of the C-4 at 1000 bhp. The Approved Engine Type Certificate No. 177 was issued on July 13, 1937 with the engine model identified as the V-1710-C4, rated for 1000 bhp at 2600 rpm at sea level on 87 octane fuel with 8.00:1 supercharger step-up gears and a Stromberg NA-F7C carburetor.[59] This describes the Type Test engine.

AEC s/n 13: Engine AC34-13 was built as a YV-1710-7(C4) to be the first flight engine for the Curtiss XP-37. In March of 1937 Allison rebuilt the engine to incorporate the new pendulum type torsional vibration damper located at the rear of the crankshaft, as well as the revision to the damper on the propeller shaft. By itself, this change increased the maximum allowable diving speed of the engine from 2850 rpm to 3380 rpm.[60] This would have been prior to the first flight of the XP-37.[61] Many primary aircraft references identify this engine as a GV-1710-7 or XV-1710-7, but more specific documents show it to be a YV-1710-7(C4).[62] It was upgraded in April/May 1938 to the 1150 bhp V-1710-11(C7) standard.[63] The engine was again overhauled during early 1941 and given a general update, but there is no reference to it receiving a Bendix Pressure carburetor nor the standardized cylinder blocks.[64] Therefore it is believed to have served out its days as a V-1710-11.

C-5: This engine was to be a carbureted engine rated at 1000 bhp for both normal and takeoff. It was being designed during the summer of 1936 in combination with the military version of the similarly rated C-6, and the 1150 bhp rated C-7, all of which were to have a new accessory housing to eliminate fuel injector piping, mount and drive.[65] No engines are known to have been built to this standard, nor was a military designation assigned.

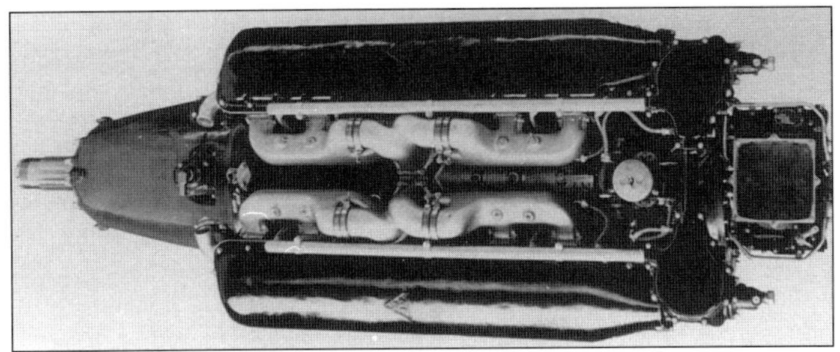

Top view of a V-1710-C6, showing the gun synchronizers to the rear of the distributors, 4-barrel carburetor, center-line location of the magneto and the new Hazen PN34054 intake manifolds. The major difference between the C-4 and C-6 was use of the dynamically balanced crankshaft developed for the 1150 bhp C-7. It is believed that this photo is of either AC34-4 or AC34-5 when rebuilt as C-6's following development testing. (Allison)

C-6: Allison issued its Specification 104-A in June 1936 defining a 1000 bhp engine that retained the supercharger gears of the C-4. This engine was the evolved form of the YV-1710-7 engine (AEC#10, AC34-8) that completed the Type Test on April 23, 1937. The C-6 was proposed to the Air Corps for use in the Curtiss XP-37 Design 80, prototype for the coming YP-37.[66] The Army evidently determined that they wanted a 1150 bhp engine for the YP-37 and that required a reduction in the supercharger ratio. This ended the military interest in the C-6, as the high compression C-7 was the 1150 bhp model and incorporated these features. The C-6 would have been procured under the military designation V-1710-7.[67]

Allison then revised the specification to create their first "commercially" offered V-1710, a design featuring a low to moderate altitude rating. This was accommodated in the revised Specification 104-B, issued in March 1937, concurrent with Wright Field's completion of their 150-hour Type Testing of the YV-1710-7(C4). The commercial engine was defined as having an internal supercharger gear ratio of 6.75:1, but retained the 6.00:1 compression ratio of the C-4. No commercial models are known to have been built, though contemporary publications do identify it as being under Approved Engine Specification No. 177, rated for 1000 bhp at 2600 rpm. At the time, access to V-1710 information and data was very carefully controlled by the Air Corps and distribution limited to "approved" aircraft manufacturers and designers. Contemporary photos showing the "Allison V-1710-C6" at the completion of the 150-hour Type Test at 1000 bhp are actually of the Air Corps YV-1710-7(C4), AC34-8, AEC #10 that had evolved to the Specification 104-A configuration by the time of the successful Type Test. The C-6 designation noted that the engine incorporated the new pendulum type of dynamic vibration damper. Many publications of the day provide engine specifics and data that is not particularly accurate or consistent with either the C-4 or C-6.

Interestingly, there were two engines built as C-6's. These were the first two engines on the YV-1710-3 contract, that is, AC34-4 (AEC#6) and AC34-5 (AEC#7). Originally these had been built respectively as the XV-1710-3(C1) Type Test engine and the XV-1710-7(C2), intended as the first flight engine for the A-11A test bed aircraft. Following declaration of its failure of the Type Test, 34-6 was rebuilt and returned to the Air Corps as a C-6. What service or function if subsequently performed has not been determined. AC34-5 was returned to Allison following its initial fitting and checkout in the A-11A. They then rebuilt it as a C-6 before delivering it to Curtiss for use in propeller testing.[68] Both engines were rated for 1000 bhp at 2600 rpm using 6.75:1 supercharger step-up ratios along with 6.00:1 compression ratio.

C-7: Only one of these engines was built as such, AEC#14, AC34-14. It was delivered to Wright Field as the XV-1710-11, and was to perform the Type Test for the 1150 bhp V-1710-11 model, intended to soon be used on production aircraft.[69] AEC#14 was actually designed in the Summer of 1936, and was equipped with a number of improved components. The major change was in improving thermal efficiency by being the first V-1710 to incorporate new cylinder blocks developing a 6.65:1 compression ratio that was to become standard on the production engines.

In the course of the engineering program done in support of this engine Ron Hazen directed that it was desirable to provide features to allow reversibility of the engine (as later done on the C-9 for the XP-38) and to eliminate the accessories drive from the reduction gear. Both of these features were soon to be characteristics of the V-1710-E/F models that were begun along with the XV-3420-1 in 1937.[70]

The C-7 was rated for 1000bhp/2600rpm Normal and 1150/2950 for Takeoff and Military. Further improvements in the induction system and manifolds [switching from PN34054 to PN36085/6] allowed achieving the C-4 rated power with the internal supercharger gear ratio reduced to 6.23:1. This had the added benefit of better matching the engine to the capabilities of the Form F-13 turbosupercharger used in the Design 80 version of the XP-37. Features of the C-7 included:

XV-1710-11(C7) Particulars when Delivered for Type Test:

Model	XV-1710-11	Valve Clearance:		Carburetor	NA-F7F3
Manuf s/n	14	Intake, in	0.010	Fuel Grade	92 Octane
Air Corps s/n	34-14	Exhaust, in	0.020	Coolant	Ethylene Glycol
Compression Ratio	6.65:1	Valve Timing, IO	44 BTC	Weight, pounds	1330 as "D1"
Supercharger Ratio	6.23:1	Inlet Closes	58 ABC	Length, in	
Reduction Gear Ratio	2.00:1	Exhaust Opens	76 BBC	Width, in	28-15/16
Propeller Shaft	No. 50	Exhaust Closes	26 ATC	Height, in	41-23/32
Date Delivered	6-28-37	Ignition, Intake	32 BTC	S/C Impeller, in	9-1/2
Contract, W535	ac-6795	Exhaust	38 BTC	Mech Efficiency,%	

Performance: 1150 bhp/2950 rpm/SL/38.6 inHgA @ 0.585 #/bhp-hr, and 1000/2600/35.85 inHgA @0.588 #/bhp-hr data taken from Type Test when running as XV-1710-13, July 1937.
Comments: Type Tested as XV-1710-13(D1) for 163-3/4 hours at Wright Field 6-28-37 to 11-19-1937.

- The first crankshaft with induction hardened main and crankpin journals.
- Forged Connecting Rods with improved grain flow.
- Nitrided Cylinder Barrels as an improvement in manufacturing process.
- Cylinder heads built to accommodate long-reach spark plugs.
- Valve and ignition timing remained as on C-4.
- First V-1710 use of either the Stromberg or Chandler-Groves non-icing, no-heat carburetors.

When Wright Field calibrated the YV-1710-9(D1) and the XV-1710-13(D1) (which were basically a V-1710-7(C4) and V-1710-11(C7)) power sections, they found that normal rated power of 1000 bhp at 2600 rpm could be achieved with 4-1/2 inHg *less* manifold pressure in the later model. This was a direct result of improvements in the induction system that allowed the reduction of the supercharger gear ratio from 8.00:1 to 6.23:1, in combination with the increase in cylinder compression ratio to 6.65:1 for better fuel economy.[71]

Following the delivery of AC34-14 to Wright Field the Air Corps converted it to become the Type Test engine for the V-1710-13(D1), rated at 1000 bhp at 2600 rpm Normal, and 1150 bhp at 2950 rpm for takeoff. This was done to meet the need for more power for the Bell XFM-1.[72]

Because of the similarity of the C-7 internal configuration to that of the V-1710-13, then flying in the Bell XFM-1, the C-7 was allowed it to go into service without a Model Test. Engine YV-1710-7(C4) [AEC#13,AC34-13], the first flight engine for the XP-37, was subsequently reworked to become a V-1710-C7 and be rated for 1150 bhp.[73]

In 1937 the Army specified this engine for use in aircraft designs submitted in response to their Proposal Circulars X-608 and X-609. Those RFPs resulted in the turbosupercharged XP-38 and XP-39 aircraft respectively.

C-8: Allison originally suggested that the C-8 would be like the V-1710-11(C7), except that the engine would not be equipped with gun synchronizers, as these would not be required on multi-engined aircraft. Wright Field rejected this proposal as they were concerned about the complications within the supply and overhaul system. Evidently, Allison prevailed. The savings of 7.7 pounds per engine along with the lower cost of components resulted in the C-8 not having gun synchronizers.[74] The major differences in the C-8, versus the C-7, were a change in the type of distributors and inclusion of provisions for propeller feathering, a feature required for twin engine aircraft applications such as the XP-38.

Additionally, valve timing was slightly revised, establishing the setting that were utilized for all subsequent production models of the V-1710. The supercharger ratio remained at 6.23:1 as in the C-7, as well as the new 6.65:1 compression ratio. Twenty of these engines were ordered as V-1710-11's for the 13 service test YP-37's per contract W535-ac-10660, though only [AC38-581, AEC#18 for the XP-38 and AC38-582, AEC#19 for the XP-37] were delivered as such.[75] The total contract cost for these 20 engines was $487,946.30.

Following the crash of the XP-38 on its delivery flight, its' damaged engines were returned to Wright Field. V-1710-11(C8) AC38-581 was held to provide spare parts, and in fact, later contributed the crankcase, crankshaft, connecting rods and pistons needed to repair the special V-1710-19 during its "Hi-Power" demonstration testing. The AC38-582 engine for the XP-37 was later upgraded during an overhaul to become a V-1710-21(C10).

C-9: This was the left-hand turning C-8 derivative engine for the XP-38, designated by the Army as the XV-1710-15. It utilized C-8 components except for those necessary to accommodate left hand rotation. The engine was purchased on a separate contract, W535-ac-10291, and only one engine (AC38-120, AEC#17) was built, it being delivered on April 4, 1938. Cost of conversion of an existing V-1710-11 or V-1710-19 to -15 left-turning configuration was given by Allison in 1938 as $2,930, putting the total cost of the engine at about $28,000. The engine was damaged in the crash landing of the XP-38 on its delivery flight. It was subsequently repaired, equipped with a propeller, and taken to Chanute Field where it was used to train mechanics in the maintenance and operation of the V-1710.

C-10: All but three of the twenty engines originally purchased as V-1710-C8's on W535-ac-10660 for the YP-37's, were delivered as C-10's. Major new features were incorporation of new torsional vibration dampers and the first use by Allison of the Bendix-Stromberg PD-12B2 two barrel "pressure carburetor", though production engines used the PD-12B4 when delivered as the V-1710-21(C10). It was rated for 1000/2600 Normal and 1150/2950 for Takeoff and Military.

The C-8 was the first model to show many of the features that were to be standard on the production C-Models. This includes the external scavenge drain on the bottom of the nose case and the crankcase with improved mounting points and strengthened to resist stresses. Exhaust sparkplug cooling was provided by a complex shrouded blast tube. (Allison)

The general appearance of the C-10 is much like the C-8, but note how the oil return lines on the front of each block are located well below the large coolant outlet nozzles. This feature was introduced on the C-10 and typifies the blocks of all subsequent V-1710 and V-3420 engines. (Allison)

By the time the V-1710-C10 model was going into development at 1150 bhp, some 3000 hours of Type Testing or equivalent development running on "C" type engines had been completed at the 1000 bhp level.[76] No Model Test had been done on the C-7/C-8, as the engine built for that purpose had been used to perform the V-1710-13(D1) Type Test. In view of this, coupled with the first use of a Bendix Stromberg pressure type carburetor and the higher compression ratio, intended to improve engine efficiency, the Air Corps required a Model Test of the V-1710-21(C10) engine. Furthermore, when the V-1710-23(D2)'s were added to the contract, it was intended to have the -21 Model Test suffice for both. The Air Corps felt that they already had sufficient running time on the V-1710-D propeller speed extension shaft.

As compared to previous models, the following major improvements were made or refined in the C-10:[77]

- Heavier crankcase diaphragms.
- Crankshaft strength and bearing oiling improved by changing oil holes.
- Improved torsional damping obtained with heavier two-node pendulum type damper weights for 4-1/2 and 6th order vibrations. Damper redesigned to eliminate eccentric loading of the pendulum weight supporting pins.[78] This change was necessary to dampen the two-node vibration at the takeofff speed of 2950 rpm.[79]
- Compression ratio of C-7/C-8 at 6.65:1 retained with improved piston design for ring belt cooling and strength.
- Heavier piston pins with improvements in design as a result of extensive endurance testing in a test fixture.
- Carburized cylinder barrels for reduced barrel and ring wear.
- Improved intake manifolds and internal supercharger permitting use of 6.23:1 blower ratio in place of 8.00:1 while obtaining higher power rating. Thus improving efficiency and fuel economy. (This change first occurred on -11/-13 engines).
- Simplified camshaft drives. Reduced cost, improved assembly, oil leakage and service.
- Improved reduction gear bearing lubrication, reducing oil flow in engine and increasing reliability.
- Simplified and strengthened reduction gear housing and accessory housing.
- Substituted Bendix pressure carburetor for float type carburetor with improved control, freedom from icing, etc.
- Propeller hydraulic oil feed improved.
- Flatted ball type valve followers in place of spherical end type.
- Numerous detail refinements for improved manufacture and durability.

Three attempts at a 150-hour Type Test occurred between February 1938 and March 2, 1939 resulting in an again strengthened lower half of the crankcase. Production of the V-1710-21(C10), and the internally similar V-1710-23(D2) engines, was authorized prior to completion of the Type Test, based on Allison agreeing to replace the upper half crankcase on any engines that failed prior to three years or 501 hours of service.[80] Problems with these components during Model Testing required continuous evolution through the final Model Test for the later V-1710-33(C15).

The lower oil pan had been made shallow to provide increased clearance for turbosupercharger installation directly below the engine, i.e. as in the P-37 installation. Location of cams along camshafts were rearranged to compensate for expansion when the cylinder blocks reached operating temperature. Also, these were the first engines to have cylinder blocks that were interchangeable with those on the newly designed "double vee" V-3420. A subsequent modification to the C-10 and early C-15's was implemented by Tech Order at overhaul to incorporate an oil jet nozzle spraying onto the rear of the Propeller Thrust Bearing in hopes that its tendency to rust on service aircraft could be eliminated.

Of the twenty engines purchased for the YP-37's, 13 were used by the 13 YP-37's aircraft. One became a spare for the XP-37. One went to the

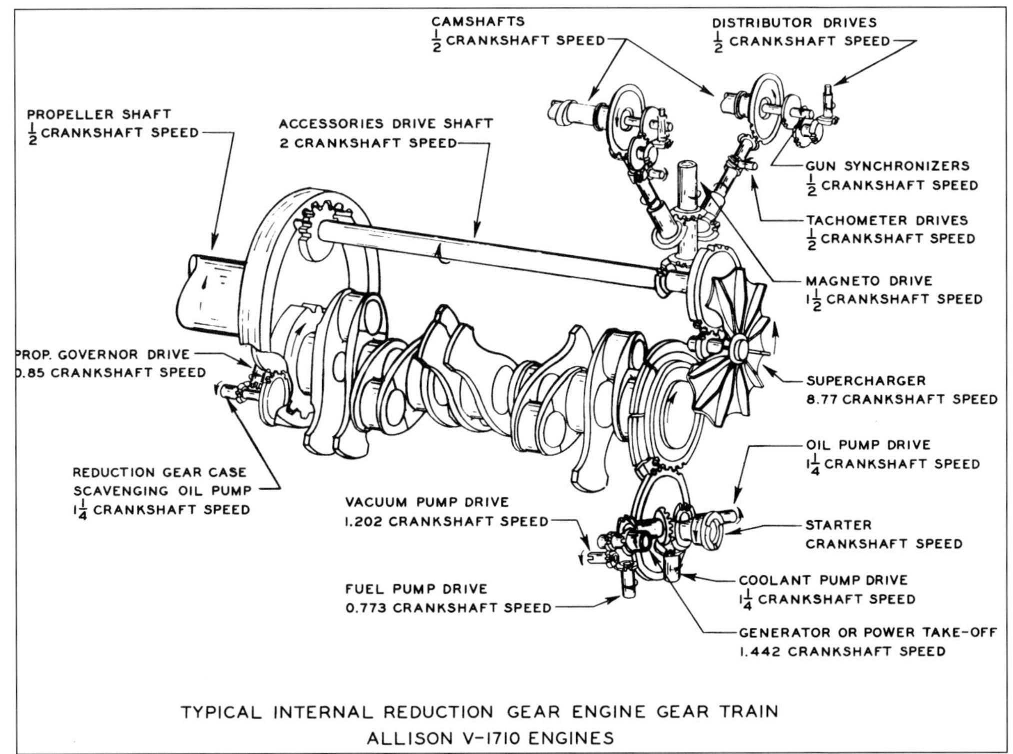

With the Army wanting a 2:1 reduction gear the "C" Model engines had the accessories driven from the inside of the reduction ring gear. Note how the magneto drive is in the center of the "V", a feature requiring it to go through the supercharger discharge pipe. Drawing is of the C-15 engine. (Allison via Hubbard)

XP-38, and another became the XV-1710-19(C13) for the XP-40. Because of the few spares left for the YP-37's, one of the "spare" V-1710-23's was delivered as a V-1710-21, and then two more engines were purchased for the Model Test (AEC#20-1 and AEC#20-2), though not given military serial numbers.[81]

Model Testing the C-10

The first Model Test engine, AEC#20-1, was tested 4-20-38 to 5-17-38 at Wright Field. The test was terminated at 110 hours when a connecting rod failed due to fatigue caused by a heavy nick on the bearing cap. The engine was damaged beyond repair.[82]

The second Model Test engine, AEC#20-2, continued the testing at Wright Field, from 10-26-38 to 11-28-38. The test was terminated after 123-1/2 hours due to a cracked lower crankcase diaphragm cap. This was a failure not seen before and was attributed to the numerous starts using a cartridge starter of a new type with higher powered cartridge that had never before been used on a V-1710.[83]

A third Model Test engine AEC#31 successfully completed the Wright Field 150-hour test program in the period 1-26-39 to 3-2-39. It confirmed the cause of the #20-2 failure by using the same crankcase, but did not use a cartridge starter. It experienced no problems. Still, Allison immediately increased the strength of the diaphragm that failed in the second Model Test on subsequent new production castings. The test did discover stress cracks at the crankcase vent holes located near the main bearing studs. It was found that venting the area was not necessary, and production immediately eliminated drilling the holes in new castings. Since most of the crankcases on the contract were already drilled, Allison guaranteed all of the crankcases for 3 years or 501 hours of operation.[84]

C-11: Allison offered their first "Altitude" rated V-1710 engine in September 1937, intended for pursuit aircraft and to be able to deliver 1175 bhp at 9,000 feet by using 8.77:1 supercharger gears and 100 octane fuel delivered through a PD-12B "dual barrel" pressure carburetor.[85] Rating was to be 1000 bhp Normal, and 1150/2950/9,000 for Takeoff and Military. The difference in this engine and the successful C-13 was the larger supercharger inlet elbow and carburetor used on the C-13. None were built.[86]

C-12: Allison offered another "Altitude" rated engine to the Air Corps on February 4, 1938. It was proposed for use by "Attack" aircraft, and was to be based on the C-10, whose 6.23:1 S/C gears would be replaced by 7.33:1 gears, giving it a capability for 1000 bhp at 4,000 feet. The upgrade would have required use of 100 octane fuel and either the NA-F7F3 or Bendix-Stromberg PD-12B carburetor. None were built.[87]

C-13: Three of these engines were actually built. These were the engine that powered the XP-40 during the 1938-39 Pursuit competition, and a pseudo C-13 for the 1939 High BHP Output Tests. The first engine on the V-1710-19 production order for the first batch of P-40's was built as a C-13, but converted to become the first C-15 prior to being accepted by the Air Corps.

The engine of primary interest was the first one, the XV-1710-19 (AC38-585?, AEC#23), built for the XP-40. The Army Materiel Division notified Allison by telegram on May 6, 1938 that they were going to procure the first engine.[88] It was the fifth engine built on the V-1710-21(C10) contract for 20 engines. Conversion involved replacing the 6.23:1 S/C gears with 8.77:1 gears and utilizing the larger PT-13B1 "three barrel" pressure carburetor and an improved supercharger inlet elbow. The PT-13B carburetor was the same as the one being used in the XV-3420-1 2300 bhp "Double Vee" engine project. The engine was delivered August 15, 1938, with the first flight in the XP-40 occurring October 14, 1938. Conversion was done under a separate contract, W535-ac-11162, for a cost of $1,712.[89]

Ratings in place for the January 1939 Pursuit Competition, when using 100 octane fuel, were:[90]

Takeoff:	1060 bhp at 2950 rpm and 42.5 inHgA
Military:	1150 bhp at 2950 rpm at 10,000 feet and 41.2 inHgA
	1000 bhp at 2950 rpm at 13,500 feet and 36.0 inHgA
Normal:	1000 bhp at 2600 rpm at 10,000 feet and 36.0 inHgA

These ratings were the same as those in place when the XP-40 first began flying in October 1938, except that the rpm had been reduced from 3000 rpm. All other values were unchanged. Interestingly, the above ratings correspond to the identical altitude power calibration curves for the V-1710-C15, but that engine was rated at a point on the curve corresponding to a lower horsepower at a higher altitude as compared to the C-13.[91]

The reduction gear case design and breather of the C-13 was such that it did not require installation of the internal oil jet for extra lubrication of the Prop Thrust Bearing required on the C-15.

With the selection of the P-40 for production the Air Corps ordered 134 aircraft and 234 V-1710-19(C13) engines to be procured with FY-39 funds. The total order for "C" models was for 776 engines, the final 542 being charged to FY-40 procurement.[92] The first engine on this order, AEC#84, was built as a V-1710-19 and was used for the Model Test. It suffered through three builds and attempts at the Model Test, during which the changes were sufficient to warrant its redesignation as the V-1710-33(C15). As a result, all of the engines delivered on this contract, W535-ac-12553, were configured as V-1710-33's. In 1939, after eight years and construction of 20 development and service test engines, along with the 59 pre-production engines for the Air Corps YP-37 and YFM-1's, this was the first V-1710 engine to receive a full production order. On or about December 1, 1939 the V-1710-19 was superseded by the V-1710-33(C15), built to Allison Specification 126-D, in the contract documents.

As of May 24, 1939 the V-1710-C13, built per Allison Specification 126-A, was released by the U.S. Government for overseas sales.[93] The first French order for 115 engines was for the V-1710-19.[94] One further V-1710-C13 was then built for the French and arrived in France during the spring of 1940, almost simultaneous with the German Wehrmacht.[95] It had been intended to use the engine in the Arsenal VG-32 prototype all-wood aircraft.[96] While the Materiel Division acknowledged that the C-13 had been released for export, they noted that in reality, the C-15 was the actual available engine, and that it should therefore also be approved for export sale. This was in November 1939.[97] On December 7, 1939 the V-1710-33 was released for export.[98] From this it is believed that the C-13 contracted for delivery to France was in reality a C-15, though it had been ordered as a C-13 and may have even carried that designation on its dataplate.

The special V-1710-19(C13) was built for the series of High BHP Output Tests. These were run in the September 1939 to January 1940 period on V-1710-21 AC38-597, configured with a C-13 induction system, but with the 8.77:1 supercharger gears replaced by 8.00:1 gears. This was done to insure that detonation would not occur due to excessive charge heating during the testing, which was run at near sea-level conditions. The tests were successful in demonstrating that the altitude rated V-1710 could safely develop a corrected output as high as 1575 bhp at 3000 rpm.[99]

C-14: This engine was described in Allison Specification #128-D as of 12-28-38. No contract resulted, though it was described as a sea-level engine rated to deliver 1150 bhp through the standard 2:1 reduction gears, PD-

This is the High BHP V-1710-19, AC38-597, when on the Wright Field rigid test stand. (Air Corps)

The highly streamlined V-1710-C15 achieved this look by the extended propeller shaft, and the use on an internal spur gear design for the reduction gear. This resulted in a prop shaft centerline only 3.172 inches above the crankshaft centerline. Many observers have wrongly concluded that the "C" model engines had epicyclic reduction gears. (Allison)

12B6 carburetor, 6.65:1 CR and 6.23:1 supercharger step-up gears. It would have used a Hydraulic Propeller Control in place of the usual Curtiss-Electric, which is likely the reason for the separate designation. The model was not produced.

C-15: These engines powered the Curtiss XP-40, P-40/B/C, *Tomahawk* Mk.I/IIA/IIB and the "castrated" (non-turbosupercharged) Lockheed P-38, their model P-322. As Altitude rated engines they all were to be operated on 100 octane fuel. The major differences, compared to the C-13, were an again revised supercharger inlet elbow, the use of the improved PT-13E1 carburetor, and introduction of the new 15 blade supercharger impeller with steel guide vanes. This change resulted in a higher altitude rating.[100] The 1940 vintage engine handbooks give a slightly different intake valve timing, IO @ 52 deg BTC, IC @ 66 deg ABC, than later, when the C-15 and all subsequent models standardized on timing of IO @ 48 deg BTC and IC @ 62 deg ABC.

The C-15 entered service with a recommended 150-hour TBO. On 2-15-1941 Allison recommended to Wright Field that results of overhauls to date justified extending ten engines to 200 hours. Results to be available following inspection and overhaul of these higher time engines would then be used to recommend further extensions of the TBO that ultimately reached 400 hours. In total, 2,550 V-1710-33(C15/C15A)'s were built. Engines built for the French and/or British prior to Lend-Lease were "commercial" sales and identified simply as V-1710-C15's and given British Military Serial Numbers. The British took over the two French orders totaling 815 engines.

The U.S. Army Air Corps received comparatively few C-15's, only 730 were ordered to support the 524 P-40's, and that order was switched to the V-1710-39(F3R) after 300 C-15's had been delivered. This change was to expedite the delivery of the improved P-40D/E *Kittyhawk*. The second major production contract, W535-ac-16323 issued in December 1940, purchased an additional 324 to be installed in the P-40B/C airframes, along with an additional 113 to be held as spares. On the other hand the British took delivery of 1,770 C-15's, including the 100 seconded to the Chinese for the AVG. In total, these engines supported 1,180 *Tomahawks* and 143 twin engined P-322's.

Developing the V-1710-C15

Improvements Prior to Model Testing the V-1710-C15

- Coolant jacket sleeves were changed from an aluminum "spinning" to stainless steel. The aluminum sleeves were found to corrode and crack after considerable use.
- Improved supercharger thrust bearing lubrication and used 8.77:1 supercharger gears used in place of 6.23:1.
- Heavier 15 blade impeller of a new design for higher altitude rating in place of the original 12 blade unit.
- Rotating steel inlet guide vanes added to impeller for higher altitude rating.
- Bendix 3-Barrel PT-13 carburetor in place of 2-Barrel PD-12 to improve rated altitude.
- Large supercharger inlet cover to accommodate the 3-Barrel carburetor.
- Modified supercharger diffuser for improved performance.
- Hollow head sodium cooled exhaust valves substituted for the old design hollow stem sodium cooled design.
- Heavier valve springs for increased intake/exhaust differential pressure.
- AN Standard sparkplugs.
- Improved radio shielding and ignition harness.
- Modified gun synchronizer drives.
- Improved construction at doweling of upper and lower crankcase.
- Standardized all threads, splines, bushings, reamed holes to ease production.
- Modified mounting brackets to accommodate Curtiss rubber mounts in P-40.[101]

Rating

Allison maintained that with the large and hurried production program for the V-1710-C15 underway, and even considering the above modifications and improvements, that the ratings for this altitude model should not be such as to increase the severity of internal stresses over those tested on the V-1710-21.[102] On March 1, 1939 they recommended that the V-1710-33 engine be rated at 930 bhp normal at 2600 rpm, and 1060 bhp at 3000 rpm, ratings giving equal severity to the V-1710-21 rated conditions. The Air Corps was not in agreement and required the 1090 bhp rating be established in Allison Specification 126 for the engine.

Model Testing the V-1710-33

This was a real ordeal for Allison and everyone involved. After the earlier V-1710-C development problems and the generally good performance of the engines then flying, particularly in the XP-40, the Model Test of the V-1710-33 should not have been difficult, and no one would have anticipated that it would take a year to complete! However, a total of five engines were used to complete this test, and by Allison's own statement, this was "far from a satisfactory history."[103] No single reason was responsible. The slightly higher rating insisted upon by the Air Corps, numerous problems in getting a vibration free test stand, engine mount, and propeller, coupled with the large number of new personnel in both testing and production, along with new suppliers of critical components each played a role.

The Model Tests not only stressed the engine and its design, but importantly the entire organization, including subcontractors and material suppliers, staff, training, workmanship and administration. Given the growth occurring in each of these arenas it should not be surprising that the testing uncovered problems. The good news is that Allison was able to resolve each, and the V-1710-33's in the field were good engines.[104]

The first Model Test engine was AEC#84-1, built as a C-13 and run at the Allison Plant 8-11-39 to 8-12-39. The engine was run on a new type of test stand not previously used with the V-1710, and drove a new Curtiss steel bladed propeller. It was also the first use of a new type of P-40 mounting bracket with rubber bushings. After only ten hours, the test had to be terminated. The reason was excessively high oil pressure found to be caused by a piece of rag having been left in the oil system. Allison rebuilt the engine with a complete set of new bearings to replace those damaged from lack of oil. In addition, a crack was found in the crankshaft roller bearing retainer, as well as a piece of lead bronze missing from the reduction gear bearing. These conditions were found to have been caused by severe high frequency horizontal vibratory forces occurring when the engine was operating at about 2520 rpm. The rubber bushings were removed and braces added to the test stand. Meanwhile improved roller bearings were procured for new production engines.

A second attempt at the Model Test was run 9-7-39 to 9-9-39, again at the Allison Plant. AEC#84-1 had been rebuilt with an improved roller bearing and a new reduction gear bearing. The test run was terminated after 22-1/2 hours as a result of a crack in the reduction gear housing. That failure was attributed to test stand and propeller vibrations causing the reduction gear nose case to crack.

A third attempt was made on 9-10-39, after replacing the reduction gear case. Nine hours into this test it was terminated due to burnt exhaust valves. These were of a new valve design, the one having hollow heads for sodium cooling. Concurrent single cylinder testing had shown that this design was not an improvement and so they reverted to the original hollow stem valve with sodium cooling prior to resuming the test.

When Allison began mass production of the V-1710 a major limitation was the shortage of adequate test cells. This is the new rigid test stand with a new C-15 Model Test Engine installed. (Allison)

C-15 Model Test engine getting a close inspection. (Allison)

At this point Model Testing of the C-13 (AEC#84-1) was terminated as the new valves would require a complete re-calibration of the engine. The engine was then rebuilt using a V-1710-23 crankcase and valves and designated as the Allison EX-7 workhorse engine. As such it ran 90 hours of endurance testing to observe the condition of the balance of the components that had been in the engine from the start. This test was run using the Curtiss mounting brackets and rubber mounts. At the end of the tests, and with over 236 hours of Model Test time on these components, they were still in good operating condition. One finding from the EX-7 was that the supercharger inlet elbow ribs were cracking, apparently due to rear end vibration. Immediately changes were made to the patterns and improved castings provided for new production engines.

A new engine, designated by Allison as their C-15 (AEC#84-2), then took over the attempt at the 150-hour Model Test, also run at the Allison Plant. This engine was identical to the first except that it reverted to the old exhaust valve design and incorporated a newly designed supercharger inlet elbow, a new design coolant pump, and stainless steel cylinder sleeves. This engine description is actually of the first V-1710-33(C15), which it was then designated. Calibration checks began on October 13, 1939. The Type Test being run during the period of November 2 through November 11, 1939. At the 50 hour point a crack was found that originated in a bolt hole in the right front mounting boss. Allison and Wright Field agreed to continuation of the test on the condition that Allison would design a stronger crankcase that would subsequently be tested through another 150-hour Type Test. The benefit of this approach was that it would get engines into the airframes coming down the line at Curtiss, even though Allison would likely have to upgrade any engines already in the field. The crack was stop drilled and the offending bolt removed. At 102-1/2 hours the crankshaft failed at the #11 crankcheek, doing considerable damage to the engine. The failure was due to the missing mounting bolt, as it had caused a concentration of loads and stresses on the diaphragm supporting the crankshaft that in turn caused an unacceptable degree of bending in the crankcheek. In a conference with the Materiel Division, the design of all crankshafts after the first fifty engines on this contract were modified to have the bore in the #6 journal (forward most) reduced by 1/8 inch, and improved chamfering and smoothing of the oil holes in the #11 crankcheek. In addition, more attention was given to the use of test stands to insure that engine mountings did not localize loads at the right front mounting point.

A fifth attempt at the Model Test was then made by another new engine AEC#84-3, identical to AEC#84-2. This attempt was made 1-26-40 to 1-29-40 and terminated after only 34 hours when the number one forked rod failed by splitting the piston pin boss, causing major damage to the engine. The cause was traced to foreign material causing a scratch in the highly stressed surface of the pin at the time it was pressed into the rod.

A third engine, AEC#150, then began anew the V-1710-33 Model Test with the intent to qualify all components of the engine except the upper crankcase and the crankshaft.[105] When these components were improved, another engine incorporating the new parts would then perform a complete Model Test to qualify the engine for unrestricted service operation. Number 150 successfully completed its share of the program, clearing the way for the final Model Test.

The sixth attempt at the Model Test used a fourth engine and started May 11, 1940. This engine incorporated improvements covering all of the defects that had shown up on previous Model Tests or in flight experience to date. The Materiel Division agreed that one more attempt would be made to qualify the engine with a Military rating of 1090 bhp, 960 bhp normal. If this was not successful then the Allison recommendation of 1040 bhp Military and 930 bhp normal would by tried. After 79 hours the crankshaft failed at the #8 crankcheek, apparently due to inadequate chamfering of the various holes in the crankcheek and adjacent journal bore.

The finally successful Model Test used a fifth engine and was begun on August 15, 1940. This test had the advantage of a much improved test stand incorporating rubber mounts that correlated well with the mounting arrangement then being used in the actual aircraft, an installation that had been relatively trouble free. The Model Test was run without special incident except for a crack in the jacket wall of the left cylinder head and some bronze flaking of the main bearings. Otherwise the engine was in very good condition. The final ratings were:

Normal:	930 bhp at 2600 rpm, sea-level to 12,800 feet.
Military:	1040 bhp at 3000 rpm, sea-level to 14,300 feet.
Takeoff:	1040 bhp at 3000 rpm, at sea-level.

Finally V-1710-33 number AC39-1125 successfully completed the 150-hour Type Test.[106] Allison then devised a "Modernization" program to cycle all of the engines delivered to date back through the factory for upgrading with the newly qualified components, all at no cost to the government. This program was a priority to all concerned and of the 228 engines effected, there were only 16 outstanding as of March 31, 1941.[107] The unmodernized V-1710-33's in squadron service had been restricted to:

Takeoff:	950 bhp at 2,770 rpm at Sea-level
Military:	950 bhp at 2,770 rpm to 8,000 feet, and
Normal:	838 bhp at 2,600 rpm to 8,000 feet

As usual, modernized engines were identified by appending an "M" to their Air Corps serial numbers, such as 39-1052M, an engine being prepared for operation in an early P-40B/C in 1994.

C-15A: Allison prepared their Specification 145-A to cover the 150 engines contracted by the Chinese. This included the 100 engines, the majority of which were constructed from off-dimension non-ferrous parts, and installed by Curtiss in the 100 airframes at delivery. These engines were diverted from French and British production contracts. Contract F-223 contributed 31, and British contract A-196 the final 69. Combined with the 50 spares, Allison identified them as V-1710-C15A's, and the total order for 150 engines was completed. These were the engines used to power the Curtiss 81-A2 *Tomahawks* for the American Volunteer Group, the *Flying Tigers*.

Allison identified these as commercial Grade "A" engines, built to military performance specifications and differed only in the interchangeability and details of subassemblies. This largely occurred due to salvaging the non-ferrous housing parts by reworking to obtain proper alignment and fits as well as using oversize studs and bushings where necessary. The end result was that when reworked, these housings had the same alignment as the Air Corps and British parts, but since they were housing parts they were not expected to normally require replacement at overhaul.[108] It is also known that the accessories installed could be quite different than those on U.S. and British military engines. The U.S. Government practice required that foreign purchasers provide their own "GFE", Government Furnished Equipment. This included, generators, starters, pumps and other engine accessories. The "A" in the designation was to identify the engines as Commercial "Grade A", indicating they were comparable to "Military" standard engines.

G-1710-C15

In the spring of 1941 the Air Corps purchased 25 C-15 engines for instructional use on contract W535-ac-18172. These were fully operational engines, but were "Grounded" as they were built from non-flight worthy parts. Deliveries were in April 1941 at a cost of $8,278 each. Since these engines were distributed to various schools it is likely that some may still exist in warehouses or museums more than fifty years later. They should *not* be considered suitable for rebuilding to power P-40 restorations.

NOTES

[1] Rickenbacker, Eddy, 1956.

[2] Corporate interest in liquid-cooled engines at Curtiss-Wright died soon after the merger of the parent firms in 1929. The Corporate focus then being on air-cooled radial engines.

[3] Wright Aeronautical Corporation Engine Specification No. R-179, 1933.

[4] Schlaifer, R. and S.D. Heron, 1950, 267. The Navy received $220,000 for its high-speed program in FY-1932, and the same amount in FY-1933.

[5] *Case History of the V-1650 Engine*, 1945. Letter of October 30, 1939, from Packard Motor Car Company to Major General H.H. Arnold, Chief of the Air Service, requesting that the Air Service reconsider them as a source of V-12 engines. They had in mind the upgrading of their earlier 4A-2500 class of engine, not the later V-1650 Merlin program.

[6] Schlaifer, R. and S.D. Heron, 1950, 258.

[7] The "Hyper" cylinder was created by S.D. Heron at Wright Field in about 1930 to investigate the claim, made by engineer H.R. Ricardo at the Britain's RAE, that "poppet" valve cylinders had reached their specific power limit, and that "sleeve" valve cylinders would be required for further advances. Heron used an air-cooled Liberty cylinder and water-cooled it, achieving 1.0 bhp/cu.in. This at a time when 0.5 bhp/cu.in. was considered high power. The Army became enthusiastic to use twelve of these cylinders in a Vee, with high temperature liquid-cooling, in an engine having a 1000 cubic inch displacement and a normal rating of 1000 bhp, with 1200 bhp for takeoff. In 1932 an agreement for the engineering and development of this cylinder, and an Army designed engine using it was reached with the Continental Motor Company. In 1934 the Army revised the dimensions of the cylinder to reduce specific power to about 0.70 bhp/cu.in. resulting in the Continental 0-1430 engine. The first such engine was not built until 1938, and passed a 50-hour development test at 1000 bhp in 1939, some seven years behind the V-1710.

Lycoming Manufacturing Company designed a cylinder similar to the Hyper on their own, and during the same time frame as the Army/Continental effort. Their cylinder was intermediate in volume between the two used by Continental, resulting in a 1234 cubic inch V-12 with individual cylinders. The Army agreed to sponsor the effort and contracted for the O-1230 in 1935, with the engine ready for testing in late 1937. The engine accomplished its 50-hour development test in March 1939 at a rating of 1000 bhp. Since the majority of the work was company funded, and since Allison was already in production with the 1150 bhp V-1710-C, Lycoming decided to abandon the O-1230 and proceed with a project able to meet the Army's desire for a much larger engine. The result was the doubled, H-2470, intended to satisfy the tentative specifications for a long-range engine issued in 1936.

In the event none of these engines were successful. The experience of Allison, that it would take 8-10 years to develop a modern engine, proved true in all cases. Furthermore, the development potential of these engines was severely constrained by the original commitment to the small cylinder size. By the time they were operating at their intended high specific power ratings, more convention engines, such as the V-1710 and Rolls-Royce V-1650 were able to operate at the same or higher specific power levels. Given their larger capacities they easily obsoleted the Hyper based engines. Schlaifer, R. and S.D. Heron, 1950, 267-283.

[8] Schlaifer, R. and S.D. Heron, 1950, 295.

[9] Mason, Francis K. 1962, 25 & 141.

[10] Engineering Section Memo Report Serial No. E-57-541-1, 1934.

[11] The Army became interested in fuel injection as a means of getting rid of the troubles experienced with contemporary carburetors, particularly icing and loss of power during negative g maneuvers. In 1926 Marvel Carburetor Co., a maker of automobile carburetors, hired M.G. Chandler to take charge of development of a fuel injection system which he had patented. In 1927 the Army ordered a single-cylinder device which Wright Field tested and declared to be at least as good and economical as a carburetor. The Army then purchased a 9-cylinder system for use on the P&W Wasp radial engine, as well as experimental systems for the Wright Conqueror and Cyclone engines. In 1928 the Marvel Carburetor Company became a part of Borg-Warner Corporation. Schlaifer, R. and S.D. Heron, 1950, 267-283.

[12] Allison Division, GMC, 1941, 6.

[13] This PN33425 manifold is for the fuel injected XV-1710-5 as of November 25, 1935, Drawing #33301, a configuration considerably different from that shown in the photo of the GV-1710-A, Build 3.

[14] Allison Drawing No.33301, 1935.

[15] TO AN 02-5AD-2, page 139 shows that compared to engines with the later streamlined manifolds, engines having Tee backfire screens lose 2 inHg MAP.

[16] *Progress on V-1710 Engine Development*, Hazen letter to Chief, Materiel Division, July 30, 1936.

[17] See Appendix 4 for details on Allison Specifications.

[18] Allison Drawing No.33700, 1936.

[19] Photo provided to author by Birch Matthews, January 1993.

[20] An interesting note is included in the test log for the XV-1710-3: "On December 3, 1935...the test was interrupted for one day to check for acceleration a master carburetor setting for the XV-1710-7 engines,..." Engineering Section Memorandum Report E-57-541-27, 1938, 11.

[21] *Progress on XV-1710-7 Engine No. 8 and Development Work for this Engine*, Allison letter to Chief, Materiel Division Wright Field, May 4, 1936.

[22] Allison Drawing No.33700, 1936.

[23] Allison Drawing No.33301, 1935.

[24] Allison contract summary compiled March 6, 1943.

[25] Engineering Section Memorandum Report, E-57-6, 1932.

[26] *Checking Reduction Gear Bearing Temperatures in Flight on V-1710 Engines*, Hazen letter to Chief, Materiel Division, September 20, 1938.

[27] *Spark Plug Performance during Type Test of XV-1710-7 Engine*, Air Corps No. 34-6, Manufacturer No. 8, Materiel Division to Allison, December 3, 1936, p.2.

[28] *Operating Instructions for the Allison V-1710-33 Engines Installed in Curtiss P-40 and P-40B Airplanes at Manufacturer's Ratings*, Wright Field Routing and Record Sheet, Maintenance Branch, April 2, 1941.

[29] Teletype TEX-T-409 from Wright Field Experimental Engineering Section to General Echols, *Pertaining to Status of Allison Engine Model Tests*, January 23, 1941. NARA RG18, File 452.8, Box 807.

[30] Wright Field Memo Report Insp-M-41-5-E, 1941.

[31] *Changes to be made in V-1710-19 Engine No. 23 at Time of Overhaul*, Allison Letter to Wright Field, February 11, 1941.

[32] Hazen, R.M. 1945 Report.

[33] Telegram from F.C. Kroeger of Allison to AAC Production Engineering Section, September 10, 1941. NARA RG18, File 452.8, Box807.

[34] Letter from Air Vice Marshal at the Royal Air Force, Cairo, dated June 25, 1941. NARA RG18, File 452.8, Box 807.

[35] The French effort got far enough along that an export license had been granted for the C-13 and C-15 engines, and one C-13 had been shipped to France prior to May 1940 for installation in the Arsenal. Given the reported date of delivery it is

likely that the actual engine was a C-13 (in compliance with the export license) but built with components from the C-15 production line. WINGS, Dec 1974, Vol. 4, No. 6, p.55.

[36] Allison Division, GMC, 1941, 7.

[37] Allison Division, GMC, 1941, 5.

[38] Engineering Section Memorandum Report, E-57-6, 1932.

[39] Engineering Section Memo Report Serial No. E-57-541-1, 1934, 2.

[40] Engineering Section Memorandum Report E-57-541-27, 1938, 6.

[41] This device failed after less than 20 hours. Cause was insufficient space in the accessory housing to provide damper springs of ample strength without radical redesign. Based upon successful experience with the drives in the airship "Shenandoah", which used friction clutches to eliminate vibration difficulties, Allison proposed use of a plate type disc clutch friction vibration damper for the XV-1710-3. This unit was built and ran the balance of the 150 hours, it resolution problem and was in excellent in excellent condition at the end of the test. By December 31, 1936 a new three plate damper was installed on engine AC34-8, set for a breakaway at 2000 in-lbf torque. It satisfactorily covered the range of operation up to 2600 rpm, however the two-node vibration of the 4-1/2 order remained a problem, in that it was sever enough to break the crankshaft in a fairly short time. Allison then designed a "tuned" damper (i.e. pendulum type) to reduce two-node vibration. This device was to be installed in Ex-2 by January 5, 1937, and installed in the AC34-8 Type Test engine when delivered 1-19-37. This cleared the engine for operation up to 2860 rpm. See: Allison Letter to Chief, Materiel Division, *Progress on V-1710 Engines*, December 31, 1936.

[42] *Progress on XV-1710-7 Engine No. 8 and Development Work for this Engine*, Ron Hazen of Allison letter to Chief, Materiel Division, Wright Field, May 4, 1936, p.2.

[43] Allison Drawing No.33700, 1936.

[44] Engineering Section Memorandum Report E-57-541-27, 1938. Testing occurred 4-19-35 to 4-13-36 on engine AC34-4, AEC#6.

[45] Allison Division, GMC, 1941, 5.

[46] Close inspection of the data plate in photos of the XV-1710-7 AC34-5 do not show an Allison model number, rather; Model XV-1710-7, Manufactures No. 7.

[47] *Progress on XV-1710-7 Engine no. 8 and Development Work for this Engine*, Allison letter to Chief, Materiel Division Wright Field, May 4, 1936.

[48] History by J.H. Hunt annotated by J.L. Goldthwaite in 1942, p.4, and Allison Division, GMC, 1941, 26.

[49] *Quotation-Allison 1710 Engines*, N.H. Gilman, Allison General Manager to Chief, Engineering Section, Wright Field, August 17, 1934.

[50] Engineering Section Memo Report, Serial No. E-57-541-9, 1935.

[51] Harold Caminez of Allison letter to Mr. H.L. Hibbard of Lockheed Aircraft, January 15, 1936.

[52] *Progress on XV-1710-7 Engine No. 8 and Development Work for this Engine*, Ron Hazen of Allison letter to Chief, Materiel Division, Wright Field, May 4, 1936, p.4.

[53] Allison Drawings No.33700 and No.33301.

[54] Notes by J.L. Goldthwaite of Allison.

[55] Wright Field Memo: Engine Disposition, transfer of YV-1710-7 #34-6 to Randolph Field, TX, 9-10-1938.

[56] Technical Report No. 4452, 1939, 10.

[57] Technical Report No. 4452, 1939.

[58] Testing by Allison, preparatory to rating the V-1710 at 1150 bhp and 2950 rpm, had identified an insufficiently damped 2-node torsional vibration problem, necessitating adoption of the pendulum type vibration damper in place of the earlier friction type dampers. It was installation of this device and its crankshaft which necessitated the Type Test penalty run.

[59] Department of Commerce, 1937.

[60] The official Operating Instructions for the YV-1710-7 in the XP-37 were issued by Wright Field on April 6, 1937, (Engineering Section Memo Report No. R-57-370-1, 1937). These were revised for the increased diving speed with a similarly titled memo dated June 10, 1937. One should not infer from the dates on these memos that they predated the actual changes in the engine, many examples can be documented where such was not the case, and no instances of the reverse are known to the author.

[61] Allison letter of March 9, 1937 to the Chief, Engineering Section, Materiel Division, Wright Field regarding rebuilding YV-1710-7 AC34-13, Mfrs. #13, to include rear end torsional balancer and revised front end damper. Total cost, including a new acceptance test, was $758.00.

[62] *Damaged Engine Parts for V-1710-13 Engine, Air Corps No. 34-11, Manufacturer's No. 15*, letter from Wright Field Chief, Procurement Section to Allison, April 1, 1938. Paragraph three in letter discusses parts necessary for modification of YV-1710-7, AC34-13, Manuf. No. 13, April 1, 1938.

[63] *Damaged Engine Parts for V-1710-13 Engine, Air Corps No. 34-11, Manufacturer's No. 15*, letter from Wright Field Chief, Procurement Section to Allison, April 1, 1938. A paragraph in this letter requests a quotation for modifying YV-1710-7 AC34-13, AEC#13, to current standards.

[64] *Changes to be Made in V-1710-11 Engine No. 13 at Time of Overhaul*, Allison letters to Ass't Chief, Materiel Division, Wright Field, February 11 and 24, 1941.

[65] *Engineering Program*, Allison Memo from Hazen to staff, July 17, 1936.

[66] Curtiss Aircraft Drawings, Microfilm Frame N119, NASM.

[67] *V-1710-27 and V-1710-29 Engines-Backfire Screen Development*, R. Hazen of Allison letter to Materiel Division, Wright Field, December 5, 1940, p.2, NARA RG 342, Backfires.

[68] Allison contract summary compiled March 6, 1943.

[69] *Allison XV-1710-11 Engine, AC34-14*, Allison letter to Materiel Division, June 2, 1937.

[70] Allison Memorandum by Ron Hazen on *Engineering Program*, July 17, 1936.

[71] Engineering Section Memo Report Serial No. E-57-1406-3, 1939.

[72] Engineering Section Memo Report Serial No. E-57-362-1, 1938.

[73] *Damaged Engine Parts for V-1710-13 Engine, Air Corps No. 34-11, Manufacturer's No. 15*, letter from Wright Field Chief, Procurement Section to Allison, April 1, 1938. A paragraph in this letter requests a quotation for modifying YV-1710-7 AC34-13, AEC#13, to current standards.

[74] *Ignition Shielding Assemblies and Gun Synchronizers on V-1710 Engines for 2 or 4 Engine Airplanes*, Allison Letter to Chief, Materiel Division, June 25, 1937, and Lt. Col. Echols response, dated August 12, 1937.

[75] Technical Order 02-5AA-2, 1939, 15.

[76] Allison Division, GMC, 1941, 10.

[77] Allison Division, GMC, 1941, 10

[78] Engineering Section Memo Report, Serial No. E-57-1406-3, 1939, 8.

[79] Engineering Section Memo Report, Serial No. E-57-1839-2, 1939.

[80] Allison Division, GMC, 1941, 12-13.

[81] Table 9-9.

[82] Allison Division, GMC, 1941, 11.

[83] Allison Division, GMC, 1941, 11.

[84] Allison Division, GMC, 1941, 12-13.

[85] Allison Specification 113-B, February 3, 1938.

[86] *Allison Engine Specification No. 113B, Allison V-1710-C11 Engine*, letter from R.M. Hazen to Chief, Materiel Division, Wright Field, February 4, 1938.

[87] *Allison Engine Specification No. 116, Allison V-1710-C12 Engine*, letter from R.M. Hazen to Chief, Materiel Division, Wright Field, February 4, 1938.

[88] *Procurement of V-1710-19*, Telegram from Materiel Division to Allison, May 6, 1938.

[89] *Conversion of V-1710-21 Engine to a V-1710-19*, Allison General Manager to Materiel Division, October 17, 1938. The cost of $1,712 provided the new gears, supercharger inlet and carburetor, but applied only if the work was done during initial assembly of the V-1710-21's then in production. On November 2, 1938 the Chief of the Wright Field Engineering Section responded that they would not be needing a second engine as ...a definite test program for the (XP-40) airplane project ...has not yet been decided upon.

[90] Technical Order 02-5A-1A, 1939.

[91] *Release for Export Sale of P-40 Type Airplane*, Inter-Office Memo, Air Corps Materiel Division, November 8, 1939. NARA RG342, RD 3479, Box 6891.

[92] *Status of Deliveries of Aircraft on 1939 Procurement Program*, October 1, 1939. RG18, File 452.1 to 452.17, Box 130 'Bulky'.

[93] *Release for Export Sale of P-40 Type Airplanes - Curtiss-Wright Corp.*, Air Corps memo dated 8-22-1939, regarding Curtiss option with France for 140 P-40 airplanes, deliveries to start July 1940 and be complete by January 1, 1941, noting that V-1710-19 to power them was released for export by the State Department as of 5-24-1939. NARA RG342, RD 3479, Box 6891.

[94] Letter from General Brett to Allison Engineering Co, November 27, 1939. NARA RG18, File452.8, Box 808.

[95] *Allison General History, 1962*, p.16.

[96] WINGS, Dec 1974, Vol. 4, No. 6, p.55, says that a V-1710-C15 had been shipped to France in May 39 for installation in the Arsenal VG-32 prototype all-wood aircraft. In fact this would have been a V-1710-C13 for it was the only V-1710 with an export license at the time! Also, Allison records show one such shipment! Expected 390 mph, never flew.

[97] Teletype Message to Materiel Division-Engineering from the Assistant Chief of the Materiel Division, November 18, 1939. NARA RG 342, RD 3479, Box 6891.

[98] *Release for Export of Allison V-1710-33 Liquid Cooled Engines*, memo from Chief of the Air Corps, December 7, 1939. NARA RG18, File 452.8, Box808.

[99] Engineering Division Memo Report, ENG-57-503-1034, 30 October 1943.

[100] Allison Division, GMC, 1941, 13.

[101] Allison Division, GMC, 1941, 13-14.

[102] Allison Division, GMC, 1941, 14.

[103] Allison Division, GMC, 1941, 21.

[104] Allison Division, GMC, 1941, 14-21.

[105] It is believed that this engine was Air Corps AC39-948.

[106] It is believed that this was Allison serial number 360.

[107] *Operating Instructions for the Allison V-1710-33 Engines Installed in Curtiss P-40 and P-40B Airplanes at Manufacturer's Ratings*, Wright Field Routing and Record Sheet, Chief Field Service Section, April 2, 1941.

[108] *Proposed V-1710-E4 Engines for Installation in Bell Airacobras for the Netherlands East Indies*, Allison letter to Chief of Air Corps, June 20, 1940. This letter recommends that the engines under consideration be constructed from non-standard parts, as was done on the 50 C-15A engines for the Chinese. NARA RG18, File 452.8, Box 807.

10

The V-1710-D Extension Shaft "Pusher" Engine

Background

Allison prepared its' first extension shaft engine design in response to a request by the Glen Martin Company in November 1934.[1] These were for the buried tractor installations in the four engine Martin Model 145-A, XB-16 and would have required propeller speed extension shafts approximately eight feet in length. The first record of a design for Bell Aircraft was a layout dated January 23, 1936, with the design substantially completed by September of that same year.[2,3] However, early concept drawings, done by Bell Chief designer Bob Woods, dating from July-September 1934 show the Allison pusher arrangement in what became the Bell multi-place fighter.[4]

his V-1710-23(D2) is one of the 39 for the Bell YFM-1 airplanes added to ie V-1710-21(C10) contract in 1939. While this was the first semblance of a roduction order, it was really only a "batch" and did not warrant setting up a ue production line. Each engine was hand built. (Allison)

The first aircraft to actually use the new Allison pusher engine was from the innovative and fledgling Bell Aircraft Company. The design was largely created by their Chief Designer Bob Woods. His was the founding product concept for the firm, and created the multi-place-fighter, or bomber-destroyer that Larry Bell was ultimately able to sell to the Air Corps.

Following an Air Corps design competition for an aircraft built to this requirement, the Bell XFM-1 was selected over the Lockheed XFM-2. The XFM-1 aircraft that resulted was loaded with advanced features, including the first high frequency alternating current electrical system, and of course, its extension shaft pusher configured and turbosupercharged Allison V-1710's.

The structure of the Bell program resulted in the Air Corps committing five of its eight available service test "YV-1710-3's" to the XFM-1. Over the six year life of the program Allison designed and/or specified six different pusher models of the engine for the X/YFM/FM. Four models were actually being built. Only one order, for a small production batch of 39 engines, was received as a result of this multi-year effort. Production of the engines for the 13 YFM-1's being completed in October 1940, and the last engine overhaul in September 1941. Allison's engineering activity on the XFM-1 extended from December 1935 to the production order for the V-1710-D2, received in January 1938. Additional development work continued through production of the D-2's, and for another 14 months. At that point, the project ended upon completion of the spare parts order.[5]

Development and Testing

Development of the extension shaft engine for the pusher really got under way with the September 22, 1936 Air Corps order for the engines for the Bell XFM-1, four for the aircraft and one for a Model Test. These engines were to be a derivative of the Type Test engine and not only different because of the extension shaft, but they were to be fuel injected and rated for 1250 bhp.

Development was paced by the problems being encountered by the V-1710-C in completing the 150-hour V-1710 Type Test at 1000 bhp. Allison

The unique and advanced features of the Bell XFM-1 propulsion system resulted in Wright Field constructing a mockup of the wing, nacelle, and fuselage. These were used to investigate the dynamics of the pusher configuration during the XV-1710-9(D1) Model Test. (Air Corps)

had designed the "D" to be the same engine, with only a revision of the coolant flow path needed to accommodate the high point of the cooling system now being at the "rear" of the engine. They also used a new reduction gear housing on the "front" or output end of the engine to accommodate the propeller speed extension shaft. Both engines used the same propeller reduction gears and shaft, as well as the same accessories and internal supercharger. The necessary rigidity and thrust support for the propeller was achieved by mounting the thrust bearing near the propeller and directly to the airframe, some five feet from the engine.

Type Testing of the pusher configuration was accomplished at Wright Field, using a complete XFM-1 nacelle/fuselage/wing segment arrangement as a test stand. No provision was made for incorporating or testing the companion turbosupercharger installation. Still, initial testing in the nacelle showed a number of problems, primarily from vibration in the structure being transmitted to the engine and propeller. Considerable work was done by both Allison and Wright Field to resolve the identified 2-node torsional vibrations inherent in this unique crankshaft-propeller system. Most of the work involved mechanical design and qualification of two plate type friction dampers. One was incorporated into the propeller drive shaft, and the other, a friction plate driven flywheel located at the opposite end of the crankshaft. Given the limited space at the rear of the crankshaft the modification required some exacting design and ingenuity. While this device was developed to satisfactorily dampen the torsional vibrations, it was complex to manufacture, bulky, and generated heat from friction due to the damping action. It was soon replaced in the V-1710-13 with the newly developed pendulum type harmonic balancer, which was not only compact, but had none of the drawbacks associated with the friction damper.

According to Bell, the XFM-1 needed at least 1250 bhp from each engine to meet their original performance objectives. Early specifications for the fuel injected version of the "D1" give this value, but it was never to be, even though Allison was working on a 1150 to 1250 bhp rating for takeoff from the outset of the project. Soon after terminating priority development of fuel injection, Allison's Chief Engineer Ron Hazen laid down his priorities for engineering development of the V-1710 in a memo dated July 17, 1936.[6] In it he establishes the priority of the carbureted "D" as #3 out of 3, pending an Army contract:

> #3. Pusher Engine. 1000 bhp Normal, 1150-1250 bhp Takeoff.
> Only such design work will be done, until a contract is obtained for this type engine, as is required to co-operate with the Bell Aircraft Company in working out necessary modifications to apply to their installation. This requires layouts of extension shaft, distributors, cooling system for rear cylinder head outlet and general study of the installation.

Given the demands being placed on Allison for engines and the necessity to complete the Type Test on the XV-1710-7, it is evident that the demand for the XFM-1 engines could only be met by adapting the current version of the engine then being Type Tested. Consequently, the evolved YV-1710-9(D1) engines were contemporaries of the YV-1710-7(C4), and similarly rated for 1000 bhp at 2600 rpm. Higher power was to require higher rpm, and the limitations of the flywheel type vibration damper and crankshaft would not allow sustained running at higher speeds. To resolve this limitation, and at the same time improve the thermal efficiency of the engine, Allison was designing the V-1710-C7, that the Air Corps later designated as the V-1710-11. The new model had to be able to run normally at 2995 rpm in order to deliver the desired 1150 bhp for Takeoff. This rating was possible only as a result of incorporating the new Allison 4-1/2 order pendulum type harmonic vibration damper. The Air Corps wanted the XFM-1 pusher engines to be similarly configured. Wright Field then became concerned about an observed 6th order harmonic vibration and took it upon themselves to redesign the damper, intending it to resolve both the 4-1/2 and 6 order harmonics. They then insisted that Allison incorporate this revision into the upgraded V-1710-D1 engines, now designated by Wright Field as the V-1710-13. Allison did the work, but would not guarantee the reliability. The Air Corps vibration damper design proved it was not up to the task. During the next attempt it caused the crankshaft to break on the XV-1710-13(D1), AC34-14/AEC#14, the Type Test engine. The original Allison designed six segment 4-1/2 order pendulum unit was then successfully fitted onto the V-1710-13 engines that had been installed in the XFM-1 not long after the first flight.

When the Air Corps ordered the thirteen service test YFM-1's, the specified engine was the V-1710-13(D1). This was soon changed to the V-1710-23(D2), an adaptation of the V-1710-21(C10) with its internal improvements, reduced supercharger step-up gear ratio, and Bendix-Stromberg pressure carburetor. This carburetor was a major improvement in fuel control as it minimized the likelihood of ice formation in the induction system

Operational History

Operationally and tactically the X/YFM-1 program was less than successful. The airframe was prone to vibration, a feature made more troublesome by the tendency of the propulsion system to vibrate in sympathy. In fact, a least one aircraft was lost due to a fire caused by a vibration ruptured oi line. Other problems had to do with the turbosupercharger operation. As best as can be determined, there was no ground running of the entire supercharged propulsion system prior to that done in the prototype XFM-1 aircraft. This is a less than ideal way to develop, tune, and test such a complex and innovative propulsion system.

The major problems had to do with controlling the turbosupercharger As conceived, the turbo was to maintain a constant sea-level manifold pressure at the inlet to the carburetor up to the critical altitude of the airplane This was to be accomplished by gradually closing the "waste gate" tha otherwise bypassed the hot exhaust gasses around the turbine. In this way, its speed could be precisely controlled to cause the connected supercharger to supply the proper air pressure and flow. If the turbo should "overspeed" the result was that the temperature of the air leaving the supercharger would be to high, with the consequence of increased chance of backfiring and/or detonation. Either condition robbed power and could be destructive to the engine. This is exactly what occurred on the first flight of the XFM-1 when its port YV-1710-9 experienced a violent backfire, damaging the engine. The turbo regulator had stuck, running the carburetor pressure to 36-37 inHgA. This in turn caused the mixture to lean-out when running at takeoff power and result in the violent backfire. Complicating the condition was the later finding that the engine tachometers were out of calibration and the engines were inadvertently running at over 3500 rpm![7] It was estimated that the damaged engine had been producing at least 1350 bhp when the backfire occurred![8] The solution to such problems was in improved control apparatus and integrated control dynamics. Concepts that were in their infancy in 1937/8, and were not really resolved until the B-17E, B-24D and P-38J went into service in 1942/43 with much improved turbosuperchargers and controls.

In 1938 there was a move by Allison and some of the aircraft manufactures to convince the Air Corps to delete the turbosupercharger from new pursuit aircraft, the YFM-1 included. Achieving a measure of success, two of the YFM-1's were stripped of their turbos during manufacture and received V-1710-41(D2A) "Altitude" rated engines in their place. These engines used the supercharger gears and carburetors from the altitude rated V-1710-33(C15). In this form they achieved the same performance and ratings as the engines in the early P-40's. In 1942, one of these aircraft was the last of the *Airacudas* to be flying.

Another priority project to improve the *Airacuda's* performance was to give it "handed" engines, as was being done for the Lockheed XP-38. By having the propellers rotating in opposite directions on a twin engined aircraft the effect of propeller torque is canceled and flight characteristics can be improved. With this in mind the Air Corps requested Allison to develop a "left-hand" turning V-1710-D engine, using components developed for the left-hand turning XV-1710-15(C9) engine for the XP-38. Such an engine would have been specified for any production Bell FM-1's. Neither the Bell FM-1, nor the proposed Allison V-1710-D left-hand models, were ever purchased.

Allison offered the "D" model to other potential customers in both tractor and pusher configurations. Only an adaptation of the D-3 tractor engine, with a completely new arrangement having a crankshaft speed extension shaft to a remote reduction gearbox was to continue. This engine became a transition design and was actually the prototype for the "E" series of crankshaft speed extension shaft engines. None of the D-3's were delivered to the Air Corps.

The "D" series had the same limitations as the "C" engines, that is, the reduction gear was limited to 1150 bhp, and could not readily be developed to higher levels for sustained operation in subsequent models. This encouraged its replacement by transitioning to the D-3 and the later "E" series of "crankshaft speed" extension shaft engines that were used on later pusher and/or "buried" engine aircraft.

Extension Drive Hardware

The extension drive shaft on the D1/D2 was a 6" O.D. tubular shaft of 0.093 inch wall thickness, 57-11/32 inches long, having integral flanges at both ends for bolting to the engine-end flexible coupling, and the other end to the propeller shaft itself. The shaft turned at propeller speed as the reduction gears were integral with the front of the engine. A mounting ring, to which a large rubber insert was vulcanized, provided flexible anchorage of the remote thrust bearing housing and mounted it to the airplane nacelle structure. The rubber mounting permitted a maximum operating misalignment of +/-1 degree between the axes of the extension shaft and the mounting plate. Two oil transfer glands in the thrust bearing housing provided for hydraulic propeller control when connected to the necessary oil supply, but were removed when a non-hydraulic propeller was used as in the case of the X/YFM-1. Standard specification 3560, grade 295, grease was used for thrust bearing lubrication.[9] Although the reduction gear case was much shorter than those on the earlier V-1710-A and V-1710-C engines, the gears, plain reduction gear bearings and concentric propeller shafts were exactly the same as used in the V-1710-C series. The plate type friction damper was also retained, although as developed it utilized a different setting and fewer friction plates. It was possible to retain the same length of two element propeller shaft as on the V-1710-C by "telescoping" it into the extension shaft. This worked to keep the dynamics of the shafting nearly the same on both the V-1710-C and "D" engines. The standard nose case extension was not needed in the pusher configuration as the propeller thrust bearing was remotely mounted to the airframe structure. The vibration damper arrangement was as previously described.

Like the V-1710-C engines, the V-1710-D's used the same firing order with the engine crankshaft and propeller both rotating clockwise, when viewed from the "anti-propeller" end of the engine, the "front" of the engine when installed as a pusher. The propeller was a three-bladed left-hand

This view of the Model Test YV-1710-9(D1) gives a good perspective of the size and position of the propeller speed extension shaft. The large thrust bearing behind the propeller mounted directly to the Bell X/YFM-1 airframe. (Allison)

Table 10-1 V-1710-"D" Series Ratings				
Model	S/C Ratio	Carburetor	Takeoff/Military bhp/rpm/alt,ft/MAP,inHgA	Normal bhp/rpm/alt,ft/MAP,inHgA
YV-1710-9(D1)	8.77:1	MC-12 Fuel Inj	1250/2800/25,000/48.6"	1000/2600/ 4,500/42.0"
YV-1710-9(D1)	8.00:1	NA-F7C1	1000/2600/25,000/41.0"	1000/2600/25,000/41.0"
V-1710-13(D1)	6.23:1	NA-F7F1	1150/2950/25,000/37.5" or 1000/2770/25,000/34.0"	880/2600/25,000/32.2"
V-1710-23(D2)	6.23:1	PD-12B6	1150/2950/25,000/37.5"	880/2600/25,000/32.7"
V-1710-41(D2A)	8.77:1	PT-13E1	1090/3000/13,200/	960/2600/12,000/32.2"
V-1710-D3	6.23:1	NA-F7L	1150/2950/25,000/37.2"	1000/2600/25,000/35.0"
V-1710-D4	8.77:1	PT-13E1	1090/3000/13,200/42.9"	960/2600/12,000/37.7"
V-1710-D5	6.23:1	PD-12G1	1150/2950/25,000/	1000/2600/25,000/

pusher type.[10] Interestingly, the data plate from V-1710-D2, AEC #79, gives the rotation to be "anti-clockwise" for both the propeller and crankshaft, suggesting a left-hand rotation for the engine. This is entirely contrary to every other piece of existing information, including the Allison manual for timing the D-2, although there is a photo of one of the YFM-1A's in which the pitch of the blade would clearly be "left-hand." It is believed that this photo has been printed from a reversed negative.[11] Photos of the XFM-1 with its right-hand engine damaged from a collapsed landing gear, clearly show that the propeller was rotating in the "right-hand" or clockwise direction when viewed from the anti-propeller end of the engine. Your author believes that the data plate statement of "anti-clockwise" was incorrectly referenced from the perspective as installed in the aircraft.

Table 10-2 Extension Shaft Pusher Engines Shipped			
Model	Number	Aircraft	Comments
X/YV-1710- 9(D1)	5	Type Test, XFM-1	Based on YV-1710-7(C4)
YV-1710-13(D1)	(4)	XFM-1	Built from earlier models, 1150 bhp similar to V-1710-11(C7), only AC34-9, 34-11and 34-12 flew.
V-1710-23(D2)	35	YFM-1, YFM-1A	Turbosupercharged, similar to V-1710-21(C10)
V-1710-41(D2A)	4	YFM-1B	Altitude rated, used XV-1710-19(C13) induction system
Total	44		

Table 10-3 V-1710-D Engine Models

D-1(-9): The Allison V-1710-D1 was developed specifically for the "pusher" installation critical to the success of the Bell XFM-1 Airacuda.

When first specified, on February 7, 1936, the D-1 was to use the Marvel Carburetor Company "Chandler Fuel Charger" port type fuel injection system from the XV-1710-5(C2). To provide the 1250 bhp needed for the Bell XFM-1 the compression ratio was to be raised from 6.0 to 7.0:1, and the supercharger step-up gear ratio of 8.77:1 from the XV-1710-1 was to be used to deliver 1250/2800/SL/48.6 inHgA. Fuel would have been 100 octane to accommodate the higher pressures. Allison configured their EX-2 "workhorse" engine per this standard and operated it complete with extension shaft beginning in May 1936. When the March 1936 decision to terminate development of the fuel injected V-1710 was made, this configuration was also abandoned.

Both Allison and the Air Corps retained their designations for the engine, but its specification was changed to be consistent with its parent, the recently defined V-1710-7(C4).

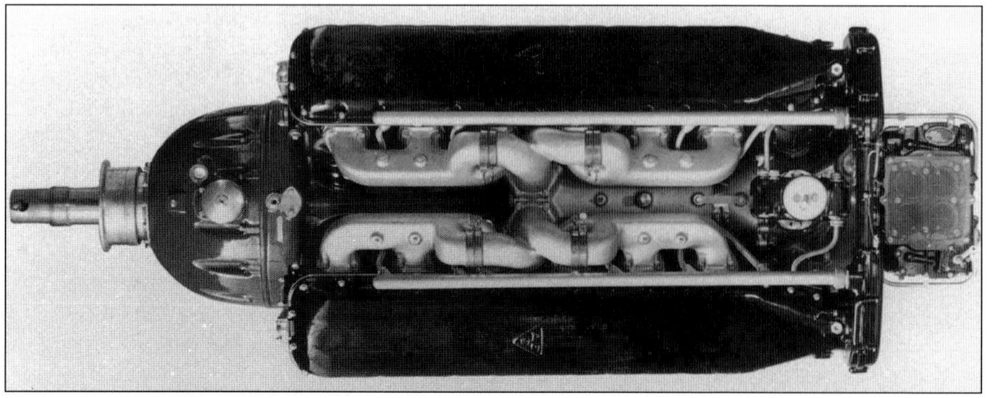

Compare this top view of YV-1710-9(D1), AEC#10, AC34-8, with the very similar view of the V-1710-C6. The only differences are the shape of the reduction gear case, detail of the propeller shaft and the use of distributors without gun synchronizers. Note the 4-barrel NA-F7C-1 carburetor and how these early ramming intake manifolds entered each branch at a right angle. This engine was identical to AC34-9, the engine damaged by backfire during the first flight of the Bell XFM-1. (Allison)

As delivered the V-1710-D1's differed physically from the C-4 in that they had an extension shaft coupling to the reduction gear in place of the streamlined nose case. They also used the coolant outlets at the rear end of the cylinder blocks, rather than at the propeller end. This was required because of the pusher installation in the tail-dragger XFM-1/YFM-1 aircraft that also necessitated a different configuration of coolant metering plugs to each cylinder in the header along the coolant jackets. This was necessary as the back end of the engine was the "high" point when installed in the tail-dragger configured pusher models of the X/YFM-1.

Five engines, from the ten YV-1710-3's procured for Service Testing under contract W535-ac-6795, were initially assigned to be built as "pushers" for the Bell XFM-1, Air Corps Project M-9-35. These were Air Corps serial numbers: 34-8, 34-9, 34-10, 34-11, 34-12. In the event, the five actually delivered were: 34-9, 34-10, 34-11, 34-12 and 34-14. Engine AC34-8, which had been intended to perform the XV-1710-9(D1) Type Test, was converted to become the final Type Test YV-1710-7(C4). Engine AC34-14 then took its place and was further updated to become the XV-1710-13(D1) Type Test engine. The V-1710-9 being qualified by association and similarity to the Type Tested YV-1710-7(C4/C6). All but AC34-14 were initially delivered by Allison as model V-1710-D1's, Air Corps model YV-1710-9, rated for takeoff at 1000 bhp. The first engine was delivered January 20, 1937 and the rest by the end of June 1937. In June 1937, prior to the first XFM-1 flight, the Air Corps desired more takeoff power for the aircraft, and on its own initiative, directed Allison to modify and uprate the four YV-1710-9(D1) engines as V-1710-13(D1)'s rated at 1150 bhp, though the first flight engines remained in the YV-1710-9 configuration. Hence, no change to the Allison model designation. Significant to this upgrading was a change in the type of harmonic vibration damper from friction to pendulum type. Furthermore, the Air Corps directed that the new pendulum unit be revised to their design, which replaced two of the 4-1/2 order pendulum weights with ones tuned for the 6th order, a change that Allison did not approve, and that subsequently led to problems during the XV-1710-13 Type Test and in operation. The YV-1710-9(D1) attempted the Type Test, installed in the XFM-1 model test stand at Wright Field, from April 13 through June 18, 1937. It was then replaced by the XV-1710-11 power section modified with the first engine's reduction gear and extension shaft, along with the relevant accessories and fittings, to become the XV-1710-13(D1). This engine was tested from June 28, 1937 through November 19, 1937. During that period it accumulated 164-3/4 hours of testing, but was unsuccessful in completing the Type Test as a consequence of failure of the Air Corps modified harmonic vibration damper.

During takeoff on the first flight of the XFM-1 the left engine, YV-1710-9 AC34-9, suffered a violent backfire, damaging the internal accessories drive shaft and housing as well as the supercharger inlet cover and intercooler. The engine continued to operate during the 20 minute flight, which ended in a successful landing executed by Lt. Ben Kelsey of the Wright Field Fighter Project Office. Subsequent experience with the early Allison's in P-40 aircraft was to show that they were prone to backfiring due to the specified valve clearance settings, but this was the first instance of the problem. The engine was removed and quickly repaired by replacing the damaged parts with ones taken from AC34-11 that Allison was then converting to a V-1710-13. Repairs to AC34-9 cost $1,048.50 for parts and $768.10 for labor, all in 1937 Depression era dollars.

The Air Corps ultimately had four engines configured as V-1710-13's, AC34-9, 34-11, 34-12 and 34-14. It is interesting to note that during this period Allison refers to the engine as a YV-1710-9, while the Army identifies it as a V-1710-13. At the time of the first XFM-1 flight the engines were equipped with the Stromberg NA-F7C1 four barrel downdraft carburetor, and when 34-9 was repaired, it retained the original 8.00:1 supercharger gears. Both items are characteristic of the YV-1710-9, not the -13, confirming that the engines initially powering the Bell XFM-1 were definitely the YV-1710-9. The right engine may have been either AC34-10 or AC34-12, which was scheduled for subsequent conversion to V-1710-11.[12] On the date of the first flight, September 1, 1937, AC34-12 was known to still be a YV-1710-9. It then became the engine to be second engine to be converted. The third engine, AC34-11 as a V-1710-13, was slightly damaged while installed in the Right Hand Nacelle when the "dog latch" on the left landing gear on the XFM-1 collapsed just after landing following the aircraft's third flight.[13] The first XV-1710-13 was AC34-14, the V-1710-13 Type Test engine.

V-1710-13's AC34-11 and AC34-12 may have been installed after the second flight of the XFM-1, a flight that had ended in a ground loop caused by a collapsed right hand landing gear. The right-hand propeller and right wing were damaged. It is possible that AC34-10 was damaged, or at least suspect, and with the V-1710-13 configured AC34-11 and AC34-12 available, the change to V-1710-13's was made.

In December 1938 the Army asked to have the "three V-1710-13's" overhauled and equipped with crankshafts and 2-node pendulum type vibration dampers as were then standard on the V-1710-23's being produced for the YFM-1's. At the time the XFM-1 was at Bell Aircraft and did not fly from September 1938 through April 1939, while being configured as the prototype YFM-1.[14] The aircraft had accumulated 46 flight hours prior to this period, and the engines probably had a great deal of additional ground running time, hence the overhauls were both warranted and desired to better represent the -23's for the coming YFM-1 model. Engine AC34-14 was badly damaged during its Type Test at Wright Field and had been Surveyed. When the XFM-1 was Surveyed at Wright Field, March 19, 1942, after 103 hours of flight, V-1710-13 engines AC34-9 and AC34-11 were installed. It had not flown since January 1940.

A problem identified with the engine installation in the XFM-1 was "charring" of the exhaust side ignition wires, unique to the pusher installation and was caused by the lack of cooling airflow around the wires during ground operation. They were in close proximity to the radiant heat from the exhaust manifolds. The only "fix" was to minimize the amount of ground running.

D-1(-13): As the V-1710-13, these engines were contemporaries of the V-1710-11(C7), though provided with the extended drive shaft and other minor changes in the accessories to accommodate the XFM-1 installation.

Improvements and new design features incorporated in the XV-1710-13, compared to the YV-1710-9 engine, were as follows:[15]

- New cylinder heads having the compression ratio increased to 6.65:1 to improve fuel economy.
- New intake manifolds (PN36085/6) and redesigned induction system that permitted a decrease in blower gear ratio from 8.0:1 to 6.23:1 and materially reduced the manifold pressure required to develop take-off and normal rated power (At 1000 bhp and 2600 rpm the difference in required manifold pressure between the two engines was 4.5 inHg, a little over 2 psi). This reduced induction temperature by about 15°F and supercharger power by about 20 horsepower.
- Cylinder heads to accommodate "long reach" spark plugs. This change materially improved the margin to detonation.
- Minor design changes in accessory drives to simplify construction or improve reliability.

- Provision for conversion to a propeller speed extension shaft model suitable for use with a tractor propeller by interchanging or replacing cylinder coolant orifices and reversing engine coolant outlet connections.

The XV-1710-11(C7) Type Test engine (AC34-14), was the first to be built in the XV-1710-13(D1) configuration, it was soon followed by the three engines remaining in the XFM-1 program, AC34-9, 34-11 and 34-12.[16] Engine AC34-11 (AEC#15) had previously attempted the YV-1710-9(D1) Type Test, but failed as a result of a broken accessories drive shaft, and then was replaced by AC34-14 and itself rebuilt as a -13.

D-2(-23): In Fiscal Year 1938, the Air Corps ordered 10 YFM-1 and 3 YFM-1A twin engined *Airacudas* and 39 V-1710-13(D1) engines to support service testing. They paid a total of $486,483.21 for the airplanes and $317,288.20 for their initial compliment of engines. An additional $335,569.00 was paid for 13 spare V-1710-23's and associated spare parts.[17] These engines were all procured by amendment to contact W535-ac-10660, which had been issued for the YP-37 engines, and joined them in production. None were delivered as V-1710-13's. The improvements, including the Bendix pressure carburetor being built into the V-1710-21's and the improved 2-node harmonic vibration damper, were also incorporated into these engines. Consequently they were all delivered as V-1710-23(D2)'s.

The change to contract W535-ac-10660 for production of the 39 V-1710-D2's was issued January 5, 1939. The last engine was shipped in October 1940, the spare parts order was completed in December 1941 and the last overhaul of a D-2 by Allison was done in September 1941.[18] The big change in this model, as compared to the V-1710-13, was the adaptation of the pressure carburetor, contemporary with the V-1710-21(C10). Otherwise it used the same V-3420 cylinder blocks, and reduced supercharger gear ratio to better match use with a turbosupercharger. The reduction gear case design of the D-2 was such that it did not require installation of the internal oil jet for extra lubrication of the Prop Thrust Bearing that was remotely mounted on the airframe.[19] A Model Test of the -23 was attempted, but not successfully completed due to problems with the crankcase. Production of the service test quantity was allowed to proceed based upon Allison's guarantee of satisfactory service of the crankcase for 500 hours.[20]

Approved operational ratings tracked those being applied to the contemporary C-10 engines, and reflect Wright Field's 1940 decision to not allow operation of early V-1710's at 2995 rpm as a result of the continuing overspeeding conditions caused by the early Curtiss-Electric propellers. As a result, the engines were not allowed to develop the rated 1150 bhp.

D-2A: Development of the V-1710-41(D2A) engine for the YFM-1B was again a fairly simple conversion, as only the supercharger gears and induction/exhaust ducting arrangements needed to be changed.[21] The 8.77:1 gears from the V-1710-C13 replaced the 6.23:1 set and allowed operating at medium altitudes without the weight, complexity and operational problems of the original turbosupercharger installation. The three barrel Bendix pressure carburetor and supercharger inlet elbow arrangement from the V-1710-C13 was to be used in place of the earlier two barrel unit as a way to improve breathing and reduce induction system pressure losses. This exercise paralleled the efforts to convert the turbosupercharged Curtiss XP-37 and Bell XP-39 into the altitude rated XP-40 and XP-39B by similar changes in carburetion, induction ducting and engine supercharger gearing. Only the four engines for the two YFM-1B aircraft were converted to this configuration, leaving no spares. Allison was paid an additional $1,690 for each of the conversions.[22]

D-3: Allison initially offered (December 1936) the type "D" extension shaft engine in a "tractor" configuration for use in the proposed Bell Model 3 single engine pursuit. As initially offered, the engine would have been rated at 1250 bhp and 2900 rpm. This would have required sea-level boosting from a turbosupercharger, the net effect being a critical altitude to 15,000 feet when using the available GE Form F-10 turbo.[23] In March 1937 Allison rerated the engine so as to not require sea-level turbo boosting. The result was 1150 bhp at 2950 rpm, though the Form F-10 was now able to deliver this rating to 20,000 feet. Three variations on this engine were offered for the 1938 X-608/X-609 Pursuit Competition, all based upon the C-7 power section. The basic D-3 was a tractor configured D-2, retaining the propeller speed extension shaft. The other two were extension shaft engines, and differed significantly from earlier models by having a *crankshaft* speed extension shaft and a remote 5:3 reduction gear able to accommodate a 25mm cannon firing through the #50 propeller hub.

The final Bell Model 3 and 4 proposals, as well as the Curtiss Design 80A (a proposed configuration of the YP-37), for the delayed 1938 Pursuit Competition invited by Circular X-609, all required an extension shaft engine and a remote reduction gear box configured such that a hollow propeller shaft was available through which a cannon could fire. The "D" series was the "extension shaft" Allison engine and Allison designed the D-3 Mod A/B models to accommodate these Bell and Curtiss requirements. The previous propeller speed extension shaft from the "pusher" arrangement could not be used, for the propeller centerline was the same as that for the crankshaft and it was not practical to fit a cannon within it. The solution was to provide a remote reduction gear with the propeller centerline far enough above the input shaft so as to provide space for the cannon. Allison Specification 110-Mod A, provided to Bell, described an engine configured as such, with a 5:3 (1.667:1) reduction gear and rated for 1150/2950/20,000 with turbosupercharger. A similar design was provided to Curtiss for their Design 80A via Allison Specification 110-Mod B, both dating from June 1937.[24]

For these engines it was necessary to revise the front nose case of the D-3 engine so that it could mount an extension shaft running at crankshaft speed. The remote reduction gear was devised such that it was of the "external spur gear" design, with the gears sized to provide ten inches of vertical offset between the shaft centerlines. In this way space was provided for the cannon. The other major design change was that the direction of rotation of the crankshaft had to be reversed to accommodate the reversal of rotation that occurs with "external spur" gears in the reduction gear box and still provide a "right-hand tractor" drive. So as to retain the same crank rotation and firing order as in previous engines, an idler gear was incorporated into the new remote reduction gear.[25] With the exception of the above noted changes, the engine was otherwise similar to the V-1710-11/13. Interestingly, the X-609 competition required the proposals to utilize the V-1710-C7 engine. The V-1710-D3 complied as it utilized the C-7 components, revised as above.[26]

The combination of all of these changes, coupled with the concurrent project to design the XV-3420-1 double vee engine, gave Allison the impetus and opportunity to create a new base line engine. The result was the V-1710-E series of extension shaft engines that used the improved components designed for the V-3420 rather than continuing to base developments on the earlier V-1710-C series.

D-4: Allison Spec 131 was issued June 1, 1939 describing this engine as a contemporary of the altitude rated V-1710-41(D2A), but equipped for a Hydromatic propeller control. None were procured.

D-5: Allison Spec 127-A, originally issued 12-22-38 and revised as of 6-1-39, describes the D-5. It was a sea-level rated engine similar to the V-1710-23(D2) except that it used the PD-12G1 pressure carburetor and was fitted for operation with a Hydromatic propeller. None were procured.

NOTES

[1] Hunt, J.H. 1942, XFM-1.

[2] Hunt, J.H. 1942, XFM-1.

[3] Hazen *Development Program,* memo of 9-14/18-36.

[4] Wood, Robert, 1934.

[5] Hunt, J.H. 1942, XFM-1.

[6] Ron Hazen, Allison Chief Engineer, Memo, *Engineering Program*, 7-17-1936.

[7] The XFM-1 was one of the first aircraft to use the new "Selsyn" type instruments for remote reading indications.

[8] *Damage to and Repair of Model YV-1710-9 Engine A.C. No. 34-9, as Result of Backfire in XFM-1 Airplane*, Hazen of Allison letter to Chief, Materiel Division, Wright Field, September 8, 1937. Also, Hunt, J.H. 1942, XFM-1.

[9] Wright Field file D52.41/16-Allison @ NASM.

[10] Curtiss Electric Controllable Pitch Propeller, Blade Design per Curtiss Dwg. No. 88996-18, Hub No. 88549. Engineering Section Memo Report, Serial No. R-57-362-1, 1938.

[11] Ethell, Jeff, 1st Quarter 1981, photo on page 63.

[12] Per Air Corps Letter of 8-18-37/8-28-37.

[13] Ethell, Jeff, 1st Quarter 1981, 61.

[14] Matthews, Birch. 1996, 47.

[15] Engineering Section Memo Report Serial No. E-57-1406-3, 1939.

[16] Allison Model Specification No. 109-C1, 1938.

[17] *Annual Report of the Chief of the Air Corps, 1938*, 10. *Sub-Title C-Maintenance and Operation of Airplanes, Depots, Stations and Organizations, Project 81-Spare Engines for New Aircraft.*

[18] Hunt, J.H. 1942, XFM-1.

[19] Allison letter, 2-24-1941.

[20] Teletype EXP-T-652 to Chief, Materiel Division, responding to General Arnold's questions on C-15, D-2, E-4, E-5, F-2R/L and F-3R engines, January 22, 1941. NARA RG18, File 452.8, Box 807.

[21] In building the -D2A and -E2A "Altitude" models from existing D-2 and E-2 engines, it was easier to stamp an "A" on the data plate than to change the dataplate and manufactures model number. Early Allison data plates did not have the Air Corps model or serial number on them, only the Allison manufactures' data. See the photo of data plate from V-1710-D2 AEC s/n 79. The Air Corps model and serial number were on a separate data plate mounted elsewhere on the crankcase.

[22] Change No.7 to Contract W535-ac-10660, December 7, 1939. NARA RG18, File 452.8, Box 808.

[23] Bell Drawing 3M001, December 1936.

[24] Curtiss Microfilm, June 1937.

[25] Bell Aircraft Preliminary Specification, Report No. 4Y003, 1937. See page 56, paragraph B-3(A).

[26] NASM Curtiss Microfilm, Roll N, Frame 121, June 1937.

11

The V-1710-E: Engines for Remote Installations

With the selection of the Bell Model 4 as the winner of the 1938 Circular X-609 single engine interceptor competition, it was necessary to provide a suitable engine for the unique installation. Although the Army had specified that X-609 proposals were to utilize the V-1710-C7, the preliminary specification for the XP-39 identifies the engine to be per Allison Specification #110-A, Model 4 (XP-39) defining the Allison V-1710-D3 "Mod A." This was effectively a re-configured V-1710-C7 equipped with an extension shaft to a remote 5:3 external type reduction gear and rated at 1150/2950/20,000', with turbosupercharger.[1] As noted earlier during the story of the XP-39, it was prudent to develop the engine from the newly designed components coming from the XV-3420-1 project rather than continuing with a version based on the earlier V-1710-C series engine. This was a good decision, for the "E" series was to see extensive development and production, being produced in many models, some capable of approaching 3000 bhp.

Under the direction of Chief Engineer Ron Hazen, the small engineering staff at Allison recognized that the V-1710-C/D engine needed a complete redesign to accomplish all of the many roles being assigned to it, and to make it easier to mass produce. Unfortunately the Army did not appropriate sufficient funds for that purpose, so Allison independently designed the V-3420 so that its parts could be modified or used directly on an improved V-1710. The result was the V-1710-E/F series of engines that powered the bulk of the Army's Pursuit aircraft during WWII.[2]

With the preliminary specification identifying the V-1710-D3 Mod A as the Bell Model 4 (XP-39) engine we have an explanation of the subsequent assignment of "E" series model numbers to the early aircraft using the extension shaft engine. Though the V-1710-E1 (the first Allison "E" engine model) was used in the Navy Bell XFL-1, that project was not under contract until nine months after the first XP-39 engine was ordered.[3] The Air Corps awarded the XP-39 contract to Bell on October 10, 1937, an action that spurred Allison to offer their E-2 for the XP-39 in Specification #115, dated November 24, 1937. Allison clearly did not want to continue development work on the then eight year old design of the V-1710-C/D.

The extension shaft configuration was a complex engineering challenge requiring the full measure of Allison ingenuity and expertise for dynamic analysis of shafting and power systems. Many others around the world had proposed such configurations over the years, particularly in the 1930s, but Allison was the only one to actually make it work to the point that it was to see sizable production.

Transmission of power through shafts is almost an art as much as it is a science. Vibration, and the potentially destructive critical speeds inherent in rotating machinery, are the bane of such systems. Usually these effects can be minimized if everything can be rigidly mounted, but in an aircraft this is nearly impossible. Instead, the power transmission system must be designed to tolerate vibration along with the flexibility of the airframe. The Bell P-39 was a rugged and comparatively rigid airframe, obviously incorporating lessons learned from Bell's earlier X/YFM-1 that was

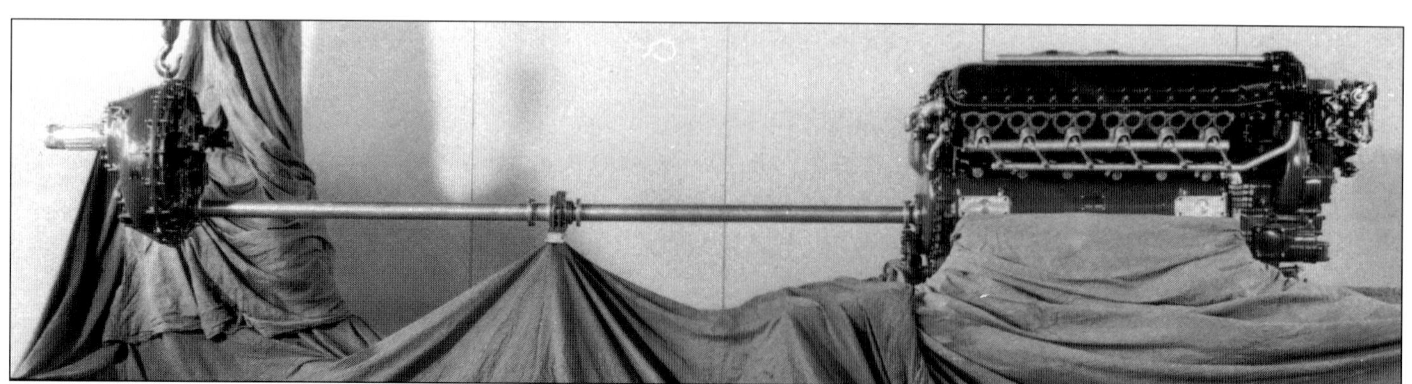

This is the first XV-1710-17(E2), Serial Number AC38-644 as delivered to Bell for installation in the XP-39. (Bell via Matthews)

Retouched photo of a production "E" model engine. Structure on top of the front cover, above the output coupling, is the crankcase breather assembly. The fitting directly below the coupling contains the forward scavenge oil pump and drain back to the crankcase. (Allison)

troubled with vibration problems. The P-39 fuselage structure was extremely rigid and provided a good foundation for the engine and extension drive. Even so, the Allison system was designed to accommodate +/-1 degree of motion or miss-alignment between the front flange of the engine and the remote reduction gear. Over the nearly ten feet involved this amounted to a total travel of 3.64 inches. Accommodation was provided by the use of splined connections between the drive shafts and couplings.

During the initial design torsional vibrations were extensively studied and considered because of the length and small diameter of the extension drive shaft. Torsional analysis gave the extension shaft good characteristics through the range of speeds normally checked, but at low speeds terrific vibrations developed in the XP-39 that resulted in broken flexible drive shafts to the accessories section.[4] A delay of about three months in the first flight of the XP-39 occurred because these vibrations in the accessories drive of the V-1710-E2 were so severe when installed in the aircraft that it would not run smoothly or safely. The effect of the vibrations were to cause erratic operation of the valves and ignition timing, and were initially attributed to the use of a plate type friction damper in the accessories drive.

The fix, that might have occurred to any automobile engineer, was to devise and install a simple hydraulic vibration damper, a shock absorber that cushioned rebound, and thereby limit the vibrations being transmitted through the short, but flexible, quill shaft driving the accessories.[5] While this device was very successful and is a primary reason for the V-1710's subsequently earned reputation as a smooth running engine, it was ineffective in solving the problem being caused by the 1/2-order vibration. Allison determined this during tests of the new damper in their EX-3 workhorse engine in early May 1939. They then determined that the low-order vibrations were being caused by lack of rigidity in the extension shafts.

At a May 22, 1938 conference in Indianapolis with Wright Field representatives on the status of the XV-1710-17, Hazen suggested the Army accept the engine in the XP-39 as it was, believing it safe to 2950 rpm. Allison would then confirm safety per the contract required 10 hour test of the workhorse V-1710-17 then on Allison torque stand. Wright Field objected as the 1/2-order vibration began at 2800 rpm, which they argued was extremely dangerous in the event of a cylinder out during takeoff or Military power conditions. Under those conditions the torsional excitation in the crankshaft system would be 1/12 of the entire indicated horsepower of the engine, far exceeding the small amount of torsional excitation (+/-1.75 degree in the range from 2800-3000 rpm) caused by the half order harmonic when all cylinders were operating normally. The engine was test run in this condition, and when one cylinder was deliberately shorted out at 2950 rpm the speed dropped to 2850 rpm and the entire engine shook so severely that it cut out due to violent shaking of the carburetor.[6]

Hazen offered that the only satisfactory fix would be to enlarge the extension shafts which would stiffen the shafts by 30 percent and raise the period of vibration approximately 400 rpm, well out of the operating range of the engine. Fabrication of new shafts was to start immediately and was expected to take 4 to 5 weeks, with another week before they would be in the airplane and permit unrestricted flying as desired by the Government. The new shafts would increase the weight by 11 to 15 pounds. The OD would be increased from 2.500 inches and 0.156 inch wall, to 2.550 inches, and the wall increased to 0.200 inches. These shafts would have the same torsional characteristics as the 3 inch diameter shafts intended for follow-on P-39's, which were to only weigh an additional 5 pounds, all in the required larger couplings and flanges. The 3 inch diameter shafts would not fit in the XP-39.

Interestingly, the urgency to complete the experimental work on the XV-1710-17 was so the test stand could be used for development and testing to support the 815 V-1710 engines contracted by the French Government. To this end the Wright Field representatives recommended obtaining the new shafts and then run the contract required 35-hour test, following which the Allison would have satisfied Government requirements on the XV-1710-17 and would be free to do other experimental work.[7] While all of this was going on, it was decided to fly the XP-39 to Langley and have the NACA put it into the wind tunnel to investigate the drag characteristics of the entire airplane.

During 1938, Allison was promoting the Altitude Rated versions of the V-1710. They also recognized the need for two-stage supercharging and began work on a mechanically driven auxiliary stage supercharger that could be attached to, and driven by gears from the accessories section. Formal notification of the Air Corps Procurement Office of this effort was given in November 1938.[8] First use of this device was intended to provide a "two-stage" engine for an improved Bell P-39, one able to regain the promise of the originally turbosupercharged interceptor. Allison, like P&W, took the approach of developing a separate "Auxiliary Stage Supercharger" that could be connected with few changes and little redesign. The Aux Stage Supercharger was driven via a power take-off coupling driven by the rugged starter gear.

Initially they drove the Auxiliary Stage through a friction clutch. This was replaced in 1942 by a hydraulic coupling that offered improved vibration isolation as well as better overall engine efficiency by using its variable speed control capability. This was important in minimizing the power consumed by the Auxiliary Stage when operating at below the "critical altitude" of the engine, and is a feature also seen on contemporary Daimler-Benz V-12 engines. While the detachable Aux Stage was a good arrangement from the perspective of manufacturing, it caused a lengthening of the engine by 18 to 22 inches. This is clearly apparent in the look of the aircraft. Their engine compartments are noticeably longer than those in single-stage equipped aircraft. Still the altitude performance was considerably improved.

The V-1710-E series engines were used in aircraft that were able to take advantage of the inherent high power capability of the engine, and coupled with the long operational life of the series, were developed to use a wide range of power settings. Table 11-1 shows the limiting values for the various models.

Table 11-1 V-1710-"E" Maximum MAP Ratings, inHgA		
Engine Model	WER Operation	Standard Emergency Operation
V-1710- 35	56	52
V-1710- 63	60	55
V-1710- 83	57	55
V-1710- 85	58.5	55
V-1710- 93	60	55
V-1710-117	61	NA
V-1710-109	61	NA
from TO AN 02-5AD-2, Section VII, Table I, May 5, 1945		

Outboard Drive Assemblies

All V-1710-35, -37, -63, and -85 engines were procured with a complete outboard drive assembly consisting of a reduction gear housing assembly, extension shaft center bearing, and propeller extension shafts. These components were originally considered a component part of the engine and were matched by serial number. Field experience found that these assemblies were mechanically satisfactory to operate for longer periods between overhauls than the basic engine. By early 1944 tech orders were issued directing that it was satisfactory to operate for longer periods between overhauls. Normal running time was set for these models was 1000 hours. If the unit was still performing satisfactorily at that time, the oil was to be changed, the oil pump removed and inspected, and if all was well another 250 hours could be run. This procedure could be repeated a second time, though in any case the assembly was to be overhauled after 1,500 hours.[9]

Table 11-4: V-1710-E Engine Models

E-1: Built to Allison Specification #112 dated 9-17-1937, the V-1710-E1 was designed as an Altitude rated engine for the Navy XFL-1 derivative of the Bell XP-39B. The Navy contract for the XV-1710-6's was dated March 3, 1939.[10] It was not the first extension shaft engine delivered, as the V-1710-D3 Mod A replacement, the XV-1710-17(E2) was first. As first proposed, Spec #112 of 9-17-1937, the E-1 was to have 5:3 reduction gearing and use an idler gear in the remote reduction gear so that the crankshaft and propeller would rotate in same direction. This was a feature of the D-3 Mod A proposed for the XP-39 as well. The extension shaft was to be of approximately 2-1/2 inches outside diameter, the engine itself was to have been a modified "standard" V-1710, that is, a V-1710-C/D power section. Collector exhaust manifolds, collecting each group of three cylinders together, were to have been used.[11] Grumman did show some interest in using this engine, for on February 14, 1938 they requested the Navy to provide specifics on a configuration using a five foot extension shaft version similar to that described for the Bell Model 3.[12] When actually constructed

This is the 1.8:1 reduction of the XV-1710-17(E2). It shows clearly the hole for the cannon through the propeller shaft, as well as the two gun synchronizers. Later units were fitted with different reduction gear ratios as well as mounting pads for the engine starter and generator. (Bell via Matthews)

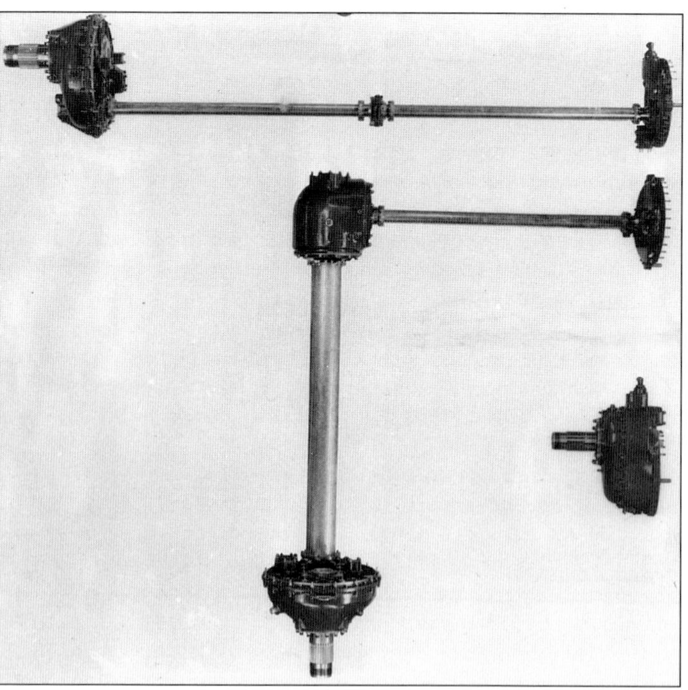

Allison built considerable flexibility into the V-1710 to enable it to be used in a wide variety of airframe configurations. The "E" model was the power section for all of these. The top view is the arrangement developed for the Bell P-39 and P-63, with crankshaft speed extension shaft driving a remote reduction gear. The angle arrangement was intended for use with the engine in the fuselage and could be used either in tractor or pusher arrangements with the gearboxes in the wings. The angle box reduced the speed 1.5:1, while the epicyclic remote box could shift gears to improve long range cruise. While tested using an "E" power unit, the application would have been V-3420 powered. The "F" external spur type reduction gear was interchangeable with the "E" drive. (Allison)

Table 11-2
V-1710-"E" Production Series Ratings

Model	Engine	Aux S/C	Carb	Takeoff[1]	Military @3000rpm	Normal @ 2600 rpm	WER @ 3000 rpm
XV-1710- 6(E1)	8.80:1	None	PD-12G1	1150/SL/46.0	1150/ 9,000/43.6[2]	1000/ 9,000/40.4	No WER Established
XV-1710-17(E2)	6.44:1	Turbo	PD-12K2	1150/SL/41.6	1150/25,000/41.6[2]	1000/25,000/	No WER Established
V-1710- 31(E2A)	8.80:1	None	PT-13B1	1150/SL/	1150/12,000/____	1000/12,000/	No WER Established
V-1710- 35(E4)	8.80:1	None	PD-12K2	1150/SL/45.5	1150/12,000/42.0	1000/10,800/37.2	1490/ 4,300/56.0 Dry
V-1710- 37(E5)	8.80:1	None	PD-12K2	1090/SL/42.9	1090/13,000/40.5	960/12,000/35.8	No WER Established
V-1710- 63(E6)	8.80:1	None	PD-12K2	1325/SL/51.0	1100/13,800/44.2	950/13,700/37.5	1580/ 2,500/60.0 Dry
V-1710- 47(E9)	7.48:1	6.85:1	PT-13E5	1325/SL/51.5	1150/21,000/49.7	1000/19,000/40.5	No WER Established
V-1710- 93(E11)	8.10:1	6.85:1	PT-13E9	1325/SL/54.0	1180/21,500/52.0	1050/10,000/43.0	1825/ SL /75.0 Wet
V-1710- 59(E12)	9.60:1	None	PD-12K2	1100/SL/46.5	1100/13,800/44.2	950/13,700/37.5	No WER Established
V-1710- 83(E18)	9.60:1	None	PD-12K2	1200/SL/52.4	1125/15,500/44.5	1000/14,000/38.8	1410/ 9,500/57.0 Dry
V-1710- 85(E19)	9.60:1	None	PD-12K6	1200/SL/50.5	1125/15,500/44.5	1000/14,000/39.2	1410/ 9,500/57.0 Dry
V-1710-117(E21)	8.10:1	7.23:1	PT-13E10	1325/SL/61.0	1000/25,000/52.0	1000/21,000/43.0	1800/24,000/76.0 Wet
V-1710-109(E22)	8.10:1	7.23:1	PT-13E15	1425/SL/59.3	1100/28,000/50.7	950/24,000/42.0	1750/3200/SL/75.0Wet
V-1710-103(E23)	8.10:1	6.85:1	PD-12K7	1325/SL/54.0	1200/22,400/51.0	1000/20,000/42.5	1820/3200/SL/ Wet
V-1710-125(E23A)	8.10:1	6.85:1	PD-12K8	1675/SL/70.0	1100/24,000/51.0	1000/20,000/42.5	1900/3200/SL/75.0Wet
V-1710-129(E23B)	8.10:1	7.23:1	PD-12K15	1675/SL/70.0	1100/29,000/51.0	1000/24,500/45.2[3]	1900/3200/SL/75.0Wet
V-1710-137(E23C)	8.10:1	7.64:1	PD-12K15	1790/SL/75.0	1150/27,500/57.0	1000/25,000/46.9	No WER Established
V-1710-E24	8.10:1	7.23:1	PD-12K7	1500/SL/	1325/3200/22,500'	1000/20,000/	None Built
V-1710-127(E27)	8.10:1	7.23:1	PD-12K15	2830/SL/100	2430/17,000/85.0	1200/26,000/	2980/11,000/100 Wet
V-1710-133(E30)	8.10:1	7.64:1	PD-12K15	1500/SL/65.5	1150/27,500/58.0	975/24,000/43.4	1800/3200/SL/75.0We
V-1710-135(E31)	9.60:1	None	PD-12K	____/SL/____	1150/ /	/ / SL /____	

1. BHP/ MAP, inHgA
2. At 2950 rpm
3. At 2700 rpm

Table 11-3
Extension Shaft Engines Shipped

Model	Number	Aircraft	Comments
XV-1710- 6(E1)	2	XFL-1	For U.S. Navy, Altitude rated
XV-1710-17(E2)	2	XP-39	Turbosupercharged, sea-level rated
V-1710- 35(E4)	2,155	P-39D/F, P-400	Alt rated, includes 810 shipped on British contracts
V-1710- 37(E5)	16	YP-39	Altitude rated
V-1710- 63(E6)	946	P-39D-2/K/L	165 shipped without reduction gear boxes
V-1710- 67(E8)	1	P-39 Proposal	Development for Lightweight P-39, Bell Model 13A
V-1710- 47(E9)	10	XP-39E, XP-63	First two-stage engine
V-1710- 93(E11)	2,554	P-63A, P-63C	Two-stage production model, 33 shipped w/o Gear boxes
V-1710- 59(E12)	25	P-39J	Early attempt at 9.6:1 supercharger gears
V-1710- 65(E16)	1	P-39 Proposal	Development engine w/Panial type s/c, not flown
V-1710- 83(E18)	1,034	P-39M	159 shipped without reduction gear boxes
V-1710- 85(E19)	9,780	P-39N/Q	1,235 shipped without reduction gear boxes
V-1710-117(E21)	2,237	P-63C	730 shipped without reduction gear boxes
V-1710-109(E22)	222	P-63E	207 shipped without reduction gear boxes, 1 w/o S/C
V-1710-103(E23)	5	XB-42	Later converted to -125(E23A) and -129(E23B)
V-1710-133(E30)	8	XB-42A, Bell L-39	1 for test, 3 for Navy L-39, 4 became -137(E23C)
V-1710-135(E31)	–	RP-63G-1	258 Converted from V-1710-115A(F31A) production.
Total	18,998		

This is XV-1710-6(E1), AEC #89, being prepared by the Navy for their comprehensive calibration tests in support of the XFL-1 flight test program, 1940. (Navy)

under the Military designation of XV-1710-6, only two such engines were built, the 88th and 89th V-1710's. Engine AEC#88 was the first flight engine for the XFL-1, while #89 performed the extensive Altitude calibrations, and after overhaul, it flew as the spare when the XFL-1 needed its engine changed during the fall of 1940.

E-2: This was the first crankshaft speed extension shaft engine to be constructed and was developed to replace the V-1710-D3 Mod A and Mod B concepts with their remote reduction gear for mounting a nose cannon. The engine used many components coming from the V-3420 program and was built to Allison Specification 115. Two engines were purchased and delivered to the Air Corps designated as model XV-1710-17. Designed as a "sea-level" engine it utilized 6.44:1 supercharger gears so as to be compatible with the turbosupercharged Bell XP-39 and its General Electric Type B-5 turbosupercharger. Both engines went through numerous modifications using different supercharger step-up gears during the evolution of the aircraft from turbosupercharged to Altitude rated. In their final configuration in the XP-39B they were effectively prototypes for the E-5 model, and used the same 8.80:1 supercharger gears. Only two of these engines were built for the Air Corps, though a third engine was configured to this standard and used by Allison as the EX-3 "workhorse" engine for in-house development and testing. The EX-3 accumulated a total of 345 hours of operation before the contract required 35 hour test was performed. During the early running, the friction damper for the accessories drive was changed to the new hydraulic type and the extension shafts were increased from 2-1/2 inch diameter and 5/32 inch wall to 2-5/8 inch diameter and 0.200 inch wall to raise the system frequency and better dampen torsional vibration. The initial form of the hydraulic damper was not able to dampen all of the vibration modes of concern. Even so, the XP-39 began its flight testing using an early version of the hydraulic damper. When the testing was completed on July 31, 1939, the hydraulic damper and stiffened extension shafts had been shown to be effective such that there were no serious resonant speeds for torsional vibration encountered from 600 to 3100 rpm.[13]

The first XV-1710-17, sn AC38-644. Note that all coolant is delivered via the lower cylinder jacket manifold i.e. no "y" fitting, and that the shield type exhaust plug blast tube was fitted. (Bell via Matthews)

E-2A: Identified by the Air Corps as the V-1710-31, these engines were actually an interim configuration of the original E-2's when they were to be equipped with the PT-13B1 carburetor and 8.80:1 supercharger gear ratio, though the workhorse engine tried 9.60:1 gears as well. Both features were similar to what was being used by the prototype altitude rated XV-1710-19(C13) engine in the XP-40. The "A" in -E2A designation identified the conversion to "Altitude" rating, an expedient way to identify the modification on the existing engine dataplate. It is believed that the engine was designated as the V-1710-31 when the Air Corps decided that the batch of thirteen YP-39's should be purchased as 12 sea-level rated Bell Model 4C's, with turbosuperchargers and V-1710-17(E3)'s, and one Bell Model 4D YP-39A that was to be altitude rated and would have used the E-2A. Priorities were such that the YP-39A was scheduled to be the first of the service test YP-39's to be delivered.[14] It then evolved that the twelve YP-39 Bell Model 12's were designed to use the altitude rated V-1710-E4, and a single YP-39A was to be turbosupercharged and powered with the V-1710-17(E3). The turbo effort was subsequently terminated and the YP-39A (Bell Model 12A) built as a Model 12, also to be powered by the E-4. No V-1710-31's were built, although the original E-2's operated in a number of altitude rated configurations as they evolved from E-2 through E-2A, to prototype E-5's. No new V-1710-31 engines were actually procured.

E-3: Twelve of the 13 YP-39's, as the Bell Model 4C, were originally intended to be turbosupercharged and were to be powered by the improved V-1710-E3, still identified by the Air Corps as the V-1710-17. With the decision to produce these aircraft with the altitude rated engine originally intended for the YP-39A, the V-1710-E4, the E-3 was terminated. Allison Specification 121-B, dated 6-1-39, described the E-3, while Spec 121-E, dated 12-18-39, was the final form of the E-5 engine. The order for sixteen of these engines was converted to the E-5 model following termination of the turbo effort.[15]

E-4: This engine was first described in Allison Spec 122, and designated as the V-1710-35 by the Air Corps and was originally to be the engine for the altitude rated Bell Model 12, YP-39A. This engine was built on a con-

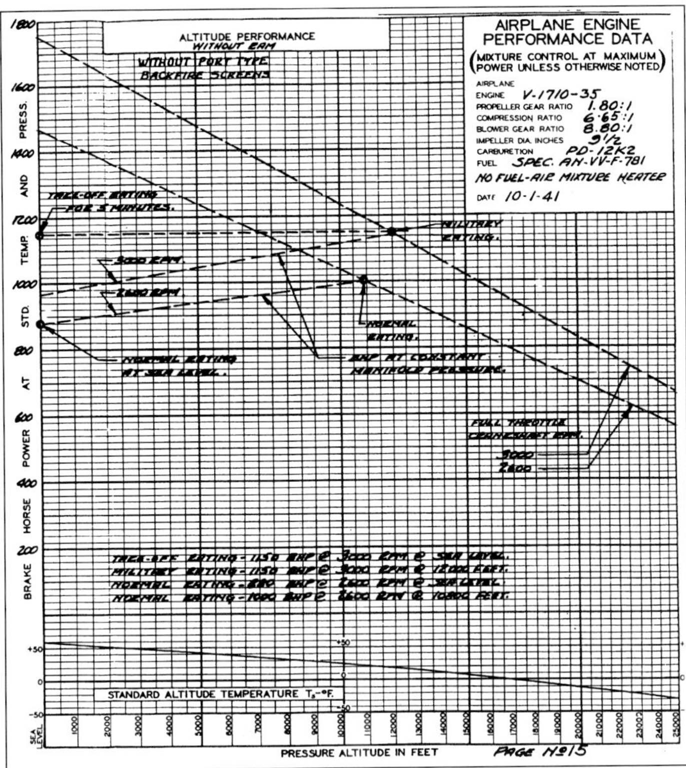

The complete performance curve for the altitude rated E-4 shows it to be a very capable mid-altitude performer. Note the capability for WER rating as high as 1750 bhp at sea-level, provided that high octane fuel was available. This curve was prepared a year before Allison proposed WER ratings to Wright Field. (Allison)

siderably strengthened crankcase and was therefore free of the power limitations that had been set by the requirement not to exceed the V-1710-C/D series Indicated horsepower equivalent to 1150 bhp at sea-level. As an altitude rated engine the E-4 was to deliver 1150 bhp at its critical altitude, which was a considerably greater IHP when the supercharger horsepower is included. This engine also introduced the "y" fitting in the coolant supply to each block, establishing the coolant piping arrangement used on all following V-1710's.

The engine experienced problems in completing the initial 150-hour Model Test at the specified 1150 bhp. Resolution required further strengthening of the crankcase, before the test was successfully completed, but not in time for the pre-production YP-39's.

As early as May 1940 Allison was authorized to begin discussions with foreign governments regarding the sale of the E-4.[16] The Model went into production December 26, 1940, though the official Model Test was not completed until January 9, 1941. Still there were several parts that were not satisfactory, including the lower crankcase casting, and the extension shaft bearings in both the engine and center support.[17] While the various improvements were being incorporated, and pending a final Model Test at 1150 bhp the E-4, along with the contemporary E-5 and F-3R engines were authorized maximum ratings of only 950 bhp at 2770 rpm.[18]

An improvement in performance was also achieved by utilizing the enlarged two-barrel PD-12K2 carburetor.[19] It subsequently became the engine for the production Bell P-39C/D and export P-400 export series. A total of 2,155 were built. In late 1942 the E-4 was one of the first engines that Allison developed for War Emergency Ratings (WER), and authorized combat operation at up to 1490 bhp with 56.0 inHgA. This 30 percent overpower rating could be maintained for up to 5 minutes at a time, and at altitudes up to 4,300 feet.

A batch of E-4 engines was proposed to be constructed for the Netherlands East Indies using non-standard non-ferrous parts, as was done with the C-15A engines provided to China for the AVG. These engines were intended for an order of 100 Bell *Airacobra* aircraft planned for delivery beginning in September 1941.[20] It appears that this transaction was not completed. If it had, the engines would have been identified on their dataplates as Allison Model V-1710-E4A, the "A" identifying them as "Commercial Grade A" engines, equivalent to military standard Model E-4, but including the reworked parts.[21]

E-5: Since the intended E-4 engines for the YP-39's were not available in time, sixteen E-5 engines were built as the V-1710-37 for the service test batch of thirteen Bell Model 12. The E-5's utilized the carburetor and supercharger gears of the E-4 engine, but they were built on the crankcase that had not been able to pass the E-4 Model Test. Consequently they were assigned the rating at which the V-1710-33(C15) production engine was being operated. Some (all?) of the engines were later Modernized with the production E-4 components and given the latter's ratings.[22] The E-5 also used the early form of accessories drive vibration damper that had double quill shafts.[23] In addition, the E-5's used a unique set of coolant metering orifices, necessitated by the use of a different coolant path from the cylinder jacket into the head.[24] This coolant path was originally the same as for the V-1710-C15 and V-1710-E2, that is, with a single supply to each block. When the E-5's were Modernized they received the "y" fittings providing two supply connections to each block, along with a set of E-4 metering orifices.

E-6: The significant feature of this V-1710-63 model was the reduction gear box ratio was changed from 1.80:1 to 2.00:1. Other general improvements over the V-1710-E4 were incorporation of larger oil pumps and an improved lubrication system, along with detail improvements to the inlet to the supercharger impeller and rotating guide vanes as used in the contemporary V-1710-F4R, provided the basis for the power ratings increases. The ignition system was not supercharged and the engine continued to use straight (97 percent) Prestone for coolant. Rated on the new Grade 100/125 fuel, the takeoff and WER ratings were increased to 1325 and 1580 bhp respectively. Originally supplied with the PD-12K2 carburetor, most were subsequently switched to the improved PD-12K6. Interestingly, 165 of the 946 E-6 engines were supplied without reduction gearboxes. This suggests the reduction gears had a longer TBO than the power sections.

This is the dataplate from V-1710-37 AEC#338, AC40-581, one of the 16 E-5 engines. It was accepted 12-18-1940. Note how the clearance for the intake valves has been restamped from 0.010 to 0.015. (Author)

E-7: This 1939 model for marine use was originally known as the Allison V-1710-H2, and rated for 1000 bhp at 2600 rpm, according to Allison Specification 135-B, dated November 13, 1939. It was the power section of the engine only, and retained the usual 6.65:1 compression ratio, but was the only V-1710 to use the 6.0:1 supercharger step-up gear option. Carburetor was the early PD-12G1 model. Water cooled exhaust manifolds were provided, weighing 90 pounds. As a direct drive engine, its weight was a comparatively low 1,175 pounds. It received no Military Model designation but at least one engine was built as EX-18 and run on the test block in early 1942.[25]

E-8: The V-1710-67, a single-stage engine using the E-6 power section, with its larger oil pumps and incorporated slight changes in the accessories drive. This was the first "E" engine to use the new 2.23:1 reduction gear, a performance enhancing feature for the proposed lightweight P-39. The Model Test engine was running in early March 1942.[26]

E-9: This was the first two-stage V-1710 to reach flight status. It incorporated the Allison designed and mechanically driven Auxiliary Stage Supercharger. Development of the two-stage V-1710 components began in 1940 and endured a long development cycle, during which a number of different supercharger drive ratios, both for the engine-stage and the Aux Stage, were tested.

Development work on the auxiliary stage for this engine dates from 1938, and seven of these engines were purchased on development contract W535-ac-19859 dated in June 1941. This followed seven months after the pioneering contract for the two-stage V-1710-F7R. One engine was for development, with two each for installation in XP-39E and XP-63 airplanes, with two engines held as spares. The engine went through considerable evolution during its development, in fact, revisions to Allison Specification No. 137 were issued through "G", all of which used 7.48:1 engine stage supercharger gears. The initial configuration of the engine drove the early design Auxiliary Stage Supercharger having a 9-1/2 inch diameter impeller driven through 8.00:1 step-up gears. As early as February 1942 the engine was running in the Allison Altitude Chamber.[27] Without the intercooler the engine weight was 1,525 pounds.[28]

Power section of the E-8 was identical to the earlier E-6, the difference being that the E-8 drove a 2.23:1 remote reduction gear. Note how the "E" engines used the same blast tube for cooling the exhaust plugs as did the "F" engines. (Allison)

The V-1710-47(E9) was Allison's first two-stage engine to be developed for flight. This is E-9 AEC#5423, AF41-30474, the photo taken 6-12-42. The Aux Stage Supercharger drive from the starter gear and universal joint coupling is plainly visible. At this point the hydraulic coupling within the Aux Stage had not been incorporated. (Allison via Hubbard)

In February 1942 development testing of a hydraulic drive for the auxiliary stage supercharger was begun. In July 1942 it was decided to replace the friction clutch in the drive with the automatic hydraulic coupling on both the F-7R and the E-9. It provided much improved control of the speed and power delivered to the Auxiliary Stage.[29] The engines being used in the flight test programs for the XP-39E and XP-63 evolved to use the 8.10:1 gears typical of the E-11 during the latter portions of their flight test programs.[30] The base line Auxiliary Stage "no-slip" drive ratio was 6.85:1. It finally evolved configured as the prototype V-1710-93(E11) and was first flown in the Bell XP-39E and early XP-63 prototypes (Project MX-90), predecessor to the production Bell XP-63A. The carburetor was the large three barrel PT-13E5 unit mounted on the inlet to the Auxiliary Stage Supercharger.

As early as September 1940 the Allison Specification No. 137 was revised as 137-A to reflect the incorporation of an intercooler.[31] The device never went into the production models of the engine. No intercooler or ADI injection were initially fitted, yet the critical altitude was a respectable 22,400 feet. The engines went through considerable development and were used to test an engine aftercooler, automatic engine controls, pistons and rings, water injection, investigation of functioning of the engine hydraulic coupling and studies of carburetion, including relocation of the carburetor from the inlet of the auxiliary stage to the inlet of the engine stage supercharger.[32] Aftercooler development for this engine was terminated in December 1943 in view of the poor characteristics of the cooler provided by the Harrison Radiator Division of General Motors and the large amount of mechanical trouble with the setup.[33]

In April 1942, 2,000 V-1710-47 engines were ordered on contract W535-ac-24557 for a like number of P-39E airplanes. In addition 270 were to be provided as spares and another 430 engines without reduction gears were to constructed. In July 1942 2,300 additional V-1710-47's were ordered for the P-63 airplanes, along with 700 spares, all on contract W535-ac-30859. This order was soon revised to the V-1710-93(E11), the developed -47, and the number to be purchased increased to 3,200 for installation, 480 spares and 1,120 spares without reduction gears. It is believed that the -47's ordered for the P-39E were canceled outright when that airplane was terminated in favor of the P-63, though it is possible that the engines were built and delivered as -93's for P-63 production.

E-10: This would have been the production version of the V-1710-6 for the Bell FL-1, had it gone into production. Features, ratings and equipment on the engine would have been essentially the same as for the E-4. Intended for the U.S. Navy per Allison Specification 143, dated April 5, 1940. No

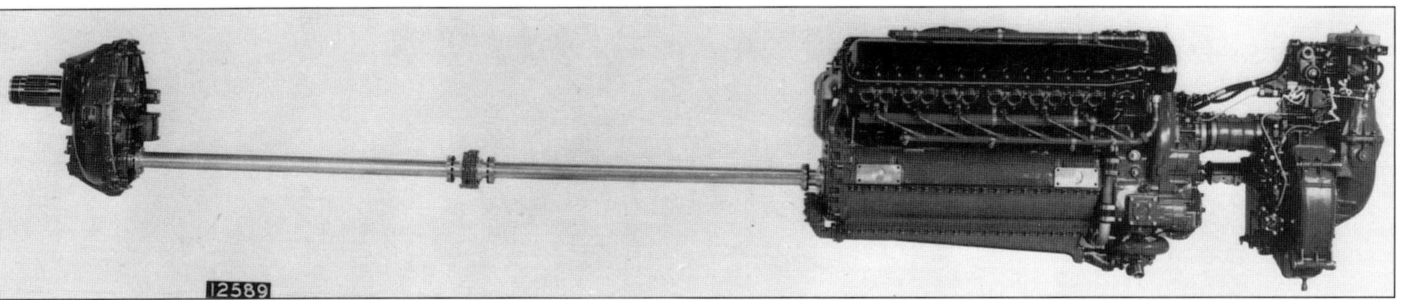

The Model Test E-11 when photographed on 5-5-43. There were few external differences between the E-9 and E-11, internally the engine benefited from a year of development and incorporated many production improvements, and of course the hydraulic drive for the Auxiliary Stage Supercharger. Gun synchronizers were mounted on the rear face of the reduction gear housing. (Allison)

contract resulted and it was intended to retain the V-1710-6 military designation.

E-11: As the V-1710-93, this was the evolved two-stage engine produced in large numbers for the Bell P-63A. As the developed E-9, it was first being designed in the Fall of 1942, but was not fully prepared for Model Testing until May 1943.[34] Many new features were incorporated, such as being equipped with an Automatic Manifold Pressure Regulator as well as being configured so that the ignition system was supercharged by the Aux Stage. It was also equipped with water injection (ADI) to enable it to both reach high altitudes and to have an impressive 1825 bhp WER rating. Two of these engine power sections were originally specified for the Douglas XA-42, later designated XB-42.

The major differences between the E-9 and E-11 engines that effected performance was the removal of both the Tee and Port backfire screens and use of the new Streamlined intake manifold gas pipes in the E-11.[35] The engine began its Model Test at Wright Field in mid-December 1942, but it was not until November 27, 1943 that the official Model Test was satisfactorily completed. At that point a considerable number of -93's had already been delivered to Bell and were flying in P-63 airplanes. Completion of the test released all of these engines for unrestricted use.[36] One reason for the extended development period was the effort required to develop water injection, and supporting improvements in pistons and rings, all in support of WER ratings for the engine. These were not used to increase the critical altitude of the engine, rather to give increased WER below 20,000 feet.[37] Allison performed the WER Rating Test in December 1943 using engine A-032554.[38]

E-12: The V-1710-59, along with the V-1710-61(F14R), were early attempts at an improved altitude rating for the single-stage engine. It was the first production engine to use 9.60:1 supercharger drive gears, on what was otherwise the E-4/E-6 engine. Critical altitude at 1150 bhp was increased to 13,800 feet from 12,000 feet for the earlier models. In an effort to improve the durability of the accessory drive gears their pressure angle was changed from 20 to 25 degrees. The supercharger impeller was improved by changing from the standard 42 to one with 47 degree inlet guide vanes. In addition, general improvements in manufacturing were incorporated, such as a nitrided crankshaft, silver-lead main and connecting rod bearings, shot blasted connecting rods, and piston pin bosses were thickened from 1/8 inch to 3/16 inch. A batch of these engines were built and installed in 25 Bell P-39J aircraft, and a companion batch of similar F-14R's were to go into 25 Curtiss P-40J's. Experience in the Bell aircraft soon found that the service life of the engines, primarily within the heavily loaded supercharger drive gears, was too short. The P-39J/P-40J projects were canceled and the engines converted back to their respective parent models.

E-13: A proposal engine only, and the intended aircraft is not known. Details are from Allison Specification 153, dated October 14, 1941, identify that it was to be a Sea-level engine for operation with a turbosupercharger with an aftercooler fitted between the cylinder banks and use an electric coolant pump for the intercooler circuit. It was to be based on the E-4/E-6 power section, but with the E-8 2.23:1 reduction gearbox. Engine stage supercharger gears were reduced to 8.10:1 to match the turbo. Engine weight was 1,485 pounds. It received no Military Model designation.

E-14: Another unique engine that would have used 7.48:1 gears in its supercharger to match the output of a two-stage turbosupercharger. The critical altitude of what would have been essentially a 3-stage supercharger configuration was a commendable 32,000 feet. Full rated power of 1400 bhp would have been available to this level. Per Allison Specification 154, dated October 22, 1941, it was to have been based upon the E-4 and was to utilize the 2.23:1 gearbox and weigh 1,445 pounds. It received no Military Model designation.

E-15: According to Allison Specification 155-A, dated October 23, 1941, this proposed engine was to have been a complex arrangement using a 8.80:1 Auxiliary Stage supercharger with intercooler feeding the engine stage that was to be driven at 7.48:1. The Aftercooler from the E-14 would have been utilized as well. Performance of this two-stage configuration was not quite as good as the turbo alternative, but still it delivered its Military rating of 1325 bhp at 30,000 feet on 100 octane fuel. A unique feature was the intended use of the comparatively large Bendix PR-58 carburetor. Power section was based upon the E-4 and was to have utilized the 2.23:1 gearbox, total weight 1,675 pounds. It received no Military Model designation.

E-16: Procured as the V-1710-65, this was to be an experimental two-stage engine equipped with a new type of Auxiliary Stage Supercharger based on using an impeller and drive of the "Panial" design, otherwise the engine was similar to the V-1710-47. The engine was never flown. The Materiel Division contracted with the Ranger Aircraft Engine Company for two of the auxiliary stage units, which were tested during the spring of 1943. Allison had considered the supercharger design at the time they were initially designing the Auxiliary Stage Supercharger and elected to not develop it due to the very complicated construction that would have been difficult to adapt for quantity production. They did consider the potential for good efficiency over a wide speed range as being of considerable value in single-stage superchargers, thereby permitting higher take-off ratings, but finally decided that their hydraulic coupling was all-in-all a better solution. When tested, the Panial Auxiliary Stage did demonstrate an improvement in efficiency that resulted in a critical altitude some 1,500-2,000 feet greater than the comparable Allison unit. They were also found to have surging problems, due to having air flow capacity too great for the engine. To properly adapt the unit, Allison would have had to develop a new two-speed drive for the Auxiliary Stage, as the Panial design incorporated internal features to accomplish the same.[39]

E-17: Allison Specification 161, dated February 11, 1942, proposed an engine that was a little less aggressive than the E-15 model in that it would have had an 8.09:1 Aux Stage Supercharger drive, with the same 7.48:1 engine stage and the Aftercooler from the E-15. Military rating would have been 1150 bhp at 24,000 feet. The carburetor was to be the PT-13E5 mounted on the inlet to the Aux Stage, and the propeller driven through the 2.23:1 gearbox, weight would have been 1,675 pounds. It received no Military Model designation.

E-18: As the single-stage V-1710-83, this engine was developed from the E-6, but with a revised accessories drive that successfully accommodated the gear loading associated with driving the supercharger at 9.60:1. The first engine was produced in August 1942 and became the first production model to successfully use this ratio.[40] Approval of the engine was based on Model Tests of the E-8, F20R and E12. When manufactured, all E-18's were rated for maximum output at 2800 rpm. Revision "B" to Allison Specification 164 changed ratings to 3000 rpm. The propeller reduction gear ratio was 2.00:1. It was also equipped with an Automatic Boost Control to reduce pilot workload. WER rating was 1410 bhp at up to 9,500 feet, with Military power of 1125 bhp available up to 15,500 feet.

E-19: This single-stage engine used the same power section as the E-18, but drove a different propeller reduction gear, the 2.23:1 unit. It became the major production version of the "E" series with 9,780 engines built. Ratings were same as for E-18.

E-20: This was a unique 1943 model based upon the E-19 and offered by Allison for use in a proposed version of the Bell Model 33, a P-63 airframe mounting a Wright 800TSB-A1 turbosupercharger in place of the usual Allison Auxiliary Stage.[41] The turbo would have been controlled by a GE Type N-1 turbo-regulator and a Packard type (Merlin-66) aftercooler was to be mounted as an intercooler. The project did not proceed, and no Military designation was assigned. It was sea-level rated for 1425 bhp to 27,000 feet using 53.5 inHgA.

E-21: As an improved V-1710-E11, the V-1710-117 saw considerable production with 2,237 produced for the Bell P-63A/C, although 730 were delivered without the remote reduction gear box. Allison performed the Model Test beginning in January 1944, using engine A-044442.[42] The Auxiliary Stage Supercharger used a slightly higher drive ratio at 7.23:1 somewhat improving its critical altitude. With ADI fluid injection it produced 1800 bhp at 24,000 feet using 76.0 inHgA and Grade 100/130 fuel, the ADI began operating at 58 inHgA. Stateside operated engines were limited to the dry takeoff rating of 55 inHgA. Mechanical improvements included use of the new 12-counterweight crankshaft and a stronger accessories case incorporating wider gears. Maximum rated rpm remained at 3000. The new crankshaft weighed 27 pounds more than the earlier unit, the result of its bore being reduced in size and the cheek shape revised.[43] The three barrel PT-13E10 carburetor, modified to incorporate a de-enrichment valve

This is the 150-hour E-22 Model Test engine, A-051920. Air from the Auxiliary Stage was delivered to the carburetor mounted on the inlet to the engine-stage supercharger, as in turbosupercharged V-1710 installations. This gave a substantial improvement in critical altitude over engines such as the E-11 and E-21 that mounted a large 3-barrel carburetor on the inlet to the Aux Stage. (Note: Allison Experimental Department Report No. A2-147, 1944.) (Allison)

for use with water injection, was installed on the inlet to the Auxiliary Stage. Coolant remained 97% Ethylene-Glycol.

E-22: The power section for the V-1710-109(E22) was the same as for the E-21, though with a completely revised connection between the discharge of the Auxiliary Stage Supercharger and the engine-stage supercharger to accommodate the carburetor, which was located as on the single-stage engines. This configuration was done to improve the efficiency of the supercharging system and resulted in increasing the critical altitude of the installation to 28,000 feet. Maximum ratings for the engine were established to take advantage of the rated 3200 rpm capable with the 12-counterweight

This is the E-22 Model Test engine during the test. In addition to having to provide all of the engine services, such as lube oil, coolant, and fuel, there is considerable instrumentation and plumbing needed to monitor engine performance and test conditions. Note the insulation blankets over the intake manifolds and valve covers to simulate a closely cowled installation. (Allison)

March 1944 found this E-23 setup for vibration testing the Mixmaster pusher prop and drive system. The engines are power sections only, suggesting that though the configuration is E-23, they are likely workhorse EX models. (Allison)

crankshaft. As a way of further improving the ability of the engine to operate at high power, the cooling system was revised to utilize greater heat removal capabilities of a pressurized mixture of 30% ethylene glycol and 70% water, the glycol acting as anti-freeze. A total of 222 E-22's were built, although only 14 were delivered configured as described above. One was delivered without the supercharger, and 207 were delivered without reduction gear boxes, as it was intended to use others already delivered. Allison began the Model Test of the engine in June 1944.[44]

There was also a V-1710-109A(E22A) model that was identical in every respect except for a larger bore in the hollow propeller shaft. The first of these was delivered in May 1945.[45] A definitive reason for the change has not been identified, but it is possible that it was done to accommodate mounting of the "pin-ball" light mounted on the nose of the RP-63's.

E-23: Allison constructed a mockup of this engine configuration as early as June 1943. The first engine was delivered to the test stand for Model Testing in March 1944, beginning a period of extensive development for the series.[46] Designated as the V-1710-103, the E-23 with its Auxiliary Stage Supercharger was adapted from the E-11 specifically for use in the first Douglas XB-42. It was built under Allison Specification 180-D. As installed in that aircraft, a pair of V-1710-103's drove the remote V-3420-B type counter-rotating propeller reduction gear box in a "pusher" configuration. Unique to the XB-42, the reduction gear ratio was 2.772:1. The engines were identified as "Unit #1" and "Unit #2", with the #1 engine on the right-hand side, as viewed when facing forward in the aircraft. From the conventional Allison perspective, the #1 power unit drive the "right hand (clockwise)" rotating propeller, though it was on the left side of the power plant apparatus as viewed from the accessories end of the engine. Unit #1 was assembled as a conventional "E" series power section, while Unit #2 was assembled to rotate as a "left-hand (counter-clockwise)" engine. If conventional Allison practice had been followed, these power sections would have been designated V-1710-E23R/L, though this was not done. Each engine and propeller was individually controllable. Both power sections used 8.10:1 supercharger gears, while Unit #1's Aux Stage was driven at 6.85:1, and Unit #2's at 6.80:1. Use of the 12-counterweight crankshaft allowed the engine to be rated at 3200 rpm, and the standard 6.65:1 compression ratio was retained. The coolant was a pressurized mixture of 30% ethylene glycol and 70% water. WER rating was 1820 bhp at sea-level, while the dry takeoff rating was 1325 bhp.

E-23A: As the V-1710-125 these engines were upgraded E-23's. While some sources report that they were the first flight engines in the second Douglas XB-42, available records do not support it. The major difference from the E-23 models was use of 7.23/7.18:1 drives for the Auxiliary Stage Superchargers. The sea-level WER rating was up to 1900 bhp for each unit, with 1675 bhp available for takeoff, both ratings being "wet.

E-23BR/L: Describes the "paired" engines in the second Douglas XB-42, which amounted to another upgrade to the E-23A and then designated by AAF as the V-1710-129R/L. The Army decided the method it had been using to designate the power sections, i.e. Unit #1 and Unit #2, was not working and reverted to a more convention R/L scheme, where R/L referred to the direction of rotation of the connected propeller. Spark retard devices were incorporated on the magnetos to aid in starting and when running at below about 1400 rpm. Allison Specification No. 194 was prepared for this engine, which was an Air Force authorized conversion of the old V-1710-103(E23) engines. The Air Force still insisted that the "oil spew" condition was in non-compliance with the contract, even though the contract applied to the V-1710-103, Allison Specification 180-D. It was not until July 1945 that an appropriate Specification 194, was developed. This model continued to have problems with "oil spew" on the left-hand engine unit when running for more than five minutes at takeoff and/or WER conditions above 75 inHgA, conditions under which each power unit could produce 1900 bhp at 3200 rpm. Power ratings were the same as for the E-23A engines, but higher step-up gear ratios of 7.23:1 and 7.18:1 for the Aux Stage Superchargers on the R/L engines respectively gave a Military rated altitude increase of 5,000 feet, to 29,000 feet.

E-23CR/L: This is the evolved form of the engines as used in the XB-42A, designated XV-1710-137R/L. Four of these engines were built from advanced E-30 power sections, the substantive differences being those associated with installation in the XB-42A and its 2.772:1 reduction gear ratio. The major change in the power sections was the reduced compression ratio of 6.00:1, as well as to incorporate numerous improvements coming from the V-1710-G series, as well as changes to the crankcase breather that allowed a dry takeoff rating of 1500 bhp at 3000 rpm from 65.5 inHgA, compared to 1425 in the E-23B.[47] Wet takeoff rating was up to 1790 bhp at 3200 rpm and 75 inHgA, although the engines were rated for 100 inHgA.[48] The drive for the Auxiliary Stage increased to 7.64:1 on the right-hand engine, and 7.59:1 on the left-hand engine. The difference being caused by the necessity of having an idler gear in the drive of the left-hand unit. Critical altitude for 1150 bhp Military power was 27,500 feet. It is believed that these engines were all produced from upgrades and modifications of previously delivered engines, i.e. no new production.

E-24: These engines were a proposal Allison made to Martin Aircraft to be used in a XB-42 type installation. They would have been a two-stage engine, and equipped with a 35 percent Aftercooler, while using the 12-counterweight crankshaft and rated for operation at 3200 rpm. Aux Stage Supercharger was to use 7.23 and 7.18:1 drive ratios as a result of the idler gear needed to keep the Auxiliary Stage Superchargers themselves interchangeable. The remote counter-rotating reduction gear would have been from the V-3420-19(B8), with 2.458:1 gears. It would have been modified to allow for independent engine operation by providing separate oil pump drives. Takeoff rating was 1500 bhp at 3200 rpm, and 1325 bhp was available at 20,000 feet when running at 3200 rpm. Allison Specification No. 181 dated August 24, 1943 described the engine. None were built.

E-25: Allison Specification 185, dated December 21, 1943, was never released. It was prepared for a conference with Bell Aircraft, but the conference finished before the specification was ready. Hazen talked Bell out of working on the installation until the F-32 engine was finalized with North American. The engine was to be an "E" version of the V-1710-119(F32R) which was the two-stage, 50 % Aftercooled engine utilizing the Speed Density carburetor and low-compression pistons. It was to have used a 2.50:1 reduction gear. It received no Military Model designation. It would have been a strong power plant, 1500 bhp for takeoff, dry, with 1900 bhp at sea-level wet, and developing Military of 1200 bhp up to 30,000 feet, all ratings at 3200 rpm on 100/130 Grade fuel.

E-26: No specifics on this model have come to light, though it was covered by Allison Specification 188. It received no Military Model designation.

E-27: In many ways this was a very unique engine, and way ahead of its time. This was the first turbo-compound, or "power feedback" engine built by any manufacturer. Based on the E-22 power section with Auxiliary Stage Supercharger, it added a General Electric CT-1 power turbine, driving the crankshaft through 5.953:1 gears. Allison built a mockup of the configuration in September 1944, followed by a series of component tests, including a complete compound engine development series run in June 1945. The first complete engine was produced in September 1945.[49] The turbine was adapted from the CH-5 turbosupercharger and driven by exhaust gasses collected by P-38 type collector manifolds. Compression ratio was reduced to 6.00:1 to accommodate very high manifold pressures up to 100 inHgA, consequently Grade 115/145 was specified. The 12-counterweight crankshaft was used allowing ratings at up to 3200 rpm. The engine was intended for the Bell XP-63H, where it would have been able to deliver a WER rating of 2980 bhp at up to 11,000 feet from 100 inHgA, using ADI injection. Considerable running of related components and complete engines was accomplished during 1944/45. The engine was able to demonstrate a 19% improvement in cruise specific fuel consumption due to the contribution of the power turbine. The aircraft never flew because the turbine was limited to inlet exhaust gas temperatures of 1725 °F, and at full power this temperature could be easily exceeded. As a consequence, Allison undertook a project of their own to design and develop a suitable power turbine with air-cooled blades, though that occurred only after the V-1710-127(E27) project was canceled.

E-28: This was a proposed adaptation of the power feedback turbine to the E-23 engine proposed for use in a XB-42 type installation with paired V-1710's. No military designation assigned.

E-29: This model is believed to have been intended for a commercial adaptation of the Douglas XB-42, the DC-8. As such, it did not require the Auxiliary Stage Supercharger and was therefore proposed as an adaptation of the E-22 derived V-1710-G3 with the new 10.25 inch diameter engine-stage supercharger impeller, driven by a two-speed gear train. Lo and Hi-speed gear ratios are believed to have been 7.48:1 and 9.60:1 respectively. Takeoff rating was to be 1620 bhp.

E-30: Based on the E-22, eight of these engines were built as military model V-1710-133. Allison began Model Testing in May 1945 using engine A-068603.[50] Compression ratio was reduced to 6.00:1 to allow the engine to be operated with 100-110 inHgA manifold pressure, on Grade 115/145 fuel with ADI. Both the 8.10:1 engine-stage and 7.64:1 Aux Stage supercharger impellers had improved rotating inlet guide vanes having a parabolic shape for increased efficiency. Five of these engine power sections were procured for the XB-42A, one used for Model Test and two sets for flight, though as installed they were designated as the V-1710-137(E23C). The other three engines were used in the Bell L-39, swept wing demonstrator conversion of a P-63, though one was operated in a P-63E where it effectively prototyped the P-63F installation and performance. Some references in the literature incorrectly identify these engines as being for the Douglas XB-42, they were not, as the E-30 was configured for use with a single rotation 2.227:1 gear box whose #60 propeller shaft was of the large bore design. Takeoff rating was 1500 bhp at 3000 rpm with 65.5 inHgA-Dry. WER rating at sea-level was 1800 bhp at 3200 rpm with 75 inHgA. Military power of 1150 bhp was available to 27,500 feet. It was

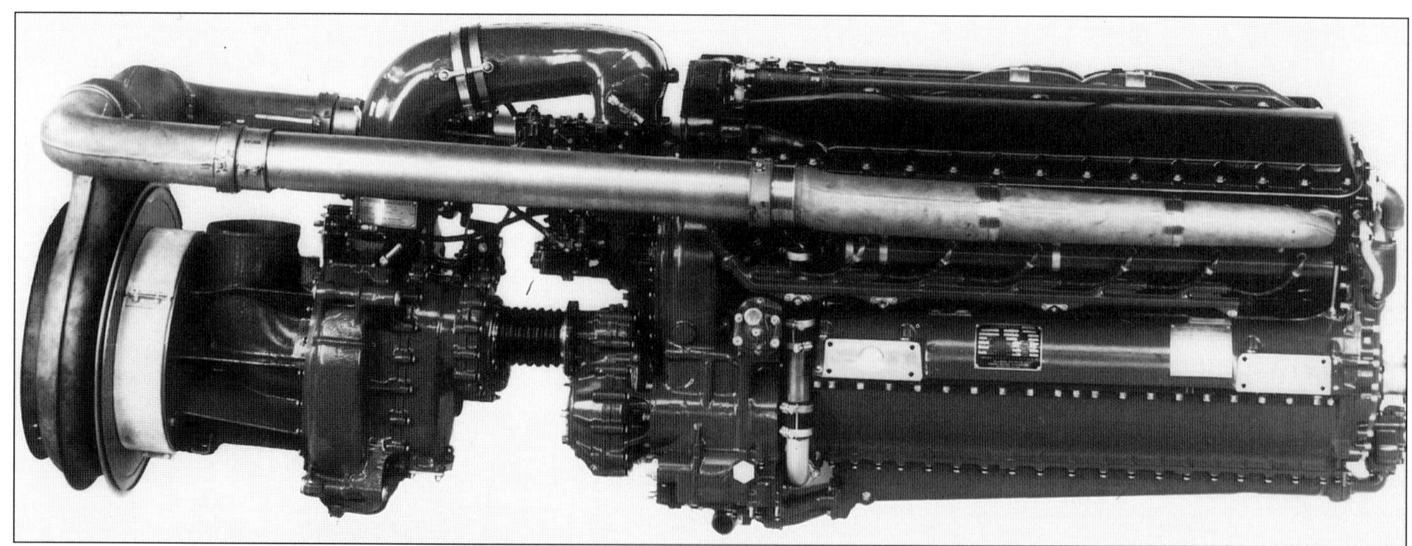

V-1710-127(E27) incorporated many internal improvements to accommodate the 3000 bhp delivered to the propeller at full power. The GE Type CT-1 power recovery turbine was supplied with 1750 °F exhaust at up to 100 inHgA. By using water injection (ADI) and running the manifolds external to the aircraft, it was hoped to keep temperatures within limits. (Allison photo Courtesy of NASM)

Coming at the end of the war, the E-30 represented the next departure point for evolution of the V-1710. In addition to being a two-stage model, it featured numerous internal improvements enabling it to be operated at manifold pressures up to 110 inHgA. (Allison)

intended to qualify the engine with a WER at 100 inHgA wet on AN-F-33R Grade 115/145 fuel, and Allison performed the WER tests in July 1946 using engine A-071878.[51] These tests may have had little or no military need, but interestingly, they were completed just prior to the 1946 National Air Races. The engines installed in the Bell P-39 *Cobra* racers incorporated many of the E-30 features. In the spring of 1946 -133 engine was converted to a V-1710-143(G6R) and test flown in the XP-51J by Allison Flight Test.[52]

E-31: This single stage engine was intended to see significant production as the V-1710-135 for the late model, altitude rated, Bell RP-63G-1, and a total of 258 of these engines were reportedly built. It is known that most (all?) were conversions of V-1710-111(F30R) and V-1710-115A(F31A) models.[53] In fact, official AAF records state that Allison Specification 182-A describes these engines, even though the specification itself says nothing about adapting for an extension shaft drive. Specification 183-A describes the V-1710-115(F31) model that was also an altitude rated engine having 9.60:1 supercharger gears. Allison lists Specification 192-A as applying to the E-31, though the summary description would suggest that it merely describes the installation of the extension shaft drive on the F-31 power section.

Four of these engines were built for the highly modified Bell P-39 "*Cobra I* and *Cobra II*" racers entered in the unlimited 1946 Thompson Race. The engines were built from V-1710-F31RA power sections with their nose cases replaced by "E" series cover plates to adapt them to the P-39. Supercharger step-up gears were 9.60:1. Running on Grade 115/145 fuel with ADI fluid mixed from 50% water, 25% Ethyl alcohol and 25% Methyl alcohol the engines would produce 2000 bhp at 3200 rpm from 82.0 inHgA at sea-level. There was no military designation.

E-32: According to Allison Specification 197 this was a V-1710-E31 configured for use as a Tank engine. In November 1947 at least one engine, A-057405, was built in this configuration. Given this serial number it is likely that the E-32 was a conversion of an earlier model.[54]

NOTES

[1] Birch Matthews letter of 3-1-90, excerpts from *Bell Model Specification (Preliminary), Model No. 4, Report No. 4Y003*, 3 June 1937, p.3.

[2] Hazen, R.M. 1945 Report, 2.

[3] Allison prepared their Specification #112 for the E-1 as of September 17, 1937.

[4] Allison Division, GMC, 1942 news release, 16.

[5] Allison Division, GMC, 1942 news release, 16.

[6] Engineering Section Memo Report, Serial No. E-57-1077-88, 1939.

[7] Engineering Section Memo Report, Serial No. E-57-1077-88, 1939.

[8] Allison Letter to Chief, Procurement Section, 11-28-1938, regarding design of a gear driven engine supercharger that can be attached as an accessory to the V-1710, thus making possible 2-stage supercharging of Allison engines.

[9] Technical Order 02-5AD-11, 1944.

[10] Per telecon with Birch Matthews, 10-15-95.

[11] *Special Features and Innovations for Bell Model 5*, Bell Report 5Y006, 1-7-1938.

[12] Grumman Letter of 2-14-38, in V-3420 NARA file.

[13] Emmick, Wm. G. 1939.

[14] *Status of Deliveries of Aircraft on 1940 Procurement Program*, October 1, 1939. RG18, File 452.1 to 452.17, Box 130 'Bulky'.

[15] Telecon w/Birch Matthews, author of Cobra, 11-2-94/dw.

[16] *Release of Information for Negotiating Foreign Sales of Allison V-1710-E4*, Specification 130-D, May 4, 1940. NARA RG18, File 452.8, Box 808.

[17] Teletype PROD-T-511 to Chief, Materiel Division, January 21, 1941. NARA RG18, File 452.8, Box 807.

[18] Teletype EXP-T-652 to Chief, Materiel Division, responding to General Arnold's questions on C-15, D-2, E-4, E-5, F-2R/L and F-3R engines, January 22, 1941. NARA RG18, File 452.8, Box 807.

[19] Allison Division, GMC, Overhaul Manual 1943, 42-43.

[20] *Netherlands PNR No. N-0138 Covering 100 Bell Airacobra Airplanes*, letter from Chief Materiel Division, June 11, 1941. NARA RG18, File 452.8, Box 807.

[21] *Proposed V-1710-E4 Engines for Installation in Bell Airacobras for the Netherland East Indies*, Allison letter to Chief of Air Corps, June 20, 1940. This letter recommends that the engines under consideration be constructed using non-standard parts, as was done on the 50 C-15A engines for the Chinese. NARA RG18, File 452.8, Box 807.

[22] Technical Order 02-5A-5, 1945.

[23] Allison Division, GMC, Overhaul Manual 1943, 42.

[24] Allison Division, GMC, Overhaul Manual 1943, 106.

[25] Allison Photolab Log, photos #05589 and 05678, Jan/Feb 1942.

[26] Allison Photolab Log, photos #05993/6008 and 06461/67, March/April 1942.

[27] Allison Photolab Log, photo #05754, Feb 1942.

[28] *Technical Data on the V-1710-47(E9) Engine*, Air Corps memo from Chief, Experimental Engineering to Chief, Materiel Division, May 22, 1941. NARA RG18, File 452.8, Box 807.

[29] Engineering Division Memo Report, Serial No. ENG-57-503-1217, 1944. The hydraulic coupling provided infinite speed control of the Auxiliary Stage by allowing more or less "slip" to occur in its drive. In this way the power required to drive the supercharger, when below the critical altitude, was reduced resulting in more economical engine operation. The coupling resembled a "Torque Converter" in a modern automobile automatic transmission except a "scoop" controlled the quantity of oil in the device, and hence the degree of slip. The first V-1710-47(E9)'s used a straight mechanical drive requiring considerable throttling of the engine when below the critical altitude.

[30] Technical Order 02-5A-5, 1945.

[31] Technical Report No. 5022, 1943, 2.

[32] Technical Report No. 5022, 1943, 18.

[33] *Progress Report on Two-Stage Allison Engine Project*, December 27, 1943. NARA RG 342, RD 3971, Box 8146.

[34] Allison Photolab Log, photos #12587/12601, May 5, 1943.

[35] *Information on Allison V-1710-E9/-E11 Engine*, AAF Memo from Chief, Experimental Engineering Section to Assistant Chief of Staff, Materiel Command, September 5, 1942. NARA RG18, File 452.8, Box 807.

[36] *150-Hour Model Test of V-1710-93 Engine*, Memo AFDSA-2E to BG B.W. Chidlaw from Chief, Power Plant & Propeller Section, Wright Field, November 27, 1943. NARA RG18, File 452.13, Box 1260.

[37] Materiel Division, Engineering Division teletype ENG-7910, September 15, 1943. NARA RG18, File 452.13, Box 1260.

[38] Allison Photolab Log, photos #16062 and 16082, Dec 22, 1943.

[39] *Contract W535-ac-25479, V-1710-65 Engine*, Allison letter No. C2E20AE from Hazen to Commanding General, AAF Materiel Command, Wright Field. May 20, 1943. NARA RG18, File 452.161-A, Box 756.

[40] Allison Photolab Log, photos #08430/43, Aug 27, 1942.

[41] Bell Drawing 33-945-001, dated 2-19-1943, via Birch Matthews.

[42] Allison Photolab Log, photos #16624/28, Jan 31, 1944.

[43] Per TO AN 02-5AD-2, p.9.

[44] Allison Photolab Log, photos #18864/79, June 10, 1944.

[45] *Daily Report of Engine Shipments*, letter from Air Technical Service Command to Allison, August 17, 1945. NARA RG342, RD3733, Box7255.

[46] Allison Photolab Log, photos #17172/89, March 15, 1944.

[47] Allison news release on Douglas XB-42 "*Mixmaster*", photo J08-11745-25652.

[48] *Approval of 100" Hg War Emergency Rating for V-1710-133 and V-1710-137 Engines*, letter from Wright Field Power Plant Laboratory to Allison, June 16, 1946. NARA RG 342, RD 3775, Box 7411.

[49] Allison Photolab Log, photos #20195, 24399 and 25135/45, Sept 1944 through Sept 1945.

[50] Allison Photolab Log, photos #23949/62, 24164/7 and 24244, May 1945.

[51] Allison Photolab Log, photos #27057/61, July 22, 1946.

[52] *Flight Test of V-1710-143 Engine in P-51J Airplane*, letter from Wright Field Power Plant Laboratory to Allison, May 31, 1946. The letter authorized converting a V-1710-133 engine to -145 for test. NARA RG342, RD3775, Box7411.

[53] Sixty-four were from F-30R production, the balance from F-31R production. *Daily Report of Engine Shipments*, letter from Air Technical Service Command to Allison, August 17, 1945. NARA RG342, RD3733, Box7255.

[54] Allison photos #30222/30234 made on November 7, 1947.

12

The V-1710-F: The "Bread & Butter" Allison

With the design of the V-3420 underway in 1938, along with the effort to incorporate its improved components into the V-1710-E series extension shaft engine, it was only natural to create a companion engine with a close coupled integral reduction gear. This model was intended to be the replacement for the "C" series that had yet to receive more than small scale pre-production orders. The new model would remove the limitations imposed by the earlier internal spur type reduction gear and weak crankcase.

In the summer of 1938 Allison reported that discussions with the various airplane companies had found a good deal of interest in the new engine on account of its improved accessory arrangement and the possibility of opposite rotation propellers on twin engine airplanes. But they wanted this engine to be built with an offset spur type reduction gear attached directly to the front of the crankcase. Allison designated this as their "F" type engine. With its higher center of thrust (8-1/4 inches above the crankshaft centerline on the 2.00:1 models) and short nose, it attracted considerable favorable comment. In a letter to the Chief of the Materiel Division at Wright Field, Allison proposed development, model and flight testing of the V-1710-F model engine to the Air Corps for a price of $45,000 for one engine, and $30,000 for the second. The new model was to weigh 1,260 pounds and offer 1200 bhp at sea-level.[1]

The Air Corps response is both interesting and important as it reflects the circumstances of the day, dominated as they were by the Great Depression. On September 16, 1938 the Division wrote to Allison that they,

"concur in the desirability of advancing the development of the V-1710-F engine in preference to the V-1710-C or -D model engines because of the numerous improvements incorporated in and greater potential capabilities of the newer model. However, as no funds are available, it is necessary to defer action on this project. Data furnished will be placed on active file for possible application in the near future."[2]

In keeping with the commitment of General Motors funding the development of the V-1710, preparatory work on the new engine continued.

The V-1710-E/F series engines were models of utilitarian design. The cylinder blocks, induction manifolds and crankshafts were interchangeable with each other, and with those used in the V-3420. An important feature to airplane designers was the availability of both "left" and "right" handed propeller rotation. This was accomplished by simply reversing the positions of several components, including the crankshaft, and installing an idler gear in the accessory drive train to cause the supercharger, camshafts, and magneto to turn in the same direction on all engines. Either the extension shaft or the integral reduction gear nose case could be installed on the front of the universal V-1710-E/F crankcase. At the rear of the en-

A typical early model V-1710-F, configured as an F-2 for use in the turbosupercharged Lockheed P-38, but built for the Allison Service School. (Allison)

gine, the accessories sections could likewise be interchanged to provide a range of supercharger drive ratios, carburetion, and even a power takeoff to drive a second or Auxiliary Stage supercharger. Configured in this manner manufacturing was comparatively simple and changing from one model to another was practical in the extreme. During the war it became Allison's practice to manufacture "E" series engines for two weeks, followed by building "F" series engines for two weeks.

With the evolution of the many V-1710-E/F models Allison practiced what today is called "downward compatibility." Continued development not only provided improved primary components for the later and higher rated models, but improvements were retrofitable into earlier models. The consequence was that engines could be improved during overhaul, while the supply of spare parts was simplified in that only the latest designs needed to be stocked. In many respects the design and production of the Allison V-1710 was the culmination of the state of the manufacturing art as developed and practiced in the United States during the first half of the 20th Century, and prior to the advent of the digital computer.

Table 12-1
Comparison of C-15 and F-2 Models

	V-1710-C15	V-1710-F2R/L
Horsepower @ TO	1040	1150
Weight, pounds	1325	1305
Cross-section, sq.ft.	5.49	5.19

The V-1710-F was introduced with a takeoff rating of 1150 bhp, but was ultimately able to deliver over 1750 bhp without further major redesign. When the V-1710-C was introduced it was applauded for being lightweight, compact, streamlined and for having a low frontal cross-sectional area. The "F" had all of these attributes as well, and even managed to improve upon the specifics. Dispensing with the streamlined reduction gear allowed shortening the engine, while not adversely effecting the suitably for streamlined aircraft installations, primarily as a result of the higher propeller thrust line. Weight was reduced as well as the engine cross section. Most important was the ability of the new external spur type reduction gear to absorb the ever increasing power output, up from the 1150 bhp maximum for the "C", to over 2300 bhp in the later V-1710-G without further major modifications.

The first aircraft to fly with the "F" was the first Lockheed YP-38. It used the "handed" V-1710-F2R/L, Army Air Corps models V-1710-27/29. The Air Corps had bought five V-1710-25(F1)'s, but these quickly evolved and were delivered as F-3R's. The F-2's were coming into service concurrent with the V-1710-33(C15) for the Curtiss P-40's. Similarities between the engines were such that they had the same induction system and backfire problems. A review of the Wright Field correspondence of the time, summer/fall 1940 and concurrent with the Battle of Britain, makes clear the sense of emergency and the necessity to resolve the introductory problems and get the P-38, P-39, P-40 and P-51 into volume production and immediate service. Fortunately the problems were largely resolved by the time of the attack on Pearl Harbor, and the V-1710 in its various forms was able to enter combat as a comparatively reliable and well developed powerplant.

Introduction of the E/F engines into production was not easy. Until production began in the new Plant 3, February 1940, all engines were produced in Plant 2, a facility built and better suited for production of prototype and experimental engines. Compounding the manufacturing challenges were the problems of qualifying the higher rated E/F engines for production and flight. This effort was being done in parallel with the shift of the C-15 manufacture to Plant 3, and the continuing Model Testing of the C-15. As of March 1940 the schedule for Model Testing was:[3]

- V-1710-33(C15) for P-40, final test to begin May 1, 1940.
- V-1710-27/29(F2R/L) for P-38, scheduled for late April 1940.
- V-1710-35(E4) for P-39, was to have started in January, but rescheduled for April/May 1940.
- V-1710-39(F3R) for NA-73/P-40D, Officially begun mid-February 1941.[4]

Much of the delay was due to the greatly increased demands of production of high quality non-ferrous castings and in proving tools and fixtures for finishing the large aluminum and magnesium parts. This problem would not be finally resolved until after production of the C-15 model was completed, July 1941. With this milestone the way was clear for all resources to be focused on production of E/F models. Another aspect of the problem was that the early E/F engines were only marginally completing the rigorous Model Tests. The engines were usually able to run the required 150-hours and deliver the required performance, but on final teardown there were problems with Prestone seepage, cracks in the crankcases, defects in the main bearings, and other minor faults.[5] The situation got to the point that in the summer of 1940 the Chief of the Materiel Division insisted that Allison redesign the offending parts before again submitting the engines for Model Testing.[6]

Production of the early "F" engines was tedious. Three F-3R's were built in October 1940, along with two F-2's. Then in November, only the two remaining F-3R's on contract ac-13841 were built. These were the engines for the first YP-38, NA-73X, XP-46 and P-40D, as well as one F-3R for the Model Test.[7] Model Testing itself was considerably delayed due to insufficient numbers of test stands and the need to incorporate various improvements. Getting started was tough.

The "F" did not go into series production until March 1941, though development and pre-production engines for the YP-38's was underway more than a year earlier.[8] The "F" became the major production series of the V-1710 program, with nearly 48,000 built in 37 basic models. During the first two and a half years of the war Allison averaged a production model change in the V-1710-E/F every 45 days. On the "F" engine the aggregate of these changes resulted in a frontal area decrease of 10 percent, a shortening of the engine by 8 to 10 inches, and reducing weight by 50 pounds. All of this was done while production rates and horsepower were substantially increased and unit costs reduced. The "F" engine got to a takeoff rating of 1425 bhp, with 1625 bhp available for WER, before the weight was increased above that of the earlier "C" model.

The evolution of the different models achieved improved performance largely because of improvements in the engine-stage supercharger and a program to incrementally strengthen the various components of the engine to support the ever higher power ratings. All "E/F" models used the 9-1/2 inch diameter, 15 blade impeller, but there was a series of changes that provided improved efficiency and provided more air at ever higher manifold pressures into the engine. Where the F-2 supercharger impeller used a fairly straight vane, relying on the supports in the inlet elbow for proper direction and swirl of the mixture into the supercharger, the F-5 improved upon this by adding curved rotating steel inducer guide vanes at the inlet. This improved the already high efficiency of the supercharger and added to the critical altitude of those models incorporating the new unit. Introduction of later refinements in the shape of the inducer vanes further improved its performance.

Supercharger gear ratios were several and varied, selection being based upon the mission of the aircraft the engine was to power. Early sea-level

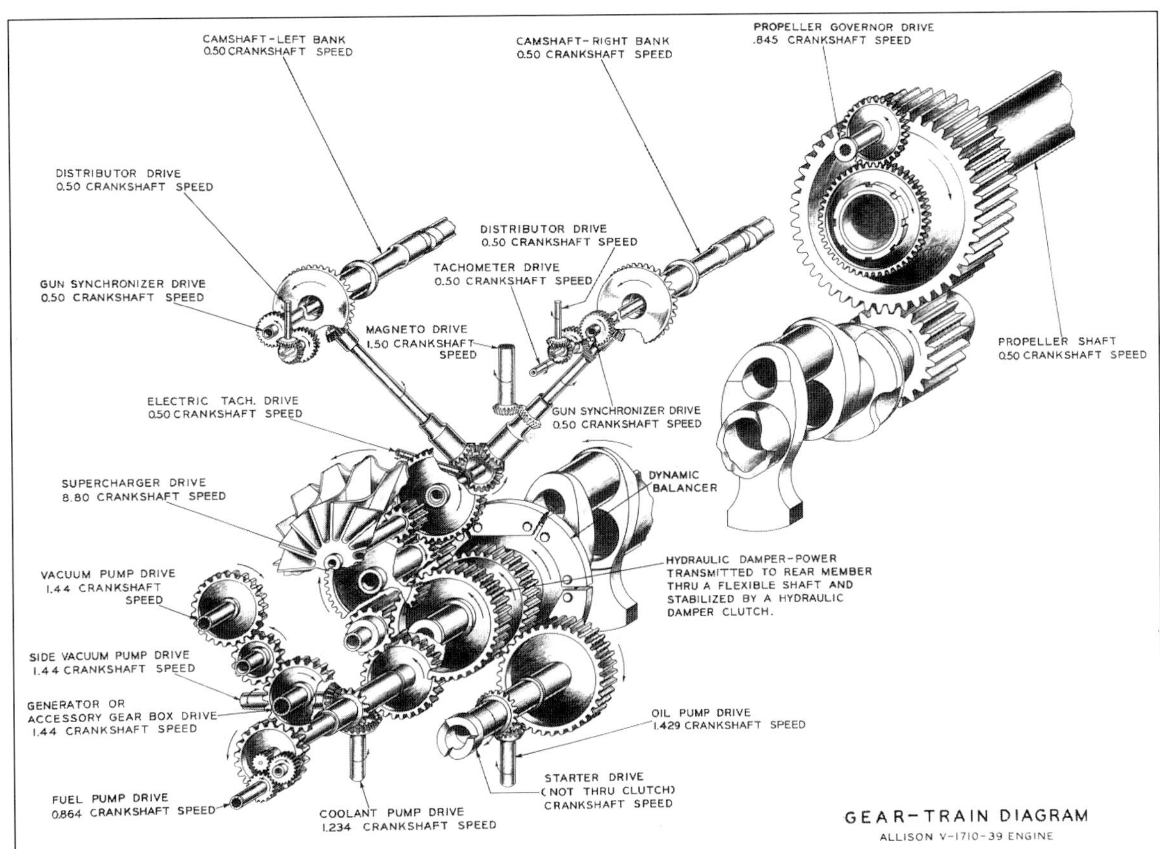

For the E/F engines Allison developed a very utilitarian and rugged gear train as shown in this drawing for the V-1710-39. (Allison via Hubbard)

rated engines used low ratios to maximize efficiency and capability of the turbosupercharger. The later P-38 engines stepped up to higher ratios, so that with the turbos still delivering sea-level pressure to the engine-stage supercharger higher manifold pressures were developed, along with a corresponding increase in power output at wide open throttle.

In the fall of 1940, the V-1710-F3R and two-stage F-7R models were the primary models being developed by Allison, though the Chief Designer and his Project Engineers had been directed to initiate a comprehensive review of the basic "E" and "F" engine in preparation for obtaining a sea-level rating of 1500 bhp.[9] This was done, resulting in a design basis for the engine at a rating of 1500/3000/SL/61.5 inHgA. The corresponding Altitude Rating, with the assumed 8.80:1 supercharger gears, was still 1150/3000/12,000/42.0 inHgA, limited by supercharger parameters, not engine strength.[10] Significantly, Allison determined that it was desirable to go to a 12 counter-weight crankshaft so that further increases in rpm and power could be achieved without increasing bearing loads. While the new crankshaft would be heavier, the resulting reduction in bearing loads allowed for a lighter overall engine than would have otherwise been required.

Within the first six months of the war Allison responded to needs from the War Theaters by demonstrating and authorizing use of War Emergency Ratings for the E/F engines. Though the official Army Air Forces channels were slow to approve their use, these ratings were soon widely used within the combat groups.

It is not clear why Allison never put a two-speed drive for the engine-stage supercharger into production for the altitude rated engines during the war. The benefits of the feature were well known and available in almost every other engine manufacturers' product line. The reason may have been that engines using the 9-1/2 inch supercharger were operating with up to 9.60:1 drive gears that at 3000 rpm which gave an impeller tip speed of about 1,200 feet per second. This was about the limit for state-of-the-art superchargers in the 1940s.[11] With these gears, the single-stage engine could deliver its Military rating up to 12,000 feet. It would have taken a considerably larger impeller and induction system to improve this very much. That was done in the later "G" series with its 10-1/4 inch supercharger, though the step-up ratio was reduced to 7.48:1 resulting in a 1,070 feet per second tip speed. The real benefit of the later redesign was to open up the entire induction system leading into the enlarged impeller. This reduced the losses in the inlet and carburetor and gave the desired improvement. At the start of the war the demand was for Allison to build the tried and true. Fundamental improvements, no matter how desirable, had to wait. The major benefit of a two-speed supercharger would have been to make more power available to the propeller at lower altitudes, though the amount would probably have been 100 bhp or less. Again, the demand must not have been there.

Many have asked why Allison did not build a two-stage "Merlin" type supercharged engine. The answer is, they did. Allison built thousands of mechanically driven two-stage supercharged engines, those with the Auxiliary Stage Supercharger, which by comparison with the two-stage Merlin were about 18 inches longer, though of similar weight. This development was the basis of the V-1710-45(F7R) and -47(E9) two-stage engines. The -45 was proposed for use in the Curtiss XP-60, but that aircraft was canceled at the start of the war and the engine was not adopted by any of the other aircraft manufacturers. The lead was then shifted to the -47 which Bell Aircraft adopted for their XP-39E and XP-63 projects. The two-stage Allison was available at the same time, with the same ratings and in a similar stage of development and reliability as the vaulted two-stage Merlin.

Table 12-2
V-1710-"F" Production Series Ratings

Engine Model	S/C Ratio	Carb	Takeoff, bhp/MAP	Normal, bhp/Alt/MAP	Military, bhp/Alt/MAP	WER, bhp/Alt/MAP	Comments
V-1710- 27/29(F2R/L)	6.44:1	PD-12K2	1150/39.4	1000/25,000/36.5	1150/25,000/39.4	None	YP-38,P-38D/E
V-1710- 39(F3R)	8.80:1	PD-12K2	1150/45.5	1000/11,000/38.5	1150/11,700/44.6	1470/SeaLev/56	P-51, P-51A
V-1710- 73(F4R)	8.80:1	PD-12K2	1325/51.0	1000/10,800/37.2	1150/12,000/42.0	1550/SeaLev/60	P-40K
V-1710- 49/53(F5R/L)	7.48:1	PD-12G1	1325/47.0	1000/25,000/37.8	1325/25,000/47.0	None	P-38F
V-1710- 51/55(F10R/L)	7.48:1	PD-12K2	1325/47.0	1100/25,000/41.0	1325/25,000/47.0	None	P-38F/G/H
V-1710- 75/77(F15R/L)	8.10:1	PD-12K7	1425/54.0	1100/27,000/45.0	1425/27,000/54.0	1600/27,000/60	P-38K,XP-60A/B
V-1710- 89/91(F17R/L)	8.10:1	PD-12K7	1425/54.0	1100/34,000/44.0	1425/24,900/54.0	1600/10,000/60	P-38H/J
V-1710- 81(F20R)	9.60:1	PD-12K6	1200/51.5	1000/14,400/38.3	1125/15,500/44.5	1360/SeaLev/57	P-51A,P-40M/N/R
V-1710- 87(F21R)	7.48:1	PD-12K7	1325/47.0	1100/ 3,000/41.0	1325/ 2,500/44.5	1500/ 5,400/52	A-36A
V-1710- 99(F26R)	9.60:1	PD-12K6	1200/51.5	955/15,700/37.0	1125/15,500/44.5	1360/SeaLev/57	P-40N
V-1710-111/113(F30R/L)	8.10:1	PD-12K8	1500/54.0	1100/30,000/44.0	1425/29,000/54.0	1600/28,700/60	P-38L/M
V-1710-115(F31R)	9.60:1	PD-12K8	1200/51.5	955/15,700/37.0	1125/15,500/44.5	1360/SeaLev/57	P-40N-40

Ratings given are on Grade 100/130 fuel, engine running at 2600 rpm "Normal", 3000 rpm for Military and WER. MAP, inHgA.

Table 12-3
"F" Model Engines Shipped

Model	Number	Aircraft	Comments
V-1710-27(F2R)	540	YP-38/P-38/A/D	Sea level rated.
V-1710-29(F2L)	540	YP-38/P-38/A/D	Sea level rated.
V-1710-39(F3R)	4,694	P-40D/E, XP-46, P-51	Altitude rated, 1,412 shipped on British contracts
V-1710-73(F4R)	1,883	P-38 #40-744, P-40K	Altitude rated.
V-1710-79(F4L)	1	P-38 #AC40-744	Believe used when turbos removed for cockpit in boom.
V-1710-49(F5R)	1,030	P-38E/F	Includes 163 shipped on British contracts
V-1710-53(F5L)	1,030	P-38E/F	Includes 163 shipped on British contracts
V-1710-45(F7R)	1	P-40	Improved model P-40, did not proceed
V-1710-51(F10R)	1,752	P-38G, P-60	Sea level rated.
V-1710-55(F10L)	1,753	P-38G	Sea level rated.
V-1710-57(F11R)	1	P-40	Used 2-speed Birmann supercharger, test only
V-1710-61(F14R)	26	P-40J	Early attempt to use 9.6:1 supercharger gears
V-1710-75(F15R)	10	P-38K, P-60	Sea level rated.
V-1710-77(F15L)	4	P-38K	Sea level rated.
V-1710-89(F17R)	5,429	P-38H/J, XB-38	Includes 2,118 shipped w/o carburetors and 8 for XB-38
V-1710-91(F17L)	5,421	P-38H/J	Includes 2,118 shipped without carburetors
V-1710-81(F20R)	4,817	P-40M, P-51A	Altitude rated.
V-1710-87(F21R)	835	A-36A	Altitude rated.
V-1710-95(F23R)	6	XP-55	Pusher installation
V-1710-F25R	(2)	XP-40Q	2 engines built, not in totals, No Military model assigned.
V-1710-99(F26R)	4,200	P-40N-20/35	Altitude rated.
V-1710-101(F27R)	2	XP-40Q-2	Two-stage
V-1710-121(F28R)	4	XP-40Q-2	Two-stage
V-1710-105(F29R)	2	P-38	Development only, not flown
V-1710-107(F29L)	2	P-38	Development only, not flown
V-1710-111(F30R)	6,344	P-38L	Includes 3,706 shipped without carburetors
V-1710-113(F30L)	6,348	P-38L	Includes 3,705 shipped without carburetors
V-1710-115(F31R)	560	P-40N-40	Altitude rated.
V-1710-115A(F31RA)	420	P-40N-40	Altitude rated.
V-1710-119(F32R)	5	XP-51J	Two-stage with aftercooler
Total	47,660		

In March 1944 Wright Field tested a V-1710-91 utilizing a intercooled and aftercooled two-stage, two-speed supercharger borrowed from a Packard built V-1650-3 two-stage Merlin. While its performance was as good as that of the Merlin, mechanically it was not a happy arrangement. The supercharger was not tightly connected to the engine, though it was directly driven from the starter drive. Three entire supercharger assemblies failed mechanically after only a few hours of testing. It was believed that the Merlin unit relied upon engine heat to maintain critical internal clearances and that the remotely mounted test installation did not meet this need. Still some interesting performance data was obtained. Where the -91 was normally rated for 1425 bhp at 3000 rpm with 54.0 inHgA at sea-level, the two-stage Merlin supercharged test engine delivered 1250 bhp at 3000 rpm with 60.0 inHgA up to an altitude of 25,700 feet. Performance very similar to that of the V-1650-3.[12] Significantly, the -91 was not normally equipped with an aftercooler. With its normal single-stage supercharger, it could not deliver more than the rated 54.0 inHgA without ADI. With turbosupercharging and ADI in place of the aftercooler, the V-1710-89/91 was capable of a WER of 2300 bhp at 3000 rpm using 90 inHgA. Using the intercoolers and turbosupercharger as in the P-38J airplane the dry WER was 1840 bhp at 3000 rpm from 75 inHgA, available up to the critical altitude of the turbo.[13]

In 1943 Allison undertook the development of an aftercooled two-stage engine, the V-1710-F32. Although the general arrangement and concept was similar to that of the Merlin, the engine utilized an Allison Auxiliary Stage Supercharger discharging directly into the conventional engine-stage supercharger. The aftercooler occupied the space normally used by the down-draught carburetor, necessitating either the use of a PT-13 carburetor on the inlet to the Auxiliary Stage or the new compact Speed Density unit mounted in the connecting duct. The choice was the new Bendix "speed density" type where engine fuel requirements were determined by a form of analog computer. No carburetor venturies were used. The benefit of this approach was reduced pressure losses in the induction system. The engine compression ratio was reduced to 6.00:1, the same as in the Merlin. This allowed the use of higher boost pressures sufficient to give the engine a War Emergency rating of 1720 bhp at 3200 rpm to an altitude of 20,700 feet, or 1900 bhp at 3200 rpm at sea-level, a benefit of having the hydraulic clutch in the drive to the Auxiliary Stage Supercharger. All-in-all, this engine would have been the match of the two-stage Merlin. It was flown in the XP-51J and XP-82A aircraft, both programs that were canceled near the end of the war.

Table 12-3: V-1710-F Engine Models

F-1R: This 1938 engine was developed to be altitude rated at 1060 bhp with a PT-13B2 carburetor and using 8.80:1 engine-stage supercharger gears, 6.65:1 compression ratio and rated on 100 Octane fuel. Designated V-1710-25 by the Air Corps, only development engines were procured and they do not show up in production listings of delivered engines. Development was concurrent with the V-1710-E1/E2, and the models differed in that the "F" has the close coupled, 2.00:1 ratio external spur gear type reduction gearbox and nose case and #50 propeller shaft. This reduction gear being interchangeable with the front cover on the "E" series of extension shaft engines. Coolant was 97% Ethylene Glycol and engine weight was 1,320 pounds. The military rating at 1150 bhp was available up to 12,000 feet. At one time, four of these model engines were specified to power the Boeing Model 333 and 333A, the aircraft that became the XB-29. Your author believes that the initial contract for these engines was converted to purchase the first five F-3R engines. The two models differed only in internal details, with the F-3R benefiting from strengthened components.

The initial "F" engine was the Altitude Rated V-1710-25(F1R) that used the PT-13B carburetor from the C-13. Note continuation of the sparkplug cooling shrouds as used on the C-10. Otherwise the engine appears much like the thousands of F Model's that followed. It shared power sections with the contemporary E-5, though featuring the compact integral 2:1 external spur type reduction gear. An initial order for five of these engines was revised to deliver them with PD-12 carburetors and internal improvements as the F-3R. (Allison)

F-2R/L: This was the first series production model of the V-1710-F. It was specifically built for the Lockheed YP-38 program in both RH and LH models as the V-1710-27/29. They went into series production in January 1941 for use by the early production series of the P-38. These were sea-level rated engines having 6.44:1 engine-stage supercharger gears and were intended for use with a turbosupercharger. The engine-stage impellers did not have rotating inlet guide vanes. The first 30 engines were built with accessories section gears having been cut with a 20 degree Pressure Angle. Subsequent engines switched to 25 degree gears in an attempt to improve their performance in the Left turning engine. The ignition systems were not supercharged and the engines used the PD-12G1 carburetor. A total of 540 of each of these models, R/L, were built. Some references relating to the first YP-38's describe the installed engines as V-1710-F3. This is believed to have been an interim reference to an F-2 having the Materiel Division designed intake manifolds with individual port backfire screens. This configuration was subsequently incorporated into the F-2 specification as a revision and the F-3 designation assigned to the first developed altitude rated engine.[14] The V-1710-27/29 completed the 150-hour Model Test on October 31, 1940, though several parts were only conditionally accepted pending redesign and subsequent qualification in the upcoming Model Tests of the V-1710-35, -37 or -39. These requirements were satisfied when the V-1710-35 completed its Model Test on January 9, 1941.[15]

Table 12-4 Takeoff Rating Comparison with V-1710-33		
	V-1710-33(C15)	V-1710-27/29 (F2R/L)
BHP	1040	1150
RPM	3000	3000
MAP @ 70 °F Inlet Air, inHgA	41.9	39.7
Mixture Temp @ 70 oF*	196	112
Friction HP**	290	240
Indicated HP	1330	1390
IMEP, psi	205	214.5

* Mixture temperature is that of the combined air with evaporated fuel mixture as it leaves the intake manifold and passes through the intake valves and into the combustion chambers. In this case the temperature is based on having 70 oF air entering the carburetor. Data from Allison Test Report A2-7, NASM file D52.41/64, p.37.
** Includes power to drive superchargers, 8.77:1 in V-1710-33, 6.44:1 in V-1710-27/29. Higher gear ratio requires more power for same rpm and airflow.

The F-2 was the first production "F", and the first to be built in both left and right turning models. First used in the Lockheed YP-38 and similar models, the engine did not require gun synchronizers and was therefore fitted with the same end-mounted distributors as the D and E model engines used. (Allison)

When the unrestricted engine operating instructions were issued in February 1941, the ratings given in Table 12-5 were established.[16]

Table 12-5 Unrestricted Ratings for V-1710-F2R/L			
	bhp/rpm/ MAP, inHgA	Fuel-Rich, gph	Fuel-Lean, gph
Takeoff and Military	1150/3000/40.3	NA	NA
Climb and High Speed	1000/2600/37.2	90	NA
Maximum Cruising	870/2280/31.4	59	NA
Desired Cruising	800/2280/29.0	50.5	46.5
Desired Cruising	750/2280/27.7	47.0	42.0

In August 1941, the Maintenance Command announced a decision to change all intake valves in V-1710's to the new sodium cooled type. The changeover in production occurred about the first of October 1941, while the valves for retrofitting into service engines did not become available until the end of November. Due to use of turbosuperchargers, it was decided to fit the new valves into the P-38 airplanes first, and some 68 airplanes had been so equipped by Pearl Harbor, December 7, 1941.[17]

F-3R: The Army Air Corps was anxious to get an altitude rated version of the new Allison "F" into the hands of the aircraft manufacturers, and into their airframes. Five F-3R's were purchased on contract W535-ac-13841 (dated January 27, 1940) to expedite Model Testing and to make the engines available to the XP-46 and XP-47 projects. Your author believes these were originally ordered as the similar V-1710-F1R model. No records have been found regarding the delivery of the V-1710-25(F1R)'s, yet it is known that one or more did exist. Other examples exist of where the government changed its purchase orders in this respect. Recipients of the five V-1710-39(F3R) engines were the North American Aviation NA-73X, Curtiss P-40D, XP-46 and XP-46A. The fifth engine was for the Model Test, which was not performed until long after the F-3R engine was flying. The Army converted its order for 776 V-1710-33's to V-1710-39's after 300 of the -33's had been delivered. The final Model Test engine actually came from that contract, W535-ac-12553. Because of required improvements to crankcases and connecting rods as a result of other Model Tests, the F-3R 150-hour test did not get underway until February 17, 1941. The test was completed with only minor problems in the cylinder head jackets and out of tolerance crankcase stud bolts.[18] The F-3R single-stage supercharger was driven by the same 8.80:1 gears as the F-1R and resulted in a Military rated ceiling of 11,700 feet at 3000 rpm. The 9-1/2 inch diameter impeller was provided with rotating steel inlet guide vanes set at an inlet angle of 42 degrees to improve efficiency over that of the supercharger in the F-2R/L. The early engines received the improved aluminum intake manifolds and backfire screen arrangements as were being used on the C-15 and F-2R/L engines. Fighter installations of the F-3R usually required that gun synchronizers be provided and these were installed in combination with specially configured distributors. The distributors and magnetos were not supercharged. The PD-12K2 carburetor was used, which had its throttle bores increased by 1/8 inch over those in the F-2R/L carburetors. Takeoff and Military ratings of 1150 bhp were eclipsed in 1942 when WER ratings were established at 1490 bhp when operating on 100 Octane fuel. A total of 4,694 F-3R engines were built.

F-4R/L: Development of this engine began in 1938 under Allison Specification 124. Given the pressure of developing other models, it benefited from numerous improvements warranting issue of a new Specification 159 in 1942. As the V-1710-73 this was a major production model beginning in 1941, with a total of 1,883 delivered for use in the Curtiss P-40K. The 8.80:1 supercharger gears of the F-3R were retained, but by rating the engine on Grade 100/130 fuel, the takeoff power was increased to 1325 bhp and the WER rating set at 1580 bhp. This could be maintained up to 2,500 feet. This was the first model to include the Automatic MAP regulator, as well as the first "F" to use 30% Ethylene Glycol and 70% water for coolant. There were no gun synchronizers.

A single XV-1710-79(F4L) was also built in 1942 for experimental purposes, then converted back to F-4R configuration at the end of the project.[19] That single engine was likely the one used on P-38 AC40-744 when its turbos were removed for installation of a second cockpit, located in the left boom.[20]

F-5R/L: This engine was the result of the AAF demand for a 1325 bhp V-1710 that was intended to improve the performance of the P-38 compared to that of the F-2R/L models. The engine-stage supercharger step-up ratio was raised from 6.44 to 7.48:1 and the impeller was fitted with the rotating steel inlet guide vanes as used on the F-3R. These improvements, along with supercharging the ignition system, resulted in the first P-38 engine to exceed 1 horsepower per pound at takeoff. It weighed 1,325 pounds. The PD-12G1 carburetor was retained, but the coolant was changed to the improved pressurized system using 30% Ethylene Glycol and 70% water. A smooth and satisfactory 150-hour Model Test was completed on the F-5R at 1225 bhp and 3000 rpm on May 13, 1941. This cleared the way to immediately begin another test, this time at the desired 1325 bhp.[21] No WER ratings were ever officially established, though 8th AF Fighter Command did authorize their pilots to use 1250 bhp from August 15, 1942. Col. Ben Kelsey was recommending 1325 bhp for up to five minutes. The Command was considering authorizing use of as much as 1450 bhp from these engines. Allison rated these V-1710-49/53 engines on Grade 100/125 fuel.

F-6R: This model was proposed by Allison for use by Curtiss in response to the 1940 Air Corps R-40C competition for a lightweight fighter. The Curtiss Design CP 40-1 proposal resembled the earlier XP-46. This was to have been the first "F" to use 9.60:1 supercharger gears. None of these engines are known to have been built. Particularly unique in the final design was the incorporation of a counter-rotating reduction gear of 1.80:1 having #40 (CCW) and #60 (CW) propeller splines, though the submittal to R-40C used a single 10 foot diameter 3-bladed propeller. The design included gun synchronizers, and with the 100 pound reduction gear would have had a total weight of just under 1,410 pounds. The engine was 20-1/8 inches longer than standard because of the reduction gear. Another observation, visible on the drawing of the CP 40-1, is that as of the date of the proposal Allison had yet to introduce the "y" fitting in the coolant supply to the cylinder blocks, a feature first incorporated on the E-4 engine. Military power of 1125 bhp would have been available to 15,000 feet.

F-7R: As the V-1710-45, this 1940 project was the first V-1710 to use the new Auxiliary Stage Supercharger and thereby become a mechanically driven two-stage V-1710. Contract W535-ac-16146, dated December 2, 1940, was issued for one experimental engine. Change No. 1 was issued in July 1942 to cover redesigning to provide a larger Auxiliary Stage Supercharger (12-3/16 inch diameter impeller replaced the earlier 9-1/2 inch unit) driven by a hydraulic coupling instead of the original friction clutch. The delivery date was changed from September 1941 to February 1943. Obviously the effort was considerably behind schedule at this point. A second contract, W535-ac-22957 was issued in December 1941 to purchase an experimental engine with aftercooler for the Classified Project No. MX-69, which was the airplane that became the Curtiss XP-60. This version with the two stage Allison never achieved flight status.[22] Allison configured their EX-39 workhorse engine as an F-7R and was running it in early 1943.[23] Other production and performance information has yet to come to light.

The engine was based on the F-3R, but with the engine-stage supercharger drive ratio reduced to 7.48:1, though in 1942 this was increased to 8.10:1. The carburetor was the large PT-13E9 installed on the inlet to the Auxiliary Stage, itself being driven by 6.85:1 gears and equipped with an automatic boost control. In anticipation of high power ratings, the cooling system was the new pressurized 30% Ethylene Glycol and 70% water. The engine operated on Grade 100/125 fuel. Apparently only two of these experimental engines were built, though they seem to have had a long life. In its initial form with the 8.10:1 gears, it also had a liquid-cooled aftercooler. Total weight was established as 1,545 pounds. Military rating was for 1150 bhp at 21,000 feet. Why this engine was not further developed has not been determined, though during most of 1943, it was under active testing by the NACA Cleveland Engine Laboratory where they worked on methods of improving its power output characteristics.[24]

F-8R: This single-stage engine was tested in the XP-40K, AC42-10219. It incorporated an improved crankshaft, bearing lubrication, and strengthened connecting rods. It was further unique in that a revised valve timing was tried. One thing that was never changed during the production life of the V-1710-E/F was the valve and ignition timing. The engine also used 9.60:1 supercharger gears and pressurized 30% Ethylene Glycol and 70% water coolant. Ratings were quite conservative at only 1150 bhp for takeoff and military as a result of using 100 Octane fuel.

F-9R: Allison originally proposed this engine to the U.S. Navy in 1940, and then again in a revised specification dated March 24, 1941. It incorporated a fighter type 2.00:1 reduction gear, and an Auxiliary Stage Supercharger using a 9-1/2 inch diameter impeller as on the early two-stage F-7R and E-9. In addition, an intercooler was provided between the supercharger stages. An interesting feature was the ability to bypass both the Auxiliary Stage and the intercooler, allowing dual ratings for the engine. With the bypass, Takeoff rating was 1325 bhp. With the second stage operating, 1125 bhp was available to 18,000 feet. Total weight was 1,510 pounds. No military designation resulted.

F-10R/L: Thousands of these V-1710-51/55 engines were built for the Lockheed P-38G/F and related models. For all intents they were the same as the F-5R/L models that they replaced on the Lockheed production lines except that they used the improved PD-12K3 carburetor, a PD-12K2 modified to accommodate being supplied with pressurized air coming from the turbo that previously had shown a tendency to lean the mixture. Allison recommended against giving these engines a WER rating due to the inadequate intercooler capacity of the early P-38 models in which they were installed.

The F-10R Model Test engine was delivered in July 1941 and its testing was largely completed that Fall. Engine AEC #11196 was prepared for the F-10L Model Test as of December 5, 1941, with its testing continuing through January 1942.[25]

F-11R: This engine was intended for the Curtiss P-53 airplane, and represented an effort to dramatically improve the efficiency of the Allison designed engine-stage supercharger then being used on all V-1710's. The V-1710-57(F11R) was the first V-1710 to use a two-speed supercharger, having the ability to shift between 6.44:1 and 8.80:1 drives. Contract W535-ac-22143 was issued for one of these engines in October 1941. The engine was to be delivered the following March at a price of $25,000. The Model Test engine, AEC #13132, was actually completed on April 2, 1942, and was unusual for a V-1710 as it used a Holly Carburetor mounted on a new inlet elbow to the supercharger.[26] Particularly unique was the supercharger impeller, which was an entirely new "mixed-flow" unit known as the "Birmann" supercharger, so named for its designer. The impeller was 10-1/4 inches in diameter and involved aspects of both centrifugal and axial flow in its design. Birmann and his firm, Turbo Engineering Company (TEC), of New Jersey was at the time designing and building turbosuperchargers with similar features for the U.S. Navy. The engine itself was otherwise similar to the F-10. References are not consistent regarding the high gear ratio. Some give it as 9.60:1, suggesting that as an experimental effort, a variety of ratios were tried. Altitude performance on Grade 100/125 fuel was improved, with Military rating of 1150 bhp being available at 16,000 feet in high blower. Had the Allison supercharger impeller been similarly increased to 10.25 inches, as was done for some models in the later "G" series, it too would have been able to produce similar performance. Evidently the improvements shown were marginal and did not justify introduction into production at the time.

F-12: Proposed sea-level rated model with single-stage, single-speed supercharger. No military designation was assigned and none were built.

F-13R/L: No military designations were issued to these models, which were considered in October 1941 as revised F-10R/L models to be built with 2.36:1 reduction gears in place of the otherwise standard 2.00:1 units. While the carburetors were expected to be the PD-12K2 unit, it is likely the PD-12K3 that had been configured for use with a turbosupercharger would have actually been used. Fuel and ratings would have continued as for the F-10R/L. The reason for changing the reduction gear ratio would likely have been to accommodate the new Hamilton-Standard high activity paddle blade hydromatic propeller, an arrangement that was actually flight tested on the F-15R/L engines in the XP-38K. None of the F-13 engines were built for the Air Force although the reduction gear was at Allison in March 1942.[27]

F-14R: This was another altitude rated development of the F-3R, designated as the V-1710-61(F14R). It was intended to power the Curtiss P-40J, and sufficient engines for a service test batch of 25 aircraft were to be produced. The Model Test engine was prepared for testing as of December 5, 1941, just prior to Pearl Harbor.[28] The bottom end of the engine was considerably strengthened and incorporated a nitrided crankshaft, shot blasted connecting rods, and piston pin bosses which were increased from 1/8 inch to 3/16 inch in thickness. All of these features later became standard on the V-1710. The rotating inlet guide vanes on the supercharger impeller were the 47 degree design, replacing the earlier 42 degree set. The carburetor remained the PD-12K2, and only the tee-type backfire screens were used. Three engines were built for development and qualification. Though none were successful at the Model Test, the third engine did complete the 150 hours. This engine was a contemporary of the V-1710-59(E12) that was powering a similar pre-production batch of 25 Bell P-39J's. Those engines suffered from poor reliability, primarily in the accessories drives, a fault also apparent on the test blocks with the V-1710-61. Consequently, the project for the early introduction of the 9.60:1 supercharger step-up gears was put into abeyance pending a redesign of the accessories section to incorporate wider gears. The engine was to be rated on Grade 100/125 fuel, that in combination with the 9.60:1 gears, limited takeoff rating to only 1100 bhp. This was achievable at 2800 rpm with only 46.5 inHgA manifold pressure. Development of this engine continued for some time, with Allison evolving the engine to incorporate the latest improvements. By May 1942 they tested it on 100 octane fuel without port type backfire screens, but for a takeoff rating of 1125 bhp from either 2800 rpm or 3000. With 70 °F carburetor air, the ratings were 1125/2800/47.81 inHgA and 1125/3000/48.68 inHgA. Fuel efficiency deteriorated slightly at the higher speed because of the greater power required to drive the supercharger and overcome internal friction. BSFC was 0.6648 at 2800 rpm and 0.6844 #/bhp-hr at 3000 rpm.[29]

F-15R/L: Several new features were incorporated into these V-1710-75/77 engines when developed in 1941. The Model Test engine was a F-15L in recognition of the earlier problems with the accessories drives in the left-turning engines. If the left-turning engine could satisfactorily complete the test then the right-turning model was qualified by association. The actual Model Test engine was A-037201 which was prepared for testing in August 1943, with testing continuing through the fall and involving at least six teardowns and rebuilds to monitor the condition of internal parts.[30] The major changes from the parent F-10R/L engines were the use of 2.36:1 reduction gears, an increase of the engine-stage supercharger step-up gears to 8.10:1, and provisions for use of Hamilton-Standard Hydromatic propellers. The Hydromatic propeller required that high pressure engine oil be used to operate the constant speed propeller and its controls. This was done by providing the proper passage ways through the hollow propeller shaft on the engine. The other unique feature was adoption of water injection (ADI), a first for the V-1710-F. It is believed that this feature was needed on the P-38K as it did not have sufficient intercooler capacity to support a WER of 1600 bhp at the altitudes where the new turbosuperchargers could lift it when equipped with the new paddle bladed propellers. A total of 19 V-1710-75's and six V-1710-77's were built. The engine never went into full series production for it was deemed that the revisions necessary to reshape the nacelles on the P-38 to accommodate the one inch larger diameter of the propeller spinner (necessary due to the higher thrust line) would have adversely effected production. The V-1710-75 was also specified, and test flown on the Curtiss XP-60A, in that installation it was fed by a GE Type B-14 turbosupercharger. That program also never got beyond the experimental stage.

F-16: This was another very unique model in that it was to be a pusher and would have had a counter-rotating propeller. As originally proposed, the purpose of the engine was to power the Curtiss XP-55 *Ascender*. The power section was contributed by the two-speed, single-stage "Birmann" supercharger of the V-1710-F11. The carburetor was the large Bendix PR-58. Takeoff rating was 1425 bhp, with a Military rating of 1250 bhp to 14,000 feet in Hi-gear. The Army elected to not continue with the project as a consequence of the demands being placed on Allison in 1942 to increase production and to provide new engines for more conventional aircraft. No military designation was assigned to the counter-rotating engine project. As a consequence, the XP-55's were fitted with a single rotation V-1710-95(F23R) configured for pusher installation.

F-17R/L: These sea-level rated engines were built as the V-1710-89/91 and were the primary production model for the Lockheed P-38, being used in the P-38H/J/L models. Derived from the F-10R/L with its 7.48:1 super

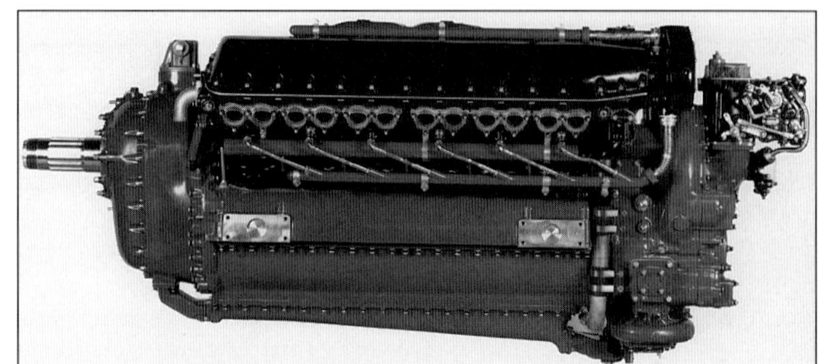

This is V-1710-F17R Model Test engine AEC#21375, photographed on 12-5-42. External differences were few, but internally it was stressed for 1600 bhp. Greater power was largely due to higher MAP produced by increasing the supercharger gears to 8.1:1 and use of Grade 130 fuel. (Allison)

charger gears, they incorporated the higher 8.10:1 engine-stage supercharger drive ratio developed for the F-15R/L, though the earlier 2.00:1 propeller reduction gears were retained. A new PD-12K7 carburetor was also required so as to provide the proper enrichment when operating at the 1600 bhp WER rating on Grade 100/130 fuel. As had become standard for sea-level P-38 engines, the ignition systems were pressurized, as was the use of 30% Ethylene Glycol and 70% water in the cooling system. These engines were rated for 1425 bhp from takeoff up to their Military rated critical altitude.

A total of three F-17R Model Test engines were used in a year long Model Test and development campaign that began in August 1942, at the same time as the similar effort on the F-15L. These engine models differed primarily only in their reduction gear assemblies, and hence provided additional basis for not testing both "R" and "L" versions of each. Changes to the engine during the test program were sufficient to warrant the Allison Photolab taking a "full series" of standard views of the evolving engine three times in August 1942, December 1942 and April 1943.[31] The last of these engines used a six counterweight crankshaft and the keystone ring and piston combination.[32] A number of test programs were run on these engines, including development of the new center manifold with venturi gas pipe and positioning bumpers. In a companion effort a series of high-power tests were run in which water injected test engines were qualified to operate at 1840 bhp and 75 inHgA. By also utilizing turbosupercharger boost at the carburetor deck, these engines demonstrated WER ratings at 2000 bhp. A total of 5,429 V-1710-89 engines were built and 5,421 V-1710-91's were delivered. In addition to the P-38, the Vega XB-38 was powered by four turbosupercharged V-1710-89's, eight of which were procured for the project.

F-17RA: This engine was identified as the V-1710-89A and was identical to the -89 except for being equipped with an automatic manifold pressure regulator.[33] Number built and usage is not known.

F-17LA: This engine was identified as the V-1710-91A and was identical to the -91 except for being equipped with an automatic manifold pressure regulator.[34] Number built and usage is not known.

F-18R: Wright Field was always interested in ways of improving the performance of aircraft and engines. In 1942, they had Allison build one V-1710-69(F18R) with Cylinder Port Fuel Injection in place of a carburetor. It is believed that the fuel metering system was "sequential" in nature, with a pulse of fuel timed for delivery with the opening of the intake valves. This would make it similar to the approach used on the GV-1710-A when modified for fuel injection in 1934. The F-18R was derived from a V-1710-51(F10R). The exercise did not lead to any production during the war, but several of Allison's post-war development proposals included this system as a necessary component to achieve very high power levels. The significant advantage of such a system is the positive assurance given that each cylinder will receive its proportional allotment of fuel. If this is not accomplished, as is typically the case in a carbureted engine, the engine must be rated based upon detonation occurring in the "leanest" cylinder. Ratings for this engine are listed as for a sea-level engine and are the same as for the F-19R, 1325 bhp from takeoff to the critical altitude. Both the F-18 and F-19 were procured on contract W535-ac-29292, approved July 9, 1942.

F-19R: In a companion test to that done on the F-18R, Allison built the V-1710-71. It was similar to the V-1710-69, but used a system for Direct Cylinder Fuel Injection. This was a feature of the German Daimler-Benz and Junkers inverted V-12's, though it is not clear just what type of injection equipment Allison used for these tests. Reference sources are inconsistent on the engine-stage supercharger gears used, some giving 8.80:1 (typical of Altitude rated V-1710's), while others way the engine was built from an F-10R, which had 7.48:1 gears. It is most likely that the source engines for the F-18 and F-19 projects were both F-10's. Ratings for this engine are listed as for a sea-level engine and are the same as for the F-18R, 1325 bhp from takeoff to the critical altitude.

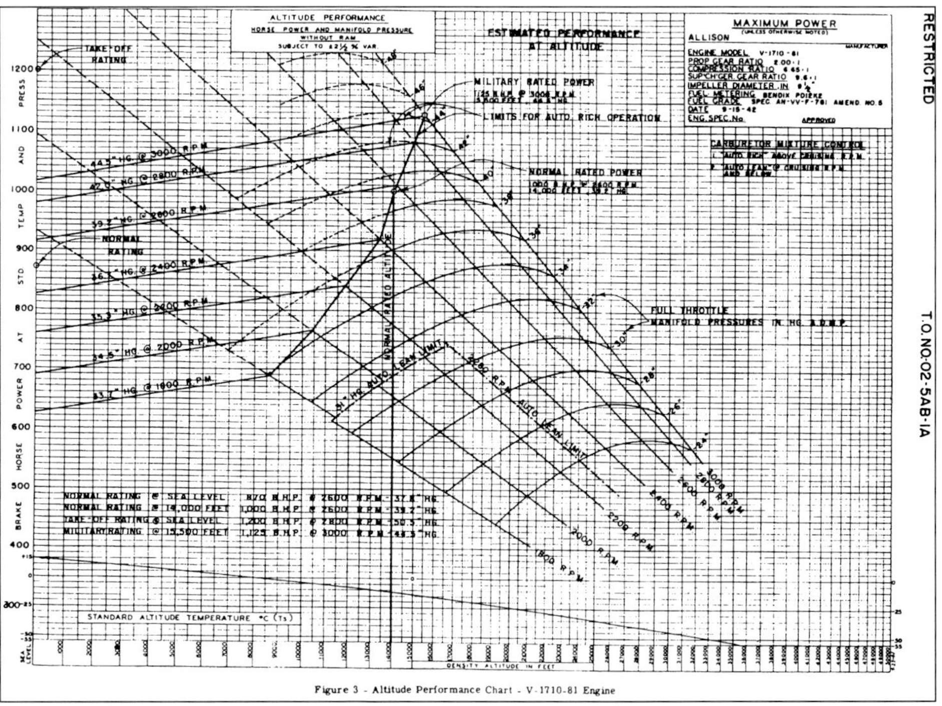

Official performance curve for F20R, though dating from prior to time AAF authorized use of WER ratings. The effect of the higher supercharger step-up gears is readily apparent by comparing to the curve for the F-21R. (AAF)

F-20R: Designated as the V-1710-81, this engine was the follow-on to the altitude rated F-4R, though with the 8.80:1 engine-stage gears replaced by 9.60:1 gears. This model was successful, where the F-14R with 9.60:1 gears had not been, primarily because of an improved accessories housing incorporating wider drive gears. A total of 4,817 of these engines were produced and used to power the North American P-51A and Curtiss P-40M/N/R aircraft, though it is believed only 2,938 went to the U.S. Army. With the new supercharger gear ratio and the resulting high power consumption by the supercharger, only 1200 bhp was allowed at takeoff, a consequence of having to contend with high ambient temperatures. The engine used the PD-12K6 carburetor and was equipped with a Automatic Manifold Pressure Regulator (Boost Control), PN42685, but did not have gun synchronizers.[35] Normal 1125 bhp was available under Military conditions at up to 15,500 feet, and the engine had a creditable WER rating of 1410 bhp at up to 9,500 feet. All ratings on AN-F-28 Grade 100/130 fuel.

There was also a V-1710-81A version, which was identical to the -81 except for having the supercharger step-up gears changed to 8.80:1.[36] It may also have been equipped with the Automatic Boost Control unit as provided on the V-1710-99(F26R). The number of engines converted to this configuration has not been established, nor has their use. Speculation is that they were used within the Continental U.S. for aircraft in training roles where the lower supercharger gears would be more compatible with the low octane fuels and lower altitude missions.

In 1942-1943 the engine was also involved in a special research program performed at the request of the Air Force by the NACA Cleveland Engine Laboratory. The goal was to reduce the wear on engine cylinders, pistons, rings, and valves while determining the limitations on current 100 octane fuel, as well as fuel likely to become available within the following two years. Investigations also considered effects of using an aftercooler and internal coolants (ADI) as ways to lower charge temperature.[37]

F-21R: With the advent of the North American A-36A and its low-level bombing and attack mission, it was important to provide an engine with performance tailored to the lower altitudes. This was accomplished by the V-1710-87(F21R), adapted from the V-1710-51(F10R) with its 7.48:1 engine-stage supercharger gears, a new PD-12K7 carburetor, and rating the engine on Grade 100/130 fuel. The result was 1325 bhp for Takeoff and a WER rating at 5,400 feet of 1500 bhp, all on a fairly conservative 52 inHgA manifold pressure. The installation required the Type E8 gun synchronizers. A total of 835 of this model being built for use by the 500 A-36A's.

The performance curve for the F-21R is the final version produced by Allison. It dates from August 1942, prior to the establishment of WER ratings. While those ratings were obtained on Grade 100/130 fuel, and the curve used Grade 100/125, the curve shows that the engine could make 1500 bhp, from 3000 rpm and 52 inHgA, only at sea-level. Since the established WER of 1500 bhp was allowed at up to 5,400 feet we can only conclude that it was achieved by running the engine well above 3000 rpm, probably near to 3200 rpm.

F-22R: According to Allison Specification 167, this engine would have used a 9.12:1 single-speed supercharger drive, though it is not clear how this would have been accomplished without a complete redesign of the accessories drive train. No contract was established and none of these engines are known to have been built.

F-23R: Six of these engines were built to support the three Curtiss XP-55 *Ascender* aircraft. As the V-1710-95 they were configured for installation as "pushers", and were equipped with an explosive device to eject the propeller should the pilot have to bail out. Based on the F-20R, and using the same 9.60:1 drive to the single-stage supercharger, it also was an altitude rated engine. Rated on Grade 100/130 fuel, 1275 bhp was available on takeoff.

F-24R: According to Allison Specification 169, prepared in the fall of 1942, this engine would have used a newly designed integral two-stage supercharger. No contract was established, nor was a military designation assigned.

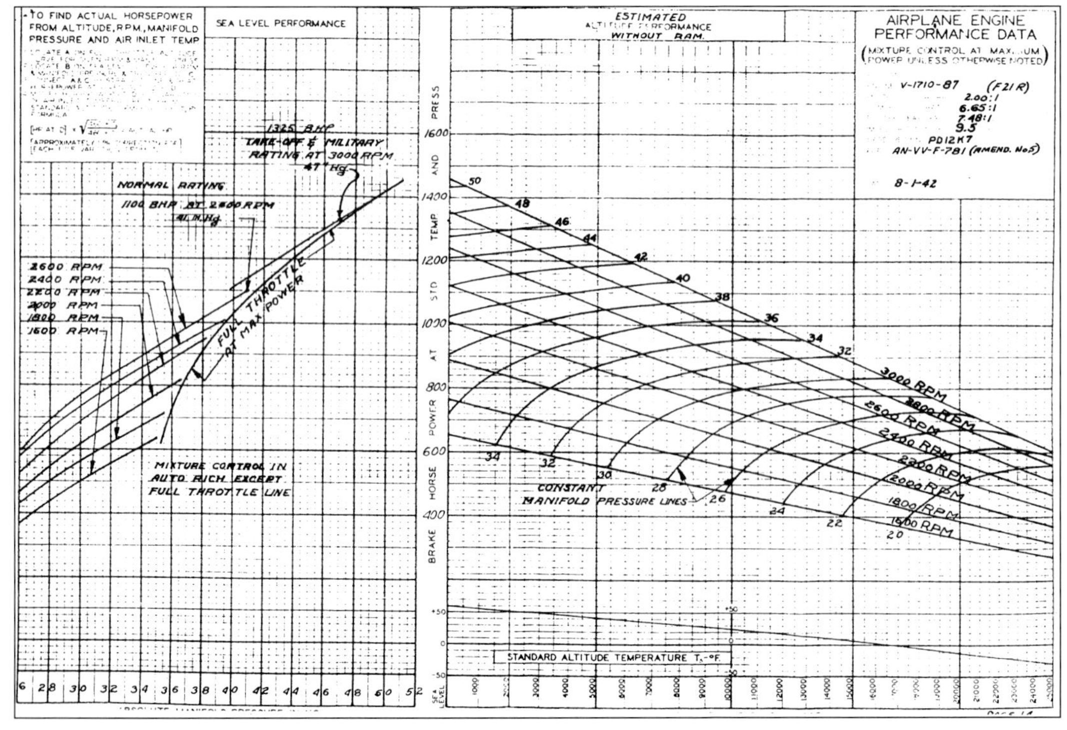

Performance curve for V-1710-87(F21R) low-level rated engine. Since the established WER of 1500 bhp was allowed at up to 5,400 feet we can only conclude that it was achieved by running the engine well above 3000 rpm, probably near to 3200 rpm. (Allison)

F-25R: Per Allison Specification 173, dated February 9, 1943, this engine was to have been equipped with the two-stage supercharger using the standard 6.85:1 driven Auxiliary Stage unit. It was intended for the XP-40Q program. In March 1943, two engines were provided by Wright Field Production Engineering Section to Curtiss-Wright, one for installation and the other as a spare.[38] Total weight was 1,520 pounds and the reduction gear was the new 2.36:1, configured for use with a Hydromatic propeller. Contract ac-22714 was apparently used to cover the project, though no military designation was assigned to the engine.

F-26R: As the V-1710-99, 4,200 of these 1943 vintage engines were built for the later blocks of the Curtiss P-40N. The engine was similar in all respects to the F-20R except that it incorporated the new Automatic Engine Control, PN43801, a more comprehensive control than the earlier Automatic Manifold Pressure (Boost) Control.[39] Performance ratings were unchanged from the earlier model. This engine had achieved a 750 hour Time Between Overhaul interval by the end of the war.

F-27R: As the V-1710-101, this two-stage engine was usually built from a single-stage F-17R (itself based upon the F-15R) for use in the advanced and improved Curtiss XP-40Q airplanes, but it required the installation of the 2.36:1 reduction gear developed for the earlier F-15R. This was done to accommodate operation of the crankshaft at up to 3200 rpm. Three of these engines were built for the XP-40Q program, with the first engine AF43-3407 ready for testing in July 1943.[40] This was one of the first engines to employ the "Keystone" piston rings, needed to minimize "blow-by" caused by ring sticking during high power, high rpm, operation. It still used the old 6 counterweight crankshaft. The carburetor was installed between the supercharger stages. Engine-stage supercharger ratios remained at 8.10:1 while the Aux Stage used the 6.85:1 gears. By November 1943, this engine with the carburetor located between the supercharger stages was being flown in the XP-40Q and obtaining 54 inHgA at altitudes up to 30,000 feet.[41] Ratings on Grade 100/130 fuel gave 1325 bhp for takeoff and allowed a WER power of 1500 bhp at up to 6,000 feet when running at 3200 rpm. Military power of 1150 bhp was available with 3000 rpm up to 20,000 feet.

F-28R/L: Based on the V-1710-101(F27R), the V-1710-121(F28R) engine was built in early 1944 and employed higher 7.23:1 drive gears for the Auxiliary Stage Supercharger that resulted in improved ratings and higher critical altitude. With these gears the performance was improved to the point that at high-altitude it equaled the Rolls-Royce RM14SM high performance two-stage Merlin.[42] It was used in the experimental Curtiss XP-40Q-2 and XP-40Q-3. The Air Force also designated a left hand model, the V-1710-123(F28L), but it is not clear that any were actually constructed. The drawing of a two-stage V-1710 in a P-38 (see Chapter 15) suggests that for a time it was considered as a competitor to a Merlin powered P-38. Only four F-28R's were built. With the higher speed into the Aux Stage Supercharger additional power could be developed. Running at 3200 rpm was possible due to incorporation of the new 12 counterweight crankshaft. WER rating was 1700 bhp at up to 26,000 feet, wet. The accompanying photo of the F-28 engine was actually the F-27R when photographed in July 1943.[43] This was acceptable given that the differences were all internal. Weight of the F-28 was 1,525 pounds compared to the 1,481 pounds of the F-27, much of the difference being due to the new 12-counterweight crankshaft.

F-29R/L: These V-1710-105/107 engines were turbocharged sea-level rated developments based on the F-15/F-17, but equipped with 12-counterweight crankshafts and used the 2.36:1 reduction gear from the F-15. Improvements in fuel control incorporated a new mechanical accelerating pump for the carburetor, the Automatic Engine Control and provided the same ratings as the F-30R/L when operating on Grade 100/130 fuel. Only two of each of these models were built, with the first being on the test stand in April 1944.[44] None are known to have flown.

F-30R/L: As the V-1710-111/113 this model powered the Lockheed P-38L and was a major production series in that 6,348 and 6,344 of each model were built respectively. Interestingly, Allison delivered about 60 percent of each model without carburetors. Evidently, the Air Force had sufficient spares for field installation. These engines were based on the F-17R/L and had the same ratings. Differences were in the use of the 12-counterweight

Differences between the F-27 and F-28 were all internal, primarily the new 12 counterweight crankshaft and higher gears for the Auxiliary Stage Supercharger. Allison used this photo of the F-27 taken on July 10, 1943 when describing the F-28. (Allison)

This is the V-1710-F30L Model Test engine, A-044441, photographed 12-20-43. Allison had found that the Accessories Section on left turning engines could be a problem, so by qualifying the F-30L they were assured that the F-30R would qualify by association. Thousands of these engines were built for late models of the Lockheed P-38. The engine featured internal improvements and the 12-counterweight crankshaft, allowing 1725 bhp at 3200 rpm. (Allison)

crankshaft and an improved PD-12K8 carburetor, though this was changed in mid-1945 to the PD-12K17 configuration. The engine was qualified by Allison for WER of 1725 bhp at 3200 rpm. The Air Force never authorized this rating for service use, instead staying with the 3000 rpm ratings. Engines built prior to V-1710-111 s/n A-063623 and V-1710-113 s/n A-036955 were originally equipped with magneto timing control mechanism. This was subsequently removed by Technical Order issued in June 1945 and replaced with the previously standard fixed magneto coupling.[45] The purpose of the timing control was to improve the starting and idle speed running characteristics of the engine.

F-31R: As the V-1710-115, 560 of these engines were built to power the final block of Curtiss P-40's, the P-40N-40. They replaced the earlier V-1710-99(F26R) used in the P-40N-20/35 Blocks. Improvements included stronger accessories housing incorporating wider and more durable supercharger drive gears, the 12-counterweight crankshaft, the improved DFLN-6 magneto, and a PD-12K8 carburetor to replace the venerable PD-12K6. These engines used the Automatic Engine Control, PN43801, that combined both propeller and boost controls into a single device. Another 420 engines were built as the V-1710-115A, which in addition to the above changes, reverted to the Automatic Manifold Pressure Regulator, PN42685.[46] Ratings on Grade 100/130 fuel remained as for the F-26R. Many of these engines were returned to Allison and converted to V-1710-135 models in the summer of 1945.[47]

F-32R: Designated as the V-1710-119, this engine was to be the Allison alternative to the two-stage Rolls-Royce Merlin, for it had two-stage mechanical supercharging and incorporated an integral aftercooler sized to remove 50 percent of the supercharger compression heat. Allison built a mockup of the engine in January 1944 and had their EX-50 workhorse engine running as a F-32R on the test stand that July.[48] Engine A-064211 was completed in January 1945 and is believed to have been dispatched to North American for installation in the XP-51J. Engine A-064213 was the Model Test engine, and went to test in May 1945.[49] The first "production" F-32R is believed to be A-064212 and was completed in April 1945, just prior to the first flight of the XP-51J, April 23, 1945. Development of the engine continued into the Fall of 1945.

One advantage of the Allison Auxiliary Stage Supercharger as used in the F-32 was its use of a variable speed hydraulic coupling to drive the impeller. When rated, or less, MAP was desired at below the critical altitude of the supercharger, the coupling would reduce the Aux Stage rpm, thereby reducing the horsepower drawn from the crankshaft. This increased the power available to the propeller, as well as the overall engine efficiency. Under WER conditions, the engine could deliver 1900 bhp at sea-level and 1720 bhp at the critical altitude of 20,700 feet. To accommodate the intended higher power levels, the engine compression ratio was reduced to 6.00:1 and ADI (water) injection was incorporated. The Auxiliary Stage Supercharger was driven by 7.64:1 step-up gears that resulted in a Military power of 1200 bhp at 30,000 feet critical altitude. The engine-stage supercharger was contained in the strengthened accessories housing that continued to use the 9-1/2 inch diameter impeller driven by 8.10:1 engine stage step-up gears as had become typical for two-stage V-1710's.

One new feature was introduced on this engine that was significant to delaying its introduction into service. That was the new Bendix "Speed Density" SD-400 computing fuel pump in place of a carburetor. It provided a whole new way to meter fuel, and proved to be troublesome to develop to the level of the engine. The 12-counterweight crankshaft was used, which allowed rating at 3200 rpm for high power when operating on Grade 100/130 fuel. As usual, on engines intended to operate at 3200 rpm 2.36:1 reduction gears were used. The March 1944 Allison Specification

The large liquid-cooled aftercooler on the V-1710-F32R sat in the location desirable for the carburetor, necessitating the use of the compact Speed Density unit. The Auxiliary Stage Supercharger used the usual 15 blade, 12-3/16 inch OD impeller. (Allison)

F-32R s/n A-064213, performing 50 hr Military Test on #7 Rigid Test Stand, August 1945. Note the use of insulated covers over the valve covers and intake manifolds to simulate conditions under the cowling, as well as a XP-51J engine mount. Close inspection shows that the engine had been given a lot of heavy running. (Allison)

184-B was based on operating on Grade 150 fuel. This gave a WER rating of 2100 bhp at 3200 rpm up to an altitude of 4,000 feet. Weight of this engine and all of its apparatus increased to a total of 1,750 pounds. A total of five of these engines were built. The engine was specified for the North American XP-51J and their initial model XP-82A, for which a F-32L version of the engine would likely have been provided. Some production lists show that the XP-82A was to have been powered by the V-1710-119/-121, suggesting that the -121 would have been the F-32L.[50] The -121 was assigned to the F-28R used in the XP-40Q project. The F-32 project began in November 1943, with the first experimental engine, the EX-50, expected to be running in March 1944, and a flight engine available by approximately May 1, 1944.

F-33R/L: Designated V-1710-139/141, these two-stage engines were offered by Allison for use in the proposed North American P-82B "Twin Mustang." North American persisted with its use of the Merlin for the Mustang derivatives and consequently these models were not developed. The basic engine features were those typical of late model two-stage engines such as the E-30 and F-32: 6.00:1 compression ratio, 8.10:1 engine-stage drive ratio and the 7.64:1 Auxiliary Stage Supercharger drive. When connecting the Aux Stage to a left-hand turning engine, it was necessary to interpose an idler gear between the starter drive and the Aux Stage. This slightly altered the overall ratio to 7.59:1 on the V-1710-141(F33L). On Grade 100/130 fuel, the ratings were the same as on the E-30, 1500 bhp at takeoff.

F-34R: According to Allison Specification 238, this engine was proposed to be a development of the F-32R, but using port type fuel injection and was to be equipped for ADI injection. No contract was established, nor was a military designation assigned.

F-35R: While no military designation was assigned, the F-35R was to have been a two-stage engine similar to the V-1710-133(E30), but equipped with an "F" type 2.36:1 reduction gear and with the carburetor installed between the supercharger stages. It continued with the 6.00:1 compression ratio and 8.10:1 engine-stage supercharger drive. The Auxiliary Stage Supercharger was to use a new drive ratio of 8.08:1. Equipped with ADI and rated on Grade 115/145 fuel the performance would have been quite outstanding: 1850 bhp at takeoff with a WER rating of 2240 bhp at 10,700 feet (wet). The dry Military rating would have been 1250 bhp at 30,200 feet.

F-36R/L: Designated as the V-1710-143/145, these two-stage engines represented another milestone in the development of the V-1710, for they were the first to adapt the 10-1/4 inch diameter engine-stage supercharger impeller. They also adopted all of the previous improvements, such as 6.00:1 compression ratio, 12-counterweight crankshaft and 30% Ethylene Glycol coolant. In addition to the entirely new accessories housing necessitated by the new supercharger, the Auxiliary Stage was further refined with a new diffuser and enlarged inlet diameter and driven by a considerably increased step-up ratio of 8.08:1. The other significant change, was the intended use of the new Bendix Fuel Metering Pump that would have provided Port Type fuel injection to each cylinder, in place of the more typical carburetor. Performance on Grade 115/145 fuel and ADI was quite outstanding with 1850 bhp for takeoff and a WER rating of 2200 bhp at up to 12,700 feet, and the Military rating was 1250 bhp at up to 32,600 feet. Allison built up at least one of these engines for testing in September 1945, which resulted in an order in October 1945, just after the war. These engines were intended to power the new North American P-82E "*Twin Mustang*" that was just then being ordered into production. At the request of the Air Force Allison was asked to build these engines from spare engines already delivered, and from spare part stocks declared surplus at the end of the war. Allison put quite an effort into securing such stocks and properly storing them while awaiting the final decisions on production.

F-37: Although this was the last model proposed for the "F" series, it would have continued the growth and diversity of the line. It was to have been a two-stage engine with all of the features of the F-36 engines, including Bendix Port Fuel Injection, while adding a 50 percent Aftercooler. On Grade 115/145 fuel and ADI, it would have been capable of 2300 bhp at sea-level, and WER of 2050 bhp at up to 23,500 feet. No military designation was assigned and none were procured.

NOTES

[1] *Model V-1710-F Type Engines*, O.T. Kreusser, Allison General Manager, letter to Chief, Materiel Division, Wright Field, August 10, 1938.

[2] *Model V-1710-F Type Engines*, letter to Allison from Chief, Engineering Section, Materiel Division, September 16, 1938.

[3] Priority Teletype PROD-T-519, from Production Engineering Section, March 12, 1940. NARA RG18, File 452.8, Box 808.

[4] Priority Teletype PROD-T-569, from Production Engineering Section to Chief of Engine Branch, Wright Field, February 11, 1941. NARA RG18, File 452.8, Box 807.

[5] *Status of Allison Engines*, Materiel Division response to questions from General Arnold, November 29, 1940. NARA RG18, File 452.8, Box 807.

[6] Letter to Allison Engineering Company from Brig. General George H. Brett, Chief, Materiel Division, AAC, July 25, 1940. NARA RG18, File 452.8, Box 808.

[7] *Allison Motor Output*, report to General Arnold by Jimmy Doolittle, Air Corps Representative at Allison, October 21, 1940. NARA RG18, File 452.8, Box 808.

[8] Hazen, R.M. 1945 Report, 4.

[9] Internal Allison Memo to Chief Designer and Design Project Engineers, 11-16-1940.

[10] Sherrick, E. 1942.

[11] This was comparable to the two-stage V-1650-3 Merlin which in hi-gear and 3000 rpm, drove its impellers at 1,070 and 1,272 feet per second.

[12] *Attempted Altitude Calibration of an Allison V-1710-91 Supercharged with a Packard Built Rolls-Royce V-1650-3 Supercharger and Aftercooler*, Wright Field file D52.41/205 Allison, March 4, 1944.

[13] Engineering Division Memo Report, Serial No. ENG-57-531-276, 1944.

[14] Allison Power Curve G-83, contained in *Preliminary Flight Test Data, YP-38, Constructor's Number 2202*, January 15, 1941. NASM File A0304210.

[15] Teletype PROD-T-511 to Chief, Materiel Division, January 21, 1941. NARA RG18, File 452.8, Box 807.

[16] *Operating Instructions for the Allison V-1710-27 & -29 Engines Installed in Lockheed YP-38 and P-38 Airplanes*, Wright Field Memo Report Serial No. PROD-M-184, February 5, 1941. NARA RG18, File 452.8, Box 807.

[17] *Sodium Valve Situation in Allison V-1710-27 and -29 Engines, Particularly in Reference to P-38 Airplanes*, letter from Production Engineering Branch to General Echols, Materiel Command, February 13, 1942. NARA RG18, File 452.8, Box 807.

[18] Teletype PROD-T-662 to Chief Materiel Division, March 13, 1941. NARA RG18, File 452.8, Box 807.

[19] *Status of Equipment Change Request, XV-1710-79(F4L)*, October 22, 1942. NARA RG18, File 452.8, Box 807.

[20] Bodie, Warren M., 1991, 204.

[21] Teletype DHQ-T-775 to Chief, Materiel Division, May 19, 1941. NARA RG18, File 452.8, Box 807.

[22] *Case History of the XP-60 Series Project*, 1945.

[23] Allison Photolab Log, photos #10800/02 and 11694, Jan 1943 and March 1943.

[24] *Improvement of Power Output Characteristics Relative to Allison V-1710-45 Engine*, letter to Commander of Materiel Command from G.W. Lewis, Director of Aeronautical Research, NACA, March 4, 1943. NARA RG18, File 452.13, Box 1260.

[25] Allison Photolab Log, photos #03580/83 and 04992/5001, July 1941 and December 1941.

[26] Allison Photolab Log, photos #03435/40, 05709/12 and 06386/91, July 1941, Feb 1942 and April 1942.

[27] Allison Photolab Log, photo #06103, March 13, 1942.

[28] Allison Photolab Log, photos #04847/56, December 5, 1941.

[29] Experimental Engineering Report EXP-M-57-503-599, 1942.

[30] Allison Photolab Log, photos #14447/56 and 15479, August 25, 1943 and November 5, 1943.

[31] The three F-17R Model Test engines were, AEC 17645, AEC 21375, and A-037963.

[32] Allison Experimental Department Report No. A2-143, 1944, 4.

[33] Technical Order 02-5A-5, 1945.

[34] Technical Order 02-5A-5, 1945.

[35] Allison Model Specification No. 163-E, 1943.

[36] Technical Order 02-5A-5, 1945.

[37] *Program for Research on the Allison V-1710-81 Engine*, letter from NACA Aircraft Engine Research Laboratory to Commanding General, AAF Materiel Center, Wright Field, January 19, 1943. NARA RG18, File 452.13, Box 1260.

[38] *XP-40Q Airplane V-1710-F25R Installation*, letter to Curtiss-Wright Corp, Buffalo, N.Y. from Fighter Branch, Production Engineering Section, Wright Field, March 25, 1943. NARA RG 342, RD 3834, Box 7636.

[39] Allison Model Specification No. 163-E, 1943.

[40] Allison Photolab Log, photos #13570/74, July 10, 1943.

[41] *Development of Increased Horsepower and Altitude for the Allison V-1710 Engines, Particularly for N.A.A. P-51 Airplane*, letter from R.M. Hazen of Allison to Commanding General, AAF Materiel Command, C1L2AE, November 2, 1943. RG18, file 452.13, Box 1260.

[42] *Development of Increased Horsepower and Altitude for the Allison V-1710 Engines, Particularly for N.A.A. P-51 Airplane*, letter from R.M. Hazen of Allison to Commanding General, AAF Materiel Command, C1L2AE, November 2, 1943. RG18, file 452.13, Box 1260.

[43] Allison Photolab Log, photo #13574, July 10, 1943 and Weight and Balance diagram of F-28 engine as of March 6, 1944.

[44] Allison Photolab Log, photo #17643, April 13, 1944.

[45] Technical Order 02-5AB-18, 1945.

[46] Technical Order 02-5A-5, 1945.

[47] *Daily Report of Engine Shipments*, letter from Air Technical Service Command to Allison, August 17, 1945. NARA RG342, RD3733, Box7255.

[48] Allison Photolab Log, photos #16409 and 19355, Jan 1944 and July 1944.

[49] Allison Photolab Log, photos #22145/47 and 24184, Jan 1945 and May 1945.

[50] *Allison's Proposed Production Schedule of V-1710 Engines*, Materiel Division, February 12, 1945. NARA RG342, RD3775, Box 7411.

13

The V-1710-G: The Ultimate V-1710

The V-1710-G was the final series of the V-1710 reciprocating engine to see production. In it Allison brought together all of its experience in design and manufacturing to produce an engine that was outstanding in practically every respect.

The only characteristic or feature that distinguishes the "G" series is that all of them had the 12-counterweight crankshaft, though this alone is not enough to distinguish the "G" series, for that crank was also used on some late model "E" and "F" series engines as well.

The "G" series was distinguished for its wide variety of features and configurations. To name a few: short nose-high centerline reduction gear or extension shaft (for P-63); with or without Auxiliary Stage Supercharger; 10-1/4 inch or 9-1/2 inch diameter Engine-stage Supercharger impellers; with or without two-speed supercharger drive; a range of engine-stage supercharger ratios, 7.48:1, 7.76:1, 8.80:1, 9.60:1; with 6.65:1, 6.50:1 and 6.00:1 compression ratios; Pressure carburetor, Speed Density "carburetor", or port fuel injection; and both left or right-hand rotation. In addition there were a number of detail improvements throughout the engines.

Production of the "G" series was quite limited by comparison to the "E" and "F" models of the V-1710. While it has often been reported that the V-1710-135(G4) was built for a production aircraft, the Bell P-63F and the RP-63G "Pinball", the only records that have come to light are for V-1710-135(E31) versions that retained the 9-1/2 inch diameter supercharger and used in Pinball aircraft. It is also known that "large numbers" of V-1710-115(F31R) engines were returned to Allison for conversion to -135 models.[1] Consequently, if 258 of these V-1710-135 engines were built, they were conversions of other models, and not "G" production. After the war, with the cancellation of the Packard contract to build V-1650 Rolls-Royce Merlin's, the Army Air Force purchased 750 V-1710-G6's for the North American P-82E/F program. As a twin engined aircraft with handed engines, the actual purchase was for 375 each of the V-1710-143/145(G6R/

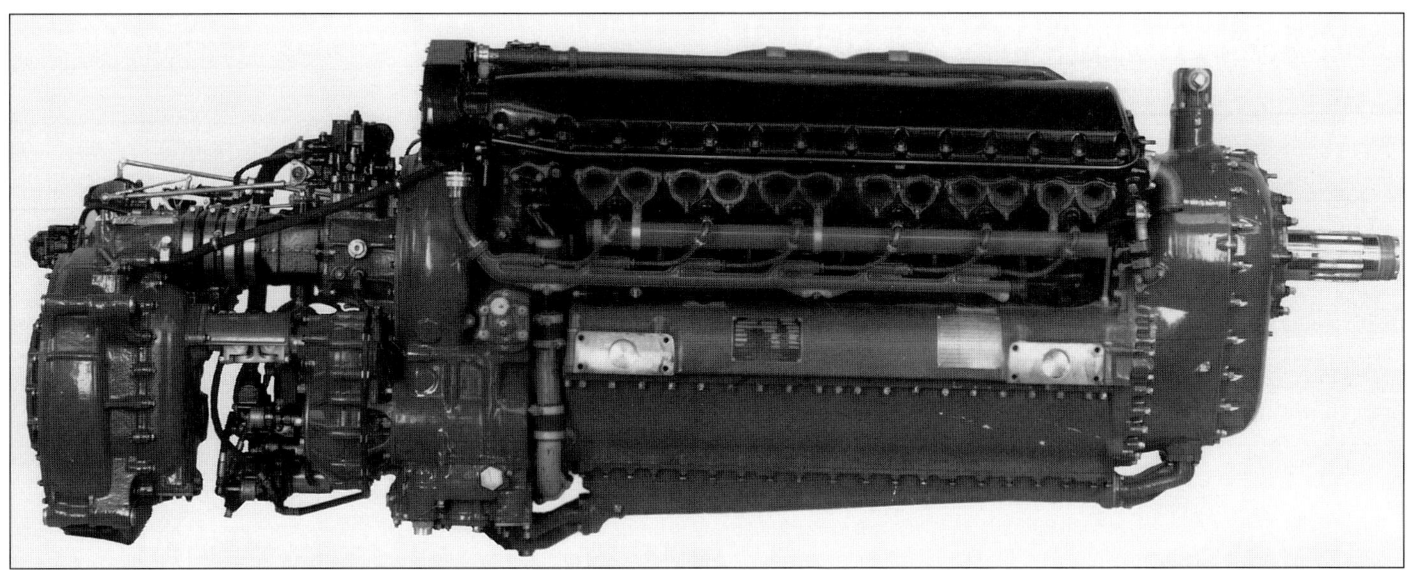

As the final V-1710 production model the V-1710-G6 was a very sophisticated two-stage engine. It incorporated the compact Speed Density carburetor, 10-1/4 inch diameter engine-stage supercharger, water injection, pressurized ignition and reduced compression ratio pistons. Notice the enlarged reduction gear case identifying the 2.36:1 gear set used in engines rated for 3200 rpm operation such as the G-6. The tall breather was to resolve the oil foaming concerns when the left turning V-1710's were at very high power. (Allison)

This is the G-6L Model Test engine, photographed 9-10-46. It completed the 150 hour test in October of 1946. Use of the Speed Density "Carburetor" in place of the earlier Pressure Carburetors removed a lot of bulk from the back of the engine. These engines used the 2.36:1 reduction gear. The larger gear raised the thrust line by one inch compared to the more typical 2.00:1 V-1710 reduction gear. (Allison)

G6L) models. When surplused in the 1950s, it was largely this batch of engines that powered the Unlimited Hydroplane Racers into the 1990s. As it was not unusual for the racers to sacrifice engine life for power, many of these engines were modified to deliver as much as 4000 bhp.

The "G" design was begun in early 1943, intended to respond to the compiled list of Wright Field desired features and improvements, while establishing a new threshold of performance for the V-1710. In this it was designed to deliver 1725 bhp at an impressive 3400 rpm. This was to be achieved by incorporating the 12-counterweight crankshaft, developed in late 1942, along with improvements in the induction flow path, supercharger, cylinder heads, as well as an improved accessories section. The first model was the V-1710-97(G1R), a single-stage sea-level rated engine, incorporating a new 10-1/4 inch diameter impeller driven through 7.48:1 gears and continuing with the standard 6.65:1 cylinder compression ratio. All power section component modifications were interchangeable with previous models of the V-1710 and V-3420 engines.[2] Key to the "G" was the enlarged single stage induction system that was able to take full advantage of a PT-13E 3-barrel carburetor.

Many of the improved features developed for the "G" were soon incorporated into late model series E/F production. In particular significance were the strengthened cylinder heads and crankcases produced using improved materials and foundry practices. But for the "G", development into a production program during the war was not to be.

The "G" represented an aggressive development target. The specific power at takeoff was 235 psi, as measured by BMEP, a comparatively moderate increase when compared to the 175 psi developed in the early V-1710-C when running at 2600 rpm for takeoff at 1000 bhp.[3] The "G" made its considerably greater power by running at much higher speeds and using high boost from the supercharger. On an Indicated MEP basis, the conditions in the cylinder where much more demanding and required minimization of inconsistencies in the distribution of mixture between the highly loaded cylinders. That the "G" was able to successfully complete the 150-hour Military Mode! Test at these conditions is a tribute to the enhanced structural strength and reliability incorporated into the V-1710 during its long development and production life.

With many of the features developed for the "G" being incorporated into E/F engines on the production lines, the G-6 went on to distinguish itself by adapting the new Bendix Speed Density Carburetor. In retrospect this was a controversial adaptation, for it is generally believed that the device was not sufficiently mature to be successful in a production. Much of the criticism of the "G" is really attributable to problems with this carburetor, which was first fitted to the V-1710-119 and continued to evolve until the SD-400-D3 model was used on production V-1710-143/145(G6R/L) engines. An objective of the "G" was to have an efficient induction system and both the Air Force and Allison elected to stay with this unit. By the end of the war, Allison had a considerable effort underway to introduce fuel injection on the V-1710 and V-3420, though the Air Force believed that it was unnecessary as they believed conventional pressure carburetors were adequate.

Allison investigated both sequential port and direct cylinder fuel injection, including building an engine of each type for experimental studies. These were 1942 projects, the XV-1710-69(F18R) and XV-1710-71(F19R), respectively with port and direct cylinder injection. Allison went on to test and study the German direct cylinder fuel injection systems. They first used a Junkers unit in mid-1944 and then used those from a Daimler-Benz DB 610 obtained after the war. These efforts resulted in proposing a fuel injection system for the V-1710-G9R/L, a design that went far enough that the Air Force designated it their V-1710-147/149, the final Military model of the V-1710. Even though the V-1710 was long out of production when these engines were being considered in 1946 and 1947, Allison continued to propose advanced models up through their V-1710-G12, and even a V-1710-H model. These engines were intended to power follow-on models of the F-82 and as such were configured to fit into the airframe with little or no structural modification. In June 1948, the Army Air Force became a fully independent arm of the military, and as the U.S. Air Force it was eager to complete the break with the past. That included a

Table 13-1
V-1710-"G" Engine Ratings

Model	Eng S/C Ratio	Eng S/C Dia, in	Aux S/C Ratio	Comp Ratio	Weight, pounds	Carb	Takeoff @ 3000 rpm	WER @ 3400 rpm
V-1710- 97(G1R)	7.48:1	10.25	Turbo	6.65:1	1500	PT-13E	1725/68.5	1725/25,000/68.5
V-1710-None(G2R/L)	7.76:1	10.25	None	6.00:1	1470	PT-13H2	1600/3200/61.7	Commercial
V-1710-None(G2R/L)	9.60:1	10.25	None	6.00:1	1470	PT-13H2	NA	Commercial
V-1710-131(G3R)-Lo	7.48:1	10.25	None	6.50:1	1475	PT-13H2	1600/3200/61.7	1360/ 3,000/___[1]
V-1710-131(G3R)-Hi	9.60:1	10.25	None	6.50:1	1475	PT-13H2	NA	1220/15,000/___[1]
V-1710-None(G4)-Lo	7.76:1	10.25	None	6.00:1	1340	PT-13H2	NA	Commercial
V-1710-None(G4)-Hi	9.60:1	10.25	None	6.00:1	1340	PT-13H2	1830/3200/	Commercial
V-1710-None(G5)	9.60:1	9.50	None	6.65:1	1500	PD-13K8	1425/__	2200/SeaLev/80
V-1710-143(G6R)	7.48:1	10.25	8.087:1	6.00:1	1595	SD-400D3	1600/74	1700/21,000/85[2]
V-1710-145(G6L)	7.48:1	10.25	8.03 :1	6.00:1	1595	SD-400D3	1600/74	1700/21,000/85[2]
V-1710-None(G8R)	7.48:1	10.25	8.087:1	6.00:1	1595	SD-400D3	1600/74	1700/21,000/85[2]
V-1710-147(G9R)	7.48:1	10.25	8.087:1	6.00:1	1625	Port Inj	1600/	____/______/__
V-1710-149(G9L)	7.48:1	10.25	8.03 :1	6.00:1	1625	Port Inj	1600/	____/______/__
V-1710-None(G10R)	2-stage	10.25	Turbo Compound	6.00:1		Port Inj	2000/__	2550/20,000/__[2]

[1] Ratings at 3000 rpm.
[2] Ratings at 3200 rpm.

Table 13-2
"G" Model Engines Shipped

Model	Number	Aircraft	Comments
V-1710-97(G1RA)	5	P-82A	Tested in P-51A
V-1710-131(G3R)	8	XC-114	4-engine C-54 Derivative
V-1710-143(G6R)	375	P-82E/F	
V-1710-145(G6L)	375	P-82E/F	
Total	763		

break with technologies of the past as well, a break that was facilitated by the rapid advancements in jet propulsion. With little interest from the customer for continuing with propeller power, and a hearty workload being assigned to Allison to produce jet engines, it should be no surprise that these final proposed models of the V-1710 were not developed.

Table 13-3: V-1710-G Engine Models

G-1R: Analytical work in support of this series was begun in late 1942, defining a model based upon use of the newly designed 12-counterweight crankshaft, and an enlarged supercharger with a 10-1/4 inch diameter impeller. The engine was designed for 1725 bhp at 3400 rpm using 68.5 inHgA manifold pressure and fed by a three barrel carburetor. Examples using either 7.48 or 9.60:1 supercharger step-up gears were considered, depending upon whether it was to be sea-level or altitude rated. As built, the engine used 6.65:1 compression ratio pistons that required AN-F-29 Grade 140 fuel. Because of the heavier crankshaft and accessories section, total engine weight reached 1,500 pounds. It is believed that only three or four of these engines were built, mostly for test and development purposes, with the first engine appearing in late 1944. As a result of this design and testing, the 12-counterweight crank became available and was used in many late model V-1710-E/F engines. The Flight Test Department first flew the engine in P-51A AF43-6012 in mid-1945, under its AAF assigned designation of V-1710-97.[4]

G-1RA: This model was similar to the G-1R, but the cylinder compression ratio was changed to 6.00:1 in an effort to improve the margin to the onset of detonation at high power settings. Several engines were built, A-065973 went onto the test blocks as a G-1RA in March 1945,[5] A-065974 went to Wright Field for propeller testing, and A-065977 to Hamilton Standard to test propellers.[6]

G-2R/L: In January 1946 Allison proposed using these engines, (derived from the F-30 with the fitting of the new two-speed supercharger and accessories section), in both left and right rotation for commercial derivatives of the Douglas XC-114/YC-116, themselves developments of the C-54 transport. They would have used the PT-13H2 carburetor, have 6.00:1 compression ratio, and two-speed (7.76:1 and 9.60:1) single-stage 10.25 inch diameter supercharger impellers. Takeoff power would have been 1600 bhp at 3200 rpm, with cruise capability of 1200 bhp from 2700 rpm at 6,000 feet in low-gear, and 1120 bhp at 2700 rpm at 15,500 feet in high. Maximum cruise fuel efficiency was a BSFC of 0.640, while normal cruise at 600 bhp was achieved at a very respectable 0.456 BSFC. Engine weight was 1,470 pounds. It is believed that none were built, though Allison did begin testing the two-speed supercharger clutch in October 1945 and continued testing it through October 1946. The engine was to incorporate a new low-tension high-frequency ignition system that differed from previous systems by having integral distributors/magnetos mounted on each cylinder bank. A mockup engine incorporating the device was built in November 1946 as well as workhorse engine EX-54 being built to develop the installation.[7]

G-2RA: This was a modified F-30R proposed to American Airlines for a liquid-cooled version of the Douglas C-54, to be known as the DC-4B.[8] Designation G-2R"A" denoted a sub-version with features specific to the requirements of American Airlines, it may also have been used to designate the "American" features.

G-3R: Eight of these engines, were military versions of the proposed commercial G-2 model, were built as V-1710-131's to power the Douglas XC-114 (C-54E derivative with lengthened fuselage), and YC-116 (similar to XC-114 but with thermal de-icing for wings). Both were contemporaries of the DC-4M "Merlinized" C-54's. They incorporated a two-speed 10.25 inch diameter engine-stage supercharger and revised compression ratio pistons for 6.50:1 to accommodate the available manifold pressure. Rated for Takeoff at 1600/3200/SL/61.7 inHgA in Lo, and with a low altitude (7.48:1) Military rating of 1360/3000/4,750 feet, and in high gear (9.60:1) produced 1220/3000/15,500 feet. The first engine, A-071870, went to test in December 1946.[9]

G-4R/L: This single-stage, two-speed (7.76:1 and 9.60:1) engine was offered for commercial use as the G-4R/L with extension shafts in the proposed Douglas DC-8 transport. The 10.25 inch diameter supercharger was used. Only the G-4 and G-5 engines were offered in configurations for operation in extension shaft installations.

G-5: This engine was intended to be similar to the G-4, though no information has come to light detailing its specifics or applications.

G-6R/L: Production for the P-82E/F made this the primary "G" series engine, with 375 delivered of each model to the AAF, as their V-1710-143/145. A two-stage engine with the new Bendix Speed Density "carburetor" and 6.00:1 compression ratio, it was capable of 2250 bhp "wet" on Grade 115/145 fuel and delivered 1600 bhp for takeoff. It used the large 10-1/4 inch diameter supercharger impeller and 7.48:1 gears in combination with 8.08 and 8.03:1 Auxiliary Stage supercharger drive ratios, right and left hand respectively. Interestingly, the original specification for the V-1710-143/145 engines was as the Allison model V-1710-F36R/L that was to have used the Bendix Fuel Metering Pump to effect Port Injection. Evidently the complexities of developing that system caused a return to the SD-400 that had been in development and testing on the V-1710-F32R for the previous three years. The Model Test engine was a G-6L that had been built from V-1710-109, s/n A-068591. It used new G-6L parts as required, and completed the 150 hours in October of 1946. The Military rating was set at 1250 bhp and 30,000 feet, a deviation from the intended 32,500 feet.[10] Allison then went on to achieve the intended altitude rating.

G-6A: This was a February 1946 proposal by Allison to investigate NACA findings that certain organic compounds in water made superior internal coolants compared to conventional water-alcohol ADI fluid. It is not clear that the project was ever undertaken.

G-7R/L: Believed to have been intended for commercial installations.

G-8R/L: These engines were similar to the G-6, but were to be equipped with integral aftercoolers suitable for use on the P-82E/F and later P-82G/H.

G-9R/L: As a further improvement in the G-6 engine, this model was similar except that "port" fuel injection (as had been intended for the F-36R/L) was to be utilized in place of the speed density carburetor. In addition low-tension/high-frequency ignition as well as an improved automatic engine control with a single lever for the pilot, a redesigned supercharger and drive was also to be provided. In February 1946, this became the last model of the V-1710 to receive a military designation, as the V-1710-147/149 respectively.[11] Ratings were to be the same as for the G-6.

G-10R/L: Another 1946 model based upon the E-30 power section and intended for future versions of the P-82, but designed so as to require no modifications to the P-82E airframe. This was a turbo-compound engine similar to the E-27 but utilizing an Allison designed air-cooled, power recovery turbine geared to the crankshaft. It also used a new two-stage integral supercharger and aftercooler to remove 50 percent of the compression heat. With the Bendix Port Type fuel injection from the G-9, the takeoff rating was 2000 bhp at 3200 rpm, and 2550 bhp was available under War Emergency conditions at 20,000 feet. A #60 spline propeller shaft was to be used as a result of the high power produced. The two-stage supercharger and auxiliary drives were repackaged into a single compact unit.

G-11R/L: Also based upon the E-30 power section, this turbo-compounded model was a proposed development of the G-10 with a change in the supercharger to an improved 5-stage design. The port type fuel injection system would have been retained. Ratings of the turbo-compound engine would have been the same as for the G-10.

G-12: In late 1947 this engine was proposed as a G-11 companion, but configured for installations utilizing the "E" type extension shaft drive system. Ratings would have been the same as for the G-11.

H-1: Yes, there was to be an "H" series too, and different from the H-1 and H-2 Marine engines designed in the 1938-1939 period. This model was to be based upon the G-10, but without the engine-stage supercharger. Instead, the Allison designed two-stage, air-cooled, power recovery turbine also drove a new two-stage supercharger. The 50 percent aftercooler was retained along with port type fuel injection.[12] While the performance ratings would not have changed from the G-10, the package should have been simpler and more efficient.

[1] *Daily Report of Engine Shipments*, letter from Air Technical Service Command to Allison, August 17, 1945. NARA RG342, RD3733, Box7255.

[2] Sherrick, E.B.1944.

[3] For comparison, the late model P&W R-1830 rated at 1350 bhp with 2700 rpm developed 210 psi BMEP.

[4] Telegram from Wright Field Power Plant Laboratory to Allison, April 12, 1945. NARA RG342, RD3775, Box 7411.

[5] Allison Photolab Log, photos #23244/53 of A-065973 3/28/45, and 25337/9 of A-067475 on 10/10/45.

[6] Telegrams in Allison V-1710 file, 452.8 at NARA RG342, RD3775, Box 7411.

[7] Allison Photolab Log, photos #27810/11 of EX-54 on 10/10/46.

[8] AERO DIGEST, January 1947, 108.

[9] Allison Photolab Log, photos #27978/87 and 28135/7 of A-071870 on 11/29/46 and 12/16/46.

[10] *V-1710-145 150-Hour Qualification Test*, Allison letter to Commanding General, Air Materiel Command, Wright Field, October 24, 1946. NARA RG342, RD 3898, Box 7861.

[11] Message from Air Corps Chief, Power Plant Laboratory, Engineering Division to R. Hazen of Allison confirming telephone conversation of 2-18-46. NARA RG 342, RD 3775, Box 7411.

[12] *Proposal for Allison Model V-1710-G10 Engines*, Allison letter to Engineering Division, Wright Field, June 6, 1946. NARA RG342, RD 3775, Box 7411.

14

The V-3420: A Double Vee for the Long Range Bomber

By the middle of the 1930s the Air Corps had begun to develop strategic plans for very long ranging bomber aircraft. World events were already causing some tension. There became a clear need for large aircraft able to carry large payloads long distances. These would require large engines, and the immediate problem was that there were no suitable large engines being planned. Furthermore, the air-cooled engine technology of the day was not particularly encouraging that they could be developed to meet the need. This concern was largely the result of experience from the one attempt in 1925 at a large air-cooled engine, the Wright Field designed, and Allison built X-4520, rated at 1200 bhp. With the anticipated early success of the liquid-cooled V-1710, Air Corps planners and engineers asked Allison to "double" the engine, by taking four of its cylinder banks and arranging them in a "X" form. In this way they hoped to quickly develop an engine having the needed power.

The 2300 bhp XV-3420-1 swung a very large propeller and created a need for all new and enlarged test facilities at both Allison and Wright Field. These delayed development testing for a considerable time. (Grantham)

Bombing strategists were pushing for what could be considered a "super bomber" – one able to carry a heavy bomb load over previously unheard of distances. In April 1934, the Air Corps issued a request for such a "Long Range Airplane Suitable for Military Purposes", the XBLR (eXperimental Bomber, Long Range), also known as Army Project "A." Boeing received an order for the XBLR-1, to be powered by four of the new 1000 bhp V-1710-3 engines. At the same time the Martin Company was authorized to begin design of their XB-16, an aircraft that was to use an extension shaft tractor version the V-1710-3. As advanced as they were, these aircraft were still severely underpowered. At the time, there were no more powerful engines then in advanced development in the U.S.

As a consequence of the delay in the V-1710 development, the X-3420 engine was also delayed. The X-3420 (itself identified as Restricted Project No. M-2-35) was identified by the Air Corps as the engine for the proposed Project "D" airplanes Restricted Projects No. M-4-35 and M-5-35.[1] These projects resulted in the Douglas XBLR-2 (XB-19) and Sikorsky XBLR-3, both intending to use four of the 1600 bhp X-3420-1's. The Sikorsky project did construct a wooden mockup, but following an Air Corps evaluation, the project was rejected in favor of the Douglas XBLR-2 in the spring of 1936.[2]

Political conditions in Europe were increasingly concerned with the aggressive re-armament of Germany then in full swing. In the U.S., with the recent arrival of the prototype Boeing Model 299 (XB-17, first flight July 28, 1935) and experimental Boeing XB-15 (first flight October 15, 1937), Army planners were already looking ahead to the next generation of heavy bombers. The operating requirement was reduced to a cliché: "10,000 pounds of bombs carried 10,000 miles."

It was clear that a lot more power was going to be required than the earlier aircraft had available. Such goals were going to require four engine bombers, with each engine being rated for at least 2000 bhp.[3] It was with

Artist rendering of V-3420-A made in 1937, prior to construction of the XV-3420-1. (Allison)

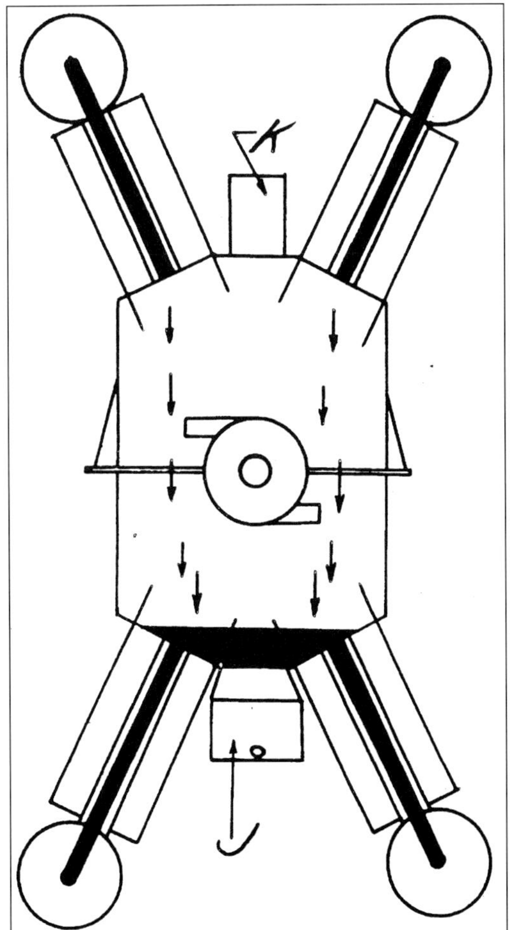

The 1918 X-24 Liberty was a 720 bhp engine that retained the 45 degrees between the banks from the Liberty-12. Features of the inverted crankcase were later used in the Allison inverted Liberty models, particularly the oil system shown here. (Note: Experimental 24-Cylinder Liberty Engine, The Bulletin of the Experimental Department, Airplane Engineering Division, U.S.A., Vol.II, No.3, December 1918, p.114-121.) (Air Service)

Accessory end of the wooden mockup of the fuel injected X-3420. The Air Corps requirement to use a single crankshaft meant use of articulated connecting rods and limited the maximum speed to 2400 rpm, and 1600 bhp. (Allison)

this in mind that the Air Corps continued to press Allison for a 2000 bhp engine.

For many years manufactures of in-line engines had taken developed V-12's and utilized their components in 4-bank "X" engines and thereby nearly doubled their output. During WW1 a Liberty had been built in this configuration. In the late 1920s, Packard had built the X-2775 from four cylinder blocks from its 2A-1500 liquid-cooled V-12. In England, Rolls-Royce was at work building the Vulture, an engine utilizing four Kestrel blocks in a "X" arrangement. And of course, Allison had experience coming from their first indigenous aircraft engine, a 24-cylinder, four bank, air-cooled eXe of 4520 cubic inches.

The Materiel Division wanted the engine to be as light as possible, and specified it be built with a single crankshaft. Allison proceeded in this fashion but was never in favor of the arrangement as they believed that a suitable crank would be too heavy, given the length of the cylinder blocks and the high power to be transmitted. Furthermore, a satisfactory "master" and "articulating" connecting rod scheme would have to be designed and developed from scratch. A task that had never been successful at the power and stress levels anticipated.

Allison began an investigation of the eXe engine at their own expense as the Air Corps was unable to provide funding at the time. Investigations reached the point where Allison was able to define the X-3420-1 engine in their Specification No. 201, dated June 12, 1935. The engine followed Air Corps requirements in that it was to use a single crankshaft, and articulated connecting rods even though this limited the rated speed of the engine to 2400 rpm, the same as that for the GV-1710-A of 1931. In this form the engine would produce 1600 bhp, weigh no more than 2,160 pounds, be equipped with 7.0:1 supercharger gears, and drive the propeller through 2.0:1 reduction gears and a single #50 spline propeller shaft. With the intent to power a long-range bomber Allison specified a 8.5:1 compression ratio to achieve high engine efficiency. The Army continued with their fascination with fuel injection and the engine was to be built with Marvel Carburetor Company Type MC-12 fuel injectors. The engine was to run on 100 octane fuel and deliver 1000 bhp at 1800 rpm for cruising at an efficiency of 0.42 #/bhp-hr, a very good fuel utilization.

At the same time, Allison offered the X-3420-3 version with exactly the same ratings and features, but driving dual or counter-rotating propeller shafts, again through 2.0:1 reduction gears.

This engine, being proposed for development at the same time the V-1710-C was continuing its struggle through Type Testing, was clearly a major distraction to the small Allison development team. The engine itself was never built, but a full size wooden mockup was. See the accompanying photo. One noticeable feature is the physical arrangement of the cylinder banks. Two vees with 60 degrees between them are clearly seen. Note that the centerlines of the vees are displaced about the single central crankshaft with 90 degrees between the inner cylinder banks. This results in different angles between the paired banks than was subsequently used in the V-3420 engines, and would of course, require a different firing order for the 24 cylinders.

Upon arrival and naming as Allison's Chief Engineer in March 1936, Ron Hazen was immediately confronted with the diversions being faced by his small staff, and particularly the problems with the Army requirement to use fuel injection rather than carburetors.

He determined that there was little hope in getting the fuel injection system to work given the present state of injector pump development. To this end he enlisted the assistance of senior management at General Motors and prevailed upon the Materiel Division to terminate the fuel injection engine models. In return, he promised to deliver a satisfactory 1000 bhp V-1710 Type Test engine in a matter of months. In September 1936, with the remaining "YV-1710-3's" released for production, and the XV-1710-7 well on its way to passing its 150-hour Type Test, the Air Corps asked Hazen if "a 1500 horsepower Allison was in the cards soon." He responded that "It would take development time to produce a 1500 horsepower engine, but I can give you a 2000 horsepower engine by doubling up on what we now have."[4]

The 1600 bhp X-3420 had effectively died with the abandonment of fuel injection. By this offer, Hazen proposed to not only better it, but to simplify the development and give the Air Corps a higher powered engine in a shorter time. By the time the concept had crystallized, he was able to base the new engine on the improved 1150 bhp V-1710-C7, and a 2300 bhp engine quickly resulted by simply pairing two 1150 bhp V-1710's, each with its own crankshaft, but mounted on a single crankcase.

On November 4, 1936 Ron Hazen directed his design team to prepare a quotation to the Air Corps on the "DV-3420 Engine."[5] He stated that the quotation was to cover the "design, [and] fabrication of one engine and 50 hours of development test[ing] of this engine at our plant." He defined the engine as follows:

> "The engine will be a two crankshaft engine using 4 banks of six cylinders each of the present V-1710 size, to develop approximately 2000 B.H.P. normal and 2300 B.H.P. take-off on 100 octane fuel. The engine will be delivered with reduction gear for a single propeller, with provision to be made for installation of reduction gear nose for double drive for oppositely rotating propellers. Provision should be made for firing small cannon through propeller shaft of single direction propeller, if feasible. Engine should be reversible with change of minimum number of parts. The general design of the V-1710 will be followed wherever possible and as many standard parts incorporated as reasonably can be. The engine is planned for 2 to 4 engine wing mount airplanes. The possibility of using this engine as a pusher should be kept in mind throughout the design."

Interestingly, Hazen notes that:

> "Design costs already incurred on the X-3420 engine are to be picked up on this item. There will be considerable design study before determining the final engine arrangement. Layouts of crankcase assembly, reduction gear assembly, accessory housing assembly, intake manifolding and engine mounting will be the major requirements."

Obviously, Allison's' arguments had finally prevailed regarding the nature of the engine, and they were soon authorized to construct a wooden mockup and perform a suitability study of the proposed V-3420 double vee engine. Contract W535-ac-9678 was authorized in May 1937 to cover the study and later to provide the first engine.

In response to the Air Corps original requirement to have the double engine use a single crankshaft, Allison conceded that it would have been lighter (2,160 pounds for the X-3420-1 versus 2,300 pounds for the XV-3420-1) but the Air Corps had not fully considered the difference in power,

In the X-3420 the single crankshaft required use of articulated connecting rods. Each pair of cylinder banks retained the 60 degree vee angle, with the inner banks 90 degrees apart, placing the centerlines of the vees 150 degrees apart.

Hazen's approach to a doubled engine featured V-1710 cylinder blocks and crankshafts. These were 12-3/4 inches apart while the angle between the vees was changed from 150 to 90 degrees.

1600 bhp vs. 2300 bhp respectively! The XV-3420-1 was the first engine to deliver 1 bhp per pound at its normal takeoff rating.

Hazen's small team was certainly spread thin with all of the programs and developments then underway. In addition to taking on the design of the DV-3420, the XV-1710-7(C3) was not yet through its Type Test, the YV-1710-7(C4) was being prepared for its December 14, 1936 first flight, the pusher engines for the Bell XFM-1 were being fabricated and prepared for Type Testing, as was the YV-1710-7(C4) engine for the XP-37. The YV-1710-C4 that ultimately passed the Type Test was yet to even be constructed. Furthermore, Allison was still working with the GM Research Laboratory on the "U" type two-cycle aircraft engine. While all of this was going on, Allison was also responding to aircraft manufactures interested in altitude rated engines. Leading this effort was Navy interest in a liquid-cooled engine for a new carrier fighter.

In March 1937, the Air Corps completed their Type Test work on the YV-1710-7(C4/C6) and returned it to Allison for the additional 116 hour penalty run. With the receipt of contract W535-ac-9678, Allison officially began work on the design of the XV-3420-1(A1), though company funded work had been underway for some time. By that September, Allison had completed the design study showing that an engine of this type could be built to handle the doubled power of the V-1710 with good performance and durability.[6] The design proposed that the crankshafts be configured such that they could be installed to turn in opposite directions from one another, a benefit for future use with counter-rotating propellers, or to accommodate handed engine installation. This feature was a cornerstone of the subsequent Allison V-3420 and V-1710-E/F/G series. Though it may have been serendipitous, it is more likely that the requirement of Lockheed's X-608 proposal (which became the XP-38 with "handed engines"), along with the numerous requests for various engine/propeller arrangements being received from the aircraft manufactures, provided the impetus.

Work on the engine progressed rapidly to the point that by January 1938 Allison had been authorized to release information on the engine to selected aircraft manufactures and airlines, as well as the Navy.[7] Boeing was then at work on a study for Pan American Airways on a V-3420 powered flying boat, and numerous other projects were in the conceptual stage by others that were interested in utilizing the engine. Much of this interest was instigated by an extended road trip and visits by Hazen to the aircraft manufacturers in 1938.

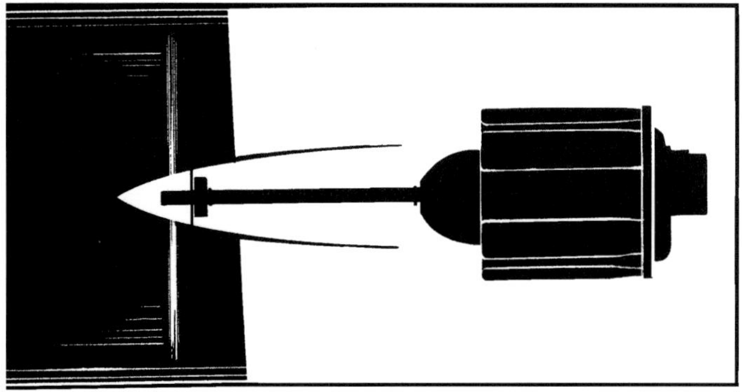

V-3420-A as a Pusher, with propeller speed extension shaft. (Allison)

The XV-3420-1(A1), AC38-119, with its 2.5:1 reduction gear, was run for the first time in April 1938. On April 18, 1938, the Materiel Division requested a quote from Allison on five V-3420's having 3.0:1 reduction gear ratios, the first engine to be for Type Testing and the next four to follow on its success. Allison offered the Type Test engine for $75,000, the others at $60,000 each.[8]

Allison was successful in utilizing the majority of the V-1710 components in the double vee engine. There were only 340 parts unique to the V-3420, accounting for 930 different pieces per engine, out of a total of 11,630 pieces in the engine.[9] Furthermore, Allison really turned on their marketing imagination with the V-3420. There was practically no arrangement of engines and/or propellers that the V-3420 was not shown to accommodate. Extension shafts, right-angle drives, counter-rotating propellers, tractors, pushers, engines buried in the fuselage, buried in the wings, anything was possible. They even offered a DV-6840 – a double, Double-Vee 3420! The DV-6840 progressed to the point that in July 1941 the Allison photolab took a complete set of photos of the setup! While none of those photos have survived, the accompanying plates give an idea of some of the considered and available arrangements. As time was to show, several of the less radical arrangements were actually built, and several flown.

Still the aircraft designers continued to ask for more. In 1939 a firm by the name of DuBois-Martin was working with Wright Field to have Allison configure the V-3420 so that it would drive a tractor propeller from

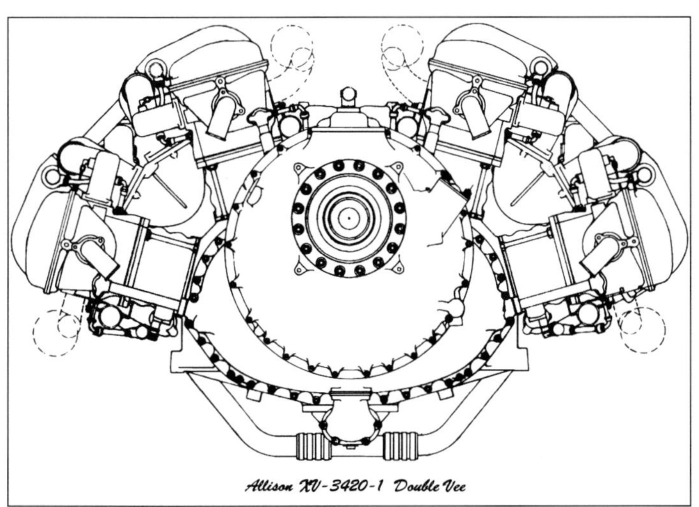

Head-on view of the 2300 bhp XV-3420-1(A1). Its design lead in the effort to redesign the V-1710 for mass production and greater growth potential. (Author)

Twin V-3420-B's with close coupled counter-rotating propellers. (Allison)

Table 14-1 Contemporary High Horsepower Engines				
	Tornado	R-2600	R-3350	V-3420
Horsepower	2350	1700	2300	2300
Weight, pounds	2,735	1,965	2,700	2,700
Number of Different Detail Parts	725	720	810	–
Number of different Assemblies	150	130	105	–
Number of standard parts	5,765	4,820	6,325	–
Total number of Part Numbers	875	850	915	–
Total number of parts per engine	10,200	8,000	10,150	11,630

the front and a pusher propeller from the rear of the engine through an extension shaft to a remote reduction gear. This tandem propeller drive was a radical departure from propeller shafting and drive arrangements originally contemplated for use with the Allison V-3420.[10]

Three years later, a comparison of contemporary 2000+ bhp high powered engines was prepared. It included the much touted Wright R-2160 Tornado, a liquid-cooled radial having seven "in-line" banks of six cylinders. The Tornado was offered to the Army in 1941 at a cost of $63,000 each for a batch of ten development engines. As you can imagine, that engine was complex by comparison with contemporary Wright engines. See Table 14-1 for details.[11]

Contract W535-ac-11328 for the six V-3420-3's was issued in June 1938, but production of the engines was contingent upon the successful completion of the 50-hour endurance test of the XV-3420-1.[12] Allison Specification 204 was issued on April 15, 1938 defining these high compression bomber engines.

Once the XV-3420-1 was running, good ideas continued to be brought forward. There were other concepts for the V-3420 that also received serious consideration. In October 1939 the Materiel Division held a conference regarding the possible conversion of the V-3420 to diesel operation.[13] Another concept proposed by Allison, was to raise the compression ratio and thereby improve the fuel efficiency to an entirely new level for a supercharged engine, as it was intended to be under 0.4 pounds per horsepower-hour. This is comparable to the 0.393 BSFC the unsupercharged and liquid-cooled *Voyager* engine attained when cruising at 52 bhp during its un-refueled around the world flight in December 1986.[14] A similar performance would have been a significant achievement for a vintage 1939 2300 bhp engine.

The interest in the V-3420 was heightened by the appearance of the XV-3420-1 in a display at the 1939 New York Worlds Fair.[15] In January 1940, Wilbur Shaw, a well known automobile racer of Indianapolis 500 fame, began discussions to obtain a V-3420 for an endeavor to break the world's automobile straight-away speed record on the Bonneville Salt Flats, Utah. The Army had no objection to his use of an engine for this purpose, providing that the engine not be taken out of the country without specific authority.[16] The V-3420 did finally make it into speed automobiles, but not until after WWII, and such projects continue today.

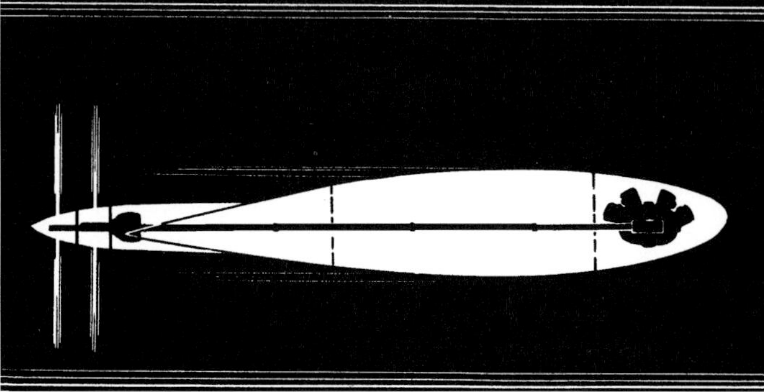

Buried in the wing, V-3420-B driving individual extension shafts to counter-rotating pusher propellers. (Allison)

Following the Worlds Fair, there was a considerable amount of interest in the engine, but the Air Corps only had the single XV-3420-1 which had been on display and they took a very hard line with writers and journalists seeking information. None beyond that on the placard at the Fair was to be released.[17]

Due to a protracted test series dominated by Materiel Division bureaucratic mentality, it was not until March 1940 that the XV-3420-1 fi-

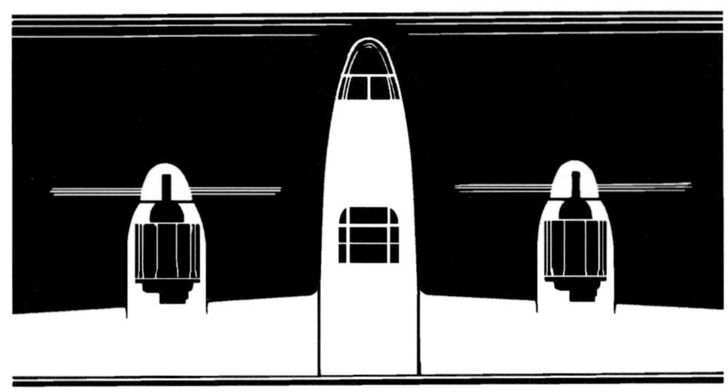

Handed V-3420-A's, as in Lockheed XP-58. (Allison)

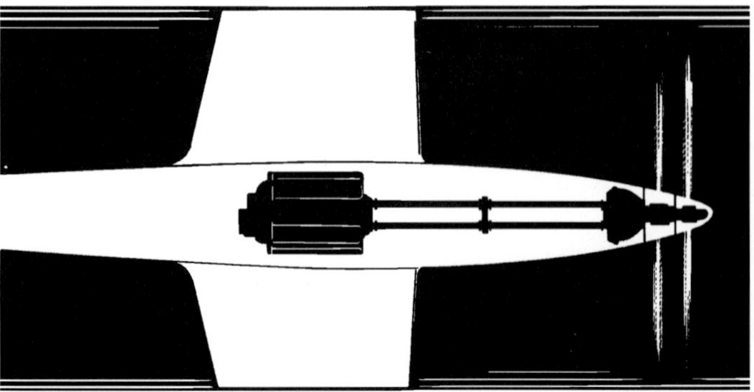

V-3420-B with extension shafts to counter-rotating reduction gears and propellers as used in Fisher P-75. (Allison)

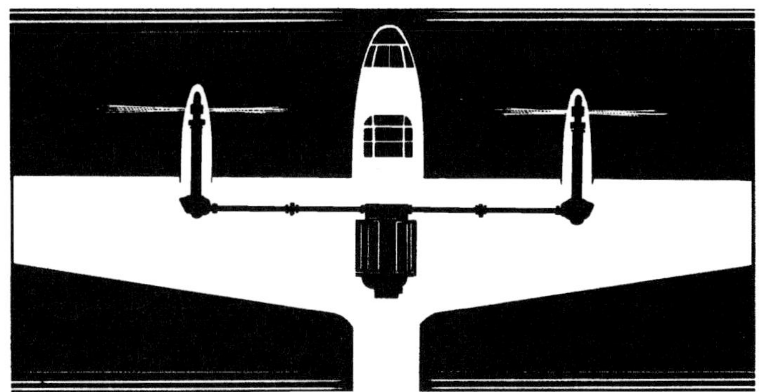

V-3420-B5 with right angle crankshaft speed drives to remote reduction gears, tractor arrangement. (Allison)

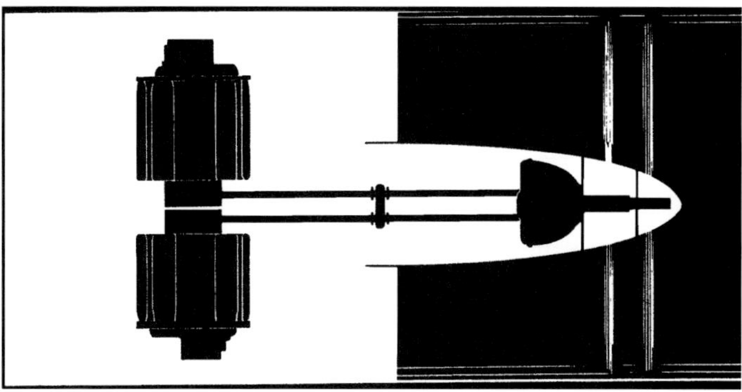

Paired V-3420-A(R/L)'s in a DV-6840 arrangement driving a counter-rotating reduction gear. (Allison)

nally began its official 50 hour test, a test critical to clearing the way for the later V-3420-3 engine to attempt a full 150-hour Type Test. In December 1939, the contract for the six XV-3420-3 engines was changed to use the same 2.50:1 reduction gear as on the XV-3420-1. This was in place of the intended 3.00:1 gear originally specified in the contract.[18] Then in September 1940 the Air Corps requested Allison to revise one of the V-3420-3's on order to become a V-3420-5, the first "B" model and the first Allison engine to have a remote dual rotation reduction gear for mounting a counter-rotating propeller. In December 1940, Allison was directed to focus their V-3420 efforts on the XV-3420-5, and to rate it at 2600 bhp for take-off using extension shafts and right angle drives. This was admittedly not the best form for the 2600 bhp engine, but it was the only engine in the power class under development at the time, and the Army needed such an engine.[19]

As a result, Allison devised two series of V-3420, the "A" and the "B". The V-3420-A series had both crankshafts turning the same direction, either right or left. The V-3420-B series crankshafts turned in opposite directions and was used in counter-rotating propeller installations. Either series could have a close coupled reduction gear or drive a remote reduction gear through extension shafts. The lessons learned in the design and testing of the V-1710-D and V-1710-E series engines had given Allison confidence in the design and operation of extension shafts and remote reduction gearing.

V-3420 Put On Hold

In the summer of 1940, Allison was expanding its engineering and production capabilities to meet the need for V-1710's to power French, British and American production fighters. Major new factory facilities were being built that required the training of thousands of new workers, many inexperienced for the work at hand. All of this was creating problems in getting sufficient V-1710's manufactured for the large numbers of Curtiss P-40, Bell P-39, and Lockheed P-38's then on order. Because these demands conflicted with the availability of qualified engineers and facilities for the necessary work on the V-3420, the Air Corps suspended all work on the V-3420 in September 1940, coincident with the peak of the Battle of Britain.[20] It was expedient to put fighters in the air immediately, and thereby prevent, or delay, the day when it would be necessary to launch very long range bombers on missions to Europe from America.

With the above action, and after completing the initial endurance runs, the Army recommended that the XV-3420-1, AC38-119, be salvaged and approved it for disposal on December 18, 1940, stating, "This engine is of no further value to the Air Corps for either development or for service

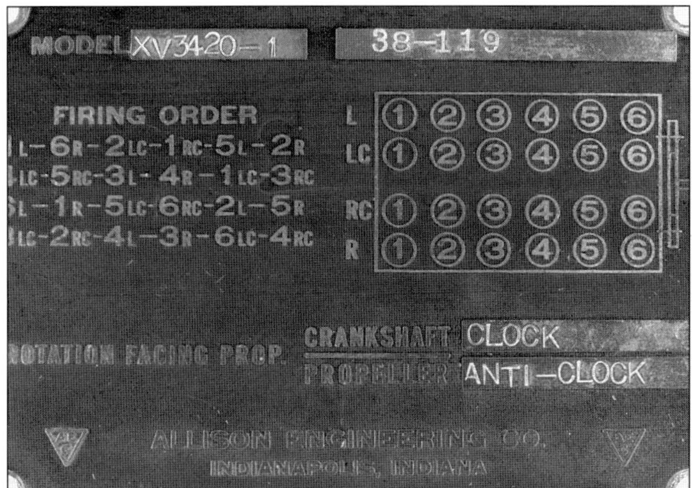

Aviation historian and writer Kevin Grantham obtained the dataplate from the XV-3420-1, AC38-119 from a junk yard in Oklahoma. As a sidenote, the single XV-1710-15(C9) left-hand engine for the Lockheed XP-38 was AC38-120, and it was Allison s/n #17. The XV-3420-1 was being built along side the first of the C-10 service test engines. (Grantham)

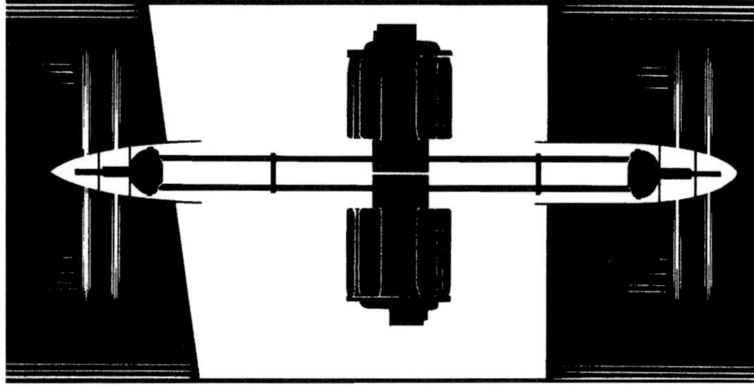

One wing of a very large aircraft with a DV-6840 set of paired V-3420-B's on each wing. Each V-3420 section driving one pusher and one tractor propeller via extension shafts. (Allison)

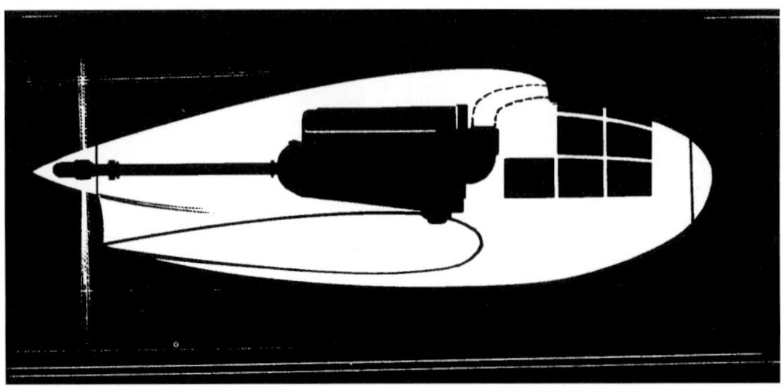

Fuselage mounted V-3420 pusher with propeller speed extension shaft, similar to arrangement with V-1710-D in Bell X/YFM-1's. (Allison)

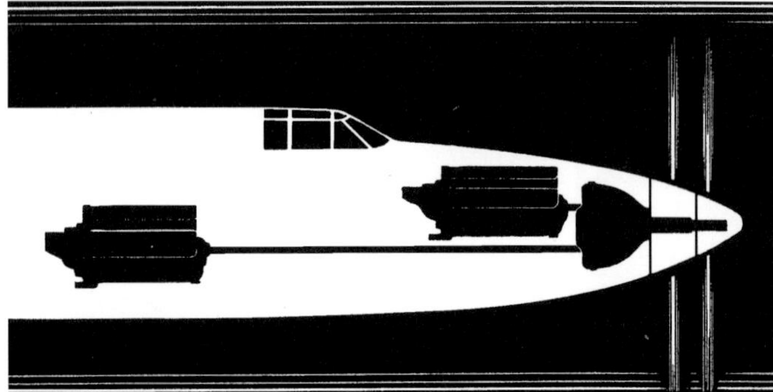

Tandem V-3420-27(A23R/L)'s in a DV-6840 configuration. This counter-rotating 3.23:1 reduction gear was actually built in 1945. (Allison)

purposes."[21] On December 27, Allison acquired the engine and had it shipped to them at Indianapolis from Wright Field. Once there, Allison performed a short test on the engine to determine its mechanical weight and balance, along with the dynamic parameters, such as "moment of inertia", that would subsequently be necessary to incorporate it into an aircraft.[22] Aviation historian Kevin Grantham subsequently came into possession of the dataplate from this engine, salvaged from a scrapyard in Oklahoma.

By February 1941 circumstances had changed and the Army requested that Allison begin work on the design of a 3000 bhp V-3420-A engine as a "standby" powerplant for installation in the Boeing B-29 Superfortress. That aircraft's Wright R-3350 was having development problems, and Allison was requested to provide a V-3420 for the B-29. The installation was to be interchangeable with the B-29's Wright R-3350.

In March 1941 Allison responded to the Wright Field Power Plant Laboratory that they could deliver a sea-level 2300 bhp V-3420 for a 150-hour Type Test within four months. This engine would incorporate changes and improvements resulting from tests of current V-1710's. In April, the Materiel Division authorized procurement of one of these engines for a price of $60,000 and requested that the engine be provided with a 2600 bhp rating. It was anticipated that this engine would be available for Type Testing in July 1941. Allison proposed production of 5 to 10 V-3420's per day as being achievable by April 1942, providing they were given sufficient priority on obtaining the necessary machine tools.[23] Interestingly, General Brett penciled a note to General Arnold that "We should have design studies made as to the possible use of this engine. Previous (information) points out that it should be a better bomber engine than pursuit." This comment suggests that at this point it had *not* been decided to use the engine as a backup to the R-3350 powered B-29. It was nearly a year later, March 3, 1942, that General Echols notified General Arnold's office that the new Model Test V-3420 had completed the 150-hour Type Test run at 2600 bhp with only minor difficulties. On his authority he planned to increase Allison facilities capacity to manufacture about 300 of these engines per month "in case it can be fit into the Heavy Bomber Program at a later date."[24] With these objectives in mind, the Air Corps ordered one V-3420-9(A11R) engine from Allison, though the contract was not signed until June 1941.

Preliminary analysis of the Boeing B-29 performance, if powered by 2450 bhp V-3420's at 25,000 feet, found that they would increase maximum speed by 13 mph over the 2200 bhp expected to be available from the R-3350. Significantly, they determined that if the gross weight of the aircraft remained unchanged, that the fuel load would be reduced by about 500 gallons, which in turn resulted in a range reduction of about 500 miles. Although the dry weight of the V-3420 proposed for this installation was similar to the 2,700 pound R-3350, the extra weight came from the liquid-cooling systems that were not present with the air-cooled R-3350. From the weight of 500 gallons of fuel, the extra weight of the V-3420 installations must be about 750 pounds for each of the four engines. It is not known to what extent credit was given for the improved fuel economy of the liquid-cooled engine, but it appears that it was not credited in this analysis. When the R-3350 powered B-29 went into combat, engine overheating was a major problem, one that required operating the engine with rich fuel mixtures to protect against overheating and engine failure. This had a very detritus impact on the load carrying capability and range of the airplane. It was some time before the engine cooling problem was resolved, largely by the introduction of direct cylinder fuel injection. This feature became available on the R-3350 late in the war.

In June 1941 a Type Test engine suitable for use in a bomber was purchased in the form of a single XV-3420-9(A11R). This engine was to be delivered to Wright Field that October. Coincident with the delivery of this engine, the U.S. was thrust into WWII and the Air Corps again directed the suspension of work on the V-3420 so that Allison could concentrate on the introduction and manufacture of the new V-1710-E/F models. By the summer of 1942, the situation regarding the engines on the B-29 had reached a crisis and Allison was put back to work on an engine for the airplane. They received an order for nine V-3420-11(A16R)'s, the first for Type Testing and the others for the XB-39, Project MX-230. These engines were to be built in the Allison experimental shop and delivered in the July to October 1942 period. The first engine was in its Allison test stand in March 1943 and the MX-230 engine nacelle was available at the same time. In October 1942, the Air Force had ordered 500 V-3420-11 engines, enough to power 100 B-39 airplanes. These engines would have been delivered in the March/July 1943 period as a result of their highest level of priority, A-1-A. Allison aggressively worked on the development of this engine and the installation for more than a year. Most of these engines were never built as the initial problems with the R-3350 had been overcome to the point that the justification for the expedited B-39 program was no longer there. At about this point, the V-3420-B was selected for the Fisher P-75 and that program began to drive the development of the engine.

By August 1941, North American Aviation was at work testing early V-3420's and determined that the engine was still having significant fuel distribution problems. Consequently, was that the engine had to be operated such that the "leanest" cylinder had enough fuel to prevent detonation, the limiting condition for design, operation, and rating of an engine. In an effort to better understand the problem, Allison constructed a translucent intake manifold and had a V-3420 running with it on the test stand in June 1942.[25] This remained a problem, to one degree or another, through the life of the V-3420 program.

Although only 157 V-3420's were built, the importance of the program should not be viewed from the production perspective. Rather, credit should be given for the design concepts developed and incorporated along with the cross-over of technology between it and the V-1710. The V-3420 was available when needed to backstop the B-29 engine program. As such it was a continuation of the original reason for the engine, to power the VLR bomber of the mid-1930s. Its availability supported the evolutionary steps in the thinking that created the Boeing B-29 and Consolidated B-32. Having the V-3420 in the late 1930s, the first 2,000+ bhp engine, gave Air Corps planners confidence to proceed with the new generation of very heavy bombers. Aircraft that would have been essential had England been lost in 1940, leaving the U.S. to face the Axis on both the east and west coasts.

As an epilog to this story, Allison proposed a fuel injected sea-level V-3420-C1 to power a follow-on to the B-29/B-39. Offered in 1945, this engine would have been a 4000 bhp alternative to power what became the Boeing B-50. Allison policy at the time had shifted to doing development only if paid for by the customer. With the wind down of the war, funds were not forthcoming and P&W was successful in getting financial support for the R-4360. That engine went on to play a major role in powering the final generation of large propeller driven bombers and transports. The V-3420 was strong enough and powerful enough to surpass the R-4360, but it was not to be.

Development of the V-3420

The V-3420, having 11,630 parts, was a complex machine, made worse when almost everything is built-up from smaller assemblies. A major objective was to keep the engine's frontal area small and total weight low, both major constraints. In this regard Allison was successful. At 11.74 square feet, the frontal area was about 2.25 times that of a V-1710F, and at 2,300 pounds the weight was about 1.8 times the F. The specific power rating of 1 bhp per pound was outstanding for the day. From the point of reliability the engine was not much more prone to failures than were any two V-1710's. The major problem with the engine was the above mentioned maldistribution of air/fuel mixture. This was somewhat surprising given that the V-1710 ramming type of streamlined intake manifolds were quite good at delivering balanced flow. The source of the problem was more likely stratification of flow as the mixture was split for delivery to the left and right Vee's of the engine.

One important consideration in the early life of the V-3420 was related to how it should be controlled in dual propeller installations. With the power available from the engine, there were really no aircraft able to use it in single propeller installations. As a result, the only viable option in 1938 when the engine was beginning testing was to use split drive shafts to separate propellers with the engine being housed in the fuselage. In March 1938, the Air Corps authorized Allison to release information on the new V-3420 to Bell Aircraft. Hazen's letter to Robert Woods, Bell's Chief Engineer, describes the XV-3420-1(A1) 2600 bhp sea-level rated tractor engine, but notes that it could be provided as a pusher and/or equipped with extension shafts operating at either crankshaft or propeller speed.[26] During the balance of the year a lively discussion transpired between Wright Field and Allison regarding the feasibility of the different types of extension shaft systems which could be developed.

The problem was to insure that each propeller was running at the same power rating under all conditions of flight. Allison suggested that the simplest approach was to gear the shafts directly to the engine. This had the advantage of insuring synchronization of the propellers and provided the needed control over torsional vibration throughout the entire engine/propeller system. The problem was that such a system required a device to insure that the pitch of both propellers was the same, with each drawing the same power. Such a device had never before been developed.[27]

Alternatively, a differential gearing system was preferred by Wright Field engineers. It would only require a propeller speed governor, matched together from each prop. The differential would then be able to deliver equal torque to each propeller and insure that both propellers were equally loaded. Allison responded that development of such a system was possible, but would be heavier and would limit the ability to use differential power from the propellers for ground maneuvering of the aircraft. Given the alternative of using conventional controls on a somewhat revolutionary drive system, or develop a more straight forward drive system and a new control to synchronize the pitch of the propellers, Allison related that, "we would much prefer to have the assignment of developing such a device (the new controller) than we would the development of a differential gearing in an engine such as the 3420."[28] The power available from the V-3420 was really ahead of its time, though Allison clearly had the resources and capability to handle it, even in 1938.

Allison did undertake a development program for a 1:1 bevel gear drive system considered for such installations. They used a V-1710-E4 test engine configured with V-1710-E2 supercharger gears to drive the new gear box through a single section of P-39 extension shafting.[29] This in turn was connected to a V-1710-D propeller speed extension shaft which drove a 2.5:1 planetary type reduction gear and mounted the propeller.[30] The combination was equivalent to one half of the system that would have been used with the intended V-3420 installation. Test work began in 1938 and was concluded in October 1940. Though the system was successful, it was never flown.[31]

While the V-3420 began as a bomber engine, it was not long before it was proposed for fighter and cargo aircraft as well. As it resulted, its major production was for the Fisher P-75 fighter, but that was years later. The first sizable effort considering the V-3420 for use in a fighter was by the Air Corps who built a 1/15 wind tunnel scale model and tested it extensively in the summer of 1939.[32]

Mechanical Features of the V-3420

The cylinder blocks were directly interchangeable with those from the V-1710-C10 and later engines. The V-3420 had a new crankcase casting that brought it all together and mounted the pair of reversible crankshafts. This

Allison test setup for the V-3420-B5 bevel gear drive was tested using the 1150 bhp workhorse V-1710-E2 in October 1940. (Allison)

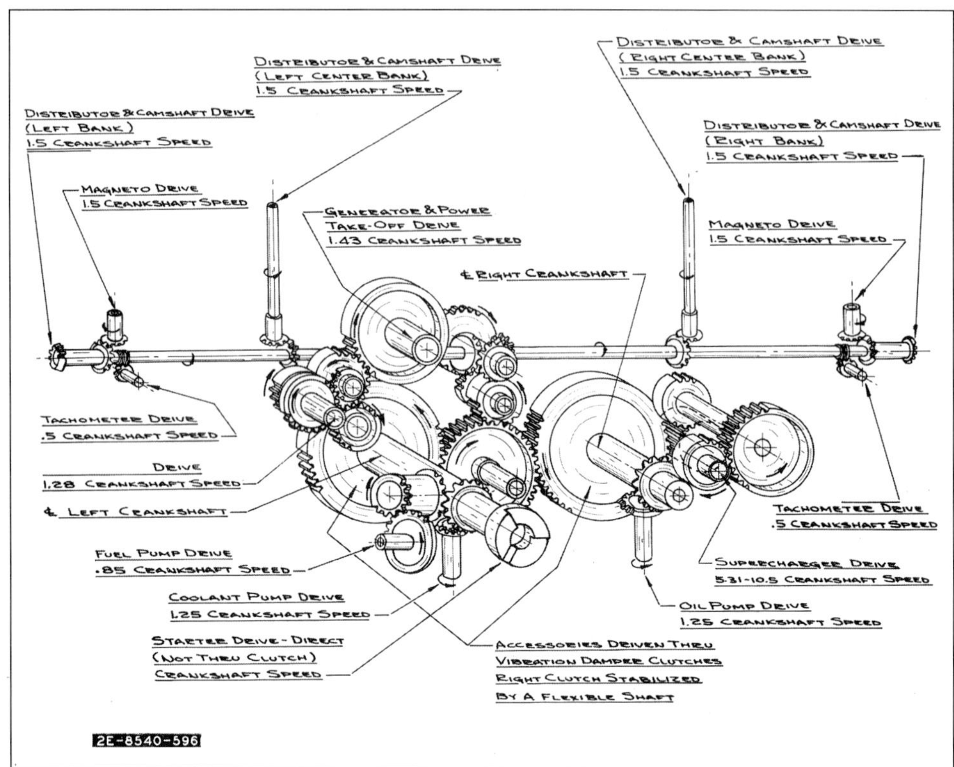

Gearing within the Accessories Section of the V-3420 shows its V-1710 heritage while having an elegance and efficient layout of its own. (Allison)

was a huge casting for the time. The crankshafts were arranged to be 6-3/8 inches on either side of the engine centerline. The center-lines of each of the pairs of 60 degree V-12's were set at an angle of 90 degrees to each other, making the outer-most blocks 150 degrees to each other, and setting the inner blocks at 30 degrees to each other. The requirements for evenly spaced power pulses were maintained by providing a cylinder at TDC on its power stroke every 30 degrees of crankshaft rotation in "A" series engines (those with same direction crankshaft rotation). The "B" series engines (with opposite rotation crankshafts) were designed to be fired every 60 degrees, with the crankshafts geared together so that they were 180 degrees out of phase.

Any V-3420 could be provided with either a direct connected reduction gear or a remote reduction gear via extension shafts similar to those developed for the V-1710-E in the Bell P-39. Usually the extension shaft configuration was used with the V-3420-B series with it's counter-rotating crankshafts. In the "B" it was important that the crankshafts be "locked" to each other to keep the engine firing order properly phased. In fact, the "B" engines were originally designed to fire the same cylinders on each crankshaft at exactly the same time. For this purpose "crankshaft timing idler gears" were provided in the gear case at the front of the engine. The arrangement was simply two interlocking pinions driven by their respective crankshafts. As a result everything was locked in step, while the cranks turned in opposite directions.

In the "A" series engines both crankshafts rotate in the same direction, either left or right. Crankshaft timing idler gears were not necessary providing a conventional single shaft reduction gear was used. This was due to having both crankshaft pinions directly engaged with the large reduction gear. This in effect "locked" the cranks together and assured proper timing and crankshaft phasing.

Engines configured as "A" models also had, as a necessity, a different ignition timing arrangement. For smooth operation it is advantageous to have many cylinders, and a corresponding number of evenly spaced power strokes. If this is not done, a large amount of potentially damaging torsional vibration is created. Consequently, the ignition pulses were staggered so as to alternate evenly between the vees in the following fashion. For right-hand propeller rotation the banks fire L-R-LC-RC, and for a left-hand propeller the banks fire L-RC-LC-R. For a right hand engine the firing order is:[33]

1L-6R-2LC-1RC-5L-2R-4LC-5RC-3L-4R-1LC-3RC-
6L-1R-5LC-6RC-2L-5R-3LC-2RC-4L-3R-6LC-4RC

Where the cylinders are numbered with number 1 being the rear most, and number 6 adjacent to the propeller. For a left hand engine the firing order is:[34]

1L-5RC-6LC-2R-5L-3RC-2LC-4R-3L-6RC-4LC-1R-
6L-2RC-1LC-5R-2L-4RC-5LC-3R-4L-1RC-3LC-6R

In the "B" series, the vees were originally timed to fire together, then Allison found it necessary to change the phasing between the two crankshafts, a parameter that allowed tuning the engine for full power. At first

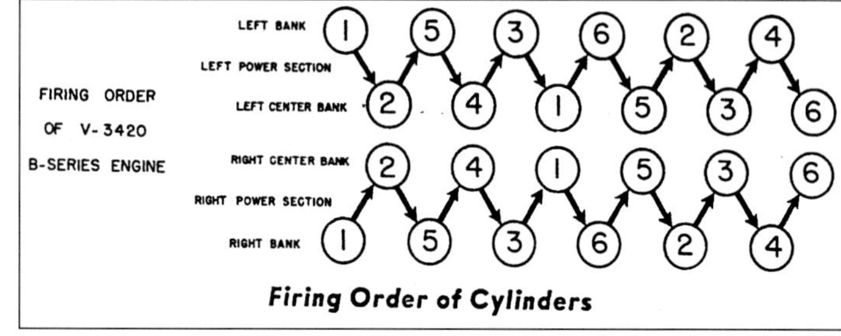

This diagram for the V-3420-23(B10) in the Fisher P-75 shows the firing order. (USAAF)

180 degrees was tried, but further experimenting resulted in setting the phasing at 150 degrees. The firing order was not changed, rather the second vee was firing about a quarter of a revolution later. The "B" firing order is:[35]

(1L,1R)-(2LC,2RC)-(5L,5R)-(4LC,4RC)-(3L,3R)-(1LC,1RC)-(6L,6R)-(5LC,5RC)-(2L,2R)-(3LC,3RC)-(4L,4R)-(6LC,6RC)

In the final development effort to get the engines in the P-75's to develop contract horsepower, it was found that by changing the "phasing" of the crankshafts from the classical 180 degrees to 150 degrees, performance guarantees could be met. This would not be enough of a change to alter the firing order, but it does mean that the cylinders no longer fire simultaneously. Rather they are separated by 30 degrees, which is the same angle as that between the inner pair of cylinder banks. While no documentation of why this was the case has come to light, it is likely that the reason had to do with smoothing flow from the single-stage supercharger into the two trunks to the intake manifolds.[36] The consequence would be that charging the cylinders would no longer be occurring "together" as shown above for the "B", but rather the flow into the manifolds would be smoothed to closer approximate the average flow.

Accessories Section and Supercharging

As done in the V-1710, the accessories section was designed to be interchangeable with future configurations and meet a variety of aircraft requirements. A single-stage, single-speed supercharger impeller, 10 inches in diameter, was situated behind and driven by the right side crankshaft. The balance of the accessories, generator, starter, pumps etc. were then mounted behind the left-hand vee. Likewise the drives were somewhat segregated, except that camshaft drives, magneto, and distributor drives remained associated with the vee they served, as on the V-1710. The rear face of each crankshaft mounted the V-1710 pendulum type vibration damper and was connected to its accessories loads by a short and flexible steel quill shaft, which itself was damped by the standard hydraulic damper to minimize single node low frequency torsional vibrations.[37] There was no mechanical connection between the crankshafts or drives at the rear of the engine. When counter-rotating propellers were used each was assured of receiving exactly the required power, usually one-half of the total, because of the crankshaft timing idler gears at the front of the engine. This balanced loads on each propeller irrespective of the loads at the accessories end of the engine.

For starting the engine, the left-hand power section crankshaft was driven by the starter, and through the crankshaft timing idler or reduction gears, caused everything else to move as well. Some later marks of the V-3420 incorporated an Auxiliary Stage Supercharger in lieu of using a turbosupercharger for maintaining sea-level power to high altitudes. The fairly large power requirements of this device were provided via a "power take-off" shaft connected to the starter gear on the left side vee crankshaft. The Auxiliary Stage Supercharger design appeared as an enlarged version of the similar unit available for the later versions of the V-1710, featuring a variable speed drive provided through a hydraulic coupling. This assured maximum efficiency without wasting power by throttling the induction system when operating at low power or at low altitudes.

The features of the V-1710-E/F accessories sections that allowed for configuring the engine for either right or left-hand rotation were also incorporated into the V-3420. These engines were configured so that the accessories rotated the same direction in all models. This was accomplished by installing an idler gear in the accessories drives on the left turning engines.[38]

Auxiliary Stage Supercharger

The V-3420-B2 was the first engine of the type to incorporate an Auxiliary Stage Supercharger. The specification for this feature dates from March 12, 1940, concurrent with the preparation of the R-40C competition entry of the McDonnell company for the airplane that later became the XP-67. The device was to be driven from the starter gear and equipped with a clutch to enable it to be mechanically engaged or disengaged depending upon operating requirements.

Like the V-1710 two-stage mechanically supercharged engines, the V-3420 used several different drive ratios for the Auxiliary Stage. Several of the models were to also include some very unique, and complex controls and devices to enable them to totally disengage the Aux Stage as a way to improve overall efficiency.

The ratios used with the V-3420 are similar to those used with V-1710, reflecting the fact that the Auxiliary Stages for both engines used impellers with an outside diameter of 12-3/16 inches. Of course, the unit on the V-3420 had larger flow passages to enable it to pass the greater airflow, and to deliver it at the same pressure to the carburetor deck.

As a consequence of having a two-stage supercharging capability, many of the Aux Stage equipped V-3420's also incorporated intercoolers. Even the first model proposed to have an Aux Stage, the V-3420-B2 of 1940 was to have had an intercooler.

V-3420 Reduction Gear Ratios

Much of the versatility of the V-3420 was in its ability to mount either an integral reduction gear or be equipped with extension shafts to remote reduction gears. Since the engine featured two separate crankshafts, it was particularly adaptable to installations mounting counter-rotating propellers. In such cases a "B" Model engine with opposite turning crankshafts would be used with each crank directly connected to a separate propeller. As a consequence of this flexibility, and continuing efforts to find applications for the V-3420, a very large number of engine models were proposed, with only about fifteen of them ever receiving military designations or purchase contracts. Many of these models were only different from others as a consequence of the type and ratio of the specified reduction gear.

Table 14-2
V-3420 Auxiliary Stage Supercharger Ratios

Aux Stage S/C Ratio	Engine Model(s)	Comments
NA	B-2	Mechanical Clutch, Proposal Only
6.96:1	A-19R, A-20R, A-21R/L, B-8, B-9	Only the B-8 in the XP-75 flew
7.38:1	B-10	P-75 Production Engine
7.84:1	B-11, B-12	Experimental Only

Table 14-3
V-3420 Reduction Gear Ratios, Proposed and Used

Red Gear Ratio	Type A or B Engines	Prop Rotation, View from Rear	Propeller Type	Size of Prop Hub(s)	Comments
2.00:1	NA	CW	Hydromatic	#50	Only for proposed single Crankshaft X-3420
2.458:1	B	CW/CCW	Electric	#60 & #80	B-4, B-8, B-10, B-11
2.50:1	A	CW	Hydromatic	#60	A – 1R, A-9R and A – 11R
2.50:1	B	CW	Electric	2ea #50	Outboard Bevel Gears to Pusher Props on B-5
2.772:1	B	CW/CCW	Electric	#60 & #80	Early B-4, also used with V-1710's on XB-42
3.00:1	A	CW	Hydromatic	#60	For proposed A-5, A-6
3.13:1	A	CW or CCW	Hydromatic	#60	w/Turbo installations, A-16, A-17, A-18, A-20
3.13:1	A	CW/CCW	NA	#60 & #80	DV-6840 Counter Rotating RG, w/Aux S/C, A-21
3.23:1	A	CW/CCW	NA	#60 & #80	DV-6840 Counter Rotating Reduction Gear
1.5&2.25	B	CW/CCW	Electric	2ea #50	Angle Drive to 2-Speed RG for McDonnell XP-67
3.80:1	A	CW	Hydromatic	#60	For proposed A-7R, none procured

When driving the large propellers needed for the high power of the single shaft V-3420 models, it was necessary to use considerably greater reduction gear ratios than typical of the V-1710. For this reason, the engines used on the XB-19A, XB-39 and XP-58 used the 3.13:1 reduction gears, while the XP-75 with its counter-rotating propellers, each driven at about half of the engine power, was able to use a 2.458:1 gear set. The selection of propeller type-be it Hydromatic or Electric-was really a question of availability. Satisfactory Hamilton Standard Hydromatic propellers were available that could handle the nominal 3000 bhp, but only the Curtiss-Electric Company had a fully qualified counter-rotating propeller available in this power range.

At the end of the war, development reached the point that a counter-rotating reduction gear was delivered for testing as a DV-6840. In this installation two V-3420 were to be installed, one above the other, so that four drive shafts would extend and connect to the reduction gear. The upper two shafts drove the #60 propeller and the lower pair the #80 propeller. The engines themselves were displaced longitudinally so as to make the entire affair fit within a large airplane. The engines would have appeared as B-8 or B-10's, but assembled so that both crankshafts in one of the V-3420 power sections would turn the same direction. In this way they were designated as V-3420-A21R/L or A-23R/L engines. The A-21R/L driven DV-6840 reduction gear was to have 3.13:1 gears, while the A-23R/L would have used 3.23:1 gears. Records fail to show whether or not the reduction gear was actually tested.

Probably the most complex reduction gear was that of the proposed McDonnell XP-67. The engine was to drive crankshaft speed extension shafts at right angles to the fuselage mounted engine. These in turn were to drive right angle reduction gear boxes mounting epicyclic gear sets. These were to be shiftable and to offer 1.50:1 or 2.25:1 overall ratios. In addition, it was proposed to incorporate a feature that would allow disengagement from the center gear box so that the propellers could windmill and drive accessories independent of the engine. This device was to be provided by

The Navy PT-8 torpedo patrol boat was powered by two 2000 bhp V-3420's, model V-3420-2(A8R/L), circa 1942. (Allison)

the aircraft manufacturer, and Allison specified that they would not be responsible in case of accidental disengagement.[39]

V-3420 Models

The V-3420 was offered for a wide range of uses and applications, spanning land, sea, and airborne installations. Quite a number of the variations were proposed, but only a few were prototyped, and fewer still were actually produced. In many respects, the V-3420 was a versatile solution to the problem of how to best power an airplane, and as a result, it was offered in many configurations and arrangements. Few practical applications were developed.

As originally conceived, the engine was to have been sea-level rated for use with a turbosupercharger. Still, as early as May 1938, Wright Field notified Allison that the Division was contemplating airplane designs using a V-3420 with an "altitude" rated built-in supercharger to obtain a high altitude rating.[40] Actual design was to await the developing experience with the XV-1710-19 altitude rated engine being proposed for the XP-40.[41] The Air Corps was working to design an airplane around the V-3420 requiring a 15,000 foot altitude rating and stated that if higher altitudes were desired turbosupercharging would be used.

As can be observed on Table 14-4, there were a lot of models of the V-3420 proposed. Many of them were never more than studies or proposals done for either the Air Force or an airframe manufacturer. Only the engines for the Fisher P-75 saw production for more than prototype aircraft.

Marine applications were considered from the beginning, with the U.S. Navy purchasing the second and third V-3420's as the two V-3420-2(A8R/L) engines used in the PT-8 fast patrol boat.[42] These were ordered in 1939, though the test program for the Model Test engine was not begun until the spring of 1942. No reasons have been found for not proceeding with production, but it is likely that aircraft production requirements took precedence. The V-3420 program was reinitiated to support the B-29 bomber backup effort.

Just as the V-1710 was doubled to create the V-3420, V-3420's could be "doubled" to create a DV-6480. In November 1946 such a proposal reached the point that Allison built a reduction gear for such an engine.[43] It was probably intended to be a competitor to engines such as the Lycoming XR-7755.

In-Service Problems

The V-3420 was reportedly a mechanically sound engine. Its one serious problem was that of obtaining a uniform mixture distribution between the two trunks coming from the engine-stage supercharger. A lot of effort was expended to rectify the problem, but it was not resolved in time for series production. It is surprising that the option of retaining the individual power sections accessories housings was not used. That approach certainly worked well for the otherwise similar Daimler-Benz Double-Vee's. Another option would have been a "balance pipe" connecting the intake trunks at the front of the engine. After the end of the war, Allison obtained a doubled Daimler-Benz DB 610 engine and studied it extensively. They appear to have been focused on that engine's direct cylinder fuel injection system, a system they were also developing for the V-3420/V-1710 at the time.

Other "Doubled" In-Line Engines

Between the world wars there were several attempts to build "four bank" liquid-cooled in-line engines, usually incorporating the cylinder banks from an existing V-12. With the exception of the 2,592 cubic inch Rolls-Royce "*Vulture*" used in the twin-engined Avro Manchester, none of these "X" configured engines went into series production. Even the "Vulture" was problematic, for it failed in service and was replaced by four Merlin's, thus creating the immortal Avro Lancaster.[44] All of these "X's" were built with a single crankshaft to save weight, and consequently required use of articulated connecting rods. This has never been a happy configuration. Instead, the "Double Vee" or "W" configuration was workable, as shown by it being the final effort to capitalize on V-12 developments and thereby find a quick and convenient way to very high powered piston engines.

Another adaptation, and attempt to double a V-12, was in the Fiat AS-6 built for the 1931 Schenider Cup Races. They coupled two AS-5 V-12's end-to-end, arranged so that the crankshafts drove concentric drive shafts to counter-rotating propellers. There was a single induction system serving all 24 cylinders, and it was the source of problems that kept the aircraft out of the 1931 race. It took two years, but these problems were overcome and the aircraft did then set the all-time sea-plane speed record.

The path blazed by the V-3420 to a doubled engine-gearing together crankshafts from two V-12's- achieved a measure of success, not only in the V-3420, but also in the Daimler-Benz DB 606, DB 610 and DB 613 engines as well.

A more sophisticated approach to doubling the "V-12" in-line was in the "H-24" type engines, such as those developed and/or attempted by Napier-Sabre, Rolls-Royce, P&W, and Lycoming. They each built various configurations of crankshafts, amounting to two "flat" twelve's constructed back-to-back. Very compact, and when finally developed by Napier, an extremely powerful engine. Had the jet engine not come on the scene, there is a good chance that such engines would have given the large radial a run for the money.

If there is a lesson in the "Double Vee" engine story, it is that there is really no easy way to make a major technological advance, say on the order of doubling output, without returning to the basics. Engineering design is a study in compromise, and bolting two engines together requires more compromise than may be justified.

The Daimler-Benz Double Vee

The German decision not to develop a large strategic bomber was probably influenced by a lack of materials and available manufacturing capacity. The need for large bombers was deferred until 1942, when the development of the "Ural Bomber" became a high priority. The result was the Heinkle He 177, an aircraft with the power from four engines squeezed into what was dimensionally only a "twin-engine" airframe. To achieve the necessary power, Daimler-Benz developed the DB 606 by combining two of their inverted, liquid-cooled, 12 cylinder, vee type DB 601 power sections on a common nose case. Although the power sections were inverted, this was a very similar configuration to the Allison V-3420 and probably owes its existence to the public display of the XV-3420-1 at the 1939 New York World's Fair.

The DB 606 was much more like a pair of Siamese V-12's, each unit having its own supercharger and accessories section, although the "off" side unit did have these features constructed in "mirror image." Cylinder capacity was 67.8 liters (4,140 in.cu.), resulting in the engine delivering 2700 bhp at takeoff. The follow-on model to the DB 601 inverted V-12 was the DB 605. With its increased ratings, it was not long before the DB 610 (with 71.5 liters or 4,364 in.cu. and 2950 bhp at takeoff) was built as a similar twin, but based on the DB 605. Both engines developed a reputation for being prone to fires. In fact, many He 177's were lost to this cause. The problem was that oil leaks would develop from the crankcases and naturally run down onto the low slung, red hot, exhaust manifolds where the oil would immediately burn. The engine also had a history of connecting rod failures which would rupture the crankcase with an oil fire immediately resulting. An interesting feature of the doubled engine was that if one power unit failed, a gear was available for "decoupling" the failed unit

Table 14-4
V-3420 Models and Ratings

MODEL	Engine S/C Ratio	Aux S/C Ratio	Carb	Red Gear Ratio	Weight, Pounds	Takeoff, bhp	MILITARY @ 3000rpm BHP/Alt/MAP,inHgA	NORMAL @ 2600 rpm BHP/ Alt/MAP,inHgA	WER @ 3000 rpm BHP/ Alt/MAP,inHgA
X-3420-1	7.000:1	None	MC-12 FI	2.00:1	2160	1600	1600/24001/SL/36.0	1000/13,250/22.5	None
XV-3420-1(A1)	6.000:1	Turbo	PT-13B3	2.50:1	2300	2300	2300[2]/25,000/	2000/25,000/	None
V-3420- 3(A2)	6.000:1	Turbo	PT-13B3	3.00:1	2350	2300	2300[2]/25,000/36.0	2000/25,000/33.5	None
V-3420- 5(B4)	6.818:1	Turbo	PR-58B1	2.458:1	5910/2	2600	2600/25,000/45.8	2100/25,000/38.6	None
V-3420- 7(B5)	6.818:1	Turbo	NA	2.50:1	3291	2600	2600/25,000/	2100/25,000/	None
V-3420- 9(A11R)	6.390:1	Turbo	PT-13E1	2.50:1	2450	2300	2300/25,000/	2000/25,000/	None
V-3420-11(A16R)	6.900:1	Turbo	PR-58B3	3.13:1	2655	2600	2600/25,000/46.2	2100/25,000/38.5	3000/28,000/51.5
V-3420-13(A16L)	6.818:1	Turbo	PR-58B3	3.13:1	2655	2600	2600/25,000/46.2	2100/25,000/38.5	3000/28,000/51.5
V-3420-15(A17R)	6.900:1	Turbo	PR-58B2	3.13:1	2630	2600	2600/25,000/	2100/25,000/	None
V-3420-17(A18R)	6.900:1	Turbo	PR-58B3	3.13:1	2655	2600	2600/25,000/46.2	2100/25,000/38.5	3000/28,000/51.5
V-3420-19(B8)	6.818:1	6.96:1	PR-58B3	2.458:1	3175	2600	2300/20,000/	2100/17,000/	/SeaLev/60.0
V-3420-21(A20R)	8.000:1	6.96:1	PR-58B2	3.13:1	2655	2600	2300/10,000/	2100/ 8,000/	None
V-3420-23(B10)	6.818:1	7.38:1	PR-58B3	2.458:1	3275	2600	2300/20,000/48.5	2100/17,000/41.0	2885/SeaLev/57.5
V-3420-25(A18L)	6.818:1	Turbo	PR-58B3	3.13:1	2655	2600	2600/25,000/46.2	2100/25,000/38.5	3000/28,000/51.5
V-3420-27(A23)[3]	8.0 & 7.8	None	PR-58B3	3.23:1	6200/2	2600	2300/10,000/	2100/ 8,000/	None
V-3420-29(B11)	6.818:1	7.84:1	PR-58B5	2.458:1	3275	2850	2300/25,000/	2100/17,000/	None
V-3420-31(A24R)	7.260:1	Turbo	PR-58B	3.13:1	2695	3000	3000/30,000/52.0	2500/30,000/45.5	None
Proposals and/or Test Only									
V-3420-A 3	9.00:1	None	PT-13B1	2.50:1	2375	2300	2300/12,000/	2000/15,500/	None
V-3420-A 4R	6.00:1	Turbo	PT-13E1	2.50:1	2330	2300	2300/25,000/	2000/25,000/	None
V-3420-A 5R	6.00:1	Turbo	PT-13E1	3.00:1	2400	2300	2300/25,000/	2000/25,000/	None
V-3420-A 6R	6.818:1	None	PT-13E1	3.00:1	2425	2300	2300/ 5,000/	2000/ 5,000/	None
V-3420-A 7R	6.818:1	None	PT-13E1	3.80:1	2550	2300	2300/ 5,000/	2000/ 5,000/	None
V-3420-A 8R/L	NA	None	NA	NA	NA	2000	2000/ SL	NA	None
V-3420-A 9R/L	6.818:1	None	PT-13E1	2.50:1	2400	2500	2500/25,000/	2050/25,000/	None
V-3420-A10R	9.50:1	None	PT-13E1	2.50:1	2450	2100	2100/15,000/	2200/15,000/	None
V-3420-A10R-A	6.4/9.5	None	PT-13E1	2.50:1	2500	2300	2200/15,000-Hi	2000/ 4,000/	None
V-3420-A10R-B	8.00:1	Turbo	PT-13E1	2.50:1	2450	2300	2300/25,000/	1875/25,000/	None
V-3420-A10R-C	8.00:1	Aux	PT-13E1	2.50:1	2725	2300	2300/10,000/	2100/25,000/	None
V-3420-A12R	7.20:1	Turbo	PT-13E1	3.00:1	2550	2600	2600/25,000/	2100/25,000/	None
V-3420-A13R	NA	NA	NA	NA	NA	NA	NA	NA	NA
V-3420-A14R	NA	NA	NA	NA	NA	NA	NA	NA	NA
V-3420-A15R/L	6.818:1	Turbo	PR-58B2	2.50:1	2530	2600	2600/25,000/	2100/25,000/	None
V-3420-A19R	6.90:1	6.96:1	PR-58B2	3.13:1	2785	2600	2300/20,000/	2100/17,000/	None
V-3420-A21R/L	6.90:1	6.96:1	PR-58	3.13:1	2785	2600	2300/20,000/	2100/17,000/	None
V-3420-A22R	7.26:1	None	PR-58B3	3.13:1	2695	2850	2850/ 4,300/	2300/ 7,500	None
V-3420-A22L	7.29:1	None	PR-58B3	3.13:1	2695	2850	2850/ 4,300/	2300/ 7,500	None
V-3420-A25	7.26:1	Turbo	NA	3.13:1	2750	3500	3300/	2600/	3600@3200rpm/ Wet
V-3420-B 1	6.00:1	Turbo	PT-13B1	2.50:1	2475	2300	2300/25,000/	2000/25,000/	None
V-3420-B 2	6.818:1	Aux	PT-13E1	1.5&2.25	2300	2600	2400/18,000/	1900/16,500/	None
V-3420-B 3	NA	NA	NA	NA	NA	NA	NA	NA	NA
V-3420-B 6	NA	NA	NA	NA	NA	NA	NA	NA	NA
V-3420-B 7	6.818:1	Turbo	PT-13	2.59:1	2955	2600	2600/25,000/	2100/25,000/	None
V-3420-B 9	6.818:1	6.96:1	PR-58	2.458:1	2975	2600	2300/20,000/	2100/17,000/	None
V-3420-B12	6.818:1	7.48:1	PR-58B5	2.458:1	3191	2850	2300/25,000/	2100/17,000/	None
V-3420-C 1	7.26:1	Turbo	Carb	3.13:1[4]	2850	4000	3600[5]/30,000/ Dry	3000[6]/30,000/ Dry	4800[5]/ / /Wet

1. Sea Level at 2400 rpm.
2. At 2950 rpm.
3. Two power sections, one with Engine-stage S/C Ratio of 7.8:1 and other with 8.0:1.
4. Also available with 3.04:1 and 3.23:1 reduction gears.
5. At 3200 rpm.
6. At 2700 rpm.

and continuing with the good "engine." This shows the degree of independence retained in the coupled power plants.[45]

Daimler-Benz also built a large capacity version of the DB 601, the DB 603, holding 44.5 liters. This engine was also paired and identified as the DB 613, quite a power plant at 89.0 liters (or 5,433 cu.in.) and delivering 3500 bhp for takeoff. Only 26 were built. There were also a number of the still later model Daimler-Benz engines which were likewise doubled, but none of them saw any more than production for development and testing.

Total production of German double vee engines was much higher than for the V-3420, but it is clear that the arrangement was no more successful. In fact, the continuing engine problems were the direct cause of the cancellation of the He 177 program after that aircraft had been placed in series production. A total of 820 DB 606's and 1,070 DB 610's were built.

V-3420 Powered Aircraft

The V-3420 played a critical role in the development and potential for many different aircraft during the late 1930s and early 1940s. It was usually the most powerful engine available for a given project, and so was often considered either in the original design concept or as a retrofit to improve aircraft performance. In this section we look at most, if not all, of the projects in which it was a consideration. Some are surprising; the engine was promoted at various times to power every type of aircraft, from fighters the size of the P-40, to cargo carriers as large as the Spruce Goose.

While the engine was often leading in the power sweepstakes, the ebb and tide of the war caused the program to vacillate between a top A-1-A priority to suspension and outright cancellation. This occurred more than once, so those intending to build an airplane around it suffered as well.

L.H. Outboard Nacelle for XB-19A as originally built for V-3420-11(A16R), dated 8-24-43. Note compartmentalization of intake. Center duct feeds radiator while duct on right feeds the turbosupercharger, and through it the engine. Note the exhaust from the turbo exiting at 4 o'clock. The two kidney shaped inlets feed cooling air into the shrouds on the exhaust collector manifolds. (Allison)

Douglas XB-19A

Douglas Aircraft Company began work on a secret experimental Project "D" bomber for the Army Air Corps on February 5, 1935. The project was initiated "in an effort to further the advancement of military aviation by investigating the maximum feasible distance into the future" (for a large bomber aircraft).[46] The project was to be built to Wright Field Type Specification X-203, which, in June 1935, directed Douglas to proceed with the design and construction of the Douglas XBLR-2 bomber.[47] The specification called for the XBLR-2 aircraft to be powered by four 1600 bhp Allison X-3420-1 engines and for it to be completed by March 31, 1938. The Air Corps also authorized Sikorsky to develop a competing design, the XBLR-1, though that aircraft was never completed.

As a result of extremely lean budgets during the Depression years, the Air Corps was unable to fund the project at a rate necessary to achieve the specified schedule. Air Corps interest in the project persisted through the long development period which accumulated many changes and modifications. Significant among them was the substitution of the new 2200 bhp Wright XR-3350-5 air-cooled engine for the X-3420-1's after that engine's termination in 1936 for want of an adequate fuel injection pump. Prior to the switch to the R-3350 the Air Corps had Wright work on an alternative power plant, the "Doubled" R-3640. This consisted of two R-

The XB-19A was powered by four 2600 bhp V-3420-11's making it the long-range bomber the Army had sought as the Allison X-3420-1 powered Douglas XBLR-2 back in 1935. The conversion from Wright R-3350 engines was done to expedite development of the power plant installation for the Boeing XB-39. (SDAM)

XB-19A when at Allison Flight Test, Weir Cook Airport. This view shows to advantage the turbo and waste gate exhausts exiting through the stainless steel heat shield. (Allison)

1820 power sections set at a 60 degree angle to each other driving a common gear box mounting counter rotating propellers, all in a "Y" configuration.[48] Apparently this was replaced by the more compact R-3350 configuration. As for the aircraft, it was given a redesignation on March 8, 1938, becoming the XB-19.

Douglas had been forced to spend much of its own money and resources on the aircraft. Because they needed staff on other critical projects, and believed that the design was now obsolete, they recommended on August 30, 1938 that the project be canceled. The Army disagreed, and the project struggled on. With Major Stanley Umstead at the controls on June 27, 1941, when the aircraft made its first flight.[49]

In this form the aircraft was tested extensively by both Douglas and Wright Field to collect data and experience needed in the design of other large aircraft then being developed. The XB-19's R-3350-5 engines were found to be prone to overheating and the performance of the aircraft was consequently constrained by the necessity of keeping the cowling cooling gills open so that the rear cylinder head temperatures could be kept within limits.

Similar problems became evident on the early R-3350 powered B-29 and B-32's, and as insurance against failure of the trouble prone R-3350's, the Army decided to develop an alternative engine installation for the Boeing B-29. The only alternative 2300 bhp class engine was the Allison V-3420, an engine whose development had been stopped in September 1940 in preference to increasing production of the V-1710. In May 1942, a change in priorities resulted in Allison being directed by the Army to restart work on the V-3420, this time with an A-1-A priority. This was followed by an order to Allison for nine V-3420-11's for the XB-39, and 500 similar V-3420's for the first 100 Boeing B-39's, the designation of the V-3420 powered B-29.[50] The immediate goal was to install the V-3420 engine in the fourth YB-29. Unfortunately, the pressure to get the B-29 "combat ready" made the needed airframe unavailable, forcing the selection of an alternate aircraft on which to develop the installation.

Wright Field studied in detail both the Curtiss C-46 and Douglas XB-19 for conversion to V-3420's for installation development. Because of balance problems caused by the forward location of the complete B-39 engine nacelle on the C-46 wing structure, the aircraft was found, unfortunately, to be unacceptable. Testing on a twin engined airframe would have greatly expedited the subsequent development effort.

The Army then commandeered the XB-19 and assigned it to the role of developing the V-3420 engine installation.[51] In this role it would adopt the installation exactly as intended for the B-29 conversion. As a side benefit, the marginal performance of the XB-19 would be greatly improved. Wright Field classified the project as "Restricted" and assigned the Classified Project No. MX-309 on September 25, 1942, contracting with the Fisher Body Division of General Motors to perform the work.[52] Selection of Fisher Body to do the work may at first seem strange, but Don Berlin, designer of the P-40, had just departed from the Curtiss P-40 program because of unhappiness with Curtiss and the Air Corps over not being allowed to install the two-stage Merlin 61 in the P-40.[53] He was then employed by the General Motors Fisher Body Aircraft Development Section.[54]

The XB-19 installation was to use four 2600 bhp V-3420-11(A16R) engines, each equipped with a GE Type CM-2 turbosupercharger, with an intercooler ahead of the engine-stage supercharger. The Type CM turbos were new technology themselves, being two-stage superchargers rated for high altitude use with engines producing a nominal 2000 bhp. The result was an aircraft intended to have a very high critical altitude of approximately 40,000 feet, though in November 1942 this requirement was revised down to 35,000 feet.[55]

Actual work to convert the XB-19 into its XB-19A configuration began in September 1942 with preparations to install the four Allison's.[56] The aircraft was delivered to Fisher at Romulus Army Air Base at Detroit in November 1942, having accumulated 147.0 hours of flight time in the hands of the Air Force, and an unknown amount by Douglas in the June to November 1941 period when they operated it.[57] It took until June 1943 for the V-3420-A16R engine to satisfactorily complete its 150-hour Type Test at Wright Field, and thereby clear the installation for flight. First flight of the then worlds largest and most powerful aircraft, the XB-19A at 160,000 pounds and 9,200 horsepower, was made in January 1944.

The airplane was flown with its original set of V-3420-11(A16R) engines for about 32 hours. Following a successful test program, the aircraft was again modified in March 1944, this time to become a cargo carrier. At this point, the engines were removed and exchanged for modernized V-3420-17(A18R) engines at Cleveland in August 1944.[58] It was then flown to Wright Field to complete propeller vibration testing and to continue performance and flight testing with the new engines.[59]

The aircraft never went into service as a freight hauler for the military. It was transferred to the new All Weather Flying Center located at Lockbourne Air Base near Columbus Ohio, and then relocated to Clinton County Army Air Base at Wilmington Ohio. On April 26, 1946, Col. Ben Kelsey, the last pilot qualified to fly the aircraft, returned it to California.[60] After a short stay, it was finally flown to Davis-Monthan Field, AZ on August 17, 1946 for storage.

One of the Douglas employees who witnessed the aircraft flying at the El Segundo factory remarked at the performance and strong, smooth sound of the V-3420 powered XB-19A.[61]

Three years later the XB-19A was scrapped. Top speed with the V-3420's was 265 mph at 20,000 feet, compared to 224 mph at 15,700 feet with the R-3350's. In reality, cooling problems with the R-3350's required cruising with the cooling gills open, which reduced actual speed to only 204 mph.[62]

Fisher XB-39

Boeing's B-29 was created in response to a Wright Field Request for Data, R-40B, issued in February 1940. The Army requested a bomber with a range of 5,333 miles with improved speed and bomb load over that of the

Under the direction of Don Berlin, Fisher modified YB-29 "Spirit of Lincoln" with four sleek new V-3420-11(A16R) engines. Each engine was rated for 2600 bhp to 40,000 feet, using the new GE Type CM two-stage turbosuperchargers. These engines replaced the original 2200 bhp Wright R-3350-21's rated to 25,000 feet when supplied by two Type B-11 turbos. (Allison)

premier existing bomber, Boeing's B-17.[63] After the war began, Boeing's B-29 took on new significance because it was the preferred representative of a new class of bomber that the Air Corps called the Very Long Range (VLR), or Very Heavy bomber (VH). Such aircraft were going to be of critical importance should Europe and the Pacific be lost. In any event, the aircraft was going to be necessary to carry the battle to the heartland of the enemy, which could conceivably be any place on the face of the earth. When the XB-29 made the first flight for the type on September 21, 1942, as well as its competitor and similarly powered Consolidated XB-32 which had flown first on September 7, 1942,[64] the war was already nearly a year old and the Allied bombing campaign had accomplished little of strategic importance. VLR's were critically needed, but both programs were in deep trouble as a result of fires and failures in their Wright R-3350 power plants. As a consequence, it was decided to proceed with a backup program to develop the only other available engine in this class, the Allison V-3420.[65] The dire circumstances can be appreciated when it is realized that Wright delivered only eight R-3350's in 1940, six in all of 1941, and by the first of August 1942, Wright had only produced an additional 18 engines. Things were not much better for the V-3420, for in 1940 only the two engines for the Navy PT-8 patrol boat were produced, followed in December 1941 by a single XV-3420-9 for testing, and then none in 1942 as the program had been on hold while V-1710 production was increased. Deliveries of V-3420's really began in January 1943, with a total of 30 being delivered during the year.

A V-3420 powered VLR Bomber project really got its start in April 1941, eight months before the war began. General Echols at Wright Field directed his staff to run some design studies with a view to starting another heavy bomber project, probably with Douglas. He was basing his directions upon conversations with O.E. Hunt of General Motors that Allison could deliver an engine for Type Testing in July 1941. Wright Field investigated the performance improvement that would result with V-3420's in either the Boeing B-29 or Consolidated B-32.[66]

In May 1942, some five months before either of the VLR bombers first flight, the General Motors Fisher Division began work for the Army on the design of the installation of the turbosupercharged V-3420-A in a B-29. The Army requested that GM design the nacelle to fit the existing firewall of the B-29. Initial design work had begun as early as February 1942 by Allison and the General Motors Product Study Group in Detroit, a group

Roll-out of the XB-39 with its new V-3420-11 engines delivering a total of 10,400 bhp. The R-3350-23 powered B-29 had 8,800 bhp.

The Fisher modified Boeing XB-39 "Spirit of Lincoln" in flight. The demand for getting the B-29's problems solved and into combat delayed the XB-39 to the point that an order for 100 of the V-3420 powered aircraft was canceled. The program then sought to use the undeveloped two-stage turbosupercharger to further increase the critical altitude to unheard of heights. This resulted in additional delays and early flights without turbos. (SDAM)

that began what was later known as the Fisher Body Aircraft Development Section.[67] The resulting aircraft was to be redesignated as the XB-39. After about 25 percent of the engineering work was done it became apparent that the promised B-29 testbed aircraft would not be available on schedule and the Army determined to substitute the severely underpowered Douglas XB-19.[68] Fisher had recently hired Don Berlin, the designer of the Curtiss P-40, and put him in charge of the design and installation of the Allison engine in the B-29 derivative.

As a Wright Field engineering development project, the "Restricted" identity of Classified Project No. MX-230 was assigned. Initially confronted with the delay in the availability of a B-29 airframe for conversion, the project was given a boost when in September 1942, the Douglas XB-19 was allocated for use as a platform to develop and demonstrate the intended Allison engine installations.

During the year 1942, the reemphasized V-3420 program successfully completed the military 150-hour Type Test of the XV-3420-9(A11R) at a Takeoff and Military Rating of 2300 bhp. This was the first American engine of any type or make to be successfully Type Tested at a rating of more than 2000 bhp. Also during the year, the V-3420-A15L ran a Model Test at a rating of 2600 bhp for both Takeoff and Military.[69]

The aircraft that finally became available for conversion as the XB-39 was actually the first of the fourteen YB-29's, YB-29-BO AC41-36954, which had been named "Spirit of Lincoln."[70] It was delivered to the Fisher plant at Cleveland Airport for installation of the Allison's in November 1943.[71,72] The engine selected for the XB-39-BO was the V-3420-11(A16R) as had been installed in the XB-19A, a model with both crankshafts rotating left, driving a single rotation right hand tractor propeller through integral reduction gears. These engines developed 2600 bhp at 3000 rpm for Takeoff and Military ratings, which with the GE Type CM turbos could go to 40,000 feet. Given the size, weight, and general complexity of the installation, a considerable amount of effort was required to design the na-

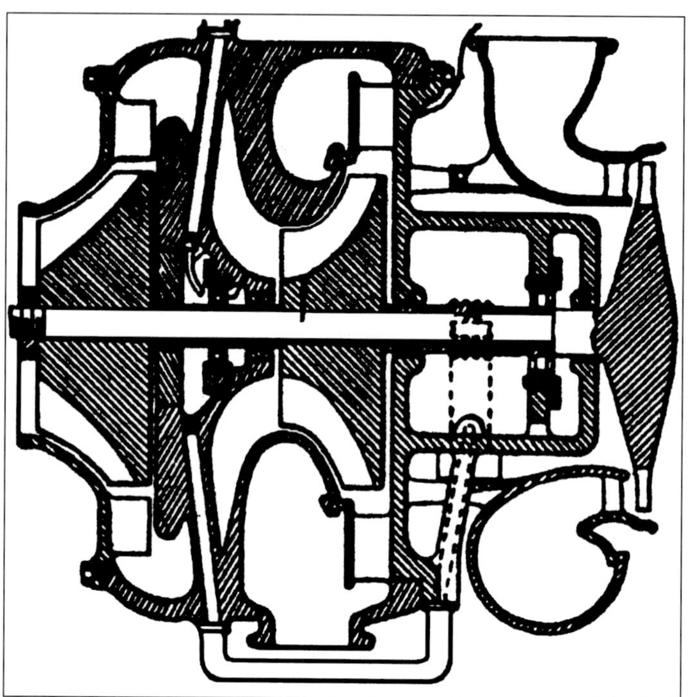

This is a patent drawing for the GE Type CHM, two-stage turbosupercharger. The CM-2 was to be the production version. (Provided by William O. Meckley, Patent holder)

This is how the advanced B-39 powered by V-3420-A25 or V-3420-C1 engines was proposed to be configured. (Allison)

celle and supporting structure for the engine, as well as to incorporate the two-stage GE Type CM-2 turbosupercharger, intercooler and connecting ducting. One difference between the B-29 and XB-39 turbosupercharger installations was the use of a single large two-stage turbo on the XB-39, where the B-29 used two 1000 hp class GE Type B-11 turbos operating in parallel to supply the quantity of air needed by the smaller R-3350 engine. The two-stage unit also allowed considerably higher critical altitudes to be obtained.

In an effort to obtain maximum performance for the XB-39, other supercharger configurations were investigated and engineered. As early as April 1943 an effort was underway to obtain U.S. Navy developed Birmann Type B mixed-flow two-stage turbosuperchargers. These were considered because they had similar performance, but weighed only 260 pounds as compared to the 468 pounds of the Type CM-2 then being used.[73] In August of 1943, the Army purchased from the U.S. Navy eight Birmann designed P-15B-3 turbos from the Turbo Engineering Corporation of New Jersey for use in the XB-19A/XB-39 projects.[74]

Critical to progressing with the engine installations was the availability of the GE Type CM-2 turbosuperchargers. As a new and complex type, GE had difficulty in meeting contract established delivery dates, and the first units were not even scheduled to go to the XB-19A until February 1944.[75]

When the "Spirit of Lincoln" arrived at Cleveland the old R-3350's were removed and preparations begun for the V-3420's. Then in a change of priorities the Army shelved work on the XB-39 in preference for expediting development of the Fisher P-75 long-range fighter.

Again the fortunes of war moved faster than the engineers were going to be able to get the P-75 into combat units. In October 1944, the airplane was canceled in preference to the new jet fighters.[76] Once more work was intensified on the XB-39, and as the first flight date drew near Wright Field decided to have the V-3420's then installed in the XB-19A and XB-39 "Modernized" so as to incorporate the latest features available for Allison engines. This included installing revised pistons with the new "Keystone" shaped rings.[77]

In July 1944 Wright Field had notified Fisher Body that changes in the tactical situation and shortages of manpower necessitated their reducing the number of types of turbosuperchargers that they were having General Electric develop. This way GE could focus on filling orders for critically needed designs then in production. As a result they proposed that the XB-39 be equipped with GE Type CH-5 turbos (a high-speed single-stage turbo) instead of the previously specified two-stage Type CM-2. The advantages were given as greatly reduced weight, improved performance and reliability, as well as ease of installation.[78]

This change in turbosuperchargers resulted in the first flights of the XB-39 being made without any turbosuperchargers installed. The table at the end of this section shows comparative results of flight testing against a similarly constrained B-29. This setback meant that the aircraft was not going to be available during the war, but Allison felt that the V-3420 might be selected for follow-on developments of the B-29 type. Allison continued with development of the engine and the installation, with Don Berlin leaving Fisher Body Division and joining the Allison Division of General Motors as Allison's Director of the Installation Engineering Section in 1945.[79]

At Cleveland, Ohio on December 9, 1944, the XB-39 made its first flight.[80,81] Performance was improved to 405 mph at 25,000 feet (compared to 365 mph at 25,000 feet from the 2200 bhp R-3350-23 powered B-29's) from the 2600 bhp Allison's. The improvement was not enough. The project was so late that it was determined not to disrupt B-29 production with a changeover to the B-39.[82]

Buried engines had always been a goal of the bomber planners and in the proposed V-3420 follow-on to the B-29 the concept dominated. (Allison)

In August 1945, Allison proposed to Wright Field its plans for future ratings available for turbosupercharged, single-stage V-3420's suitable for the B-39. The engine was being developed to truly impressive performance levels utilizing features being demonstrated on the new V-1710-G series engines. As an example, the proposed fuel injected V-3420-C1, when run on Grade 145 fuel and water/alcohol injection, would produce 4800 bhp, with a dry weight of 2,850 pounds.[83] Allison planned to develop this engine in a post-war program that would have cost $3,600,000 and would have required two years before production engines would have been available.[84]

Unfortunately for the V-3420, it was Allison policy and practice after the War not to undertake any development not specifically funded by the customer, in this case the Air Force.[85] A follow-on model of the Boeing B-29 was built, the Boeing B-50A. It used the 3500 bhp P&W R-4360-35, 28 cylinder air-cooled radial engine, which had also been in development throughout the war. Had Allison pursued the V-3420-C program, that aircraft may well have been powered by Allison rather than P&W. The R-4360, weighing 3,490 pounds, was used extensively by both commercial and military aircraft in the post-war period. Later development models of the R-4360 (specifically the "VDT" which used compound turbos) never reached the 4300 bhp of the V-3420.[86]

It appears that extension shaft drives would have been necessary to maintain aircraft balance in the advanced B-39 model. (Allison)

Table 14-5 Comparison of B-29 and XB-39 Performance (As Flight Tested at 10,000' Without Turbos)				
Weight, pounds	**XB-29** 100,000		**XB-39** 100,000	
	Air Speed, mph	Power, bhp	Air Speed, mph	Power, bhp
Military Power	331	2350	346	2600
Normal Power	317	2120	320	2080
Auto-Lean Power	268	1450	277	1407
		Miles per Gallon Comparison		
	True Air Speed, mph	B-29 mpg	XB-39 mpg	Percent Improvement
	320	0.288	0.324	12.5
	280	0.492	0.576	17.1
	240	0.660	0.714	8.2
	220	0.693	0.745	7.5
	215	0.696	na	na

As seen Table 14-5, the XB-39 improvements in speed due to the extra power provided by the un-turbosupercharged V-3420-A16R's were not significant enough to justify the full-scale development of the B-39. The improvement in "mileage" or fuel efficiency was considerable, and mostly due to the reduced drag of the liquid-cooled engine installation.

Martin B-26 Bomber

Very little information is available on this concept airplane, but this artist's rendition dates from 1941 and clearly shows a B-26 bomber powered by a pair of V-3420's. Details as to which model have not been found.

Hughes HFB-1 *"Spruce Goose"*

In the spring of 1943 a Colonel R.C. Wilson of the AAF in the Pentagon passed on to a Hughes-Kaiser subcontractor, Fleetwings, a concern about being able to secure the P&W R-3420 Wasp Major engines intended for the cargo seaplane (500 passenger HFB-1, also known as the H-4 and/or HK-1 plywood Flying Boat). Wilson's office had no authority to communicate directly with Mr. Hughes on the HK-1 project, so he was working through intermediaries and passed his message that, "the Allison 3420 engine is more readily available than the X-Wasp and that it was rated at 2600 horsepower and has every prospect of an increased horsepower rating."[87] This message evidently did reach Mr. Hughes, eliciting a reaction from General Motors.

When the General Motors Headquarters in Detroit heard about Hughes interest in using eight V-3420 engines to power his airplane, they told the Allison Los Angeles Field Office, "not to encourage use of the 3420." Detroit considered Hughes rather unstable and simply did not want to get involved.[88]

One of the reasons the H-4 did not fly until after the war was the lack of suitable engines. Had Hughes been able to use V-3420-31's, which weighed 2,695 pounds and were available in 1944 rated for 3000 bhp, the story of the *Spruce Goose* might have been considerably different. These engines provided the same power as the later 3000 bhp R-4360-4A's actually used on the aircraft, but those engines were much heavier at 3,410 pounds each.

Curtiss-Wright C-46

In 1938, lots of effort was being expended on finding aircraft to use the V-3420. This effort was not only coming from Allison, but from within the airframe industry, and encouraged by the Air Corps and others as they became aware of the 2300 bhp engine.

One firm requesting detailed information on the engine was the Curtiss-Wright Corporation's St. Louis Airplane Division. In June 1938 they stated

Repowering the Martin B-26 with a pair of V-3420's would have made a most attractive airplane, and would have given it outstanding performance. (Allison)

Table 14-6 Performance of a V-3420 Powered Curtiss C-46			
	P&W R-2800-5	V-3420-11 w/Turbo	V-3420-C174a
High Speed (Military), mph/bhp/alt	270/3000/14,000'	324/5200/14,000' 354/5200/25,000'	320/4810/14,000' 333/4600/21,000'
High Speed (Normal bhp)	264/2900/13,000'	297/4200/13,000'	305/4200/13,000'
Service Ceiling, feet	24,600	37,500	32,500
Max Gross Weight, Pounds	45,000	50,586	48,507
Takeoff Distance, feet	1,950	1,975	1,790

their wish to consider the use of the Allison V-3420 for proposed military and commercial airplane designs.[89] Given the nature of the projects the St. Louis Division was working on at the time, this would have been the first consideration of the engine for use on what ultimately became the C-46, an aircraft powered by early models of the P&W R-2800.

With the 1942 rush to get the V-3420 powered XB-39 into the air, it was necessary to find a suitable airframe able to provide inflight test data on the proposed engine/turbosupercharger installation. A study of the C-46 was begun in January 1943. Two different engines were considered, a turbosupercharged V-3420-11, and a two-stage supercharged development model, the V-3420-C174a.[90] The results were that the turbosupercharged installation was not practical as the amount of equipment designed into the XB-39 nacelle resulted in moving the aircraft center of gravity nearly four feet forward (the propeller disk would have been 56.8 inches forward). This was believed too much for safe operation of the airframe.

The two-stage engine was able to utilize an entirely new design nacelle, one that maintained the maximum cross-section of the previous R-2800's, but was incompatible with the XB-29. Balance was not so badly impacted, moving forward only 19 inches.[91] This was considered a feasible installation, but the Air Force was convinced that it wanted a turbosupercharged installation and decided to use the Douglas XB-19 as the test bed instead. This decision resulted in a lot of delay, for the project was at least twice as big as the effort that would have been required to modify the C-46.

Performance of the souped-up C-46 would have been impressive.

This kind of performance improvement was not lost upon the Curtiss-Wright company. They were under some pressure as a result of the service conditions and demands of flying the "Hump" from India into China. This extremely dangerous route was hard on the limited cargo carriers of the day, the C-46, C-47 and C-87. The very high operating altitudes and heavy operating weights were taking a heavy toll on aircraft and crew. In the Spring of 1943, the Air Corps was facilitating efforts by Curtiss-Wright to install either the Allison V-3420-C174a two-stage, or V-3420-11 turbosupercharged engine in the C-46. By that fall, such an installation was being considered for a C-46 follow-on.[92]

Lockheed XP2V-1 Conversion

The Wright R-3350 powered Lockheed-Vega "*Neptune*" land based patrol bomber (Lockheed Model 26) had a long and distinguished career. Design work on this replacement for the Lockheed *Hudson* and *Ventura* patrol bombers began in 1941, just prior to Pearl Harbor. With the focus on getting current production types to the front, work was not authorized by the Navy until February 19, 1943, when they ordered two XP2V-1 prototypes (to be powered by 2300 bhp R-3350-8's), though the contract for them was not issued until April 4, 1944. First flight of the type occurred on May 17, 1945.[93] Early in its active development phase Lockheed seriously considered converting the airplane to V-3420 power. While the reasons the installation never went into production are not known, the long-time Navy aversion to liquid-cooled engines may have again held sway.

CONVERSION INSTALLATION
ALLISON V-3420

This installation of the V-3420-A22R/L in the XP2V-1 involved the minimum changes from the original R-3350 engines. While compact it did not exploit the low drag advantages of the inline engine. (B. Matthews)

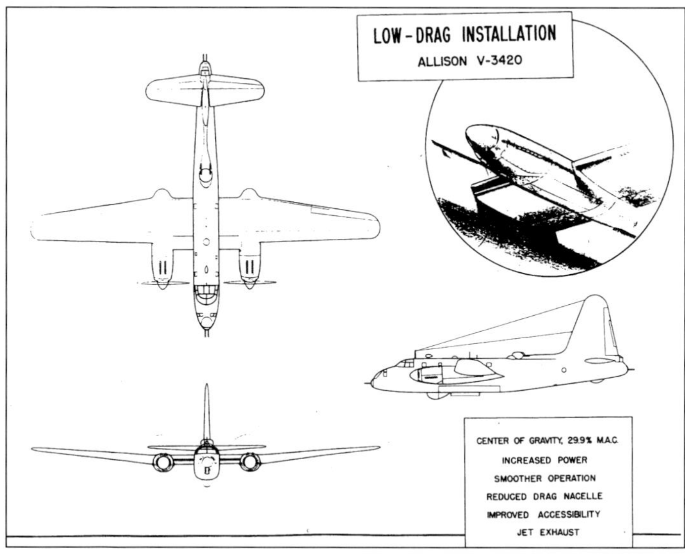

This 3-view shows the way the new streamlined nacelles would have looked on the Lockheed XP2V-1. (B. Matthews)

Two different versions of the conversion to V-3420 power were proposed in a detailed study by Lockheed in July 1944.[94] In the first, the R-3350 cross-section at the firewall was retained. In the second a "low-drag" installation was achieved by changing the landing gear stowage so that the main wheels laid flat in the nacelle. This required relocating the "chin" mounted radiators.

Both airplanes were to have been powered by Allison V-3420-A22R/L single-stage, single-speed engines having 3.13:1 propeller reduction gears. In either case the propeller was to have been a four bladed 17'-0" Curtiss-Electric unit rather than the XP2V-1 hydraulic prop.

The basic conversion retained the R-3350 firewall and would have accomplished engine cooling using a large "chin" radiator which was to be fed cooling air through a streamlined inlet fashioned much like that on the original Lockheed Model 22 proposal of 1937. This inlet fed both the radiator and carburetor. Oil cooling was accomplished using a radiator installed behind an extended leading edge, the outlet air flow, and consequently the oil temperature, being controlled by a fowler flap on the upper surface of the wing. The entire installation was 92.125 inches long, 1.375 inches longer than the R-3350 installation, and resulted in the thrust line being 13.625 inches higher. Even so, the minimum 12 inch propeller to fuselage clearance declared by the Navy was not met in these installations as the larger prop required by the more powerful engines resulted in only 5.1 inches of clearance.

The reasons for proposing the conversion to the low-drag installation were first an increase in power, smoother operation, a reduced drag nacelle with improved accessibility and considerable thrust from using jet exhausts. In this case the engine extended 90.75 inches from firewall to propeller, though the firewall and propeller were moved forward 1.75 inches as compared to the XP2V-1. The thrust line was 8 inches higher than on the R-3350 powered XP2V-1. Carburetor air was obtained from streamlined inlets on either side of the cowling, just behind the propeller spinner. With the elimination of the chin mounted radiator for this installation, both the coolant and oil radiators were placed in the extended leading edge of the wing, between the engine and fuselage.

The selected V-3420-A22R/L engines were originally defined by Allison in June 1943, as 2600 bhp sea-level engines with 6.90:1 supercharger gears and intended for the turbosupercharged Lockheed XP-58. The specification was updated on July 20, 1944 to describe them as altitude rated engines for the proposed Allison powered XP2V-1 conversion. In this form, they developed 2850 bhp from 3000 rpm at Takeoff and up to 4,300 feet, while being rated for continuous operation at 2300 bhp with 2700 rpm at 7,500 feet.[95]

Lockheed XP-58 *"Chain Lightning"*

It is sometimes baffling as to why there were three such different, but similar Lockheed airplanes as the P-38, XP-49 and XP-58. Your author believes the reasons go back to one man, General Olds. In a hand written 1939 memo to Col. Spaatz he states:[96]

"• Lockheed P-38 type to replace all pursuits *if* such types can withstand terminal velocity dives *and* carry one rear gunner. References: Bristol 2-seater fighters shot down more aircraft of all types on the WWI front than any other pursuit squadrons during the latter months of the war. Both German and British are now swinging to 2-engine fighters for speed, fire power, and arc of fire.

• Limit radius of action of light, undefended bombers to that of P-38 types for necessary protection.

• Increase fire power, (fuel) tank security and armor on medium bombers and self defense when cover cannot be taken by formations in weather and delete requirement for long range escort which is nothing more or less than another bomber carrying guns instead of bombs. (Note: If escort types are considered inseparable they should be organized integrally as a 4th squadron in each Bombardment Group of the striking force.)"

The design of the XP-58 certainly provided the exact features demanded by General Olds. It was a P-38 "type", had a rear gunner, offered long range escort as a fighter, had heavy firepower and was able to withstand terminal velocity dives. But there is more, including the unique way the aircraft was procured.

When finally built the XP-58 was powered by modernized V-3420-11/13 engines with GE Type XE turbos. With 3000 bhp available from each engine at 28,000 feet the 40,000 pound airplane was capable of 430 mph. (SDAM)

Coming out of the lean between wars period and the Depression, the Army Commanders were extremely value conscious and able to get some excellent bargains. One example is how they finessed new airplanes they wanted out of the airframe manufacturers. In the case of the XP-58, Lockheed originally agreed to provide it under a $1.00 change order to the XP-49 contract. In exchange for the Government agreed to release the P-38 for export.[97] This is what is meant when it is stated "the airplane was procured as a result of foreign sales agreements."

In its original concept the XP-58 was to have been a derivative of the XP-49, and also powered by 1700 bhp Continental IV-1430 engines. As such, the XP-58 would essentially have been an improved P-38 tailored for high altitude, medium range escort duties.[98]

When Lockheed ran into compressibility troubles on the P-38, they became worried about similar difficulties on the XP-58 and suggested the elimination of the turbosuperchargers and redesign as a low altitude ground support airplane with its 75 mm cannon. The technical staff of the Materiel Division agreed, but a November 1942 conference in Washington DC on pursuit aircraft, argued that the high altitude version of the XP-58 represented an excellent and immediate answer to the Pursuit Directorate's requirements for a bombardment destroyer type of airplane. As such, it was not envisioned as an aircraft which would have to be employed in terminal velocity dives, and hence the compressibility troubles would not be so serious. As a result the Pursuit Board recommendation was to go ahead with procurement of two airplanes as in their original concept.[99] During the subsequent protracted design and development period, the XP-58 was assigned practically all of the promised high powered engines of the period. When it finally flew, it was powered by Allison V-3420's.

Engine Selection

When the XP-58 project was started in early 1940 the aircraft was to be an improved P-38 powered by two Continental IV-1440 engines. In May of 1940 Wright Field decided the aircraft should be two-place and use the new Continental IV-1430 engines instead. In July 1940, it was realized that as a heavily armed two-place, the aircraft would be inferior to the Lockheed XP-49, which was to be a P-38 type powered by either the P&W X-1800 "H-2600" Type liquid-cooled engine (1850 bhp), or the Wright R-2160 "Tornado" liquid-cooled radial engine. It was therefore logical, in Army parlance, to switch the engine installations between the two aircraft as they were both still in the early design stages.

In October 1940 Lockheed was notified that the P&W X-1800 engine had been canceled so that P&W could concentrate on production and development of their air-cooled radial engines, particularly the R-4360. The Army the requested that Lockheed study the possibility of installing other engines, the "Tornado" R-2160, Lycoming H-2470, and P&W R-2800 being suggested candidates. The Tornado configuration was selected for development that December. In this form, work continued on the design and construction of two prototypes until early 1943, when it became clear that the R-2160 was not going to become a reality.

Allison V-3420's were then suggested, giving the added benefit of a total of some 1300 bhp more than previously available, though gross weight did increase and offset some of the gain. With the engine change to V-3420-11/13's, and the adoption of four 37 mm cannon for armament, the first aircraft (AF41-2670) was specified to be available to fly in August 1943. Due to the priority need for Lockheed to reassign its personnel to the task of installing leading edge fuel tanks on P-38J's, work on the XP-58's was delayed. The first aircraft became available for engine ground running in January 1944, the results of which showed unacceptable overheating problems. In April 1944, the second aircraft was canceled in deference to the promising jet powered Lockheed XP-80 which was then absorbing all available Lockheed resources.

Still, modernized V-3420's were installed in the first aircraft and it made its first flight, from Burbank, California on "D-Day", June 6, 1944 with a landing at Muroc Flight Test Base (now Edwards AFB), after a 50 minute flight.

After twenty-five flights the contractor's test program was complete and the aircraft was delivered to Wright Field that October. Considerable trouble had been experienced with torching from the large GE Type XE turbosupercharger exhaust which resulted in scorching the right hand rudder and aft boom section.

During a high speed run in August 1944, engine trouble developed which necessitated a replacement engine. The experience at Wright Field was not particularly good and only a few flights were made, with the major problems being with the aircraft's hydraulic system. In May 1945, the aircraft was placed in "Class 26", to be used for ground instructional purposes only.

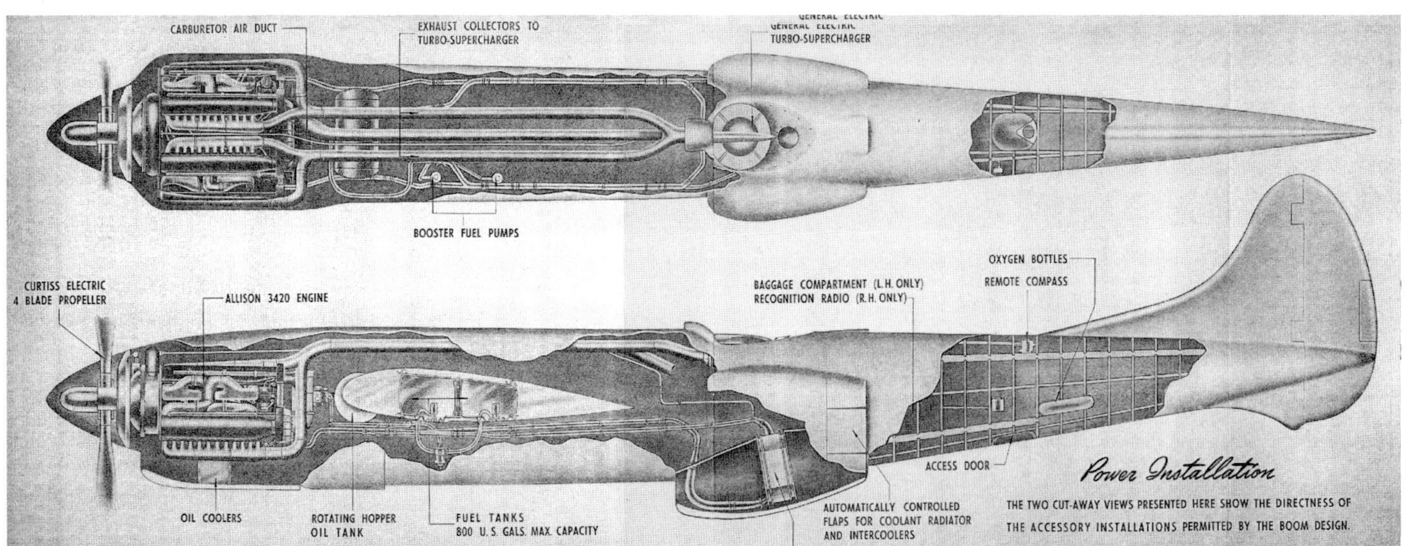

Internal layout of the engine, turbo, and systems in each nacelle of the XP-58 was quite different than on the similar appearing P-38. (USAAF)

Had the airplane gone into production, it would have been powered by V-3420-17/25(A18R/L) engines and fed air from GE Type E-2 turbosuperchargers. These engines had the same ratings as the A-16's and differed only in internal details.

Engine Installation and Cooling

The general arrangement of the engine, systems, and turbosupercharger were as an in the P-38, but everything was proportionally larger. A single turbo was used with each engine, with the experimental General Electric Type XE first installed, a large single-stage unit rated for use with engines in the 2201-3000 bhp range. This was soon replaced by the GE Type E-2, a model configured specifically for use in the XP-58 and able to support the V-3420's to a critical altitude of 25,000 feet. The manifolds connecting the engine to the turbo were designed to handle 325 pounds per minute of 1800 degree exhaust gasses from each engine, all at a differential pressure of 20 inHg. The effect of the shrouds and slip stream passing the ducting was sufficient to reduce the temperature of the gasses to less than 1450 degrees by the time they reached the turbo. The exhaust manifolds were made from 0.049 inch thick stainless steel and weighed 250 pounds.[100]

One subtle difference from the P-38 was the location of the turbo and intercoolers. In the XP-58 these were located along with the radiators in compartments midway back on the booms. The radiator scoops and exit chutes were located on the bottom of the boom while the intercooler radiators were on the sides, the location of the radiators on the P-38. The turbo was mounted directly above the radiators, making for a compact, though busy, section of the booms.

Each engine, sea-level rated for 2600 bhp, required 300 pounds per minute of oil to move 300 hp of heat to the oil coolers, while 530 gpm of coolant was required to reject 860 hp. The pressurized coolant system operated at no more than 15 psi. The engines had a WER rating of 3000 bhp each, and the cooling systems were able to handle this amount of power. Early flight testing found that the cooling air control flaps were marginal in performing their function, though satisfactory cooling was obtained. This was attributed by the Flight Test Engineers to the coolant radiators having been designed for the considerably greater amount of heat that the liquid-cooled Wright R-2160 *Tornado* would have produced.[101]

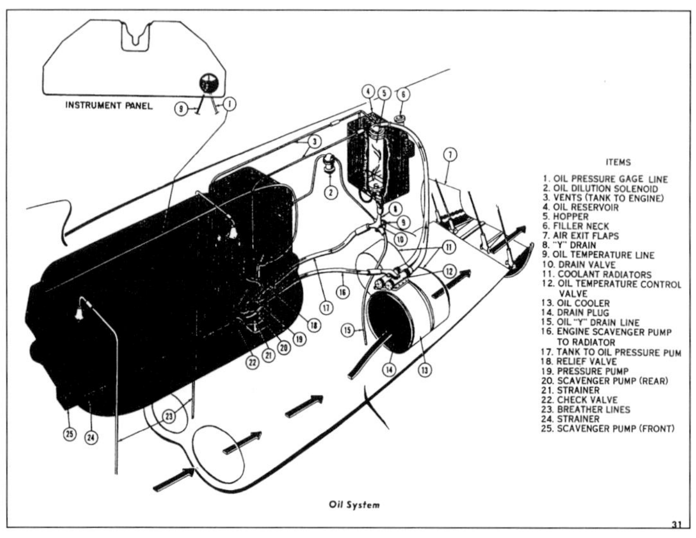

P-75A Oil cooling system. (USAAF)

General Motors Fisher XP-75 *"Eagle"*

William S. Knudsen, past president of GM and head of the Government procurement and production programs, believed that traditional airframe production methods needed to be considerably revised with an emphasis on maximizing the interchangability of major parts and components. He devoted much time and energy to the advancement of this concept and deserves credit for significant improvements in aircraft manufacturing techniques that resulted. Probably the most radical application of his theories was embodied in the Fisher XP-75.[102]

With Donovan R. Berlin employed by General Motor's Fisher Body Division, Fisher submitted a proposal to the Air Corps Materiel Division in September 1942 for a heavily-armed high-performance interceptor. It was to be a hybrid aircraft built around a new and heavy fuselage incorporating the V-3420. The wings, empennage, landing gear, and other systems came from aircraft then in mass production. The engine to be used would be the first of the V-3420-B's, that is, the model with counter-rotating crankshafts. Furthermore it was to be fitted with extension shafts to a remote counter-rotating reduction gear installed in the aircraft's nose, like the P-39. Contract W535-ac-33962 was awarded to Fisher on October 1, 1942 covering two prototypes, designated XP-75 (aircraft AF43-46950 and 43-46951). Delivery was to be in six months, with the aircraft specified to achieve 389 mph at sea-level and 434 at 20,000 feet, along with an initial 4,200 feet per minute rate of climb enabling it to achieve 20,000 feet in 5.5 minutes and an absolute ceiling of 39,000 feet. It would have been quite a performer given its 18,210 pound maximum gross weight.

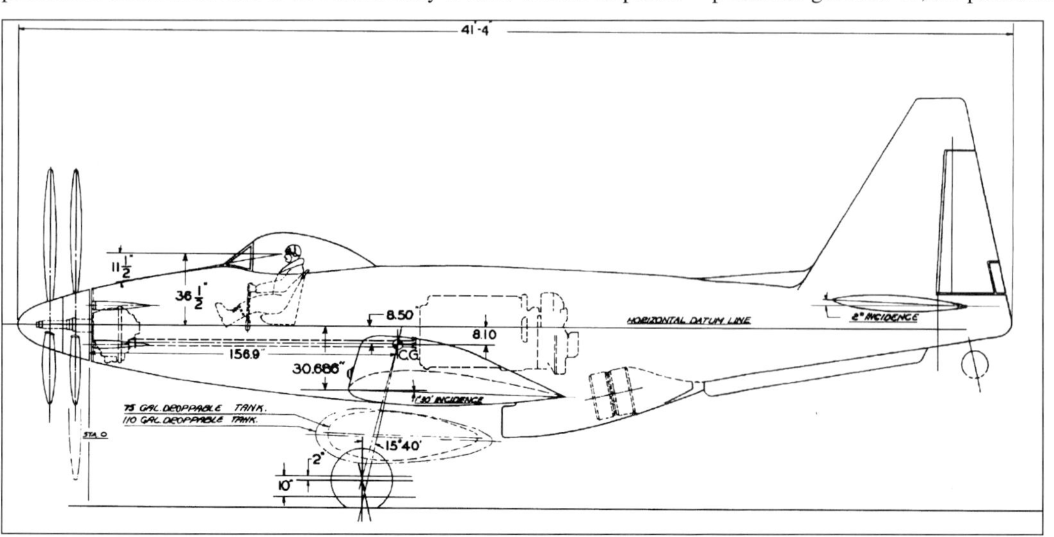

This is the P-75A production version of the Eagle and powered by the 2850 bhp two-stage intercooled V-3420-23(B10). At its 20,000 foot critical altitude 2300 bhp was available. Although the original P-75 concept was to use components from other production airplanes mounted on a new fuselage, the P-75A was an entirely new airplane. (Allison via Hubbard)

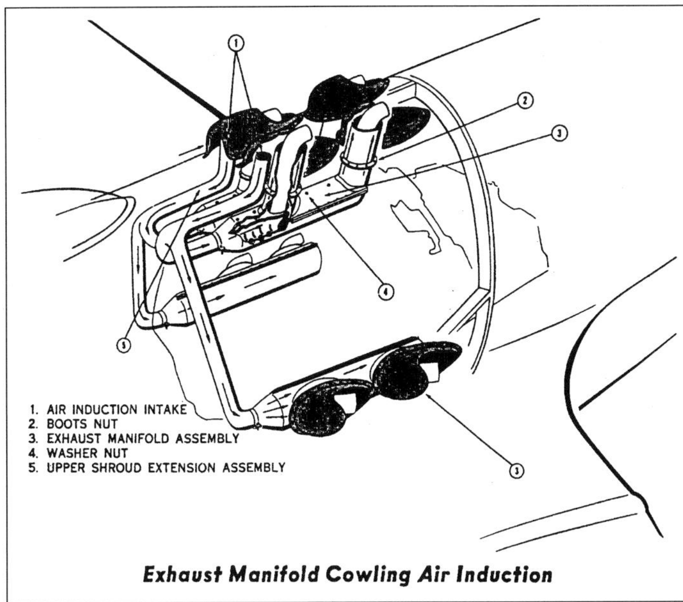

The exhaust system on the P-75A was a challenge simply because there were so many ports on the V-3420. Collectors routed the hot exhausts from each bank through cooling shrouds to exits on the fuselage. (USAAF)

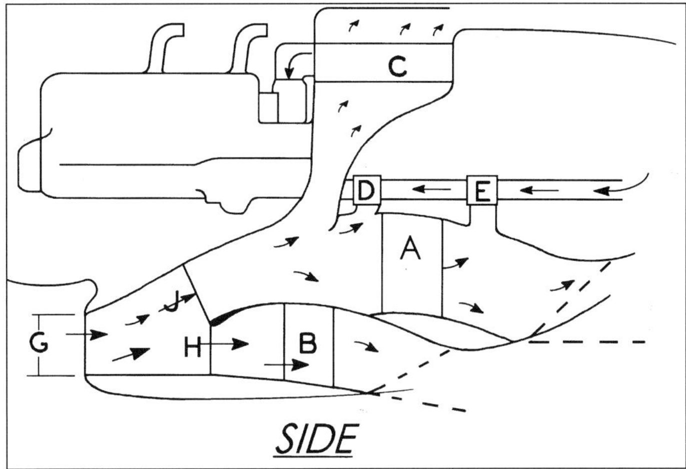

Airflow through the radiators, oil coolers and air cooled intercooler made for a lot of internal ducting in the P-75A. "A", two 16 inch diameter coolant radiators, "B", two 16 inch diameter oil radiators, "C" one 50% effective Intercooler, "D" is carburetor cold air intake, "E" is carburetor hot air intake, "G" is 465 square inch inlet, "H" is 30 % of air to oil coolers. Cooling systems designed for 2850 bhp condition, 19,000 BTU/min to oil, 39,500 BTU/min to coolant. The XP-75 prototypes had suffered from inadequate engine cooling. (Fisher via Hubbard)

The Army Materiel Division's view on the XP-75 was that it was an attempt to solve the problems incidental to the installation of the V-3420 in a pursuit airplane, a concept they had been pursuing since 1939. While General Echols was at Wright Field as Assistant Chief of the Materiel Division, the Division tried to interest several of the existing pursuit manufactures in an airplane mounting the V-3420. However, apparently none of them were interested, saying, in effect, that it was not a pursuit engine and would never be of any use as such. However, when Mr. Don Berlin left Curtiss and went with Fisher Body Division, he made some preliminary studies and determined that an airplane could be built for something between 12,000 and 12,500 pounds gross weight with an excellent initial rate of climb using the two-stage version of the engine. Due to Mr. Berlin's personal drive, the project began rapidly moving forward by using parts of existing airplanes to the maximum extent in order to get the first airplane flying within the deadlines date of six months.[103]

On November 17, 1943, the first flight of the XP-75 took place, powered by the V-3420-B8. Concurrently, the first "production" V-3420 was shipped.[104] It looked like the V-3420 program was finally really going somewhere.

Allison had a fair amount of trouble in getting the V-3420-B engine in the XP-75 to deliver its full power. The reason was recognized early in that the V-3420 used a single engine driven supercharger with two trunks, one leading to each pair of cylinder banks or vee. This resulted in a uneven flow to the cylinder banks under certain conditions, and was finally resolved by changing the phasing between the two crankshafts. An earlier effort to resolve the problems focused on developing a suitable port type fuel injection system. This had been an objective of the Materiel Command for some time, as they had intended to use it on the V-1710. By the fall of 1943 it was determined that the V-1710 was operating fine with its carburetor and ramming manifolds. As a result the fuel injection project was then focused on the V-3420, the goal being to resolve the much more complex distribution problem. They believed that chances of fuel injection resolving the V-3420 problems were pretty good.[105]

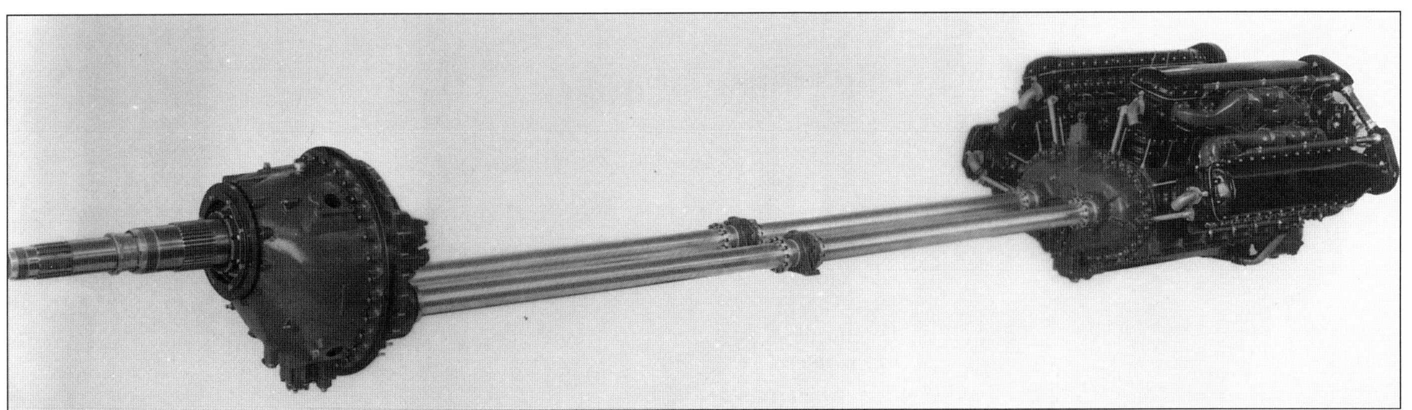

The V-3420-B4 was built as Allison workhorse engine EX-41 and photographed 4-1-43. It was a single-stage engine, the first configured for the XP-75. (Allison)

The program was troubled by changing mission requirements from the Army Air Force, each of which required redesign and delay. Finally on June 7, 1944 an order for 2,500 P-75A-1-GC "*Eagle*" escort fighters, to be produced at Cleveland, was placed by contract W535-ac-41011. On October 6, 1944, even with production finally underway, the *Eagle* became one of the first end-of-the-war cancellations, as it was obvious that the P-75A's would not be available in sufficient quantities to get into service before the end of hostilities. Consequently, the production contract was canceled, although Fisher was authorized to complete up to 20 of the aircraft already on the line. Total finally built, of all models, was only 14. At the close of the program, the third production P-75A (AF44-44551) with a two-stage, intercooled V-3420-23(B10) was bailed to Allison for further flight testing and development work of the V-3420.[106]

Allison went on to install their V-3420-B12 engine in AF44-44551, which was an intercooled engine similar to the B-10, but offering War Emergency ratings of 3150 bhp at sea-level, and 3050 bhp at 18,400 feet, both using 3000 rpm. The goal of the project was development of the engine, not the airframe.[107]

One of the concerns about liquid-cooled aircraft has always been the entire weight of the installation. The specifics of the P-75A may be interesting, if not instructive. The Power Plant Group weight breakdown is given in Table 14-7.

All of this was for the P-75A-1-GC aircraft that, with the Allison V-3420-23(B10), had an Empty Weight of 11,574.0 pounds.[108]

Table 14-7
P-75A Engine Installation

Engine Weight as Installed		3,191.0
Engine Accessories		221.7
Power Plant Controls		44.3
Propellers		843.9
Starting System		55.2
Cooling System		544.0
* Radiators & Shutters		244.1
* Liquid	268.5	
* Piping	31.4	
Lubricating System:		217.7
* Tanks and Protection	136.4	
* Piping etc.	81.3	
Fuel System		802.8
* Tanks and Protection	629.6	
* Piping etc.	173.2	
TOTAL, pounds		5,920.6

The second production XP-75A. Performance of the first XP-75A was below specifications so this airplane was flown to Moffett Field, California and tested in the full scale NACA wind tunnel. The third airplane, (AF44-44551) was provided to Allison on a no-cost bailment contract for ten hours of flight testing. (Allison)

Table 14-8
Double Vee Model V-3420 Engines Shipped

Model	Number	Aircraft	Comments
V-3420- 1(A1)	1	Type Test	Development contract W535-ac-09678
V-3420- 2(A8R/L)	2	PT-8 Boat	Navy PT Boat project
V-3420- 9(A11R)	1	XB-19A Model Test	Development engine
V-3420-11(A16R)	17	XB-19A ,XB-39, XP-58	Flight engines. 500 ordered for the B-39 were Canx.
V-3420-13(A16L)	3	XP-58	Flight engines
V-3420-15(A17R)	1	XB-39	Development and test only
V-3420-17(A18R)	7	XB-19A ,XB-39	Flight engines
V-3420-25(A18L)	4	XB-19A	Flight engines
V-3420-21(A20R)	1	XP-75	Intended for XP-75, not flown
V-3420- 5(B4)	4	XP-75	Intended for XP-75, not flown
V-3420- 7(B5)	1	XP-67 or FM-1?	Fuselage mounted, bevel gear drives to twin props.
V-3420-19(B8)	12	XP-75	Aircraft 43-46950 & 43-46951
V-3420-23(B10)	101	P-75, P-75A	Aircraft 44-32161/6 and 44-44549/554
V-3420-29(B11)	2	P-75	Development only
V-3420-B12		P-75A-1GC	Engine devel only, A/C44-44551 Not in total built
Total	157		37 "A" Models and 120 "B" Models

Table 14-9 V-3420 Engine Models

The following describes each model of the V-3420 as designated by Allison. Many of these models were only conceptual, or existed as paper proposals to aircraft manufacturers interested in preparing a proposal to a likely customer. Engines that actually existed were usually purchased by the Army and issued a series model number.

X-3420-1: This engine was to be of particular importance to the Air Corps and its long range bomber program, intended as it was for the Project "D" airplane. The engine itself was identified as Restricted Project M-2-35, and defined by Allison Specification 201 dated June 12, 1935.[109] The engine was to use four cylinder banks from the fuel injected V-1710-5(C2), mounted on a crankcase and connected by master and articulated rods to a single crankshaft. The angle between the inner banks of cylinders was to be 90 degrees, while 60 degrees was retained between each of the outer pairs of banks. The result was 150 degrees between the centerlines of the vees. Compression ratio was 8.5:1 to provide high efficiency when powering long range bombers. A single-stage 11.0 inch diameter supercharger impeller, driven by 7.00:1 step-up gears, was provided for the altitude rated engine. Two Marvel MC-12 direct cylinder fuel injectors, as being developed for the V-1710-5/9, were to be used. Allison fuel injection induction manifolds and air throttle were to be used to control air into the cylinders. Takeoff power was to be 1600 bhp at 2400 rpm, limited by the articulated rods and crankshaft arrangement. 1000 bhp was available for cruising at 13,500 feet, where fuel consumption was to be less than 0.42 pounds per bhp per hour using 100 octane fuel. A single right hand turning #50 spline propeller shaft was fitted and driven by a 2.0:1 reduction gear, the design of which appears to have featured a large plain bearing. Total weight was estimated at 2,160 pounds. Only a wooden mockup of the complete engine was constructed.

X-3420-3: This engine was defined by Allison Specification 202 dated June 13, 1935. It was to be identical in every respect to the X-3420-1 except that it was to be equipped with a 2.0:1 counter-rotating reduction gear. Total weight was estimated at 2,200 pounds.

The following are the two-crankshaft V-3420's as outlined by Ron Hazen in 1936. The Allison V-3420-A series engines were defined by having the two crankshafts rotating in the same direction. As usually applied they tended to be used in installations with integral, single rotation, reduction gearboxes, but this was not a necessary distinguishing feature. The "B" series engines had crankshafts that turned in opposite directions.

V-3420-A1: This was the single XV-3420-1(A1), AC38-119. Its design incorporated the many lessons learned in development and testing of the early V-1710 models and was also designed to improve manufacturability. Such features as the reversible crankshafts, to accommodate right/left rotation, and counter-rotating propellers, were incorporated. A new crankcase and Accessories section was designed. Though laid out to incorporate features similar to those of the V-1710-E/F, it was sized for the V-3420 and included a 10.0 inch diameter single-stage engine driven supercharger. The physical arrangement of the cylinder banks differed from that in the earlier X-3420 in that there was 90 degrees between the centerlines of the vees, the X-3420 used 150 degrees. For its size and rating it was little larger than the "two" V-1710's which it was intended to combine. Total weight was 2,300 pounds, was 1,690 sq.in., that is, 11.74 sq.ft.[110] Conceptually, it was a single-stage engine with 6.00:1 step-up gears intended for turbo-supercharging. The engine used a 2.50:1 reduction gear ratio and drove a single right-hand propeller. Compression ratio was unique at 7.25:1.

View of underside of fuel injected X-3420 wooden mockup engine. (See: Photographs of the Allison X-3420, Project M-2-35, March 11, 1936. Wright Field File D52.41/6-Allison.) (Air Corps)

The XV-3420-1 design was formally begun in November 1936, with construction under experimental contract W535-ac-09678, issued in March 1937. The engine was constructed and running in April 1938, but due to a somewhat short sighted policy from a development standpoint, a number of minor items were used to delay acceptance by the Air Corps until December 30, 1938. Consequently, the 50-hour development test to be performed at Wright Field was delayed to the point where most of the interest was lost. The test was further delayed by minor failures in the accessory housing gear train bearings and again by a faulty cotter pin which came loose and caused a rod end failure, seriously damaging the engine. After rebuild, another failure of the accessory drive gear bearings occurred at 20 hours, and at forty hours the Government furnished Schwartz wood propeller lost 6 to 8 inches from one blade tip, shaking the engine badly. When torn down on June 10, 1940, the center crankcase was found cracked. By this time, the V-1710 production program was so enlarged and all consuming that the Army directed that development of the V-3420 be suspended.[111]

The XV-3420-1 had an impressive crossection, though total frontal area was only a little greater than that of two V-1710's. Specific weight was much less, being the first engine to achieve 1 bhp/pound. (Allison)

A-2: Designated as the V-3420-3, Allison was requested on April 18, 1938 to quote on five of these engines. They were to have 3.00:1 reduction gears and SAE #60 propeller shafts. The first engine was to be for Type Testing, the following four were to be produced pending completion of the Type Test. Allison offered the engines under their Specification 204 at a price of $60,000 each. In June 1938, contract W535-ac-11328 was issued for six XV-3420-3(A2)'s. They retained the 7.25:1 compression ratio of the XV-3420-1, 6.00:1 step-up gears and PT-13B1 carburetor. They were sea-level rated to 25,000 feet with 2300 bhp at 2950 rpm from only 36 inHgA, and a weight of 2,350 pounds. In December 1939, the contract for the six engines was changed to use the same 2.50:1 reduction gear as on the XV-3420-1 and redesignating the six engines as V-3420-1's. This was in place of the intended 3.00:1 gear originally specified in the contract, and saved the government $90,000.[112] These engines were never delivered as V-3420-3's. Contract changes were issued and they were built as the first V-3420-5(B4) and V-3420-7(B5)'s to Allison Specifications 216 and 217 respectively.

A-3: This was a 1938/1939 commercial offering of the V-3420 to the major aircraft manufacturing firms of the day. Single rotation 2.50:1 reduction gear, 2,375 pounds. These were Altitude rated engines using a 9.00:1 supercharger step-up gear and rated for 2300 bhp at 3000 rpm. Two different Specifications were prepared for what appears to be the same engine, probably to accommodate the installation needs of two different manufactures. None were contracted.

A-4R: This was a 1939 commercial offering of a sea-level V-3420 offered to the major aircraft manufacturing firms. It used the improved PT-13E1 carburetor and standardized on the V-1710 6.65:1 compression ratio and used 6.00:1 step-up gears. A single rotation 2.50:1 reduction gear, 2,350 pound dry weight offered 2300 bhp at 3000 rpm up to 25,000 feet. None were contracted.

A-5R: This 1939 proposal described a sea-level engine with a single rotation 3.00:1 reduction gear and 2,400 pound dry weight. Power remained the normal 2300 bhp. None were contracted.

A-6R: This was a 1939 proposal of an Altitude rated version of the V-3420-A5R. Supercharger gears were changed to 6.82:1 and weight increased to 2,425 pounds for the single rotation 3.00:1 reduction gear engine. Power was 2300 bhp to 5,000 feet. None were contracted.

A-7R: This 1939 proposal was for a development of the V-3420-A6R with a revised single rotation 3.80:1 reduction gear, 2,550 pound dry weight. Ratings were as for the A-6R. None were contracted.

A-8R/L: Designated V-3420-2, one engine of each rotation was procured under Navy Contract 68427 for marine service in 1939. Rated power was 2000 bhp for the engines which were for use in the PT-8 torpedo boat. The V-3420-2(A8L) Model Test engine, AEC #173, stood for its official portraits in May 1942.[113] It is not clear if this was before of after the test work.

A-8RLW: A commercial V-3420 made from a rebuilt V-3420-A8 for the race boat of driver Gar Wood. No specifications prepared.

A-9R/L: No military designation for this experimental engine, but Allison Specification 212 described it as a turbosupercharged engine that was submitted in response to Army Request for Data R-40A, October 1939. The request was seeking 4000-5000 bhp engines for installation in pursuits and medium bombers. The single engine would have been rated for 2500 bhp up to 25,000 feet using two GE Type B-1 turbosuperchargers, intercoolers, and controls that were expected to weigh 600 pounds per engine. Frontal area for the engine was 1690 square inches, nearly 12 square feet! None were contracted.

A-10R/L: No military designation, but Allison Specification 213 described this turbosupercharged engine and was also submitted in response to Army Request for Data R-40A, October 1939. The engine was to be used in an aircraft proposed for the Circular R-40C Pursuit Competition. It would have used a 2.50:1 reduction gear, with a weight of 2,450 pounds. Allison offered a "basic" and three optional versions of this engine. None were contracted.

- The "basic" was an Altitude rated single-speed, single-stage 9.50:1 supercharger ratio, 2,450 pounds, 2100 bhp for takeoff and 2200 bhp at 15,000 feet;
- Option "A" was Altitude rated with a two-speed (6.40:1 and 9.50:1) single-stage supercharger ratio, 2500 pounds, 2300 bhp for takeoff, 2300 bhp to 4,000 feet in Lo and 2200 bhp at 15,000 feet in Hi-gear;
- Option "B" was Sea-Level rated with its engine-stage supercharger having 8.00:1 step-up gears, and used a Type C-5 Turbosupercharger and two intercoolers with ducting which added 560 pounds to the basic 2,450 pound dry weight. 2300 bhp was available from takeoff up to the 25,000 foot critical altitude of the turbo;
- Option "C" was a single-speed, two-stage supercharged design which is believed to have been constructed so that the first stage could be "clutched" out at altitudes below 10,000 feet. Two intercoolers, ducting and controls added 290 pounds to the engine dry weight for a total of 2,725 pounds. 2300 bhp was available for takeoff, with 2100 bhp available at the critical altitude of 25,000 feet.

A-11R: Designated as the XV-3420-9, this engine was similar to the XV-3420-1. Only one engine was built, Manufactures No. 5425 (AAF41-50951), which was used for Type Testing. It retained the 2.50:1 reduction gear, and drove the single-stage supercharger at 6.39:1. On Grade 100/130 fuel, this 1941 engine was rated for 2300 bhp when turbosupercharged up to 25,000 feet. This was the prototype engine for the XB-19A/XB-39 projects, purchased on contract W535-ac-19402, dated June 21, 1941. Allison had the engine ready to begin testing on December 2, 1941, a week before Pearl Harbor.[114] Allison also operated its EX-29 workhorse engine configured as an A-11R. Wright Field performed the Type Test between February 12, 1942 and June 16, 1942. At 75:30 hours into the test, the left rear reduction gear pinion bearing failed. Allison rebuilt the engine and Wright Field continued the 150-hour endurance test, including a 75:30 penalty run without further major incident. The test resulted in the recommendation to approve the engine as a Type (Materiel Center letter to Allison of July 6, 1942 granted Type approval for the V-3420 engine). It was also recommended that Allison continue work on mixture distribution and spark-plug performance in the V-3420, noting that improvements were considered essential prior to flight use.[115]

A-12R: Sea-level rated for 2600 bhp up to 10,000 feet, with 2100 bhp at 25,000 feet. Single-stage supercharger step-up gears were 7.20:1. It was to have 3.00:1 reduction gear and weigh 2,550 pounds. No military designation. None were contracted.

A-13: A proposed development that would have been unique in having a two-speed propeller reduction gear. No other specifics available as Allison never prepared a specification for the model. None were contracted.

A-14R/L: Another engine intended for marine use. No other specifics available as Allison never prepared a specification for the model. None were contracted.

A-15R/L: No military designation. A Sea-level rated Model Test engine having 6.82:1 supercharger gears and 2.50:1 reduction gears while weighing 2,530 pounds was built. The engine delivered 2600 bhp up to 25,000 feet, the carburetor was the large PR-58B2. Allison built the A-15L Model Test engine and identified it as their workhorse EX-22A. This engine was first put on test in January 1942 and continued as such at least through all of 1943. Test records show that it was torn-down for inspection at least 24 times during the two years of testing.[116]

A-16R: Designated as the V-3420-11, this engine was similar to the V-3420-9(A11R), but with the supercharger step-up gears increased to 6.90:1.

V-3420-A15L Model Test engine, EX-22A, was the focus of an 18 month test program. This photo taken at start of program, 1-27-42. (Allison)

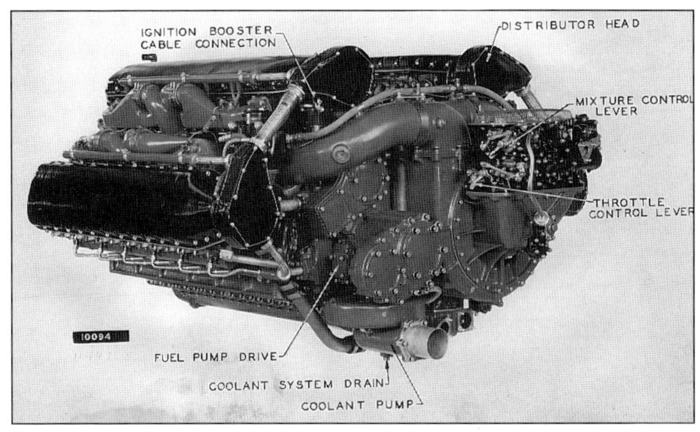

As the first V-3420 slated for production the A-16 received more extensive development. These photos were annotated in anticipation of use in technical orders introducing the engine supporting service introduction. The packaging of accessories and supercharger shows a high degree of thought and organization. (Allison via Hubbard)

It used the large Bendix PR-58B3 carburetor and the reduction gear ratio increased to 3.13:1. It was sea-level rated on Grade 100/130 fuel at 2600 bhp up to 25,000 feet, with a WER rating of 3000 bhp to 28,000 feet. Dry weight was 2,655 pounds. Seventeen of these engines, including one for development, were built in 1942 for the Douglas XB-19A, Boeing/Fisher XB-39 and Lockheed XP-58. When the decision to develop the B-39 was made in June 1942 nine of these engines were ordered on contract W535-ac-29321, for delivery in July/October 1942. On October 28, 1942, Allison was issued contract W535-ac-30124 for 500 V-3420-11 engines, enough to power 100 B-39's. These engines were to be delivered in the March/July 1943 period, but were canceled with the termination of the B-39 program before production began.

A-16L: Designated as the V-3420-13, this was the left-turning companion to the V-3420-11, specifically intended for use in the Lockheed XP-58. Because of the use of an idler gear to cause the accessories to rotate the same direction as in the A-16R, the resulting supercharger ratio was 6.818:1, otherwise all particulars and performance were as in the V-3420-A16R. Four of these engines were built.

Viewed from the top a V-3420 was a lot of machinery. The space between the power sections accommodated the engine beathers, and importantly, the exhaust stacks for the inboard banks. The intent to use the V-3420-A16 in high altitude aircraft is clear as the supercharger discharge pressure is piped directly to the magnetos. (Allison via Hubbard)

This July 1943 photo gives a good idea of weight, distribution, and size of the V-3420-A18R used for turbosupercharged installations. (Allison via Hubbard)

This July 1943 photo is of a mockup of the V-3420-A19, fitting a wooden Auxiliary Stage Supercharger to an existing V-3420-A. The assembly would have been quite compact for a 2600 bhp engine. (Allison via Hubbard)

A-17R: This sea-level engine was a proposed development of the V-3420-11 and was designated V-3420-15. It differed by having the highest compression ratio of any Allison gasoline engine, 7.25:1, along with the different carburetor, the PR-58B2. All other parameters were as for the V-3420-11, including ratings. The higher compression ratio may have been intended to improve the efficiency, particularly under cruise conditions, but the predicted BSFC remained practically identical to the earlier engine. Not flown.

A-18R: Designated V-3420-17, this was an improved V-3420-11 intended for the XP-58. It was proposed in 1942 and was approved for production by the AAF in February 1943, but never flown. These engines were the first "A" Models not to use the 2.50:1 reduction gear, instead using a 3.13:1 reduction gear. Allison tested this gear on a workhorse engine for 25 hours at both 2600 and 2800 rpm without distress even though the change from 2.50 to 3.13:1 gears resulted in an 18 percent increase in torque reaction imposed on the crankcase.[117] Seven of these engines were built, they incorporated the PR-58B3 carburetor and other internal engine improvements. Sea-level ratings, with 2600 bhp available, were unchanged from the earlier model.

A-18L: Designated V-3420-25, this was an improved V-3420-13 left-turning engine intended for the XP-58, but believed not to have flown. Four of these engines were built. They incorporated an improved PR-58B3 carburetor and other improved details of internal engine components. Ratings were unchanged from the earlier model.

A-19R: This was a 1943 proposal for a two-stage mechanically supercharged engine driving a single propeller through an integral 3.13:1 reduction gear. The usual single-stage 10.0 inch impeller was driven through 6.90:1 step-up gears while the Auxiliary Stage was driven through its hydraulic coupling with a 6.96:1 step-up. No military designation was assigned. Total weight would have been 2,830 pounds. Takeoff power was 2600 bhp at 3000 rpm with a 2300 bhp Military rating at 20,000 feet. Allison assembled a mockup of the engine, but it is believed that none were actually operated.[118]

A-20R: A March 1943 specification of an Altitude rated two-stage engine based on the Sea-Level V-3420-17(A18R) and designated V-3420-21, the important difference was the increased engine-stage supercharger of 8.00:1 and the use an Auxiliary Stage Supercharger driven through 6.96:1 step-up gears. It could be constructed for either right or left rotation. Initially intended to power the Fisher XP-75, but was never flown. It was rated for 2600 bhp at takeoff, with 2300 bhp available up to an altitude of 10,000 feet. This arrangement was considered early in 1943 during the conceptual design process for the XP-75 and assumed fitting a comparatively large diameter, single rotation propeller. Such an arrangement would have required a tall landing gear as well as having likely caused directional and control problems with an aircraft that needed to minimize propeller torque. For this reason the XP-75 and all subsequent P-75's were propelled by counter-rotating propellers, the forte of the Allison V-3420-B series.

A-21R/L: This model described one of the two power sections in a two-stage DV-6840 installation, "Double V-3420's." No military model number was assigned. Auxiliary Stage drive ratio was 6.96:1, and the engine-stage supercharger ratio was 6.90:1, with PR-58 carburetors mounted in the duct between the Auxiliary and Engine Stage superchargers. The power sections would have been those from the V-3420-19(B8), resulting in four input shafts to drive the counter-rotating #60 and #80 propellers shafts through a 3.13:1 gear train. Engines were to be installed in "tandem", with the front engine connected to the reduction gearbox by extension shafts approximately 3 feet long, and the rear power section connected via extension shafts approximately 25 feet long. This configuration was designated as an "A" Model since both crankshafts in either power section "turned in the same direction", though those in the front power section turned opposite those in the rear section. The entire assemblage would have weighed 6,300 pounds. Each power section was Altitude rated for 2600 bhp for Takeoff and had a Military rating of 2300 bhp at 20,000 feet. One can only speculate on the aircraft which evidently required this power and configuration, though interestingly, in November 1946 Allison had the reduction gear for the V-6840 setup for testing.[119]

A-22R/L: This was an engine proposed to Lockheed in 1943 for the XP-58, but not flown. Similar to the V-3420-11/13, but with the PR-58B3 carburetor and other changes in structural details. Intended for turbosupercharger installations and sea-level rated at 2600 bhp for takeoff up to 25,000 feet. WER rating was 3000 bhp at up to 28,000 feet for a maximum of five minutes, all on Grade 100/130 fuel. A total of seven of these engines were built, probably one for Model Testing and three each R/L for the XP-58 aircraft. In July 1944 Allison revised the specification, this time describing an Altitude rated engine for a proposed low-drag installation to replace the Wright R-3350-8 engines in Lockheed's new XP2V-1 *Neptune* land-based patrol bomber. The primary change in the engine was use of 7.26:1 supercharger drive gears in place of the 6.90:1 gears in the sea-level engine, along with the increased Takeoff and Military ratings. Both engines used the 3.13:1 integral reduction gears, and for the converted XP2V-1, would have turned four bladed 17'-0" Curtiss Electric propellers. The new engine was rated for 2850 bhp at 3000 rpm for Takeoff, and up to 4,300 feet. Continuous rating at 2700 rpm was for 2300 bhp at up to 7,500 feet. Of the seven A-22's believed to have been built, it is not clear to which configuration they were actually constructed. Retention of the A-22 designation for two quite different engines is not surprising, as neither was ever purchased as a Military model. Instead, the A-22 was the "Lockheed" engine, and as the customers needs changed, so did the configuration and performance ratings.

A-23R/L: Designated as the V-3420-27, this was an Altitude rated configuration in another "Double V-3420" DV-6840 arrangement. Designed in November 1943, this time using a V-3420-17(A18R) and V-3420-25(A18L). One power section was mounted ahead of the other in a tandem arrangement so that each could drive extension shafts to a remote reduction gear box fitted with counter-rotating propeller shafts as in the A-21. Crankshafts and extension shafts rotated CCW on the upper unit and drive the #60 propeller shaft clockwise, it also had supercharger step-up gears of 8.00:1. Crankshafts and extension shafts rotated CW on the lower unit and drove the #80 propeller shaft counter-clockwise. Its supercharger step-up gears were unique at 7.80:1. Propellers were independent of each other, as was done for the XB-42 *Mixmaster*. Each power section was Altitude rated for 2600 bhp at takeoff with the remote reduction gear ratio of 3.23:1. Total weight of the installation was to be 6,200 pounds. It was not flown.

A-24R: Designated as the V-3420-31, this was a sea-level rated engine similar to the V-3420-17(A18R), but with an increase in the engine-stage supercharger ratio to 7.26:1 and higher ratings. On Grade 100/130 fuel 3000 bhp was available for takeoff up to 30,000 feet. Weight for the November 1944 proposal to the Military would have been 2,695 pounds, while the August 1945 commercial offering weight had increased to 2,740 pounds as a result of general improvements. This was a concept engine only. Never flown.

A-25: This sea-level engine was proposed in August 1945 for the Allison powered follow-on to the Boeing B-29/XB-39, the aircraft which became the P&W R-4360 powered B-50. It was to be similar to the V-3420-17(A18R), except for adapting the engine-stage supercharger ratio of 7.26:1 as proposed for the V-3420-A24R/L and utilized a reduced compression ratio of 6.00:1 and using 30% Glycol-70% Water coolant and Grade 100/130 fuel.[120] It was also to be provided with ADI for use during Takeoff and WER conditions. The engine produced 3500 bhp from 3200 rpm for Takeoff, and 3600 bhp for WER wet, both conditions requiring that the turbo provide 35.2 inHgA to the carburetor deck. Dry Military rating was 3300 bhp up to the critical altitude of the turbosupercharger, intended to be well over 30,000 feet, all on Grade 100/130 fuel.

Ratings for the V-3420A25	
Takeoff & Military	3500 bhp at 3200 rpm, wet
Normal Power	2600 bhp at 2700 rpm, dry
War Emergency	3600 bhp at 3200 rpm, wet

Allison V-3420-B series engines were defined by having the two crankshafts rotating in opposite directions. As usually applied they tended to be used in installations with remote reduction gearboxes, but this was not a necessary distinguishing feature.

B-1: Allison Specification 205 was issued in October 1938 describing this as a Sea-level engine using the PT-13B1 carburetor and having 7.25:1 compression ratio and 6.00:1 supercharger step-up gears. It was to drive a 2.50:1 SAE #60 remote reduction gear driven by a common, crankshaft speed, extension shaft. This engine was the only example of such a configuration, and it did not result in a contract or development. Total weight, with the extension shaft, would have been 2,475 pounds. Performance would have been 2300 bhp up to the critical altitude of the turbo, 25,000 feet. It is not known what aircraft would have used the arrangement. None were contracted.

B-2: This model was proposed as an Altitude rated engine to power the first aircraft designed by McDonnell Aircraft, and was submitted to the Air Corps for consideration as McDonnell's entry in the FY-40 Circular R-40C competition. They would have mounted the V-3420 engine in the fuselage and driven pusher props aft of the wings via extension shafts and right-angle gear drives to two-speed planetary type remote reduction gears. The overall reduction gear ratio could be shifted between 1.50:1 and 2.50:1 to optimize performance over a wide range of conditions. Propellers were to have been mounted on SAE #50 shafts. Shafting and gear boxes in total weighed 1485 pounds. Allison offered a "clutched" Auxiliary Stage Supercharger which would have resulted in 2150 bhp at 17,000 feet, or if intercooled, 2250 bhp at 18,000 feet. Basic performance with the Aux Stage de-clutched was 2450 bhp for Takeoff and up to 5,000 feet.

McDonnell responded with a design that incorporated the Allison gearing and shafting, but required that it could be used interchangeably with the V-3420, Pratt & Whitney X-1800 or H-3130, or Wright Tornado engines.[121] The Army rejected their Circular R-40C proposal, which McDonnell subsequently revised into the much different McDonnell XP-67. None were contracted.

B-3: No contract was issued for this engine which was proposed in the Circular R-40D competition. Installation would have utilized two V-3420-B power sections in a DV-6840 configuration. None were contracted, though the Air Corps did issue a short form contract to Allison in December 1940 covering the purchase of engineering data and a study of a long range bombardment engine of 4000-5000 bhp.[122]

B-4: Defined in Allison Specification 216 in August 1940, this was the first "B" series engine to receive a contract and military designation, as the V-3420-5. The contract was W535-ac-11328 which had been originally issued for the V-3420-3. The V-3420-5 and V-3420-7's took the place of the V-3420-3. This was a single-stage sea-level rated engine for use with either a turbosupercharger or with an Auxiliary Stage Supercharger. The engine drove a remote reduction gearbox. Engine-stage supercharger ratio was 6.82:1 while the reduction gearbox ran 2.458:1 gears. With the hold on V-3420 development implemented in 1941 by the Army, this Model languished until needed in 1943 for the Fisher XP-75. Allison built their EX-41 workhorse engine as a B-4 and had it ready for testing in early April 1943.[123] Revision "C" to the Allison specification was issued in May 1943 and noted a change in the reduction gear ratio to 2.772:1, established a

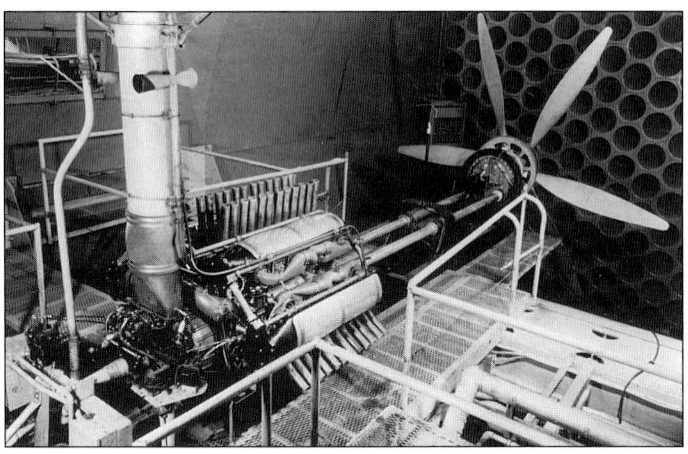

After extensive Model Testing the B-4 was used by GM's Aeroproducts Division to develop counter-rotating propellers for the XP-75. (Allison via Jim Doll)

This is the B-10 Model Test engine, A-045799-3, photographed on 7-27-44. Auxiliary Stage Supercharger is not installed. (Allison via Jim Doll)

turbo installation, and upgraded performance to that available on Grade 130 fuel, 2600 bhp at 3000 rpm up to 25,000 feet. Weight was 2,955 pounds. To expedite the XP-75 program this engine was used by Allison to accomplish the 50-hour approval test at 2600 bhp, it then went on and successfully completed a 7-1/2 hour WER test at approximately 3050 bhp and 60 inHgA, without water injection.[124] Following these tests this same engine was used for a simulated 150-hour Model Test. At 129 hours the test stand mount for the reduction gear failed, wrecking the propellers. There were no failures in either the engine or reduction gear box.[125] This testing established confidence in the mechanical reliability of the coming XP-75 engines.

B-5: Designated V-3420-7, this sea-level rated power section was similar to the V-3420-5(B4) except that 90 degree engine mounted 1:1 bevel type gear boxes drove extension shafts to another set of remote bevel type gear boxes, to drive in-turn remote 2.50:1 reduction gear boxes mounting SAE #50 pusher propellers. The entire assemblage weighed 3,640 pounds and delivered 2600 bhp up to 25,000 feet. This engine is believed to have been initially considered in a design study of a V-3420 powered Bell YFM-1 follow-on aircraft. The contract was W535-ac-11328, the contract that had been originally issued for the V-3420-3 was revised in September 1940 to describe the two-stage V-3420-7. The V-3420-5 and V-3420-7(B5)'s took the place of the V-3420-3, though early development was shelved in preference of focusing on production of V-1710's needed in fighter aircraft. Work on the V-3420 series did not resume until June 1942, after thousands of V-1710's had been delivered.[126] In an interesting twist, the Allison Photolab took a full series of this engine at the end of the war, August 1945.[127] It is not clear what the intended application might have been.

B-6: Proposed engine, which would have driven a single propeller. No other information has come to light on its specifics. None were contracted.

B-7: The power section would have been similar to the B-4 and sea-level rated with 6.82:1 supercharger gears and extension shafts driving counter-clockwise reduction gears 2.59:1. The front propeller would have turned clockwise and mounted an SAE#50 propeller, while the rear propeller turned counter-clockwise and mounted a SAE #70 propeller. The intended application has not been determined, but 2600 bhp was available to 25,000 feet. Total weight was to be 2,955 pounds. No Allison Specification was prepared and none were contracted.

B-8: Designated as the V-3420-19, twenty five of these two-stage engines were ultimately built for use by the Fisher XP-75. When originally specified in 1942 the engine was to be based on the B-4 and equipped with an Auxiliary Stage Supercharger which was driven via 6.96:1 step-up gears. In March 1943 the specification was revised and based on the B-5 power section while the counter-rotating propeller shafts were changed from a #60 and #80 to a #50 and #70.At the same time the requirement for gun synchronizers was deleted. The 2.458:1 reduction gear ratio and 6.82:1 engine-stage supercharger gears were not changed. Rated on Grade 100/130 fuel for 2600 bhp at takeoff, and 2300 bhp Military at up to 20,000 feet for a weight of 3,175 pounds. Allison completed the first B-8 in October 1943, and in March 1944 the Model Test -19 was delivered to Wright Field.[128] Not only had the delivery from Allison been slow, the Wright Field Engine Laboratory was slower in providing the needed test stand and facilities to support the testing. The engine then performed the 150-hour Model Test at 2600 bhp, and in the process qualified the similar B-10 model as well.[129] It then went on to perform the Dry (without water injection) WER rating test at 60 inHgA, approximately 2900 bhp.[130] Following successful Dry WER testing the engine was to be fitted with water injection and undertake testing to establish a WER rating of 3400 or 3500 bhp for both the B-8 and B-10 engines.[131]

B-9: This engine was to be similar to the V-3420-B8, but with the counter-rotating reduction gear directly coupled to the power section, i.e. without extension shafts. Weight was to be 2,975 pounds. Performance as for the B-8.

B-10: Designated as the V-3420-23, this engine was similar to the V-3420-19 except that it was equipped with the improved 12-counterweight crankshafts and the Auxiliary Stage Supercharger ratio was increased to 7.38:1. Weight was 3,275 pounds. Engine-stage supercharger ratio remained 6.82:1, and the carburetor was the Bendix PR-58B3. A total of 101 of these engines were built and intended for the Fisher XP-75 and P-75A-1-GC. Takeoff rating on Grade 100/130 fuel was 2600 bhp, with 2300 bhp available at up to 20,000 feet, and a sea-level WER of 2885 bhp on 57.5 inHgA. In addition Allison ran Continuous Power tests to failure on this engine. The first exceeded 40 hours at 2800 bhp, and the second was to be run at 3000 bhp.[132] Results have not been located. The first B-10 was ready for testing in March 1944, and it appears that three engines were used in the Model Testing program. The last of these was A-045799-3, apparently they all carried the same contract serial number, suggesting that the first two engines probably were damaged beyond repair in the course of testing. This particular engine began testing in late July 1944.[133]

B-11: In August 1944 this engine was designated V-3420-29. It was similar to the V-3420-23(B10) except that the coolant was changed to a mixture of 30 percent water and 70 percent glycol and the carburetor was changed to the PR-58B5. The Auxiliary Stage Supercharger ratio was also increased to 7.84:1. Engine weight remained at 3,275 pounds even though improvements were incorporated which included heavier connecting rods and different rod bearings. The engine was on an Allison dynomoter in February 1945.[134] Though the engine was not flown, takeoff rating was increased to 2850 bhp, and a Military rating of 2300 bhp was available at up to 25,000 feet on Grade 100/130 fuel.

B-12: This engine was a development of the V-3420-B11 having a 50 percent air-to-air intercooler. The result was an improvement in ratings, though 2850 bhp was still available for takeoff, and 2400 bhp up to 25,000 feet. WER was 3150 bhp at sea-level. This engine was not given a military designation, but was flown in the #3 P-75A, AF44-44551. Dry weight was 3,191 pounds.

C-1: This model was to be the result of an Allison post-war proposed 18 month development program. The V-3420-C1 would have been a single-stage, turbosupercharged with intake port fuel injection, having 6.00:1 compression ratio and using 30% Glycol-70% Water coolant and Grade 115/145 fuel. Allison intended to replace the original 10 inch diameter engine-stage supercharger impeller with one of 12-3/16 inches in diameter. Consideration was being given to an Auxiliary Stage Supercharger instead of the turbo, in which case its diameter would have been increased from 11 inches to about 17 inches. Allison recommended using two BH1 Turbos with each engine as the present CH5 turbos, which had 17.0 sq.in. nozzle area, but were inadequate for ratings above about 3200 bhp. They forced the engine backpressure into the range of 50 to 55 inHgA. The BH1 had effective nozzle area of 13.0 sq.in., so parallel units dramatically reduced backpressure.[135]

Ratings for the V-3420-C1

Takeoff & Military	4000 bhp at 3200 rpm, wet
Takeoff & Military	3500 bhp at 3200 rpm, dry
Normal Power	3000 bhp at 2700 rpm, dry
War Emergency	4800 bhp at 3200 rpm, wet

Allison's post-war policy of not undertaking developments unless sponsored and paid for by the Government caused this engine to not be selected for the aircraft that became the Boeing B-50. Instead the P&W R-4360 was used, even though it did not produce as much power. Had the "C" series V-3420 been developed they could well have gone on to power the many aircraft such as the Boeing 377 *Stratocruiser* and Convair B-36. The engine certainly had the capability.

NOTES

[1] *Power Plants for Project "A", Restricted Project No. M-3-35, and Project "D", Restricted Projects Nos. M-4-35 and M-5-35*, Air Corps Materiel Division letter to Chief of the Air Corps, November 9, 1935. NARA RG18, Entry 167D, Box 46.

[2] Jones, Lloyd S. 1962, 59-61.

[3] Kelsey, Benjamin S. B.G. USAF (Ret.) 1982, 88.

[4] Allison Division, GMC, 1942 news release, 15.

[5] Allison Memorandum to Cruzan, Goldthwaite, Reynolds, McCowall and Buttner, *Quotation on DV-3420 Engine*, November 4, 1936.

[6] Hazen, R.M. 1941 Draft, 5.

[7] Allison letter of January 27, 1938. NARA RG342, V-3420 file.

[8] Materiel Division conference at Allison, April 18, 1938. NARA RG342, V-3420 file.

[9] Engineering Section Memo Report, E-57-1077-66, 1938.

[10] Wright Field Experimental Engineering Section letter to DuBois-Martin Airplane Co., November 8, 1939. NARA.

[11] *Case History of the R-2160 Engine*, 1945.

[12] Hazen, R.M. 1945, 75-79.

[13] Conference with Mr. J.W. Nelson regarding Diesel Engines, Wright Field Engineering Section Memo Report DM-57-503-23, October 20, 1939.

[14] Norris, Jack 1988.

[15] Telegram to Engineering Section, Wright Field, April 22, 1939.

[16] *Domestic Sale (W. Shaw) [of V-3420]*, Air Corps memo to General Brett, Chief, Materiel Division, January 25, 1940.

[17] Chief of the Air Corps Routing and Record Sheet of May 15, 1939. NARA RG18, File 452.8, Box 808.

[18] *Contract W535-ac-11328, XV-3420-3 Engines*, memo to Chief of Air Corps, December 12, 1939. NARA RG18, File 452.8, Box 808.

[19] Teletype from Experimental Engineering Section at Wright Field, December 18, 1940. NARA RG18, File 452.8, Box 807.

[20] Hazen, R.M. 1941 Draft, 5.

[21] *Allison XV-3420-1 Engine, Serial Number 38-119*, memo from Chief, Experimental Engineering Section to The Accountable Officer, Wright Field, December 18, 1940.

[22] Allison Photolab Log, photo #04060 was taken on 9/23/41 of the engine rigged for the moment of inertia measurements.

[23] *Allison X-3420 Engine*, memo for General Arnold from General O.P. Echols, Materiel Division, March 28, 1941. NARA RG18, File 452.8, Box 807.

[24] *Allison X-3420 Engine*, memo for Colonel Beebe of General Arnold's office, from General O. P. Echols, Materiel Division, March 3, 1942. NARA RG18, File 452.8, Box 807.

[25] Allison Photolab Log, photo #07211 taken June 10, 1942.

[26] Hazen letter to Mr. Robert J. Woods, Chief Engineer at Bell Aircraft, March 10, 1938. V-3420 file at NARA.

[27] *Propeller Drives From a Single Engine*, Wright Field Engineering Section Memorandum Report Serial No. Prop-51-638, August 31, 1938.

[28] *Bevel Gear Drives for Model V-3420 Single Engine Installation*, Hazen letter to Chief, Materiel Division, Wright Field, November 8, 1938.

[29] Allison letter to Chief, Materiel Division, Wright Field, April 12, 1939.

[30] Allison letter to Chief, Materiel Division, Wright Field, March 22, 1939.

[31] *50-Hour Contractor's Test of the Bevel Gear Drive #39500*, AAC Contract No. W535-ac-13017, Allison Test Department Report No. D2-1, October 1, 1940.

[32] *Test of 1/15 Scale Model Materiel Division Design No. 348 Proposed Pursuit Airplane, One V-3420 Allison Engine - Five-Foot Wind Tunnel Test No. 228*, Air Corps Technical Report, Serial No. 4462, referenced in letter to Allison from Director, Army Aeronautical Museum, September 20, 1939.

[33] Looking foreword, from the anti-prop end of the engine, "L" is the left bank and "R" the right most. "LC" is the left-center bank and "RC" is the right-center bank.
[34] Operators Manual, Allison Engine Installations, 1st Edition.

[35] NASM microfiche B0002900: *Allison V-3420*

[36] Matthews, Birch. 1996, 187.

[37] Technical Order 02-5AH-2, 1949, 17.
[38] NASM Microfiche B0002900: *Allison V-3420*.

[39] Allison Model Specification No. 214,1940.

[40] *Allison V-3420 Engine-High Altitude with Geared Supercharger*, Air Corps letter to Allison, May 23, 1938. Copy in V-3420 files at NARA.

[41] *Allison V-3420 Engine-High Altitude with Geared Supercharger*, Air Corps letter to Allison, June 15, 1938. Copy in V-3420 files at NARA.

[42] Allison Model Specification No. 211, 1939.

[43] Allison photos #27918/27922 made on November 7, 1946.

[44] *FLIGHT, 1954*, 12.

[45] von Gersdorff, Kyrill and Kurt Grasmann, 1985.

[46] Francillon, Rene J. 1979, 319-324.

[47] The XBLR designation meant eXperimental Bomber-Long Range.

[48] *Power Plants for Project "A", Restricted Project No. M-3-35, and Project "D", Restricted Projects Nos. M-4-35 and M-5-35*, Air Corps Materiel Division letter to Chief of the Air Corps, November 9, 1935. NARA RG18, Entry 167D, Box 46.

[49] Goss, SMS USAF(Ret) Dan, 1987, 14-18.

[50] Contract W535-ac-29321 for 9 V-3420-11 experimental engines, June 1942. Contract W535-ac-30124 for 500 V-3420-11 production engines, October 28, 1942.

[51] AAC Letter, Subject: *Model XB-19A Airplane (XB-19 Airplane with Allison V-3420-11 Engines)*. From Chief, Technical Air Staff, Wright Field to Chief, Experimental Engineering Section, Wright Field, dated September 25, 1942. Letter establishes the "Restricted" MX-309 Project. NARA RG 342, RD 3766, Box 7366.

[52] Wright Field letter of 9-25-42, NARA RG 342, RD 3763, Box 7366.

[53] Shamburger, Page and Joe Christy, 1972, 20.

[54] Hazen, R.M. 1941 Draft, 6.

[55] Letter from F.O. Carroll, Brig. General USA, Chief, Experimental Engineering Section to General Electric directing a revision to the Type CM turbos diffusers to allow rating the V-3420-11 with a Military rating of 2600bhp/3000rpm/35,000'. Delivery of the first four units, to go into the XB-19A, were scheduled for 2-20-1943. Dated, 11-30-1942. NARA RG 342, RD 3761, Box 7359.

[56] The XB-19, AC38-471, was originally conceived as the XBLR-2, and was intended to be powered by 1,600 hp Allison X-3420-1's. It was completed in May 1941 with Wright 2200 hp R-3350-5's, first flight 6-27-41.

[57] *XB-19A #38-471 History Card.*

[58] Ron Hazen, article on V-1710, 1945, p.7.

[59] *Allison V-3420 Engine Installation in B-29 and XB-19A Airplanes*, memo for the Deputy Chief of Air Staff from Wright Field, September 1, 1944. NARA RG18, File 452.13-L, Box 756.

[60] Goss, SMS USAF(Ret) Dan, 1987, 14-18.

[61] Comments of Everett Smith, Sacramento, CA, October 26, 1995.

[62] Francillon, Rene J. 1979, 319-324.

[63] Four designs were submitted to Wright Field in response to RFD R-40B, with the Boeing Model 345 being the most revolutionary. It was also the first to receive a military designation, XB-29. Two others were which were not to make it off of the drawing boards were the Lockheed XB-30, a R-3350-13 powered bomber version of the *Constellation* transport, and the Douglas XB-31 (to be powered by R-4360's, canceled in late 1941 after B-29 was ordered into production). The Air Corps did later authorize the Consolidated XB-32 (used the same R-3350's as B-29) as a backup to the B-29 program, though it only reached combat in limited numbers and then at the very end of the war. See LeMay, General Curtis E. and Bill Yenne, 1989, 35.

[64] Harding, Stephen and James I. Long, 1983, 5.

[65] The XB-19 was originally powered by R-3350-5's, each rated at 2000 bhp for takeoff. Both the XB-29 and XB-32 used the 2200 bhp Wright R-3350-13 engine. The backup Allison V-3420-11(A16R) was rated for 2600 bhp.
[66] Teletype, General Echols to Assistant Chief, Materiel Division, April 7, 1941 and response, dated April 9, 1941.

[67] Hazen, R.M. 1945.

[68] Hazen, R.M. 1945, 75-79.

[69] Allison "Firsts" as of 1942, from J.L. Goldthwaite files at Allison.

[70] Bowers, Peter M. 1966, 290.

[71] Hazen, R.M. 1945.

[72] Birdsall, Steve 1980, 35.

[73] *Experimental Installation, Birmann Type B Turbosuperchargers in XB-19A*, Inter-Office Memo, Chief, Power Plant Laboratory to Chief, Engineering Division, Wright Field, April 6, 1943.

[74] Referenced letter to the Bureau of Aeronautics, *"Contract NOa(s)681- Eight P-15B-3 Units for installation in the XB-19A Airplanes"*, August 28, 1943.

[75] GE Letter to Commanding General AAF Materiel Command, Wright Field, *Delivery Schedule for GE Type CM-2-A1 for the XB-19A and XB-39 Airplanes*, February 2, 1944. NARA RG 342, RD 3761, Box 7359.

[76] Hazen, R.M. 1945.

[77] Ron Hazen of Allison letter to Commanding General, AAF Materiel Command, Wright Field, *Transmittal of Engine Flight Charts for V-3420 "A" Series Engines - 11, -13, 17, -25*, May 25, 1944.

[78] Wright Field Chief, Power Plant Laboratory letter to Mr. Don Berlin of the Fisher Body Division, GMC, dated July 13, 1944.

[79] Sobey, A.J. 1988 Interview. In 1945 the Fisher Aircraft Development activity was closed and some of the airframe types transferred to Allison to form the Installation Engineering Section, and to expand the Allison flight test activity. Don Berlin was among the group transferred to Indianapolis.

[80] Hazen, R.M. 1945.

[81] O'Brian, Don, 1950.

[82] Bowers, Peter M. 1966, 290.

[83] Don R. Berlin, Allison Director of Installation Engineering Section, letter to Director, Air Technical Service Command, Wright Field, Subject: *Future Ratings for Allison Model V-3420 Single-Stage, Turbosupercharged Engines*, August 23, 1945.

[84] R.M. Hazen, Allison Chief Engineer, letter to Director, Air Technical Service Command, Wright Field, Subject: *Estimated Development Costs of V-3420 Engines for Possible Application to B-39 Airplane*, October 15, 1945.

[85] I think you have to go back to the basic philosophy that Detroit had and that was "We'll build anything that the government wants, as long as they pay for it." That was pretty strict! Kline, John 1994.

[86] P&W Engine Designations, 1956 Edition, page G-12.

[87] Letter from Col. R.C. Wilson to Samuel H. Husbands, Defense Supplies Corp, June 30, 1943. NARA RG18, File 452.13, 1260.

[88] Letter from Donald P. Frankel, Allison Field Engineer, LA Office, to Wm C. Campbell, Director of Communications, Allison Gas Turbine Division, GMC, April 2, 1990.

[89] Curtiss-Wright Corporation, St. Louis Airplane Division letter to Chief, Materiel Division, Wright Field, June 24, 1938.

[90] This is not a new designation system for the V-3420, rather C174a was the number given the proposed engine's performance curve and used as a designator in the preliminary study.

[91] *Installation Study of Allison Engines, V-3420-C174a Two-Stage and V-3420-11 Turbo-Supercharged, in the Curtiss-Wright C-46 Airplane*, Engineering Division Memorandum Report Serial No. EXP-M-51/4412-1-5, April 10, 1943.

[92] Letter from Chief, Aircraft Laboratory, Engineering Division to Curtiss-Wright Corporation, Buffalo, NY, dated October 4, 1943. NARA RG 342, RD3761, Box 7359.

[93] Francillon, Rene J. 1982, 258-259.

[94] Lockheed Preliminary Design drawing PD106, *Installation of Allison V-3420 in New Nacelle for XP2V-1*, 7-17-1944.

[95] *Allison Model Specification No. 233-A*, July 20, 1944.

[96] NARA RG18, Entry 293B, Box 249.

[97] CONFIDENTIAL *XP-58 Airplane Description*, July 1, 1940. NARA RG18, Entry 293B, Box 244.

[98] Technical Report No. 5489, 1946.

[99] Experimental Engineering Section, CONFIDENTIAL Memorandum, 1942, 4.

[100] *Engine Exhaust System, Tentative Specification XP-58*, Lockheed Report No. 2345, Appendix B, July 28, 1941.

[101] Pitkin, J. 1944, page 18.1400.

[102] *General Motors and the Aviation Industry*, draft of March 6, 1957, p.13.

[103] Experimental Engineering Section, CONFIDENTIAL Memorandum, 1942, 6.

[104] O'Brian, Don, 1950.

[105] *Fuel Injection—Allison Engines*, letter by Command of General Arnold to Commanding General, Materiel Command, September 23, 1943. NARA RG18, File 452.13, Box 1260.

[106] *P-75 Eagle: America's WW-II Mystery Fighter*, Air Classics, V.16, N7, July 1980, Robert L. Trimble, p.56-65.

[107] Fisher Body Detroit Division, Aircraft Development Section, Report No. X-249, 1943, 107.

[108] *Group Weight Statement for P-75A-1-GC*, December 1, 1944, contained in Fisher Body Detroit Division, Aircraft Development Section, Report No. X-249, 1943.

[109] *Power Plants for Project "A", Restricted Project No. M-3-35, and Project "D", Restricted Projects Nos. M-4-35 and M-5-35*, Air Corps Materiel Division letter to Chief of the Air Corps, November 9, 1935. NARA RG18, Entry 167D, Box 46.

[110] *Supplemental Report to Committee Report on Request for Data R40-A*, Wright Field, Oct. 21, 1939. Competition between P&W X-1800, Wright Tornado, Continental XH-2860, and Allison V-3420-A9/A10. Competition for >1800 bhp engine for pursuit aircraft.

[111] Allison Division, GMC, 1941, 8-9.

[112] *Contract W535-ac-11328, XV-3420-3 Engines*, memo to Chief of Air Corps, December 12, 1939. NARA RG18, File 452.8, Box 808.

[113] Allison Photolab Log, photos #06828/37 of AEC#173 were taken May 16, 1942.

[114] Allison Photolab Log, photos #04824/33 of AEC#05425 were taken December 2, 1941.

[115] Experimental Engineering Report EXP-M-57-503-663, 1942.

[116] Allison Photolab Log has many entries for the EX-22A.

[117] *Development Test of 3.13:1 Reduction Gear Assembly on V-3420 Engine—Allison No. EX-40*, Hazen of Allison letter to Commanding General, Materiel Center, Wright Field, January 19, 1943. NARA RG18, File 452.13, Box 1260.

[118] Allison photo log, photos #11870/75, March 18, 1943.

[119] Allison photo log, photos #27918/27922, November 7, 1946.

[120] *Future Ratings for Allison Model V-3420 Single Stage Turbo-Supercharged Engines*, Allison letter from R. Hazen to Director Air Technical Service Command, USAAF, Wright Field, August 3, 1945.

[121] *Allison 3420 engine, and gearing and shafting for same can be used interchangeably with 1800, 3130 and Tornado engines*, letter from J.S. McDonnell to Materiel Division, Wright Field, December 12, 1939.

[122] *Short Form Contract w535 ac-17366*, letter of authorization from General H.H. Arnold to Assistant Secretary of War, December 21, 1940. NARA RG18, File 452.8, Box 807.

[123] Allison Photolab log, photos #12837/40, April 1, 1943.

[124] *V-3420 Engine Test Status*, memo to Major General O.P. Echols, March 3, 1944. NARA RG18, File 452.13-M, Box 756.

[125] *Summary of Testing on V-3420-B Type Engines at the Allison Plant*, Allison Chief Engineer letter to Materiel Division, #X1E2BE, May 2, 1944. NARA RG18, File 452.13, Box 1260.

[126] Ron Hazen, article on V-1710, 1945, p.4.

[127] Allison Photolab log, photos #24748/50, August 8, 1945.

[128] Allison Photolab log, photos #15023/32, October 5, 1943.

[129] *Progress Report on XP-75 Airplane*, report to General Echols, February 26, 1944. NARA RG18, File 452.13-M, Box 756.

[130] *Summary of Testing on V-3420-B Type Engines at the Allison Plant*, Allison Chief Engineer letter to Materiel Division, #X1E2BE, May 2, 1944. NARA RG18, File 452.13, Box 1260.

[131] *V-3420 Engine Test Status*, memo to Major General O. P. Echols, March 3, 1944. NARA RG18, File 452.13-M, Box 756.

[132] *Summary of Testing on V-3420-B Type Engines at the Allison Plant*, Allison Chief Engineer letter to Materiel Division, #X1E2BE, May 2, 1944. NARA RG18, File 452.13, Box 1260.

[133] Allison Photolab log, photos #19584/93, July 27, 1944.

[134] Allison Photolab log, photo #22650, February 22, 1945.

[135] *Future Ratings for Allison Model V-3420 Single Stage Turbo-Supercharged Engines*, Allison letter from R. Hazen to Director Air Technical Service Command, USAAF, Wright Field, August 3, 1945.

15

The Rolls-Royce Merlin V-1650 vs the Allison V-1710

The Controversy

Allison's V-1710 and the Rolls-Royce Merlin were both on the side of the Allies during WWII. Discussions between knowledgeable individuals about these two seemingly similar engines tend to quickly become polarized. Which was best, and what were the failings of the Allison, are usually the focus of the arguments. In this chapter we will look at the similarities and differences in these two engines with the intent to provide a factual basis for the discussions that are sure to continue.

Both engines were high temperature liquid-cooled 60 degree V-12's of similar capacity, with the Allison displacing 1710 and the Merlin 1650 cubic inches. Both were developed during the 1930s and were available just in time for the coming war. Each represented the best in engineering design, materials and manufacturing practiced in their respective companies, and for that matter, their countries. Such similarities become the basis for competition in the marketplace and naturally each developed its own cadre of ardent supporters, particularly among those who flew them into combat.

Throughout the war Allison was a frequent advertiser in popular periodicals and technical journals. (Authors collection)

Background

In the Spring of 1940, long before U.S. entry into the war, the British Purchasing Commission was given entry to U.S. war materiel manufacturers and allowed to purchase armaments on a "cash and carry" basis. They rapidly met with practically every U.S. firm seeking such business, particularly those in the aircraft industry. As an example, the British contracted with North American Aviation on May 29, 1940 for a new fighter, the NA-73, later to become known as the *Mustang*.

With the Nazi engulfment of Western Europe in the spring of 1940 provoking a flurry of military preparedness in Washington, Henry Ford, Sr., a renowned pacifist, proclaimed on May 28, 1940, that the Ford Motor Company stood ready to "swing into a production of a thousand airplanes of standard design a day."

Significantly, that same day his former employee, William S. Knudsen, then President of General Motors and who had directed that firm in outclassing the Ford Motor Co. for more than a decade, had just been appointed Commissioner for Industrial Production by President Roosevelt.[1] Ford had said that they would produce the thousand planes a day "without meddling by Government agencies." Ford's son Edsel then traveled to Washington to discuss production programs with Knudsen, who proposed that Ford should begin with engines, rather than airplanes.

Concurrently the British Purchasing Commission had approached Washington for help with the urgent production of 6,000 Rolls-Royce Merlin engines needed for the coming Battle of Britain. Knudsen proposed that Ford take on the job. Henry Ford agreed and had Edsel publicly announce the Rolls-Royce project.[2]

The two-stage Merlin, such as this Packard Built V-1650-7, is a handsome engine that is being well maintained and is likely to be flown well into the 21st Century in privately owned P-51 Mustangs. (Author)

Rolls-Royce advertisements during the war were usually in British periodicals, where they took every opportunity to promote their engines. (Authors collection)

But then Ford's cantankerous nature appeared. Lord Beaverbrook, Minister for War Production in London, hailed the Ford announcement as a major step forward in the British war effort. Within hours Knudsen received a call from Edsel Ford telling him that "we can't make those motors for the British." Knudsen flew to Detroit and talked to Ford directly, who said he was willing to make the motors for Britain if his contract was channeled through the U.S. Government, but he would not sign a contract directly with the British as it was against his principles to provide war materials directly to a foreign belligerent.[3] Knudsen attempted to talk Henry into a compromise, and mentioned that he had told President Roosevelt of the deal, and how it had pleased the President. That ended whatever chance there might have been of salvaging the deal, for Ford was no fan of Roosevelt. Three days after announcing the deal to build the Merlin in the U.S., Edsel had to announce that Ford would not be building the British engine.[4]

Adding to the atmosphere surrounding these events was the Emmons Board Report, issued in June 1940. It took the Air Corps to task for only planning on "one type of liquid-cooled" pursuit engine, meaning the Allison V-1710. In response, it is probable that the firms already planning to use the V-1710 were asked to consider producing similar aircraft powered by the Merlin.[5,6]

Concurrent with these events at Ford, the U.S. Advisory Commission of the Council of National Defense and the British Purchasing Commission, had determined that the requirements for 1000 bhp liquid-cooled engines could not be satisfied by the Allison Company alone. Dr. Mead, Chairman of the President's Liaison Committee, informed General Arnold, Chief of the Air Corps, of the British desire to build the Merlin in the U.S. On June 17, 1940 General Arnold responded to Dr. Mead that the Air Corps would collaborate in production of the Rolls-Royce engine *provided* that the Army would receive engines for 3,000 airplanes *over and above* those then included in existing production programs (emphasis added).

On June 19 General Brett sent an *Extra Priority* teletype message to General Arnold recommending that he immediately request that the Materiel Division be permitted to place the just received pattern Merlin on the test block for an observation run. He felt that this was essential as they were concerned that the engine would be able to meet Air Corps test standards. With the pending order for 2-3,000 engines being considered, the recommendation was that the run was necessary before signing the contract.[7]

Things were happening fast, but it was a cooperative atmosphere and the necessary requirements were met. The end result was that Arnold finessed 3,000 extra airplanes for the Air Corps, these being in addition to the 18,000 plane program that had been authorized by Congress that spring. When these 3,000 engines were added to the 6,000 required for the British there was a critical need to find a suitable organization able to manufacture the complex Merlin.[8]

Following the false start with the Ford Motor Company, Knudsen then met with the Packard Motor Company on June 24, 1940 and requested they consider producing 9,000 Merlin's.[9] Packard was a good choice, for through the 1920s they had produced a number of large V-12 aircraft engines of their own design, and were still producing a derivative marine V-12 then being used to power Navy PT boats. Packard had earlier sought Air Corps support for returning to aircraft engine manufacturing. Their entreaty was probably remembered when a manufacturer was sought. Packard began work on June 27, 1940, using the Rolls-Royce drawings that had already been delivered to Ford. It took until September 13, 1940 to complete the formal contracts, one for the 6,000 British Merlin XX engines and a separate Air Corps Contract for 3,000 similar single-stage V-1650-1 engines for the U.S.[10]

Single-Stage, two-speed Merlin XX. These were the model contracted by the British in the U.S. (Rolls-Royce via Graham White)

The Packard built V-1650-1 Merlin was a single-stage, two-speed engine differing from the Merlin XX only in the reduction gear ratio and propeller shaft. This engine is equipped with the British prop shaft, identifying it as a Packard Merlin XX. (Rolls-Royce via G. White)

The primary objective for building the Merlin in the U.S. was to obtain an off-shore supplier to the Empire aircraft building program. That program was intended to carry on the fight were England to be lost, or alternatively, to provide the engines needed for production of British four engine bombers. While the bulk of Packard production was committed to the British Empire, for use of U.S. manufacturing capacity, an additional 50 percent of production was reserved for allocation to the U.S. Air Corps. In addition to Packard, the British had established three Rolls-Royce operated factories as well as a major production factory operated by Ford Motor Co. in Manchester, England, all building the Merlin.[11]

With the September 1940 contracts for the single-stage Packard Merlin, the Air Corps had nearly three times as many Merlin's on order as it had V-1710's, and all of the Allison's were scheduled for delivery by January 1, 1941.[12] This situation evened out somewhat in December when the U.S. ordered an additional 3,691 V-1710's to be built in five different models. Clearly any commitment to the Allison by the U.S. was pending ordering of the aircraft that were to use it. The Merlin was ordered before identifying the U.S. aircraft it was to power. The first Packard built Merlin, a V-1650-1, was run in early 1941, with deliveries beginning that October, some 15 months after Packard began the project.

By the late summer of 1940, world events were moving at a rapid pace. The Battle of Britain had just been fought to a standstill, Europe was firmly in the hands of the Axis, and ominous threats were being monitored in the Pacific. In the U.S. the Air Corps was looking for ways to use its windfall Merlin's. By October 1940, Lockheed had completed a performance study of a Merlin powered P-38 and Curtiss was doing similar work on a Merlin powered P-40.

Introduction of the Merlin

The Curtiss P-40, which had been the program to launch mass production of the V-1710, was also the initial recipient of the U.S. Merlin allocation. To expedite introduction of the Merlin into an airframe, P-40D AC40-360 was modified with a Rolls-Royce built Merlin 28 and began flight testing on June 30, 1941. Production models of the Packard V-1650-1 were not available until January 1942.[13] Interestingly, other than simply being a suitable airframe for the available Merlin's, one of the primary reasons given for the Production Board to insist on this installation was to free Allison V-1710's for use in critically needed Lockheed P-38's and Bell P-63's.[14] The Army Air Force felt that the Allison powered Curtiss P-40N was superior to the Packard Merlin powered P-40, but because of the time necessary to convert Packard to all two-stage Merlin production, some 3,500 single-stage V-1650-1 engines would be available. In order to utilize these U.S. allocated Merlins, the Curtiss Aircraft Company was directed to produce the Merlin powered P-40F and follow-on P-40L aircraft.

Table 15-1
US Liquid-Cooled Engine Orders As Of October 1940

	For B & F	For Air Corps	Total
Allison V-1710	4,315	1,050	5,365
Packard-Built Merlin	6,000	3,000	9,000

By March 1943 the various Air Forces were having problems obtaining sufficient spare engines and parts for their V-1650-1 Merlin powered P-40's, most of which were at the front in combat theaters. As a measure of

As the first Warhawk model, the P-40F was visibly different from the P-40D/E Kittyhawk by the lack of the downdraft carburetor scoop for the Allison engine, and slightly deeper chin cowl to house the larger radiators and duct to the updraft carburetor for the V-1650-1 Merlin. The updraught carburetor arrangement caused considerable trouble in the Mediterranean Theater, the result of not having an air filter. The dust damaged the engines, causing sever shortages of spares and ultimately necessitating refitting with V-1710's as the P-40R. This is aircraft AF41-13997, the 397th in the series. (Curtiss via SDAM)

The U.S. Army was slow to realize the winner they had in the North American Mustang. The Allison Mustangs, P-51/A, F-6A, and A-36, were intended for Army Cooperation and ground support, a low and mid-altitude mission. Typifying these models is P-51A AF43-6012, being flown by Allison Chief Pilot "Pinky" Grimes. (Allison)

the problem, the British allowed 600 of their single-stage engines to be converted to spares and parts to support the P-40F/L aircraft already in the field. The original program had specified 12 percent spare parts; actual experience was that 35-50 percent were required.[15]

With the advent of Lend-Lease, and then the entry of the U.S. into the war, the Army took over the contracting for Merlin's going to the British. As a result the British were concerned about modifications that Packard might introduce, modifications that might adversely effect the interchangability of the engine and/or its parts with British built examples. Consequently, as Packard was approaching completion of the initial contracts for the 9,000 engines[16] they had a clause inserted in the contract that prevented the Packard Motor Car Company from making any alterations in the design of the engine, or in the quality of material or workmanship, without getting prior approval of Rolls-Royce Limited through their local representatives.[17] This arrangement worked well, and Packard did make a number of improvements and changes in the engine during the production run, but they were always coordinated with the local Rolls-Royce representatives.

Another effect of Lend-Lease was to provide aircraft manufacturers a chance to practice their art. An example being North American Aviation committing to a new All-American fighter, the NA-73 Mustang. The engine specified by the British for all of these early models was the Allison V-1710-F3R. The whole idea was to utilize American production capacity, and a U.S. built Merlin was still only a concept in the minds of a few at the time the NA-73 was ordered. It was more than a year later, July 7, 1941, before the Air Corps made its first Mustang purchase, and even that was specifically as Defense Aid, and intended for the British. Most of this batch of 150 aircraft was delivered through Lend-Lease to Britain, though following Pearl Harbor, 55 were commandeered for modification to photo reconnaissance configuration, becoming the U.S. Army's F-6A and known as the *Apache* until July 1942. Significantly, two more aircraft from this batch were retained by NAA for conversion to Merlin power as the XP-51B (XP-78).[18]

This is the Rolls-Royce Merlin 60. Featuring Hooker's innovative two-stage supercharger, and incorporating a liquid-cooled aftercooler, it promised excellent high altitude performance. Packard built the similar Merlin 61 for the British, and the companion V-1650-3 for the U.S. (G. White)

Improved Models of the Merlin

There were really two Merlin engines: those with single-stage and those with two-stage integral superchargers. While the two-stage engine became famous in the later Spitfire and Mustang airplanes, it was like other engineering developments and required a considerable gestation before it was able to enter series production.

General Arnold responded to a news item on the new Spitfire, with its two-stage Rolls-Royce Merlin 61 high altitude engine in December 1942. He stated "it is going to be an excellent high altitude, high performance engine…(but) is still in the experimental stage and there are still several bugs. The Spitfire with its new engine will not be any better than our P-38. There are a lot of experimental airplanes with experimental engines that look very good on paper, but once they are built and actually test flown, it is usually another story. So until the airplanes and engines are actually in production and out on the flying line, we cannot decide one way or another about them."[19]

It was March 1943 before the two-stage Merlin V-1650-3 could be introduced into limited production, and then only at a rate of one engine per day. This model was a major improvement in that it gave a critical altitude approaching that of a turbosupercharged engine.

As an interesting sidebar to the two-stage Merlin story, back in May 1941 the U.S. Office of Production Management responded to General Echols, Chief of the Materiel Division, regarding the feasibility of such an engine. Their stated opinion was that "the Merlin engine borders so closely on obsolescence that it would not be economical to spend time and money trying to improve it...our best bet is to concentrate on the development of the Allison, Continental, Lycoming, and Wright Tornado."[20] Concurrently, there was considerable concern within U.S. production circles that the British were dumping a "second-rate" engine upon the industry while they focused on building the 2000 bhp Napier Sabre. General Echols was of the opinion that the Sabre was too complicated to be put into production by people other than those already familiar with it, but believed that the two-stage Merlin was similar enough to warrant trial by an American manufacture.[21]

When Packard was finally in production of the two-stage Merlin, it proved to be an engine that offered quite an advantage in the right airframe. That airframe was found to be the North American P-51 Mustang. When used in the models P-51B/C/D/H/K, both the two-stage Merlin and the aircraft became legends.

Baseline Engines

The Rolls-Royce Merlin and the Allison V-1710, contemporary, though clearly independent designs, had a lot in common. They were high temperature liquid-cooled, 60 degree V-12's with a piston stroke of six inches and having 10 inch long connecting rods (the primary parameter that defines the frontal cross-section area of a vee type engine). The Merlin, with its' 5.40 inch bore, displaced 1650 cu.in. compared to the Allison's 1710 cu.in., defined by its 5.50 inch bore. The result was a somewhat trivial four percent difference in total displacement.

A parameter of considerable importance to the aircraft designer is the cross-section area of the engine. The early Merlin's, through the Mk XX and similar V-1650-1, had a cross-section of 5.85 square feet.[22] Within a reasonable range, the arguments on frontal area are fairly academic, for when the engines were fully "dressed", with the various tanks, plumbing, cooling, exhaust stacks and air inlets needed in a full scale installation, the whole assembly becomes important. It was the width of the pilots shoulders, and the need for him to be sitting high enough to see over the nose of a V-12 powered aircraft when in flight, that determined the airplane cross-sectional area. In a radial engined airplane, the frontal area was a considerably different issue, and the comparison between radials and in-line V-12's was usually won by the in-line. The differences shown below are actually fairly minor with respect to the aircraft they were installed in.

Table 15-2
Frontal Area Comparison

Frontal Area, Ft^2	Description
5.85	Merlin X, RM2SM, 1320 bhp on 100 octane for takeoff.
5.85	Merlin XX, V-1650-1, 1385 bhp
5.91	The early Allison V-1710-C7 , 1150 bhp
5.49	for V-1710-C15, 1090 bhp
5.19	for V-1710-F, 1150 bhp in 1941, 1600 bhp in 1944

There was one physical difference between the two engines which did limit power and effect performance. That divergence occurred back in 1937 when Allison increased the V-1710 compression ratio from 6.00:1 to 6.65:1 as a means to improve fuel economy. With both engines rated on the same fuel, the V-1710 reached detonation limits at an Indicated Horsepower about 10 percent below that of the Merlin. The larger capacity of the V-1710 cut this difference approximately in half, in terms of net shaft horsepower.[23] Still it was a real limitation. In their late model two-stage high BMEP engines, Allison reduced the compression ratio back to 6.00:1.

Both engines were able to run at fairly high speeds due to their heavily counterbalanced crankshafts. While it may not be apparent that an even-firing 60 degree V-12 would require counterbalancing, it is necessary to minimize the dynamic loads and couples on the supporting bearings during high speed running. In this regard, the Merlin incorporated eight counterweights while the V-1710-C used six. Allison made these heavier beginning with the V-1710-E/F, and they were subsequently further increased in weight and number when the 12-counterweight crankshaft was introduced in late model E/F and subsequent V-1710-G models built after 1942.

In comparing the engines it is also necessary to understand that both were developed extensively during the war. Consequently, comparisons need to be specific as to period, theater of use, and the assigned mission. What results should be an understanding that there was an appropriate mission for each engine, just as there was for each aircraft, and that both engines were constrained and guided in their development by separate corporate and military policies, as well as the circumstances of the day.

It should be apparent that these engines were more alike than they were different. As such, they set into motion a series of evolutionary steps involving joint projects and ventures which 50 years after WWII cumulated with the merging of the Allison and Rolls-Royce companies.

The following comparison looks at the differences, the strengths, and the reasons for the controversy over which one was the best engine, or in the eyes of some, what was wrong with the Allison.

Early Development of the Merlin

The first water-cooled 1000 bhp Rolls-Royce Merlin was built as their P.V.12 (Private Venture 12 cylinder) in the autumn of 1933. It departed from previous Rolls-Royce V-12 practice by having a crankcase with integral cylinder blocks and detachable heads. The reduction gear was of a new type having double cut helical gear teeth. Following their earlier Kestrel practice the engine incorporated a single-stage, single-speed supercharger with a forged aluminum impeller and rotating steel inlet guide vanes. Testing soon showed that the reduction gear gave trouble and the crankcase frequently cracked. This could have been easily cured in a conventional configuration, but the lack of removable cylinder blocks made future field repairs and maintenance both difficult and expensive. The engine was then redesigned to use the earlier Kestrel type removable cylinder blocks and a more conventional external spur type reduction gear.

The P.V.12, and follow-on Merlin Mk I engines, were built with "ramp head" or "semi-penthouse" shaped combustion chambers. In this form the engine first flew at the Rolls-Royce Hucknall flight development center in a Hawker Horsley, December 1935, a year before the V-1710 made its first

This is an early Merlin with ramp head combustion chambers. It was a single-stage, single-speed engine similar to this one that made the first Merlin flight in a Hawker Horsley, December 1935. The fundamental features of the later Merlin are quite clear in this early model. (G. White)

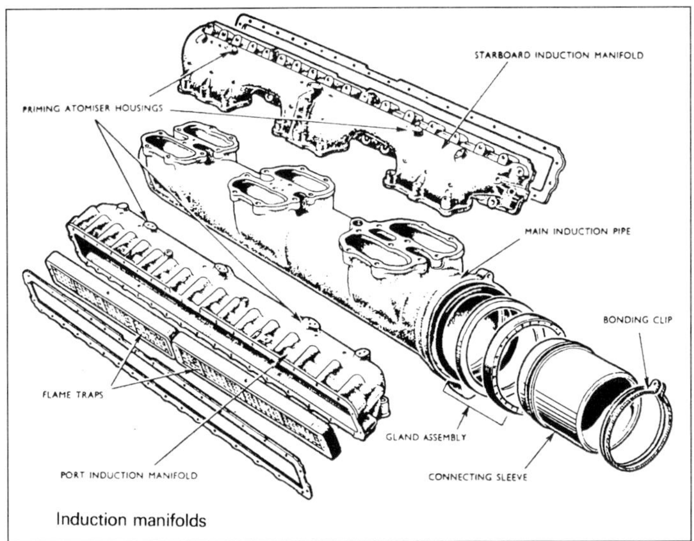

Induction manifolds

This was the layout of the Merlin gas pipe for mixture delivery to the cylinder banks. While Allison used a "ramming" design to help pack mixture into the cylinder, the Merlin plenum served as a reservoir of pressurized mixture. Each bank was served by a smaller plenum that fitted the rectangular flame traps to limit backfires. These had to be modified to incorporate a dividing flange at mid span, a feature to contain a backfire within a single quadrant of three cylinders. Although the Rolls-Royce and Allison intake manifolds were quite different, they went through similar phases and troubles during development. (Courtesy of Rolls-Royce Heritage Trust)

flight.[24] Although two 100 hour tests were completed, it was found that in the full scale engine, the ramp head was not satisfactory. Neither performance nor reliability goals were being achieved. Major problems involved cracking between the combustion chamber and the surrounding cooling water spaces.[25] The fix was to revert to a single piece Kestrel type cylinder block having a single plane head. This was a setback to Rolls-Royce in that the new design allowed the possibility of direct leakage of combustion gasses and/or coolant, should there be any problem at the cylinder head joint. This design first appeared in the Merlin Mk II, which in June 1937 completed 100 hours flying in the Horsley in a 6-1/2 day period, and as such was released for production.[26] A third approach to the head/block joint problem was to use a two-piece design. Such a design was prepared, though the press of production and the coming war delayed its introduction until the Packard built Merlin XX went into production in 1941.[27]

The Merlin was switched to straight ethylene glycol as its coolant in 1935, increasing coolant outlet temperature to 135 degrees Centigrade (275 degrees Fahrenheit). The decision to abandon "water" cooling was made following a visit by Rolls-Royce engineers to Wright Field in early 1935.[28] To improve effectiveness of cooling of the engine internals, the coolant mixture was soon changed to 70 percent water and 30 percent glycol, operating at 15 psig to raise the allowable coolant temperatures to the desired 135°C. This approach was important in reducing cylinder head temperatures and thereby allow the engine to operate at high boost pressures. The glycol was retained solely for its anti-freeze characteristics.

When the early Merlin entered squadron service, the first major trouble to occur was backfiring. The cure was to fit flame traps into the induction pipes as a field modification. The flame traps, or backfire screens, themselves proved to be a problem, just as on the contemporary V-1710. Vibration would cause the Merlin screen frames to fail and the foils to come adrift. These would then enter the cylinders or get caught in the valves and could lead directly to a backfire. Like the Allison, the Merlin also had the problem of clogging flame traps due to fuel dye, dirt, and oil. One change that had to be introduced was to weld a blank into the manifolds so as to divide each induction manifold into groups of three cylinders. This was done to prevent a backfire from extending to all cylinders on one bank, an event that could be particularly troublesome. When engines began to be operated at very high boost pressure, the flame traps were not sufficient to prevent really violent backfiring. Good maintenance and high quality spark plugs were essential to prevent backfiring. The flame traps alone could not suppress it.[29]

U.S. Assessment of the Early Merlin and V-1710

In July 1937, just after the Allison YV-1710-C6 had completed its 150-hour Type Test at 1000 bhp, the Wright Field Engineering Section compared the Allison to just released performance data on the contemporary Merlin II. They compared the Merlin II to the V-1710-C7, then being prepared to be Type Tested at 1150 bhp, and determined that, "...the Allison 'with exhaust driven supercharger' will deliver more power output at all altitudes than the Merlin by a margin of 35 to 150 bhp." The installation weight of the V-1710 and supercharger was about 100 pounds greater than that of the single-stage Merlin. Their comparison had the Merlin's' takeoff power at 900 bhp, and the V-1710 at 1150 bhp. They concluded by stating, "It is believed that this difference in takeoff rating is of very great importance for engines intended for bombardment applications."[30] This most telling statement reveals the thinking of the day regarding the intended mission of the V-1710. Of course, the V-1710 engine had just begun flight testing in the turbosupercharged Curtiss XP-37 and, except for the Bell XFM-1, no other orders existed, although earlier it was planned to use it in the Boeing XB-15 and Martin XB-16 multi-engined bombers.

As early as April 1938 Allison was seeking ways to further increase the power of the V-1710. The initial proposals were based upon achieving higher power output by using a turbosupercharger to boost the manifold pressure at the inlet to the carburetor.[31] The Materiel Division "appreciated" Allison's proposal, but chose to not support it at the time.[32] This may well have been because of the press of other priorities, or possibly the lack of funds, or even the lack of insight.

In August 1939, Wright Field prepared an engineering study on how to increase the power output of the V-1710, based upon their reading of performance data on the Merlin contained in the April 26, 1939 issue of the British aviation publication *The Aeroplane*. The study identified two approaches for increasing the output of the V-1710. The first, preferred by the Air Corps Technical Executive, would result in 1350 bhp for takeoff,

The Merlin III (RM.1S) differed from the Merlin II mainly in having the new universal propeller shaft. Ratings in Spitfires and Hurricane were the same for both models. (Courtesy of Rolls-Royce)

and use a turbosupercharger to maintain this power to altitude. This was determined to be the easier path. The second method would have resulted in an engine similar to the Rolls-Royce RM2SM Merlin described in the article, and would have required Allison to develop a two-speed supercharger. This approach required an engine capable of considerably more Indicated Horsepower and, would have meant strengthening the engine for mechanical and thermal stresses, as well as a completely revised supercharger and accessories section. The resulting engine was planned to deliver 1200 bhp at an altitude of 16,750 feet.

To implement this improvement program, Wright Field suggested Allison be contracted to build a V-1710-19 for the Air Corps. The Air Corps would then test in progressive steps of +50 bhp, beginning at 1150 bhp, until the engine failed. The contract incentive for Allison provided a progressive payment based upon the power achieved by the engine. If 1500 bhp was achieved, Allison would receive a 175 percent payment of the contract price.[33] This test program was successfully concluded using a V-1710-19 type engine which incorporated 8.0:1 supercharger gears instead of the usual 8.77:1, as sufficient Manifold Air Pressure (MAP) was available due to the low elevation of the test site. The tests were run between September 1939 and January 1940, and demonstrated 1575 bhp when running at 3000 rpm and 58.4 inHgA.[34] The engine did not fail.

Evidently General Arnold did not care for the measured engineering approach suggested by his staff at the August 1939 juncture. He had a reputation for being quite direct and his action in this instance was to send a letter to the then President of General Motors, William S. Knudsen, notifying Allison of his view of the V-1710 and its horsepower potential in light of "recently obtained" Merlin performance data. In his September 1939 letter he states:

> "The basic fact which it is desired to point out in connection with these figures on the Merlin engine are that this engine *today* is capable of being rated at 1320 horsepower; that it already has designed for it, and available for installation, a two-speed geared supercharger; that in that respect it is superior to the present Allison engine; and that unless extensive development work is undertaken at once by the Allison Engineering Company, there is very little possibility that our Allison engine will be able to match the potential horsepower available in any of the more powerful modern engines.[35]

This was certainly a cheaper way of getting Allison to undertake an aggressive development program without the necessity of Government funding.

While Allison may have had the resources to commit to an aggressive development program in 1938, things were now different and they were not having it easy in 1939. They were busy:

- Delivering the 1150 bhp V-1710-C10 for the Curtiss YP-37's,
- Delivering the 1150 bhp V-1710-D2 for the Bell YFM-1's,
- Type Testing the altitude rated V-1710-C15 for the Curtiss P-40,
- Flying the V-1710-E2 in the new Bell XP-39,
- Building the first V-1710-F1,
- Model Testing the XV-3420-1 while building the V-3420-3's,
- Developing the Auxiliary Stage supercharger for the V-1710-E/F.

In addition, in May 1939 they had begun construction work on Plant 3, which was to become the major V-1710 manufacturing facility. They certainly had their hands full, and employment was just beginning to escalate to support the orders for engines which had been placed by the French and British. Remember, it was only in April 1939 that the U.S. Army had placed its first production order and contract for series production models, a total of 837 engines in five models.

In response to the General's concerns, Allison did refocus its energies. General Motors provided, and/or transferred considerable staff, equipment and support to augment and expand the base at Allison. These efforts ultimately produced a lot of engines, but little additional rated horsepower was coming from the early engines. It was not until combat experience

Table 15-3
Comparable Single-Stage Merlin and V-1710 Engines

	Rated Output, bhp	Condition	Impeller OD, in	S/C Ratio	Tip Spd, fps	Boost, psig	MAP, inHgA	Crit Alt, ft	Wt, lbs	bhp/#	Comments
Merlin II/III	1,030	TO	10.25	8.58:1	1,151	+6.25	42.6	16,250	1335	1.08	87 Oct, 1939
Merlin II/III	1,440	WER	10.25	8.58:1	1,151	+16	62.5	5,500	1335	1.54	100 Octane, 1940
Merlin Sprint	2,160	WER+	NA	NA	NA	+27	84.9	SL	1400	0.87	3200 rpm, Est. wt
Merlin XX-Lo	1,260	NA	10.25	8.15:1	1,093	+14	58.6	12,250	1450	0.81	
Merlin XX-Hi	1,175	WER	10.25	9.49:1	1,273	+16	62.6	21,000	1450	0.90	
V-1650-1 -Lo	1,385	WER	10.25	8.15:1	1,093	+12	54.3	8,600	1535	0.81	March 1941
V-1650-1 -Hi	1,250	WER	10.25	9.49:1	1,273	+12	54.3	15,600	1535	1.14	March 1941
Merlin 50M	1,585	WER	9.50	9.09:1	1,130	+18	66.7	2,750	1385	0.81	"Cropped" S/C
V-1710-19(Hi-bhp)	1,575	Test	9.50	8.00:1	995	+14	58.4	SL	1325	1.14	100+Octane, 1939
V-1710-33(C15)	1,040	TO	9.50	8.77:1	1,091	+5.9	42.0	SL	1325	1.19	100 Octane, 1941
V-1710-33	1,090	Mil	9.50	8.77:1	1,091	+4.4	38.9	13,200	1325	0.78	100 Octane, 1941
V-1710-39(F3R)	1,150	TO/Mil	9.50	8.80:1	1,094	+7.2	44.6	11,700	1310	0.82	Grade 100, 1941
V-1710-39(F3R)	1,490	WER	9.50	8.80:1	1,094	+12.8	56.0	4,300	1310	0.88	Grade 125, 1942
V-1710-81(F17R)	1,125	TO/Mil	9.50	9.60:1	1,194	+7.1	44.5	15,500	1352	1.14	Grade 130, 1943
V-1710-81	1,410	WER	9.50	9.60:1	1,194	+13.3	57.0	9,500	1352	0.83	Grade 130, 1943

* Rated at 3000 rpm on 100/130 Grade Fuel, 1940 on.
Note: In September 1939 the RAF made 100 Octane fuel the Service standard.

was reviewed in early 1942, soon after U.S. entry into the war, that Allison began testing and authorizing operation of its engines with War Emergency Ratings. These had the desired effect of delivering the horsepower equivalent to, or greater than that provided by the comparable Merlin's.[36] In the interval between 1939 and 1942 Allison had incorporated numerous changes and improvements within the engine that allowed it to be successfully operated at the considerably higher ratings.

Type Testing the Packard Built Merlin

The Air Corps performed the 150-hour Type Test on the V-1650-1 at Wright Field from August 12 to November 1, 1941. The test consisted of calibration, torsional vibration checks, endurance running, teardown inspection at the completion of the endurance running, and a 20 hour penalty run.

At teardown inspection after the initial 150 hours the following parts were found to be in an unsatisfactory condition: Three pistons were cracked, chrome plating on the cam follower had cracked or chipped off, the impeller shaft had worn enough to allow the impeller to engage the supercharger case, copper rivets in the clutch plates of both high-gear supercharger clutches had sheared off, and the spline coupling for the supercharger drive had cracked.

As a demonstration of the critical military needs of the day the Air Corps concluded:

> It is recommended that the V-1650-1 engine be put into service at the foregoing ratings and that on the basis of dive tests the minimum manifold pressure for diving be limited to 20 inHgA. Since several parts were found to be in an unsatisfactory condition during and at the conclusion of the Type Test, it is recommended that steps be taken by the manufacturer to correct these and that a Model Test be run on a production engine after correction.[37]

Approved Military ratings for the 1,535 pound engine were:

Low-Gear, 1235 bhp at 3000 rpm to 12,000 feet, and
Hi-Gear, 1120 bhp at 3000 rpm to 19,000 feet,
Takeoff Power, 1300 bhp at 3000 rpm.

During peacetime when the V-1710 was going through Type and Model Testing, many of those tests were declared failures or as unsatisfactory based on findings far fewer in number or significance as those occurring to the V-1650-1.

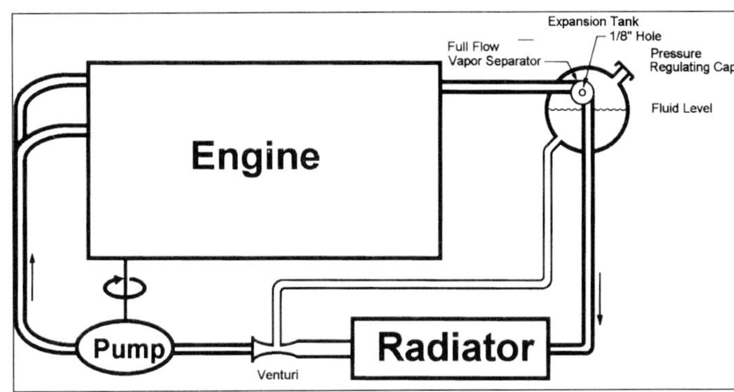

North American revised the British plumbing with this arrangement for the Merlin P-51. It had the advantage of raising the system pressure enough to cause vapor bubbles to be condensed by the flowing coolant. (Author)

From a historical perspective, the Type Test was performed by Packard Manufacturers engine No. A-4, Air Corps No. 41-46522. Its Type Testing was immediately followed by the first preliminary running of a Packard Built two-stage Merlin 61. For that testing, Packard modified V-1650-1 s/n A-9 to incorporate Model 61 components supplied by Rolls-Royce, including the latest type Merlin 28 pistons. A preliminary series of runs on the engine that was to be the V-1650-3 had accumulated 3 hour and 14 minutes by December 23, 1941. The weight of this engine was 1,680 pounds.[38] The first Packard built two-stage supercharger was installed on an engine May 23, 1942, but the first production V-1650-3 was not shipped until April 17, 1943.[39]

Engine Cooling

One important difference between the engines was the way in which internal heat was removed. Both engines used similar plumbing arrangements to deliver oil and coolant to the cylinders, blocks and moving components, but the relative quantities and roles were different. As is apparent in the following tables, the Allison was a slightly more efficient engine. Note the lower total cooling load. The most interesting aspect, though is how this heat was removed. As shown in the table, the V-1710 used the coolant to remove about 75 percent of the heat, while in the Merlin the figure was about 90 percent. In both cases, the balance was removed by the oil, primarily that being "splashed" from the connecting rods into the crankcase. Several dramatic examples of this basic difference between the engines have occurred when modern operators of aircraft powered by V-1710's have attempted to use oil coolers intended for use with a Merlin.

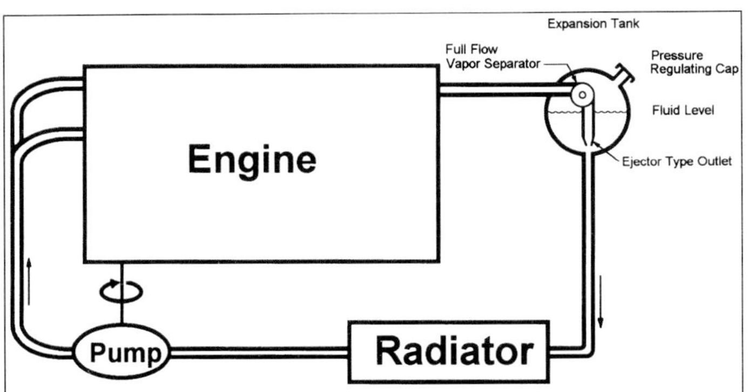

British standard vapor-phase cooling system as used by Rolls-Royce. This system can handle large quantities of vapor but has the disadvantage that the pump intake pressure is low because of the pressure drop through the radiator, and the pump is subject to flow starvation. (Author)

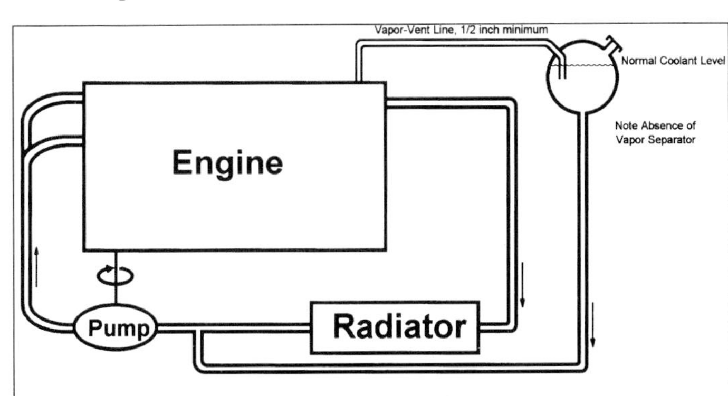

The U.S. Army/Allison handbook liquid-phase cooling system has full system pressure at the pump inlet and only liquid circulating in the system. Vapor coming from the heads is vented directly to the reservoir. (Author)

Table 15-4 Cooling Loads of V-1710 and Merlin						
Engine Model	Rating, bhp to Prop/rpm/Alt	Total Cooling Req'd, hp	Heat to Coolant/Oil, percent	Aux Stage Heat to Oil, hp	Coolant Mixture	Comments
V-1710- 33(C15)	1090/3000/13,200	510	80.4/19.6	None	97% Glycol	Altitude Rated
V-1710- 81(F20R)	1200/3000/T.O.	585	73.5/26.5	None	30/70% Water	Altitude Rated
V-1710-111(F30R)	1500/3000/T.O.	630	74.6/25.4	None	30/70% Water	Turbosupercharged
V-1710-119(F32R)	1500/3200/T.O.	725	75.9/24.1	55	30/70% Water	120 hp to Aftercooler
V-1650-1 Merlin	1300/3000/T.O.	699	89.3/10.7	None	30/70% Water	Low-Gear, Spec AM 1035
V-1650-3 Merlin	1400/3000/T.O.	777	86.4/13.6	None	30/70% Water	160 hp to Aftercooler
Cooling loads are expressed in horsepower. T.O. is Takeoff						

Another fundamental difference has to do with the way in which the coolant is used to absorb its share of the cooling load. Rolls-Royce opted for a design that allowed a considerable amount of water vapor (steam) to be produced in the cylinder heads. This minimized the amount and weight of coolant that had to be carried, but also reduced the margin to overheating. This was dramatically demonstrated in a test done at Wright Field with simulated coolant leaks in similar Allison and Merlin powered P-40's. The Allison tolerated a loss of most of its coolant with no adverse effects, while the Merlin seized after a loss of only a few gallons of coolant when the coolant began boiling and high pressure steam blocked the flow of coolant into the engine. In Chapter 16 we look at the details of how the V-1710 cylinders were cooled, and consider the effect of varying coolant mixtures and flow rates through the engine.

Supercharging

Where the V-1710 was developed specifically to be mated with a turbosupercharger, Rolls-Royce had determined not to attempt to use the feature. Instead they developed "ejector" type jet thrust exhaust stacks as a way to recover a portion of the energy otherwise being wasted with the hot exhaust gasses. Such stacks provide a significant amount of propulsive thrust to a high speed aircraft. Consequently, they are a very effective and light weight power augmenting device in the proper installation.

Allison, without much fanfare, developed its own gear driven superchargers for the V-1710. Initial credit for this capability goes to the General Electric Company, for they had been having Allison fabricate prototype turbosupercharger compressors during the 1920s. Allison's willingness to take on the design of its own superchargers was in deference to the practice of the other major U.S. aircraft engine manufactures, most of whom,

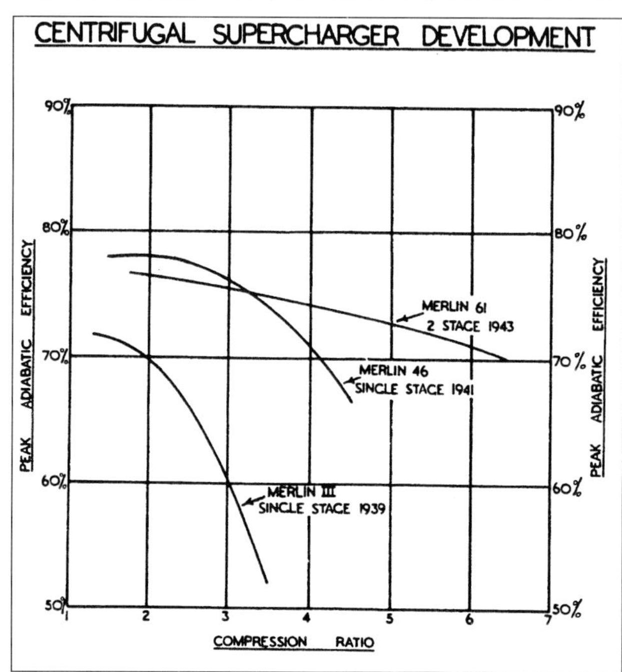

This figure shows the dramatic improvements in supercharger efficiency Hooker was able to achieve in the single-stage Merlin. Likewise, the two-stage Merlin was a formidable achievement, maintaining 70 percent efficiency to a compression ratio of 6.4:1. Such a ratio would provide sea-level manifold pressure up to an altitude of nearly 43,000 feet. (See: Gas Turbines For Aircraft Propulsion, Slide #7, Dr. S. Hooker, Rolls Royce Limited, Barnoldswick, April 1945, NEAM Box 331.) (NEAM)

Table 15-5 Related Cooling System Parameters							
Engine Model	Rating, bhp to Prop/rpm/Alt,ft	Max Oil Flow, LBM/min	Max Oil Temp, F	Oil Temp Rise, F	Max Coolant Flow, gpm	Max Coolant Temp, F	Coolant Temp Rise, F
V-1710- 33(C15)	1090/3000/13,200	125	185	66.6	215	250	14.7
V-1710- 81(F20R)	1200/3000/T.O.	160	185	80.6	250	250	13.3
V-1710-111(F30R)	1500/3000/T.O.	180	203	72.5	265	250	13.7
V-1710-119(F32R)	1500/3200/T.O.	175	203	81.5	225	250	13.9
V-1650-1 Merlin	1300/3000/T.O.	115	194	53.7	150	250	25.1
V-1650-3 Merlin	1400/3000/T.O.	120	194	72.0	150	250	25.4

until at least 1942, utilized GE designed and built superchargers. Whether by luck or design, the early Allison superchargers were more efficient than those being used on other American engines, as well as those on the Merlin prior to the Merlin XX.[40]

Rolls-Royce took a measured approach to its superchargers, building largely on the experience from highly boosting their Schneider Cup "R" racing engines of 1929 and 1931. Their early supercharging efforts were based upon the work of Mr. James Ellor, who had come to Rolls-Royce from the Royal Aircraft Establishment in the late 1920s where he had done pioneering work in the RAE supercharger development program. It was he who was responsible for the large supercharger that was the heart of the "R."[41]

In 1940, just prior to the Battle of Britain, Stanley J. Hooker, then new to Rolls-Royce, became involved in the mathematical treatment of the design and performance of superchargers and quickly suggested simple changes in the shapes of the supercharger vanes, diffuser and intake elbow which resulted in the much improved Merlin XX and Merlin 45 series engines. These engines were rapidly introduced on British aircraft soon after the Battle of Britain, and with them, assured air-superiority over the home islands for a considerable period.

Building on this experience Hooker conceived a 2-stage arrangement of two existing Rolls-Royce superchargers as an expedient to meet a Ministry of Defense requirement for an engine to power a high altitude bomber. This engine went on to become the important Merlin 60 series, widely used in Spitfires, Mosquitoes, and Mustangs.

Investigations into cooling of the cylinders and heads, done by the NACA during the war, suggests an answer to the question of why, though attempted several times, the single-stage Merlin was found to not be amenable to turbosupercharging.[42] Certainly the Merlin cylinder could tolerate high manifold pressures, as demonstrated by wartime ratings that reached 67 inHgA dry and 90 inHgA with ADI in the later two-stage engines. It is believed that the increase in backpressure, caused by the turbo, increased combustion chamber wall and/or exhaust valve temperatures to the point where detonation would occur. The fact that the Merlin used only half of the coolant flow (125 gpm in typical engines), as compared to the V-1710, meant that there was less margin before critical temperatures in the cylinder would be reached.

Rating the Engines

As mentioned elsewhere, a reciprocating engine is simply a very precise air pump, one that then burns a proportional quantity of fuel to produce power. To a first approximation, it is the amount of air processed by the engine that determines the power, not the amount of fuel. The amount of power the fuel produces in the engine is known as the Indicated Horse Power, IHP. The power available from the engine to drive the propeller is defined as the Brake Horse Power, BHP, and is the power remaining after the engine friction is overcome, along with the power consumed by the engine accessories, including the supercharger. With all else being equal, the IHP is the maximum power available to the engine and is usually established as a consequence of the grade of fuel and the shape and cooling of the engine combustion chamber shape being used.

In developing the Merlin, Hooker established that the engine produced 10.5 IHP per pound of air consumed per minute. Since this factor is primarily established by the efficiency of the combustion process in a given engine, it should remain nearly constant for different models of the same engine. For the V-1710-F engine Allison established that the similar parameter was 10.6 IHP/LBM/min.[43] These engines were fundamentally more alike than they were different, though the V-1710 used a 6.65:1 compression ratio while the Merlin was fixed at 6.00:1.

Reviewing Table 15-3, it is not immediately clear why the single-stage Merlin could develop both more power and more manifold pressure, and achieve higher critical altitudes than the V-1710. The combinations of rotor diameter and supercharger drive step-up ratios gave almost identical tip speeds to both engines, meaning that the supercharger pressure ratios should be quite similar. Yet there was a difference in the outlet or resulting manifold pressure produced by the superchargers. This means that it must have been the pressure at the supercharger inlet, just downstream from the wide open throttle and ahead of the supercharger impeller, where the differences were occurring. An investigation of the differences in the flow areas at the single-stage supercharger inlets shows that the Allison area was about 25 percent *less* than that of the Merlin. When flowing the quantity of air needed to match the Merlin XX, the Allison would have had over 1 inHg of extra pressure loss, not including any losses upstream of the throttle.

The sea-level rated Allison's used the Stromberg PD-12K carburetor, which had 24.4 square inch throat area and matched the supercharger inlet area. The early altitude rated engines used the PT-13 carburetors, with 41.3 square inch throat area narrowing down to feed the original size inlet to the supercharger. When Packard built the single-stage Merlin V-1650-1 (identical to the Merlin XX, except with American accessories), they used the PD-16 carburetor with 38.3 square inches of flow area. The later two-stage Merlin V-1650-7 used the PD-18 carburetor, with 46.4 square inches of throat area. The difference between the flow areas, and the resulting higher losses in the pressure available at the supercharger inlet was the major reason for the disparity in altitude performance of the single-stage altitude rated Allison.

This conclusion is confirmed by the Wright Field Engineering Section in a report produced in February 1938 that considering options available for *Altitude Ratings of V-1710 Engines equipped with Gear Driven Superchargers*.[44] This report was prepared at the time that Wright Field, Allison and Curtiss were all looking for alternatives to the complexity of turbosupercharging, particularly in fighter or pursuit type aircraft. This report notes that the Merlin II, rated at 16,000 feet with 6-1/4 psi (13 inHg Boost),

> "...has a very much enlarged induction system to handle the large quantity of low density air passing into the carburetor and up to the supercharger inlet. To obtain the same performance with an Allison engine no doubt the supercharger inlet system would have to be materially enlarged..."

As a near-term alternative, Allison suggested fitting the PT-13 carburetor and could thereby obtain a 12,000 foot critical altitude rating, though it would still be 110 bhp less than the Merlin at 16,000 feet. Allison also proposed development of a 14,000 foot critical altitude engine with a completely redesigned supercharger and induction system. They estimated that it would require 1-1/2 to 2 years to complete. At 16,000 feet this engine would have exceeded the Merlin III output on 87 Octane fuel by 60 bhp. It is no surprise that this approach did not proceed given the circumstances of the day and the intent to use the turbosupercharger to boost flow through the V-1710 induction system. The Army planners clearly stated that if higher altitudes were required, then turbosuperchargers would be used.[45]

The early single-stage Merlin III centrifugal supercharger impeller had 16 radial vanes, was 10-1/4 inches in diameter, and discharged through a 12 vane diffuser. At the time this supercharger, at 62-65 percent efficient, was believed by Rolls-Royce to be the most efficient in the world. Alternatively, the Allison V-1710-C15 and V-1710-E/F used a 15 blade impeller

Availability of 100 octane fuel allowed rerating the single-stage, single-speed 1030 bhp Merlin III such as this for 1440 bhp, up to 5,500 feet using 62.5 inHgA. This was a major factor contributing to the Spitfires and Hurricanes winning the Battle of Britain in 1940. (G. White)

that was 9-1/2 inches in diameter, discharged through a 6 vane diffuser, and in the early V-1710-E/F engines, was also 62-65 percent efficient.

The single parameter of impeller diameter is essential in describing the potential performance and capability of a centrifugal supercharger or compressor. When multiplied by the rotational speed of the impeller it establishes the speed at the perimeter of the wheel, the impeller tip speed. This speed determines the amount of kinetic energy available to be imparted to the air which is being compressed. The more energy imparted to the air, the more pressure rise the supercharger can theoretically develop. Since it is this pressure that establishes how much air can be pumped into the engine cylinders, it then is the parameter that defines the power the engine will produce. This is the most important parameter in establishing the rated altitude and power capability of a supercharged engine. The intake ducting and pressure losses ahead of the supercharger inlet are therefore extremely critical to the altitude performance of the engine as they are multiplied by the supercharger pressure ratio. This ratio can approach 4:1 on single-stage superchargers when running at high engine speeds.

As shown in Table 15-3, both the early Merlin and V-1710 had very similar impeller tip speeds, yet the single-stage Merlin was rated for greater MAP and power. According to Robert Schlaifer of Harvard University in his epic 1950 tome *Development of Aircraft Engines*, the V-1710-C15 and contemporary Merlin III, that powered the early Hurricanes and Spitfires, were roughly equal in supercharger pressure ratio and efficiency. He attributes the reason for the difference in maximum power to the V-1710-C15 simply having lower mechanical strength.[46] Your author disagrees with this attribution. As can be seen in Table 15-3, both of these early engines were of similar weight and rated power *prior* to 100 octane fuel becoming available in Britain. The aggressive engine rating practices of Rolls-Royce and the RAE allowed them to immediately rerate the Battle of Britain era Merlin III to what the U.S. Military later called War Emergency Ratings. The V-1710-C15 was never rated by such methods due to the fact that it was long out of production when such ratings were first allowed by the Army Air Forces and Wright Field in late 1942. The comparable engines at that point were the V-1710-F3R and Merlin XX/45. The lightest of all of these engines, and the one that exceeded the WER performance of all of the single-stage Merlin's, except for the special low-level rated "cropped" Merlin 50M, was the V-1710-39(F3R). Again, according to Schlaifer, after the introduction of the V-1710-39 in 1941, the maximum power of the Allison, disregarding the altitude at which this power was obtained, remained quite competitive, and even exceeded, that of the various models of the Packard-built Merlin.

The real reason for the difference in ratings was due to Army policies and practices on rating and use of their engines. The U.S. Army was *very* conservative in this regard, focusing on reliability over performance. The misconception on comparable power was further aggravated by the Wright Field practice of not including horsepower ratings in its technical manuals, unless it was the value associated with "takeoff." Army Tech Orders only provided the "not-to-exceed" MAP during WER operating conditions, so comparable horsepower for the V-1710 and Merlin was not widely known. The British on the other hand, rated their engines so that takeoff and WER values were similar, and published the values. Another factor was the differences in the Type and Model Testing requirements. Both the U.S. and British required a test regime that simulated actual flight cycles and covered the range of conditions from idle to Takeoff with sustained periods in excess of 90 percent of full Military power. A major difference was the Rolls-Royce/British practice of only requiring 100 hours of such running to qualify, where the U.S. had required 150 hours of the V-1710 from the beginning.[47] Remembering how the early Allison's often times were within only a few hours of satisfying this requirement when encountering untimely failures at first seems unfair. However the result was not, for the Allison enjoyed a reputation for considerable reliability. When Packard began supplying Merlin's to the U.S. Army, those engines were required to meet the standard U.S. 150-hour Type Test.

The primary criticism of the V-1710 was of its comparatively limited high altitude capability. This can be attributed to the "apples vs. oranges" comparison of single-stage Allison's against the later two-stage Merlin used in the follow-on models of the North American Aviation P-51 Mustang. In any case, in 1943 Allison began work on what became the "G" engine, whose major feature was an enlarged induction system that was fully capable of matching the altitude performance of the Merlin.

Two-Stage Supercharging

Need for a Two-Stage Engine

Stanley Hooker of Rolls-Royce began work on designing a two-stage supercharger for the Merlin in the spring of 1940. The purpose was to meet the requirements of a very high flying British bomber project. Consideration was once again given to adapting the turbosupercharger, but Rolls-Royce was reticent to lose the significant exhaust jet thrust and simplicity

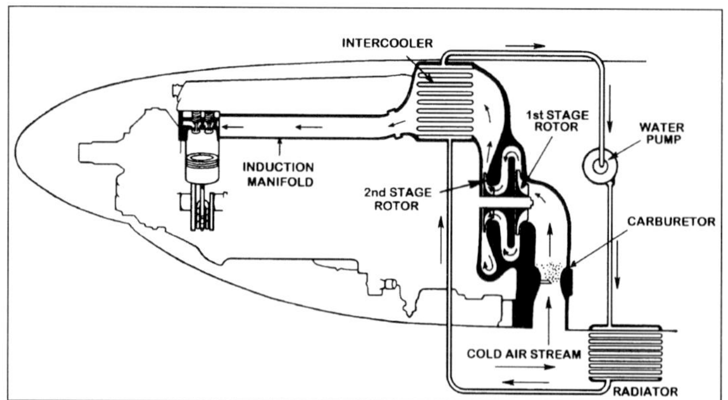

This diagram shows the path and how the mixture was cooled from about 400 oF coming from the supercharger, to approximately 180 oF leaving the intercooler in the two-stage Merlin. By using a liquid-cooling loop independent of the engine cooling system it was possible to achieve a lower mean temperature in the cooler. It also made for a very compact cooler, essential in the limited space on the engine and under the cowling of an airplane.

The tightly cowled Merlin in the P-51D was a very efficient use of space. Close observation of this Merlin, modified for racing at Reno, shows that the aftercooler has been removed and a Merlin 20 type pipe used, but fitted for copious amounts of ADI. (Author)

of their lightweight ejector exhaust stacks. By ingeniously coupling the developed supercharger from the abandoned Vulture engine, that was designed to support 1000 bhp at 30,000 feet, to the existing Merlin 45 supercharger, Hooker was able to provide a Merlin capable of 1000 bhp at 30,000 feet.[48] It was necessary to retain the two-speed drive, and also introduce an aftercooler to reduce charge temperatures to tolerable levels. This engine was quickly adapted for the Spitfire as the Merlin 61 and was test flying in 1941. Even so, it took some 15 months before the first production engine was delivered, and even then production was slow as a number of refinements were incorporated into the engine and supercharger to provide a suitable level of reliability and control.

When the Rolls-Royce Merlin is generally discussed in modern times it is usually with this two-stage engine in mind. Most believe this was the "*Merlin*", and of course it was the engine that, when fitted into the North American P-51 "*Merlin*" *Mustang*, made it the high altitude, long-range world beater that is remembered today.

Technical Details

A consequence of Hookers' approach was to nearly double the required power to drive the supercharger when operating at low altitudes. This was due to the high supercharger pressure rise necessary for 1000 bhp at 30,000 feet. The major effect of this additional power was to heat the air discharging from the supercharger some 205 Centigrade degrees (369 Fahrenheit degrees) above the inlet air temperature. At the resulting high temperature, and with the available fuels, it would be impossible to not incur damaging detonation. Rolls-Royce then designed a compact air-to-water aftercooler which removed 40 to 50 percent of the supercharger induced temperature rise. This solved the problem, insuring that the mixture temperature into the engine cylinders never exceeded 212 degrees Fahrenheit, though it added considerably to the engine weight.

There are methods other than aftercooling available to compensate for the consequences of the hot mixture. These include use of internal coolants, such as enriching the fuel/air mixture, injection of ADI or the use of a

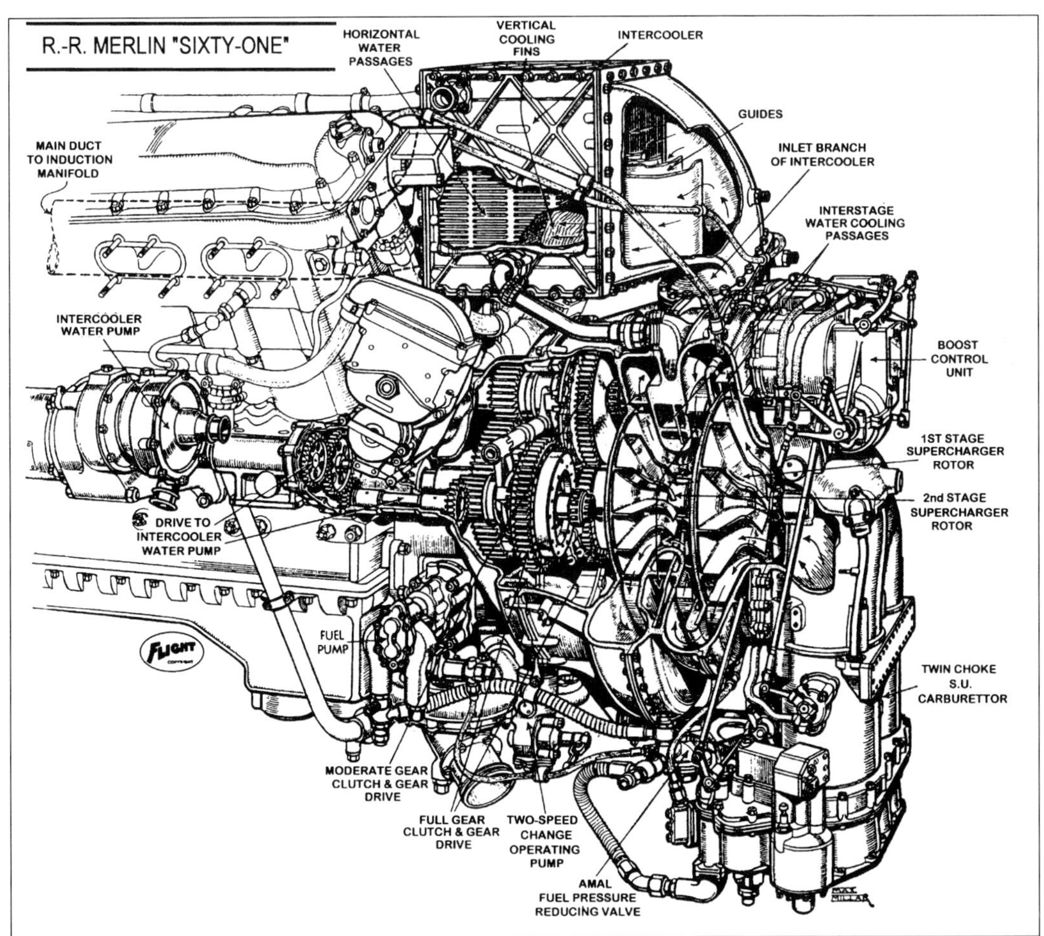

In December 1942 the British magazine FLIGHT published this detailed drawing of the Rolls-Royce Merlin 61 for all the world to see. The complexity and ingenuity of the design is obvious. It both exemplified and challenged the expertise of Rolls-Royce to develop and field the engine in the middle of the war.

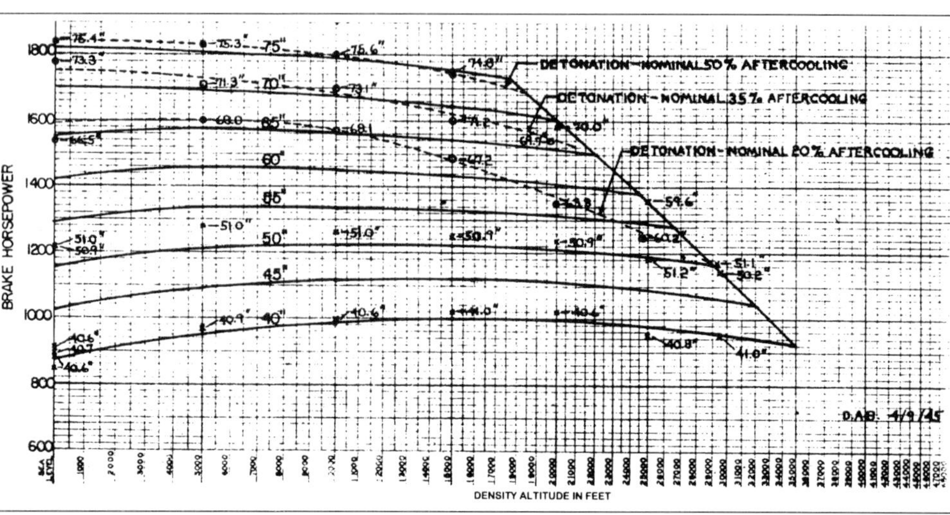

The Allison "Merlin", i.e. 2-stage with intercooling, V-1710-119 (F32R) test results show how the amount of intercooling effected detonation limited power. With 50% intercooling, the same as provided for the Merlin, the engine delivered 1700 bhp at 17,500 feet, from 75 inHgA. The usual Military rating at 1150 bhp was available up to 30,000 feet, from 50 inHgA. (See: Allison Experimental Department Report, C5-22, Scoop Position Test For Setting Controls For V-1710-F32 Engine, April 14, 1945. NARA RG343, RD3775, Box 7411.)

higher grade fuel. Some or all of these methods were used at various times. During the Battle of Britain the introduction of 100 Octane fuel allowed the early single-stage Merlin to run safely at a considerably higher manifold pressure, or correspondingly, carry the original power to a higher altitude. Better fuels were developed during the war, with Rolls-Royce ultimately operating a two-stage RM14SM Merlin at 30 pounds boost (91 inHgA) producing 2300 bhp using 160 octane fuel doped with extra TEL.[49] Using 160 Grade fuel in the Merlin 66 or similar V-1650-3 allowed detonation free operation up to 25 pounds boost (81 inHgA), equivalent to 2000 bhp in low-blower gear and 1700 bhp in hi-blower. A single squadron of fighters would use about 10,000 gallons of this fuel per week, whereas total UK production was about 70,000 gallons per week in the fall of 1943. Fuel clearly made a difference, but in these cases the outstanding performance was being made possible by the aftercooler incorporated into the two-stage Merlin. The normal ratings for the Merlin were on 130 Grade fuel, on which the Merlin operated quite flexibly. A comment by the U.S. Assistant Military Air Attaché in London, notes that, "The Rolls people have overcome the problem of higher octane fuel by greater intercooling."[50]

This is the V-1710-93(E11) Model Test engine, AEC #32381, photographed May 5, 1943. Testing was extensive and involved several engines over a period of about a year, well after the engine was in service. Rated for 1325 bhp at Takeoff from 54 inHgA and 1825 bhp from 75 inHgA WER/Wet. Military power of 1150 bhp was available up to 22,400 feet. (Allison)

On November 28, 1938 Allison presented its plans for developing a two-stage supercharger for the V-1710 to the Air Corps. The Air Corps was not encouraging and little priority or interest was forthcoming. Still Allison did continue to layout the "E" and "F" engines with a mind to accommodating an Auxiliary Stage Supercharger at a later date.

Alternatively, the two-stage V-1710 is hardly known and seldom receives comment even though no less than 36 models were built or designed and at least 5,905 were built. As discussed in Chapters 1 and 16, the Allison approach to a two-stage engine was considerably different than what Rolls-Royce applied to the Merlin, though it was introduced into production at the same time. Allison used a separate "Auxiliary Stage" supercharger, which gave similar performance, except that production models did not incorporate an aftercooler. As a consequence, Allison developed ADI coolant injection to suppress detonation at WER power levels in its 2-stage engines. The other drawback to the device was its arrangement as an "appendage" to the engine. The result was an engine considerably longer than the two-stage Merlin, and it was this feature that kept the Allison two-stage engine from being considered for early installation in the P-51B/C/D model Mustangs. The two-stage V-1710-F32R was later used in the P-51J and P-82E/F airplanes. The two-stage engine programs of both firms began in earnest at about the same time and the highly supercharged engines were available in similar numbers, and at the same time, mid-1943.[51]

The first two-stage V-1710 was the F-7R, which had both an Auxiliary Stage Supercharger and an Aftercooler. When the development emphasis shifted to the E-9 engine for the XP-63, the aftercooler was abandoned, though one was mounted in an E-9 equipped XP-39E for awhile. A major problem with the aftercooler was the complex ducting needed to properly fit the Auxiliary Stage Supercharger, carburetor and aftercooler in place. This ultimately required the development of the Speed Density "carb", which required a protracted development period and delayed the engine until after the war. Alternatively, Allison could have returned to their earlier practice of placing the carburetor ahead of the Auxiliary Stage Supercharger, but they were reticent to lose the altitude advantage gained by locating the carburetor between the supercharger stages.

In place of an aftercooler Allison opted for water injection (ADI) as an expedient way to cool the charge. It is likely this was done to minimize the weight of the installation for the ADI fluid would not have to be carried for the full duration of a mission while the aftercooler would. Although the follow-on E-11 engines used in most of the P-63's were equipped for water

Table 15-6
Comparable Two-Stage Merlin and V-1710 Engines

	WER Output, bhp	Impeller Eng/1st OD, in	S/C Gear Ratios	Boost, psig	MAP, inHgA	Critical Altitude, feet	Weight, pounds	Comments
Merlin 61	1560 Lo-Gear	10.1/11.5	6.391:1	+15	60.5	12,000	1640	Spitfire IX,1942
Merlin 61	1370 Hi-Gear	10.1/11.5	8.03:1	+15	60.5	24,000	1640	Spitfire IX,1942
Merlin 266	1705 Lo-Gear	10.1/12.0	5.79:1	+18	66.6	5,750	1645	Spitfire XVI 1943
Merlin 266	1580 Hi-Gear	10.1/12.0	7.06:1	+18	66.6	16,000	1645	Spitfire XVI 1943
V-1650-3	1600 Lo-Gear	10.1/12.0	6.391:1	+18.25	67.0	11,800	1690	P-51B/C, 1943
V-1650-3	1330 Hi-Gear	10.1/12.0	8.095:1	+18.25	67.0	23,000	1690	P-51B/C, 1943
V-1650-7	1720 Lo-Gear	10.1/12.0	5.802:1	+18.25	67.0	6,200	1690	P-51D/K, 1944
V-1650-7	1505 Hi-Gear	10.1/12.0	7.349:1	+18.25	67.0	19,300	1690	P-51D/K, 1944
V-1650-9	1830 Lo-Gear	10.1/12.0	6.391:1	+24.60	80.0	Sea-Lev	1745	P-51H Wet,Dry as -3
V-1650-9	1490 Hi-Gear	10.1/12.0	8.095:1	+15.27	61.0	13,750	1745	P-51H Dry, 1944
			Eng/Aux[1]					
V-1710-93(E11)	1825 WER	9.5/12.25	8.1/6.85	+22.15	75.0	SeaLev	1620	Wet, P-63A/C, 1942
V-1710-93(E11)	1180 Mil	9.5/12.25	8.1/6.85	+10.85	52.0	21,500	1620	Dry, P-63A/C, 1942
V-1710-117(E21)	1800 WER	9.5/12.25	8.1/7.23	+22.64	76.0	24,000	1660	Wet, P-63A/C, 1943
V-1710-117(E21)	1100 Mil	9.5/12.25	8.1/7.23	+10.85	52.0	25,000	1660	Dry, P-63A/C, 1943
V-1710-119(F32R)	1900 WER	9.5/12.25	8.1/7.64	+23.62	78.0	SeaLev	1750	Wet, P-51J, 1944/5
V-1710-119(F32R)	1200 Mil	9.5/12.25	8.1/7.64	+10.85	52.0	30,000	1750	Dry, P-51J, 6:1CR
V-1710-143/5(G6)	1600 Take-off	10.25/12.25	7.5/8.09	+21.66	74.0	SeaLev	1595	Wet, P-82E, 1946
V-1710-143/5(G6)	2250 WER	10.25/12.25	7.5/8.09	+34.43	100.0	SeaLev	1595	Wet[2], P-82E, 6:1CR
V-1710-143/5(G6)	1700 WER	10.25/12.25	7.5/8.09	+27.06	85.0	21,000	1595	Wet, P-82E, 6:1CR
V-1710-143/5(G6)	1250 Mil	10.25/12.25	7.5/8.09	NA	NA	32,700	1595	Dry, P-82E, 6:1CR

1. Allison Aux Stage S/C hydraulic coupling optimized drive ratio as needed for MAP.
2. This rating never authorized for Squadron use, required Grade 150 fuel.

injection, the system was usually not in service. As a consequence the engines operated with considerably reduced maximum manifold pressure, 60 inHgA Dry instead of 75 inHgA Wet. At these limits the engine was sensitive to fuel Grade, and it remained for the F-32R engine to introduce a fully capable aftercooler.

The only wartime mass production application of the two-stage V-1710 was in the Bell P-63, almost all of which were committed to Lend-Lease and used primarily in Russia. Little has been heard of them in contemporary accounts, though it has been reported that the Russians liked them.

One interesting sidelight, General Echols of Wright Field desired to have a two-stage Merlin tried out in the Bell P-63.[52] To this end, the designation V-1650-5 was assigned, and a mockup of the engine was built by Packard, using a V-1710-E output shaft for connection to the standard P-63 reduction gearbox. Since the Air Force had no interest in the P-63 for its own use, and the two-stage Merlin's were critically needed for the P-51B aircraft, the project never took wing.

When the two-stage Merlin was installed in the P-51B it replaced the single-stage Allison V-1710-81 used in the P-51A through there were other changes made throughout the aircraft and its systems. The -81 engine was rated for takeoff and Military power of 1125 bhp at 3000 rpm up to 15,500 feet, though its WER rating was 1410 bhp, all for a weight of 1,352 pounds. The P-51B was powered by the Packard built V-1650-3 Merlin, rated in low-gear for 1600 bhp at 3000 rpm from 67 inHgA up to an altitude of 11,800 feet. In high-gear it developed 1330 bhp at 3000 rpm from 67 inHgA up to 23,000 feet, with a weight of 1,690 pounds, 338 pounds more than

Continental was slow in getting the Merlin into production, but they did build 797 V-1650-7 Merlin's for the U.S. Army Air Forces. This photo emphasizes the sophisticated induction system, with its two-stage liquid-cooled compressor wheels and large aftercooler. (Continental, via G. White)

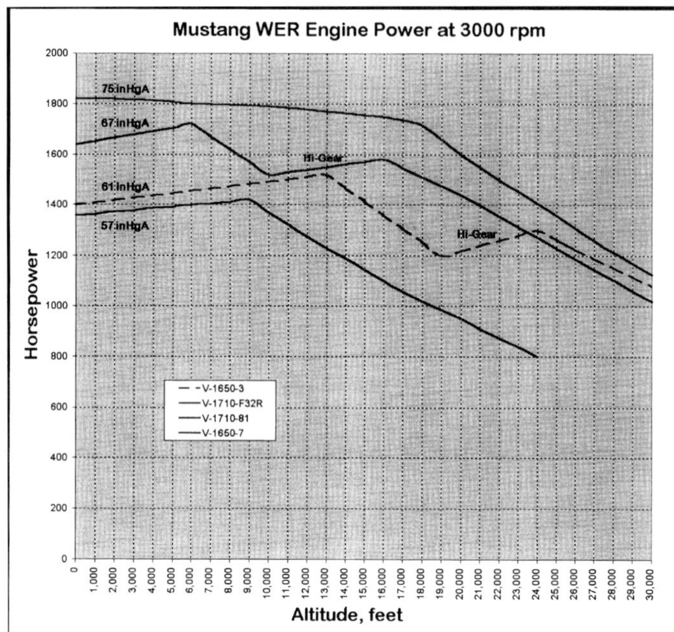

These curves bracket the WER power available to the Mustang from the engines powering the various models. The V-1710-F32R curve clearly shows the advantage of having a continuously variable drive to the first-stage supercharger. The V-1650 Merlin curves show the improvement gained by having even a two-speed supercharger drive. The V-1710-81 is representative of single-stage, single-speed supercharged engines. They make a lot of power for a given MAP, but their altitude capability is limited. Once the supercharger is fixed it was the addition of intercooling that gave these two-stage engines their outstanding performance.

Table 15-7
Contracts for Packard Built Merlin's

Contract	Date	Customer	Number Ordered	Models
A-787	9-13-40	Britain	6,000	Merlin 28, 29, 31
W535-ac-15678	9-13-40	Air Corps	3,000	V-1650-1
-15678 Supp #6	6-31-42	Britain	14,000	Merlin 28
-15678 Supp #6	6-31-42	US Army	34	V-1650-1
-15678 Supp #38	6-24-43	US Army	27	V-1650-1
-15678 Supp #38	6-24-43	US Army	13,325	V-1650-3
-15678 Supp #38	6-24-43	Britain	5,892	Merlin 68
-15678 Supp #38	6-24-43	Britain	2,508	Merlin 69
-15678 Supp #52	3-2-44	Britain	4,500	Merlin 38
-15678 Supp #76	2-8-44	Britain	24,197	Merlin 68
-15678 Supp #76	2-8-44	Britain	2,503	Merlin 69
TOTAL ORDERED			75,986	

Table 15-8
Production of the License Built Merlin Engine

	Packard Single-Stage Merlin 28/29/31/33 38/224/225/266P and V-1650-1	Packard Two-Stage Merlin 68/69 V-1650-3/7	Continental Two-Stage V-1650-7	Total
1941	45	0	0	45
1942	7,251	5	0	7,256
1943	12,292	2,792	0	15,084
1944	7,171	15,798	22	22,991
1945	NA	NA	775	13,346
Total	26,759+	18,595+	797	58,722

From U.S. Department of Commerce

the Allison it replaced. At these high power settings, the fuel economy was terrible, and for that reason, along with mechanical and cooling limitations, the pilot had to limit operating either engine at these extremes to periods of 5 minutes or less.

A direct comparison between the engines and Mustang models is simply not fair. Each was entirely different, as they were intended for entirely different missions and by design were powered by different engines. Certainly the high altitude capability of the clean P-51 airframe was realized by the installation of the two-stage Merlin, but just as fortuitously was the availability of the mission. At the time, the Army Air Force had developed a critical need for a long range high altitude escort fighter, and the *Merlin Mustang* was able to meet the need in a exemplary fashion. The long range came from cruising at high altitude (available with the two-stage Merlin) and use of external fuel tanks.

Table 15-9
Cost of Comparable Airplanes and Engines

Airplane	Engine	Cost of Engine, $	GFE, $	Total Airplane, $
P-40B	V-1710-33(C15)	16,555	7,393	51,205
P-40E	V-1710-39(F3R)	16,855	6,541	51,878
P-40F	V-1650-1	21,016	8,341	59,967
XP-40Q-3 #571	V-1710-121(F28R)	10,503	8,082	45,582
P-51	V-1710-39(F3R)	12,914	12,918	61,783
P-51A	V-1710-39(F3R)	10,907	11,590	49,889
XP-51B	V-1650-3*	12,914	12,918	61,783
P-51D	V-1650-7	17,558	8,654	50,619
P-63C	V-1710-117(E21)	13,160	11,689	53,813

* British Engine provided from Defense Aid, without Royalty payment.

Packard-Built Merlin Production

Reviewing Packard production of the Merlin definitely shows that the program was for the benefit of the British. Most of the single-stage engines were utilized in either Canadian manufacture of the four engined Lancaster bomber and twin engine Mosquito or similar British built aircraft. British two-stage Merlin's were also shipped to England and installed in late marks of the Spitfire as well. Because of the

Table 15-10
Allison V-1710 and Packard V-1650 Cost and Production

	1941	1942	1943	1944	1945	1946	Wartime Totals
V-1650 Cost	$20,185	$21,016	$16,919	$17,555	$17,558	-	-
V-1710 Cost	$16,131	$12,239	$11,268	$10,561	$13,176	$26,824	-
V-1650 Production	49	7,251	12,295	22,969	12,571	-	55,135
V-1710 Production	6,447	14,905	21,063	20,191	5,487	10	68,103

insistence that the Packard and Rolls-Royce produced engines not only be interchangeable, but have interchangeable parts, the Packard Built Merlin's were readily issued for use throughout the RAF.

The contracts issued to Packard for production of the Merlin are listed in Table 15-7.[53]

A much greater variety of models were actually built by Packard, but these occurred as changes to the original quantities ordered. For example, only enough V-1650-3 engines were built to power and support the 1,988 P-51B and 1,750 P-51C aircraft. North American then switched to the P-51D and P-51K models totaling 9,802 aircraft that were all powered by the V-1650-7 engine. These were followed by another 555 P-51H lightweight Mustangs with V-1650-9 engines. There were a total of 12,622 Merlin powered P-51 and P-82 aircraft, for which the Air Force took delivery of 13,325 two-stage engines, all delivered under Supplement #38 to the original contract.

V-1650-3/7/9 and Merlin 60 series were all two-stage engines. At the end of the war production was terminated following engine number 55,523. It is not clear why Packard supplied production figures do not agree with those published by the U.S. Department of Commerce, which lists a total of 58,722 U.S. built Merlin's compared to Packard and Continental figures that combine for 56,320.

The only other use for the U.S. allocation of Merlin engines was in the 2,011 Curtiss P-40's that were powered by the V-1650-1 Merlin. This aircraft consumed all 3,000 of the first allocation of Air Corps engines and at least 600 more as spares. Even so, service conditions and reliability problems due to the consumption of available spare parts and resulted in hundreds of these airplanes being re-engined with V-1710's as P-40R's. Still, in the final analysis of the procurement contracts the U.S. share of Packard Merlin's was somewhat less than 22 percent, compared to the original allocation requirement of one-third of production.

Cost of the two engines is believed to have differed primarily by the need to pay royalty on the Merlin, though the amount has not been validated. Examples are given in Table 15-9. Much of the variation in cost is due to the extent of production history and costs that had previously been recovered, as well as cost of radios and propellers and other items included in GFE.

Benefit of production rates and improvements in manufacturing efficiency were apparent on both engines. The numbers make for interesting consideration.[54]

Practical Differences-Fuel Economy

An important consideration with respect to range of an aircraft is fuel consumption during cruise conditions. The V-1650-7, similar to the V-1650-3 but with slightly lower supercharger gear ratios, was used in the major production Mustang, the P-51D. In cruise at 16,500 feet it delivered 520 bhp at 2000 rpm from 27 inHgA, while burning 48 gallons of fuel per hour. This is a BSFC of 0.60 pounds per horsepower per hour. By comparison, the contemporary (two-stage via turbosupercharging) Allison V-1710-111/113 used in the P-38L, cruising at 15,000 feet with 525 bhp from 1600 rpm and using 31 inHgA, consumed 0.45 pounds of fuel per horsepower per hour. The conditions are not quite the same, but close enough to see that the Merlin required more fuel. This was due to the lower compression ration and the power required to drive the two-stage supercharger, even when the engine was throttled back, and in low gear. The difference in fuel consumption (efficiency) between the two-stage Merlin and the turbosupercharged V-1710 is over 33 percent, which is entirely due to the P-38's turbosuperchargers providing the bulk of the power required to supercharge the intake air at altitude.

The recommended method for cruising a turbosupercharged aircraft like the P-38 is to fully open the throttle and control the turbo speed to

Table 15-11
Merlin and V-1710 Powered P-38 & P-51D Aircraft, Cruise Conditions

Aircraft	Engine Model	Altitude, feet	Engine rpm	MAP, inHgA	Speed, mph	Total Fuel, gph	Mileage, mpg	BSFC, #/bhp-hr	Aircraft Drag, hp
P-51D	V-1650-7	30,000	2250	FT-Hi	348	62	5.61	0.60	606
P-38L	V-1710-111/3	30,000	1900	27	285	77	3.70	0.45	1004
P-38H	V-1710-89/91	25,000	1700	26	272	64	4.25	0.45	835
P-38H	V-1710-89/91	20,000	1600	26	254	59	4.31	0.45	769
P-38L	V-1710-111/3	20,000	1650	27	248	66	3.76	0.45	861
P-51D	V-1650-7	20,000	2100	FT-Lo	316	57	5.54	0.54	619
Merlin P-38	V-1650-19/21	20,000	2100	FT-Lo	254	71	3.59	0.54	769
Merlin P-38	V-1650-19/21	25,000	2250	FT-Hi	272	85	3.18	0.60	835
T-S/C'd P-51	V-1710-G1	30,000	2000	30	350	48	7.29	0.45	620

Lockheed assumed the Merlin P-38 would have the lower drag typical of the P-38G/H.
FT-Lo, Full Throttle in Lo-Gear. FT-Hi, Full Throttle in Hi-Gear.

provide the desired MAP and airflow to the engine. Engine speed is controlled by the propeller, so during slow speed economy cruise the airflow through the engine is actually quite low. The result is that the single-stage engine driven supercharger is requiring very little power. In the case of a P-38 cruising with about 525 bhp from each engine at 16,500 feet, the turbo is providing compression work in the range of 30-40 bhp though this can reach 300 hp during combat conditions.

The P-38 had two engines, and thus required twice the fuel, but remember that in economical cruise, speeds and drag are much reduced. Table 15-9 compares the P-38 and P-51D in their maximum long-range cruise configurations, with only wing racks installed. First of all, it is quite clear that the P-51 was a *very* clean aircraft, able to cruise 50 to 75 mph faster at altitude on the same total fuel burn as the P-38. This strength of the Mustang was a consequence of the airframe, not the engine.[55] This will become quite apparent when we consider the performance of the proposed "Merlin P-38."

The cruising "mileage", (in miles per gallon) of the Merlin Mustang at 20,000 feet was 5.5 mpg, but would have been only 3.6 mpg for the "Merlin P-38." This is about the same as the turbosupercharged P-38L at 3.8 mpg, but not nearly as good as the P-38H at 4.3 mpg. If really long-range cruise had been desired the right combination would have been a turbosupercharged P-51. Such an aircraft would have had a phenomenal 30 percent increase in range to 7.3 mpg at 30,000 feet when flying 350 mph.

The truth in this exercise is that both of these aircraft and engines were excellent when used so as to exploit their strengths. We were indeed fortunate to have adequate numbers of them when they were needed.

Overhaul Life

The actual service life of a particular engine is not always that intended by the powers to be. In general both the Merlin and V-1710 engines were "mature", that is sufficiently developed and built to quality standards that allowed the various models to achieve their intended service life. British experience with their Merlin's is fairly well established, showing that the 240-300 hour expected life of the engine in 1939 had by 1945 improved to 300-500 hours, depending upon application. Engines in various fighter aircraft ranged from 300-360 hours, bomber aircraft ran 360-420 hours and transport aircraft 480-500 hours before overhaul. Experience from 1942 onward was that about 35 percent of the engines achieved these levels, and that the average engine achieved about 60 percent of the intended rated life.[56] By contemporary standards this performance was quite satisfactory. It is known that one V-1650-3 used in a stateside training aircraft exceeded 1,000 hours with only routine servicing before being removed for overhaul.[57]

The first YV-1710-7 in the A-11A demonstrated the intended 300 hour operating life, though the first batch of V-1710-33(C15)'s were scheduled for removal for overhaul at 150 hours. Accelerated service testing of the early P-40's soon established 200 hours before overhaul, and then began steady progress to raise the allowable service life. With V-1710's powering fighters, the scheduled overhaul interval by 1942 was 400 hours. While statistical information similar to that referenced for the British Merlin's has yet to be found, it appears from surveying overhaul plates on existing engines that many V-1710's made it through at least one overhaul, with 250-400 hours having been logged. *Tale Spins*, a wartime house publication of the Allison Service Department, shows numerous P-40s and P-38's in training service that in 1943 were achieving the then allowed maximum of 750 hours on their engines. One P-38H, based at Muskogee Army Air Field in Oklahoma is shown after compiling 750:16 hours on its original pair of V-1710-F17's.[58] During 1944 all V-1710's were authorized 480 hour TBO's, with a maximum of 1,000 hours allowed under certain conditions.[59]

Another aspect of the similarities and differences between the engines relates the labor-hours needed to overhaul the engines, along with the average operating hours being accumulated between overhauls. Comparable data was collected only near the end of the war, and then only for overhauls of engines within the Continental U.S. For perspective radial engine data is included in Table 15-12.[60]

"Sprint" Engines

As mentioned earlier, in August 1939, the Wright Field Power Plant Laboratory investigated the published performance of the Merlin RM2SM. As the single-stage two-speed Merlin II it was rated for 1320 bhp at takeoff and 1145 bhp at 16,750 feet when running at 3000 rpm. From this performance, Wright Field concluded that the Merlin had a "potential" power at sea-level of 2040 bhp if it had the mechanical strength and fuel to enable it to be run at full boost.[61]

This was a fairly accurate forecast of the performance Rolls-Royce secretly demonstrated with the Merlin III "Sprint" engine that they were preparing for an attempt on the World Speed Record in the specially prepared N17 Spitfire. In May 1938 that engine was developing a BMEP of 323.9 psi when running at 3200 rpm and developing 2160 bhp from 84.9 inHgA manifold pressure (27 psig boost). This required use of a special fuel "cocktail."[62] Rolls-Royce often relied on this experience as they worked to further develop the Merlin over the years, for it gave them considerable confidence in the fundamental ability of the Merlin to take manifold pressure and make power. In the fall of 1939, Wright Field determined that the V-1710-33, just then going into production for the Curtiss P-40, was capable of 1760 bhp at 3000 rpm if run on a similar basis, though at the assumed lower rpm, the BMEP was considerably less at 271.6 psi.[63]

As of August 1939 Wright Field had established a requirement for the V-1710 to deliver 1350 bhp at takeoff, and by use of the turbosupercharger, maintain that rating to altitude. The plan was to use the basic V-1710-F5, with its single-speed low gear ratio supercharger rated for 1150 bhp, and then to use turbo boost at the carburetor inlet to achieve the desired 1350 bhp rating. As an alternative, consideration was given to adopt the Merlin

Table 15-12
Flying Hours and Labor to Overhaul

	V-1650	V-1710	Wright R-1820	P&W R-1830	P&W R-2800
1945 1st Qtr					
Labor Hours	251	134	104	147	242
Flying Hours	302	362	591	580	469
1945 2nd Qtr					
Labor Hours	259	153	93	141	241
Flying Hours	200	387	609	562	500
Primary Use	Pursuit	Pursuit	Bomber	Bomber	Pursuit

approach-to develop an enlarged two-speed, single-stage, gear driven supercharger. A significant consideration in the final determination was that this later path would result in an engine with a much higher Indicated Horse Power (IHP), meaning that the entire engine would likely have to be strengthened.[64]

What actually occurred was a combination of both approaches. The Altitude Rated XV-1710-19(C13) and V-1710-33(C15) had been built to use the larger PT-13 carburetor, giving Military ratings of 1090 bhp at up to 10,000 and 13,200 feet respectively on 100 octane. The V-1710-49/53(F5R/L) retained the smaller PD-12 carburetor but were able to deliver the requisite 1325 bhp for takeoff and up to 25,000 feet when boosted with a turbo. Once the press of getting production up to the requirements of the war was behind them, Allison began to work on the V-1710-G series. This model had an enlarged 10-1/4 inch diameter supercharger impeller and feature 3200 rpm ratings. In early 1944 the V-1710-131(G3) was tested, configured much like the Merlin XX, even to the point of using the same impeller diameter along with a two-speed gear drive and with weight increased to 1,475 pounds. For takeoff it matched the single-stage Merlin's with 1600 bhp at 3200 rpm and 1220 bhp at 15,500 feet when running at 3000 rpm. Opening up the induction system was the major improvement.

A further series of tests were run by Wright Field to establish War Emergency Ratings for the V-1710-89/91(F17R/L) sea-level engines. These tests used a complete turbosupercharged setup and deviated from normal practice in that the turbo was operated to deliver pressure greater than sea-level to the carburetor. The two superchargers in series could develop considerably higher manifold pressure. One early series of tests was successful in establishing a rating, using the standard 7-1/2 hour WER test, of 2000 bhp at 3000 rpm from 75 inHgA, equivalent to a BMEP of 308.7 psi. This test was run in March 1944 and used Grade 150 fuel, and engine weight was 1,350 pounds.[65]

In the final production models of these engines the two-stage V-1710-G6 was rated for 2250 bhp at 3200 rpm and 100 inHgA with ADI while the advanced RM14SM Merlin was approved for 2080 bhp at 3000 rpm from 25 psi Boost (80.8 inHgA) on Grade 150 fuel. Not bad for engines originally intended and designed for rating at 1000 bhp.

A Hybrid V-1710/Merlin

In early 1944, the Engineering Division at Wright Field attempted to operate an Allison V-1710-91 engine supercharged by a two-stage, two-speed supercharger and aftercooler taken from a Packard built V-1650-3 Merlin. The lash-up was for engineering investigation and was not intended to be installed in an aircraft. Considerable difficulties were encountered. In fact, three complete supercharger assemblies were consumed in the project, all due to internal failures. The problem was attributed to the fact that the assemblies were mounted remote from the engine and thereby did not benefit from internal heat that was evidently necessary to properly maintain internal clearances for the bearings in the Merlin unit.[66] Still, some interesting data was obtained. The Allison engine with the Merlin supercharger and the PD-18A carburetor arrangement gave a hi-blower critical altitude of 25,700 feet, practically identical to the 25,800 feet capability of the V-1650-3. At this condition and 3000 rpm, the V-1710 delivered 1205 bhp from 60 inHgA.[67] The comparable rating for the V-1650-3 at its high-blower critical altitude of 25,800 feet was 1210 bhp at 3000 rpm with 61 inHgA.

Valve Gear

One area where the V-1710 did improve upon the Merlin was in the design of the valve actuating gear. When designed in 1930, the V-1710 incorporated earlier Allison experience with roller actuated rocker arms, while the Merlin continued Rolls-Royces' earlier practice of having the cams in direct contact with hardened steel "slipper" pads on the rocker arms. These were a continuing source of problems caused by the "scuffing" or rubbing action every time the cam actuated the rocker. These remain a problem component for Mustang pilots 50 years after the last Merlin was built. In fact, most of the serviceable spare rockers have long since been consumed. Consequently, in 1994 an entrepreneur made available a set of roller actuated rockers for the still flying Merlin's, at a cost of $30,200 per engine set![68] The original Packard built Merlin engines cost the government about $20,000 each toward the end of the run, including the royalty payment to Rolls-Royce.

A Merlin Powered P-38

This is great topic for discussion during the cocktail hour. It was so during the war and so it remains, more than 50 years after the event. Everybody seems to have an opinion on how the Merlin would have "improved" the Lockheed P-38. It did not happen, and we will discuss the factual aspects of studies done at the time.

First, this was not a one-time idea, for it was considered several times during the war years, and even got to the point were a P-38 was delivered to the Rolls-Royce Hucknall Flight Test Center and marked for conversion to single-stage Merlin XX engines. The turbos and intercoolers would have been removed. It appears that this installation was never effected, but that is getting ahead of the story.[69]

To go back, the earliest record of considering a Merlin powered P-38 was a Lockheed study done in October 1940, just following the signing of the September 1940 contract by the British Purchasing Commission with Packard to license build 9,000 Rolls-Royce Merlin XX's.[70]

Important to our investigation are the circumstances of the time as they relate to the agreement with the U.S. to receive one-third of the Packard built engines. The Army was therefore looking for suitable airframes in which to use it's allocation of engines. Interestingly, at the time this was the largest order for pursuit aircraft engines then on the books for the Army, and it overshadowed the orders that had been placed with Allison for the V-1710 by more than a factor of three.

Lockheed was at the time heavily committed to gearing up to deliver on their large backlog of orders for the P-38. This included 80 for the U.S. and 667 similar P-322's for Britain. In addition, Lockheed was then increasing production of their Lockheed Hudson and Ventura bombers, mostly again for Britain. Concurrent with the date of the Merlin study, Lockheed was just beginning to deliver the 13 V-1710-27/29 powered YP-38's, so it was a busy time for the growing engineering and production departments.[71] Consequently, it is likely that Lockheed was not particularly eager for another redesign of the P-38, but it was considered, probably because of the importance and interests of their two major customers.[72]

Merlin XX Powered P-38

As shown in Table 15-13, the Merlin XX offered similar, and in some instances improved, performance in comparison with the then current YP-38 powered by the V-1710-F2.[73] The subsequent Allison powered P-38D/E models could not match the performance of the YP-38 because of numerous changes and Military requirements, such as incorporation of armor plating and self-sealing fuel tanks. By this measure it can be inferred that a combat ready version of the Merlin XX powered P-38 would have maintained a small performance edge over comparable early V-1710 models. If so, then why did the Army not press on and put the aircraft into production?

Initial deliveries of single-stage Packard built Merlin's did not begin until early 1942, just after U.S. entry into WWII and concurrent with delivery of the first of the Allison powered P-322-B's. At about this same

Table 15-13 Performance of a Merlin XX P-38		
	Merlin XX	V-1710-F2
Data source	Rpt 2036	YP-38
Takeoff bhp	1280/3000	1150/3000
Critical Altitude Military bhp	1170	1150
Critical Altitude, feet	21,000	25,000
Critical Altitude Max Speed, mph	431	405
P-38 Weight, pounds	14,500	14,348
Sea Level Rate of Climb, Ft/min	3,160	
Sea Level Max Speed, mph	354	
High Speed Cruise bhp/rpm	875/2650	1000/2600
High Speed Cruise mph/Alt, ft	393/20,000	/25,000
High Speed Cruise Fuel Use, #/bhp-hr	0.485	
Range at High Speed Cruise, miles/gal	2.78	310
Service Ceiling, feet	38,100	38,000
Engine, with Turbo & Ducts, pounds	1,430	1,590
Normal Range, miles	640	650
Normal Fuel Capacity, gallons	230	210

time, the two-stage Merlin was entering limited production in England and it was decided to shift as much of Packard's effort to this model as possible. This in turn further limited the near term availability of the Merlin XX, and made any interest in the two-stage engine by Lockheed uncertain due to Packard's production capacity being assigned to the Spitfire and Mustang.[74] Consequently, as the longest ranging escort fighter of the period the available Allison powered P-38 models were selected for continued and uninterrupted development and production.

Merlin 61 Powered P-38

With the introduction of the two-stage Merlin 61 in Britain, and the desire by many in the U.S. to have Packard switch to its production, it was only natural that Lockheed consider a two-stage Merlin powered P-38. A comprehensive study of such an aircraft was prepared by Lockheed in June 1942.[75] The comparison was done to a turbosupercharged Allison V-1710-F17 powered airframe incorporating core type intercoolers, as introduced on the P-38J model.

These engines developed approximately the same maximum power, but attained their altitude ratings with different types of supercharging. The assumed airframe was a modification of the P-38F, including complete armament and identical useful loads. The changes to the airframe were to accommodate the different cooling systems associated with removing the turbos and wing leading edge intercoolers when the Merlin

Table 15-14 Engines For a Merlin 61 P-38		
	Allison F-17	R-R Merlin 61
Supercharger, Integral	7.48:1	6.39:1 & 8.03:1
Exhaust System	Turbo	Jet Stacks
Engine Power Ratings, bhp		
Military Power, rpm	3000	3000
Sea Level	1425	1375
13,000, ft.	1425	1520
19,500, ft.	1425	1255
24,000, ft.	1425	1300
27,000, ft.	1425	1195
Takeoff Power, rpm	3000	3000
Horsepower	1425	1300
Normal Rated Power, rpm	2600	2850
Sea Level	1100	1080
16,700, ft.	1100	1240
22,400, ft.	1100	1035
27,700, ft.	900	1080
For purposes of the performance analysis, thrust from optimally designed ejector exhausts was added to these values.		

Table 15-15 Performance of a Merlin 61 Powered P-38				
	Allison V-1710-F17		R-R Merlin 61	
	Military	Normal	Military	Normal
Altitude for Maximum Speed , ft	27,000	25,000	27,300	30,200
Maximum, Speed, mph	418	395	423	403
Sea Level	360	326	343	326
5,000, ft	365	342	365	345
10,000, ft	382	358	386	364
15,000, ft	396	374	406	381
20,000, ft	408	386	406	395
25,000, ft	416	395	414	388
30,000, ft	414	383	419	402
Absolute Ceiling, feet	42,300	38,500	42,300	41,200
Service Ceiling, feet	41,600	37,800	41,600	40,400
Climb in 5 minutes, feet	17,800	-	17,800	-
Climb to 20,000 feet, minutes	6.2	-	5.9	-
to 25,000 feet, minutes	8.7	-	8.4	-
to 30,000 feet, minutes	12.2	-	11.8	-
Distance to Takeoff of 50ft Obstacle, ft.	1,640	-	1,770	-

Table 15-16
Advanced V-1710 and Merlin Powered P-38's

	Standard P-38J	Advanced Allison P-38	Advanced Allison P-38	Merlin P-38
War Emergency Power, bhp	1600	1725	2000Wet	2000Dry
WER Engine, rpm	3000	3200	3400	NA
Engine/Propeller Gear Ratio	2.00:1	2.36	2.36	2.36
Propeller Diameter, ft	11.5	11.5	12.5	12.5
Turbosupercharger, GE Type	B-33	B-39	B-39	Jet Exh
Fuel, Grade	130	140	140	Special
Propeller Activity Factor	89.3	89.3	110	110
Propeller Weight Increase, lbs	0	0	51	51
Per Engine Weight Increase,lbs	0	0	45	NA
P-38 Operational Weight, lbs	16,200	16,200	17,250	16,500
Increase in Max Speed, mph	0	12	16	38
Increase in Max Climb, fpm	0	490	850	1,300
Impact on Maneuverability, %	0	0?	-5	-1
Maximum Sea Level Speed, mph	356	364	382	398
Maximum Speed @ 30,000', mph	436	448	452	468
Max Climb in 5 min., feet	18,700	21,600	23,000	25,600
Time to Climb to 30,000',min.	8.7	7.2	6.2	6.2
Absolute Ceiling, feet	43,900	43,900	43,700	43,900

was installed. For purposes of proper balance it would have been necessary to locate both the coolant and aftercooler radiators aft on the nacelle booms.

Performance calculations took into consideration the changes in weight and added drag effects of the revised radiators, as well as the gains from ejector exhausts and the loss in propulsion efficiency with increased altitude and compressibility.

Not included in the performance table are the differences in range due to the considerable difference in specific fuel consumption between the two installations. The Allison was considerably more efficient than the Rolls-Royce in this regard because of the difference of power required to drive the geared superchargers. Fuel consumption data in the engine specifications showed the Merlin to require approximately 5 percent more fuel in low-blower and 10 percent more in high-blower.

The Lockheed report concluded that the installations were very similar in empty weight, useful load and maximum performance. The one outstanding advantage of the Allison was in range at high altitude. Given that it was not known if or when the Merlin 61 might be available, and that there was little performance advantage in the Merlin P-38, coupled with the ready availability of the V-1710 and GE turbo's it should not be surprising that Lockheed chose to stay with Allison engine and deliver airplanes having superior range.

Advanced Merlin Powered P-38

In 1944, with a supply of two-stage Packard Merlin's available to the U.S. Army, Kelly Johnson of Lockheed once again had their Phil Colman investigate the benefits of adapting the Merlin to the P-38. The Allison engine considered in the Advanced P-38 study appears to have been the single-stage V-1710-G1, which was rated for 1725 bhp and normal operation at up to 3400 rpm using AN-F-29 Grade 140 fuel. The 2.36:1 reduction gears would have required a revised cowl line for the P-38 nacelle, a feature that would have been the same as that in the one-off P-38K. It used both these same gears and the proposed high activity propellers. To achieve the 2000 bhp rating with the Allison it would have been necessary to provide water/ADI internal cooling, which was the primary reason for the increase in operating weight of the aircraft. The two new 30 gallon ADI tanks weighed 50 pounds, and a 15 minute supply of ADI added 500 pounds to the takeoff weight.[76]

With the installation of the two-stage Merlins the turbosuperchargers would have been removed, along with their supporting systems and intercoolers. Of course the aftercooled two-stage Merlins weighed more than the V-1710's they would have been replacing, but to a degree they would have been able to compensate for this with the total of 250 thrust-horsepower which the short ejector exhaust stacks would have generated when the aircraft was flying with full power, resulting in 450 mph at 24,000 feet. Lockheed assumed that the two-stage Merlin P-38 would have had the same low-drag characteristics as the streamlined P-38G, which was free of installation drag as its intercoolers were located in the wing.

With the tabulated improvements in performance expected of the two-stage Merlin powered P-38, the question is why did they not proceed? Several answers have been suggested, the most likely being a combination of them all:

- As shown with the decision not to place the much improved P-38K in production, demand for P-38's in 1943/44 was such that any further interruption in production to introduce a new model was out of the question.
- The high demand for the new two-stage Merlin in 1943/44 was such that engine production capacity was simply not available. The second U.S. source for two-stage Merlin's, the Continental Engine Co., was literally years behind in producing the engine.
- While not investigated in the Lockheed performance study, a major factor in the success of the Allison powered P-38 was its ability to fly extremely long missions because of the high efficiency of its turbosupercharged engines. This advantage would have been lost in a two-stage Merlin powered P-38. The critical mission of the day, in both Europe and the Pacific, was long-range high altitude escort.
- It has been charged that politics played a role, that the highest levels of Government directed that all efforts to supplant the Allison engine in the P-38 were to stop. If true, the right personalities were

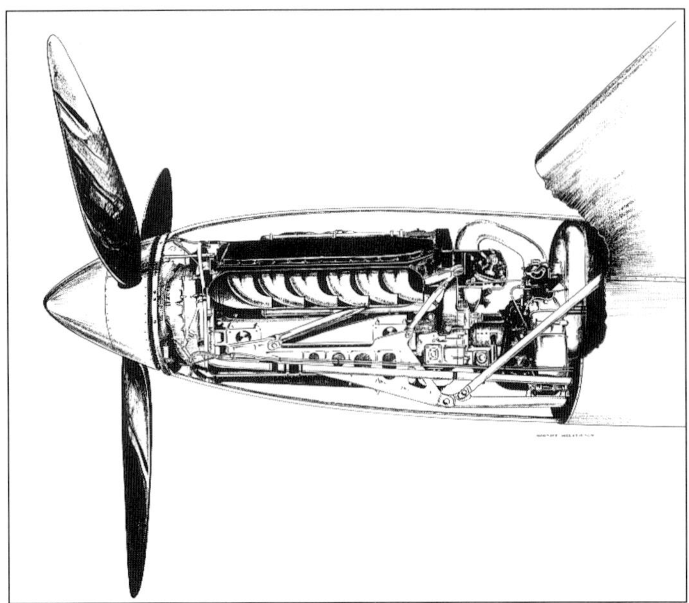

This sketch shows a two-stage V-1710-F28R/L in a very sleek P-38 nacelle. The turbo would have been removed as well as the intercooler. ADI injection would have been needed to limit detonation at very high power. (Allison)

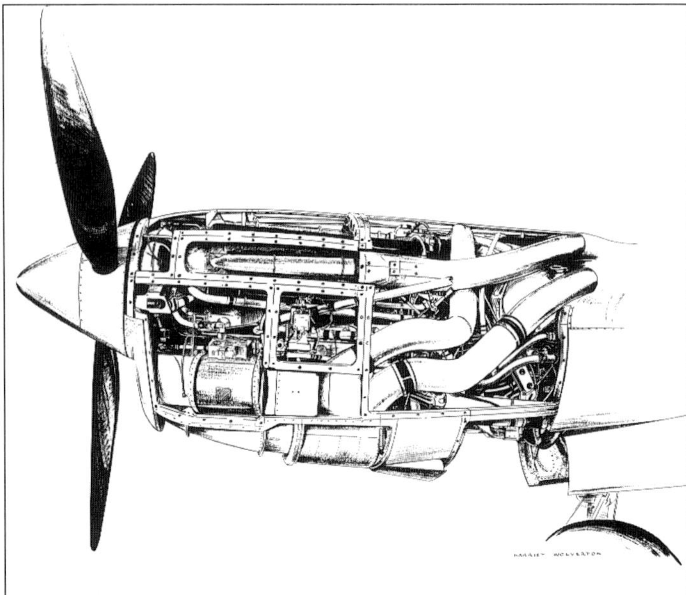

This is the standard nacelle arrangement used with V-1710-F17R/L in the turbosupercharged P-38J. It would have been greatly simplified in the two-stage adaptation. (Allison)

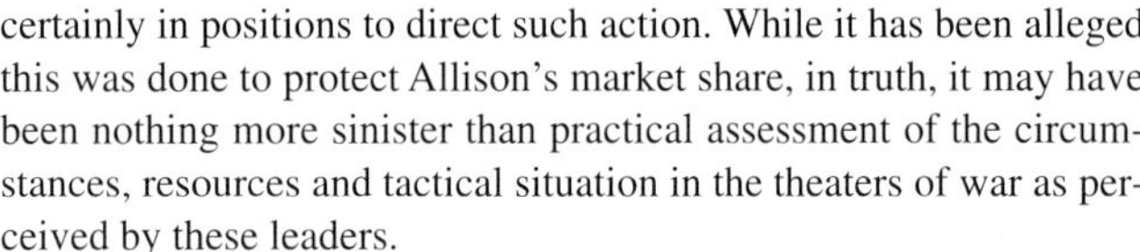

certainly in positions to direct such action. While it has been alleged this was done to protect Allison's market share, in truth, it may have been nothing more sinister than practical assessment of the circumstances, resources and tactical situation in the theaters of war as perceived by these leaders.

- By the summer of 1944 War Production planners were already cutting back and canceling production of aircraft and engines that would have otherwise become surplus with the coming end of the war. Allison production figures certainly show this effect, along with the knowledge that Packard production of the Merlin would end concurrent with cessation of hostilities.

In December 1944 General Arnold concurred with AAF Technical Service Command at Wright Field that the studies and efforts to install V-1650-3 engines in a P-38 be abandoned.[77] So ended the Merlin P-38 projects.

Alternative Allison Two-Stage Powered P-38

All of the talk and interest in a two-stage Merlin powered P-38 caused Allison to consider a two-stage mechanically supercharged version of the P-38 as well. Allison artist Harriet Wolverton made the above sketches in the summer of 1943. The new turbosupercharged V-1710-F17 installation for the P-38J is offered for comparison with the sleek elongated nacelle that was proposed to house the two-stage V-1710. The engine appears to be the V-1710-F28R/L with the "between the stages" carburetor as used in the V-1710-101(F27R). The engine did not incorporate either an intercooler or an aftercooler. Water injection would likely have been included.

Packard Training School

In February 1942 Packard was asked by the AAF Technical Training Command to arrange to give familiarization training to AAF officers and enlisted men on the Merlin engine. Training began on March 15, 1942 for the first class of 18 non-commissioned officers. The capacity of the school was increased to where it could handle 300-500 personnel each month in classes of 90 students. This was later raised to 720 students per month by operating on a three shift schedule. On average, the trainees received about 140 hours of specialized training on the operation and maintenance of the Merlin. With the end of the war approaching, the school was reduced to half scale in May 1944, and continued to be scaled back until it was terminated on November 30, 1944.[78]

The 3,352 trainees attending the school were as follows:

AAF Service Personnel	2,594
AAF Civilian Mechanics	308
AAF Air Service Command	75
Packard Field Service	375

The scope of the Packard School was considerable smaller than the 101,000 trained by the Allison Service School.

Summary

The Allison V-1710 and the Rolls-Royce V-1650 liquid-cooled V-12 aircraft engines were really very many different engines. The active production period for each extended for nearly 15 years. A large number of variations of the engines were built to meet the needs of military tacticians related to the key rolls both engines played during the war. Both engines were needed, and while one particular model or another was preferred for a given mission, not every airplane, and not every engine, should be expected to perform every mission with equal superiority or effectiveness.

The V-1710 was designed to be a two-stage engine, albeit, with the first stage being a turbosupercharger. When pressed into service as a single-stage power plant it should not be a surprise that it would be "altitude" limited. Power ratings were quite similar once the U.S. military allowed WER ratings, the V-1710 was able to equal or exceed comparable Merlin performance.

Although the V-1710 had a four year headstart in design and testing, the Government funded Merlin benefited from the experienced Rolls-Royce development team. They had the engine ready and with it the Battle of Britain was won. At that point in time, the V-1710 had just gone into series production. Furthermore, Rolls-Royce was able to grow the capabilities of the engine with their masterful arrangement of superchargers and aftercooling to provide a compact two-stage engine that was able to establish air superiority over both Europe and Japan. Allison was able to match this performance, but was never able to get its most advanced engines into production combat aircraft during the war.

Which was the better engine? For what mission? In which airplane? At what altitude? Were you sitting behind it? When?

Both engines have been acknowledged as the best of the arts of manufacturing and engineering in their respective countries of origin at the time. Furthermore, both are a credit to the inspiration and commitment of the men and women who provided them when they were really needed.

NOTES

[1] In anticipation of the need to mobilize US Industrial Production capacity Knudsen was recruited to head this aspect of the job for the nation. He was in this position early in 1940 and one of his early actions was to aid the British Purchasing Commission in obtaining a U.S. manufacture for the R-R Merlin engine. He personally appealed to Henry Ford Sr. in an attempt to have Ford build the Merlin in the US as they were doing in England. When Ford reneged, he was then instrumental in getting Packard to produce the Merlin. To his credit, his General Motors background did not interfere with what the country needed.

[2] Lacey, Robert, 1986, 378-389.

[3] Interestingly, the at this time the Ford Motor Companies of Britain and Germany were producing cars, trucks and armaments flat out for their respective nations. This included production of the Rolls-Royce Merlin!

[4] Terminating the Merlin project was a problem for Ford, and in an attempt to salvage some public face, the Ford Motor Co. began studies on July 9, 1940 for a 1650 cubic inch aluminum V-12 to be rated at 2000 bhp with a built-in turbosupercharger. After spending some $2 million of their own money Ford was successful in selling the engine to the Army for use in the new Sherman tank. It was built as a naturally asperated 450 bhp V-8, though a few 700 bhp V-12's were later built. It was quite successful and 25,741 were built.

[5] Bodie, Warren M. 1994, 113.

[6] This is only speculation by your author, but we do know that at least Curtiss and Lockheed performed such studies soon after Packard was contracted to manufacture Merlin's. Curtiss did go on to build Merlin powered models of the P-40.

[7] Extra Priority Teletype from General Brett to General Arnold, June 19, 1940. NARA RG18, File 452.8, Box 808.

[8] *Allison Engine Situation*, 1941.

[9] Gruenhagen, Robert W. 1980, 31.

[10] Packard Motor Car Company, 1945.

[11] Harvey-Bailey, Alec, 1984, 12.

[12] Allison Engineering Company O. T. Kreusser, General Manager, letter to Brig. General G.H. Brett, Chief of Materiel Division, December 12, 1939. NARA RG 18, File 452.8, Box 808.

[13] Bowers, Peter M. 1979, 485.

[14] *Case History of the V-1650 Engine*, 1945. Letter to Production Division, Wright Field from Air Force Chief of Staff, February 27, 1943. It says "P-63", though your author believes it meant "P-39".

[15] *Case History of the V-1650 Engine*, 1945. Transcript of Conversation between Colonel Sessums and Colonel Johnson, March 11, 1943.

[16] This first contract for the 9,000 single-stage Merlin's was completed about January 1, 1943, 2-1/2 years after being signed.

[17] *Rolls-Royce Merlin Engines Built by Packard Motor Car Company*, letter from Director General, British Air Commission to Brigadier General B.W. Chidlaw, Assistant Chief of Staff, Army Air Forces, December 3, 1942. NARA RG18, File 452.13, Box 1260.

[18] Gruenhagen, Robert W. 1980, 195.

[19] Memorandum for Mr. Swope: *News Item Reference Spitfire Engines*, by H.H. Arnold, December 18, 1942. H.H. Arnold manuscript, Roll 165, Library of Congress.

[20] *Case History of the V-1650 Engine*, 1945. *Merlin Engine Improvement*, letter from Office of Production Management, Washington, D.C. to Brig. General O. P. Echols, Chief, Materiel Division, May 28, 1941.

[21] *Case History of the V-1650 Engine*, 1945.

[22] Packard Motor Car Company, 1945.

[23] Schlaifer, R. and S.D. Heron, 1950, 306-307.

[24] Harvey-Bailey, Alec, 1982, 31.

[25] It is interesting to note how similar the problems in developing the Merlin were to those experienced by Allison in developing the early V-1710. The difference in the time required is clearly in the amount of government financial support, and commitment to keeping Rolls-Royce in a position of leadership.

[26] Harvey-Bailey, Alec, 1984, 9.

[27] Harvey-Bailey, Alec and Michael Evans, 1984, 102-104.

[28] Nockolds, Harold, 158.

[29] Harvey-Bailey, Alec, 1984, 36, 60-61.

[30] Engineering Section Memo Report, Serial No. E-57-1643-1, 1937.

[31] *Proposed Turbo Supercharger Tests and Higher Power Outputs of Allison V-1710 Engines*, Letter from Allison to Chief, Materiel Division, April 25, 1938.

[32] *Proposed Turbo Supercharger Tests and Higher Power Outputs of Allison V-1710 Engines*, Letter from Chief, Materiel Division to Allison, June 1, 1938.

[33] Engineering Section Memo Report Serial No. E-57-4272-1, 1939.

[34] Engineering Division Memo Report, ENG-57-503-1034, 30 October 1943.

[35] *Case History of the V-1650 Engine*, 1945, frame 1868. *Relative Horsepower Output of Present Allison Engines and the Rolls-Royce Merlin Engine*, H.H. Arnold Major General, Chief of the Air Corps to Mr. Wm. Knudsen, President, General Motors Corp, September 11, 1939 letter.

[36] Schlaifer, R. and S.D. Heron, 1950.

[37] Technical Report No. 4761, 1942.

[38] *Preliminary Testing of Engine No. A-9 Converted to Merlin 61 Type*, Packard report No. 29, December 23, 1941. NARA RG342, RD 3466, Box 6842.

[39] Packard Motor Car Company, 1945, 12.

[40] Low supercharger efficiency was a direct result of the fact that the GE units were adaptations of the superchargers they had developed for use with their turbosuperchargers. Since there was a large excess of power available from the exhaust gas driving the turbo, it was not necessary to have a particularly efficient compressor. Furthermore, since the turbosupercharger required use of an intercooler between the turbo-compressor and the carburetor/engine-stage supercharger, the consequence of an inefficient compressor (which results in high temperature induction air) was not really a problem. Finally, radial engines with superchargers were intended for installations requiring very little manifold pressure (boost), for the fuels of the 1930's simply could not tolerate high manifold pressures in hot air-cooled cylinders. In fact, the inefficiency of these early superchargers was to a degree useful, for the heat they produced improved the evaporation of the fuel and therefore the uniformity of distribution to the cylinders in the radial engine. In this circumstance Allison was wise to develop their own supercharger.

[41] Hooker, Sir Stanley, 1984, 31.

[42] Bodie, Warren M., 1991, xvi.

[43] Merlin indicated specific mixture consumption was 10.5 IHP/LBM-AIR/MIN. For the XV-1710-6 at 3000 rpm it was 10.22 IHP/LBM-AIR/MIN, for the V-1710-G1 @ 3400 RPM, 8.1:1 supercharger, 6.65:1 Compression Ratio, 9.5" diameter supercharger of 65% efficiency, 1725 BHP, 68.5 inHgA, value @ 2140 IHP is 10.03 IHP/LBM-AIR/MIN. Differences are due to differences in supercharger efficiency. V-1710-F with 8.8:1 supercharger of 70% efficiency had 10.62 IHP/LBM-AIR/MIN, 1500 bhp, 61.5 inHgA at 3000 rpm.

[44] Engineering Section Memo Report, Serial No. E-57-1860-1, 1938.

[45] *Allison V-3420 Engine-High Altitude with Geared Supercharger,* Air Corps letter to Allison, June 15, 1938. Copy in V-3420 files at NARA.

[46] Schlaifer, R. and S.D. Heron, 1950, 306-307.

[47] Redsell, Arthur, 1996, 18-22.

[48] Hooker, Sir Stanley, 1984, 52-56.

[49] *Estimated P-51B Mustang Performance with Improved Fuel (160 Grade)*, memo to file regarding conference at R-R Derby on October 23, 1943. NARA RG 18, File 452.1, Box 698.

[50] Letter from Assistant Military Air Attaché in London to Major General O. P. Echols, Materiel Command, June 28, 1943. NARA RG 18, File 452.13-M, Box 758.

[51] The two-stage Allison was first run in the spring of 1941, but this was only the beginning of the development of the drive and controls. Rolls-Royce had a two-stage Merlin 60 bomber engine in production in November 1941, though limited production of the two-stage Merlin 61 for fighters did not begin until March 1942, and it was not available for production Mustangs until 1943. Schlaifer, R. and S.D. Heron, 1950, 309.

[52] *Case History of the V-1650 Engine*, 1945. Note to Col. Barber.

[53] Packard Motor Car Company, 1945, 1-3.

[54] Table 68—*Average Unit Cost of AAF Delivered Engines*, and Table 69—*Factory Deliveries of All Military Aircraft Engines, By Plant*, from AAF Statistical Digest, 1946.

[55] The Germans much admired the low-drag Mustang. One of their technical writers said, "A comparison of flight measurements shows quite unmistakably that the Mustang is far superior aerodynamically to all other airplanes and that it maintains this superiority in spite of its considerably greater wing area." German data, taken at a lift coefficient of 0.2 revealed the following wing profile drag coefficients:

He 177	0.0109
Bf 109B	0.0101
Ju 288	0.0102
FW 190	0.0089
Mustang	0.0072

From H.H. Arnold Manuscript, Roll 194, Von Karman Report, Library of Congress.

[56] Harvey-Bailey, Alec, 1984, 92.

[57] V-1650-3 No. V-301579, installed in P-51B AF43-12259, 11-25-1943, removed 8-2-1944 for having exceeded maximum allowable time before overhaul. Total time, 1009:15 hours. Packard Motor Car Company, 1945.

[58] *Tale Spins*, Volume II, Number 6, published in 1943.

[59] *Price Adjustment Board Report-1944*, 15.

[60] Table 103—*Average Man-Hours per Major Airplane Engine Overhaul in ConUS*, and Table 104—*Average Operating Hours of Airplane Engines at Overhaul in ConUS*, from AAF Statistical Digest, 1946.

[61] Engineering Section Memo Report Serial No. E-57-4272-1, 1939.

[62] Schlaifer, R. and S.D. Heron, 1950, 222.

[63] Engineering Division Memo Report, ENG-57-503-1034, 30 October 1943.

[64] Engineering Section Memo Report, Serial No. E-57-4272-1, 1939.

[65] Engineering Division Memo Report, Serial No. ENG-57-531-267, 1944.

[66] Rolls-Royce reported that tail bearing failures on the two-stage supercharger in the Merlin 66 with the Bendix carburetor, the same as in the V-1650-7, when operated in low temperatures. The British SU carburetor provided an oil hotspot for the bearing and did not appear to have the problem. Harvey-Bailey, Alec, 1984, 45.

[67] Engineering Division Memo Report, Serial No. ENG-57-503-1166, 1944.

[68] The improved valve train is offered by Roush Technologies. Cox, Jack 1994, 29.

[69] Correspondence from Dave Birch, Archivist of the Rolls-Royce Heritage Trust to Graham White, January 28, 1993.

[70] Packard Motor Car Company, 1945, 1-3.

[71] An interesting observation is that some of the performance curves in the Lockheed report actually are labeled Model YP-38, with the "Y" having been erased.

[72] For perspective on the task of introducing a new model, work on the P-322-B *Lightning* had begun in April 1940, but the first P-322-B (Allison powered) was not ready for delivery until January 1942.

[73] Posner, E.C. 1940.

[74] A similar dilemma was faced at Curtiss which resulted in P-40 designer Don Berlin leaving the firm when they were denied access to the two-stage Merlin.

[75] Colman, Philip A. 1942.

[76] Colman, Philip A. 1944.

[77] *Installation of Packard-Merlin V-1650-3 Engines in a P-38 Airplane*, letter commanded by General Arnold to Director, AAF Air Technical Service Command, Wright Field, December 18, 1944. NARA RG18, File 452.13-N, Box 756.

[78] Packard Motor Car Company, 1945, 16.

16

Engineering the V-1710

Allison started as an experimental and engineering "job" shop. Their work being specialty projects and high performance equipment in support of "Indy 500" racers. This explains their location in Speedway, Indiana, near Indianapolis. This background not only developed expert engineering skills, but since they built and manufactured what they designed, they evolved a highly skilled group of mechanics, machinists and craftsmen able to fabricate and construct advanced designs.

A lot of their specialty work featured power transmission, shafting and gear boxes. When one studies photos of the special airship propeller drives built for the U.S. Navy, you come away impressed with the quality of the work and the practicality of the design. In the subsequent development of the V-1710-B, with its ability to rapidly reverse direction of rotation, one can appreciate the magnitude of the endeavor and the elegance of their engineering solution to the mechanical design challenges.

Engineering designs of this caliber are not seat-of-the-pants affairs. Careful and measured analysis is necessary before committing the design to metal, and then a systematic development and test program must be followed to insure that every component and requirement is fully demonstrated and satisfactory before the race or before flight. The Allison Engineering Company exemplified this approach to their products, the V-1710 being the case in point. Each of their engines, even the earliest of them, was preceded by a comprehensive engineering design study. Every casting, bolt, bearing, gear and shaft was carefully analyzed to meet stringent, self imposed, design standards and criteria. Individual components were then carefully tested in the laboratory and on the test bench to validate the design and the functioning of the parts. All of this was done using hand calculations and graphical analytical techniques, many of which are lost on modern engineers. That these methods worked and were appropriate

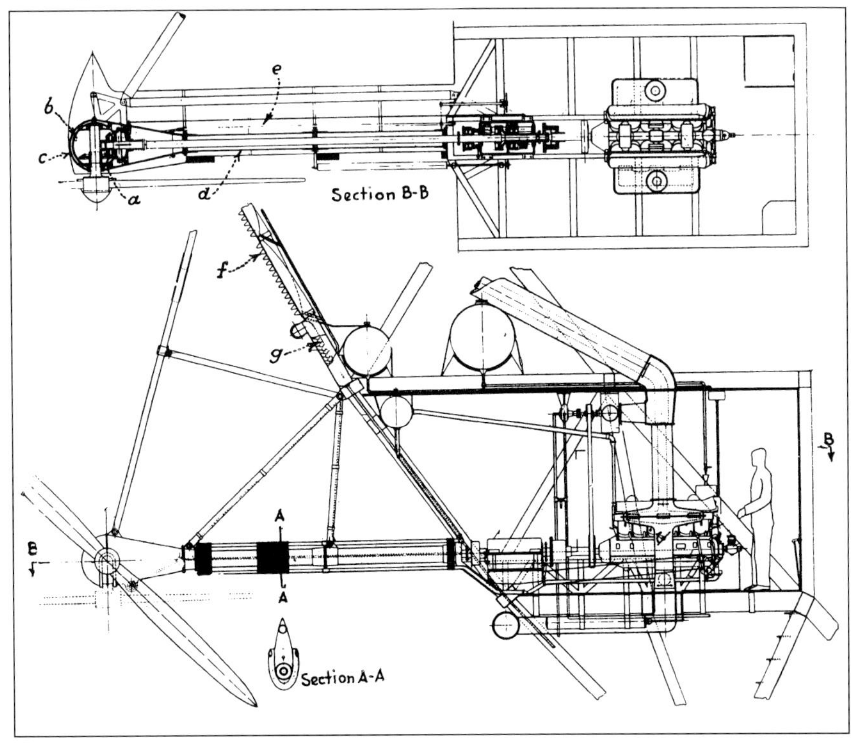

This drawing shows the arrangement of the Allison extension drives and swiveling gear heads built for the Navy's largest airships. The reversible V-1710-4 was intended to replace the German Maybach engines shown. (Air Service)

was demonstrated in the reliability and accomplishments of the V-1710 during WWII.

This is not to say that the V-1710 engine did not suffer problems in service. The major problems are discussed elsewhere in this text, though upon review, many of the early manufacturing problems were not caused by the engineering of the engine or its components. Rather they can be attributed to the need to increase production to unheard-of levels in a very short time, using, for the most part, an unsophisticated labor force recruited off of Mid-western farms. Such problems did not persist for long. Allison instituted new quality control systems and training of its personnel to the point that supply and quality were not factors in safe flight or in the fielding of adequate numbers of competitive combat aircraft. At its peak, Allison employed approximately 23,000 people, a number that does not count employment by suppliers and sub-contractors which were major contributors and probably doubled the number of people working directly on the program.

In addition to training its own manufacturing staff in the technical skills required to produce such a complex engine, Allison also trained over 101,000 people to operate and maintain its engines.[1] This is nearly 1.5 people trained by the Allison Service School for every engine produced. All-in-all, a huge undertaking and accomplishment. Still, while all of this led to the ultimate success of the V-1710 program, when we look at the V-1710's early days none of this was in place, and the demands and expectations upon Allison and the V-1710 were enormous.

Engine Development, Testing and Improvement

The best experience and engineering does not guarantee success in an endeavor such as developing a new engine, particularly if it is an aircraft engine where low weight and high performance are viewed as mutually exclusive. Such an engine is not "invented." Rather it starts from a concept, often times a "Patentable" one, but its success comes about only when the concept is refined by rigorous and extensive development and testing. Since this is likely to extend over a protracted period, it is also necessary to have the financial and political resources necessary to obtain and support the endeavor.

EX-Workhorse Engines

Allison performed extensive analytical and design work prior to constructing engines and starting testing. Testing for contract purposes was usually carried out by the AAC/AAF at their Wright Field facilities. Still Allison performed extensive component, single cylinder, and full scale testing on its engines. A long series of "workhorse" engines were utilized to perform first-of-a-kind tests and to prove new concepts and components prior to receiving Wright Field funding or authorization to integrate into production. These engines were designated with an "EX-" prefix, with at least 55 of them used over the course of development and support of the V-1710. Each of these engines was built, tested and rebuilt numerous times as needed by the assigned test programs. EX-11 did at least 59 cycles, and EX-14 more than 57. It was not unusual for an engine to built up as a different model, depending upon the need at the time, so the table just shows how things were at some point in the life the engine. It is likely that few of these engines retained much more than a dataplate as an original component throughout their lives.

There appears to have been little consistency in the ownership of these workhorse engines. Many, probably most, were paid for by the government, but not as "delivered" items. The various test reports and performance dates were the contractual deliverables, not the actual engine. These EX- engines are not included in production totals.

Engines and/or major components were often painted with a stencil identifying the "EX-" engine to which they belonged. The data plates also remain a mystery as to the exact nomenclature used, but it is believed that the engine model was often shown, at least on some of the later engines, as "V-1710-FR." Significantly the Allison model number was not identified. Instead a detailed build-up package was prepared by Engineering Test describing the component part numbers that would be utilized for the purpose intended for a particular engine buildup or test.

Not long after Ron Hazen took over as Chief Engineer in 1936 the records begin to mention test work Allison was doing using their own "house" engine, the EX-1. It was used to test any number of design changes, often using substandard parts in an effort to define the proper specifications for components. In addition to fundamental engine components like

Table 16-1
Partial List of Allison Workhorse Engines

Test Engine	Configuration	Comments
EX-1	V-1710-C	Component testing
EX-2	V-1710-C2	Fuel Injected testing as XV-1710-5 during 1936
EX-3	V-1710-E2	Used to develop extension shaft system
EX-8	V-1710-C15	C-15 development engine
EX-13	V-1710-F3R	Ran qualifications prior to 1st Mustang Flight, Oct 1940
EX-22A	V-3420-A15L	Tested Jan 42 through Nov 43, went through at least 24 teardowns
EX-23	V-1710-F3R	Tested during 1941
EX-29	V-3420-A11R	Tested during December 1941
EX-39	V-1710-F7R	Being tested through Jan 1943
EX-41	V-3420-B4	Being tested in April 1943
EX-46	V-1710-F32R	Two-stage with Aftercooler, for XP-51J
EX-50	V-1710-F32R	Being tested in July 1944, for XP-51J
EX-54	V-1710-G2R	Being tested in October 1946
IX-1	I-16	Allison's first Jet engine, from GE I-16 (Whittle) drawings, Jan 1945
IX-2	I-16	
IX-3	J33	Allison's first model of the GE designed J33, July 1945

Table 16-2
V-1710 Technology Milestones

Major Technology/Improvements	Introduced On:	Date
Basic Engine: 60° V-12, 5.5"Bore, 6.0"Stroke, 1710 cu.in., Sea-level Rated, 8.25" S/C, Geared, 5.8:1 CR.	GV-1710-A(A1)	1929/30
Reversible Airship Engine	XV-1710-4(B1R)	1932
9-1/2 " Dia Supercharger & 6 Counterweight Crankshaft	XV-1710-1(A2)	1932
Direct Fuel Injection	XV-1710-5	1934/5
Hazen "Raming" type Intake Manifold	XV-1710-7(C3)	1936
6.65:1 Compression Ratio	XV-1710-11(C7)	1936
Pendulum Type Vibration Damper	V-1710-7/11(C4/C7)	1936
Design of Auxiliary Stage S/C	V-1710-F7/E9	1938
Bendix Pressure Carburetor	V-1710-21/23(C10/D2)	1938
Design of Double Vee XV-3420-1	XV-3420-1(A1)	1938
Universal Cylinder Block	V-3420, V-1710-C10/D2	1938
Reversible 6-cwt Crankshaft	V-3420, V-1710-E2/F1	1938
First Altitude Rating	XV-1710-19(C13)	1938
Remote Reduction Gearbox	V-1710-D3/E2	1937
Hydraulic Type Vibration Damper	V-1710-17(E2)	1939
Integral External Spur-Type Gearbox	V-1710-25(F1)	1939
RH/LH Engine Rotation by Assembly	V-1710-27/29(F2R/L)	1939
Auxiliary Stage S/C w/Hydraulic Drive	V-1710-E9, F7	1942
War Emergency Ratings on Grade 100	E4, E6, F3, F4	1942
2-Stage Engine Production	V-1710-93(E11)	1942
Anti-Detonant Injection (ADI)	V-1710-93(E11)	1942
12-cwt Crankshaft, 3200 rpm Ratings	V-1710-E21/F27/G1	1942
Interstage Carburetion	V-1710-E22	1944
Aftercooler on 2-Stage Engine	V-1710-119(F32R)	1944
Bendix Speed Density Carburetor	V-1710-119(F32R)	1944
Keystone Piston Rings	E11, E19, F17, F26	1943
Feedback Turbine/Turbocompound	V-1710-127(E27)	1944
Reduced Compression Ratio, 6.00:1	V-1710-E30,G6	1945/6
10-1/4 inch Diameter Supercharger	V-1710-G1	1942

crankshafts, cylinder banks, and pistons, important information was also gathered on carburetor settings, coolant pump performance, selection of supercharger gears, and the ultimate horsepower capability of the design.

The EX-2 was being operated in 1936 as the fuel injection development engine, configured similar to a XV-1710-5, and at least for a while, with the extension shaft for the D series. In this form it was intended to represent the engine proposed for the Bell FM-1 where fuel injection improved fuel distribution would allow detonation free operation up to the necessary 1250 bhp. At the end of the fuel injection program, it was reconfigured as a carbureted "C" model and continued an active test life.

The EX-3 was the test engine used to develop the extension shaft configuration for the V-1710-E in the Bell P-39 program.[2]

Many of these engines went through many build-up/teardown cycles. (Note that with the introduction of the jet engine, a similar designation system for them was used, the IX- series for the GE "Type I Supercharger" derived jet.)

The following table identifies the major technical changes and features introduced on the engine as it went from concept and Type Testing on through development. Much of this development was initiated to provide additional capability, such as more power at higher altitudes. Following this table, the text of this chapter explains the engineering and sometimes the tactical background of these major innovations and improvements in the steadily improved and capable V-1710.

Supercharging the V-1710

By the end of the 1920s, the concept of having an integral supercharger in an engine was just coming into practicality. While many of the early installations looked like superchargers, that is, they used a gear driven centrifugal impeller, their actual function was to improve the distribution of the fuel/air mixture to the cylinders. In fact, contemporary accounts refer to these devices as "rotary induction" systems, reflecting that they did little if anything to boost manifold pressure or power. Rolls-Royce in England was one of the first to build an engine strong enough to take advantage of increasing the impeller speed to the point that additional mixture and manifold pressure could be used to dramatically increase the sea-level power output. This was demonstrated by their highly boosted 1929 and 1931 Schneider Cup "R" race engines.

One of the major reasons for not developing the "supercharging" capability on standard engines was simply that engines could not be safely operated at high specific power levels with the low octane fuels of the day. [See below, *Effect of Fuel on Performance Ratings*]. Furthermore, with the exception of the development of the turbosupercharger by General Electric, there was and had been no systematic development within the U.S. of supercharging, its theory, or the apparatus necessary to achieve it.

A reason for the original Air Corps interest in the new Allison engine was that they did not believe that air-cooled radial engines could be adequately supercharged. This was because of the high cylinder head temperatures inherent in air-cooled engines of the day. With senior Air Corps Brass intellectually committed to high altitude bombers, lack of an engine suitable for such an aircraft was a real problem. This was a major influence on their desire to support the development of the V-1710 liquid-cooled engine.

From the outset Allison and the Air Corps both intended the liquid-cooled V-1710 to be a "sea-level" rated engine. That is, the throttle would be wide open at sea-level for maximum power. Altitude capability would be provided by boosting ambient air to sea-level pressure by a turbosupercharger that was being developed for the Air Corps by the General Electric Company.

With the turbo always providing sea-level pressure, it was not necessary to use the engine-stage supercharger to develop more than the intended original design power of 750 bhp on an engine with a displacement of 1710 cubic inches. Consequently, the GV-1710-A used a 8-1/4 inch diameter impeller having partially shrouded blades. It was wholly designed by Allison, which was a non-conventional approach as other manufactures of the day were obtaining their integral superchargers from General Electric.

The supercharger in the XV-1710-1(A2) was increased to 9-1/2 inches in diameter, from 8-1/4 inches, but otherwise similar with straight blades and a vaneless diffuser. The enlarged supercharger was to handle the larger volume of air needed for rating at 1000 bhp. Shown here is the entire Accessories section, including carburetor, supercharger, coolant and oil pumps, and a mounting pad for the starter. (Allison)

Early testing found that the impeller blades failed as a consequence of the high speeds and resulting "tuning fork" vibrations of the blades. Allison corrected this by extending the back face of the impeller to the periphery, creating a "half-shrouded" configuration. This basic design was retained throughout the production life of the V-1710 series.

With the introduction of the V-1710-C series engines, the Air Corps specified that the engine would be designed to operate at 1000 bhp and 2600 rpm. To accomplish this without increasing the capacity of the engine beyond the original 1710 cubic inches, it was necessary to operate the engine with some positive manifold pressure. This in turn required a larger capacity supercharger. As a result, the diameter of the impeller was increased to 9-1/2 inches, along with other changes to allow the passage of the necessarily larger airflow.

A study of the fundamentals of centrifugal compressor theory quickly shows that the diameter of the impeller is the most important parameter in the design and performance of this type of supercharger. The diameter, coupled with the rpm of the impeller, not only determines the stresses within the material from which it is constructed, but the amount of power required to drive it. The "square" of a parameter known as "tip speed", which is the product of the diameter times the rpm, defines the power needed from the drive, as well as defining the pressure rise the supercharger will develop. It should now be apparent why superchargers turning 20,000 rpm consume large amounts of power at high engine speeds.

As the fuel/air mixture passes through the supercharger impeller its velocity is increased to a value near that of the rim of the impeller. It is not at all unusual to find velocities approaching the speed of sound. This high velocity mixture must then be decelerated in the supercharger's "diffuser", a process that converts the high kinetic energy of the mixture into static pressure, which is then useful to the engine as "manifold pressure." This is actually where the "compression" occurs, and as mandated by the laws of physics, results in a considerable increase in the temperature of the mixture. On engines in which fuel is injected ahead of the supercharger, a fuel/air mixture is being compressed by the supercharger which provides a sizable side benefit by contributing the heat necessary to evaporate the fuel. Some 40 degrees Fahrenheit of cooling results, making the mixture more dense and reducing the likelihood of detonation.

More than any other component in an engine, the supercharger benefits from the artistic and analytic skills of the engineer. An inefficient design requires more horsepower, and any inefficiency is power that shows up in the mixture as undesirable heat. Such heat, if above a threshold temperature determined by the quality of the fuel being used and the physical design of the combustion chambers, will immediately cause detonation. If not quickly corrected, detonation will rapidly break the pistons. The Allison V-1710 9-1/2 inch diameter supercharger was one of the most efficient superchargers of the period. It was used in a variety of arrangements, both with and without rotating inlet guide vanes, and sometimes with different diffusers to tailor it to the needs of various installations. These factors defined different efficiencies for the superchargers used on different models.

As an aircraft climbs to higher altitudes, the ambient air pressure decreases with the result that the supercharger must work harder to maintain the same discharge pressure. Another way of seeing this is to observe that the pressure ratio across the supercharger must increase with altitude. If the supercharger speed is not changed, the only way to accomplish this is to further open the throttle. When the throttle is wide open the supercharger is doing all that it can, and such an "altitude rated" engine is now at its

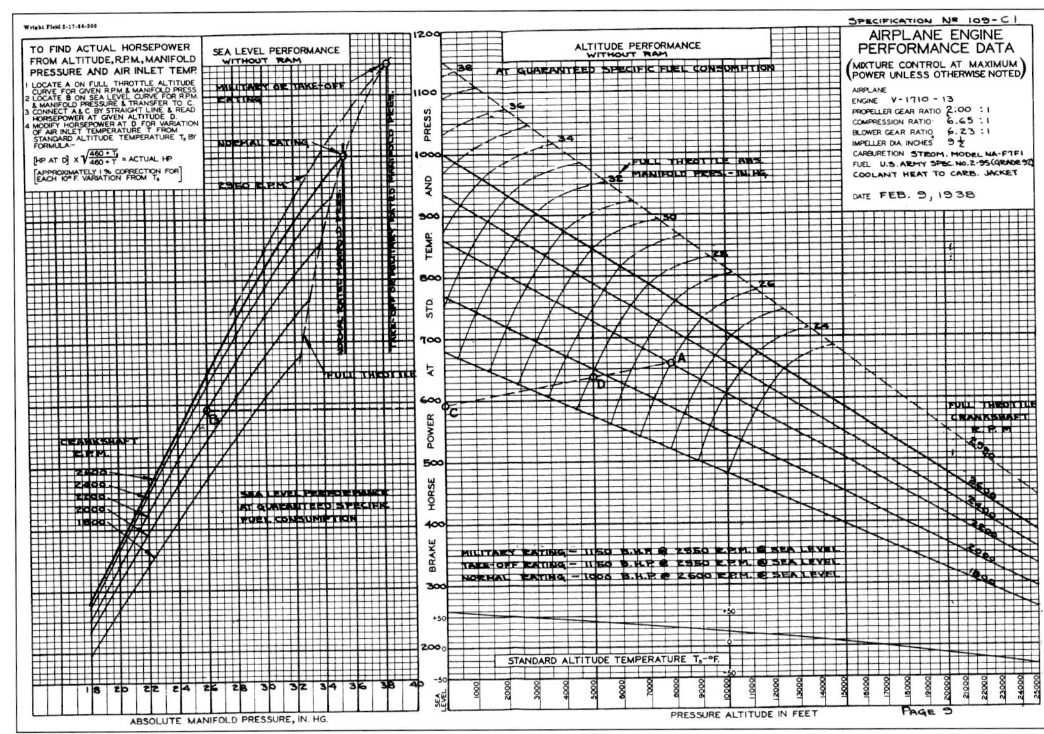

This is the complete performance curve for the V-1710-13(D1), typical for any "Altitude" rated engine. When the D-1 was "Sea-Level" rated for installation in the Bell XFM-1 the effect was to maintain constant power capability to the critical altitude of the turbosupercharger. The curves define bhp as related to rpm, MAP, and altitude. (Allison Spec No. 109-C1)

"critical altitude." While all of this is occurring, the engine power is assumed to remain constant since the airflow through the engine, as observed by a constant manifold pressure, is being maintained by the supercharger.

Reference to Figure 16-1 shows that as altitude increases, the power required to drive the supercharger also increases. This is a reflection of the extra work necessary because of the increased pressure ratio demanded of the supercharger. A second consideration is the efficiency of the supercharger. It is not constant, and is defined by the excellence of the particular machine doing the supercharging. Often the efficiency degrades as the pressure ratio and airflow increase above the normally rated conditions. The centrifugal superchargers in the Allison V-1710 were about as efficient as could be had during the period, and in some instances considerably better. Still, the best efficiency observed was only in the high 60 percent range, with off-design performance dropping into the 50's. Fifty years of further development of this type of supercharger has made dramatic improvements, adding as much as 20 points to these values. The benefits are obvious. Figure 16-1 gives the horsepower required based upon an airflow of one pound per second. The Allison V-1710 would normally require from 6,000 to 12,000 pounds per hour, depending upon the desired power. Doubling the horsepower values in the figure is equivalent to 7,200 pounds per hour, tripling is 10,800 pounds per hour. Delivery of 10,800 pounds/hr for a Military rating at 24,000 feet, with the supercharger at 60 percent efficient would require 225 hp from the drive. This is power not available to the propeller. To lessen confusion, Allison usually defined Military ratings as 1150 bhp, and determined the altitude at which this occurred. This needs to be kept in mind when comparing V-1710 ratings to other engines that typically defined Military ratings at a selected altitude and let the power be the variable.

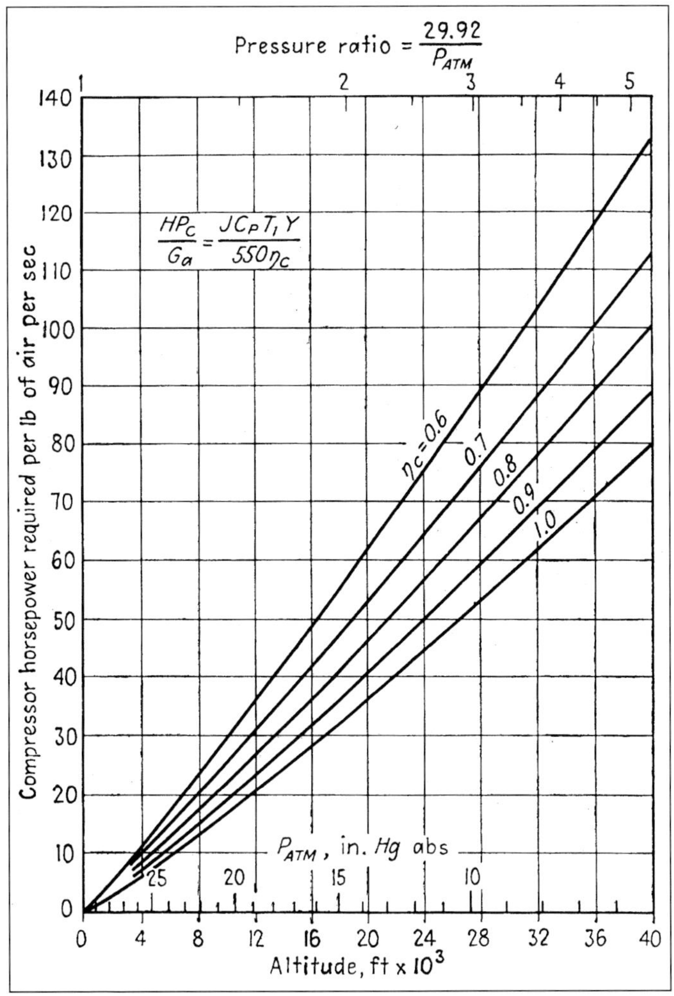

RIGHT: Supercharger Horsepower is effected by altitude, efficiency and air flow into engine. Higher altitudes require higher pressure ratio from supercharger, which in turn demands more horsepower. Improved efficiency reduces the required drive power. N_c is Efficiency Factor. (Liston, Power Plants for Aircraft, 1953)

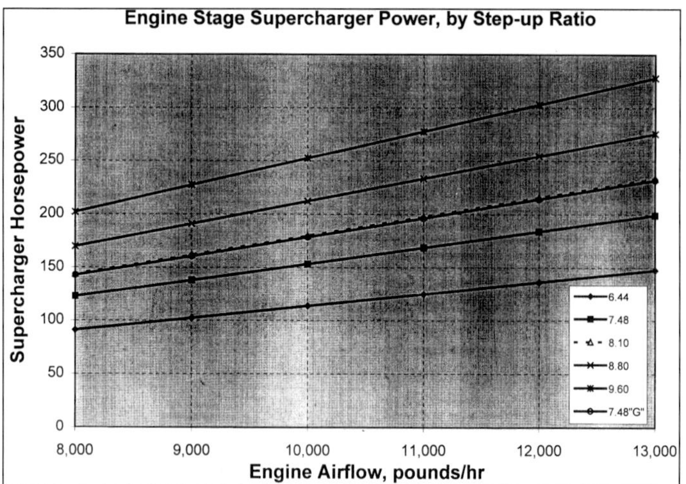

Figure 16-1 Engine Stage Supercharger input power shown for different step-up gear ratios.

As a consequence of the power demands of the supercharger, the available power to the propeller is considerably reduced at altitude on altitude rated engines. This was because the engine is limited by detonation considerations, defined by fuel grade and mixture temperature, to the "indicated" horsepower input to the crankshaft. Propeller power, observed as brake horsepower, is what remains after friction and supercharger requirements are met.

This situation also applies to the Auxiliary Stage Supercharger. As a two-stage engine, the altitude advantage provided by the Aux Stage was due to it being able to provide sea-level air pressure to the engine to a considerably higher altitude. While this required more drive power, the benefit was that greater brake horsepower could be carried to a higher altitude. The elegance of the Allison hydraulic drive coupling between the engine and the Aux Stage was that it demanded only the amount of power that could be used effectively by the Aux Stage.

It was Allison's' practice to tailor the supercharger step-up drive gears to the particular service or mission intended for the engine. The objective always being to minimize the power taken from the engine to drive the supercharger. The V-1710-F2 engines powering the early turbosupercharged Lockheed P-38's, were designed to deliver 1150 bhp from sea-level up to the critical altitude of the turbosupercharger. The turbosupercharger was designed to maintain an outlet pressure equivalent to sea-level up to the point where the rpm of the turbine could no longer be safely increased. Typically this was about 25,000 feet for the single-stage GE turbo's. As such, the V-1710-F2 did not require much internal "supercharging" by the engine-stage supercharger, and the selected step-up gears provided an increase in impeller speed of 6.44:1 over that of the crankshaft. This minimized the amount of supercharging "heat" added to the mixture while at the same time minimized the losses otherwise incurred in throttling the engine. The P-38's Pilots Manual instructs the pilot to operate his engines with the throttle full open, and to control power by adjusting propeller rpm and manifold pressure by using turbo boost. This maximized the overall efficiency and was a prime reason for the P-38s' long range cruise capability.

When it became necessary to provide engines with higher output, Allison used a higher step-up gear ratio, such as the 7.48:1 in the 1325 bhp V-1710-F5 engines for the P-38F. The higher power came from the increased manifold pressure now available at the same engine speed. In making the conversion to "altitude" rated engines the same approach was used. The major difference being that the higher step-up gears required the engine to be "throttled" for operation at altitudes below the "rated", or "critical" altitude. This is inefficient, and consequently such engines do not show fuel consumption rates as good as those typical of sea-level rated engines.

The final altitude rated models of the V-1710-F engine were using 9.6:1 supercharger drive gears. At 3000 rpm and WER ratings, these engines were able to develop 57 inHgA manifold pressure and 1410 bhp, on available Grade 100/130 fuel. None of the Allison single-stage production engines used water-alcohol injection.

Figure 16-1 shows the way that the power required to drive the supercharger dramatically increases for the higher step-up gear ratios. Experience with this class of engines found that approximately 10 IHP would be produced per pound of air flowing through the engine. Expressed as 0.167 IHP/LBM-hr it provides an indication of the amount of power the engine would actually produce over the range of available airflow's. Brake-horsepower at the propeller would of course require that the power being consumed by internal friction and the supercharger would also have be accounted for, as would the effects of intake and exhaust pressures "pumping" the crankshaft. Assuming that the pumping effects offset each other, and acknowledging that the V-1710 at 3000 rpm used about 147 hp to overcome the internal friction of the engine and power all of the internal systems, pumps and drives (excepting the supercharger compression work) then fixes a case for an example.

If we assume 10,000 LBM/hr of air passing through the engine, about 1672 IHP could be produced, which after adjustment for friction and supercharger work, results in 1285 bhp from a 9.60:1 equipped engine, and 1415 bhp in a 6.44:1 equipped engine. This of course assumes that the 6.44:1 equipped engine could provide the 10,000 LBM/hr to the cylinders, in fact this was just about the limit of its capabilities. In addition to showing the full range of step-up ratios used with the 9-1/2 inch diameter impeller, Figure 16-1 also shows the horsepower required for the 10-1/4 inch diameter impeller used with some of the "G" series engines, all with the 7.48:1 gears. Interestingly, at 3000 rpm this unit requires the same amount of power as does the "F" engine 9-1/2 inch impeller when driven through 8.10:1 step-up gears. This is primarily due to both impellers operating at nearly the same "tip-speeds" when at 3000 rpm. Of course the "G" was rated for 3200 rpm operation, where its tip-speed was proportionally higher, which in turn requires more power. At the higher rpm the 10-1/4 inch supercharger provided a 2.8:1 pressure ratio, compared to 2.4:1 at the lower speed. In this regard it was equivalent to the pressure ratio developed by the 9-1/2 impeller driven by 8.80:1 step-up gears, though other dimensional changes allow it to pass a greater quantity of mixture.

With the introduction of the V-1710-G series, Allison developed a 10-1/4 inch diameter version of its supercharger, though not all "G" model engines used the larger unit. Combined with the increase in allowed rated speed of the engine to 3200 rpm, this supercharger was capable of very high manifold pressures, and correspondingly high power. The net effect was that most operation was with the engine "throttled", unless WER power was being demanded, all of which required the use of ADI for operation at above Military power.

Engine-Stage Supercharger Gear Ratios

The early development V-1710's utilized the engine-stage supercharger to both insure a uniform fuel/air mixture and to develop the full rated power of the engine at sea-level conditions. Since the engine was intended to operate in series with a turbosupercharger there was no need to further boost manifold pressure to deliver rated power. As the aircraft climbed to altitude, the turbo would continue to provide the engine with mixture at sea-level density. Furthermore, with 1710 cubic inches of capacity the en-

Table 16-3
Available V-1710 Supercharger Gear Ratios

Number of Teeth		V-1710-A/C/D	V-1710-E/F/G
Drive	Driven	Ratios	Ratios
50	20	5.00:1	6.00:1
51	19	5.37:1	6.44:1*
52	18	5.78:1	6.93:1
53	17	6.23:1*	7.48:1*
54	16	6.75:1	8.10:1*
55	15	7.33:1	8.80:1*
56	14	8.00:1*	9.60:1*
57	13	8.77:1*	10.52:1
58	12	9.67:1	11.60:1

* Indicates ratios commonalty used in V-1710's.

gine was able to produce the required 1000 bhp at a speed of only 2600 rpm with a modest amount of boost, about 40 inHgA, or some 5 psi above sea-level static pressure.

As improvements were made in the engine induction system during development, Allison engineers would change the supercharger gear ratio to reduce the amount of work required to drive the supercharger and thereby generally reduce stresses and improve the overall engine efficiency. Consequently there were a number of ratios used, from 8.77:1 on the XV-1710-1(A2) down to 6.23:1 in the V-1710-21/23(C10/D2), all being rated to deliver 1000 bhp at 2600 rpm. This lower ratio was very fortuitous in the V-1710-21/23 engines, since with their turbosuperchargers providing pressurized air from the intercoolers, the engines were able to produce full power with little danger of backfire or detonation, and therefore were suitable for rating up to 1150 bhp on the available 92 octane fuel.

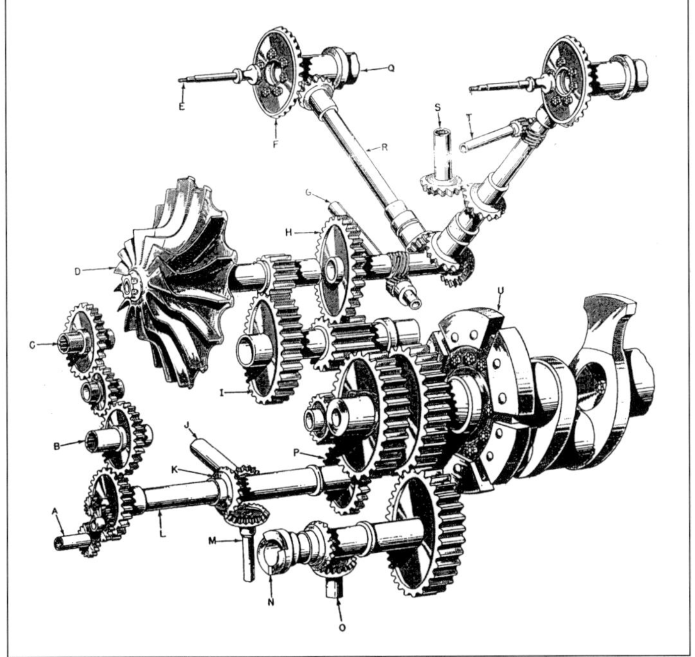

In the V-1710-E/F/G, accessories including the supercharger, were driven from the rear of the crankshaft, via the hydraulic vibration damper at a fixed 2.4:1 step-up ratio as the input to the supercharger gears. Different ratios were used at "I" to give the desired speed to the S/C impeller "O". (Allison)

When the altitude rated versions of the V-1710-C were introduced, the approach was to return to the 8.77:1 supercharger gear ratio used in the XV-1710-1. In this case, the numerous improvements throughout the induction system provided a much higher "boost" pressure than the engine required to produce maximum power at sea-level, so it was necessary to limit the pressure and power, by "throttling" the engine. As the aircraft climbed, the pilot could continue to open the throttle and thereby regain the power otherwise lost as a result of the reduced air density at altitude. With the higher supercharger gear ratios in the altitude engines, more work was required to drive the supercharger. This is why the XV-1710-19, with its 8.77:1 supercharger gears in the XP-40, could only deliver 1060 bhp, while the V-1710-21 (6.23:1 s/c ratio) from which it was derived, delivered 1150 bhp under the same conditions. The crankshaft was doing the same amount of work in both engines, but the supercharger was absorbing the rest of the power in the XP-40. Still there was a sizable benefit, for the 1060 bhp could be maintained to the altitude at which the throttle was fully opened, this point being defined as the "Critical Altitude" of the engine. The aircraft can still climb above this altitude, but the power is reduced by the inability of the engine to draw in enough air for full power.

Considering the available supercharger ratios, the V-1710-A/B/C/D accessory drive operated at a fixed input ratio of 2:1, while the V-1710-E/F/G accessories drive came via a gear driven by the crankshaft resulted in a fixed speed increase of 2.40:1. This was established by the ratio of the number of teeth on the driven side of the hydraulic vibration damper as compared to the number of teeth on the supercharger drive idler gear, this ratio is common to all of the V-1710-E/F/G models. Specific supercharger final speeds for all V-1710's are then determined by the final step-up gears in the impeller drive.

The Auxiliary Stage Supercharger

With the developing interest in removing the turbosupercharger from the XP-37, XP-39, and XFM-1 in 1938/9, it was clear to all involved (the Air Corps, Bell Aircraft, Curtiss-Wright and Allison) that an airplane with a two-stage supercharger was still going to be needed. In November 1938, Allison began efforts to develop such a unit, though they chose a fairly unique approach, that being a separate Auxiliary Stage supercharger that could be attached to the production engine with little modification or revision.[3]

It was intended to provide a critical altitude similar to turbosupercharged engines, about 25,000 feet. To handle the large volumes of low density air a high altitudes it was necessary to have a physically larger impeller than the engine-stage unit. This was to compensate for the low air density at 25,000 feet, which meant that the unit was going to have to draw in 2.23 times as much air (by volume), because of the low density, and compress it by about 2.7 times in order to deliver sea-level conditions to the engine-stage supercharger. This was about the limit for the technology for the centrifugal superchargers of the day. In fact Allison had at first attempted to use a 9-1/2 inch diameter impeller with enlarged passages, but this was soon revised to improve the altitude capability.

In its final design, Allison selected a 12-3/16 inch diameter impeller with 15 radial vanes, and drove it by connecting to the crankshaft through the engine starter gear. This provided a robust drive able to deliver the comparatively large 350 hp sometimes needed by the Auxiliary Stage. Interestingly, the impeller was designed to turn the opposite direction of the engine-stage supercharger. As with the engine-stage drives, a range of step-up ratios were used as dictated by the installation and aircraft mission. In its initial form, on the V-1710-45(F7), the drive for the Auxiliary Stage was an all mechanical single speed affair with a friction clutch for vibration damping.[4]

What was unique with the developed drive was the method for insulating the supercharger from the torsional vibrations inherent in a direct drive from the crankshaft. In 1942 Allison fitted a small hydraulic clutch, much like that later used in automatic automobile transmissions, to drive the Auxiliary Stage. This provided another significant benefit in that it could be operated to allow a significant amount of "slip" and thereby provide an infinitely variable control of the speed of the Auxiliary Stage impeller. The unit speed was as controllable as that of a turbosupercharger. The benefit being that the engine throttle could be efficiently maintained wide open so that manifold pressure and power were controlled by the speed of the Auxiliary Stage. This minimized the amount of compression heating in the induction system while at the same time reduced the power required to drive the Auxiliary Stage. Quite a neat setup.

The first application of the Auxiliary Stage Supercharger was on the V-1710-45(F7R), intended for an early version of the Curtiss XP-60 Pursuit project (MX-69). While the engine was built and tested, it soon took a second position behind the V-1710-47(E9) being developed for the Bell XP-39E/XP-63 aircraft when the Curtiss project was canceled.

One idiosyncrasy of the installation on the P-63A was that Allison controlled the degree of slip, and thereby the speed of the Aux Stage drive, by controlling Absolute Total pressure at the carburetor. This resulted in the minimum slip, maximum rpm, condition not being reached until the airplane was well above the full throttle "critical" altitude. It was estimated that there was a performance improvement of a few mph, or a higher critical altitude available.[5] This may well be one of the reasons for relocating the carburetor to "between the stages" on the later V-1710-109 engines used in the P-63D/E.

The P-63A/C installations, with the V-1710-93 and V-1710-117 respectively, had the carburetor mounted on the intake elbow to the Auxiliary Stage. As before, the fuel was still injected into the eye of the engine-stage supercharger impeller. This arrangement was found to cause an unnecessary reduction in the critical altitude, by about 3,000 feet, due to the pressure drop naturally occurring through the carburetor venturis that restricted the flow into the Auxiliary Stage. Later models such as the V-1710-109 mounted the carburetor on the inlet elbow to the engine-stage supercharger as was done in the turbosupercharged installations. The effect was to improve performance. On the V-1710-G6 engines, built after the war for the North American P-82's, this arrangement was further streamlined by the use of the Bendix Speed Density "carburetor", which was really an early analog computer that mechanically measured engine speed, atmospheric pressure and then calculated air density from manifold pressure and temperature. The proper amount of fuel was then metered to maintain the desired power and fuel/air ratio.

With the introduction of the Auxiliary Stage Supercharger, along with the usual controls for engine mixture, rpm, and manifold pressure, the pilot would be very busy. Given the demands of combat and the need for the pilot to focus outside the cockpit, Allison developed a single control to manage all of these parameters, as well as the additional demands of controlling use of ADI injection and the Aux Stage. This was their Automatic Engine Control.

The Control was calibrated so that the engine was never operated closer than 2 inHg to the limiting conditions for detonation, irrespective of the altitude, air temperature, engine speed, aircraft speed, or demanded power. As constructed, the differences due to changes in temperature between summer and winter conditions at the critical altitude resulted in a gain in performance of approximately 5 inHg at sea-level and 11 inHg at altitude. The complexity of the calculations necessary to select the proper limiting MAP to protect against detonation over a range such as this would have been impossible for any person to do in combat. Relieved of such

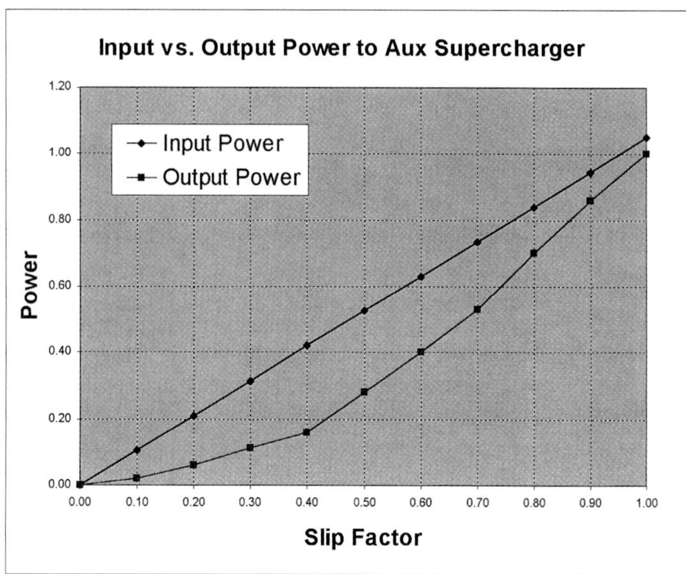

Variation of Auxiliary Stage Supercharger input power with hydraulic coupling slip. (From: Dolza, John 1945, 491-496.)

requirements, it meant that the pilot could confidently maneuver his aircraft without worry about damaging his engine.[6]

This type of control was found to be very sensitive to engine speed, requiring that the propeller governor had to be stable and accurate to within +30 rpm, as excessive speed could quickly exceed the detonation margin. Short duration overspeeding, in the range of 20-30 seconds was not harmful since the Auxiliary State fluid coupling would not fill fast enough to overboost in such cases.

The cooling loads given in Table 16-4 are the maximum values airframe designers needed to accommodate two-stage engines. For the Auxiliary Supercharger, these values correspond to the condition of 50 percent slip, which only occurred when the engine was at less than rated conditions. The cooler load is based upon 25 percent of the maximum power required by the supercharger when at full power at the critical altitude. For example, at full power the V-1710-143 Auxiliary Stage Supercharger would require 472 hp at the critical altitude, and the hydraulic coupling would be consuming another 4% or about 20 horsepower. Total drive power taken from the engine crankshaft would then be about 490 bhp, enabling the engine to operate at its full WER power of 2250 bhp under these conditions.

A feature which never found its way into production V-1710's equipped with the Auxiliary Stage Supercharger was an intercooler or aftercooler, though the device was a part of the design of the two-stage engines from the beginning. Examples were flight tested through the war. Consequently

Table 16-4
Two-Stage Engine Design Cooling Loads

	Rated Power, bhp	Coolant Radiator, hp	Engine Oil Cooler, hp	Aux S/C Oil Cooler, hp
V-1710- 93(E11)	1325	430	155	45
V-1710-117(E21)	1325	445	170	50
V-1710-119(F32R)	1500	460	175	55
V-1710-143(G6R)	1600	605	187	118

Table 16-5
Engine Models, Use of Auxiliary Stage Drive Ratios

Ratios	6.85:1	7.18:1	7.23:1	7.59:1	7.64:1	8.03:1	8.087:1
Eng Rotation	Right	Left	Right	Left	Right	Left	Right
V-1710-E	5	2	6	1	3	-	-
V-1710-F	3	1	1	1	2	1	3
V-1710-G	-	-	-	-	-	2	2

the engines without aftercoolers were sensitive to the conditions that caused backfiring and detonation for they were operating with manifold temperatures very near this limit.[7] Maximum, or War Emergency Rated power levels requiring more than about 60 inHgA were only available through the use of ADI.[8] The similarly configured V-3420-B10, used in the Fisher P-75A, was also equipped with an Auxiliary Stage Supercharger, though sized to supply the doubled quantity of air. This engine was provided with an intercooler, a feature which should have greatly improved the overall operation of the installation.

Nearly ten percent of the V-1710's built were equipped with Auxiliary Stage Superchargers-at least 5,905 engines. Furthermore, about one-third of the models of the engine were so equipped. The breakdown of those with known ratios is given in Table 16-5.

Table 16-5 shows that only four different drive ratios were used. The left turning engines had an idler gear in the drive that caused a reduction of about five percent in the overall ratio, but the Aux Stages were identical.

Interestingly, it was a development of the GE Form F-8 turbo which was later redesignated as the Type B turbosupercharger. General Electric Type B turbos flew not only on the Allison powered Lockheed P-38s, but the Boeing B-17s, Consolidated B-24s, Boeing B-29s, and almost every other turbosupercharged production aircraft, except for the Republic P-47. That aircraft used the larger, though otherwise similar, Type C turbo. The Type B turbo ultimately was developed to support engines operating at up to 1800 hp and with critical altitudes of up to 33,000 feet. The original impeller was retained through all Type B models.

Auxiliary Stage Supercharger Drive Gear Ratios

The Auxiliary Stage Supercharger was driven via a gear off of the starter shaft. This resulting "power take-off" then drove through a universal joint to accommodate any miss-alignment between the engine accessories section and the Auxiliary Stage. The drive connected into a small hydraulic clutch, or torque converter, much like those used on modern day automobile automatic transmissions. Control of the output speed of the torque converter was by a "scoop" which controlled the quantity of oil within the device. When filled with oil, the input and output sections of the clutch were effectively "locked" together and both turned at the speed of the input shaft. By "scooping" out more oil than was being supplied, the output shaft was allowed to slip, and as a result would rotate at some speed less than that of the input shaft. In this way, the Auxiliary Stage impeller speed was infinitely controllable. This was particularly advantageous in an engine without intercooling between the Aux Stage and engine-stage superchargers for it meant that the Aux Stage would be heating the induction air only the minimum amount, consistent with the desired power level and altitude of operation. Furthermore, the amount of power required by the Aux Stage was always minimized, and engine efficiency maximized by this approach.

In the search for ever higher aircraft performance and altitude capability, Allison provided several different maximum step-up ratios within the Auxiliary Stage housing itself. The selection of a ratio for a specific aircraft was determined by its mission, but a good example is provided by the development of the Douglas XB-42/XB-42A aircraft, which used several different models of the V-1710, each characterized by adaptation of a different drive ratio to the Auxiliary Stage Supercharger.

The V-1710 Auxiliary Stage Supercharger was driven off the rugged starter gear via a universal coupling into a hydraulic torque converter "H" that functioned as a variable speed control. Different step-up ratios "D":"B" were used in different engines to match the mission of the aircraft. (Allison)

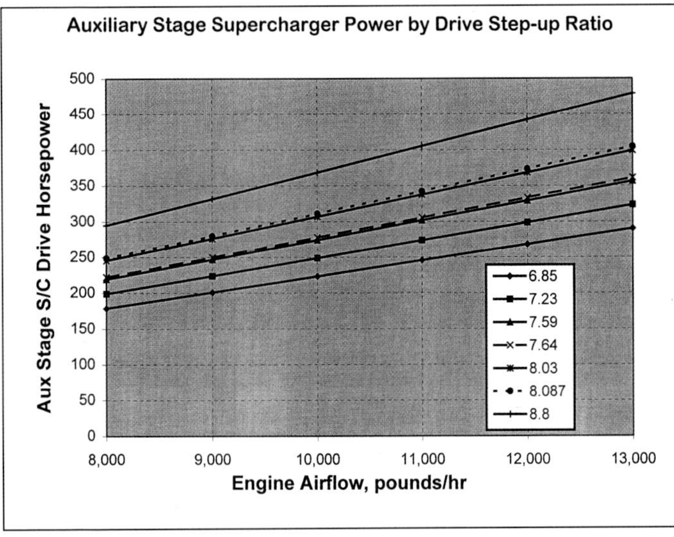

Figure 16-2 Auxiliary Stage Supercharger input power, shown for different step-up gear ratios at minimum slip.

Table 16-6
Models with Auxiliary Stage Superchargers

Engine Model	Aux S/C Ratio	Engine S/C Ratio	Number Built	Date	Military Rating, bhp/rpm/Alt,ft/MAP	Comments
V-1710- 45(F7R)	6.85:1	7.48:1	1	1940	1150/3000/21,000/49.7	Experimental Only
V-1710- 47(E9)	6.85:1	7.48:1	3+	1940	1150/3000/21,000/49.7	XP-39E, XP-63, XP-63A, E-11 Prototype
V-1710- 65(E16)	7.23:1	8.10:1	1	1940	1150/3000/25,800/	Experimental Only
V-1710- 93(E11)	6.85:1	8.10:1	2,521	1940	1150/3000/22,400/51.5	P-63A and P-63C
V-1710-101(F27R)	6.85:1	8.10:1	3	1943	1150/3000/22,400/~50	XP-40Q
V-1710-103(E23)	6.85:1	8.10:1	5	1943	1200/3000/22,400/~51	XB-42 during first six months of flight
V-1710-109(E22)	7.23:1	8.10:1	222	1943	1100/3000/30,000/50.0	P-63D, P-63E
V-1710-109A(E22A)	7.23:1	8.10:1	128	1944	1100/3000/30,000/50.0	P-63D or P-63E
V-1710-117(E21)	7.23:1	8.10:1	2,237	1943	1100/3000/25,000/51.5	P-63A and early P-63C
V-1710-119(F32R)	7.64:1	8.10:1	5	1943	1200/3200/30,000/52.0	XP-51J, XP-82A(Canceled)
V-1710-121(F28R)	7.23:1	8.10:1	4	1943	1100/3200/28,000/	XP-40Q, Advanced P-38
V-1710-123(F28L)	7.23:1	8.10:1	None	1943	1100/3200/28,000/	Advanced P-38
V-1710-125(E23A)	6.85:1	8.10:1	2+	1945	1100/3000/24,000/51.0	XB-42 #43-50225, Early Flights
V-1710-127(E27)	7.23:1	8.10:1	1	1944	2430/3200/17,000/85.0	XP-63H Turbo-Compound
V-1710-129L(E23BL)	7.18:1	8.10:1	2+	1945	1100/3000/29,000/49.5	XB-42 #43-50225
V-1710-129R(E23BR)	7.23:1	8.10:1	2+	1945	1100/3000/29,000/49.5	XB-42 #43-50225
V-1710-133L(E30L)	7.59:1	8.10:1	None	1945	1150/3000/27,500/58.0	Proposal Only
V-1710-133R(E30R)	7.64:1	8.10:1	8	1945	1150/3000/27,500/58.0	Navy L-39, P-63F-1
V-1710-137L(E23CL)	7.59:1	8.10:1	2+	1945	1150/3000/27,500/57.0	XB-42A, model converted from E-30
V-1710-137R(E23CR)	7.64:1	8.10:1	2+	1945	1150/3000/27,500/57.0	XB-42A, model converted from E-30
V-1710-139(F33R)	7.64:1	8.10:1	None	1945	1150/3000/27,500/57.0	Proposed for Production P-82A
V-1710-141(F33L)	7.59:1	8.10:1	None	1945	1150/3000/27,500/57.0	Proposed for Production P-82A
V-1710-F35R	8.087:1	8.10:1	None	1945	1250/3200/30,200/	Proposal Only
V-1710-143(F36R)	8.80:1	7.48:1	None	1945	1250/3200/32,600/	P-82E
V-1710-145(F36L)	8.80:1	7.48:1	None	1945	1250/3200/32,600/	P-82E
V-1710-F37R	8.087:1	8.80:1	None	1945	1600/3200/30,000/	Based on E-30 but with Port Fuel Injection
V-1710-143(G6R)	8.087:1	7.48:1	375	1945	1250/3200/32,700/68.0	P-82E
V-1710-145(G6L)	8.03:1	7.48:1	375	1945	1250/3200/32,700/68.0	P-82E
V-1710-G8R/L	8.087:1	7.48:1	None	1946	1250/3200/32,700/	G-6 but equipped with Aftercooler
V-1710-147(G9R)	8.087:1	7.48:1	None	1946	1250/3200/32,700/68.0	G-6 but with Port Fuel Injection
V-1710-149(G9L)	8.03:1	7.48:1	None	1946	1250/3200/32,700/68.0	G-6 but with Port Fuel Injection
V-3420-A19R	6.96:1	6.90:1	None	1943	2300/3000/20,000/	Alternate configuration for XP-75
V-3420-19(B8)	6.96:1	6.82:1	12	1942	2300/3000/20,000/	XP-75
V-3420-B9	6.96:1	6.82:1	None	1943	2300/3000/20,000/	Proposal
V-3420-23(B10)	7.38:1	6.82:1	101	1943	2300/3000/20,000/48.8	XP-75-1, P-75A-1
V-3420-29(B11)	7.84:1	6.82:1	2	1944	2300/3000/25,000/50.5	Intended for XP-75, not flown
V-3420-B12	7.84:1	6.82:1	1+	1944	2400/3000/25,000/	Tested in XP-75, as B-11 w/50% Intercooling

Figure 16-2 shows the horsepower required to drive the Auxiliary Stage Supercharger when running with minimum slip, that is with the impeller at its maximum speed. The power input to the Auxiliary Stage, was not proportional at reduced loads on the Auxiliary Stage. Rather, when the control was at 50 percent slip, the load on the drive was not only that due to the supercharger, but also included an additional 25 percent of the design maximum supercharger load. This extra power was lost to the oil in the hydraulic clutch as friction, and is the major reason for Allison Specifications requiring so much cooling capacity for the Auxiliary Stage oil. Fortunately, for maximum power at the critical altitude, the power to drive the coupling at minimum slip was only about four percent of the supercharger input.[9]

With Allison's penchant for standardizing manufacturing and maximizing the interchangability of assemblies, the Auxiliary Stage superchargers were all built to rotate the same direction, though opposite to that of the engine-stage supercharger. This meant that in a "left-hand" rotating engine installations, it was necessary to interpose an idler gear in the power take-off drive coming from the engine accessories section. This idler in turn caused a slight reduction in the actual overall drive step-up ratio, as seen by the Auxiliary Stage impeller. While the difference was not enough to effect the performance of the engine, it is the reason why tabulations of the step-up ratios for engines with Auxiliary Stage Superchargers will show different values for left-hand engines. The Auxiliary Stages themselves were identical.

There were differences between models of the Auxiliary Stage units. Most of the early units were equipped with a PT-13 carburetor mounted on the inlet to the Auxiliary Stage, though the fuel was still injected into the engine-stage supercharger. Another change was that the diffuser would be tailored to the specific installation, giving different performance for otherwise similar appearing models.

No less than 22 Army Air Force models of the V-1710-E/F/G, and three models of the V-3420, utilized the Auxiliary Stage supercharger. Most of these were fairly late models. Of the 17 models that were actually purchased or contracted by the Air Force, starting from the V-1710-117, only two were not equipped as two-stage engines utilizing the Auxiliary Stage Supercharger. Although major production was only for the Bell P-63A/C and North American P-82E/F series, a number of experimental aircraft used the engine. These all came too late to play a role in the war, and all were soon eclipsed by the capabilities of the new jet powered aircraft. Still the two-stage Allison was a very capable high altitude engine.

Turbosupercharging the V-1710

By January 30, 1933, Wright Field had already requested a proposal from General Electric to begin work on a turbosupercharger suitable for the Allison V-1710 at a rating of 1000 bhp. On March 10, 1933, the Materiel Division requested the General Electric Company to convert one of their Form F-7 turbosuperchargers for use, rather than the earlier proposal to convert one of the 90 production Form F-2F units then being manufactured. The Form F-2F was the largest turbo then being built and was used on the 700 bhp liquid-cooled Curtiss-Wright V-1570-55 *Conqueror* being installed in the Curtiss XP-6F. General Electric responded within ten days, offering a new model, their Form F-8, to be delivered within five months at a cost of $10,006. The new unit was to be similar to the Form F-7 design, but required a larger impeller (12-1/4 inch diameter), nozzle box and turbine. The unit was to be able to maintain 30 inches of mercury (sea-level pressure) at all altitudes up to 20,000 feet using a speed of approximately 19,500 rpm.[10]

With this exchange of requirements and proposals, a fundamental design concept was established: the effect of a turbosupercharger on the engine backpressure should be no greater than the pressure being developed by the turbo compressor. That is, both the engine intake and exhaust should always "believe" the engine is at sea-level. Had the Form F-2F been adopted as the basic unit, the backpressure on the engine would have been 1-1/2 times sea-level, an amount which would have had a detrimental effect on overall engine performance. The Wright Field staff led in establishing this basis for the design and use of the turbosupercharger.

A considerable amount of work is required to compress the thin air at high altitudes to sea-level pressure. This work raises the temperature to levels which are incompatible with developing high power in the engine. To this end it is usual to provide an "intercooler" to remove at least 50 percent of the heat of compression before passing the compressed air to the carburetor and engine-stage supercharger. All WW II era turbosupercharged installations incorporated this device. It was found in practice to be undesirable to "over" cool the supercharged air, for this would have required an excessively heavy intercooler, greater internal pressure losses to friction, and usually resulted in lead fouling of the sparkplugs.

Since nearly 50 percent of the fuel energy remains in the engine exhaust gasses as pressure and temperature, the turbos had more than enough energy available to provide the 150 to 300 hp required to drive their compressors. GE therefore did not find it necessary to develop particularly high efficiency in its turbosuperchargers, with the result that the Allison engine-stage supercharger was the more efficient of the two. These results were appropriate for the purposes to which the devices were being used.

The V-1710-A/B versions of the engine used a 3:2 internal spur type reduction gear riding on a 10 inch OD plain bearing. Observe the excellent machine work on these components and how the external gear just in front of the bearing drove the accessories via a shaft mounted in the bulged housing. (Allison)

Reduction Gear Assembly

From the very beginning of the V-1710 in 1929, the engine was intended to run at high crankshaft speeds, mandating that a reduction gear would be incorporated. Allison came to the design of this portion of the engine with considerable experience, for they had been building reduction gears since early in WWI for a wide variety of engines designed by others. This included designs by the Army engineers at McCook Field along with series manufacture of complete reduction gear assemblies for the Packard A-1500 and A-2500 water-cooled V-12's.

The original concept for the V-1710 deviated considerably from the previous units, which had been either epicyclic or external spur gear designs. With their considerable expertise with steel backed plain bearings, Allison elected to incorporate one into the reduction gear of the VG-1710, which was then the unit used on the GV-1710-A and V-1710-B engines for the Navy. These gears were of 3:2 ratio, well suited to the fixed pitch propellers of the day, and the maximum engine rated speed of 2400 rpm.

Beginning with the contract discussions between Wright Field and Allison that defined the XV-1710-1(A2), and on through the V-1710-C/D series, Wright Field was not enamored with the overhung crankshaft pinion design of the internally driven spur type reduction gear. This was a design feature looked upon as having questionable reliability and one that would cause problems for the crankshaft. Allison claimed the feature provided weight-saving construction, though Wright Field believed there was a reasonable chance that the saved weight might need to be added in the crankshaft or crankcase to prevent failures such as those occurred on two crankshafts in the earlier Navy GV-1710-A engine, both of which were believed to have been induced by this cantilever feature.

Wright Field had similar concerns about the even larger 14 inch diameter plain bearing for the 2:1 reduction gear. It had a rotative speed which would go as high as 1820 rpm in dives (6,670 feet per minute sliding velocity), and being of such short axial length (2-1/2 inches) that it seemed impossible to adequately lubricate without supplying a relatively large quantity of oil, most of which would flow unused from the ends of the bearing. Indeed such must have been the case, for it was stated by Allison that the oil flow to this bearing was five gallons per minute, and it was found by test that the total flow through the entire engine did not exceed fifteen gallons per minute.[11] The similar construction in the GV-1710-A engine had used a smaller 10 inch diameter bearing for its 3:2 reduction gear which operated at a maximum sliding velocity of 4,890 feet per minute. Since the 14 inch bearing must, because of its size, be distorted under operating loads, Wright Field believed that it also seemed possible that there might result excessive local loads even though conventionally computed (average) loads were very low.

These questions were principally in mind because of the intention to operate the engine at speed and power levels in excess of any there-to-fore developed by this type of engine. Consequently, during the operation and testing of many of the early engines, thermocouples were built into the plain bearing which were closely monitored. Any sudden increase in temperature was cause for immediate shutdown, or landing.

As early as February 1935, Allison responded to a Wright Field request to consider replacing the large plain bearing with a conventional (ball or roller type) reduction gear for the XV-1710-3/5. This would have required a major redesign with new upper and lower half crankcases, complete with main bearings and studs, as well as a new crankshaft with a revised front end to take the new reduction gear and reduction gear bearings. A new pinion gear as well as a new propeller shaft and reduction gear assembly would also have been required. Complicating the arrangement was the need for a new accessory drive gear and drive shaft. The proposed design featured an outboard bearing for the crankshaft gear so that the crankcheek adjacent to the reduction gear would not be subjected to bending loads. It also eliminated the necessity for the heavy roller type thrust bearing on the front of the crankshaft. The design allowed a significant reduction in oil flow to the reduction gear which then allowed the main engine pump to meet all oil requirements with margin to spare. The redesign resulted in an estimated weight reduction of 12 pounds in overall engine weight.[12] The project was shelved.

Allison too became concerned about the overhung crankshaft pinion on the production V-1710-33(C15). When the RAF cabled Allison in August 1941 for permission to operate their V-1710-C15s at rated Takeoff manifold pressure for five minutes of combat only, but at up to the critical altitude of the engine, Allison was emphatic in rejecting the request "...on account of extreme overload on the overhung pinion [in the Reduction Gear] at the higher power involved."[13] To insure the best possible performance from the assembly, Allison produced the crankshaft and reduction gear for the C-15 series engines as a matched pair. At the time of manufacture the parts were hand fitted and given identical serial numbers. During subsequent overhauls it was necessary to keep the reduction gear pinion, all pinion gears, and the crankshaft together as a matched set for reinstallation in the original engine.[14]

This redesign was deferred while the basic V-1710-C was developed and worked its way through the Type Test. With the growing needs for greatly increasing power, and the general redesign of components initiated by the V-3420 project, Allison took the step to incorporate these new features and components into the V-1710-E/F series of engines.

Propeller Reduction Gear Ratios

The purpose of a propeller reduction gear is to match the optimum speed of the engine to the optimum speed of the propeller. Up until the late 1920s or early 1930s, most propellers were fixed pitch, meaning that engine rpm was the primary control of power available to the pilot. This greatly compromised performance, for when maximum thrust is needed at takeoff, with the airplane at a very low speed, the propeller should be at its maximum speed. Conversely, when the airplane is cruising at high speed it is desirable to have the propeller take a larger "bite", otherwise the engine speed must be reduced. In every case the speed and diameter of the propeller are limited by the poor efficiency and noise that results if the propeller blade tips are allowed to approach the speed of sound. Consequently, the selection of the reduction gear ratio is a compromise. With such being the case, the GV-1710-A was designed with a 3:2 engine to propeller gear ratio.

The Army always intended to operate the V-1710-C series engines with the newly available variable pitch propellers. With this device the propeller can be operated over a wide range of conditions at near its maximum efficiency. Propeller efficiency is very important. A good propeller can be 80 to 85 percent efficient, meaning that "only" 200 to 150 hp is being "wasted" or lost by an engine operating at 1000 bhp. Operating with a fixed pitch propeller can result in efficiencies as low as 60 percent under some conditions. Wasting 400 horsepower could have devastating results under the wrong conditions. The entire V-1710-C/D series of engines were built with a 2:1 reduction gear ratio. Significantly, the change from 3:2 to 2:1 required a major redesign and resizing of the entire reduction gear assembly.

In the GV-1710-A the "ring" gear was constructed to ride within a 10 inch diameter plain bearing, with the straight cut driven teeth on the inside diameter of the ring. Such an arrangement is described as an "internal spur type" reduction gear. The driving spur, or pinion gear, was integral with the crankshaft. The auxiliary drive shaft took its power from a separate gear cut into the *exterior* of the ring, it ran at twice crankshaft speed.[15]

For the V-1710-C the ring gear had to be of a larger diameter to provide the increase in the ratio to 4:2 (2:1), meaning that the crankshaft would now turn four revolutions to every two propeller revolutions. The ring gear was therefore increased to 14 inches in diameter, resulting in a very large bearing surface and high sliding velocities. Alternatively, it was not practical to reduce the diameter of the crankshaft pinion by a proportional amount as the contact loads on the teeth would have been too high. This time the accessory drive was taken off of the *inside* of the ring gear by a small gear running at twice crankshaft speed. Interestingly, by changing its drive from "outside" to "inside", the direction of rotation was also reversed, as was the direction of rotation of the various accessories, such as the supercharger. This drive provided the power and synchronization to the magneto, distributors, and camshafts, as well as power to the supercharger impeller. Its ratio, in combination with the fixed centers, or distance between the supercharger impeller shaft and the accessories drive shaft, defined the available supercharger step-up ratios, such as: 6.23:1, 8.77:1.

By the time the V-1710-C15 engines were entering service, the reduction gears were satisfactory performers, although there were considerable problems during initial operations with rusting of the propeller thrust bearing. This was due to the way in which the internal engine airflow and crankcase venting breathers were arranged. The net effect was that ambient air was drawn in through the thrust bearing, which did not receive sufficient oil to protect it from corrosion. A number of innovative field fixes were implemented while the factory developed and retrofit a special spray nozzle to augment the supply of oil to the bearing.

At a rating of 1150 bhp, the V-1710-C reduction gear was the limiting feature of the engine. With this in mind, and following at least one effort to design a convention type of reduction gear behind them, Allison abandoned the plain bearing reduction gear when they designed the XV-3420-1 and V-1710-E/F engines in 1938.

With the introduction of the V-1710-E/F series engines, and the adoption of the external spur type reduction gears, much more flexibility for adjusting the reduction gear ratio was available to the aircraft designer. No longer was it necessary to change the diameter of the ring gear, and its supporting plain bearing in order to change the ratio. Rather, by exchanging the crankshaft pinion and providing a matching reduction gear, the change could be effected. Furthermore, within limits of overall diameters and spacing, it was a comparatively simple change to implement. This was made somewhat simpler in the "F" series since the pinion was an integral component within the reduction gear, driven by an internally splined coupling bolted directly to the front flange of the crankshaft. Consequently, the reduction gear ratios used with these engines vary widely, though they were usually consistent for a particular aircraft type.

Bell Aircraft initially specified a unique 1.8:1 gear ratio for the V-1710-E and its remote external spur type reduction gear for their P-39. This was subsequently changed to a more typical 2.0:1 when it was found that the prop was sometimes entering high speed stalls. For the otherwise similar V-1710-F and its integral external spur gears, the Air Corps wanted to continue with the 2.00:1 ratio. There was little that was remarkable about these reduction gears other than their reliability and versatility. The entire assembly was fabricated by the Cadillac Division of General Motors, and typified the quality reputation earned by Cadillac.

As the variety of V-1710 models expanded to meet the ever changing requirements of the military and aircraft manufacturers, Allison was tasked to develop several versions of these reduction gears, each having a different gear ratio. In order to insure reliability Allison designed these gear sets to have similar maximum loading on the meshing gear teeth. As a result, each ratio was defined by a unique center-to-center distance, that is, the distance between the centerlines of the crankshaft and propeller. These units are described below.

The P-38K used the V-1710-75/77 engines, which were adapted for high altitude operation and used a new Hamilton-Standard Hydromatic propeller. Optimum use of this prop required a different reduction gear ratio, hence the new model engine. As a consequence of adopting the 2.36:1 reduction gear with its one inch higher propeller centerline, Lockheed would have had to again (this would have been the fourth such production change) completely revise the cowling and forward nacelle of the P-38. This was the reason given for Lockheed's decision not to adopt the V-1710-75/77 for P-38 production.

Table 16-7
Reduction Gear Ratio Usage

Usage	Crankshaft to Propeller Ratio
P-39B/D/E/F/J/Q	1.80:1
P-39K/L/M/N/Q	2.00:1
P-38E/F/H/J/L	2.00:1
P-40D/E	2.00:1
P-51/P-51A/A-36A	2.00:1
P-39M/N/Q-1	2.23:1
P-63A/C	2.23:1
P-38K	2.36:1
P-40Q	2.36:1
P-51J	2.36:1
P-82E/F	2.36:1

Reduction Gear Centerline-To-Centerline Distances

1.80:1 Remote reduction gear for V-1710-E Models, with 10.0 inches center-to-center, crank to prop. Pinion has 25 teeth, Gear has 45 teeth, 25 degree pressure angle.[16] No. 60 propeller shaft bored with 3-3/4 inch hole for cannon. This ratio was only used by the early Bell P-39 models, and was changed in the major production models because it was found to cause the propeller to stall under certain conditions because of the high propeller rotational speed.[17]

2.00:1 For V-1710-C/D, has 3.172 inches center-to-center, crank to prop, No. 50 propeller shaft**.** Incidentally, many aviation historians and writers persist in incorrectly identifying this reduction gear as an "epicyclic" or "planetary" gear design. It was not. This perception is probably due to the comparatively close centerlines of the propeller and crankshaft in the "long nose." They were only 3.172 inches apart due to the use of the "internal" spur gear arrangement, but it is clearly an error and needs to be rectified in numerous accounts.[18]

2.00:1 For V-1710-E extension shaft engine. Hollow No. 60 propeller shaft for cannon.

2.00:1 For V-1710-F, 8.25 inches center-to-center, crank to prop**.** Pinion has 22 teeth, Gear had 44 teeth, both with 25 degree pressure angle.[19] No. 50 propeller shaft.

2.23:1 For V-1710-E, Hollow No. 60 propeller shaft for cannon. Actual ratio was 2.227:1.

2.36:1 For V-1710-F/G Models, 9-1/4 inches center-to-center, crank to prop.[20] This unit resulted from the higher rated rpm available with the new 12-counterweight crankshafts. Full utilization of the extra power required no increase in the effective rpm of the propeller, thus the new ratio was needed. Fitted with a No. 50 propeller shaft.

Engine Vibration

The V-1710 was well known for it's smooth operation and sweet sound. This is a result of the fundamentals of the Allison design and the attention to design details intended to minimize stresses within the engine. As a test, Allison intentionally ran their EX-2 extension shaft workhorse engine at the critical speed, without dampers. The point of maximum 2-node vibration stress occurs at the #6 main bearing. The crankshaft broke at this point after 55 minutes of operation.[21] Control of such vibration forces is essential. In this section we look and the evolution of V-1710 vibration control devices, devices that are essential in high powered engines.

First a few basic fundamentals: A rotating shaft becomes dynamically unstable due to resonance when its rotational speed corresponds to the natural frequency of lateral vibration in the shaft. In the discussion of harmonic vibrations it is common to identify a "critical" speed as the speed at which vibration amplitudes are noticeably severe. These critical speeds are known as "natural frequencies", and are often expressed in "cycles per second", or in modern terminology as "Hertz."

Natural frequencies exist for a number of modes. The first "mode" is known as the fundamental, and is characterized by having a single "node" or zero amplitude location within the system, as well as a steadily increasing amplitude away from the node. Higher modes typically do not generate deflections with as large of amplitudes, and they are of necessity observed only at still higher frequencies.

Only the first three modes were of concern in the V-1710, and their natural frequencies were nominally at 25.6 cps, 47.3 cps and 278 cps.[22] In

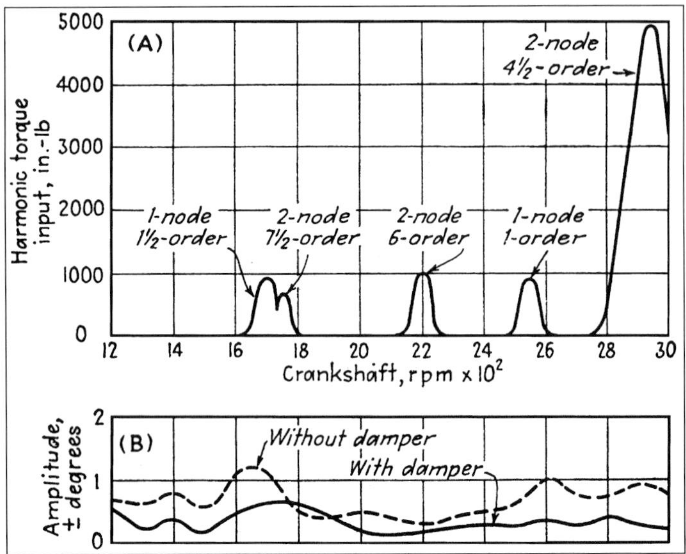

Figure 16-3 For the undamped V-1710-C, the upper curve shows that not all critical frequencies contain enough energy to require damping. The lower curve shows the necessity and effectiveness of the damper on keeping vibration amplitudes below +/- 1 degree. (Hazen and Montieth in SAE Transactions, Vol 43, No2, Aug. 1938.)

the early Allison V-1710's the single-node frequencies varied between 30 and 160 cycles per second, and the two-node frequencies in the range of 160 to 350 cycles per second. When the engine running at 3000 rpm the corresponding fundamental frequency is 50 cycles per second.

Of particular concern in an engine such as the V-1710 is the physical location of the "nodes" associated with each of the modes of vibration. At the location of the node there is essentially no transverse deflection of the vibrating element, consequently the stresses due to the vibration are at their maximum. These are candidate sites for material failure and are a major concern to the designer. For the V-1710-E/F/G the locations were:

1st Mode: Reduction Gear Coupling
2nd Mode: Quill Shaft and Propeller
3rd Mode: Quill Shaft, No. 4 Main Journal and Propeller

Every force generated within the engine can be compounded, and appear as a vibration, when it coincides with other forces or vibrations within the system. The primary cause of these intermittent and periodic forces is the frequency at which the cylinders fire, and the firing order. When all of the combinations are investigated a large number of the possible forces cancel each other. The designer then evaluates the remaining forces existing at each of the fundamental frequencies and determines which result in critical or resonant vibrations that might be destructive. These frequencies are simply multiples of the fundamental or natural frequencies of the system. The multiplier is known as the "order" of the vibration, and for the V-1710 were interesting over the range of 1/2, 1, 1-1/2, 2,..., 7-1/2.

The engine designer must also be concerned about how much vibratory energy is available in each of the harmonic orders, and for each of the natural frequencies. Significantly, many are so weak that internal engine

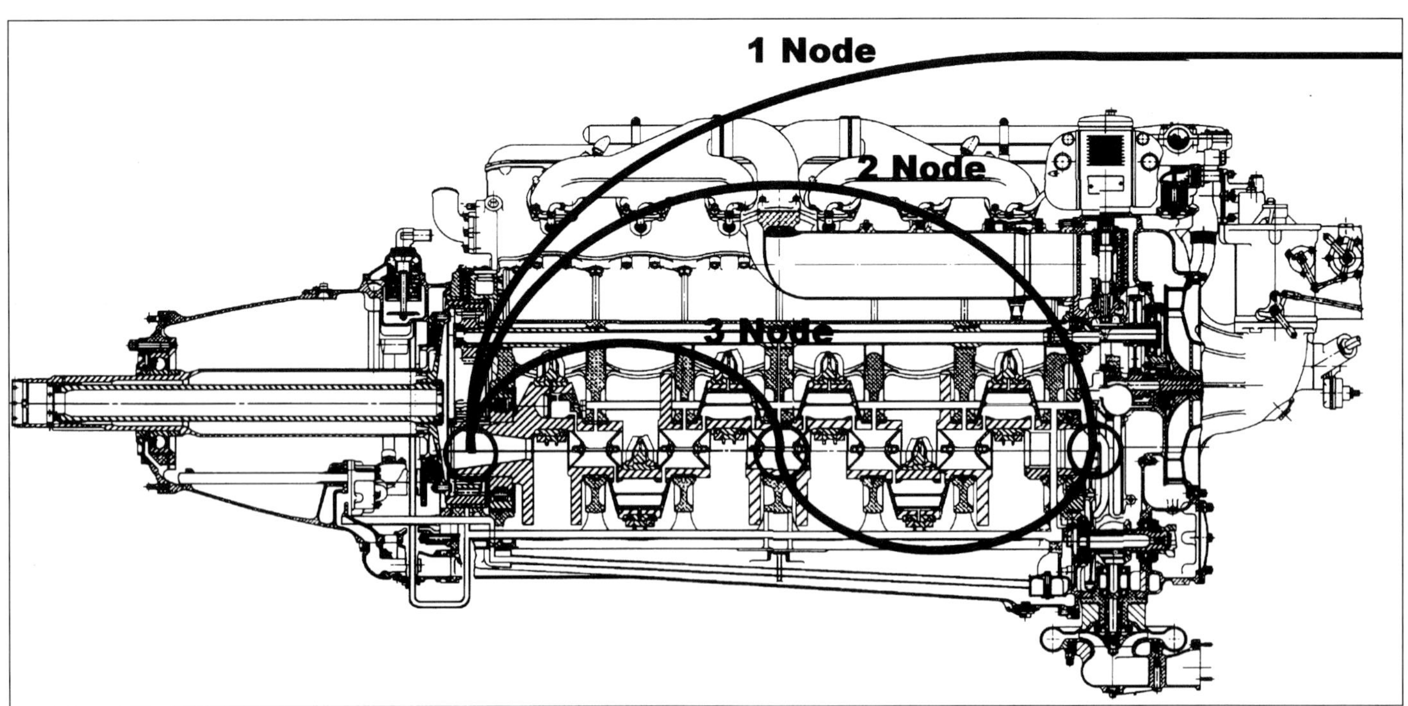

Torsional vibrations run rampant through the long flexible crankshaft in a supercharged V-12. This drawing shows the physical locations within the engine where the fundamental modes have zero amplitude nodes, the locations where vibration induced stresses are maximum. Damping a particular mode is most effective if the damping force is applied at a node or anti-node. Arrangement of the dampers was different between the C/D and E/F/G engines primarily because of how the integral supercharger was driven, and consequently where its damping inertia was applied. In the C/D (as shown) the pendulum damper was assigned the 2 and 3 node vibrations while the spring drive quill shaft and supercharger damped the single node. This was revised on the later models where the accessories were driven from the rear of the crankshaft through the hydraulic damper. It resolved the 1 and 2 node vibrations while the pendulum damper handled the 3 node. (Author)

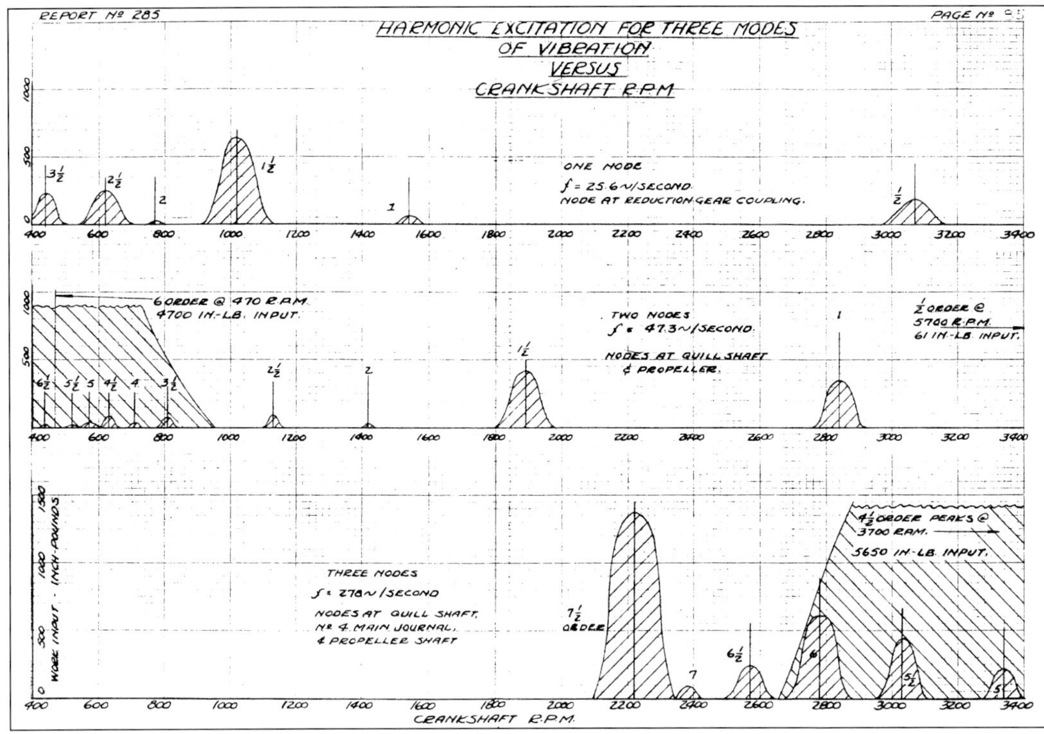

The undamped vibration spectrum for the V-1710-E2 was considerably different than that characteristic of the "C" model engines and required different mitigation measures. The important consideration is the amount of energy contained within a given harmonic, and the likelihood that the engine would be operating within its range. The combination of the Dynamic Pendulum Absorber along with the Hydraulic Damper smoothed the V-1710 over a very wide range and protected against damaging harmonics in very diverse installations. (From: Montieth, O.V. 1939, 66.) (Allison)

friction is able to completely dampen them, while others are outside of the normal range of operation to the point that they can be ignored. Still some, such as the small amplitude 3-node 4-1/2 and 7-1/2 order harmonics contain enough energy to break crankshafts and must be damped. Others, such as the 2-node 6th-order, while nearly as energetic as the damaging 3-node 4-1/2 order, occur at such a low engine speeds (470 rpm) that the engine can easily accommodate them during the short and infrequent occasions when it passes through that critical speed.

The Engine: Allison believed an engine with a six-throw crankshaft was particularly attractive for use in an aircraft as it was the shortest type to give inherent smoothness due to natural balance of the reciprocating and centrifugal forces as well as associated force couples. It also provided suitable overlapping of power impulses, and the ease of manifolding for good mixture distribution. When incorporated into a 12-cylinder engine it provided a low ratio of maximum to average torque, low specific weight due to the efficient utilization of crankcase and crankshaft structure, and in the Vee arrangement, a shape easily adaptable into a streamlined aircraft fuselage.[23]

In the view of Allison's Chief Engineer Ron Hazen, practically every step in the development of the reciprocating aircraft engine tended to make the torsional vibration problem more severe. Engines prior to the V-1710, with their low speed and direct drive to a heavy wooden propeller, resulted in engines free of resonant vibrations across nearly all of their admittedly narrow operating speed range. With the introduction of such features as reduction gears, constant speed light weight metal propellers, and a wide range of operating rpm's, the result was a requirement for smooth operation from as low as 1000 rpm up to 3500 rpm during dives. Improvements in fuel, and the adoption of increased compression ratios, resulted in increased mean effective pressure within the cylinders, which added to the amplitude of vibrations and the need to mitigate two-node vibrations.

Significant to the consequences of vibration is the amount of torsional energy associated with each harmonic being input to the system. The resulting vibration amplitudes are generally proportional to this input. Fortunately, the damping of vibrations is greatly aided by piston, ring and bearing friction of the lubricated parts, and was quite effective in resolving the higher frequency vibrations, say those in the range of 100 cycles per second and above. Still, the dominant harmonics do express themselves and must be resolved at the critical speeds within the operating range. Allison found for the V-1710, that for torsional inputs as large as 5000 in-lb, severe vibrations could extend over a range of 400 rpm. The Air Corps usually required that torsional vibrations not exceed +/- 1 degree of twist in an excited shaft. Since the torque values are affected by indicated mean effective pressure, compression ratio, firing order and reciprocating masses, the various changes and improvements incorporated over the years all required constant attention to their effects on vibration.

With the acquired body of knowledge about vibration and its causes Allison was able to incorporate specific mitigating features into the engine. The approach was first to minimize the generation of the potentially damaging harmonic torque's, and secondly, to provide devices which were able to effectively neutralize those remaining vibrations that would have otherwise effected the reliability and performance of the various engine models. These effects were accomplished by the careful selection of the materials, and their rigidities, so that most of the naturally occurring harmonics resided outside of the usual operating range of the engine. Consequently only the relatively few remaining vibrations then had to be damped.

Torsional Vibration Dampers: In order to reduce or eliminate damaging vibration, the amplitude of the harmonic torque must be dissipated either as heat in a friction damper, or offset by action of a dynamic balancer. In the evolution of the V-1710 Allison applied both methods to resolve specific requirements of various models.

The **friction damper** they initially used was a multiple-disc type having alternate steel and bronze plates. One set of plates was connected to the vibrating member and the other set to a stabilizing or flywheel mass, such as the supercharger impeller, propeller, or a flywheel. A supply of oil was

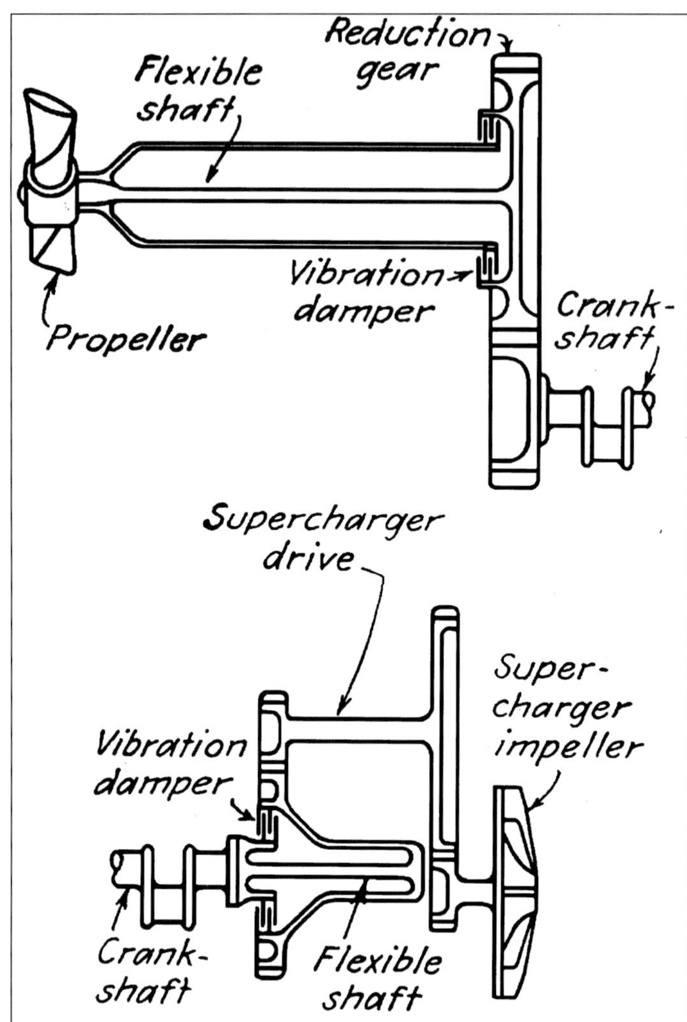

The upper figure shows how Allison incorporated the friction damper into the propeller shaft for the V-1710-C/D. The lower drawing is similar to the 1- and 2-node friction damper used at the rear of the crankshaft on the first build of the XV-1710-17(E2) prior to the first flight of the XP-39. (Hazen and Montieth in SAE Transactions, Vol 43, No2, Aug. 1938.)

required to lubricate the plates in the damper and to remove the heat generated when dissipating vibrational energy. The advantage of this type of damper is that it is generally effective on any kind of vibration as soon as the vibrating member begins to vibrate above a predetermined threshold. Alternatively, this type of damper is not entirely satisfactory, as some level of torsional vibration must exist in order to have energy absorption. For this reason the damper needs to be located at the point of maximum torsional oscillation, which is normally at the front of the crankshaft for single-node and the front and rear of the crankshaft for two and three-node vibrations. A complication is the important requirement to attach one set of plates to a mass of considerable rotating inertia.

Allison successfully incorporated such a device into the propeller shaft beginning with the final form of the XV-1710-3(C1) engine. The heavy mass of the propeller was carried by the stiff outer shaft while the drive torque was transmitted through a very flexible inner shaft. The two shafts were connected at the rear through the friction damper. Since the inner shaft was flexible compared to the crankshaft, the amplitude at the friction damper was high enough to give good damping of single-node vibrations without excessive amplitude at the rear of the shaft. The damper was also at an effective point for two-node vibration since it was near the anti-node, which in flexible systems, was found to be a better location that at the rear of the crankshaft. However, a friction damper is not particularly effective for two-node vibrations since the threshold stresses in the crankshaft become high for even comparatively small angular amplitudes.

For the design and development of the new E/F engines, the vibration characteristics were completely reassessed. The torsional system of the first, second, and third modes of vibration was analyzed, and the severity of vibration of the various harmonics, the resonant speeds, the approximate spread of the vibration, the natural frequency of vibration, and the location of the nodes. Damping the vibrations was achieved by use of both a hydraulic and a dynamic pendulum vibration damper. The hydraulic damper mitigated all the harmonics of the first and second modes of vibration while the dynamic damper was designed to mitigate the 7-1/2 and 4-1/2 orders of the third mode of vibration.[24]

Dynamic dampers, which had been in use on automobiles for a number of years, were a late addition to aircraft engines. This was a result of the early direct drive aircraft engines operating over a fairly narrow speed range, in contrast to automobile engines which drove through a gearbox and required a wide speed range. Allison recognized from the outset that their lightweight geared, high-speed, high-powered engine was going to require fairly sophisticated vibration damping. In order to improve upon the friction damper, Allison was quick to adopt the new pendulum-type damper so successfully worked out for radial aircraft engines by E.S. Taylor and R. Chilton in 1935.[25]

Like the friction damper, the dynamic pendulum damper should be located in the crankshaft system where it is subjected to maximum vibration amplitudes. If the crankshaft system is designed so that the middle node of three-node vibrations is approximately in the center of the crankshaft, then the damper may be located on either end of the crankshaft. The pendulum damper has the advantage that it can be tuned to almost any harmonic order, rather than a particular vibration frequency. When more than one resonant speed occurs within the operating range of an engine, the weights can be mounted such that more than one harmonic order can be damped at the various critical speeds.

In the late 1930s when Allison was designing the V-1710-E/F and V-3420 engines, the analytical state-of-the-art relating to the dynamics of the

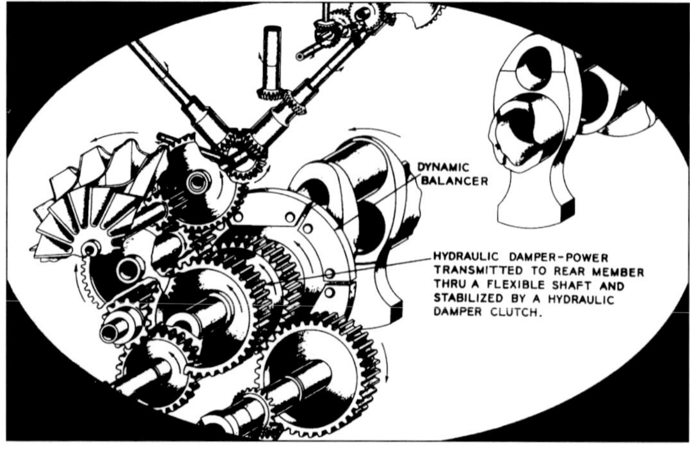

The pendulum damper was mounted directly to the aft end of the crankshaft. The pendulum weights were supported by two pins and free to oscillate when excited by torsional vibrations. In the process they opposed, and greatly diminished, the vibrational energy exciting the crankshaft.

engine and its components was in its infancy. Strict mathematical analysis of the complex engine-crankshaft-propeller system was nearly impossible, and many of the needed analytical techniques had not been developed. As a result, Allison engineers used considerable insight, along with graphical analyses techniques, to direct their investigations and testing to define and resolve the complex dynamics occurring within the engine models. They advanced their knowledge and understanding by devising special test apparatus, such as electronic detonation sensors and torsiometer equipment, for use in running engines. During the war, they also constructed an electrical analog of the entire drive system that functioned as an analog computer and provided considerable information as to the dynamic performance of the mechanical components as a consequence of proposed changes and revisions to components.[26]

Amplitude curves for both single-node and two-node vibrations in the V-1710-C are shown in Figure 16-3. Also included are torsiometer test results showing the effectiveness of the pendulum damper in reducing the amplitude of the vibrations. Allison's success in managing vibrations within the V-1710 was demonstrated in the evolution from the V-1710-C to the crankshaft speed extension shaft version of the V-1710-D3 Mod A. This engine was built as a prototype of the V-1710-E series, but with a 56 inch crankshaft speed extension shaft to a remote reduction gear. Vibration testing found that there was no change in either the single-node or two-node frequencies in the damped engines.

With the introduction of the V-1710-E/F engines vibration management was completely revised. After first attempting to use a plate type damper in the accessories drive train, (needed for friction to dampen the first two modes of vibration) they went on to devise an all new hydraulic damper. Like the plate damper it was mounted on the rear of the crankshaft where it used the high rotational inertia of the supercharger impeller as the stabilizing element. The angular vibration energy was dissipated as heat by "pumping" small quantities of oil through metering orifices between the vanes in the damper. The third mode vibrations were then assigned to the pendulum absorber mounted at the rear of the crankshaft.

Evolution of the Damper System The vibration problems encountered as the V-1710 was upgraded, and/or configured for different installations, were not unexpected by Allison, but each case required Allison to anticipate the need or provide a timely resolution. The reason for not adopting more comprehensive solutions at earlier times was simply the requirement to keep the weight of the engine as low as practical while meeting Army specifications. For example, engines using the stiffer 12-counterweight crankshaft incurred a 27 pound weight increase as compared to the 6-counterweight unit it replaced.

The first V-1710, the GV-1710-A, was built with a simple statically balanced crankshaft, that is, one without counterweights. The throws on the crank were in the conventional symmetrical arrangement and spaced every 120 degrees. With this arrangement, the normal speed of the engine was set at 2400 rpm and was limited during overspeeding to 2600 rpm as a consequence of both the torsional vibrations and the bearing loads at the higher speeds. This limitation was revised beginning with the long-nose XV-1710-1 and V-1710-C series engines, all of which were equipped with a crankshaft having six counterweights designed to offset half of the dynamic forces on the bearings. The counterweights were made from SAE 1020 steel and electric resistance welded to the forged steel crankshaft prior to final machining and heat treatment. Consequently, the early V-1710-C engines could be rated for normal running at 2600 rpm due to the reduced dynamic bearing loads.

When the engine was redesigned as the 1000 bhp XV-1710-3, and required to be run at up to 2800 rpm, a major torsional vibration zone was entered which presented a serious design challenge. To limit these vibrations, plate type vibration dampers were incorporated into both the rear of the crankshaft and into the propeller shaft at the front of the crankshaft. Use of these types of friction damper was less than optimum in that it required careful preloading of their spring packs, and then when "slipping" at the critical speeds, considerable friction heat could be developed.

When the need came to increase output to 1150 bhp, the rated speed had to increase into the previously prohibited 2950-3000 rpm range. It was no longer feasible to simply counter-balance the crankshaft, for at these conditions torsional vibrations rather than dynamic forces were the limiting factor. To achieve this output, it was necessary to reduce these vibrations, a task that was beyond the limits of the friction type dampers Allison had been successfully employing in drive systems for ten years. At this point the pendulum damper was introduced.

A summary of the evolution of the damping features provided by Allison is interesting and informative, as it shows the degree of sophistication incorporated into the engine over the years. The major evolutionary steps are outlined below:

- The GV-1710-A(A1) used a long and flexible "quill" shaft, driven by the reduction gear and made of spring steel to provide single-node damping and a smooth drive to the supercharger and accessories. In this regard the supercharger acted to smooth the torsional vibrations otherwise passing throughout the engine. Significantly, the drive was taken from the reduction gear mounted at the front of the crankshaft.
- The XV-1710-1(A2) introduced the six-counterweighted crankshaft and retained the long flexible quill shaft in the supercharger drive. As a result the early models could run at up to 2800 rpm, at which point the 2-node 4-1/2 order became a limitation.
- During extensive development testing of the XV-1710-3(C1) it was found that several changes had to be made to accommodate vibration. First, the propeller shaft showed a resonant 1-node 1-1/2 order vibration at 2400 rpm, as well as a 6-1/2 order 2-node vibration of +/- 1/2 degree occurring at a frequency of 220 cycles per second. In June 1935 Allison decided to provide a larger propeller shaft, thereby intending to move its fundamental frequency to be above the planned operating speed range. This change resulted in replacing the original SAE #40 sized propeller shaft with the SAE #50, which then became standard on all subsequent V-1710's, except for the "E" with its hollow SAE #60 shaft to accommodate a cannon. The new shaft moved the single node 1-1/2 order vibration to 2900 rpm and shifted the high frequency 2-node vibration to 2100 rpm, but gave a dangerously high +/-1 degree of crankshaft twist. Allison then decided to incorporate vibration damping devices into the design. The result was the double propeller shaft with friction damper, as well as a spring type harmonic vibration damper at the rear of the crankshaft. The resulting crankshaft was identified as PN33302, and fitted the EX-46 harmonic vibration damper, which was of the spring driven flywheel type, intended to cancel the 2-node oscillation at 2100 rpm. This crankshaft, along with the new double propeller shaft, was installed in the XV-1710-3 (Build 4) engine in August 1935. The new propeller shaft was successful in moving the single node vibration to below 1700 rpm where it was no longer of concern. This was a two concentric shaft arrangement where the outer shaft carried the weight of the propeller, while the drive power was transmitted through the flexible center shaft. The friction plates between the shafts were mounted on the front of the reduction gear. Tests of this arrangement showed that the one-node resonance was now at 1650 rpm and two-node at 2000 rpm, both having been very nearly canceled in amplitude. After 14-1/2 hours

of operation the EX-46 harmonic damper failed. As there was insufficient space for a stronger damper Allison again had to reconsider their entire approach to damping vibration within the engine. Based upon the success of their drives in the airship "Shenandoah", on which they had used friction clutches to eliminate its vibration problems, a disc clutch friction damper, PN33115, was developed. With this damper mounted on the rear of the mating PN33685 crankshaft, both the one and two-node vibrations were reduced to acceptable levels.[27]

- The XV-1710-7(C3), and the first build of the YV-1710-7(C4) Type Test engine, both used the PN33685 crankshaft and PN33115 disc type friction damper, along with the friction damped two shaft propeller shaft. The vibration response and conditions were initially believed to be satisfactory, as they had been demonstrated in the XV-1710-3 (Build 4).
- The V-1710-9(D1) pusher engine was designed as a contemporary of the YV-1710-7(C4), though it was to use the PN34600 crankshaft, which differed mainly in requiring less lubricating oil flow to the bearings.[28] The same damping system was used, though testing had determined that the propeller drive friction damper breakaway preload setting, at 4800 in-lb, was too high. While the nose case shapes of the C/D were entirely different they used the same propeller shafts. On the V-1710-D it simply telescoped into the extension shaft.
- During this period Allison was also working on what was to become the XV-1710-11(C7), the first 1150 bhp rated engine. This rating required it to operate at up to 2950 rpm. For operating at this speed and power Allison determined that, because of the 2-node 4-1/2 order vibration, it would be necessary to install a dynamic damper of the new pendulum type. The approach selected was to adapt the "Dynamic Pendulum Vibration Absorber", a device recently invented by E.S. Taylor, and intended to resolve serious vibration problems inherent in radial engines. As adopted by Allison's J.L. Goldthwaite, the device was to have six 4-1/2 order segments, and be effective for the 3-node vibrations. With this in mind they developed the PN36000 crankshaft, which incorporated six PN34758 bifilar pendulum weights mounted on twelve PN34749 suspension pins, and defined the tuning of the PN36022 damper for the 4-1/2 order.[29]
- At 116 hours into the Type Test of the YV-1710-7(C4) the crankshaft was found to be cracked at the No. 4 (center) journal. Consequently, its original friction damper equipped PN33685 crankshaft was replaced by a PN36000 crankshaft having the new pendulum damper and six 4-1/2 order weights as intended for the 1150 bhp engines. In this configuration the engine soon successfully completed the 150-hour Type Test, as well as the additional 116 hours of penalty running on the new crankshaft and damper.
- With the need to provide more power for the Bell XFM-1, the Air Corps decided to have Allison uprate the YV-1710-9(D1) engines to become 1150 bhp rated YV-1710-13(D1)'s. Since the YV-1710-9's had been built to the final configuration of the Type Test YV-1710-7(C4/C6) they already incorporated the PN36000 crankshaft and pendulum type harmonic vibration damper. During the Type Test of the XV-1710-13 the Air Corps Wright Field became concerned about the 6th order vibration and exchanged two of the weight segments and pins for ones tuned for it. Allison objected to this change and refused to guarantee the engine. They believed that the remaining 4-1/2 order segments were too light to sufficiently dampen the 4-1/2 order energy. Allison was proved correct when the damper hub was damaged when the four 4-1/2 order weights were unable to adequately dampen the input energy. Following repair to the damper, the YV-1710-13 Type Test engine, AC34-14, completed the test and was rebuilt by Allison to again incorporate the six 4-1/2 order segments. The engine then ran a further 13-3/4 hours at 1000 bhp and 2600 rpm at which point is was significantly damaged when the crankshaft (PN36000) failed at both the #3 and #4 crankpins. Analysis found fatigue cracks originating at the crankpin oil holes, which had acted as stress risers. The extensive running with only four 4-1/2 order weights had caused the excessive two-node stresses to be concentrated at these mid-position locations. The ultimate fix was to increase the weight of the 4-1/2 order segments which were tuned to minimize 1st and 2nd mode torsional vibrations.
- As a consequence of these experiences the YV-1710-13(D1)'s in the Bell XFM-1, as well as all of the subsequent V-1710-C and V-1710-D series engines, was fitted with similar 2-node 4-1/2 order pendulum type dampers. With installation of the device it became practical to revise the loading on the friction type damper used in the V-1710-C propeller drive from 4,800 in-lbf to 2,000 in-lbf by reducing the number of friction plates from seven to three.[30]
- With the complete redesign of the engine for the V-1710-E/F series, the vibration management system on the whole engine was again entirely revised. Given that the "E" engine was to drive its propeller through a long crankshaft speed extension shaft, it was necessary that all of the damping now be accomplished at the rear of the crankshaft. This was because in the new configuration there was no room for installing a damper on the output shaft to the propeller for either the "E" or "F". At the same time, there was no longer a way to have both the "E" and "F" models be able to share power sections, and also retain the long quill shaft to drive the accessories and supercharger as done in the "C" and "D" engines. The solution was to design a short, but still flexible quill shaft, to be driven from the rear of the crankshaft through a friction damper. In this arrangement the supercharger was to provide the stabilizing inertia for damping the 1 and 2-node vibrations using the new plate type friction damper. Three-node vibrations were assigned to the pendulum damper.
- When the V-1710-17(E2) powered Bell XP-39 began ground running in February 1939 in preparation for its first flight, a serious torsional vibration problem was identified that took several months for Allison to resolve. It was found that when mounted in the comparably flexible airframe, 1st and 2nd order vibrations were passing into the accessories section and causing the magneto and camshafts to see excessive vibration amplitudes that resulted in the extremely rough running and inability to reach rated engine speed. The problem was believed to be due to the use of a clutch plate type friction damper for first and second modes, and the fact that it passes vibrations below the torque setting of the spring pack.[31] As a fix, Allison designed and perfected a hydraulic type friction damper able to dampen the high amplitude 1 and 2-node vibrations before the passed into the accessories section. The resulting device defined the vibration control system for the tens of thousands of V-1710 and V-3420's in the coming production run. Even so, Allison was chagrined when the new device was found to be ineffective in resolving the problem in the XP-39.[32] They then traced thc vibration to be due to excessive torsional flexibility of the extension shafts to the remote 1.8:1 reduction gear. These were stiffened by 30 percent when the aircraft was rebuilt as the XP-39B and resolved the problem.[33] The final units were 2-5/8 inch OD, with 0.200 inch walls, but were not available until the aircraft was reconfigured as the XP-39B.[34]
- The hydraulic friction damper in the E/F was found to be ineffective in damping three node vibrations since the amplitudes were so small. Still considerable vibratory energy was being input at the higher

speeds so Allison revised the pendulum damper and tuned it specifically for the 4-1/2 and 7-1/2 order 3-node vibrations. In the V-1710-17 the damper pendulum segments weighed 1.18 pounds and 0.88 pounds respectively. Three segments of each were used, with the 4-1/2 order being able to accommodate 5,650 in-lb and the 7-1/2 order 1,370 in-lb.[35] As the V-1710-E/F was upgraded over the production run these basic weights were used to counter-balance vibratory energy, though the amount of angular displacement varied with the specific installation and power level.

- The 12-counterweight crankshaft was designed in 1942 to offset more of the dynamic loads on the bearings, and thereby improve the reliability and power capability of the basic engine. Subsequently it was determined that this crank was suitable for normal running at up to 3200 rpm. Engines with this rating were late model E/F's and those in the "G" series. As a consequence of this higher speed it was found that the left bank camshaft was also generating excessive torsional vibrations. To counter this Allison added a small pendulum type absorber on the left bank camshaft in the G-6 engines. This provided the desired smoothing and reduction in torsional vibrations induced by the uniform opening and closing of the valves. It was not necessary to further dampen the right side as the magneto drive had the desired effect.

The high speed "G" model engines incorporated a dynamic balancer (11), with three weight segments (3) on the left side camshaft (1) to dampen torsional vibrations. The right camshaft was dampened by the magneto drive (12). The camshafts and other accessories, including the supercharger, were driven off of the hydraulic damper, which was driven by the dynamically balanced crankshaft (9). (Allison)

The Allison V-1710 earned a reputation as a very smooth running engine. In this regard, the inherent advantages of a V-12 were accentuated by the attention to smooth functioning by proper and adequate damping of the various accessories and propeller system loads and stresses. Over its design life the engine operating speed was increased from 2400 to 3200 rpm, and power from 750 bhp to over 3000 bhp. This could not have been done without the advanced design, analyses, and manufacturing skills Allison brought to the program.

Working of the Dynamic Pendulum Vibration Absorber

In 1935 E.S. Taylor conceived, and put into practice, a damper to manage the serious torsional vibrations being encountered in geared radial aircraft-engine-propeller systems. His device was known as the "Pendulous Vibration Absorber."[36] When Allison's J.L. Goldthwaite became aware of the device he saw the possibility of a simple, comparatively light-weight and

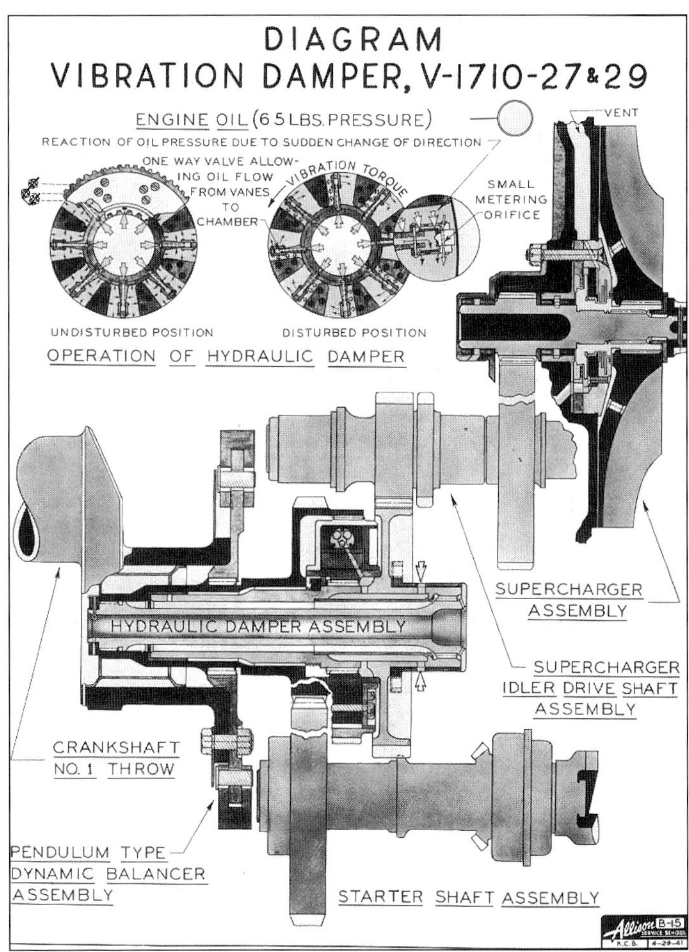

Allison developed this hydraulic damper to remove the 1st and 2nd mode vibrations from the engine. Note how it drove the supercharger and accessories and was directly connected to the rear of the crankshaft, just behind the pendulum damper used to compensate for 3rd mode vibrations. The hydraulic vibration damper was not Allison's first choice, rather it was an expedient attempt to solve torsional vibration problems in the XP-39 and get that airplane flying. The device was extremely effective and contributed greatly to the reputation for smooth running of the V-1710. (Allison Service School)

friction-less damper which could overcome the difficulties inherent with the plate type friction damper. He was able to adapt the concept into the precious little space available within the V-1710 at the rear of the crankshaft. This was a timely occurrence as he soon used it to resolve a critical vibration problem inherent in the uprating the engine to 1150 bhp at 2950 rpm. When the YV-1710-7(C4) Type Test engine was found to have a cracked crankshaft, due to a manufacturing defect at the point of maximum 3-node torsional vibration, the pendulum absorber was quickly incorporated by adapting PN36000 crankshaft as a replacement. With this crankshaft the engine became the first to successfully complete the 150-hour Type Test at 1000 bhp or more.[37]

Allison referred to this device as the "harmonic vibration damper" and it functioned by having six small weights mounted on the rear flange of the crankshaft such that they were free to pivot, that is, to swing like pendulums. The dynamic pendulum damper has the advantage that it can be tuned to almost any harmonic order, rather than just a particular frequency of vibration. When more than one resonant speed occurs within the operating range of an engine the weights on a dynamic damper can be tuned to the various harmonic orders necessary to match the critical speeds. The resulting device was extremely successful in controlling the torsional vibrations in the crankshaft, and was directly responsible for allowing the later engines to operate satisfactorily at up to 3500 rpm.[38]

The following discussion describes the finally developed unit which was used on all production V-1710-E/F/G and V-3420 engines:

> "To dampen three node vibrations a dynamic pendulum absorber was used.[39] The device was installed at the rear of the crankshaft and drives the supercharger and accessories, while the pendulum damper was tuned for the 4-1/2 and 7-1/2 orders which were most sever for the 3-node vibrations. The 4-1/2 order peaks at 3700 rpm, which was outside the normal operating range, but its energy extended down to 2800 rpm and contained such a large amount of energy that it required dampening. The 7-1/2 order had a large peak at 2200 rpm, and therefore needed to be damped, as this was a speed commonly used during cruising flight.
>
> "When the pendulum absorber is revolving in the crankcase at speeds other than the critical speed, the speed at which the exciting frequency equals the natural frequency of the system, the weights are at rest or are oscillating at very small amplitudes relative to the carrier. At a critical speed, the damper weights are caused to oscillate in opposition to the undesired vibrational energy. This oscillation causes the pin to roll at high speeds on the surface of the damper hole.[40]
>
> "The dynamic pendulum absorbers consists of six bifilar suspended pendulum weights, each supported by two 3/8 inch diameter rollers which bear in oversize holes in both the pendulum weight and the hub or carrier. The hub is bolted to the anti-prop end of the crankshaft. Three pendulum weights are tuned to dampen the 7-1/2 order and three are tuned for 4-1/2 order three node vibrations.
>
> "The dynamic pendulum absorber theory is that under the action of centrifugal force and harmonic torque excitation, the bifilar mounted pendulum weight will respond as a simple pendulum. The harmonic torque induces oscillatory motion in the pendulum weights, producing a reaction torque, which at the condition of equilibrium, is equal and opposite to the exciting harmonic torque. The number of free oscillations of the pendulum weights per revolution of the crankshaft depends only upon the geometry of the pendulum and not upon the angular velocity or rpm of the crankshaft.
>
> "With the bifilar suspension system, every particle of the pendulum weights are constrained to move in a circular path and since the weight has parallel motion when displaced, it will behave as a simple pendulum of length equal to the difference in the diameters of the holes and rollers. As the rollers have a rolling motion in the holes when the weight is displaced the bearing is nearly frictionless.
>
> "The mass of the absorber should be such that the amplitude of the pendulum is not too large. Large amplitudes are not recommended because the natural frequency (in terms of crankshaft speed) of a dynamic absorber decreases appreciably after an amplitude of 20 degrees is exceeded. Although the absorber can be tuned for a particular amplitude greater than 20 degrees, there is a loss of efficiency at other amplitudes.
>
> "Each 4-1/2 order weight weights 1.178 pounds and is tuned for resonance at 3700 rpm with an amplitude of +/- 15.5 degrees, at which point it offsets a harmonic torque in the crankshaft of 2428 in-lb, 202 ft-lb.
>
> "Each 7-1/2 order weight weights 0.871 pounds and is tuned for resonance at 2220 rpm with an amplitude of +/- 12.5 degrees, at which point it offsets a harmonic torque in the crankshaft of 1994 in-lb, 166 ft-lb."[41]

In some aircraft installations it was found that the pins supporting the 4-1/2 order weights were galling the mating surfaces of the carrier. A number of fixes were tried, such as grit blasting and lead plating of the pins. Little improvement was noted. The NACA then investigated the effect of increasing the pin diameter by 33 percent, to approximately 1/2 inch, as a way of increasing the surface area and thereby spreading the contact loads. As a consequence of the theory of operation this could be accommodated by properly increasing the diameter of the holes in the carrier and weights and not change the damped critical speed. This scheme worked extremely well, though the production run of the V-1710 was nearing its end, so the modification was not incorporated during the war.[42] When the V-1710-143/145(G6R/L) engines were built after the war, Allison did incorporate larger sized pins in the dynamic balancer.

Vibration Management in the V-3420

The V-3420 used the same crankshafts as the V-1710, along with the same dynamic and pendulum dampers. The only change was that the 3rd Mode Pendulum Damper was fitted with only 7-1/2 order weights. The hydraulic damper continued to be used for damping the 1 and 2 modes. With this arrangement the V-3420-9 was able to successfully complete the Type Test for the V-3420. Crankshaft vibration at takeoff rpm was a maximum of 0.8 degrees, while the limit was 1 degree.[43] This was very smooth considering that the engine had inherent mixture distribution problems causing uneven power distribution between cylinders.

Bearing and Crankcase Loads

With the ever increasing need to develop more of power, it was necessary to increase manifold pressure to the degree allowed by the available fuels, and then to further increase the speed of the engine. Higher crankshaft speed by itself increases the reciprocating loads on the bearings, independent of the power, but increased power is produced by higher average gas pressure within the cylinders. The combination of the two results in the total bearing loads and defines the support required from the crankcase. Table 16-8 shows the loads developed for two typical versions of the V-1710 operating at their limiting power and speed ratings. The benefits of the 12 counterweight crankshaft are apparent in the fact that the mean bending moment on the crankcase, even at the much higher rating of the V-1710-G1, is less than that for the V-1710-F with the six counterweight crankshaft.

Table 16-8
Summary of Usage of Vibration Dampers

Model	At Propeller Reduction Gear	At Rear End of Crankshaft	At Accessories Drive	At Left Bank Camshaft
V-1710-A1	Quill Shaft to S/C	None	Long Quill Shaft	None
V-1710-B	TBD	TBD	Long Quill Shaft	None
V-1710-A2	Quill Shaft to S/C	None	Long Quill Shaft	None
V-1710-C1	Quill Shaft to S/C	None	Long Quill Shaft	None
V-1710-C1/C2/C3	7-plate Friction *	Frict-Flywheel	Long Quill Shaft	None
V-1710-C4/C7	3-plate Friction **	Pendulum	Long Quill Shaft	None
V-1710-C8/C10/C15	3-plate Friction **	Pendulum	Long Quill Shaft	None
V-1710-D1(9)	3-plate Friction	Frict-Flywheel	Long Quill Shaft	None
V-1710-D1(13)/D2	3-plate Friction	Pendulum	Long Quill Shaft	None
V-1710-E2 (Build 1)	None	Pendulum	Friction Plate Damper	None
V-1710-E2 (Build 2)	None	Pendulum	Hydraulic Damper	None
V-1710-E	None	Pendulum	Hydraulic Damper	None
V-1710-F	None	Pendulum	Hydraulic Damper	None
V-1710-G	None	Pendulum	Hydraulic Damper	Pendulum
V-3420-A	None	Pendulum, 7-1/2 Order, 3rd Mode	Hyd Damper, 1 & 2 Modes	None
V-3420-B	None	Pendulum	Hydraulic Damper	None

* Friction damper clutches set for 4800 in-lbf.
** Friction damper clutches set for 2000 in-lbf

Although the six counterweight crankshaft in the V-1710-E/F engines enjoyed about a ten percent reduction in stresses as compared to the six counterweight crankshaft in the earlier V-1710-C series, the 12 counterweight crankshaft was another significant advance. It was used is some late model V-1710-E and V-1710-F engines, but was a central feature in the "G" series. The net weight increase amounted to 27 pounds, the extra metal being in the counterweights and in thickened journals, accomplished by drilling smaller lightening bores only two inches inside diameter. Each counterweight exceeded the moment of the counterbalanced crankpin by 43 percent. As a result, the 7,732 pound maximum centrifugal force developed by the connecting rod big ends at each crankpin was partially balanced by an opposite 3,086 pound centrifugal force developed by the counterweights adjacent to the crankpin. More weight could have been incorporated into the counterweights so as to completely balance these dynamic forces, but at considerable weight gain for the entire engine. As shown below, the bearing loads remained acceptable and the entire crankcase load and resulting bending was reduced. Engineering is a series of tradeoffs. Good engineering includes loading materials and components up to their capability, consistent with desired lifetimes, and in this example, Allison excelled.

Crankshaft Overspeed Rating

The Air Corps required that at the end of each Type or Model Test the engine be tested for its ability to accept overspeed, such as that which might be expected during a power dive or due to an improperly operating propeller governor. Since these were expected to be infrequent occurrences the engine had only to sustain the overspeed for 30 seconds, but the test had to be repeated ten or more times, as defined in the particular contract documents.

The "C" series of engines were rated for overspeeds of up to 3600 rpm, a fairly robust 20 percent over the normally "rated" speed. With the introduction of the "E" and "F" model engines, this capability was increased to 4100 rpm, even though the engines were still rated at 3000 rpm for normal operation. With the later 12-counterweight crankshaft this capability was further improved to be in excess of 4400 rpm.[44]

Conversion of V-1710-E/F and -G Engines from RH to LH Rotation

The twin engined XP-38 required "handed" engines, that is the propellers were to turn in opposite directions. Providing the feature required a fair amount of engineering and modifications costing $2,900 in 1938 to convert a right hand V-1710-C8 into a left hand V-1710-15(C9). At about the

Table 16-9
Crankcase Flexure Loads

	V-1710-F 1500 bhp, 6 cwt Crank	V-1710-F 3000 rpm, 12 cwt Crank	V-1710-G1 1725 bhp/3400 rpm	V-1710-G1 Overspeed, 3500 rpm
Max Vertical Bending Moment, in-lbf	178,000	151,500	172,100	182,372
Max Horizontal Bending Moment, in-lbf	132,000	106,000	104,200	110,420
Mean Bending Moment, in-lbf	150,108	123,583	132,204	140,095

Table 16-10 Loads On Bearings At Maximum Rpm And Power			
	V-1710-F 1150 bhp, 3000 rpm	V-1710-F 1500 bhp, 3000 rpm	V-1710-G1 1725 bhp/3400 rpm and Max Crank Angle
Crankpin & Connecting Rod			
Max Bearing Load, lbf	15,100	16,250	20,500 @ 562.5 degrees
Mean Bearing Load, lbf	11,122	11,772	14,676
Max Bearing Pressure, psi	2,618	2,817	3,554
Mean Bearing Pressure, psi	1,928	2,041	2,544
Blade Connecting Rod			
Max Bearing Load, lbf	13,540	19,750	18,060 @ 15 degrees
Mean Bearing Load, lbf	5,331	5,838	7,311
Max Bearing Pressure, psi	4,056	5,956	5,446
Mean Bearing Pressure, psi	1,608	1,761	2,205
End Main Bearings, #1 & 7			
Max Bearing Load, lbf	9,970	10,450	8,730 @ 555 degrees
Mean Bearing Load, lbf	7,628	7,881	6,207
Max Bearing Pressure, psi	1,883	1,974	1,778
Mean Bearing Pressure, psi	1,441	1,488	1,264
Rubbing Velocity, ft/sec	49.05	49.05	55.59
Intermediate Main Bearings, #2,3,5&6			
Max Bearing Load, lbf	10,820	12,250	15,050 @ 195 degrees
Mean Bearing Load, lbf	7,429	8,238	8,120
Max Bearing Pressure, psi	2,483	2,811	3,066
Mean Bearing Pressure, psi	1,705	1,890	1,654
Center Main Bearing, #4			
Max Bearing Load, lbf	14,400	15,260	16,020 @ 187.5 degrees
Mean Bearing Load, lbf	10,060	10,252	10,542
Max Bearing Pressure, psi	1,932	2,047	2,231
Mean Bearing Pressure, psi	1,350	1,376	1,468
0 degrees Crank Angle is defined as TDC of #1 LH Cylinder.			

same time the Air Corps tasked Allison to develop a left-handed pusher version of the V-1710-D2, a project which was not completed. With the order for the YP-38's in hand, it was apparent that a large number of "left-hand" engines would be required. Consequently, it was expedient to use the components coming from the V-3420 program to simplify the conversion. With this in mind, the V-1710-E/F were designed from the outset to be easy to convert, if necessary even in the field.

The original design of the V-1710-F2R/L only required reassembly of the parts already in the engine, plus the use of a properly handed starter dog and R/L turning starter. The procedure to convert a V-1710-27(F2R) engine to a V-1710-29(F2L) engine was as follows:[45]

1. Turn the crankshaft end for end, retaining the connecting rods in their usual positions. This results in the forked rods being installed in the right bank and the blade rods in the left bank. Remount the damper hub and reduction gear coupling so that they remain in their usual front/rear locations.

2. Rewire the distributor heads to provide for the different firing order:

Right Hand Rotation 1L-2R-5L-4R-3L-1R-6L-5R-2L-3R-4L-6R
Left Hand Rotation 1L-6R-5L-2R-3L-4R-6L-1R-2L-5R-4L-3R

The firing order of each cylinder block is the same in either case, namely 1-5-3-6-2-4, however the phase relationship between the banks has been changed as a result of the reverse rotation of the crankshaft. By this expedient, there is no change in the valve mechanism (since the accessories continue to rotate in the same direction) or intake manifolding.

3. Reassembly of the Accessories Housing

a. Turn the starter shaft oil pump bevel drive gear end-for-end. This reverses the direction of rotation of the oil pump relative to the crankshaft, but since the crankshaft rotation has been reversed, rotation is the same relative to the housing.

b. Turn the generator and pump drive shaft gear end-for-end. This engages the generator and pump drive shaft directly with the gear off the end of the crankshaft instead of through an idler gear. The idler gear is moved to an upper "dummy" shaft as described below for the supercharger drive. Reverse rotation of this shaft in combination with reverse rotation of the crankshaft retains the direction of rotation of all accessories driven from this shaft, including the vacuum pump, fuel pump and generator.

c. Install a left hand rotation starter dog.

d. The same direction of rotation of the supercharger impeller is obtained with the reverse rotation of the crankshaft by incorporation of an additional gear in the supercharger gear train. The supercharger drive idler gear is turned end-for-end. The accessory drive idler gear is moved up from the lower dummy shaft to an upper dummy shaft and is thereby interposed as an additional idler in the supercharger drive train. When the accessory drive idler gear is installed on the lower dummy shaft (right-hand engine) it drives the generator and pump drives. In the left-hand engine the generator and pump drives are driven directly from the end of the crankshaft.

4. The Reduction Gear Assembly

a. The oil nozzle which provides a jet of oil to the reduction gears is reversed from one side of the gears to the other to account for the reversal of rotation.

b. The scavenge oil pump in the reduction gear is also reassembled by changing the positioning of one casting in the pump to enable suitable operation when reversed.

Difficulties with the Left Hand Rotation V-1710-29 Engine

Operation of the early V-1710-29(F2L) at its rated 1150 bhp and 3000 rpm for extended periods resulted in severe galling, pitting, and even cracking of the contact surfaces of the teeth on the supercharger drive idler and mating accessory drive idler gears. In an effort to resolve the problem, the pressure angle on the gear teeth was first changed from 20 degrees to 25 degrees. Though an improvement, it was not sufficient to resolve the problem. Still, the 25 degree design was put into service as the production standard for all engines after the first thirty pairs of V-1710-27 and V-1710-29's.

The problem was finally determined to be that, with the gears relocated for left-hand operation, the plane of the gear driven from the accessory drive idler was being displaced from the plane of the driving gear. This offset in driving planes resulted in a slight tilting or cocking of the accessory drive idler gear. When coupled with the small size of the supercharger gears the result was unsatisfactory performance. This was not a problem on the right-hand engines, as their accessory drive idler gear only had to pass a minimal amount of power to drive the generator, fuel and vacuum pumps, where on the left-hand engines, it drove the supercharger with its much higher power requirements. The supercharger requires about ten times as much power as do all of the other accessories together.

Simple changes to the gear face angles and quality of the teeth were not able to satisfactorily resolve the problem. The ultimate solution was to provide wider gears unique to the left-hand engines, along with a redesigned accessory case to accommodate them in engines rated for the higher power levels. The simple expedient of reversing the position of the gears as a way to accommodate left-hand operation was lost, but at a small cost for a major improvement in reliability.

As a consequence of the left turning engines seeing somewhat more severe loading with the accessories section it was common for such engines to perform the Model Test for representative series, such as the V-1710-F17R/L and V-1710-G6R/L.

Chemical-Cooling

Water-cooled engines of the day used a vented water cooling system that was seriously limited in performance at high altitude, was necessity to keep the water temperature below 160°F at 20,000 feet, otherwise the coolant all boiled away. Curtiss engineers, encouraged by the Army and Navy, got the idea of using ethylene glycol instead of water in the radiator and thereby allow the radiator temperatures to run up to 300°F. It was expected that they could then appreciably increase the power output at altitude to at least the standard sea-level rating of 550 bhp. Although the V-1570 was an excellent 550 bhp water-cooled engine, it was a complete failure when it was attempted to use ethylene glycol as a high temperature coolant.[46]

Another advantage attributed to Ethylene Glycol as a coolant was an anticipated reduction in radiator size because it could remain a liquid at high temperatures. The Air Corps had sponsored a number of tests using this coolant in the latest models of the Curtiss-Wright 650 bhp V-1570 Conqueror. The Conqueror cylinder featured a closed end steel cylinder liner in mechanical contact with the aluminum head. This was not entirely successful at the intended 300°F coolant outlet temperature because differences in thermal expansion would cause a gap and result in hot spots. Furthermore, the type of joints used to isolate the cooling passages were designed for water cooling and were not able to deal with glycol's tendency to leak through even the smallest passages.

The Air Corps believed that for their future bombers to operate at stratosphere altitudes it would be necessary to use turbosupercharged liquid-cooled engines using high temperature coolant. This was because air-cooled cylinder cooling limitations prevented them from delivering sufficient specific power under the conditions encountered in operating at such altitudes. Although they were constrained by monetary limits caused by the Depression, they still followed closely the Navy V-1710 program, and soon found funds sufficient to obtain an engine of their own.

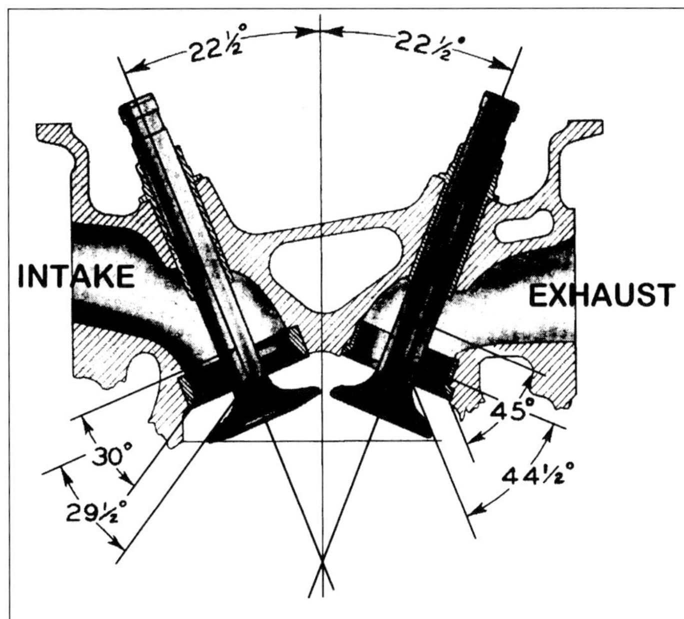

Allison developed a pentagon shaped combustion chamber with two inlet and two exhaust valves, the combination giving excellent flow conditions for gasses in and out of the cylinder. Note the smooth and direct flow paths for gas and the large coolant passages integral with the head. Generous amounts of coolant were directed around the hot exhaust valves. All of the valves incorporated liquid metal (sodium) to aid in transferring heat from the very hot valve heads into the shank, and from there into the aluminum head casting. (Allison)

Cylinder Block Details[47]

Norm Gilman began the design of what became the V-1710 with a critical view of previous attempts at high temperature liquid cooling. His approach incorporated features that were successfully retained through the life of the V-1710. Figure 16-4 shows several of these features to advantage, and all are discussed below.

Critical to the efficiency and ability to make power in an engine is its capability to "breath." At very slow engine speeds an engine should be capable of drawing in its full swept volume, i.e. displacement, on ever other revolution. As rpm increases this becomes much more difficult, and engines without superchargers are hard pressed to aspirate even 80 percent of their capacity. Consequently, power is reduced. A supercharger overcomes this effect by pumping high density mixture into the cylinder, but even so, it is critical that the pathways be as free and open as possible. Any unnecessary restriction or change in direction of flow reduces the ability of the engine to produce power. The V-1710 has a particularly good induction system that not only is supercharged, but takes advantage of the pulsing nature of the flow to pack additional mixture into the cylinder before the compression stroke begins. These same considerations apply to the exhaust side of the cylinder as well, and here again Allison provided a smooth and direct path for the exiting of the hot exhaust.

The Allison cylinder is of the "wet sleeve" type, meaning that the piston rides in a thin steel cylinder liner (barrel), the exterior of which is in direct contact and wetted by the coolant. To enhance heat transfer, Gilman utilized a ribbed exterior of the liner and filled an adjacent thin metal sleeve to control flow and insure that the coolant flow adjacent to the liner was turbulent for maximum cooling effectiveness. Early development testing in the XV-1710-B showed that the liners were too thin to resist the naturally occurring side loads developed by the pistons during the "thrust" portion of the power stroke, and these had to be made slightly heavier. Likewise, Type Testing of the V-1710-C found that the thin sleeve used to control coolant flow was prone to cracking in high time engines, so it was changed from aluminum to stainless steel.

Figure 16-4 Note the cooling paths as used on all engines from the V-1710-E4 on. The lower manifold orifices in the cylinder block jacket metered coolant to each cylinder liner, where it then passed into the head to flow out the port on the front of the head. A stainless steel sleeve separated the metered flow from the stagnate coolant filling the jacket. The upper rear supply port had been blanked off on the C-3 through C-15, E-2 and E-5 engines. On subsequent engines it was used to provide additional coolant directly into the head and around the valves, necessary to allow WER ratings of the engines going into combat. The oil supply to the valve gear also provided some cooling. (Allison)

The all important gas-tight seal between the cylinder liner and the aluminum head/cylinder block was accomplished by an interference shrink fit between the liner and the head. The lower joint between the cylinder barrel and the block went through considerable evolution during the life of the engine. At various times, copper and/or steel gaskets or washers were used between the mating surfaces, though in the late model engines this was eliminated in favor of the simplicity of a direct metal to metal joint. The whole assembly was held together, and made leaktight by the force developed when the compression nuts on the bottom of each of the six sleeves was torqued to the required 2,100 pound-feet. This action generated some 20 tons of force on each joint.

The aluminum cylinder blocks were cast with a wide corrugated cross-section, the purpose being to accommodate the stresses induced when the steel barrel expanded in the aluminum block. Note that the explosion forces generated within the cylinders were not carried by the liners or the block. Rather, long studs connected the supporting crankcase to the heads, and when torqued, carried these loads down into the crankcase.

The coolant flow path received a considerable amount of engineering design and attention during the development life of the engine. The pre-Hazen engines used cylinder blocks having two coolant inlets, the main one connecting high on the rear of the block, just under the distributor, and a second smaller supply connecting to the supply manifold integral with the coolant jacket at the base of the block. This manifold was drilled and fitted with metering orifices which insured the proper quantity of coolant was directed onto each cylinder liner. Photos of these early engines clearly show how the coolant supply pipes, coming from the below engine coolant pump, form a "y" and split the flow to the two inlets. These blocks also featured the coolant outlet on the front face of the block.

The Hazen blocks, first used on the XV-1710-7(C3), blocked off the rear upper supply port while the lower manifold to each cylinder was enlarged to distribute all of the coolant from the pump. One advantage of this arrangement was that it was now easy to use the engine in either a tractor or pusher installation, for either of the upper front, or rear coolant connections could be used as the outlet. The installation determining which was the higher, and that port being selected for connection to the radiator while the other was blanked off. Of course, it was necessary to reverse the order of the metering orifices in the supply manifold to assure each cylinder received its share of coolant when the engine was installed in a pusher like the Bell XFM-1. The single supply to each block was the arrangement used when the V-1710-33(C15) engines were manufactured.

This arrangement was quite adequate for the 1150 bhp rated engines for which Hazen intended it, but his genius, or foresight was further appreciated when it became necessary to increase cooling for the much higher ratings later approved for the V-1710-E/F engines. All that was needed was to increase the capacity of the coolant pump and reinstitute the "y" in the supply pipe, thereby reconnecting the upper inlet and dramatically improving the cooling of the cylinder head. As originally developed the V-1710-17(E2) and V-1710-37(E5) did not have the dual coolant connections, though when the E-5 went through the Modernization program the "y" fitting was incorporated. This same upgrade was incorporated into "C" engines at overhaul. This marked the transition point and all subsequent V-1710-E/F/G engines included the feature, though it is believed that the later models utilized a considerably enlarged upper connection, thereby allowing an increase in the total coolant flow.

As the coolant circulated up past the cylinder barrel a fairly uniform temperature was maintained, insuring good conditions for cooling the pis-

Rear-quarter of V-1710-33(C15), AEC #155, clearly shows the unique casting supporting the 3-barrel carburetor and fitting to the supercharger inlet. Though difficult to clearly make out, note that the coolant fitting on the rear of the cylinder head was not fitted when this model was originally built. While such modernization may not be historically correct, it is entirely proper in the interest of safe flight. All V-1710's were continually given the latest components during overhaul. That they were retro-fittable is a tribute to Allison. Modern builders assure greater reliability by incorporating later model parts and features. (Author)

ton and rings. On the engines with the "y" connection a separate supply was provided direct to the back end of the head. Flow entered at the rear of the block and effectively provided a reservoir of coolant surrounding each of the valve stems and covering the combustion chambers. Coolant exiting the liner sleeve region also passed through these passages, all leading to a single discharge port at the front of the block. Most of the heating of the coolant occurred in the head, where passageways directed it close to the valve seats and valve stems. Production V-1710's had all four valves equipped with internal sodium coolant, the result being a considerable amount of heat to be removed from the head, though the valves operated in the severe conditions with little duress. In all, a very clean arrangement and one that saw few leaks and little thermal distress in service.

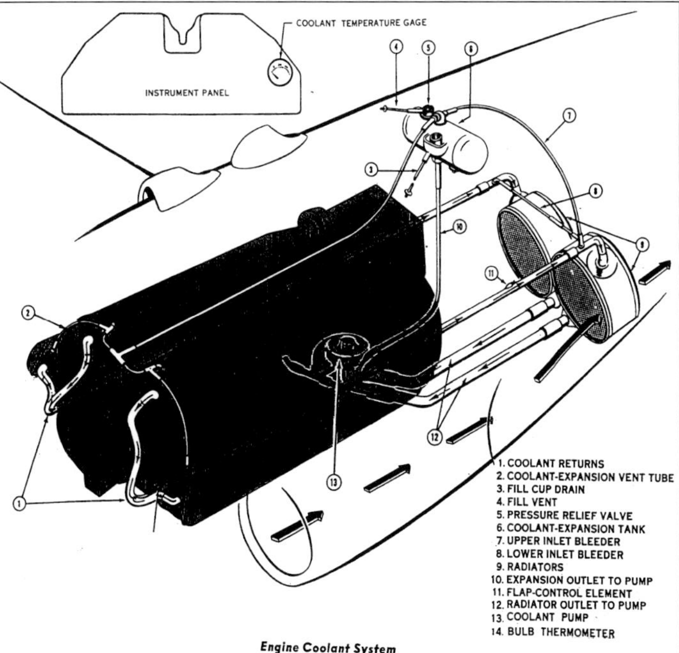

Typical of a wide range of cooling systems for the V-1710 was that used for the V-3420-B10 in the large P-75A. It was compact and well designed to insure adequate cooling under all operating conditions. It addition it incorporated features for thrust recovery in the heated air. (USAAF)

When ready to install, the cylinder block was a leak-tight assembly that could be completely leak tested to insure that it would remain so during operation.

Engine Cooling Systems

Early in the development of liquid-cooled engines it was recognized that water was the preferred coolant, due to its very high specific heat and latent heat of evaporation. However, when coupled with the demand for low radiator drag, i.e. a small radiator cross-sectional area, water at atmospheric pressure was out of the question. Many attempts were made at using some form of evaporative cooling, but this was quite troublesome and was found to be totally impractical in a military aircraft. The feature was demonstrated by some very innovative systems which were developed using a large portion of the surface of the aircraft constructed to function as a steam condenser. An excellent example being the upper wing-surface radiators fitted to the Curtiss V-1400 powered Curtiss R3C-2 Schneider Cup racers of 1925.

In view of these limitations, in the 1920s Wright Field pursued use of chemical coolants having high boiling points, and ethylene glycol was found to be particularly ideal for the purpose. Not only could it remove large quantities of heat by operating at high temperatures, but it operated at comparatively low pressure, while also possessing inherent anti-freeze capabilities. Consequently, the Allison V-1710 and the contemporary Rolls-Royce Merlin, were both developed to operate on nearly pure ethylene glycol, known by the trade name of "Prestone" in the U.S. Should the coolant be diluted with water by more than about 5 percent, its cooling capabilities at low pressure were greatly diminished.

From this work the Army had originally intended to operate its Prestone cooled engines at 300 °F. Initial operation of the XV-1710-1, which was designed to accommodate both Prestone and the 300 °F temperature, resulted in an interesting finding. The high temperature was causing the amount of heat absorbed by the engine oil to increase to the point where the oil cooler would have to be as large as the radiator. When Allison took everything into consideration it was found that 250 degrees was a more realistic temperature, and all subsequent V-1710's operated at this temperature. Even with its advantages, Prestone cooling was still not as effective as water cooling, and it still caused numerous problems within the entire aircraft cooling system due to its tendency to "creep" through joints and other small passages. By comparison, water was easier to control.

With this knowledge, Rolls-Royce in 1938-39 began to advocate use of pressurized water cooling.[48] In such a system the pressure forces the water to remain a liquid, preventing boiling and steam formation, even at temperatures in the 250 to 300 degree range. They proposed adding 30 percent ethylene glycol to the mixture, but only to function as an anti-freeze. The solution remains liquid down to -15 °C.

Rolls-Royce had originally switched their Merlin to chemical cooling after finding out about its advantages, and its use in the Allison during a visit to Wright Field in April 1935. They were then the first to switch to pressurized water cooling for production installations, with Allison following in 1942. Allison also specified a similar 70 percent water, 30 percent ethylene glycol mixture, and both manufactures requiring something less than one percent corrosion inhibitors in the mixture as well. With the institution of pressurized water cooling it was again possible to allow in-

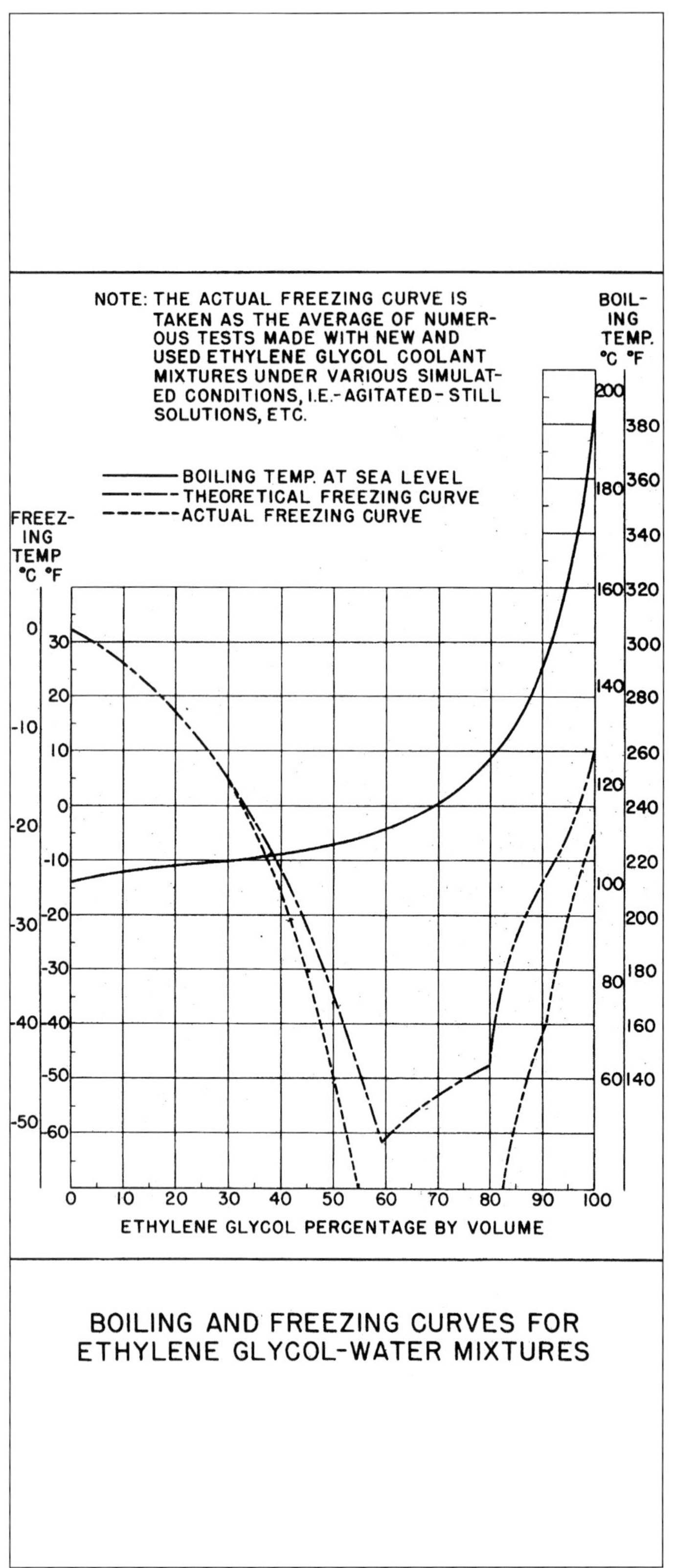

Both the boiling and freezing points are dramatically effected by the amount of water in the mixture. (Allison)

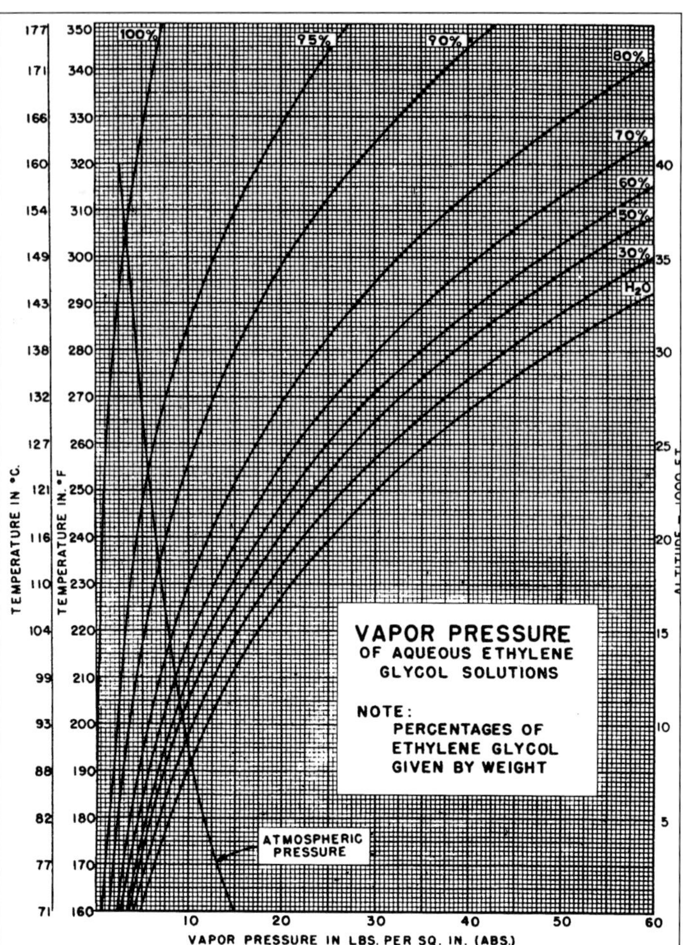

Pressure needed to keep from boiling varies with concentration of water and is strongly affected by altitude. (Allison)

creasing the power output of the V-1710, as the water was particularly effective in increasing cooling of the hot spots in the cylinder heads around the valves. This can be seen in tables of engine model versus type of cooling.

Another difference in the cooling concepts between Allison and Rolls-Royce was Allison's practice of reliance on liquid-phase convective heat transfer while Rolls used predominantly vapor-phase cooling, meaning that local boiling would be allowed at hot-spots. The result was believed to provide a lighter and perhaps more effective cooling system for the Rolls engines.[49] Changing from one type of coolant to the other was not a simple matter of draining one and filling with the other. The plumbing itself had to be changed to accommodate the relative amounts of vapor being produced and the proper location of system components such as the radiator, expansion tank, vapor separator and connections to the coolant pump.

Collaboration of the effectiveness of Allison's switch to pressurized water in its coolant system, was provided by the NACA-LMAL laboratory at Langley Field, Virginia, in 1944.[50] In the course of their work on investigating the cooling requirements of the V-3420, they found that in general, the reduction in engine-head temperatures in changing from 100 percent glycol, or 80-20 glycol-water mixtures to either 30-70 or 0-100 Prestone-water mixtures were preferable in the V-1710.

Allison identified two locations within the engine for which cooling was of critical importance. These were the piston crown and between the

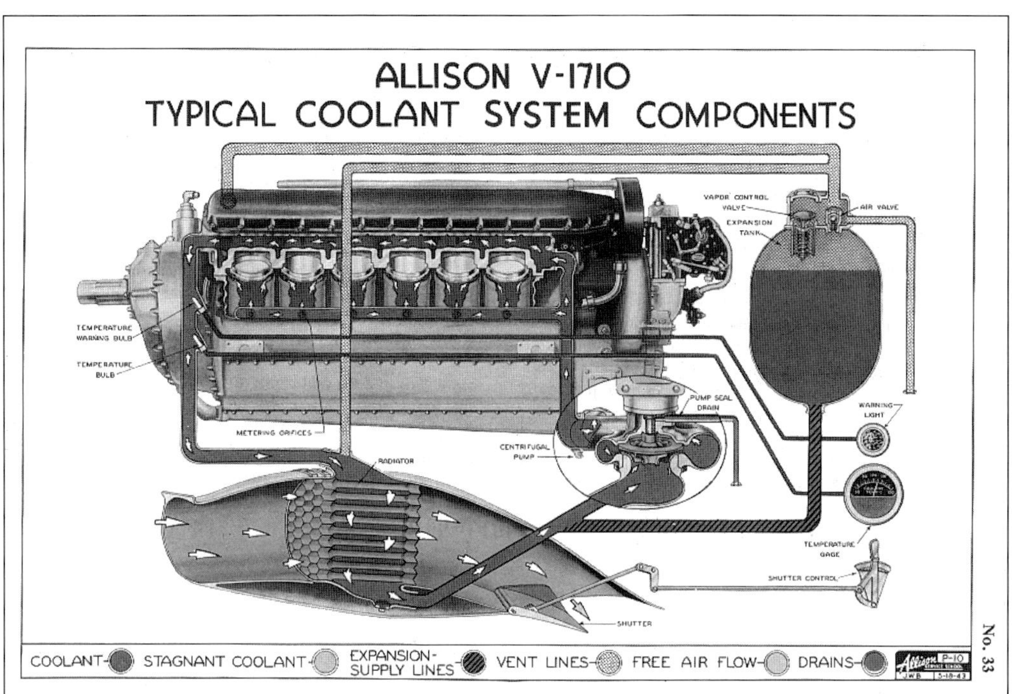

This is the typical cooling system when nearly pure ethylene glycol was the coolant. When the 30/70 glycol/water mixture was introduced the various manufacturers made many detail changes in this system. (TO 30-5A-1)

exhaust valves in the cylinder head. They successfully limited the piston-crown temperature by controlling the amount of coolant provided to the cylinder barrels. The switch to water based mixtures was effective for improved cooling of the cylinder head, as some vapor production was then allowed. The nature of vapor production in pressurized water is that the temperature remains constant as long as liquid is present, a very desirable feature within the heads.

With the one-time "standard" 97-3 glycol-water mixture, between the exhaust valve temperatures in the V-1710 were about 512°F.[51] With the change to 30-70 mixtures the this temperature was reduced to about 456°F. Pure water, the best coolant, only offered a reduction to about 444°F, all of this using 35 psig at the cylinder block outlet.[52] It should be clear why the change in coolant mixtures was adopted, for it provided a wider margin to detonation for engines being operated at very high Military and WER power ratings.

In 1941 Allison ran a series of tests on a "Dummy" V-1710-E/F driven by a dynamometer, and instrumented to determine the effects of using different designs of main bearings. At 3000 rpm it was found that the crankshaft bearings consumed about 24 hp in friction, resulting in heat that was absorbed by the lubricating oil. When combined with a complete set of connecting rods and pistons, the friction heat into the oil was about 71 hp and the main bearings were running at about 258 °F. As a part of the test oil pressure was increased by 10 psi, which resulted in decreasing bearing temperatures by only 3 degrees.[53] All of this needs to be viewed in light of the design heat load on the early V-1710-E/F models of about 155 hp. Clearly the oil was being used to provide a portion of the overall cooling needed within the engine as the total heat load to the oil was twice that due to friction of the mechanical parts. The bulk of the cooling provided by the oil would be that due to the oil spraying onto the cylinder walls.

As an example of the cooling requirements of one of these engines, the V-1710-81(F20R). When operating at take-off conditions of 1200 bhp and 3000 rpm, was specified to have no more than 430 hp absorbed by the 250 gpm of coolant, and the concurrent heat removed by the engine oil was to be less than 155 hp into 160 pounds per minute oil flow.[54] There are no "rules" about the split of cooling to be accomplished between the coolant and oil. Both systems remove a considerable amount of waste heat from the engine, and are critical to satisfactory performance. The following table provides some insight to the variation of cooling loads, which are dominated by the engine power level. Recognize that the values in these tables come from the various model specifications, and represent the maximums, that is, the values to which the aircraft installation engineer would design the cooling systems. In actual operation, the various temperature rises and cooling loads were usually somewhat less, say 15-30 percent, even at the same power levels.

Of particular interest is the considerable difference between how heat was removed from the V-1710 and Merlin engines. Clearly, Rolls-Royce committed more of the cooling load to the coolant, while Allison allocated

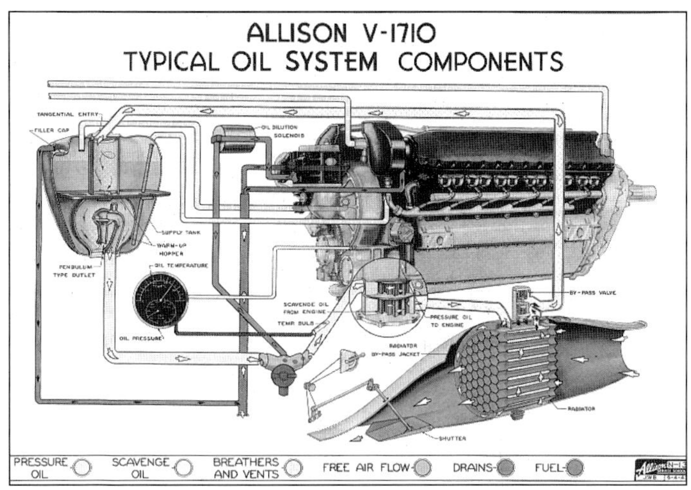

The engine oil supply or "pressure" pump delivered clean and cooled oil to the engine. Hot used oil collected in sumps in the lower crankcase. Scavenge pumps in the front and rear of the crankcase pumped the hot oil through the oil coolers and into the reservoir. (TO 30-5A-1)

Table 16-11 Cooling Loads, in Terms of Propeller Power						
Engine Model	Rating, bhp to Prop/rpm/Alt	Total Cooling Req'd, hp	Heat to Coolant/Oil, percent	Aux Stage Heat to Oil, hp	Coolant	Comments
V-1710- 11(C8)	1150/2950/T.O.	485	84.5/15.5	None	97% Glycol	Turbosupercharged
V-1710- 33(C15)	1090/3000/13,200	510	80.4/19.6	None	97% Glycol	Altitude Rated
V-1710- 67(E8)	1150/3000/T.O.	570	75.4/24.6	None	97% Glycol	Altitude Rated
V-1710- 81(F20R)	1200/3000/T.O.	585	73.5/26.5	None	30/70% Water	Altitude Rated
V-1710-109(E22)	1425/3000/T.O.	645	73.6/26.4	55	30/70% Water	No Aftercooler
V-1710-111(F30R)	1500/3000/T.O.	630	74.6/25.4	None	30/70% Water	Turbosupercharged
V-1710-119(F32R)	1500/3200/T.O.	725	75.9/24.1	55	30/70% Water	120 hp to Aftercooler
V-1710-143(G6R)	1600/3200/S.L.	792	76.4/23.6	117	30/70% Water	No Aftercooler
V-1710-143(G6R)	1100/2700/S.L.	575	76.5/23.5	48	30/70% Water	No Aftercooler
V-1650-1 Merlin	1300/3000/T.O.	699	89.3/10.7	None	30/70% Water	Low-Gear, Spec AM 1035
V-1650-3 Merlin	1400/3000/T.O.	777	86.4/13.6	None	30/70% Water	160 hp to Aftercooler

Cooling loads are expressed in horsepower, to convert to BTU/minute multiply by 42.42. Specific heat of oil is taken as 0.504 BTU/LBM-F. T.O. is Takeoff, S.L. is Sea Level.

more to the oil system. This was accomplished by the considerably different rates of coolant and oil flow selected by the two different designers. The Merlin used less coolant flow, nominally 60 percent of the Allison, and as a result saw nearly twice the temperature rise as in the V-1710. We can speculate that the reason the V-1710 had a fairly large amount of heat absorbed by its oil was because of the larger quantity of oil being supplied. The two-stage Merlin and the later V-1710's have similar temperature rise in their oil, which should be expected as the heat was coming from contact with the cylinder walls. The additional heat to the V-1710 oil resulted from the greater flow, this "extra" oil being thrown about in the crankcase where it not only lubricated the cylinder walls, but absorbed a considerable amount of heat in the process.

Another important observation is the amount of cooling required of the Allison Auxiliary Stage Supercharger oil. This was largely a result of the inefficiency of the hydraulic coupling when operating at altitudes below the "critical" altitude, which was the "no-slip" condition. Still the single-stage V-1710's generated about 100 bhp less total cooling load than the equivalent single-stage Merlin. The direct reason for such being the 10 percent higher compression ratio than the Merlin.

One most interesting experiments, which illustrated the differences between the liquid/vapor phase approaches to cooling a V-12 aircraft engine, was demonstrated in 1941 by Curtiss in a comparison of similar P-40's, one a P-40F equipped with a V-1650-1 Merlin, the other with the V-1710 Allison, and both cooled by the then standard, nearly pure Prestone.[55] On a cool day, the airplanes were lashed to the ground so that powers in the cruising range could be drawn from the engines while coolant was intermittently drained from the systems. The Allison cooling system had a capacity of 15 gallons. As coolant was drained from the system the temperatures stabilized at progressively higher values, starting in the neighborhood of 250 °F, and when 15 gallons had been drained, the temperatures were stable around 300 °F and the test was discontinued. With the Roll-Royce powered plane, which had a capacity of 12 gallons, the first few

Table 16-12 Cooling System Parameters							
Engine Model	Rating, bhp to Prop/rpm/Alt,ft	Oil Flow, LBM/min	Max Oil Temp, F	Oil Temp Rise, F	Max Coolant Flow, gpm Temp, F	Max Coolant F	Coolant Temp Rise,
V-1710- 11(C8)	1150/2950/T.O.	110	185	57.5	NA	250	NA
V-1710- 33(C15)	1090/3000/13,200	125	185	67.4	215	250	14.7
V-1710- 67(E8)	1150/3000/T.O.	NA	NA	NA	NA	NA	NA
V-1710- 81(F20R)	1200/3000/T.O.	160	185	81.6	250	250	13.3
V-1710-109(E22)	1425/3000/T.O.	170	203	84.4	265	250	13.8
V-1710-111(F30R)	1500/3000/T.O.	180	203	74.9	265	250	13.7
V-1710-119(F32R)	1500/3200/T.O.	175	203	84.1	225	250	13.9
V-1710-143(G6R)	1600/3200/S.L.	202	203	78.0	180	250	19.0
V-1710-143(G6R)	1100/2700/S.L.	NA	NA	NA	NA	250	NA
V-1650-1 Merlin	1300/3000/T.O.	115	194	54.8	150	250	25.1
V-1650-3 Merlin	1400/3000/T.O.	120	194	74.2	150	250	25.4

gallons of coolant drained made no difference in the coolant temperatures until a point was reached where draining about two additional quarts of coolant caused the temperatures to rise so fast that the engine operator was unable to stop the engine before it failed. The experience demonstrated that decreasing the flow of coolant in an engine causes little change except an increase in the amount of vapor-phase cooling occurring until the system ultimately vapor-locks and coolant flow stops, with catastrophic results. The well known vulnerability of liquid-cooled aircraft to otherwise minor combat damage is well documented. This experiment suggests that the Allison approach provided a margin of safety over that typical of a Merlin powered installation.

When Allison introduced pressurized water cooling in 1942, with 30 percent Prestone as anti-freeze, the nominal operating pressure for these systems was maintained by the pressure cap within the range of 16 to 20 psig.[56] At these pressures water would remain a liquid, and not boil, at up to 260-265 °F, providing protection up to an altitude of 26,000 feet.

The early systems specified a minimum of 94 percent glycol, giving a boiling point of about 335 °F at sea-level, and operating at a pressure of 3-5 psi above atmospheric under all flight conditions. Cooling systems in turbosupercharged airplanes had another problem in that the glycol would boil at atmospheric pressure if above 25,000 feet. It was for this reason that the desired level of glycol was established at 97 percent in the P-38, determined by a sea-level boiling point of approximately 365°F.[57]

When the post-war V-1710-G6 engines were qualified for operation in the F-82E/F airplanes on duty in Alaska, the engine was able to operate on a special Winter mixture of 70 percent Ethylene Glycol and 30 percent water, although the maximum allowable outlet temperature was reduced to 230 degrees.[58] This was necessary because of the extremely low temperatures that could be incurred. Otherwise these engines used the normal 30/70 mixture.

Engine Efficiency

Common practice expresses the efficiency of a mechanical drive engine in terms of "pounds of fuel used per hour per brake-horsepower." Known as Brake Specific Horsepower, the more efficient the engine, the less fuel is consumed per horsepower per hour. That is, "less" is "better."

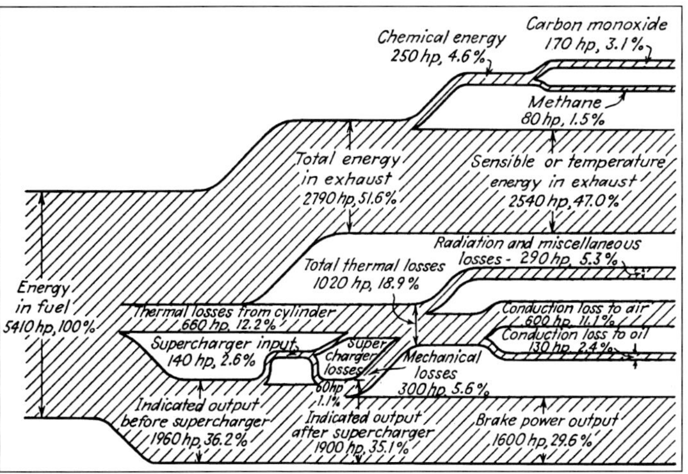

This graphic traces the use and waste of energy contained within the fuel. The calculations were originally done for an air-cooled engine, a liquid-cooled engine has all of the same concerns, only the proportions of heat in each path would be somewhat different. (Adapted from Liston, Joseph, 1953, Figure 4-65, p.226, which obtained the diagram from Engine Compounding for Power and Efficiency, Pierce and Walsh, SAE Quarterly Transactions, Vol.2, No.2, April 1948, p.317.)

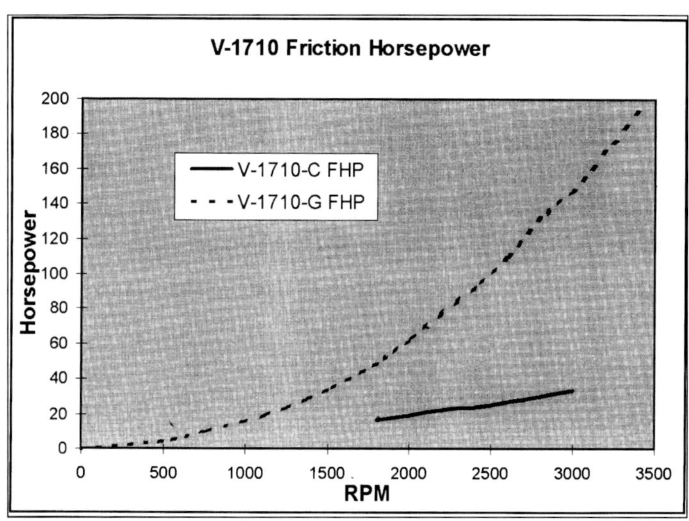

V-1710-C and V-1710-G had quite different levels of internal friction, due largely to the different supercharger gear ratios, wider gears, stiffer valve springs, larger pumps and stronger piston rings in the "G". Measurements made without blades on supercharger.

The fuel consumed must provide the desired power to the propeller and in addition, overcome friction and produce enough power to drive all of the engine accessories, under the conditions of engine speed and altitude. Consequently the fuel is actually producing what is defined as "indicated" horsepower, which is the power that would exist if the engine were friction free and the accessories, such as the supercharger and generator, did not have to be driven. The available useful power is then the "brake" horsepower. For an engine such as the V-1710, at 3400 rpm, the friction horsepower can approach 200 hp, the supercharger power being dependent upon the engine supercharger gear ratio, and the position of the throttle. At maximum throttle with 9.6:1 gears, the supercharger alone can consume well over 200 hp. On the other hand, full throttle on an engine with 6.44:1 gears requires only about 50 hp to drive the supercharger. Since this power derives from fuel which does not directly provide power to the propeller, it is obvious that the engine with the lower supercharger gears will be the more efficient, even at the same net horsepower to the propeller.

Other factors effect efficiency as well. Of particular importance is the turbosupercharger, if one is used. There is an adverse effect of the turbo in that it causes "back pressure" on the engine. This means that on the normal "exhaust" stroke, the piston must actually "pump" or "push" much of the exhaust gas out of cylinder. This takes power away from the propeller. The positive advantage of the turbosupercharger is that its' compressor delivers high pressure air to the integral engine-stage supercharger. This in turn allows a lower gear ratio to be used for the engine stage, thus making more power available to the propeller, and increases efficiency. In the early development of the General Electric turbosupercharger it was realized that the best "balance" was for the turbosupercharger discharge pressure to be similar to the backpressure required to drive the turbo. Consequently, most GE turbos were designed to deliver "sea-level" air pressure to the engine carburetor, hence the engine intake and exhaust cycles never knew that they were not operating under their normal sea-level conditions. Since nearly half of the energy supplied by the fuel exits the engine as heat in the exhaust gases, the turbo was able to perform its function without adversely effecting the engine performance or efficiency. Its' benefit was that the turbo allowed use of less power by the engine-stage supercharger. These effects are quite apparent in the comparison of the efficiencies of "altitude" and "sea-level" engines.

Table 16-13
Part Throttle Engine Performance

	Dis-placement, cu.in	S/C Gear Ratio	Cruise Power, bhp	% of Takeoff Power	Fuel Burn, #/bhp/hr	Fuel Efficiency, %
Liberty 12-A	1650	None	300/1200/SL	71.4	0.525	25.1
Voyager/Teledyne 200	200	None	52/2220/10,600	67.5	0.393	33.5
V-1650-1 Merlin, 1-Stage	1650	8.15-Lo	648/1855/	49.8	0.450	29.3
V-1650-7 Merlin, 2-stage	1650	7.35-Hi	520/2000/16,500	34.9	0.554	23.8
V-1710-C15	1710	8.77	715/2280/10,000	68.8	0.493	26.7
V-1710-F2, w/Turbo	1710	6.44	670/2280/25,000	58.3	0.377	35.6
V-1710-F20R	1710	9.60	750/2300/14,000	62.5	0.579	22.8
V-1710-F30, w/Turbo	1710	8.10	795/2300/37,800	53.0	0.465	28.4
V-1710-G6, 2-Stage	1710	7.48	700/2000/20,500	43.8	0.430	30.7

One of the reasons for the initial Air Corps interest in the liquid-cooled Allison V-1710 was its inherent capacity for high specific power and its adaptability for turbosupercharging. This was in comparison to the alternative air-cooled radial engines of the 1930s that required an extra measure of fuel to be used as an "internal coolant", particularly when at high power levels. This of course adversely effects their fuel efficiency.

During the development of the V-1710-C Allison elected to increase the engines' mechanical compression ratio from 6.00:1 to 6.65:1 for the expressed purpose of increasing the engine efficiency. By this change alone, the theoretical efficiency of the engine was increased by about 4.2 percent, a non-trivial amount when considering long-range cruise applications. The 6.65:1 compression ratio was used by all subsequent production V-1710's, except for the late model two-stage V-1710-G's which used 6.00:1 or 6.50:1 pistons as a way to minimize the likelihood of detonation during high manifold pressure, high power, operation.

The highest efficiency in a given reciprocating engines occurs when the engine is operating at part load, as internal friction and supercharger loads are minimized and the carburetor can be adjusted for lean running. If it is a two-stage installation then it also helps if there is a variable speed geared or turbo-supercharger ahead of the engine stage.

When throttled back for cruise conditions the V-1710 was quite efficient. It was this feature, coupled with proper operating technique, which gave the P-38 it's long-range capability. As detailed in the Allison engine Technical Orders, and later taught by Charles Lindbergh to airmen in the South Pacific, proper cruise control involved operating at reduced engine speed, but with the engine throttle "wide-open" (thereby minimizing pressure losses in the induction system), and maintaining desired engine manifold pressure and power with the turbo. This shifted the burden of compressing the air to the turbo which had plenty of energy available. As a side benefit the turbo was operating at near its maximum rpm, so there was little "turbo-lag" in getting it to accelerate to provide extra power in an emergency. All the pilot had to do to get combat power was to increase engine rpm with the propeller control. For very long cruise conditions, some carburetors could be set to "lean", which further reduced the amount of fuel being consumed, although at the "cost" of increasing the temperature of the exhaust gasses. This was not a limiting condition for the V-1710 with its sodium cooled valves.

High Output Development of the V-1710

The "1000 bhp" phase of the V-1710 effectively ended on September 3, 1936 when Allison received approval from the Air Corps for that rating at both normal and maximum conditions. As a result of this approval the Air Corps released the remaining engines then under contract for fabrication, the official notification being received on September 22, 1936.[59] Not one to sit idyll by, Ron Hazen quickly prepared a memo to Norm Gilman, Allison General Manager, outlining a plan for future development of the V-1710. Gilman received and approved the program on September 18, 1936. It is with the authorization of this program that the development of the practical V-1710 got underway. While 1000 bhp was an aggressive goal in 1930, by 1936 Lt. Ben Kelsey of the Air Corps Fighter Projects Office was seeking at least 1600 bhp for the advanced aircraft that he had in mind.[60]

In Hazen's program, he acknowledges that the design modifications necessary for the pusher engine with extension shaft to the propeller (the V-1710-D1) had been completed and five such engines released for production. Furthermore, design work for an 1150 bhp engine able to meet 1937 Type Specifications (this would be the V-1710-C7) was practically complete and development work in support of that goal would be complete by the end of October 1936. Although he believed that the then current staffing level was practically a minimum for carrying on any program on the V-1710, he proposed that the following program should be accomplished as rapidly as possible, but without increasing personnel.

1. High Compression Ratio Pistons. Having already achieved the 1150 bhp level, he thought it was "more important to obtain better fuel economy in the near future than to attempt higher power outputs." He proposed developing pistons with 7.5 and 8.0:1 compression ratios, to be tested using 100 octane fuel.
2. Cylinder Head Design. Based upon his experience with the U-engine and other projects, it appeared probable that an appreciable gain in both power and fuel economy was possible by modification of the combustion chamber shape and sparkplug location. To this end design studies and single cylinder tests were directed.
3. Reduction Gear. Although the present reduction gear had gone through the Type Test, he believed "it is ... perhaps the most questionable item in the V-1710 design." He proposed, "studying other types with a view of obtaining a lighter and cheaper design, and one with less friction loss and oil heating."
4. Accessory Housing. The design had gone through a number of previous revisions and was facing still more. Hazen wanted a new accessory arrangement that not only would meet future requirements, but that would facilitate reversibility (LH/RH operation), be simpler to machine and with reduced cost of production.
5. Steel Cylinder Construction. Knowing that still higher power was going to be demanded from the V-1710, and with the increasing difficulty of obtaining reliable aluminum castings able to meet these demands, he proposed that an all-steel cylinder and head be devel-

Table 16-14
Growth in Specific Power Ratings

bhp	bhp/cu.in.	Weight, lbm	bhp/lbm	Fuel Grade	Year	Engine Model
750	0.439	1035	0.725	87	1931	GV-1710-A(A1) Build #2
1000	0.585	1265	0.803	92	1937	V-1710-7(C4)
1090	0.636	1325	0.813	100	1939	V-1710-33(C15)
1150	0.673	1305	0.881	100	1940	V-1710-27/29(F2R/L)
1325	0.775	1345	0.985	100/125	1941	V-1710-49/53(F5R/L)
1425	0.833	1345	1.059	100/125	1942	V-1710-51/55(F10R/L)
1580 WER	0.924-Dry	1435	1.101	100/125	1941	V-1710-63(E6)
1600 WER	0.936-Dry	1350	1.185	100/130	1941	V-1710-89/91(F17R/L)
1600 WER	0.936-Dry	1395	1.147	100/130	1943	V-1710-111/113(F30R/L)
1900 WER	1.111-Wet	1811	1.049	100/130	1945	V-1710-129(E23B)*
2980 WER	1.743-Wet	1983	1.503	115/145	1944	V-1710-127(E27)*
2250 WER	1.316-Wet	1595	1.411	115/145	1946	V-1710-143/145(G6R/L)*

* 2-stage supercharged, WER ratings with ADI. V-1710-129 high weight due to remote installation and 4000 bhp rated reduction gear box.

oped. He believed this was practical based upon Allison's experience with the brazed steel U-cylinder and related GM experience. In the event, improved foundry practices solved the aluminum casting problem and little effort was expended in the pursuit of an all-steel cylinder for the V-1710.

Hazen's program provided the foundation for the design of the V-3420 and the 1150 bhp V-1710-E/F models that evolved from it.

By the fall of 1940, the 1150 bhp V-1710-F3R, E-4 and F-7R models were the primary models being developed by Allison, though the Chief Designer and his Project Engineers had been directed to initiate a comprehensive review of the basic E/F engine in preparation for obtaining a sea-level rating of 1500 bhp.[61] This was done, resulting in an assessment of the engines' design basis for rating at 1500/3000/SL/61.5 inHgA. The corresponding Altitude Rating, with the assumed 8.80:1 supercharger gears, was still 1150/3000/12,000/42.0 inHgA, limited by supercharger parameters, not engine strength.[62] Significantly, Allison determined that it was desirable to go to a 12 counter-weight crankshaft so that further increases in rpm and power could be achieved without producing additional loads on the main bearings. While the new crankshaft would be heavier, the resulting reduction in bearing loads allowed for a lighter overall engine than would have otherwise been possible at the higher power levels.

Suggested components for possible improvement in the 1940 review were:[63]

- Cylinder heads, to improve strength of the top deck as a beam, along with the transfer of loads to the hold down studs. In addition, coolant flow distribution within the coolant jacket was to be investigated, as well as the gas flow through and shape of the ports, intake inserts, camshaft timing and the use of salt cooling for the intake valves.
- A strengthened crankshaft, with more generous fillets, nitriding, shot blasting, and the improved balance available with increased counterweighting.
- Connecting rods, pistons, and the crankcase were to be similarly strengthened and refined in detail.
- The efficiency of the supercharger was to be generally investigated and improved along with the development of a two-stage unit and improved controls for the supercharger, particularly in two-stage and turbo installations.

By 1942 Allison was able to show dramatic improvements in the Military and War Emergency ratings of the various Model Tested V-1710s.[64] They continued to develop the engine, and even more power was available with the introduction of War Emergency Ratings, a rating authorized only after a 7-1/2 hour run at the desired level, a very severe test.

As is apparent from Table 16-14, a key to these improvements in specific ratings of the V-1710 during the war was the availability of much improved fuels, while detail improvements throughout the engine provided mechanical reliability. In addition, the later two-stage models adapted ADI injection which allowed significantly increased manifold pressure in engines which were not otherwise intercooled.

Most impressive is the increase in power, without a proportional increase in engine weight. To the credit of Allison and their subcontractors, they were able to significantly improve the strength and durability of engine components while increasing production. The V-1710 was a comparatively light weight engine, but it matched or beat the competition on the basis of specific performance.

The High Powered V-1710 Test Project

On August 29, 1939, three days before the action by Germany that started WWII, General Arnold directed engineering investigations and tests be conducted for the purpose of increasing the power output of the Allison engine. The first priority was to increase the power of the engine in the P-40 as soon as practicable. The program was to be expedited, with the first action to be a test of a V-1710 at 50 bhp increments until the maximum output or failure was encountered. The engine was to be similar to the V-1710-33, but with 8.0:1 supercharger gears to limit detonation when testing at sea-level.[65]

In response, the Wright Field Power Plant Laboratory ran a series of tests to determine the mechanical power output capability of the altitude rated V-1710. This was in the period of September 1939 through January 1940. These tests may well have been motivated by a demand from Bell Aircraft for 1350 bhp which was required if their XP-39B was to demonstrate the contract specified 400 mph. Removal of the turbo from the XP-

39 and its redevelopment as the XP-39B had been ordered just prior to this testing.[66]

The engine used was a V-1710-21 (AC38-597) modified to XV-1710-19(C13) configuration and delivered to the Wright Field Power Plant Laboratory on September 20, 1939.[67] It had been modified with the installation of the PT-13B carburetor, along with heavier valve springs (to eliminate valve surge), improved sodium cooled exhaust valves, a modified dynamic balancer with six 4-1/2 order weights to reduce the torsional vibrations occurring in the vicinity of 3000 rpm, along with steel inlet guide vanes to the supercharger impeller, and cylinder head coolant outlet elbows reworked to provide increased coolant flow. The other deviation from the V-1710-19 specification was the use of 8.0:1 supercharger gears instead of the standard 8.77:1 gears. This was done to reduce the mixture temperature and thereby reduce the likelihood of damaging detonation during the testing. Since the testing was to be done at near sea-level, the slower speed supercharger would have adequate capacity to develop the requisite manifold pressures for the intended high power levels. The testing was to start at the rated 1150 bhp, and 3000 rpm, and increase in increments of 50 bhp after operating for five minutes at each level. Testing was to continue in this fashion until either a predetermined limit was reached, including maximum power or failure. Routine teardowns of the engine were to occur at frequent intervals so that the condition of the internals could be assessed prior to any failure.

The testing of this pre-production "C" series engine, the 37th V-1710, reached a level of 1460 bhp at 56.8 inHgA with a specific fuel consumption of 0.606 pounds of fuel per horsepower per hour. When corrected to standard conditions this was equivalent to 1575 bhp. The fuel used was standard 92 Octane Specification AN-9528 (unleaded) plus 6.0 ml tetraethyl lead per gallon, which probably resulted in a rating of about 100 Octane.

At the first teardown, after 80 minutes of high power running, a 5 inch crack was found in the center main web in the upper half of the crankcase. Replacement parts, including the crankcase, main bearings, crankshaft, connecting rods, and conn-rod bearings, were taken from V-1710-11, AC38-581. This was the V-1710-11 which had been damaged in the crash of the XP-38 the previous February. These parts functioned without failure for an additional 13.5 hours of high power testing. Testing was finally terminated due to distortion and burning of valves which resulted in engine backfiring.[68]

From a mechanical standpoint, the testing proved the basic engine components for operation at higher power levels, and identified improvements that were required in the valves and valve seats. Furthermore, it was established that in high power operation, the V-1710 should be equipped with a "colder" heat range on its' spark plugs. These improvements were subsequently incorporated into all V-1710's.

In late 1943 Wright Field began a further series of tests to determine the limiting factors in operating the V-1710 at even higher power levels. This program was in support of the performance and reliability development being done for the two-stage engine program. Those tests are summarized here:

- V-1710-89(F17R) began a 7-1/2 hour War Emergency test at 65 inHgA at 3000 rpm using AN-F-28 Amendment 2 fuel (Standard 100/130 Grade) on December 16, 1943. The engine failed after 45 minutes due to pre-ignition (detonation) caused by use of standard issue sparkplugs. Engine was Glycol cooled.[69]
- V-1710-91(F17L) began a 7-1/2 hour War Emergency test at 65 inHgA at 3000 rpm in early November 1943 using AN-F-28 Amendment 2 fuel. The engine was equipped with "Cammed" pistons to improve the "fit" between the piston and the thrust face of the cylinder wall. The test was split into portions using pressurized water coolant and pressurized 70 percent Glycol and 30 percent Water. This coolant mixture did not allow detonation free operation at as high a manifold pressure as did water alone. The reason was the better cylinder head cooling obtained with water.[70]
- The highest powered 7-1/2 hour War Emergency test was accomplished on a V-1710-91(F17L), beginning March 1, 1944 at Wright Field. Fuel was 104/150 Grade fuel, and 10 inHg of "ram" provided to the carburetor by the laboratory blower. Turbo backpressure was simulated at 45 inHgA, and the engine successfully completed the test at a rating of 2000 bhp, 3000 rpm and 75 inHgA. Engine weight was 1,350 pounds and the coolant was the usual Glycol. The P-38J turbosupercharged arrangement was not used, as this test was a step proceeding running of the actual turbosupercharged test.[71]
- V-1710-91(F17L) began a 7-1/2 hour War Emergency Rating test at 75 inHgA at 3000 rpm in March 1944. Test installation was to mock-up that of the Lockheed P-38J, and included the GE Type B-33 turbosupercharger and an intercooler. The PD-12K7 carburetor had to have its jets enlarged to provide sufficient fuel for a 0.10 pounds/horsepower-hour BSFC required to eliminate detonation. Fuel was Grade 44-1 rated at 104/150 PN. The engine produced 1880 bhp at 3000 rpm with 77.0 inHgA. Turbo backpressure was 51 inHgA and the "ram" from the turbo into the carburetor was 42.2 inHgA, somewhat greater than the usual carburetor deck pressure for sea-level rated (i.e. turbosupercharged) engines of 30 inHgA. After 4-1/2 hours at high power a piston failed ruining the engine. A retest at 70 inHgA was recommended.[72]
- In late March 1944 another V-1710-91(F17L) attempted a 7-1/2 hour War Emergency test, this time at 70 inHgA. The same turbosupercharged arrangement as in the P-38J was used. This time the test was successful at a rating of 1700 bhp, 3000 rpm and 70 inHgA.[73]

Still later Allison undertook a program to develop the Turbocompound engine. Such an engine was configured in the proposed V-1710-G10 series to be optimized for cruise power and maximum efficiency. As such rated Military power (Dry) was anticipated to be 2250 bhp to 28,500 feet, with WER of 2550 bhp (Wet) to 20,000 feet and Continuous Cruise at 1700 bhp to 30,000 feet. Furthermore, if needed, the engine could have been optimized for peak output in the range of 3000 bhp.[74]

As a continuation of this program Allison ran some experimental Triptane tests on their 1500 bhp rated engines and got 2400, 2500, 2600 and 2700 bhp from them. Based on a takeoff ratings, the specific weight of the V-1710-C was 1.21 pounds per horsepower. With the first V-1710-F engines this was improved to 1.13 lb/hp, and on the last V-1710-F model the weight averaged 0.91 lb/hp. Allison was the first and only manufacturer of aircraft engines in the world to rate its piston engines at less than one pound of weight per horsepower.[75]

Boosting Power with ADI

With aircraft engines the practice is to establish their limiting ratings by determining the engine speed and power at which detonation can just be detected within the engine on a given fuel. If such detonation is allowed to persist the engine can be expected to fail catastrophically. Increasing the octane rating, or the Performance Number, of the fuel will allow additional power, but this avenue is limited by the cost, practicably and safety of the available fuels. Alternatively, it has been found that injection of a number of fluids and compounds along with the fuel/air mixture can significantly suppress the onset of detonation.

The Allison V-1710 was designed to be used in aircraft with turbosuperchargers. Such installations have intercoolers for the purpose of cooling the supercharged air coming from the turbo, and before passing it through the engine. A supercharger not only increases the pressure of the air, but also its temperature. As a result of this compression, followed by that done by the rising piston, there is a limit to the manifold pressure that can be developed without encountering detonation. Since the developed temperature rise in the cylinder is always a multiple of the inlet mixture temperature, there becomes a very practical limit to the maximum manifold pressure that can be tolerated. With 100/130 rated fuel, this limit was about 70 inHgA MAP in the V-1710's with 6.65:1 compression ratio pistons.

Allison V-1710's were for the most part used in aircraft which did not have intercoolers or aftercoolers. As for production aircraft, only the turbosupercharged P-38's had intercoolers, the power from the single-stage P-39's, P-40's, P-51's, and two-stage P-63's were all limited by the effects of this heat of compression. In the P-63 this was particularly acute, for it used the two-stage mechanically driven Auxiliary Stage supercharger. It was for this reason that the P-63's were fitted with ADI injection, a first for a production aircraft. It was to be used whenever War Emergency power was required. Later on, this feature was incorporated into most of the aircraft using Allison two-stage engines.

Anti-Detonate-Injection (ADI) fluid is the common designation of fluids used for this purpose. They work on the principle of being an internal coolant, that is, the ADI fluid evaporates and absorbs heat that would otherwise cause higher density within the cylinder and as a consequence, increase the pressure and temperature of the compressed mixture. The result is detonation and the exploding or burning of the mixture in an uncontrolled manner. Such burning can be very damaging to a high powered engine. Probably the most practical of all such fluids is simply water, preferably distilled water. Testing has shown that mixing alcohol with the water improves the effect at high power, while it also serves as an anti-freeze when the aircraft is operating in the frigid temperatures at high altitudes.

In the 1938-1940 period the NACA did a series of tests investigating the parameters that effected the onset of detonation.[76] They concluded that these were, in no particular order:

- Compression Ratio.
- Inlet Air Temperature.
- Inlet Air Pressure.
- Spark Advance.
- Fuel/Air Ratio.
- Size and Shape of the Combustion Chamber.
- Temperature of the Combustion Chamber and Cylinder Walls.
- Engine Speed.

The significant finding from this work was in quantifying the relationship between the variables that result in detonation. They confirmed that there is a fuel specific threshold where the density and temperature of the fuel/air mixture will detonate. This threshold is somewhat different for each engine and engine speed, and is primarily determined by the characteristics of the specific fuel being burned.

Furthermore, the NACA developed relationships that established this critical density and temperature as a function of first, the mechanical compression ratio, and then the additional compressive heating done on the charge as a result of the mechanical compression. As might be expected, these relationships act as multipliers on the initial density and temperature of the mixture as it enters the cylinder. Consequently, the temperature and pressure of the supercharged mixture in the intake manifold are critical in defining the conditions which initiate detonation.

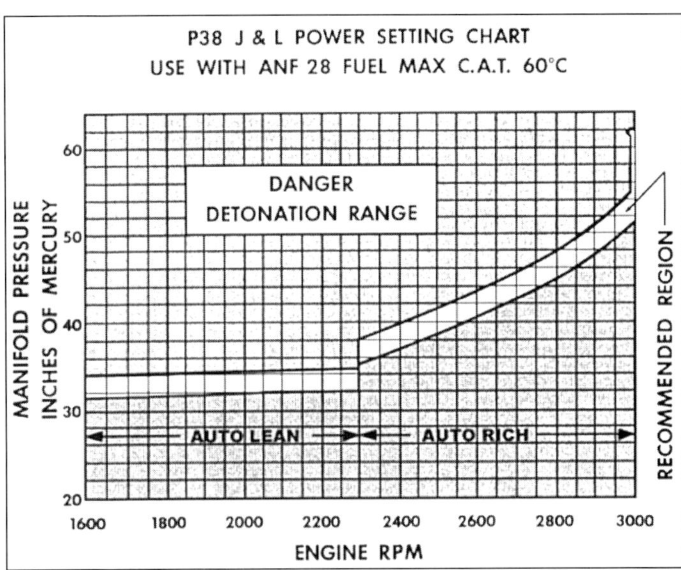

P-38J/L pilots were instructed to operate their F-17/F-30 engines within this MAP/RPM band so as to avoid detonation when operating on standard AN-F-28 fuel. The automatic engine controls provided by Allison were a great assist. (AAF Manual 51-127-1, 1945, 72.)

In their investigations into the use of ADI, the NACA established that it is the cooling effect of the evaporating fluid, primarily occurring within the induction system and *prior* to entry into the cylinder, that effectively delays detonation.[77] On the V-1710 the fuel was sprayed into the eye of the engine-stage supercharger. This not only provided excellent mixing, but maximum cooling of the mixture due to evaporation of the fuel. Similarly, it was found that for best results, ADI should be injected at the same location. Spraying it directly into the cylinder intake ports considerably reduced its effectiveness in reaching high power levels before incurring detonation.

They also found that pure water was the most effective ADI fluid at power levels up to about 1500 bhp, levels that were within the capability of the engine without ADI providing the inlet air temperature was 60 degrees Fahrenheit or less. For occasions when higher power was desired, it was best to use a water/alcohol mixture, usually a 50:50 blend. Denatured ethyl alcohol (ethanol) was used for the referenced series of tests, but similar results were realized using methanol.

The quantities of internal coolant required increase in a fairly linear manner with the desired power level. For example, when a V-1710 with 6.0:1 compression ratio pistons was tested using 100/130 Grade fuel, it developed 2310 bhp, with 0.079 fuel/air ratio and 60 °F inlet air, while requiring 663 pounds per hour of water-ethanol ADI to suppress detonation. This is a ratio of 0.45 pounds of ADI per pound of fuel. Without this ADI, the induction air temperature would have been about 230 °F, and only 1500 bhp could be developed from the limiting 63 inHgA manifold pressure. With ADI, the induction air was cooled to 120 °F and 2310 bhp resulted from the 88.5 inHgA manifold pressure that could now be tolerated.

During the above mentioned test program it was established that the subject V-1710, with a 9.5 inch diameter engine-stage supercharger impeller and 9.6:1 supercharger drive ratio, required the inlet air to the carburetor be boosted to above atmospheric sea-level pressure if manifold pressures greater than about 70 inHgA at 3000 rpm were to be achieved.[78] This

Table 16-15 WER Performance with ADI Injection						
	Engine Stage Ratio	Aux Stage Ratio	Carburetor Location	Aux S/C Weight, lbs	Military Critical Altitude, Ft.	Wet, bhp/alt,ft/MAP
V-1710- 93(E11)	8.10:1	6.85:1	Inlet to Aux Stage	175	22,400	1825/SL/75*
V-1710-109(E22)	8.10:1	7.23:1	Inlet to Engine Stg	125	28,000	1750/SL/75
V-1710-117(E21)	8.10:1	7.23:1	Inlet to Aux Stage	175	30,000	1800/24,000/76
V-1710-133(E30)	8.10:1	7.64:1	Inlet to Engine Stg		27,500	2200/SL/80-85
V-1710-143/5(G6R/L)	7.48:1	8.08/8.03:1	None in air path	125	32,700	1700/21,000/85

* -93 operating at 3000 rpm. All others at 3200 rpm.
-109 had higher critical altitude than otherwise similar -117. Lighter weight of Aux Stage due to not having large PT carb installed. Per TO AN 02-5AD-2, p.30, p.133, p.176-181 for performance charts.

again points to the limitations of the induction system, particularly when fitted with a two-barrel PD-12 type carburetor. When the later V-1710-G engines were introduced, with their 10-1/4 inch impellers and rated for 3200 rpm, this limit was raised to about 100 inHgA, at which point 2250 bhp could be delivered, wet.[79]

The use of ADI was not unique to Allison. During the summer of 1943 the War Department determined that all front line Pursuits should have the feature, and announced to Theater Commanders that water injection equipment would start on new production airplanes on the schedule shown in Table 16-16.[80]

Table 16-16
Aircraft to be Equipped with ADI

Series	The First of:
P-47	August 1943
P-63	September 1943
P-38	October 1943
P-61	October 1943
P-39	November 1943
P-40	November 1943
P-51	To Be Determined

The feature provided approximately 400 additional horsepower, and a 15 minute supply of ADI fluid was to be carried. Only the P-47, P-63 and P-61 were equipped with water injection, for production of the P-39 and P-40 were soon reduced to only those needed to support the training program. When the P-51 was reengined with the two-stage Merlin with its integral aftercooler, ADI was not needed. Interestingly, post-war Merlin powered racing Mustangs have all incorporated the feature. It is critical to not only the ability to make the extreme power needed, it is the feature that gives some hope of enough engine life to make it through a race.

Engine Efficiency with ADI

When one views a "carburetor curve" it is apparent that as power is increased to "high power" settings, that the fuel flow is increased proportionally. Much of this fuel is being wasted, for the resulting mixture is "rich"; in fact, all of the oxygen has been burned out of the mixture. This extra fuel is simply acting as an internal coolant, providing a degree of protection against detonation. It can be considered as a very expensive form of ADI. This is obvious from the extremely poor efficiencies shown by engines running at takeoff power, particularly those that are air-cooled.

Since ADI fluid replaces the need to use fuel as an internal coolant, there is an immediate improvement in efficiency as measured by fuel consumption. Of course ADI fluid is consumed in the process, and it has to be carried aloft as well, but there is a considerable difference in cost and engine performance. As a result of the need to reduce fuel flow when using ADI, manufactures provide a "derichment" valve on their carburetors which reduced fuel flow in proportion to the amount of ADI required to produce the desired power. Table 16-17 gives an example of wet and dry performance.

Effect of Fuel on Performance Rating

The first systematic investigations into aircraft engine "knock" began in the early 1920s. One of the early and important findings was that there were wide differences between fuels depending upon where the crude oil itself came from. It was found that some fuels were much preferred to others for use in aircraft. As a result, much of the testing of the day suggested that it was the chemical composition of the fuel that made the difference. But still, composition alone was shown to be insufficient to determine the ability of a fuel to resist knock. In 1926 it was found that "iso-octane" was excellent in resisting knock. Likewise a petroleum compound known as "normal heptane" was exceedingly poor in this regard. By 1929, knowledge of fuel composition, and the growing dissatisfaction with the quality and consistency of available fuels in use by military and commercial interests, led away from chemical composition being the determination of fuel suitability for aircraft use.

Table 16-17
V-3420-C1 Engine at Military Power, Grade 145 Fuel

	ADI, #/hr	BHP	RPM	BSFC, #/hp/hr	Thermal Eff, %	Heat Rejection, BTU/min Coolant	Heat Rejection, BTU/min Oil	Fuel Burn in 15 min, Pounds
Wet	600	4000	3200	0.590	23.3	35,500	12,900	590
Dry	0	3600	3200	0.740	18.6	44,700	13,440	666

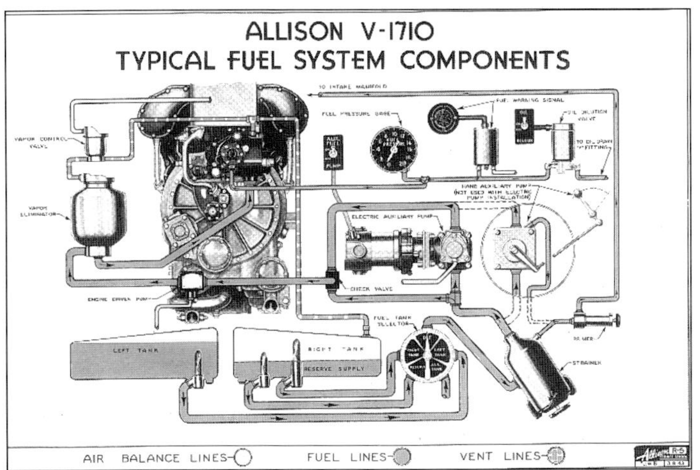

Each aircraft fuel system was unique for the needs of the particular airplane. Still they included these fundamental parts. (TO 30-5A-1)

In 1930 the "octane" rating scale for aviation fuel was introduced. It worked on the basis of comparing the knock tendency of an unknown fuel, operating in a standardized liquid-cooled test engine, to that of a particular reference blend of "iso-octane" and "normal heptane." The ratio of these two components then established the "octane" number for the fuel.

100 Octane Fuel

By the mid-1930s 100 octane fuel was available in limited quantities, primarily for testing and special purposes, though in 1934 100,000 gallons were procured for delivery to Hamilton Field for practical test.[81] Only just prior to the Battle of Britain, in the summer of 1940, did 100 octane fuel become available in tanker load quantities.

Testing also established that fuel performance could only be satisfactorily quantified by properly testing in an actual engine.[82] For perspective, when expressed in modern terminology, testing at that time found that fuel meeting then current Military Specifications varied in performance over a range of Performance Numbers from a low of 30 to a high of 50.

Prior to Pearl Harbor, available Grade 100 fuel had been almost entirely produced by blending gasoline, directly distilled from crude oil (straight-run gasoline), and doped with various high Performance Number blending agents. These were produced by means of expensive and complicated refining processes. Immediately after Pearl Harbor, these blending agents were in critical short supply because of the surge in demand for high performance aviation gasoline. The refiners responded by substituting catalytically cracked (cat cracked) gasoline for the previous straight-run commodity. Cat cracked gasoline had been in use prior to this time as an automobile motor fuel, but U.S. authorities had deemed it undesirable for aviation use due to the large amount (20 to 30 percent) of "aromatics" inherent in the product. These aromatics had the benefit of significantly improving the fuels resistance to knock, but they introduced a number of problems in systems designed for the previous "paraffinic" family of fuel blends. Particularly, American self-sealing (rubber lined) fuel tanks were unsatisfactory when exposed to aromatic fuel blends, as were most other rubber and natural materials then being used in U.S. aircraft fuel and carburetion systems.[83]

Cat cracked fuel was a life saver for the U.S. war effort, and the Allies to whom it was supplied. The American public was quickly placed on severe gas rationing so as to free up adequate quantities of aromatic fuels for the military services. Alternatively, the U.S. could have expanded its production capacity for paraffinic blending agents, but these would have required time and a considerable capital investment. It also would have consumed a huge amount of steel for new refineries, steel that was then being directed to the critically needed Liberty ship and destroyer construction programs for the Navy.[84]

Lean/Rich Performance Ratings

Prior to Pearl Harbor the U.S. had given little attention to the "rich mixture" performance of aviation fuels. This was probably a consequence of U.S. aviation's exclusive use of Paraffinic based fuels (produced primarily from domestic crude oil fields). Such fuels tend to exhibit little performance variation between lean/rich mixtures. Alternatively, Britain and Germany utilized a wide variety of crude oil sources and did not have an aversion to aromatics in their fuels. Consequently they had paid much greater attention to the performance of fuels for use in rich mixture conditions. As a result of these endemic variations almost all engine manufactures had developed their engines to different levels of performance, determined by the limits of the fuel available to them. American engine manufactures had taken advantage of this effect and, with the concurrence of the Military Services, usually rated their engine on selected Grade 91 and Grade 100 fuel.

After entry into the war, it became apparent that some sources of both grades of fuel were inferior to the fuels which had been used for engine development and rating. As a result of tests, it was found that the F3 Method did not satisfactorily determine rich mixture performance and that engines had really been developed on Grades 91/96 and 100/120, whereas actual service fuel might be as low as 91/91 and 100/100 respectively.[85] As a result, the ASTM F4 Method was quickly developed and adopted to establish the rich rating of fuel.[86]

The F4 Method for rich rating was quite a different procedure in that it involved supercharging the fuel rating test engine. Furthermore, it required more extensive data collection and analysis than needed for the F3 Method, which was still used for lean ratings.

Grading of fuel performance, as originally developed using the octane scale, could only rate fuels up to 100 octane. It was now necessary to have a standard for higher relative values and for this purpose the "Performance Number" scale was established. Like the "octane" scale, it makes use of reference fuels and "bracketing", but in this case they are pure iso-octane doped with a proscribed amount of Tetra Ethyl Lead, TEL.[87] The PN which results then is an index that approximates relative engine performance based on brake and friction horsepower developed by two different fuels, one of which is the reference fuel.[88]

Engine Performance With Different Fuels

An approximate and useful relationship for engine performance can be developed from observing the supercharged absolute manifold pressure at the onset of engine knock. It was found that in the CFR fuel rating test engine, for manifold pressures greater than about 24 inHgA, that the manifold pressure, minus seven inches, was proportional to the fuel PN required to give knock-free operation at each manifold pressure level.

As an example, if an engine that operates just knock-free at 36 inHgA on Grade 91/96 fuel is to be operated on Grade 100/130, and it is desired to reset the engine controls to take advantage of the additional fuel capability, the new limiting manifold pressure can be determined. Grade 91/96 fuel has a rich Performance Number of 88, while the rich PN for Grade 100/130 is 130.[89] Assuming the same fuel/air ratio for the two conditions, the result is:[90]

$$((36-7)*130/88) + 7 = 42.8 + 7 = 50 \text{ inHgA}$$

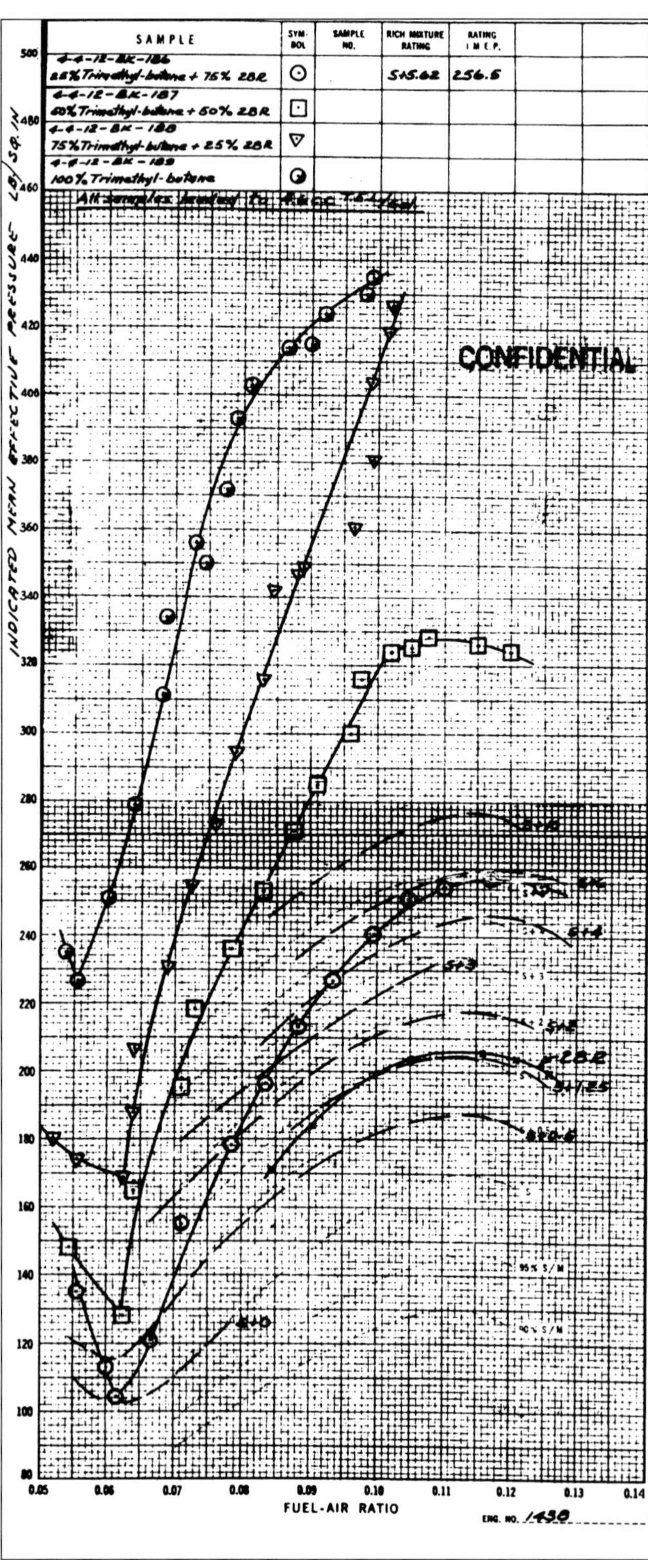

Allison regularly tested fuels being used in its engines. These curves show the effect of Triptane when blended with standard AN-F-28R fuel, doped with 4.6 cc of TEL/gallon. With straight Triptane, able to run detonation free at 430 IMEP, a V-1710 at 3000 rpm would develop 2785 ihp, a V-1710-G1 running at 3400 rpm, would be able to develop 3157 ihp. Adjusting for friction and supercharger loads would reduce these values to about 2400 and 2650 bhp respectively. (Allison via Hubbard)

Which means approximately a 25 percent increase in the allowable bhp of the engine, though this would only be true if the manifold temperatures in the two cases were similar.

The effect of improved fuel is dramatically shown with the increase in MAP and power between the F-2 and F-5 models, used in the same aircraft and with the same intercoolers. By the above formula, the approximate MAP allowed by the increased fuel rating from Grade 100 to Grade 125 is:

$$((39.4-7)*125/100) + 7 = 47.5 \text{ inHgA}$$

This is very close to the actual 47.0 inHgA rating specified for the F-5 engines.

Adding TEL to gasoline increases its resistance to knock and consequently its Octane or Performance Number. Unfortunately the effects are not linear, meaning that a disproportionate amount of TEL must be added to achieve a given improvement in PN. The further away from PN 100, the less the effect. For example, 1.28 cc of TEL per gallon of iso-octane is required to achieve 130 PN, doubling the amount of TEL produces a PN of only 143.

Meeting Fuel Quantity Requirements

In November 1941, the military revised its specifications to allow an increase from 3.0 to 4.0 cc of TEL in 100 PN fuel. This allowed lower grade stocks to be used, and in effect increased the output of suitable Grade 100 fuel by 25 percent. In mid-1943 this specification was again increased, this time to 4.6 cc, which increased fuel supplies by another 7 percent.[91] These production increases came about because they allowed producing greater amounts of straight-run gasoline from the available crude oil, though that gasoline was coming from the refinery at lower PN. By increasing the amount of TEL which could be added, the rating was maintained at the Grade 100/130 level required for the bulk of Allied combat aircraft.

Specifications for gasoline used by commercial airlines in the late 1940s did not allow more than 4.0 cc of TEL, except in Grade 115/145, which allowed up to 4.6 cc per gallon. For a number of reasons, it was found that fuel should never be doped with more than 6.0 cc of TEL per gallon, which results in 161 PN. The super fuels of the late 1940s, such as Triptane, can also be doped with TEL, and can in this way give lean/rich performance numbers equivalent to Grade 200/300. Production of such fuels is extremely expensive and at best, the U.S. production capacity was only a few hundred gallons per day.[92]

Fuels for Allison Engines

The early Type Test V-1710-C engines were all developed and rated on 92 octane fuel per Army Specification 2-95. The first engine to be rated for operation on 100 Octane fuel was the first Altitude rated engine, the XV-1710-19(C13) for the Curtiss XP-40. This was necessary because of the higher pressure ratio developed by the increased supercharger drive ratio, which caused both higher manifold pressures and temperatures. The 100 Octane fuel allowed the full mechanical capability of the engine to be utilized without being limited by knock (detonation), where 92 Octane would have been limiting. Allison continued to rate its early engines depending upon whether or not they were sea-level or altitude rated, the former on 91 or 92 Octane, the latter on 100 Octane.

With the 1941 introduction of the V-1710-F5 series of sea-level engines for the Lockheed P-38, Allison began to rate engines using the lean/rich Grade 100/125 PN fuel to specification AN-F-27. However, because of quality and supply problems, they used only Grade 125 for rating engines prior to the V-1710-89/91(F17R/L), which began deliveries in 1943.

Table 16-18 Typical V-1710 Takeoff Ratings versus Fuel Grade				
Engine Model	Year	S/C Ratio	Fuel Grade, PN	Takeoff, bhp/rpm/MAP,inHgA
V-1710-11/15(C8/C9)	1938	6.23:1	92 Octane	1150/2950/37.5
V-1710-27/29(F2R/L)	1940	6.44:1	100	1150/3000/39.4
V-1710-49/53(F5R/L)	1941	7.48:1	125	1325/3000/47.0
V-1710-89/91(F17R/L)	1943	8.10:1	130	1425/3000/54.0
V-1710-143/5(G6R/L)	1945	7.48:1	145	1600/3200/74.0

These, and all V-1710's delivered through the end of the war were rated on Grade 100/130 fuel produced to military specification AN-F-28. The V-1710-G series, and a few late V-1710-E/F, engines were developed and rated on Grade 115/145 fuel (specification AN-F-33), but none of these saw service during the war.

Through most of the war, the aircraft being flown within the Continental U.S. in various training and liaison roles were operated on 91 octane fuel. This was a fuel which was not suitable for combat. The reason was simply the critical fuel supply situation. It was standard practice to retard the fixed timing of the magnetos in such service as a way to protect them from detonation damage, particularly during take-off. For the V-1710, the revised timing was 29 degrees BTC for exhaust and 23 degrees BTC for the intake plugs.[93] At the same time, the maximum manifold pressures were reduced by 5 inHgA, meaning somewhat reduced power was available.[94] Since the aircraft were not usually operating at maximum gross weights, the restriction against full use of Military and WER ratings was not a significant constraint on the quality of training provided. While official Air Force technical orders required resetting the timing on any engine being sent to foreign theaters, a number of engines with retarded timing reached England in 1944 and were installed in P-38's without having been reset for full power operation on Grade 100/130 fuel. The Materiel Command took immediate action to insure this situation was rectified.[95]

Accessories section of V-1710-37(E5), AEC #338. (Author)

Accessories Section

Allison continually refined the design of the V-1710 to improve its manufacturability. A key to their success was the fundamental building block nature of the engine, consisting of a universal power section which could be fitted in the front with a variety of nose cases, and in the rear with an even larger number of accessories sections. In every instance, the important gear ratios, carburetors, superchargers, and arrangement of auxiliary accessories could be altered to meet the needs of aircraft designers and missions.

Two-stage engines with the Auxiliary Stage Supercharger were considerably more complex around the accessories section. This is a V-1710-143(G6R). (Author)

The accessories section received a considerable amount of design attention during the long production run. Major changes were necessary to allow use of wider gears internal to the section. These were incorporated to handle the greater torque required to drive some of the later accessories, as well as to improve reliability. Some of the important, but subtle differences between models are shown on the accompanying photographs.

Accessories section of a V-1710-81(F20R). Note similarities with E-5 accessories section. (Author)

This is a late model V-1710-135(E31), s/n A-057314. These single-stage engines were converted from other models for use in the RP-63 program. Color scheme is non-standard, though attractive and easy to keep clean when used in a P-63 restoration. (Author)

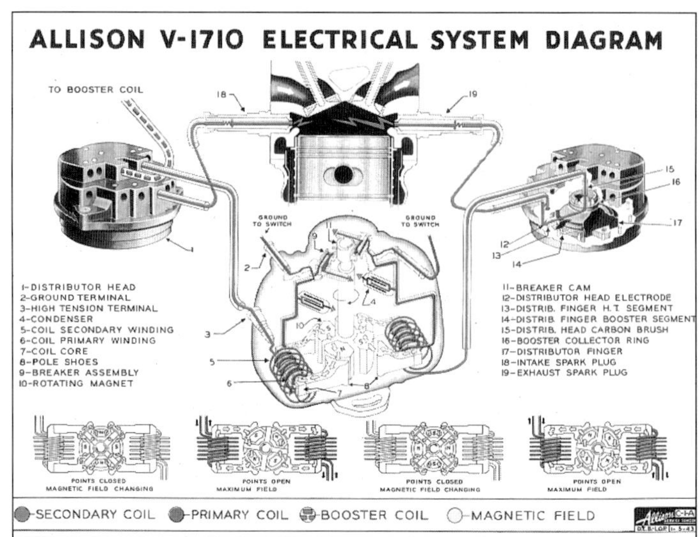

The V-1710 ignition system used a single fixed timing magneto with dual distributors providing sparks to dual plugs in cylinders. The 4-pole magneto turned at 1-1/2 times crankshaft speed, forming six sparks for every turn of the crankshaft. (TO 30-5A-1)

Ignition

As with contemporary aircraft engines, the V-1710 was equipped with dual ignition, that is, two sparkplugs in each cylinder. Each plug was fired by a different distributor, one for all of the "exhaust" plugs and the other for the "intake" plugs. The plugs were arranged opposite each other, on the cylinder centerline, and at right angles to the block. A single Type DFLN6 Scintilla double, fixed timing, "4-pole" magneto was driven via the accessories drive at 1-1/2 times crankshaft speed. Consequently, the magneto formed six sparks for every crankshaft revolution. Since the engine is a "4-cycle", each cylinder must be fired on every other revolution. This arrangement provides the necessary twelve sparks for every two revolutions.

Practically all production V-1710's, except for some of the late V-1710-E/F and G's, operated with "fixed" timing of the magneto, that is, the timing of the spark was 28 degrees BTC on the Intake Plugs and 34 degrees BTC for the Exhaust Plugs, although the V-1710-C/D's did use slightly different ignition timing settings. When introduced the retarding type magnetos were effective in easing starting and idling, and resolved complaints that Wright Field had been voicing for years. The spark timing was automatically advanced as the engine speed was increased so that for 1400 rpm and above the normal fully advanced fixed timing was in effect.[96]

The distributors direct the spark formed in the magneto to the appropriate cylinder consistent with the firing order of the engine. Mechanically they are driven from the rear of each camshaft, and two types were used. For installations requiring gun synchronizing, a distributor was used that incorporated the device. Such distributors can be identified by the horizontal orientation of the distributor cover, which allows the gun synchronizer to mount on the camshaft centerline. On engines used in installations which did not require gun synchronizers the distributor cover is vertically oriented and mounted directly onto the rear of the camshaft. For sea-level rated engines intended to operate at very high altitudes, provisions were made for "pressurizing" the distributors with induction air taken from ahead of the fuel injection nozzle. This air had been pressurized to near "sea-level" conditions by either the aircraft's' turbosupercharger or the Auxiliary Stage Supercharger. This feature significantly reduced the tendency for ignition "shorts" and the spark jumping to ground during operation at high altitudes.

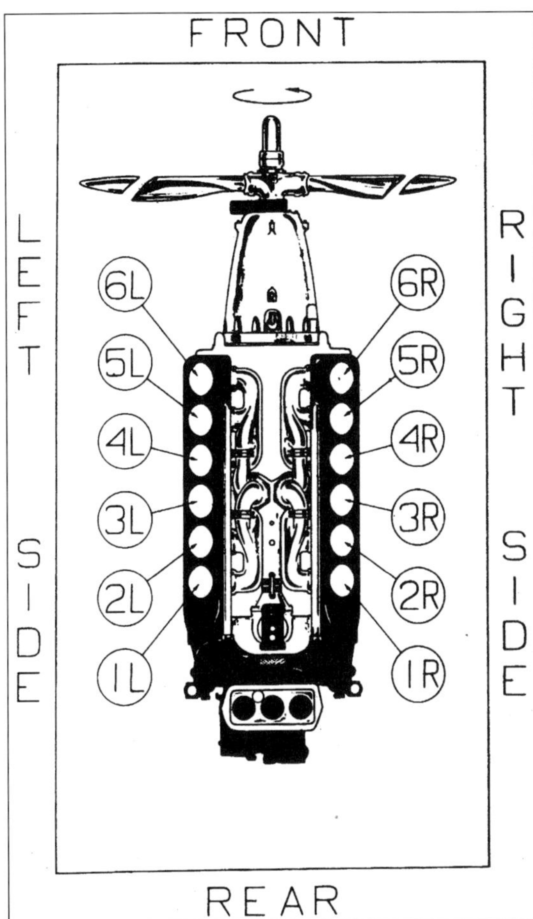

As shown on this C-15, Allison used the U.S. standard nomenclature for identifying the cylinders of the V-1710. The Packard Built Rolls-Royce Merlin numbers from the front. (Allison via Hubbard)

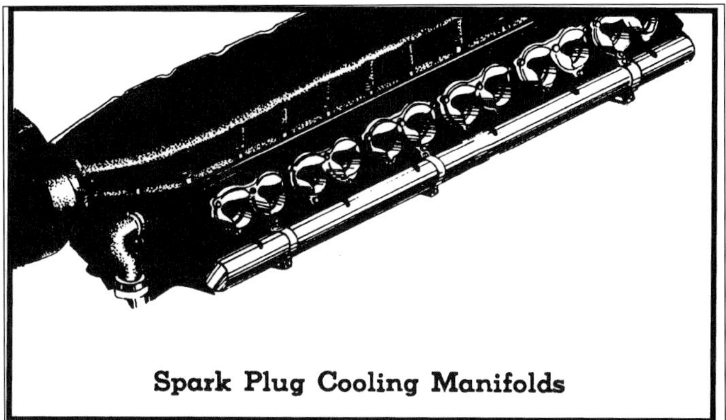

Blast tubes delivered slipstream ram air through slots directly onto the shank of each exhaust sparkplug. This was important in the tightly cowled installations where radiant heat from the exhaust pipes would otherwise soon damage the plugs and/or heat them to the point where they would cause pre-ignition. (USAAF)

The single-stage engines were not equipped with pressurized ignitions, and still operated quite well to high altitudes. Tests were run on Wright Field P-51A AF43-6009, which was operated at maximum altitudes of 30,000 to 31,500 feet for 15 hours, and above 15,000 feet for 26 hours without ignition lead failures. This was with ignition harnesses that had accumulated 293 hours of service on the airplane.[97]

Several different arrangements were used for shielding the ignition wires. Early engines used formed tubes to bundle the six wires going to a bank of intake or exhaust plugs. Appropriate fittings and braided shields were used to continue the shielding to the sparkplug body and onto the engine block. Late model engines used a more economical shield made from two thin steel sheetmetal stamped halves formed and then crimped together. Tightly cowled engines presented a problem for the ignition wires adjacent to the exhaust plugs, particularly on engines with collector type exhaust manifolds necessary for turbosupercharged installations. The problem was burning of the wires due to the high temperature environment and radiant heat. The solution was the incorporation of a "blast tube" which directed a flow of ram-air, taken from the propeller slipstream, onto each of the sparkplugs. Even more trying was the installation of the V-1710-D pusher engines in the Bell X/YFM-1, for there was no "slipsteam" to provide cooling during ground operations, which of necessity, had to be limited. Even with the blast tube, the exhaust heat in the vicinity of the "exhaust" plugs made it necessary to frequently replace them.

Concurrent with the development of the early V-1710, the design, construction, and manufacture of sparkplugs was also undergoing an evolution. Early on, plugs were often of two piece manufacture and it was necessary to frequently disassemble them for cleaning. Leakage of the high pressure gasses was a frequent malady, which, for the most part, was resolved when the sealed ceramic insulator was introduced in the 1930s. With the V-1710 and its high cylinder pressures and thermal loads, duty on the spark plugs was particularly severe and demanding. In modern parlance we would specify a "cold" plug for such service, meaning that the design of the insulator and center electrode would have the ability to rapidly transfer heat to the cylinder head and thereby remain "cool." This was essential, for a "hot" plug could become incandescent and cause pre-ignition and even detonation. Either condition being fatal to the engine.

Ignition distributors for the V-1710 look remarkably like those used in automobiles. Each set of plugs is fired by its own distributor. The RH unit fires all Exhaust plugs while the LH unit fires the Intake plugs. (Author)

John Kline, the Allison Tech Rep to Bell Aircraft for much of the XP-39 and XFL-1 flight testing programs, tells of an early XFL-1 flight which nearly resulted in the loss of the aircraft at Patuxent River, the Naval Air

Test Center located on Chesapeake Bay. On the takeoff roll, the aircraft was unable to develop full power and the pilot elected to abort, unfortunately ending up in a ditch at the end of the runway. His investigation centered on the plugs, which when pulled were heavily fouled. In response to questions, it was determined that the aircraft had been on the "compass rose" the previous day for some 3-4 hours, all with the engine running at idle and low power. As could have been anticipated, a high performance engine operating on leaded fuel will foul the plugs, and it can occur in only minutes. Good operating practice was that the pilot would "clear the plugs" frequently during extended periods of low power operation, such as cruise or during ground operation. "Clearing" simply required increasing engine rpm and power for a few minutes, which would "heat" the "cold" plugs and burn-off any accumulation of lead or fuel additives.[98]

Another example of the attention to spark plugs and their care, comes from Don Wright, who was the Allison Tech Rep at Ladd Field, located outside of Fairbanks, Alaska, where the P-39's were being turned over to the Russians under Lend-Lease.[99] While these were all brand new aircraft, they each had several hours of acceptance testing as well as the delivery flight time on them. This could amount to 30 to 50 hours total time. The Russians insisted that each aircraft, being given to them under Lend-Lease, be in pristine condition, and complete in every detail at the time they accepted it.[100] To accomplish this, an assembly line was setup in a hangar and the aircraft processed accordingly. Included was the requirement by the local Russian Air Force representative to install a "new" set of sparkplugs. As can be imagined, with every engine requiring 24 plugs, it would not take long to exhaust the available supply, particularly given the fact that Alaska was literally at the end of the supply line. Furthermore, at that time each plug cost about $5.00, which would equate to about $40.00 in the 1990s, so a complete set of plugs would be worth about $1000. Facing this dilemma, as well as the prospect of backing up deliveries, Don went into Fairbanks and found an old spark plug cleaner/sandblaster, which he was able to get the owner to loan him. With this device setup in a back room, the plugs on arriving aircraft would be ceremoniously removed and "discarded", then "new" plugs installed. Each having been taken from carefully husbanded sparkplug boxes. Properly cleaned and gapped, the plugs provided like-new performance.

With the introduction of long-range escort missions it was necessary to cruise at low engine rpm, but with the throttle wide open. In this condition, the amount of supercharging heat added to the mixture was minimal, and when the outside air temperature was low the result was lead fowling of the sparkplugs. Instructions were issued to pilots to "clear" their plugs by occasionally increasing rpm, but this remained an ever present problem. Late in the war it was studied, and when the P-82E was introduced, resulted in a cautionary note to the pilot to maintain at least 60 degC (140 °F) in the intake manifold to preclude the problem. Should the plugs foul, any attempt to increase power could quickly lead to detonation and subsequent engine failure.[101]

The final evolution of the V-1710 ignition system was to be fitted to the late V-1710-G2 model, offered to airlines to power a version of the Douglas DC-4B. This would have been a low-tension (low-voltage), high-frequency ignition system built around an integral distributor/magneto, one of which would have been fitted to the camshaft on each cylinder bank. Such a device would have greatly simplified the construction and layout of the accessories section of the engine.

Spark Plugs

High performance engines have always been hard on spark plugs, and probably no other component within the engine is more critical to the proper functioning of the engine. Forming a proper spark at the critical moment, in the high pressure and hot conditions within the cylinder, is a difficult requirement. This was particularly true with the comparatively low powered magneto type ignition systems used on aircraft engines of the period.

The first V-1710's used "two-piece" plugs that had changed little since WWI. These could be removed and disassembled for cleaning and reuse, but were also prone to gas leakage and other maladies. By the time of the P-40, single piece porcelain plugs were in general use, and the Air Corps even required Allison to guarantee the plugs on some of the service test engines to last for 500 hours. Even later plugs would not be able to meet such a requirement.

Good crew chiefs were able to avoid reliability and operational problems for their engines by paying attention to the plugs. Both antidotal and experimental evidence was developed that showed that plugs needed to be cleaned and/or replaced, and properly gapped, after any operation at WER power. The British ran a series of tests on a V-1710-F3R in early 1944 that showed that lead from the high octane fuel would bridge the electrodes while a glaze would often deposit on the insulator during high power running. One interesting finding was that "cyclic" running was more likely to cause these faults than straight forward endurance running. They also determined that the first plug to fail was invariably in the No. 1 right bank exhaust position. By increasing the plug gap, the incidence of plug failure was postponed. Another finding was that the AC-LS86 plugs gave better performance than British twin-electrode types, both in flight and on the test bench.[102]

Significantly the British followed up their work on fouling of plugs in the F-3R with an investigation of the amount of TEL being delivered to each cylinder. They found that there was considerable variation between cylinders, with the No.1 Right Bank cylinder receiving over twice the amount of TEL than average, and that the average cylinders were receiving about 15 percent less TEL than provided in the supplied fuel. No wonder the plugs in the No. 1 right bank cylinder tended to become fouled with lead. It was due to this type of problem that the "Madam Queen" venturi gas intake pipe was introduced by Allison.[103]

The usual spark plugs were Champion C35S or C34S. Alternatively, AC-LS86 plugs could be used.

Valve and Ignition Timing

Synchronizing and timing the opening and closing of the valves is absolutely critical to the power and efficiency of an engine. This is apparent when it is recognized that an engine is simply an air measuring machine, and that when a proportional amount of fuel is mixed with the air and burned, a portion of the heat released is available to be recovered as useful mechanical work. The engine simply processes the air and provides a machine to recover some of the power available in the hot combustion gasses.

Ideally, combustion should occur at the point of peak pressure in the cylinder, that is when the piston is at Top-Dead-Center, TDC. Likewise it is important that the maximum amount of air be drawn into the cylinder during the "intake" stroke, just prior to the beginning of the "compression" stroke, which is necessary to create the peak pressure at TDC. Considering the mechanics involved in the 4-stroke cycle, it would appear that the intake valves should be open for 180 degrees of crankshaft rotation, followed by 360 degrees of rotation to accommodate the compression and power strokes, (all with the valves closed), and then the exhaust valves should be opened for the remaining 180 degrees of crankshaft rotation, during which the spent exhaust gasses are pushed out of the cylinder in preparation for beginning the next cycle. Valve timing as just described would only be effective only for a *very* slow running engine. As the speed of the engine increases, the flow of the fuel/air mixture into the cylinders

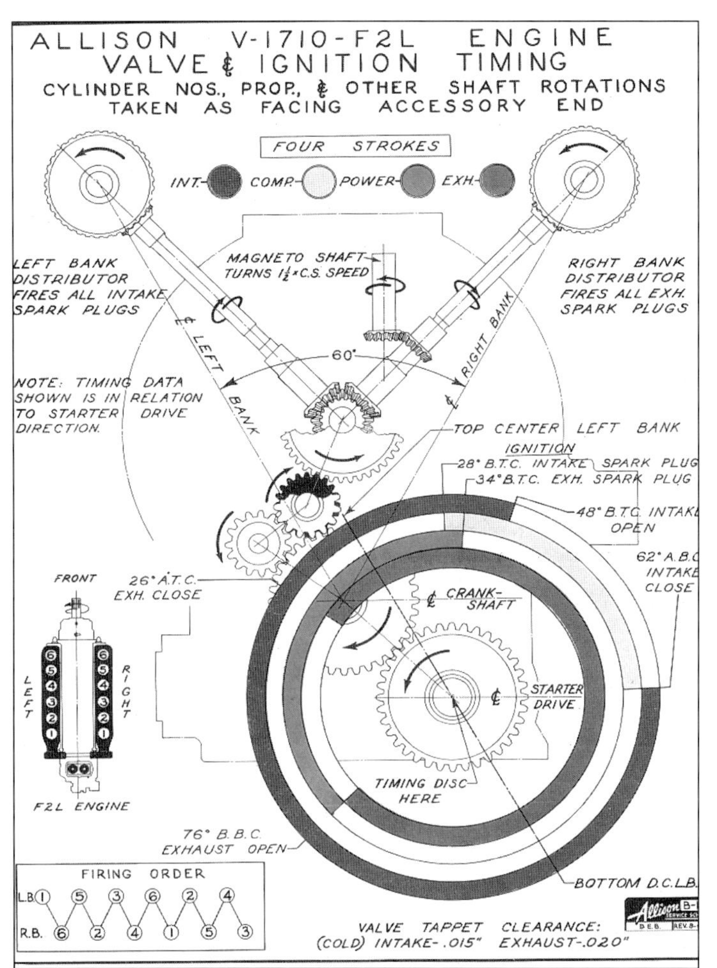

Specifics for timing the LH turning V-1710-F/G engines are explained on this graphic. The RH turning C-15 was timed as above for its crankshaft turned the same way. (TO 30-5A-1)

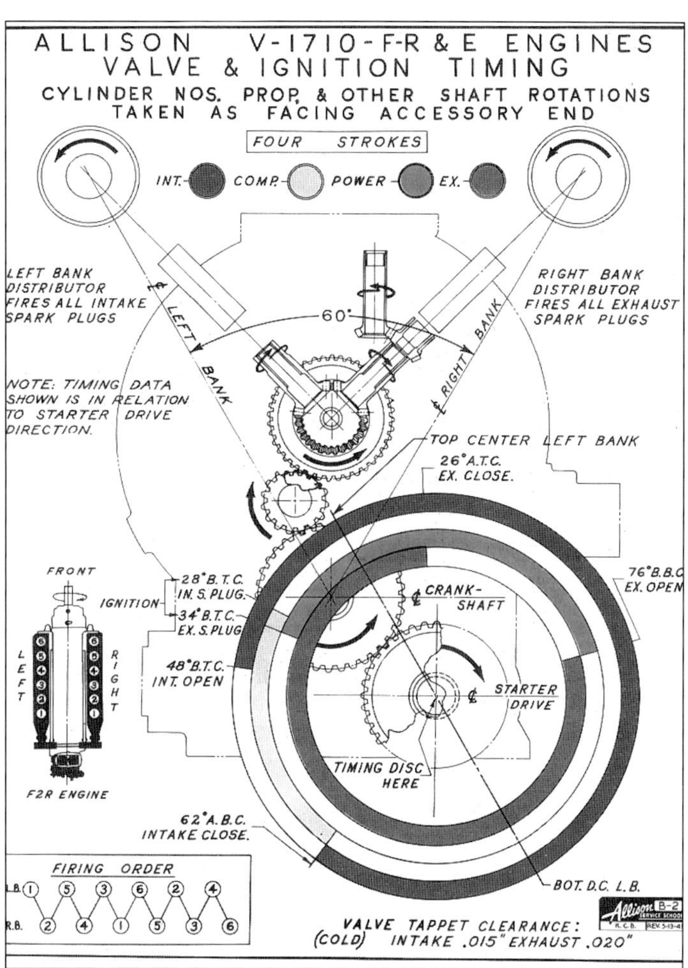

The RH turning engines were timed in a similar fashion as the LH models. (TO 30-5A-1)

begins to look more like a continuous flow, and it is both prudent and necessary to modify the valve timing "ideal."

The GV-1710-A was a comparatively slow running engine, rated at 2400 rpm. While it had an integral supercharger, very little manifold pressure was necessary to produce its rated 750 horsepower. Consequently, Allison followed conventional design practices and initially selected a "symmetrical" pattern of valve timing. The intake was opened 15 degrees BTC (Before-Top-Center) and remained open for 250 degrees of the crankshaft rotation, closing at 55 degrees After-Bottom-Center, ABC. Meanwhile the exhaust valves opened at 55 degrees BBC (Before-Bottom-Center) and closed at 15 degrees ATC (After-Top-Center), also remaining open for a total of 255 degrees. Implicit in this cycle was a period of "overlap" during which *both* the exhaust and intake valves were open, of some 30 degrees, the period from when the intake opens at 15 degrees BTC, through 15 degrees ATC when the exhaust valve closes.

Valve overlap is important as some of the cool pressurized mixture in the intake manifold is allowed into the cylinder while the piston at near TDC on the exhaust stroke and pushes the residual spent hot gasses out into the exhaust manifold. Since both sets of valves are only slightly open at this position in their stroke, very little of the fresh mixture is lost directly to the exhaust. As a consequence of this sequencing, the fresh charge that is then drawn into the cylinder during the intake stroke is greater than it would otherwise be since it is not diluted by residual exhaust gasses. To prevent the hot gasses from initiating a backfire into the induction system, it is necessary that the manifold pressure be greater than the exhaust pressure, or that there be enough flow in through the partially open intake valve so that temperatures are held low enough to prevent the ignition of the fresh charge.

Allison did both an analytical study and experimental work on the first V-1710-C1 to convince themselves that further benefits were available if both the duration of valve "open", and the amount of overlap between intake and exhaust valves was extended.[104] This was first incorporated in the XV-1710-3(C1) which utilized an intake duration of 258 degrees, with an exhaust duration of 266 degrees and 44 degrees of overlap. The successful Type Test engine, the YV-1710-7(C4) operating at 1000 bhp and 2600 rpm, was even more aggressive with its intake duration of 282 degrees and exhaust duration of 278 degrees. It also had the overlap increased to 70 degrees.

With the introduction of the V-1710-11(C8) Allison established the valve timing that would remain a standard for all subsequent production models of the V-1710 and V-3420, except on the C-15 which delayed opening the intake valve by 4 degrees. Standard intake duration was 290 degrees, the exhaust duration 282 degrees with an overlap of 74 degrees.

Two intake and two exhaust valves in each cylinder are operated through "Y" rockers from a single cam lobe. (Author)

The advantages of having a universal valve, and similarly ignition timing, over the entire family of engines should be obvious. While there may have been some reasons for trying more aggressive timing in an all out drive for power, the compromises did not justify it. For most of an engine's operating life it will be in a part load cruising condition where fuel economy and reliable operation are the critical considerations. The extra peak power needed in combat is actually limited by either the detonation characteristics of the fuel or the mechanical strength of the engine. Since the V-1710 was able to operate up to the detonation limits of available fuels, there was really no benefit to be gained by adopting a more aggressive valve timing.

There was also no need to change valve timing on engines which were operated with a turbosupercharger since the turbo was intended not to increase the back pressure on the engine above sea-level pressure. Consequently, the turbo had little, if any, effect on the conditions within the cylinder. This was not the case for the "next generation" of recip engines that Allison began actively developing late in W.W.II, the "Turbo-Compound" engines. In this configuration, a turbine is used to recover power available in the exhaust gasses and feed it directly back into the crankshaft. Had this engine completed development and become a commercial product, it well may have had revised valve timing. In its development form as the V-1710-127(E27) it retained the standard timing and was able to develop as much as 2825 bhp at 3000 rpm, as well as 2975 bhp at 3200 rpm and 100 inHgA.

Carburetors

During the V-1710 development period, the engines were equipped with various models of the Stromberg NA-F7 float type carburetor (the "F" denoting 4-barrels and the 7 the size of the barrels). These gave satisfactory performance at up to 1150 bhp, although there were susceptible to negative-g cutout and icing and required jacketing with hot coolant to insure adequate evaporation of the fuel. This latter occurrence was the reason for the loss of the XP-38 on its delivery flight. It had not been thought that icing would be a problem in a turbosupercharged installation, but events showed this not to be the case.

By the time the YP-37's and YFM-1's were being delivered in 1939, the new Bendix-Stromberg pressure carburetor was available and incorporated. This feature alone defined the C-10 and D-2 models from the otherwise similar C-8/9 and D-1 engines. The pressure carburetor was sometimes referred to as a "fuel injection" carburetor, and in many respects it functioned much like a single-point or throttle body fuel injection system used on modern automobiles. The "carburetor" portion of the device was really only a airflow measuring device, combined with the throttles. Controls were also incorporated that set the selected mixture ratio and metered the proportional quantity of fuel needed by the engine. That fuel was then injected under pressure into the eye of the engine-stage supercharger. In this way, excellent mixing with the air was achieved and a portion of the supercharger compression heat negated by the evaporation of the fuel. Various models of this carburetor became standard on nearly all of the large engines produced in the U.S. during the war.

The carburetor was available in two general forms, identified by Bendix as being either a "PD" (two-barrel) or "PT" (three-barrel) arrangement. Obviously the three-barrel units could accommodate a greater flow and would be used whenever a larger volumetric flowrate was needed. Late in the war a "PR" model having a single "Rectangular" flow section was introduced for some models of very powerful engines such as the V-3420, R-3350 and R-4360, as well as a few V-1710's intended for high altitude.

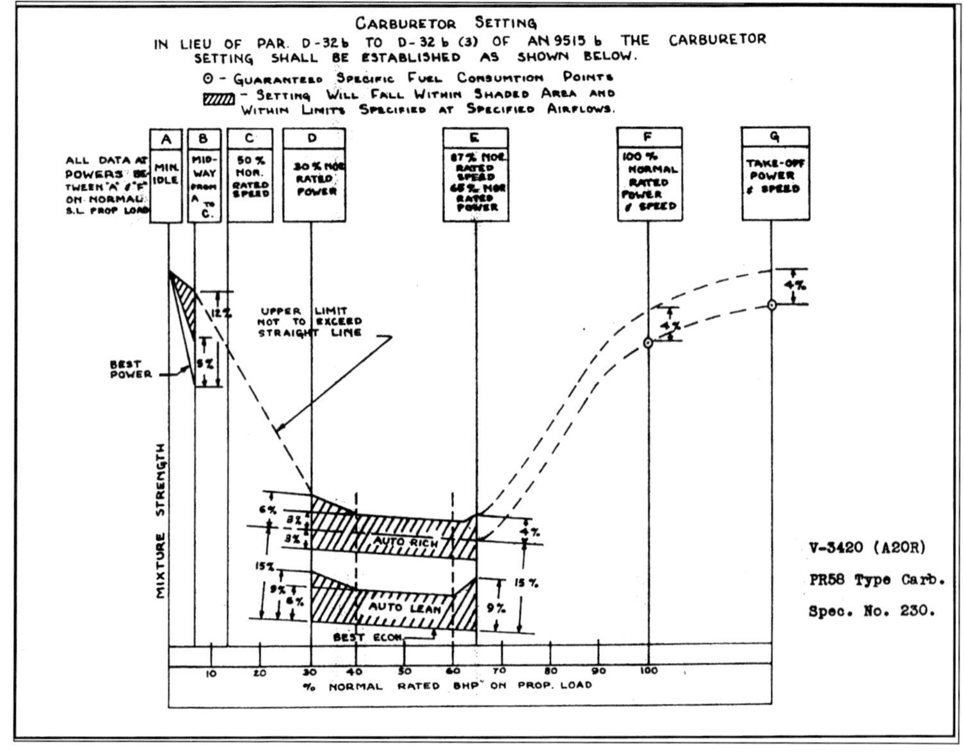

Carburetor curves all have this general shape and these features. The reason for cruising at reduced rpm and power should be apparent, high power dramatically increases use of fuel. The carburetor is designed to automatically provide the proper ratio of fuel to meet the specific requirements over a wide range of conditions. (Allison)

Table 16-19
Representative Carburetors

Model	Description
PD-12G1	Total Throat Area 22.838 sq.in., with a baseline diameter of 1-1/16 inches for the "G" models.
PD-12K2	Barrel diameter is 1/4 inch larger than on the PD-12G1 giving a total throat area of 24.353 sq.in.. The accelerator pump mounted remotely on S/C inlet cover. Had 3-1/4 inch diameter venturis.
PD-12K6	Has accelerator pump mounted remotely on S/C inlet cover.
PD-12K15	Uses mechanical type accelerator pump mounted on carburetor body.
PT-13B1	Carburetor for the XV-3420-1, also used on the XV-1710-19 Altitude Rated engine for the XP-40. Calculated flow area is 41.32 sq.in
PT-13E9	Flow area of 41.32 sq.in., uses mechanical type accelerator pump mounted on carburetor body.
PT-13E10	Differs from the PT-13E9 in that it incorporates a derichment valve for use with water injection and a mechanical type accelerator pump mounted on carburetor body.
PD-18A1	Used on Packard-Built Rolls-Royce V-1650-3 two-stage Merlin. Throttle bores were 5-1/2 inches in diameter. Flow area 47.52 sq.in.
PR-58B3	"Rectangular" cross-section of 58 sq.in. Used on many of the V-3420 double vee engines.

The various PD/PT carburetors were further identified by a dash number and letter, as well as a suffix. The number defined the number of 1/4 inch increments over a baseline diameter for the model, and therefore the flow area for the venturis. In this way it can be seen why the 1090 bhp altitude rated V-1710-C15 for the P-40 elected to use the PT-13 rather than the PD-12B used on the sea-level rated 1150 bhp V-1710-C10 from which it was derived. The C-15 was able to deliver its rating at an altitude of 13,200 feet simply because of the larger volume of its carburetor and induction system compared to the C-10 at sea-level. Actual airflow through the two engines was essentially identical at these conditions.

Figure 16-4a As with the Bendix Pressure type carburetors the Speed Density unit injected the fuel directly into the eye of the engine-stage supercharger. Throttles were still necessary to control the flow of air into the engine. (USAF TM 52-12)

Allison used PD carburetors on all of its turbosupercharged (sea-level) engines, while the C-15 altitude rated engines used the PT. When the two-stage V-1710 with the Auxiliary Stage Supercharger was introduced it used the PT carburetor as well. In all of these installations the carburetor was mounted in a "down-draft" position directly above the inlet to the supercharger, though on the two-stage it was upon the inlet to the Aux Stage. With the carburetor ahead of the Aux Stage the pressure drop though the carburetor impeded the capacity of the Aux Stage supercharger and Allison introduced a "Between Stage" arrangement. This improved performance, adding 1,500 to 2,500 feet to the critical altitude, as well as 7-12 mph to the airplane. With the introduction of the E/F altitude rated engines Allison reverted to the PD carburetors, though models with slightly larger throttle bores. The altitude performance was maintained by increasing the engine-stage supercharger gear ratio.

The suffix on the designation noted the detailed configuration, setting or modification state of the carburetor. In many cases these details were specific to a particular aircraft model.

Following the evolution from "float" to "pressure" carburetors, the next technological improvement in the device was to dispense with it entirely. This was done with the introduction of Speed Density "carbure-

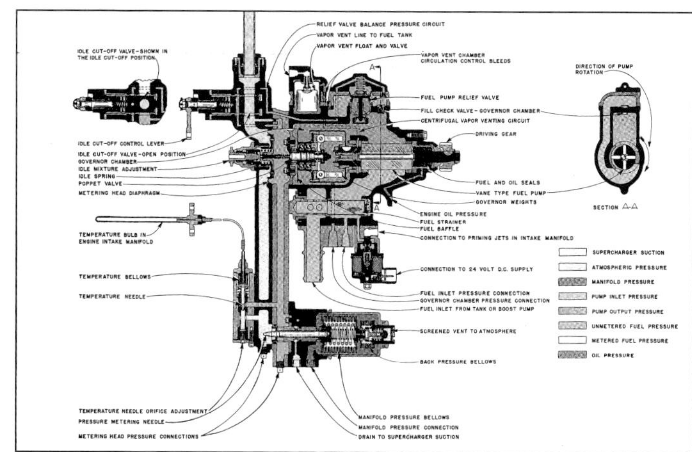

Figure 16-4b. This is the heart of the Bendix Speed Density carburetor. Observe how the engine temperature and pressure signals were combined with rpm to control the pressure and flow of fuel in proportion to the airflow into the engine. (USAF TM 52-12)

tors." These were not actually carburetors, but rather a fuel metering device that used engine rpm, manifold pressure and temperature in combination with local atmospheric pressure and temperature to calculate the airflow through the engine. By combining this information with desired mixture ratio, the result was the amount of fuel required in proportion to the flow of air. Allison adopted this unit on the aftercooled V-1710-F32R as a way to improve mixture control over a wide range of power and altitude while eliminating the undesirable restrictions to flow within the induction system caused by venturis and metering jets. Figure 16-4 shows how the Bendix SD-400 unit operated.

Because of the pre-production nature of the Bendix unit, Allison became quite involved in its development. Following promising but generally unsatisfactory operation of the SD-400-A model, an improved but experimental SD-400-B1 unit was extensively tested in 1945. Lessons learned were incorporated into a considerably improved SD-400-D1, a version of which went on to be used in the V-1710-G6 production engine.

Problems were encountered in the device as a result of having rotating parts and ball bearings operating while submerged in gasoline. New types of bearing lubrication and diaphragm seals were necessary to resolve seizures that would in turn stop the engine. The very necessary temperature metering device required redesign to minimize damage during handling and to improve its durability. The possibility of failure of the fuel pump drive shaft and governor (eliminating all fuel flow and stopping the engine) necessitated incorporating features for having an electric fuel pump able to provide a constant flow of fuel sufficient to "get home" safely as a backup.

With the introduction of the SD-400-D2, a derichment valve was incorporated to allow use with water-alcohol injection systems, as well a mechanical accelerating pump to improve transient operation. Difficulty in the development of the SD-400 effectively prevented the two-stage Allison from getting into the production version of the light-weight Mustang near the end of the war.

Intake Manifolds for the V-1710

A reciprocating engine is simply a pump working on a combustible mixture of fuel and air. The whole question of engine performance then relates to how well the air can be delivered to the cylinders where the combustion process occurs. The capability of the intake manifolds to uniformly deliver mixture to the many cylinders has a lot to do with the ratings, performance, and even the reliability of the engine. After a shaky start the V-1710 was developed into an engine with particularly good intake manifolds and breathing.

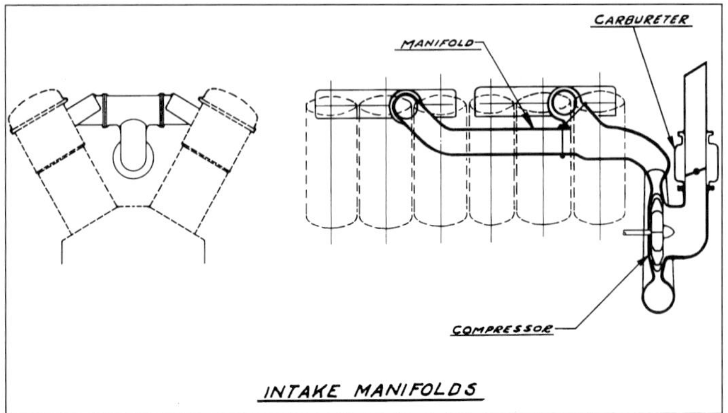

Intake Manifolds on the GV-1710A and XV-1710-1(A2) were fairly simple but became a source of considerable trouble during the development of the early engine due to the sever maldistribution of mixture. Although these manifolds were revised several times on the pre-Hazen engines, this general arrangement was retained. (Author)

All V-1710 engines used intake manifolds that grouped quadrants of three cylinders. In the early GV-1710-A, XV-1710-1, XV-1710-3 and XV-1710-7(C2) these were fed from straight tees directly off of a central trunk running in the center of the Vee, between the cylinder banks. When Ron Hazen arrived at Allison in 1936, his assessment of the engine determined that many of its problems were due to the wide spread in the quantity and richness of mixture being delivered to the various cylinders. This was due in no small part to the significantly different distances the mixture had to travel through the manifolds to reach the connected cylinders. Hazen's major contribution to the engine was to conceive and design the tuned "ramming" manifolds that approximately equalized the distances to each cylinder while utilizing the pulsating nature of the flow to create pressure waves to pack more mixture into the cylinders. The new manifold reduced the variation in power from each cylinder from the previous +/-10 percent to less than +/-2 percent. It was with these new manifolds that the YV-1710-C4/C6 was finally able to complete the 150-hour Type Test in 1937.[105]

Completing the Type Test did not end development of the intake system. While the new Hazen manifolds were decidedly better, a further improvement was introduced in the C-10 engine that smoothed the flow by eliminating the sharp right angle turn as the mixture entered the quadrant. Next there were the changes brought about by the backfiring incidents occurring during the introduction of the P-40 *Tomahawk*. This resulted in strengthening the manifolds by making them of aluminum, rather than the original magnesium. Allison also introduced a new tee fitting, the "divided tee" (PN40041), that featured a continuation of the dividing web down to the surface of the backfire screen, effectively isolating a backfire to a single quadrant. The old PN34162 tee used through Model Testing of the V-1710-C15.[106] Still, Wright Field did not believe the backfire problem had been solved so they directed redesign of the manifold to incorporate backfire screens at the inlet ports to the cylinders. The entire issue was finally resolved when Allison introduced an improved Tee design with oval cross-section that maintained constant velocity through the tee sections and on through the distribution quadrants. With these "streamlined" manifolds it was possible to remove both the cylinder port backfire screens and the screens that had been installed in the "tee" to straighten the flow and distribute it evenly between the banks. These new manifolds were Accelerated Service tested in P-38's during the summer of 1942 with favorable results. On August 31, 1942, Allison was informed to proceed at once to install them on all "F" engines in production and service.[107] On September 17, 1942 Allison was notified to utilize these "Streamlined" or "Constant Velocity" intake manifolds on all engines and that all backfire screens, both those in the manifold tee and the intake port backfire screens were to be removed.[108]

Another feature which was incorporated was known within Allison circles as the "Madam Queen" gas intake pipe, the large pipe connecting to the distribution tees ahead of the ramming manifolds. The device was first used in an attempt to resolve the uneven fuel distribution occurring in the XV-3420-1 engine, in December 1939.[109] It was unsuccessful in that application, but was subsequently successful in resolving a V-1710 problem occurring during low-power cruising at very high altitudes. Even with supercharging, during low power cruise the velocities within the trunk connecting the engine-stage supercharger to the distribution tee would drop comparatively low, and with the low temperatures typical of high altitude, fuel was condensing and collecting as liquid in the bottom of the trunk. This in turn could cause fires and a number of other problems. The solution was quite elegant. What was done was to take a "boost venturi" similar to those used in a carburetor to measure airflow, and install it in the

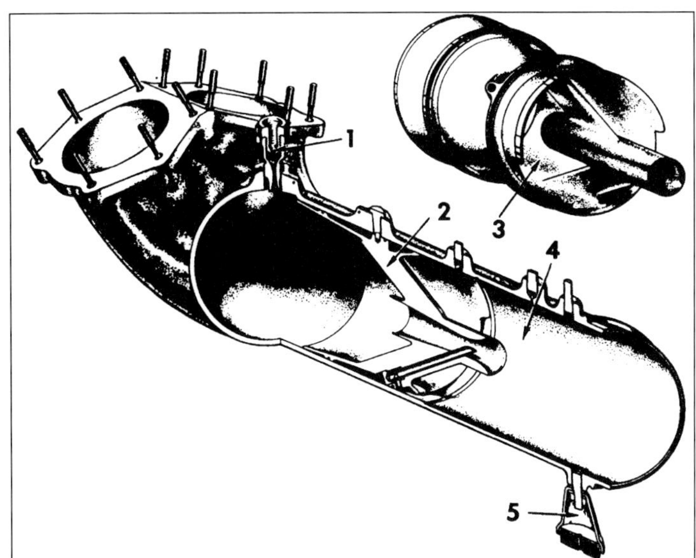

The Madam Queen gas intake pipe incorporated a boost venturi connected to the floor of the pipe and was able to draw up and reatomize any fuel that collected in the pipe. "Bumpers" were fitted on the cylinder blocks to insure proper location. The streamlined passages into the tee fittings allowed deleting the backfire screens. 1-Idle Fuel Injector, 2/3-Venturi Assembly, 4-Gas Intake Pipe, 5-Pipe Support. (Allison Service School)

intake pipe. The low pressure created within the venturi was then connected to a small tube positioned just above the floor of the trunk. Should any fuel collect in the trunk it was immediately drawn up the tube and reatomized back into the airstream. This type of gas intake pipe can be identified by the five bosses on the upper side of the pipe. Earlier models have only three bosses.[110]

For historical purposes, the following summary of induction manifold part numbers is given to delineate the major design changes.

Original "Log" Type Manifolds: The distribution manifold incorporated two T-adapters supplying the front six and the rear six cylinders respectively. Each branch of a "T" then fed three adjacent cylinders through a long narrow slot. During development, these manifolds were enlarged at least twice to better accommodate higher power levels. When the XV-1710-1(A2) was tested it was with PN19420 manifolds, though the units finally installed are PN19408/9.[111]

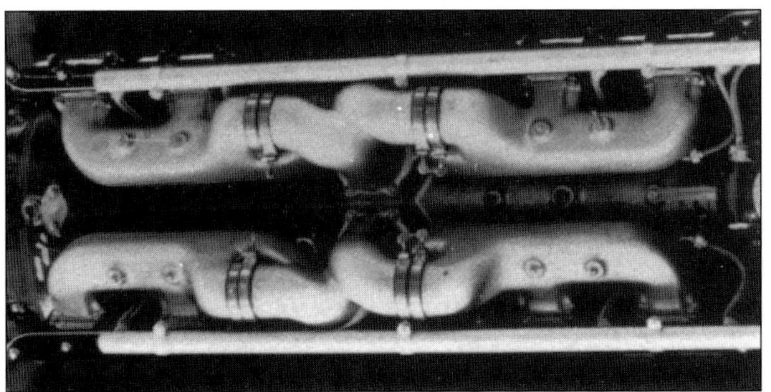

The initial PN34550 ramming manifolds by Hazen are typified by this installation on the V-1710-C6. Note how the tees are blended together and mount on top of the backfire screen. The connection to each quadrant is at nearly right angles and was found to cause unwarranted pressure loss. This setup was used on the C4/C6/C7/C8/C9/D1 and V-3420-A1.

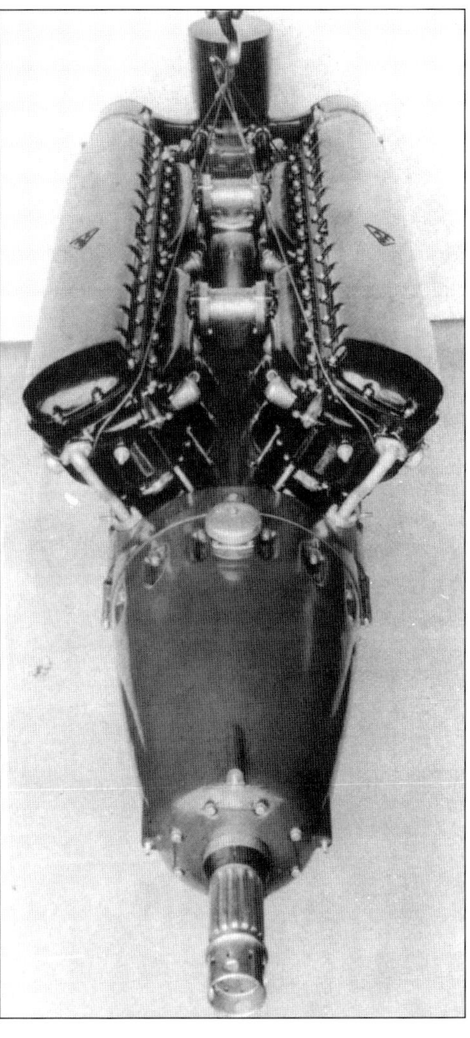

This early view of the XV-1710-1(A2) clearly shows the trunk and tee manifolds and how they fitted with a single chamber spanning each group of three cylinders. (Air Corps)

Individual Port/Plenum Manifolds: These manifolds were first used on Build 3 of the XV-1710-3(C1), the first engine to incorporate cylinder blocks having individual intake ports, a feature needed to facilitate the intended introduction of port type fuel injection. Photos of the XV-1710-7(C2) show these PN33550 manifolds. Testing in early 1936 showed they contributed to the serious maldistribution of mixture, +/- 10 percent between cylinders. This was a major cause of the reliability problems delaying the Type Test at 1000 bhp.

Hazen "Ramming" Manifolds: Ron Hazen led the development of a new manifold intended to take advantage of the mixture flow inertia and resolve the maldistribution problem. These manifolds were able to deliver mixture with only +/- 2 percent differences between cylinders. They also incorporated backfire screens, located in the tee coming from the supercharger delivery trunk, to insure even distribution between the banks. These manifolds were made of magnesium and assigned PN34054 and only used on engines through the C-9 and D-1.[112]

Improved Ramming Manifolds: When the 1150 bhp C-10 and D-2 were introduced they were fitted with an improved manifold to smooth the transition into each quadrant. The result was the PN36085/6 manifold, initially fabricated from magnesium, but during the trauma of the P-40 backfire period, was strengthened and made fireproof by fabricating in aluminum.

Divided Tee: These tee-fittings were original equipment on engines in the second production contract, W535-ac-16323, and were retrofit onto all existing service engines. By way of example, the V-1710-27/29 used PN40041 intake manifold tees, in which the dividing vane came down to the top surface of the PN40043 backfire screen, a feature intended to reduce the volume of charge that could be ignited by a backfire from any one cylinder by separating the branches to each bank. The V-1710-33 and foreign sales V-1710-C15 used PN34162 tees, in which the dividing vane terminated about 9/16 inch above the PN36930 backfire screen. At the same time the new Manifold Tees were introduced and retrofit onto engines in the field the new aluminum PN36085/6 manifold quadrants were installed. These manifolds retained the same part number, though the suffix was changed to "BB", as PN36085/6 BB. These changes coincided with the changeover in production contracts to W535-ac-16323.

Port Screen Manifolds: Based on continuing backfire events, Wright Field still was not satisfied that Allison's actions would resolve the problem. At their insistence, a new PN40772/3 "wide flange" aluminum manifold that fit a backfire screen at each intake port was introduced in May 1941. It was to be used in combination with the PN40041 Tee and a backfire screen. Concurrently, Allison was successful in getting the Materiel Division to agree to reset the clearance on all intake valves from 0.010 to 0.015 inches. Tech Order 02-5A-15 was issued in October 1941 to insure that all engines in the field had been properly modified.

A consequence of the double backfire screens was to require rerating of the engines, not a desirable condition. Wright Field issued new performance curves accounting for the backfire screens and revised rated rpm (to minimize overspeeding due to Curtiss-Electric governors) to 2800 rpm in August 1941.

The port screens were installed on engines AF40-504 through 42-33581 at time of manufacture. Beginning with engine AAF42-33582 a PN41726 spacer was installed in place of the port backfire screen.[113] Subsequently many of the PN40772/3 manifolds were made serviceable by inserting the spacer. A port so modified is shown in an adjacent photo. Consequently both salvaged PN40772/3 and the earlier PN36085/6 manifolds were used in combination with the new streamlined tees.

The insert to smooth the flow path into the cylinder when the port type backfire screens were deleted is evident in this photo of a PN40772 "wide flange" manifold. (Author)

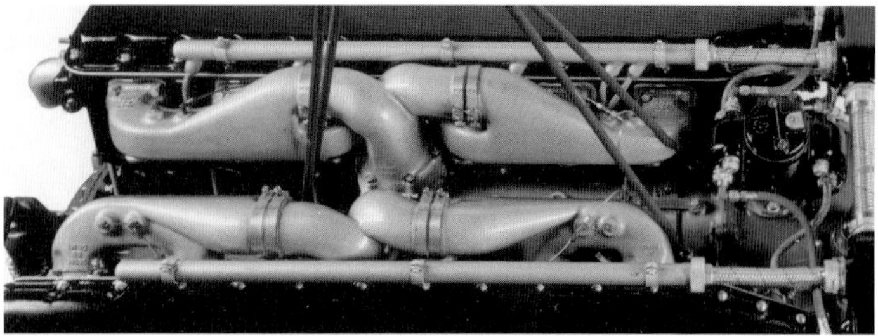

This is the first two-stage V-1710-47(E9), AEC#5421, AC41-30474. Close examination shows PN40041 Divided Tees with backfire screens and PN38085/6 manifolds (without port screens). When this photo was taken in June 1942 the "Military Standard" V-1710 was to be equipped with both tee and port backfire screens. As shown the performance would be improved without the port screens. (Allison)

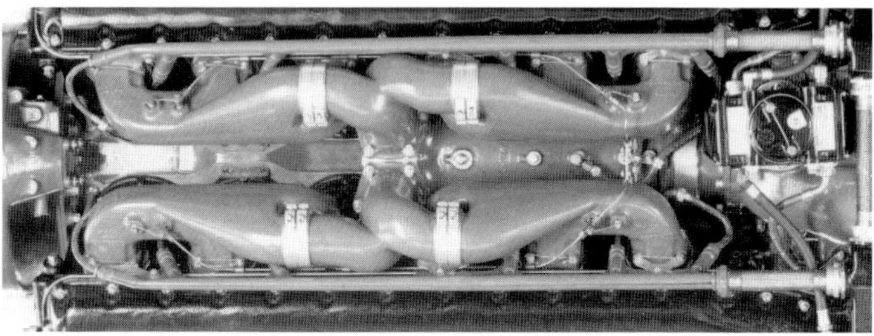

Intake Manifolds as fitted on the 1944 vintage V-1710-119(F32R) with aftercooler were PN36085/6DD. Note use of venturi type gas intake pipe and how magneto was offset on the right side of the gas pipe. (Allison via J. Wetzel provided by G. White)

Streamlined Manifold Tees: By June 1942 Allison had developed the new Streamlined gas pipe and tee that allowed removing the tee backfire screens. With the new components, and without screens, a test engine had endured over 2,200 backfire events without serious consequences. In addition, a service wide inspection done per Tech Order 02-5A-20 in March 1942 had found that the earlier problems with valve seat damage had been resolved by resetting the intake valve clearance. Consequently, the back-

This photo shows the gas intake pipe and PN40041 tee from P-38F "Glacier Girl" (upper units, with backfire screens) in comparison to the constant velocity pipe and tee (PN42790). Note that in both cases the dividing web inside the tee comes down to the mating surface. Both setups would connect to either the PN36085/6 or PN40772/3 3-port manifold castings. (Author)

SECTION III
SPECIFIC OPERATING INSTRUCTIONS

RESTRICTED

ENGINE: V-1710-73 P-40K

DATE:

CONDITION	FUEL PRESSURE LB/IN.^{2}	OIL PRESSURE LB/IN.^{2}	OIL TEMP. °C	COOLANT TEMP. °C
DESIRED	12-16	60-70	60-80	105-115
MAXIMUM	16	85	95	125
MINIMUM	12	55		85
IDLING	9	15		

MAX. PERMISSIBLE ENGINE OVER SPEED: 3120 R.P.M.

MAX. ALLOWABLE OIL CONSUMPTION AT:

NORMAL RATED POWER 13.3 QTS./HR.
MAXIMUM CRUISING 10 QTS./HR.
MINIMUM SPECIFIC FUEL FLOW 5 - 7 QTS./HR.

Fuel Specification No. AN-VV-F-781 (Amend. 5)

FUEL GRADE S-1 + 1.0 OCTANE

⊕ OPERATING CONDITION	HORSE POWER	R.P.M.	MAN. PRESS. (IN. HG)	PRESSURE ALTITUDE (IN FEET)	BLOWER CONTROL POSITION	USE LOW BLOWER BELOW	MIXTURE CONTROL POSITION	MIN. F/A RATIO	FUEL FLOW GAL/HR	MAX. CYL. HD. TEMP. °C	REMARKS
TAKE-OFF	1325	3000	51.0	Sea Level	-	-	Auto Rich	-	155	-	5 Minute Operation Only
WAR EMERGENCY	1550	3000	60.0	Sea Level	-	-	Auto Rich	-	174	-	5 Minute Operation Only
MILITARY RATED POWER	1150	3000	44.6	10700 11700*	-	-	Auto Rich	-	132	-	15 Minute Operation Only
⊙ NORMAL RATED POWER (100%)	1000	2600	38.5	9900 10800*	-	-	Auto Rich	-	105	-	*With streamline manifolds. Other altitude figures are for tee backfire screens installed.
MAX. CRUISING (75%)	750	2280	30.8	9900 10800*	-	-	Auto Rich	-	65	-	-
DESIRED CRUISE (67%)	670	2280	28.2	9900 10800*	-	-	Auto Rich Auto Lean	-	56 50	-	-
DESIRED CRUISE (60%)	600	2190	26.0	9900 10800*	-	-	Auto Rich Auto Lean	-	50 45	-	-
CRUISE FOR MIN. SPECIFIC FUEL FLOW	375 420 460 490 450	1950 1950 1950 1950 1950	22.5 22.5 22.5 22.5 22.5	Sea Level 5000 10000 15000 20000	-	-	Auto Lean Auto Lean Auto Lean Auto Lean Auto Lean	-	28 32 35 37 34	-	-

REVISED 12-1-41

⊕ REFER TO T.O. NO. 00-10 FOR DEFINITION OF EACH OPERATING CONDITION ⊙ MAXIMUM PERMISSIBLE CONTINUOUS HORSE POWER

T.O. NO. 02-5AB-1A

RESTRICTED

Performance Improvement with Streamlined Manifolds compared to the earlier engines with backfire screens in the manifold tees. An additional 900 to 1,000 feet in rated altitude was available in engines equipped with the new Streamline manifolds. Neither of these engines were using port backfire screens at the time these ratings were issued. (USAAF)

fire episodes were no longer a concern. On this basis the Materiel Division was willing to install the new gas pipe and tee and at the same time remove the port backfire screens.

Elimination of the screens added 2,400 feet in critical altitude and 7 to 10 mph in top speed of the P-39 and P-40.[114] Note that this gain in critical altitude is a combination of engine performance and ram effects due to aircraft speed. Table 16-20 shows a page from the Operation Instructions, with a nominal 1,000 foot gain in critical altitude, this of course being established for the "average" airplane.[115]

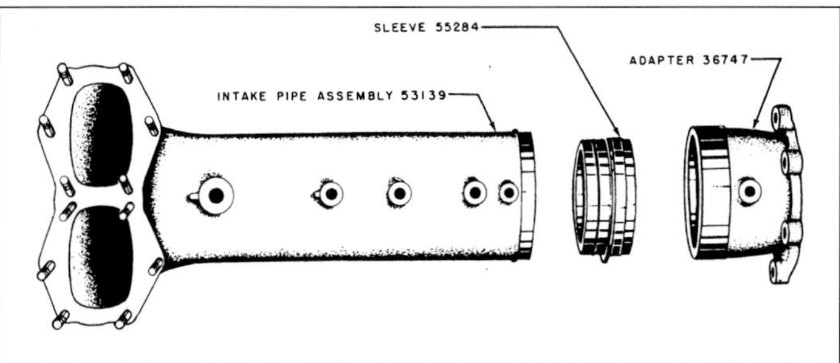

This view of the late model gas intake pipe shows the constant velocity oval passages that mated with a matching tee and eliminated the need for backfire screens. (Army Air Forces)

The new divided tee, PN42790, was distinguished by having an oval section for mounting to the matching center manifold trunk. The intake manifold sections mating to the three cylinder segments could be either the old PN36085/6 manifolds, or the PN40772/3 manifolds with a sleeve inserted in place of the port screens to provide a smooth passageway into the cylinder.[116]

Still, it was not until September 1942 that Wright Field concurred in the use of this manifold on turbosupercharged airplanes. They had extensively tested twelve engines, including V-1710-27, -29, -49 and -53's in P-38F's for about 50 hours each and reported no backfires. On this basis the streamlined manifolds were approved for installation and use on all V-1710's.[117] This testing confirmed that the original cause of the backfiring was the tight intake valve clearance setting, coupled with the overspeeding due to poor propeller governor performance.

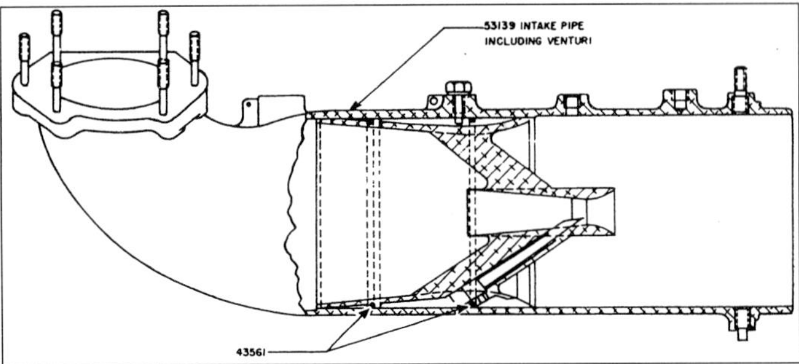

For reasons known only to the participants this somewhat sophisticated gas intake pipe was known as the "Madam Queen". It was particularly effective in reatomizing any gasoline that condensed or had not fully evaporated coming out of the supercharger. (Army Air Forces)

Center Venturi Intake Pipe The "Madam Queen" venturi-type center gas intake pipe was developed and ultimately retrofit onto all V-1710's from the F-5 models onward. The purpose of the device was to improve fuel distribution between the banks and cylinders by insuring that the fuel was completely evaporated before it entered the cylinder ports. As can be seen on the accompanying figure, a type of "boost" venturi was fitted into the gas pipe coming from the supercharger. Within the device was a tube connecting to the bottom of the pipe and arranged to pull any collected liquid fuel and reatomize it. In 1945 the initial 1943 PN43710 version used on production engines was replaced by an improved PN53139 assembly that replaced whatever early pipe was installed, as well as being used in production of most V-1710-111/113 engines. The new assembly also was equipped with "bumpers" mounted on the cylinder blocks to insure the pipe maintained proper position within the vee of the engine. It was also made of a more durable material so as to resist chemical attack by the fuel.[118] Madam Queen gas intake pipes replaced the PN42791 trunk and was installed on the engine models beginning in 1943 as listed in Table 16-20.[119]

Table 16-20
Use of the Madam Queen Intake Pipe

Model	Engine Serial Number	Date
	PN43710	
V-1710-89	A-038549	November 30, 1943
V-1710-91	A-039385	November 29, 1943
V-1710-85	A-032627	November 24, 1943
V-1710-99	A-041632	December 1, 1943
V-1710-93	All Engines	1943/4
	PN53139	
V-1710-111	A-063845 to A-067413	1944/5
V-1710-113	A-064156 to A-067969	1944/5
And on all subsequent production		

Upgrading the thousands of V-1710's in the field with the improved manifolds would have been a major undertaking, so was usually accomplished during major overhauls at a centralized depot. Specifics for installation of the constant velocity gas intake pipes and tees were issued in July 1943 in T.O. 02-5A-46, with the new parts installed at overhaul.[120]

A subsequent improvement was made in February 1944 to rework or replace the gas intake pipe to improve the split line seal. Included were two other important changes, addition of a support for the gas intake pipe and bumpers placed to improve alignment and minimize movement of the manifold tees in the event of a backfire. These changes were simply external sheet metal fittings mounted on existing bolts and clamps.[121]

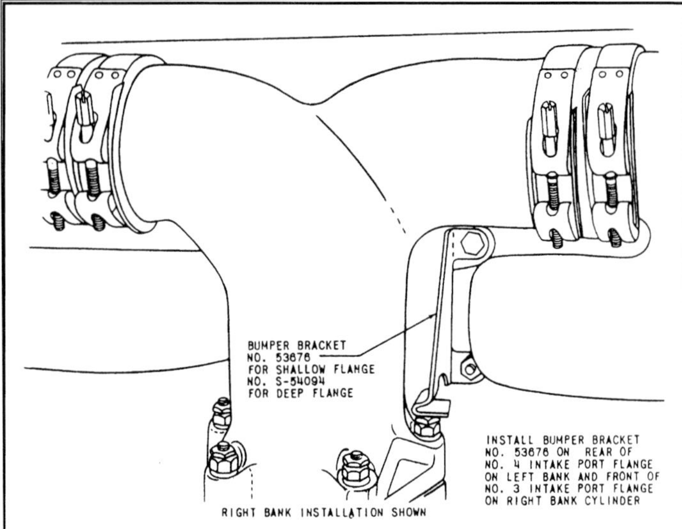

These simple brackets were effective in preventing excessive deflection of the gas intake pipe and manifold tees in the event of a violent backfire. This reduced the likelihood that a manifold would break and lead to a fire. (TO 02-5A-64)

Exhaust Manifolds and Ejector Stacks

American practice regarding the engine as a component of the aircraft was quite different from the example established by Rolls-Royce in England. Rolls was very involved in the integration of the engine and its various system within the airframe, while U.S. practice was that the engine was provided by the Government to the airframe manufacture, who was to use it to meet his performance guarantees. Typifying this approach, Allison supplied an adapter plate to connect the exhaust manifolds or stacks to the engine. Since each cylinder had two exhaust ports, six plates were required for each bank of cylinders. It was then up to the airframe manufacture to weld his exhaust system to the plates, and bolt them to the heads.

Airframe manufacturers were responsible for designing the exhaust stacks and welding them to the Allison supplied mounting flanges. This photo details the "12 port" , low drag exhaust stacks seen on each side of many late model Bell P-39's. (Author)

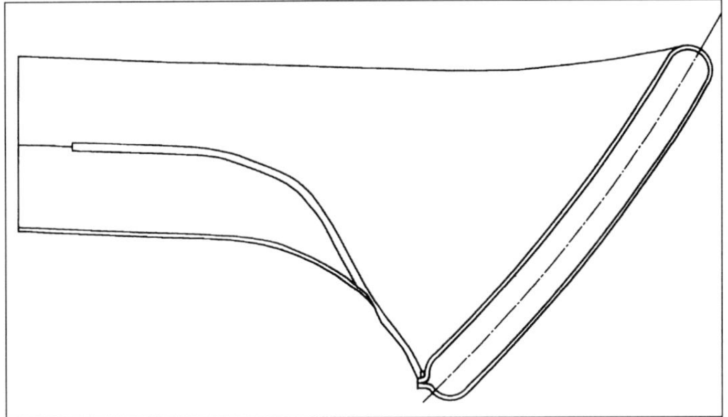

This "Fishtail" exhaust stack was designed by Curtiss and used on many V-1710-F powered P-40's. Fabricated from sheetmetal the formed stack was welded to the Allison provided adapter plate. Each assembly covered the two ports of a single cylinder. (Curtiss)

In turbosupercharged installations it was necessary to provide fairly complex manifolding that would remain leak-tight while containing 1800°F exhaust gasses at up to 50 psia! One such system which proved quite adequate, was the one on the Lockheed P-38. With few changes it was adapted to other aircraft and test programs and satisfactorily operated at very high pressures, such as the V-1710-127(E27) turbo-compound developed for the Bell XP-63H.

The Rolls-Royce and German RLM research organizations led the way in demonstrating that simple "ejector" type exhaust stacks could lend substantial thrust to a high speed fighter while adding almost no weight or complexity to the installation. By 1940, most U.S. manufactures were aware of these benefits and had access to NACA research into how to properly design such devices to maximize the benefits and minimize adverse backpressure on the engine.

As a result, simple straight cut ports, angled back along the longitudinal axis of the airframe to provide forward thrust, were often used. Another style was the "fish tail" design, which had the advantage of somewhat reducing the flames issuing from the stacks. These were particularly necessary if the pilot was to retain any night vision. While made in many shapes and sizes, they were all welded to the plates supplied by Allison.

Fishtail exhaust stacks on a P-40E. Even when made from tough material, the high temperatures, vibration, and sonic velocities requires careful inspection and maintenance to insure there are no leaks into the engine compartment. Note how the cooling air for the blast tube below the spark plugs is connected through an elbow. (Author)

Allison found that with the introduction of WER ratings, their previous recommendations on the size of the ejector jet nozzle was inadequate to pass the additional flow without creating excessive back pressure and potentially causing detonation. They had previously recommended that stack areas of 3.2 to 3.6 square inches per cylinder were adequate for up to 1325 bhp at takeoff and 1150 bhp up to 12,000 feet. At the new ratings they recommended 4.2 to 4.4 square inches per cylinder. As an expedient, Allison suggested that field forces could simply saw off the ends of the stacks to achieve the desired area.

Examples can be found with almost every type of aircraft having quite different types of ejectors installed. One of the most effective was the low drag, hi-thrust stacks used by Bell on many of their late model P-39/P-63's. These had 12 ports on each side of the engine/fuselage. This reduced the cross-sectional area of the aircraft, and its consequential drag, while maintaining the desired jet thrust and efficiency.

Turbo-Compounding the V-1710

Early in 1939 the Materiel Division at Wright Field asked Allison to undertake an experimental program to improve the performance of the internal combustion engine by turbo-compounding. This was to be accomplished by routing the exhaust gasses through a turbine, which in turn would be connected to the engine crankshaft, thereby increasing the power and efficiency of the engine. They proposed that the Allison inline engine, with its four-valve combustion chamber, would be most suitable to their scheme and purpose.[122]

The normal V-1710 was constructed so that each of the two exhaust valves in every cylinder had its own exhaust port. What the Materiel Division wanted Allison to do was to revise the cam and exhaust rocker arrangement such that one port would be connected to the power recovery turbine and the other ported directly to atmosphere. In operation the valve connected to the turbine manifold would open first and release the very high pressure exhaust gasses, on the order of 1000 psi, until the pressure was down to 15 or 25 psig, at which point the first valve would be closed and the valve porting to atmosphere would be opened to complete blowdown of the cylinder. Fairly complex, but workable, though operating conditions for the turbine would have been quite severe. This arrangement was not adopted at the time, but it obviously got people thinking. "Good" ideas have a hard time dying, and this approach to turbo-compounding resurfaced in September 1945. At that time, the NACA Aircraft Engine Research Laboratory in Cleveland undertook procurement of a set of suitably configured camshafts that were already available from the Air Technical Service Command at Wright Field.[123] Other than producing a study and report, this project does not appear to have progressed, certainly not to flight hardware.

By 1944 Allison had its mass production problems under control and reliable engines were being delivered in large numbers. Allison engineers were then able to turn their attention to the demands for improving the performance of Allison powered aircraft. The earlier interest in turbo-compounding was re-investigated.

Military tactics at the time required improved takeoff performance, to carry ever greater quantities of fuel for the long range escort missions, and at the same time provide more power at ever higher altitudes. It was well known that recovery of the nearly 50 percent of fuel energy wasted in the exhaust gases could go a long way in meeting these goals. In 1944 Allison launched a major effort to improve the V-1710 and satisfy the demands from the operating theaters. The path selected was to develop a "turbo-compound" engine. This was an ambitious development as no one had ever flown such a machine, or even demonstrated a practical test cell setup.

The project went through several development and test phases. Component parts were subjected to extensive development testing under conditions similar to those anticipated for the final V-1710-127(E27) turbo-compound engine. The actual engine utilized components from the latest Allison standard, the V-1710-E30, which was rated for operation at 3200 rpm, with high manifold pressures and able to maintain sustained high power. Its major component differences from earlier models involved using the 12 counterweight crankshaft and pistons having reduced compression ratio of 6.00:1 instead of the standard of 6.65:1.

The concept followed Allison's basic design approach of interfacing stand-alone components with minimum redesign of each, and thereby imposing minimum disruption of the production line. The engine was configured as an "E" model, for driving a remote reduction gear box, had integral engine-stage supercharging, and was mechanically coupled to an Auxiliary Stage Supercharger. The configuration had a two-stage breathing ca-

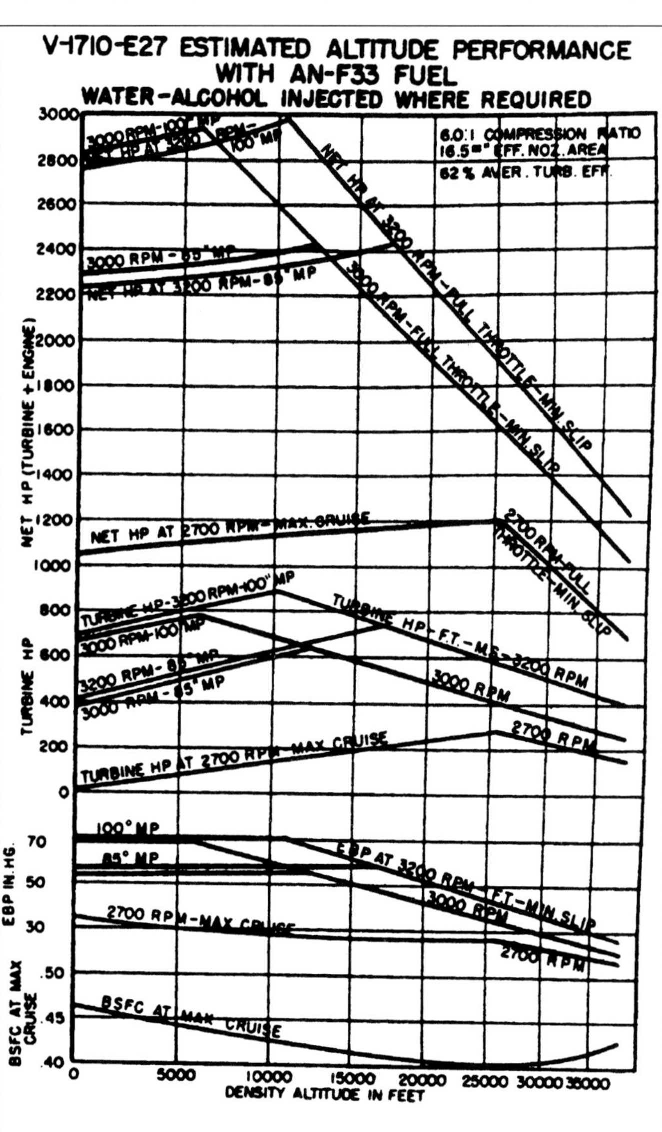

In the turbo-compound Allison brought all of their high power developments together. These curves are based upon actual development tests of both components and full scale engines. (NACA)

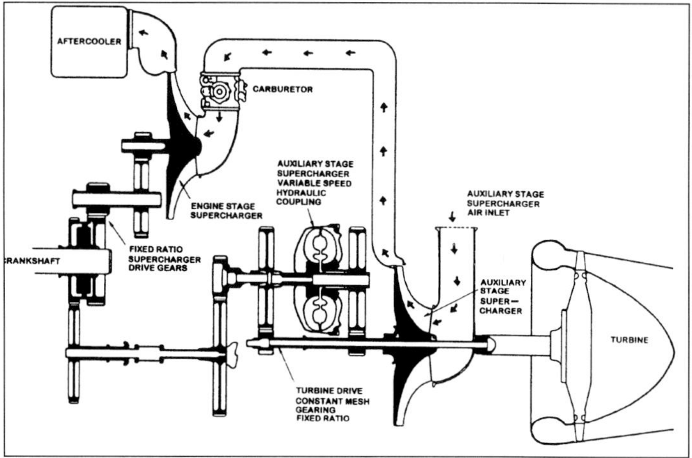

The 890 bhp turbine in the E-27 drove through a hollow Auxiliary Stage drive shaft, connecting to the crankshaft via a fixed 5.953:1 ratio. The Aux Stage was driven in the usual way through the hydraulic coupling. (NACA)

pability and a high critical altitude, all in support of the installation in the new Bell XP-63H. The Auxiliary Supercharger was direct connected to the power recovery exhaust gas turbine, itself derived from the General Electric Type CH-5 turbosupercharger used in late model 2800 bhp P-47's and coming 3000 bhp R-4360 powered bombers.

The turbine was built by GE as their CT-1 power turbine. It differed from the turbine in a standard Type CH-5 in that the nozzles and buckets were arranged to drive the wheel in the opposite direction. It had a nozzle flow area of 16.5 square inches and operated with an efficiency of about 62 percent. No waste gate was provided as it was intended to operate the turbine at all times and the sizing of the nozzle area defined the maximum turbine power and engine backpressure. GE set the maximum allowed exhaust gas temperature into the turbine at 1725 degrees Fahrenheit, which was somewhat higher than the nominal 1600 degrees seen in normal turbo operations and was clearly at the upper limit for the design and then available turbine materials. This temperature constraint was the "Achilles Heel" of the project.

It was found by testing, that with the engine at 3200 rpm, and at 11,000 feet, with water injection and 100 inHgA manifold pressure, the turbine caused a back pressure of 70 inHgA while developing 890 bhp. At this manifold pressure, and with the turbine coupled to the engine, 2980 bhp was delivered to the propeller. The turbine was connected to the engine through 5.953:1 gears contained in the Auxiliary Stage Supercharger gearbox. The connecting shaft was concentric with the Aux Stage impeller, but the impeller continued to be driven through its standard hydraulic coupling. In this way, the Aux Stage only consumed the amount of power necessary to provide the demanded pressure at the inlet to the engine-stage supercharger, minimizing the temperature of the air inducted by the engine and helping to reduce the threat of detonation.

Ducting of the hot exhaust gases from the engine to the power turbine was accomplished by using standard late model P-38 manifolds and fittings. These were suitable up to the maximum back pressures, which could reach 100 inHgA, nearly 50 pounds per square inch.

The one-off Bell XP-63H, intended to test the engine in flight began as a P-63E and was modified to accommodate the turbo-compound Allison engine installation. The fuselage was not lengthened, as the P-63 series had already been "stretched" to accommodate the "two-stage" Allison's. At first viewing it appears that the installation was somewhat crude as the exhaust manifolds were simply run external to the fuselage. In fact, this was an attempt to accomplish a degree of exhaust gas cooling prior to passing them through the temperature limited turbine. P-38 type collector manifolds were bolted directly to the V-1710 cylinder blocks, and covered with a simple sheet metal shroud, open on both ends, which extended into the airstream and continued to just past the rear of the cylinder blocks. From this point on, the circular pipes were external to the fuselage back to the point where they entered through a circular cut-out, and were connected to the turbine nozzle box, situated on the centerline of the aircraft.

The turbine exhaust was then collected by a shroud and ducted to a 12 inch diameter ventral exit, again on the fuselage centerline. This required the elimination of the ventral stabilizing fin normally seen on late model P-63's, hence the profile of the XP-63H resembled the earlier P-63A and C models. At maximum power and altitude this exhaust duct configuration should have been able to contribute a further 100-200 pounds of net propulsive thrust, though no figures have turned up. The only other significant modification was the relocation of the oil tank, though it appears to have remained near its original position.

All of this suggests that the XP-63H may have had a fairly "aft" center of gravity, as the V-1710-127 was expected to weigh 1,950 pounds, dry. The usual engine for the P-63E, the V-1710-109, weighed in at 1,660 pounds, dry. The CT-1 assembly alone probably weighed about 150 pounds, and was located quite far aft. Possibly, the ventral fin removal assisted in restoring balance along with ballast weight in the nose.

Table 16-21
V-1710-127(E27) Ratings

Rating	BHP/RPM/Altitude, ft/MAP	Comments
War Emergency	3090/3200/28,000/100 inHgA	Wet
Military	2320/3200/28,000/ 85 inHgA	Dry
Normal	1740/3000/33,000/ 52.5inHgA	Dry
Maximum Cruise	1340/2700/26,000/	Dry, BSFC 0.365 #/hp-hr

With all of this, the V-1710-127(E27), was quite a performer. Allison's proposed guaranteed performance, on AN-F-33 grade 115/145 fuel and equipped with an indigenous air-cooled turbine able to operate at 1950 °F is given in Table 16-21.

Lean Fuel/Air mixtures are known to cause high exhaust gas temperatures. Allison found that even the normal "rich" mixture ratio of 0.08:1 pounds of fuel per pound of air, and water injection sufficient to depress detonation at 100 inHgA, would result in 1825 °F exhaust gases.

They believed that a power turbine able to operate with 1950 °F gas would have to be developed to realize the promise of the turbo-compound. In 1946, Allison believed the effort could be better directed to establishing their role in the field of jet propulsion. Consequently the program was terminated and the V-1710-127, and its XP-63H testbed, never flew.

Prior to terminating the program Allison proposed the V-1710-G10R/L series of turbo-compound engines. These units would have utilized an air-cooled turbine of Allison's own design. At the same time, Allison proposed the V-1710-H series, which would have removed the engine-stage supercharger as well as the Auxiliary Stage Supercharger and provided an Allison designed two-stage supercharger driven by a two-stage air cooled turbine. This would have offered un-compromised engine efficiency, as well as extremely high power levels. In fact such an arrangement was subsequently developed by Pratt & Whitney for the R-4360-VDT, intended to power the proposed Consolidated B-36C series of very heavy long-range aircraft.

In June 1945, with the war winding down, resources were available to look into new technical fields. Compound engines were one of the selected fields investigated by the Army Development Engineering Office. Their report showed that for a year, the General Electric Company had been working on a compound gas turbine to be used on the P&W R-4360-VDT. The Allison V-1710-127 project for the P-63 had been under contract for six months and was showing progress. In addition to this project, Allison was considering the possibility of compounding the V-3420 as well. Pratt & Whitney was not under contract, but was cooperating with GE on the R-4360 project and had submitted a proposal to for a turbo-compound R-2800-E type engine. The Studebaker Corporation was discussing compounding their new H-9350 with General Electric, as was the Lycoming Division, who felt that their liquid-cooled R-7755 was ready for the feature. Wright Aeronautical had discussed the concept with Wright Field and the Air Forces, but the bulk of their effort had been directed at obtaining the power necessary to drive cooling fans, with no provision for altitude performance of the engine. Wright was expected to enter into contracts with the Navy Bureau of Aeronautics on that project. In general, turbocompounding had been very much delayed by work on jet and turbine engines, as reflected in its status at General Electric where it was number eleven on their priority list, the same as being in "inactive" status. The Air Force did not intend to terminate any of the pending projects and was taking measures to accelerate the program within the limitations of their priority status which was set at "3A" for both the General Electric and Allison efforts.[124]

In 1948, Wright Aeronautical Corporation began testing their turbo-compound R-3350. This configuration, with its inherent high efficiency, went on to become the mainstay of the final generation of piston powered airliners and bomber aircraft such as the Douglas DC-7, Lockheed Constellation, and Lockheed P-2 Neptune. To accommodate the high exhaust gas temperatures, Wright developed air-cooled blading for their power turbine. It was a good approach, but one several years behind the Allison program that developed the first turbo-compound aircraft engine, the V-1710-127.

Meeting the Needs of the Aircraft Designers

Even with the size and complexity of being a Division of General Motors Corporation, Allison appears to have been a cooperative participant with the many aircraft manufacturers and customers for its engines. This is not to say that there was not criticism of Allison, for there appear to have been cycles of displeasure voiced by the Army, usually dealing with slowness to respond with a particular engine model, production schedule, or incorporation of some improvement desired by the military. Many of these have been investigated in the course of compiling the record for this book with the result being that Allison should be given credit for responding in an aggressive fashion whenever an issue got to the level requiring management involvement. In general, legions of issues never reached this level because of the can-do attitude and resources available to Allison representatives in the field and support from the factory.

In many technical areas, Allison was far ahead of the Army. This would include: abandoning fuel injection (1936), adopting twin crankshafts for the V-3420 (1937), the need for altitude rated engines (1938), need to develop two-stage mechanical supercharging (1938), resolution of the 1940 backfire crisis, rating engines for WER (1942), removal of backfire screens (1942), use of ADI (1942), developing Turbo-Compounding (1943), and finding an application for the aftercooled two-stage (1944).

Criticism by the Materiel Division was usually related to the factors created by the extraordinary demands to increase production rates on en-

Table 16-22
Major Allison Models

	V-1710	V-3420
AAF Models	74	14
Navy Models	3	1
Allison Models		
A	2	25
B	2	12
C	16	1
D	6	
E	32	
F	37	
G	12	
H	2	
Total	110	38

Brookley Field Army Air Base in Mobile, Alabama was one of many facilities around the world where the Army setup engine overhaul facilities. Production line methods were used to expedite overhauls and to control quality. (SDAM)

gines that had yet to complete development. This was particularly acute during the 1940-42 production of the C-15, E-4, F-2R/L and F-3R.

Allison seems to have been quite responsive to the interests of the aircraft designers. They provided at least 148 different model specifications, many of which went through a number of revisions as the various engine and aircraft designs evolved. These are tabulated, by engine model in Table 16-22.

The military models were designated only after a contract was put in place, even though, in some instances no engines were delivered. The Allison model count does not include sub-variants such as R/L turning, or those of a conceptual nature that never resulted in a specification.

Providing such a long list of engines was a demanding undertaking on Allison's part, but a natural consequence of working to meet the demand for their products. Given the building block nature of the V-1710 design the scope of these models is probably wider than that demanded of any other engine manufacturer. Having the ability to accommodate buried installations, fuselage mounted engines, remote propellers, doubled power sections, and a wide range of supercharging options gave them an engine that could be adapted to almost any conceivable application or aircraft.

Engine Overhaul

When the YV-1710-7 was first flown in the A-11A, one objective was to operate the engine through a complete overhaul cycle, intended to be set at 300 hours. The first engine did this successfully. As a result, the Air Corps decided to establish the introductory service life for the engine at 150 hours. This was the limit for all of the early engines, including the V-1710-21 and -23 in the Curtiss YP-37 and Bell YFM-1 respectively. When the Curtiss P-40 began service powered by the early V-1710-33, it also was operating the engines only 150 hours before overhaul. Based upon the condition of the early engines at overhaul, Allison recommended in early 1941 that the interval be lengthened to 200 hours on engines which had been "Modernized." The life of the engine was consistently lengthened until established at the intended 300 hours.

When the V-1710-E/F engines were introduced and well established in production the allowed interval between overhauls was continually raised until, by the end of the war, many engines in Training command were operated to the maximum of 1,000 hours before overhaul.

John Cartwright, a retired Allison employee who had been trained at the Allison V-1710 Service School when it was held at the old Antler's Hotel in downtown Indianapolis, went on to be a crew chief on P-38's in the 8th and 15th Air Forces. He relates how his was the first P-38 in the Fighter Group in which both V-1710's went to the then time limit of 300 hours without prior removal for cause. He remembers that they had few problems with the V-1710, other than being shot-up, or shot-down.[125] By the end of the War, engines in combat theaters were going 750 hours before overhaul.

The following is taken from the engine record of V-1710-85, AAF A-034850. This engine was used in a stateside training regime, and gives an idea of the intensity of operations as Squadrons worked-up for assignment to combat theaters:

- Accepted by the AAF, October 27, 1943
- Installed in P-39Q-15-BE, AAF No. 44-2354, and delivered by Bell Aircraft on 11-2-1943. Total time 1:35 hours.
- Received by the 392nd Fighter Squadron, Santa Rosa, CA, 11-7-1943. (The 392nd FS then went on to distinguish itself in the 9th Air Force, where they flew the Republic P-47 as a component of the 367th Fighter Group.)
- Flown to Sacramento Air Depot (SAD) with 355:35 hours Total Time, several tech order improvements incorporated while at the depot. In 108 days with 392nd the aircraft and engine had averaged 3:20 hours of flight per day.
- Aircraft and engine assigned to 432nd Base Unit at Portland AAB, Oregon on April 30, 1944.
- Flown with 432nd until July 5, 1944 when engine was removed for having exceeded maximum allowed flight time before overhaul. Total time at removal was 615:05 hours. During 65 days with the 432nd the engine had averaged 3:55 hours of flight per day.
- Shipped to Sacramento Air Service Center for overhaul on July 7, 1944. Overhaul completed August 26, 1944.
- Shipped to AAF storage on March 18, 1946.
- This engine is still flying in 1997!

Overhaul Process

Overhaul of engines involved a complete inspection upon arrival at the overhaul facility, which was usually a military depot, and in some instances located in the combat theaters. It was not unusual to find evidence of battle and/or crash damage, as well as missing accessories. Cannibalization of

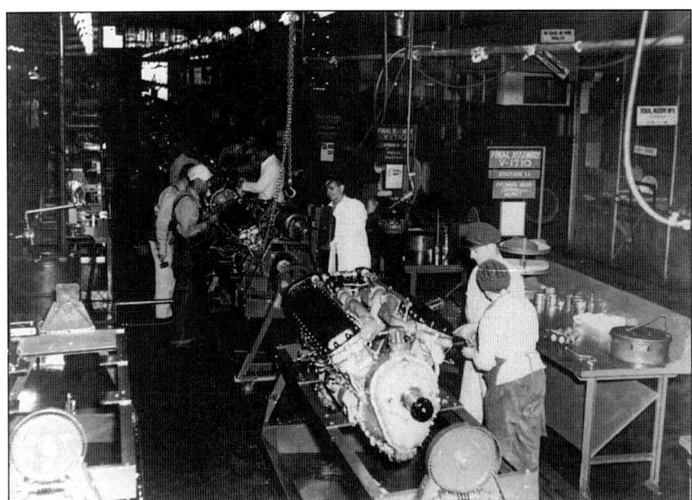

V-1710-F engines are nearing the final assembly at the Brookley Field AAB Overhaul facility in Mobile Alabama. This facility also overhauled R-2600 engines. (SDAM)

engines going into overhaul was a time honored practice of crew chiefs working to keep their own birds flying. Typical missing or worn out accessories were items such as generators, carburetors, magnetos, as well as miscellaneous bits and pieces.

After inspection, the engines were disassembled into their major sections, which were then broken down into component parts. Each part was then inspected for wear and compliance with pre-established dimensional limits. Highly stressed steel parts were magnafluxed to identify flaws or cracks not otherwise detectable. Minor defects could be removed, including those on bearing surfaces of connecting rods and crankshafts. Exacting repair procedures were used to rework such components to prepare them for reuse. Allison provided oversized bearings, pistons and cylinder liners in standard sizes to allow almost every engine to be returned to service. The important requirement was to keep the parts of a given engine together, and if pistons or liners had to be replaced, the final assembly had to match very tight weight limits to insure that the completed engine would run smoothly and not overstress the already heavily loaded components.

As an example of the close tolerances required, after regrinding and polishing the journals of a crankshaft, the seven main bearings had to be no more than 0.012 inches out of line. The twelve pistons, with a normal weight of 3.720 to 3.750 pounds, had to match within 0.040 pounds (18 grams), and the twelve piston pins had to all be of the same type. Each piston and connecting rod was marked as to the cylinder in which it had originally been installed, and they were to remain in that position.[126] With the introduction of the 12-counterweight crankshaft, the tolerance for piston weight variation was opened up to 0.050 pounds.[127]

The overhauled engines were reassembled using all new gaskets and rubber parts, as well as having renewed electrical systems and accessories. After passing their full complement of inspections, and a run-in at up to full rated power, they were "pickled" to protect from corrosion and exposure, then placed in shipping crates for issue to operational units.

During the war, the Army setup a number of engine overhaul facilities. Most were at the Air Depots within the Continental U.S. in locations such as Sacramento, Fairfield, Middletown, and San Antonio, but many operating theaters operated their own facilities as well. As a consequence, there appears to have been some concern in the quality of the overhauled engines returned to the combat commands. In June 1944, a report from General Eaker of the Mediterranean Theater went to General Arnold noting that overhauled engines were not giving the operating hours equivalent to new engines. General Arnold was necessarily concerned about the allegation and immediately initiated an investigation.

General Eaker responded that further investigation within the Mediterranean Theater found that the average time on Allison engines at removal from P-38's was 212 hours, and that the earlier suggestion of a problem with overhauled engines was incorrect. The investigation determined that as of May 1944, flying time at overhaul of the P&W R-1830 (Consolidated B-24) was 236 hours, while the Wright R-1820 (Boeing B-17) was operating 199 hours and the R-3350 (Boeing B-29), 116 hours. The report tells that *averages* are not the proper way to assess the life of an engine, rather a statistical approach as used for mortality determination was more appropriate. On this basis, it was determined that 50% of the R-3350's were still running at 200 hours in Theater, and that within the Continental U.S. approximately 49% had flown over 300 hours. The study validated the original assumption that the average flying time of the R-3350 would be at least 200 hours.[128] The final conclusion and response to the Mediterranean Theater was that overhauled Allison engines being shipped to them were giving more than satisfactory performance.[129]

It may be hard for some to consider 200 or 300 hours a satisfactory overhaul life from the perspective of the 1990s when it is common for engines on commercial jet airplanes to stay on the airplane for tens of thousands of hours. Better than any other factor this may be the best measure of how far aviation technology has come in the last fifty years.

Pistons and Piston Pins

As could be expected with an engine that went through so many models and power upgrades, the V-1710 used several different pistons and piston pins. On the GV-1710-A, the piston top was dished, much like those in the later Rolls-Royce Merlin's. The usual production V-1710 used a standard flat topped piston which gave the desired 6.65:1 compression ratio. Table 16-23 shows the more typical applications.[130, 131]

By mid-1944, the overhaul facilities had been directed to remove and dispose of Type I and Type II pistons as well as the Type A light piston pins.[132] At the same time, they were directed to install only the PN53567

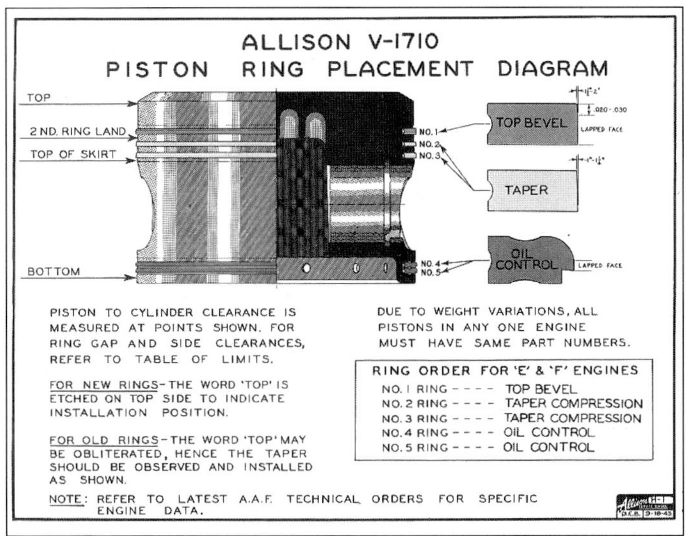

Typical piston and piston ring configuration of early V-1710 models. The later "Keystone" rings were similarly installed, only their cross-sectional shape was different. (Allison Service School)

Table 16-23
Usage of Standard Size Pistons

Type	Piston Part No.	Piston Max Wt, lbs	Piston Min Wt, lbs	Used in V-1710 Models	Oversize Pistons, inches OD
I	36380	3.695	3.665	-27-29-33-35-37-39	Not available
II	41380	3.730	3.700	-27-29-35-39-49-53-63	Not available
III	42380	3.750	3.720	-35-39-49-53-51-55-63-73-81-83-85-87	+0.010", +0.020"
IV	43567	3.750	3.720	-51-55-81-85-89-91-93	+0.010", +0.020"
V	53567			-89-91-93-111-113 and Subsequent E/F/G's[4]	

1.All pistons in any one engine must be of the same basic part number.
2.A later part number piston will never be replaced by an earlier PN.
3.Oversized pistons were slightly heavier, could be used with standard pistons provided the maximum weight variation within the engine was less than 0.040 pounds, although late model engine having the 12-counterweight crankshaft were able to accommodate a maximum weight variation of up to 0.050 pounds.
4."G" models using these part numbers were the few with 6.65:1 compression ratio.

"Tapered" or "Keystone" top ring in the V-1710-89/91, -111/113 and later engines. These were the engines rated for 1600 bhp and higher.

One interesting enterprise on the part of Allison was the "cammed" piston. This was used in some very high power development engines, and involved grinding a contour into the perimeter surface of the piston. The purpose was to accommodate the effect of high side loads normally directed at the cylinder liner and improve piston cooling by increasing the contact surface between the piston and cylinder wall. The piston was therefore shaped (cammed) to match the deflected cylinder wall when subjected to high side thrust. The pistons were modified by machining to reduce the diameter along the piston pin axis.[133] Test experience with these pistons confirmed an advantage to this design, particularly in the original pistons with the rectangular piston ring groove.

The pistons in the 6.0:1 compression ratio "G" engines differed by having a shorter height above the piston pin sufficient to change the nominal combustion chamber clearance volume from 421 cc to 467 cc. This was accomplished by reducing the height of the piston deck by a little less than 1/8th inch (approximately 0.118 inches).

Another development incorporated into the post-war V-1710-G6 engines was to slightly reduce the diameter of the piston. The purpose being to prevent scuffing or sticking of the piston when operating at very high power where the resulting high temperatures caused the aluminum piston diameter to increase more than the steel liner. This in turn resulted in somewhat poor sealing of the rings when not at WER power levels, and meant excessive amounts of oil in the combustion chambers. Consequently, these engines often had excessive "blow-by" and/or fouled spark plugs in normal operation.

Table 16-24
Piston Pins

Type	Pin Weight	Pin Inside Diameter, in
A	Light	1-3/32
B	Medium	1.000
C	Heavy	15/16

The light weight Type A pin should not be used in any engine with rated power above 1150 bhp.
All pins in any one engine must be of the same type.
Type A pins used in: V-1710-F2R/L, F3R, E4.

Piston Rings

The V-1710 piston always used three compression rings, located above the piston pin, and two oil control rings, both located in the same groove on the piston skirt below the piston pin. Each ring had its perimeter cut and shaped to provide a gas tight seal between the piston and cylinder wall under the varied conditions seen in high powered service.

In creating the two-stage mechanically supercharged engine Allison was pushing the Indicated Horsepower to considerably higher levels than was required in a turbosupercharged engine able to deliver the same shaft horsepower. One consequence of this was a temporary restriction issued by directive in October 1943 that limited all two-stage engines to no more than 45 inHgA manifold pressure. What was occurring was that at about 60 inHgA, piston rings were sticking. This allowed a considerable amount of combustion gas to blow-by the pistons and then be vented from the crankcase. Of course this was destructive to the engine. Allison responded with the "keystone" piston ring, which was used in a specially shaped top grove on the piston. These resolved the problem sufficiently to restore the intended engine ratings.[134] One of the causes of the sticking problem was the change in fuel formulation allowed by later amendments to AN-F-28 Grade 100/130. Resolution took most of the fall of 1943 to study and ef-

Table 16-25
Implementation of Keystone Piston Rings

Model	From Serial Number	Date in 1943
V-1710-89	A-038116	October 9
V-1710-91	A-038993	October 11
V-1710-85	A-041869	November 2
V-1710-99	A-041226	November 3
V-1710-93	A-032437	October 14

And on all subsequent production

fect. Production of the new Keystone rings and pistons was somewhat later.[135]

The improved "Keystone" or tapered top ring, had a wedge shaped cross-section, with the top and bottom surfaces angled toward each other at 15 degrees.[136] It completed development in mid-1943 and was introduced into production on the following models, beginning with the engine serial number listed in Table 16-25.[137]

Some modern builders of high performance V-1710's do not consider any of the original piston rings to have been very good. These builders each have their own secret formulas, but a typical arrangement might be to replace the two top compression rings with ones having steeper "keystone" angles. The third ring groove is recut to accept a wider, but square, land to accommodate a modern oil control ring, one borrowed for a current production diesel engine. The oil ring groove at the bottom of the skirt is left empty, as it is no longer needed for oil control. Leaving it open reduces friction while increasing oil flow to the piston pin.

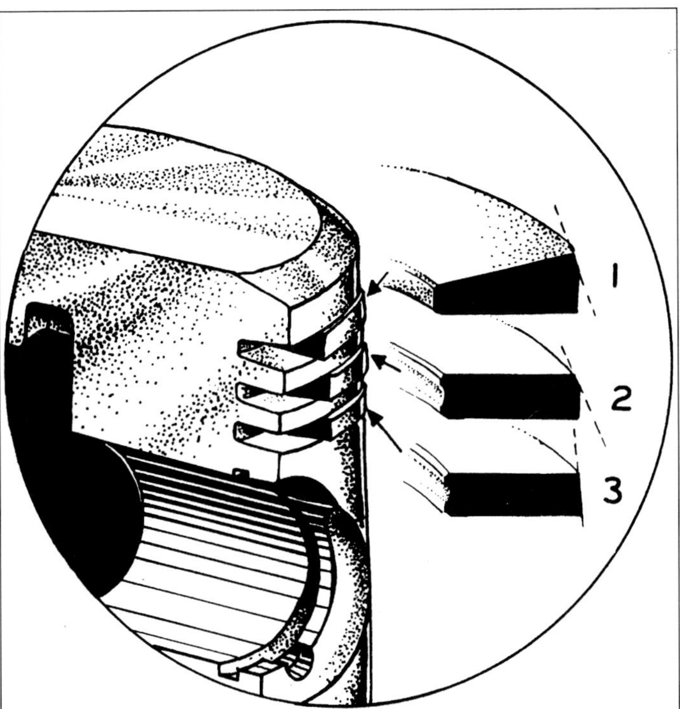

The keystone piston ring in the top compression groove was introduced to allow the higher BMEP developed by two-stage V-1710's. Rings are a small but very important component in an engine. Note the subtle differences in shape of each ring. (Allison)

NOTES

[1] May 1946 Allison receives plaque from AAF for training 101,000 on V-1710 during war.

[2] Schmid, J.C. 1940, 2.

[3] Allison Letter to Chief, Procurement Section, 11-28-1938, regarding design of a gear driven engine supercharger that can be attached as an accessory to the V-1710, thus making possible 2-stage supercharging of Allison engines.

[4] Engineering Division Memo Report, Serial No. ENG-57-503-1217, 1944, 3.

[5] Matthews, Birch. 1996, 200.

[6] Dolza, John 1945, 491-496.

[7] Induction temperatures either too high or too low were a problem. According to TO AN 01-60JJA-1, *Pilots Manual for the North American P-82E*, with the high powered V-1710-143/145(G6L/R), there was a serious danger of plug fouling if the Manifold Temperature dropped below 60 °C (140 °F), and the Maximum allowed temperature was 175 °C (347 °F). This was of course on Grade 115/145 fuel, which was not available in the combat theaters during the war. Grade 100/130, the usual WW II combat fuel, would have allowed a maximum temperature in the range of 250 °F, at which point detonation would have been encountered.

[8] The water-alcohol mixture is known as Anti-Detonate-Injection (ADI) fluid, and acts as an internal coolant. This limits the temperature of the mixture, and thereby prevents destructive detonation within the combustion chambers. Although engines with this feature were rated at up to 2250 bhp, actual power produced by the crankshaft was higher by the amount required to drive the superchargers. In such cases both the engine-stage and Auxiliary Stage superchargers were consuming over 300 bhp each. Obviously Allison had gone a long way to strengthen the engine to be able to reliably produce such power from the engine, though the increase itself was simply due to the increase in manifold pressure developed by the supercharger.

[9] Dolza, John 1945, 491-496.

[10] *Turbo-supercharger for 1000 HP Engine*, Letters between Materiel Division and General Electric, file Superchargers, 1933, contained in Record Group 342, NARA.

[11] Air Corps Technical Report No. 4038, 1937, 49.

[12] *Proposed V-1710 Engine Reduction Gear Bearing Redesign*, Caminez of Allison Letter to Captain Powers, Power Plant Branch, February 11, 1935. NARA RG 342, V-1710 1932/1935.

[13] *Cable Regarding Re-Rating Allison V-1710-C15 Engine*, received August 28, 1941 from British Air Commission. Wright Field file D52.41/81 Allison.

[14] Technical Order 02-5A-27, 1943.

[15] Caminez, H., Report No. 27, 1932, 12.

[16] Montieth, O.V. 1939, 15.

[17] *Some Comments on American Fighter Planes*, SECRET report by Lt. Col. Thomas Hitchcock, AAF Air Attaché, London, February 2, 1943, p.4. NARA RG18, File 452.13, Box 756.

[18] Drawing 33700 (XV-1710-5/7) shows dimension between center lines to be 3.172". This is confirmed by Allison Report No. A2-7, 1941.

[19] Sherrick, E. 1942, 291.

[20] Allison Specification C172-C.

[21] Buttner, H.J. 1937.

[22] Montieth, O.V. 1939.

[23] Hazen, R.M. and O.V. Montieth, 1938, 335-341.

[24] Montieth, O.V. 1939, 5.

[25] Taylor, E.S. 1936, 81-89.

[26] Stewart, Hugh B. 1946, 238-245.

[27] Engineering Section Memorandum Report E-57-541-27, 1938.

[28] *V-1710 Engine Crankshaft*, Allison letter to Chief, Materiel Division, December 11, 1936.

[29] Technical Report No. 4452, 1939. Also, Allison letter to Chief, Engineering Section, Wright Field, March 9, 1937.

[30] Rebuilding of two V-1710-13 engines with current production two node torsional vibration dampers on crankshaft, Allison letter to Materiel Division, August 16, 1938.

[31] Montieth, O.V. 1939, Rev 0 and 1.

[32] *Development of Tests of the V-1710-17 and XV-3420-1 Engines*, Wright Field Chief, Power Plant Branch report of May 3, 1939 notes that they were notified that "preliminary tests of the new hydraulic vibration damper in the V-1710-17 engine showed no improvement in torsional vibration characteristics. The contractor, therefore, is considering further design changes to improve characteristics. NARA File RG342, RD3367, File452.8 Allison Engines, Box 6449.

[33] Engineering Section Memo Report, Serial No. E-57-1077-88, 1939.

[34] Emmick, Wm. G. 1939.

[35] Montieth, O.V. 1939.

[36] Taylor, Charles Fayette, 1960.

[37] At 116-1/2 hours into the YV-1710-7(C4) Type Test, the crankshaft (PN 34600) was found to be almost completely cracked through. The failure was attributed to fatigue stresses from an insufficiently damped two-node torsional vibration encountered during the required overspeed running, compounded by an extraordinary condition of surface hardening in the #4 journal. The replacement crankshaft was from the YV-1710-13, and of the new design (PN36000) which incorporated the six-weight pendulum type dynamic balancer (PN34747 Hub, PN34748 Weights) tuned for 4-1/2 order, two-node, torsional vibrations. With this addition, it was also necessary to change the propeller shaft friction damper was changed from the seven friction plate to three friction plate design. See: Technical Report No. 4452, 1939.

[38] Hazen, R.M. and O.V. Montieth, 1938, 335-341.

[39] Montieth, O.V. 1939, 66.

[40] Meyer, Andre' J., Jr.1945.

[41] Sherrick, E. 1942, 270.

[42] Meyer, Andre' J., Jr.1945.

[43] Experimental Engineering Memo Report, Serial No. EXP-M-57-503-655, 1942.

[44] *Allison "Firsts" as of 1942*, from J.L. Goldthwaite files at Allison.

[45] Inter-Office Memo, *Conversion of V-1710-27 to V-1710-29*, Chief Experimental Engineering Section, Wright Field, June 30, 1941.

[46] Nordenholt, G.F. 1941, 218-223.

[47] Technical Order 30-5A-1, 1944.

[48] Hives, E.W. and F. Ll. Smith of Rolls-Royce Ltd, 1940, 106-117.

[49] *Conference with Allison Representatives on the Tests of the V-3420 Engine in the XB-39 Nacelle*, CONFIDENTIAL Memorandum for Chief of Research, NACA-LMAL, Langley Field, VA, July 27, 1944. NARA RG 342, RD 3761, Box 7359.

[50] *Conference with Allison Representatives on the Tests of the V-3420 Engine in the XB-39 Nacelle*, CONFIDENTIAL Memorandum for Chief of Research, NACA-LMAL, Langley Field, VA, July 27, 1944. NARA RG 342, RD 3761, Box 7359.

[51] Presentation of glycol-water mixtures is in terms of "Volume Percent".
[52] Povolny, John H. and Louis J. Chelko, 1945, Figure 26(b).

[53] *Crankshaft and Connecting Rod Bearing Performance in Dive Test Dummy Engine*, Allison Test Engineer, John Hubbard, Jan/Feb 1941.

[54] Allison Model Specification No.163-E, 1943.

[55] *Conference with Allison Representatives on the Tests of the V-3420 Engine in the XB-39 Nacelle*, CONFIDENTIAL Memorandum for Chief of Research, NACA-LMAL, Langley Field, VA, July 27, 1944. NARA RG 342, RD 3761, Box 7359.

[56] Materiel Command Memorandum Report SN ENG-57-503-1173, March 15, 1944.

[57] *Operators Manual for Allison Engine Installations*, Allison Division of GMC, Fifth Edition, Sept. 15, 1943.

[58] Allison Model Specification No. 245-B, 1946.

[59] Hazen Memorandum to Gilman, *Future Development Work on V-1710 Engine*, September 18, 1936.

[60] Ethell, Jeff, 1st Quarter 1981, 69 and 80.

[61] Internal Allison Memo to Chief Designer and Design Project Engineers, 11-16-1940.

[62] Sherrick, E. 1942.

[63] Internal Allison Memo to Chief Designer and Design Project Engineers, 11-16-1940.

[64] Allison "Firsts" as of 1942, from J.L. Goldthwaite files at Allison.

[65] *Allison Engine*, letter from Major General H.H. Arnold, Chief of the Air Corps, to Chief of Materiel Division at Wright Field, August 29, 1939. NARA RG18, File 452.8, Box 808.

[66] Telecon with Birch Matthews, 11-18-1994.

[67] *Progress Report on High Output Tests of Modified V-1710-19 Engine*, report from Wright Field to Chief of the Air Corps, September 26, 1939. NARA RG18, File 452.8, Box 808.

[68] Engineering Division Memo Report, ENG-57-503-1034, 30 October 1943.

[69] Allison Experimental Department Report No. A2-143, 1944.

[70] Engineering Division Memo Report, Serial No. ENG-57-503-1173, 15 March 1944.

[71] Engineering Division Memo Report, Serial No. ENG-57-531-267, 1944.

[72] Engineering Division Memo Report, Serial No. ENG-57-531-294, 16 May 1944.

[73] Engineering Division Memo Report, Serial No. ENG-57-531-299, 12 May 1944.

[74] Allison letter of June 6, 1946, NARA RG342, RD3775, Box 7411.

[75] Hazen, R.M. 1945 Report. Report of Allison Chief Engineer, 1945.

[76] Lee, Dana W. 1940.

[77] Nelson, R. Lee, Myron L. Harries, and Rinildo J. Brun, 1945.

[78] It was also necessary to reduce the diameter of the pistons by 0.008 inch from their standard dimensions in order to assure satisfactory operation at the high power levels developed during the tests. The reduced diameter compensated for the increase in piston diameter due to the high temperatures encountered during the high powered runs.

[79] V-1710-G6 modified to run in *Cobra II* during the 1948 Thompson Race. Used 115/145 PN fuel delivered through a PR-58 carburetor along with 8.80:1 supercharger gears and running at 3200 rpm. Result was 2250 bhp at sea-level from 100 inHgA, with ADI.

[80] Arnold, General H.H., August 19, 1943.

[81] *Annual Report of the Chief of the Air Corps, 1934,* 39.

[82] Technical Order 06-5-4, 1950, 84-85.

[83] Technical Order 06-5-4, 1950, 80.

[84] To produce one gallon a day of Grade 100/130 during WWII, a plant cost of about $35 was involved as well as a need for about 120 pounds of mostly steel. Technical Order 06-5-4, 1950, 81.

[85] The new rating system used Performance Number in place of "Octane" ratings and was presented for Lean/Rich operating conditions. Rich values were used to achieve maximum performance for Takeoff, Military and WER.
[86] Technical Order 06-5-4, 1950, 87.

[87] Tetra-Ethyl-Lead (TEL), a proprietary compound developed by the Ethyl Corporation, which when added to petroleum fuel increased its resistance to knock or detonate.

[88] Technical Order 06-5-4, 1950, 89.

[89] See Table 2 for Performance Number vs Octane Number. Technical Order 06-5-4, 1950, 94.

[90] Technical Order 06-5-4, 1950, 90.

[91] Technical Order 06-5-4, 1950, 91.

[92] Technical Order 06-5-4, 1950, 82.

[93] Technical Order 02-5A-55, 1943.

[94] *P-40 Pilots Training Manual.*

[95] *Timing on V-1710 Engines Used in P-38 Airplanes in England*, memo to Major General O. P. Echols, April 27, 1944. NARA RG18, File 452.13, Box 1260.

[96] Technical Order 02-5AD-2, 1945, 115.

[97] *Test of Ventilated Ignition System on Allison Engines at High Altitude*, memo to General Arnold from Chief Engineering Division, Wright Field, November 20, 1943. NARA RG18, File 452.13, Box 1260.

[98] Kline, John. 1994.

[99] Deliveries of aircraft to the Russians began at Ladd Field in August 1942 and continued to the end of the war. A total of 7,983 aircraft were delivered, with 148 in 1942, 2,662 in 1943, 3,164 in 1944 and 2,009 in 1945. The bulk of these deliveries were Bell P-39 and P-63's, although P-40's, A-20's, B-25's, C-47's and AT-6's were included as well. Cohen, Stan, 1981, 45-46.

[100] During 1944 the Russians were complaining that they were sometimes being given overhauled engines as spares, and they insisted that they receive only new engines. In a similar way they followed the life of their Lend-Lease aircraft, for example, P-40K-5 AAC No. 42-9807 crashed in the Soviet Union in the summer of 1943 after flying a total of 61 hours since delivery in September 1942. The cause was determined to be attributable to a missing crankshaft oil transfer tube in engine AC NO. 42-99098, and therefore they insisted that their account be credited for the cost of the airplane. Letter from Government Purchasing Commission of the Soviet Union in the USA to the International Section U.S. War Department, September 13, 1943. NARA RG18, File 452.13, Box 1260.

[101] See Caution noted on Gages shown in *P-82E-1 Pilots Manual.*

[102] *Spark Plug Fouling with Lead*, letter from R.B. Mansell, Air Vice-Marshal, R.A.F. to War Department, Washington, D. C., May 23, 1944. NARA RG18, File 452.13, Box 1260.

[103] *Lead Distribution-Allison Engine*, letter from British Air Commission, to War Department, Washington, D. C., July 24, 1944. NARA RG18, File 452.13, Box 756.

[104] *Timing Analysis*, Allison Report No. 224.

[105] Hazen, R.M. 1941, 488-500.

[106] Allison Report No. A2-7, 1941.

[107] *Results of Test of Constant Velocity Manifold on P-38 Installations*, Teletype Message from Wright Field Production Engineering Section, to Engineering Division, August 31, 1942.

[108] *Streamlined or Constant Velocity Intake Manifolds for Allison V-1710 Engines*, Materiel Command letter to Major General C.P. Echols, September 17, 1942.

[109] *Progress Report No. 10, Allison XV-3420-1*, Contract W535-ac-9676, Period Ending December 15, 1939. NASM Wright Field file D52.41/25-Allison.

[110] *Allison Service School Handbook, Supplement for Auxiliary Stage Supercharger*, page 4.

[111] Leyes, Rick 1997.

[112] Goldthwaite, J.L., notes compiled 6-10-45.

[113] Technical Order 02-5AB-4B, 1942, 46.

[114] Letter from Col. T. A. Sims, Assistant Technical Executive to Major General O. P. Echols, Commanding General, Materiel Command, June 25, 1942.

[115] Technical Order 02-5AB-1A, 1942.

[116] Sherrick, E. 1942, Fig. 128.

[117] *Streamlined or Constant Velocity Intake Manifolds for Allison V-1710 Engines*, letter from Materiel Center Technical Executive to Major General O. P. Echols, AAF Materiel Command, September 17, 1942. NARA RG18, File 452.8, Box 807.

[118] Technical Order 02-5A-42, 1945.

[119] *Introduction Into Production*, listing, NARA RG 342, RD 3774.

[120] *Overhaul compliance list for V-1710-85, A-034850*, accomplished at Sacramento Air Service Center, 8-26-1944.

[121] Technical Order 02-5A-64, 1944.

[122] *Compound Engine*, letter to Allison from Chief, Materiel Division, February 3, 1939. NARA RG342, Allison V-1710, Experimental Project.

[123] *Camshafts for V-1710 Engine*, ATSC letter dated September 7, 1945. NARA RG342, RD 3971, Box 8146.

[124] *Compound Engine Projects*, letter report from Chief, Propulsion & Accessories, Engineering Division at Wright Field to Commanding General, AAF, June 23, 1945. NARA RG342, RD3774, Box 7409.

[125] *Cartwright, John.*

[126] Technical Order 30-5A-1, 1944, Section 10, Page 4.

[127] Technical Order 02-5A-9, 1944.

[128] *Flying Time at Overhauls—R-3350, R-1820 and R-1830 Engines*, memo for Major General O.P. Echols, July 17, 1944. NARA RG18, File 452.13, Box 756.

[129] *Overhaul of Engines in the Mediterranean Allied Air Forces*, letter from BG Patrick W. Timberlake, Deputy Chief of Air Staff to Materiel & Services Command, August 11, 1944. NARA RG18, File 452.13-L, Box 756.

[130] Allison Division, GMC, 1943 Overhaul Manual, 76.

[131] Technical Order 30-5A-1, 1944, Section 10-14.

[132] Technical Order 02-5A-9, 1944.

[133] Materiel Command Memo Report Serial No. ENG-57- 503-1173, 1944.
[134] Priority teletype messages of October 6 and 12, 1943 from Wright Field Production Engineering Section to Curtiss-Wright, Buffalo, N.Y. NARA RG 342, RD 3834, Box 7636.

[135] *Development of Increased Horsepower and Altitude for the Allison V-1710 Engines (Particularly for N.A.A. P-51 Airplane)*, Allison letter C1L2AE from Hazen to Commanding General, AAF Materiel Command, November 2, 1943. NARA RG18, File 452.13, Box 1260.

[136] Technical Order 30-5A-1, 1944, Section 10, page 21.

[137] *Introduction into production*, listing, NARA, RG 342, RD 3774.

17

Manufacturing the Allison

Perspective on the significance of the Allison contribution to WWII comes from Eddie Rickenbacker who noted in a 1956 *U.S. News and World Report* that, "From 1940 to 1945 Allison turned out 70,000 engines – 60 percent of all of the engines supplied for U.S. fighter aircraft during the war!"[1]

Preparing for War

During the 1930s, when the V-1710-C was being worked through Type Testing, the pre-W.W.I facilities of Plant 1 were adequate for the hand building, and rebuilding, of the few engines involved. In January 1938, when the order for 51 engines to power the YP-37's and YFM-1's was received, the expanded facilities provided by Plant 2 were more than adequate for the production task.[2] All of this changed when the U.S. Government woke up to the worsening European situation, and recognized the need for large quantities of modern high powered fighter engines. Within the country the V-1710 was the only such engine even remotely ready for production.

Anticipating the acceptance of the V-1710 for service, Allison had undertaken the construction of a new production facility, Plant 2, which was started in 1936, and dedicated on Decoration Day, May 30, 1937, only a month after the V-1710 completed its Type Test.[3] The plant was built to have a production capacity of perhaps 100 engines per year.[4] This of course was before Munich, and the sudden recognition by the western world of the rise of Hitler and the threat he represented. Plant 2 continued as the primary production facility until two years later. In 1939, and again on Decoration Day. Immediately following the annual Indianapolis Speedway 500 Mile Race, General Motors hosted the ground breaking for Allison Plant 3 at the nearby site of an old pear orchard.[5]

This facility was funded and built by General Motors in response to an appeal by then Assistant Secretary of War Louis Johnson to William S. Knudsen, then President of General Motors, for a plant to mass produce the V-1710. General Motors made the commitment even though there was no assurance from the government that contracts would be forthcoming for any more than about 900 engines, intended for the recently ordered Curtiss P-40's as well as initial P-38 and P-39 production. The commitment required GM to invest another $5,970,000 in manufacturing facilities, which was in addition to the $3,800,000 it had invested in Allison to date, and occurred even though Allison had produced a nearly unbroken string of unprofitable years since being purchased by GM ten years earlier.[6] This was a testimonial of faith, for with nearly ten million Depression era dollars invested General Motors had yet to recover, through net earnings, an amount equal to its initial $592,000 investment in Allison.[7] This investment was recognized and appreciated by the principals in the Air Corps. As General Kelsey related in his book, "...certainly if it had not been for the substantial investment by General Motors the Allison V-1710 would not have become available in 1936 as a 1000 bhp engine.[8]

Increasing Production Capacity

By December 31, 1936, with the Consolidated/Bell A-11A having just begun flight testing, and with just three of the ten "YV-1710-3's" having been delivered, the Air Corps had paid the Allison Engineering Company $419,423.70. Additional related costs incurred by the Air Corps made their investment a grand total of $506,361.19. These costs included the cost to modify the A-11A, Air Corps direct labor, material, travel and overhead.[9] For the eight engines delivered through 1936 Allison had been paid by the government a total of only $277,950.[10] During 1938 additional orders were placed for the engines to power the service test Curtiss YP-37 and Bell YFM-1, but those engines were not delivered until 1939, and the Air Corps never advanced funds; they only paid for engines delivered and satisfying test specifications. Still General Motors continued to support the program, and even provided funds to build and equip Plant 2 so that the service test engines could be fabricated and a moderate production capability established.

On Saturday April 15, 1939, the War Department contacted Allison on behalf of the French and asked to obtain one V-1710-19 engine, and inquired how long would it take them to reach a production capacity of 300 engines per month.[11] The response quickly involved General Arnold, who was informed that there was only one V-1710-19, the XV-1710-19 then installed in the XP-40. Furthermore, the U.S. expansion program was going to require 1,200 V-1710's, of which 409 were assigned to early batches of P-40 and P-38 airplanes, and the planned Allison Plant 3 was only intended to reach 75 engines per month at maximum capacity.[12] The response to the War Department then noted the existing U.S. "expansion" plan and what Allison would be able to do to meet the French request. This proposal was actually implemented, and was a confirming basis for the Allison Division breaking ground for Plant 3 on Decoration Day, May 30, 1939.

When the production order finally came in from the Air Corps, in June 1939, it was for 837 engines.[13] Unfortunately, from a production stand-

Table 17-1 Allison Capability to Increase Production, 1939		
Month	US Expansion Plan, Engines/mo	Proposed Response to French, Engines/mo
1	0	0
2	1	1
3	0	0
4	2	2
5	3	3
6	5	35
7	15	35
8	20	35
9	25	60
10	30	75
11	40	150
12	60	150
13	75	150
14	75	300

point, this included four models of the V-1710, each differing basically from any other.[14] A fifth model was added in December 1939 with a separate contract for 81 additional engines for the P-39C.[15,16]

Getting Production Started, and Other Problems

Allison was constructing Plant 3, Model Testing the V-1710-19, recruiting and training its production staff and committed to a rigorous production schedule as 1939 drew to an end. All of this was made even more critical following the September 1, 1939 Nazi invasion of Poland, the event marking the beginning of WWII. In this situation, things quickly became acrimonious. As early as February 7, 1940 the Chief of the Air Corps notified the Assistant Secretary of War stating that Allison had defaulted on the contract for delivery of engines for the Curtiss P-40. The Secretary then wrote to Mr. Knudsen, President of General Motors, calling attention to the default in delivery and requesting a new delivery schedule. This was provided, and emphasized that Allison expected to be on their original delivery schedule, and to have made up the delinquent deliveries by September 15, 1940.[17]

This was an ambitious plan and commitment, for Allison had only delivered two engines in January and four in February. What Knudsen knew was the standard and extent of resources of the General Motors Company that he was directing to the effort. He had raided the entire automobile manufacturing system for their best people, at all levels, and moved them unceremoniously to Indianapolis, where they stayed for the duration.

At the same time the engine was having problems. In the late Fall of 1939, the Air Corps decided to shift from the C-13 to the similar, but improved V-1710-C15. This engine was having considerable trouble with crankshafts and crankcases in attempting to complete the rigorous Model Test at the contract specified 1090 bhp. In March 1940 the crankshaft was redesigned, and then in late April it was decided to change the lubrication system due to connecting rod bearing failures that occurred during acceptance tests of the Curtiss P-40.

Allison argued that the altitude rated V-1710-33(C15) should not be rated for 1090 bhp, as the power required to drive its larger supercharger resulted in overstressing the engine. They believed that 1040 bhp was the more appropriate figure. Finally in June 1940 it was decided by the Air Corps to accept the Allison recommendation and proceed with unrestricted production of the engine, although it was to be limited to 950 bhp until both the Accelerated Service Test on six P-40's and the Model Test at 1040 bhp were successfully completed.

By the promised September 15, 1940 date, the delivery of V-1710-33's was up to the promised schedule and Curtiss production was no longer being constrained. Allison was at the same time bringing through the modifications needed to establish the full 1040 bhp rating.[18]

Not only the early development, but the series production of the V-1710 was made possible only by the financial support and long range perspective of General Motors. This they did in the face of considerable adversity. By 1938 the direct development costs had exceeded military contract payments by nearly a million dollars. Through the period of engineering, design, prototype production and development, all work had been done by Allison, without either interference or technical help from General Motors. When the problems of war production arose from late 1938 onwards, all of the immense resources of General Motors in manpower, production skill, organizational genius, and research were brought into full play.

Table 17-2 Allison Production Rate	
Production Rate at End of Year	Engines per Month
1938	2
1939	8
1940	300
1941	1100
1942	1400
1943	2100
1944	2100
1945	None

Table 17-3 Production Buildup				
	Floor Space, sq.ft.	Personnel	Engines Delivered	Comments
1938	93,598	512	14	
1939	404,652	1,213	46	
1940	1,221,660	7,347	1,175	342 to US, 833 to British
All figures as of December 31 of the year. From: Doolittle, Major J.H. 1941.				

Table 17-4
Growth of Allison

Year	End of Year No. Employees	Number of Bearings	No. of Engines Ordered	Number of Engines Delivered per Dept of Commerce	Number of Engines Delivered per Allison
1913	20	None	NA	NA	NA
1917	100	None	250	NA	2
1919	350	None	NA	NA	250
1927	200	NA	NA	NA	NA
1928	NA	NA	NA	NA	NA
1929	108	329,149	NA	NA	NA
1930	101	NA	NA	NA	NA
1931	113	NA	1	NA	0
1932	115	NA	0	NA	1
1933	163	54,199	5	NA	2
1934	150	174,228	11	NA	1
1935	252	NA	0	NA	3
1936	256	NA	0	NA	2
1937	322	474,737	22	NA	7
1938	530	NA	43	NA	12 V-1710's
					1 XV-3420-1
1939	786	NA	1,053	NA	48
1940	4,303	NA	7,891	1,141 V-1710's	
				2 V-3420's	1,153 w/342 to AAC
1941	9,673	NA	NA	6,447 V-1710's and 1 V-3420	6,433
1942	14,323	NA	NA	14,905	15,319
1943	23,019 (Max)	NA	NA	21,063 V-1710's	
				30 V-3420's	20,350 V-1710's
				31 V-3420's	
1944	17,940	NA	NA	20,191 V-1710's	
				112 V-3420's	20,191 V-1710's
				112 V-3420's	
1945	9,512	NA	NA	5,486 V-1710's	5,492 V-1710's
				10 V-3420's	610 J33's
				10 V-3420's	610 J33's
1946	4,969	NA	NA	NA	17 V-1710's
					905 J33's
1947	6,928	NA	NA	NA	603 V-1710's
					1,614 J33 & J35's
1948	6,946	NA	NA	NA	145 V-1710's
					1,102 Jets
1949	8,282	NA	NA	NA	V-1710 Spares
TOTALS		10 Million			

• This table does not include Workhorse (Test) Engines, School Engines or "Equivalent" Engines from Spares.
• 1933 Orders were for 3 ea V-1710-4(B)'s, one XV-1710-1, and one XV-1710-3, also the U-250 s/n 34-.
• 1934 Orders were for 10 ea YV-1710-3's.
• 1937 Orders were for 20 ea V-1710-11(CB)'s for 13 YP-37 aircraft, one XV-1710-15(C9) for the XP-38, and it is believed, the single XV-3420-1.
• 1938 Orders were for 39 V-1710-23(D2)'s for the 13 YFM-1's, two V-1710-17(E2)'s for the Bell XP-39, and possibly the two XV-1710-6(E1)'s for the Bell XFM-1.
• In June 1939 the U.S. Army released its first production contract that ultimately delivered 969 engines.

Table 17-5a
Deliveries of Allison Engines, by Model

Year	Total No. Del	No. This Model	Model	Application
1932	1	1	GV-1710-A(A1)	50-hr Navy Development Test at 750 bhp
1933	2	1	XV-1710-1(A2)	50-hr Army Development Test at 1000 bhp
		1	V-1710-4(B1R)	150-hr Navy Reversible, Type Test at 650 bhp
1934	1	1	XV-1710-3(C1)	150-hr Army Type Test at 1000 bhp
1935	3	2	V-1710-4(B2R)	Pre-production Reversible Dirigible engines.
		1	XV-1710-7(C2)	XV-1710-7(C2), for initial fitup in A-11A.
1936	2	1	XV-1710-7(C3)	150-hr Army Type Test at 1000 bhp.
		1	YV-1710-7(C4)	First Flight engine for A-11A
1937	7	1	XV-1710-9(D1) to	150-hr XFM-1 "Pusher" Type Test, converted and completed "C"
		-	YV-1710-7(C4)	Type Test
		4	YV-1710-9(D1)	Flight engines intended for Bell XFM-1
		1	YV-1710-7(C4)	First Flight engine for Curtiss XP-37
		1	XV-1710-11(C7)	Development Test of 1150 bhp rated engine
1938	14	2*	V-1710-11(C8)	RH engine for XP-38, Curtiss XP-37/Design 80
		1	XV-1710-15(C9)	LH engine for Lockheed XP-38
		8	V-1710-21(C10)	Engines for Curtiss YP-37's
		1*	XV-1710-19(C13)	First Flight engine for Curtiss XP-40
		1	XV-1710-17(E2)	First Flight engine for Bell XP-39
		1	XV-3420-1(A1)	50-hr Army Development Test at 2300 bhp
1939	46	12	V-1710-21(C10)	Curtiss YP-37's, includes 2 for Type Testing.
		1	XV-1710-17(E2)	Spare engine for Bell XP-39
		29	V-1710-23(D2)	Engines for Bell YFM-1's
		2	V-1710-41(D2A)	Engines for Bell YFM-1B's
		1	V-1710-33(C15)	For Curtiss P-40
		1	XV-1710-6(E1)	For Navy Bell XFL-1
1940	1178	1	XV-1710-6(E1)	Spare, Bell XFL-1, used for Altitude Tests
		1	V-1710-C13	Exported to France, just before its fall
		299	V-1710-33(C15)	For U.S. Army Curtiss P-40 production
		16	V-1710-37(E5)	For the 13 Bell YP-39's
		4	V-1710-27(F2R)	For Lockheed YP-38's
		4	V-1710-29(F2L)	For Lockheed YP-38's
		3	V-1710-35(E4)	For Bell P-39 production
		2	V-1710-39(F3R)	For Curtiss P-40D and NA-73X
		3	V-1710-39(F3R)	For Curtiss XP-46, XP-46A, and Model Test
		1	V-3420-A8R	For Navy PT-8 Torpedo Boat, RH turning
		1	V-3420-A8L	For Navy PT-8 Torpedo Boat, LH turning
		2	V-1710-41(D2A)	For Bell YFM-1B's
		6	V-1710-23(D2)	For Bell YFM-1's
		833	V-1710-C15	For Curtiss Tomahawk 81A (British)
1941		?	V-1710-C15	For Curtiss Tomahawk 81A and Lockheed P-322's (British)
		50	V-1710-C15A	For China and AVG
		?	V-1710-27(F2R)	For Lockheed YP-38's, P-38D/E
		?	V-1710-29(F2L)	For Lockheed YP-38's, P-38D/E
		?	V-1710-35(E4)	For Bell P-39 Production
		?	V-1710-39(F3R)	For Curtiss P-40D/E, P-51 Mustang
		?	V-1710-49(F5R)	For Lockheed P-38F
		?	V-1710-53(F5L)	For Lockheed P-38F
		?	V-1710-51(F10R)	For Lockheed P-38F/G/H
		?	V-1710-55(F10L)	For Lockheed P-38F/G/H

Table 17-5b
Deliveries of Allison Engines, by Model

Year	Total No. Del	No. This Model	Model	Application
		29	V-1710-61(F14)	For Curtiss P-40J, 3 used for Model Test
		25	V-1710-59(E12)	For Bell P-39J
1942		?	V-1710-35(E4)	For Bell P-39 Production
		?	V-1710-39(F3R)	For Curtiss P-40D/E, P-51A Mustang
		?	V-1710-49(F5R)	For Lockheed P-38F
		?	V-1710-53(F5L)	For Lockheed P-38F
		?	V-1710-51(F10R)	For Lockheed P-38F/G/H
		?	V-1710-55(F10L)	For Lockheed P-38F/G/H
		?	V-1710-47(E9)	For Bell XP-39E and Development, 2-Stage S/C
		1883	V-1710-73(F4R)	For Curtiss P-40K
		?	V-1710-63(E6)	Bell P-39D-2, P-39K, P-39L
		<5	V-1710-75(F15R)	Curtiss P-60A/B
		?	V-1710-75(F15R)	Lockheed P-38K
		835	V-1710-87(F21R)	North American A-36A Invader
		1034	V-1710-83(E18)	Bell P-39L/M/N/Q
		?	V-1710-67(E8)	Bell P-39M
		?	V-1710-89(F17R)	For Lockheed P-38H/J, Vega XB-38
		?	V-1710-91(F17L)	For Lockheed P-38H/J
		?	V-1710-81(F20R)	North American P-51A
		?	V-1710-93(E11)	Bell P-63A/C, 2-Stage Supercharger
		?	V-1710-85(E19)	Bell P-39M/N/Q-1
1943		?	V-1710-51/55(F10R/L)	For Lockheed P-38F/G/H
		?	V-1710-47(E9)	For Bell XP-39E and Development, 2-Stage S/C
		?	V-1710-63(E6)	Bell P-39D-2, P-39K, P-39L
		?	V-1710-75/77(F15R/L)	Lockheed P-38K, Curtiss P-60A/B
		?	V-1710-89(F17R)	For Lockheed P-38H/J, Vega XB-38
		?	V-1710-91(F17L)	For Lockheed P-38H/J
		?	V-1710-81(F20R)	North American P-51A "Mustang"
		?	V-1710-93(E11)	Bell P-63A/C, 2-Stage Supercharger
		?	V-1710-85(E19)	Bell P-39M/N/Q-1
		1	V-1710-65(E16)	For development of Planiol Aux S/C, not flown
		1	V-1710-45(F7R)	Two-stage S/C w/Aftercooler, Experimental P-40
		Few	V-1710-F25R	XP-40Q Program, 2-Stage Supercharger
		6	V-1710-95(F23R)	Curtiss XP-55 "Ascender"
		31	V-3420-	Fisher P-75
1944		10	V-1710-103 & 137's	Douglas XB-42/XB-42A "Mixmaster"
		4	V-3420-11/13	Lockheed XP-58 "Chain Lightning"
		108	V-3420-23(B10)	Fisher P-75
1945		?	V-3420-23(B10)	Fisher P-75
1946		2	V-1710-143/5(G6R/L)	North American P-82E
		3	V-1710-133(E30)	Bell L-39 Swept Wing Test Aircraft
1947		603	V-1710-143/5(G6R/L)	North American P-82E/F
		1,614	J33 and J35	Jet engines
1948		145	V-1710-143/5(G6R/L)	North American P-82E/F
		1,102	J33 and J35	Jet engines

* These two V-1710-11(C8)'s and the XV-1710-19(C13) were diverted from the batch of 20 V-1710-21(C10)'s purchased for the YP-37's. Two additional C-10's (AEC#20 and #20-1) were built to perform the ModelTest, though neither received Air Corps serial numbers. It is possible that another C-10 was purchased to replace the diverted XV-1710-19, though no records have surfaced to prove it. There may be some double counting the production of the V-1710-21's.

Table 17-6
Wartime Engine Shipments

	Jan	Feb	Mar	Apr	May	June	July	Aug	Sept	Oct	Nov	Dec	Total
1940													
V-1710	3	7	6	7	12	29	60	70	230	285	176	256	1141
V-3420	0	0	0	0	0	1	0	0	1	0	0	0	2
R-1820, Wright	85	222	149	202	160	211	171	225	187	255	227	231	2325
R-3350, Wright	0	3	0	0	0	0	2	0	0	1	2	0	8
R-2800, P&W	0	0	1	0	1	0	0	1	1	2	1	8	15
1941													
V-1710	134	300	313	237	403	404	513	709	725	765	845	1101	6447
V-3420	0	0	0	0	0	0	0	0	0	0	0	1	1
V-1650	0	0	0	0	0	0	0	4	4	5	10	26	49
R-1820, Wright	317	269	359	382	320	342	430	402	472	444	424	526	4687
R-3350, Wright	3	1	0	0	0	0	0	0	0	0	1	1	6
R-2800, P&W	13	30	63	95	113	103	131	118	103	161	272	264	1466
1942													
V-1710	1101	1041	1177	1151	1203	1252	1265	1326	1330	1379	1301	1379	14905
V-3420	0	0	0	0	0	0	0	0	0	0	0	0	0
V-1650	109	149	333	505	602	702	801	800	800	800	800	850	7251
R-1820, Wright	552	564	829	738	803	818	776	987	884	974	993	928	9846
R-3350, Wright	1	1	2	0	1	7	6	13	5	11	12	7	66
R-2800, P&W	269	358	379	406	459	515	501	516	449	520	531	526	5429
1943													
V-1710	1430	1346	1452	1551	1700	1925	2020	2105	2084	1936	1514	2000	21063
V-3420	2	4	4	2	2	1	0	5	3	4	1	2	30
V-1650	850	864	615	606	1222	1002	1146	964	1203	1290	1268	1265	12295
R-1820, Wright	867	881	1007	874	965	764	837	537	660	840	688	776	9696
R-3350, Wright	19	11	29	10	21	14	22	31	102	169	233	256	917
R-2800, P&W	531	464	624	479	361	524	744	842	485	842	882	811	7589
R-4360, P&W	0	0	0	0	0	0	0	0	2	2	1	3	8
1944													
V-1710	2002	2007	2101	2101	2101	1900	1702	1971	1190	1366	1073	677	20191
V-3420	5	2	2	6	1	2	11	11	13	28	17	14	112
V-1650	1564	1490	2000	1905	1275	2239	2268	2754	2027	1950	1861	1636	22969
R-1820, Wright	814	830	682	616	844	615	648	589	601	383	326	331	7279
R-3350, Wright	291	336	366	389	334	417	454	504	557	544	420	656	5268
R-2800, P&W	894	1013	1012	994	971	1101	836	1021	800	1067	942	914	11565
R-4360, P&W	2	5	3	1	4	4	1	2	1	0	1	0	25
1945													
V-1710	1103	904	1309	1145	828	156	3	21	16	0	1	0	5486
V-3420	4	0	1	0	1	2	0	1	1	0	0	0	10
Jets, Allison	0	1	0	6	10	27	111	158	128	104	65	0	610
V-1650	2190	1708	1777	1034	1936	1191	1502	812	382	5	34	0	12571
R-1820, Wright	297	275	279	245	311	277	130	77	99	37	2	0	2029
R-3350, Wright	706	714	871	977	1070	1134	1068	489	263	254	12	4	7562
R-2800, P&W	1087	1134	1271	1032	1185	1225	1169	565	131	97	58	22	8976
R-4360, P&W	2	6	9	10	5	25	17	4	4	5	16	7	110

Production numbers for similar rated radial and Packard V-1650 Merlin engines are provided for perspective. Second source production of engines is not included.

Table 17-7
Wartime Engine Shipments-Summary

	1940	1941	1942	1943	1944	1945	Totals
V-1710, Allison	1,141	6,447	14,905	21,063	20,191	5,486	69,233
V-3420, Allison	2	1	0	30	112	10	145
Jets, Allison	0	0	0	0	0	610	610
V-1650, Packard	0	49	7,251	12,295	22,969	12,571	55,135
R-1820, Wright	2,325	4,687	9,846	9,696	7,279	2,029	35,862
R-3350, Wright	8	6	66	917	5,268	7,562	13,827
R-2800, P&W	15	1,466	5,429	7,589	11,565	8,976	35,026
R-4360, P&W	0	0	0	8	25	110	143

Note: Production deliveries to U.S. Military during war. Note that discrepancies between Allison provided numbers on Table 17-1 and Department of Commerce numbers above have not been resolved. Both sources are creditable, though Allison numbers may include special test engines and in some cases redeliveries of engines as different models. Not included are second sources of radial engines. Packard records show 55,523 Merlin engines built.

Manufacturing the V-1710

That the engine was suitable for mass production is attributable to its General Motors heritage, and the practical skills and approach of the original Allison design and development team. Of course many individuals on the team, including Allison's Chief Engineer Ron Hazen, had previously served in other GM Divisions and roles, and many had received advanced education at the General Motors Institute. The team was well versed in the requirements and techniques for mass production.

The mechanics at Allison were a capable and innovative lot. Hand building each of the development and pre-production engines called upon all of their skills. Don Wright, who worked on the production line when the last of the engines for the YP-37's and YFM-1's were being built relates that,

> "We had a guy who repaired everything, a welder. He was a guy that would take a file and heat treat it for you so that you could make a scrapper out of it. He was a master at it. That's why I say every place you went, there was a guy that was an expert, and when I first went there they were scrapping all of the bearings to fit, then we started to line-bore them. Originally, every bearing in there had 100 percent Persian Blue on it when it went into the engine, that's what makes them run, but oh there were so many things that I saw that just amazed me. They'd take a crankshaft and build it up, Arni Beckwith particularly, with rods, the whole thing, and then go back with an engine graph and etch the numbers on the bearings etc. If you would touch one of those rods, it's gone! Never would he touch one of them. We've had quite a change from that time. We didn't even have a time clock, you just showed up and went to work, and then you went home!"[19]

John Kline adds as how, "We didn't have any security guards or anything else around that place either. It was a family."

Putting the Allison Into Mass Production

One month before receiving the first Army quantity order, and two months before declaration of war in Europe, General Motors broke ground on Plant 3. It was to be 390,000 square feet of factory, office, testing space, and located in Speedway. The concept of the Allison production schedule, with each succeeding Axis aggression in Europe, was revised upward by the Army to the point that when construction on Plant 3 finally ended in June 1943 Allison had built, and was using, three million square feet of plant space for the V-1710. The new Plant 3 anticipated the coming U.S. involvement in the war and was planned and built without windows and incorporated other "blackout" features. This was three years before the U.S. was drawn into the war.

Putting a modern high-powered, custom-built, precision engine into mass production presented many challenges:[20]

Table 17-8
Allison's Early Manufacturing Plants and Facilities

Plant	Description/Use
1	Jim Allison's original facility in Speedway.
2	Begun in 1936, used for Engineering Development and Testing during war.
3	Begun in May 1939, this was the major V-1710 production facility. Production began in February 1940.
4	Completed in 1943, part of the Plant 3 complex.
5	Largest facility, built by Defense Plant Corp. Construction begun June 1942, first delivery in February 1943.
6	Leased space which housed Service Training School during WWII. Vacated in 1945.
7	Located at 21st St. and Northwestern Ave, from 1950.
8	Became Research and Engineering Center, 1954.
9	
10	Allison Flight Test at Weir Cook Municipal Airport, Indianapolis

- The V-1710 contained over 7,000 pieces.
- Rough casting and blanks were to be analyzed before acceptance, inspected individually after each machining process, and again after test of each engine. One-fifth of all Allison employees were engaged as inspectors.
- All highly stressed steel parts [crankshafts, connecting rods], and all rotating and reciprocating parts were required to be "magnafluxed" to insure their integrity, both before and after machining.
- Sixty-two different metals or alloys were specified for use in each engine. All of these were soon to be placed high on the Wartime list of critical metals.
- Many parts required precision machining to the highest standards known. Involute gear profiles and some splines had to be machined to 0.0001 inch tolerance, crankshaft pins and main journals to 0.0005 inch.

A typical V-1710, the V-1710-F3R, had 7,161 parts by actual count, but only about 700 "piece" or different parts.[21] The competing Rolls-Royce Merlin/Packard V-1650 had about 11,000 parts, with some 4,500 "piece" parts. There were about 1,800 "piece" parts in a typical American radial engine.[22,23] By 1943, only 140 of the 700 V-1710 "piece" parts were being machined by Allison at Indianapolis, with 109 suppliers providing raw materials, semi-finished and finished parts from more than 100 plants located in 60 cities. Cadillac, Chevrolet, Delco-Remy, and eight other divisions of General Motors contributed significantly to production of the V-1710.[24]

General Motors tapped its far-flung divisions for managerial talent to tool up for the precision work, for plant expansion, for personnel training, for material procurement, and for subcontracting. It was not necessary to go outside the General Motors organization for the "know how" that was required.[25]

Typical of the good fortune of having experienced manufacturing people available for the expansion was with the arrival of Mr. Settle, who took on the position of Manager of Materiel Procurement in 1940. With strategic materials being increasingly difficult to obtain, he heard about a stock of 2,000 tons of high quality steel that was considered unusable because of its shape. He was able to acquire the unusual square bars with round corners which ranged in length from 2 to 8 feet, and at a good price. He then had all of this steel rolled to sizes useable in the Allison engine, an effort that required hauling it from one steel mill to another for the various sizing and heat treatments, but by obtaining this stock prior to the freezing of vital materials by the Government, he had provided Allison with the cushion to be sure they could meet early production schedules. This reserve was exhausted in 1942, by which time the priority system was working well. With Allison's high priority and ability to find substitute material when needed, they had few raw material supply problems. During the course of the war Allison was spending about $1,000,000 a day for materials and labor. Exclusive of payroll, it is estimated that Allison spent about one billion dollars during the war, an awesome amount for the time.[26]

At the beginning Allison management had some misgivings about the raw labor force that would be available in the Indiana region. More than half of the men, whose average age was 22, came from nearby Indiana, Ohio, and Illinois communities. They were former soda-jerkers, high school boys who had never worked before, white collar workers, and a few former WPA workers. From June 1939 to December 1941 personnel increased from somewhat over 500 to nearly 10,000.

The personnel department, as it hired 100 men a day from all walks of Middle Western life, developed a technique of hand-picking likely candidates for the precision work. An applicant's character was analyzed, his associations and background checked, but his appearance was ignored once he got a clean bill of health on a thorough physical examination. Allison management impressed itself with the small percentage of "wrong guesses" on its selection of new workers. The experience indicated that the Middle West offered above average men for mechanical training. As it turned out, of the people Allison hired during the War, 10 percent were skilled, 20 percent semi-skilled and the rest were hardly trained at all. The ingenuity of the American people and their aptitude for mechanical things is what won the war.[27]

The defense program migratory worker problem was the gravest personnel problem Allison faced, and it arose from the excellence of Allison training. Many employees lived in the Indianapolis area on the "hook." As soon as they heard of a plant opening nearer their home town, they were prone to leave, because with Allison training they were sure of a job anywhere. At the 11,000 worker level, and while hiring about 100 men per day, Allison had a turnover of 170 men a month.

The nucleus of 400 or more old-time Allison master mechanics who had custom-built the twenty different engines through the eight-year development period made an invaluable "faculty" for the education of the thousands of new trainees. They willingly passed along their mechanical skills and techniques as fast as the trainees could learn them.

As Allison tooled up and trained personnel, management and workers developed many new manufacturing techniques. The hand micrometer, long regarded as the symbol of precision work, was not sensitive enough to check the closest limits to which parts had to be machined. New automatic gauges were devised. Available materials were adapted wherever possible, as well as recent metallurgical research and developments for new methods of processing and heat treatment. Shot blasting was adopted to toughen the "skin" of highly stressed parts such as connecting rods. Nitriding, it was discovered, gave greater strength and durability to the crankshaft, allowing it to reliably deliver the coming demands for increases in power.

Although Allison had always used subcontractors for specialty components, such as the large and complex crankcase, accessories and intake manifold castings, they had always performed the final machining on the parts themselves. With the need to significantly increase production it was necessary for Allison to develop a network of suppliers and subcontractors to provide the large quantities of finished high quality parts.

Subcontracting, as done in the automobile industry, was expanded as a way of obtaining a flow of raw, semi-finished and finished parts, with the Cadillac and Delco-Remy Divisions of General Motors making the major contributions. They were so closely integrated in the Allison organization that their production departments operated virtually as units of the Allison shop.[28] Cadillac, with its precision machining and manufacturing background, made some 250 Allison parts, including the crankshafts, connecting rods and reduction gear assemblies. Delco-Remy provided the aluminum and magnesium castings, along with 75 different machined parts. Other GM Divisions were subcontractors as well. These included Chevrolet, New Departure, Hyatt Bearing, Delco Products, Packard Electric, A.C. Spark Plug, Antioch Foundry, Harrison Radiators, and Inland. Outside corporations included another 93 suppliers of raw materials, semi-finished, and finished parts.

In February 1940, engine production began from Plant 3, with the first engine, a V-1710-C15, being delivered in June 1940.[29] Within the next three months, Allison produced almost twice as many engines as had been built during the entire 10 years of the development and preproduction period. Production during 1940 and 1941 continued to be stepped up, as the V-1710 was specified to power ever more new Army Pursuit planes. In October 1940, Allison delivered the 500th V-1710, and by the end of the year a total of 1,153 engines had been delivered. The maximum intended production rate for the plant, 1,000 engines per month, was reached in December 1941. With Plant 3 in production, Plant 1 was turned over entirely to bearing manufacture.[30] Still, the effort was not without problems and controversy.

On July 1, 1940 Jimmy Doolittle had been recalled to active duty and assigned by Major General H. Arnold, Chief of the Army Air Forces, as the Assistant District Supervisor of the Central Air Corps Procurement District, with duty as the Army Air Corps representative at the Allison Indianapolis Plant. He completed his assignment in November 1940 and submitted a report to Arnold making the following points:[31]

- Upon arrival, the major engine problems were varying from crankshaft and bearing failures to coolant pump and Prestone leaks. (The V-1710-C15 was the major production model at the time).
- He felt that the Allison liquid-cooled engines would not take the punishment and overloads that could be imposed on older air-cooled engine designs. He recommended that until their "temperamentally" could be corrected, pilots should be cautioned to operate the Allison engines within strict limitations. (This was the period when the early P-40's were having their backfire problems and unmodernized engines performance ratings were temporarily restricted.)
- He stated that the problems with Allison were largely managerial. He noted that many personnel changes had taken place and how certain top managers were more interested in quantity than quality. He gave credit to C.E. Hunt (V.P. Engineering for GM) for straightening out the mess, and that progress was being made toward turning out reliable engines due to "his knowledge, skill, patience, honesty, and never-failing cooperation."
- The most disturbing single production problem was dirt. Failures prior to assembly and during engine run-ins were due to sand from castings, metal chips from machine operations, emery dust, magnaflux powder, dirt from new construction, and dirt introduced during engine run-ups. Doolittle felt the problem was due to attitudes which had to be changed among the foremen and supervisors. Corrections caused much delay, but were finally, albeit reluctantly, accomplished.
- On the positive side, Doolittle noted that future engines, if redesigned as planned [V-1710-E/F's were entering prototype production], should develop 1,500 hp, but would entail a substantial increase in weight and alteration of the cylinder head, induction system, and exhaust system to increase strength and volumetric efficiency.

His Summary:

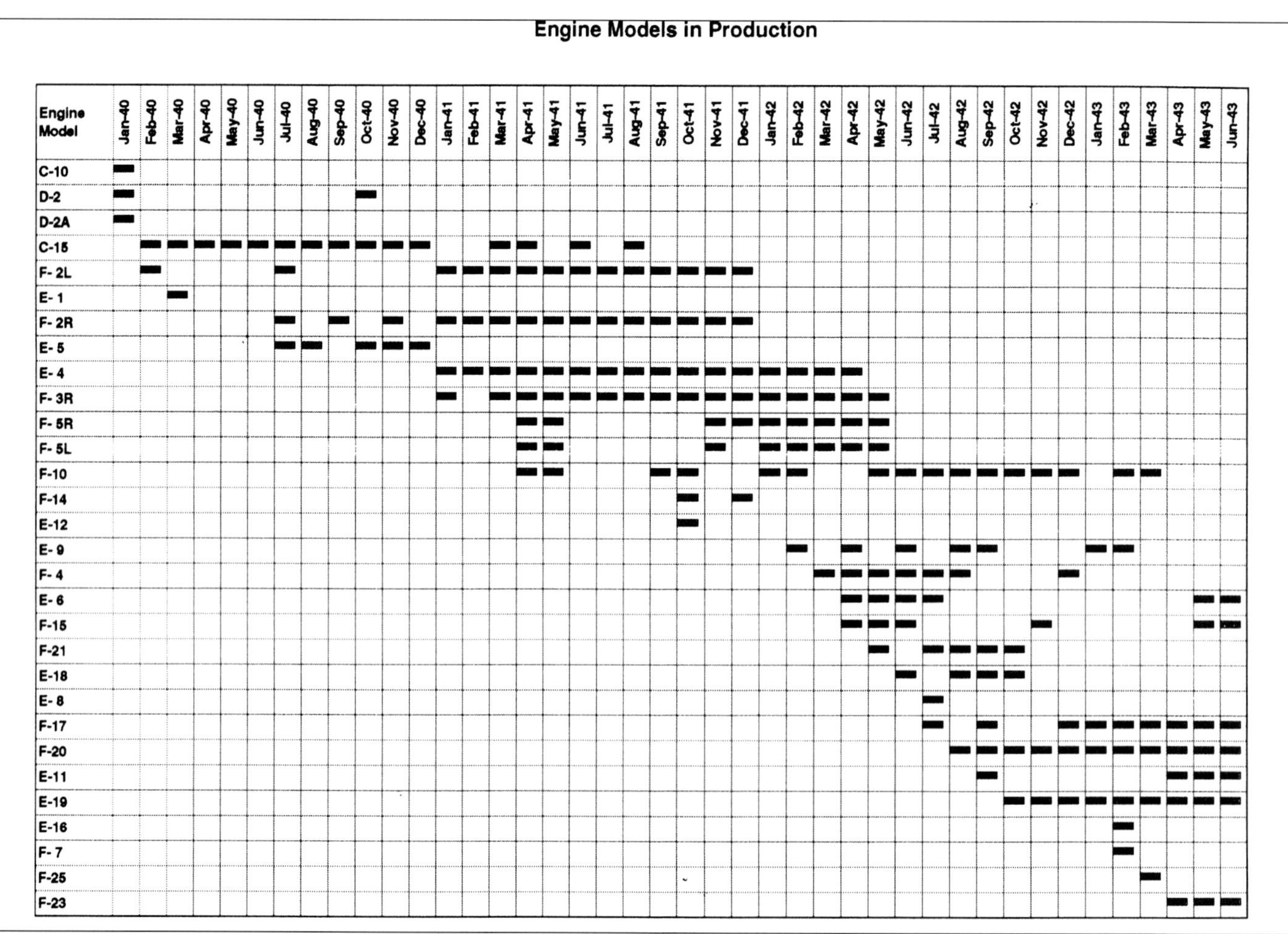

This chart shows the variety of engine models being manufactured and delivered early in the war, and when they were in production. In many cases only a few examples of each model were being built in any given month. During 1942 a new model entered production every month. Workhorse and development engines of these models often preceded these dates by many months, or even years. (Allison)

- No organization, unaccustomed to making aeronautical equipment, can produce a satisfactory product until they acquire the "aviation viewpoint" and starts working to aeronautical instead of commercial standards.
- The bigger and more successful an organization (i.e. General Motors), the harder it is for them to realize this.

It needs to be kept in mind that more engines were delivered during 1940 than were on order January 1, 1940. Furthermore, actual deliveries to the Air Corps of only 342 engines during the year were due to a contract change reducing the quantity of C-15 engines from 776 to 300 and substitution of the new F-3R engine, a model not yet developed and involved a contract change delaying deliveries. All of the requested C-15 engines were delivered during the year. Another complication was the change in specifications which resulted in converting the Sea-Level E-2 engines for the Bell YP-39's into Altitude Rated E-5 engines, a model which had yet to be completely developed and resulted in delaying their delivery. This performance would hardly seem to merit the criticism that has been made.[32]

Obviously the problems identified by Major Doolittle were resolved, for the V-1710 went on to make an outstanding record for itself while absolving each of his stated criticisms. Sizable quantities were produced. In fact by October 1941, the 5,000th V-1710 had been built.[33] By December 1941, Allison production had exceeded the factory's design goal of 1,000 engines per month and was delivering 1,100 completed engines per month.[34] The attack on Pearl Harbor completely rearranged production objectives with the result that the only "schedule" became the maximum number of engines that could be built, assembled, and tested each month.[35] That peak was reached in December 1943 when 2,600 engines were produced, though peak employment had occurred in October 1943 when 23,019 workers, of which 31.4 percent were women, were on the Allison payroll.[36,37] Some references identify a different peak employment figure, that of 20,703 workers. It is believed that the difference is that the first figure represents total employment within the Allison Division, there being 20,703 employed on the V-1710 and another 2,316 producing Allison sleeve bearings.

From the "automobeelers" came the uniquely American principle of mechanical simplification, which they aggressively applied to the V-1710. Born in the experience that made the sewing machine and the automobile mass consumer products, their efforts resulted in making the Allison V-1710 as American as ham and eggs. This can be seen by comparing the reasons for the difference in the parts count between the V-1710 and otherwise similar European liquid-cooled engines. For example, piece parts in one small sub-assembly were reduced from 38 to 3 by simply casting the part whole rather than bolting it together. Additionally, this resulted in a part that could be machined with greater accuracy and offering improved durability and significantly reduced production cost. The benefits of reduced production cost were passed largely on to the U.S. Government as purchaser. During the course of the war, the unit engine price dropped to one-third that of the engines powering the XP-38, XP-39 and XP-40.

Allison considered the implications of this degree of standardization to be important, suggesting that the V-1710 was easier to maintain than a European engine (British, German, Italian, Russian, or Japanese copies) with many times the number of parts. They believed this resulted in a reduction in the amount of maintenance and overhaul required for the engines, which in turn could be equated to greater fighting hours available for a squadron of fighting ships. In some way, this had to contribute to overcoming some of the numerical disadvantage in the number of available aircraft in the combat theaters. This certainly was the experience of the V-1710 powered *Tomahawks* of the *Flying Tigers,* as they cannibalized aircraft and engines to maintain their limited fighting strength.

The Manufacturing Process

Manufacturing the actual engines was a complex and exacting process. It not only required the coordination of an extensive network of subcontractors and material suppliers, but Allison reserved for itself many of the most exacting fabrication steps and processes.

An aircraft engine necessarily operates with many of its components stressed to the full extent of their ability to carry load. It was in exploitation of this factor that allowed Allison to be the first to achieve production engines able to deliver over one horsepower for each pound of engine weight. As an example, in the case of ferrous (steel) parts, many of which were forgings, it required the removal of a lot of surplus material. A 160 pound forging was supplied from which a 36 pound finished propeller shaft resulted. Although the finished propeller shaft appears to have been formed from a drawn tube with a flange, strength demanded physical properties which could only be achieved by using a single high quality forging. Similarly, the 120 pound, 6-counterweight crankshaft began as a 620 pound forging.

Because each part of the engine must be almost perfect, inspection was as important as fabrication. Of every ten men in the plant, two were inspectors, either Allison, British or U.S. Army. Every part was inspected at each stage of its progress from receiving room through the manufacturing, assembly, and testing. The completed engine represented successfully passed some 70,000 inspection hurdles.

The tolerances for the various parts were carefully worked out as a balance of efficiency in production and necessity in operation. On the crankcase and cylinder heads, which were nearly four feet long, certain tolerances had to be held to within +/-0.005 inches. One of the reasons for such control was to accommodate the growth of the engine, about 1/8 inch in length, from cold start to normal running temperatures.

Complicating the manufacture of components were the stresses imposed during assembly. As an example, the aluminum cylinder heads were each fitted with six steel cylinder sleeves and 24 valve inserts, all shrunk fit into place. The head was then bolted to the aluminum cylinder jacket and the assembly then made leaktight by tightening a compression nut on each cylinder liner to 2,100 lb-ft torque. To accommodate the resulting distortion due to assembly pre-stresses and to insure a perfectly flat cylinder head mating surface, Allison machined the cylinder heads in a fixture that "bowed" them so that about 0.009 inches of concavity was induced in the bottom surface, and a similar amount of convexity in the top surface. When assembled, the assembly would be perfectly straight.[38]

Once cylinder heads were rough machined, they were given an anodizing treatment as a protective coating. The next machining operations were to +/-0.005 inch tolerance and prepared the heads for shrink fitting the valve-seat and cylinder barrel inserts. These were accomplished by heating the heads in an oven to 470-490 °F, while the inserts were cooled to about -40 or -50 °F by immersion in a bath of kerosene and dry ice.[39] The intended interference fit between the cylinder barrels and the cylinder head was 0.018 to 0.023 inches. Heating the head caused the holes in the casting to increase about 0.028 inches, while the chilled barrels contracted about 0.005 inches. The resulting gap allowed a loose fit and insured that the six barrels would properly register in their seats. A hydraulic press was used to position and hold all six liners in position while the temperatures equilibrated and the liners were permanently captured by the head.[40]

This method, as used by Allison, was quite different from that typically used by manufactures of radial engine cylinders. They were obligated to use a shrink fit as well, but also needed a threaded connection to insure adequate capture. The major reason for the success of the Allison joint was that liquid-cooling limited cylinder head temperatures to something around 450 °F. Air-cooled cylinder heads could go well above this during high power operation, and they could not tolerate the possibility that the faster growing aluminum head would come loose from the steel cylinder liner.

After assembly of the cylinder head, the final operation was grinding and honing of the cylinder barrels to the final dimensions and surface finish, all of which was quite exacting at +0.002, -0.000 inches.

Once assembled, every engine was given a "green run." Performance during this seven hour test was checked constantly to guard against developing problems or failure. Even if the performance was perfect, the engine would then be completely torn down, each part cleaned and carefully inspected. All of the parts from an engine were kept together, and the steel parts magnafluxed. Once all of the parts were cleared, the engine would process through "final assembly" with a dedicated five man team, and their own inspector. The engine then returned to the test beds where it received a two hour run prior to final acceptance. At this point the engines were prepared for storage, safety wired, and all external units checked, along with a final check of valve clearance and timing.[41]

According to Don Wright, who worked in production and testing before beginning his career in Allison's Service Department,

> "once production got into full swing and confidence in the engine was established, Allison instituted a sampling system for the "green run" and teardown. They had started out with the idea that they would have crews [do the engine build], the way we'd worked for years. We found out pretty soon that that wasn't the best way to do it. So they set it up that one guy built up the crankshaft and put it in the crankcase, move it down, the next guy would put the pistons on, then the banks went on and so forth." "Now it would get up to the end here and it was to be either a right or left-hand engine. And for some reason the people had difficulty figuring out which way the engine was going to run. And they were finding out when they got to the test stand that a lot of them were wrong. So they had, I don't know how many engines that were timed to run in the wrong direction. Well if you think you've ever heard a cannon go off, you should hear one of these. Well at any rate, they started bringing those back in, and because I had been with those other guys, you just didn't miss that if you worked for Marty Becker or Stockdale [guys who had been longtime Allison technicians in the early days], you just didn't do that, you did it right, and you did it right the first time! And you'd better not miss. So they decided to setup a station and set me on the station to time all of the engines coming off either the final or green production lines. I'm doing this, I'm doing pretty good, doing about 28 engines a day."[42]

Attention to Detail

The V-1710 is a complex machine. Manufacturing, assembly, and maintenance all required exacting and consistent application of procedures. For example, the procedure for tightening cylinder head bolts is always specified by the manufacturer, and trained engine mechanics know they must follow the tightening sequence and torque values as specified. In the case of the V-1710, the life of the main journal bearings is directly affected by the manner in which the hold down nuts are tightened. Incorrect tightening results in distortion of the crankcase and main journal bearing bores. Similarly, crankcase distortion also results if the rear cover, i.e. Accessories Section, is removed from the engine and the four rear cylinder holddown nuts are not first relaxed.[43] Skill, technique, and knowledge were essential to the success of the engine mechanic, and to the mission.

Demands on Production

As the V-1710 began its service life, and as the demands of combat began to be felt, Allison continued, without letup its program of engine development and improvement. A variety of sources provided basis for these changes: lessons learned from their extensive program of abuse to engines in the testing rooms and through suggestions for improvements being returned from the military and Allison field representatives. The 1000 bhp engine of 1937 was upgraded and rated for 1150 bhp in 1938, and accommodated a variety of diverse altitude ratings required for the various aircraft. All of this led to improvements and changes in the design and bits and pieces associated with increasing the power, performance, durability and service life of the engine. As fast as an improvement could be proved, it was ordered into production, all without stopping the production lines. According to Ron Hazen, the watch cry was, "the enemy won't wait a year for model changes," so for two and a half years there was a "model change" in production averaging every 45 days.[44] In fact, during the major changeover from the V-1710-C15 to the considerably different V-1710-E/F, production rates were maintained at over 60 percent of the previous peak level during the month that it took to effect the changeover.[45]

Once the production process was well underway, Allison introduced a tactic to further boost production, and at the same time meet the demand for increased numbers and quality of the new V-1710-E and -F models. By the simple expedient of building only one basic model, either the "E" or the "F", for a two week period, they were able to minimize the confusion of mixing models on the production line, or the complexity and expense of maintaining parallel lines.

Improving the Breed

As the production program matured, the needs of engineering, production, and the customer were constantly changing. By 1944 it was clear that demand for the V-1710 had peaked and that not only did Allison need to begin to look for new products and markets for the coming post-war period, they also needed to further improve the V-1710 and its profit potential. Starting in March 1944, a very large number of detail part improvements were placed into production, involving more than half of the total parts in the engine. A number of these improvements were made available as kits for modification of older models already produced and in the field. Some of the parts affected were:[46]

- Twelve counterweight crankshaft in place of six, permitting higher power and engine speed with greater reliability throughout the engine and bearings, along with a reduction in vibration.
- Fatigue strength of connecting rods improved by shot-peening.
- Exhaust valves provided with nichrome head for longer life at higher power.

These and other improvements resulted in a significant increase in the overhaul life on all Allison engines, from a previous high of 480 hours maximum to 1,000 hours under certain conditions.

The V-1710 power output was significantly increased over the production life of the engine. Specifically, the takeoff rating was increased from 1000 bhp to 1600 bhp, while it earned a combat rating of up to 2300 bhp. While a large part of the increase was due to improved design and attention to detail, advances in metallurgy, development of water-alcohol injection along with improvements in fuels provided the biggest boost, it is also important to recognize the contribution made by the manufacturing standards and improved techniques Allison and its suppliers developed for

Table 17-9
Manufacturing the Cylinder Block

- Receive Head Casting and inspect.
- Rough mill top face 0.010-0.012 in. convex in longitudinal plane, bottom face 0.010-0.012 concave. Inspect.
- Rough mill outside flanges, intake pads, oil drain pad and both ends of head.
- Finish mill oil and coolant pads.
- Rough mill camshaft bearing and sides of bearing brackets.
- Through drill and ream 14 stud holes.
- Mill oil drain slot.
- Rough bore combustion chamber diameter.
- Drill six holes in each intake and exhaust boss, drill six spark plug holes on each side.
- Profile front and rear end flanges.
- Cut out excess stock in coolant cavities, between cylinder bores and exhaust cavities.
- Anodize cylinder head and prepare for painting.
- Profile cylinder head combustion chambers.
- Mill 22.5 deg flat angle in dome.
- Tap drill and counterdrill 34 holes for studs in bottom. Rough spot-face 7 hold-down stud holes and bosses.
- Semi-finish tap stud holes in bottom of head and 10 holes for jacket sleeve bolts. Rough-tap 6 holes in top of head. Inspection.
- On both intake and exhaust sides, rough bore intake valve insert holes, drill and bore intake valve guide holes, end mill area at 6 holes.
- Form six intake chambers between valve guide and valve insert holes.
- Form six exhaust chambers.
- Finish bore exhaust and intake valve holes, semi-finish valve guide holes. inspection.
- Tap drill 26 holes on intake side, 34 holes on exhaust side, and 12 spark plug holes.
- Semi-finish spot-face 6 intake and 6 exhaust valve guide bosses to tolerance of +/-0.005 inches from bottom of insert holes.
- Radius spark plug holes on inside; cut out excess stock below valve seats; polish out cutter marks around intake and exhaust holes, combustion chamber and dome. Clean and inspect.
- Pressure test for leaks with Prestone at 40 psi, head heated to 280 °F.
- Use oven, heat head to 470-480 °F, hand assemble 12 intake and 12 exhaust inserts in head. Inspect.
- Chamfer ends of camshaft bearing bosses.
- Finish hand tap stud holes in head.
- Finish mill bottom face 0.008-0.009 in. concave.
- Semi-finish mill top face of cylinder head.
- Finish bore cylinder barrel holes +0.001, -0.002 in. on diameter and +/-0.003 in. on depth.
- Lap bottom face. Clean and inspect cylinder bores and domes.
- Heat cylinder head to 480 °F. Press chilled cylinder barrels into place. Inspect.
- Finish tap 12 spark plug bushing holes. Assemble 12 spark plug bushings with copper washers.
- Assemble 34 studs on bottom of head. Assemble 10 cylinder stud sleeves in bottom of head. Clean and inspect.
- Assemble miscellaneous parts, including jacket and sleeve assembly to head and barrel assembly.
- Tighten six cylinder barrel nuts to 2,100 lb-ft torque using hydraulic nut tightener.
- Pressure test with Prestone for leaks at 250 °F.
- Lap bottom side of jacket, clean and inspect.
- Place assembly into machining block, torque 12 holddown bolts to 600 lb-ft, for succeeding machining operations.
- Finish mill top of cylinder head to 0.001 in.
- Finish mill and chamfer 13 camshaft bearing bosses. Tap drill dowel pin holes, 24 holes in camshaft bearing bosses, 10 in head cover holes; drill 22 flange holes in top of head.
- Semi-finish tap 24 holes in camshaft bearing bosses and 10 cylinder head cover screw bushings. Finish tap same.
- Drill vent holes in camshaft bearing stud holes.
- Lap top of cylinder head, wash, inspect.
- Finish mill distributor flange face and camshaft location bearing bosses.
- Rough and finish bore camshaft bearings. Inspect.
- Remove machining fixture.
- Finish diamond bore valve guide holes to 0.0005 in. on diameter.
- Spot-face hold-down stud hole bosses in topface. Inspect.
- Finish mill 6 intake and 6 exhaust pads, along with coolant pads and bosses.
- Drill and ream 7 dowel holes in camshaft bearings. Finish mill camshaft drive pad.
- Face off spark plug inserts flush with recess in head.
- Clean, inspect and paint with two coats of black enamel.
- Press in 24 valve stem guide bushings.
- Semi-finish ream valve guide bushings.
- Recut intake valve seats for 30 deg angle and proper width.
- Finish hand tap 4 holes in camshaft drive pad; 4 holes in distributor flange end; 3 holes each in front and rear coolant pads; 24 holes in intake manifold; 30 holes in exhaust manifold pads. Inspect.
- Assemble 12 brass inserts in intake pad, 9 others, and 22 steel inserts on top face.
- Retap spark plug bushings and inserts.
- Grind and lap valve seats to gage.
- Finish grind 6 cylinder barrels to +0.002, -0.000 inches.
- Hone cylinder barrels to +0.002, -0.000 inches.
- Finish ream valve guide holes.
- Inspect barrels for size and finish. Finish lap 14 hold-down stud bosses, and all other finished surfaces to remove all scratches.
- Assembly of various studs, pipe plugs, safety wires, cotter pins.
- Wash, Final Inspection, Army Inspection.

fabricating the engine. The basic engine components, dimensions, size, and weight remained largely unchanged, while the ratings were continually increased.

A major 1944 improvement was the introduction of water-alcohol injection on two-stage supercharged engines, particularly those in the Bell P-63. Allison provided kits with the necessary parts to control the ADI fluid as well as to adjust the carburetors for proper functioning during water injection use on their V-1710-E11 engines.[47] The technology allowing this technique for dramatically increasing power was first developed and applied to production engines by Allison.

Based upon Takeoff ratings alone, the specific weight was 1.24 pounds per horsepower for the V-1710-C, and 1.33 pounds per horsepower for the first V-1710-F engine. The last V-1710-F weighed 0.91 pounds per horsepower. The "F" rating was up to 1425 bhp takeoff and 1625 bhp WER before Allison had to increase the weight over that for the V-1710-C15. Allison was the first aircraft engine manufacturer in the world to operate its engines at a weight of less than one pound per horsepower. Weight was controlled by efforts at overall simplification of piece parts and the substitution of magnesium for aluminum and aluminum for steel wherever possible. It got to the point where Allison was developing more horsepower than the airplane designers could use without a marked change in the airplanes they were powering.[48]

Spare Parts Manufacture

Just as it is necessary to have spare engines available to support the number of aircraft built and being operated, it is also necessary to have sufficient spare engine parts to support the overhaul and operation of the engines in the field. The following table shows the number of equivalent engines which could have been assembled from the quantity of parts actually produced.[49] It also suggests that the V-1710 was a fairly economical engine to overhaul, in that from this table it appears that approximately ten to twenty percent spare parts were required to both support the service life of the engines and stock the supply system. This of course includes the quantities of spares provided to support planned upgrades and improvements in engines as they passed through the various overhaul facilities.

Table 17-10
Production, Including Spares

Year	Equivalent Number of Engines Built	Engines Shipped
1940	1,153	1,153
1941	7,150	6,433
1942	18,500	15,319
1943	23,950	20,350
1944	22,700	20,190

Supporting the Allison engine at the far corners of the world was itself a world class challenge, often made more difficult by the Army. Early in 1942 a crisis developed in Panama where they were locally overhauling Allison engines in support of the fighters stationed there to protect the Canal. The local command sent a report through channels that there was a serious shortage of Allison parts needed to support the overhauls. The Air Service command then requested detailed lists of parts needed and quickly supplied seven transport loads of parts, taking them from the overhaul facilities at Middletown and Fairfield Air Depots. These parts were sufficient to overhaul 75 engines, though the quantity of parts requisitioned would have been sufficient to overhaul 700 engines. The Panama overhaul facility had the capacity to process 15 engines a month. Evidently someone was frustrated in attempts to get parts, so ordered far more than needed, hoping to get the needed quantity. Such hoarding of parts was detrimental to the entire effort. The Materiel Division was routinely challenged by the practice.[50]

Technical Advances in Manufacturing

The V-1710 was a complex machine to manufacture. When it became necessary to maximize production for the war it became necessary to find ways to build the engine utilizing, for the most part, unskilled workers. This was done by giving the training needed to qualify them to perform only specifically assigned tasks. Furthermore, it was necessary to improve the productivity, by minimizing wastage and economizing on the use of labor and material. At the same time the continuing need to raise the output of the V-1710 required that the quality of individual parts be improved to carry the additional loads. Using the great manufacturing experience of General Motors some very innovative advances were made.

A particularly important improvement was the speeding of aluminum foundry methods done during the war. Early in the war Allison found existing sources were inadequate to supply the intricate aluminum castings which were required to carry the very high loads in the V-1710. By coincidence, it was found that at Antioch College, in Yellow Springs, Ohio near Dayton, there was a foundry which had found a way to cast art objects to

Table 17-11
Advances Due to Increased Knowledge in Design and Production

	Liberty-12, 1918	Allison V-1710, 1943
Takeoff horsepower	400 bhp @ 1700 rpm	1425 bhp @ 3000 rpm
Weight per bhp, pounds	2.3	0.93
Displacement, cubic inches	1650	1710
Reduction Gear Ratio	Direct Drive	2.0:1
Bore and Stroke, inches	5.0 x 7.0	5.5 x 6.0
Time Between Overhauls, hours	50	500+
Crankshaft Weight, pounds	105	95
Crankshaft Weight/horsepower	0.26	0.065
Crankshaft Steel tensile strength, psi	135,000	140,000
Crankshaft Counterweights	None	6
Inspection	Visual	Magnetic

exceedingly close dimensions. General Motors bought the foundry and hired the teacher and his wife who had developed the process to work on application of their technique to high strength aluminum castings. Their work supplemented the two foundries set up at Anderson and Bedford, Indiana under the operation of the Delco-Remy Division of GMC to supply aluminum castings for the V-1710. These castings proved to be infinitely better than any others available anywhere in the world, and at less than half the cost. At the close of the war Allison purchased the Bedford Foundry and General Motors sold the Antioch Foundry back to the professor, his wife, and their associates. General Motors continued to pay a royalty on all castings made which involved their unique process.[51]

In this way forms were duplicated to an accuracy that eliminated the previously slow process of contour machining and hand fitting. Such techniques, once developed, significantly accelerated production and quality. The result of all of this really became apparent in the V-1710-E/F models, the crankcase of which, compared to the V-1710-C, required about 10 percent fewer operations to manufacture, weighed less and was much more rugged. Development and use of permanent molds for the intricate castings of components, such as the crankcase, reduced the cost of production and raised the endurance limit for engine operation by at least fifty percent.[52] One of the first examples of the Antioch process was the crankcase for the V-3420-9 Type Test engine. The initial Alcoa cast crankcase was damaged early in the test and the engine was rebuilt in the spring of 1942 with one cast by the Antioch process. This unit became the first V-3420 to successfully complete the 150-hour Type Test.[53]

Achieving these aggressive production goals would have been impossible without the closest cooperation between the aircraft and automobile industries and the U.S. Army Air Corps. The mass production ideas of the "automobeelers" were agreeably reconciled to the precision requirements of the aircraft industry, and there was an effective blending of abilities. As the customer, the Air Corps, worked to improve communications between the squadrons and Wright Field, as well as maintain a strong presence in the factory. Everything was focused on getting quality engines out the door.

The Table 17-11 provides a comparison of the extent of advances made from the early Liberty-12 through the V-1710. While it might be said there is no comparison, the truth is that considerable advances were made to allow an engine of similar size as the Liberty-12 to produce 3-1/2 times as much power, and with at least three times the reliability.

Highly loaded components like the crankshaft were able to carry the higher power levels because of features such as counterbalancing, which significantly reduces the reciprocating loads reaching the bearings. This reduces the deflection of the crankcase and consequently reduces the amount of bending forced onto the crankshaft itself. Of course there is an increase in the weight of the crankshaft due to the integral counterweights, but as shown in Table 17-11, this can more than offset the alternative. Improved manufacturing technique and procedure also played an important role. An example is the effect of shot peening the crankshaft shown in Figure 17-1 and the improvement in ability to tolerate stress as compared to that of the originally untreated crankshaft. This was further improved when the nitrided crankshaft was introduced early in 1942.[54] As long as rated loads were not exceeded, these crankshafts would safely operate forever.

Allison Personnel and Pay

As has been noted earlier, the growth in employment at Allison in support of the V-1710 and the war effort was nothing short of phenomenal. Table 17-12 provides some detail during the buildup to maximum production. The pay rates are of some interest fifty years after the fact. Recognize that these are the average rates paid to men, many of which were considered skilled craftsmen and would have been at near the top of their respective crafts.

Subcontracting and Subcontractors

Allison always used subcontractors in the production of its various engines, but this practice was developed to the level of art during the war. Subcontractors were used extensively. At the peak of production activity there were more than 300 subcontractors building pieces or sub-assemblies for Allison. Allison shied away from subcontracts on the West Coast as they considered it too far away from their base of operations.[55] As a result, almost all suppliers and subcontractors were located between the Mississippi River and the East Coast.

These 300 contractors provided specific "piece parts." This doesn't include suppliers of raw materials and generally standard "nuts and bolts" type items. If these vendors were included, some 1,250 outside suppliers would be listed to account for the $80,000,000 Allison was spending with them each year. Although other General Motors Corporation Divisions were major participants in the program, their comparable billings amounted to considerably less at approximately $6,000,000.[56]

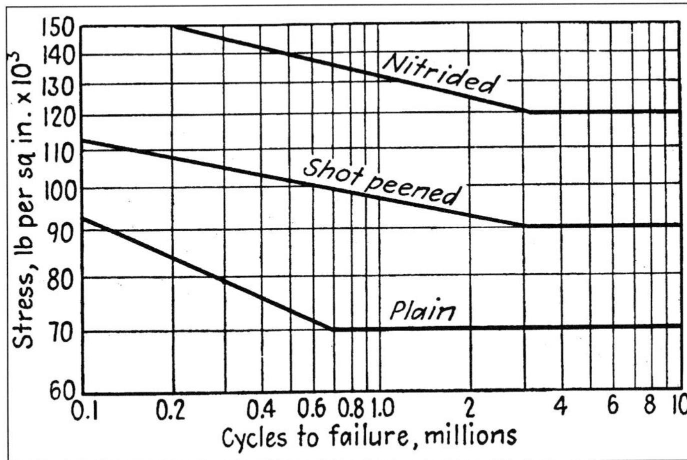

Figure 17-1. The endurance of highly stressed parts like the crankshaft were greatly improved by the new techniques of shot-peening and Nitriding the finished parts. (Liston, Joseph, 1953, Figure 8-30, 508.)

Table 17-12
Growth in Employment Since Date of Expansion, July 1, 1939

	Working Employees	Hourly Base Average Rate, $/hr	Monthly Earnings, All Emp.
July 1939	672	.7271	$160.29
December 1939	1,138	.7929	$170.08
January 1940	1,311	.7645	$186.98
June 1940	3,603	.7518	$181.98
December 1940	7,241	.8140	$194.27
May 1941	8,766	.9421	$221.14
December 1941	12,348	.9070	$225.15
May 1942	13,633	.9754	$228.92
December 1942	16,865	1.0079	$243.30
May 1943	18,993	.9980	$263.18
October 1943	21,115	1.0047	$243.93

Significant in this regard was the contribution of the Cadillac Motor Car Division of the parent General Motors Corporation. Cadillac had developed a long history as the producer of the highest quality automobiles in the U.S. As such, they were an obvious candidate when considering the conversion from automobile production to wartime military engine products. In fact, other automobile manufacturers made similarly successful forays into aircraft engine production, such as Packard with its' production of the V-1650 Merlin, along with Ford, Buick, Chevrolet and Studebaker manufacturing Pratt & Whitney radials. All would agree that the transition was not painless, for the demands of components in aircraft service requires extremely high tolerances and attention to detail. Still, the strength of the auto manufacturers was in their ability to mass produce vehicles with interchangeable parts. Interchangeability requires precision, and this experience is what made their transition to aircraft engine production successful.

Cadillac took on the mass production of the most critical components in the V-1710. They produced the crankshaft, camshafts, connecting rods, piston pins, the entire reduction gearbox, and the rotating steel supercharger rotor vanes. Every component was a complex and exacting production challenge.

To bring these components into large scale mass production, Cadillac introduced numerous innovations that had the effect of reducing material consumption and scrap rates, while at the same time reduced the man-hours required to produce a part. For example, the finished camshaft weight was 6.9 pounds, though Cadillac started with a 26 pound forging. Previously, Allison had machined the part from a 79 pound section of bar stock. Additionally, Cadillac introduced new processes such as honing the final finish on the inside of the hollow camshaft once the hole had been precision bored. In this way, they produced a finished part free of any scratches or tool marks, any of which would likely lead to its failure when stressed in the high powered engine.

As the power ratings on the V-1710 steadily increased during its production life, it was necessary for Cadillac to improve its methods and processes in ways that gave improved reliability at the higher stress levels, without incurring significant increases in size or weight of the parts. The manufacture of the polished connecting rods required 86 different operations, compared to an automobile rod which needed only 29. The V-1710 rods were then shotblasted to impart the necessary surface conditions in the parts to prevent fatigue failures. New manufacturing processes were also justified based on the scale of production. For example, where Allison had milled the slot in the "forked" connecting rod, Cadillac was able to devise a "broaching" process with which two broaches, cut the slot in just three simple passes. A sizable savings in production time resulted.[57]

During the course of the V-1710 production program it often fell to Allison to aid its subcontractors with materials, priority, technical assistance, as well as capital, to insure that deliveries would be as specified and on time. By the end of production it was estimated that, of the complete engine, one-third of the engine was built by each: Allison, other GM Divisions, and outside sub-contractors.[58]

Table 17-13
Engine Price by Year

Year	Price per Engine	Number of Engines Delivered
1932	$75,000	1, GV-1710-A
1933	$25,000	3, V-1710-4's
1934	$38,000	2, XV-1710-1 and XV-1710-3
1935	$27,500	3, YV-1710-3's
1936	$27,500	2, YV-1710-3's
1937	$28,442	7
1938	$26,560	12
1939	$26,283	48
1940	$22,860	1,153
1941	$18,540	6,433
1942	$14,000	15,319
1943	$12,000	20,350 V-1710's
	-	31 V-3420's
1944	$9,500	
	$30,278[76]	21,381 V-1710's
		112 V-3420's
1945	$9,304	5,492 V-1710's
	-	10 V-3420's
	$55,917	610 J33's
1946	$13,500	17 V-1710's
	$20,800	905 J33's
1947	$23,329	603 V-1710's
	-	1,614 J33 & J35's
1948	$23,329	145 V-1710's
	-	1,102 Jets

* Air Corps actually paid Allison $70,656.57 total for the XV-1710-1 and XV-1710-3.

*The ten YV-1710-3 engines purchased on contract W535-ac-06795 totaled approximately $300,000 by time of delivery.

Price of the Engine

Mass production has been shown to be the vehicle to reduce the costs of an item, and this was certainly true for the V-1710, as shown in Table 17-13.[59]

Production Summary

In all, Allison states that 70,033 V-1710 engines were built, from #1 on August 13, 1931 to s/n A-074125 on June 2, 1948. The engine which qualified at 1000 bhp in 1937 remained basically the same engine that produced, in its final configuration, 1600 bhp for takeoff and had an emergency WER rating of up to 2300 bhp.[60] These numbers do not include equivalent engines provided as spare parts, "workhorse" engines, or those delivered from non-flightworthy parts as "school engines."

Of the 70,000 liquid-cooled V-1710 Allison engines, there were some 55 models produced. Of these, 29 had been perfected and applications specified such that they were placed in mass production. Of the balance, all but five had reached the stage of flight testing during the war. In addition to the V-1710, there were thirteen models of the V-3420 that were built for flight testing. Furthermore, there were several models of both types of engines specifically modified and built for use as marine engines. These accomplishments emanated from an engineering department that, including blueprint boys, was staffed by only 25 people when the V-1710-C/E/F were being placed in production. At the end of the war the engineering department had about 1,000 people fully employed, with about half of the engineers having come to Allison from other General Motors Divisions to serve during the emergency.[61]

Growing the production organization, while at the same time delivering ever increasing quantities of more powerful and reliable engines, was itself a prodigious accomplishment. For example, in the 28 months following the introduction into production of the V-1710-E/F, 18 individual models were produced, while the production rate increased from 300 per month to over 2000 per month. Every new model represented an improvement in

Table 17-14
General Motors Delivery of War Materiel

Product Class	Total, Billion $
Military Trucks	2.09
Allison Engines	1.04
P&W Engines	1.35
Jet Engines	0.03
Aircraft and Sub-Assemblies	1.31
Aircraft parts and Propellers	1.13
Tanks and Armored Cars	2.00
Marine Diesels	1.35
Guns, mounts and controls	1.15
Ammunition	0.47
Other	0.41
Total	12.30

at least one or more feature resulting in improved performance of the aircraft in which they were installed.[62]

As large as the Allison war effort was, it was only one of a number of significant contributions made by General Motors during the war. In terms of dollars paid to GM for delivery of war materials, the effort was only about 8 percent of the total. Surprisingly, licensed production by GM of P&W radial engines was an even larger program, as was the production of diesel engines, largely for submarines. Given this comparatively small role to the whole of the corporation, the commitment of resources to make the Allison program successful is reflective of vision and credit to the shareholders and management of General Motors.

Production of the V-3420

The V-3420 suffered an on-again/off-again existence, depending upon the direction of the war and the needs or demands of various aircraft strategic policy and other engine development programs. For example, early in the war, it's development was entirely stopped to allow Allison to focus on production of V-1710's for fighters. Then early in 1942, the Wright R-3350, powering the Boeing B-29 and Consolidated B-32 Very Heavy Bombers, was in serious trouble and the V-3420 development program again became a top priority. Allison was asked to expedite development and production of engines for the B-39 version of the Boeing B-29. An order for 500 engines was received, only to soon be canceled when that need again waned. The V-3420 then languished until the critical need for long-range fighters developed in 1943, resulting in the development of the Fisher P-75. Problems with that aircraft led to constant revisions in the production schedules for its engine, though it did garner the bulk of V-3420 production. Originally the V-3420 for the P-75 was scheduled to enter production at a rate of 5 per month beginning in February 1944, and reach 350 per month by November 1944.[63] Only about 157 V-3420's of all models, were built.

Army-Navy "E" Production Award

In 1906, the U.S. Navy inaugurated the system of awarding an "E" for excellence to those units of the Navy which displayed exceptional loyalty, devotion, and service beyond the call of duty.

Following Pearl Harbor, the need for a tremendous increase in war production brought home the realization that the men and women of American industry were partners with the fighting forces in the great struggle for human freedom. Thus was born the Army-Navy Production Award, a joint recognition by the Army and Navy of exceptional performance and patriotism on the production front.

Final selection and awarding of the Army-Navy "E" Award was made by the Army and Navy Boards for Production Awards. The award was evidenced in two forms:

1. The privilege of flying an Army-Navy "E" Award burgee above the selected plant for a period of six months.
2. The presentation of a special silver "E" insignia pin, for permanent personal possession, to each man and woman as evidence of outstanding service to the armed forces and to the country.

Production performance was reviewed by the Board every six months to determine whether continuation of the high honor was justified. If so, a service star was added to the burgee for every half-year of continuing outstanding service. The award became a source of pride and an incentive to carry on the patriotic service – to "keep the flag flying" until Victory was won. Allison men and women received the award for the first time on November 5, 1942.[64]

Allison won the "E" award four times:

1st Army-Navy "E" won 11-5-42, 1,379 engines per month.
2nd Army-Navy "E" won 2-16-44, 2,009 engines per month.
3rd Army-Navy "E" won 9-16-44, 1,982 engines per month.
4th Army-Navy "E" won 4-21-45, 1,310 engines per month.[65]

The Army and Navy jointly awarded this burgee, to be displayed for six months, signifying "Exceptional" performance and patriotism the production front. Allison won the award four times. (Authors collection)

The End

Soon following "V-E Day" the War Production Division notified General Arnold that, as a result of revisions for P-63 and P-38 airplane delivery schedules, deliveries of V-1710-111, -113, -117, -109 and -109A engines were in excess of requirements. Therefore, on May 28, 1945 they had taken action to discontinue manufacture of the listed engines. As these types constituted the current production engines (excepting experimental or developmental models) this action effectively terminated reciprocating engine production at Allison. Included in the order was notice of forthcoming direction to dispose of tools, jigs, dies and fixtures for manufacturing the V-1710.[66] With this notice production of the V-1710 was ended, even though the war in the Pacific was not over.

Jet engine production at Allison went into high gear, filling much of the void caused by termination of the V-1710. Then "V-J Day", August 14, 1945, signaled the start of a two-day holiday and celebration for the workers at Allison. It was marked by a triumphant and relieved note of Victory. For many who had come to Allison as a patriotic duty, or because their peacetime jobs had been displaced, it signaled the opportunity to return to the work or career of their choice – their temporary duty having been most satisfactorily accomplished. For many other thousands, who had trained and committed themselves for a lifetime in heavy industry, future stern challenges lay ahead, the size and order of which could not be visualized when work resumed following the victory celebration.[67]

And Then a Few More

After the war the Army Air Forces rapidly downsized, but there was still a critical need for long-range fighters. The new jets simply did not have the range to escort B-29's. For this reason the North American P-82E/F *Twin Mustang* went into limited production, and Allison received an order for 750 V-1710-G6 engines to power them.

These engines did not enter series production until 1947, long after the production lines for the V-1710 had been closed down and dismantled. In fact the G-6 engines were not even built in Plant 3, the facility original built to produce the V-1710. Complicating production was direction from the Air Force to utilize previously surplused parts, and even components from assembled earlier model engines in the G-6's. Actual production took about 18 months to complete, with the final 150-hour Model Test completed on November 6, 1947, at a combat rating of 2300 bhp.[68] The last engine, serial number A-074125, was assembled in December 1947, then weighed out and delivered from Plant 5 in January 1948.[69]

In the period from 1945 through 1948, Allison Engineering and Marketing continued to conceive and propose follow-on V-1710 and V-3420 engines for both military and commercial projects and aircraft. The military versions would have been quite exotic, with features such as direct cylinder fuel injection, intercooling, and turbo-compounding. While there were numerous proposals, and Allison model numbers designated, none made it into the test cell. The revolution of the turbojet had leap-frogged the evolution of the reciprocating engine, and even very sophisticated models of the V-1710 were not able to stay in the race.

Allison Service School

The Allison Service School was a part of the Allison manufacturing program. In fact it was even identified within Allison as Plant 6. The following discussion is excerpted from the Official *History of Army Air Forces Training Detachment, Allison Division GMC,* by John W. Varley, Captain, USAAF.[70]

From the forward, penned by a Captain Goulet:

"Allisons have from the start always shown the keenest interest in urging themselves on toward doing their utmost to further the war effort. In addition to the furtherance of the war effort Allison was vitally interested in seeing that their product operated at its maximum efficiency by training the Army personnel in proper maintenance methods. Thousands of men have graduated and gone forth to do their job toward winning W.W.II. I feel that I can state without fear of contradiction that the Tenth Army Air Forces Technical Training Detachment, Allison Division, GMC, has contributed to placing the Allied Air Forces in the strategically position they are now in (October 1943)."

The following is a summary of the beginning of the Allison Service School and gives a picture of developments prior to the establishment of a school for Army Air Force personnel.

Training a Workforce

Following receipt of the 1939 production order for engines to power the P-40 it became necessary to expand the number of people knowledgeable of the engine. To familiarize mechanics in the field with this new liquid-cooled engine, men were drawn from the Allison Experimental Department to comprise the first of the "Field Service" crew. At that time, factory employees started the training with only six students-operating in one room at Plant Number Three [which was under construction at the time]. Equipment was salvaged from factory rejections to build one complete engine for use in class demonstration and in July 1940, the first class was started.

As the European war grew, Allison equipped planes were sent into action around the world and the Allied Nations asked for instruction on Allison Engines. Programs were designed to effectively meet these demands. More room for the school soon became necessary, with classes now numbering in the fifties. Three rooms were set aside in Plant Number Two, and the original room at Plant Number Three was used for test stand run-up instruction.

After December 7, 1941, the Air Forces grew by leaps and bounds. How to train the additional thousands was a growing problem. Mobile training units were started as a way to carry information directly to the field.

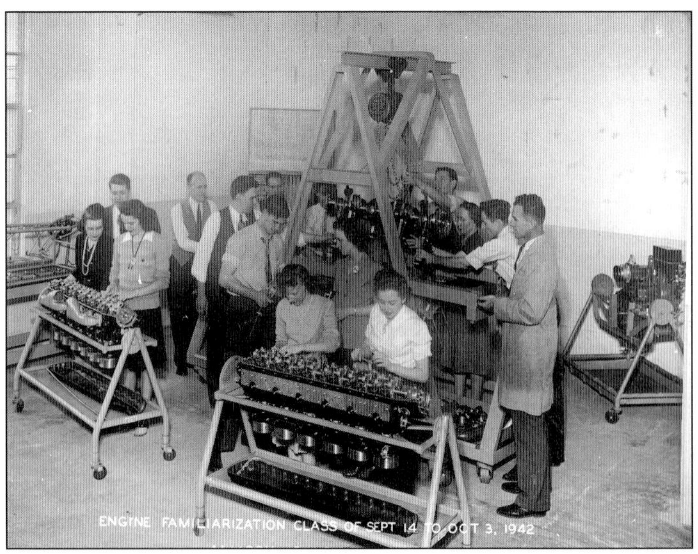

The Allison Service School taught both Servicemen and Plant workers the intricacies of the V-1710 and how to manufacture, operate, and maintain it. (Allison)

By June 1941, so many Allison Engines were powering planes for the fighter commands that up-to-date on-location instruction of mechanics was the only way to be sure of maximum performance. A group of four instructors from the Service School became the first "Mobile Unit" men and began Field Training on the East and West coasts. It was soon apparent that this form of training was most effective and that personnel requiring instruction could be reached and trained in a short period of time. Units sent into the various war areas were named "Fire Squads."

Mobil Unit equipment varied, depending on the specific training job. Cutaway engines, assemblies, parts, and large chests called Instructional Aid Kits, containing slide films, charts, handbooks, parts books, and timing devices were all used. As each Mobile Unit was organized, a new kit was assembled and packed, ready for shipment to any area for the use of traveling instructors on a moment's notice.

Semi-trailer Army trucks or cargo airplanes carried these Mobile Units to fields all over the United States. There were also two in foreign areas. Each unit was supervised by an Army Officer. The Allison instructor from the airplane plant and an air instructor with six or seven enlisted personnel traveled with each unit. Upon arrival at the designated field, they reported to the Commanding Officer. The technical staff of Army instructors aided in their own particular field, each a specialist in one subject, giving the trainee every advantage of up-to-the-minute information.

Crew chiefs and commanding officers found these Mobile Units indispensable in training men that otherwise would not be able to take advantage of training. Half-day courses were arranged on those fields for those "on the line", taking half of the group in the morning and the other half in the afternoon, without stopping routine work.

In the two short years following the original group of four men starting out, Mobile Units had trained thousands of men at the fields and depots comprising the facilities of the U.S. Army Air Forces devoted to pursuit and interceptor airplane training and operation. While these Mobile Units served to carry instruction on the Allison engine to personnel wherever they were located, the army needed and demanded more. Thus, officials of the Army and Allison Division, GMC, planned and established a school for AAF personnel.

The Allison Service School came about from a meeting between Army and Allison officials held March 26, 1942. The Army desired to send men to the Allison engine factory for the purpose of studying the V-1710 engine, and their consequent development into Air Force instructors. As a result of this meeting, twelve men were to be ordered to report weekly for a 28 day course of instruction [although only a 22 day course was offered prior to November 16, 1943].

On April 13, 1942, on 24-hour notice, one hundred enlisted men arrived from Keesler Field, Mississippi. While it is true that so many were not expected, and that there were no accommodations available for housing and feeding, immediate action by officials of Allison soon found the men located in the Antlers Hotel on Meridian Street in downtown Indianapolis, and ready to begin their work toward becoming instructors on the Allison engine.

It was imperative to have some kind of school set up which would be large enough to accommodate this number of students. An emergency arrangement was made at the Antlers Hotel for training these students. Working all day Sunday and far into the night, the various engine models were installed in the ballroom to comprise a student laboratory and lecture room. The dining room became the mess hall, and one floor of hotel rooms were reserved for sleeping accommodations. The school at Plant Number Two continued with civilians and special classes, and the new addition provided a total school floor space of approximately 16,500 square feet.

It was obvious that a hotel and makeshift arrangements for teaching facilities were not practical and did not tend toward efficient and effective training. On orders from Washington, an officer was dispatched to investigate other quarters. After protracted negotiations with Indiana Central College, the men's dormitory was to be used for quarters and the actual school location was to be in the gymnasium and several other rooms. This arrangement was projected as a temporary program lasting approximately ninety days, the space not being available until May 24, 1942, at which time the school was moved to the College.

The new location provided a total floor space of 24,875 square feet. More laboratory space was available and programs became more adequate. A fuselage area with one P-39 airplane was introduced at the Plant Number Two school where each student was instructed in the installation and ground tests of an engine in a running plane. Safety hazards barred a fuselage area at the college, and trainees had to be transported for this part of the training.

Knowing that the use of the College could only be temporary, a third site for the school was sought. The Army produced plans for the construction of barracks to provide housing and messing facilities adjacent to various factories where the enlisted men were to be trained.

The school was officially activated on May 15, 1942, under the command of Captain Leonard Smith. The activation orders designated the unit as the *Army Air Forces Training Detachment, Allison Division, GMC*. On October 15, 1942 the detachment was relocated from Indiana Central College in Indianapolis, five miles southwest to the Maywood and Mars Hill suburbs, where Allison officials had negotiated a more efficient and suitable school location. This location was an old Hercules Paper Box Company factory site, covering 27 acres and 22 buildings, the main structure being a brick building 500 feet long, 50 feet wide, and 3 stories high. In June 1942, all the service school classes then located in Plant Number Two were moved into this area and temporarily occupied adjacent frame buildings while the main building was converted into a school. With practically no construction except for partitions, an excellent set of facilities was developed. With 56,200 square feet for training, each phase was covered on every part of engine maintenance in a minimum amount of time.

Construction of barracks and other necessary facilities was begun on September 1, 1942 by the Indianapolis Area Engineers. A total of 30 buildings were to be constructed, the breakdown being somewhat interesting:

2	Mess Halls, 208 man capacity each
2	Latrines
1	Day Room
1	Administration Building
1	Supply Building
23	Barracks, 34 man capacity each

Only one Mess Hall, six barracks, one latrine, the Day Room and the Administration Building were actually built.

At the time of the move into these facilities, Acting First Sargent Henry E. Ostapa presented the following picture of the school upon arrival:

> "The AAFTD school is located about five miles southwest of Indianapolis, on the east side of Holt Road, and about one quarter mile from Stout Field Army Air Base. The Allison Service School, Plant Number Six, and its adjoining grounds are to the south and farm land borders the area on the east and north. The area consists of about twenty acres of land and ten buildings."

From the beginning of the Service School, cutaways of the engine and all the subassemblies played a prominent role in instruction. Because of various limitations of other devices, the Visual Aids Department developed a Lucite model of the accessories gear train. This was the first time Lucite had been used for instructional purposes.

A timing model was produced for laboratory demonstration on how to time the engine. The complete timing procedure could be carried out on the model, thus visually training the student before he individually worked on an engine.

By May 1942, the first Service School Handbooks were published. The handbook was designed to be used as a general training guide and contained 146 pages of information including 183 illustrations concerning proper procedure for overhauling, servicing, and maintaining the V-1710 engine. In addition, there were 42 full color charts incorporated throughout the text on installations, coolant, fuel, electrical and lubrication systems, timing charts, cutaway drawings, gear trains, and wiring diagrams. The first publication of 10,000 copies of the Handbook, estimated to be a year's supply, lasted three and a half months. By July 1943, 26,000 copies had been made, not including thousands of copies made by the Air Force and issued as a Technical Order [TO No. 30-5A-1].

With the class entering the school on November 16, 1943, the course was changed from 22 days to 28, which enabled the School to add further periods of practical training on live engines installed in the different types of Army planes powered by Allison engines. The new course devoted about 71 percent of the hours to laboratory practice, with hands-on engine work, about 11 percent more than previous. Important to being able to increase the hands-on time was the acquisition of a number of Class 26 airplanes to form a flight line in a field adjoining the school buildings. All of the planes were equipped with V-1710's in operating condition and were used for practical training in third echelon maintenance work and troubleshooting.

Procurement of these planes had been difficult and, upon arrival, they were usually in such a damaged condition that extensive repair work had to be undertaken to render them suitable for school use. The successful development of the use of these Class 26 airplanes represented a very real achievement on the part of the Allison Service School. The instruction given in the 28 day program was considerably more practical and effective than that given earlier.

One new feature of the 28 day program was the addition of a study of the Auxiliary Stage Supercharger. Engines equipped with the unit [the V-1710-93(E11)] had gone into production and the longer program made it possible to include it in the course of study, thus keeping students informed of an important recent development.

The methods of instruction used in the school were based upon the five natural steps of teaching as defined by the Army: definition of purpose, detailed outline, demonstration, discussion, and drill. These steps were applied according to a very carefully planned program that had been planned and designed in its minutest details and thoroughly tested before any instruction was given. The planning had been done at a conference attended by the representatives of the AAF, Allison Division, and specialists in training. After this conference had determined the objectives and subject matter of the course, experts in training designed the course, preparing a detailed outline of each subject. Thus designed, the course was tested on a group, and revised to correct errors and deficiencies. Then it was put into operation.

This procedure was intended to eliminate any use of trial and error methods in teaching. Such methods were never employed in the Allison Service School, and changes in the program were made only when improved equipment or alterations in the subject matter of the course rendered them desirable.[71]

One serious problem faced by the school in the Fall of 1943 was in obtaining qualified instructors. Many of the most experienced instructors were being constantly lost to the draft. It was impossible to replace them with men of equal qualifications. However, new instructors were being trained all of the time. Another problem was the arrival of many students with insufficient fundamental training to take full advantage of the instruction being offered. Allison supported a faculty of 66 instructors.[72]

The morale of both the students and permanent party enlisted men remained at a high level throughout. The detachment services, supply, dispensary, post exchange and the like were efficiently operated and contributed to the maintenance of a favorable environment for the successful achievement of the training mission of the detachment. Inspections resulted in generally favorable evaluations of the program with the usual findings of minor irregularities and deficiencies, primarily of an administrative nature.

Representing Allison during the early days of developing the school was Mr. A.C. Hazen, Director of Training, Allison Service School.[73] He was interviewed by Captain Varley on December 27, 1943, and the following is taken from that interview.

> Q: Why did Allison starting training AAF personnel?
>
> A: All manufactures of war products were told by General Arnold that they would have to train Army personnel because he felt that the factory where the product was being manufactured was best equipped to train the men who were to use those products.
>
> Q: What was the basis for the course of instruction given to the first class?
>
> A: The instruction was based on the experience of men who had taken the training previously and also on the limitations of the facilities available.
>
> Q: How was the 22 day course curriculum developed?
>
> A: Three representatives from each of four groups got together at the Flint, Michigan facilities of the General Motors Institute. They spent 5-1/2 weeks, beginning in May 1942, designing the course of instruction that was to be given by the Allison Division in the 22 day course. The four groups represented were:
>
> U.S. Army Air Corps
> Engineering Department, Allison Division
> Service School Department, Allison Division
> General Motors Institute, Training Specialists

During the first year the school was in operation, the usual student load was about 100. By the end of 1943 there were usually 200 students enrolled. During the Christmas holiday in 1943, there were 250. When the 28 day class schedule was fully implemented, the enrollment was maintained at 200. This meant that a new class of 40 students was entering every week. Previously, under the 22 day schedule, each class started with 50 new students.

Students were usually enlisted men, enrolled in the course which was officially titled, *Airplane Engine Mechanic Course (MOS 762), Special-Allison V-1710 Engine*. About one percent of the Army students were officers. In addition, civilians, including both men and women were trained along with the military men.[74]

After the War, the massive training program by Allison was no longer needed and in October 1945 Allison gave up the facilities at what had been its "Plant 6." In recognition for the contribution Allison made in training its far-flung forces, Allison received a plaque in May 1946 from AAF for having trained 101,000 men on the V-1710 during war.[75] This is almost 1.5 trainees per engine built. Almost all of these trainees were men.

Table 17-15 Class 26 Airplanes Used by Allison Service School		
Model	Air Corps No.	Date Received
Class 26 Airplanes in use as of July 7, 1943		
P-39	41-6821	4-16-42
P-39D-1	41-38314	10-7-42
P-39D-1	41-28374	11-23-42
P-38F-5LO	42-12617	9-30-42
Additional Airplanes obtained July 7 to Nov 16, 1943		
P-39N-1-BE	42-9345	4-12-43
P-39N-1-BE	42-9299	6-18-43
P-38	40-756	10-22-42
P-40	39-158	11-12-43
Additional Airplanes obtained Nov 16 to Dec 31, 1943		
P-40C	41-13382	11-4-43
P-39D-1	41-28384	10-9-42
Additional Airplanes obtained 12-31-43 to Mar 1, 1944		
F-4A-1-LO	41-2379	1-13-44
P-40E	40-449	1-26-44

"Tech Reps" - Allison in the Field

The manufacturers responsibility for his products does not end at the shipping dock. For many reasons, including self interest, it is important that a manufacture institute a support system for products in the field, particularly if those products are complex and/or have unique service requirements. In the case of the V-1710, the task was even more daunting, for the engine was being used by some 15 foreign air forces and being operated on every inhabited continent in the world. At the peak of the war Allison had 400 field service representatives scattered around the world to check on the operation and performance of their engines.[76]

Given Allison's neophyte position in mass production and the wide ranging need for product support, their support and commitment to the forces using their engines is commendable. It is likely that their many years of close workings with the Materiel Division at Wright Field had instilled the benefits and necessity to support the military customer with its needs for operating and testing the special equipment Allison was providing. As a result, Allison had its Technical Representatives in-place at the various aircraft manufactures to aid in the technical support and maintenance of the engines which were being installed and tested in new aircraft. From the first, this was occurring. For example, the Allison representative T.S. McCrae was with the A-11A during its accelerated flight testing of the YV-1710-7(C-4). In fact, he even attempted to wrangle a flight in the two-seat aircraft, claiming that he needed the flight to get first hand observations of the engines performance.

Allison Tech Reps developed a reputation for being "get-it-done" kind of guys. They were committed specialists who brought with them extensive knowledge and a willingness to work hard right along-side the troops. In Chapter 4 we discussed the key role of Arne Butteberg, Tye Lett, and Teh Chang Koo, who were Allison's representatives with the *Flying Tigers* in China. They were there to meet Chennault's troops when they arrived in Burma, and they were there long after the AVG was inducted into the U.S. Army Air Force. These men not only supported their Allison's, but provided their technical expertise on much of the rest of the Curtiss *Tomahawk* and its equipment, for there was no one else around with even a smattering of relevant expertise. The record of devotion to the troops and their mission continued throughout the war and was a standard every Allison Tech Rep worked to maintain.

The key to the Tech Reps success was knowledge and expertise. Many had long and varied backgrounds in aviation, while others were barely out of school. What they all shared was the training and experience gained in the Allison factory, the test beds, and on the flight lines. Their stories are poignant, funny, and tragic. Sometimes all of the above.

As an example of innovation, required even when "at home" in the "ConUS", John Kline, Allison Installation Engineer on the XP-39, relates how:

> "Back in those days, anytime you were more than 50 miles away from the home office you were automatically an expert, no one to challenge you!" This both created and demanded ingenuity. He relates an occasion when he and Maury Rose were up at Bell when the XP-39 was being put together and they had to re-time an engine. Well typical, I didn't have any of the equipment, like a fasten-on timing dial, and of course we had to have it done "yesterday." So "Maury", I said, "you can plan on working late with me." So he said, "How are we going to do this?" I said, "There's only one way that we can. We don't have a dial indicator so we'll layout everything on the spinner of the prop!" Maury swore, "It can't be done!" Well we did it and it worked out just fine! You just had to remember to figure your pencil marks to account for the 9:5 reduction gear ratio."[77]

Allison Tech Reps earned a reputation for solving problems, sometimes brought about by nothing more than their own curiosity. Fortunately, they felt empowered to deal with all aspects of the engine, including how it was being used and maintained. Dick Askren started early with Allison and tells a story about a problem that was common in the P-39.

He relates as to:

> "How this was one of those real unimportant things. When we had modification programs to fix something, wing flaps, engine, etc., little problems would often get ignored. For a long time we had been finding that the lower left stud on the [reduction] gear box, where it was mounted to the airframe structure, [was being] broken off. And a lot of people thought this was a real disaster, broken parts and all. One time, over at Plant 6, I got my fill of broken lower left studs, so I got Scotty Mervin, who was an instructor in the [Allison Service] School, well we had some captive P-39's there for guys to runup. I said to Scotty, run it up, and I got up inside there [in the cannon compartment] and that quick, I could tell what was happening to those studs, but nobody had ever done this [before]. I was standing in the wheel well looking at the back side of the gearbox. There was one part of the structure, which when you would accelerate or decelerate, would twist that angle plate and that would break off the stud!"[78]

As a follow-up to Dick's observation, Allison's Experimental Department did a systematic load test on a P-39. The test was done by John Hubbard in November 1942. Analytically, the ring of studs attaching the reduction gear to the airframe were considerably stronger than needed to simply carry the torque due to the design load of 1070 bhp. The expectation was that torsional deflections of the forward fuselage might be causing the studs to break. The test apparatus was interesting, for it used an airplane with the entire drive train, except for having only a crankshaft mounted in the en-

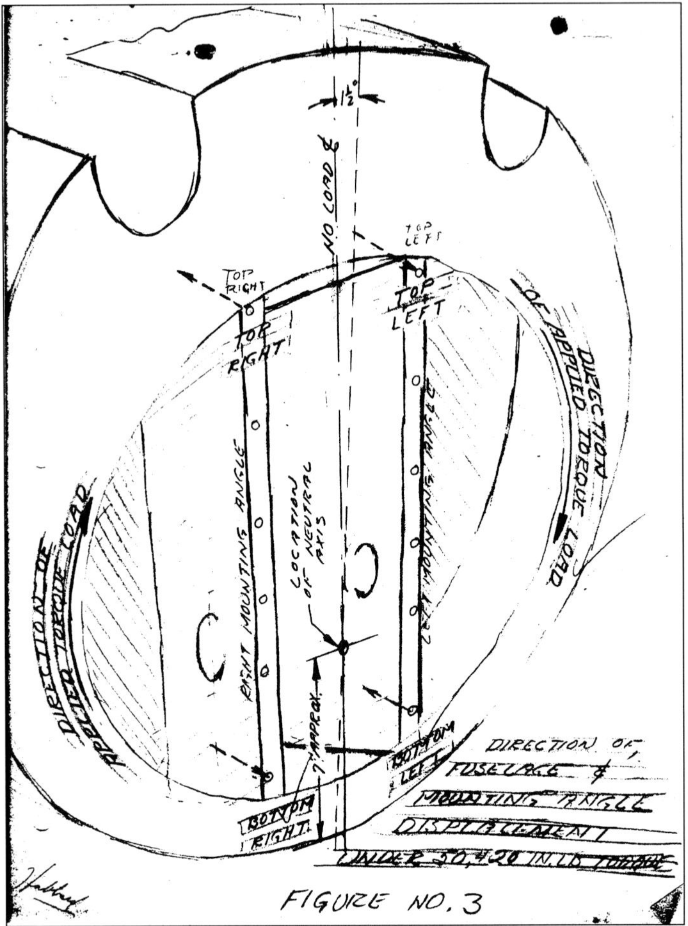

This 1942 sketch by Allison's John Hubbard shows the stress patterns and deflection in the nose of the P-39 that was causing the failure of mounting bolts for the remote reduction gear. (Hubbard)

gine, and it was rigidly attached to a plate to keep it from turning when the torque was applied at the propeller. What was found was that not only the lower left, but also the upper right stud were both being loaded to well above their safe load carrying capacity when acted upon by fluctuating loads. This also meant that the rear of the reduction gearbox was carrying loads that correctly should have been accommodated by the airframe. With the full torque equivalent to 1070 bhp, it was found that the bulkhead at the front of the fuselage twisted 1-1/2 degrees from its neutral position. The recommended fix was to strengthen the structure supporting the gearbox.[79] The Tech Reps took the information to heart, and one report from a front line overseas squadron reported that they had resolved the problem by tightening all studs to the same torque, then recheck them for tightness every 25 hours. They had 24 airplanes with over 150 hours apiece, and had only found three broken studs after applying the procedure.[80]

Spark plugs have always been a problem, though not so bad with modern high energy ignition systems. These were not available during W.W.II. So it is no wonder that the Tech Reps have numerous tales to tell of dealing with these critically important little plugs. Don Wright tells of a state-side base that was having a particularly bad time with engines that were rough running, cutting out and that sort of thing.

> "They said something is wrong, this engine won't work. So I happened to be there in the hangar where they were building up these engines to put into the P-38's, and I noticed these women (working on the plugs). They had this round thing filled with this gunk, and you were supposed to put a little bit of it on the first thread on the spark plug, but they were dipping it [the plug] down in there and turning it around! You know what the results were, so at the risk of being thrown out completely, I said, "Do you know that you guys can't do that!" And they told me that they had always done it that way! I went to the Engineering Officer and told him, "You know you've got a problem with fouled plugs?" "Oh yes, all kinds of fouled plugs!" I can tell you what's wrong, (but) it's up to you to correct, I'm out of it, but you need to watch how they put that anti-seize on the spark plugs![81]

John Kline tells of a similar problem that occurred on the early P-40 production line at Curtiss:

> "I had instructed those guys, you take the shipping plugs [out], clean out the threads of all that gunk which has been sprayed into the cylinder, [and] don't screw the new plug in there until its been cleaned. Well when we got them out at the airport, and on the first runup they fouled the plugs and had to change the whole set. With that many plugs, and they were worth about $5.00 apiece at that time, so I had a pretty good idea about how to get around this. Pete Jansen was the factory manager there at Curtiss, he was a tough old Dane or Swede or whatever, I got hold of him and I said you're complaining about these sparkplugs and you guys are the ones who are the problem! I said I'll show you. I took him out there. Well he put a stop to that. Before I had to do that I told Charlie Place, the Air Corps inspector at the airport, he [was the one who] accepted airplanes or not, I told him what was going on. He said, "I'll take care of that!" He shutdown the flight line, no more airplanes are leaving here until that's fixed. He was a big help to me. Of course it took a few steaks and a lot of whisky. Anyway that finally fixed the problem."[82]

Life as an Allison Tech Rep may have involved a lot of interesting work in interesting places, but often under trying conditions and with a great deal of uncertainty. Don Wright went around the World on his various assignments from Allison during the war. How he got to the CBI is the story that follows:

> "After I had come back from Alaska [supporting Lend-Lease P-39's and P-63's going to Russia] I went to this outfit at Goldsboro N.C. The CO [Commanding Officer], was a guy with a slouch hat, and another guy [who] was named Alison, [was] the Engineering Officer. I trained the pilots for engine operation of the P-51A, I talked to them for about two weeks. So then I went to leave and I told the CO, "I guess I'm through here, they all shook hands and said, "well, we'll see you later." I didn't know anything about that. Well I went back to Indianapolis and about 2-3 weeks later I got orders from Mines Field saying, "Report to Miami, FL."
>
> Take your time, but go from Miami to Puerto Rico then to South America, Ascension Island, The Gold Cost of Africa, across there to [Egypt], and finally I end up in Karachi, India. This occurred about the first of the year, 1944. These are kind of strange orders.
>
> [I was following that outfit, the 1st Composite Group, from Goldsboro, and] anytime they had a plane down for an engine, I would just stay there, fix that one, then move on to the next place. I finally got in there and found out what these guys did, they had P-51A's deck loaded on the [USS] *Saratoga* that they brought to Karachi, the processing station, and they had gliders, and they had C-47's as support

aircraft, but most were P-51's and A-36 types. I found out a little bit later when these guys all arrived they were on their way to Burma to work with Stillwell.

I was their Allison Representative, though I hadn't known that before. They also had a North American representative and several other guys as well. It turned out that this guy was Phil Cochran [Col. Philip Cochran, CO of the 1st Air Commando Group, who was in the comics as "Flip Corkin"] and they had Johnny Alison the Exec Officer and the guy who was the Engineering Officer, Capt Jennett. Anyhow, we had to process all those engines and airplanes through and then they decided they are going to takeoff and the next thing you know they are all gone and they decided to leave me behind. Then I found out they said, "well they didn't have any engine problems so they would send for me if they needed anything." So I thought, "Boy, this is the greatest thing."

So then I started working with the CBI, General Oliver sent down some orders and I went over to a landing field in "Bolier", where they were processing Chinese pilots. And I don't think you have anything to fear about somebody being modern in China and in those countries. We'd have 60 of them come in, they'd wash-out maybe 20, 15 would kill themselves in flying accidents. We had these old P-40's, and the last day, when they graduated, was a great day. We'd get in these revetments, they would sound an "Alert", and everybody had to scramble. They'd come out of there all going every direction you could think of trying to get into the air and then they'd run into each other. I saw one of them, he was sailing along, there was an Indian driving a big truck, and on the back he's got a roller and water tank to spray and keep the dust down. This guy ran the P-40 right up on that big roller and it had this sharks head painted on it, the Indian turned around and looked out the window and saw this sharks' head, opened his door and took off himself! The truck was still moving. The collision knocked the prop and gear box off the engine, and the engine was still running. I'm standing over in the hangar, and I said to the guy with me, "hey look at that would ya, you just can't stop those Allison engines!"

They would have an awful time. Then they would give them the best P-40's you could get. They came over deck loaded in crates, and would be erected in Karachi, all of the American pilots coming back, ferry pilots and so-forth, would just drool over these brand new P-40's, M and N series, late models and they couldn't have one, they had to fight with something else, then we would give them to these Chinese and away they'd go boom-boom-bang all over the country, it was quite a sight I tell you!

I'll never forget, one time they told me "why don't you teach these guys something?", so I said I'd teach them how to check the Cuno strainer. So someone said that they would have to first tell them how to take off the radiators, they hang down below there, so I showed them how to do that and I put it all back together. So I went out there the following day. By the time I got there, here were all these guys sitting there, they had torn every radiator and Cuno filter out, sitting there with big smiles on their faces, and there wasn't an airplane you could fly!"[83]

It is evident that Allison's Tech Reps took the view that their engines were a part of an aircraft's propulsion system. Just because a problem was elsewhere in the aircraft was not cause for them to ignore it. As a case in point, Don Wright tells about the technique needed to properly rig the turbosuperchargers on the twin engined Lockheed P-38 so that the pilot would get the same power from both engines:

"One of the things that I always marveled at, and you had to do it in order to understand it, was how you set the turbos on the P-38 to cut in at the same time, because the linkage setup in the P-38 was such that it was a lot further to one engine than it was to the other. So you had to know a little bit about how much play or slack there was going to be in the pulleys and so forth that ran back to the turbos, and then you had to set them accordingly. I decided that this was a job that I'd learn how to do there at the Experimental Flight [at Lockheed's Burbank factory], and I finally did. I got to the place were I could compensate for this and that and every other thing and it turned out pretty good. I went up to Santa Rosa, CA. They had P-38's in flights, Red, White and Blue, and most every one of them had the problem that the turbos didn't cut in at the same time. One was always leading and giving more horsepower than the other, so I made a mistake. I said I'd go out and fix one of these for you, so I went down to this hangar and rigged one of them up. The next thing I know, they've [the Army] given me a jeep and I'm suppose to travel around to all of these places and set these turbosuperchargers. I said, you know, that really is not my job here."[84] Dick Askren, another Allison engineer and Tech Rep, commented, and "Suddenly, you're Mr. Turbo!"

The Allison Tech Reps played a critical role in the success of the V-1710 and in its diverse employments around the world. To them should go much of the credit for giving Allison a persona and demystifying the bureaucracy of a large and growing enterprise.

Allison Flight Test

It was usual for each of the major engine manufacturers to have their own flight test operation for the purpose of refining and testing of engines under actual conditions of flight. Allison was comparatively late in developing this capability. Rolls-Royce, Pratt & Whitney, and Wright already had extensive company operated flight test capability when Allison established its operation in the fall of 1941 at Weir Cook Airport, Indianapolis.

The operation was set up under the direction of Harry Carsher, then head of the Installation Engineering Department. He had been a design engineer on the Ford Trimotor and had come to Allison with a considerable aviation background.

Allison Flight Test Department took time out to stand for a photo. Chief Pilot "Pinky" Grimes is in the center, in leather jacket. The well worn Duze fasteners on the P-51A cowling attest to use of the airplane in many engine test programs. (Allison)

Initially, Allison had no facilities at Weir Cook, but they were able to temporarily borrow the CAA hangar. From there they moved to the Roscoe Turner hanger before finally getting facilities on the south side of the airport for their own, identified as Plant 10.

The initial staff consisted of Jim Knott, a mechanic named Ray Freeze, Pinky Grimes the test pilot, and Horace Roberts. Their first assignment was to work on the crankcase breathers, which at that time were creating quite a lot of problems as changes in pressure, altitude, or speed often caused the engines to spew oil directly out of the crankcase breathers.[85] Their initial compliment of aircraft consisted of one P-40 on a bailment contract.

Major Jimmy Doolittle, then assigned to Allison as the government representative, allowed them to use his P-40E for tests. General Arnold had assigned it to him to commute to the headquarters for the Central Procurement District in Detroit.

Later, Allison acquired a great many different airplanes and offered a full scale engine test program to investigate major changes coming from engineering and production, as well as working with the airplane manufacturers to conduct engine related tests that would otherwise impact their own test schedules. Allison had a good reputation with its customers for helping create good power plant installations.[86]

NOTES

[1] Rickenbacker, Eddy, 1956.

[2] On January 5, 1938 an order was received for 51 engines, the largest single order received by Allison up to that time. While issued on January 5, 1938, it consolidated previous authority to purchase engines for the 13 YP-37 and thirteen twin-engined YFM-1's, along with 7 and 13 spares respectively, for a total of 59 engines. The balance of the batch were attributed to 1937 orders. *Allison History*, 1962, 15.

[3] Hazen, R.M. 1945 Report, 2.

[4] Goldthwaite, John L. 1950, 5.

[5] Hazen, R.M. 1945 Report.

[6] Allison Division, GMC. 1950, 21.

[7] General Motors Corporation. 1957, 24.

[8] Kelsey, Benjamin S. B.G. USAF (Ret.) 1982, 31.

[9] Engineering Section Memo Report, E-57-680-14, 1937.

[10] Allison Division, GMC. 1950, 18.

[11] Teletype from General Brett to General Arnold, 4-17-1939. NARA RG 342, RD3480, Box 6896.

[12] On Dec 8, 1939 the French ordered 115, and an additional 700 C-15's in Feb 1940, later converted to the British. On May 25, 1940 the British ordered 3570, in 3 models. *Allison History*, 1962, 15. Allison Historical Dates by Don O'Brien says 3,500 ordered. Check of contract numbers shows only 3500, but in 5 models. Author.

[13] The number of engines on the original contract were 837, which was subsequently increased by 132 to provide engines for 66 P-38's. There was no increase in cost to the Government as the considerable gains in efficiency Allison was enjoying due to the concurrent orders for the British reduced unit cost.

[14] The 837 engines ordered, by model were:

V-1710-19(C13)	776
V-1710-27(F2R)	23
V-1710-29(F2L)	22
V-1710-17(E2)	16

Before delivery the -19 was changed to -33(C15), and after 300 were delivered the balance was shifted to -39(F3R). In November 1939 81 V-1710-35's, a fifth model, were added under a separate contract. In September 1940 66 each of the -27/29 were added to the contract at no cost, a consequence of production efficiencies coming from the large British orders.

[15] Allison Division, GMC, 1941, 2. The contract was W535-ac-13420, November 1939.

[16] December 14, 1940, U.S. orders 3,691 V-1710's, in 5 different models. Allison Division, GMC, 1941, 3.

[17] *Allison Engine Situation*, 1941.

[18] *Allison Engine Situation*, 1941.

[19] Wright, Donald F. 1994.

[20] Allison Division, GMC, 1942 news release, 9-14.

[21] One of the early advantages of the jet engine was its inherent simplicity, only "one" moving part. As developed by the mid-1990s, a front-line military jet engine has more than 30,000 parts, 1,300 of them rotating. Mechanical Engineering, December 1995, p.60.

[22] Allison Engine, The, 1942.

[23] *Aeroplane, The,* February 27, 1942, 248-249.

[24] Allison Engine, The, 1942.

[25] Allison Engine, The, 1942.

[26] *Settle R.E., Report of*, Allison Material Procurement Manager, 1946.

[27] Zigmunt, Joan E. 1994.

[28] Fleming, Roger, 1957, 8.

[29] Allison Division, GMC, 1941, 2.

[30] Allison Division, GMC, 1941, 4.

[31] Doolittle, General James H. "Jimmy", 1992, 200-201.

[32] Allison Division, GMC, 1941, 2-3.

[33] O'Brian, Don, 1950.

[34] Knott, James E. 1968, 3.

[35] General Motors Corporation. 1957, 27.

[36] O'Brian, Don, 1950.

[37] Hazen, R.M. 1945 Report, 3.

[38] *American Machinist*, March 5, 1942, 165-177.

[39] In 1943 a change was made to a special refrigerated box rather than the somewhat dangerous kerosene mixture.

[40] *American Machinist*, May 11, 1944, 101-102.

[41] *American Machinist*, March 5, 1942, 168.

[42] Wright, Donald F. 1994.

[43] *Tightening of Cylinder Hold Down Nuts—All V-1710 Series*, N.Z.A.P. 110/C-3, Sept 4, 1944. Wright Field file D52.41/183-Allison.

[44] Hazen, R.M. 1945 Report, 5.

[45] Allison Division, GMC, 1942 news release, 14.

[46] *Price Adjustment Board Report-1944*, 14.

[47] *Price Adjustment Board Report-1944*, 15.

[48] Hazen, R.M. 1945 Report, 6.

[49] *Price Adjustment Board Report-1944*, 13.

[50] *Shortage of Parts for Panama*, memo to Chief, Air Service Command from Wright Field, March 10, 1942. NARA RG18, File 452.8, Box 807.

[51] Allison Division, GMC 1950, 9.

[52] *Allison "Firsts" as of 1942*, from J.L. Goldthwaite files at Allison.

[53] Experimental Engineering Report EXP-M-57-503-663, 1942.

[54] When the V-3420-9 Type Test engine was rebuilt in the spring of 1942 it was equipped with two nitrided crankshafts. The successful completion of the Type Test then qualified them for general V-1710 and V-3420 use. Experimental Engineering Report EXP-M-57-503-663, 1942.

[55] Settle, R.E., Report of Allison Material Procurement Manager, 1946.

[56] Allison Division, GMC 1950, 35-39.

[57] AERO DIGEST, January 1943, 191-198.

[58] *Allison History*, 1962, 17.

[59] Allison Division, GMC 1950, Appendix.

[60] *Allison History*, 1962, 17 and 23.

[61] Hazen, R.M. 1945 Report, 7.

[62] *Price Adjustment Board Report-1944*, 9.

[63] *Price Adjustment Board Report-1944*, 11.

[64] Presentation Brochure, 1942.

[65] Provided by Joan Zigmunt, April 1998, augmented by *Historical Notes*, compiled by Don O'Brien, Allison Public Relations. Available at NASM, file B1001025. See Department of Commerce, CAA Report *US Military Aircraft Acceptances 1940-1945* for production figures.

[66] *Termination of V-1710 Type Aircraft Engines*, letter to Commanding General, AAF, May 28, 1945. NARA RG 342, RD3775, Box7411.

[67] *Allison History*, 1962, 17.

[68] O'Brian, Don, 1950.

[69] *AllisoNews*, Vol. XIII, No. 1, Commemorative issue, July 4, 1953, p.8.

[70] Official History of Army Air Forces Training Detachment, Allison Division GMC, through July 7, 1943, prepared for the Second District Technical Training Command, Army Air Forces, by John W. Varley, Captain, AAC Historical Officer and Lloyd J. Goulet, Captain, Air Corps Commanding Officer. Microfilm roll A2455 Index 1833, 234.021 7/39-12-41 through 234.027 7/43-2-44, frame 1104 through 1432, from HQ USAFHRC, Maxwell AFB AL 36112.

[71] Things change slowly in the Military. Your author was a student in the USAF course at Chanute AFB, IL, for Aircraft Maintenance Officers in 1965. These methods and regime were still the order of the day, unchanged from the time when the Allison Service School prototyped them.

[72] Ron Hazen, article on V-1710, 1945, p.4.

[73] Not related to Mr. R. Hazen, Allison Chief Engineer.

[74] Ron Hazen, article on V-1710, 1945, p.4.

[75] O'Brian, Don, 1950.

[76] Ron Hazen, article on V-1710, 1945, p.4.

[77] Kline, John 1994.

[78] Askren, R.W. "Dick" 1994.

[79] Hubbard, John D. 1942. Mr. Hubbard has contributed significantly to this history of the V-1710 by graciously providing considerable technical material, file H13.

[80] Tale Spins, September 1943, 10.

[81] Wright, Donald F. 1994.

[82] Kline, John 1994.

[83] Wright, Donald F. 1994.

[84] Wright, Donald F. 1994.

[85] Knott, Jim, 1989.

[86] Knott, Jim, 1989.

18

Epilogue: Post-War Racing and Soldiering On

In the quest for speed, there are only two avenues available: reduce the drag of the vehicle or put in more propulsive power. While every racer intends to improve both areas, there are physical constraints in reducing drag below a threshold defined by the vehicle. Alternatively, the benefits of additional power are practically unending, though it is a game of diminishing returns, for power consumed in drag varies as the cube of the speed. It takes a disproportionate amount of power to produce an increment of speed. Still this is an essential element in the quest for speed, and any serious racer must obtain the maximum amount of power he can afford.

How to "Soup-up" an Engine

The options available to an engine builder are no different than those considered, applied, and developed by Allison engineers in their quest to upgrade the V-1710. The only difference is the relative weighting of performance requirements.

Allison's goals were to provide maximum power and efficiency at the maximum altitude consistent with optimum engine life, and all at minimum installed weight. These are the same for the racer, but they are often willing to accept reduced engine life, sometimes by factors of 100-1000, in exchange for an increase in power by a factor of 1.5 to 2!

An engine makes power in proportion to how much air it can be pump. By running faster and pumping higher density air, more air and fuel goes through the engine and greater power results. But there are limits. These are set by the ability of the fuel to resist destructive detonation. To this end, the builder/racer must control the processes that work to the detriment of the engine.

Superchargers, pistons, piston pins, combustion chambers, valves, and gears all have physical limitations of strength, internal aerodynamics, or heat that limit performance. To a certain extent the builder can, by proper attention to detail, improve or maximize the capability of each of these components in turn. Once done, then the operational parameters must be controlled. These include mixture temperature and pressure, fuel, valve and ignition timing, and the engine cooling and lubrication oil systems.

Optimum delivery of mixture to the cylinders is where the real magic occurs. Serious power requires that a comparatively low temperature high density mixture be supplied to the cylinders. This can, and has, been achieved by many techniques on highly boosted engines. This includes aftercooling, ADI, Nitros Oxide and/or use of fuels such as Methanol that substantially cool the mixture as they evaporate. There are tradeoffs in applying any of these approaches and it is how they are implemented that provides the "racers edge."

Building a V-1710 for Racing

For starters, it is important to begin with the best parts available. This means a core engine in good physical condition, and then fitted with a 12 counterweight crankshaft and heavy duty pistons. Depending upon overall design of the modified engine, pistons with compression ratios ranging from 6:1 to 12:1 may be fitted.

From this point individual builder preferences come into play. Many builders have learned from experience ways to improve piston sealing and will modify the ring grooves in the pistons to accept modern rings better suited to running at very high power levels. Other preferences such as strengthened quill shafts and supercharger step-up gear ratios are determined by experience and the intended use of the engine.

For extreme power, the induction system will often be considerably revised. The original ramming manifolds become limiting at something over 2300 bhp and most builders intending to exceed this level opt for fuel injection and a plenum type intake manifold. Interestingly, many of these take the form of those intended by Allison for the fuel injected XV-1710-5(C3) engine of 1936.

All of these high power engines run at speeds considerably above the nominal 3000 rpm specified by Allison. Given a sufficient oil supply they can easily run at 4400 rpm, but then other factors become important. These include the considerably increased power demanded by the supercharger and the problem of valve "float" due to insufficient valve spring force to hold the valves closed against extreme manifold pressures and "bounce" caused by the high speeds. The usual response is to provide stiffer valve springs.

Some builders have resolved the limitations of the quill shaft drive into the accessories and supercharger by removing the engine stage supercharger. In its place turbos are used, and often in some quite unique arrangements. Examples have been seen where each bank has its own turbo, fed exhaust gasses via P-38 type collector manifolds. All of these high power setups require either ADI or Methanol fuel, and in some cases both. Certainly the duty was severe and the life of these marvelous engines limited at best.

Cobra I was based upon a P-39Q-10 and flown by Jack Woolams. Originally built with a V-1710-85 giving 1200 hp at 3000 rpm and 57 inHgA, for the 1946 Thompson both Cobras were given modified V-1710-135 single stage engines able to develop 2200 bhp at 3400 rpm with 85 inHgA and ADI. (Bowers via Roland Harper)

Cobra I and Cobra II

Unlimited Air Racing, which had been deferred during W.W.II, resumed in 1946 with the unlimited Thompson Race being the premier event. Whereas most pre-war Thompson competitors had been either exotic homebuilts or one-off manufacturers prototypes, the 1946 was a race between recently surplused high performance fighters, mostly P-51D *Mustangs* and P-38 *Lightning's*.

These aircraft were for the most part prepared for the race by demilitarizing to remove the armament, armor plate, and radios, as well as the bullet proof fuel tanks in favor of sealing the wing structure to make it function as a tank. Power for these aircraft was usually the stock engine slightly modified to allow it to run at War Emergency Power, hopefully throughout the race.

Carrying this concept one step further were three Bell Aircraft test pilots, Tex Johnston, Jack Woolams, and Chalmers "Slick" Goodlin, who at the time were on the team preparing the Bell XS-1 rocket plane for its first supersonic flight. Their concept was to take the smallest airframe available and stuff the largest engine they could find into it. As it happened, Bell had just purchased two surplus Bell P-39Q-10-BE aircraft, for $750 each, for the sole purpose of obtaining their propellers. This was a cost effective way of proceeding with the L-39 swept wing P-63 derivative test program. The pilots were able to convince company President Larry Bell to let them purchase each aircraft for one dollar, and race the two aircraft in the 1946 Thompson. Bell approved the concept, but insisted that the effort be done under the contractual banner of a company entirely separate of Bell Aircraft.[1]

A classic period photo if there ever was one. Tex Johnston is shown with Cobra II after he piloted it to victory in the 1946 Thompson. (Roland Harper)

The original V-1710-85 engines were removed and replaced by V-1710-135(E31) engines. Four of these engines were obtained from government surplus, as they had originally been built for the Bell RP-63G "Pinball" aircraft where they delivered 1425 bhp at 3000 rpm for Takeoff, but at low altitude could produce 2200 bhp at 3400 rpm with 85 inHgA manifold pressure and water-alcohol injection. The E-31 engine did not have an Auxiliary Stage Supercharger, for none was needed for the low altitude mission. It also retained the 9.5 inch diameter supercharger impeller and was driven by the original 9.60:1 step-up gears, though Bill Wise (Allison Division engineering service representative at Bell) replaced the 6.65:1 pistons with new 6.00:1 pistons having specially shaped crowns to improve the resistance to detonation.[2]

Under the supervision of Bell's Jack Berner, two of these engines were completely torn down and rebuilt to improve their performance when operating continuously at maximum power. This was done by installing the new pistons, along with improvements in the oil system for the accessories section.

The aircraft were further modified to provide an enlarged engine induction air scoop and to incorporate a seventy-five gallon water-alcohol tank into the nose cannon bay to serve the P-63 type ADI system needed for the high powered E-31. They also enlarged the oil system from the stock 8 gallons to 13 gallons to insure that there would be sufficient oil for the long Thompson when racing at sustained high power settings. A new oil cooler was mounted on the aircraft centerline, just below the cockpit, with a flow through air scoop fashioned and provided with a shutter on the exit to regulate air flow and control oil temperature. It was still found necessary to use an anti-foaming additive from Sohio (Standard Oil of Ohio) as a means to minimize the severe oil "frothing" condition that developed when the engine was running at continuous high power. To handle all of the available power, the 4-bladed propellers taken from the two P-63's being used in the L-39 project were used. The combination gave tremendous torque reaction on the small aircraft, and contributed some exciting times for pilots and spectators alike during the races. Additional fixed trim tabs were added to the rudder to counter the torque and all of the fabric covered moveable surfaces on the aircraft were reskined with metal to solve a significant "ballooning" and control problem encountered at high speeds.

Cobra I and II qualified for the 1946 Thompson at 392.7 and 409.1 mph respectively, both establishing new records. *Cobra I* had been detonating on the qualifying run and it was decided to return the airplane to Niagara Falls and install one of the spare E-31 engines. Unfortunately, after completing the engine change, Jack Woolams took the aircraft out over Lake Ontario for a very high speed run and fatally crashed. While never completely explained, it was believed that the Plexiglas used to replace the heavy bullet-proof glass windshield failed due to the high speed and the twisting of the airframe caused by the high torque reaction from the high power being produced.[3]

The windshield on *Cobra II* (Race 84, NX92848, AF42-20869) was immediately strengthened and it went on to win the 1946 Thompson and establish a new speed record of 373.9 mph. Tex Johnston ran the first three laps at 3400 rpm and 85 inHgA, after which he had established sufficient lead that he throttled back to 62 inHgA for the balance of the 48 minutes 8 seconds it took him to complete the 300 mile race.

In 1947, *Cobra II* was again a contestant, but this time things had changed. The racer was now owned by Rollin Stewart, and Jay Demming had signed on as its pilot. Due to the press of time to prepare for the race, no changes were made to either the engine or airframe from those of the previous year. In spite of the old Plexiglas rupturing during his qualifying run, Demming was able to qualify for the 4th position at 386.8 mph. Competition was tougher this year with the entry of two F2G *Corsairs*, powered by huge P&W R-4360's, and provided with trick fuel and water injection.

Cobra II held in and made it a good race, capturing 3rd place, with the final speeds being 396.1 and 390.1 for the Corsairs and 389.8 for Demming in *Cobra II*.

With Rollin Stewart as owner, and a year in which to prepare, the plan was to come back with a much more competitive *Cobra II* for the 1948 Thompson. Allison's Don Nolan was charged with the responsibility of developing the engine installation for 1948.

The old racing V-1710-135 was removed and the last of the four original E-31's modified to incorporate Allison's latest improvements, including installation of the new induction and accessories section with the 10-1/4 inch diameter impeller (but not including the Auxiliary Stage Supercharger) from the V-1710-G6. Nolan and his Chief Mechanic Ray Freese, installed 8.80:1 engine-stage supercharger gears in place of the 7.48:1 gears standard in the G-6.[4]

Flight tests were conducted on several fuels, principally mixtures of menthol, benzine, and acetone, which required complete recarburization to obtain proper fuel/air ratios. During this program *Cobra II* was able to exceed 438 mph from the 2850 bhp being developed when using 115 inHgA and 3350 rpm.[5] Comparative runs were also made using Shell "Triptane"

Race #30 was a P-63C-5, NX63231, piloted by Charles Tucker and was usually the fastest P-63 in the various post war races. In the 1946 Bendix and 1947 Tinnerman he ran a fairly stock E-21 engine that retained the Auxiliary Stage Supercharger. In 1948, running on Shell Triptane at 100 inHgA and 3200 rpm, but without ADI, the engine blew. For 1949 the engine was considerably modified. The Aux S/C was removed and 8.8:1 engine stage gears installed, along with a large PD-18 carburetor in place of the PT-13E. Enough ADI was provided to run the entire race at full power, 1800 bhp from 76 inHgA. (Note: Telecon with Birch Matthews, 4-23-95.) The modified engine could develop as much power as the two-stage, but on less manifold pressure and with improved reliability. For the1949 Sohio and Thompson he posted a qualifying speed of 393.3 mph, reaching 420 mph on the straightaways. Along the way three feet had been clipped from each wing. (Roland Harper)

(2,4,4, tri-methyl pentane) and water injection.[6] Since performance was similar, and the quantity of fuel reduced, the decision was made to use the Shell fuel for a guarantee of sufficient fuel quantity for the long race.[7]

With the engine modified, it was reinstalled along with a stock 3-bladed propeller to improve directional stability. In addition, a number of other airframe improvements were accomplished. *Cobra II* won the 1948 Thompson pole (against the F2G *Corsairs*, *Mustangs* and P-63's) with a qualifying speed of 418.3 mph.

Things did not come out so well in the race. *Cobra II* ended up lapping the entire field, including the Corsairs which had blown their intake ducts, but on about the 12th lap of the twenty scheduled, the engine began cutting out on the backstretch, only to come on strong for the remainder of the circuit. Because of the confusion, the pilot Charles Brown, an Allison Division test pilot, lost track of the number of laps and quit after 19. At that point his average speed was 392.4 mph, some 10 mph faster than the eventual winning P-51.

It turned out that the problem was due to a triangular 10 inch piece of cowling behind the engine that had vibrated off, allowing hot exhaust gasses to be drawn in. Unfortunately, this was directly in line with an asbestos insulated fuel line, but the heat was sufficient to cause vapor-lock, even with 35 psig fuel boost pressure. No wonder the engine would cut out. After the race and with the cowl repaired, the aircraft refueled, and everything was ready to go racing again.

But the *Cobra II* never raced again. It sat-out the 1949 race, which was the last race of the era. An effort was made in the mid-1960s to rebuild it as *Cobra III* and race it at Reno. On a test flight following extensive airframe modifications, and again using a 4-bladed propeller, control problems developed and the aircraft was abandoned and totally destroyed in the resulting crash.[8]

Other Racers

Several P-63's have been modified for racing over the years, along with a few other Allison powered aircraft. One of these is the sweet sounding P-51A belonging to the Planes of Fame museum in Chino, California. For several years they campaigned it in the Bronze division at Reno. They never attempted to push the rare airplane, but still it was able to put on a good show.

Another perennial favorite is Lefty Gardner's P-38L, which has had the turbos and intercoolers removed and Model 322 cowlings installed. He is a steady competitor in the Bronze Heats at Reno, and puts on an awesome display of aerobatics between races.

Lefty Gardner has raced and shown N13Y for decades. The airplane is a P-38L, though the turbos and intercoolers have been removed allowing the streamlined early P-38 cowlings to be used. (Author)

Table 18-1
Allison Powered Warbirds

Type	Number Extant	On Rebuild	Flyable
Bell P-39	43	8	2
Bell P-63	17	3	4
Curtiss P-40	80	24	19
Lockheed P-38	27	5	8
North American P-51A/A-36	3+	3+	3+
North American P-82E/F	0	0	0
Yak-3UA	10	10	3+

Soldiering On

With the Warbird movement gaining momentum in the 1990s, aircraft types thought to be non-existent are being found and brought through restorations once thought impossible. Rarity and money are the essential ingredients for some of these recoveries and restorations.

A recent compilation in *Warbirds Worldwide* lists some interesting numbers regarding the potential for Allison engined aircraft restorations.[9]

Given that a fair number of these aircraft are in museums and are unlikely to ever be made flyable, it is still gratifying to see that there are a number of aircraft being rebuilt to flyable status. Particularly interesting is that a number of these are unique models once thought to have been completely lost, such as the early P-40 *Tomahawk*, Lockheed P-38F, and Allison powered North American P-51A/A-36. There is plenty of interest, and a fair supply of parts, both of which will be needed to keep the Allison V-1710 flying well into the 21st Century.

Unlimited Hydroplane Racing

This was really the afterlife for the Allison V-1710. The boat racers preferred the late model Allison and its 12-counterweight crankshaft over the Rolls-Royce Merlin. This was because of its ruggedness and availability. Each team had its own formula and secrets. While some did little more than bolt in stock engines, many teams fielded a complete set of overhaul tools and aggressively modified nearly every component in an effort to extract every bit of performance.

The preferred engines were those in the 100 series, as they benefited from the many durability upgrades and incorporated the heavily balanced 12 counterweight crankshafts able to run to 4500 rpm. Compression ratios were often raised by milling the heads, or by obtaining special pistons. Many used proprietary supercharger gears as high as 12:1 in order to extract additional manifold pressure from the supercharger, this in addition to often running the engines at well over 4000 rpm. Of course special fuels were obligatory for such operation, as was one form or another of ADI or nitro.[10]

While the Hazen streamlined intake manifolds were quite adequate for up to about 2300 bhp, at higher levels they presented a considerable restriction. Many teams went to various types of fuel injection systems, usually drawing from a large plenum placed between the banks, much like the arrangement used on Indy race engines of today. Others deleted the Allison supercharger entirely and utilized turbosuperchargers in their own, sometimes quite elaborate configurations. Interestingly, most of these would use P-38 type exhaust headers to collect and pass the hot exhaust to the turbos which were of modern design rather than the GE type used during the war.

Modified engines for the Hydros would routinely get 3200 to 3300 bhp with 80-100 inHgA at 4100 to 4200 rpm. At these conditions, it was

Miss Great Lakes, hydroplane U4, powered by a V-1710-E configured engine on startup is typical of the early Allison powered boats. (Allison)

necessary to replace the standard Allison valve springs, rated for 120 LBF, with custom springs rated at 220 LBF. The stock Allison sodium cooled valves were trouble free and the stock cams were usually retained.

One problem faced by the boat racers was the dreaded broken "quill" shaft. Things had never changed, and the reasons for Allison to have incorporated the device to both dampen torsional vibration and protect the drive gears, were still present. In their pursuit of speed, and when running at well over 4000 rpm, the engine could rapidly accelerate to much higher speeds should the boats prop pop out of the water. When this occurred, the supercharger was already consuming something on the order of 700 to 1000 bhp! It would not take many such excursions before the quill shaft would fail, and everything would become very quiet. This was one of the reasons for going to multistage turbosupercharging, for there were no longer any gears in the supercharger drive system.

Few, if any, of these modified engines were ever run on a suitable dynamometer able to really measure the power output. Still, knowing the parameters such as manifold pressure and engine speed it is entirely likely that many of the boat engines were producing in excess of 4000 bhp during important races. Not bad for an engine originally designed for 1000 bhp!

How about a Double Vee 3420 Dragster? Quad Al sported two V-3420-B engines that gave a total of 6840 cubic inches of displacement and even in stock form would deliver over 6000 horsepower! This is a monster 'hiding' behind that VW Bonnet! (Allison)

Allison at the Dragstrip

In 1965 Jim Lytle built "*Quad Al*", a four-wheel drive dragster powered by two V-3420's! It had a total of 6840 cubic inches, 96 sparkplugs, 384 valves, and produced upwards of 8000 bhp. "*Quad Al*" earned a spot in the *Guinness Book of World Records* as the "most powerful piston-engine car ever."[11]

The other arena where the V-1710 has made a showing is in Tractor Pulling. There are some really elaborate tractors used in these events, and the Allison is the heart of many of them.

It is in Tractor Pulling that the rule book and instruction manuals were really tossed out. Since the "pulls" last only for seconds, many pullers dispensed with coolant and ran "dry sleeve" cylinder liners. It was found that a liner made for a certain model of the Cummings Diesel could be made to fit and live for a short while without coolant.[12]

These engines had everything done to them that the boat racers did and more. One engine is known to have had three turbosuperchargers arranged in series, so that each was amplifying the pressure delivered by the others. Such an installation would have to run on alcohol, and/or gasoline with *lots* of ADI injection fluid, or as in some cases, nitrous oxide. How much power? Certainly 4,000 bhp, and probably a lot more.

Where did they all Go?

At the end of the war there were thousands of V-1710's in crates, and many thousands more still mounted on airplanes destined for the smelter. These were removed before melting the airplane down because many of the unique metals in an engine would contaminate the desired aluminum. Most of these engines were themselves torn down, parts segregated, and then melted down.

Stories abound about docks full of V-1710's in their crates that had been sold to junk dealers, who then had the task of breaking apart the crates and taking the engines alone and dumping them in the cargo holds of ships for shipment overseas as scrap.

Still many survived. The Allison's went to work on farms and in factories around the country, often pumping water or doing other mundane chores. As late as the 1980s, one of the hydroplane teams tells of stopping at a freeway rest stop in Nebraska and in the quiet of the prairie hearing the distinct note of a V-1710. Taking the next exit, and following the sound they found a farmer using a V-1710 for pumping water. He had two more still in their crates, as spares. A deal was struck, and the hydro team picked up the three engines on their way back home. The farmer then switched to diesel power.

The most desirable engines were the "late" models, like the V-1710-111 from the P-38L, and any others that had the late connecting rods and 12 counterweight crankshafts, particularly the V-1710-143/145(G6R/L) engines taken from the P-82's when they were scrapped. These were the engines preferred by the hydroplane racers and tractor pullers. Many teams modified their engines extensively, and ran them really hard. As a result, none of these engines, including used and/or modified parts, can be considered for powering restored airplanes.

One of the most surprising incidents to befall the dwindling supplies of V-1710s occurred during the Hunt Brothers "silver boom" of the 1970s. The Hunt's succeeded in running the commodity price of silver to unheard of levels. Many people who had been holding V-1710's for a long time, and seeing little future for them, proceeded to break them apart and take out the main and connecting rod bearings. Only to melt them down for their silver.[13] They turned a quick profit, but nothing like what would be coming with the resurgence of the Warbirds. Today there is a growing demand for V-1710's, most recently accelerated by the production of 20 Russian built Yak-3UA's, all powered by the V-1710 since the Russian V-12 was neither available or considered satisfactory for modern operation.

NOTES

[1] Matthews, Birch J. 1993, 185-198.

[2] Matthews, Birch J. 1996, 297.

[3] As noted earlier, Allison tested a P-39 in November 1942 and determined that the fuselage twisted 1.5 degrees when torque equivalent to 1070 bhp at 3000 rpm was applied through a 2.23:1 reduction gear. With *Cobra II* at 2850 bhp and 3350 rpm the fuselage would have twisted about 3.6 degrees. It is no wonder that the Plexiglas and cowling panels were popping out.

[4] Letter from Allison Director of Public Relations to Birch J. Matthews, August 19, 1963, transmitting information provided by Mr. Don Nolan.

[5] Huntington, Roger, 1989, 156.

[6] This fuel is reported to offer a "Octane" or PN rating of 200/300! Shell offered it to any of the 1948 teams for $1.75 a gallon, which probably did not cover but a fraction of its actual cost. A byproduct of using the fuel was that it was so slow burning that spark timing became a critical factor. Both of the 1948 F2G Corsairs blamed their backfiring and consequential loss of the race to this factor and the backfiring it caused. Huntington, Roger, 1989, 143.

[7] Methanol weighs 6.75 pounds per gallon and requires a fuel/air ratio of about 0.17:1 while it provides a Lower Heating Value of only 7,658 BTU/pound. Grade 100/130 gasoline weighs 5.75#/gal and gives best power with a fuel/air ratio of about 0.115:1, while providing 19,300 BTU/pound (LHV). Triptane is more dense, similar to gasoline. Since horsepower demands the same number of BTU's be provided by either fuel, about 2.2 times as many gallons of methanol as gasoline will be required to produce the same amount of power. In a long race one becomes concerned about having enough fuel.

[8] Matthews, Birch J. 1993, 236-241. Crash occurred on August 10, 1968.

[9] Warbirds Worldwide, *Warbird Watch*, Journal 38, Fall 1995, p.8.

[10] Boushea, Al 1995.

[11] Stone, Matthew L. 1993, 22-23.

[12] Boushea, Al 1995.

[13] Babin, Gus 1995.

Appendix 1: Allison V-1710 Serial Numbers

Throughout this book your author has attempted to rigorously follow the engine model nomenclature as used by the principals at the time. The following is provided as a key to understanding how it all worked.

Military and Manufacturers Serial Numbers

The Allison Engineering Company began their series of manufacturers numbers for the V-1710 with AEC 1, the GV-1710-A engine for the U.S. Navy. Subsequently, each engine delivered to a purchaser was likewise issued a sequential "AEC" number determined by the order in which it was begun. Purchasers-the Navy, Air Corps or British-then assigned a separate serial number unique to their own methods. For the Navy Bureau of Aeronautics, this was simply the next number reflecting the total number of engines they had purchased from all sources.

The Air Corps/Army Air Forces was the major buyer of V-1710's. Their scheme identified the Fiscal Year providing the purchasing funds, followed by a number that is generally a sequential number reflecting the number of engines being procured during that Fiscal Year from all manufacturers. In FY-1934, the Air Corps bought the U-250 engine from Allison and gave it the number AC34-1. They then ordered the single XV-1710-3 as AC34-4, and followed with the order for ten YV-1710-3 pre-production engines that were assigned serial numbers AC34-5 through AC34-14. Engines AC34-2 and AC34-3 were not Allison built engines.

This scheme was continued by the Army through FY-1943, June 30, 1943. For reasons yet to be fully determined, though obviously much more practical, the Army determined to adopt a serial number that was equivalent to the Allison manufacturers number. At this point Allison added an "A" prefix to their manufacturers number, continued the series without interruption. The Army number was identical. Even so, rules were made to be broken. There must have been a number of factors at work. By example, Table A1-1 shows V-1710-93(E11)s built in mid-1943, with no consistency.

The likely explanation is that contract W535-ac-24557 was the major production order and the later contract W535-ac-30859 was for a preproduction batch of E-11s, giving them earlier Allison s/n's and production dates. Similar anomalies can be found with other models. For example, with the V-1710-39(F3R).

Another consequence of the integration of Allison and Army serial numbers was the complete loss of association with engine types. No longer did the Allison serial number refer solely to the V-1710. The V-3420 engines were interspersed within the production sequence and received serial numbers accordingly. From this point on there is no way to tell how many of a type were built solely by the serial number. This became even more the case when Allison began manufacturing jet engines, for they received serial numbers within the existing sequence as well. From 1943 on, it may also be that blocks of serial numbers were assigned to particular models, such as the V-1710-111/113(F30R/L)'s for the Lockheed P-38. As a result, the serial number no longer related to the production sequence. Rather, it was a contract artifact.

In this same way, the serial number no longer meant the total number of engines delivered. Instead it became the total number ordered. This resulted when production orders were canceled as the end of the war came in sight. It is believed that 69,233 V-1710's were actually built, along with 145 V-3420's, and 610 jet engines by the end of 1945. These were followed by production of 650 V-1710-143/145(G6R/L) engines for the P-82E/F airplanes. The final production V-1710-143/145 was serial number A-074125, delivered in January 1947. Even so, we also know that V-1710-145 A-074136 was being held as a spare by the Air Force in June 1948.

Table A1-1
V-1710-93(E11) Engine Serials

Allison Serial Number	Army Serial Number	Contract	Date Delivered
28731	43-1311	W535-ac-30859	May 1943
32413	42-101055	W535-ac-24557	Summer 1943
A-032433	A-032433	W535-ac-24557	Fall 1943
A-032441	A-032441	W535-ac-24557	Fall 1943

Table A1-2
British Engine Serials-Examples

Allison Serial Number	Air Ministry Serial Number	Comments
510	A 199666	British Type Test Engine
854	A 199967	N.Africa,15 Bullet Holes, kept running
899	A 200012	Flying Tiger P-8127, in California 1996
963	A 200076	To Flying Tigers
964	A 199977	To Flying Tigers
2879	A 201293	On Display At Allison Factory, 1996
2998 to 3032+	None	Flying Tigers-Spare Engines
NA	A 200983	C-15 in AAC P-40 AC39-289 @ McChord Field
NA	A 205964	F-3R in AAC P-40E AC40-548
6587	A 206891	F-3R in AAC P-40E AC41-5534 @ Langley Field

British Serial Numbers

It would not be fair to have something as fundamental as a numbering scheme that would not also be confusing or ambiguous. When the British Air Ministry began placing orders with Allison they also assigned "A" numbers to each engine. As a result, dataplates on these "Commercial" engines show only the Allison model number, such as V-1710-C15, along with the Allison serial number and the one assigned by the Air Ministry. The confusion can occur since everyone ended up using an "A" prefix. For examples see Table A1-2.

The first 100 engines built from non-standard parts for the American Volunteer Group, Flying Tigers, were taken directly from the original British and French contracts. All received Air Ministry serials. The 50 spare engines purchased directly by the Chinese are only identified by their Allison serials as they had no association with British purchase or production.

Also apparent in the numbers is that once Lend-Lease went into effect, it appears that the U.S. Government took possession of the remaining British engines and used them without assigning their own serial numbers. Note the examples of both C-15 and F-3R engines installed in aircraft delivered to the U.S. Army.

V-3420 Serial Numbers

The first double engine, the XV-3420-1 did not receive an Allison serial number, rather it was identified by its Air Corps serial, 38-119. This appears to have been an anomaly, for even the Allison built air-cooled X-4520 was given "Manufacturers Number 1", as well as the Air Corps AC25-521. With this exception it appears that all of the rest of the double engines did receive both Allison and Army serials as explained above.

Allison Workhorse Engines

As a way to expedite development of models and features for their engines Allison built and maintained a number of engines that they referred to as "workhorse" engines. All of these engines were identified by the prefix EX-. It is believed that Allison owned these engines outright, though there are many examples of testing being done on them that was paid for by the customer.

It was not unusual for an EX- engine to be built-up at different times as a different model, including switching between "E" and "F" configurations. Some of these engines were real workhorses, enduring more than 50 build-up/tear-down cycles. At lease 54 EX- engines have been identified, and it is known that some of them went through at least two incarnations in addition to many builds. Included in this list were both V-1710 and V-3420 examples, though none of these engines are included in the engine production figures.

Details of a designation system for these engines have not been determined, but it does appear that some show on their dataplate the Model as V-1710-FR. That is, they do not identify the sub-model. A word of caution: anyone coming across an engine with such a data plate should not consider using it in an airplane. There is no telling what trials and tribulations it has gone through in a past life.

When the jet engines came along, it was necessary to have them available for testing as well. Allison identified these workhorse engines as IX-1, IX-2, IX-3, etc. The first two were their first copies of the General Electric Type I-16 "Supercharger", and the IX-3 was Allison's first example of the GE Type I-40 jet engine. That model subsequently was designated as the J33, of which Allison was to become the primary manufacturer.

School Engines

For a variety of uses in training, both static and for ground running, the Air Forces purchased a number of School Engines. The dataplates on these engines show the serial number to be as S.E. 369, a V-1710-51(F10R), or S.E. 582, a V-1710-85(E19). These engines are complete in every detail, but were constructed from non-flight worthy components. A number of these engines are known to be on display in various museums, and at least one was used during the filming of Tora-Tora-Tora. It is believed that there were not "582" Allison School Engines. Rather, these were purchased sequentially by the Army from a number of engine manufacturers to support the various training programs during the war. Certainly there were more than 100 V-1710's built as School Engines. These are not included in the total production numbers.

Another example was the Army's original purchase of C-15 engines for training. These were identified as G-1710-C15 models and were built from non-flight worthy parts, but intended to be operable for use in training mechanics on maintenance and operation of the engine. Such engines are not suitable for rebuilding to use in a flight worthy airplane.

Table A1-3
V-3420 Serials-Examples

Allison Serial Number	Air Forces Serial Number	Engine Model	Comments
None	38-119	XV-3420-1(A1)	First V-3420 Engine
EX-41	NA	V-3420-5(B4)	Allison Workhorse Engine
173	NA	V-3420-A8L	Navy PT-8 Boat
4525	41-50951	V-3420-9(A11R)	Type Test, November 1941
A-029879	A-029879	V-3420-13(A16L)	Flew in XP-58
A-032655	A-032655	V-3420-19(B8)	Flew in XP-75
A-045799	A-045799	V-3420-23(B10)	Model Test Engine for XP-75
A-059831	A-059831	V-3420-17(A18R)	Used for Jet Engine Compressor Tests

Appendix 2: Engine Ratings

An aircraft engine must perform over a very wide range of altitudes, air temperatures, and aircraft speeds. Even at full throttle, each condition defines a different amount of power. As a result, there can be confusion as to exactly how much power it is producing. Furthermore, given the long period of production of the V-1710 and the many different installations, there is little apparent consistency in the ratings assigned to the engine. In fact, it was not so arbitrary. Allison was instrumental in creating a number of the fundamental definitions of engine performance that were utilized by the military and various engine manufacturers of the period.

The V-1710 was originally intended by both the Air Corps and Allison to be a sea-level engine. As such, the performance on the test bed and that at altitude were, for all intents, the same. Previously, very few military aircraft had internally supercharged engines and could routinely exceed 10,000 feet. And if they did, they were not expected to demonstrate outstanding performance at high altitude. This pretty well defined aviation prior to 1940 and in such a world, a universally accepted method for altitude rating was not needed.

When Allison got the Air Corps to allow the altitude rated V-1710-C13 in the 1938 Pursuit Competition, it became necessary to determine an appropriate rating strategy. Allison described the issues in a March 1, 1939 letter to Wright Field regarding options for definition the ratings for the coming altitude rated V-1710-F1. Everyone agreed that at altitude, with reduced atmospheric temperature and pressure, maximum power was desired. The question was how to define a corresponding rating for Takeoff and performance at intervening altitudes. For the F-1, it's Military rating was to be 1150 bhp with 3000 rpm at 12,000 feet.

Air Corps practice had been to require this same power to be delivered when tested on their sea-level test stands. Allison argued that this was inappropriate for a highly supercharged altitude rated engine. It would be penalized since it would be operating with a considerably higher ihp at sea-level. A consequence would likely be damage resulting from excessive mixture temperatures due to such operation at conditions it would never see at altitude. Their alternative proposal was to define Takeoff power as being a value 40 percent above the difference between the power obtained at sea-level when operating at the altitude rated manifold pressure. They believed this would give a reasonable balance of manifold pressure, mixture temperature, and power in consideration of the severity of testing.

The performance curves shown throughout this book reflect on the Allison recommendations that were finally adopted as the standard method for establishing ratings for Altitude engines. It was Allison practice to define 1150 bhp for the Military rating of their altitude engines, the variable was the critical altitude. By changing supercharger gear ratios, and/or adding an Auxiliary Supercharger Stage, they were able to select the altitude at which this power was delivered to match the aircraft mission. Over the course of the production run, various models of the single stage V-1710 were rated for Military power of 1150 bhp at altitudes from sea-level up to 15,500 feet. Some of the two-stage engines delivered 1150 bhp at up to 27,500 feet.

It is important to note that a centrifugal supercharger, as used in the V-1710 and contemporary engines, delivers air flow and produces pressure in proportion to the rpm of its impeller. For this reason a series of straight lines are drawn on the performance chart at typical engine speeds. Since engine power is proportional to airflow, these lines define the power available at each engine speed, which of necessity decreases as the altitude increases and the air thins.

The "rated" power for an altitude engine then becomes somewhat arbitrary, hence the typical use of 1150 bhp. Even so, for every rpm and power on this curve, there is a corresponding manifold pressure necessary to deliver the requisite quantity of air. As the airplane descends, and if rpm is held constant, the higher density of the incoming air will cause MAP to increase, and with it ihp. To prevent mechanical damage to the engine due to detonation, Allison recommended that the manifold pressure equivalent to the rated power at altitude not be exceeded. Lines of constant MAP show a decreasing amount of power available from the engine as the airplane descends. It is for this reason that an airplane will exhibit its maximum speed at the critical altitude of its engine.

In the case of the V-1710-F1, this approach resulted in a sea-level power of only 1000 bhp. Everyone-Allison, the aircraft manufacturers, and the military-agreed that it was not viable or necessary to reduce the available Takeoff power to such an extent. Allison's initial proposal was to restore 40 percent of the reduction, relying on the comparatively brief periods when takeoff power was required and the inherent strength of the engine to assure reliable operation.

In one way or another, manufacturers and Air Forces around the world adopted similar approaches to rating their engines. Use of two-speed superchargers and/or ADI were ways that the low altitude ratings could be otherwise adjusted, but they still had to contend with same considerations.

Table A2-1
Altitude Ratings for Early Single Stage Engines

Model	Takeoff	Military	Comments
V-1710-19(C13)	1060/2950/SL/42.5	1150/2950/10,000/41.2	XP-40
V-1710-33(C15)	1040/3000/SL/42.0	1040/3000/14,300/38.0	P-40 Tomahawk
V-1710-33(C15)	NA	1090/3000/13,200/38.9	Air Corps Demanded, 1939
V-1710-33(C15)	1040/3000/SL/42.0	1040/3000/13,200/	As Model Tested, 1940
V-1710-33(C15)	1040/2800/SL/44.2	1040/3000/12,800/40.0	July '41, w/Port BF Screens
V-1710-25(F1R)	1060/3000/SL/42.0	1150/3000/12,000/38.0	Proposed, 1939
V-1710-25(F1R)	1150/3000/SL/45.5	1150/3000/12,000/44.6	Adopted, 1940
V-1710-39(F3R)	1150/3000/SL/45.5	1150/3000/11,700/44.6	Rated, 1940
V-1710-73(F4R)	1325/3000/SL/51.0	1150/3000/12,000/42.0	Rated, 1941

All of these models used 8.8:1 supercharger gears and show the various rating strategies in effect as they went into service. Note that conflicting MAP's are given in references due to whether or not backfire screens were being used.

Although the Air Corps adopted the Allison approach to rating, when it came to its first application [the V-1710-33(C15)], selection of the numerical values became a major issue. Allison was willing to rate the C-15 for 1090 bhp, as limited by ihp considerations, at up to an altitude of 11,800 feet. The Air Corps wanted the 1090 bhp, but at a higher altitude where Allison was only willing to support 1040 bhp. The Air Corps insisted and, as a result, went through four different Model Test engines before finally accomplishing a successful Model Test on a fifth engine at the recommended 1040 bhp. While it might seem that 50 hp would not be so significant, the real issue was the ihp and how much power had to be delivered to the supercharger to produce the additional power at the higher altitude.

With the continuing demand for power, Allison soon revised their approach and allowed the same power at Takeoff as for the Military altitude rating. This was accomplished by additional strengthening of the engine during the development of the first V-1710-E/F engines, and is the reason why there were delays in getting the altitude rated V-1710-35(E4) and V-1710-39(F3R) through their respective Model Tests in 1940/41. When the second generation of these engines [V-1710-63(E6) and V-1710-73(F4R)] became available, they were rated for Takeoff at 1325 bhp and followed the constant MAP line up to their "Wide Open Throttle" altitude. Still, the "Military" rating usually remained 1150 bhp.

Appendix 3: Engine Data by Allison Model

Allison Model	Military Model	Number Built	Comp Ratio	Weight, lbs	S/C OD, inches	S/C Gear Lo-Ratio	S/C Gear Hi-Ratio	Aux S/C OD, inches	Aux S/C Gear Ratio	Carburetor Model	Propeller Gear Ratio & Shaft Size
GV-1710-A1 Build 1	V-1710- 2	(1)	5.88:1	1026	8.25	7.33:1	none	none	none	NA-U8J	3:2#40
GV-1710-A1 Build 2	V-1710- 2	1	5.85:1	1035	8.25	8.00:1	none	none	none	NA-U8F, Downdraft	3:2#40
V-1710-A 2	V-1710- 1(XV)	1	5.85:1	1160	9.50	8.77:1	none	none	none	NA-F7B	2.00:1#40
V-1710-B 1R	V-1710- 4(XV)	1	7.00:1	1162	none	none	none	none	none	Two NA-Y6E	3:2#40
V-1710-B 2R	V-1710- 4	2	7.00:1	1160	none	none	none	none	none	Two NA-Y6E	3:2, Flanged
V-1710-C 1 Bld 1	V-1710- 3(XV)	1	5.75:1	1160	9.50	8.77:1	none	none	none	NA-F7C	2.00:1#50
V-1710-C 1 Bld 2	V-1710- 3(XV)	(1)	5.90:1	1160	9.50	8.00:1	none	none	none	NA-F7C	2.00:1#50
V-1710-C 1 Bld 3	V-1710- 3(XV)	(1)	6.05:1	1214	9.50	8.00:1	none	none	none	NA-F7C	2.00:1#50
V-1710-C 1 Bld 4	V-1710- 3(XV)	(1)	6.05:1	1214	9.50	8.00:1	none	none	none	NA-F7C	2.00:1#50
V-1710-C 2	V-1710- 5(YV)	(1)	6.00:1	1290	9.50	8.00:1	none	none	none	MC-12 Fuel Inj	2.00:1#50
V-1710-C 2	V-1710- 7(XV)	1	6.05:1	1214	9.50	8.00:1	none	none	none	NA-F7C	2.00:1#50
V-1710-C 3	V-1710- 5(YV)	0	6.00:1	1290	9.50	8.00:1	none	none	none	MC-12 Fuel Inj	2.00:1#50
V-1710-C 3	V-1710- 7(XV)	1	6.00:1	1245	9.50	8.00:1	none	none	none	NA-F7C	2.00:1#50
V-1710-C 4	V-1710- 7(YV)	3	6.00:1	1269	9.50	8.00:1	none	none	none	NA-F7C	2.00:1#50
V-1710-C 5	V-1710-None	0	na	na	9.50	na	none	none	none	na	2.00:1#50
V-1710-C 6	V-1710- 7(YV)	(3)	6.00:1	1280	9.50	8.00:1	none	none	none	NA-F7L	2.00:1#50
V-1710-C 6	V-1710-Comm	0	6.00:1	1280	9.50	6.75:1	none	none	none	NA-F7L	2.00:1#50
V-1710-C 7	V-1710- 11(XV)	1+(1)	6.65:1	1310	9.50	6.23:1	none	none	none	NA-F7F3?	2.00:1#50
V-1710-C 8	V-1710- 11	2	6.65:1	1310	9.50	6.23:1	none	none	none	NA-F7F3	2.00:1#50
V-1710-C 9	V-1710- 15	1	6.65:1	1305	9.50	6.23:1	none	none	Turbo	NA-F7F3	2.00:1#50
V-1710-C10	V-1710- 21	<20	6.65:1	1310	9.50	6.23:1	none	none	none	PD-12B4/6	2.00:1#50
V-1710-C11	V-1710-None	0	na	1310	9.50	8.77:1	none	none	none	PD-12B1	2.00:1#50
V-1710-C12	V-1710-None	0	6.65:1	1310	9.50	7.33:1	none	none	none	NA-F7F3	2.00:1#50
V-1710-C13	V-1710- 19(XV)	1+(2)	6.65:1	1320	9.50	8.77:1	none	none	none	PT-13B1	2.00:1#50
V-1710-C14	V-1710-None	0	6.65:1	1325	9.50	6.23:1	none	none	None	PD-12B6	2.00:1#50
V-1710-C15	V-1710- 33	2,550	6.65:1	1325	9.50	8.77:1	none	none	none	PT-13E1	2.00:1#50
V-1710-C15A	V-1710-None	50	6.65:1	1325	9.50	8.77:1	none	none	none	PT-13E1	2.00:1#50
V-1710-D 1	V-1710- 9(YV)	(5)	6.00:1	1316	9.50	8.00:1	none	none	none	NA-F7C1	2.00:1#50
V-1710-D 1	V-1710- 9(YV)	0	7.00:1	1300	9.50	8.77:1	none	none	Turbo	MC-12 Fuel Inj	2.00:1#50
V-1710-D 1	V-1710- 9(YV)	(1)	7.00:1	1290	9.50	8.00:1	none	none	none	MC-12 Fuel Inj	2.00:1#50
V-1710-D 1	V-1710- 13	3	6.65:1	1330	9.50	6.23:1	none	none	Turbo	NA-F7F1	2.00:1#50
V-1710-D 2	V-1710- 23	39	6.65:1	1350	9.50	6.23:1	none	none	Turbo	PD-12B6	2.00:1#50
V-1710-D 2	V-1710- 23	0	6.65:1	1350	9.50	6.23:1	none	none	none	NA-F7F3	2.00:1#50
V-1710-D 2A	V-1710- 41	(4)	6.65:1	1350	9.50	8.77:1	none	none	none	PT-13E1	2.00:1#50
V-1710-D 3	V-1710-None	0	6.65:1	1335	9.50	6.23:1	none	none	Turbo	NA-F7L	2.00:1#50
V-1710-D 3(A)	V-1710-17(XV)	0	6.65:1	1280	9.50	6.23:1	none	none	Turbo	NA-F7L	1.67:1#50
V-1710-D 3(B)	V-1710-None	1?	6.65:1	1285	9.50	6.23:1	none	none	Turbo	NA-F7L	1.67:1#50
V-1710-D 4	V-1710-None	0	6.65:1	1375	9.50	8.77:1	none	none	none	PT-13E1	2.00:1#50
V-1710-D 5	V-1710-None	0	6.65:1	1325	9.50	6.23:1	none	none	Turbo	PD-12G1	2.00:1#50
V-1710-E 1	V-1710- 6(XV)	2	6.65:1	1364	9.50	8.80:1	none	none	none	PD-12G1	1.80:1#60
V-1710-E 2	V-1710- 17	2	6.65:1	1350	9.50	6.44:1	none	none	Turbo	PD-12B4	1.80:1#60
V-1710-E 2A	V-1710- 31	(2)	6.65:1	1360	9.50	8.80:1	none	none	none	PT-13B1	1.80:1#60
V-1710-E 3	V-1710- 17	(16)	6.65:1	1375	9.50	na	none	none	none	PD-12	1.80:1#60
V-1710-E 4	V-1710- 35	2,155	6.65:1	1375	9.50	8.80:1	none	none	none	PD-12K2	1.80:1#60
V-1710-E 5	V-1710- 37	16	6.65:1	1375	9.50	8.80:1	none	none	none	PD-12K2	1.80:1#60
V-1710-E 6	V-1710- 63	946	6.65:1	1435	9.50	8.80:1	none	none	none	PD-12K2/K6	2.00:1#60
V-1710-E 7	V-1710-None	2	6.65:1	1175	9.50	6.00:1	none	none	none	PD-12G1	Direct
V-1710-E 8	V-1710- 67	1	6.65:1	1445	9.50	8.80:1	none	none	none		2.00:1#60
V-1710-E 8	V-1710- 67	1	6.65:1	1445	9.50	8.80:1	none	none	none	PD-12K6	2.227:1#60
V-1710-E 9	V-1710- 47	3?	6.65:1	1595	9.50	7.48:1	none	12.1875	6.85:1	PT-13E5	2.227:1#60
V-1710-E10	V-1710- 6	None	6.65:1	<1400	9.50	8.80:1	none	none	none	PD-12K2	1.80:1 #60
V-1710-E11	V-1710- 93	2,521	6.65:1	1613	9.50	8.10:1	none	12.1875	6.85:1	PT-13E9	2.227:1#60
V-1710-E12R	V-1710- 59	25	6.65:1	1345	9.50	9.60:1	none	none	none	PD-12K2	1.80:1 #60
V-1710-E13	V-1710-None	1?	6.65:1	1485	9.50	8.10:1	none	none	Turbo	PD-12K2	2.227:1 #60
V-1710-E14	V-1710-None	na	6.65:1	1445	9.50	7.48:1	none	none	2Stg Tur	PD-12K2	2.227:1 #60
V-1710-E15	V-1710-None	?	6.65:1	1675	9.50	7.48:1	none	12.00	8.80	PR-58	2.227:1 #60

Allison Model	Military Model	Number Built	Comp Ratio	Weight, lbs	S/C OD, inches	S/C Gear Lo-Ratio	S/C Gear Hi-Ratio	Aux S/C OD, inches	Aux S/C Gear Ratio	Carburetor Model	Propeller Gear Ratio & Shaft Size
V-1710-E16	V-1710- 65	1	6.65:1	1660	9.50	8.10:1	none	12.1875	7.23:1	na	2.227:1 #60
V-1710-E17	V-1710-None	?	6.65:1	1645	9.50	7.48:1	none	12.1875	8.087:1	PT-13E5	2.227:1 #60
V-1710-E18	V-1710- 83	1,034	6.65:1	1435	9.50	9.60:1	none	none	none	PD-12K6	2.00:1#60
V-1710-E19	V-1710- 85	9,780	6.65:1	1452	9.50	9.60:1	none	none	none	PD-12K6	2.227:1#60
V-1710-E20	V-1710-None	0	na	1450	9.50	8.10:1	none	na	turbo	PD-12K7	2.227:1#60
V-1710-E21	V-1710-117	2,237	6.65:1	1660	9.50	8.10:1	none	12.1875	7.23:1	PT-13E-10	2.227:1#60
V-1710-E22	V-1710-109	222	6.65:1	1660	9.50	8.10:1	none	12.1875	7.23:1	PD-12K15	2.227:1#60
V-1710-E22A	V-1710-109A	128	6.65:1	1660	9.50	8.10:1	none	12.1875	7.23:1	PD-12K15	2.227:1#60
V-1710-E23 Unit#1	V-1710-103 Unit#1	5	6.65:1	1800	9.50	8.10:1	none	12.1875	6.85:1	PD-12K7	2.772:1#50
V-1710-E23 Unit#2	V-1710-103 Unit#2	5	6.65:1	1800	9.50	8.10:1	none	12.1875	6.80:1	PD-12K7	2.772:1#70
V-1710-E23A Unit#1	V-1710-125 Unit#1	2+	6.65:1	1811	9.50	8.10:1	none	12.1875	6.85:1	PD-12K8	2.772:1#50
V-1710-E23A Unit#2	V-1710-125 Unit#2	2+	6.65:1	1811	9.50	8.10:1	none	12.1875	6.85:1	PD-12K8	2.772:1#70
V-1710-E23BL	V-1710-129L	2+	6.65:1	1811	9.50	8.10:1	none	12.1875	7.18:1	PD-12K15	2.772:1#70
V-1710-E23BR	V-1710-129R	2+	6.65:1	1811	9.50	8.10:1	none	12.1875	7.23:1	PD-12K15	2.772:1#50
V-1710-E23CL	V-1710-137L	2+	6.00:1	1811	9.50	8.10:1	none	12.1875	7.59:1	PD-12K15	2.772:1#70
V-1710-E23CR	V-1710-137R	2+	6.00:1	1811	9.50	8.10:1	none	12.1875	7.64:1	PD-12K15	2.772:1#50
V-1710-E24 Unit#1	V-1710-None	2+	6.65:1	1810	9.50	8.10:1	none	12.1875	7.23:1	PD-12K7,w/ADI	2.458:1#50
V-1710-E24 Unit#2	V-1710-None	2+	6.65:1	1810	9.50	8.10:1	none	12.1875	7.18:1	PD-12K7,w/ADI	2.458:1#70
V-1710-E25	V-1710-None	3+	6.00:1	TBD	9.50	8.10:1	none	12.1875	7.64:1	SD-400	2.478:1#60
V-1710-E26	V-1710-None	na	na	na	na	na	na	na	na	na	na
V-1710-E27	V-1710-127	1	6.00:1	1983	9.50	8.10:1	none	12.1875	7.23:1	PD-12K15	2.48:1#60
V-1710-E28	V-1710-None	na	na	na	na	na	na	na	na	na	na
V-1710-E29	V-1710-None	0?	na	na	10.25?	7.48?	9.60?	na	na	na	na
V-1710-E30	V-1710-133	8	6.00:1	1660	9.50	8.10:1	none	12.1875	7.64:1	PD-12K15	2.227:1#60
V-1710-E31	V-1710-135	258	6.65:1	1500	9.50	9.60:1	none	none	none	PD-12K-8?	2.227:1 #60
V-1710-E31	V-1710-Racer	4	6.00:1		9.50	9.60:1	none	none	na	na	na
V-1710-E32	V-1710-None	na	na	na	na	na	na	na	na	na	na
V-1710-F 1R	V-1710- 25	none?	6.65:1	1320	9.50	8.80:1	none	none	none	PT-13B2	2.00:1#50
V-1710-F 2L	V-1710- 29	540	6.65:1	1305	9.50	6.44:1	none	none	Turbo	PD-12G1	2.00:1#50
V-1710-F 2R	V-1710- 27	540	6.65:1	1305	9.50	6.44:1	none	none	Turbo	PD-12G1	2.00:1#50
V-1710-F 3R	V-1710- 39	4,694	6.65:1	1310	9.50	8.80:1	none	none	none	PD-12K2	2.00:1#50
V-1710-F 4	V-1710-None	0	6.65:1	1290	9.50	6.44:1	none	none	Turbo	PD-12B6	2.00:1#50
V-1710-F 4L	V-1710- 79	1	6.65:1	1345	9.50	8.80:1	none	none	none	na	2.00:1#50
V-1710-F 4R	V-1710- 73	1,883	6.65:1	1345	9.50	8.80:1	none	none	none	PD-12K2	2.00:1#50
V-1710-F 5L	V-1710- 53	1,030	6.65:1	<1345	9.50	7.48:1	none	none	Turbo	PD-12G1	2.00:1#50
V-1710-F 5R	V-1710- 49	1,030	6.65:1	<1345	9.50	7.48:1	none	none	none	PD-12G1	2.00:1#50
V-1710-F 6R	V-1710-None	None	6.65:1	<1410	9.50	9.60:1	na	none	none	PD-12K2	1.8:1,#40<
V-1710-F 7R	V-1710- 45	1	6.65:1	1515	9.50	7.48:1	none	12.1875	6.85:1	PT-13E9	2.36:1#50
V-1710-F 8	V-1710- 43	1	6.65:1	1310	9.50	9.60:1	none	none	none	na	2.00:1#50
V-1710-F 9R	V-1710-None	?	6.65:1	1410*	9.50	7.48:1	none	9.50	8.00:1	PD-12K2	2.00:1 #50
V-1710-F10L	V-1710- 55	1,753	6.65:1	1345	9.50	7.48:1	none	none	Turbo	PD-12K3	2.00:1#50
V-1710-F10R	V-1710- 51	1,752	6.65:1	1345	9.50	7.48:1	none	none	Turbo	PD-12K3	2.00:1#50
V-1710-F11R	V-1710- 57	1	6.65:1	1360	10.25	6.44:1	8.80:1	none	none	na	2.00:1#50
V-1710-F12	V-1710-None	0	na	na	na	na	na	na	na	na	na
V-1710-F13L	V-1710-None	0	6.65:1	1345	9.50	7.48:1	none	none	none	PD-12K2	2.36:1#50
V-1710-F13R	V-1710-None	0	6.65:1	1345	9.50	7.48:1	none	none	none	PD-12K2	2.36:1#50
V-1710-F14R	V-1710- 61	26	6.65:1	1320	9.50	9.60:1	none	none	none	PD-12K2	2.00:1#50
V-1710-F14R	V-1710- 61	3-Test	6.65:1	1320	9.50	9.60:1	none	none	none	PD-12K2	2.00:1#50
V-1710-F15L	V-1710- 77	6	6.65:1	1365	9.50	8.10:1	none	none	Turbo	PD-12K7	2.36:1#50
V-1710-F15R	V-1710- 75	19	6.65:1	1365	9.50	8.10:1	none	none	Turbo	PD-12K7, wet	2.36:1#50
V-1710-F16	V-1710-None	0	6.65:1	1375	10.25	6.44:1	8.80:1	none	none	PR-58	2.00:1#50
V-1710-F17L	V-1710- 91	5,421	6.65:1	1350	9.50	8.10:1	none	none	Turbo	PD-12K7	2.00:1#50
V-1710-F17L	V-1710- 91A	(5,421)	6.65:1	1350	9.50	8.10:1	none	none	Turbo	PD-12K7	2.00:1#50
V-1710-F17R	V-1710- 89	5,429	6.65:1	1350	9.50	8.10:1	none	none	Turbo	PD-12K7	2.00:1#50
V-1710-F17R	V-1710- 89A	(5,429)	6.65:1	1350	9.50	8.10:1	none	none	Turbo	PD-12K7	2.00:1#50
V-1710-F18R	V-1710- 69	1	6.65:1	1350	9.50	7.48:1	none	none	none	Cyl Port Inject	2.00:1#50
V-1710-F19R	V-1710- 71	1	6.65:1	1350	9.50	7.48:1	none	none	none	Direct Cyl Inj	2.00:1#50
V-1710-F20R	V-1710- 81	4,817	6.65:1	1352	9.50	9.60:1	none	none	none	PD-12K6	2.00:1#50
V-1710-F20R	V-1710- 81A		6.65:1	1352	9.50	9.60:1	none	none	none	PD-12K6	2.00:1#50
V-1710-F21R	V-1710- 87	835	6.65:1	1353	9.50	7.48:1	none	none	none	PD-12K7	2.00:1#50
V-1710-F22R	V-1710-None	0	na	na	na	na	na	na	na	na	na

Allison Model	Military Model	Number Built	Comp Ratio	Weight, lbs	S/C OD, inches	S/C Gear Lo-Ratio	S/C Gear Hi-Ratio	Aux S/C OD, inches	Aux S/C Gear Ratio	Carburetor Model	Propeller Gear Ratio & Shaft Size
V-1710-F23R	V-1710- 95	6	6.65:1	1355	9.50	9.60:1	none	none	none	PD-12K6	2.00:1#50
V-1710-F24R	V-1710-None	0	na	na	na	na	na	na	na	na	na
V-1710-F25R	V-1710-None	0	6.65:1	1520	9.50	8.10:1	none	12.1875	6.85:1	PT-13E9	2.36:1#50
V-1710-F26R	V-1710- 99	4,200	6.65:1	1355	9.50	9.60:1	none	none	none	PD-12K6	2.00:1#50
V-1710-F26R	V-1710- 99A	4,200	6.65:1	1355	9.50	8.80:1	none	none	none	PD-12K6	2.00:1#50
V-1710-F27R	V-1710-101	3	6.65:1	1520	9.50	8.10:1	none	12.1875	6.85:1	PD-12K7	2.36:1#50
V-1710-F28L	V-1710-123	None?	6.65:1	1555	9.50	8.10:1	none	12.1875	7.18:1	PD-12K12	2.36:1#50
V-1710-F28R	V-1710-121	4	6.65:1	1555	9.50	8.10:1	none	12.1875	7.23:1	PD-12K12	2.36:1#50
V-1710-F29L	V-1710-107	2	6.65:1	1430	9.50	8.10:1	none	none	Turbo	PD-12K8	2.36:1#50
V-1710-F29R	V-1710-105	2	6.65:1	1430	9.50	8.10:1	none	none	Turbo	PD-12K8	2.36:1#50
V-1710-F30L	V-1710-113	6,344	6.65:1	1395	9.50	8.10:1	none	none	Turbo	PD-12K8	2.00:1#50
V-1710-F30R	V-1710-111	6,348	6.65:1	1395	9.50	8.10:1	none	none	Turbo	PD-12K8	2.00:1#50
V-1710-F31R	V-1710-115	560	6.65:1	1385	9.50	9.60:1	none	none	none	PD-12K8	2.00:1#50
V-1710-F31RA	V-1710-115A	420	6.65:1	1385	9.50	9.60:1	none	none	none	PD-12K8	2.00:1#50
V-1710-F32R	V-1710-119	5	6.00:1	1750	9.50	8.10:1	none	12.1875	7.64:1	SD-400	2.36:1#50
V-1710-F33L	V-1710-141	0	6.00:1	1560	9.50	8.10:1	none	12.1875	7.59:1	PD-12K15	2.36:1#50
V-1710-F33R	V-1710-139	0	6.00:1	1560	9.50	8.10:1	none	12.1875	7.64:1	PD-12K15	2.36:1#50
V-1710-F34	V-1710-None	na	na	na	na	na	na	na	na	na	na
V-1710-F35R	V-1710-None	0	6.00:1	na	9.50	8.10:1	none	12.1875	8.087:1		2.36:1#50
V-1710-F36L	V-1710-145	0	6.00:1	na	10.25	7.48:1	none	12.1875	8.03:1	Bendix Port Inj	2.36:1#50
V-1710-F36R	V-1710-143	0	6.00:1	na	10.25	7.48:1	none	12.1875	8.087:1	Bendix Port Inj	2.36:1#50
V-1710-F37	V-1710-None	0	6.00:1	na	10.25	8.80:1	none	12.1875	8.087:1	Bendix Port Inj	2.36:1#50
V-1710-G 1R	V-1710- 97	5	6.65:1	1500	10.25	7.48:1	none	none	none	PT-13E	2.36:1#50
V-1710-G 1RA	V-1710- 97A?	5	6.00:1	1500	10.25	9.60:1	none	none	none	PT-13E	2.36:1#50
V-1710-G 2R/L	V-1710-None	na	6.00:1	1470	10.25	7.76:1	9.60:1	none	none	PT-13H2	2.36:1#50
V-1710-G 2RA	V-1710-Comm	?	6.00:1	1475	10.25	7.76:1	9.60:1	none	none	PT-13H2	2.36:1#50
V-1710-G 3R	V-1710-131	8	6.50:1	1475	10.25	7.48:1	9.60:1	none	none	PT-13H2	2.36:1#50
V-1710-G 3R	V-1710-Comm	?	6.50:1	1475	10.25	7.48:1	9.60:1	none	none	PT-13H2?	2.36:1#50
V-1710-G 4L	V-1710-Comm	0	6.00:1	1340	10.25	7.76:1	9.60:1	none	none	PT-13H2	Various:1#50
V-1710-G 4R	V-1710-Comm	0	6.00:1	1340	10.25	7.76:1	9.60:1	none	none	PT-13H2	Various:1#50
V-1710-G 5	V-1710-None	0	6.65:1	1500	9.50	9.60:1	none	none	none	PD-13K-8	2.227:1
V-1710-G 6	V-1710-None	(2)?	6.00:1		10.25	8.80:1	none	none	Removed	PR-58	2.36:1#50
V-1710-G 6A	V-1710-None	0	6.00:1	1595	10.25	7.48:1	none	12.1875	8.087:1	SD-400-D3	2.36:1#50
V-1710-G 6L	V-1710-145	375	6.00:1	1595	10.25	7.48:1	none	12.1875	8.03:1	SD-400-D3	2.36:1#50
V-1710-G 6R	V-1710-143	375	6.00:1	1595	10.25	7.48:1	none	12.1875	8.087:1	SD-400-D3	2.36:1#50
V-1710-G 7R/L	V-1710-Comm	0	na	na	na	na	na	na	na	na	na
V-1710-G 8R/L	V-1710-None	0	6.00:1	1595	10.25	7.48:1	none	12.1875	8.087:1	SD-400-D3	2.36:1
V-1710-G 9L	V-1710-149	0?	6.00:1	1625	10.25	7.48:1	none	12.1875	8.03:1	Port Fuel Inj	2.36:1#50
V-1710-G 9R	V-1710-147	0?	6.00:1	1625	10.25	7.48:1	none	12.1875	8.087:1	Port Fuel Inj	2.36:1#50
V-1710-G10L	V-1710-None	0	6.00:1	na	2-stg	na	na	Integral	na	Port Injection	2.36:1#60
V-1710-G10R	V-1710-None	0	6.00:1	na	2-stg	na	na	Integral	na	Port Injection	2.36:1#60
V-1710-G11	V-1710-None	0	6.00:1	na	2-stg	na	na	Integral	na	Port Injection	2.36:1#60
V-1710-G12	V-1710-None	0	6.00:1	na	2-stg	na	na	Integral	na	Port Injection	2.36:1#60
V-1710-H 1	V-1710-None	0	6.00:1	na	External	none	none	none	na	Port Injection	2.36:1#60
V-1710-H1	V-1710-Marine	2?	6.65:1	1515	9.50	TBD	none	none	none	PD-12B4	2.50:1
V-1710-H2	V-1710-Marine	2?	6.65:1	1175	9.50	6.00:1	none	none	none	PD-12G1	Direct
V-3420-A 1	V-3420- 1(XV)	1	7.25:1	2300	10.0	6.00:1	none	none	Turbo	PT-13B1	2.50:1#60
V-3420-A 2	V-3420- 3(XV)	1	7.25:1	2350	10.0	6.00:1	none	none	Turbo	PT-13B1	3.00:1#60
V-3420-A 3	V-3420-Comm		6.65:1	2375	10.0	9.00:1	none	none	Turbo	PT-13B1	2.50:1#60
V-3420-A 4R	V-3420-Comm	0	6.65:1	2350	10.0	6.00:1	none	none	Turbo	PT-13E1	2.50:1#60
V-3420-A 5R	V-3420-None	0	6.65:1	2400	10.0	6.00:1	none	none	Turbo	PT-13E1	3.00:1#60
V-3420-A 6R	V-3420-None	0	6.65:1	2425	10.0	6.82:1	none	none	none	PT-13E1	3.00:1#60
V-3420-A 7R	V-3420-None	0	6.65:1	2550	10.0	6.82:1	none	none	none	PT-13E1	3.80:1#60
V-3420-A 8R/L	V-3420-None	2	na	na	na	na	none	none	none	na	na
V-3420-A 8RLW	V-3420-Comm	?	na	na	na	na	none	none	none	na	na
V-3420-A 9R	V-3420-None	0	6.65:1	2400	10.0	6.82:1	none	none	none	PT-13E1	2.50:1#60
V-3420-A 9R/L	V-3420-None	0	6.65:1			Yes	none	none	Turbo	na	na
V-3420-A10R	V-3420-None	0	6.65:1	2450	10.0	9.50:1	none	none	none	PT-13E1	2.50:1#60
V-3420-A10R-"A"	V-3420-None	?	6.65:1	2500	10.0	6.40:1	9.50:1	none	none	PT-13E1	2.50:1#60
V-3420-A10R-"B"	V-3420-None	?	6.65:1	2450	10.0	8.00:1	none	none	Turbo	PT-13E1	2.50:1#60
V-3420-A10R-"C"	V-3420-None	?	6.65:1	2725	10.0	8.00:1	none	Yes	Aux	PT-13E1	2.50:1#60

Allison Model	Military Model	Number Built	Comp Ratio	Weight, lbs	S/C OD, inches	S/C Gear Lo-Ratio	S/C Gear Hi-Ratio	Aux S/C OD, inches	Aux S/C Gear Ratio	Carburetor Model	Propeller Gear Ratio & Shaft Size
V-3420-A11R	V-3420- 9	1	6.65:1	2450	10.0	6.39:1	none	none	Turbo	PT-13E1	2.50:1#60
V-3420-A12R	V-3420-None	0	6.65:1	2550	10.0	7.20:1	none	none	Turbo	PT-13E1	3.00:1#60
V-3420-A13	V-3420-	0	na	na	na	na	na	na	na	na	na
V-3420-A14R/L	V-3420-	0	na	na	na	na	na	na	na	na	na
V-3420-A15L	V-3420-None	(1)	6.65:1	2530	10.0	6.82:1	none	none	Turbo	PR-58B2	2.50:1#60
V-3420-A15R	V-3420-None	(1)	6.65:1	2530	10.0	6.90:1	none	none	Turbo	PR-58B2	2.50:1#60
V-3420-A16L	V-3420-13	4	6.65:1	2655	10.0	6.818:1	none	none	Turbo	PR-58B3	3.13:1#60
V-3420-A16R	V-3420-11	17	6.65:1	2655	10.0	6.90:1	none	none	Turbo	PR-58B2	3.13:1#60
V-3420-A16R	V-3420-11	4	6.65:1	2655	10.0	6.90:1	none	none	Turbo	PR-58B3	3.13:1#60
V-3420-A17R	V-3420-15	?	7.25:1	2630	10.0	6.90:1	none	none	Turbo	PR-58B2	3.13:1#60
V-3420-A18L	V-3420-25	4	6.65:1	2655	10.0	6.818:	none	none	Turbo	PR-58B3	3.13:1#60
V-3420-A18R	V-3420-17	?	6.65:1	2655	10.0	6.90:1	none	none	Turbo	PR-58B3	3.13:1#60
V-3420-A19R	V-3420-None	0	6.65:1	2785	10.0	6.90:1	none	12.1875	6.96:1	PR-58B2	3.13:1#60
V-3420-A20R	V-3420-21	1	6.65:1	2655	10.0	8.00:1	none	12.1875	6.96:1	PR-58B2	3.13:1#60
V-3420-A21R/L	V-3420-None	?	6.65:1	3150	10.0	6.90:1	none	12.1875	6.96:1	PR-58	3.13:1#60P
V-3420-A22L	V-3420-None	?	6.65:1	2695	10.0	7.29:1	none	none	none	PR-58B3	3.13:1#60A
V-3420-A22R	V-3420-None	7?	6.65:1	2695	10.0	7.26:1	none	none	none	PR-58B3	3.13:1#60A
V-3420-A22R/L	V-3420-None	7	6.65:1	2655	10.0	6.90:1	none	none	Turbo	PR-58B3	3.13:1#60
V-3420-A23R/L	V-3420-27	1	6.65:1	3100	10.0	8.00:1	none	none	none	PR-58B3	3.23:1#60 P
V-3420-A24R	V-3420-31	0	6.65:1	2695	10.0	7.26:1	none	none	Turbo	PR-58B	3.13:1#60A
V-3420-A24R	V-3420-Comm	0	6.65:1	2695	10.0	7.26:1	none	none	Turbo	PR-58B	3.13:1#60A
V-3420-A25	V-3420-Comm	0	6.00:1	2750	10.0	7.26:1	Turbo	none	Turbo	Carb	3.13:1#60
V-3420-B 1	V-3420-None	?	7.25:1	2475	10.0	6.00:1	none	none	Turbo	PT-13B1	2.50:1#60
V-3420-B 2	V-3420-None	0	6.65:1	2300	10.0	6.82:1	none	Yes	Yes	PT-13E1	1.5:1& 2.25:1
V-3420-B 3	V-3420-	0	na								na
V-3420-B 4	V-3420- 5	4	6.65:1	2955	10.0	6.82:1	none	none	Turbo	PR-58B1	2.458:1#60P
V-3420-B 5	V-3420- 7	1	6.65:1	3640	10.0	6.82:1	none	none	Turbo	PR-58B2	2.50:1#60P
V-3420-B 6	V-3420-	0	na	na	na	na	na	na	na	na	na
V-3420-B 7	V-3420-None	0	6.65:1	2955	10.0	6.82:1	none	none	Turbo	PT-13	2.59:1#50F
V-3420-B 8	V-3420-19	12	6.65:1	3175	10.0	6.82:1	none	12.1875	6.96:1	PR-58B2	2.458:1#50F
V-3420-B 9	V-3420-None	0	6.65:1	2975	10.0	6.82:1	none	12.1875	6.96:1	PR-58	2.458:1#50F
V-3420-B10	V-3420-23	101	6.65:1	3275	10.0	6.82:1	none	12.1875	7.38	PR-58B3	2.458:1#50F
V-3420-B11	V-3420-29	2	6.65:1	3275	10.0	6.82:1	none	12.1875	7.84:1	PR-58B5	2.458:1#50F
V-3420-B12	V-3420-None	0	na	3191	10.0	6.82:1	none	none	7.84:1	PR-58	2.458:1#50F
V-3420-C 1	V-3420-None	0	6.00:1	2850	10.0	7.26:1	Turbo	none	Turbo	Carb	3.13:1*
X-3420-1	X-3420- 1	0	8.50:1	2160	11.0	7.00:1	none	none	none	MC-12 Fuel Inj	2.00:1#50
X-3420-3	X-3420- 3	0	8.50:1	2200	11.0	7.00:1	none	none	none	MC-12 Fuel Inj	2.00:1
DV-6840-A1	DV-6840-None	0	6.65:1	6750	10.0	6.82:1	none	none	Turbo	2 ea PT-13E1	___:1#60P

Appendix 4: Engine Performance by Military Model

Military Model	Allison Model	Takeoff bhp	Takeoff rpm	Takeoff MAP, inHgA	Military Rating, bhp/rpm/Alt,ft/MAP	War Emergency Rating, bhp/rpm/Alt,ft/MAP	Octane or Grade	Coolant Mixture
V-1710- 1(XV)	V-1710-A 2	800	2400	43.5	1010/2600/SL/43.5	None	92	97% EthyleneGlycol
V-1710- 2	GV-1710-A1 Build 1	650	2400	na	None	None	80	97% EthyleneGlycol
V-1710- 2	GV-1710-A1 Build 2	750	2400	34	None	None	87	97% EthyleneGlycol
V-1710- 3(XV)	V-1710-C 1 Bld 1	1000	2645	42.4	1000/2650/SL	None	92	97% EthyleneGlycol
V-1710- 3(XV)	V-1710-C 1 Bld 2	1000	2650	41.35	1000/2650/SL	None	92	97% EthyleneGlycol
V-1710- 3(XV)	V-1710-C 1 Bld 3	1000	2650	42.2	1000/2650/SL	None	92	97% EthyleneGlycol
V-1710- 3(XV)	V-1710-C 1 Bld 4	1000	2650	43.0	1000/2650/SL	None	92	97% EthyleneGlycol
V-1710- 4	V-1710-B 2R	690	2400	-2.26	None	None	83	97% EthyleneGlycol
V-1710- 4(XV)	V-1710-B 1R	690	2400	-2.26	None	None	83	97% EthyleneGlycol
V-1710- 5(YV)	V-1710-C 2	1000	2600	44.0	1000/2600/SL/44.0	na	100	97% EthyleneGlycol
V-1710- 5(YV)	V-1710-C 3	1000	2600	44.0	1000/2600/SL/44.0	na	100	97% EthyleneGlycol
V-1710- 6	V-1710-E10	1150	3000	na	1150/3000/12,000/__	None	100	97% EthyleneGlycol
V-1710- 6(XV)	V-1710-E 1	1150	3000	46.0	1150/3000/10,000/	None	100	97% Ethylene Glycol
V-1710- 7(XV)	V-1710-C 2	1066	2600	44.0	1000/2600/SL/44	na	92	97% EthyleneGlycol
V-1710- 7(XV)	V-1710-C 3	1064	2600	40.35	1000/2600/SL/41.0	1123/2800/SL/42.7	92	97% EthyleneGlycol
V-1710- 7(YV)	V-1710-C 4	1046	2600	40.68	1000/2600/SL/41.0	1123/2800/SL/42.7	92	97% EthyleneGlycol
V-1710- 7(YV)	V-1710-C 6	1000	2600	40.0	1000/2600/SL/40.0	None	92	97% EthyleneGlycol
V-1710- 9(YV)	V-1710-D 1	1000	2600	40.46	1000/2600/SL/40.46	None	92	97% EthyleneGlycol
V-1710- 9(YV)	V-1710-D 1	1250	2800	48.6	1250/2800/Turbo/48.6	None	100	97% EthyleneGlycol
V-1710- 9(YV)	V-1710-D 1	1250	2800	na	1100/2600/	1250/2800/	100	97% EthyleneGlycol
V-1710- 11	V-1710-C 8	1150	2950	37.5	1000/2770/25,000/33.4	1150/2950/25000/37.5	91	Prestone w/4%H2O
V-1710- 11(XV)	V-1710-C 7	1150	2950	37.5	1000/2770/25,000/33.4	1150/2950/25000/37.5	91	Prestone w/4%H2O
V-1710- 13(XV)	V-1710-D 1	1150	2950	38.9	1150/2950/25,000/38.9	none	92	97% EthyleneGlycol
V-1710- 15(XV)	V-1710-C 9	1150	2950	38.4	1150/2950/25,000/38.4	None	92	97% EthyleneGlycol
V-1710- 17(XV)	V-1710-E 2	1150	2950	37.7	1150/2950/25,000/41.6	None	92	97% Ethylene Glycol
V-1710- 17	V-1710-E 3	1150	2950	41.6	1150/2950/25,000/41.6	None	91	97% Ethylene Glycol
V-1710- 17(XV)	V-1710-D 3(A)	1150	2950	37.2	1150/2950/20,000/37.2	None	92	Prestone w/4%H2O
V-1710- 19(XV)	V-1710-C13	1060	2950	42.5	1000/2950/13,500/41.2	1150/2950/10,000/41.2	100	97% EthyleneGlycol
V-1710- 21	V-1710-C10	1150	2950	37.7	1000/2770/25,000/33.4	1150/2950/SL/37.7	92	Prestone w/4%H2O
V-1710- 23	V-1710-D 2	1150	2950	37.7	1150/2950/25,000/37.5	None	92	Prestone w/4%H2O
V-1710- 23	V-1710-D 2	1150	2950	38.9	1150/2950/25,000/	none	92	97% EthyleneGlycol
V-1710- 25	V-1710-F 1R	1150	3000	na	1150/3000/12,000/44.6	None	100	97% Ethylene Glycol
V-1710- 27	V-1710-F 2R	1150	3000	39.4	1150/3000/25,000/39.4	None	92	97% Ethylene Glycol
V-1710- 29	V-1710-F 2L	1150	3000	39.4	1150/3000/25000/39.4	None	92	97% Ethylene Glycol
V-1710- 31	V-1710-E 2A	1150	2950	41.6	1150/3000/12,000/	None	91	97% Ethylene Glycol
V-1710- 33	V-1710-C15	1040	3000	42.0	1040/3000/14,300/41.0?	1090/3000/13,200/38.9	100	97% Ethylene Glycol
V-1710- 35	V-1710-E 4	1150	3000	45.5	1150/3000/12,000/42.0	1490/3000/ 4,300/56.0	100	97% Ethylene Glycol
V-1710- 37	V-1710-E 5	1090	3000	na	1090/3000/13,300/40.5	None	100	97% Ethylene Glycol
V-1710- 39	V-1710-F 3R	1150	3000	45.5	1150/3000/11,700/44.6	1490/3000/ 4,300/56.0*	100	97% Ethylene Glycol
V-1710- 41	V-1710-D 2A	1090	3000	na	1090/3000/13,200/	na	100	97% EthyleneGlycol
V-1710- 43	V-1710-F 8	1150	3000	na	1150/3000/12,000/	na	100	30% Glycol-70% H2O
V-1710- 45	V-1710-F 7R	1325	3000	51.5	1150/3000/21,000/49.7	None	100	30% Glycol-70% H2O
V-1710- 47	V-1710-E 9	1325	3000	54.0	1150/3000/21,000/49.7	None	100/125	30% Glycol-70% H2O
V-1710- 49	V-1710-F 5R	1325	3000	47.0	1325/3000/25,000/47.0	None	100/125	30% Glycol-70% H2O
V-1710- 51	V-1710-F10R	1325	3000	47.5	1325/3000/25,000/47.0	1450/3000/SL/51.0	100/125?	30% Glycol-70% H2O
V-1710- 53	V-1710-F 5L	1325	3000	47	1325/3000/25,000/47.0	None	100/125	30% Glycol-70% H2O
V-1710- 55	V-1710-F10L	1325	3000	47.5	1325/3000/25,000/47.0	1450/3000/SL/51.0	100/125	30% Glycol-70% H2O
V-1710- 57	V-1710-F11R	1325	3000	na	1150/3000/16,000/HiBlo	1225/3000/5000/	100/125	30% Glycol-70% H2O
V-1710- 59	V-1710-E12R	1100	2800	46.5	1100/3000/13,800/44.2	None	100/125	97% EthyleneGlycol
V-1710- 61	V-1710-F14R	1100	2800	46.5	1100/3000/13,800/44.2	None	100/125	97% EthyleneGlycol
V-1710- 63	V-1710-E 6	1325	3000	51.0	1150/3000/11,800/42.0	1580/3000/ 2,500/60.0	100/125	97% Ethylene Glycol
V-1710- 65	V-1710-E16	1325	3000	na	1150/3000/25,800/	NA	100/130	30%Ethylene Glycol?
V-1710- 67	V-1710-E 8	1150	3000	45.5	1150/3000/11,800/42.0	na	100/125	97% Ethylene Glycol
V-1710- 69	V-1710-F18R	1325	3000	na	1325/3000/25,000/	na	100/125	30% Glycol-70% H2O
V-1710- 71	V-1710-F19R	1325	3000	na	1325/3000/25,000/	na	100/125	30% Glycol-70% H2O

Military Model	Allison Model	Takeoff bhp	Takeoff rpm	Takeoff MAP, inHgA	Military Rating, bhp/rpm/Alt,ft/MAP	War Emergency Rating, bhp/rpm/Alt,ft/MAP	Octane or Grade	Coolant Mixture
V-1710- 73	V-1710-F 4R	1325	3000	51.0	1150/3000/12,000/42.0	1580/3000/2,500/60.0	100/125	30% Glycol-70% H2O
V-1710- 75	V-1710-F15R	1425	3000	54.0	1425/3000/27,000/54.0	1600/3000/ /	100/125	30% Glycol-70% H2O
V-1710- 77	V-1710-F15L	1425	3000	54.0	1425/3000/27,000/54.0	1600/3000/ /	100/125	30% Glycol-70% H2O
V-1710- 79	V-1710-F 4L	1150	3000	na	1150/3000/11,800/	na	100/130	30% Glycol-70% H2O
V-1710- 81	V-1710-F20R	1200	3000	51.5	1125/3000/15,500/44.5	1410/3000/9,500/57.0	100/130	30% Glycol-70% H2O
V-1710- 81A	V-1710-F20R	1200	3000	51.5	na	na	100/130	30% Glycol-70% H2O
V-1710- 83	V-1710-E18	1200	3000	51.5	1125/3000/15,500/44.5	1410/3000/ 9,500/57.0	100/130	97% EthyleneGlycol
V-1710- 85	V-1710-E19	1200	3000	51.5	1125/3000/15,500/44.5	1410/3000/ 9,500/57.0	100/130	97% EthyleneGlycol
V-1710- 87	V-1710-F21R	1325	3000	47.0	1325/3000/ 2,500/47.0	1500/3000/ 5,400/52	100/130	30% Glycol-70% H2O
V-1710- 89	V-1710-F17R	1425	3000	54.0	1425/3000/24,900/54.0	1600/3000/10000/60.0	100/130	30% Glycol-70% H2O
V-1710- 89A	V-1710-F17R	1425	3000	54.0	1425/3000/24,900/54.0	1600/3000/10000/60.0	100/130	30% Glycol-70% H2O
V-1710- 91	V-1710-F17L	1425	3000	54.0	1425/3000/24,900/54.0	1600/3000/10,000/60.0	100/130	30% Glycol-70% H2O
V-1710- 91A	V-1710-F17L	1425	3000	54.0	1425/3000/24,900/54.0	1600/3000/10,000/60.0	100/130	30% Glycol-70% H2O
V-1710- 93	V-1710-E11	1325	3000	54.0	1150/3000/22,400/51.5	1825/3000/SL/75.0Wet	100/130	97% EthyleneGlycol
V-1710- 95	V-1710-F23R	1275	3000	54.3	1125/3000/15,500/44.5	None	100/130	30% Glycol-70% H2O
V-1710- 97	V-1710-G 1R	1725	3400	68.5	1725/3400/25,000/68.5	1725/3400/25,000/68.5	140	30% Glycol-70% H2O
V-1710- 97A	V-1710-G 1RA	1725	3400	68.5	1725/3400/25,000/68.5	1725/3400/25,000/68.5	140	30% Glycol-70% H2O
V-1710- 99	V-1710-F26R	1200	3000	51.5	1125/3000/15,500/44.5	1360/3000/SL/57.0	100/130	30% Glycol-70% H2O
V-1710- 99A	V-1710-F26R	na	3000	na	1125/3000/	1360/3000/SL	100/130	30% Glycol-70% H2O
V-1710-101	V-1710-F27R	1325	3000	na	1150/3000/22,400/~50.0	1500/3200/ 6,000/60.0	100/130	95% Glycol
V-1710-103 Unit#1	V-1710-E23 Unit#1	1325	3000	54.0	1200/3000/22,400/~51.0	1820/3200/SL/__	100/130	30% Glycol-70% H2O
V-1710-103 Unit#2	V-1710-E23 Unit#2	1325	3000	54.0	1200/3000/22,400/~51.0	1820/3200/SL/__	100/130	30% Glycol-70% H2O
V-1710-105	V-1710-F29R	1500	3200	57.0	1500/3200/30,000/57.0	1650/3200/SL/__	100/130	95% Glycol
V-1710-107	V-1710-F29L	1500	3200	57.0	1500/3200/30,000/57.0	1650/3200/SL/__	100/130	95% Glycol
V-1710-109	V-1710-E22	1425	3000	59.3	1100/3000/30,000/50.0	1835/3200/SL/76.0Wet	100/130	100%Ethylene Glycol
V-1710-109A	V-1710-E22A	1425	3000	59.3	1100/3000/30,000/50.0	1835/3200/SL/76.0Wet	100/130	100%Ethylene Glycol
V-1710-111	V-1710-F30R	1500	3000	54	1425/3000/29,000/54.0	1600/3000/28,700/60.0	100/130	30% Glycol-70% H2O
V-1710-113	V-1710-F30L	1500	3000	54	1425/3000/29,000/54.0	1600/3000/28,700/60.0	100/130	30% Glycol-70% H2O
V-1710-115	V-1710-F31R	1200	3000	na	1125/3000/15,500/44.5	1360/3000/SL/57.0	100/130	30% Glycol-70% H2O
V-1710-115A	V-1710-F31RA	1200	3000	51.5	1125/3000/15,500/44.5	1360/3000/SL/57.0	100/130	30% Glycol-70% H2O
V-1710-117	V-1710-E21	1325	3000	55.0	1100/3000/25,000/51.5	1800/3000/24,000/76.0	100/130	97%Ethylene Glycol
V-1710-119	V-1710-F32R	1500	3200	62.5	1200/3200/30,000/52.0	1900/3200/SL/78.0	100/130	30% Eth Gly/70% H2O
V-1710-121	V-1710-F28R	1425	3000	na	1100/3200/28,000/__Dry	1700/3200/26,000/__Wet	100/130	30% Glycol-70% H2O
V-1710-123	V-1710-F28L	1425	3000	na	1100/3200/28,000/__Dry	1700/3200/26,000/__Wet	100/130	30% Glycol-70% H2O
V-1710-125 Unit#1	V-1710-E23A Unit#1	1675	3200	70Wet	1100/3000/24,000/51.0	1900/ 3200 /SL/75.0Wet	100/130	30% Glycol-70% H2O
V-1710-125 Unit#2	V-1710-E23A Unit#2	1675	3200	70Wet	1100/3000/24,000/51.0	1900/ 3200 /SL/75.0Wet	100/130	30% Glycol-70% H2O
V-1710-127	V-1710-E27	2825	3000	100	2430/3200/17,000/85.0	2980/3200/11,000/100.0	115/145	30% Glycol-70% H2O
V-1710-129L	V-1710-E23BL	1675	3200	70.0	1100/3000/29,000/49.5	1900/ 3200 /SL/75.0Wet	100/130	30% Glycol-70% H2O
V-1710-129R	V-1710-E23BR	1675	3200	70.0	1100/3000/29,000/49.5	1900/ 3200 /SL/75.0Wet	100/130	30% Glycol-70% H2O
V-1710-131	V-1710-G 3R	1600	3200	61.7	1220/3000/15,500/52Hi	None	100/130	30% Glycol-70% H2O
V-1710-133	V-1710-E30	1500	3000	65.5	1150/3000/27,500/58.0	1800/ 3200 /SL/75.0	100/130	30% Glycol-70% H2O
V-1710-135	V-1710-E31	1200	3000	na	1125/3000/15,000/	1360/3000/SL/	100/130	na
V-1710-137L	V-1710-E23CL	1790	3200	75.0	1150/3000/27,500/57.0	None	100/130	30% Glycol-70% H2O
V-1710-137R	V-1710-E23CR	1790	3200	75.0	1150/3000/27,500/57.0	None	100/130	30% Glycol-70% H2O
V-1710-139	V-1710-F33R	1500	3000	65.5	1150/3000/27,500/57.0	None	100/130	30% Glycol-70% H2O
V-1710-141	V-1710-F33L	1500	3000	65.5	1150/3000/27,500/57.0	None	100/130	30% Glycol-70% H2O
V-1710-143	V-1710-F36R	1850	3200	na	1250/3200/32,600/Dry	2200/3200/12,700/Wet	115/145	30% Glycol-70% H2O
V-1710-143	V-1710-G 6R	1600	3200	74	1250/3200/32,700/68	2250/3200//100.0 wet	115/145	30% Glycol-70% H2O
V-1710-145	V-1710-F36L	1850	3200	na	1250/3200/32,600/Dry	2200/3200/12,700/Wet	115/145	30% Glycol-70% H2O
V-1710-145	V-1710-G 6L	1600	3200	74	1250/3200/32,700/68	2250/3200//100.0 wet	115/145	30% Glycol-70% H2O
V-1710-147	V-1710-G 9R	1600	3200	74	1250/3200/32,700/68	2250/3200//100.0 wet	115/145	30% Glycol-70% H2O
V-1710-149	V-1710-G 9L	1600	3200	74	1250/3200/32,700/68	2250/3200//100.0 wet	115/145	30% Glycol-70% H2O
V-1710-Comm	V-1710-C 6	1000	2600	35.0	1000/2600/SL/35.0	NA	92	97% EthyleneGlycol
V-1710-Comm	V-1710-G 2RA	1600	3000	na	1120/2700/15,500/__Hi	None	100/130	30% Glycol-70% H2O
V-1710-Comm	V-1710-G 3R	1600	3200	61.7	1215/3000/18,000/__Hi	1350/3000/ 6,500/__Lo	100/130	30% Glycol-70% H2O
V-1710-Comm	V-1710-G 4L	1830	3200	Wet	1225/3000/17,000/__Hi	None	100/130	30% Glycol-70% H2O
V-1710-Comm	V-1710-G 4R	1830	3200	Wet	1225/3000/17,000/__Hi	None	100/130	30% Glycol-70% H2O
V-1710-Comm	V-1710-G 7R/L	na	na	na	na	na	na	na
V-1710-Marine	V-1710-H1	1150	2950	TBD	1150/2950/SL/	None	92	97% Ethylene Glycol
V-1710-Marine	V-1710-H2	1000	2600	TBD	1000/2600/SL/	None	na	97% Ethylene Glycol
V-1710-None	V-1710-C 5	1000	2600	42.4	na	na	87	97% EthyleneGlycol

Military Model	Allison Model	Takeoff bhp	Takeoff rpm	Takeoff MAP, inHgA	Military Rating, bhp/rpm/Alt,ft/MAP	War Emergency Rating, bhp/rpm/Alt,ft/MAP	Octane or Grade	Coolant Mixture
V-1710-None	V-1710-C11	1150	2950	39.0	1175/2950/ 9,000/39.0	None	100	97% EthyleneGlycol
V-1710-None	V-1710-C12	1130	2950	na	1150/2950/ 4,000/	None	92	97% EthyleneGlycol
V-1710-None	V-1710-C14	1150	2950	37.7	1150/2950/20,000/37.7	None	92	97% EthyleneGlycol
V-1710-None	V-1710-C15A	1040	3000	41.0	1040/3000/14,300/	1090/3000/13200/38.9	100	97% Ethylene Glycol
V-1710-None	V-1710-D 3	1150	2950	37.2	1150/2950/20,000/37.2	None	92	Prestone w/4%H2O
V-1710-None	V-1710-D 3(B)	1150	2950	37.2	1150/2950/20,000/37.2	None	92	Prestone w/4%H2O
V-1710-None	V-1710-D 4	1090	3000	42.9	1090/3000/13,200/42.9	None	100	97% Ethylene Glycol
V-1710-None	V-1710-D 5	1150	3000	na	1150/3000/25,000/	None	100	97% Ethylene Glycol
V-1710-None	V-1710-E 7	1000	2600	TBD	1000/2600/SL/	None	na	97% Ethylene Glycol
V-1710-None	V-1710-E13	1300	3000	na	1150/3000/25,000/	None	na	na
V-1710-None	V-1710-E14	1400	3000	na	1400/3000/32,000/__	None	na	na
V-1710-None	V-1710-E15	1325	3000	na	1325/3000/30,000/__	None	na	na
V-1710-None	V-1710-E17	1325	3000	na	1125/3000/24,000/	None	na	na
V-1710-None	V-1710-E20	1425	3000	53.5	1425/3000/27,000/53.5	na	na	na
V-1710-None	V-1710-E24 Unit#1	1500	3200	Dry	1325/3200/20,000/__	1800/3200/SL/75Wet	100/130	30% Glycol-70% H2O
V-1710-None	V-1710-E24 Unit#2	1500	3200	Dry	1325/3200/20,000/__	1800/3200/SL/75Wet	100/130	30% Glycol-70% H2O
V-1710-None	V-1710-E25	1500	3200	62.5	1200/3200/30,000/52.0	1900/3200/SL/78.0	100/130	30% Glycol-70% H2O
V-1710-None	V-1710-E26	na	na	na	na	na	na	30% Glycol-70% H2O
V-1710-None	V-1710-E28	na	na	na	na	na	na	na
V-1710-None	V-1710-E29	1620?	na	na	na	na	na	na
V-1710-None	V-1710-E32	na	na	na	na	na	na	na
V-1710-None	V-1710-F 4	1200	3000	na	1200/3000/25,000/	None	100	97% EthyleneGlycol
V-1710-None	V-1710-F 6R	1100	3000	na	1125/3000/15,000/__	None	na	na
V-1710-None	V-1710-F 9R	1325	3000	na	1125/3000/18,000/__*	None	100	na
V-1710-None	V-1710-F12	na	na	na	na	na	na	na
V-1710-None	V-1710-F13L	1325	3000	47	1325/3000/25,000/47.0	na	100/125	30% Glycol-70% H2O
V-1710-None	V-1710-F13R	1325	3000	47	1325/3000/25,000/47.0	na	100/125	30% Glycol-70% H2O
V-1710-None	V-1710-F16	1425	3000	na	1300/3000/ 2,600/__-Lo	None	na	30% Glycol-70% H2O
V-1710-None	V-1710-F22R	na	na	na	na	na	na	na
V-1710-None	V-1710-F24R	na	na	na	na	na	na	na
V-1710-None	V-1710-F25R	1510	na	61.0	1150/3000/22,400/50.0	1360/3000/17,250/60.0	130	na
V-1710-None	V-1710-F34	na	na	na	na	na	na	na
V-1710-None	V-1710-F35R	1850	3200	na	1250/3200/30,200/Dry	2240/3200/10,700/Wet	115/145	30% Glycol-70% H2O
V-1710-None	V-1710-F37	2000	3200	Wet	1600/3200/30,000/Dry	2050/3200/23,500/Wet	115/145	30% Glycol-70% H2O
V-1710-None	V-1710-G 2R/L	1600	3200	na	1200/2700/ 6,000/__	None	100/130	30% Glycol-70% H2O
V-1710-None	V-1710-G 5	1425	3000	na	1125/3000/15,000/	na	100/130	30% Glycol-70% H2O
V-1710-None	V-1710-G 6	na	3200	na	na	2250/3200/SL/100	na	na
V-1710-None	V-1710-G 6A	1600	3200	74	1250/3200/32700/	/3200/ /100.0 wet	115/145	30% Glycol-70% H2O
V-1710-None	V-1710-G 8R/L		3200	na	1250/3200/32700/	/3200/ /100.0 wet	115/145	30% Glycol-70% H2O
V-1710-None	V-1710-G10L	2000	3200	na	2250/3200/28,500/Dry	2550/3200/20,000/Wet	115/145	30% Glycol-70% H2O
V-1710-None	V-1710-G10R	2000	3200	na	2250/3200/28,500/Dry	2550/3200/20,000/Wet	115/145	30% Glycol-70% H2O
V-1710-None	V-1710-G11	2000	3200	na	2250/3200/28,500/Dry	2550/3200/20,000/Wet	115/145	30% Glycol-70% H2O
V-1710-None	V-1710-G12	2000	3200	na	2250/3200/28,500/Dry	2550/3200/20,000/Wet	115/145	30% Glycol-70% H2O
V-1710-None	V-1710-H 1	2000+	3200	na	2250+/3200/28,500/Dry	2550+/3200/20,000/Wet	115/145	30% Glycol-70% H2O
V-1710-Racer	V-1710-E31	2000	3200	82	na	2200/3400/SL/85.0Wet	145	na
V-3420- 1(XV)	V-3420-A 1	2300	3000	na	2300/3000/25,000/__	None	100	97% EthyleneGlycol
V-3420- 3(XV)	V-3420-A 2	2300	2950	36.0	2300/2950/25,000/36.0	None	100	97% EthyleneGlycol
V-3420- 5	V-3420-B 4	2600	3000	45.8	2600/3000/25,000/45.8	None	130	97% EthyleneGlycol
V-3420- 7	V-3420-B 5	2600	3000	na	2600/3000/25,000/45.8	None	130	97% EthyleneGlycol
V-3420- 9	V-3420-A11R	2300	3000	na	2300/3000/10,000/__	None	100	97% EthyleneGlycol
V-3420-11	V-3420-A16R	2600	3000	46.2	2600/3000/25,000/46.2	3000/3000/28,000/51.5	100/130	97% EthyleneGlycol
V-3420-13	V-3420-A16L	2600	3000	45.8	2600/3000/25,000/46.2	3000/3000/28000/51.5	100	97% EthyleneGlycol
V-3420-15	V-3420-A17R	2600	3000	na	2600/3000/25,000/	None	100	97% EthyleneGlycol
V-3420-17	V-3420-A18R	2600	3000	46.2	2600/3000/25,000/45.8	3000/3000/28000/51.5	130	97% EthyleneGlycol
V-3420-19	V-3420-B 8	2600	3000	na	2300/3000/20,000/	____/3000/SL/60.0	100/130	97% EthyleneGlycol
V-3420-21	V-3420-A20R	2600	3000	na	2300/3000/10,000/	None	130	97% EthyleneGlycol
V-3420-23	V-3420-B10	2600	3000	50.5	2300/3000/20,000/48.8	2885/3000/SL/57.5 inHg	na	na
V-3420-25	V-3420-A18L	2600	3000	46.2	2600/3000/25000/46.2	3000/3000/28000/51.5	na	na
V-3420-27	V-3420-A23R/L	2600	3000	na	2300/3000/10,000/__	None	100/130	na
V-3420-29	V-3420-B11	2850	3000	56.5	2300/3000/25,000/50.5	None	130	30% Glycol-70% H2O
V-3420-Comm	V-3420-A 3	2300	3000	na	2300/3000/12,000/__	None	100	97% EthyleneGlycol

Military Model	Allison Model	Takeoff bhp	Takeoff rpm	Takeoff MAP, inHgA	Military Rating, bhp/rpm/Alt,ft/MAP	War Emergency Rating, bhp/rpm/Alt,ft/MAP	Octane or Grade	Coolant Mixture
V-3420-Comm	V-3420-A 4R	2300	3000	na	2300/3000/25,000/__	None	100	97% EthyleneGlycol
V-3420-Comm	V-3420-A 8RLW	na	na	na	na	na	na	na
V-3420-Comm	V-3420-A24R	3000	3000	52.0	3000/3000/30,000/52.0	None	130	30% Glycol-70% H2O
V-3420-Comm	V-3420-A25	3600	3200	na	3300/3200/___/__ Dry	3600/3200/___/__ Wet*	100/130	30% Glycol-70% H2O
V-3420-None	V-3420-A 5R	2300	3000	na	2300/3000/25,000/__	None	100	97% EthyleneGlycol
V-3420-None	V-3420-A 6R	2300	3000	na	2300/3000/ 5,000/__	None	100	97% EthyleneGlycol
V-3420-None	V-3420-A 7R	2300	3000	na	2300/3000/ 5,000/__	None	100	97% EthyleneGlycol
V-3420-None	V-3420-A 9R/L	2500	3000	na	2500/3000/25,000/__	None	100	97% EthyleneGlycol
V-3420-None	V-3420-A10R	2100	3000	na	2200/3000/15,000/__	None	100+	97% EthyleneGlycol
V-3420-None	V-3420-A10R-"A"	2300	3000	na	2300/3000/ 4,000/__Lo	None	100+	97% EthyleneGlycol
V-3420-None	V-3420-A10R-"B"	2300	3000	na	2300/3000/10,000/__	None	100+	97% Ethylene Glycol
V-3420-None	V-3420-A10R-"C"	2300	3000	na	2300/3000/ 10,000/__	None	100+	97% Ethylene Glycol
V-3420-None	V-3420-A12R	2600	3000	na	2600/3000/25,000/__	None	100	97% Ethylene Glycol
V-3420-None	V-3420-A13	na	na	na	na	na	na	na
V-3420-None	V-3420-A14R/L	na	na	na	na	na	na	na
V-3420-None	V-3420-A15L	2600	3000	na	2600/3000/25,000/__	None	100	97% Ethylene Glycol
V-3420-None	V-3420-A15R	2600	3000	na	2600/3000/25,000/__	None	100	97% Ethylene Glycol
V-3420-None	V-3420-A19R	2600	3000	na	2300/3000/20,000/	None	130	97% Ethylene Glycol
V-3420-None	V-3420-A21R/L	2850	3000	na	2300/3000/20,000/	None	130	97% EthyleneGlycol
V-3420-None	V-3420-A22L	2850	3000	na	2850/3000/ 4,300/	None	130	30% Glycol-70% H2O
V-3420-None	V-3420-A22R	2850	3000	na	2850/3000/ 4,300/	None	130	30% Glycol-70% H2O
V-3420-None	V-3420-A22R/L	2600	3000	46.2	2600/3000/25000/46.2	3000/3000/28000/51.5	100/130	30% Glycol-70% H2O
V-3420-None	V-3420-B 1	2300	3000	na	2300/3000/25,000/__	None	100	97% EthyleneGlycol
V-3420-None	V-3420-B 2	2600	3000	na	2400/3000/18,000/__	None	100+	97% EthyleneGlycol
V-3420-None	V-3420-B 3	na	na	na	na	na	na	na
V-3420-None	V-3420-B 6	na	na	na	na	na	na	na
V-3420-None	V-3420-B 7	2600	3000	na	2600/3000/25,000/	None	100	97% EthyleneGlycol
V-3420-None	V-3420-B 9	2600	3000	na	2300/3000/20,000/	None	130	97% EthyleneGlycol
V-3420-None	V-3420-B12	2850	3000	na	2850&2400/3000/SL&25k'	3150&3050/3000/SL&18.4	na	na
V-3420-None	V-3420-C 1	4000	3200	na	3600/3200/30000/__ Dry	4800/3200/___/__Wet**	115/145	30% Glycol-70%H2O
X-3420- 1	X-3420-1	1600	2400	36.0	1600/2400/SL/36.0	None	X-100	na
X-3420- 3	X-3420-3	1600	2400	36.0	1600/2400/SL/36.0	None	X-100	na
DV-6840-None	DV-6840-A1	5000	3000	na	5000/3000/25,000/__	None	100+	97% EthyleneGlycol

Appendix 5: Allison Engineering Company Specifications

These specifications were used by Allison to control the configuration, features and accessories specific to any given engine. The Air Corps was not particularity interested in them, other than for reference as they relied upon their own contracts for engine specifics. Furthermore, the Air Corps only assigned a model number when they had decided to purchase an engine. Even so, some designations were applied and no engine was ever delivered, usually because events overtook the specific need or development and the effort was applied to a later model. These specifications were used to communicate with the aircraft design and engineering firms, providing them with the particulars and performance necessary to the aircraft design process. Only if the Air Corps purchased the aircraft, and needed the engine would a contract be developed and the appropriate model numbers be assigned.

Allison assigned the 100-series to the V-1710, and the 200-series to the four bank 3420 engines. When over one hundred models of the V-1710 were proposed they began taking the next number available in the 200 series. Not all of the hundred series were used for V-1710, exceptions including the Model 501 turboprop engine under specification 196.

Some of the early Specifications can be a little confusing:

AEC Spec 100 None
AEC Spec 101 XV-1710-7(C2) Engine
AEC Spec 102 XV-1710-5(C2) Engine
AEC Spec 103 V-1710-D1 Engine for XFM-1, 2-7-36.
AEC Spec 104 V-1710-C4, YV-1710-7 Engine completing Type Test, 4-37.
AEC Spec 105 V-1710-C7(-11) Engine as of 12-23-36.
AEC Spec 105-A V-1710-C7 Engine, January 25, 1937.
AEC Spec 105-B V-1710-C7 Engine, May 1937.
AEC Spec 105-F V-1710-C8 Engine, Revised 9-29-37.
AEC Spec 105-L V-1710-C10 Engine

It can be seen that Allison viewed the -C7/-C8/-C10 as being evolution's of the same engine. The C-8 had different valve timing and a propeller feathering capability, but otherwise was the same as the C-7. The C-10 had internal improvements and different part number cylinder blocks, but all had the same higher compression ratio. So the changes could be either minor or major, and retain the same Spec number, though with an appropriate suffix. Another factor that comes up, is that the specifications were not routinely updated, once production of a given model was underway. Updates usually occurred only when a matter of contract significance was involved. An example being that when WER ratings were introduced for engines then in production, the Specifications do not mention the fact or show the revised performance.

Military Model numbers were assigned only when an actual contract to procure an engine was issued. Normally this would be long after Allison would have assigned their Manufacturers Model number and descriptive specification. The Military was not particularly interested in either, as their contracts provided all of the specification they needed.

Allison Spec No.	Allison Model	Military Model	Year	Description and Comments
None	GV-1710-A1 Build 1	V-1710- 2	1930	Allison VG-1710, Predates use of Allison Engine Specification Numbers.
None	GV-1710-A1 Build 2	V-1710- 2	1931	Allison VG-1710, Predates use of Allison Engine Specification Numbers.
None	V-1710-A 2	V-1710- 1(XV)	1931/33	Allison Engine Specifications, NONE ASSIGNED.
None	V-1710-B 1R	V-1710- 4(XV)	1933	Reversible Model Test Engine, Predates use of Allison Engine Specifications.
None	V-1710-B 2R	V-1710- 4	1935	First two Reversible production engines, Predates use of Allison Engine Specifications.
None	V-1710-C 1 Bld 1	V-1710- 3(XV)	1934	Retroactively known as V-1710-C1. Allison Engine Specifications, NONE ASSIGNED.
None	V-1710-C 1 Bld 2	V-1710- 3(XV)	1935	Retroactively known as V-1710-C1. Allison Engine Specifications, NONE ASSIGNED.
None	V-1710-C 1 Bld 3	V-1710- 3(XV)	1935	Retroactively known as V-1710-C1. Allison Engine Specifications, NONE ASSIGNED.
None	V-1710-C 1 Bld 4	V-1710- 3(XV)	1935	Retroactively known as V-1710-C1. Allison Engine Specifications, NONE ASSIGNED.
None	V-1710-E 2A	V-1710- 31	1939	Altitude Rated Intrim configuration of E-2 as they evolved into E-5 prototypes.
101	V-1710-C 2	V-1710- 7(XV)	1935	XV-1710-7(C2), 6.0:1CR, 8.0:1 S/C, 2:1 RG, 1245 pounds, 1000/2600/SL/44", 12/10/35."
102	V-1710-C 2	V-1710- 5(YV)	1935	YV-1710-5(C?), XV-1710-7(C2) w/MC-12 Fuel Inj, 6.0:1CR, 8.0:1 S/C, 2:1 RG, 1300 #s, 1000/2600/SL/44", 12/10/35."
102	V-1710-C 3	V-1710- 5(YV)	1936	YV-1710-5(C?), XV-1710-7(C2) w/MC-12 Fuel Inj, 6.0:1CR, 8.0:1 S/C, 2:1 RG, 1300 #s, 1000/2600/SL/44", 12/10/35."
102	V-1710-D 1	V-1710- 9(YV)	1935	YV-1710-5(C?), XV-1710-7(C2) w/MC-12 Fuel Inj, 6.0:1CR, 8.0:1 S/C, 2:1 RG, 1300 #s, 1000/2600/SL/44", 12/10/35." 0."
103	V-1710-D 1	V-1710- 9(YV)	1936	YV-1710-9(D1), XV-1710-7 w/MC-12 Fuel Inj, 7.0:1CR, 8.77:1 S/C, 2:1 RG, 1300 #s, 1250/2800/SL/48.7, 2/7/36.
103-	V-1710-C 3	V-1710- 7(XV)	1936	XV-1710-7(C3), 1st w/Hazen manifolds, Attempted Type Test, NA-F7C Carb, 6.00:1CR, 8.00:1 S/C, 2:1 RG, 1245# , 1064/ 2600/SL/40.35,"
103-	V-1710-C 5	V-1710-None	1936	Contemporary of C-6 and C-7.
103-	V-1710-D 1	V-1710- 9(YV)	1936	YV-1710-D1, 6.00:1CR, 8.00:1 S/C, 2:1 RG, 1316 pounds, 1000/2600/SL/40.46," 2/19/36. @6/27/95.
104-A	V-1710-C 6	V-1710- 7(YV)	1936	V-1710-C6, 6.0:1CR, 8.00:1 S/C, 2:1 RG, 1260 #, 1000/2600/SL/40," 6/23/36 Proposed to Mil, Replaced 3-24-37.
104-B	V-1710-C 6	V-1710-Comm	1937	V-1710-C6, 6.0:1CR, 6.75:1 S/C, 2:1 RG, <1290 pounds, 1000/2600/SL/40," 3/24/37.
105-B	V-1710-C 7	V-1710- 11(XV)	1936	AEC List 11/30/94, says Spec 109, no contract issued. Likely that this was the engine specified for X-608 & X-609 aircraft.
105-E	V-1710-C 7	V-1710- 11(XV)	1936	V-1710-C7 1150 bhp engine, per AEC Spec 105-E dated as of 7-23-37, original spec dated 1-25-1937.
105-F/H	V-1710-C 8	V-1710- 11	1937	1150 bhp engine, 105-H for XP-38 per AEC Ref List 12/23/41.
105-L/K-1	V-1710-C10	V-1710- 21	1938	V-1710-C10, PD-12B4 Carb, 6.65:1CR, 6.23:1 S/C, 2:1 RG, 1310 pounds, 1150/2950/SL/37.7," 6/27/38.
106	V-1710-D 1	V-1710- 9	1936	1000 bhp @ 2600 rpm, V-1710-D1 Pusher engines, per Dwg #34405, as of 12/7/36.
107	V-1710-D 1(Ext)	V-1710- 9	1936	Spec for Prop Speed Extension Drive Assembly for V-1710-D1 Pusher engines, as of 12/7/36.
108	V-1710-C 4	V-1710- 7(YV)	1936	YV-1710-7(C4), 1st to complete Type Test 4-23-37, NA-F7C Carb, 6.00:1CR, 8.00:1 S/C, 2:1 RG, 1269# , 1000/2600/SL/ 41.0," 12/23/36.
109-C1	V-1710-D 1	V-1710- 13	1937	V-1710-13(D1), NA-F7F1 Carb, 6.65:1CR, 6.23:1 S/C, 2:1 RG, 1330 pounds, 1150/2950/SL/38.9, 2/9/38.
109-E	V-1710-D 1	V-1710- 13	1937	1150 bhp V-1710-D3 RH Pusher for Bell XFM-1, Project M-9-35, 6-25-1937 revised 12-17-37.
109-F	V-1710-D 2	V-1710- 23	1937	V-1710-13(D2), NA-F7F3 Carb, 6.65:1CR, 6.23:1 S/C, 2:1 RG, 1330 pounds, 1150/2950/SL/38.9, 2/9/38.
110	V-1710-D 3	V-1710-None	1937	V-1710-D3 Tractor w/Prop Speed Extension for Bell Model 3, NA-F7L Carb, 6.65:1CR, 6.23:1 S/C, 2:1 RG, 1335 pounds, 1150/2950/SL/37.2," 3/29/37.
110-A	V-1710-D 3(A)	V-1710-17(XV)	1937	V-1710-D3, Crank speed extension to remote 5:3 RG for Bell X-609 Model 3/4 proposals, 1280 pounds, 1150/2950/SL/37.2, 5/18/37.
110-B	V-1710-D 3(B)	V-1710-None	1937	V-1710-D3, 65.5 Crank spd extension to remote 2:1 RG for Curtiss Model 80A, 1285 pounds, 1150/2950/SL/37.2", 5/18/ 37.
111-D	V-1710-C 9	V-1710- 15	1937	V-1710-15(C9), NA-F7F3 Carb, 6.65:1CR, 6.23:1 S/C, 2:1 RG, 1305 pounds, 1150/2950/SL/38.4, 3/25/38.
112-E	V-1710-E 1	V-1710- 6(XV)	1937	XV-1710-6(E1), PD-12G1Carb, 6.65:1CR, 8.80:1 S/C, 1.80:1 RG, <1375 pounds, 1150/3000/10,000/__", 100 Oct, 10/10/ 39.
113-B	V-1710-C11	V-1710-None	1938	V-1710-C11, PD-12B1 Carb, 6.65:1CR, 8.77:1 S/C, 2:1 RG, 1310 pounds, 1150/2950/ 9,000/39, 100 Oct, 9/21/37.
114-C	V-1710-D 2	V-1710- 23	1938	V-1710-D2, LH Pusher, PD-12B6 Carb, 6.65:1CR, 6.23:1 S/C, 2:1 RG, 1350 pounds, 1150/3000/SL/37.5, 7/6/39.
114-D	V-1710-D 2A	V-1710- 41	1940	Revision of V-1710-23 to V-1710-41(D2A) for YFM-1B, Contract W535-ac-10660.
115-F	V-1710-E 2	V-1710- 17	1937	V-1710-17(E2), 6.65:1CR, 6.44:1 S/C, 1.80:1 RG, <1350 pounds, 1150/2950/25,000/37.7, 92 Oct, 10/10/39.
116	V-1710-C12	V-1710-None	1938	V-1710-C12, NA-F7F3 Carb, 6.65:1CR, 7.33:1 S/C, 2:1 RG, 1310 pounds, 1150/2950/ 4,000/__", 92 Oct, 1/21/38.
117-D	V-1710-C13	V-1710- 19(XV)	1938	V-1710-19(C13), PT-13B1 Carb, 6.65:1CR, 8.77:1 S/C, 2:1 RG, 1320 pounds, 1150/2950/10,000/42.5, 100 Oct, 4/4/39.
118-A	V-1710-H1	V-1710-Marine	1938	V-1710-H1, 6.65:1CR, TBD:1 S/C, 2.50:1 RG, 1515 pounds, 1150/2950/SL/__", 92 Oct, 10/10/39.
119-D	V-1710-F 2R	V-1710- 27	1938	V-1710-27(F2R), 6.65:1CR, 6.00:1 S/C, 2.0:1 RG, 1305 pounds, 1150/3000/25,000/__", 92 Oct, 12/5/39.
120-C	V-1710-F 1R	V-1710- 25	1938	V-1710-25(F1R), PT-13B2 Carb, 6.65:1CR, 8.80:1 S/C, 2:1 RG, 1320 pounds, 1150/3000/12,000/44.6?, 2/10/39.
121-A	V-1710-E 3	V-1710- 17	1938	V-1710-17(E3), PD-12__ Carb, 6.65:1CR, ___:1 S/C,__:1 RG, 1375 pounds, ___/3000/__"__/__"_, Sea Level rated.
121-B	V-1710-E 3	V-1710- 17	1939	Engines to be purchased on W535-ac-12553, changed to V-1710-E5's. Spec'd 12/18/39 per AEC 11/30/94.
121-E	V-1710-E 5	V-1710- 37	1939	V-1710-37(E5), PD-12G1 Carb, 6.65:1CR, 8.80:1 S/C, 2:1 RG, 1375 pounds, 1090/3000/13,300/42.9, 12/18/39.
122	V-1710-E 4	V-1710-None	1938	V-1710-E w/1.8:1 RGR and RH & LH Prop (Counter-rotating?), per Index of AEC Specs.
123-E	V-1710-F 3R	V-1710- 39	1938	V-1710-39(F3R), PD-12K2 Carb, 6.65:1CR, 8.80:1 S/C, 2:1 RG, 1310 pounds, 1150/3000/12,000/42.0, 9/20/40, Altitude Rated.
124	V-1710-F 4	V-1710-None	1938	V-1710-F4R/L, PD-12B6 Carb, 6.65:1CR, 6.44:1 S/C, 2.0:1 RG, 1290#, 1200/3000/25,000/__", 100 Oct, 12/19/39, Sea Level rated.
125-A	V-1710-C10	V-1710- 21	1938	V-1710-C10, PD-12B6 Carb, 6.65:1CR, 6.23:1 S/C, 2:1 RG, 1320 pounds, 1150/2950/SL/37.7, 12/28/38.
126-D	V-1710-C15	V-1710- 33	1938	V-1710-33(C15), PT-12E1 Carb, 6.65:1CR, 8.77:1 S/C, 2:1 RG, 1325 pounds, 1150/2950/13,200/38.9, 12/5/38.

127-A	V-1710-D 5	V-1710-None	1938	V-1710-D5, PD-12G1 Carb, 6.65:1CR, 6.23:1 S/C, 2:1 RG, 1360 pounds, 1150/2950/25,000/__", 6/1/39.
128-D	V-1710-C14	V-1710-None	1938	V-1710-C14, PD-12B6 Carb, 6.65:1CR, 6.23:1 S/C, 2:1 RG, 1325 pounds, 1150/2950/20,000/37.7, 12/28/38.
129	V-1710-E 3	V-1710-None	1939	V-1710-E Model w/1.80:1 RGR and RH Prop per Index of AEC Engine Specifications.
130-E	V-1710-E 4	V-1710- 35	1938	V-1710-35(E4), PD-12K2 Carb, 6.65:1CR, 8.80:1 S/C, 1.80:1 RG, 1375#, 1150/3000/12,000/42.0, 100 Oct, 9/29/41.
131	V-1710-D 4	V-1710-None	1939	V-1710-D4, Hydromatic Prop, PT-13E1 Carb, 6.65:1CR, 8.77:1 S/C, 2:1 RG, 1375#, 1090/3000/13,200/42.9, 6/1/39.
132-B	V-1710-F 2L	V-1710- 29	1939	V-1710-29(F2L), 6.65:1CR, 6.00:1 S/C, 2.0:1 RG, 1305 pounds, 1150/3000/25,000/__", 92 Oct, 12/5/39.
133-F	V-1710-F 5L	V-1710- 53	1939	V-1710-53(F5L), PD-12G1 Carb, 6.65:1CR, 7.48:1 S/C, 2.0:1 RG, 1345#, 1150/3000/25,000/47, 125 Oct, 9/26/41.
133-F	V-1710-F 5R	V-1710- 49	1939	V-1710-49(F5R), PD-12G1 Carb, 6.65:1CR, 7.48:1 S/C, 2.0:1 RG, 1345#, 1150/3000/25,000/47, 125 Oct, 9/26/41.
134	V-1710-E 6	V-1710- 63	1941	Improved -E4 w/higher ratings, Douglas Requirements per Index of AEC Specs, 12/23/41.
135	V-1710-E 7	V-1710-None	1939	V-1710-E7, PD-12G1 Carb, 6.65:1CR, 6.00:1 S/C, Direct Drive, 1175#, 1000/2600/SL/__", ___ Oct, 11/13/39.
135	V-1710-H2	V-1710-Marine	1939	V-1710-E7, PD-12G1 Carb, 6.65:1CR, 6.00:1 S/C, Direct Drive, 1175#, 1000/2600/SL/__", ___ Oct, 11/13/39.
136-B	V-1710-E 8	V-1710- 67	1941	As listed in Index of AEC Engine Specifications, 12/23/41.
137-G	V-1710-E 9	V-1710- 47	1940	V-1710-47(E9), PT-13E5 Carb, 6.65:1CR, 7.48:1 S/C, 6.85:1 Aux, 2.23:1 RG, 1595#, 1150/3000/21,000/49.7, 125Oct, 7/8/42.
138	V-1710-F 6R	V-1710-None	1940	V-1710-F6R, PD-12K2 Carb, 6.65:1CR, 9.60:1 S/C, 1.80:1Dual Counter-rotating RG, 1410#, 1125/3000/15,000/_", 130 Oct, 3/12/40.
139-E	V-1710-F 7R	V-1710- 45	1940	V-1710-F7R, PT-13E9 Carb, 6.65:1CR, 7.48:1 S/C, 6.85:1Aux S/C, 2.36:1RG, 1515#, 1125/3000/15,000/_", 100 Oct, 3/12/40.
140	V-1710-F 8	V-1710- 43	1940	2.00:1 RGR, no contract per Index of AEC Engine Specifications.
141-B	V-1710-F 9R	V-1710-None	1940	V-1710-F9R, PD-12K2, 6.65:1CR, 7.48:1S/C, could bypass 8.00:1 AuxS/C & Intercooler, 2.00:1RG, 1510#, 1125/3000/18,000, 100 Oct, 6/6/41.
142-C	V-1710-E11	V-1710- 93	1940	V-1710-93(E11), PT-13E9 Carb, 6.65:1CR, 8.1:1 S/C, 6.85:1 Aux, 2.23:1 RG, 1620#, 1150/3000/22,400/50.0, 130Oct, 1/16/43.
143	V-1710-E10	V-1710- 6	1940	V-1710-6(E10), PD-12K2Carb, 6.65:1CR, 8.80:1 S/C, 1.80:1 RG, <1400#, 1150/3000/12,000/__", 100 Oct, 10/10/39.
144	V-1710-C10	V-1710-Comm	1940	Commercial version of Military V-1710-C10 per Index of AEC Engine Specs 11/30/94.
145-A	V-1710-C15A	V-1710-None	1940	Chinese Contract for 150 engines for AVG, all 150 were C-15A's with unique accessories, many hand built.
146-C	V-1710-F10L	V-1710- 55	1941	V-1710-55(F10L), PD-12K3 Carb, 6.65:1CR, 7.48:1 S/C, 2.0:1 RG, 1345#, 1325/3000/25,000/47, 125 Oct, 2/5/42.
146-C	V-1710-F10R	V-1710- 51	1941	V-1710-51(F10R), PD-12K3 Carb, 6.65:1CR, 7.48:1 S/C, 2.0:1 RG, 1345#, 1325/3000/25,000/47, 125 Oct, 2/5/42.
147	V-1710-F11R	V-1710- 57	1941	Spec 12/01/41 per Index of AEC Engine Specifications 11/30/94.
148	V-1710-F12	V-1710-None	1941	No contract, Sea-Level rated w/single speed S/C, Spec 6/17/41 per Index of AEC Specs 11/30/94.
149	V-1710-F13L	V-1710-None	1941	Basically same as V-1710-F10L except 2.36:1RGR, Spec 10/06/41 per Index of AEC Engine Specifications 11/30/94.
149	V-1710-F13R	V-1710-None	1941	Basically same as V-1710-F10R except 2.36:1RGR, Spec 10/06/41 per Index of AEC Engine Specifications 11/30/94.
150	V-1710-E12R	V-1710- 59	1941	Engines built on Contract W535-ac-16323 per Index of AEC Engine Specifications, 12/23/41.
151	V-1710-F14R	V-1710- 61	1941	Basic V-1710-F3R w/9.6:1 S/C ratio per Index of AEC Engine Specifications 11/30/94.
152-F	V-1710-F15L	V-1710- 77	1941	V-1710-77(F15L), PD-12K7 Carb, 6.65:1CR, 8.1:1 S/C, 2.36:1 RG, 1365#, 1425/3000/27,000/54", 125 Oct, 1/16/43.
152-F	V-1710-F15R	V-1710- 75	1941	V-1710-75(F15R), PD-12K7 Carb, 6.65:1CR, 8.1:1 S/C, 2.36:1 RG, 1365#, 1425/3000/27,000/54", 125 Oct, 1/16/43.
153	V-1710-E13	V-1710-None	1941	V-1710-E13, PD-12K2 Carb, 6.65:1CR, 8.1:1 S/C, 2.227:1 RG, Turbo w/Aftclr, 1485#, 1150/3000/25,000/, ___ Oct, 10/14/41.
154	V-1710-E14	V-1710-None	1941	V-1710-E14, PD-12K2 Carb, 6.65:1CR, 7.48:1 S/C, 2.227:1 RG, w/2-stageTurbo S/C, 1445#, 1100/3000/32,000/, ___ Oct, 10/22/41.
155-A	V-1710-E15	V-1710-None	1941	V-1710-E15, PR-58 Carb, 6.65:1CR, 7.48:1 S/C, 2.227:1 RG, 8.8:1AuxS/C w/Intercool +Aftclr, 1675#, 1325/3000/30,000/, ____Oct, 10/23/41.
156-C	V-1710-E 6	V-1710- 63	1941	Improved V-1710-35(E4) w/PD-12K2 Carb, 6.65:1CR, 8.80:1 S/C, 2.00:1 RG, 1435#, 1150/3000/12,000/42.0, 100 Oct, 6/24/42.
157-B	V-1710-E 8	V-1710- 67	1941	Improved V-1710-35(E4) w/PD-12K2 Carb, 6.65:1CR, 8.80:1 S/C, 2.227:1 RG, 1445#, 1150/3000/12,000/42.0, 100 Oct, 6/24/42.
158	V-1710-E16	V-1710- 65	1942	Basic V-1710-E for experimental use, no performance guarantees per Index of AEC Engine Specifications.
159-?	V-1710-F 4L	V-1710- 79	1942	May not have had a Spec due to being an experimental project.
159-B	V-1710-F 4R	V-1710- 73	1942	V-1710-73(F4R) w/PD-12K2 Carb, 6.65:1CR, 8.80:1 S/C, 2.00:1 RG, 1345#, 1150/3000/12,000/42.0, 100 Oct, 1/22/42.
160	V-1710-F16	V-1710-None	1942	V-1710-F16, 2-spd 10.25 Birmann S/C, PR58 Carb, 6.65:1CR, 6.44 & 8.80:1 S/C, 2.00:1 RG, 1375#, 1250/3000/14,000/__, 100 Oct, 6/24/42.
161	V-1710-E17	V-1710-None	1942	V-1710-E17, PT-13E5 Carb, 6.65:1CR, 7.48:1 S/C, 8.09:1 Aux, w/Aftercooler, 2.23:1 RG, 1645#, 1125/3000/24,000/__", ___Oct, 2/11/42.
162-E	V-1710-F17L	V-1710- 91	1941	V-1710-91(F17L), PD-12K7 Carb, 6.65:1CR, 8.10:1 S/C, 2.0:1 RG, 1350#, 1425/3000/25,000/54", 130 Oct, 11/30/42. @6/27/95.
162-E	V-1710-F17L	V-1710- 91A	1944?	V-1710-91(F17L), PD-12K7 Carb, 6.65:1CR, 8.10:1 S/C, 2.0:1 RG, 1350#, 1425/3000/25,000/54", 130 Oct, 11/30/42. w/Auto MAP Reg.
162-E	V-1710-F17R	V-1710- 89	1941	V-1710-89(F17R), PD-12K7 Carb, 6.65:1CR, 8.10:1 S/C, 2.0:1 RG, 1350#, 1425/3000/25,000/54", 130 Oct, 11/30/42. @6/27/95.
162-E	V-1710-F17R	V-1710- 89A	1944?	V-1710-89A(F17R), PD-12K7 Carb, 6.65:1CR, 8.10:1 S/C, 2.0:1 RG, 1350#, 1425/3000/25,000/54", 130 Oct, w/Auto MAP Reg.
163-	V-1710-F20R	V-1710- 81A	1943	Similar to -81, except equipped with automatic Boost Control as in V-1710-99.
163-E	V-1710-F20R	V-1710- 81	1942	V-1710-81(F20R), PD-12K6 Carb, 6.65:1CR, 9.60:1 S/C, 2.0:1 RG, 1352#, 1410/3000/ 9,500/57.0, 130 Oct, 3/31/43.

164-B	V-1710-E18	V-1710- 83	1942	V-1710-83(E18) w/PD-12K6 Carb, 6.65:1CR, 9.60:1 S/C, 2.00:1 RG, 1435#, 1410/3000/ 9,500/57.0, 130 Oct, 11/2/42.
165-B	V-1710-E19	V-1710- 85	1942	V-1710-85(E19) w/PD-12K6 Carb, 6.65:1CR, 9.60:1 S/C, 2.23:1 RG, 1435#, 1410/3000/ 9,500/57.0, 130 Oct, 11/2/42.
166-B	V-1710-F21R	V-1710- 87	1942	V-1710-87(F20R) w/PD-12K7 Carb, 6.65:1CR, 7.48:1 S/C, 2.00:1 RG, 1353#, 1500/3000/ 5,400/52.0, 130 Oct, 8/1/42.
167	V-1710-F22R	V-1710-None	1942	No contract, single speed S/C w/9.12:1 gear per Index of AEC Engine Specifications 11/30/94.
168	V-1710-F23R	V-1710- 95	1942	V-1710-95(F23R) w/PD-12K6 Carb, 6.65:1CR, 9.60:1 S/C, 2.00:1 RG, 1355#, 1125/3000/15,500/44.5, 130 Oct, 11/3/42.
169	V-1710-F24R	V-1710-None	1942	With integral two-stage S/C per Index of AEC Engine Specifications 11/30/94.
170	V-1710-G 1R	V-1710- 97	1942	V-1710-97(G1R), PD-13E Carb, 6.65:1CR, 7.48:1 & 10.25 S/C, 2.00:1 RG, 1500#, 1725/3400/25,000/68.5", 140 Oct, 12/11/42.
170-	V-1710-G 1RA	V-1710- 97A?	1943/44	Similar to V-1710-G1R but with 6.00:1 Compression Ratio and 9.60:1 S/C gears, for Altitude Rating, per Index of AEC Engine Specs 11/30/94.
171-A	V-1710-E20	V-1710-None	1943	V-1710-E20, PD-12K7 Carb, 6.65:1CR, 8.10:1 S/C w/T-S/C & Intercooler, 2.23:1 RG, 1450#, 1425/3000/27,000/53.5, ___ Oct, 1/1/43.
172-C(C)	V-1710-G 2R/L	V-1710-None	1946	V-1710-G2R/L, PT-13H2 Carb, 6.00:1CR, 2-Spd S/C 7.76:1& 9.6:1 w/10.25 Imp, 2.36:1 RG, 1470#, 1600/3200/SL/__", 130 Oct, 5/17/46.
173	V-1710-F25R	V-1710-None	1943	V-1710-F25R, PT-13E9 Carb, 6.65:1CR, 8.10:1 S/C, 2.36:1 RG, 1520#, 1360/3000/17,250/60.0, 130 Oct, 2/9/43.
174-B	V-1710-F26R	V-1710- 99	1943	V-1710-99(F26R), PD-12K6 Carb, 6.65:1CR, 9.60:1 S/C, 2.0:1 RG, 1355#, 1425/3000/25,000/54", 130 Oct, 11/30/42.
174-B	V-1710-F26R	V-1710- 99A	1944	V-1710-99A(F26R), PD-12K6 Carb, 6.65:1CR, 8.80:1 S/C, 2.0:1 RG, 1355#, ____/3000/__",000/__"_.
175	V-1710-F27R	V-1710-101	1943	V-1710-101(F27R), PD-12K7 Carb, 6.65:1CR, 8.10:1 S/C, 2.36:1 RG, 1520#, 1500/3200/ 6,000/60, 130 Oct, 9/30/43.
176	V-1710-F28L	V-1710-123	1944	V-1710-123(F28L), PD-12K12 Carb, 6.65:1CR, 8.10:1 S/C, 2.36:1 RG, 1555#, 1700/3200/26,000/__", 130 Oct, 1/22/44.
176	V-1710-F28R	V-1710-121	1944	V-1710-121(F28R), PD-12K12 Carb, 6.65:1CR, 8.10:1 S/C, 2.36:1 RG, 1555#, 1700/3200/26,000/__", 130 Oct, 1/22/44.
177-A	V-1710-F29L	V-1710-107	1943	V-1710-107(F29L), PD-12K8 Carb, 6.65:1CR, 8.10:1 S/C, 2.36:1 RG, 1520#, 1500/3200/30,000/57, 130 Oct, 9/20/43.
177-A	V-1710-F29R	V-1710-105	1943	V-1710-105(F29R), PD-12K8 Carb, 6.65:1CR, 8.10:1 S/C, 2.36:1 RG, 1520#, 1500/3200/30,000/57, 130 Oct, 9/20/43.
178-A	V-1710-E21	V-1710-117	1943	V-1710-117(E21), PT-13E10 Carb, 6.65:1CR, 8.1:1 S/C, 7.23:1 Aux, 2.23:1 RG, 1660#, 1100/3000/25,000/52.0, 130Oct, 11/12/43.
179-E	V-1710-E22	V-1710-109	1943	V-1710-109(E22), PD-12K15 Carb, 6.65:1CR, 8.1:1 S/C, 7.23:1 Aux, 2.23:1 RG, 1660#, 1100/3000/28,000/__", 130Oct, 7/13/44.
179-E	V-1710-E22A	V-1710-109A	1944	V-1710-109A(E22A), PD-12K15 Carb, 6.65:1CR, 8.1:1 S/C, 7.23:1 Aux, 2.23:1 RG, 1660#, 1100/3000/28,000/__", 130Oct, 7/13/44.
180-D	V-1710-E23 Unit#1	V-1710-103 Unit#1	1943	V-1710-103(E23) Unit#1, PD-12K7 Carb, 6.65:1CR, 8.1:1 S/C, 6.85:1 Aux, 2.772:1#50 RG, 1800#, 1100/3000/28,000/__", 130Oct, 7/13/44.
180-D	V-1710-E23 Unit#2	V-1710-103 Unit#2	1943	V-1710-103(E23) Unit#2, PD-12K7 Carb, 6.65:1CR, 8.1:1 S/C, 6.80:1 Aux, 2.772:1#70 RG, 1800#, 1100/3000/28,000/__", 130Oct, 7/13/44.
181-A	V-1710-E24 Unit#1	V-1710-None	1943	V-1710-E24 Unit#1, PD-12K7 Carb, 6.65:1CR, 8.1:1 S/C, 7.23:1 Aux w/AftClr, 2.458:1#50 RG, 1810#, 1325/3200/20,000/__, 130Oct, 8/24/43.
181-A	V-1710-E24 Unit#2	V-1710-None	1943	V-1710-E24 Unit#2, PD-12K7 Carb, 6.65:1CR, 8.1:1 S/C, 7.18:1 Aux w/AftClr, 2.458:1#50 RG, 1810#, 1325/3200/20,000/__, 130Oct, 8/24/43.
182-F	V-1710-F30L	V-1710-113	1943	V-1710-113(F30L), PD-12K8 Carb, 6.65:1CR, 8.10:1 S/C, 2.0:1 RG, 1395#, 1500/3000/30,000/56.5", 130 Oct, 9/28/44.
182-F	V-1710-F30R	V-1710-111	1943	V-1710-111(F30R), PD-12K8 Carb, 6.65:1CR, 8.10:1 S/C, 2.0:1 RG, 1395#, 1500/3000/30,000/56.5", 130 Oct, 9/28/44.
183-A	V-1710-F31R	V-1710-115	1943	V-1710-115(F31R), PD-12K8 Carb, 6.65:1CR, 9.60:1 S/C, 2.0:1 RG, 1385#, 1360/3000/SL/57, 130 Oct, 12/20/43.
183-A	V-1710-F31RA	V-1710-115A	1943	V-1710-115A(F31R), PD-12K8 Carb, 6.65:1CR, 9.60:1 S/C, 2.0:1 RG, 1385#, 1360/3000/SL/57, 130 Oct, 12/20/43.
184-C	V-1710-F32R	V-1710-119	1943	V-1710-119(F32R), SD-400 Carb, 6.00:1CR, 8.10:1 S/C, 7.64:1 Aux S/C, AftClr, 2.36:1 RG, 1750#, 1900/3200/SL/78, 130 Oct, 4/20/45.
185	V-1710-E25	V-1710-None	1943	V-1710-E25, SD-400 Carb, 6.00:1CR, 8.10:1 S/C, 7.64:1 Aux S/C, AftClr, 2.36:1 RG, 1750#, 1900/3200/SL/78, 130 Oct, 4/20/45.
186-B	V-1710-E27	V-1710-127	1944	Turbine feedback V-1710-E27, PD-12K15, 6.00:1CR, 8.10:1 S/C, 7.23:1 Aux S/C, 2.48:1 RG, 1983#, 2980/3200/11,000/100, 145 Oct, 7/13/44.
187	V-1710-E28	V-1710-None	1944	Turbine feedback adapted to V-1710-E23 per Index of AEC Engine Specifications.
188	V-1710-E26	V-1710-None	1944	No information per Index of AEC Engine Specifications.
189	V-1710-E29	V-1710-None	1945	Same as V-1710-E22 except adapted to Douglas Requirements per Index of AEC Engine Specs.
190-A	V-1710-E30	V-1710-133	1945	V-1710-133(E30), PD-12K15 Carb, 6.00:1CR, 8.1:1 S/C, 7.64:1 Aux, 2.23:1 RG, 1660#, 1150/3000/27,500/58.0, 130Oct, 5/21/45.
191-B	V-1710-G 3R	V-1710-131	1945	V-1710-131(G3R), PT-13H2 Carb, 6.50:1CR, 2-Spd S/C 7.48:1& 9.6:1 w/10.25 Imp, 2.36:1 RG, 1475#, 1600/3200/SL/61.7", 130 Oct, 11/15/46.
192-A	V-1710-E31	V-1710-135	1946	V-1710-F31RA engines converted to "E" type engines, per Index of AEC Engine Specs.
192-A	V-1710-E31	V-1710-Racer	1946	V-1710-F31RA engines converted to "E" type engines, per Index of AEC Engine Specs.
193	V-1710-E23A Unit#1	V-1710-125 Unit#1	1945	V-1710-125(E23A) Unit#1, PD-12K8 Carb w/H2OInj, 6.65:1CR, 8.1:1 S/C, 6.85:1 Aux, 2.772:1#50 RG, 1811#, 1100/3000/28,000/51", 130Oct, 4/17/45.
193	V-1710-E23A Unit#2	V-1710-125 Unit#2	1945	V-1710-125(E23A) Unit#2, PD-12K8 Carb w/H2OInj, 6.65:1CR, 8.1:1 S/C, 6.80:1 Aux, 2.772:1#70 RG, 1811#, 1100/3000/28,000/51", 130Oct, 4/17/45.
194	V-1710-E23BL	V-1710-129L	1945	V-1710-129L(E23BL), PD-12K15 Carb w/H2OInj, 6.65:1CR, 8.1:1 S/C, 7.18:1 Aux, 2.772:1#70 RG, 1811#, 1100/3000/28,000/51", 130Oct, 4/18/45.
194	V-1710-E23BR	V-1710-129R	1945	V-1710-129R(E23BR), PD-12K15 Carb w/H2OInj, 6.65:1CR, 8.1:1 S/C, 7.23:1 Aux, 2.772:1#50 RG, 1811#, 1100/3000/28,000/51", 130Oct, 4/18/45.

197	V-1710-E32	V-1710-None	1945	V-1710-E31 converted to Tank Engine, per Index of AEC Engine Specifications.
198	V-1710-F33L	V-1710-141	1945	V-1710-141(F33L), PD-12K15 Carb w/H2OInj, 6.00:1CR, 8.1:1 S/C, 7.59:1 Aux, 2.36:1#50 RG, 1560#, 1150/3000/27,500/57, 130Oct, 8/6/45.
198	V-1710-F33R	V-1710-139	1945	V-1710-139(F33R), PD-12K15 Carb w/H2OInj, 6.00:1CR, 8.1:1 S/C, 7.64:1 Aux, 2.36:1#50 RG, 1560#, 1150/3000/27,500/57, 130Oct, 8/6/45.
199-A	V-1710-E23CL	V-1710-137L	1945	V-1710-137L(E23CL), PD-12K15 Carb w/H2OInj, 6.00:1CR, 8.1:1 S/C, 7.59:1 Aux, 2.772:1#70 RG, 1811#, 1150/3000/27,500/57, 130Oct, 7/26/45.
199-A	V-1710-E23CR	V-1710-137R	1945	V-1710-137R(E23CR), PD-12K15 Carb w/H2OInj, 6.00:1CR, 8.1:1 S/C, 7.64:1 Aux, 2.772:1#50 RG, 1811#, 1150/3000/27,500/57, 130Oct, 7/26/45.
201	X-3420-1	X-3420- 1	1935	X-3420-1, MC-12 Fuel Inj, 8.50:1CR, 7.0:1 S/C, 2:1 RG, 2160 pounds, 1600/2400/SL/36, Direct cylinder fuel injected, 6/12/35. @6/27/95.
202	X-3420-3	X-3420- 3	1935	X-3420-3, MC-12 Fuel Inj, 8.50:1CR, 7.0:1 S/C, Dual 2:1 RG, 2200 pounds, 1600/2400/SL/36, Direct cylinder fuel in jected, 6/12/35.
203-F	V-3420-A 1	V-3420- 1(XV)	1938	XV-3420-1(A1), PT-13B1 Carb, 7.25:1CR, 6.00:1 S/C, 2.50:1 RG, 2300 pounds, 2300/3000/25,000/__", 100 Oct, 10/11/38.
204	V-3420-A 2	V-3420- 3(XV)	1938	XV-3420-3(A2), PT-13B1 Carb, 7.25:1CR, 6.00:1 S/C, 3.00:1 RG, 2350 pounds, 2300/2950/25,000/36.0, 100 Oct, 4-15-38.
205	V-3420-B 1	V-3420-None	1938	V-3420-B1, PT-13B1 Carb, 7.25:1CR, 6.00:1 S/C, 2.50:1 RG, 2475# w/Ext Shaft, 2300/3000/25,000/__", 100 Oct, 10/10/38.
206-A	V-3420-A 3	V-3420-Comm	1939	V-3420-A3, PT-13B1 Carb, 6.65:1CR, 9.00:1 S/C, 2.50:1 RG, 2375#, 2300/3000/12,000/__", 100 Oct, 6-6-39.
206-A	V-3420-A 4R	V-3420-Comm	1938	V-3420-A4R, PT-13E1 Carb, 6.65:1CR, 6.00:1 S/C, 2.50:1 RG, 2350#, 2300/3000/25,000/__", 100 Oct, 6-6-39.
208	V-3420-A 5R	V-3420-None	1939	V-3420-A5R, PT-13E1 Carb, 6.65:1CR, 6.00:1 S/C, 3.00:1 RG, 2400 pounds, 2300/3000/25,000/__", 100 Oct, 6-12-39.
209	V-3420-A 6R	V-3420-None	1939	V-3420-A6R, PT-13E1 Carb, 6.65:1CR, 6.82:1 S/C, 3.00:1 RG, 2425 pounds, 2300/3000/ 5,000/__", 100 Oct, 6-13-39.
210	V-3420-A 7R	V-3420-None	1939	V-3420-A7R, PT-13E1 Carb, 6.65:1CR, 6.82:1 S/C, 3.80:1 RG, 2550 pounds, 2300/3000/ 5,000/__", 100 Oct, 6-28-39.
211	V-3420-A 8R/L	V-3420-None	1939	One each LH and RH turning 2000 bhp engine delivered to Navy in 1940 for PT-8 Boat, Navy Contract 68427 for Marine Service.
212	V-3420-A 9R	V-3420-None	1939	V-3420-A9R, PT-13E1 Carb, 6.65:1CR, 6.82:1 S/C, 2.50:1 RG, 2400 pounds, 2500/3000/25,000/__", 100 Oct, 10-9-39.
213	V-3420-A10R	V-3420-None	1939	V-3420-A10R, PT-13E1 Carb, 6.65:1CR, 9.5:1 S/C, 2.50:1 RG, 2450 pounds, 2200/3000/15,000/__", 100+ Oct, 10-10-39.
213	V-3420-A10R-"A"	V-3420-None	1939	V-3420-A10R-Opt A, PT-13E1 Carb, 6.65:1CR, 6.4:1 & 9.5:1 1-stage S/C, 2.50:1 RG, 2500 pounds, 2200/3000/15,000/__", 100+ Oct, 10-10-39.
213	V-3420-A10R-"B"	V-3420-None	1939	V-3420-A10R-Opt B, PT-13E1 Carb, 6.65:1CR, 8.0:1 1-stage S/C + T-S/C, 2.50:1 RG, 2450 pounds, 2300/3000/10,000/__", 100+ Oct, 10-10-39.
213	V-3420-A10R-"C"	V-3420-None	1939	V-3420-A10R-Opt C, PT-13E1 Carb, 6.65:1CR, 8.0:1 2-stage S/C, 2.50:1 RG, 2725 pounds, 2300/3000/10,000/__", 100+ Oct, 10-10-39.
214	V-3420-B 2	V-3420-None	1940	V-3420-B2, PT-13E1 Carb, 6.65:1CR, 6.82:1 S/C, For 1939 R40-C Competition. 2-spd remote RG, 2300 pounds, 2450/3000/5,000/__", 100+ Oct, 3-12-40.
215	DV-6840-A1	DV-6840-None	1940	DV-6840-A1, PT-13E1 Carbs, 6.65:1CR, 6.82:1 S/C, 2-spd remote RG, 6750 pounds, 5000/3000/25,000/__", 100+ Oct, 5-21-40.
215	V-3420-B 3	V-3420-	1940	No Contract, For Contract R40-D Competition. Consists of two V-3420 power sections.
216-C	V-3420-B 4	V-3420- 5	1943	V-3420-B4, PR-58B3 Carb, 6.65:1CR, 6.82:1 S/C, 2.458:1 remote RG, 2955 pounds, 2600/3000/25,000/__", 130 Oct, 5-18-43.
217	V-3420-B 5	V-3420- 7	1940	V-3420-B5, PR-58B2 Carb, 6.65:1CR, 6.82:1 S/C, 2.50:1 remote RG, 3640 pounds, 2600/3000/25,000/__", 130 Oct, 5-18-43.
217-A	V-3420-B 5	V-3420- 7	1942	Contract ac-11328. RGR 2.5:1, 3640#, Spec 5/19/42.
218	V-3420-A11R	V-3420- 9	1941	V-3420-A11R, PT-13E1 Carb, 6.65:1CR, 6.39:1 1-stage S/C + T-S/C, 2.50:1 RG, 2450 pounds, 2300/3000/25,000/__", 100 Oct, 4-9-41.
219	V-3420-B 6	V-3420-	1941	Built for a single propeller.
220	V-3420-A12R	V-3420-None	1941	V-3420-A12R, PT-13E1 Carb, 6.65:1CR, 7.20:1 1-stage S/C + T-S/C, 3.00:1 RG, 2550 pounds, 2600/3000/25,000/__", 100 Oct, 4-30-41.
221	V-3420-A13	V-3420-	1941	Utilizes a two speed propeller.
222	V-3420-A14R/L	V-3420-	1941	Boat engine with CW-CCW rotation.
223	V-3420-B 7	V-3420-None	1941	V-3420-B7, PT-13 Carb, 6.65:1CR, 6.82:1 S/C, 2.59:1 Counter Rotating remote RG, 2955 pounds, 2600/3000/25,000/__", 100 Oct, 8/22/41.
224-A	V-3420-A15L	V-3420-None	1942	V-3420-A15L, PR-58B2 Carb, 6.65:1CR, 6.82:1 1-stage S/C + T-S/C, 2.50:1 RG, 2530 pounds, 2600/3000/25,000/__", 100 Oct, 5-5-42.
224-A	V-3420-A15R	V-3420-None	1942	V-3420-A15R, PR-58B2 Carb, 6.65:1CR, 6.90:1 1-stage S/C + T-S/C, 2.50:1 RG, 2530 pounds, 2600/3000/25,000/__", 100 Oct, 5-5-42.
225-C	V-3420-A16L	V-3420-13	1942	V-3420-A16L, PR-58B2 Carb, 6.65:1CR, 6.82:1 1-stage S/C + T-S/C, 3.13:1 RG, 2655 pounds, 2600/3000/25,000/__", 100 Oct, 2-26-43.
225-C	V-3420-A16R	V-3420-11	1942	V-3420-A16R, PR-58B2 Carb, 6.65:1CR, 6.90:1 1-stage S/C + T-S/C, 3.13:1 RG, 2655 pounds, 2600/3000/25,000/__", 100 Oct, 2-26-43.
226	V-3420-A17R	V-3420-15	1942	V-3420-A17R, PR-58B2 Carb, 7.25:1CR, 6.90:1 1-stage S/C + T-S/C, 3.13:1 RG, 2630 pounds, 2600/3000/25,000/__", 100 Oct, 2-26-43.
227-A	V-3420-A18L	V-3420-25	1943	RGR 3.13:1, 2655#, Spec 3/02/43.
227-A	V-3420-A18R	V-3420-17	1942	V-3420-A18R, PR-58B3 Carb, 6.65:1CR, 6.90:1 1-stage S/C + T-S/C, 3.13:1 RG, 2655 pounds, 2600/3000/25,000/__", 130

				Oct, 3-2-43.
228-A	V-3420-B 8	V-3420-19	1942	V-3420-B8, PR-58B2 Carb, 6.65:1CR, 6.82:1 S/C, 6.96:1 Aux S/C, 2.458:1CrtR remote RG, 3175 pounds, 2300/3000/20,000/__, 130 Oct, 3/12/43.
229	V-3420-A19R	V-3420-None	1943	V-3420-A19R, PR-58 Carb, 6.65:1CR, 6.90:1 S/C, 6.96:1 Aux S/C, 3.13:1 RG, 2830 pounds, 2300/3000/20,000/__", 130 Oct, 3/19/43.
230	V-3420-A20R	V-3420-21	1943	V-3420-A20R, PR-58 Carb, 6.65:1CR, 8.00:1 S/C, 3.13:1 RG, 2655 pounds, 2300/3000/10,000/__", 130 Oct, 3/10/43.
231	V-3420-B 9	V-3420-None	1943	V-3420-B9, PR-58 Carb, 6.65:1CR, 6.82:1 S/C, 6.96:1 Aux S/C, 2.458:1 CtrR integral RG, 2975 pounds, 2300/3000/20,000/__, 130 Oct, 3/9/43.
232	V-3420-A21R/L	V-3420-None	1943	V-3420-A21R, 2 each, PR-58 Carb, 6.65:1CR, 6.90:1 S/C, 6.96:1 Aux S/C, CrtR 3.13:1 RG, 3150 pounds, 2300/3000/22,000/__, 130 Oct, 6/12/43.
233-A	V-3420-A22L	V-3420-None	1944	V-3420-A22L, PR-58B3 Carb, 6.65:1CR, 7.29:1 S/C, 3.13:1 RG, 2695 pounds, 2850/3000/ 4,300/__", 130 Oct, 7/20/44.
233-A	V-3420-A22R	V-3420-None	1944	V-3420-A22R, PR-58B3 Carb, 6.65:1CR, 7.26:1 S/C, 3.13:1 RG, 2695 pounds, 2850/3000/ 4,300/__", 130 Oct, 7/20/44.
233-D	V-3420-A22R/L	V-3420-	1943	Same as -A18R w/exceptions, RGR 3.13:1, 2655#, Spec 7/20/44.
234-C	V-3420-B10	V-3420-23	1943	V-3420-B10, PR-58B3 Carb, 6.65:1CR, 6.82:1 S/C, 7.38:1 Aux S/C, 2.458:1CrtR remote RG, 3275 pounds, 2300/3000/20,000/48.8, 130 Oct, 4/25/44.
235	V-3420-A23R/L	V-3420-27	1943	V-3420-A23, Two V-3420's, PR-58B3 Carb, 6.65:1CR, 8.0:1& 7.8:1 S/C, 3.23:1 RG, 3100# each, 2600/3000/10,000/__", 130 Oct, 11/27/43.
236	V-3420-B11	V-3420-29	1944	V-3420-B11, PR-58B5 Carb, 6.65:1CR, 6.82:1 S/C, 7.84:1 Aux S/C, 2.458:1CrtR remote RG, 3275 pounds, 2300/3000/25,000/50.5, 130Oct, 8/7/44.
237	V-3420-A24R	V-3420-31	1944	V-3420-A24R, PR-58B Carb, 6.65:1CR, 7.26:1 1-stage S/C + T-S/C, 3.13:1 RG, 2695 pounds, 2600/3000/25,000/__", 130 Oct, 11/30/44.
238	V-1710-F34	V-1710-None	1945	V-1710-F32R w/provisions for fuel & H2O/Alcohol Inj, Spec 8/03/45 per AEC Specs Index 11/30/94.
239-A(C)	V-1710-G 3R	V-1710-Comm	1945	V-1710-G3R, PT-13H2? Carb, 6.50:1CR, 2-Spd S/C 7.48:1& 9.6:1 w/10.25 Imp, 2.36:1 RG, 1475#, 1600/3200/SL/61.7", 130 Oct, 11/13/45.
240	V-3420-A25	V-3420-Comm	1945	Concept engine, proposed for B-29/XB-39, 8-10-1945.
241	V-3420-C 1	V-3420-None	1945	1945 Proposal for Installation in B-29 A/C.
242- (C)	V-3420-A24R	V-3420-Comm	1945	V-3420-A24R, PR-58B Carb, 6.65:1CR, 7.26:1 1-stage S/C + T-S/C, 3.13:1 RG, 2740 pounds, 3000/3000/25,000+/__", 130 Oct, 8/21/45.
245-A	V-1710-F36L	V-1710-145	1945	Original designation of what became the V-1710-G6L.
245-A	V-1710-F36R	V-1710-143	1945	Original designation of what became the V-1710-G6R.
245-C	V-1710-G 6L	V-1710-145	1945	V-1710-145(G6L), SD-400-D3 w/H2OInj, 6.00:1CR, 7.48:1 S/C, 8.03:1 Aux, 2.36:1#50 RG, 1595#, 1250/3200/30,000/68, 145Oct, 12/3/46.
245-C	V-1710-G 6R	V-1710-143	1945	V-1710-143(G6R), SD-400-D3 w/H2OInj, 6.00:1CR, 7.48:1 S/C, 8.087:1 Aux, 2.36:1#50 RG, 1595#, 1250/3200/30,000/68, 145Oct, 12/3/46.
246	V-1710-G 7R/L	V-1710-Comm	1946	Per Index of AEC Specs 11/30/94.
247- (C)	V-1710-G 4L	V-1710-Comm	1945	V-1710-G4L, ExtShaft, PT-13H2 Carb, 6.00:1CR, 2-Spd S/C 7.76:1& 9.60:1 w/10.25 Imp, __:1 RG, 1340#, 1830/3200/SL/_Wet, 130 Oct, 12/31/45.
247- (C)	V-1710-G 4R	V-1710-Comm	1945	V-1710-G4R, ExtShaft, PT-13H2 Carb, 6.00:1CR, 2-Spd S/C 7.76:1& 9.60:1 w/10.25 Imp, __:1 RG, 1340#, 1830/3200/SL/_Wet, 130 Oct, 12/31/45.
251-A(C)	V-1710-G 2RA	V-1710-Comm	1946	V-1710-G2RA, PT-13H2 Carb, 6.00:1CR, 2-Spd S/C 7.76:1& 9.60:1 w/10.25 Imp, 2.36:1 RG, 1475#, 1200/2700/SL/_", 130 Oct, 1/25/46.
252	V-1710-G 8R/L	V-1710-None	1946	Integral Supercharger w/Aftercooler, Spec 2/04/46 per Index of AEC Engine Specs 11/30/94.
254	V-1710-G 9L	V-1710-149	1946	V-1710-149(G9L), Port Fuel Inj, 6.00:1CR, 7.48:1 S/C, 8.03:1 Aux, 2.36:1#50 RG, 1625#, 1250/3200/30,000/68, 145Oct, 2/18/46.
254	V-1710-G 9R	V-1710-147	1946	V-1710-147(G9R), Port Fuel Inj, 6.00:1CR, 7.48:1 S/C, 8.087:1 Aux, 2.36:1#50 RG, 1625#, 1250/3200/30,000/68, 145Oct, 2/18/46.
255	V-1710-G10L	V-1710-None	1946	Turbo-compound w/Integral auxiliary drives and two-stage supercharger, Spec 5/15/46 per Index of AEC Specs 11/30/94.
255	V-1710-G10R	V-1710-None	1946	Turbo-compound w/Integral auxiliary drives and two-stage supercharger, Spec 5/15/46 per Index of AEC Specs 11/30/94.
269-A	V-1710-G11	V-1710-None	1946	Turbo-compound w/6.0:1 CR, integral accessory drives and five-stage S/C, Spec 11/14/47 per AEC Index 11/30/94.
270	V-1710-G12	V-1710-None	1946	Same as -G11 turbo-compound but with E-Type shaft to Reduct Gear, Spec 11/14/47 per AEC Specs Index 11/30/94.
None?	V-1710-F18R	V-1710- 69	1942	Experimental Model similar to F-10R w/Port Injection, per Index of AEC Engine Specifications 11/30/94.
None?	V-1710-F19R	V-1710- 71	1942	Similar to V-1710-F10R but with Cyl Head fuel Injection per Index of AEC Engine Specs 11/30/94.
None?	V-1710-F35R	V-1710-None	1945	Variable speed S/C per Index of AEC Engine Specs 11/30/94.
None?	V-1710-F37	V-1710-None	1945	Like F-36 but uses Bendix Fuel Metering Pump and Port Injection. Large S/C and 50% Aftercooler.
None?	V-1710-G 5	V-1710-None	1945	Similar to G-4 w/exceptions, per Index of AEC Engine Specifications 11/30/94.
None?	V-1710-G 6	V-1710-None	1946	na
None?	V-1710-G 6A	V-1710-None	1946	na
None?	V-1710-H 1	V-1710-None	1946	V-1710-H1 2550+ bhp, turbo-compound w/separate 2-stage S/C and 2-stage air-cooled turbine, 50% Aftercooler and Port Injection, 1947.
None?	V-3420-A	V-3420- 3	na	na
None?	V-3420-A 8RLW	V-3420-Comm	na	Rebuilt A-8R/L for Gar Wood Boat, no Specs.
None?	V-3420-B12	V-3420-None	1944	Same as B-11 but with 50% Intercooling.

Appendix 6: Allison Powered Aircraft, First Flight Dates

Aircraft Model	Tail Number	1st Flight Date	Total Time, hrs	Disposition
A-11A	33-208	12/14/36	311	First flight testbed for V-1710. Class 26 at Chanute Field, IL 3/40.
XP-37	37-375	3/1/37	162.0	Class 26 at Chanute Field, IL 9/41.
XFM-1	36-351	9/1/37	103	Lt. Ben Kelsey pilot on 1st flight. Class 26, 8-40.
XP-40	38-010	10/1/38	433.2	Last flight Oct 1942 at Wright Field. Surveyed 8/31/44.
YP-37	38-472	1/24/39	168.5	Class 26 at Wichita Falls, TX 1/13/42.
XP-38	37-457	1/27/39	11:51	Crashed at Mitchell Field, NY, 2-11-39.
XP-39	38-326	4/6/39	9	Transfered to NACA Langley in June 1939 with 9 hrs TT.
YP-37	38-473	4/28/39	181.6	Class 26 11/7/41 at Chanute AAF, IL.
YP-37	38-480	8/25/39	171.2	Class 26 at Chanute, 11/7/41.
YP-37	38-474	10/26/39	180.3	Class 26 8/4/42.
YP-37	38-475	11/7/39	212.0	Class 26 11/21/41 at Lowry Field, CO.
YP-37	38-476	11/10/39	98.7	Class 26 at Biloxi, Miss, 1/2/42.
YP-37	38-477	11/10/39	129.7	Class 26 at Ladd Field, AK, 5/29/41.
YP-37	38-478	11/14/39	182.7	Class 26 at Biloxi, Miss, 1/13/42.
YP-37	38-479	11/15/39	196.7	Class 26 at Chanute AAF, IL, 11/7/41.
YP-37	38-481	11/22/39	279.3	Accepted from contractor 11/22/39. Class 26 2/3/42.
YP-37	38-482	11/22/39	182.4	Accepted from contractor 11/22/39, Class 26 1/5/42 @ Biloxi, Miss.
XP-39B	38-326	11/25/39	87.2	Wrecked as XP-39B in Aug. 1940.
YP-37	38-483	11/27/39	116.5	Class 26 11/7/41 at Chanute AAF, Il.
YP-37	38-484	11/30/39	88.1	Class 26 11/6/41 at Shepard AAF.
P-40	39-156	4/4/40	NA	First production P-40, achieved 366.7 mph at 15,000 feet on 1090 bhp.
XFL-1	BuNo1588	6/13/40	NA	Land Fill
YP-39	40-027	9/13/40	NA	Wrecked near Buffalo NY 10-12-40, LG would not extend.
YP-38	39-689	9/17/40	23.9	Wrecked 11-4-41 at Glendale CA.
XP-51	NX19998	10/26/40	3:20	NA-73X, engine AEC #301, wrecked 11/20/40 due to fuel starvation. Repaired and flown until 7/15/41.
YP-38	39-690	1/25/41	234.4	Wrecked 12-9-41
XP-46A	40-3054	2/15/41		Delivered to the Air Corps on September 12, 1941.
YP-38	39-692	3/28/41	114.2	Class 26
XP-51	AG345	4/16/41		1st RAF Production airframe, retained by North American.
YP-38	39-694	4/23/41	114.7	Wrecked 11-6-41 at McClellan Field, CA when left engine failed on takeoff. Class 26
YP-38	39-695	4/29/41	31.3	Wrecked 7-23-41at Alpane Michigan due to engine failure. Class 26
YP-38	39-696	4/30/41	75.5	Class 26 on 7-27-42
YP-38	39-697	5/3/41	181.3	Class 26 on 1-5-43
YP-38	39-698	5/14/41	56.3	Class 26 on 1-5-43
YP-38	39-699	5/22/41	31.8	Wrecked near Atlanta, Michigan 6-23-41.
YP-38	39-700	5/24/41	150.8	Class 26, 7-27-42
YP-38	39-701	5/29/41	170.7	Class 26, 1-5-43.
XP-40F	40-360	6/30/41		Modified P-40D rebuilt with Rolls-Royce Merlin 28 of 1300 bhp. Model 87-B3.
XP-46	40-3053	9/22/41		Powered by V-1710-39(F3R). First flight date may be delivery date.
XP-39E	41-19501	2/21/42	13:55	Crashed 3-26-42.
XP-39E	41-19502	4/4/42	>100	This was first to fly with 2-stage E9 engine.
XP-63	41-19511	12/7/42		Crashed Jan 1943.
XP-63	41-19512	2/5/43		Crashed May 1943.
P-63A	42-68861	4/26/43		First of 1,725 aircraft, ordered Sept 29, 1942 with V-1710-93(E11).
XP-63A	42-78015	4/27/43		Third prototype aircraft, ordered June 1942 with V-1710-93(E11).
XB-38	41-2401	5/19/43	12	Crashed 6-15-43 due to fire in #3 nacelle from leaking exhaust manifold.
XP-55	42-78845	7/19/43		First flight, at Scott Field, IL. Crashed 11-15-43 during stall testing.
XP-75	43-46950	11/17/43		
XP-55	42-78846	1/9/44		
XP-55	42-78847	4/25/44		
XB-42 No.1	43-50224	5/6/44	144:25	NASM
XP-58	41-2670	6/6/44	<50	Class 26, May 1945, Wright Field.
XB-42 No.2	43-50225	8/1/44	118:20	Crashed 12-16-45 near Bolling Field, Washington DC due to fuel exhaustion.
XP-75A	44-4549	9/15/44		1st Production P-75 Eagle. 404 mph @ 32,000' w/2300 bhp. Crashed 10-10-44
XP-51J	44-76027	4/23/45		
P-82E	46-255	2/17/47	NA	First block of 100, AF46-255 to AF46-354
XB-42A	43-50224	5/27/47	18:20	NASM

Appendix 7: Partial List of V-1710 Related Military Technical Orders

Technical Orders:
Tech Orders Series 01- applicable to aircraft.
Tech Orders Series 02- applicable to engines.
Tech Orders Series 03- applicable to accessories.

Note: Dates given for specific TO's are not necessarily the date of the first issue, as no comprehensive listing is available. Many of the TO's were revised and issued numerous times.

TO 01-25C-54	Required removal of the supercharger oil seal vent and routing it to the engine compartment. April 1941.
TO 01-25C-55	Relocated the two crankcase breathers, hoped to improve (reduce) thrust bearing and propeller shaft corrosion. April 1941.
TO 02-5-5	Allison – Model Designations with Army Air Forces and British Equivalents and Technical Order Compliance on British Numbers, January 3, 1945.
TO 02-5A-1	Operation & Flight Instructions – V-1710-11, -13, -21 and -23 Engines
TO 02-5A-1A	Preliminary Operation Instructions for V-1710-19, January 16, 1939.
TO 02-5A-6	Issued as a Technical Radiogram, Feb 14, 1941, directed inspection of backfire screens on V-1710-33.
TO 02-5A-8	Inspection and Reworking of Propeller Shaft Reduction Gearing – V-1710-11, -21, -23, -33 and -41 Engines, September 18, 1940, revised February 25, 1941. Wright Field file D52.41/48, Allison.
TO 02-5A-9	Interchangeability of Pistons and Piston Pins – V-1710 Series, July 28, 1944.
TO 02-5A-10	Improved lube of Propeller Thrust Bearings in V-1710-19/-21/-33 by installing external oil tube assembly and spray nozzle. Not required on V-1710-11 engines. Allison issued TO requirement January 8, 1940.
TO 02-5A-12	Issued to retard propeller thrust bearing corrosion by application of soft grade grease, Spec 3560, on the inner and outer propeller shafts, May 1941.
TO 02-5A-13	Issued to cause clearance on all V-1710 intake valves to be reset from 0.010 to 0.015 inches. May 1941.
TO 02-5A-14	Correct Backfire Screens for Installation in Intake Manifolds, 1941.
TO 02-5A-15	Directs the replacement of intake manifolds and installation of individual port backfire screens in V-1710-27/29 and -33 engines, June 1941. Parts to implement this modification were shipped by air to the Philippines, Hawaii and Canal Zone. Allison requested that the old manifolds be returned to them as they were to be used on Foreign contract engines, thus freeing manufacturing capacity for producing the new manifolds required by the Air Corps.
TO 02-5A-20	Inspection of Valves, Valve Mechanism and Backfire Screens – V-1710 Series, March 6, 1942. D52.41/78 Allison, NASM, Backfire file.
TO 02-5A-29	Inspection, Cleaning and Replacement of Distributor Drive Housing Vents – V-1710 Series, Nov 28, 1943.
TO 02-5A-34	Modification of booster coil lead wire, 4-12-43.
TO 02-5A-39	Replacement of distributor drive assemblies and inspect bearings, 12-28-43.
TO 02-5A-43	Rework cylinder head coolant outlet elbows, 5-3-43.
TO 02-5A-46	Install constant velocity gas intake pipe and tees, 7-15-43.
TO 02-5A-50	Increase carburetor fuel inlet pressure, 3-27-43.
TO 02-5A-55	Retarded Ignition Timing For Flight With 91 Octane Fuel, Spec No. AN-F-26, V-1710 Series, 8-13-43.
TO 02-5A-56	Replacement of gas intake pipe tie strap, 9-2-43.
TO 02-5A-57	Rework ignition tube assembly, 10-13-43.
TO 02-5A-59	Installation of directional priming jets, 10-21-43.

TO 02-5A-60	Instructions for use of 100 octane fuel, 5-24-43.
TO 02-5A-61	Installation of 1750cc enrichment jet in carburetor, 10-27-43.
TO 02-5A-63	Aligning And Anchoring Cylinder Jacket Coolant Sleeves – V-1710 Series, 3-13-44.
TO 02-5A-67	Proper Safetying Of Main Lever Clamping Screw On Automatic Engine Controls, V-1710 Series, 2-21-44.
TO 02-5A-68	Installing and Pinning of Camshaft Drive Shaft Bushings, V-1710 Series, March 15, 1944.
TO 02-5A-71	Installation of rubber bushing on distributor end of magneto, 7-29-44.
TO 02-5A-72	Replacing Center Gas Intake Pipe And Installing Gas Intake Pipe Sleeve, New Manifold Hoses, And Split Sleeves To Improve Fuel Distribution – V-1710-49,-51,-53,-55,-81,-81A,-83,-85,-89,-89A,-91,-91A,-99,-99A,-111, -113, -115 and -115A, 8-18-45.
TO 02-5A-73	Enlarge vent holes in spray nozzle gaskets, 8-15-44.
TO 02-5A-75	Identification and Installation of Neoprene Coolant Hose – V-1710 Series, October 20, 1944.
TO 02-5A-76	Break Edges of Thrust Face Holes of Supercharger Impeller Locating Bearing Cage – V-1710 Series, October 20, 1944.
TO 02-5A-77	Installation of Cotter Pin in Rocker Arm Bracket Studs – V-1710 Series, October 25, 1944.
TO 02-5A-78	Lubrication of Exterior Linkages, Automatic Manifold Pressure Regulators, and Auxiliary Stage Supercharger Controls – V-1710-81,- 83,-85,-89,-91,-93,-99,-111,-113,-115,-109 and -117, Dec 12, 1944.
TO 02-5A-79	Air-Maze Oil Filters, Installation, Inspection and Maintenance – V-1710-49,-51,-53,-55,-81,-81A,-83,-85,-89,-89A,-91,-91A,-93,-99,-99A,-111, -113, -115, -115A and -117, June 4, 1945.
TO 02-5A-81	Proper Tightening and Safetying of Pressure Selector Cam Lever Clamp Screw on Manifold Pressure Reg No. 42685 – V-1710-81,-81A,-83,-85,-89A,-91A,-99,-99A,-111, -113, -115 and -115A, May 24, 1945.
TO 02-5A-83	Replacement of Valve Springs and Tightening of Rocker Arm Bracket Stud Nuts – V-1710-109, -109A, -111, -113, -115, 115A, -117 and -135, August 27, 1945.
TO 02-5AA-1	Operation and Flight Instructions V-1710-21 and Associated Models.
TO 02-5AA-2	Handbook of Service Instructions for V-1710-11, -21, -23. April 25, 1939, revised 10-10-1941. Engines had backfire screens and incorporated intake valve clearance specified as 0.015 inches.
TO 02-5AB-2	Handbook of Service Instruction for V-1710-27, -29, -39, -49, -51, -53 and -55 Engines, Revised 5-5-42.
TO 02-5AB-3	Overhaul Instructions for V-1710-27, -29, -39, -49, -51, -53, -55, -73, -81, -87, -89, and -91 Aircraft Engines, December 1, 1943.
TO 02-5AB-10	Removal of Automatic Manifold Pressure Regulators – V-1710-89 and -91. Purpose: P-38 airplanes with these engines do not require Automatic MAP regulators. TO required removal from spare engines only, and then only at time of installation. D52.41/72 and D52.41/192 Allison, NASM, Backfire file.
TO 02-5AB-11	Replacing Reduction Gear Pinion Plug Rivets – V-1710-27,-29,-39,-49,-51,-53,-55,-73,-81,-87,-89,-91,-99,-111, -113, and -115, January 10, 1945.
TO 02-5AB-16	Replacement of Reduction Gear Pinion Bearing Thrust Washer Positioning Dowel – V-1710-27,-29,-39,-49,-51,-53,-55,-73,-81,-91,-99,-111, -113, and -115, January 26, 1945.
TO 02-5AB-17	Adjustment of Automatic Manifold Pressure Regulators – V-1710-89A. -91A, 111, and -113, Feb 2, 1945.
TO 02-5AB-18	Removal of Magneto Timing Control Mechanism – V-1710-111 and -113, June 8, 1945.
TO 02-5AB-19	Replacing Steel Oil Drain Tube of Automatic Manifold Pressure Regulator – V-1710-111, -113, and -115, April 21, 1945.
TO 02-5AC-6	Directs resetting intake valve clearance to 0.015 inches and inspection of Valves, Valve Mechanism and Backfire Screens in V-1710-33. Results of the monitoring requirements of this TO found that, in the period 4-3-41 to 5-17-41, 43 engines from contract W535-ac-12553 had burned screens while only 3 engines from contract W535-ac-16363 were so impacted. The first 300 engines [all V-1710-33(C15)'s]on W535-ac-12553 did not originally have divided tee manifolds, all of the subsequent engines did. The relative numbers of engines in service at this juncture is not known.
TO 02-5AC-9	Directs the installation of aluminum intake manifolds and individual port backfire screens on V-1710-33 engines. Manifold PN40772 and PN40773, Backfire Screen, intake port PN40840 (also known as 41D8959). Applied to 550 engines then in service. Caused engines to be re-rated due to increase in pressure losses across the new screens. All production capacity allocated to retrofit program, Allison did not expect to be able to incorporate into new engines until week of July 14, 1941. There is no reason to believe that this modification was ever accomplished on engines built for British contracts, or applied to those used by the AVG.
TO 02-5AC-12	Replacement of Oil Pump Assembly, V-1710-33. Purpose: To improve the operating characteristics of the oil pump assembly and reduce the tendency of the pressure pump to air lock, particularly at high altitudes. PN37425 to be replaced by PN40125. This TO applied to early engines, those produced on contract W535-ac-12553. Removed pumps were to be returned to Allison for rework within 10 days to the improved configuration, and PN. D52.41/85 Allison, NASM.TO 02-5A-8 Inspection and Reworking of Propeller Shaft Reduction Gearing, V-1710-11, -21, -23, -33 and -41 Engines. Manufacture of PN34051 made it difficult to maintain accurate gear tooth forms, the condition causes burring or galling of the gear teeth of the crankshaft reduction gear pinion, PN34050 and the accessory drive shaft, gear PN36150, resulting in excessive wear on the internal gear teeth in the main reduction gear, PN34051. This TO required removal of the reduction gear on any overhauled engine, or engine in which any of these gears had been replaced, and inspection for the likely damaged condition. If the teeth were found galled or burred, then they were to be hand stoned, the engine run for 1:15 hrs, and reinspected. This process was to continue until the condition was removed. D52.41/48 Allison, NASM.
TO 02-5AH-0	Operation Maintenance and Overhaul Handbook for Allison V-1710-"E" Type Engines.
TO 02-5AH-2	Handbook of Service Instructions for Models V-1710-143 and -145 Aircraft Engines.
TO 02-5AH-3	Overhaul Instructions for Models V-1710-143 and -145 Aircraft Engines.

Appendix 8: Selected Wright Field Contracts

Selected contracts related to Allison engines, issued by Wright Field Materiel Division.

W535-ac-05592 To Allison for XV-1710-1(A2), AC33-42, March 7, 1932.
W535-ac-06192 To Allison for U-250 2-cyl, 2-cycle engine AC34-1.
W535-ac-06551 To Allison for XV-1710-3(C1), AC34-4, 1934?
W535-ac-06795 To Allison for service test quantity of 10 YV-1710-3's, 6-15-34, flight engines released for production 9-22-36, last engine shipped and order completed 6-25-37.
W535-ac-07949 To Bell Aircraft for installation of XV-1710-7(C2) in A-11A, 1936
W535-ac-08892 To Seversky for 77 P-35's, cost $1,636,250. Per 1936 Pursuit competition.
W535-ac-09678 To Allison for XV-3420-1 AC38-119, March 1937.
W535-ac-09974 To Lockheed for XP-38 (AC37-457, w/C-7/C-9) at a cost of $163,000, 6-23-37.
W535-ac-10291 To Allison for XV-1710-15(C9) AC38-120, 1938
W535-ac-10341 To Bell Aircraft for XP-39 (AC38-326), October 7, 1937.
W535-ac-10535 To Curtiss-Wright for 13 YP-37's, 1938
W535-ac-10660 To Allison for 20 V-1710-11's for YP-37's, 1-5-38 amended to add 39 V-1710-13(D1)'s for YFM-1's, last shipped as a D-2, Oct 1940.
W535-ac-10830 To Allison for first XV-1710-17(E2) AC38-644, 1938.
W535-ac-11122 To Bell Aircraft for 13 YFM-1's, 1938.
W535-ac-11162 To Allison for conversion of V-1710-11(C8) to XV-1710-19(C13) for XP-40.
W535-ac-11279 To Allison for second XV-1710-17(E2) AC38-931.
W535-ac-11328 To Allison for six V-3420's, March 1938.
W535-ac-12414 To Curtiss-Wright for 524 Curtiss P-40's, 4-26-39.
W535-ac-12523 To Lockheed Aircraft for 13 YP-38's, $2,180,725, ordered 4-27-1939.
W535-ac-12553 To Allison for 837 (132 later added for total of 969) V-1710-27/29/33/37/39's, FY-39/40.
W535-ac-12632 To Wright Aero Corp. for R-2160 component tests, June 1939.
W535-ac-12635 To Bell Aircraft for 12 YP-39's and one YP-39A, $1,073,445, 1939.
W535-ac-13709 To Wright Aero Corp. for first R-2160 Tornado Engine, Jan 1940.
W535-ac-15678 To Packard to build 3,000 Rolls-Royce Merlin V-1650-1 V-12 Engines.
W535-ac-15850 To Republic for 773 R-2800 powered P-47B aircraft, 9-13-40.
W535-ac-16323 To Allison for 3691 V-1710-27/29/33/35/39/49/51/53/59/61's
W535-ac-19438 To WAC for ten development R-2160 Engines, Nov 1940.
W535-ac-21623 To Allison for 8,982 V-1710-49/51/55/63/73/75/81/83/85/89's
W535-ac-33362 To Fisher Body Co. to install V-3420's in Douglas XB-19A, 1942.

Appendix 9: Differences Between the V-1710-C15 and V-1710-F2R Engines

Description of the Engines

(1) Model and Rating: Allison V-1710-F2R, Air Corps V-1710-27
Military: 1150 bhp at 3000 rpm at sea-level
Normal: 1000 bhp at 3000 rpm at sea-level
Take-off: 1150 bhp at 3000 rpm
Rated Over-speed: 3120 rpm

(2) The Allison V-1710-F2R engine is a 12-cylinder, 60 degree V-type, high temperature (250 °F) liquid cooled (ethylene glycol) engine of 5-1/2 inch bore by 6 inch stroke, 1710 cubic inches piston displacement. The cylinders are arranged in two blocks of six cylinders each, in line. Each cylinder block is a complete unit, consisting of a single aluminum casting for the combustion chambers, six steel barrels, and a cast aluminum jacket. There are four valves per cylinder, two intake and two exhaust, which are operated by a single overhead camshaft and rocker arms. The heat-treated steel cylinder barrels and the valve seats are shrunk into machined recesses in the head, and the bronze valve guides are pressed into the heads. The exhaust valves are sodium-cooled, and their seats as well as the valve insert seats are stellite faced.

(3) The crankcase consists of upper and lower aluminum alloy castings which split on the horizontal crankshaft centerline. The bearing caps are integral with the lower half. The crankcase decks, on which the cylinder blocks are seated on solid copper gaskets, have 14 long hold-down studs for each block. These studs transmit gas loads directly to the crankcase.

(4) The crankshaft is a conventional six-throw, partially counterbalanced type, all surfaces of which are machined and lapped or polished. The crankshaft journals are hollow and fitted with removable aluminum alloy plugs. The counter-weights are welded to the forged shaft, providing a dependable and easily produced design. Each end of the shaft has a nine-bolt flange, which provides a mount at the rear for the dynamic balancer hub, into which is splined the outer member of the hydraulic vibration damper unit. On the front flange is mounted a female gear tooth coupling for driving the reduction gear pinion. The crankshaft is supported in seven steel-backed, copper-lead lined main bearings.

(5) The drive for the overhead camshafts is provided by a gear train which starts at the rear end of the crankshaft and drives through a hydraulic vibration damper; then, in order: an idler gear, a combination spur and bevel gear, and inclined drive shafts to bevel gears at the rear end of the crankshaft of each cylinder head.

(6) The reduction gearing, together with the propeller shaft, is encased in a short nose section consisting of two aluminum alloy castings which are stud mounted to the front face of the crankcases. The reduction gear is an external spur gear mounted by bolts to a flange on the propeller shaft. The propeller shaft is supported at the front by the thrust bearing, and at the rear by a large roller bearing. The pinion is mounted in two plain steel-backed copper lead lined bearings and is splined to and driven by a loose gear tooth coupling which isolates the crankshaft from any bending load caused by gear reaction. The propeller shaft thrust line is located 8-1/4 inches above the crankshaft centerline, which puts it roughly in the geometrical center of the frontal area of the engine. The front scavenging oil pump is located in the reduction gear housing, and is driven by a "tongue and groove" coupling from the reduction gear pinion. The pad for mounting the propeller governor is at the upper rear of the housing, so located that the governor is located between the two front cylinders. Provision is made in the governor oil system to accommodate both hydraulic and electric constant speed propellers.

Oil for lubrication of the reduction gear assembly is put into the rear housing through a fitting which pilots into the front end of the main oil line in the upper crankcase. The oil is distributed from this fitting, through drilled passages, to the steel-backed, lead bronze lined pinion bearings, the governor drive shaft bushings, the governor mounting pad and to a triple jet oil nozzle which sprays oil onto the gear teeth as they come out of mesh.

(7) The pistons are machined from aluminum alloy forgings. Each piston is equipped with three compression rings above the piston pin, and two oil rings in a single groove near the bottom of t he skirt. The hardened, ground and lapped piston pins float in both piston and rod bushing, and are located endwise in the piston by spring steel snap rings. The under side of the piston head is grid ribbed for strength and as an aid to heat transfer.

(8) The accessories housing is mounted directly on the rear of the crankcase and is a magnesium alloy casting. It contains the centrifugal supercharger and the drives for the coolant pump, camshafts, fuel pump, two vacuum pumps or hydraulic pumps, oil pump, starter, and generator. The supercharger impeller is 9-1/2 inches in diameter and is driven at 6.44 times the crankshaft speed. The supercharger is driven by a gear train which starts with a spur gear on the outside of the hydraulic damper inner shaft, which is flexibly connected to the rear end of the crankshaft through a quill shaft and the hydraulic vibration damper. This gear drives an intermediate shaft, on the end of which is the final drive gear which drives the impeller pinion shaft. The impeller shaft runs in two steel backed floating bushings which are faced with lead bronze inside and outside. The inertia of the rapidly rotating supercharger impeller is utilized to maintain constant speed on the inner member, or "paddle", of the hydraulic damper. The single node low frequency[1] torsional vibratory motion of the outer member of the hydraulic damper and that of the rear end of the crankshaft, to which the outer member is splined) is effectively damped by oil pressure built up in the hydraulic damper by virtue of the differential motion of the inner and outer members. The torsionally flexible quill shaft connecting the inner and outer members serves to limit the travel of the hydraulic damper and to hold the two members in proper mechanical relation. A take-off from this same gear train drives the camshafts and magneto.

The starter is located on the rear face of the accessories housing, at the lower right had corner, and it turns the crankshaft through the spur gear on the outer member of the hydraulic damper, which is rigidly splined to the crankshaft.

A jack shaft in the lower left section of the housing, driven from the inner member of the hydraulic damper, drives the coolant pump through a bevel gear, and the fuel pump, generator, and rear vacuum pump (or hydraulic pump) are driven through a train of gears extending upward to the left alongside the supercharger inlet elbow, this drive originating at the rear end of the above mentioned jack shaft. A second vacuum pump (or hydraulic pump) is driven by a bevel gear meshing with the coolant pump bevel gear. The oil pump is mounted on the bottom of the housing, and is bevel gear driven from the starter shaft. A spiral gear electric tachometer drive is taken off from the camshaft drive train in the center of the housing, the drive shaft extending to the mounting pad on the left side of the housing. Another tachometer drive is furnished on the right had camshaft drive shaft, this also being a spiral gear drive.

(10) A single Bendix-Stromberg two barrel PD-12G1 down draft pressure injection type carburetor with automatic mixture control is mounted on a separate supercharger inlet elbow casting which has the diffuser vanes cast integrally with it. The injection nozzle is mounted on the elbow just below the carburetor, and fuel is sprayed from it directly into the air entering the supercharger impeller. The fuel air mixture passes through the impeller and diffuser vanes into a volute scroll which discharges into the "ramming" type manifolds. At the point in the manifold system where the air flow splits up into the two banks, two circular shaped screens are furnished which are for the purpose of stopping backfire flame travel.

(11) Ignition is supplied by one fixed timing high-tension Scintilla double magneto. The current from the magneto is distributed to the spark plugs through two separate camshaft driven high-tension distributors. Two spark plugs are used for each cylinder. When the engine is enclosed in a tight cowl, a blast of cooling air for the exhaust spark plugs and elbows is let into aluminum tubing manifolds from a source of pressure.

(12) Engine Weight:
Basic engine, including integral supercharger drive mechanism, propeller reduction gears, coolant pump and piping on the engine, engine lubrication system oil pumps, starter connections including starter dog, tachometer drives, fuel pump drive, generator drive, vacuum pump drives, propeller governor drive, and all piping and controls between engine parts:

Item	Weight
Bare Engine Weight	1,193.87 lbs.
Carburetor -	PD-12G1 33.52
Carburetor Screen and Gaskets -	PD-12G1 0.89
Hot Spot, without Connecting Pipes	None
Magneto	12.08
Ignition Distributors and Cable Assembly	24.00
Spark Plugs: 12 BG-LS321; 12 Aero LS3AD	5.69
Priming System on Engine (36905)	0.53
Exhaust Flanges and Gaskets (Army)	8.56
Cooling Air Deflectors and Baffles	None
Accessory Drive Covers	2.23
Total Dry Weight of Engine	1,281.4

(13) Comparison of V-1710-27 and V-1710-33 Engines.
Due to the necessity of evolving a design to eliminate as far as possible all "bottlenecks" to future power rating increases, the V-1710-27 and -29 engines represented a major redesign which could hardly be called a redesign of a previous model, excepting, of course, that all past experience with previous models was fully exploited. The following comparisons describe the detail differences in the V-1710-27 (Allison V-1710-F2R) and the V-1710-33 (Allison V-1710-C15).

(13.a) Reduction Gear Assembly
The -27 reduction gear represented a major change from the -33 type in that the -27 employed a pair of 2:1 external spur gears instead of the 2:1 internal spur gears as used in the -33. The -27 pinion was mounted in two plain bearings in the rear half of the reduction housing, and was driven by the crankshaft through a loosely fitted splined coupling, which design completely isolated the crankshaft from the bending load caused by gear reaction that was present in the -33, which had the pinion splined directly on the end of the crankshaft and had no "outboard" bearing for the pinion.

The propeller shaft of the -27 was a single shaft much shorter and stiffer than the double, telescoped shaft of the -33, and the torsional vibration damping friction clutch of the -33 was eliminated in the -27, due to the widely different torsional characteristics.

The -33 used an extremely long, one piece housing, in which the propeller shaft center was 3.172 inches above the crankshaft centerline, while the -27 used a much shorter and more compact two piece housing in which the propeller shaft was located 8.25 inches above the crankshaft centerline. In the -27, all reduction gear assembly parts were contained in the reduction gear housing, while in the -33, the large plain bearing for the internal ring gear (eliminated in the -27) was mounted in the front end of the crankcase.

The propeller governor mounting pad on the -27 was located on the upper rear wall of the reduction gear housing in such a position that the governor was mounted horizontally in the vee between the number 6 cylinders, and was spur gear driven from the rear end of the propeller shaft. The -33 governor pad was located near the rear end of the reduction gear housing so that the governor was located in front of, and at approximately the same angle as the right hand cylinder block, which location necessitated a bevel gear and shaft drive from the reduction gear scavenge oil pump.

The reduction gear scavenge oil pump on the -27 was mounted coaxial to the reduction gear pinion and was driven by a "Tongue and Groove" drive from the pinion, while the -33 reduction gear pump was driven through a pair of spur gears from the propeller shaft.

The -27 reduction gears were cooled and lubricated by three oil jets from a nozzle so located as to direct the jets onto the teeth as they came out of mesh, while the -33 gears did not have direct oiling.

The reduction gear breathers were identical on both models.

(13.b) Crankcase

The crankcase of both models consisted of two aluminum alloy castings which were split at the crankshaft centerline and were doweled together by hollow dowels through which the main bearing studs passed.

The -27 crankcase had a circular pilot and mounting flange at the front end for the reduction gear assembly, while the -33 case carried a large diameter plain bearing, for the internal reduction gear, in the front end. The rear flanges for attaching the accessories drive housings were essentially the same in both models. The cross sectional form of both crankcases were essentially the same.

The major structural difference in the -27 and -33 crankcases was in the main bearing cap sections of the lower cases, which in the -33 were shaped like the conventional loose cap (but were integral with the case) while the cap sections in the -27 were carried the full depth of the lower case, the main bearings stud nuts being mounted in counterbores in the bottom face of the case. This necessitated considerably longer main bearing studs in the -27 than in the -33.

The -27 had seven plain main bearings, while the -33 used a large roller bearing in the number 7 position, the other six being identical to those of the -27.

The -33 utilized an accessories drive shaft which was mounted inside the upper crankcase immediately below the apex of the two cylinder decks, and extended the full length of the case. The -27 did not have this shaft, and utilized the same space for the main oil line, from which vertical drilled holes fed oil to the main bearings. The -33 had two oil lines, the main oil line being located to the right side of the upper case, and the auxiliary line, to the large reduction gear bearing, at the right side of the lower case. The oil to the reduction gear on the -27 was taken directly from the front end of the main oil line.

Both models used fourteen long cylinder hold down studs threaded directly into the cylinder decks of the upper crankcase.

The -33 upper crankcase had two vertically drilled bolt holes at each corner for engine mount bolts. The -27 also had vertical bolt holes similarly located, and in addition, four rectangular pads with vertical faces, parallel to the longitudinal centerline, each having four stud holes and a bored pilot hole for engine mount brackets, were available for mounting the engine.

(13.c) Crankshaft

The crankpin O.D., main journal O.D. and the crankcheek contour and thickness were dimensionally the same on both models, excepting that on the -27 shaft, all twelve crankcheeks were the same thickness, while the -33 shaft had a considerably thicker front cheek to withstand the combined torsional and bending loads due to the gear reaction from the overhung reduction gear pinion. The -27 crankpin and main journal I.D. were 1/8 inch smaller (making the load carrying path 1/8 inch thicker) than those of the -33.

The -27 shaft had flanges at each end for the attachment of the tuned pendulum type dynamic vibration damper hub at the rear, and the internal gear tooth coupling member for driving the reduction gear pinion at the front. The end flanges were identical, each having nine bolt holes for attachment. The shaft is assembled end-for-end in the crankcase for the -29 engine.

The dynamic damper hub had a central internal spline for driving the outer member of the hydraulic vibration damper.

The -33 shaft had splines at both ends for the mounting of the reduction gear pinion at the front and the starter and pump drive gear and dynamic damper at the rear.

Main journal and crankpin oil plugs were similar in both models.

(13.d) Connecting Rods

The -27 and -33 rods differed only at the big ends. The -27 rods were strengthened by the use of bolts with full round heads having tapered seats, with screw driver slots to hold the bolts from turning at assembly. The use of these bolts allowed the elimination of the notches which were milled in the -33 rods to hold the bolt locking lugs. The elimination of these notches strengthened the rods at a fairly critical point. (It is planned to standardize on yet further improved rods, part number 37883 and 37884, on the -E4(-35), F2R(-27), F2L(-29) and F3R(-39) engines for production).

(13.e) Accessories Drive Housing Assembly

The -27 accessories drive housing casting is a completely different casting than that of the -33. All accessories drives were relocated to give a more compact grouping of accessories and engine pumps. The -27 supercharger driving train (described in paragraph 9 above) was arranged so that the inertia of the impeller could be utilized as a force for stabilizing the rear end of the crankshaft torsionally, while the -33 supercharger was driven from the reduction gear by a long, flexible shaft extending the length of the crankcase, and had no connection with the rear end of the crankshaft.

The -27 engine was a ground boosted sea-level engine, equipped with a 9-1/2 inch diameter supercharger impeller driven at 6.44 times crankshaft speed, while the -33 was an altitude rated engine in which the 9-1/2 inch diameter impeller was driven at 8.77 times crankshaft speed. The -27 impeller did not use a rotating guide vane such as was used on the -33.

On the -27, advantage was taken of the space below and to the left side of the supercharger inlet elbow to mount accessories in the space which was not thus utilized on the -33.

The -27 hydraulic damper was not used on the -33.

In the -27, the vertical magneto drive was moved over to the right of center and was driven from the right hand camshaft drive shaft, while the -33 magneto drive was located on the vertical centerline. This necessitated a streamlined housing for the magneto drive shaft, passing across the supercharger scroll outlet on the -33, but on the -27 the scroll was led around the left side of the magneto drive.

The generator drive on the -27 was located in the gear train in the supercharger inlet elbow casting, parallel to the crankshaft centerline, while on the -33, it was located at the lower left corner of the accessories drive housing and was at a 15 degree angle above the crankshaft centerline, being bevel gear driven as compared to the spur gear drive in the -27.

The vacuum pump drive, which on the -33 was located forward of the generator drive, was located immediately above the generator and in the same spur gear train as the generator, on the -27. In addition, the -27 had a second vacuum pump (or hydraulic pump) drive located near the bottom of the housing on the left side, which was driven by a bevel gear drive from the coolant pump.

The fuel pump drive on the -27 was located below the generator drive, and in the same spur gear train, while on the -33 it was located just forward of and below the generator.

The -27 starter drive was located on the rear face of the housing, at the lower right hand corner, while the -33 starter drive was on the engine centerline directly below the supercharger inlet elbow.

The -27 oil pump was mounted in an adapter located on the bottom face of the housing, to the right of the engine centerline. Its drive shaft was vertical, and it was bevel gear driven from the starter shaft. The -33 oil pump was mounted on the right side of the housing near the bottom, its drive shaft being horizontal. It was bevel gear driven from the starter shaft.

On the -27, all scavenge and pressure passages connecting the pump to the accessories housing, and the inlet and outlet connections to and from the oil tank, were located in an oil pump adapter casting, while these passages were in the oil pump body castings on the -33.

The -27 and -33 oil pumps differed mainly in that the -27 pump gears were mounted in round bodies which fitted into a bore in the pump adapter, while the -33 pump bodies were mounted directly on the accessories housing.

The method of oil pressure regulation was different in the two models in that, on the -33, the pressure relief valve was operated by pressure from a line on the outlet side of the Cuno oil strainer, while on the -27, is was operated by oil pump outlet pressure (on inlet side of Cuno).

The -27 used an oil pressure operated automatic Cuno oil filter, while the -33 used a manually turned Cuno filter. The filters were located at practically the same point on both models.

The coolant pump on the -27 was located on the bottom face of the accessories drive housing to the left of center, while the -33 pump was nearly the same location, being on the centerline of the engine.

(13.g) Coolant Pump

The coolant pumps differed mainly in that the -27 had a vertical inlet from the bottom and a single scroll discharge pipe with a tee branching off into two branches leading to the cylinder blocks, while the -33 pump had a horizontal inlet and two discharge branches taken directly from the scroll.

The -27 pump packing was a spring loaded non-adjustable Crane type seal consisting of a spring which loaded a plastic carbon disc against a disc of bakelite in such manner as to also force it radially inwardly against the shaft, the complete assembly turning with the shaft and eliminating friction between the shaft and the plastic seal. The -33 seal consisted of a formed plastic metallic packing in which the shaft ran, the packing being adjusted by pressure obtained by displacing a plastic material by means of a screw. This plastic material was forced in between the packing glands which in turn exerted pressure on the metallic packing, in which the shaft turned.

(13.h) Carburetor

The -27 used the Bendix-Stromberg model PD-12G1 two barrel carburetor, while the -33 used the Bendix-Stromberg model PT-13E1 three barrel carburetor.

(13.i) Intake Manifold Tees and Backfire Screens

The -27 used the PN40041 intake manifold tees, in which the dividing vane is brought down to the top surface of the PN40043 backfire screen in order to reduce the volume of charge which could be ignited by a backfire from any one cylinder. The -33 used the PN34162 tees, in which the dividing vane terminated about 9/16 inch above the backfire screen PN36930.

(13.j) Distributor Drives

The distributor drives were considerably different, due to the fact that no gun synchronizers were required on the -27. On the -27, the distributor finger drive shafts were co-axially driven from the rear ends of the camshafts, while on the -33, the coaxial shafts drive the vertical distributor finger shafts through bevel gears, and the gun synchronizer drives were extended from the coaxial shafts.

(13.k) Timing

The ignition timing on the -27 was:
Intake plugs 28 degrees B.T.C.,
Exhaust plugs 34 degrees B.T.C.,

while on the -33 it was 29 degrees and 35 degrees respectively.

The valve timing on both models was as follows:

Intake Opens 52 degrees B.T.C.
Intake Closes 66 degrees A.B.C.
Exhaust Opens 76 degrees B.B.C.
Exhaust Closes 26 degrees A.T.C.

Reference: Excerpted from NASM file ALLISON D52.41/64: Allison Division of GMC, Test Department Report No. A2-7: 150-hour Model Test, Air Corps Specification #AN-9502, on the Allison V-1710-F2R Aircraft Engine, M'F'R's #305, Air Corps Model V-1710-27, Serial # 39-1131, Contract No. W535-ac-12553, January 23, 1941.

[1] As noted in Tech Order AN 02-5AH-2.

Appendix 10: Manufacturing Allison Steel Backed Bearings

That the Allison firm was able to survive the untimely death of Jim Allison and the economic Depression of the 1930s was largely due to their ongoing business manufacturing steel backed plain bearings that they provided to engine manufacturers around the world. The experience gained in this endeavor, including large scale production, provided a basis for the coming expansion into engine development and manufacturing. We will now take a look at how the bearing came about, how it evolved, and how it was produced.

Allison Steel Backed Bearings

Simply stated, Allison went into the bearing business as a matter of necessity when the Government decided to modify several thousand Liberty-12 aircraft engines left in supply depots following the end of WWI. The intent of these modifications was to add reliability and life to the engines, though there was also a 30-50 bhp increase in power as a result of the new bearings allowing the rated speed of the Liberty to be raised to 1800 rpm.[1]

The Liberty was originally equipped with conventional bronze shell babbitt lined bearings, which were definitely one of the engine's weak features, though standard for the day. This was largely due to the differences in thermal expansion between the steel connecting rods and the bronze bearing that required large clearances and resulted in a setup which was not tolerant of heavy bearing loading.

Air Corps engineers at the Fairfield Air Depot (located near Dayton, Ohio) were experimenting with a bearing having a steel back and babbitt lining, but they were apparently stumped on how to provide a suitable bearing surface for the blade rod in the blade and fork assembly to ride upon. Allison engineers were consulted on the problem. The Government had approached Allison with a proposal that, if this weakness in the Liberty could be overcome they were prepared to contract for modifications of several thousand engines.

Allison engineers were of the opinion that the only satisfactory connecting rod bearing for the Liberty would be one made with a steel back to which a suitable bearing bronze could be applied in some manner to obtain a bearing surface for the blade rod. Such an arrangement would require an indestructible bond between the bronze and steel. Experiments were started on this basis.

As devised by Allison for the Liberty, the steel shell supported a layer bronze, which was then babbitt lined in the conventional manner.[2] The design was complicated by the dissimilar requirements to have a relatively hard bronze to support the blade rod, and a much softer material needed to cushion the inner diameter of the "forked" rod riding on the hardened crank-pin journal. The Air Corps was pleased with the early results and ordered 100 sets of these new bearings.

Manufacture of the bearing followed a trial and error period to find a way to consistently produce high quality bearings. The original method was to machine the steel shell and then to weld an enlarged, but thin steel sleeve or jacket around the outside such that one side was left open. With the shell laying on its end such that the opening was on top, a ring of cast bronze was then set on the opening. When placed in a furnace the bronze would melt and run down into the space, displacing borax flux which had been filling the space. Allison soon learned that they needed to quickly

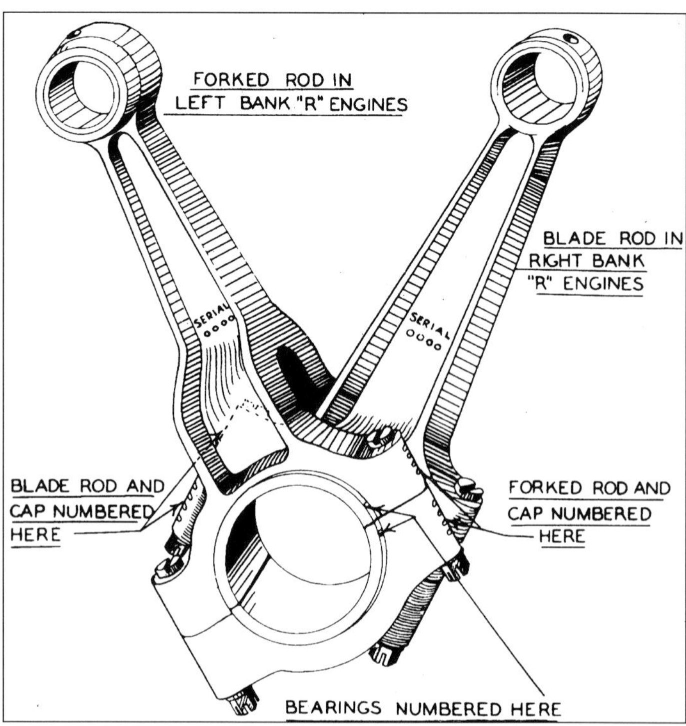

Blade and fork connecting rods as used in V-1710 and V-3420 engines. (Allison)

quench the bronze to assure the bond by controlling grain size as well as controlling the distribution of the lead, which otherwise tended to settle out at the lower side of the bearing shell. For the original Liberty bearings a conventional babbitt was used as the material to support the steel shell surrounding the crankpin. A similar approach was used to cast the inner bearing surface. The follow-up steps were to then machine away the thin steel jackets and finish the bearing surfaces.

To prevent the separation and settling of the lead during melting the previously cast lead-bronze rings were replaced by a crucible of molten lead-bronze which was removed from the furnace just before pouring, and stirred violently to get the lead evenly dispersed. With this method, followed by a rapid quench to provide a fine grain structure and tightly adhering bronze, there were very few visible lead particles in the final product. The final development of the Liberty bearing manufacturing process required understanding and standardizing all of the many variables in the process, each being specific to the size, loading and usage intended for the bearing.

After the Liberty program Allison continued to work on the process and next developed an improved bearing having a lead-bronze inner surface as well. This allowed still higher loadings at even higher temperatures.[3] With these new bearings the way was paved for the rapid development of high-powered engines that occurred over the next twenty years.

Following the initial production of connecting rod bearings for the Liberty, Allison went on to provide Master Rod bearings for Wright radial engines, and shortly thereafter, complete bearing sets for the V-12 Curtiss Conqueror. Other manufacturers were soon equipping their engines with the bearings as well. The Wright T-3 Marine engine, Packard aero and marine engines along with the Pratt & Whitney aircraft engines being early examples. General Motors also used them in its Winton Diesels.[4] Quantities were not large, for when production was begun in 1927 there were only a total of 1,400 aircraft engines built in all of the U.S.! It was one of these engines, a 1927 built Wright J-5, that was specially fitted with Allison steel-backed bearings just before it that powered Lindbergh's *Spirit of St. Louis* on its epic solo flight across the Atlantic to Paris, and into history in May 1927.[5]

Allison also provided the bearings to foreign manufactures, with fairly large numbers going to the French Hispano-Suiza Company. Rolls-Royce in England attempted to develop a similar bearing on their own, but soon gave up the effort in favor of purchasing manufacturing rights and paying a royalty to Allison for every bearing they subsequently manufactured.[6] As a consequence of the new bearing business and the impetus given aviation by Lindbergh's crossing of the Atlantic, Allison saw its sales increase from barely $200,000 in 1925, passing $425,000 in 1927 and approach $675,000 in 1928.[7] The depression caused sales to drop off, from 329,149 bearing units in 1929 to 54,199 in 1933. By 1934 volume was up to 174,228 units and in 1937 reached 474,737.[8]

In 1932 Allison used this experience to develop an economical steel-backed bronze-lined bearing for automobile engines. The bronze was cast onto thin steel plates that were then cut into strips and formed into bearing shells. After developing the process through the pilot-line stage the program was transferred to the AC Spark Plug Company of GMC who produced the bearings. These were used in some non-General Motors automobiles for some time.[9,10]

Beginning in 1935 Allison began experiments with silver lined bearings, and these were soon being provided to Pratt & Whitney. Early production was done using the same process as used to hot-cast the lead-bronze bearings. Continuing work on electroplating the silver wearing surface was eventually successful and became the predominant method of manufacture.

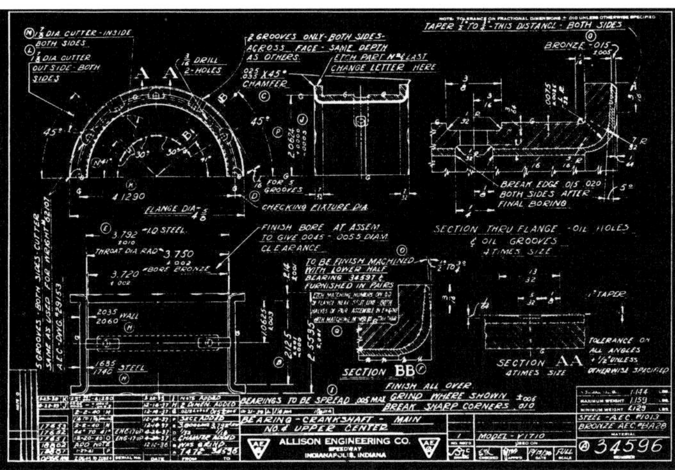

This production drawing of the upper half of the main crankshaft thrust bearing, located in the center of the engine at position #4, gives an idea of the details necessary to manufacture an acceptable bearing. Note that this design began in December 1936, during the period of the Hazen redesign. Changes made 4-26-37 applied to engine #17 and up. AEC#17 was the single XV-1710-15(C9) for the XP-38. This drawing was still effective as of 3-23-42. (Allison via Hubbard)

Bearings for the V-1710

Work on improving the bearings, crankshaft and connecting rods as a system continued throughout the war years. Advanced developments and experimental results obtained by Allison in early 1941 are interesting as they defined the direction future production components were going to follow. These results were determined during "dive" (over-speed) testing done on the Indianapolis V-1710 test beds primarily to resolve a number of early bearing troubles being encountered in the field.

Allison was experiencing an undesirable number of bearing rejections of their standard steel backed bronze bearings being used in the V-1710-33(C15). Considering the trend toward higher power and the likelihood of needing even higher crankshaft speeds in the future, they sought a substitute bearing material which would offer fewer rejections, yet carry the increasing loads due to higher power and speed. Bearing failure in the field was believed by Allison to be caused by over-speeding which was attributed to lack of proper automatic or manual propeller pitch control. As a consequence they desired to provide additional over-speed margin as a matter of safety. The tests investigated the effects of extreme loading and difficult lubricating conditions on the crankshaft bearings as affected by various designs and materials. Testing of each combination was pursued until failure occurred. The conclusions are interesting and significant:[11]

> Steel backed, silver-lead plated connecting rod and main bearings running on a nitrided crankshaft gave the maximum speed reached without bearing failure, 4100 rpm.
>
> Lead-plated silver rod bearings were superior to bearings with only silver plating.
>
> Silver-coated, lead-plated and silver-coated steel connecting rod bearings were superior to the steel backed lead plated, bronze connecting rod bearings from the standpoint of seizure under over-speed conditions.
>
> When failures occurred, they took place either at the No. 2 or No. 5 connecting rod bearing and were the result of insufficient oil supply to that position. The silver bearings required less oil than the

bronze type and the straight groove main bearings allowed more oil to go to the connecting rods than the spiral grooved main bearings.

On a related matter, Allison built and delivered all of the V-1710-33's on contract W535-ac-12553, using a costly and time consuming process that involved line-boring the bronze main bearings while installed in the crankcase. Lead plated, pre-fitted copper-lead bearings were first used on the W535-ac-16323 contracted V-1710-33's.[12] With the introduction of the considerably changed and strengthened E/F engine crankcases it was believed that deflection of the main bearing support webs would be considerably reduced. To that end an experimental program was run early in 1941 to investigate a new manufacturing procedure which relied upon interchangeable bearings designed to not require line-boring.

The results were quite satisfying. It was found that crankshaft speeds of 3500 rpm could be routinely achieved without encountering excessive main bearing temperatures. Various combinations of bearings were used, including the early type of steel-backed bronze, leaded with tin plating (incorporated the "straight" type oil groove), as well as steel backed bronze having copper and lead plating and incorporating the "spiral" oil groove.[13] On the basis of these tests the interchangeable main bearing design was introduced into production, greatly simplifying the manufacture and overhaul of the subsequent V-1710-E/F/G engines. These also used a silver plated bearing rather than lead plating.

When the silver-lead bearings were introduced into production early in 1942 there were a number of introductory problems. The bearing surfaces were necessarily quite soft, and consequently susceptible to damage by abrasive materials such as sand and/or any metal chips circulated by the engine oil system. This became somewhat of a crisis for those units operating in dusty or sandy areas. Fortunately most of the problems appeared stateside during the period when new fighter squadrons were working up and training before their overseas deployments.

The problems were traced to a lack of proper pre-oiling and flushing of the oil systems on airplanes following engine changes. It was also found to be essential to replace the oil coolers following an internal engine failure, for otherwise they became a source of chips or sand and caused failure of the replacement engine. Many of the failures were occurring in P-39 airplanes, where it was determined that the original oil tank was impossible to completely flush to remove chips or sand. A redesign of the tank was necessary to eliminate the internal traps causing the problem.[14]

The sand was getting into the engine through the induction system, and for this reason sand filters were soon incorporated into aircraft coming from the production lines. At the same time work was underway to improve the effectiveness of filters in the lubricating system. As a consequence of the lessons learned, design changes and improved procedures, the silver plated bearings went on to give very satisfactory performance in not only the V-1710 but similarly equipped Merlin and radial engines as well.

Bearing Production

With the onset of war in 1939 the demand for bearings outstripped production capability and Allison personnel helped train and establish the AC Spark Plug Division of GM as a second source of the bearings. One consequence of the war was that supplies of foreign industrial silver were soon exhausted and it became necessary to substitute silver from the U.S. Treasury, an action ordered by the War Production Board. The cost of bearings went up accordingly, for the price of silver had been about 36 cents per ounce, where using Treasury or domestic silver cost 71 cents per ounce. During the war, from 1941 through 1945, the silver disbursed in bearings would have produced 13,955,000 U.S. silver dollars, a total of 11,993,000 troy ounces, nearly 412 tons.

While the silver lined bearings contributed significantly to the operating life of the wartime V-1710, for a fair number of the engines that survived the war they caused their demise. During the 1970s the Hunt Brothers of Texas were able to get control of the silver commodity market and succeeded in running the price of silver to unheard of levels. People who had been holding V-1710's for a long time, and seeing little other future for them, proceeded to break them apart and take out the main and connecting rod bearings. Only to melt them down for the silver!

In total, nearly 10,000,000 bearings were supplied by Allison for the war, with Allison using about 4.5 million of them for the V-1710. During the course of the war it was found that the number of spare bearings which were being ordered was more than required to support the engines in service. This was the direct result of the durability of the bearings being in excess of expectations by military planners. By the end of 1943 Allison was providing the highest grade sleeve bearings, which were not only being used in Allison engines, but by Packard in building Rolls-Royce Merlin's, Pratt & Whitney, Wright, Jacobs and Electro-Motive.[15] The V-1650 *Merlin* powering the North American *Mustang* achieved its acclaim for power and reliability riding on Allison bearings!

NOTES

[1] Fleming, Roger 1950, 1.

[2] Goldthwaite, John L. 1950, 1.

[3] Goldthwaite, John L. 1950, 1.

[4] Fleming, Roger 1950, 18.

[5] Fleming, Roger 1950, 2.

[6] Fleming, Roger 1950, 25. Packard Built Rolls-Royce V-1650 Merlin's of WWII fame also used the Allison bearing.

[7] Fleming, Roger 1950, 12.

[8] Fleming, Roger 1950.

[9] Fleming, Roger 1950, notes, 1.

[10] Fleming, Roger 1950, notes, 4.

[11] Hubbard, John D. 1941.

[12] McCrae, T.S. 1941, 8-9.

[13] Hubbard, John D. Report No. C9-6, 1941.

[14] *Main Bearing Failures, V-1710 Engines, and Sand Contamination of Lubricating Oil*, Wright Field memo to Commanding General, AAF Materiel Command from Chief, Experimental Engineering Section, June 1, 1942. NARA RG18, File 452.8, Box 807.

[15] *Allison Division Operating Report*, as of December 1, 1943.

Appendix 11: Marine V-1710's

With the availability of the high powered engine from Allison it was only natural for it to be considered for marine use. In the early days Allison was eager for any market opportunity, and the result was that several models were offered. With the demands of building up for wartime airplane production all of the marine projects were shelved, including the V-3420-A8R/L project for the Navy PT-8. Later in the war, when early model P-38 engines were effectively surplus after having been overhauled, a program was put together by Gar Wood to convert a number of them to marine use.

After the war the potential of the V-1710 for marine propulsion was exploited by the Unlimited Hydroplane racers. They collected hundreds of surplus V-1710's, preferably the -100 series models, and modified them to produce as much as 4000 bhp in racers. The resulting "Thunder Boats" were just that, exciting racers that achieved unheard of speeds on water.

Allison designated their early marine engines as the "H" series. Although the use of the designation in 1938 suggests that a "G" series had also been anticipated, no early documentation has been found. The first use of "G" otherwise dates from 1942. Since extensive use of the earlier "H" designation had been mostly forgotten, Allison did use it in proposals of advanced aircraft V-1710's offered in 1946.

Marine H-1: This 1938 project provided an "E" series power section with a 2.5:1 reduction gear weighing 235 pounds for Marine use, with total weight at 1,515 pounds. In addition, a 60 pound salt-water cooling pump and an additional 90 pounds of water cooled exhaust manifolds were required. Maximum power was 1150 bhp at 2950 rpm. An "E" series designation was not assigned, nor was a military designation given. Whether or not any engines were delivered to this standard has not been determined.

Marine H-2: In 1939 Allison offered an engine also known as the V-1710-E7, having 6.00:1 supercharger step-up gears, but without a reduction gear. Total weight was down to 1,175 pounds, and rated for 1000 bhp at 2600 rpm. An additional 90 pounds of water cooled exhaust manifolds were required. A military designation was not given. Whether or not any engines were delivered to this standard has not been determined.

Marine

While the details have not come to light, there were several efforts during the middle of the war to adapt the V-1710 to boat projects. At one point the well known boat racer Gar Wood, and his GarWood Industries in Algonac, Michigan, was under contract to design and test a Marine conversion, using either new or used engines. He was to develop three sets of paired left and right turning engines. Models considered were the F-5R/L, F-10R/L and F-17R/L coming from the P-38 aircraft.[1]

[1] Requisition No. T-1981 from Marine Design Branch, AAF Marine Section, for shipment of six V-1710-F engines to GarWood Industries, Algonac, MI, November 17, 1943. NARA RG18, File 452.13, Box 1260.

Bibliography

Bibliography

Archives

Allison Engine Company, Indianapolis, Indiana.

Hoover Institute, Stanford U., Stanford, CA.

Kirtland AFB, New Mexico. Personal Operations Records-Air Force.

Library of Congress, Manuscript Collection, Washington DC.

NARA, National Archives and Records Agency, Washington DC.

NASM, National Air and Space Museum, Smithsonian Institution, Washington DC.

NEAM, New England Air Museum.

SDAM, San Diego Aerospace Museum, San Diego, CA.

Stanford University, Engineering Library, Stanford, CA. NACA Reports and period periodicals

USAFM, US Air Force Museum, Wright Patterson AFB, Ohio.

Allison Publications/Releases

Allison Division, GMC, 1942 news release. Condensed History of the Allison Engine.

Allison Division, GMC, Overhaul Manual 1943. Allison Overhaul Manual for V-1710-"E" Type Engines, 2nd Edition August 1, 1943.

O'Brian, Don, 1950. Historical Dates, Allison Public Relations, 1950. Available at NASM, file B1001025.

Tale Spins, September 1943. Published by Allison Service Department, Volume II, Number 6.

Allison Technical Reports

Allison Experimental Department Report No. A2-143, 1944. 7-1/2 Hour War Emergency Rating Test at 65 inHgA Manifold Pressure on V-1710-F17R Engine A-038577, June 10, 1944. Wright Field file D52.41/202-Allison.

Allison Experimental Department Report No. A2-147, 1944. 150 Hour Test on the Allison V-1710-E22 Aircraft Engine, Manufacturer's No. A-051920, Army Air Forces Model V-1710-109, August 21, 1944. Wright Field file D52.41/201-Allison.

Allison Report No. A2-7, 1941. 150 Hour Model Test on Allison V-1710-27 Aircraft Engine, Manuf s/n 305, AAC s/n 39-1131, Contract W535-ac-12553, January 23, 1941. File D52.41/64-Allison at NASM.

Allison Service Engine Report No. A2-6, 1940. Engineering Inspection, Allison Engine #301, V-1710-F3R, A.C. Engine #40-4395, Model V-1710-39, December 5, 1940. Wright Field File: D52.41/54-Allison.

Buttner, H.J. 1937. On Condition of Parts Affected by Loosening of Dynamic Damper #36022 on Model V-1710-D1 Engine #14 After 98 Hours on Type Test at Wright Field, Allison Engineering Co. Test Memo Report, November 15, 1937. Wright Field file, D52.41/16-Allison.

Caminez, H., Report No. 26, 1930. Stress Analysis of Allison VG-1710 Engine Rated 750 H.P. at 2400 R.P.M., Allison Engineering Company Report No. 26, September 24, 1930, 72 pages.

Caminez, H., Report No. 27, 1932. Allison GV-1710-A Engine Data, Allison Engineering Company Report No. 27, June 3, 1932.

Caminez, H., Report No. 28, 1930. Analysis of Valve Cam and Springs, Allison VG-1710 Engine, Report No.28, 33 pages, September 26, 1930.

Caminez, H., Report No. 43, 1932. Description of Allison GV-1710-A Engine, Report No. 43, June 3, 1932, Allison Engineering Co.

Caminez, Harold and F.N.M. Brown. 1935. Large Engine Development, paper delivered to the 1935 ASME National Technical Aeronautic Meeting-Aeronautics Division, St. Louis.

Emmick, Wm. G. 1939. Allison Report No. 286, Thirty-Five Hour Development Test of the Allison V-1710-17 Engine, August 15, 1939. Wright Field file D52.41/18-Allison.

Hubbard, John D. 1941. Durability of Main and Connecting Rod Bearings Affected by Material, Lubrication and Load, Allison Test Department Report, Test Orders 298 & 490, Testing occurred 3-4-41 through 8-23-41.

Hubbard, John D. 1942. Allison Experimental Department Report No. I4-3, Effect of Torsional Deflection of P-39 Airplane Nose Structure on Movement of Reduction Gear Mounting Angles and Stress in Bottom Left Angle Stud, November 23, 1942.

Hubbard, John D. Report No. C9-6, 1941. Performance of Modified Interchangeable "E" and "F" Main and Connecting Rod Bearing Installations on 3500 rpm Overspeed Tests, Allison Test Department Report No. C9-6, written 3-4-1941.

McCrae, T.S. 1941. Difficulties Encountered in Allison Engines, Particularly Fire and Backfiring, T.S. McCrae, Allison Assistant Chief Engineer to Materiel Division, Wright Field, June 17, 1941.

Montieth, O.V. 1934. Tests of Allison V-1710-4 Engine in Accordance with Navy Specification E-4G from May 23, 1934 to June 25, 1934, Allison Engineering Company Report #71, July 3, 1934. From NASM microfilm B0002220.

Montieth, O.V. 1939. Allison Report No. 285, Detail Design Analysis and Torsional Vibration Analysis of the Allison Model V-1710-E2 Aircraft Engine, August 15, 1939.

Schmid, J.C. 1940. V-1710-E and F Dive Test to Determine the Advisability of Decreasing Oil Flow by Decreasing Main Bearing Clearance, Allison Experimental Department Report No. 311, 1-6-1940, File H-10.

Sherrick, E.B. 1942. Allison Engineering Department Report No. 369, Analysis of Crankshaft and Connecting Rod Bearings and Crankcase Flexure of the Allison V-1710-F Engine 1500 bhp 3000 rpm with Twelve Counterweight Crankshaft #43900, 11-25-1942. Copy at NASM.

Sherrick, E.B.1944. Allison Engineering Department Report No. 379, Design Analysis of the Allison V-1710-G1R Engine 1725 bhp 3400 rpm Power Section Components, 1-1-1944. Copy at NASM.

Air Service Information Circulars

Air Service Information Circular No.143, 1920. Report on Performance and Design of Five Representative Geared Aviation Engines, Engineering Division, Vol. II, No. 143, McCook Field, June 7, 1920.

Air Service Information Circular No.551, 1926. First 50-Hour Test of the Air-Cooled Liberty Engine, Vol. VI, No. 551, January 25, 1926.

Books

Althoff, William F. 1990. *Sky Ships: A History of the Airship in the US Navy*, New York: Orion Books.

Angelucci, Enzo with Peter Bowers, 1985. *The American Fighter*, New York: Orion Books

Ascani, Maj. General Fred J. and John M. Fitzpatrick, 1993. *Taking Some Friends for a Ride, an article in Test Flying at Old Wright Field, From the Piston Engine to Jet Power*, Chilstrom, Ken and Penn Leary, editors, Omaha, Nebr.: Westchester House.

Beauchamp, Gerry, To be published in 1995. *Curtiss Hawk 75.*

Bentele, Max, 1991. *Engine Revolutions: The Autobiography of Max Bentele*, SAE Historical Series, Warrendale, PA: SAE Inc.

Birch, David, 1987. *Rolls-Royce and the Mustang*, RRHT Historical Series No. 9, Derby, England: Rolls-Royce Heritage Trust.

Birdsall, Steve 1980. *Saga of the Superfortress*, Garden City: Doubleday.

Bodie, Warren M. 1991. *The Lockheed P-38 Lightning, The Definitive Story of Lockheed's P-38 Fighter*, Hiawassee, Georgia: Widewing Publications.

Bodie, Warren M. 1994. *Republic's P-47 Thunderbolt, From Seversky to Victory*, Hiawassee, Georgia: Widewing Publications.

Bowers, Peter M. 1966. *Boeing Aircraft Since 1916*, New York: Funk & Wagnalls.

Bowers, Peter M. 1979. *Curtiss Aircraft 1907-1947*, Annapolis, Maryland: Naval Institute Press

Boyne, Walter J. 1984. *de Havilland DH-4, From Flaming Coffin to Living Legend*. Washington DC: Smithsonian Institution Press.

Byttebier, Hugo T. 1972. *The Curtiss D-12 Aero Engine*, Smithsonian Annals of Flight Number 7, Washington DC: Smithsonian.

Caidin, Martin. 1960. *Black Thursday*, New York: Elsevier-Dutton

Chilstrom, Ken and Penn Leary, editors. 1993. *Test Flying at Old Wright Field, From the Piston Engine to Jet Power*. Omaha, Nebr.: Westchester House

Cohen, Stan, 1981. *The Forgotten War*, Missoula, Montana: Pictorial Histories.

Dickey, Philip S. III. 1968. *The Liberty Engine 1918-1942*, Smithsonian Annals of Flight, Vol. 1, No. 3. Washington DC: Smithsonian Institution Press.

Doolittle, J.H. with Carroll V. Glines, 1992. *I Could Never Be So Lucky Again, Autobiography of General James H. "Jimmy" Doolittle*. New York: Bantam Books.

Ford, Daniel, 1991. *Flying Tigers*, Washington and London: Smithsonian Institution.

Francillon, Rene J. 1979. *Douglas XP-48*, McDonnell Douglas Aircraft Since 1920, London: Putnam.

Francillon, Rene J. 1982. *Lockheed Aircraft Since 1913*. London: Putnam & Co.

Green, William and Gordon Swanborough, 1977. *US Army Air Force Fighters*, Part 1, New York: Arco.

Green, William. 1971. *Augsburg Eagle, The Story of the Messerschmitt 109*, New York: Doubleday

Green, William. 1979. *Warplanes of the Third Reich, The*, Garden City, NY: Doubleday.

Gruenhagen, Robert W. 1980. *Mustang, The Story of the P-51 Fighter*, Revised Edition, New York: Arco Publishing.

Harding, Stephen and James I. Long, 1983. *Dominator, The Story of the Consolidated B-32 Bomber*, Missoula, Montana: Pictorial Histories.

Hardy, M.J. 1979. *The North American Mustang, The Story of the Perfect Pursuit Plane, P-51*, New York: Arco Publishing Co.

Harvey-Bailey, Alec and Dave Piggott, 1993. *The Merlin 100 Series-The Ultimate Military Development*, RRHT Historical Series No. 19, Derby, England: Rolls-Royce Heritage Trust.

Harvey-Bailey, Alec and Michael Evans, 1984. *Rolls-Royce: the Pursuit of Excellence*, RRHT Historical Series No. 3, Derby, England: Rolls-Royce Heritage Trust.

Harvey-Bailey, Alec, 1982. *Rolls-Royce: the Formative Years, 1906-1939*, RRHT Historical Series No. 1, Derby, England: Rolls-Royce Heritage Trust.

Harvey-Bailey, Alec, 1984. *Merlin in Perspective-the Combat Years*, The, RRHT Historical Series No. 2, , 2nd Edition, Derby, England: Rolls-Royce Heritage Trust.

Hickman, Ivan, 1990. *Operation Pinball, The USAAF's secret aerial gunner program in WWII*, Osceola, Wisconsin: Motorbooks International.

Hooker, Sir Stanley, 1984. *Not Much of an Engineer, Autobiography of Sir Stanley Hooker*, Shrewsbury, England:Airlife.

Huntington, Roger, 1989. *Thompson Trophy Racers*, Osceola, WI: Motorbooks.

Johnson, C.L."Kelly" with Maggie Smith, 1985. *Kelly, More than My Share of It All*, Washington DC: Smithsonian Institution Press.

Jones, Lloyd S. 1962. *U.S. Bombers, B1-B70*, Los Angeles: Aero Publishers.

Kelsey, Benjamin S. B.G. USAF (Ret.) 1982. *The Dragon's Teeth, The Creation of United States Air Power for World War II*. Washington DC: Smithsonian Press.

Lacey, Robert, 1986. *Ford, The Men and the Machine*, Boston-Toronto: Little, Brown and Company.

LeMay, General Curtis E. and Bill Yenne, 1989. *Superfortress*, New York: Berkley.

Losonsky, Frank S. and Terry M. Losonsky, 1996. *Flying Tiger-A Crew Chief's Story*, Atglen, PA: Schiffer Military/Aviation History.

Mason, Francis K. 1962. *The Hawker Hurricane*, MacDonald Aircraft Monographs, Garden City, NY: Doubleday & Co.

Matthews, Birch J. 1993. *Wet Wings & Drop Tanks*, Atglen, PA: Schiffer.

Matthews, Birch J. 1996. *Cobra! Bell Aircraft Corporation 1934-1946*, Atglen, PA: Schiffer Military/Aviation History.

Mitchell, Rick, 1992. *Airacobra Advantage: The Flying Cannon*, Missoula, Montana: Pictorial Histories.

Nockolds, Harold. *Magic Of A Name*, The, 3rd Edition, Henley-on-Thames, Oxfordshire: G.T. Foulis, Distributed in US by Motorbooks International.

Norris, Jack 1988. *Teledyne Continental Motors Liquid-Cooled 200 HTCC BHP vs. Fuel Consumption Curve* No. 88037, updated 3-10-88, from Voyager-The World Flight, Northridge CA:Norris.

Page', Victor W. 1929. *Modern Aviation Engines*, Volume Two, New York: Norman W. Henley Publishing Co.

Rickenbacker, Edward V. 1967. *Rickenbacker, an autobiography of Edward V. Rickenbacker*, Englewood Cliffs: Prentice-Hall

Robertson, Bruce, 1961. *Spitfire-The Story of a Famous Fighter*, Letchworth, Herts: Harleyford

Schlaifer, R. and S.D. Heron, 1950. *Development of Aircraft Engines and Development of Aviation Fuels*, Graduate School of Business Administration, Boston: Harvard University.

Scott, Robert L., Colonel, Air Corps, U.S. Army, 1944. *God is my Co-Pilot*, New York: Charles Scribner's.

Shamburger, Page and Joe Christy, 1972. *The Curtiss Hawks*, Kalamazoo, Michigan: Wolverine Press.

Sloan, Alfred P., Jr. 1964. *My Years with General Motors*, Garden City, New York: Doubleday.

Sonnenburg, Paul, and William Schoneberger. 1990. *Allison-Power of Excellence, 1915-1990*. Malibu: Coastline

Taylor, Charles Fayette, 1960. *The Internal-Combustion Engine in Theory and Practice*, Volume II, Cambridge, Massachusetts: M.I.T. Press.

Thomas, Lowell and Edward Jablonski, 1976. *Doolittle, A Biography*, New York: A Da Capo Baperback.

Thorner, Robert H. 1946. *Aircraft Carburetion*, New York: Wiley & Sons.

USAF Historical Advisory Committee, 1975. *Encyclopedia of U.S. Air Force Aircraft and Missile Systems*, Volume 1, Post-World War II Fighters, 1945-1973, Washington DC: GPO.

von Gersdorff, Kyrill and Kurt Grasmann, 1985. *Die deutsche Luftfahrt, Flugmotoren und Strahltriebwerke*, Koblenz: Bernard & Graefe Verlag.

Wagner, Ray, 1960. *American Combat Planes*, Hanover House.

Wagner, Ray, 1990. *Mustang Designer, Edgar Schmued and the Development of the P-51*, New York: Orion.

Drawings

Allison Drawing No.33301, 1935. XV-1710-5 Installation Drawing, for engines AC34-5 and AC34-6, AEC #7 and #8, NASM Microfilm frame #AF366 dated 11-25-1935.

Allison Drawing No.33700, 1936. XV-1710-7 Installation Drawing, for engines AC34-5 and AC34-6, AEC #7 and #8, NASM Microfilm frame #AF393, dated February 21, 1936.

Bell Drawing 3M001, December 1936. Secret Drawing: Proposed Interceptor Pursuit Airplane, Model 3, December 4, 1936. NARA, Box 49 3W2/6/18/F.

Bell Drawing 3M001, March 1936. Inboard Profile of Bell Model 3, March 1936. NARA RG18, Entry 167D, Box 49.

Curtiss drawing SK-1090, 1937. Curtiss Microfilm Roll L, Frame 97, dated 12-2-1937, NASM Archives.

Curtiss Microfilm, June 1937. NASM Archives, Roll N, Frame 121.

Wood, Robert, 1934. Preliminary Multi-Place Fighter, 9-19-1934.

Government Documents

AAF Statistical Digest, 1946. H.H. Arnold manuscript collection, Roll 186, Library of Congress.

Airplane Engineering Division, 1918, Vol.1, No. 2. Liberty Epicyclic Reduction Gear-Design of Reduction Gearing for Use with Liberty Engines-Comparison with Rolls-Royce Unit, a report by Research Department, Chief Engineers Office, Bulletin of the Experimental Department, Airplane Engineering Division, 1918, Vol.1, No. 2. Stanford University W87.30/3.

Airplane Engineering Division, 1918, Vol.1, No. 3, 108-116. The Bulletin of the Airplane Engineering Department, U.S. Army, McCook Field, Vol.I, No.3, August 1918.

Airplane Engineering Division, January 1919. Supercharging Progress in the United States, The Bulletin of the Experimental Department, Airplane Engineering Division, U.S.A., Vol.II, No.4.

Allison Engine Situation, 1941. Arnold, H.H., Major General, Chief of the Air Corps memorandum Facts on the Allison Engine Situation, to Mr. R.A. Lovett, Assistant Secretary of War for Air, May 14, 1941. NARA RG18, File 452.8, Box 807.

Allison Engines Under Wartime Conditions, 1942. Headquarters European Theater of Operations, U.S. Army, Technical Report, September 16, 1942. NARA, RG 342, RD 3774.

Annual Report of the Chief of the Air Corps, 1928. H.H. Arnold manuscript, Roll 181, Library of Congress.

Annual Report of the Chief of the Air Corps, 1930. H.H. Arnold manuscript, Roll 181, Library of Congress.

Annual Report of the Chief of the Air Corps, 1932. H.H. Arnold manuscript, Roll 181, Library of Congress.

Annual Report of the Chief of the Air Corps, 1934. H.H. Arnold manuscript, Roll 182, Library of Congress.

Annual Report of the Chief of the Air Corps, 1938.. U.S. Army, Wright Field, August 27, 1938.

Arnold, General H.H., August 19, 1943. Memorandum from General Arnold: Personal Letters to Theatre Senior Air Commanders. H.H. Arnold manuscript, Roll 131, Library of Congress.

Department of Commerce, 1940-1945. U.S. Military Aircraft Acceptances, 1940-1945, U.S. Department of Commerce, Civil Aeronautics Administration.

Department of Commerce, 1946. Aircraft, Engine and Propeller Production, U.S. Military Aircraft Acceptances 1940-1945, U.S. Department of Commerce, Civil Aeronautics Administration, 1946.

Doolittle, Major J.H. 1941. Allison Engines, a report compiled at the request of Chief of the Air Corps, Major General H.H. Arnold by Major J.H. Doolittle, Asst. Supervisor, Central Procurement District, Detroit, MI, January 19, 1941. NARA RG18, File 452.8, Box 807.

Forecast of Production of Planes and Engines for Week Ending July 26, 1918. H.H. Arnold manuscript, Roll 197, Library of Congress.

Price Adjustment Board Report-1944. Allison Division, GMC.

Sanwald, G.L. 1940. Calibration Of Allison XV-1710-6 Engine, Navy Bureau of Aeronautics, Report AEL-711, December 23, 1940.

Silverstein, Abe and F.R. Nickle, 1939. Tests of the XP-39 Airplane in the NACA Full-Scale Tunnel, September 27, 1939, Confidential Memorandum Report. NARA RG18, Entry 293B, Box 254.

Summary Report of French, British and Chinese Contracts, 1942. Orders from French, British and Chinese Contracts for AIRPLANE ENGINES-CONTRACTUAL DELIVERIES of Quantities Ordered in US by British Empire, January 31, 1942. NARA RG18, Entry 293B, Box 238-Allison.

Superchargers. NARA RG 342. Letters between Materiel Division and General Electric on Superchargers, 1933, contained in Record Group 342, NARA.

Type Designations, Army Aircraft Engines, Second Edition, 1936. Compiled by Materiel Division, Wright Field, May 1936. NARA RG18, Entry 167D, File 452.8, Box 49.

Manuscript Collections

Chennault Papers, Hoover Institution, Stanford, CA.

Arnold, H.H. Manuscript Collection. Library of Congress.

NACA Documents

15th Annual Report of the NACA, 1929.

16th Annual Report of the NACA, 1930.

Lee, Dana W. 1940. The Effects of Engine Speed and Mixture Temperature on the Knocking Characteristics of Several Fuels, Technical Note No. 767, April 2, 1940.

Meyer, Andre' J., Jr.1945. Elimination of Galling of Pendulum-Vibration Dampers Used in Aircraft Engines, NACA Advance Restricted Report ARR No. E5G31.

Nelson, R. Lee, Myron L. Harries, and Rinildo J. Brun, 1945. Effect of Internal Coolants on the Knock-Limited Performance of a Liquid-Cooled Multi-Cylinder Aircraft Engine With a Compression Ratio of 6.0:1, Memo Report No. E5F30, June 30, 1945.

Povolny, John H. and Louis J. Chelko, 1945. Cylinder-Head Temperatures and Coolant Heat Rejection of a Multicylinder, Liquid-Cooled Engine of 1710 Cubic Inch Displacement, NACA Technical Note No. 1606.

Wartime Report A-78, 1944. Flying Qualities of a High-Speed Bomber with a Dual Pusher Propeller Aft of the Empennage, originally issued in November 1944 as Memorandum Report A4K04.

Pamphlets

Allison Motors, 1920. The Allison Twelve Marine Engine, a product brochure believed to have been published in 1920.

Allison Engine, The, 1942. Allison Division Stockholders Bulletin for 1942.

Presentation Brochure, 1942. Presentation of ARMY-NAVY PRODUCTION AWARD to Allison Division, GMC, November 5, 1942.

Periodicals

Allen, Grant. 1969. "Vega XB-38," *Aero Album*, Volume 5, Spring 1969.

Beauchamp, Gerry, 1977. "Futuristic Hawks, A Story about the Curtiss X/YP-37 Aircraft," *AAHS Journal*, Spring 1977, Vol.22, No.1.

Bourdon, M.W.,1944. "Trends in the Development of Aircraft Engines, report on Sir Roy Feddens' Wilbur Wright Memorial Lecture before the Royal Aeronautical Society," *Automotive and Aviation Industries*, 24-25/116-120, September 15, 1944.

Bowers, Peter M. 1974. "Around the World in 175 Days," *Wings*, Vol. 4, No 6, December.

Boyne, Walt. 1973. "The First, and the Only, The Douglas XB-42/42A/43," *Airpower*, Vol.3, No.5, Sept. 1973.

Carter, Dusty, Spring 1988. "Vultee "P-38"," *AAHS Journal*, Spring 1988.

Christy, Joe, 1971. "The Convoluted Quest for Fighter Engines," *Air Trails*, Military Aircraft.

Dolza, John 1945. Allison Division GMC, "Coordination of Supercharger Speed to Manifold Pressure," *SAE Journal*, Vol.53, No.8, August 1945.

Echols, Maj. Gen O. P. 1941. "Materiel Division," *Flying and Popular Aviation*, September.

Ethell, Jeff, 1st Quarter 1981. "The Story of Aviation Pioneer Benjamin Kelsey," *Aviation Quarterly*, Vol.7, No.1.

Flight , 1954. "The Two Rs, 1904-1954, A Commemorative History of Rolls-Royce Aero Engines," reprint, May 7, 1954.

Frank, Gerhardt W. 1929. "High-Temperature Liquid-Cooling, S.A.E. Journal," Vol. XXV, No.4, Frank, ME, Materiel Division Air Corps, October.

Gilman, N.H. 1927. "The Air Cooled Liberty Engine, Chief Engineer, Allison Engineering Company," *Aviation*, December 19, 1927.

Goldthwaite, J.L. 1932. "Shaft Drives for Airship and Airplane Propellers," J.L. Goldthwaite, Assistant Chief Engineer, Allison Engineering Company, Presented at the Sixth National Aeronautic Meeting of the ASME, June 6/8, 1932. Printed in *ASME* Transactions AER-55-17.

Goss, SMS USAF(Ret) Dan, 1987. "Douglas XB-19," U.S. Air Force Museum *Friends Bulletin*, Vol.12, No.2.

Hazen, R.M. 1941 Draft. "Report of R.M. Hazen, Allison Chief Engineer," for *SAE Journal* article, Nov 1941.

Hazen, R.M. 1941. "The Allison Aircraft Engine Development," *SAE Journal* (Transactions), Vol.49, No.5, November 1941.

Hazen, R.M. 1945. "Development of the Allison 3420, Chief Engineer, Allison Division of GM Corp.," *Aero Digest,* October 1, 1945.

Hazen, R.M. and O.V. Montieth, 1938. "Torsional Vibration of In-Line Aircraft Engines," Allison Engineering Co., *SAE Journal*, Vol. 43, No. 2, August 1938.

Hives, E.W. and F. Ll. Smith of Rolls-Royce Ltd, 1940. "High-Output Aircraft Engines," *S.A.E. Journal* (Transactions), Vol. 46, No. 3, March 1940.

Knott, James E. 1968. "World War II Hero-The Allison V-1710," by James E. Knott, Manager Allison Division of General Motors, for *Aerospace Historian* Magazine, manuscript from late 1960s.

Liston, Joseph, 1953. Data from *SAE* Transactions, Vol.51 (1943), p.422, tabulated as Table 8-3 in Power Plants for Aircraft.

Ludwig, Paul A. 1996. "Hap Arnold's Ghost Fighters," *Wings*, Volume 26, No.4, August 1996.

Martin, Bob, 1993. "Pretty Bird! The Bell XFL-1 Airabonita," *Journal, American Aviation Historical Society*, Fall 1993.

Matthews, Birch J. 1993. "Cobra...," *AAHS Journal*, Volume 8, Number 3, Fall 1963.

Mitchell, Kent A. 1992. "The One and Only Fairchild XC-31 Cargo Plane," *AAHS Journal*, Volume 37, Number 1, Spring 1992.

Mormillo, Frank B. 1982. "The 'Planes of Fame' P-51A," *Aircraft Illustrated*, February 1982.

Nordenholt, G.F. 1941, "The Allison Engine, A record of its development from an idea to mass production," *Product Engineering*, May 1941. NASM File B1001025.

Norton, William 1981. "Liberty Aircraft Engines," *Air Classics*, Vol.17, No.3., March.

Raflo, Diane 1994. "Allison Moving Forward," *Turbomachinery International*, (July/August).

Redsell, Arthur, 1996, 18-22. "Memories, Part Two," Rolls-Royce Heritage Trust Archive, Number 43, Volume 14, Issue 3, 1996.

Rickenbacker, Eddy, 1956. "Rickenbacker Says: Big Industry Gives U.S. The Edge Over Russia," an interview with Eddy Rickenbacker, *U.S. News & World Report*, January 6, 1956.

Stewart, Hugh B. 1946. "Electrical Model for Investigation of Crankshaft Torsional Vibrations in In-Line Engines," Allison Division GMC, *S.A.E. Journal* (Transactions), Vol 54, No. 5, May 1946.

Taylor, E.S. 1936. "Eliminating Crankshaft Torsional Vibration in the Radial Aircraft Engine," *S.A.E.* Transactions, Vol. 31, March 1936.

Tydon, Walter, 1997. "XP-40 Marsupial Coolant System," The, *AAHS Journal*, Summer 1997, Vol.42, No.2. Tydon was Chief Engineer on the XP-40.

Vandenberg, Maj. Hoyt, 1941. "Pursuit," *Flying and Popular Aviation*, September 1941.

Waag, Robert J. 1973. "Kingcobra, America's Great Giveaway Fighter," *Airpower*, Vol. 3, No.5, September 1973.

Wood, Carlos. 1947. "Design and Development of the Douglas XB-42," Carlos Wood, Chief, Preliminary Design, Douglas Aircraft Company, Inc., *Aeronautical Engineering Review*, January 1947.

Betts, Ed, 1997. "Post-WWII Airliners," Circa 1945-1949. *AAHS Journal*, Summer 1997, Vol.42, No.2.

Hazen, R.M. 1933. "6-Cylinder 'U' Type Radial Engine Weight Summary Estimate," 3-28-1933.

Aeroplane, The, February 27, 1942. "The Rolls-Royce Merlin XX Motor,"

American Machinist, March 5, 1942. "Allison Set for Peak Production, Armament Section."

Lett, Tye M. Jr. December 1942. "We Learned War Maintenance With the AVG," Overseas Representative, Allison Division of GM, Aviation, December, 1942.

Aero Digest, January 1943. "How Cadillac Mass-Produces Allison Engine Parts to Close Tolerances."

American Machinist, May 11, 1944. "Shrink-Fit Assembly of Cylinders Is Done on a Special Press."

Journal of Military Aviation, January/February 1987. "Combat Report: Double Trouble, A Look at the Twin Mustang," Vol.1, No. 2.

Cox, Jack 1994. EAA World, in *Sport Aviation*, October 1994.

Stone, Matthew L. 1993. "Allison Wonderman," *Autoweek*, December 27, 1993.

Specifications

Allison Engineering Company, 1920 Specification. Allison Twelve 400 Horsepower Gasoline Marine Motor Specification, Indianapolis, Indiana, no date, probably 1920.

Allison Model Specification No. 109-C1, 1938. Model V-1710-13(D1), issued 3-22-1937, revised 2-9-1938.

Allison Model Specification No. 110, 1937. Model V-1710-D3 Geared Tractor Aircraft Engine, Issued March 29, 1937, revised May 20, 1937.

Allison Model Specification No. 163-E, 1943. Model V-1710-81(F20R), revision March 31, 1943.

Allison Model Specification No. 198, 1945. Model V-1710-139/141(F33R/L), August 6, 1945.

Allison Model Specification No. 211, 1939. Model V-3420-A8R/L, October 9, 1939.

Allison Model Specification No. 214,1940. Model V-3420-B2, March 12, 1940.

Allison Model Specification No. 217, 1940. Model V-3420-7(B5), August 6, 1940.

Allison Model Specification No. 230, 1943. Model V-3420-A20R, March 10, 1943.

Allison Model Specification No. 245-B, 1946. Model V-1710-143/145, issued 10-8-1945, revised 12-3-45.

Allison V-1410 and VG-1410 Engine Specification, Commercial, no date, probably 1927.

Allison V-1650 Engine Specification, no date, probably 1927.

Bell Aircraft Preliminary Specification, Report No. 1Y012, 1936. "Airplane – Multiplace Fighter Detail Specification", Report No. 1Y012, 7 March 1936.

Bell Aircraft Preliminary Specification, Report No. 4Y003, 1937. Model Specification (Preliminary) for Interceptor Pursuit Airplane, Bell Model No. 4, June 3, 1937.

Bell Aircraft Report No. 33-947-006, 1944. Model Specification for Single Engine Training Fighter Airplane, Model RP-63C-2-BE, November 15, 1944, courtesy of Birch Matthews.

Department of Commerce, 1937. Bureau of Air Commerce, Washington, APPROVED ENGINE SPECIFICATION NO. 177, file E-99, Allison, V-1710-C4, July 13, 1937.

Fisher Body Detroit Division, Aircraft Development Section, Report No. X-249, 1943. Model Specification P-75A-1-GC, Single Engine, Interceptor and/or Long Range Fighter Aircraft, July 8, 1943, with revisions through November 7, 1944, Contract No. W535-ac-41011.

North American Aviation Report No. NA-8033, 1944. Model Specification for the Model XP-51J Airplane, N.A.A. Model NA-105B, Fighter-Offensive, Contract W535-ac-37857, Engineering Department Report, January 1, 1944.

Wright Aeronautical Corporation Engine Specification No. R-179, 1933. WAC Engine Model SGIV-1800 (Reduction Gear 7:5), 750 bhp to 12,000 Feet Altitude, Liquid-Cooled, March 16, 1933. NARA RG18, Entry 167D, Box 49.

Technical Orders

AAF Manual 51-127-1, 1945. Pilot Training Manual for the P-38 Lightning, August 1, 1945.

Technical Order 00-25-30, 1945. Unit Costs of Aircraft and Engines, entry for P-75A, August 1, 1945.

Technical Order 01-25CF-1, 1943. Pilots Flight Operating Instructions, P-40D and P-40E Airplanes, February 25, 1943.

Technical Order 01-60JJA-2, 1947. P-82E Airplane Erection and Maintenance Instructions, North American Aviation, January 10, 1947.

Technical Order 02-5A-01A, 1939. Preliminary Operation Instructions – Allison V-1710-19, January 16, 1939, Wright Field file D52.41/14-Allison.

Technical Order 02-5A-05, 1945. Allison-Model Designations, January 3, 1945.

Technical Order 02-5A-09, 1944. Interchangeability of Pistons and Piston Pins—V-1710 Series, July 28, 1944.

Technical Order 02-5A-27, 1943. Inspection and Repair of Crankshaft-V-1710 Series, April 14, 1943.

Technical Order 02-5A-42, 1945. Allison-Replacing Center Gas Intake Pipe and Installing Gas Intake Pipe Sleeve, New Manifold Hoses, and Split Sleeves, May 24, 1945, revised August 18, 1945.

Technical Order 02-5A-55, 1943. Retarded Ignition Timing for Flight with 91 Octane Fuel, Specification No. AN-F-26, V-1710 Series, August 13, 1943.

Technical Order 02-5A-64, 1944. Rework or Replacing the Gas Intake Pipe to Improve Split Line Seal—V-1710 Series, February 15, 1944.

Technical Order 02-5AA-2, 1939. Handbook of Service Instructions for the Model V-1710-21 Engines and Associated Models, April 25, 1939.

Technical Order 02-5AB-18, 1945. Allison-Removal of Magneto Timing Control Mechanism, V-1710-111 and -113, June 8, 1945.

Technical Order 02-5AB-1A, 1942. Handbook of Operating Instructions, V-1710-39, -73, -81 Aircraft Engines, December 20, 1942. NASM Wright Field file D52.41/134-Allison.

Technical Order 02-5AB-4B, 1942. Parts Catalog, Model V-1710-39 Engine and Associated Models, August 25, 1942. NASM Wright Field file D52.41/131-Allison.

Technical Order 02-5AD-11, 1944. Interchanging, Identification, and Overhaul of Outboard Drive Assemblies-V-1710-35, -37, -63, -83, and -85, February 1, 1944.

Technical Order 02-5AD-2, 1945. Service Instructions for Models V-1710-35, -63, -83, -85, -93, -109, -117 Aircraft Engines, May 5, 1945.

Technical Order 02-5AH-2, 1949. Handbook of Service Instructions for Models V-1710-143 and -145 Aircraft Engines, Revised February 17, 1949.

Technical Order 06-5-4, 1950. Aviation Fuels and Their Effects on Engine Performance, USAF, NAVAER-06-5-501.

Technical Order 30-5A-1, 1944. Information Guide: Allison V-1710 Engines, January 1, 1944.

Unpublished Material

Aeronautical Engine Laboratory, 1935. Study on All Aspects of Supercharging, Naval Aircraft Factory, Philadelphia, PA, March 29, 1935.

Allison Division, GMC 1941. Company Development, a history of the V-1710 prepared by Allison January 15, 1941.

Allison Division, GMC 1950. Allison-Its Products And Their Relationship to the General Motors Product Line, Indianapolis.

Allison Division, GMC 1962 Draft. Allison General History.

Allison Division, GMC early 1940s. Allison Early History, internal manuscript prepared in early 1940s.

Allison History, 1962. Under cover letter to Miss Alberta Pemberton, July 8, 1962.

Bell Report 1Y023, 1937. XFM-1 Performance Estimate Based on NACA Wind Tunnel Tests of the Half Size Model and Allison 1000 and 1150 bhp Engines, Bell Report 1Y023, September 10, 1937.

Bell Report 1Y027, 1937. Actual Balance Statement, Model XFM-1 Multi-Place Fighter Airplane, October 28, 1937.

Colman, Philip A. 1942. Study of P-38 With Allison F-17 and Rolls-Royce 61 Engines, Lockheed Report No. 2726, Aerodynamics Group, Lockheed Aircraft Corp., June 9, 1942.

Colman, Philip A. 1944. P-38 Performance Comparison, Allison and Rolls-Royce Engines, Lockheed Report No. 4598, Aerodynamics Group, Lockheed Aircraft Corp., Approved by C. L. Johnson, February 9, 1944.

Cruzans, Mr. 1947. From a speech by Mr. Cruzans, What is Allison Doing in the Bearing Business? Allison Bearing Division.

Fleming, Roger 1950. Notes on Allison, GMC History, internal Allison memo for use in Allison-Its Products And Their Relationship to the General Motors Product Line, Public Relations, February 1, 1950.

Fleming, Roger, 1957. Memo, Internal Allison History, Public Relations.

General Motors Corporation. 1957. General Motors and the Aviation Industry, a report developed by GM in 1957.

Goldthwaite, John L. 1950. Notes on Allison, G.M. History for use in Allison-Its Products And Their Relationship to the General Motors Product Line, February 1, 1950.

Hazen, R.M. 1943. Effect of Exhaust Back Pressure on War Emergency Ratings,, Letter from Allison Chief Engineer to Commanding General AAF, Materiel Center, Wright Field, January 13, 1943. NARA RG 342, RD 3774.

Hazen, R.M. 1945 Report. Report of Allison Chief Engineer, 1945.

Hazen, R.M. 1957. Aircraft Engine Information, data provided by R.M. Hazen, Technical Assistant to General Manager, for use in Allison History prepared in 1957. Letter of July 24, 1957.

Hunt, J.H. 1942, A-11A. Allison internal report, A-11A History, via J.L. Goldthwaithe.

Hunt, J.H. 1942, XFM-1. Allison internal report, Allison XFM-1 History, with comments by J.L. Goldthwaithe.

Jahnke, R.L. 1942. Zone Office Report No. 2, Preliminary Report on Change of Performance Ratings on Airplanes in England, R.L. Jahnke Allison Zone Manager in British Isles to Mr. W.C. Gould, Allison Service Manager, October 14, 1942.

Lett, Tye M. Jr. 1942. Ground Crewing the 'Flying Tigers, Overseas Representative of the Allison Division, General Motors, 1941-42.

Leyes, Rick 1997. Curator for Aero Propulsion, NASM letter to Author, Personal inspection of XV-1710-1, AEC #2, February 12, 1997.

Marion County Clerk's Office, 1915. Miscellaneous Record Book 88, page 375, Indiana.

Millikan, Clark B. 1940. California Institute of Technology Report Number 284, A Comparative Wind Tunnel Test on Two Alternative Wings for the North American NA-X73 Airplane, Graduate Aeronautical Laboratories, September 20, 1940.

Packard Motor Car Company, 1945. Chronological Development of the Packard-Built Rolls-Royce Merlin Engines, Section II, Aircraft Engineering Department.

Pitkin, J. 1944. XP-58 Power Plant, V-3420 Installation, Lockheed Report No. 4594, August 8, 1944.

Posner, E.C. 1940. Model P-38 with Merlin XX Engines, Performance Calculations, Lockheed Report No. 2036, Aerodynamics Group, Lockheed Aircraft Corp., October 24, 1940.

Wagner, Ray 1991. Great Aviation Failures: Curtiss P-46, by Ray Wagner, Archivist, San Diego Aerospace Museum, 1991.

White, R.J. 1938. Determination of Drag and Cooling from Tests on 1/3 Scale Model of Nacelle and Tail Boom, Model 22, Lockheed Report No. 1187, August 15, 1938.

Unpublished Interviews

Askren, R.W. "Dick" 1994. Interview with "Dick" Askren, Engineer involved with the early design of V-1710, and the Service Department, by author. Tape recording. Indianapolis, Indiana, November 28, 1994.

Atkinson, Robert P. 1989. Outstanding Memories of Allison Retirees- Robert P. Atkinson, August 29, 1989.

Babin, Gus, 1995. Allison engine builder, interviewed in Apache Springs, AZ, September 1995.

Bosler, T.C. 1952. Letter to O.T. Kreusser, Allison Historical Notes-Allison Phase of 500-Mile Race History, August 28, 1952.

Boushea, Al 1995. Discussion with Al Boushea, engine mechanic and V-1710 builder,Septermber 24, 1995.

Cartwright, John, 1989. Outstanding Memories of Allison Retires-John Cartwright, Allison employee.

Kline, John 1994. Interview with John Kline, Allison Tech Rep (Retired), by author. Tape recording. Indianapolis, Indiana, November 28, 1994. Kline was Allison Installation Engineer on XP-37, XP-38, XP-39, XP-40 and NA-73X.

Knott, Jim, 1989. Recollections of Jim Knott, Allison Retiree, January 1989.

Sobey, A.J. 1988. RECOLLECTIONS of ALLISON, one of a series obtained by Allison Division, 9-30-88.

Worden, Harold W. 1962, Letter to Allison Public Relations Office, July 8, 1962.

Wright, Donald F. 1994. Interview with Don Wright, retired Allison Service Department, by author. Tape recording. Indianapolis, Indiana, November 28, 1994.

Zigmunt, Joan E. 1994. Interview with Joan E. Zigmunt, retired from Allison Public Relations Department, by author. Tape recording. Indianapolis, Indiana, November 28, 1994.

Wright Field Letters, Memos, Reports

Brett, General George H. July 25, 1940. Letter from, Chief of Materiel Division, to Allison Engineering Company, July 25, 1940. NARA RG18, File 452.8, Box 808.

Case History of the B-42 Airplane Project, 1948. Compiled by Historical Office, Air Materiel Command, Wright-Patterson AFB, January 1948. Copy presented in American Aviation Historical Society Journal, Summer 1992.

Case History of the F-82E, F and G, 1951. Compiled by Martin J. Miller, Jr., USAF/AMC, Jan. 1951, AFHRC 202.1-49.

Case History of the R-2160 Engine, 1945. Historical Division, Air Technical Service Command Wright Field, June 1945. Copy from Maxwell AFB, Microfilm Roll #A2073.

Case History of the V-1650 Engine, 1945. Compiled by the Historical Office, Air Technical Service Command, Wright Field, December 1945. Microfilm roll #A2073.

Case History of the XP-60 Series Project, 1945. Historical Study No. 45, Historical Office, Materiel Command, Wright Field.

Engineering Division Memo Report, Serial No. ENG-57- 503-1034, 30 October 1943. High B.H.P. Output Tests on an Allison V-1710-19 Engine, Tested Sept 22, 1939 to Jan 11, 1940. AAF Materiel Command, Wright Field file: D52.41/186-Allison at NASM.

Engineering Division Memo Report, Serial No. ENG-57- 503-1166, 1944. Attempted Altitude Calibration of an Allison V-1710-91 Engine Supercharged with a Packard Rolls-Royce V-1650-3 Supercharger and Aftercooler, AAF Materiel Command, March 4, 1944. Wright Field file D52.41/205-Allison, at NASM.

Engineering Division Memo Report, Serial No. ENG-57- 503-1173, 15 March 1944.

Engineering Division Memo Report, Serial No. ENG-57- 503-1217, 1944. First and Second Attempted Model Tests of the Allison V-1710-93 Engine, AAF Materiel Command, June 3, 1944. Wright Field File D52.41/211-Allison, at NASM.

Engineering Division Memo Report, Serial No. ENG-57- 531-267, 1944. Preliminary 7-1/2 Hour War Emergency Rating Test of the Allison V-1710-91 Engine on Grade 104/150 Fuel and AC-EX433 Spark Plugs, AAF Materiel Command, March 27, 1944.

Engineering Division Memo Report, Serial No. ENG-57- 531-276, 1944. Attempted 7-1/2 hour War Emergency Rating of Allison V-1710-89 Engine as Installed in P-38J Airplane Operated on Grade 104/150 Fuel and AC 433M Spark Plugs, April 22, 1944.

Engineering Division Memo Report, Serial No. ENG-57- 531-294, 16 May 1944.

Engineering Division Memo Report, Serial No. ENG-57- 531-299, 1944. 7-1/2 hour War Emergency Rating of V-1710-91 on Grade 104/150 Fuel and C34S Spark Plugs, May 12, 1944.

Engineering Section Memo Report, Serial No. E-57- 6, 1932. Conference with Messsrs. N.H. Gilman and H. Caminez of the Allison Engineering Company, Reason Type V-1710 Engine, Nov. 3, 1932, Wright Field File: D52.41/2-Allison, NASM.

Engineering Section Memo Report, Serial No. E-57- 240-4, 1934. Adaptation of Fuel Injector Discharge Nozzles to Manifolding for V-1710-2 Engine, July 11, 1934. Wright Field file D52.41/4-Allison.

Engineering Section Memo Report, Serial No. E-57- 250-9, 1934. Conference at Materiel Division June 21, 1934 on Allison XV-1710-1 Engine, Wright Field Materiel Division, July 12, 1934.

Engineering Section Memo Report, Serial No. E-57- 362-1, 1938. Failure of the XV-1710-13 Engine, Air Corps No. 34-14, Materiel Division, January 27, 1938.

Engineering Section Memo Report, Serial No. E-57- 541- 1, 1934. Conference with Mr. H. Caminez of Allison Engineering Company on Allison XV-1710-3 Engine, Held July 11, 1934. Wright Field, July 20, 1934.

Engineering Section Memo Report, Serial No. E-57- 541- 9, 1935. Conference with Mr. Caminez of Allison Engineering Company, Regarding V-1710 Type Engines, February 19, 1935, dated March 2, 1935.

Engineering Section Memo Report, Serial No. E-57- 541-27, 1938. 150-Hour Type Test of Allison XV-1710-3 Engine, September 10, 1938.

Engineering Section Memo Report, Serial No. E-57- 680-14, 1937. Summary of Air Corps Activity and Costs in Connection with Development of Allison V-1710 Type Engines, January 4, 1937.

Engineering Section Memo Report, Serial No. E-57-1077-66, 1938. Conference with Representatives of Allison Engineering Company on V-3420 Engines and Cabin Superchargers, Wright Field, by Opie Chenoweth, April 18, 1938. Copy in V-3420 files at NARA.

Engineering Section Memo Report, Serial No. E-57-1077-88, 1939. V-1710-17 Engine, May 25, 1939, Contract W535-ac-10830. NARA RG18, File 452.8, Box 808.

Engineering Section Memo Report, Serial No. E-57-1406-3, 1939. Attempted Type Test of Allison YV-1710-9 and XV-1710-13 Engines, Materiel Division, January 28, 1939, Wright Field file D52.41/16-Allison.

Engineering Section Memo Report, Serial No. E-57-1643-1, 1937. Comparison of Rolls-Royce "Merlin" and Allison V-1710 Engines, July 2, 1937.

Engineering Section Memo Report, Serial No. E-57-1839-2, 1939. First Attempted Type Test of Allison V-1710-21 Engine, Wright Field, January 28, 1939. Wright Field file D52.41/15-Allison.

Engineering Section Memo Report, Serial No. E-57-1860-1, 1938. February 24, 1938.

Engineering Section Memo Report, Serial No. E-57-4272-1, 1939. Engineering Study on Means of Increasing the Power Output of the Allison V-1710 Engine, Materiel Division, written by Opie Chenoweth of the Wright Field Power Plant Laboratory, August 8, 1939. NARA RG18, File 452.8, Box 808.

Engineering Section Memo Report, Serial No. R-57- 311-9, 1938. Coolant Level Inspection and Overhaul Time of Allison V-1710-7 Engine in A-11A Airplane, Materiel Division, July 18, 1938.

Engineering Section Memo Report, Serial No. R-57- 362-1, 1938. Failure of the V-1710-13 Engine, Air Corps No. 34-14, January 27, 1938.

Engineering Section Memo Report, Serial No. R-57- 370-1, 1937. Operating Instructions for the Allison YV-1710-7 Engine in the XP-37 Airplane were issued by Wright Field, April 6, 1937.

Experimental Engineering Report EXP-M-57- 503-126, 1940. Allison V-1710-15 Engine for Aircraft Technical School at Chanute Field, May 9, 1940.

Experimental Engineering Report EXP-M-57- 503-599, 1942. Third Attempted 150 Hour Type Test on V-1710-61, May 18, 1942. Wright Field file D52.14/114-Allison.

Experimental Engineering Report EXP-M-57- 503-655, 1942. Torsional Vibration Survey of Allison V-3420-9 Engine, July 6, 1942. Wright Field file D52.41/122-Allison.

Experimental Engineering Report EXP-M-57- 503-663, 1942. Type Test of V-3420-9, July 9, 1942. Wright Field file D52.41/121-Allison.

Experimental Engineering Section, CONFIDENTIAL Memorandum, 1942. A Review of the Experimental Airplane Program, for General Vanaman from Chief, November 6, 1942.

Materiel Command Memo Report Serial No. ENG-57- 503-1173, 1944. Short High Power Endurance Test of V-1710-81 No. A-034356 Modified with Cammed Pistons and a Pressurized Water Coolant System. AAF, March 15, 1944.

Materiel Command Memo Report, Serial No. ENG-57- 503-848, 1943. Tentative specification outline, dated April 24, 1943, per Progress Report No.1 on 7-1/2 hour Approval Test for WER of V-1710-93(E11), December 9, 1943.

Materiel Division Memo Report, Serial No. EXP-M-57- 535-3, 1941. Calibration of Allison V-1710-21 Engine, AC#38-582, with Four Different Exhaust Stacks, Air Corps, January 29, 1941. NARA RG 342, file: V-1710 Jan-Feb '41.

Project MX-467, 1945. Compound Engine Projects, letter from Engineering Division to Commanding General AAF, June 23, 1945. NARA RG342, RD3774, Box 7409.

Technical Bulletin No.46, 1926. US Army, Aircraft Development Section.

Technical Report No. 3585, 1932. Performance Test of Allison X-4520 Engine, Air Corps January 25, 1932. Copy from Bill Lewis via Mick Jefferies

Technical Report No. 4038, 1937. Performance Characteristics and 50-Hour Development Test at 800 bhp of Allison XV-1710-1 Engine, A.C. No. 33-42, Mfr's No.2, Air Corps September 8, 1937. Wright Field File: D00.12W/4038, NASM.

Technical Report No. 4069, 1935. 50-Hour Development Test of Allison XV-1710-3 Engine, Air Corps, March 4, 1935. Wright Field file D00.12W/4069.

Technical Report No. 4452, 1939 150-Hour Type Test of Allison YV-1710-7 Engine, Air Corps No. 34-8, Manufacturer's No. 10, by C.E. Mines, Associate Mechanical Engineer, Air Corps, April 6, 1939. Wright Field File D00.12W/4452, copy at NASM.

Technical Report No. 4761, 1942. 150-Hour Type Test of V-1650-1 Engine, April 18, 1942. NASM File B0028380.

Technical Report No. 5022, 1943. Final Report of Development, Procurement, Performance and Acceptance of XP-39E Airplanes, Army Air Forces, September 21, 1943. Wright Field file D00.12W/5022.

Technical Report No. 5306, 1945. Final Report of Development, Procurement, Performance and Acceptance of the XP-55 Airplane, AAF, October 22, 1945.

Technical Report No. 5489, 1946. Final Report of the Procurement, Inspection, Testing, and Acceptance of the Two Lockheed XP-58 Airplanes, Army Air Forces Technical Report No.5489, April 30, 1946. NASM file A0305500.

Technical Report No. 5504, 1946. Final Report of the Development, Testing, and Acceptance of the Republic XP-47J Airplane, Army Air Forces, June 26, 1946. HQ USAFHRC Microfilm Roll 30703, frame 0274.

Technical Report No. 5653, 1947. Final Report on the Procurement, Inspection, Testing, and Acceptance of North American Model XP-51F, G & J Airplanes, USAF, December 4, 1947. Microfilm Roll 30703.

Technical Report No. 5673, 1948. Final Report of the Procurement, Inspection, Testing, and Acceptance of North American XP-82 and XP-82A Airplanes, Army Air Forces Technical Report No.5673, February 18, 1948. HQ USAFHRC Microfilm Roll 30703.

Wright Field file D52.1/1422-Curtiss, 1940. Airplanes-Curtiss YP-37 (Fighter) Operation Instructions, revised September 6, 1940, available at NASM.

Wright Field Memo Report Insp-M-41-5-E, 1941. Ratings on Allison V-1710 Engines, January 25, 1941.

Wright Field Report No. TSEST-A2, 1946. Aircraft Characteristics, Production and Experimental, April 1, 1946.

Glossary

Air Corps

In 1925 the President commissioned the Morrow Board to study Federal policy for the furtherance of aeronautical progress. The result was a recommendation of a ten year appropriation and development plan. In 1926 Congress passed the Army Five Year Aviation Program, the Navy Five Year Aviation Program, and a Civil Aeronautics Act. In addition to authorizing increases in planes and personnel, the Army program established the Air Corps, thereby giving a measure of stature to Army aviation, though not the separate air force desired by many.[1]

BHP

Brake Horsepower, the amount of power delivered at the output shaft of an engine. It is less than the IHP because of the need to overcome engine friction and provide power to the supercharger and accessories.

Boost

The British use this term to describe the pressure of the mixture in the induction system of an engine, e.g. +12 psig. It is calibrated as "gauge" pressure, meaning that pressure above atmospheric. The usual units are pounds per square inch. To convert to absolute Manifold Air Pressure, as typically used by the Americans, it is necessary to multiply by 2.04 and add sea-level static pressure of 29.92 inHgA. +12 psig equates to 54.4 inHgA. The drawback to using "Boost" to describe engine operation is that it is not directly proportional to the engine power, and during low power operations may even be reported as "negative."

BMEP

Brake Mean Effective Pressure, a convenient parameter for comparing performance of different engines as it is the average pressure driving the piston. Comparisons overcome the differences in rpm, cylinder size, and number of cylinders.

Carburetors

Bendix-Stromberg developed the pressure injection carburetor in the mid-1930s. The device was far superior to earlier "float type" carburetors then in use. Its' significant feature was that it metered pressurized fuel which was then usually sprayed directly into the eye of the supercharger impeller, which had the effect of insuring optimum mixing with the air and provided excellent mixture distribution. This method of injecting the fuel also minimized the likelihood of ice formation within the induction system since the cooling effect of fuel evaporation occurred within the supercharger, which was heating the air while compressing it. The "float" was also eliminated, and along with it the tendency of float type carburetors to cutoff the flow of fuel during "negative-g" flight maneuvers. It was still configured much like a conventional carburetor as far as the airflow metering and induction pressure losses were concerned. It was not until the development of the Speed Density "Carburetors late in the war that such losses were eliminated. The Speed Density unit was actually a form of analog computer which used throttle position, engine rpm, manifold pressure, manifold temperature and exhaust pressure to schedule the proper quantity of fuel.

From the YP-37 and YFM-1 on, Allison used two types of carburetors, one a "dual-barrel" unit and the other a "three-barrel." The number of "barrels" referring to the number of circular passages, throttle plates and venturis used in the carburetor body. These were designated as "PD-" and "PT-" respectively, the "P" describing the unit as a "pressure" type carburetor. The number "12" or "13" is the number of "quarter inches" the diameter is over a minimum of 1-5/16 inches for the PD-12K, and 1-3/16 inches for the PD-12G models.[2] Combining all of these parameters gives a throat area for the PD-12G1 of 22.832 square inches, for the PD-12K of 24.353 square inches, and for the PT-13, 41.316 square inches, though in no case was all of this area was available for airflow since each barrel had a "boost venturi" located within it. The addition of a letter suffix to the designation describes a major modification to the unit, further numerical suffixes define specific configurations of jets and other settings necessary in particular installations.[3]

The PD-12K carburetors were entirely satisfactory for the V-1710 at sea-level ratings up to about 1425 hp. The PT-13 was first used on the XV-3420-1, but when it was desired to create an "altitude rated" engine for the XP-40 this unit was adapted as it caused less obstruction and significantly improved the critical altitude of the aircraft. This general strategy seems to have been applied throughout the production life of the V-1710. On the two-stage engines which had the carburetor on the inlet to the Auxiliary Stage supercharger the PT-13E was used, though on similar installations with the carburetor on the inlet to the engine-stage supercharger the PD-12K was used. The later V-3420's utilized the PR-58 carburetors, here the designation describes a "pressure" type carburetor with a "rectangular" shaped flow passage, and 58 square inches of flow area. These carburetors were satisfactory for engines producing 3000 bhp, and were also used on other large engines such as the Pratt & Whitney R-4360's, and in more recent times, on unlimited racing V-12's. This gives an idea of the power developed by unlimited racing engines.

The PD-12D8 carburetor was original equipment on most V-1710-111/113(F30R/L) engines. In February 1945 production engines were equipped with PD-12K17, which was identical to the earlier unit except for the improved air density compensation provided by a new two ply bellows automatic mixture control unit which was vented to boost venturi suction instead of being vented to the carburetor deck pressure.[4]

Coolant: AN-E-2 Ethylene Glycol "Prestone[tm]"

Commercially pure (CH_2OH-CH_2OH) with 2.5% by weight Triethanolamine Phosphate corrosion inhibitor. The fluid is hydroscopic, meaning that it will absorb moisture from the air if left in open containers. It also expands about 10 percent when the engine reaches operating temperatures, requiring care in providing adequate expansion space in the provided expansion tank. Only in an emergency was automotive ethylene glycol anti-freeze to be used, for it contained soluble oil and other undesirable additives.[5] Allison Specification #125 (1941 era) for Ethylene Glycol coolant specified water was not to exceed 6 percent by weight, and that the Corrosion Inhibitor was optional, and to be Sodium Nitrite, $NaNO_2$ in the amount of 0.02 pounds per gallon if used.

By 1944 the Army standardized on four mixtures of coolant, which included NaMBT (sodium mercaptobenzothiazole) as an inhibitor, these were:[6]

Type	Mixture %Glycol/%Water	Freezing Temp, F	Boiling Temp at 20,000 ft, F
A	100/0	-5	256
B	80/20	-52	210
C	70/30	-65	198
D	30/70	+5	180

In turn the engine manufactures specified which mixtures were to be used in their engines:

Manufacturer	Winter, below 4 F	Summer, above 4 F
Allison	B (C)	A (D)
Packard	C	D

In 1943 Allison began rating its engines using Type D coolant in pressurized systems, switching to Type C coolant for extreme cold weather operations.[7] This greatly improved the cooling in the cylinder heads and allowed operating at higher rated power because of the considerably better heat removal characteristics of water.

Critical Altitude
Defined as the altitude where, with the throttle is wide open, it is no longer possible to maintain rated horsepower. A sea-level rated engine achieves this condition at sea-level and is therefore dependent upon either a turbosupercharger or Auxiliary Stage Supercharger to maintain rated power to critical altitude as defined by this device.

Detonation
When conditions in an engine cylinder reach the point necessary for spontaneous ignition, in essence an explosion, rather than the smooth burning expected as a result of proper spark ignition. Extremely high pressures are then developed along with heat which "hammer" the pistons. Very rapidly holes can be pounded through pistons, often with catastrophic results for the whole engine. Assuming the engine and cylinders are not defective, the way to control detonation is to either reduce power, increase the octane rating of the fuel, or to inject an internal coolant to reduce the pressure within the cylinder. Water, and/or Water/Methanol are typical Anti-Detonant-Injection (ADI) fluids used for this purpose. Common usage of terms, describes conditions or ratings were ADI is being used, as "wet." These are usually the maximum power conditions.

FHP
Friction Horsepower, the power lost in overcoming the friction in bearings and sliding surfaces such as the pistons on the cylinder walls. This power varies with the square of the engine rpm, and contributes significantly to the heat that must be removed from the engine by the coolant.

Fuel Grades
The following designations refer to Army/Navy standard fuel specifications.

AN-F-22 Aviation Gasoline
Aviation Grade 62 aircraft engine fuel.[8]

AN-F-23 Aviation Gasoline
Aviation Grade 73 aircraft engine fuel.[9]

AN-F-24 Aviation Gasoline
Aviation Grade 80 aircraft engine fuel.[10]

AN-F-25 Aviation Gasoline
Aviation Grade 87 aircraft engine fuel.[11]

AN-F-26 Aviation Gasoline
Aviation Grade 91/96 aircraft engine fuel.[12]

AN-F-27 Aviation Gasoline
Believe this is 100/125 grade fuel, used early in war per fuel specification AN-VV-F-781, dated Sept 26, 1940. This was a 100/125 Octane fuel which was replaced by AN-F-28 as of December 23, 1942.[13] It also went through several "Amendments", such as Amendment 5 that was specified for use in P-38H airplanes per TO 01-75FF-1.

AN-F-28 Aviation Gasoline
During WWII this was the usual premium fuel. It was rated with a performance number of 100/130, for the lean/rich, or normally aspirated/supercharged operating conditions. According to NACA Technical Note No. 996, issued in 1945, the density of this fuel was 0.72 g/ml and typically included 4.6 milliliters of "ethyl fluid", or "tetra-ethyl-lead", $Pb(C_2H_5)_4$ per gallon of mixture. The gross heat of combustion is 20,000 BTU/LBM, while the net heat of combustion is 18,700 BTU/LBM. The difference is due to the heat lost through the formation of water during the combustion process. Since the water is lost to the environment as a vapor during the exhaust process, that energy is not available to any engine.

AN-F-28R is reported by Hubbard to be "green" colored, 4.5 cc TEL/gal.

AN-F-28 Amend#2 is reported by Hubbard to by "blue" colored.

AN-F-29 Aviation Gasoline
Grade 140 fuel available in limited quantities near the end of the war.

AN-F-33 Aviation Gasoline
This was the premium fuel, and only became available late in the war. It was only to be used when truly high performance was required. Its Performance Numbers were 115/145, and such fuel was usually required anytime it was intended to operate an engine with 70 inHgA or more manifold pressure, whether or not water injection was being used. Often referred to as "Grade 145."

Fuel: Trimethyl-butane
A special fuel known as "Triptane" which has very good "rich" performance at high manifold pressures. When doped with 4.6 cc TEL/gallon, allowable IMEP of 400 psi available at 0.082 Fuel/Air ratio. Reported that this is equivalent Grade PN 200/300. This compares to about 210 psi for AN-F-28 fuel, neither case using ADI.

Fighter Aircraft
A mission designation describing a class of airplanes, usually capable of high performance and armed for offensive and/or defensive missions and capable of besting similar aircraft challenging the mission.

FY-39
The U.S. Government functions financially on a Fiscal Year basis. In the 1930s and 40's the calendar period covered was July 1 through June 30 of the following year. Aircraft and engines ordered by the Air Corps were identified with a serial number prefixed by the fiscal year of the funds used for the purchase. Contracts for purchases may extend over several fiscal years, for example the first 200 Curtiss P-40's were procured on contract W535-ac-12414, with 134 aircraft delivered with serial numbers AC39-156/289 and the final 66 with serial numbers AC40-292/357. The contract ultimately procured 524 P-40 and P-40B/C aircraft. The year in which the aircraft or engine was delivered is not reflected in the serial number, only the Fiscal Year of the funds used.

IHP
Indicated Horsepower, the total amount of mechanical power extracted from the fuel burned in an engine. The reference parameter for determining the mechanical efficiency of an engine.

Lend-Lease
US Senate passed the Lend-Lease bill empowering the President to give "all-out" aid to Britain or any other countries opposing the Axis, March 8, 1941.

Manifold Air Pressure (MAP), inHgA
The American way of describing the pressure of the mixture in the induction system of an engine, e.g. 61 inHgA, read as "inches Mercury, Absolute." This is an "absolute" pressure and is not effected by altitude effects. As such it is proportional to the density of the mixture, assuming constant mixture temperatures, and changes in MAP are directly proportional to power. To convert the measurement to gauge pressure or "Boost", divide the value by 2.04 and subtract the value of standard sea-level absolute pressure, 14.7 pounds per square inch. 61 inHgA equates to +15.2 psig Boost.

Mixture Temperature
This is the temperature of the combined air with evaporated fuel mixture as it leaves the intake manifold and passes through the intake valves and into the combustion chambers. It is a critical parameter since the piston compresses the mixture by the

multiple of the compression ratio. At some mixture temperature the compression will elevate the temperature to the level at which "detonation" occurs. Use of an "internal coolant", which included excess fuel or ADI fluid, can be used to reduce the mixture temperature. ADI also has the effect of slowing the rate of burning.

Model Test/Type Test of Engines
These designations were used by Wright Field to qualify an engine for military service. The Type Test was done on the first of the "Type", such as the V-1710-7(C4/C6) that qualified the V-1710 for flight and use by the Army. Model Tests were run on each sub-series of a Type, such as the V-1710-33(C15), prior to its introduction to service. Both tests were normally for 150 hours, with differing amounts of time at each power level specified. There were also "qualification" tests used for some new models, particularly if the engine was needed by the airframe manufacturer. These tests were usually for either 35 or 50 hours, and were often run on a Workhorse engine, rather than an engine built on the relevant service production contact.

Preliminary Flight Test, 50-Hour
The actual definition of such a test was more than just running the engine for 50 hours. The following is an example of the test specified for the V-1710-125 in October 1944.[14]

Object of the Test: To gain a preliminary flight rating for the V-1710-125 engine employing Water-Alcohol injection and operating at 75"Hg Manifold Pressure and 3200 Crankshaft RPM.

(1) 15 hours of alternate periods of:

A. 5 minutes at: 3200 rpm and 1740 bhp per power control chart. ADCMP shall not drop below 75" Hg.

B. 10 minutes at: 1600 rpm and 570 bhp with inlet air pressure of 29.9+/-0.3 inHgA. (ADCMP is Absolute Dry Center Manifold Pressure)

(2) 12-1/2 hours of alternate periods of:

A. 5 minutes at: 3200 rpm and 1740 bhp per power control chart. ADCMP shall not drop below 75" Hg.

B. 10 minutes at: 2280 rpm and 787 bhp with inlet air pressure of 29.9+/-0.3 inHgA.

(3) 22-1/2 hours continuous operation at: 2280 rpm and 787 bhp with inlet air pressure of 29.9+/-0.3 inHgA.

Pursuit Aircraft
Army Air Corps designation of fighter type airplanes dating from WWI. Designation F- for "Fighter" was introduced at the time the Army Air Forces became the U.S. Air Force, 1948.

Sniffle Valve
With the introduction of pressurized cooling systems it became necessary to insure that the coolant was not allowed to boil, a problem of considerable proportions in an aircraft producing high power and coolant temperatures and needing to operate over a wide range of ambient pressures. Since the boiling point for a given coolant mixture is determined by the absolute pressure of the coolant, it was necessary to have a device able to compensate for the changes in atmospheric pressure as the aircraft changed altitude. This was the Sniffle Valve, which was located on the low pressure side of the cooling system, usually on the header tank. Typical setpoints for the valve were 16 psia, though this would be increased for coolant mixtures with boiling points below the engine operating temperature of 250 °F. This would include the Army Standard Winter Coolant which was 80% Glycol and boiled in the range of 236-246 °F.[15]

U.S. Army Air Forces
This arm of the Army was created on June 20, 1941 by Army Regulation 95-5, which was a redesignation of the Army Air Corps. There were several subsequent steps in the restructuring, with Air Corps not officially ending until at least 1942. The AAF became a separate force in 1948, as the U.S. Air Force.

W535-ac-
Government purchase contracts were numbered for allocation management and tracking. The W535- prefix identifies a contract as a Wright Field Materiel Division procurement. The -ac- identifies the funds were expended in support of an Air Corps program or project. Suffix numbers were assigned sequentially without regard to the scope or nature of the acquisition. It made no difference whether they were for aircraft, engines or accessory items. From a historical perspective contract numbers add a chronological order to when actions were consummated with a contract.

NOTES

[1] Kelsey, Benjamin S. B.G. USAF (Ret.) 1982, 32.

[2] Tech Order AN 02-5AB-2, p.22.

[3] Thorner, Robert H. 1946, 218-286.

[4] *PD-12K17 Carburetor for V-1710-111/113(F30R/L) Engines*, Allison letter to Director Air Technical Service Command, Wright Field, February 22, 1945.

[5] *Operators Manual for Allison Engine Installations*, Allison Division of GMC, Fifth Edition, Sept. 15, 1943, pages 6-7 and 94-96.

[6] *Aircraft Engine Coolants—Ethylene Glycol*, T.O. 24-25-1, April 10, 1944.

[7] Technical Order 01-60JJA-2, 1947, 110-111.

[8] Fuels table, see Hubbard notebook H5.

[9] Fuels table, see Hubbard notebook H5.

[10] Fuels table, see Hubbard notebook H5.

[11] Fuels table, see Hubbard notebook H5.

[12] Fuels table, see Hubbard notebook H5.

[13] Allison Model Specification No.163-E, 1943.

[14] Report of Complete Verification Test, Q.T.R. Serial No. Allison-44-10, *50 Hour Preliminary Flight Rating Test, of Engine V-1710-125 Manufacturer's Serial No. A-042209*, 7 October 1944.

[15] Pitkin, J. 1944, page 18.14004.

Index

COBRA! THE BELL AIRCRAFT CORPORATION 1934-1946

Birch Matthews

COBRA! is a fully documented history of Bell Aircraft Corporation and their piston engine fighters built during the Great Depression and through World War II.

Size: 8 1/2" x 11"
over 700 b/w and color photographs, line drawings,
432 pages, hard cover
ISBN: 0-88740-911-3 $59.95